Carla W.
MONTGOMERY
Professor Emerita
Northern Illinois University

ENVIRONMENTAL GEOLOGY

eighth edition

McGraw-Hill Higher Education

Boston Burr Ridge, IL Dubuque, IA New York San Francisco St. Louis
Bangkok Bogotá Caracas Kuala Lumpur Lisbon London Madrid Mexico City
Milan Montreal New Delhi Santiago Seoul Singapore Sydney Taipei Toronto

The McGraw·Hill Companies

ENVIRONMENTAL GEOLOGY, EIGHTH EDITION

This book is printed on recycled, acid-free paper containing 10% postconsumer waste.

1 2 3 4 5 6 7 8 9 0 QPD/QPD 0 9 8 7

ISBN 978–0–07–282691–3
MHID 0–07–282691–6

Publisher: *Thomas D. Timp*
Executive Editor: *Margaret J. Kemp*
Senior Developmental Editor: *Margaret B. Horn*
Senior Marketing Manager: *Lisa Nicks*
Project Manager: *Joyce Watters*
Senior Production Supervisor: *Sherry L. Kane*
Associate Design Coordinator: *Brenda A. Rolwes*
Cover Design: *Studio Montage, St. Louis, Missouri*
Lead Photo Research Coordinator: *Carrie K. Burger*
Photo Research: *Jerry Marshall*
Supplement Producer: *Mary Jane Lampe*
Compositor: *Laserwords Private Limited*
Typeface: 10/12 Times Roman
Printer: *Quebecor World Dubuque, IA*
Cover Photo: *Horizontal axis wind turbines at San Gorgonio Pass, California. © The McGraw-Hill Companies Inc./Doug Sherman Photographer.*
Photos not credited on page provided by author, Carla W. Montgomery.

Library of Congress Cataloging-in-Publication Data

Montgomery, Carla W., 1951-
Environmental geology / Carla W. Montgomery. — 8th ed.
p. cm.
Includes index.
ISBN 978–0–07–282691–3--- ISBN 0–07–282691–6 (hard copy : alk. paper) 1. Environmental geology. I. Title.

QE38.M66 2008
550—dc22

2007026271

www.mhhe.com

Carla W.
MONTGOMERY
Professor Emerita
Northern Illinois University

ENVIRONMENTAL GEOLOGY

eighth edition

McGraw-Hill Higher Education

Boston Burr Ridge, IL Dubuque, IA New York San Francisco St. Louis
Bangkok Bogotá Caracas Kuala Lumpur Lisbon London Madrid Mexico City
Milan Montreal New Delhi Santiago Seoul Singapore Sydney Taipei Toronto

The McGraw-Hill Companies

ENVIRONMENTAL GEOLOGY, EIGHTH EDITION

Published by McGraw-Hill, a business unit of The McGraw-Hill Companies, Inc., 1221 Avenue of the Americas, New York, NY 10020.

This book is printed on recycled, acid-free paper containing 10% postconsumer waste.

1 2 3 4 5 6 7 8 9 0 QPD/QPD 0 9 8 7

ISBN 978–0–07–282691–3
MHID 0–07–282691–6

Publisher: *Thomas D. Timp*
Executive Editor: *Margaret J. Kemp*
Senior Developmental Editor: *Margaret B. Horn*
Senior Marketing Manager: *Lisa Nicks*
Project Manager: *Joyce Watters*
Senior Production Supervisor: *Sherry L. Kane*
Associate Design Coordinator: *Brenda A. Rolwes*
Cover Design: *Studio Montage, St. Louis, Missouri*
Lead Photo Research Coordinator: *Carrie K. Burger*
Photo Research: *Jerry Marshall*
Supplement Producer: *Mary Jane Lampe*
Compositor: *Laserwords Private Limited*
Typeface: 10/12 Times Roman
Printer: *Quebecor World Dubuque, IA*
Cover Photo: *Horizontal axis wind turbines at San Gorgonio Pass, California. © The McGraw-Hill Companies Inc./Doug Sherman Photographer.*
Photos not credited on page provided by author, Carla W. Montgomery.

Library of Congress Cataloging-in-Publication Data

Montgomery, Carla W., 1951-
Environmental geology / Carla W. Montgomery. — 8th ed.
p. cm.
Includes index.
ISBN 978–0–07–282691–3--- ISBN 0–07–282691–6 (hard copy : alk. paper) 1. Environmental geology. I. Title.

QE38.M66 2008
550—dc22

2007026271

www.mhhe.com

Environmental Geology is affectionately dedicated to the memory of Ed Jaffe, whose confidence in an unknown author made the first edition possible.

–CWM–

CONTENTS

SECTION ONE Foundations

SECTION TWO Internal Processes

SECTION THREE Surface Processes

SECTION FOUR Resources

SECTION FIVE Waste Disposal, Pollution, and Health

SECTION SIX Other Related Topics

PREFACE

ABOUT THE COURSE

Environmental Geology Is Geology Applied to Living

The *environment* is the sum of all the features and conditions surrounding an organism that may influence it. An individual's physical environment encompasses rocks and soil, air and water, such factors as light and temperature, and other organisms. One's social environment might include a network of family and friends, a particular political system, and a set of social customs that affect one's behavior.

Geology is the study of the earth. Because the earth provides the basic physical environment in which we live, all of geology might in one sense be regarded as environmental geology. However, the term *environmental geology* is usually restricted to refer particularly to geology as it relates directly to human activities, and that is the focus of this book. Environmental geology is geology applied to living. We will examine how geologic processes and hazards influence human activities (and sometimes the reverse), the geologic aspects of pollution and waste-disposal problems, and several other topics.

Why Study Environmental Geology?

One reason for studying environmental geology might simply be curiosity about the way the earth works, about the *how* and *why* of natural phenomena. Another reason is that we are increasingly faced with environmental problems to be solved and decisions to be made, and in many cases, an understanding of one or more geologic processes is essential to finding an appropriate solution.

Of course, many environmental problems cannot be fully assessed and solved using geologic data alone. The problems vary widely in size and in complexity. In a specific instance, data from other branches of science (such as biology, chemistry, or ecology), as well as economics, politics, social priorities, and so on may have to be taken into account. Because a variety of considerations may influence the choice of a solution, there is frequently disagreement about which solution is "best." Our personal choices will often depend strongly on our beliefs about which considerations are most important.

ABOUT THE BOOK

An introductory text cannot explore all aspects of environmental concerns. Here, the emphasis is on the physical constraints imposed on human activities by the geologic processes that have shaped and are still shaping our natural environment. In a real sense, these are the most basic, inescapable constraints; we cannot, for instance, use a resource that is not there, or build a secure home or a safe dam on land that is fundamentally unstable. Geology, then, is a logical place to start in developing an understanding of many environmental issues. The principal aim of this book is to present the reader with a broad overview of environmental geology. Because geology does not exist in a vacuum, however, the text introduces related considerations from outside geology to clarify other ramifications of the subjects discussed. Likewise, the present does not exist in isolation from the past and future; occasionally, the text looks both at how the earth developed into its present condition and where matters seem to be moving for the future. It is hoped that this knowledge will provide the reader with a useful foundation for discussing and evaluating specific environmental issues, as well as for developing ideas about how the problems should be solved.

ORGANIZATION

The book starts with some background information: a brief outline of earth's development to the present, and a look at one major reason why environmental problems today are so pressing—the large and rapidly growing human population. This is followed by a short discussion of the basic materials of geology—rocks and minerals—and some of their physical properties, which introduces a number of basic terms and concepts that are used in later chapters.

The next several chapters treat individual processes in detail. Some of these are large-scale processes, which may involve motions and forces in the earth hundreds of kilometers below the surface, and may lead to dramatic, often catastrophic events like earthquakes and volcanic eruptions. Other processes—such as the flow of rivers and glaciers or the blowing of the wind—occur only near the earth's surface, altering the landscape and occasionally causing their own special problems. In some cases, geologic processes can be modified, deliberately or accidentally; in others, human activities must be adjusted to natural realities. The section on surface processes concludes with a chapter on climate, which connects or affects a number of the surface processes described earlier in Section Three.

A subject of increasing current concern is the availability of resources. A series of five chapters deals with water resources, soil, minerals, and energy, the rates at which they are being consumed, probable amounts remaining, and projections of future availability and use. In the case of energy resources, we consider both those sources extensively used in the past and new sources that may or may not successfully replace them in the future.

Increasing population and increasing resource consumption lead to an increasing volume of waste to be disposed of; thoughtless or inappropriate waste disposal, in turn, commonly creates increasing pollution. Three chapters examine the interrelated problems of air and water pollution and the strategies available for the disposal of various kinds of wastes. The introduction to this section presents some related concepts from the field of geomedicine, linking geochemistry and health.

The final two chapters deal with a more diverse assortment of subjects. Environmental problems spawn laws intended to solve them; chapter 19 looks briefly at a sampling of laws and international agreements related to geologic matters discussed earlier in the book, and some of the problems with such laws and accords. Chapter 20 examines geologic constraints on construction schemes and the broader issue of trying to determine the optimum use(s) for particular parcels of land—matters that become more pressing as population growth pushes more people to live in marginal places.

Of course, the complex interrelationships among geologic processes and features mean that any subdivision into chapter-sized pieces is somewhat arbitrary, and different instructors may prefer different sequences or groupings (streams and ground water together, for example). An effort has been made to design chapters so that they can be resequenced in such ways without great difficulty.

New and Updated Content

Environmental geology is, by its very nature, a dynamic field in which new issues continue to arise and old ones to evolve. This is reflected in changes to this edition:

- **Chapter 10: Climate—Past, Present, and Future** is a new chapter that includes topics such as oceanic thermohaline circulation and climate feedbacks.
- **New boxed readings** include the Sumatran earthquakes and tsunami, Hurricane Katrina, and the mercury poisoning at Minamata, Japan.
- **New topics** include the Volcanic Explosivity Index, geological carbon sequestration, and E85.

For a more complete description of new/updated material in this edition, visit the Online Learning Center at www.mhhe.com/montgomery8e.

Features Designed for the Student

This text is intended for an introductory-level college course. It does not assume any prior exposure to geology or college-level mathematics or science courses. The metric system is used throughout, except where other units are conventional within a discipline. (For the convenience of students not yet "fluent" in metric units, a conversion table is included on the inside back cover, and in some cases, metric equivalents in English units are included within the text.)

Each chapter opens with an introduction that sets the stage for the material to follow. In the course of the chapter, important terms and concepts are identified by boldface type, and these terms are collected as "Key Terms and Concepts" at the end of the chapter for quick review. Many chapters include actual case histories or specific examples. The Glossary includes both these boldface terms and the additional, italicized terms that many chapters contain.

Every chapter concludes with review questions and exercises, which allow students to test their comprehension and apply their knowledge, followed by suggested readings and pertinent references.

Readings and Examples Connect to Life Experiences

Each chapter includes one or more boxed readings relating to the chapter material. Some involve a situation, problem, or application that might be encountered in everyday life. Others offer additional case histories or relevant examples. The tone is occasionally light, but the underlying issues are nonetheless real. (While some boxes were inspired by actual events, and include specific factual information, all of the characters quoted, and their interactions, are wholly fictitious.)

NetNotes Link Readers to Timely, In-depth Resources

The "NetNotes" at the end of each chapter offer modest collections of Internet sites that provide additional information and/or images relevant to the chapter content. These should prove useful to both students and instructors. An effort has been made to concentrate on sites with material at an appropriate level for the book's intended audience and also on sites likely to be relatively stable in the very fluid world of the Internet (government agencies, educational institutions, or professional association sites).

Appendices Provide Background and Reference Tools

Relative to the length of time we have been on earth, humans have had a disproportionate impact on this planet. Appendix A explores the concept of geologic time and its measurement and looks at the rates of geologic and other processes by way of putting human activities in temporal perspective. Appendix B gives an introduction to topographic and geologic maps and satellite and other kinds of imagery, highlighting some new techniques for examining the earth. Appendix C provides short reference keys to aid in rock and mineral identification, and the inside back cover includes units of measurement and conversion factors.

ACKNOWLEDGMENTS

A great many people have contributed to the development of one or another edition of this book. Portions of the manuscript of the first edition were read by Colin Booth, Lynn A. Brant, Arthur H. Brownlow, Ira A. Furlong, David Huntley, John F. Looney, Jr., Robert A. Matthews, and George H. Shaw, and the entire book was reviewed by Richard A. Marston and Donald J. Thompson. The second edition was enhanced through suggestions from William N. Mode, Laura L. Sanders, Jeffrey J. Gryta, Martin Reiter, Robert D. Hall, Robert B. Furlong, David Gust, and Stephen B. Harper; the third was refined with the assistance of Susan M. Cashman, Robert B. Furlong, William N. Mode, Frank Hanna, Laura L. Sanders, Paul Nelson, and Michael A. Velbel; the fourth was strengthened through the input of reviewers Pascal de Caprariis, James Cotter, John Vitek, Paul Schroeder, Steven Lund, Barbara Ruff, Gordon Love, Herbert Adams, Michael McKinney, Thomas E. Hendrix, Clifford Thurber, Ali Tabidian, Dru Germanoski, and Randall Scott Babcock. The fifth edition was improved thanks to reviews by Barbara L. Ruff, Michael Whitsett, John F. Hildenbrand, Alvin S. Konigsberg, Vernon P. Scott, Jim Stimson, Kevin Cole, Gilbert N. Hanson, Doreen Zaback, and Ann E. Homes, the sixth edition, by careful reviews from Ray Beiersdorfer, Ellin Beltz, William B. N. Berry, Paul Bierman, W. B. Clapham, Jr., Ralph K. Davis, Brian E. Lock, Gregory Hancock, Syed E. Hasan, Scott W. Keyes, Jason W. Kelsey, John F. Looney Jr., Christine Massey, Steve Mattox, William N. Mode, William A. Newman, Clair R. Ossian, David L. Ozsvath, Alfred H. Pekarek, Paul H. Reitan, and Don Rimstidt; and improvements in the seventh edition were inspired by reviewers Thomas J. Algaeo, Ernest H. Carlson, Douglas Crowe, Richard A. Flory, Hari P. Garbharran, Daniel Horns, Ernst H. Kastning, Abraham Lerman, Mark Lord, Lee Ann Munk, June A. Oberdorfer, Assad I. Panah, James S. Reichard, Frederick J. Rich, Jennifer Rivers Coombs, Richard Sleezer, and Michael S. Smith. This eighth edition has benefited from careful reading and astute suggestions by Richard Aurisano, Wharton County Junior College; Thomas B. Boving, University of Rhode Island; Ernest H. Carlson, Kent State University; Elizabeth Catlos, Oklahoma State University; Dennis DeMets, University of Wisconsin; Hailiang Dong, Miami University; Alexander Gates, Rutgers University; Chad Heinzel, Minot State University; Edward Kohut, University of Delaware; Richard McGehee, Austin Community College; Marguerite Moloney, Pensacola Junior College; Lee Slater, Rutgers University; Dan Vaughn, Southern Illinois University, Carbondale; and additional comments by Nathan Yee, Rutgers University.

The thoughtful suggestions of all of the foregoing individuals, and many other users who have informally offered additional advice, have substantially improved the text, and their help is most gratefully acknowledged. If, as one recent reviewer commented, the text "just keeps getting better," a large share of the credit certainly belongs to the reviewers. Any remaining shortcomings are, of course, my own responsibility.

M. Dalecheck, C. Edwards, I. Hopkins, and J. McGregor at the USGS Photo Library in Denver provided invaluable assistance with the photo research over the years. The encouragement of a number of my colleagues—paticularly Colin Booth, Ron C. Flemal, Donald M. Davidson, Jr., R. Kaufmann, and Eugene C. Perry, Jr.—was a special help during the development of the first edition. The ongoing support and interest of fellow author, deanly colleague, and ecologist Jerrold H. Zar helped immensely to make several revision cycles survivable. Thanks are also due to the several thousand environmental geology students I have taught, many of whom in the early years suggested that I write a text, and whose classes collectively have provided a testing ground for many aspects of the presentations herein.

My family has been supportive of this undertaking from the inception of the first edition. A very special vote of appreciation goes to my husband Warren—ever-patient sounding board, occasional photographer and field assistant—in whose life this book has so often loomed so large.

Last, but assuredly not least, I express my deep gratitude to the entire McGraw-Hill book team for their enthusiasm, professionalism, and just plain hard work, without which successful completion of each subsequent edition of this book would have been impossible.

Carla W. Montgomery

SUPPLEMENTS

For the Student

Online Learning Center (http://www.mhhe.com/montgomery 8)

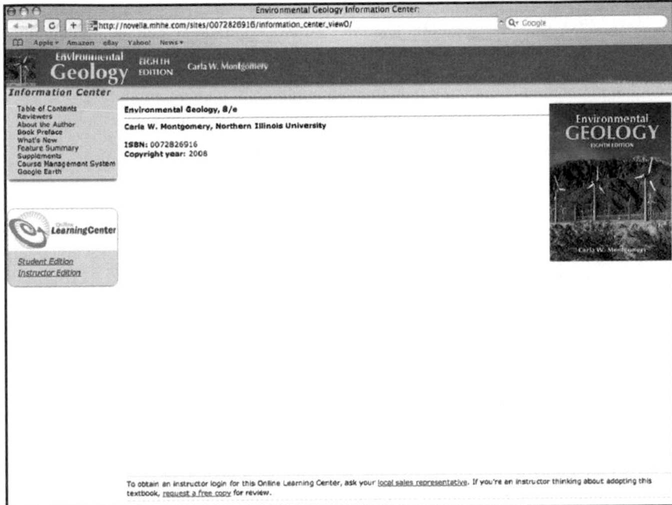

Student Resources—Everything You Need in One Place:

- Chapter Summary
- Student Study Guide with review questions and answers
- Practice quizzes
- Additional study tools
- Guest essays
- Interactive maps
- Animations of geologic concepts
- Environmental timeline
- Geologic timeline
- Career information

For the Instructor

Online learning Center (http://www.mhhe.com/montgomery 8)

In addition to **all of the student resources** on the Online Learning Center, the Instructor Edition has these assets:

- Answers to end-of-chapter review questions
- Google Earth worksheets
- PowerPoint Lecture Outlines for every chapter
- CPS by eInstruction
- Access to McGraw-Hill's Presentation Center

Presentation Center

Accessed through the Online Learning Center, the **Presentation Center** is an online digital library containing assets such as art, photos, tables, PowerPoints, animations, and other media types that can be used to create customized lectures, visually enhanced tests and quizzes, compelling course websites, or attractive printed support materials.

Instructor's Testing and Resource CD-ROM

This cross-platform CD-ROM provides a computerized test bank utilizing software to quickly create customized exams. The user-friendly program allows instructors to sort questions by format; edit existing questions or add new ones; and scramble questions for multiple versions of the same test. ISBN-13: 978-007-332419-7 (ISBN-10: 007-332419-1)

PACKAGING OPPORTUNITIES

McGraw-Hill offers packaging opportunities that provide students with valuable course-related materials at a substantial cost savings. Instructors, ask your McGraw-Hill sales representative for information on discounts and special ISBNs for ordering a package that contains one or more of the following:

Laboratory Exercises in Environmental Geology by Blatt

These lab exercises focus on a broad range of environmental issues and also cover social issues relevant to environmental concerns and the effects of human interaction in geologic processes. ISBN 0-697-28288-0

Laboratory Manual for Physical Geology, Sixth Edition, by Jones/Jones

Seventeen exercises on geological methods, incorporating examples of the scientific method, are offered in this lab manual. It also includes "In Greater Depth" problems, a more challenging probe into certain issues that are more quantitative in nature and require in-depth, critical thinking. These problems are unique to this type of manual. ISBN-13: 978-0-07-305091-1 (ISBN-10: 007-305091-1)

Physical Geology Laboratory Manual, Thirteenth Edition, by Zumberge et al.

This manual features thirty exercises in which students study earth materials, geologic interpretation of topographic maps, aerial photographs and earth satellite imagery, structural geology, and plate tectonics and related phenomena. ISBN-13: 978-0-07-298861-1 (ISBN 007-298861-4)

Exploring Environmental Science with GIS by Stewart et al.

This short book provides a set of exercises aimed at students and instructors who are new to GIS but are familiar with the Windows operating system. The exercises focus on improving analytical skills, understanding spatial relationships, and understanding the nature and structure of environmental data. Because the software used is distributed free of charge, this text is appropriate for courses and schools that are not yet ready to commit to the expense and time involved in acquiring other GIS packages. ISBN-13: 978-007-297744-8 (ISBN-10: 007-297744-2)

Annual Editions: Environment by Allen

This small text is a compilation of current articles that explore the global environment, the world's population, energy, the biosphere, natural resources, and pollution. ISBN-13: 978-007-351544-1 (ISBN-10: 007-351544-2)

Taking Sides: Clashing Views on Controversial Environmental Issues, Twelfth Edition, by Easton

The ***Taking Sides*** volume presents the arguments of leading environmentalists, scientists, and policymakers. The issues reflect a variety of viewpoints and are staged as pro and con debates. Topics are organized around four core areas: general philosophical and political issues, the environment and technology, disposing of wastes, and the environment and the future. ISBN-13: 978-007-351443-7 (ISBN-10: 007-351443-8)

Sources: Notable Selections in Environmental Studies, Second Edition, by Goldfarb

This volume brings together primary source selections of enduring intellectual value—classic articles, book excerpts, and research studies—that have shaped environmental studies and our contemporary understanding of it. The book includes carefully edited selections for the works of the most distinguished environmental observers, past and present. Selections are organized topically around the following major areas of study: energy, environmental degradation, population issues and the environment, human health and the environment, and environment and society. ISBN-13: 978-007-303186-6 (ISBN-10: 007-303186-0)

CUSTOM PUBLISHING

Did you know that you can design your own text or laboratory manual using any McGraw-Hill text *and* your personal materials to create a custom product that correlates specifically to your syllabus and course goals? Talk to your McGraw-Hill sales representative about this creative option.

Presentation Center

The Presentation Center, accessed through the Online Learning Center at www.mhhe.com/montgomery8, is an indispensable online digital library that allows instructors to create customized course tools such as classroom lectures and presentations, testing materials, or online course communications.

Every Illustration from the Book

Every piece of line art and every photo, already in PowerPoint format, are at your disposal.

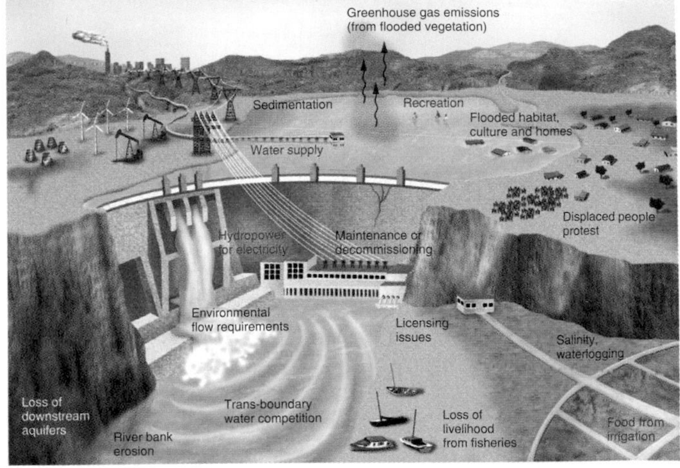

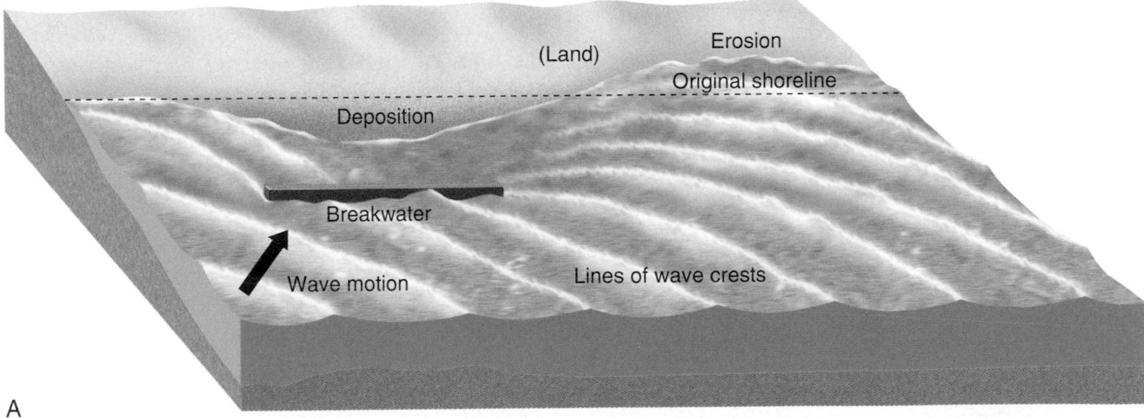

A

B

Table Library

In addition to the hundreds of illustrations and photos, every table in the textbook is available in ready-to-use digital files.

PowerPoint Lecture Outlines

A customized presentation combining art and photos from the text with instructor-written lecture notes is available for each chapter in the book.

Active Art

These special art pieces consist of key environmental geology illustrations that are converted to a format that allows you to break the art down into core elements, and then group the various pieces and create customized images. This is especially helpful with difficult concepts because they can be explained to students step-by-step.

Global Base Maps

Frustrated that your students can't identify cities/states/countries beyond their immediate surroundings? In response to the many professors asking for more maps, we've developed 22 base maps for all world regions and major subregions. These maps are offered in four versions: black-and-white and full color, both with labels and without labels. All of these choices allow you flexibility in planning class activities, quizzing opportunities, study tools, and PowerPoint enhancements.

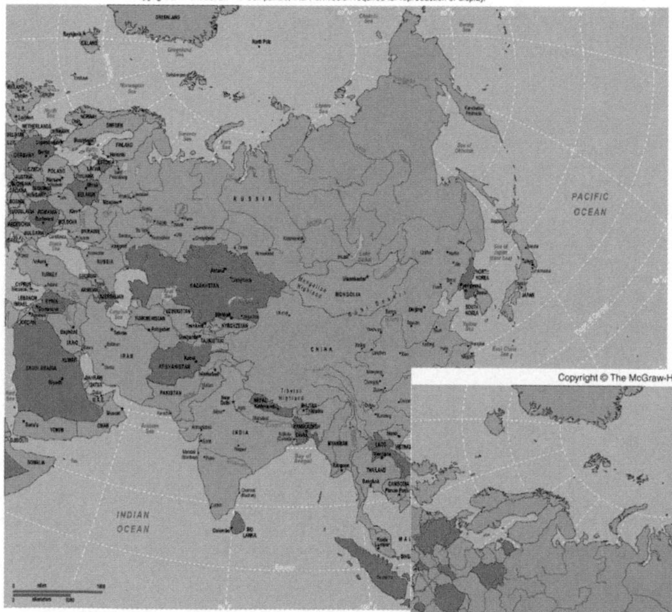

Copyright © The McGraw-Hill Companies, Inc. Permission required for reproduction or display.

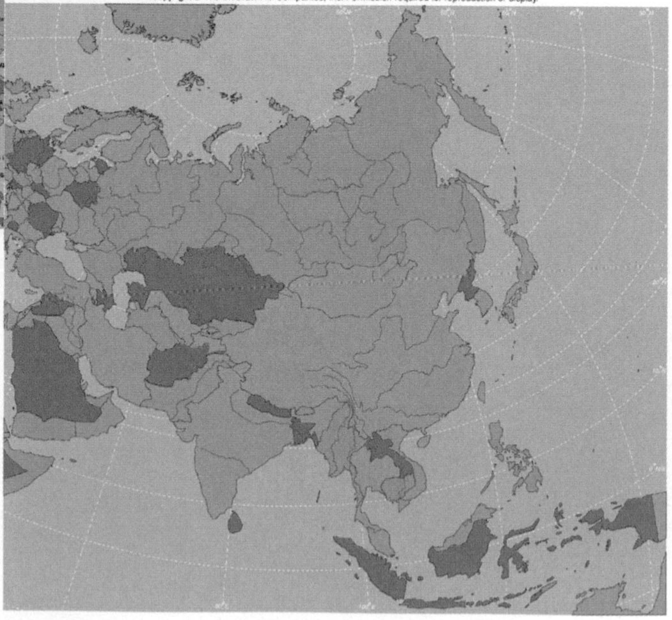

Animations

One hundred and sixty-five full-color animations illustrating many different concepts covered in the study of environmental science and geology are available for your use.

Additional Photo Library

Over 500 full-color photographs *additional* to those already in the textbook are included in the Presentation Center. Chosen for their specific contribution to topics in environmental science and in geology, these photos are searchable by content and will add interest and contextual support to your lectures.

Access to Other McGraw-Hill Titles!

The Presentation Center library includes thousands of assets from many McGraw-Hill titles. This ever-growing resource gives instructors the power to utilize assets specific to an adopted textbook as well as content from all other books in the library. The Presentation Center's dynamic search engine allows you to explore by discipline, course, textbook chapter, asset type, or keyword. Simply browse, select, and download the files you need to build engaging course materials. All assets are copyrighted by McGraw-Hill Higher Education but can be used by instructors for classroom purposes.

SECTION ONE

FOUNDATIONS

A sense of historical perspective helps us to appreciate current problems and to anticipate future ones. Many modern environmental problems, such as acid rain and groundwater pollution, have come upon us very recently. Others, such as the hazards posed by earthquakes, volcanoes, and landslides, have always been with us. Recognition of the impact of natural hazards is worldwide. The 1990s were designated the United Nations Decade for Natural Disaster Reduction (though ironically, both the number of catastrophic events, and their toll in dollars and lives, soared to all-time highs in that period). In recent years, we have seen unprecedented international cooperation on issues such as preservation of the ozone layer.

Chapter 1 briefly summarizes major stages in the earth's development and allows us to begin to see where current and future human activities fit in. It provides us with some information about the solar system to help the reader judge the degree to which other planets might provide solutions to such problems as lack of resources and living space. It also introduces the concept of cyclicity in natural processes, and points out that the interrelationships among natural processes may be complex.

The size and growth of earth's human population bear strongly on the ways and extent to which geology and people interact, which is what environmental geology is all about. In fact, many of our problems are as acute as they are simply because of the sheer number of people who now live on the earth, either overall or in particular locations. This will be evident in later discussions of resources, pollution, and waste disposal. Again, the global impact of the issues is underscored by the fact that in 1992, more than 170 nations came together in Rio de Janeiro for the United Nations Conference on Environment and Development, to address such issues as global climate change, sustainable development, and environmental protection. The fact that follow-up discussions are ongoing reflects the complexity of the issues and the practical challenges of resolving them.

It is difficult to talk for long about geology without discussing rocks and minerals, the stuff of which the earth is made. Chapter 2 introduces these materials and some of their basic properties. Specific physical and chemical properties of rocks and soils are important in considering such diverse topics as resource identification and recovery, waste disposal, assessment of volcanic or landslide hazards, weathering processes and soil formation, and others.

Geology provides us land to live on, resources with which to build and manufacture, energy resources—and even scenery and recreation. Delicate Arch, Arches National Park. (Note people on horizon left of arch for scale.)

1 An Overview of Our Planetary Environment

About five billion years ago, out of a swirling mass of gas and dust, evolved a system of varied planets hurtling around a nuclear-powered star—our solar system. One of these planets, and one only, gave rise to complex life-forms. Over time, a tremendous diversity of life-forms and ecological systems developed, while the planet, too, evolved and changed, its interior churning, its landmasses shifting, its surface constantly being reshaped. Within the last several million years, the diversity of life on earth has included humans, increasingly competing for space and survival on the planet's surface. With the control over one's surroundings made possible by the combination of intelligence and manual dexterity, humans have found most of the land on the planet inhabitable; they have learned to use not only plant and animal resources, but minerals, fuels, and other geologic materials; in some respects, humans have even learned to modify natural processes that inconvenience or threaten them. As we have learned how to study our planet in systematic ways, we have developed an ever-increasing understanding of the complex nature of the processes shaping, and the problems posed by, our geological environment. **Environmental geology** explores the many and varied interactions between humans and that geologic environment.

As the human population grows, it becomes increasingly difficult for that population to survive on the resources and land remaining, to avoid those hazards that cannot be controlled, and to refrain from making irreversible and undesirable changes in environmental systems. The urgency of perfecting our understanding, not only of natural processes but also of our impact on the planet, is becoming more and more apparent, and has motivated increased international cooperation and dialogue on environmental issues. (However, while nations may readily agree on *what* the problematic issues are, agreement on *solutions* is often much harder to achieve!)

Night view of earth shows most urbanized areas by concentrations of lights. Image by Craig Mayhew and Robert Simmon, NASA GSFC, based on DMSP data

EARTH IN SPACE AND TIME

The Early Solar System

In recent decades, scientists have been able to construct an ever-clearer picture of the origins of the solar system and, before that, of the universe itself. Most astronomers now accept some sort of "Big Bang" as the origin of today's universe. At that time, enormous quantities of matter were created and flung violently apart across an ever-larger volume of space. The time of the Big Bang can be estimated in several ways. Perhaps the most direct is the back-calculation of the universe's expansion to its apparent beginning. Other methods depend on astrophysical models of creation of the elements or the rate of evolution of different types of stars. Most age estimates overlap in the range of 15 to 20 billion years, although a few researchers suggest an age closer to 10 billion years. (Some also suggest an oscillating universe, which eventually stops expanding and collapses back to another "Big Bang," and so on in cyclic fashion. In such a case, the estimated "age of the universe" is of the *last* "Big Bang" before the present.)

Stars formed from the debris of the Big Bang, as locally high concentrations of mass were collected together by gravity, and some became large and dense enough that energy-releasing atomic reactions were set off deep within them. Stars are not permanent objects. They are constantly losing energy and mass as they burn their nuclear fuel. The mass of material that initially formed the star determines how rapidly the star burns; some stars burned out billions of years ago, while others are probably forming now from the original matter of the universe mixed with the debris of older stars.

Our sun and its system of circling planets, including the earth, are believed to have formed from a rotating cloud of gas and dust, some of it debris from older stars (figure 1.1). Most of the mass of the cloud coalesced to form the sun, which became a star and began to "shine," or release light energy, when its interior became so dense and hot from the crushing effects of its own gravity that nuclear reactions were triggered inside it. Meanwhile, dust condensed from the gases remaining in the flattened cloud disk rotating around the young sun. The dust clumped into planets, the formation of which was essentially complete over 4½ billion years ago.

The Planets

The compositions of the planets formed depended largely on how near they were to the hot sun (figure 1.2). The planets formed nearest to the sun contained mainly metallic iron and a few minerals with very high melting temperatures, with little water or gas. Somewhat farther out, where temperatures were lower, the developing planets incorporated much larger amounts of lower-temperature minerals, including some that contain water locked within their crystal structures. (This later made it possible for the earth to have liquid water at its surface.) Still farther from the sun, temperatures were so low that nearly all of the materials in the original gas cloud condensed—even materials like methane and ammonia, which are gases at normal earth surface temperatures and pressures.

The result was a series of planets with a variety of compositions, most quite different from that of earth. This is confirmed by observations and measurements of the planets. For example, the planetary densities listed in table 1.1 are consistent with a higher metal and rock content in the four planets closest to the sun and a much larger proportion of ice and gas in the planets

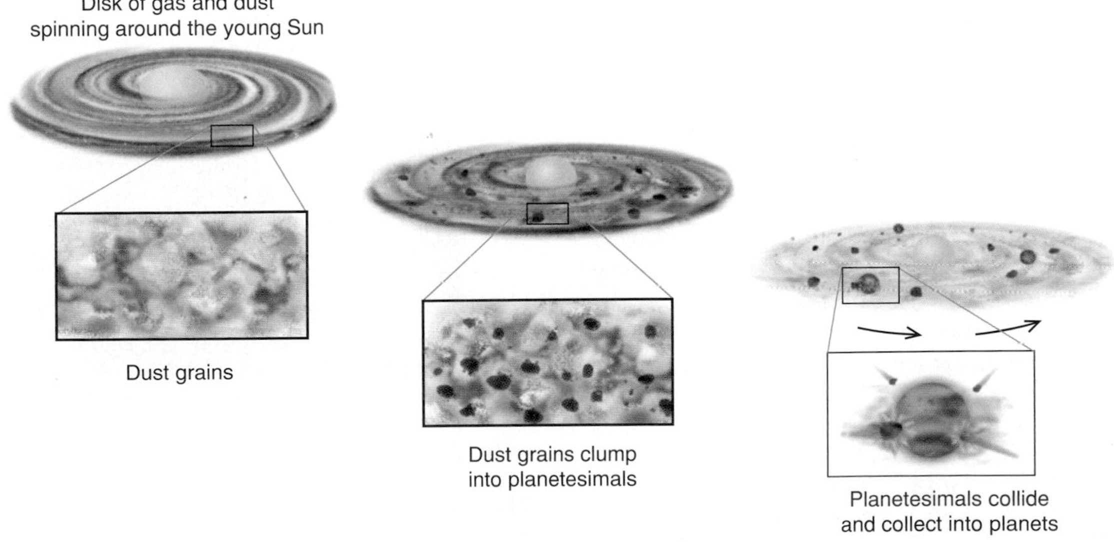

FIGURE 1.1 Our solar system formed as dust condensed from the gaseous nebula, then clumped together to make planets.

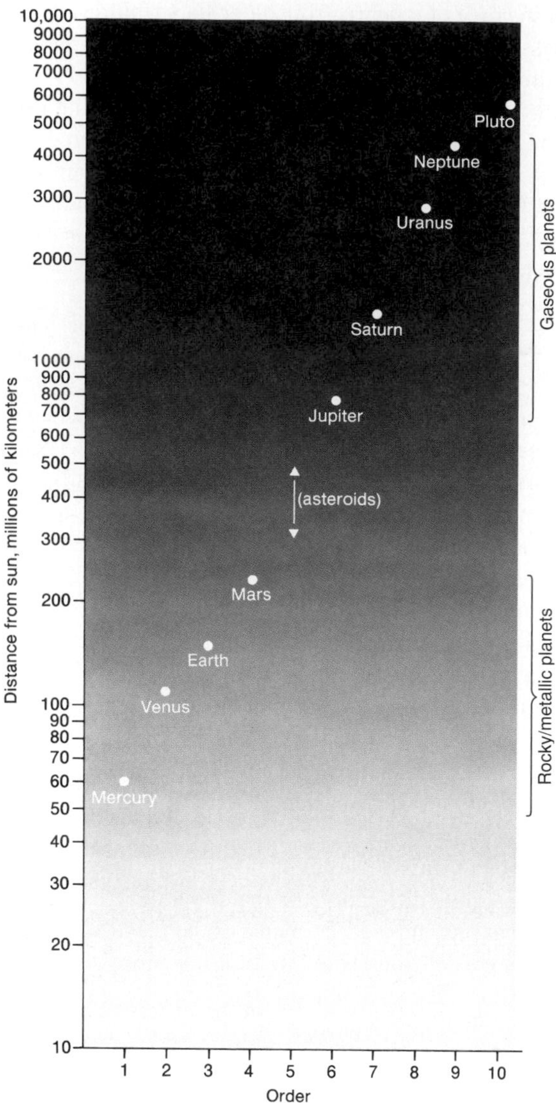

FIGURE 1.2 As this graph shows, the spacing of the planets' orbits exhibits a geometric regularity. Note that the distance scale is logarithmic—outer planets are *much* farther out, proportionately, and formed at much colder temperatures.

farther from the sun (see also figure 1.3). These differences should be kept in mind when it is proposed that other planets could be mined for needed minerals. Both the basic chemistry of these other bodies and the kinds of ore-forming or other resource-forming processes that might occur on them would differ considerably from those on earth, and may not have led to products we would find useful. (This is leaving aside any questions of the economics or technical practicability of such mining activities!) In addition, our principal current energy sources required living organisms to form, and so far, no such life-forms have been found on other planets or moons. Venus—close to Earth in space, similar in size and density—shows marked differences: Its dense, cloudy atmosphere is thick with carbon dioxide, producing planetary surface temperatures hot enough to melt lead through runaway greenhouse-effect heating (see chapter 10). Mars would likewise be inhospitable, as will be seen later in the chapter.

FIGURE 1.3 The planets of the solar system vary markedly in both composition and physical properties. For example, Mercury (A) is rocky, iron-rich, dry, and pockmarked with craters; Jupiter (B) is a huge gas ball, with no solid surface at all, and dozens of moons of ice and rock that circle it to mimic the solar system in miniature. Note also the very different sizes of the two planets.

Photographs courtesy NSSDC Goddard Space Flight Center

TABLE 1.1

Some Basic Data on the Planets

Planet	Mean Distance from Sun (millions of km)	Mean Temperature (°C)	Equatorial Diameter, Relative to Earth	Density* (g/cu. cm)	
Mercury	58	167	0.38	5.4	Predominantly rocky/metal planets
Venus	108	464	0.95	5.2	
Earth	150	15	1.00	5.5	
Mars	228	−65	0.53	3.9	
Jupiter	778	−110	11.19	1.3	Gaseous planets
Saturn	1427	−140	9.41	0.7	
Uranus	2870	−195	4.06	1.3	
Neptune	4479	−200	3.88	1.6	
Pluto	5900	−225	0.23	1.8	

Source: Data from NASA.

*No other planets have been extensively sampled to determine their compositions directly, though we have some data on their surfaces. Their approximate bulk compositions are inferred from the assumed starting composition of the solar nebula and the planets' densities. For example, the higher densities of the inner planets reflect a significant iron content and relatively little gas. Pluto, anomalous in composition and orbit, may in fact be a stray body captured into its current position long after the planets formed.

Earth, Then and Now

The earth has changed continuously since its formation, undergoing some particularly profound changes in its early history. The early earth was very different from what it is today, lacking the modern oceans and atmosphere and having a much different surface from its present one, probably more closely resembling the barren, cratered surface of the moon. The planet was heated by the impact of the colliding dust particles and meteorites as they came together to form the earth, and by the energy release from decay of the small amounts of several naturally radioactive elements that the earth contains.

These heat sources combined to raise the earth's internal temperature enough that parts of it, perhaps eventually most of it, melted, although it was probably never molten all at once. Dense materials, like metallic iron, would have tended to sink toward the middle of the earth. As cooling progressed, lighter, low-density minerals crystallized and floated out toward the surface. The eventual result was an earth differentiated into several major compositional zones: the central **core;** the surrounding **mantle;** and a thin **crust** at the surface (see figure 1.4). The process was complete before 4 billion years ago.

Although only the crust and a few bits of uppermost mantle that are carried up into the crust by volcanic activity can be sampled and analyzed directly, we nevertheless have a good deal of information on the composition of the earth's interior. First, scientists can estimate from analyses of stars the starting composition of the cloud from which the solar system formed. Geologists can also infer aspects of the earth's bulk composition from analyses of certain meteorites believed to have formed at the same time as, and under conditions similar to, the earth. Geophysical data demonstrate that the earth's interior is zoned and also provide information on the densities of the different layers within the earth, which further limits their possible compositions. These and other kinds of data indicate that the earth's core is made up mostly of iron, with some nickel and a few minor elements; the outer core is molten, the inner core solid. The mantle consists mainly of iron, magnesium, silicon, and oxygen combined in varying proportions in several different minerals. The earth's crust is much more varied in composition and very different chemically from the average composition of the earth (see table 1.2). Crust and uppermost mantle together form a somewhat brittle shell around the earth. As is evident from table 1.2, many of the metals we have come to prize as resources are relatively uncommon elements in the crust.

The heating and subsequent differentiation of the early earth led to another important result: formation of the atmosphere and oceans. Many minerals that had contained water or gases in their crystals released them during the heating and melting, and as the earth's surface cooled, the water could condense to form the oceans. Without this abundant surface water, which in the solar system is unique to earth, most life as we know it could not exist. The oceans filled basins, while the continents, buoyant because of their lower-density rocks and minerals, stood above the sea surface. At first, the continents were barren of life.

The earth's early atmosphere was quite different from the modern one, aside from the effects of modern pollution. The first atmosphere had little or no free oxygen in it. It probably consisted dominantly of nitrogen and carbon dioxide (the gas most commonly released from volcanoes, aside from water) with minor amounts of such gases as methane, ammonia, and various sulfur gases. Humans could not have survived in this early atmosphere. Oxygen-breathing life of any kind could not exist before the single-celled blue-green algae appeared in large numbers to modify the atmosphere. Their remains are found in rocks as much as several billion years

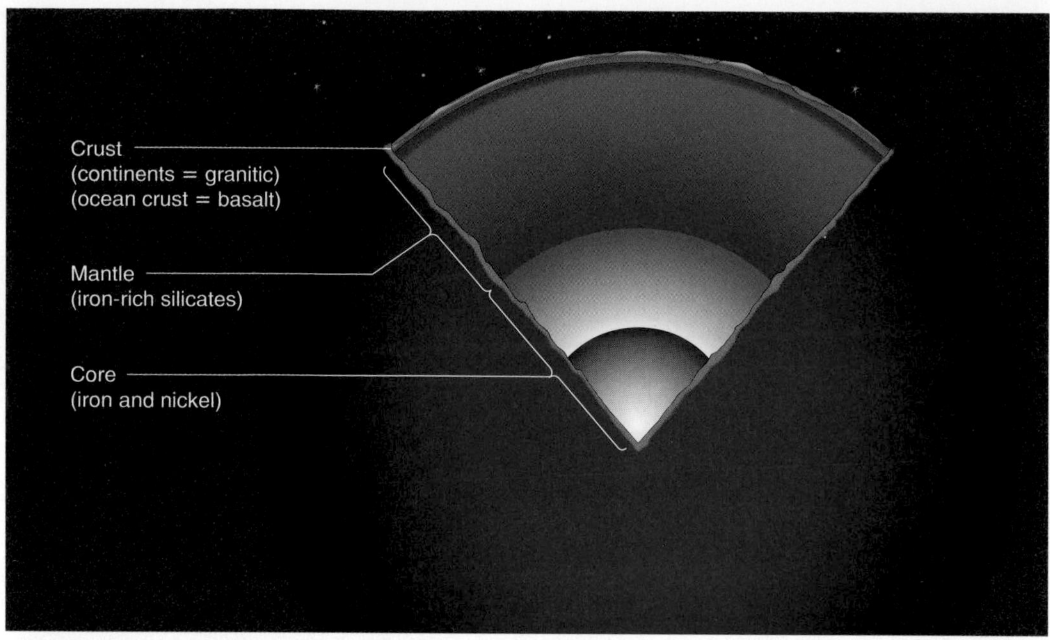

FIGURE 1.4 A chemically differentiated earth. The core consists mostly of iron; the outer part is molten. The mantle, the largest zone, is made up primarily of ferromagnesian silicates (see chapter 2) and, at great depths, of oxides of iron, magnesium, and silicon. The crust (not drawn to scale, but exaggerated vertically in order to be visible at this scale) forms a thin skin around the earth. Oceanic crust, which forms the sea floor, has a composition somewhat like that of the mantle, but is richer in silicon. Continental crust is both thicker and less dense. It rises above the oceans and contains more light minerals rich in calcium, sodium, potassium, and aluminum.

old. They manufacture food by photosynthesis, using sunlight for energy, consuming carbon dioxide, and releasing oxygen as a by-product. In time, enough oxygen accumulated that the atmosphere could support oxygen-breathing organisms.

Life on Earth

The rock record shows when different plant and animal groups appeared. Some are represented schematically in figure 1.5. The earliest creatures left very few remains because they had no hard skeletons, teeth, shells, or other hard parts that could be preserved in rocks. The first multicelled oxygen-breathing creatures probably developed about 1 billion years ago, after oxygen in the atmosphere was well established. By about 600 million years ago, marine animals with shells had become widespread.

The development of organisms with hard parts—shells, bones, teeth, and so on—greatly increased the number of preserved animal remains in the rock record; consequently, biological developments since that time are far better understood. Dry land was still barren of large plants or animals half a billion years ago. In rocks about 500 million years old is the first evidence of animals with backbones—the fish—and soon thereafter, early land plants developed, before 400 million years ago. Insects appeared approximately 300 million years ago. Later, reptiles and amphibians moved onto the continents. The dinosaurs appeared about 200 million years ago and the first mammals at nearly the same time. Warm-blooded animals took to the air with the development of birds about 150 million years ago, and by 100 million years ago, both birds and mammals were well established.

Such information has current applications. Certain energy sources have been formed from plant or animal remains.

TABLE 1.2

Most Common Chemical Elements in the Earth

Whole Earth		Crust	
Element	**Weight Percent**	**Element**	**Weight Percent**
Iron	32.4	Oxygen	46.6
Oxygen	29.9	Silicon	27.7
Silicon	15.5	Aluminum	8.1
Magnesium	14.5	Iron	5.0
Sulfur	2.1	Calcium	3.6
Nickel	2.0	Sodium	2.8
Calcium	1.6	Potassium	2.6
Aluminum	1.3	Magnesium	2.1
(All others, total)	.7	(All others, total)	1.5

(Compositions cited are averages of several independent estimates.)

Knowing the times at which particular groups of organisms appeared and flourished is helpful in assessing the probable amounts of these energy sources available and in concentrating the search for these fuels on rocks of appropriate ages.

On a time scale of billions of years, human beings have just arrived. The most primitive human-type remains are no more than 3 to 4 million years old, and modern, rational humans (*Homo sapiens*) developed only about half a million

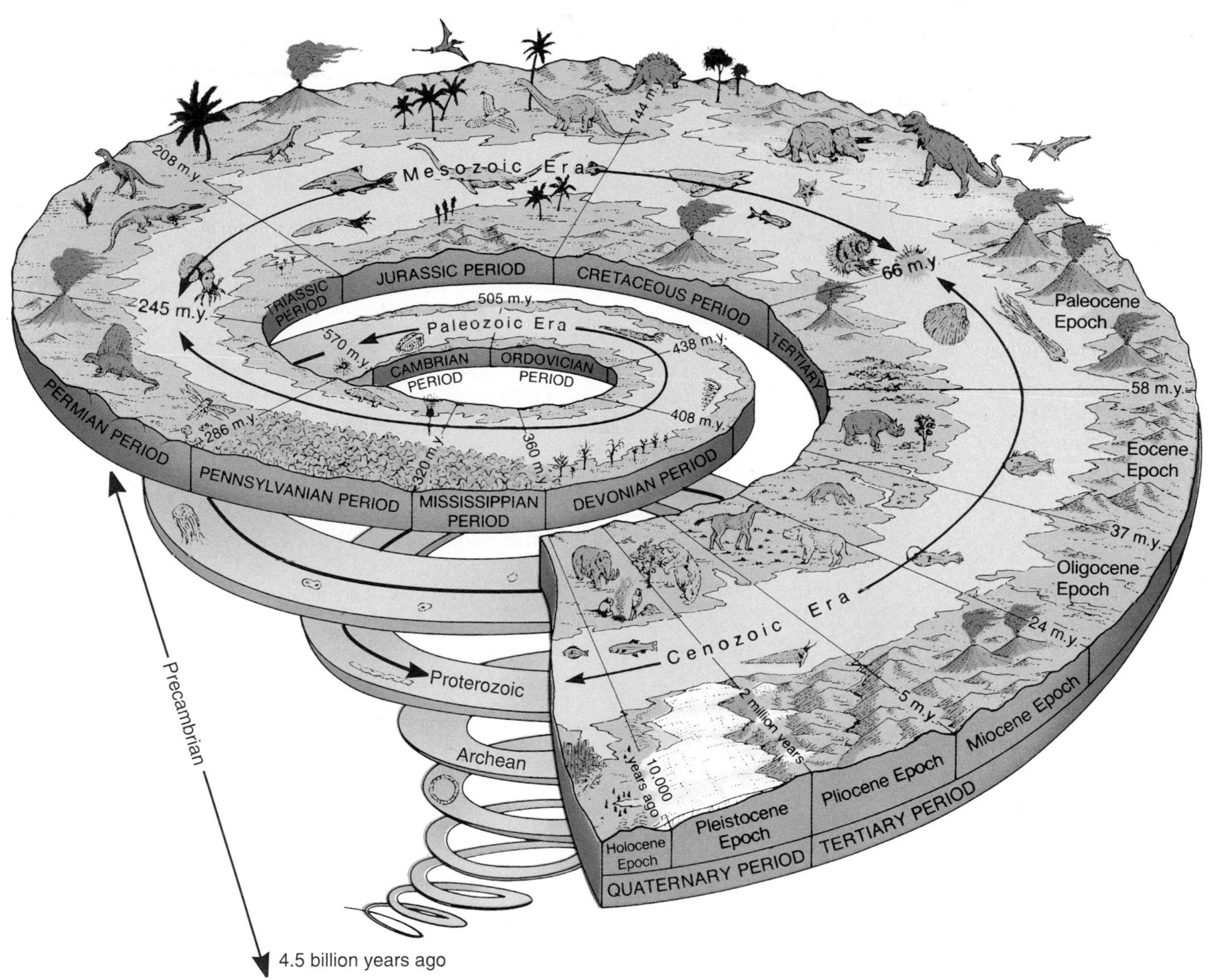

FIGURE 1.5 The "geologic spiral": Important plant and animal groups appear where they first occurred in significant numbers. If earth's whole history were equated to a 24-hour day, modern thinking humans (*Homo sapiens*) would have arrived on the scene just about ten seconds ago.
Source: Modified after U.S. Geological Survey publication Geologic Time

years ago. Half a million years may sound like a long time, and it is if compared to a single human lifetime. In a geologic sense, though, it is a very short time. If we equate the whole of earth's history to a 24-hour day, then shelled organisms appeared only about three hours ago; fish, about 2 hours and 40 minutes ago; land plants, two hours ago; birds, about 45 minutes ago—and *Homo sapiens* has been around for just the last ten *seconds*. Nevertheless, we humans have had an enormous impact on the earth, at least at its surface, an impact far out of proportion to the length of time we have occupied it. Our impact is likely to continue to increase rapidly as the population does likewise.

GEOLOGY, PAST AND PRESENT

Two centuries ago, geology was mainly a descriptive science involving careful observation of natural processes and their products. The subject has become both more quantitative and more interdisciplinary through time. Modern geoscientists draw on the principles of chemistry to interpret the compositions of geologic materials, apply the laws of physics to explain these materials' physical properties and behavior, use the biological sciences to develop an understanding of ancient life-forms, and rely on engineering principles to design safe structures in the presence of geologic hazards. The emphasis on the "why," rather than just the "what," has increased.

The Geologic Perspective

Geologic observations now are combined with laboratory experiments, careful measurements, and calculations to develop theories of how natural processes operate. Geology is especially challenging because of the disparity between the scientist's

TABLE 1.3

Some Representative Geologic-Process Rates

Process	Occurs Over a Time Span of About This Magnitude
Rising and falling of tides	1 day
"Drift" of a continent by 2–3 centimeters (about 1 inch)	1 year
Accumulation of energy between large earthquakes on a major fault zone	10–100 years
Rebound (rising) by 1 meter of a continent depressed by ice sheets during the Ice Age	100 years
Flow of heat through 1 meter of rock	1000 years
Deposition of 1 centimeter of fine sediment on the deep-sea floor	1000–10,000 years
Ice sheet advance and retreat during an ice age	10,000–100,000 years
Life span of a small volcano	100,000 years
Life span of a large volcanic center	1–10 million years
Creation of an ocean basin such as the Atlantic	100 million years
Duration of a major mountain-building episode	100 million years
History of life on earth	Over 3 billion years

laboratory and nature's. In the research laboratory, conditions of temperature and pressure, as well as the flow of chemicals into or out of the system under study, can be carefully controlled. One then knows just what has gone into creating the product of the experiment. In nature, the geoscientist is often confronted only with the results of the "experiment" and must deduce the starting materials and processes involved.

Another complicating factor is time. The laboratory scientist must work on a time scale of hours, months, years, or, at most, decades. Natural geologic processes may take a million or a billion years to achieve a particular result, by stages too slow even to be detected in a human lifetime (table 1.3). This understanding may be one of the most significant contributions of early geoscientists: the recognition of the vast length of geologic history, sometimes described as "deep time." The qualitative and quantitative tools for sorting out geologic events and putting dates on them are outlined in appendix A. For now, it is useful to bear in mind that the immensity of geologic time can make it difficult to arrive at a full understanding of how geologic processes operated in the past from observations made on a human timescale. It dictates caution, too, as we try to project, from a few years' data on global changes associated with human activities, all of the long-range impacts we may be causing.

Also, the laboratory scientist may conduct a series of experiments on the same materials, but the experiments can be stopped and those materials examined after each stage. Over the vast spans of geologic time, a given mass of earth material may have been transformed a half-dozen times or more, under different conditions each time. The history of the rock that ultimately results may be very difficult to decipher from the end product alone.

Geology and the Scientific Method

The **scientific method** is a means of discovering basic scientific principles. One begins with a set of observations and/or a body of data, based on measurements of natural phenomena or on experiments. One or more **hypotheses** are formulated to explain the observations or data. A hypothesis can take many forms, ranging from a general conceptual framework or model describing the functioning of a natural system, to a very precise mathematical formula relating several kinds of numerical data. What all hypotheses have in common is that they must all be susceptible to testing and, particularly, to *falsification.* The idea is not simply to look for evidence to support a hypothesis, but to examine relevant evidence with the understanding that it may show a hypothesis to be wrong.

In the classical conception of the scientific method, one uses a hypothesis to make a set of predictions. Then one devises and conducts experiments to test each hypothesis, to determine whether experimental results agree with predictions based on the hypothesis. If they do, the hypothesis gains credibility. If not, if the results are unexpected, the hypothesis must be modified to account for the new data as well as the old or, perhaps, discarded altogether. Several cycles of modifying and retesting hypotheses may be required before a hypothesis that is consistent with all the observations and experiments that one can conceive is achieved. A hypothesis that is repeatedly supported by new experiments advances in time to the status of a **theory,** a generally accepted explanation for a set of data or observations.

Much confusion can arise from the fact that in casual conversation, people often use the term *theory* for what might better be called a hypothesis, or even just an educated guess. ("So, what's your theory?" one character in a TV mystery show may ask another, even when they've barely looked at the first evidence.) Thus people may assume that a scientist describing a theory is simply telling a plausible story to explain some data. A scientific theory, however, is a very well-tested model with a very substantial and convincing body of evidence that supports it. A hypothesis may be advanced by just one individual; a theory has survived the challenge of extensive testing to merit acceptance by many, often most, experts in a field.

The classical scientific method is not strictly applicable to many geologic phenomena because of the difficulty of experimenting with natural systems. For example, one may be able to conduct experiments on a single rock, but not to construct a whole volcano in the laboratory. In such cases, hypotheses are often tested entirely through further observations and modified as necessary until they accommodate all the relevant observations (or discarded when they cannot be reconciled with new data). This broader conception of the scientific method is well illustrated by the development of the theory of plate tectonics, discussed in chapter 3. "Continental drift" was once seen as a wildly implausible idea, advanced by an eccentric few, but in the latter half of the twentieth century, many kinds of evidence were found to be explained consistently and well by movement of plates—including continents—over earth's surface. Still, the details of plate tectonics continue to be refined by further studies. Even a well-established theory may ultimately be proved incorrect. In the case of geology, complete rejection of an older theory has most often been caused by the development of new analytical or observational techniques, which make available wholly new kinds of data that were unknown at the time the original theory was formulated.

FIGURE 1.7 Mount Pinatubo in eruption, June 1991.
Photograph by R. Batalon, USAF, courtesy NOAA

The Motivation to Find Answers

In spite of the difficulties inherent in trying to explain geologic phenomena, the search for explanations goes on, spurred not only by the basic quest for knowledge, but also by the practical problems posed by geologic hazards, the need for resources, and concerns about possible global-scale human impacts, such as ozone destruction and global warming.

The hazards may create the most dramatic scenes and headlines, the most abrupt consequences: The 1989 Loma Prieta (California) earthquake caused more than $5 billion in damage; the 1995 Kobe (Japan) earthquake (figure 1.6), similar in magnitude to Loma Prieta, caused over 5200 deaths and about $100 billion in property damage; the 2004 Sumatran earthquake claimed nearly 300,000 lives. The 18 May 1980 eruption of Mount St. Helens took even the scientists monitoring the volcano by surprise, and the 1991 eruption of Mount Pinatubo in the Philippines (figure 1.7) not only devastated local residents but caught the attention of the world through a marked decline in 1992 summer temperatures. Efforts are underway to learn to predict earthquakes and volcanic eruptions reliably enough to save lives, if not property. Likewise, improved understanding of stream dynamics and more prudent land use can together reduce the damages from flooding (figure 1.8), which amount in the United States to over $1 billion a year and the loss of dozens of lives annually. Landslides

FIGURE 1.6 One of many buildings damaged in the 1995 earthquake in Kobe.
Photograph by R. Hutchinson, courtesy NOAA/National Geophysical Data Center

FIGURE 1.8 A major river like the Mississippi floods when a large part of the area that it drains is waterlogged by more rain or snowmelt than can be carried away in the channel. Such floods—like that in summer 1993, shown here drenching Jefferson City, Missouri—can be correspondingly long-lasting. Over millennia, the stream builds a floodplain into which the excess water naturally flows; we build there at our own risk.
Photograph by Mike Wright, courtesy Missouri Department of Transportation

FIGURE 1.9 Slope failure on a California hillside undercuts homes.
Photograph by J. T. McGill, USGS Photo Library, Denver, CO.

and other slope and ground failures (figure 1.9) take a similar toll in property damage, which could be reduced by more attention to slope stability and improved engineering practices. It is not only the more dramatic hazards that are costly: on average, the cost of structural damage from unstable soils each year approximately equals the combined costs of landslides, earthquakes, and flood damages in this country.

Our demand for resources of all kinds continues to grow and so do the consequences of resource use. In the United States, average per-capita water use is 1500 gallons per day; in many places, groundwater supplies upon which we have come to rely heavily are being measurably depleted. Worldwide, water-resource disputes between nations are increasing. As we mine more extensively for mineral resources, we face the problem of how to minimize associated damage to the mined lands (figure 1.10). The grounding of the *Exxon Valdez* in 1989, dumping 11 million gallons of oil into Prince William Sound, Alaska, was a reminder of the negative consequences of petroleum exploration, just as the 1991 war in Kuwait, and the later invasion of Iraq, were reminders of U.S. dependence on imported oil.

As we consume more resources, we create more waste. In the United States, total waste generation is estimated at close to 300 million tons per year—or more than a ton per person. Careless waste disposal, in turn, leads to pollution. The Environmental Protection Agency continues to identify toxic-waste disposal sites in urgent need of cleanup; by 2000, over 1500 so-called priority sites had been identified. Cleanup costs per site have risen to over $30 million, and the projected total costs to remediate these sites alone is over $1 trillion. As fossil fuels are burned, carbon dioxide in the atmosphere rises, and modelers of global climate strive to understand what that may do to global temperatures, weather, and agriculture decades in the future.

These are just a few of the kinds of issues that geologists play a key role in addressing.

FIGURE 1.10 Unreclaimed hillside disturbed by open-pit mining, Elko County, Nevada, is subject to landslides (lower right) and rapid erosion and weathering, contributing to water pollution.
Photograph courtesy USGS Photo Library, Denver, CO.

Wheels Within Wheels: Earth Cycles and Systems

The earth is a dynamic, constantly changing planet—its crust shifting to build mountains; lava spewing out of its warm interior; ice and water and windblown sand and gravity reshaping its surface, over and over. Some changes proceed in one direction only: for example, the earth has been cooling progressively since its formation, though considerable heat remains in its interior. Many of the processes, however, are cyclic in nature.

Consider, for example, such basic materials as water or rocks. Streams drain into oceans and would soon run dry if not replenished; but water evaporates from oceans, to make the rain and snow that feed the streams to keep them flowing. This describes just a part of the *hydrologic* (water) *cycle,* explored more fully in chapters 6 and 11. Rocks, despite their appearance of permanence in the short term of a human life, participate in the *rock cycle* (chapters 2 and 3). The kinds of evolutionary paths rocks may follow through this cycle are many, but consider this illustration: A volcano's lava (figure 1.11) hardens into rock; the rock is weathered into sand and dissolved chemicals; the debris, deposited in an ocean basin, is solidified into a new rock of quite different type; and some of that new rock may be carried into the mantle via plate tectonics, to be melted into a new lava. The time frame over which this process occurs is generally much longer than that over which water cycles through atmosphere and oceans, but the principle is similar. The Appalachian or Rocky Mountains as we see them today are not as they formed, tens or hundreds of millions of years ago; they are much eroded from their original height by water and ice, and, in turn, contain rocks formed in water-filled basins

FIGURE 1.11 Bit by bit, lava flows like this one on Kilauea have built the Hawaiian Islands.

and deserts from material eroded from more-ancient mountains before them (figure 1.12).

Chemicals, too, cycle through the environment. The carbon dioxide that we exhale into the atmosphere is taken up by plants through photosynthesis, and when we eat those plants for food energy, we release CO_2 again. The same exhaled CO_2 may also dissolve in rainwater to make carbonic acid that dissolves continental rock; the weathering products may wash into the ocean, where dissolved carbonate then contributes to the formation of carbonate shells and carbonate rocks in the ocean basins; those rocks may later be exposed and weathered by rain, releasing CO_2 back into the atmosphere or dissolved carbonate into streams that carry it back to the ocean. The cycling of chemicals and materials in the environment may be complex, as we will see in later chapters.

Furthermore, these processes and cycles are often interrelated, and seemingly local actions can have distant consequences. We dam a river to create a reservoir as a source of irrigation water and hydroelectric power, inadvertently trapping stream-borne sediment at the same time; downstream, patterns of erosion and deposition in the stream channel change, and at the coast, where the stream pours into the ocean, coastal erosion of beaches increases because a part of their sediment supply, from the stream, has been cut off. The volcano that erupts the lava to make the volcanic rock also releases water vapor into the atmosphere, and sulfur-rich gases and dust that influence the amount of sunlight reaching earth's surface to heat it, which, in turn, can alter the extent of evaporation and resultant rainfall, which will affect the intensity of landscape erosion and weathering of rocks by water. . . . So although we divide the great

FIGURE 1.12 Rocks tell a story of constant change. The sandstones of Zion National Park preserve ancient windswept dunes, made of sand eroded from older rocks, deeply buried and solidified into new rock, then uplifted to erode again.

variety and complexity of geologic processes and phenomena into more manageable, chapter-sized units for purposes of discussion, it is important to keep such interrelationships in mind. And superimposed on, influenced by, and subject to all these natural processes are humans and human activities.

NATURE AND RATE OF POPULATION GROWTH

Animal populations, as well as primitive human populations, are generally quite limited both in the areas that they can occupy and in the extent to which they can grow. They must live near food and water sources. The climate must be one to which they can adapt. Predators, accidents, and disease take a toll. If the population grows too large, disease and competition for food are particularly likely to cut it back to sustainable levels.

The human population grew relatively slowly for hundreds of thousands of years. Not until the middle of the nineteenth century did the world population reach 1 billion. However, by then, a number of factors had combined to accelerate the rate of population increase. The second, third, and fourth billion were reached far more quickly; the world population is now over 6 billion and is expected to rise to over 9 billion by 2050 (figure 1.13).

Humans are no longer constrained to live only where conditions are ideal. We can build habitable quarters even in extreme climates, using heaters or air conditioners to bring the temperature to a level we can tolerate. In addition, people need not live where food can be grown or harvested or where there is abundant fresh water: The food and water can be transported, instead, to where the people choose to live.

Growth Rates: Causes and Consequences

Population growth occurs when new individuals are added to the population faster than existing individuals are removed from it. On a global scale, the population increases when its birthrate exceeds its death rate. In assessing an individual nation's or region's rate of population change, immigration and emigration must also be taken into account. Improvements in nutrition and health care typically increase life expectancies, decrease mortality rates, and thus increase the rate of population growth. (Worldwide, life expectancy is about 67 years and is expected to increase to nearly 75 years by the year 2050.) Increased use of birth-control or family-planning methods reduces birthrates and, therefore, also reduces the rate of population growth; in fact, a population could begin to decrease if birthrates were severely restricted.

The sharply rising rate of population growth over the past few centuries, as shown in figure 1.13, can be viewed another way. It took until about A.D. 1830 for the world's population to reach 1 billion. The population climbed to 2 billion in the next 100 years, to 3 billion in 30 more years, and so on. It has taken less and less time to add each successive billion people, as ever more people contribute to the population growth and individuals live longer. The last billion people were added to the world's population within less than a decade.

There are wide differences in growth rates among regions (table 1.4; figure 1.13). The reasons for this are many. Religious or social values may cause larger or smaller families to be regarded as desirable in particular regions or cultures. High levels of economic development are commonly

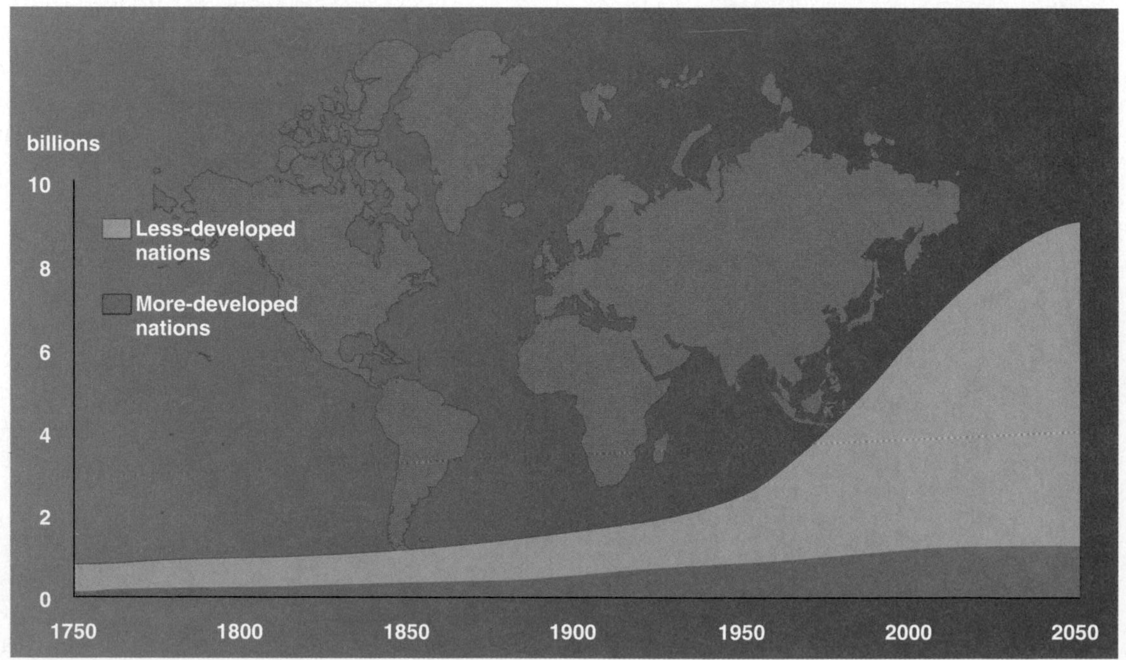

FIGURE 1.13 World population, currently over 6 billion, is projected to reach over 9 billion by 2050. Most of that population increase will occur in less-developed countries.

Sources: Data from United Nations Population Division and Population Reference Bureau, 2002

TABLE 1.4

World and Regional Population Growth and Projections (in millions)

Year	World	North America	Latin America and Caribbean	Africa	Europe	Asia	Oceania
1950	2520	172	167	221	547	1402	13
2005	6477	329	559	906	730	3921	33
2050 (projected)	9262	457	805	1969	660	5325	46
Growth rate (%/year)	1.2	0.6	1.6	2.3	−0.1	1.3	1.0
Doubling time (years)	58	117	41	30	*	54	70

*Not applicable; population declining.

Source: "United Nations World Population Estimates and Projections" (http://www.popin.org/pop1998/) and "2005 World Population Data Sheet," Population Reference Bureau, 2005.

associated with reduced rates of population growth; conversely, low levels of development are often associated with rapid population growth. The impact of improved education, which may accompany economic development, can vary: It may lead to better nutrition, prenatal and child care, and thus to increased growth rates, but it may also lead to increased or more effective practice of family-planning methods, thereby reducing growth rates. A few governments of nations with large and rapidly growing populations have considered encouraging or mandating family planning; India and the People's Republic of China have taken active measures.

A relatively new factor that strongly affects population growth in some less-developed nations is AIDS. In more developed countries, less than 1% of the population aged 15 to 49 may be afflicted; the prevalence in some African nations is over 30%. In Botswana, where AIDS prevalence is close to 40%, life expectancy is down to 35 years. How this population will change over time depends greatly on how effectively the AIDS epidemic is controlled.

Even when the population growth rate is constant, the number of individuals added per unit of time increases over time. This is called **exponential growth,** a concept to which we will return when discussing resources in section four. The effect of exponential growth is similar to interest compounding. If one invests $100 at 10% per year, compounded annually, then, after one year, $10 interest is credited, for a new balance of $110. At the end of the second year, the interest is $11 (10% of $110), and the new balance is $121. By the end of the tenth year (assuming no withdrawals), the interest for the year is $23.58, but the interest *rate* has not increased. Similarly with a population of 1 million growing at 5% per year: In the first year, 50,000 persons are added; in the tenth year, the population grows by 77,566 persons. The result is that a graph of population versus time steepens over time, even at a constant growth rate. If the growth rate itself also increases, the curve rises still more sharply.

For many mineral and fuel resources, consumption has been growing very rapidly, even more rapidly than the population. The effects of exponential increases in resource demand are like the effects of exponential population growth (figure 1.14). If demand increases by 2% per year, it will double not in 50 years, but in 35. A demand increase of 5% per year leads to a doubling in demand in 14 years, and a tenfold increase in demand in 47 years! In other words, a prediction of how soon mineral or fuel supplies will be used up is very sensitive to the assumed rate of change of demand. Even if population is no longer growing exponentially, consumption of many resources is.

Growth Rate and Doubling Time

Another way to look at the rapidity of world population growth is to consider the expected **doubling time,** the length of time required for a population to double in size. Doubling time (D) in years may be estimated from growth rate (G), expressed in percent per year, using the simple relationship $D = 70/G$, which is derived from the equation for exponential growth (see "For Further Thought" question 3 at the chapter's end). The higher the growth rate, the shorter the doubling time of the population (see again table 1.4). By region, the most rapidly growing segment of the population today is that of Africa. Its population, estimated at 906 million in 2005, was growing at about 2.3% per year. The largest segment of the population, that of Asia, is increasing at 1.3% per year, and since the nearly 4 billion people there represent well over half of the world's total population, this leads to a relatively high global average growth rate. Europe's population has begun to decline slightly, but Europe contains only about 10% of the world's population. Thus, the fastest growth in general is taking place in the largest segments of the population.

The average worldwide population growth rate is about 1.2% per year. This may sound moderate, but it corresponds to a relatively short doubling time of about 58 years. At that, the present population growth rate actually represents a decline from nearly 2% per year in the mid-1960s. The United Nations projects that growth rates will continue to decrease gradually over the next several decades, probably as a result of scarcer

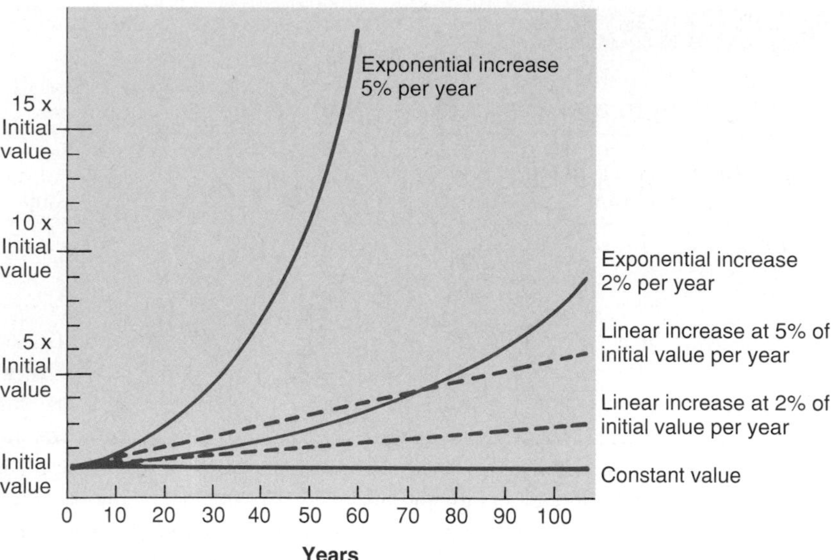

FIGURE 1.14 Graphical comparison of the effects of linear and exponential growth, whether on consumption of minerals, fuels, water, and other consumable commodities, or population.

resources coupled with increased population-control efforts (and, perhaps, AIDS). However, a decreasing *growth rate* is not at all the same as a decreasing population. Depending upon projected fertility rates, estimates of world population in the year 2050 can vary by several billion. Using a medium fertility rate, the Population Reference Bureau projects a 2050 world population of over 9.2 billion. Figure 1.15 illustrates how those people will be distributed by region, considering differential growth rates from place to place.

Even breaking the world down into regions of continental scale masks a number of dramatic individual-country cases. Discussion of these, and of their political, economic, and cultural implications, is well beyond the scope of this chapter, but consider the following: While the population of Europe is nearly stable, declining only slightly overall, in many parts of northern and eastern Europe, precipitous declines are occurring. By 2050, the populations of Russia, Ukraine, and Bulgaria are expected to drop by 23, 29, and 34 percent, respectively.

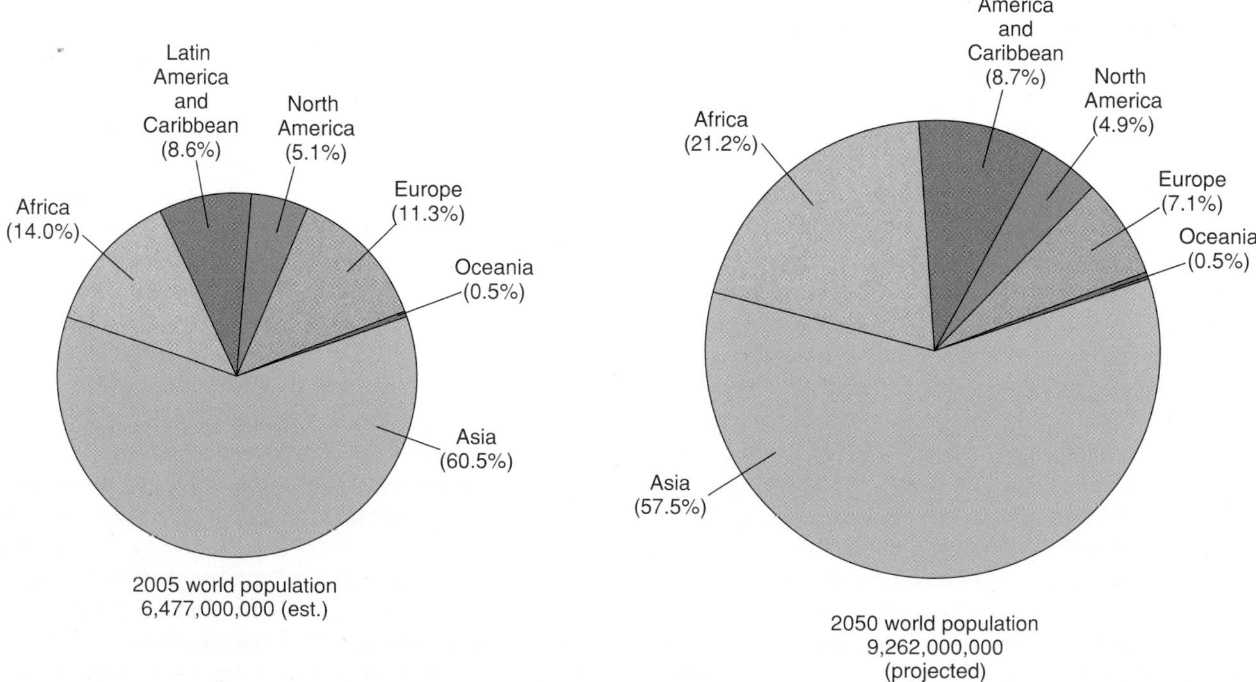

FIGURE 1.15 Population distribution by region in 2005 with projections to the year 2050. Size of circle reflects relative total population. The most dramatic changes in proportion are the relative growth of the population of Africa and decline of population in Europe. Data from table 1.4.

Conversely, parts of the Middle East are experiencing explosive population growth, with projected increases by 2025 of 31% in Lebanon, 91% in Syria, 121% in Iraq, and 197% in the Palestinian Territory. The demographics differ widely between countries, too—for example, in Japan, only 14% of the population is below age 16, with 20% age 65 or older; in Tanzania, 45% of people are under 16 and only 3% age 65 or over. Thus, different nations face different challenges. Where rapid population growth meets scarcity of resources, problems arise.

IMPACTS OF THE HUMAN POPULATION

The problems posed by a rapidly growing world population have historically been discussed most often in the context of food: that is, how to produce sufficient food and distribute it effectively to prevent the starvation of millions in overcrowded countries or in countries with minimal agricultural development. This is a particularly visible and immediate problem, but it is also only one facet of the population challenge.

Farmland and Food Supply

Whether or not the earth can support five billion people, or eight, or eleven, is uncertain. In part, it depends on the quality of life, the level of technological development, and other standards that societies wish to maintain. Yet even when considering the most basic factors, such as food, it is unclear just how many people the earth can sustain. Projections about the adequacy of food production, for example, require far more information than just the number of people to be fed and the amount of available land. The total arable land (land suitable for cultivation) in the world has been estimated at 7.9 billion acres, or about 1.2 acres per person of the present population. The major limitation on this figure is availability of water, either as rainfall or through irrigation. Further considerations relating to the nature of the soil include the soil's fertility, water-holding capacity, and general suitability for farming. Soil character varies tremendously, and its productivity is similarly variable, as any farmer can attest. Moreover, farmland can deteriorate through loss of nutrients and by wholesale erosion of topsoil, and this degradation must be considered when making production predictions.

There is also the question of what crops can or should be grown. Today, this is often a matter of preference or personal taste, particularly in farmland-rich (and energy- and water-rich) nations. The world's people are not now always being fed in the most resource-efficient ways. To produce one ton of corn requires about 250,000 gallons of water; a ton of wheat, 375,000 gallons; a ton of rice, 1,000,000 gallons; a ton of beef, 7,500,000 gallons. Some new high-yield crop varieties may require irrigation, whereas native varieties did not. The total irrigated acreage in the world has more than doubled in three decades. However, water resources are dwindling in many places; in the future, the water costs of food production will have to be taken into greater account.

Genetic engineering is now making important contributions to food production, as varieties are selectively developed for high yield, disease resistance, and other desirable qualities. These advances have led some to declare that fears of global food shortages are no longer warranted, even if the population grows by several billion more. However, two concerns remain: One, poor nations already struggling to feed their people may be least able to afford the higher-priced designer seed or specially developed animal strains to benefit from these advances. Second, as many small farms using many, genetically diverse strains of food crops are replaced by vast areas planted with a single, new, superior variety, there is the potential for devastating losses if a new pest or disease to which that one strain is vulnerable enters the picture.

Food production as practiced in the United States is also a very energy-intensive business. The farming is heavily mechanized, and much of the resulting food is extensively processed, stored, and prepared in various ways requiring considerable energy. The products are elaborately packaged and often transported long distances to the consumer. Exporting the same production methods to poor, heavily populated countries short on energy and the capital to buy fuel, as well as food, may be neither possible nor practical. Even if possible, it would substantially increase the world's energy demands.

Population and Nonfood Resources

Food is at least a renewable resource. Within a human life span, many crops can be planted and harvested from the same land and many generations of food animals raised. By contrast, the supplies of many of the resources considered in later chapters—minerals, fuels, even land itself—are finite. There is only so much oil to burn, rich ore to exploit, and suitable land on which to live and grow food. When these resources are exhausted, alternatives will have to be found or people will have to do without.

The earth's supply of many such materials is severely limited, especially considering the rates at which these resources are presently being used. Many could be effectively exhausted within decades, yet most people in the world are consuming very little in the way of minerals or energy. Current consumption is strongly concentrated in a few highly industrialized societies. Per-capita consumption of most mineral and energy resources is higher in the United States than in any other nation. For the world population to maintain a standard of living comparable to that of the United States, mineral production would have to increase about fourfold, on the average. There are neither the recognized resources nor the production capability to maintain that level of consumption for long, and the problem becomes more acute as the population grows.

Some scholars believe that we are already on the verge of exceeding the earth's **carrying capacity,** its ability to sustain its population at a basic, healthy, moderately comfortable standard of living. Whether or not this is the case depends, in

Box 1.1 A Case of Population Density: Reality in Round Numbers

Peter dashed into the office of Shawkat, the graduate assistant in his biology class, dropped into a chair, and pulled out the last problem set to ask about. "Man, that traffic was *awful*—and there's no place to park on-campus, either, once you get here! There are just *way* too many people in this town. And it seems that way everywhere these days."

"It could be worse, you know," replied Shawkat. "Sure, you have some crowded cities in the United States, but there's also lots of land to spread out onto if you want more space. In my country, Bangladesh, it's a lot more crowded all over. These days, we have about 2600 people per square mile of the whole country."

"But that doesn't sound so terrible," objected Peter, reaching for his calculator. "Let's see, a square mile is 5280 times 5280 . . . divide by 2600 . . . and you've got 10,700 square feet, give or take, per person. A hundred feet by a hundred and ten, roughly—sounds pretty roomy, really."

"Ah, but you're counting every scrap of land as if you could live on it, and not need it for anything else, either. About 3/4 of our land is used just for crops to feed the people. And it takes land for roads and schools and offices and factories, not to mention rivers and mountains and forests . . ." Shawkat broke off, noticing that Peter still looked unconvinced. "Okay, let me put it another way," he suggested. "It's as if you took the population of the United States and stuffed everybody into the state of Arizona, and they not only had to live there, but work and grow all their food and do everything else in just that little space."

"I guess I see your point," Peter conceded. "But it's just a temporary problem. We've landed men on the moon, there's a space station now—I'll bet that in 50 years we'll be living on the Moon or even Mars! Then all we have to do is ship the extra people there. Plenty of room!"

"Plenty of *room,* I'll grant you. But even if we assume those colonies could somehow be self-supporting, what does it take to get the people there? It's been estimated that we'll have 9 billion people on earth by the year 2050. Let's suppose that somehow we get population growth rates worldwide as low as 0.1% per year by then, and want to hold earth's population steady. That's 9 *million* people a year to ship into space. Nearly 50 years after the first space flights, and we still put only a few astronauts on each ship. Even if we packed a hundred people onto every flight, that's 90,000 launches a year, or about 250 a day. Hard to imagine that kind of launch capacity worldwide—to say nothing of the cost! Besides, the countries with the fastest-growing populations, who face the worst problems, are hardly the ones with the resources to launch rockets on that scale."

"So," concluded Shawkat with a grin, "looks like we're stuck solving the problems here. Shall we get back to that homework you've been avoiding discussing? You may be just the person who's going to figure out how to feed all those billions!"

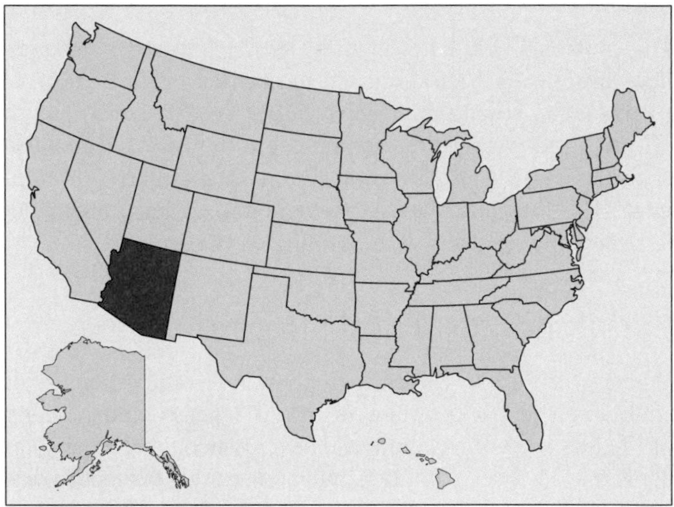

FIGURE 1 Imagine trying to compress the United States into an area just the size of Arizona, allowing room not just for the people but for all of their activities—farming, manufacturing, recreation, everything—and you have a situation like that of the most densely populated nations.

part, on what one considers an acceptable standard of living. Estimates of sustainable world population made over the last few decades range from under 7 billion to over 100 billion persons. The wide range is attributable to considerable variations in model assumptions. (For instance, the upper ranges include such assumptions as very high productivity of farmland, comparable to present record levels, with no land reserved for recreation or urbanization.) If the lower estimates of carrying capacity are accurate, this means that, even if we could somehow halt population growth within the next half-century, no possible redistribution of resources could guarantee everyone that basic standard of living. Certainly, given global resource availability, even the present world population could not enjoy the kind of high-consumption lifestyle to which the average resident of the United States has become accustomed.

It is true that, up to a point, the increased demand for minerals, fuels, and other materials associated with an increase in population tends to raise prices and promote exploration for these materials. The short-term result can be an apparent increase in the resources' availability, as more exploration leads to discoveries of more oil fields, ore bodies, and so on. However, the quantity of each of these resources is finite. The larger and more rapidly growing the world population, and the faster its level of development and standard of living rise, the more rapidly limited resources are consumed, and the sooner those resources will be exhausted.

Land is clearly a basic resource. Six, nine, or 100 billion people must be *put* somewhere. Already, the global average population density is about 47 persons per square kilometer of land surface (120 persons per square mile), and that is counting *all* the land surface, including jungles, deserts, and mountain ranges, except for the Antarctic continent. The ratio of people to readily inhabitable land is plainly much higher. Land is also needed for manufacturing, energy production, transportation, and a variety of other uses. Large numbers of people consuming vast quantities of materials generate vast quantities of wastes. Some of these wastes can be recovered and recycled, but others cannot. It is essential to find places to put the latter, and ways to isolate the harmful materials from contact with the growing population. This effort claims still more land and, often, resources. All of this is why land-use planning—making the most of every bit of land available—is becoming

FIGURE 1.16 The landslide hazard to these structures in Rio de Janeiro is obvious, but space for building here is limited.
Photograph © Will and Deni McIntyre/Getty Images

FIGURE 1.17 Even where safer land is abundant, people may choose to settle in hazardous—but scenic—places, such as barrier islands. South Dade County, Florida.
Photograph © Alan Schein Photography/Corbis

increasingly important. At present, it is too often true that the ever-growing population settles in areas that are demonstrably unsafe or in which the possible problems are imperfectly known (figures 1.16 and 1.17).

Uneven Distribution of People and Resources

Even if global carrying capacity were ample in principle, that of an individual region may not be. None of the resources—livable land, arable land, energy, minerals, or water—is uniformly distributed over the earth. Neither is the population (see figure 1.18). In 2005, the population density in Singapore was about 17,950 persons per square mile; in Australia, 7.

Many of the most densely populated countries are resource-poor. In some cases, a few countries control the major share of one resource. Oil is a well-known example, but there are many others. Thus, economic and political complications enter into the question of resource adequacy. Just because one nation controls enough of some commodity to supply all the world's needs does not necessarily mean that the country will choose to share its resource wealth or to distribute it at modest cost to other nations. Some resources, like land, are simply not transportable and therefore cannot readily be shared. Some of the complexities of global resource distribution will be highlighted in subsequent chapters.

Disruption of Natural Systems

Natural systems tend toward a balance or equilibrium among opposing factors or forces. When one factor changes, compensating changes occur in response. If the disruption of the system is relatively small and temporary, the system may, in time, return to its original condition, and evidence of the disturbance will disappear. For example, a coastal storm may wash away beach vegetation and destroy colonies of marine organisms living in a tidal flat, but when the storm has passed, new organisms will start to move back into the area, and new grasses will take root in the dunes. The violent eruption of a volcano like Mount Pinatubo may spew ash and gases high into the atmosphere, partially blocking sunlight and causing the earth to cool, but within a few years, the ash will have settled back to the ground, and normal temperatures will be restored. Dead leaves falling into a lake provide food for the microorganisms that within weeks or months will break the leaves down and eliminate them.

This is not to say that permanent changes never occur in natural systems. The size of a river channel reflects the maximum amount of water it normally carries. If long-term climatic or other conditions change so that the volume of water regularly reaching the stream increases, the larger quantity of water will, in time, carve out a correspondingly larger channel to accommodate it. The soil carried downhill by a landslide certainly does not begin moving back upslope after the landslide is over; the face of the land is irreversibly changed. Even so, a hillside forest uprooted and destroyed by the slide may, within decades, be replaced by new growth in the soil newly deposited at the bottom of the hill.

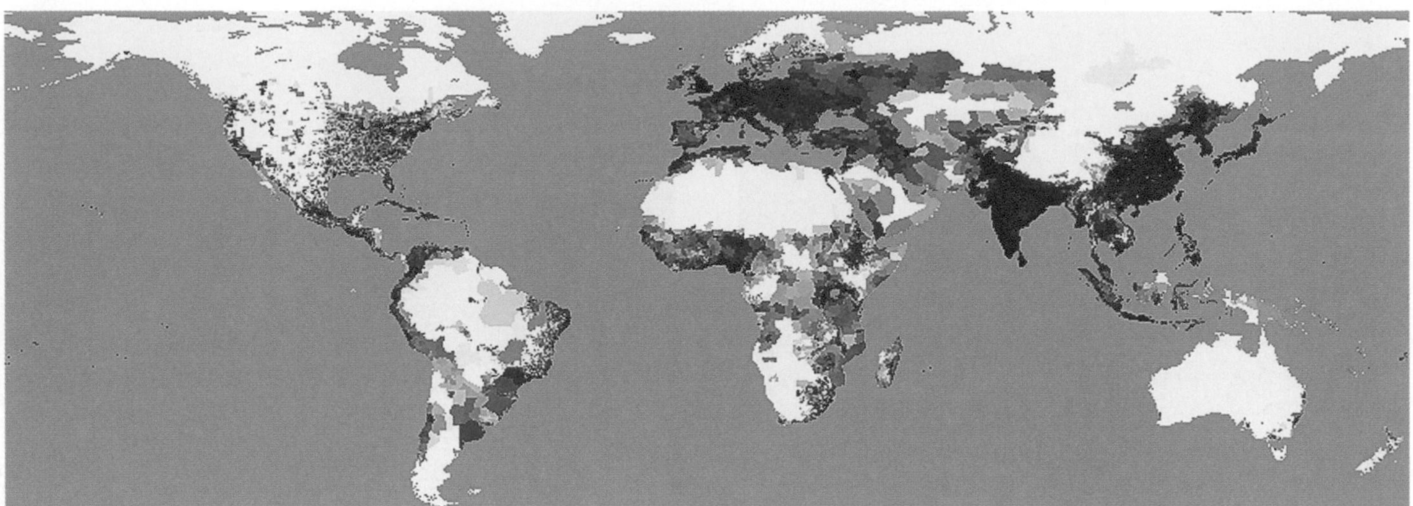

FIGURE 1.18 Global population density; the darker the shading, the higher the population density. Compare with the chapter-opener image, and note that overall population density on the one hand, and urbanization/development on the other, are not necessarily closely correlated.

From Waldo Tobler, Uwe Deichmann, Jon Gottsegen, and Kelly Maloy in Technical Report TR–95–6. Copyright © 1995 National Center for Geographic Information and Analysis, Santa Barbara, CA. Reprinted by permission

Human activities can cause or accelerate permanent changes in natural systems. The impact of humans on the global environment is broadly proportional to the size of the population, as well as to the level of technological development achieved. This can be illustrated especially easily in the context of pollution. The smoke from one campfire pollutes only the air in the immediate vicinity; by the time that smoke is dispersed through the atmosphere, its global impact is negligible. The collective smoke from a century and a half of increasingly industrialized society, on the other hand, has caused measurable increases in several atmospheric pollutants worldwide, and these pollutants continue to pour into the air from many sources. It was once assumed that the seemingly vast oceans would be an inexhaustible "sink" for any extra CO_2 that we might generate by burning fossil fuels, but decades of steadily climbing atmospheric CO_2 levels have proven that in this sense, at least, the oceans are not as large as we thought. Likewise, six people carelessly dumping wastes into the ocean would not appreciably pollute that huge volume of water. The prospect of 6 *billion* people doing the same thing, however, is quite another matter.

Is Extraterrestrial Colonization a Solution?

In recent decades, space travel has become feasible, if not yet commonplace. Martian landers have explored more of that planet's surface, and there is talk of sending astronauts to the moon again. This has inspired suggestions that space travel could alleviate many of the problems related to earth's evergrowing population. Colonization of other planets would indeed provide more land for people to live on, in the abstract; we could mine other planets' ores to supply our needs; giant solar-energy collectors in space could solve our energy needs; wastes, such as radioactive wastes, could be propelled into space away from the earth; and so on. Unfortunately, the weight of evidence suggests that this simply cannot be done on the scale needed to solve major environmental problems, given existing population pressures, and the time factor.

A major objection to colonization is that none of the other planets or other bodies in the solar system is really suitable for human habitation. It might be possible on a few such bodies to build small, controlled environments within which a few people could live, perhaps starting with a permanent outpost on earth's moon. Examination of the dust-dry, barren, airless lunar surface (figure 1.19) makes clear how constrained

FIGURE 1.19 The lunar surface is not an environment in which humans could live outside protective structures, with high energy and resource costs. Geologist/astronaut Harrison Schmitt, lunar module commander of the Apollo 17 mission.

Photograph courtesy NASA

A

B

FIGURE 1.20 (A) Mars shares many surface features with the earth, including canyons, dunes, slumps, and stream channels. However, there is currently no surface water, life, or breathable atmosphere on the planet. (B) Photograph of Mars Lander 2 site.
Photographs courtesy NASA

life on the moon would be. On most planets or moons, even limited habitation would be impossible. Any such outposts would almost certainly have to be constructed using resources from earth. As noted earlier in this chapter, the other planets all have different compositions from the earth's, most of them very different. The types of processes that have shaped them since they formed have also differed. It is quite unlikely that suitable ores or fuels could be found on them in any quantity. Mars is the least hostile alternative environmentally, and its surface looks not too forbidding (figure 1.20). But it is cold; there is no liquid water, and only a little surface water ice near the south pole; it has not so much as a blade of grass by way of vegetation (a flurry of excitement over "life on Mars" died down when features interpreted as trace fossils of microorganisms could not be shown conclusively to be so); and we cannot breathe its atmosphere. Even if the living conditions elsewhere in the solar system were more hospitable, the cost to transport any significant number of people to another planet, and the quantity of earth's materials needed to support them, would be prohibitive. It looks, in short, as though we will have to solve our environmental problems right here. And time is of the essence: Every hour, now, the world population increases by nearly *10,000 people!*

SUMMARY

The solar system formed over 4½ billion years ago. The earth is unique among the planets in its chemical composition, abundant surface water, and oxygen-rich atmosphere. The earth passed through a major period of internal differentiation early in its history, which led to the formation of the atmosphere and the oceans. Earth's surface features have continued to change throughout the last 4+ billion years, through a series of processes that are often cyclical in nature, and commonly interrelated. The oldest rocks in which remains of simple organisms are recognized are more than 3 billion years old. The earliest plants were responsible for the development of free oxygen in the atmosphere, which, in turn, made it possible for oxygen-breathing animals to survive. Human-type remains are unknown in rocks over 3 to 4 million years of age. In a geologic sense, therefore, human beings are quite a new addition to the earth's cast of characters, but they have had a very large impact. Geology, in turn, can have an equally large impact on us.

The world population, now about 6½ billion, might be over 9 billion by the year 2050. Even our present population cannot entirely be supported at the level customary in the more developed countries, given the limitations of land and resources. Extraterrestrial resources cannot realistically be expected to contribute substantially to a solution of this problem.

Key Terms and Concepts

carrying capacity 15
core 5
crust 5
doubling time 13
environmental geology 2
exponential growth 13
hypothesis 8
mantle 5
scientific method 8
theory 8

Exercises

Questions for Review

1. Describe the process by which the solar system is believed to have formed, and explain why it led to planets of different compositions, even though the planets formed simultaneously.
2. How old is the solar system? How recently have human beings come to influence the physical environment?
3. Explain how the newly formed earth differed from the earth we know today.
4. What kinds of information are used to determine the internal composition of the earth?
5. How were the earth's atmosphere and oceans formed?
6. What are the differences among facts, scientific hypotheses, and scientific theories?
7. Many earth materials are transformed through processes that are cyclical in nature. Describe one example.
8. The size of the earth's human population directly affects the severity of many environmental problems. Explain this idea in the context of (a) resources and (b) pollution.
9. If earth's population has already exceeded the planet's carrying capacity, what are the implications for achieving a comfortable standard of living worldwide? Explain.
10. What is the world's present population, to the nearest billion? How do recent population growth rates (over the last few centuries) compare with earlier times? Why?
11. Explain the concept of doubling time. How has population doubling time been changing through history? What is the approximate doubling time of the world's population at present?
12. What regions of the world currently have the fastest rates of population growth? The slowest?
13. Briefly evaluate the feasibility of space colonization as a means of alleviating land and natural-resource shortages.

For Further Thought

1. Investigate the geologic history of some particular part of the country. (A good starting point might be the state geological survey(s) in the region of interest.) Was the area ever an ocean basin, a desert, subject to volcanic activity? How long ago was it first inhabited by humans?
2. The urgency of population problems can be emphasized by calculating such "population absurdities" as the time at which there will be one person per square meter or per square foot of land and the time at which the weight of people will exceed the weight of the earth. Try calculating these "population absurdities" by using the world population projections from table 1.4 and the following data concerning the earth:

   Mass 5976×10^{21} kg*

   Land surface area (approx.) 149,000,000 sq. km

   Average weight of human body (approx.) 75 kg

   Or follow up on Peter and Shawkat's discussion by estimating the costs of the colonist-export process they began to consider—in money or resources—by researching typical costs of current space missions.
3. Derive the relation between doubling time and growth rate, starting from the exponential-growth relation

   $$N = N_0 \, e^{(G/100) \cdot t}$$

   where N = the growing quantity (number of people, tons of iron ore consumed annually, dollars in a bank account, or whatever); N_0 is the initial quantity at the start of the time period of interest; G is growth rate expressed in percent per year, and "percent" means "parts per hundred"; and t is the time in years over which growth occurs. Keep in mind that doubling time, D, is the length of time required for N to double.
4. Select a single country or small set of related countries; examine recent and projected population growth rates in detail, including the factors contributing to the growth rates and trends in those rates. Compare with similar information for the United States or Canada.

*This number is so large that it has been expressed in scientific notation, in terms of powers of 10. It is equal to 5976 with twenty-one zeroes after it. For comparison, the land surface area could also have been written as 149×10^6 sq. km, or $149{,}000 \times 10^3$ sq. km, and so on.

Suggested Readings/References

Abramovitz, J. N. 2001. Averting unnatural disasters. Chap. 7 in *State of the world 2001.* New York: W.W. Norton & Co.

Badash, L. 1989. The age-of-the-earth debate. *Scientific American* 261 (August): 90–96.

Board on Natural Disasters. 1999. Mitigation emerges as a major strategy for reducing losses caused by natural disasters. *Science* 284 (18 June): 1943–47.

Creath, W. B. 1996. *Home buyers' guide to geologic hazards.* Arvada, CO: Amer. Inst. of Prof. Geologists.

Ewert, J. W., and Harpel, C. J. 2004. In harm's way: Population and volcanic risk. *Geotimes* (April): 14–23.

Greeley, R. 1985. *Planetary landscapes.* London: Allen & Unwin.

Lutz, W., O'Neill, B. K., and Scherbov, S. 2003. Earth's population at a turning point. *Science* 299 (28 March): 1991–92.

Parfit, M. 1998. Living with natural hazards. *National Geographic* (July): 2–35.

Population Reference Bureau. 2005. *2005 World Population Data Sheet.* Washington, D.C.: PRB. Also available online http://www.prb.org/pdf05/05WorldDataSheet_Eng.pdf

Russell, M. 2006. First life. *American Scientist* 94 (January–February): 32–39.

Simons, E. L. 1989. Human origins. *Science* 245 (September): 1343–50.

Simpson, S. 2003. Questioning the oldest signs of life. *Scientific American* (April): 70–77.

United Nations Department of Economic and Social Affairs. 2003. *World population prospects: The 2002 revision.* New York: United Nations.

Vondrak, R. R., and Crider, D. H. 2003. Ice at the lunar poles. *American Scientist* 91 (July–August): 322–29.

World Resources Institute. 2000. *World Resources 2000–2001.* New York: Oxford University Press. Also accessible online.

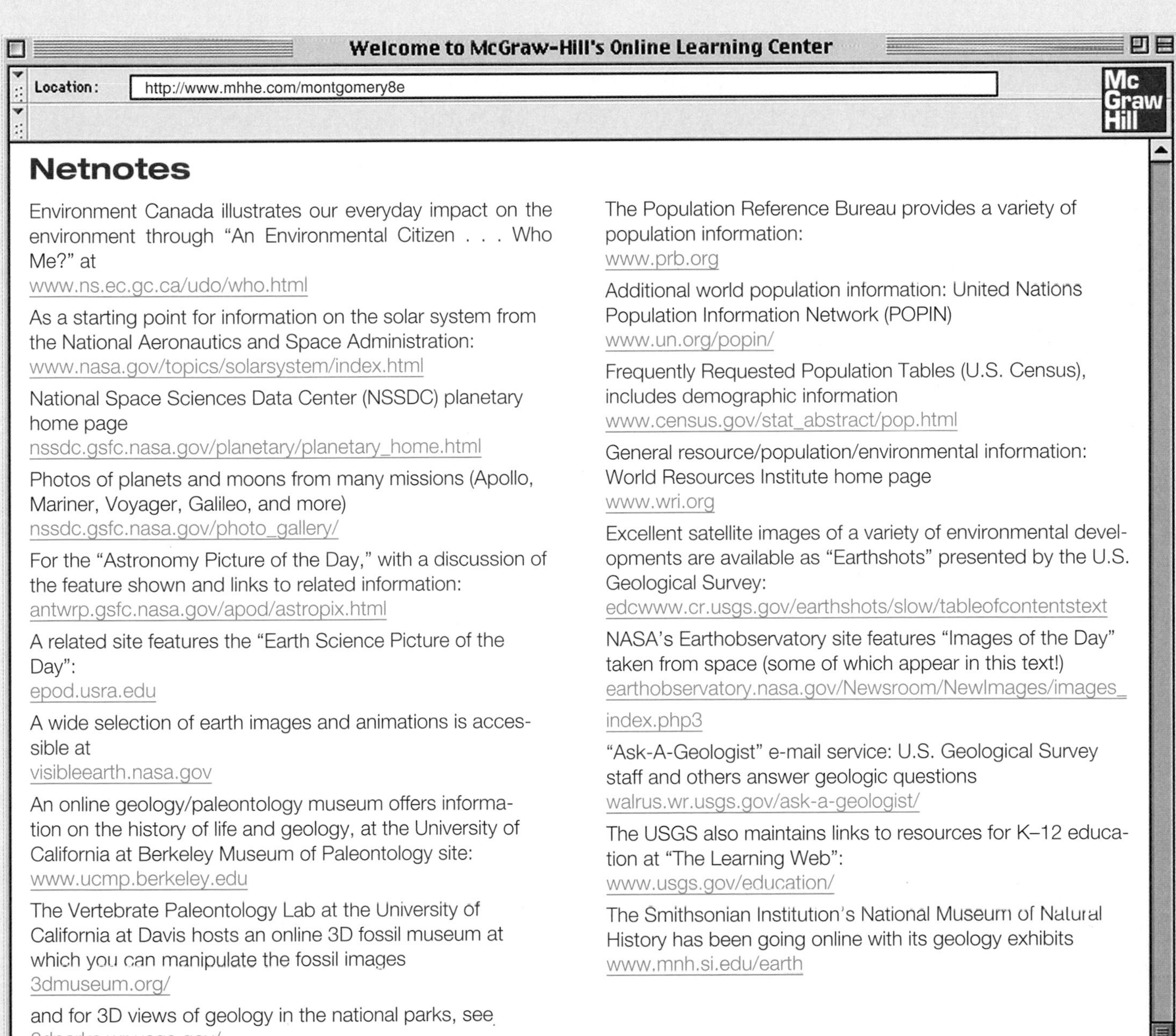

Netnotes

Environment Canada illustrates our everyday impact on the environment through "An Environmental Citizen . . . Who Me?" at
www.ns.ec.gc.ca/udo/who.html

As a starting point for information on the solar system from the National Aeronautics and Space Administration:
www.nasa.gov/topics/solarsystem/index.html

National Space Sciences Data Center (NSSDC) planetary home page
nssdc.gsfc.nasa.gov/planetary/planetary_home.html

Photos of planets and moons from many missions (Apollo, Mariner, Voyager, Galileo, and more)
nssdc.gsfc.nasa.gov/photo_gallery/

For the "Astronomy Picture of the Day," with a discussion of the feature shown and links to related information:
antwrp.gsfc.nasa.gov/apod/astropix.html

A related site features the "Earth Science Picture of the Day":
epod.usra.edu

A wide selection of earth images and animations is accessible at
visibleearth.nasa.gov

An online geology/paleontology museum offers information on the history of life and geology, at the University of California at Berkeley Museum of Paleontology site:
www.ucmp.berkeley.edu

The Vertebrate Paleontology Lab at the University of California at Davis hosts an online 3D fossil museum at which you can manipulate the fossil images
3dmuseum.org/

and for 3D views of geology in the national parks, see
3dparks.wr.usgs.gov/

The Population Reference Bureau provides a variety of population information:
www.prb.org

Additional world population information: United Nations Population Information Network (POPIN)
www.un.org/popin/

Frequently Requested Population Tables (U.S. Census), includes demographic information
www.census.gov/stat_abstract/pop.html

General resource/population/environmental information: World Resources Institute home page
www.wri.org

Excellent satellite images of a variety of environmental developments are available as "Earthshots" presented by the U.S. Geological Survey:
edcwww.cr.usgs.gov/earthshots/slow/tableofcontentstext

NASA's Earthobservatory site features "Images of the Day" taken from space (some of which appear in this text!)
earthobservatory.nasa.gov/Newsroom/NewImages/images_index.php3

"Ask-A-Geologist" e-mail service: U.S. Geological Survey staff and others answer geologic questions
walrus.wr.usgs.gov/ask-a-geologist/

The USGS also maintains links to resources for K–12 education at "The Learning Web":
www.usgs.gov/education/

The Smithsonian Institution's National Museum of Natural History has been going online with its geology exhibits
www.mnh.si.edu/earth

2 Rocks and Minerals—A First Look

Considering the limited number of chemical elements in nature, the variety of substances found on earth and the diversity of their physical properties are astonishing. The same is true even of just the rocks and minerals. Most of these are made up of an even smaller subset of elements, yet they are very diverse in color, texture, and other properties. The differences in the physical properties of rocks, minerals, and soils determine their suitability for different purposes—extraction of water or of metals, construction, manufacturing, waste disposal, agriculture, and other uses. For this reason, it is helpful to understand something of the nature of these materials. Also, each rock contains clues to the kinds of processes that formed it and to the geologic setting where it is likely to be found. For example, we will see that the nature of a volcano's rocks may indicate what hazards it presents to us; our search for new ores or fuels is often guided by an understanding of the specialized geologic environments in which they occur.

Even the simple cubic crystals of halite (common table salt, sodium chloride) can be beautiful.
Photograph © The McGraw-Hill Companies Inc./Doug Sherman, photographer

ATOMS, ELEMENTS, ISOTOPES, IONS, AND COMPOUNDS

Atomic Structure

All natural and most synthetic substances on earth are made from the ninety naturally occurring chemical elements. An **atom** is the smallest particle into which an element can be divided while still retaining the chemical characteristics of that element (see figure 2.1). The **nucleus,** at the center of the atom, contains one or more particles with a positive electrical charge (**protons**) and usually some particles of similar mass that have no charge (**neutrons**). Circling the nucleus are the negatively charged **electrons.**

The number of protons in the nucleus determines what chemical element that atom is. Every atom of hydrogen contains one proton in its nucleus; every oxygen atom contains eight protons; every carbon atom, six; every iron atom, twenty-six; and so on. The characteristic number of protons is the **atomic number** of the element.

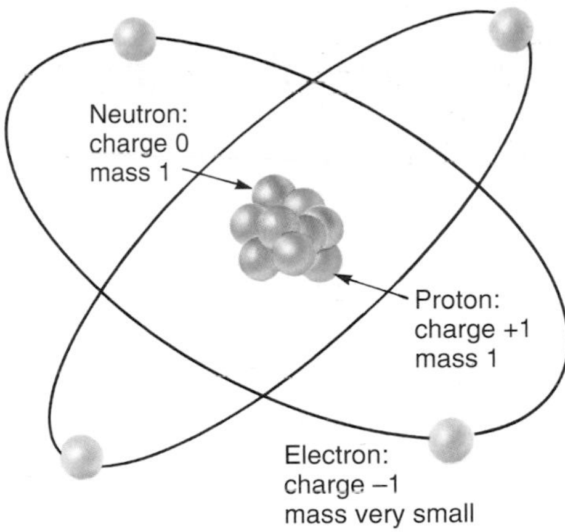

FIGURE 2.1 Schematic drawing of atomic structure (greatly enlarged and simplified). The nucleus, containing protons (here represented by blue spheres) and neutrons (peach), is actually only about 1/1000th the overall size of the atom; the nucleus is surrounded by the electrons (pale yellow).

Elements and Isotopes

With the exception of the simplest hydrogen atoms, all nuclei contain neutrons, and the number of neutrons is similar to or somewhat greater than the number of protons. The number of neutrons in atoms of one element is not always the same. The sum of the number of protons and the number of neutrons in a nucleus is the atom's **atomic mass number.** Atoms of a given element with different atomic mass numbers—in other words, atoms with the same number of protons but different numbers of neutrons—are distinct **isotopes** of that element. Some elements have only a single isotope, while others may have ten or more. (The reasons for these phenomena involve principles of nuclear physics and the nature of the processes by which the elements are produced in the interiors of stars, and we will not go into them here!)

For most applications, we are concerned only with the elements involved, not with specific isotopes. When a particular isotope is to be designated, this is done by naming the element (which, by definition, specifies the atomic number, or number of protons) and the atomic mass number (protons plus neutrons). Carbon, for example, has three natural isotopes. By far the most abundant is carbon-12, the isotope with six neutrons in the nucleus in addition to the six protons common to all carbon atoms. The two rarer isotopes are carbon-13 (six protons plus seven neutrons) and carbon-14 (six protons plus eight neutrons). Chemically, all behave alike. The human body cannot, for instance, distinguish between sugar containing carbon-12 and sugar containing carbon-13.

Other differences between isotopes may, however, make a particular isotope useful for some special purpose. Carbon-14 is naturally radioactive, which means that, over a period of time, carbon-14 nuclei decay at a known rate; this fact allows carbon-14 to be used to date materials containing carbon. (See appendix A.) Differences in the properties of two uranium isotopes are important in understanding nuclear power options: only one of the two common uranium isotopes is suitable for use as reactor fuel, and must be extracted and concentrated from natural uranium as it occurs in uranium ore. The fact that radioactive elements will inexorably decay—releasing energy—at their own fixed, constant rates is part of what makes radioactive-waste disposal such a challenging problem, because no chemical or physical treatment can make those waste isotopes nonradioactive and inert.

Ions

In an electrically neutral atom, the number of protons and the number of electrons are the same; the negative charge of one electron just equals the positive charge of one proton. Most atoms, however, can gain or lose some electrons. When this happens, the atom has a positive or negative electrical charge and is called an **ion.** If it loses electrons, it becomes positively charged, since the number of protons then exceeds the number of electrons. If the atom gains electrons, the ion has a negative electrical charge. Positively and negatively charged ions are called, respectively, **cations** and **anions.** Both solids and liquids are, overall, electrically neutral, with the total positive and negative charges of cations and anions balanced. Moreover, free ions do not exist in solids; cations and anions are bonded together. In a solution, however, individual ions may exist and move independently. Many minerals break down into ions as they dissolve in water. Individual ions may then be taken up by plants as nutrients or react with other materials. The concentration of hydrogen ions (pH) determines a solution's acidity.

Compounds

The electrical attraction between oppositely charged ions can cause them to become chemically bonded together. This is called **ionic bonding.** Bonds between atoms may also form if the atoms share electrons. This is **covalent bonding.** Whatever

Box 2.1 The Periodic Table

Some idea of the probable chemical behavior of elements can be gained from a knowledge of the **periodic table** (figure 1). The Russian scientist Dmitri Mendeleyev first observed that certain groups of elements showed similar chemical properties, which seemed to be related in a regular way to their atomic numbers. At the time (1869) that Mendeleyev published the first periodic table, in which elements were arranged so as to reflect these similarities of behavior, not all the elements had even been discovered, so there were some gaps. The addition of elements identified later confirmed the basic concept. In fact, some of the missing elements were found more easily because their properties could to some extent be anticipated from their expected positions in the periodic table.

We now can relate the periodicity of chemical behavior to the electronic structures of the elements. For example, those elements in the first column, known as the alkali metals, have one electron in the outermost shell of the neutral atom. Thus, they all tend to form cations of +1 charge by losing that odd electron. Outermost electron shells become increasingly full from left to right across a row. The next-to-last column, the halogens, are those elements lacking only one electron in the outermost shell, and they thus tend to gain one electron to form anions of charge −1. In the right-hand column are the inert gases, whose neutral atoms contain all fully filled electron shells.

The elements that occur naturally on earth have now been known for decades. Additional new elements must be created, not simply discovered, because these very heavy elements, with atomic numbers above 92 (uranium), are too unstable to have lasted from the early days of the solar system to the present. Some, like plutonium, are by-products of nuclear-reactor operation; others are created by nuclear physicists who, in the process, learn more about atomic structure and stability.

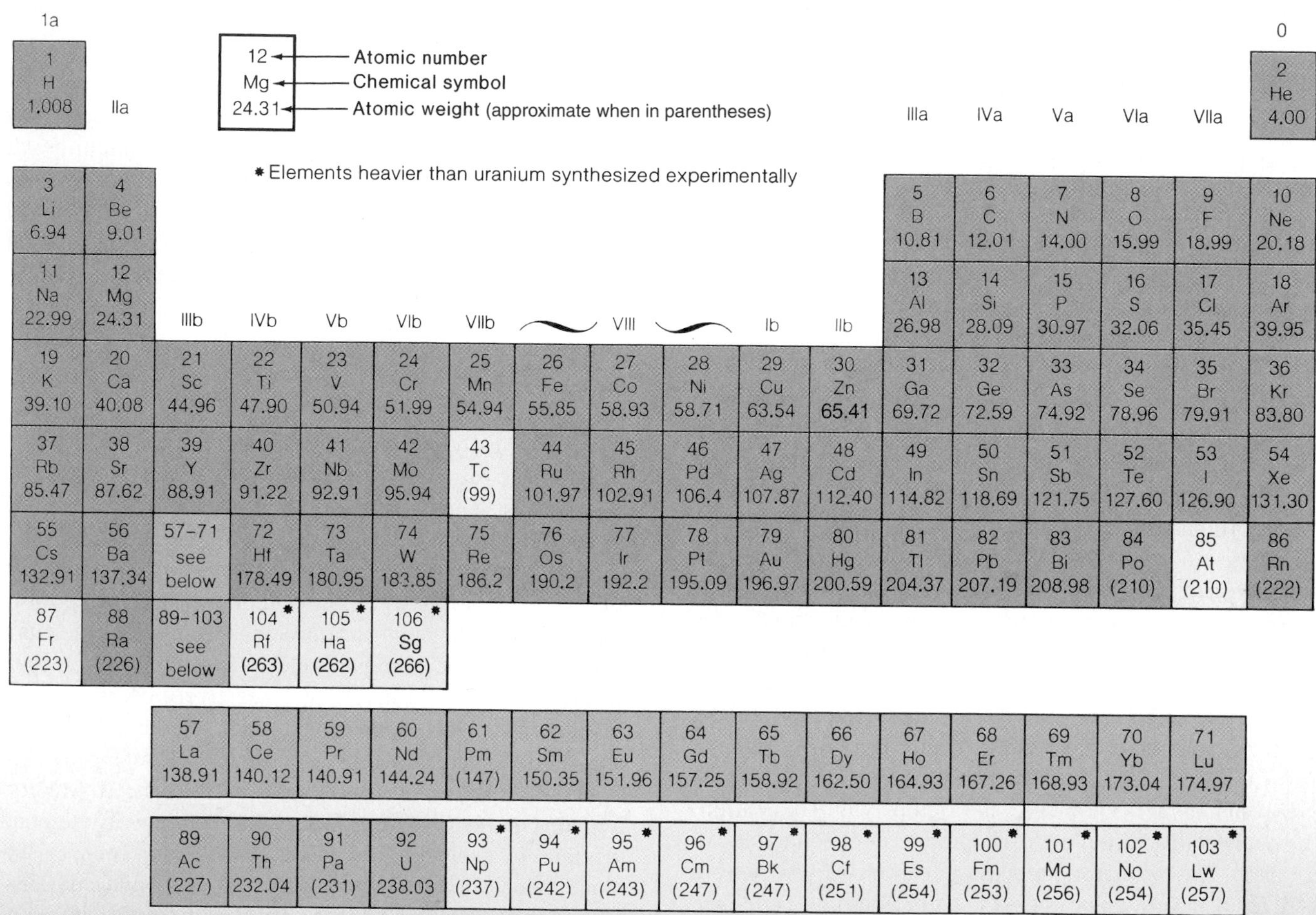

Key: 12 = Atomic number; Mg = Chemical symbol; 24.31 = Atomic weight (approximate when in parentheses)

* Elements heavier than uranium synthesized experimentally

1a	IIa	IIIb	IVb	Vb	VIb	VIIb	VIII	VIII	VIII	Ib	IIb	IIIa	IVa	Va	VIa	VIIa	0
1 H 1.008																	2 He 4.00
3 Li 6.94	4 Be 9.01											5 B 10.81	6 C 12.01	7 N 14.00	8 O 15.99	9 F 18.99	10 Ne 20.18
11 Na 22.99	12 Mg 24.31											13 Al 26.98	14 Si 28.09	15 P 30.97	16 S 32.06	17 Cl 35.45	18 Ar 39.95
19 K 39.10	20 Ca 40.08	21 Sc 44.96	22 Ti 47.90	23 V 50.94	24 Cr 51.99	25 Mn 54.94	26 Fe 55.85	27 Co 58.93	28 Ni 58.71	29 Cu 63.54	30 Zn 65.41	31 Ga 69.72	32 Ge 72.59	33 As 74.92	34 Se 78.96	35 Br 79.91	36 Kr 83.80
37 Rb 85.47	38 Sr 87.62	39 Y 88.91	40 Zr 91.22	41 Nb 92.91	42 Mo 95.94	43 Tc (99)	44 Ru 101.97	45 Rh 102.91	46 Pd 106.4	47 Ag 107.87	48 Cd 112.40	49 In 114.82	50 Sn 118.69	51 Sb 121.75	52 Te 127.60	53 I 126.90	54 Xe 131.30
55 Cs 132.91	56 Ba 137.34	57–71 see below	72 Hf 178.49	73 Ta 180.95	74 W 183.85	75 Re 186.2	76 Os 190.2	77 Ir 192.2	78 Pt 195.09	79 Au 196.97	80 Hg 200.59	81 Tl 204.37	82 Pb 207.19	83 Bi 208.98	84 Po (210)	85 At (210)	86 Rn (222)
87 Fr (223)	88 Ra (226)	89–103 see below	104* Rf (263)	105* Ha (262)	106* Sg (266)												

57 La 138.91	58 Ce 140.12	59 Pr 140.91	60 Nd 144.24	61 Pm (147)	62 Sm 150.35	63 Eu 151.96	64 Gd 157.25	65 Tb 158.92	66 Dy 162.50	67 Ho 164.93	68 Er 167.26	69 Tm 168.93	70 Yb 173.04	71 Lu 174.97
89 Ac (227)	90 Th 232.04	91 Pa (231)	92 U 238.03	93* Np (237)	94* Pu (242)	95* Am (243)	96* Cm (247)	97* Bk (247)	98* Cf (251)	99* Es (254)	100* Fm (253)	101* Md (256)	102* No (254)	103* Lw (257)

FIGURE 1 The periodic table.

the nature of the bonding, when atoms or ions of different elements combine in this way, they form **compounds.** A compound is a chemical combination of two or more chemical elements, in particular proportions, that has a distinct set of physical properties, usually very different from those of any of the individual elements in it. Consider, as a simple example, common table salt (sodium chloride). Sodium is a silver metal, and chlorine is a greenish gas that is poisonous in large doses. When equal numbers of sodium and chlorine atoms combine to form table salt, or sodium chloride, the resulting compound forms colorless crystals that do not resemble either of the component elements.

MINERALS—GENERAL

Minerals Defined

A **mineral** is a naturally occurring, inorganic, solid element or compound with a definite chemical composition and a regular internal crystal structure. *Naturally occurring,* as distinguished from synthetic, means that minerals do not include the thousands of chemical substances invented by humans. *Inorganic,* in this context, means not produced solely by living organisms or by biological processes. That minerals must be *solid* means that the ice of a glacier is a mineral, but liquid water is not. Chemically, minerals may consist either of one element—like diamonds, which are pure carbon—or they may be compounds of two or more elements. Some mineral compositions are very complex, consisting of ten elements or more. Minerals have a definite chemical composition or a compositional range within which they fall. The presence of certain elements in certain proportions is one of the identifying characteristics of each mineral. Finally, minerals are crystalline, at least on the microscopic scale. **Crystalline** materials are solids in which the atoms or ions are arranged in regular, repeating patterns (figure 2.2). These patterns may not be apparent to the naked eye, but most solid compounds are crystalline, and their crystal structures can

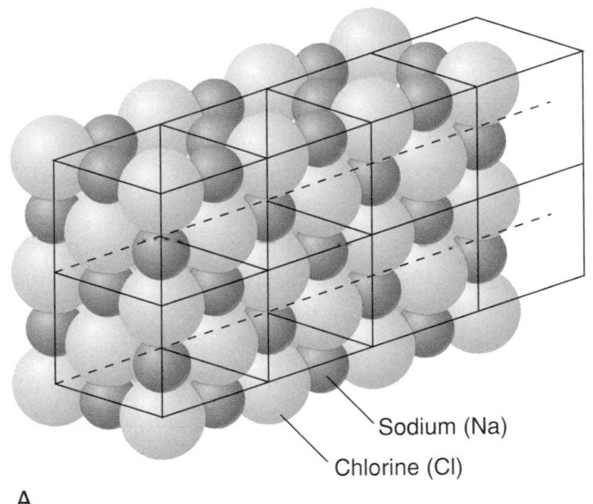

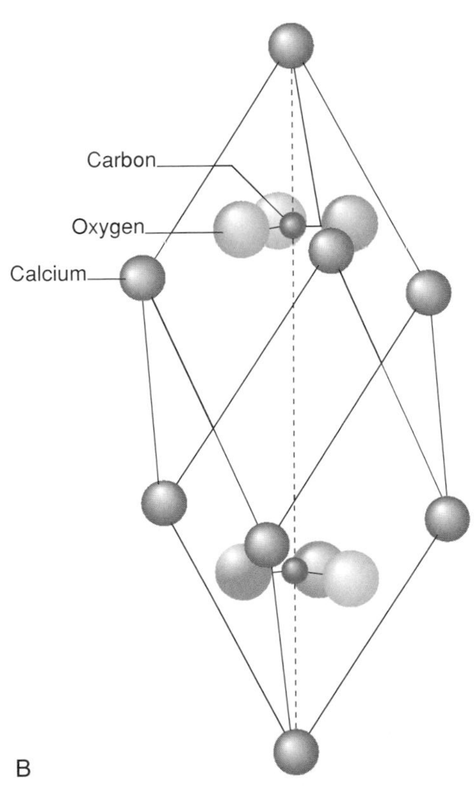

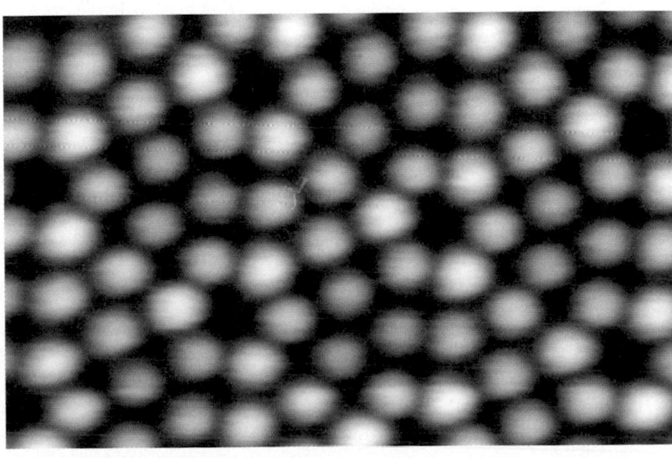

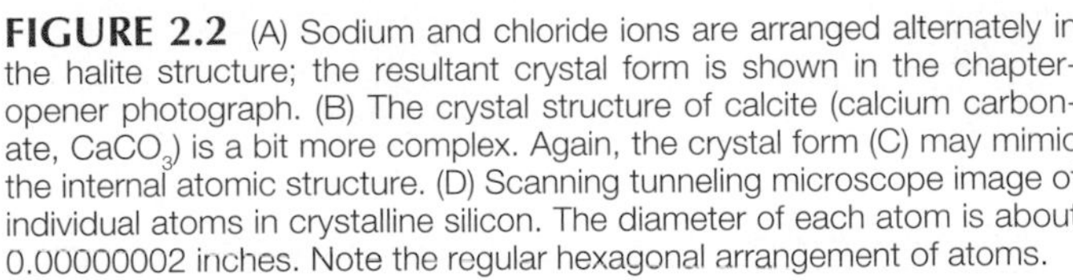

FIGURE 2.2 (A) Sodium and chloride ions are arranged alternately in the halite structure; the resultant crystal form is shown in the chapter-opener photograph. (B) The crystal structure of calcite (calcium carbonate, $CaCO_3$) is a bit more complex. Again, the crystal form (C) may mimic the internal atomic structure. (D) Scanning tunneling microscope image of individual atoms in crystalline silicon. The diameter of each atom is about 0.00000002 inches. Note the regular hexagonal arrangement of atoms.

Photograph (C) © Doug Sherman/Geofile. (D) Image courtesy Jennifer MacLeod and Alastair McLean, Queen's University, Canada

be recognized and studied using X rays and other techniques. Examples of noncrystalline solids include glass (discussed later in the chapter) and plastic.

Identifying Characteristics of Minerals

The two fundamental characteristics of a mineral that together distinguish it from all other minerals are its chemical composition and its crystal structure. No two minerals are identical in both respects, though they may be the same in one. For example, diamond and graphite (the "lead" in a lead pencil) are chemically the same—both are made up of pure carbon. Their physical properties, however, are vastly different because of the differences in their internal crystalline structures. In a diamond, each carbon atom is firmly bonded to every adjacent carbon atom in every direction. In graphite, the carbon atoms are bonded strongly in two dimensions into sheets, but the sheets are only weakly held together in the third dimension. Diamond is clear, colorless, and very hard, and a jeweler can cut it into beautiful precious gemstones. Graphite is black, opaque, and soft, and its sheets of carbon atoms tend to slide apart as the weak bonds between them are broken.

A mineral's composition and crystal structure can usually be determined only by using sophisticated laboratory equipment. When a mineral has formed large crystals with well-developed shapes, a trained mineralogist may be able to infer some characteristics of its internal atomic arrangement because crystals' shapes are controlled by and related to this atomic structure, but most mineral samples do not show large symmetric crystal forms by which they can be recognized with the naked eye. Moreover, many minerals share similar external forms (figure 2.3). No one can look at a mineral and know its chemical composition without first recognizing what mineral it is. Thus, when scientific instruments are not at hand, mineral identification must be based on a variety of other physical properties that in some way reflect the mineral's composition and structure. These other properties are often what make the mineral commercially valuable. However, they are rarely unique to one mineral and often are deceptive. A few examples of such properties follow.

Other Physical Properties of Minerals

Perhaps surprisingly, color is often not a reliable guide to mineral identification. While some minerals always appear the same color, many vary from specimen to specimen. Variation in color is usually due to the presence of small amounts of chemical impurities in the mineral that have nothing to do with the mineral's basic characteristic composition, and such variation is especially common when the pure mineral is light-colored or colorless. The very common mineral quartz, for instance, is colorless in its pure form. However, quartz also occurs in other colors, among them rose pink, golden yellow, smoky brown, purple (amethyst), and milky white. Clearly, quartz cannot always be recognized by its color, or lack of it.

Another example is the mineral corundum (figure 2.4), a simple compound of aluminum and oxygen. In pure form, it too is colorless, and quite hard, which makes it a good abrasive. It is often used for the grit on sandpaper. Yet a little color from trace impurities not only disguises corundum, it can transform this utilitarian material into highly prized gems: Blue-tinted corundum is also known as sapphire, and red corundum is ruby. The color of a mineral can vary within a single crystal. Even when the color shown by a mineral sample is the true color of the pure mineral, it is probably not unique. There are

A

B

FIGURE 2.3 Many minerals may share the same external crystal form; (A) galena (PbS) and (B) fluorite (CaF_2) form cubes, as do halite (chapter-opener photo) and pyrite (figure 2.9).

Photographs © The McGraw-Hill Companies Inc./Doug Sherman, photographer

FIGURE 2.4 The many colors of these corundum gemstones—including ruby and sapphire—illustrate why color is a poor guide in mineral identification.
Photograph by Dane A. Penland, © 1992 Smithsonian Institution

TABLE 2.1

The Mohs Hardness Scale

Mineral	Assigned Hardness
talc	1
gypsum	2
calcite	3
fluorite	4
apatite	5
orthoclase	6
quartz	7
topaz	8
corundum	9
diamond	10

For comparison, the approximate hardnesses of some common objects, measured on the same scale, are: fingernail, 2½; copper penny, 3; glass, 5 to 6; pocketknife blade, 5 to 6.

more than 3500 minerals, so there are usually many of any one particular color.

Hardness, the ability to resist scratching, is another easily measured physical property that can help to identify a mineral, although it usually does not uniquely identify the mineral. Classically, hardness is measured on the Mohs hardness scale (table 2.1), in which ten common minerals are arranged in order of hardness. Unknown minerals are assigned a hardness on the basis of which minerals they can scratch and which minerals scratch them. A mineral that scratches gypsum and is scratched by calcite is assigned a hardness of 2½ (the hardness of an average fingernail). Because a diamond is the hardest natural substance known on earth, and corundum the second-hardest mineral, these minerals might be identifiable from their hardnesses. Among the thousands of "softer" (more readily scratched) minerals, however, there are many of any particular hardness, just as there are many of any particular color.

Not only is the external form of crystals related to their internal structure; so is **cleavage,** a distinctive way some minerals may break up when struck. Some simply crumble or shatter into irregular fragments. Others, however, break cleanly in certain preferred directions that correspond to planes of weak bonding in the crystal. There may be only one direction in which a mineral shows good cleavage (as with mica, discussed later in the chapter), or there may be two or three directions of good cleavage. Cleavage surfaces are characteristically shiny (figure 2.5).

A

B

FIGURE 2.5 Another relationship between structures and physical properties is cleavage. Because of their internal crystalline structures, many minerals break apart preferentially in certain directions. (A) Halite has the cubic structure shown in figure 2.2A, and breaks into cubic or rectangular pieces, cleaving parallel to the crystal faces. (B) Fluorite also forms cubic crystals, but cleaves into octahedral fragments, breaking along different planes of its internal structure.
Photograph (A) © The McGraw-Hill Companies Inc./Bob Coyle, photographer; (B) © The McGraw-Hill Companies Inc./Doug Sherman, photographer

A number of other physical properties may individually be common to many minerals. *Luster* describes the appearance of the surfaces—glassy, metallic, pearly, etc. Some minerals are noticeably denser than most; a few are magnetic. Only by considering a whole set of such nonunique properties as color, hardness, cleavage, density, and others can a mineral be identified without complex instruments.

Unique or not, the physical properties arising from minerals' compositions and crystal structures are often what give minerals value from a human perspective—the slickness of talc (main ingredient of talcum powder), the malleability and conductivity of copper, the durability of diamond, and the rich colors of tourmaline gemstones are all examples. Some minerals have several useful properties: table salt (halite), a necessary nutrient, also imparts a taste we find pleasant, dissolves readily to flavor liquids but is soft enough not to damage our teeth if we munch on crystals of it sprinkled on solid food, and will serve as a food preservative in high concentrations, among other helpful qualities.

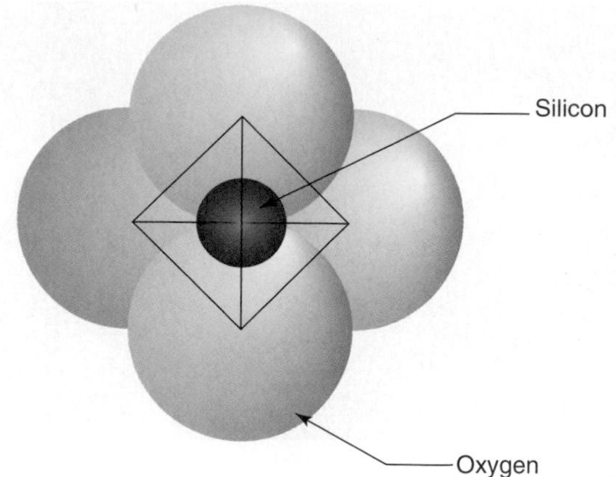

FIGURE 2.6 The basic silica tetrahedron, building block of all silicate minerals.

TYPES OF MINERALS

As was indicated earlier, minerals can be grouped or subdivided on the basis of their two fundamental characteristics—composition and crystal structure. In this section, we will briefly review some of the basic mineral groups. A comprehensive survey of minerals is well beyond the scope of this book, and the interested reader should refer to standard mineralogy texts for more information. A summary of physical properties of selected minerals is found in appendix C.

Silicates

In chapter 1, we noted that the two most common elements in the earth's crust are silicon and oxygen. It comes as no surprise, therefore, that by far the largest group of minerals is the **silicate** group, all of which are compounds containing silicon and oxygen, and most of which contain other elements as well. Because this group of minerals is so large, it is subdivided on the basis of crystal structure, by the ways in which the silicon and oxygen atoms are linked together. The basic building block of all silicates is a tetrahedral arrangement of four oxygen atoms (anions) around the much smaller silicon cation (figure 2.6). In different silicate minerals, these *silica tetrahedra* may be linked into chains, sheets, or three-dimensional frameworks by the sharing of oxygen atoms. Some of the physical properties of silicates and other minerals are closely related to their crystal structures (see figure 2.7). In general, however, we need not go into the structural classes of the silicates in detail. It is more useful to mention briefly a few of the more common, geologically important silicate minerals.

While not the most common, *quartz* is probably the best-known silicate. Compositionally, it is the simplest, containing only silicon and oxygen. It is a framework silicate, with silica tetrahedra linked in three dimensions, which helps make it relatively hard and weathering-resistant. Quartz is found in a variety of rocks and soils. Commercially, the most common use of pure quartz is in the manufacture of glass, which also consists mostly of silicon and oxygen. Quartz-rich sand and gravel are used in very large quantities in construction.

The most abundant group of minerals in the crust is a set of chemically similar minerals known collectively as the *feldspars.* They are composed of silicon, oxygen, aluminum, and either sodium, potassium, or calcium, or some combination of these three. Again, logically enough, these common minerals are made from elements abundant in the crust. They are used extensively in the manufacture of ceramics.

Iron and magnesium are also among the more common elements in the crust and are therefore found in many silicate minerals. **Ferromagnesian** is the general term used to describe those silicates—usually dark-colored (black, brown, or green)—that contain iron and/or magnesium. These silicates may or may not contain other elements also. In general, ferromagnesian minerals weather relatively readily. Rocks containing a high proportion of ferromagnesian minerals, then, also tend to weather easily, which is an important consideration in construction. Individual ferromagnesian minerals may be important in particular contexts. *Olivine,* a simple ferromagnesian mineral, is a major constituent of earth's mantle; gem-quality olivines from mantle-derived volcanic rocks are the semiprecious gem *peridot. Asbestos* is not a single mineral, but a general term for various minerals that occur in a form that can be separated into needles or thin, strong fibers; however, the several different asbestos minerals are all ferromagnesian silicates.

Like the feldspars, the *micas* are another group of several silicate minerals with similar physical properties, compositions, and crystal structures. Micas are sheet silicates, built on an atomic

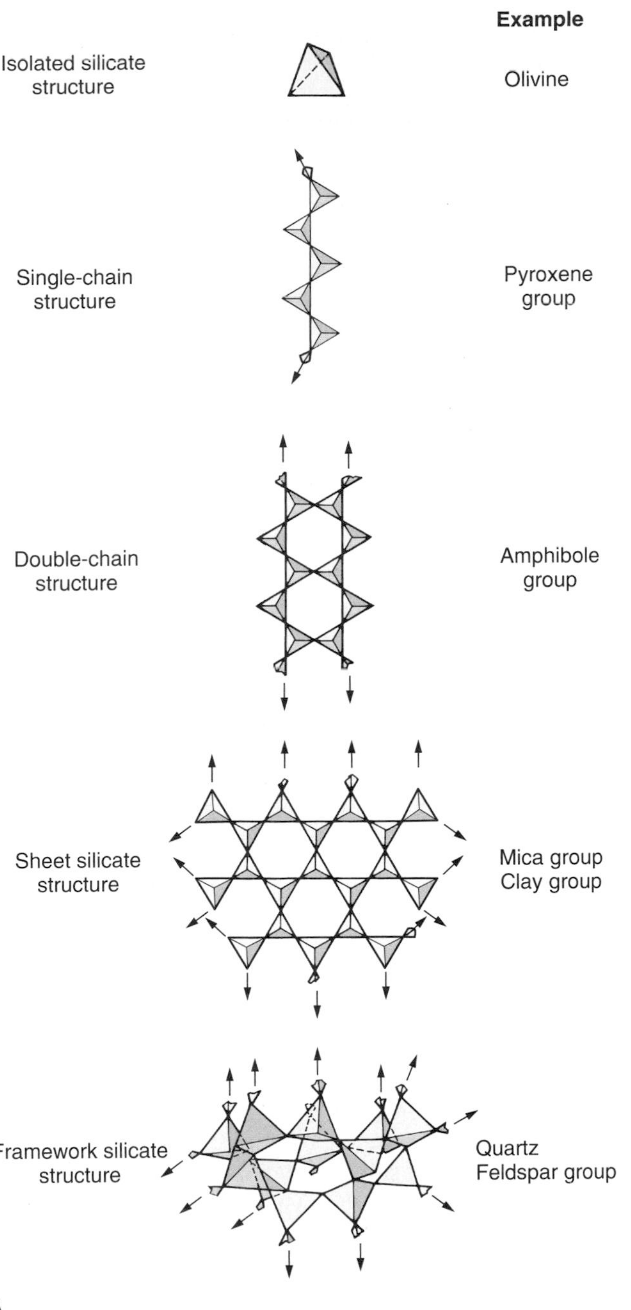

FIGURE 2.7 (A) Silica tetrahedra link together by sharing oxygen atoms (at the corners of the tetrahedra) to form a variety of structures. (Pyroxenes and amphiboles are among the ferromagnesian silicates, described in the text.) (B) The ability of mica to cleave into thin sheets reflects the ease of separating the sheets of tetrahedra in the crystal.
Photograph (B) © The McGraw-Hill Companies Inc./Bob Coyle, photographer

scale of stacked-up sheets of linked silicon and oxygen atoms. Because the bonds between sheets are relatively weak, the sheets can easily be broken apart (figure 2.7B).

Clays are another family within the sheet silicates; in clays, the sheets tend to slide past each other, a characteristic that contributes to the slippery feel of many clays and related minerals. Clays are somewhat unusual among the silicates in that their structures can absorb or lose water, depending on how wet conditions are. Absorbed water may increase the slippery tendencies of the clays. Also, some clays expand as they soak up water and shrink as they dry out. A soil rich in these "expansive clays" is a very unstable base for a building, as we will see in later chapters. On the other hand, clays also have important uses, especially in making ceramics and in building materials. Other clays are useful as lubricants in the muds used to cool the drill bits in oil-drilling rigs.

A sampling of the variety of silicates is shown in figure 2.8.

Nonsilicates

Just as the silicates, by definition, all contain silicon plus oxygen as part of their chemical compositions, each nonsilicate mineral group is defined by some chemical constituent or characteristic that all members of the group have in common. Most often, the common component is the same negatively charged ion or group of atoms. Discussion of some of the nonsilicate mineral groups with examples of common or familiar members of each follows. See also table 2.2.

The **carbonates** all contain carbon and oxygen combined in the proportions of one atom of carbon to three atoms of oxygen (written CO_3). The carbonate minerals all dissolve relatively easily, particularly in acids, and the oceans contain a great deal of dissolved carbonate. Geologically, the most important, most abundant carbonate mineral is calcite, which is calcium carbonate. Precipitation of calcium carbonate from seawater is a major process by which marine rocks are formed (see the discussion under "Sediments and Sedimentary Rocks" later in this chapter). Another common carbonate mineral is dolomite, which contains both calcium and magnesium in approximately equal proportions. Carbonates may contain many other elements—iron,

A

B

C

D

FIGURE 2.8 A collection of silicates: (A) olivine; (B) feldspar; (C) hornblende, a variety of amphibole; and (D) quartz.
Photograph (A) © Albert J. Copley/Visuals Unlimited. (B & C) © The Natural History Museum, London

manganese, or lead, for example. The limestone and marble we use extensively for building and sculpture consist mainly of carbonates, generally calcite; calcite is also an important ingredient in cement.

The **sulfates** all contain sulfur and oxygen in the ratio of 1:4 (SO_4). A calcium sulfate—gypsum—is the most important, for it is both relatively abundant and commercially useful, particularly as the major constituent in plaster of paris. Sulfates of many other elements, including barium, lead, and strontium, are also found.

When sulfur is present without oxygen, the resultant minerals are called **sulfides.** A common and well-known sulfide mineral is the iron sulfide *pyrite.* Pyrite (figure 2.9) has also been called "fool's gold" because its metallic golden color often deceived early gold miners and prospectors into thinking they had struck it rich. Pyrite is not a commercial source of iron because there are richer ores of this metal. Nonetheless, sulfides comprise many economically important metallic ore minerals. An example that may be familiar is the lead sulfide mineral *galena,* which often forms in silver-colored cubes (figure 2.3A). The rich lead ore deposits near Galena, Illinois, gave the town its name. Sulfides of copper, zinc, and numerous other metals may also form valuable ore deposits (see chapter 13). Sulfides may also be problematic: when pyrite associated with coal is exposed by strip-mining and weathered, the result is sulfuric acid in runoff water from the mine.

Minerals containing just one or more metals combined with oxygen, and lacking the other elements necessary to classify them as silicates, sulfates, carbonates, and so forth, are the **oxides.** Iron combines with oxygen in different proportions to form more than one oxide mineral. One of these, magnetite, is, as its name suggests, magnetic, which is relatively unusual among minerals. Magnetic rocks rich in magnetite were known as lodestone in ancient times and were used as navigational aids like today's more compact compasses. Another iron oxide, hematite, may sometimes be silvery black but often has a red color and gives a reddish tint to many soils. Iron oxides on Mars's surface are responsible for that planet's orange hue. Many other oxide minerals also exist, including corundum, the aluminum oxide mineral mentioned earlier.

Native elements, as shown in table 2.2, are even simpler chemically than the other nonsilicates. Native elements are minerals that consist of a single chemical element, and the minerals' names are usually the same as the corresponding elements. Not all elements can be found, even rarely, as native elements. However, some of our most highly prized materials, such as gold, silver, and platinum, often occur as native elements. Diamond and graphite are both examples of native

TABLE 2.2

Some Nonsilicate Mineral Groups*

Group	Compositional Characteristic	Examples
carbonates	metal(s) plus carbonate (1 carbon + 3 oxygen ions, CO_3)	calcite (calcium carbonate, $CaCO_3$) dolomite (calcium-magnesium carbonate, $CaMg(CO_3)_2$)
sulfates	metal(s) plus sulfate (1 sulfur + 4 oxygen ions, SO_4)	gypsum (calcium sulfate, with water, $CaSO_4 \cdot 2H_2O$) barite (barium sulfate, $BaSO_4$)
sulfides	metal(s) plus sulfur, without oxygen	pyrite (iron disulfide, FeS_2) galena (lead sulfide, PbS) cinnabar (mercury sulfide, HgS)
oxides	metal(s) plus oxygen	magnetite (iron oxide, Fe_3O_4) hematite (ferric iron oxide, Fe_2O_3) corundum (aluminum oxide, Al_2O_3) spinel (magnesium-aluminum oxide, $MgAl_2O_4$)
hydroxides	metal(s) plus hydroxyl (1 oxygen + 1 hydrogen ion, OH)	gibbsite (aluminum hydroxide, $Al(OH)_3$; found in aluminum ore) brucite (magnesium hydroxide, $Mg(OH)_2$; one ore of magnesium)
halides	metal(s) plus halogen element (fluorine, chlorine, bromine, or iodine)	halite (sodium chloride, NaCl) fluorite (calcium fluoride, CaF_2)
native elements	mineral consists of a single chemical element	gold (Au), silver (Ag), copper (Cu), sulfur (S), graphite (carbon, C)

*Other groups exist, and some complex minerals contain components of several groups (carbonate and hydroxyl groups, for example).

FIGURE 2.9 Pyrite ("fool's gold"), an iron sulfide (FeS_2), a very common sulfide mineral.
Photograph © The McGraw-Hill Companies, Inc./Doug Sherman, photographer

carbon. Sulfur may occur as a native element, either with or without associated sulfide minerals. Some of the richest copper ores contain native copper. Other metals that may occur as native elements include tin, iron, and antimony.

ROCKS

A **rock** is a solid, cohesive aggregate of one or more minerals, or mineral materials (for example, volcanic glass, discussed later). This means that a rock consists of many individual mineral grains (crystals)—not necessarily all of the same mineral—or crystals plus glass, which are firmly held together in a solid mass. Because the many mineral grains of beach sand fall apart when handled, sand is not a rock, although, in time, sand grains may become cemented together to form a rock. The properties of rocks are important in determining their suitability for particular applications, such as for construction materials or for the base of a building foundation. Each rock also contains within it a record of at least a part of its history, in the nature of its minerals and in the way the mineral grains fit together. The three broad categories of rocks—*igneous, sedimentary,* and *metamorphic*—are distinguished by the processes of their formation. However, they are also linked, over time, by the *rock cycle.*

Box 2.2 Asbestos—A Risky Business?

It was a spectacular fall day in the Midwest, the harvest well underway, the sun bright and warm. But Keisha and Kim were staring glumly at century-old Tower Hall, which had been undergoing renovation for more than two years.

"It may be gorgeous when they finish," observed Keisha, "but meanwhile, it's a nuisance—my econ class is meeting in a chem lab this term because we don't have the Tower Hall classrooms back yet."

"I heard they're already way over budget, as well as way behind schedule," replied Kim. "Something about running into a lot of asbestos in the walls and around the pipes. I guess they don't have a lot of choice; the law says that once they've disturbed it, they have to get rid of it, because asbestos causes cancer, or something."

"Well, I certainly don't want that," responded Keisha. "But I still wish they'd hurry up and finish!"

What may be most ironic about this situation is that it could be argued that much asbestos abatement mandated under current U.S. law is unnecessary, an overreaction based on a simplistic view of perceived dangers.

Asbestos is not a single mineral. The term describes a number of ferromagnesian silicates that have in common a needlelike or fibrous crystal form. Half a dozen different minerals are classified as "asbestos" in the regulations of OSHA (the Occupational Safety and Health Administration). It is certainly true that regular occupational exposure to asbestos (incurred before government regulations sharply reduced such exposure) has been associated with later increased incidence of mesothelioma (a type of lung cancer) and other respiratory diseases, including a scarring of the lungs called asbestosis. The risks can extend to families of asbestos workers, exposed secondhand to fibers carried home on clothes, and to residents of communities where asbestos processing occurred. However, the kinds of diseases caused by asbestos can be caused (or aggravated) by other agents and activities (such as smoking) too, and generally develop long after exposure. Even for workers in asbestos mining and manufacturing industries, the link between asbestos exposure and lung disease is far from direct. For setting exposure limits for the general public, it has been assumed that the risk from low, incidental exposure to asbestos can be extrapolated linearly from the identified effects of high occupational exposures, while there is actually no evidence that this yields an accurate estimate of the risks from low exposures.

Moreover, studies have clearly shown that different asbestos minerals present substantially different degrees of risk. Mesothelioma is associated particularly with occupational exposure to the amphibole *crocidolite,* "blue asbestos," and asbestosis with another amphibole, *amosite,* "grey asbestos," mined almost exclusively in South Africa. The serpentine mineral *chrysotile,* "white asbestos" (figure 2), which represents about 95% of the asbestos used in the United States, appears to be the least hazardous, even at the old occupational-exposure levels, and to pose *no* significant health threat at the much lower concentrations found in the air of buildings such as schools—even those with damaged, exposed asbestos-containing materials.

Finally, keep in mind that asbestos minerals are naturally present in the environment already. By one estimate, an adult outdoors would normally inhale, on average, nearly *4000 asbestos fibers a day.* Meanwhile, billions of dollars a year are spent on asbestos abatement in U.S. buildings. Is this an effective use of resources? What further information would you want in order to evaluate the risks more precisely?

The interested reader is referred to Gunter (1994) as a starting point for investigation.

FIGURE 1 Renovation of older buildings is often complicated by asbestos-abatement requirements.

FIGURE 2 Chrysotile asbestos, or "white asbestos," accounts for about 95% of asbestos mined and used in the U.S.
© The McGraw-Hill Companies Inc./Bob Coyle, photographer

In chapter 1, we noted that the earth is a constantly changing body. Mountains come and go; seas advance and retreat over the faces of continents; surface processes and processes occurring deep in the crust or mantle are constantly altering the planet. One aspect of this continual change is that rocks, too, are always subject to change. We do not have a single sample of rock that has remained unchanged since the earth formed. Therefore, when we describe a rock as being of a particular type, or give it a specific rock name, it is important to realize that we are really describing the form it has most recently taken, the results of the most recent processes acting on it.

For example, as will be described in the following section, an *igneous rock* is one crystallized from molten material, formed at high temperatures. But virtually any rock can be melted, and in the process, its previous characteristics may be obliterated. So when the melt crystallizes, we have an igneous rock—but what was it before? A *sedimentary rock* is formed from sediment, debris of preexisting rocks deposited at low temperatures at the earth's surface. But the sediment, in turn, is derived by weathering—physical and chemical breakdown of rocks—and a given deposit of sediment may include bits of many different rocks. (Look closely at beach sand or river gravel.) The essence of the concept of the **rock cycle,** explored more fully at the end of this chapter, is that rocks, far from being the permanent objects we may imagine them to be, are continually being changed by geological processes. Over the vast span of geologic time, billions of years, a given bit of material may have been subject to many, many changes and may have been part of many different rocks. The following sections examine in more detail the various types of rock-forming processes involved and some of the rocks they form.

Igneous Rocks

At high enough temperatures, rocks and minerals can melt. **Magma** is the name given to naturally occurring hot, molten rock material. Silicates are the most common minerals, so magmas are usually rich in silica. They also contain some dissolved water and gases and generally have some solid crystals suspended in the melt. An **igneous** rock is a rock formed by the solidification and crystallization of a cooling magma. (*Igneous* is derived from the Latin term *ignis,* meaning "fire.")

Because temperatures significantly higher than those of the earth's surface are required to melt silicates, magmas form at some depth below the surface. The molten material may or may not reach the surface before it cools enough to crystallize and solidify. The depth at which a magma crystallizes will affect how rapidly it cools and the sizes of the mineral grains in the resultant rock.

If a magma remains well below the surface during cooling, it cools relatively slowly, insulated by overlying rock and soil. It may take hundreds of thousands of years or more to crystallize completely. Under these conditions, the crystals have ample time to form and to grow very large, and the rock eventually formed has mineral grains large enough to be seen individually with the naked eye. A rock formed in this way is a **plutonic** igneous rock. (The name is derived from Pluto, the Greek god of the lower world.) *Granite* is probably the most widely known example of a plutonic rock (figure 2.10A). Compositionally, a typical granite consists principally of quartz and feldspars, and it usually contains some ferromagnesian minerals or other silicates. The proportions and compositions of these constituent minerals may vary, but all granites show the coarse, interlocking crystals characteristic of a plutonic rock. Much of the mass of the continents consists of granite or of rock of granitic composition.

A magma that flows out on the earth's surface while still wholly or partly molten is called **lava.** Lava is a common product of volcanic eruptions, and the term **volcanic** is given to an igneous rock formed at or close to the earth's surface. Magmas that crystallize very near the surface cool more rapidly. There is less time during crystallization for large crystals to form from the magma, so volcanic rocks are typically fine-grained, with most crystals too small to be distinguished with the naked eye. In extreme cases, where cooling occurs very fast, even tiny crystals may not form before the magma solidifies. The resulting clear, noncrystalline solid is a natural **glass** (figure 2.10B). The most common volcanic rock is *basalt,* a dark rock rich in ferromagnesian minerals and feldspar (figure 2.10C). The ocean floor consists largely of basalt. Occasionally, a melt begins to crystallize slowly at depth, growing some large crystals, and then is subjected to rapid cooling (following a volcanic eruption, for instance). This results in coarse crystals in a fine-grained groundmass, a *porphyry* (figure 2.10D).

Though there are fundamental similarities in the origins of magmas and volcanic rocks, there are practical differences too. Differences in the chemical compositions of magmas lead to differences in their physical properties, with magmas richer in silica tending to be more viscous. This, in turn, produces differences in the behavior of the volcanoes from which the magmas erupt, explaining why the Hawaiian volcano Kilauea erupts so quietly that tourists can almost walk up and touch the lava flows in safety, while Mount St. Helens and the Philippine volcano Pinatubo are prone to violent, sudden, and devastating explosions. Relationships among magma origins, magma types, and volcanoes' eruptive styles will be explored further in chapter 5.

Regardless of the details of their compositions or cooling histories, all igneous rocks have some textural characteristics in common. If they are crystalline, their crystals, large or small, are tightly interlocking or intergrown (unless they are formed from loose material such as volcanic ash). If glass is present,

FIGURE 2.10 Igneous rocks, crystallized from melts. (A) Granite, a plutonic rock; this sample is the granite of Yosemite National Park. (B) Obsidian (volcanic glass). (C) Basalt, a volcanic rock. (D) Porphyry, an igneous rock with coarse crystals in a finer matrix.

(A) Photograph by N. K. Huber, USGS Photo Library, Denver, CO; (B), (C), and (D) photographs © The McGraw-Hill Companies Inc./Bob Coyle, photographer

crystals tend to be embedded in or closely surrounded by the glass. The individual crystals tend to be angular in shape, not rounded. There is usually little pore space, little empty volume that could be occupied by such fluids as water. Structurally, most igneous rocks are relatively strong unless they have been fractured, broken, or weathered.

Sediments and Sedimentary Rocks

At the lower end of the spectrum of rock-formation temperatures are the **sedimentary** rocks. **Sediments** are loose, unconsolidated accumulations of mineral or rock particles that have been transported by wind, water, or ice, or shifted under the influence of gravity, and redeposited. Beach sand is a kind of sediment; so is the mud on a river bottom. Soil is a mixture of mineral sediment and organic matter. Most sediments originate, directly or indirectly, through the weathering of preexisting rocks—either by physical breakup into finer and finer fragments, or by solution, followed by precipitation of crystals out of solution. The physical properties of sediments and soils bear on a broad range of environmental problems, from the stability of slopes and building foundations, to selection of optimal waste-disposal sites, to how readily water drains away after a rainstorm and how likely that rain is to produce a flood.

When sediments are compacted or cemented together into a solid cohesive mass, they become sedimentary rocks. The set of processes by which sediments are transformed into rock is collectively described as **lithification** (from the Greek word *lithos,* meaning "stone"). The resulting rock is generally more compact and denser, as well as more cohesive, than the original sediment. Sedimentary rocks are formed at or near the earth's surface, at temperatures close to ordinary surface temperatures. They are subdivided into two groups—clastic and chemical.

Clastic sedimentary rocks (from the Greek word *klastos,* meaning "broken") are formed from the products of the mechanical breakup of other rocks. Natural processes continually attack rocks exposed at the surface. Rain and waves pound them, windblown dust scrapes them, frost and tree roots crack them—these and other processes are all part of the physical weathering of rocks. In consequence, rocks are broken up into smaller and smaller pieces and ultimately, perhaps, into individual mineral grains. The resultant rock and mineral fragments may be transported by wind, water, or ice, and accumulate as sediments in streams, lakes, oceans, deserts, or soils. Later geologic processes can cause these sediments to become lithified. Burial under the weight of more sediments may pack the loose particles so tightly that they hold firmly together in a cohesive mass. Except with very fine-grained sediments, however, compaction alone is rarely enough to transform the sediment into rock. Water seeping slowly through rocks and sediments underground also carries dissolved minerals, which may precipitate out of solution to stick the sediment particles together with a natural mineral cement.

Clastic sedimentary rocks are most often named on the basis of the average size of the particles that form the rock. *Sandstone,* for instance, is a rock composed of sand-sized sediment particles, $\frac{1}{16}$ to 2 millimeters (0.002 to 0.08 inches) in diameter. *Shale* is made up of finer-grained sediments, and the individual grains cannot be seen in the rock with the naked eye. *Conglomerate* is a relatively coarse-grained rock, with fragments above 2 mm (0.08 inches) in diameter, and sometimes much larger. Regardless of grain size, clastic sedimentary rocks tend to have, relatively, considerable pore space between grains. This is a logical consequence of the way in which these rocks form, by the piling up of preexisting rock and mineral grains. Also, as sediment particles are transported by water or other agents, they may become more rounded and thus not pack together very tightly or interlock as do the mineral grains in an igneous rock. Many clastic sedimentary rocks are therefore not particularly strong structurally, unless they have been extensively cemented.

Chemical sedimentary rocks form not from mechanical breakup and transport of fragments, but from crystals formed by precipitation or growth from solution. A common example is *limestone,* composed mostly of calcite (calcium carbonate). The chemical sediment that makes limestone may be deposited from fresh or salt water; under favorable chemical conditions, thick limestone beds, perhaps hundreds of meters thick, may form. Another example of a chemical sedimentary rock is *rock salt,* made up of the mineral halite, which is the mineral name for ordinary table salt (sodium chloride). A salt deposit may form when a body of salt water is isolated from an ocean and dries up.

Some chemical sediments have a large biological contribution. For example, many organisms living in water have shells or skeletons made of calcium carbonate or of silica (chemically equivalent to quartz). The materials of these shells or skeletons are drawn from the water in which the organisms grow. In areas where great numbers of such creatures live and die, the "hard parts"—the shells or skeletons—may pile up on the bottom, eventually to be buried and lithified. A sequence of sedimentary rocks may include layers of **organic sediments,** carbon-rich remains of living organisms; *coal* is an important example.

Gravity plays a role in the formation of all sedimentary rocks. Mechanically broken-up bits of materials accumulate when the wind or water is moving too weakly to overcome gravity and move the sediments; repeated cycles of transport and deposition can pile up, layer by layer, a great thickness of sediment. Minerals crystallized from solution, or the shells of dead organisms, tend to settle out of the water under the force of gravity, and again, in time, layer on layer of sediment can build up. Layering, then, is a very common feature of sedimentary rocks and is frequently one way in which their sedimentary origins can be identified. Figure 2.11 shows several kinds of sedimentary rocks. The photograph at the start of this section is also of sedimentary rock (sandstone).

Sedimentary rocks can yield information about the settings in which the sediments were deposited. For example, the energetic surf of a beach washes away fine muds and leaves coarser sands and gravels; sandstone may, in turn, reflect the presence of an ancient beach. Distribution of glacier-deposited sediments of various ages contributes to understanding not only of global climate change but of continental drift (see chapter 3).

Metamorphic Rocks

The name *metamorphic* comes from the Greek for "changed form." A **metamorphic** rock is one that has formed from another, preexisting rock that was subjected to heat and/or

A

D

B

E

C

F

FIGURE 2.11 Sedimentary rocks, formed at low temperatures. (A) Limestone. (B) Shale. (C) Sandstone (enlarged to show the rounded sand grains). (D) Conglomerate, a coarser-grained rock similar to sandstone; note that many of the fragments here are rocks, not individual mineral grains. (E) Coal seams (dark layers) in a sequence of sedimentary rocks exposed in a sea cliff in southern Alaska. (F) The Grand Canyon is cut into sedimentary rocks; the "shelves" are formed by more weathering-resistant layers.

Photograph (A) by I. J. Witkind, USGS Photo Library, Denver, CO. (B) © The McGraw-Hill Companies Inc./Bob Coyle, photographer. (C) © The McGraw-Hill Companies Inc./Doug Sherman, photographer. (D) © The McGraw-Hill Companies Inc./John A. Karachewski, photographer

pressure. The temperatures required to form metamorphic rocks are not as high as magmatic temperatures. Significant changes can occur in a rock at temperatures well below melting. Heat and pressure commonly cause the minerals in the rock to recrystallize. The original minerals may form larger crystals that perhaps interlock more tightly than before. Also, some minerals may break down completely, while new minerals form under the new temperature and pressure conditions. Pressure may cause the rock to become deformed—compressed, stretched, folded, or compacted. All of this occurs while the rock is still solid.

The sources of the elevated pressures and temperatures of metamorphism are many. An important source of pressure is simply burial under many kilometers of overlying rock. The weight of the overlying rock can put great pressure on the rocks below. One source of elevated temperatures is the fact that temperatures increase with depth in the earth. In general, in most places, crustal temperatures increase at the rate of about 30°C per kilometer of depth (close to 60°F per mile)—which is one reason deep mines have to be air-conditioned! Deep in the crust, rocks are subjected to enough heat and pressure to show the deformation and recrystallization characteristic of metamorphism. Another heat source is a cooling magma. When hot magma formed at depth rises to shallower levels in the crust, it heats the adjacent, cooler rocks, and they may be metamorphosed; this is **contact metamorphism.** Metamorphism can also result from the stresses and heating to which rocks are subject during mountain-building or platetectonic movement. Such metamorphism on a large scale, not localized around a magma body, is **regional metamorphism.**

Any kind of preexisting rock can be metamorphosed. Some names of metamorphic rocks reflect their compositions and suggest what the earlier rock may have been. *Marble* is metamorphosed limestone in which the individual calcite grains have recrystallized and become tightly interlocking. The remaining sedimentary layering that the limestone once showed may be folded and deformed in the process, if not completely obliterated by the recrystallization. *Quartzite* is a quartz-rich metamorphic rock, often formed from a very quartz-rich sandstone. The quartz crystals are typically much more tightly interlocked in the quartzite, and the quartzite is a more compact, denser, and stronger rock than the original sandstone.

Other metamorphic rock names describe the characteristic texture of the rock, regardless of its composition. Sometimes the pressure of metamorphism is not uniform in all directions; rocks may be compressed or stretched in a particular direction (*directed stress*). When you stamp on an aluminum can before tossing it in a recycling bin, you are technically subjecting it to a directed stress—vertical compression—and it flattens in that direction in response. In a rock subjected to directed stress, minerals that form elongated or platy crystals may line up parallel to each other. The resultant texture is described as **foliation,** from the Latin for "leaf" (as in the parallel leaves, or pages, of a book). *Slate* is a metamorphosed shale that has developed foliation under stress. The resulting rock tends to break along the foliation planes, parallel to the alignment of those minerals, and this characteristic makes it easy to break up into slabs for flagstones. The same characteristic is observed in *schist,* a coarser-grained, mica-rich metamorphic rock in which the mica flakes are similarly oriented. The presence of foliation can cause planes of structural weakness in the rock, affecting how it weathers and whether it is prone to slope failure or landslides. In other metamorphic rocks, different minerals may be concentrated in irregular bands, often with darker, ferromagnesian-rich bands alternating with light bands rich in feldspar and quartz, producing a striped appearance. Such a rock is called a *gneiss* (pronounced "nice"). Because such terms as *schist* and *gneiss* are purely textural, the rock name can be modified by adding key features of the rock composition: "biotite-garnet schist," "granitic gneiss," and so on. Several examples of metamorphic rocks are illustrated in figure 2.12.

The Rock Cycle

It should be evident from the descriptions of the major rock types and how they form that rocks of any type can be transformed into rocks of another type or into another distinct rock of the same general type through the appropriate geologic processes. A sandstone may be weathered until it breaks up; its fragments may then be transported, redeposited, and lithified to form another sedimentary rock. It might instead be deeply buried, heated, and compressed, which could transform it into the metamorphic rock quartzite; or it could be heated until some or all of it melted, and from that melt, an igneous rock could be formed. Likewise, a schist could be broken up into small bits, forming a sediment that might eventually become sedimentary rock; more-intense metamorphism could transform it into a gneiss; or extremely high temperatures could melt it to produce a magma from which a granite could crystallize. Crustal rocks can be carried into the mantle and melted; fresh magma cools and crystallizes to form new rock; erosion and weathering processes constantly chip away at the surface. Note that the appearance (texture) of a rock can offer a good first clue to the conditions under which it (last) formed. A more comprehensive view of the links among different rock types is shown in generalized form in figure 2.13. In chapter 3, we will look at the rock cycle again, but in a somewhat different way, in the context of plate tectonics. Most interactions of people with the rock cycle involve the sedimentary and volcanic components of the cycle.

A B C D E

FIGURE 2.12 Metamorphic rocks have undergone mineralogical, chemical, and/or structural change. (A) Marble, metamorphosed limestone. (B) Quartzite, metamorphosed sandstone. (C) Slate. (D) Schist. (E) Gneiss.

(B) © The McGraw-Hill Companies Inc./Bob Coyle, photographer, (C) photograph by P. D. Rowley, courtesy USGS Photo Library, Denver, CO.

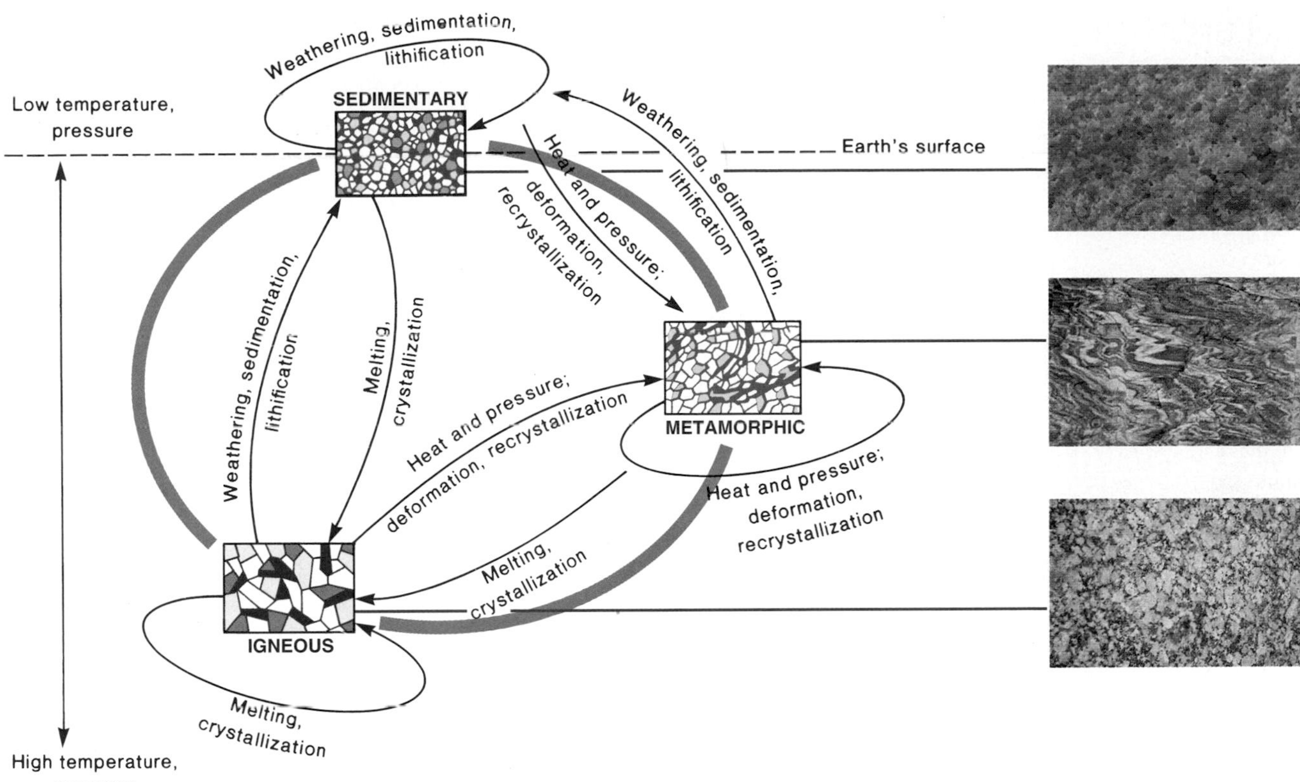

FIGURE 2.13 The rock cycle—a schematic view. Basically, a variety of geologic processes can transform any rock into a new rock of the same or a different class. The geologic environment is not static; it is constantly changing. The full picture, too, is more complex: any type of rock may be found at the earth's surface, and weathered; though the melts that form volcanic rocks are created at depth, the melts crystallize into rock near the surface; and so on.
Igneous rock photograph by N. K. Huber, USGS Photo Library, Denver, CO; other photographs © The McGraw-Hill Companies Inc./Doug Sherman, photographer

SUMMARY

The smallest possible unit of a chemical element is an atom. Isotopes are atoms of the same element that differ in atomic mass number; chemically, they are indistinguishable. Atoms may become electrically charged ions through the gain or loss of electrons. When two or more elements combine chemically in fixed proportions, they form a compound.

Minerals are naturally occurring inorganic solids, each of which is characterized by a particular composition and internal crystalline structure. They may be compounds or single elements. By far the most abundant minerals in the earth's crust and mantle are the silicates. These can be subdivided into groups on the basis of their crystal structures, by the ways in which the silicon and oxygen atoms are arranged. The nonsilicate minerals are generally grouped on the basis of common chemical characteristics.

Rocks are cohesive solids formed from rock or mineral grains or glass. The way in which rocks form determines how they are classified into one of three major groups: igneous rocks, formed from magma; sedimentary rocks, formed from low-temperature accumulations of particles or by precipitation from solution; and metamorphic rocks, formed from pre-existing rocks through the application of heat and pressure. Through time, geologic processes acting on older rocks change them into new and different ones so that, in a sense, all kinds of rocks are interrelated. This concept is the essence of the rock cycle.

Key Terms and Concepts

Exercises

Questions for Review

1. Briefly define the following terms: *ion, isotope, compound, mineral,* and *rock.*
2. What two properties uniquely define a particular mineral?
3. Give the distinctive chemical characteristics of each of the following mineral groups: silicates, carbonates, sulfides, oxides, and native elements.
4. What is an igneous rock? How do volcanic and plutonic rocks differ in texture? Why?
5. What are the two principal classes of sedimentary rocks?
6. Describe how a granite might be transformed into a sedimentary rock.
7. Name several possible sources of the heat or pressure that can cause metamorphism. What kinds of physical changes occur in the rock as a result?
8. What is the rock cycle?

For Further Thought

1. It is generally true that the more ancient rocks are less widely found and that their histories are more difficult to interpret. Consider why this should be so in the context of the rock cycle.
2. What kinds of rocks underlie your region of the country? (Your local geological survey could assist in providing the information.) What does this tell about the region's history? Are there currently any significant geologic hazards in the area? Any identified resources?

Suggested Readings/References

Berry, L. G., Mason, B., and Dietrich, R. V. 1983. *Mineralogy.* 2d ed. San Francisco: W. H. Freeman.

Building stones of our nation's capital. 1998. U.S. Geological Survey.

Dana, J. D. 1985. *Manual of mineralogy.* 20th ed. Revised by C. S. Hurlbut, Jr. and C. Klein. New York: John Wiley & Sons.

Dictionary of geology and mineralogy. 2d ed. 2003. New York: McGraw-Hill.

Ehlers, E. G., and Blatt, H. 1982. *Petrology: Igneous, sedimentary, and metamorphic.* San Francisco: W. H. Freeman.

Gunter, M. E. 1994. Asbestos as a metaphor for teaching risk perception. *Journal of Geological Education* 42: 17–24.

———. 1999. Quartz—The most abundant mineral in the earth's crust and a human carcinogen? *Journal of Geoscience Education* 47: 341–49.

Hibbard, M. J. 2001. *Mineralogy: A geologist's point of view.* New York: McGraw-Hill.

Manutchehr-Danal, M. 2005. *Dictionary of gems and gemology.* 2d ed. New York: Springer-Verlag.

O'Donoghue, M. 1976. *VNR color dictionary of minerals and gemstones.* New York: Van Nostrand Reinhold.

Perkins, D. 2002. *Mineralogy.* 2d ed. New York: Prentice Hall.

Raymond, L. A. 2002. *Petrology: The story of igneous, sedimentary, and metamorphic rocks.* New York: McGraw-Hill.

Sen, G. 2001. *Earth's materials: Rocks and minerals.* New York: Prentice Hall.

Thorpe, R. 2000. *Field guide to igneous rocks and metamorphic rocks set.* New York: John Wiley & Sons.

U.S. Geological Survey. 1998. *Building stones of our nation's capital.* Washington, D.C. U.S. Government Printing Office. Also available online: pubs.usgs.gov/gip/stones

Welcome to McGraw-Hill's Online Learning Center

Location: http://www.mhhe.com/montgomery8e

Mc Graw Hill

NetNotes

A listing of state, U.S., provincial, and various international geological surveys and associations
www.lib.berkeley.edu/EART/surveys.html

Information from and about the U.S. Geological Survey:
geology.usgs.gov

The USGS photo library has thousands of images online:
libraryphoto.er.usgs.gov/

In particular, see "Minerals close-up" at
wrgis.wr.usgs.gov/docs/parks/rxmin/mineral.html

The USGS also has an online booklet, "Natural Gemstones":
pubs.usgs.gov/gip/gemstones/

The U.S. Naval Research Laboratory has put many crystal structures online at
cst-www.nrl.navy.mil/lattice

The Mineralogical Society of America's "Higher Education Connection," with links to other sites dealing with minerals:
www.minsocam.org/

A site with extensive information about, and images of, a great variety of minerals is the Mineralogy Database:
webmineral.com

Links to images of rocks and minerals in the Smithsonian Institution Gem and Mineral Collection of the National Museum of Natural History is found at
www.nmnh.si.edu/minsci/images/gallery/gallery.htm

Pictures of microscope views of thin slices of igneous and metamorphic rocks, which often reveal interesting textural details, are online at
www.geolab.unc.edu/Petunia/IgMetAtlas/mainmenu.html

"The Mineral Gallery," a collection of images of and information about many minerals:
www.theimage.com/mineral/minerals1.html

A site of the University of Calgary provides links to many sites with information about mineral and rock properties and genesis:
www.geo.ucalgary.ca/~tmg/Research/thermo_links.html

A site featuring "Minerals and Gems on Stamps" from many countries, Afghanistan to Zimbabwe, is
stampmin.home.att.net/

SECTION TWO

INTERNAL PROCESSES

The processes that shape the earth can be broadly divided into "internal processes" and "surface processes." In this section, we will explore the causes and effects of the former. While internal processes produce effects at the earth's surface—earthquakes and volcanic eruptions are examples—they are "internal" in the sense that they are mainly caused or driven, directly or indirectly, by the heat of the earth's interior. The surface processes, by contrast, are driven by the external heat of the sun, which supplies the energy to drive the winds and power the hydrologic cycle, as well as by gravity. Some of the earth's internal heat is left over from this planet's accretion and differentiation, and more heat is continually produced by the decay of radioactive elements in the earth.

Plate tectonics, described in detail in chapter 3, provides the conceptual framework for understanding many aspects of earthquakes and volcanoes, which are discussed in chapters 4 and 5. We will see that, far from being haphazard occurrences, most of these phenomena occur in predictable zones on the earth. They are the logical consequences of the shifting of the earth's crust and the resultant creation and destruction of the crust and underlying uppermost mantle rocks. Knowledge of plate tectonics is not only useful for anticipating volcanic and earthquake hazards, either. It is also an important aid to understanding how and where many kinds of mineral deposits are formed, which, in turn, is helpful in the search for new supplies of limited mineral resources.

Folding and faulting resulting from the collision of continental landmasses. Cook Inlet, Alaska. Photograph by N. J. Silberling, USGS Photo Library, Denver, CO.

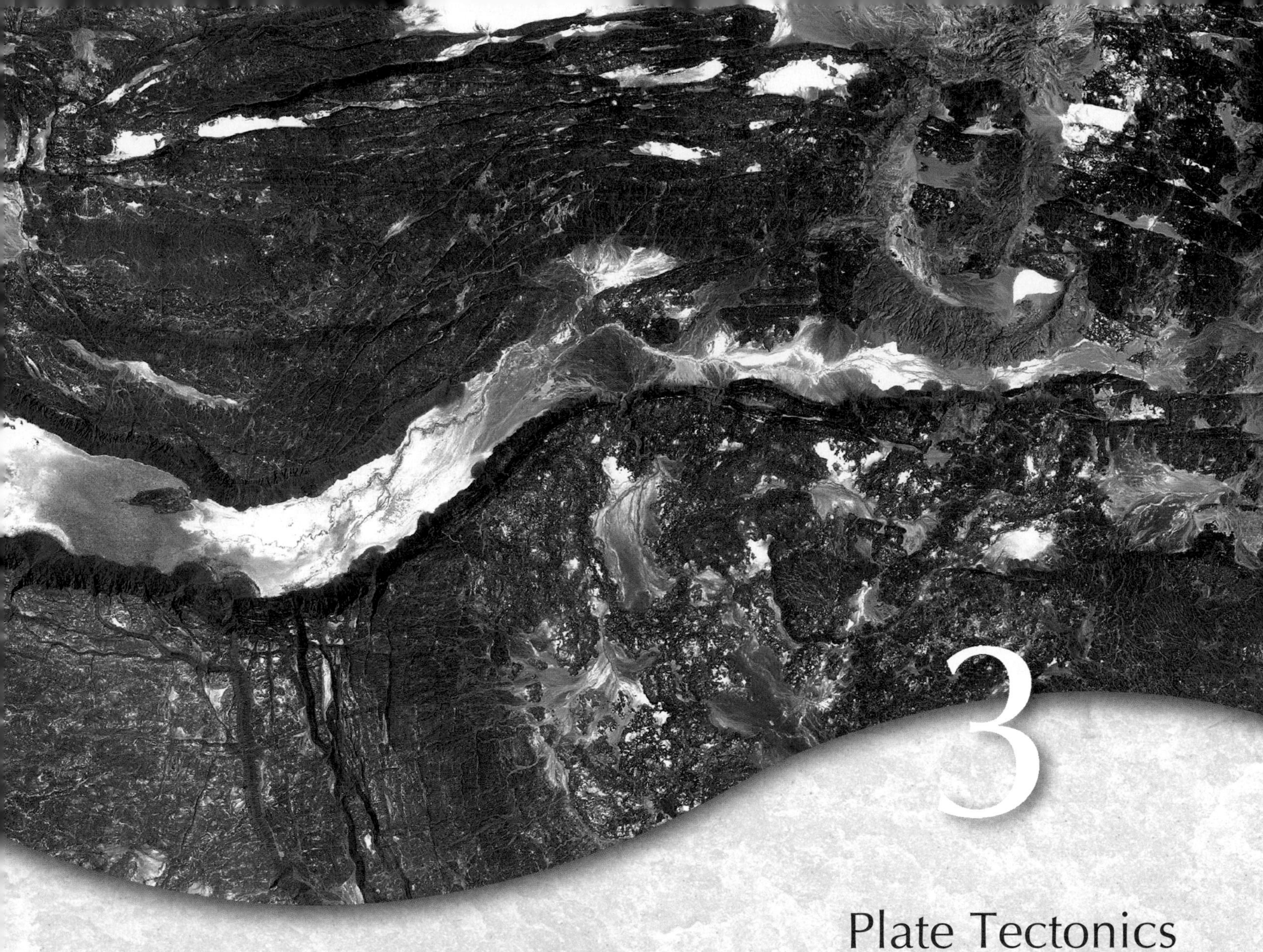

3

Plate Tectonics

Several centuries ago, observers looking at global maps noticed the similarity in outline of the eastern coast of South America and the western coast of Africa (figure 3.1). In 1855, Antonio Snider went so far as to publish a sketch showing how the two continents could fit together, jigsaw-puzzle fashion. Such reconstructions gave rise to the bold suggestion that perhaps these continents had once been part of the same landmass, which had later broken up.

This concept of **continental drift** had an especially vocal champion in Alfred Wegener, who began to publish his ideas in 1912 and continued to do so for nearly two decades. Several other prominent scientists found the idea intriguing. However, most people, scientists and nonscientists alike, had difficulty visualizing how something as massive as a continent could possibly "drift" around on a solid earth, or why it should do so. In other words, no mechanism for moving continents was apparent. There were obvious physical objections to solid continents plowing through solid ocean basins, and no evidence of the expected resultant damage in crushed and shattered rock on continents or sea floor. A great many reputable scientists either scoffed at the idea, or, at best, politely ignored it.

As it turns out, the supporting evidence was simply undiscovered or unrecognized at the time. Beginning in the 1960s (thirty years after Wegener's death), data of many different kinds began to accumulate that indicated that the continents have indeed moved. Continental drift turned out to be just one aspect of a broader theory known as *plate tectonics,* which has evolved over the last several decades. **Tectonics** is the study of large-scale movement and deformation of the earth's outer layers. **Plate tectonics** relates such deformation to the existence and movement of rigid "plates" over a weak, plastic layer in the earth's upper mantle.

In the Afar region of Ethiopia, three plates move apart, rending the continental crust. The large valley at the center formed when blocks of crust dropped into a rift thus formed; more cracks due to rifting appear at bottom.
Image courtesy NASA/GSFC/METI/ERSDAC/JAROS, and U.S./Japan ASTER Science Team

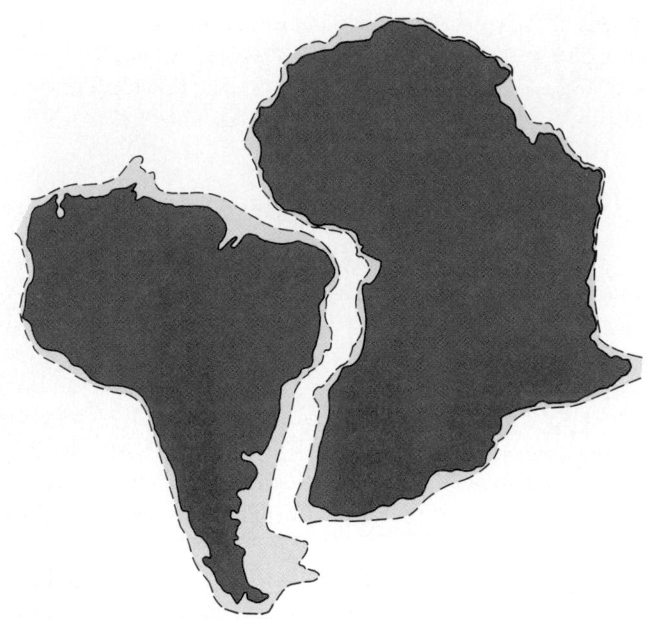

FIGURE 3.1 The jigsaw-puzzle fit of South America and Africa suggests that they were once joined together and subsequently separated by continental drift. (Blue shaded area is the continental shelf, the shallowly submerged border of a continent.)

PLATE TECTONICS—UNDERLYING CONCEPTS

One of the many obstacles to accepting the concept of continental drift was imagining solid continents moving over solid earth. However, as already noted, the earth is not rigidly solid from the surface to the center of the core. In fact, a plastic, locally partly molten zone lies relatively close to the surface so that, in a sense, a thin, solid shell of rock floats on a semisolid layer below. The existence of plates, and the occurrence of earthquakes in them, reflect the way rocks respond to stress.

Stress and Strain in Geologic Materials

An object is under **stress** when force is being applied to it. The stress may be **compressive,** tending to squeeze or compress the object, or it may be **tensile,** tending to pull the object apart. A **shearing stress** is one that tends to cause different parts of the object to move in different directions across a plane or to slide past one another, as when a deck of cards is spread out on a tabletop by a sideways sweep of the hand.

Strain is deformation resulting from stress. It may be either temporary or permanent, depending on the amount and type of stress and on the physical properties of the material. If **elastic deformation** occurs, the amount of deformation is proportional to the stress applied, and the material returns to its original size and shape when the stress is removed. A gently stretched rubber band or squeezed tennis ball shows elastic behavior. Rocks, too, may behave elastically, although much greater stress is needed to produce detectable strain. Once the **elastic limit** of a material is reached, the material may go through a phase of **plastic deformation** with increased stress. During this stage, relatively small added stresses yield large corresponding strains, and the changes are permanent: the material does not return to its original size and shape after removal of the stress. A glassblower, an artist shaping clay, a carpenter fitting caulk into cracks around a window, and a blacksmith shaping a bar of hot iron into a horseshoe are all making use of the plastic behavior of materials. In rocks, folds result from plastic deformation (figure 3.2A). A **ductile** material is one that can undergo extensive plastic deformation without breaking.

If stress is increased sufficiently, most solids eventually break, or **rupture.** In **brittle** materials, rupture may occur before there is any plastic deformation. Brittle behavior is characteristic of most rocks at near-surface conditions, with low temperature and low confining (surrounding) pressures. It leads to faults and fractures (figure 3.2B). Graphically, brittle behavior is illustrated by line A in figure 3.3. At greater depths,

A

B

FIGURE 3.2 (A) Folding such as this occurs while rocks are deeper in the crust, at elevated temperatures and confining pressure. (B) This rock suffered brittle failure under stress, which produced a small fault. (Uneven thicknesses of layers at bottom suggest previous plastic deformation.)

(A) Photograph by M. R. Mudge, (B) by W. B. Hamilton, both USGS Photo Library, Denver, CO.

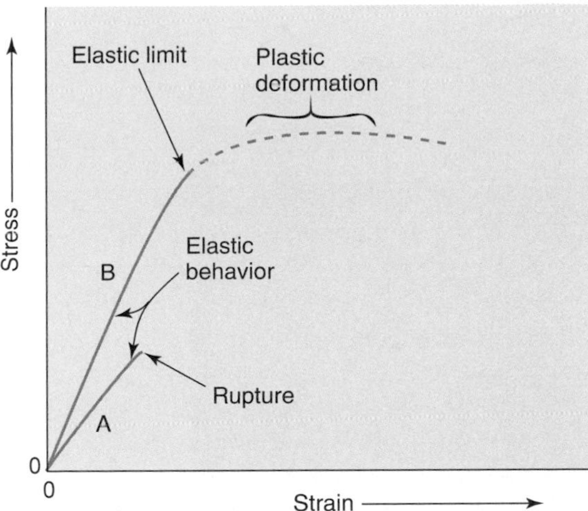

FIGURE 3.3 A stress-strain diagram. Elastic materials deform in proportion to the stress applied (straight-line segments). A very brittle material (curve A) may rupture before the elastic limit is reached. Other materials subjected to stresses above their elastic limits deform plastically until rupture (curve B).

where temperatures are higher and rocks are confined and, in a sense, supported by surrounding rocks at high pressure, rocks may be ductile (dashed section of curve B in figure 3.3). The effect of temperature can be seen in the behavior of cold and hot glass. A rod of cold glass is brittle and snaps under stress before appreciable plastic deformation occurs, while a warmed glass rod may be bent and twisted without breaking.

The physical behavior of a rock, then, is affected by external factors, such as temperature or confining pressure, as well as by the intrinsic characteristics of the rock. Also, rocks respond differently to different types of stress. Most are far stronger under compression than under tension; a given rock may fracture under a tensile stress only one-tenth as high as the compressive stress required to break it at the same pressure and temperature conditions. Consequently, the term *strength* has no single simple meaning when applied to rocks unless all of these variables are specified. Even when they are, rocks in their natural settings do not always exhibit the same behavior as carefully selected samples tested in the laboratory. The natural samples may be weakened by fracturing or weathering, for example. Several rock types of quite different properties may be present at the same site. Considerations such as these complicate the work of the geological engineer.

Time is also a factor in the physical behavior of a rock. Materials may respond differently to given stresses, depending on the rate, the period of time over which the stress is applied. (This can be seen on a human timescale in the phenomenon of fatigue: a machine component or structural material may fail under stresses well below its rupture strength if the stress has been applied repeatedly over a period of time.) A ball of putty will bounce elastically when dropped on a hard surface but deform plastically if pulled or squeezed more slowly. Rocks can likewise respond differently depending on how stress is applied, showing elastic behavior if stressed suddenly, as by passing seismic waves from an earthquake, but ductile behavior in response to prolonged stress, as from the weight of overlying rocks or the slow movements of plates.

Lithosphere and Asthenosphere

The earth's crust and uppermost mantle are solid, somewhat brittle and elastic. This outer solid layer is called the **lithosphere,** from the Greek word *lithos,* meaning "rock." The lithosphere varies in thickness from place to place on the earth. It is thinnest underneath the oceans, where it extends to a depth of about 50 kilometers (about 30 miles). The lithosphere under the continents is both thicker on average than is oceanic lithosphere, and more variable in thickness, extending in places to about 250 kilometers (over 150 miles) beneath the highest mountain ranges.

The layer below the lithosphere is the **asthenosphere,** which derives its name from the Greek word *asthenes,* meaning "without strength." The asthenosphere extends to an average depth of about 300 kilometers (close to 200 miles) in the mantle. Its lack of strength or rigidity results from a combination of high temperatures and moderate confining pressures that allows the rock to flow plastically under stress. Below the asthenosphere, as pressures increase faster than temperatures with depth, the mantle again becomes more rigid and elastic.

The asthenosphere was discovered by studying the behavior of seismic waves from earthquakes. Its presence makes the concept of continental drift more plausible. The continents need not scrape across or plow through solid rock; instead, they can be pictured as sliding over a softened, deformable layer underneath the lithospheric plates. The relationships among crust, mantle, lithosphere, and asthenosphere are illustrated in figure 3.4.

Recognition of the existence of the plastic asthenosphere made plate motions more plausible, but it did not prove that they had occurred. Much additional information had to be accumulated before the rates and directions of plate movements could be documented and before most scientists would accept the concept of plate tectonics.

Locating Plate Boundaries

The distribution of earthquakes and volcanic eruptions indicates that these phenomena are far from uniformly distributed over the earth (figures 4.6, 5.4). They are, for the most part, concentrated in belts or linear chains. This is consistent with the idea that the rigid shell of lithosphere is cracked in places, broken up into pieces, or plates. The volcanoes and earthquakes are concentrated at the boundaries of these lithospheric plates, where plates jostle or scrape against each other. (The effect is somewhat like ice floes on an arctic sea: most of the grinding and crushing of ice and the spurting-up of water from below occur at the edges of the blocks of ice, while their solid central portions are relatively undisturbed.) Fewer than a dozen very

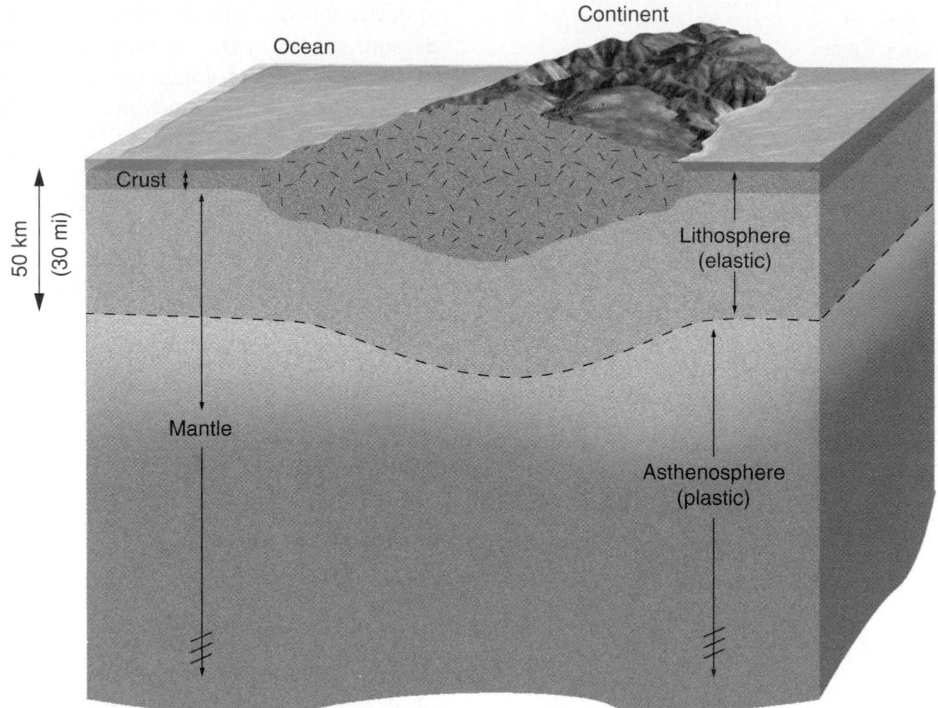

FIGURE 3.4 The outer zones of the earth (not to scale). The terms *crust* and *mantle* have compositional implications (see chapter 1); *lithosphere* and *asthenosphere* describe physical properties (the former, elastic and more rigid; the latter, plastic). The lithosphere includes the crust and uppermost mantle. The asthenosphere lies entirely within the upper mantle. Below it, the rest of the mantle is more rigidly solid again.

large lithospheric plates have been identified; as research continues, many smaller ones have been recognized in addition.

PLATE TECTONICS—ACCUMULATING EVIDENCE

The continental "jigsaw puzzle" was not the only evidence Wegener sought to explain via continental drift. Continental rocks form under a wide range of conditions and yield varied kinds of information. Sedimentary rocks, for example, may preserve evidence of the ancient climate of the time and place in which the sediments were deposited. Such evidence shows that the climate in many places has varied widely through time. We find evidence of extensive glaciation in places now located in the tropics, in parts of Australia, southern Africa, and South America (figure 3.5). There are desert sand deposits in the rocks of regions that now have moist, temperate climates and the remains of jungle plants in now-cool places. There are coal deposits in Antarctica, even though coal deposits form from the remains of a lush growth of land plants. These observations, some of which were first made centuries ago, cannot be explained as the result of global climatic changes, for the continents do not all show the same warming or cooling trends at the same time. However, climate is to a great extent a function of latitude, which strongly influences surface temperatures: Conditions are generally warmer near the equator and colder near the poles. Dramatic shifts in an individual continent's climate might result from changes in its latitude in the course of continental drift.

Sedimentary rocks also preserve fossil remains of ancient life. Some plants and animals, long extinct, seem to have lived only in a few very restricted areas, which now are widely separated geographically on different continents (figure 3.6). One example is the fossil plant *Glossopteris,* remains of which are found in limited areas of widely separated lands including India, southern Africa, and even Antarctica. The fossils of a small reptile, *Mesosaurus,* are similarly dispersed across two continents. The distribution of these fossils was, in fact, part of the evidence Wegener cited to support his continental-drift hypothesis. It is difficult to imagine one distinctive variety of animal or plant developing simultaneously in two or more small areas thousands of kilometers apart, or somehow migrating over vast expanses of ocean. (Besides, *Glossopteris* was a land plant; *Mesosaurus,* a freshwater animal.) The idea of continental drift provided a way out of the quandary. The organism may have lived in a single, geographically restricted area, and the rocks in which its remains are now found subsequently were separated and moved in several different directions by drifting continents.

Continental reassembly could be refined using details of continental geology—rock types, rock ages, fossils, ore deposits, mountain ranges, and so on. If two now-separate continents were once part of the same landmass, then the geologic features presently found at the margin of one should have counterparts

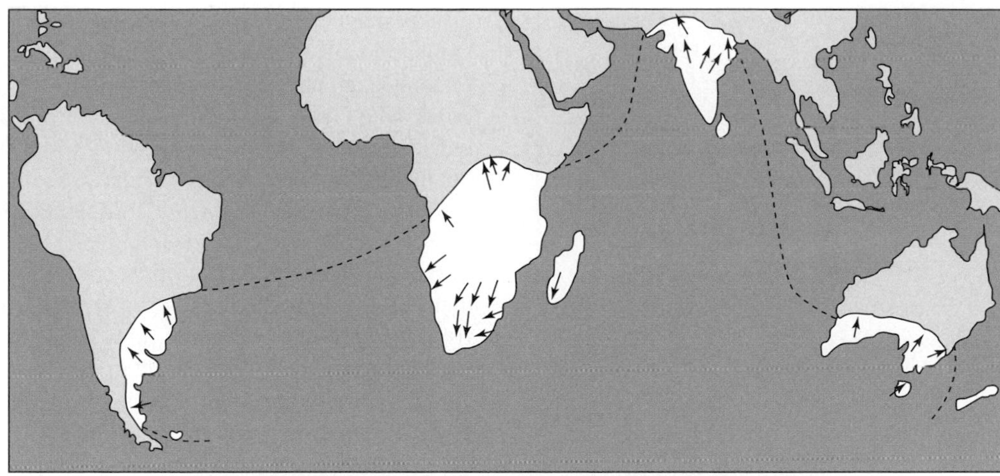

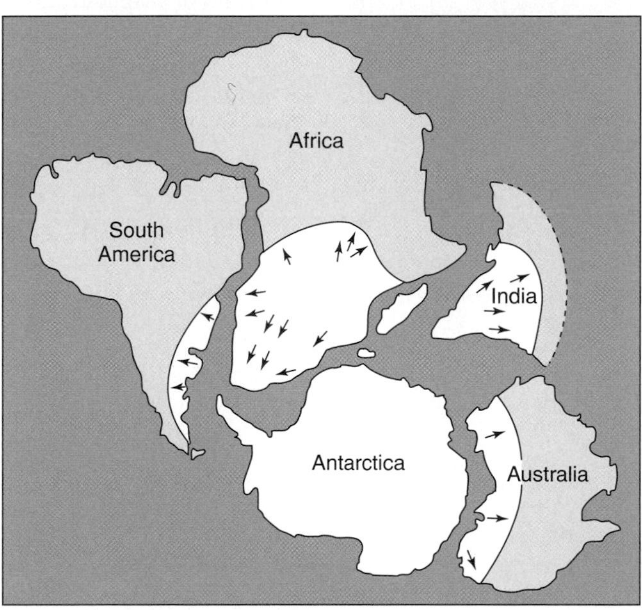

FIGURE 3.5 Glacial deposits across the southern continents suggest past juxtaposition. Arrows indicate directions of ice flow.
After Arthur Holmes, Principles of Physical Geology, *2d ed., Ronald Press, New York, NY, 1965*

at the corresponding edge of the other. In short, the geology should match when the puzzle pieces are reassembled. And it typically does—for example, the Appalachian Mountains of North America continue into the counterpart Caledonides of Greenland and the British Isles. Still, without a demonstrable mechanism for shifting continents around, continental drift left most geologists unconvinced.

The Topography of the Sea Floor

Those who first speculated about possible drifting continents could not examine the sea floor for any relevant evidence. Once topographic maps of the sea floor became available, intriguing features were revealed (figure 3.7). For example, beneath the Atlantic Ocean is an obvious ridge, running north-south about halfway between North and South America to the west, and Europe and Africa to the east. Similar ridges are found on the floors of other oceans, as seen in the same figure. They rise as high as 2 km (1.3 mi.) above the surrounding plains. In other places, frequently along the margins of continents, there are trenches several kilometers deep. It remained for scientists studying the ages and magnetic properties of seafloor rocks, as described below, to demonstrate the significance of the ocean ridges and trenches to plate tectonics.

Magnetism in Rocks—General

The rocks of the ocean floors are rich in ferromagnesian minerals, and such minerals are abundant in many rocks on the continents as well. Most iron-bearing minerals are at least

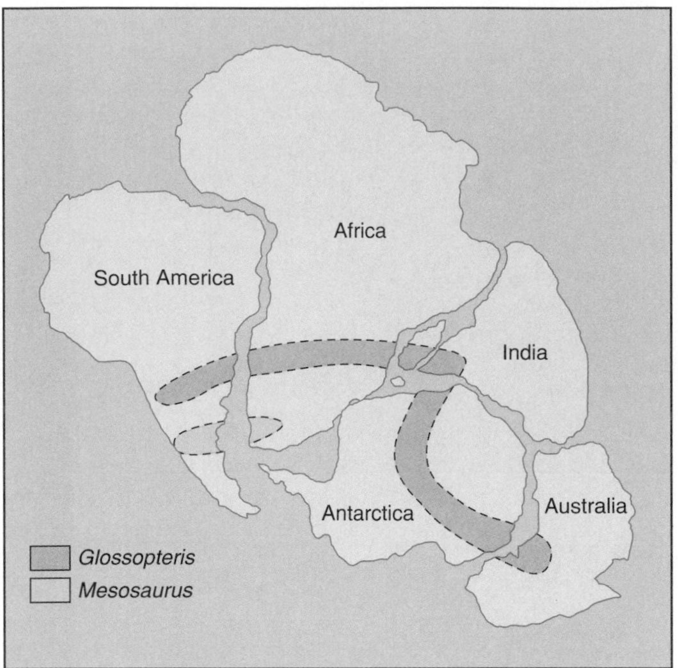

FIGURE 3.6 Curious distribution of fossil remains of the fern *Glossopteris* and reptile *Mesosaurus* could be explained most easily by continental drift.

weakly magnetic at surface temperatures. Each magnetic mineral has a **Curie temperature,** the temperature below which it remains magnetic, but above which it loses its magnetic properties. The Curie temperature varies from mineral to mineral, but it is always below the mineral's melting temperature. A hot magma is therefore not magnetic, but as it cools and solidifies, and iron-bearing magnetic minerals (particularly magnetite) crystallize from it, those magnetic crystals tend to line up in the same direction. Like tiny compass needles, they align themselves parallel to the lines of force of the earth's magnetic field, which run north-south, and they point to the magnetic north pole (figure 3.8). They retain their internal magnetic orientation unless they are heated again. This is the basis for the study of **paleomagnetism,** "fossil magnetism" in rocks.

Magnetic north, however, has not always coincided with its present position. In the early 1900s, scientists investigating the direction of magnetization of a sequence of volcanic rocks in France discovered some flows that appeared to be magnetized in the opposite direction from the rest: their magnetic minerals pointed south instead of north. Confirmation of this discovery in many places around the world led to the suggestion in the late 1920s that the earth's magnetic field had "flipped," or reversed

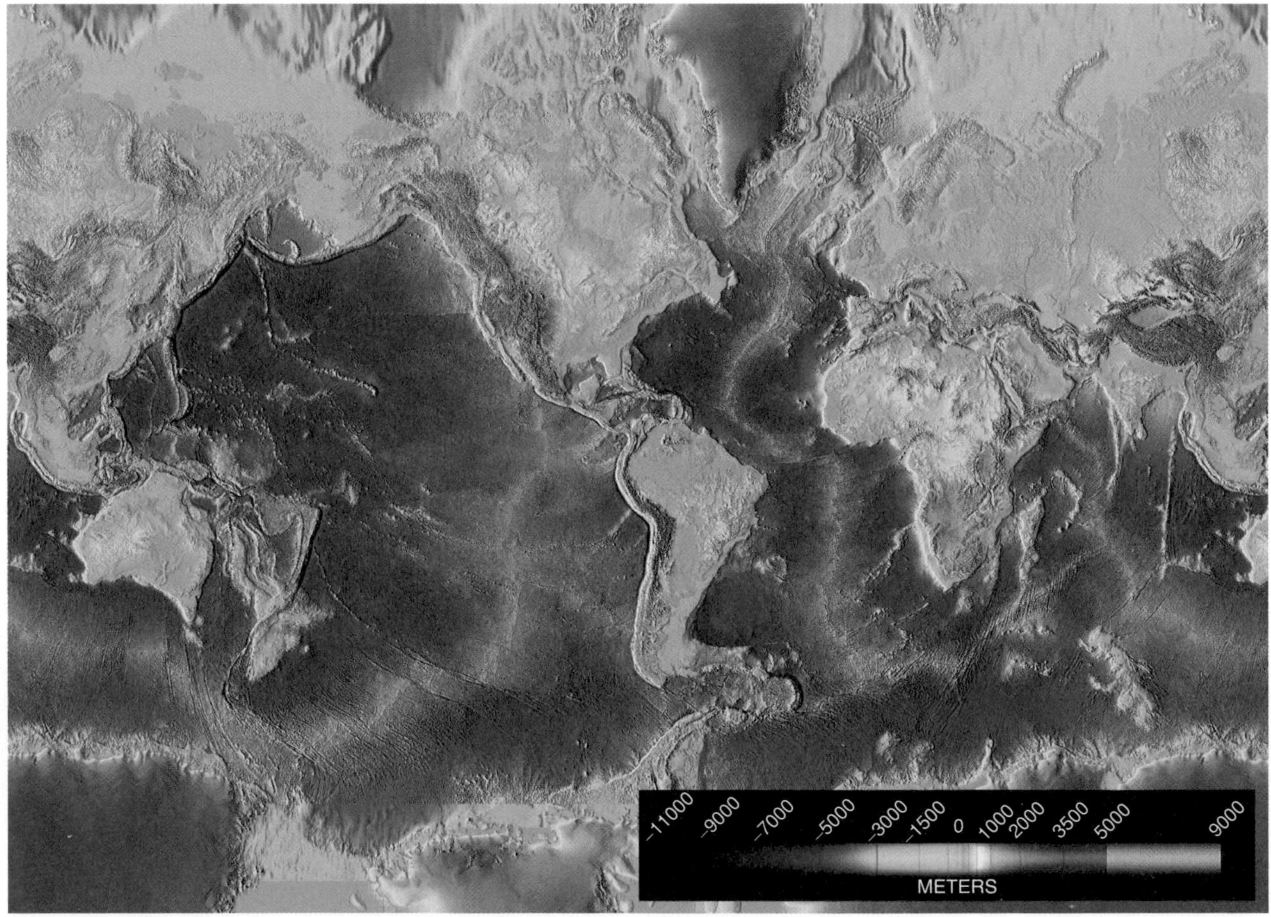

FIGURE 3.7 Global relief map of the world, including seafloor topography. Note ridges and trenches on the sea floor.
Source: Image courtesy of NOAA/National Geophysical Data Center

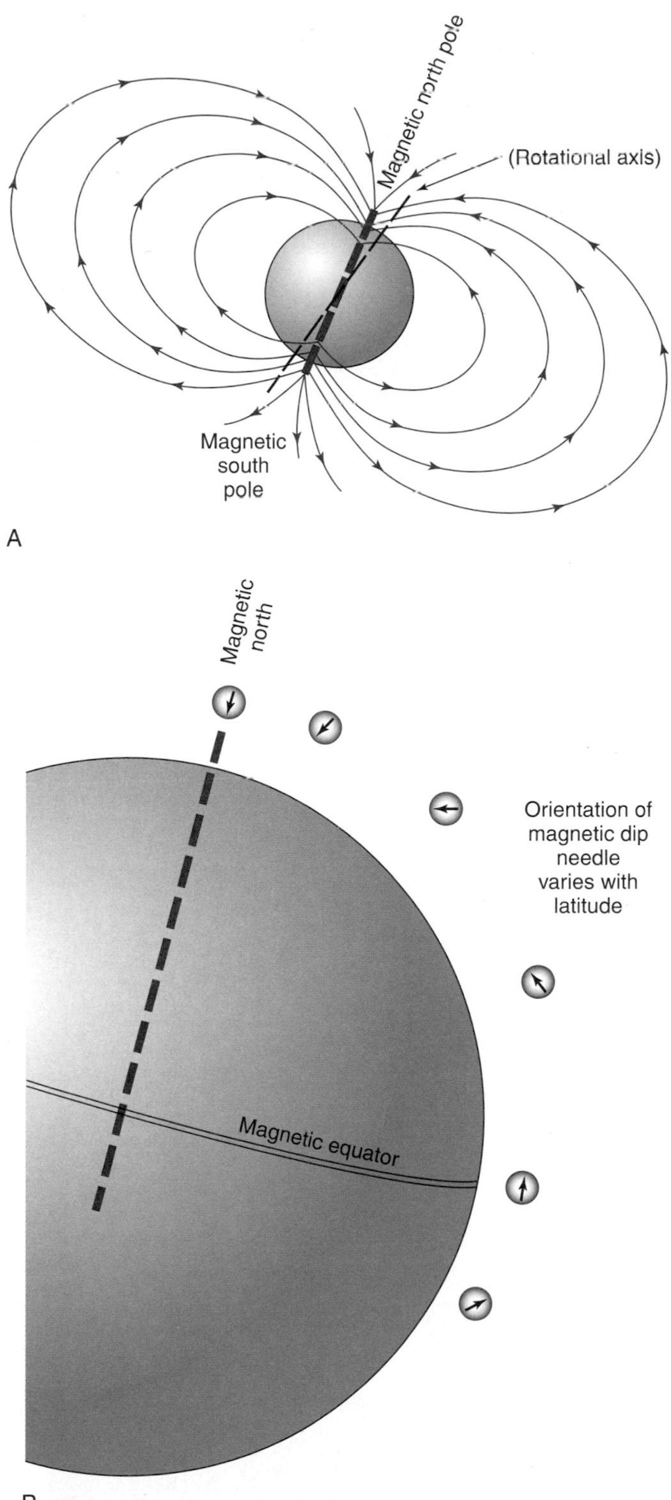

FIGURE 3.8 (A) Lines of force of earth's magnetic field. (B) A magnetized needle—or magnetic mineral's magnetism—will dip more steeply nearer the poles, just as the lines of force of the earth's field do.

polarity; that is, that the north and south poles had switched places. During the time those surprising rocks had crystallized, a compass needle would have pointed to the magnetic south pole, not north.

Today, the phenomenon of magnetic reversals is well documented. Rocks crystallizing at times when the earth's field was in the same orientation as it is at present are said to be normally magnetized; rocks crystallizing when the field was oriented the opposite way are described as reversely magnetized. Over the history of the earth, the magnetic field has reversed many times, at variable intervals. Sometimes, the polarity remained constant for millions of years before reversing, while other reversals are separated by only a few tens of thousands of years. Through the combined use of magnetic measurements and age determinations on the magnetized rocks, geologists have been able to reconstruct the reversal history of the earth's magnetic field in some detail.

The explanation for magnetic reversals must be related to the origin of the magnetic field. The outer core is a metallic fluid, consisting mainly of iron. Motions in an electrically conducting fluid can generate a magnetic field, and this is believed to be the origin of the earth's field. (The simple presence of iron in the core is not enough to account for the magnetic field, as core temperatures are far above the Curie temperature of iron.) Perturbations or changes in the fluid motions, then, could account for reversals of the field. The details of the reversal process remain to be determined.

Paleomagnetism and Seafloor Spreading

The ocean floor is made up largely of basalt, a volcanic rock rich in ferromagnesian minerals. During the 1950s, the first large-scale surveys of the magnetic properties of the sea floor produced an entirely unexpected result. As they tracked across a ridge, they recorded alternately stronger and weaker magnetism below.

At first, this seemed so incredible that it was assumed that the instruments or measurements were faulty. However, other studies consistently obtained the same results. For several years, geoscientists strove to find a convincing explanation for these startling observations.

Then, in 1963, an elegant explanation was proposed by the team of F. J. Vine and D. H. Matthews, and independently by L. W. Morley. The magnetic stripes could be explained as a result of alternating bands of normally and reversely magnetized rocks on the sea floor. Adding the magnetism of normally magnetized rocks to earth's current field produced a little stronger magnetization; reversely magnetized rocks would counter the earth's (much stronger) field somewhat, resulting in apparently lower net magnetic strength.

These bands or "stripes" of normally and reversely magnetized rocks were also parallel to, and symmetrically arranged on either side of, the seafloor ridges. Thus was born the concept of **seafloor spreading,** the moving apart of lithospheric plates at the ocean ridges. If the oceanic lithosphere splits and plates move apart, a rift in the lithosphere begins to open. But the result is not a 50-kilometer-deep crack. As the plates are pulled apart, asthenosphere can flow upward. The resulting release of pressure allows extensive melting to occur in the asthenosphere, and the magma rises up into the rift to form new sea floor. As the magma cools and solidifies to form new basaltic rock, that rock becomes magnetized in the prevailing direction of the earth's magnetic field. If the plates continue to move apart, the new rock will also split and part, making way for more magma to form still younger rock, and so on.

If the polarity of the earth's magnetic field reverses during the course of seafloor spreading, the rocks formed after a reversal are polarized oppositely from those formed before it. The ocean floor is a continuous sequence of basalts formed over tens or hundreds of millions of years, during which time there have been dozens of polarity reversals. The basalts of the sea floor have acted as a sort of magnetic tape recorder throughout that time, preserving a record of polarity reversals in the alternating bands of normally and reversely magnetized rocks. The process is illustrated schematically in figure 3.9.

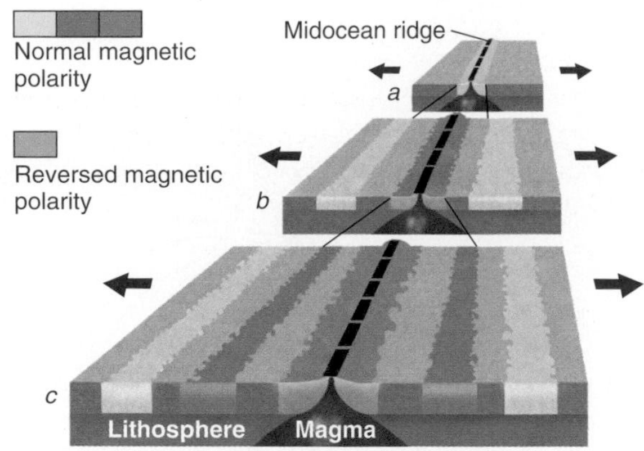

FIGURE 3.9 Seafloor spreading leaves a trail of bands of normally and reversely magnetized rocks on the sea floor. Here, (*a*) represents about 5 million years ago; (*b*) 2 to 3 million years ago; (*c*) the present.
From USGS publication "This Dynamic Earth" by W. J. Kious and R. I. Tilling

Age of the Ocean Floor

The ages of seafloor basalts themselves lend further support to this model of seafloor spreading. Specially designed research ships can sample sediment from the deep-sea floor and drill into the basalt beneath.

The time at which an igneous rock, such as basalt, crystallized from its magma can be determined by methods described in appendix A. When this is done for many samples of seafloor basalt, a pattern emerges. The rocks of the sea floor are youngest close to the ocean ridges and become progressively older the farther away they are from the ridges on either side (see figure 3.10). Like the magnetic stripes, the age pattern is

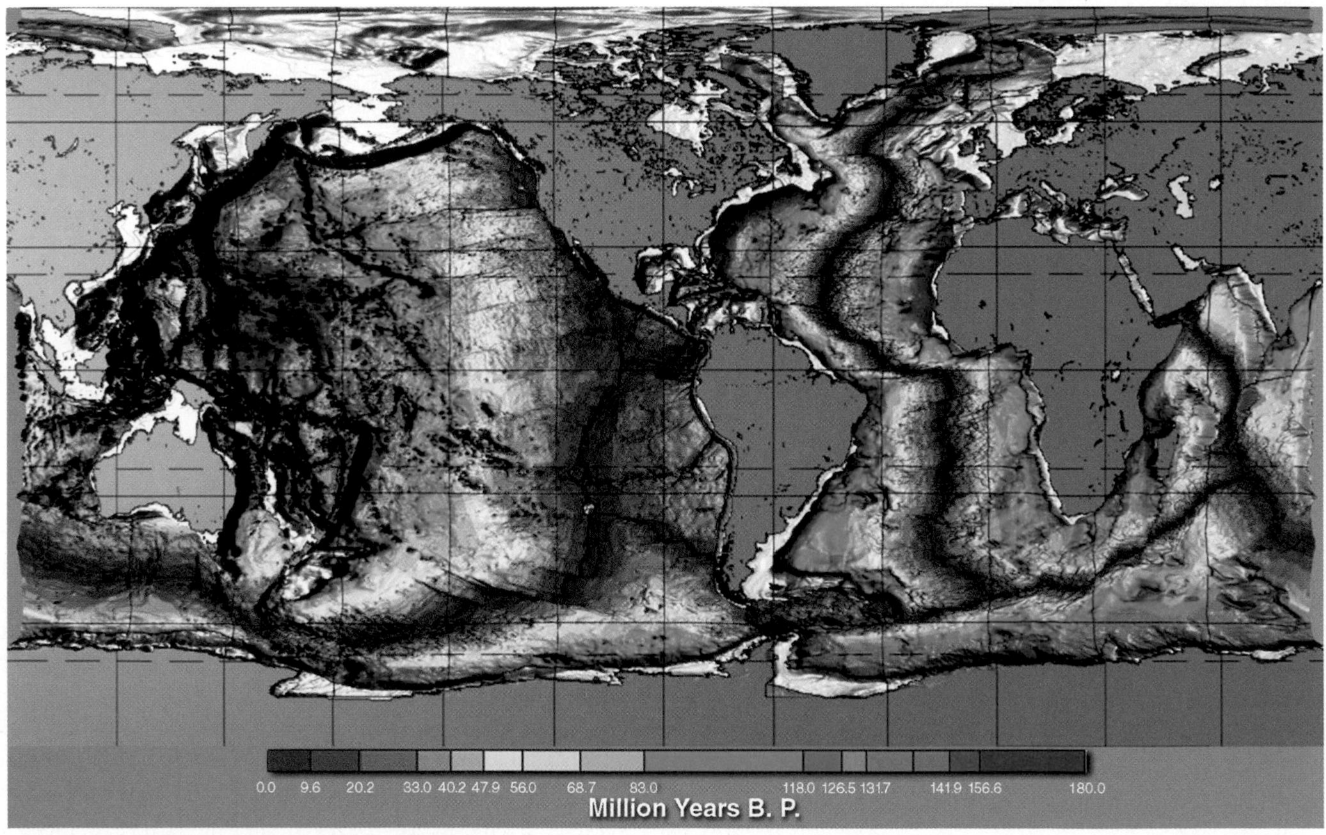

FIGURE 3.10 Age distribution of the sea floor superimposed on a shaded relief map. ("B.P." means "before present"). Note relative spreading rates of Mid-Atlantic Ridge (slower) and East Pacific Rise (faster), shown by wider color bands in the Pacific.
Source: Marine Geology and Geophysics Division of the NOAA National Geophysical Data Center

symmetric across each ridge. As seafloor spreading progresses, previously formed rocks are continually spread apart and moved farther from the ridge, while fresh magma rises from the asthenosphere to form new lithosphere at the ridge. The oldest rocks recovered from the sea floor, well away from the ridges, are about 200 million years old.

Polar-Wander Curves

Evidence for plate movements does not come only from the sea floor. For reasons outlined later in the chapter, much older rocks are preserved on the continents than in the ocean—some continental samples are over 4 billion years old—so longer periods of earth history can be investigated through continental rocks. Studies of paleomagnetic orientations of continental rocks can span many hundreds of millions of years and yield quite complex data.

The lines of force of earth's magnetic field not only run north-south, they vary in dip with latitude: vertical at the magnetic poles, horizontal at the equator, at varying intermediate dips in between, as was shown in figure 3.8B. When the orientation of magnetic minerals in a rock is determined in three dimensions, one can thus determine not only the direction of magnetic north, but (magnetic) latitude as well, or how far removed that region was from magnetic north at the time the rock's magnetism assumed its orientation.

Magnetized rocks of different ages on a single continent may point to very different apparent magnetic pole positions. The magnetic north and south poles may not simply be reversed, but may be rotated or tilted from the present magnetic north and south. When the directions of magnetization and latitudes of many rocks of various ages from one continent are determined and plotted on a map, it appears that the magnetic poles have meandered far over the surface of the earth—*if* the position of the continent is assumed to have been fixed on the earth throughout time. The resulting curve, showing the apparent movement of the magnetic pole relative to the continent as a function of time, is called the **polar-wander curve** for that continent (figure 3.11). We know now, however, that it isn't the poles that have "wandered."

From their discovery, polar-wander curves were puzzling because there are good geophysical reasons to believe that the earth's magnetic poles should remain close to the geographic (rotational) poles. (In particular, the fluid motions in the outer core that cause the magnetic field could be expected to be strongly influenced by the earth's rotation.) Additionally, the polar-wander curves for different continents do not match. Rocks of exactly the same age from two different continents may seem to point to two very different magnetic poles (figure 3.12). This confusion can be eliminated, however, if it is assumed that the magnetic poles have always remained close to the geographic poles but that the *continents* have moved and rotated. The polar-wander curves then provide a way to map the directions in which the continents have moved through time, relative to the stationary poles and relative to each other.

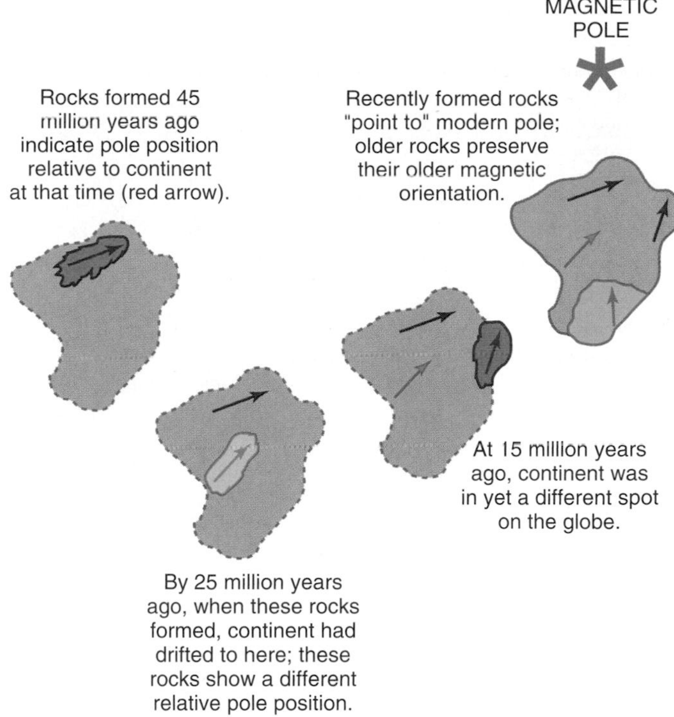

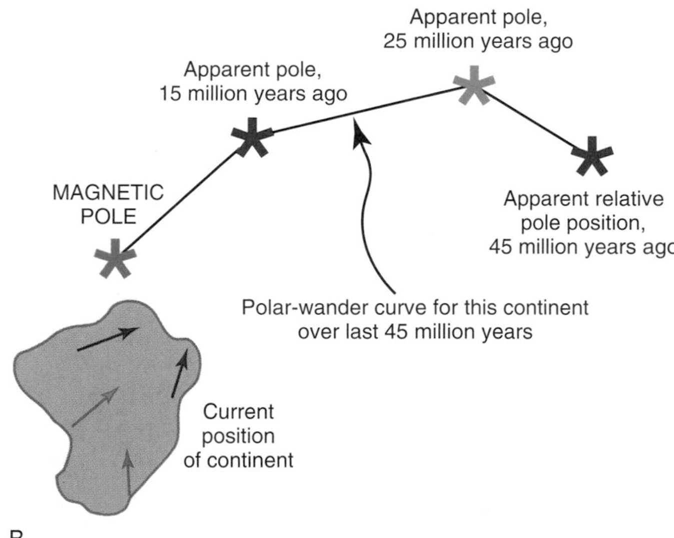

FIGURE 3.11 "Polar-wander curves" actually reflect wandering continents. (A) As rocks crystallize, their magnetic minerals align with the contemporary magnetic field. But continental drift changes the relative position of continent and magnetic pole over time. (B) Assuming a stationary continent, the shifting relative pole positions suggest "polar wander."

Efforts to reconstruct the ancient locations and arrangements of the continents have been relatively successful, at least for the not-too-distant geologic past. It has been shown, for example, that a little more than 200 million years ago, there was a single, great supercontinent that has been named *Pangaea*, from the Greek for "all lands." The present seafloor spreading ridges are the lithospheric scars of the subsequent breakup of

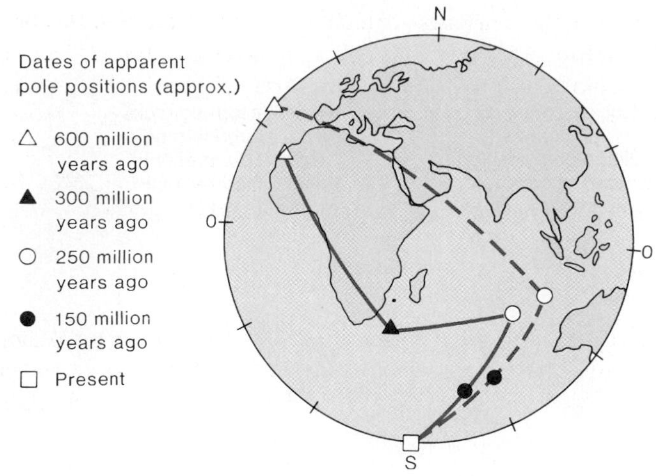

FIGURE 3.12 Examples of polar-wander curves. The apparent position of the magnetic pole relative to the continent is plotted for rocks of different ages, and these data points are connected to form the curve for that landmass. The present positions of the continents are shown for reference. Polar-wander curves for Africa (solid line) and Arabia (dashed line) suggest that these landmasses moved quite independently up until about 250 million years ago, when the polar-wander curves converge.

After M. W. McElhinny, Paleomagnetism and Plate Tectonics. *Copyright @ 1973 by Cambridge University Press, New York, NY. Reprinted by permission*

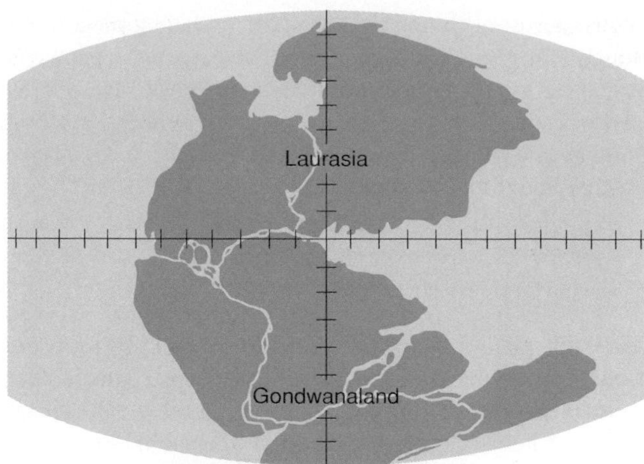

A 200 million years ago

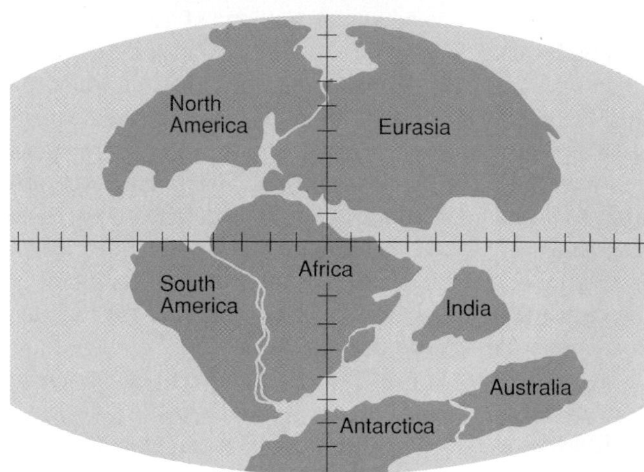

B 100 million years ago

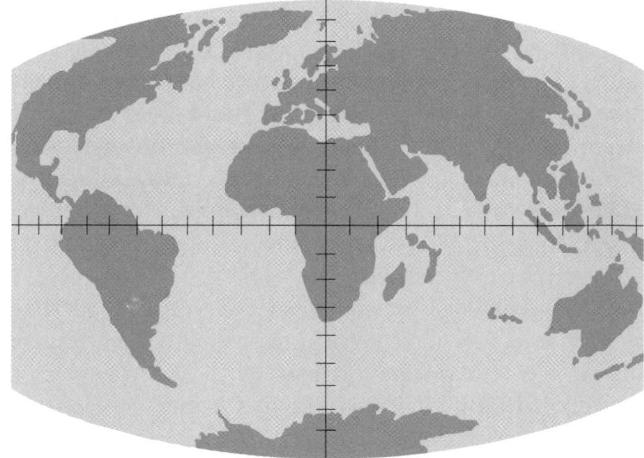

C Today

FIGURE 3.13 Reconstructed plate movements during the last 200 million years. *Laurasia* and *Gondwanaland* are names given to the northern and southern portions of Pangaea, respectively. Although these changes occur very slowly in terms of a human lifetime, they illustrate the magnitude of some of the natural forces to which we must adjust. (A) 200 million years ago. (B) 100 million years ago. (C) Today.

After R. S. Dietz and J. C. Holden, "Reconstruction of Pangaea," Journal of Geophysical Research, *75:4939–4956, 1970, copyright by the American Geophysical Union*

Pangaea (figure 3.13). However, they are not the only kind of boundary found between plates.

TYPES OF PLATE BOUNDARIES

Different things happen at the boundaries between lithospheric plates, depending on the relative motions of the plates and on whether continental or oceanic lithosphere is at the edge of each plate.

Divergent Plate Boundaries

At a **divergent plate boundary,** lithospheric plates move apart. The release of pressure facilitates some melting in the asthenosphere, and magma wells up from the asthenosphere, its passage made easier by deep fractures formed by the tensional stresses. A great deal of volcanic activity thus occurs at divergent plate boundaries. In addition, the pulling-apart of the plates of lithosphere results in earthquakes along these boundaries.

Seafloor spreading ridges are the most common type of divergent boundary worldwide, and we have already noted the formation of new oceanic lithosphere at these ridges. As seawater circulates through this fresh, hot lithosphere, it is heated and reacts with the rock, becoming metal-rich. As it gushes back out of the sea floor, cooling and reacting with cold seawater, it may precipitate potentially valuable mineral deposits. Economically unprofitable to mine now, they may nevertheless prove useful resources for the future.

Continents can be rifted apart, too, and in fact most ocean basins are believed to have originated through continental rifting. The process may be initiated either by tensional forces pulling the plates apart, or by rising hot asthenosphere along the rift zone. In the early stages of continental rifting, volcanoes may erupt along the rift, or great flows of basaltic lava may pour out through the fissures in the continent. If the rifting continues, a new ocean basin will eventually form between the pieces of the continent. This is happening now in east Africa: The easternmost part of the continent is being rifted apart from the rest, and an ocean will one day separate the pieces. In the Afar region of Ethiopia, shown in the chapter-opening satellite image, this rift zone meets two others at what is called a *triple junction.* Along the other two branches—which have formed the Red Sea and the Gulf of Aden—continental separation has proceeded to the point of formation of oceanic lithosphere. Figure 3.14 illustrates the progression from continental rift zone to ocean basin. More-limited continental rifting, now believed to have halted, created the zone of weakness in central North America that includes the New Madrid Fault Zone, still a seismic hazard.

Convergent Plate Boundaries

At a **convergent plate boundary,** as the name indicates, plates are moving toward each other. Just what happens depends on what sort of lithosphere is at the leading edge of each plate

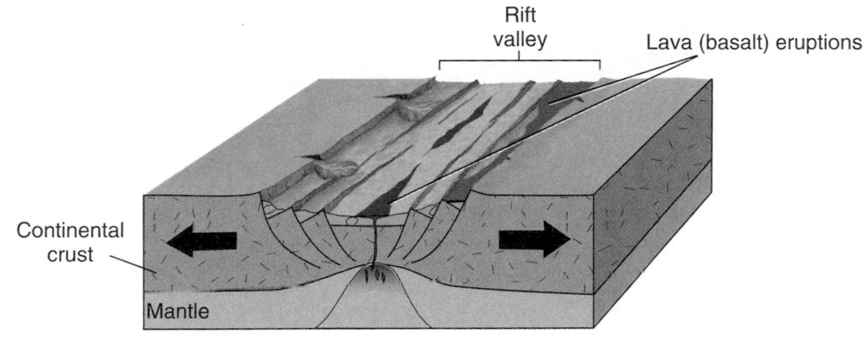

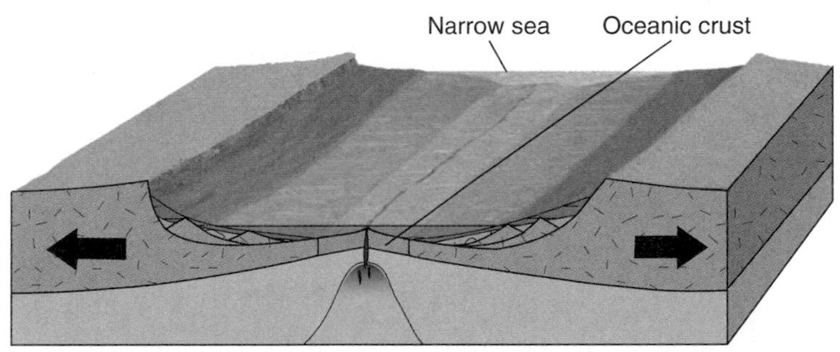

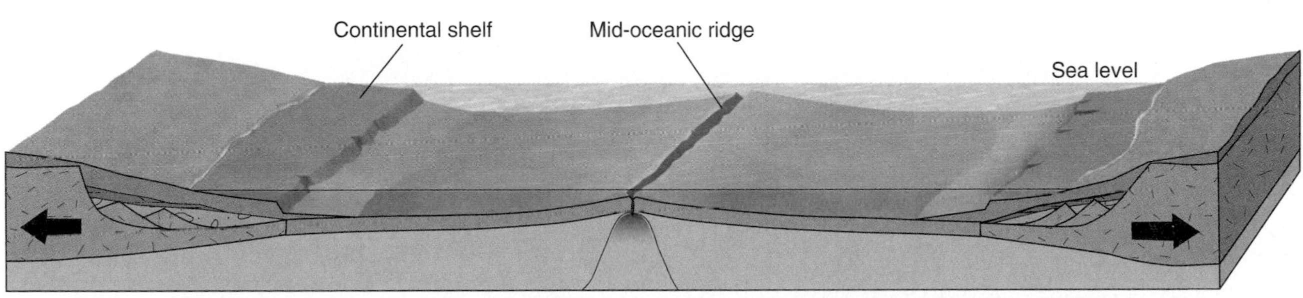

FIGURE 3.14 As continental rifting begins, crust is stretched, thinned, and fractured. Eventually, the continental pieces are fully separated, and oceanic crust is formed between them. The ocean basin widens as divergence continues.

(figure 3.15). Continental lithosphere is relatively low in density and is therefore buoyant with respect to the dense, iron-rich mantle, so it tends to "float" on the asthenosphere. Oceanic lithosphere is more similar in density to the underlying asthenosphere, so it is more easily forced down into the asthenosphere as plates move together.

Most commonly, oceanic lithosphere is at the leading edge of one or both of the converging plates. One plate of oceanic lithosphere may be pushed under the other plate and descend into the asthenosphere. This type of plate boundary, where one plate is carried down below (subducted beneath) another, is called a **subduction zone** (see figure 3.15A,B). The nature of

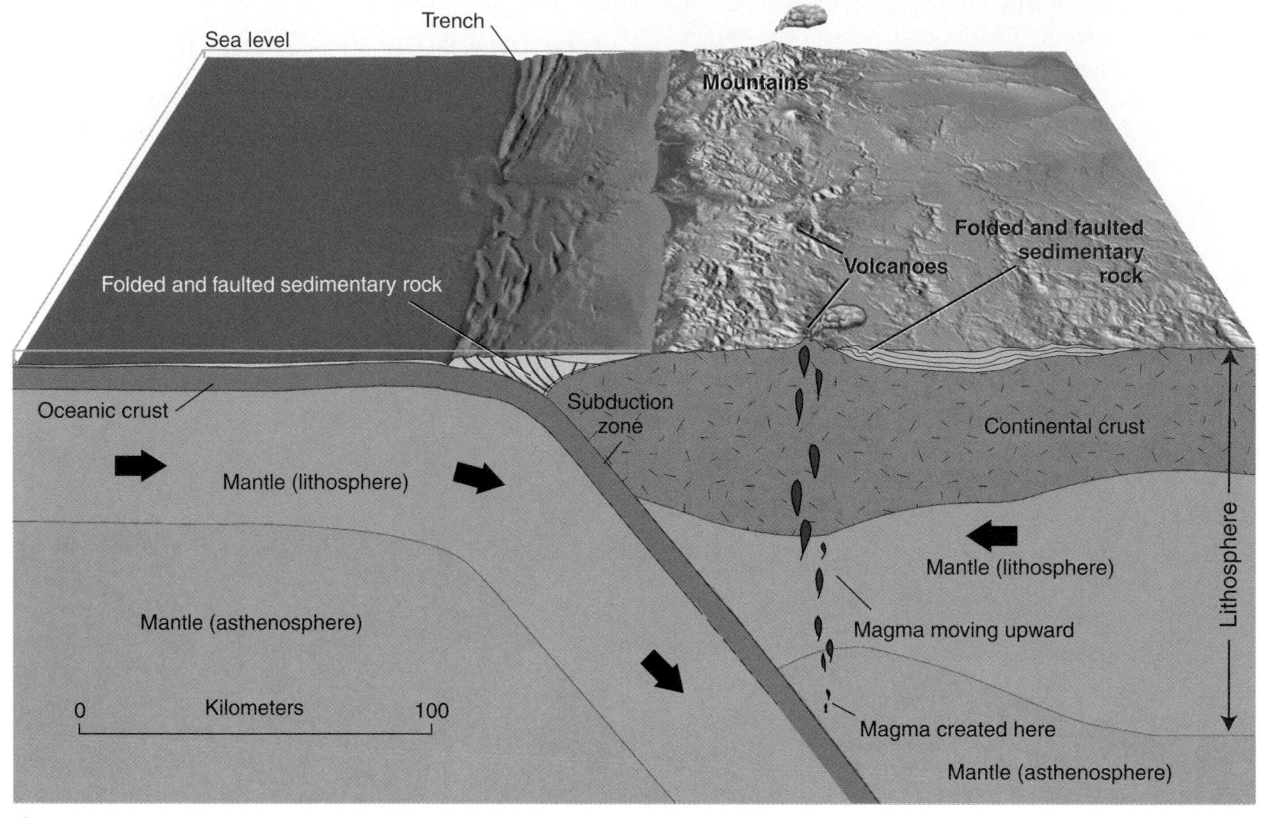

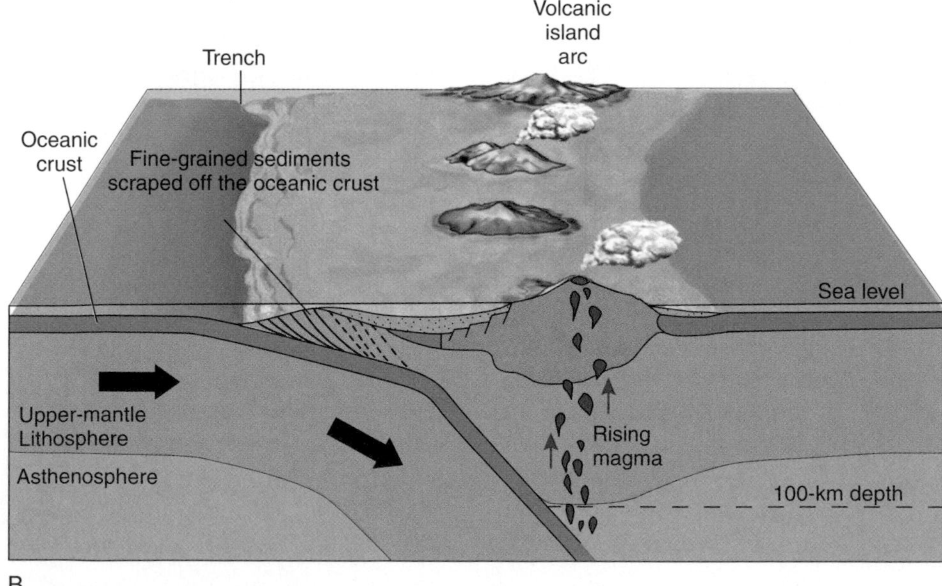

FIGURE 3.15 (A) At ocean-continent convergence, sea floor is consumed, and volcanoes form on the overriding continent. (B) Volcanic activity at ocean-ocean convergence creates a string of volcanic islands. (C) Sooner or later, a continental mass on the subducting plate meets a continent on the overriding plate, and the resulting collision creates a great thickness of continental lithosphere, as in the Himalayas (D).

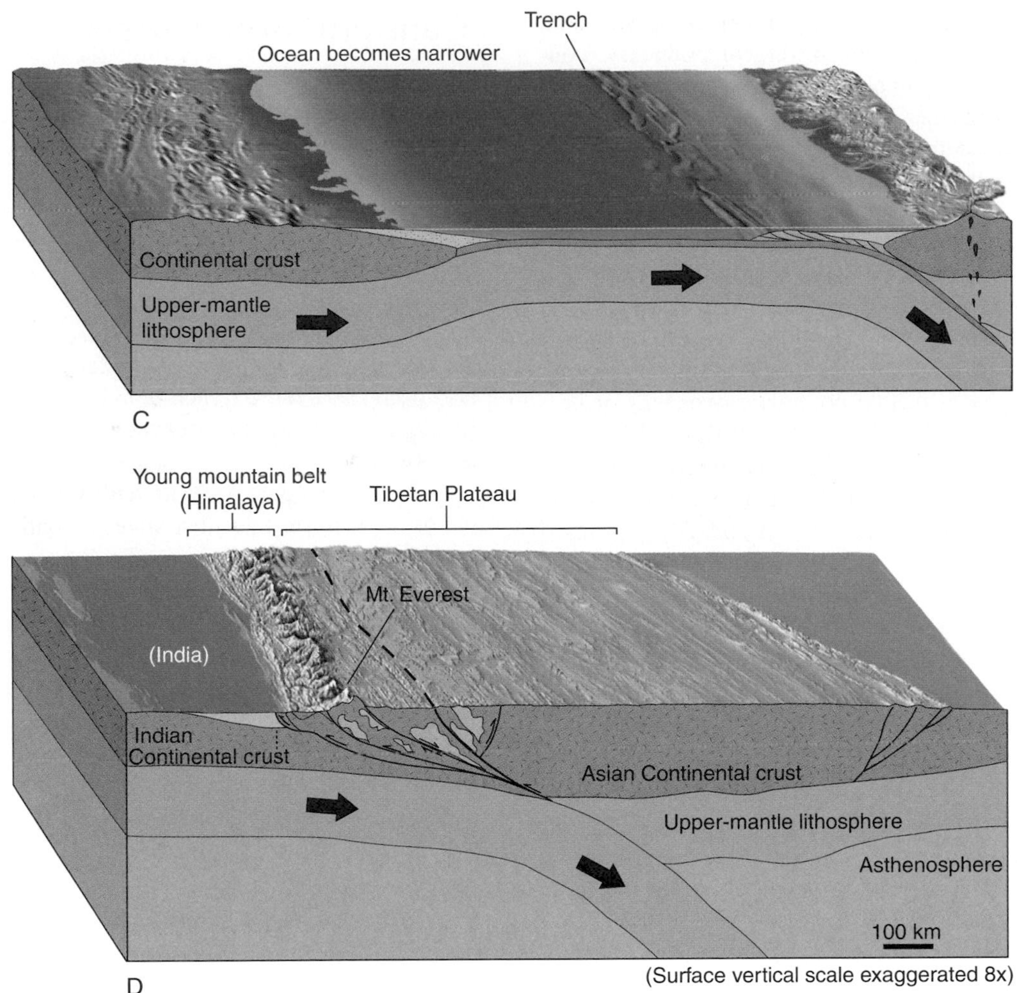

FIGURE 3.15 *Continued*

subduction zones can be demonstrated in many ways, a key one of which involves earthquake depths, as described in chapter 4.

The subduction zones of the world balance the seafloor equation. If new oceanic lithosphere is constantly being created at spreading ridges, an equal amount must be destroyed somewhere, or the earth would simply keep getting bigger. This excess sea floor is consumed in subduction zones. The subducted plate is heated by the hot asthenosphere. Fluids are released from it into the overlying mantle; some of it may become hot enough to melt; the dense, cold residue eventually breaks off and sinks deeper into the mantle. Meanwhile, at the spreading ridges, other melts rise, cool, and crystallize to make new sea floor. So, in a sense, the oceanic lithosphere is constantly being recycled, which explains why few very ancient seafloor rocks are known. Only rarely is a bit of sea floor caught up in a continent during convergence and preserved. Most often, the sea floor in a zone of convergence is subducted and destroyed. The buoyant continents are not so easily reworked in this way; hence, very old rocks may be preserved on them. Because all continents are part of moving plates, sooner or later they all are inevitably transported to a convergent boundary, as leading oceanic lithosphere is consumed.

Subduction zones are, geologically, very active places. Sediments eroded from the continents may accumulate in the trench formed by the down-going plate, and some of these sediments may be carried down into the asthenosphere to contribute to melts being produced there, as described further in chapter 5. Volcanoes form where molten material rises up through the overlying plate to the surface. At an ocean-ocean convergence, the result is commonly a line of volcanic islands, an **island arc** (see figure 3.15B). The great stresses involved in convergence and subduction give rise to numerous earthquakes. The bulk of mountain building, and its associated volcanic and earthquake activity, is related to subduction zones. Parts of the world near or above modern subduction zones are therefore prone to both volcanic and earthquake activity. These include the Andes region of South America, western Central America, parts of the northwest United States and Canada, the Aleutian Islands, China, Japan, and much of the rim of the Pacific Ocean basin.

If there is also continental lithosphere on the plate being subducted at an ocean-continent convergent boundary, consumption of the subducting plate will eventually bring the continental masses together (figure 3.15C). Two landmasses collide, crumple, and deform. One may partially override the other, but the buoyancy of continental lithosphere ensures that neither sinks deep into the mantle, and a very large thickness of continent may result. Earthquakes are frequent during active collision as a consequence of the large stresses involved in the process. The extreme height of the Himalaya Mountains is attributed to just this sort of continent-continent collision. India was not always a part of the Asian continent. Paleomagnetic evidence indicates that it drifted northward over tens of millions of years until it "ran into" Asia (recall figure 3.13), and the Himalayas were built up in this collision (figure 3.15D). Earlier, the ancestral Appalachian Mountains were built in the same way, as Africa and North America converged prior to the breakup of Pangaea. In fact, many major mountain ranges worldwide represent sites of sustained plate convergence in the past, and much of the western portion of North America consists of bits of continental lithosphere "pasted onto" the continent in this way (figure 3.16). This process accounts for the juxtaposition of rocks that are quite different in age and geology.

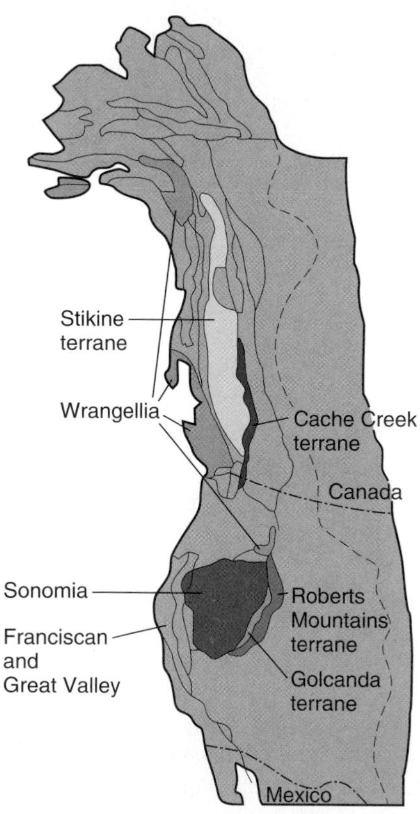

FIGURE 3.16 A *terrane* is a region of rocks that share a common history, distinguished from nearby, but genetically unrelated, rocks. Geologists studying western North America have identified and named many different terranes, some of which are shown here.
After U.S. Geological Survey Open-File Map 83–716

Transform Boundaries

The actual structure of a spreading ridge is more complex than a single, straight crack. A close look at a mid-ocean spreading ridge reveals that it is not a continuous rift thousands of kilometers long. Rather, ridges consist of many short segments slightly offset from one another. The offset area is a special kind of fault, or break in the lithosphere, known as a **transform fault** (see figure 3.17A). The opposite sides of a transform fault belong to two different plates, and these are moving in opposite directions. As the plates scrape past each other, earthquakes occur along the transform fault. Transform faults may also occur between a trench (subduction zone) and a spreading ridge, or between two trenches (figure 3.17B), but these are less common.

The famous San Andreas fault in California is an example of a transform fault that slices a continent sitting along a spreading ridge. The East Pacific Rise, a seafloor spreading ridge off the northwestern coast of North America, disappears under the edge of the continent, to reappear farther south in the Gulf of California. The San Andreas is the transform fault between the subduction zone and the spreading ridge. Most of North America is part of the North American Plate. The thin strip of California on the west side of the San Andreas fault, however, is moving northwest with the Pacific Plate. The stress resulting from the shearing displacement across the fault leads to ongoing earthquake activity.

HOW FAR, HOW FAST, HOW LONG, HOW COME?

Geologic and topographic information together allow us to identify the locations and nature of the major plate boundaries at present (figure 3.18). Ongoing questions include the reasons that plates move, and for how much of geologic history plate-tectonic processes have operated.

Past Motions, Present Velocities

Rates and directions of plate movement have been determined in a variety of ways. As previously discussed, polar-wander curves from continental rocks can be used to determine how the continents have drifted. Seafloor spreading is another way of determining plate movement. The direction of seafloor spreading is usually obvious: away from the ridge. Rates of seafloor spreading can be found very simply by dating rocks at different distances from the spreading ridge and dividing the distance moved by the rock's age (the time it has taken to move that distance from the ridge at which it formed). For example, if a 10-million-year-old piece of sea floor is collected at a distance of 100 kilometers from the ridge, this represents an average

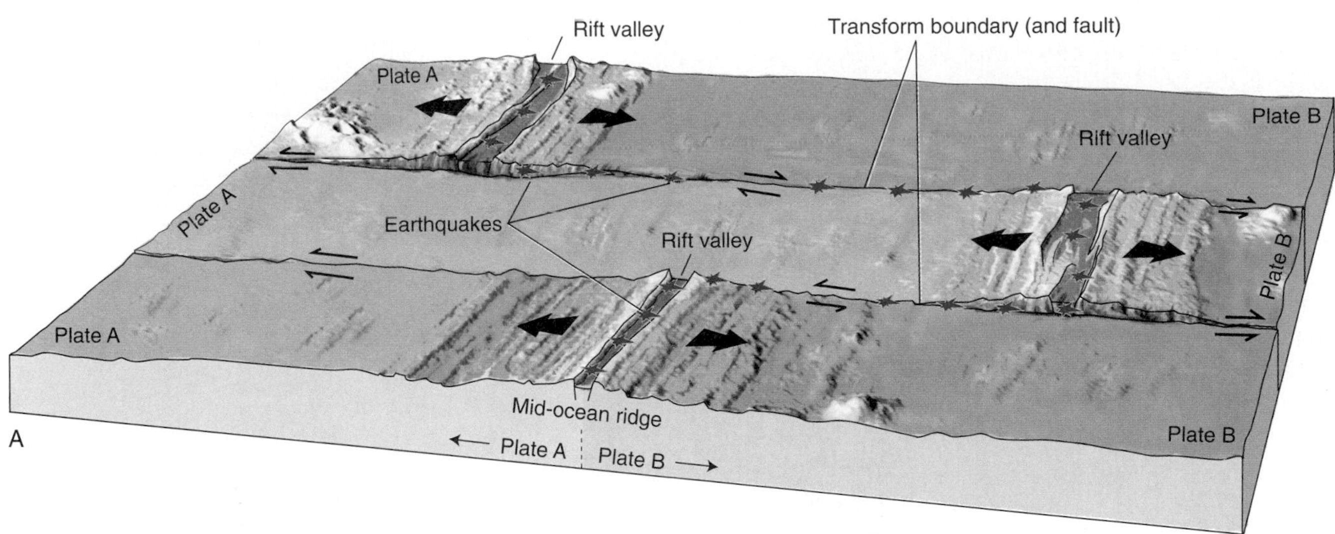

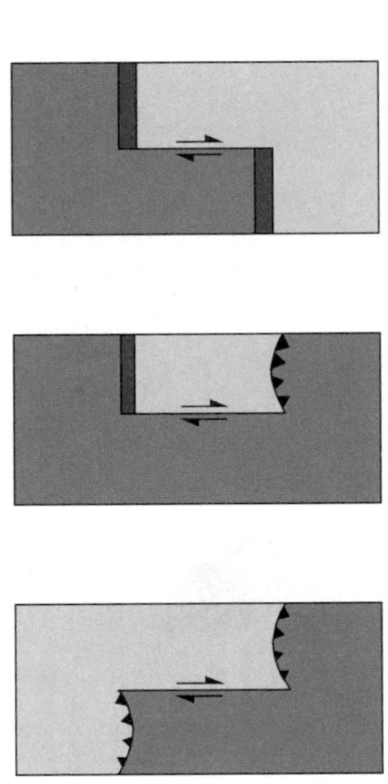

FIGURE 3.17 (A) Seafloor spreading-ridge segments are offset by transform faults. Red asterisks show where earthquakes occur in the ridge system. (B) Types of transform boundaries. Red segments are spreading ridges; "sawteeth" point to the overriding plate at a subduction boundary.

rate of movement over that time of 100 kilometers/10 million years. If we convert that to units that are easier to visualize, it works out to about 1 centimeter (a little less than half an inch) per year.

Another way to monitor rates and directions of plate movement is by using mantle **hot spots.** These are isolated areas of volcanic activity usually not associated with plate boundaries. They are attributed to rising columns of warm mantle material (*plumes*), perhaps originating at the base of the mantle. Reduction in pressure as the plume rises can lead to partial melting, and the resultant magma can rise up through the overlying plate to create a volcano. If we assume that mantle hot spots remain fixed in position while the lithospheric plates move over them, the result should be a string of volcanoes of differing ages with the youngest closest to the hot spot. (There is evidence that from time to time, churning of mantle material can shift the plumes, but in many cases, the data are consistent with essentially stationary hot spots over millions of years.)

A good example can be seen in the north Pacific Ocean (see figure 3.19). A topographic map shows a V-shaped chain of volcanic islands and submerged volcanoes. When rocks from these volcanoes are dated, they show a progression of ages, from about 75 million years at the northwestern end of the chain, to about 40 million years at the bend, through progressively younger islands to the still-active volcanoes of the island of Hawaii, located at the eastern end of the Hawaiian Island group. From the distances and age differences between pairs of points, we can again determine the rate of plate motion. For instance, Midway Island and Hawaii are about 2700 kilometers (1700 miles) apart. The volcanoes of Midway were active about 25 million years ago. Over the last 25 million years, then, the Pacific Plate has moved over the mantle hot spot at an average rate of 2700 kilometers/25 million years, or about 11 centimeters (4.3 inches) per year. The orientation of the volcanic chain shows the direction of plate movement—west-northwest. The kink in the chain at about 40 million years ago indicates that the direction of movement of the Pacific Plate changed at that time.

Hot spots occur under continents as well as beneath ocean basins (Yellowstone Park's geothermal features are powered by a hot spot, for example). They are, however, somewhat easier to detect in oceanic regions, perhaps because the associated magmas can more readily work their way up through the thinner oceanic lithosphere to build visible volcanoes.

Past plate motions can be quantified using other methods, too—for example, considering widths of strips of sea floor produced over different periods of time (recall figure 3.10),

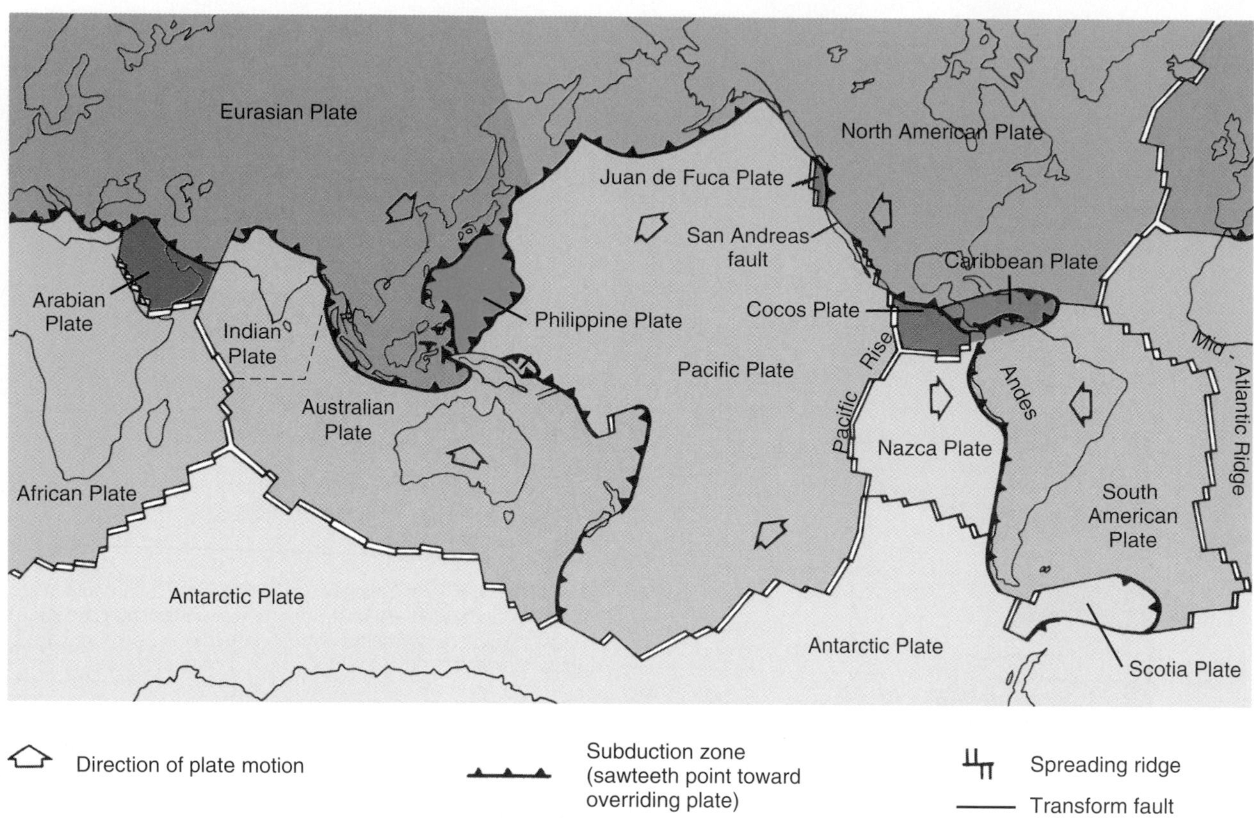

FIGURE 3.18 Principal world lithospheric plates. Arrows denote approximate relative directions of plate movement. The nature of the recently recognized boundary between the Indian and Australian plates is still being determined.
Source: After W. Hamilton, U.S. Geological Survey

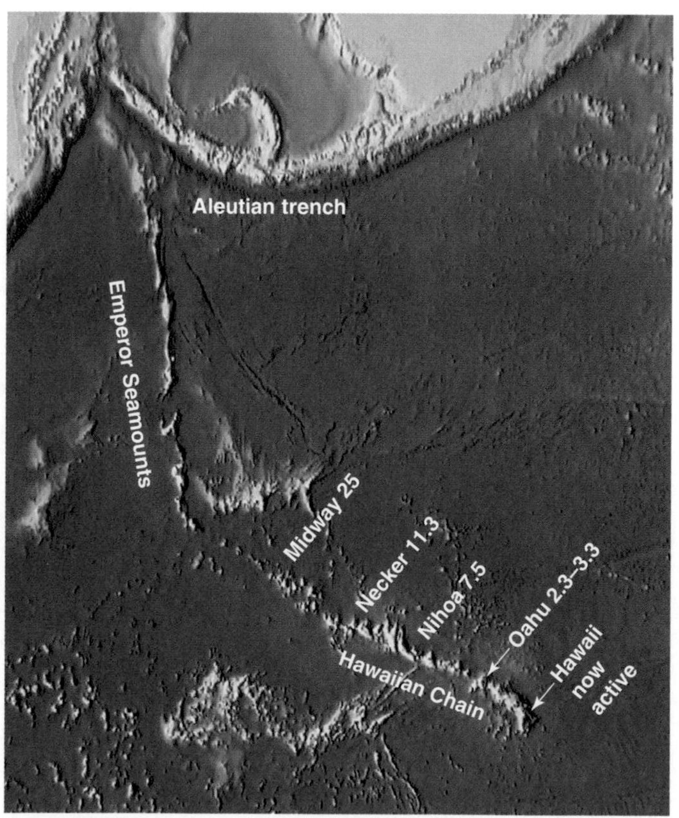

FIGURE 3.19 The Hawaiian Islands and other volcanoes in a chain formed over a hot spot. Numbers indicate ages of volcanic eruptions, in millions of years. Movement of the Pacific Plate has carried the older volcanoes far from the hot spot, now under the active volcanic island of Hawaii.

or using points of known age and position along a continent's polar-wander curve. Satellite-based technology now allows direct measurement of modern plate movement (Box 3.1 "Technology Meets Plate Tectonics"). Looking at many such determinations from all over the world, geologists find that average rates of plate motion are 2 to 3 centimeters (about 1 inch) per year. In a few places, movement at rates up to about 10 centimeters per year is observed, and, elsewhere, rates may be slower, but a few centimeters per year is typical. This seemingly trivial amount of motion does add up through geologic time. Movement of 2 centimeters per year for 100 million years means a shift of 2000 kilometers, or about 1250 miles!

Why Do Plates Move?

A driving force for plate tectonics has not been definitely identified. For many years, the most widely accepted explanation was that the plates were moved by large **convection cells**

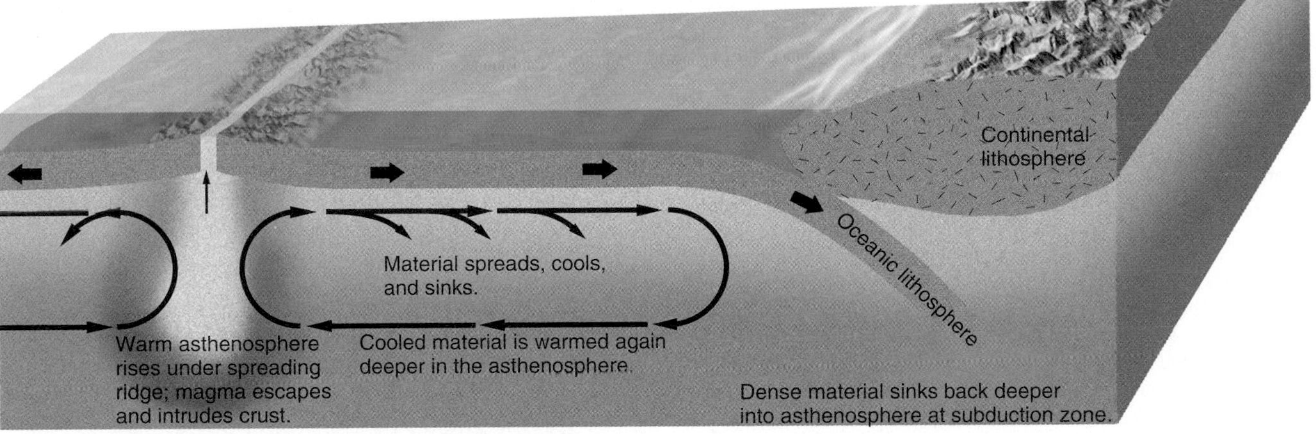

FIGURE 3.20 One possible driving force behind plate tectonics is slow convection in the partly molten asthenosphere. Alternatively, plate motions may drive asthenospheric circulation.

slowly churning in the plastic asthenosphere (figure 3.20). According to this model, hot material rises at the spreading ridges; some magma escapes to form new lithosphere, but the rest of the rising asthenospheric material spreads out sideways beneath the lithosphere, slowly cooling in the process. As it flows outward, it drags the overlying lithosphere outward with it, thus continuing to open the ridges. When it cools, the flowing material becomes dense enough to sink back deeper into the asthenosphere—for example, under subduction zones.

The existence of convection cells in the asthenosphere has not been proven definitively. There is some question, too, whether flowing asthenosphere could exercise enough drag on the lithosphere above to propel it laterally and force it into convergence. One alternative explanation, for which considerable evidence has accumulated, is that the weight of the dense, down-going slab of lithosphere in the subduction zone pulls the rest of the trailing plate along with it, opening up the spreading ridges so magma can ooze upward. Mantle convection might then result from lithospheric drag, not the reverse. This is described as the "slab-pull" model. Another model involves slabs of lithosphere "sliding off" the topographic highs formed at spreading ridges and rift zones by rising warm asthenospheric mantle, and perhaps dragging some asthenosphere along laterally ("ridge-push"). Many geoscientists now believe that a combination of these mechanisms is responsible for plate motion; and there may be contributions from other mechanisms not yet considered. That lithospheric motions and flow in the asthenosphere are coupled seems likely, but which one may drive the other is still unclear.

History of Plate Tectonics

How long plate-tectonic processes have been active is also not entirely clear. The magnetic stripes characteristic of seafloor spreading are apparent over even the oldest, 200-million-year-old ocean floor. From continental rocks, apparent polar-wander curves going back more than a billion years can be reconstructed, although the relative scarcity of undisturbed ancient rocks makes such efforts more difficult for the earth's earliest history. It seems clear that the continents have been shifting in position over the earth's surface for at least two billion years, though not necessarily at the same rates or with exactly the same results as at present. Plate-tectonic processes have long played a major role in shaping the earth and are likely to continue doing so for the foreseeable future.

PLATE TECTONICS AND THE ROCK CYCLE

In chapter 2, we noted that all rocks may be considered related by the concept of the rock cycle. We can also look at the rock cycle in a plate-tectonic context, as illustrated in figure 3.21. New igneous rocks form from magmas rising out of the asthenosphere at spreading ridges or in subduction zones. The heat radiated by the cooling magmas can cause metamorphism, with recrystallization at an elevated temperature changing the texture and/or the mineralogy of the surrounding rocks. Some of these surrounding rocks may themselves melt to form new igneous rocks. The forces of plate collision at convergent margins also contribute to metamorphism by increasing the pressures acting on the rocks. Weathering and erosion on the continents wear down preexisting rocks of all kinds into sediment. Much of this sediment is eventually transported to the edges of the continents, where it is deposited in deep basins. Through burial under more layers of sediment, it may become solidified into sedimentary rock. Sedimentary rocks, in turn, may be metamorphosed or even melted by the stresses and the igneous activity at the plate margins. Some of these sedimentary or metamorphic materials may also be carried down with subducted oceanic lithosphere, to be melted and eventually recycled as igneous rock. Plate-tectonic activity thus plays a large role in the formation of new rocks from old that proceeds continually on the earth.

Box 3.1 Technology Meets Plate Tectonics

By the latter part of the twentieth century, plate-tectonic theory was firmly established, well tested by a variety of data. Still, newer investigative techniques continue to evolve, and they provide ever more information by which to test and refine our understanding.

One example has been *seismic tomography.* This technique uses variations in the speed of seismic waves from large earthquakes to identify more- and less-dense regions in the earth's interior, often corresponding to warmer and cooler material (figure 1). Among other applications, this technique has helped in locating mantle plumes and in exploring relationships between mantle temperatures and plate motions.

Modern high-precision global positioning system (GPS) technology allows direct measurement of the relative motions of many points on the earth's surface (figure 2). This has not only enabled direct measurement of plate motions; it has also revealed differential movements within single plates, which can indicate stress buildup that could be leading to an earthquake.

Multibeam sonar bathymetry (figure 3) is providing more-precise topographic information about the sea floor, which can be used to enhance our understanding of underwater plate boundaries. And just recently, dredging of samples from under the Arctic polar ice, itself a technological advance, has revealed startling information about a newly recognized class of "ultraslow" spreading ridges: that at these ridges, chunks of mantle rock are pushed up to the seafloor surface. Such surprises continue to drive scientific exploration and discovery.

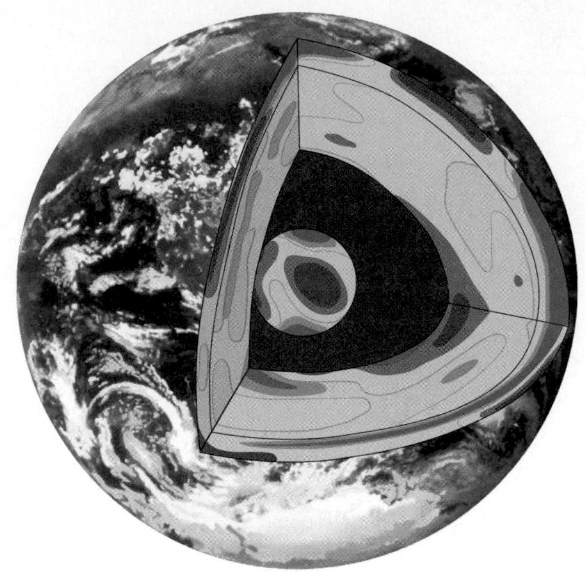

FIGURE 1 Cutaway view of seismic-velocity variations in the earth's interior. Blue (colder) regions have above-average velocities; red-brown (warmer), below-average. *Reprinted with permission from A. M. Dziewonski, "Global Images of the Earth's Interior," Science 236:37, April 13, 1987. Copyright @ 1987 American Association for the Advancement of Science*

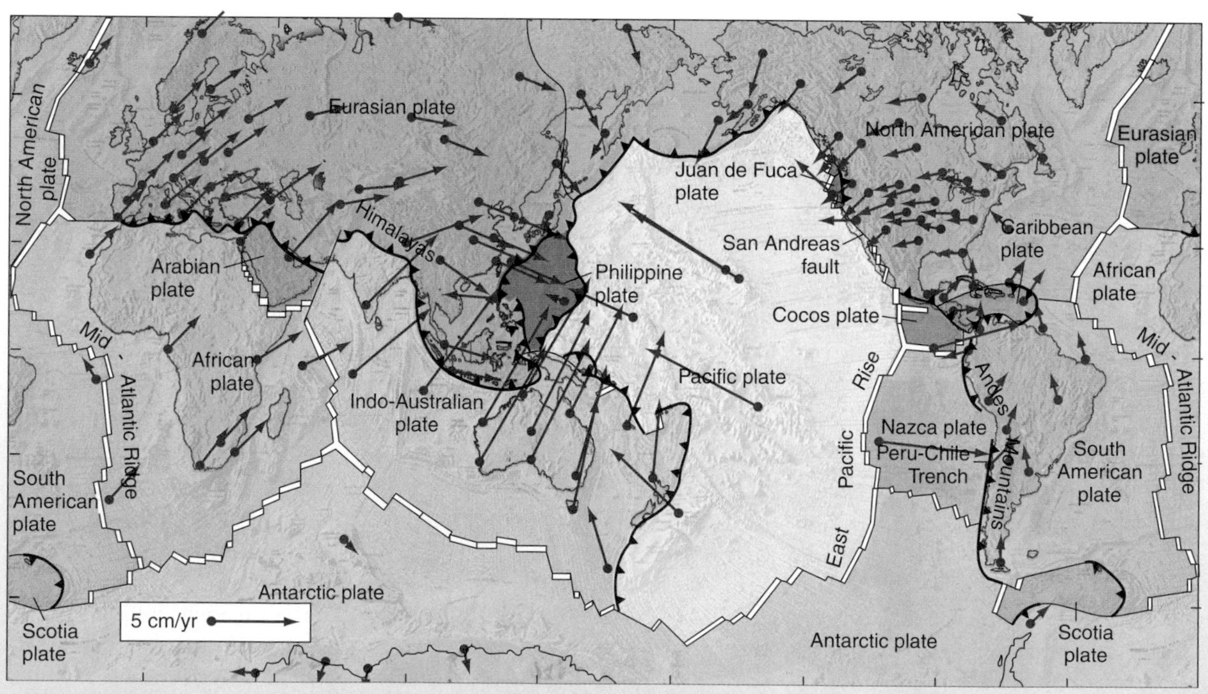

Divergent boundary

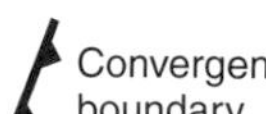

Convergent boundary

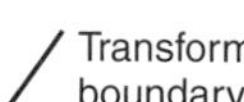

Transform boundary

FIGURE 2 Motion of points on the surface is now measurable directly by satellite using GPS technology. *Image courtesy NASA*

(Box 3.1 Continued)

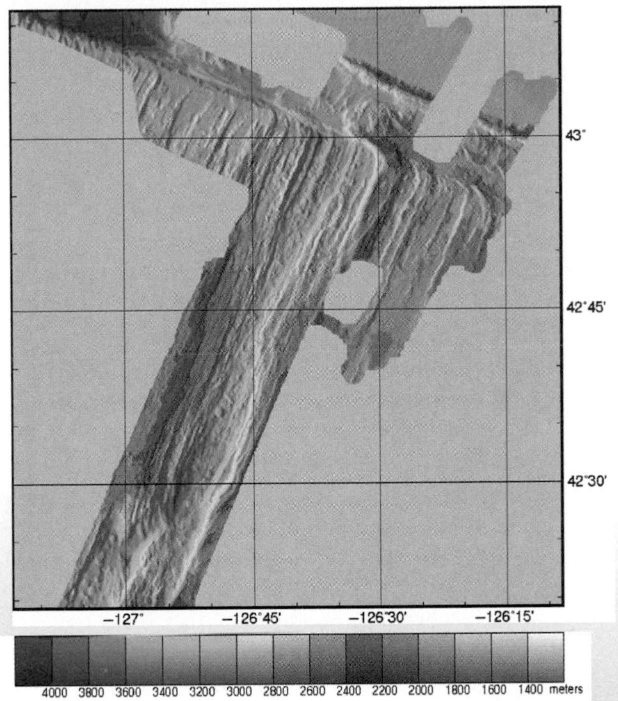

FIGURE 3 Multibeam sonar image of section of the Juan de Fuca Ridge in the northeastern Pacific Ocean, showing the rift valley, and adjacent transform fault (top of image).

Image created by Chris Keeley, Lamont-Doherty Observatory of Columbia University, Palisades NY

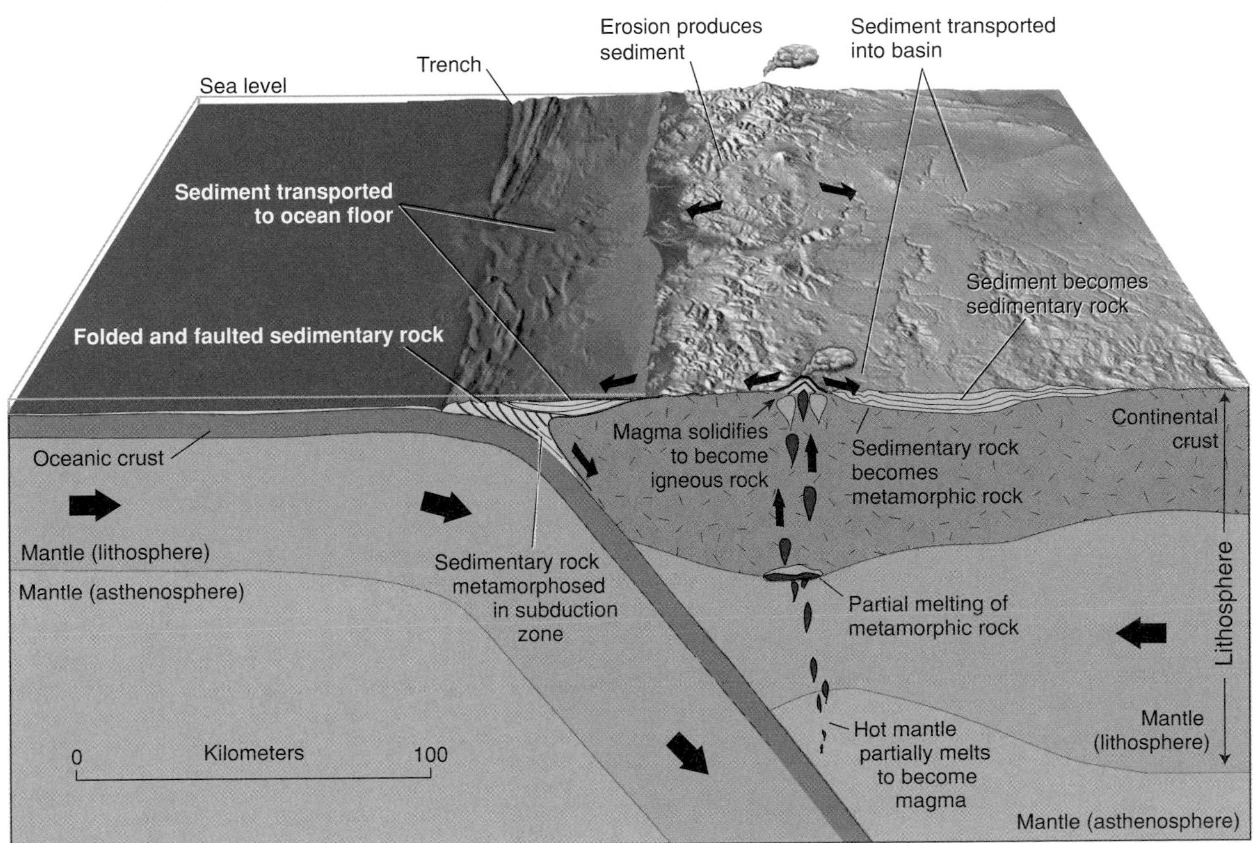

FIGURE 3.21 An ocean-continent convergence zone illustrates some of the many ways in which plate-tectonic processes transform rocks: Continents erode, producing sediment to make sedimentary rocks; asthenosphere melts to make magma to form new igneous rock; compression and magmatic heating metamorphose or even melt rocks; and so on in a cycle of continuous change.

SUMMARY

Rocks subjected to stress may behave either elastically or plastically. At low temperatures and confining pressures, they are more rigid, elastic, and often brittle. At higher temperatures and pressures, or in response to stress applied gradually over long periods, they tend more toward plastic behavior.

The outermost solid layer of the earth is the 50- to 100-kilometer-thick lithosphere, which is broken up into a series of rigid plates. The lithosphere is underlain by a plastic layer of the mantle, the asthenosphere, over which the plates can move. This plate motion gives rise to earthquakes and volcanic activity at the plate boundaries. At seafloor spreading ridges, which are divergent plate boundaries, new sea floor is created from magma rising from the asthenosphere. The sea floor moves in conveyor belt fashion, ultimately to be destroyed in subduction zones, a type of convergent plate boundary, where the sea floor is carried down into the asthenosphere and eventually remelted. Magma rises up through the overriding plate to form volcanoes above the subduction zone. Where continents ride the leading edges of converging plates, continent-continent collision may build high mountain ranges.

Evidence for seafloor spreading includes the age distribution of seafloor rocks, and the magnetic stripes on the ocean floor. Continental drift can be demonstrated by such means as polar-wander curves, fossil distribution among different continents, and evidence of ancient climates revealed in the rock record, which can show changes in a continent's latitude. Past supercontinents can be reconstructed by fitting together modern continental margins, matching similar geologic features from continent to continent, and correlating polar-wander curves from different landmasses.

Present rates of plate movement average a few centimeters a year. One possible driving force for plate tectonics is slow convection in the asthenosphere; another is gravity pulling cold, dense lithosphere down into the asthenosphere, dragging the asthenosphere along. Plate-tectonic processes appear to have been more or less active for much of the earth's history. They play an integral part in the rock cycle—building continents to weather into sediments, carrying rock and sediment into the warm mantle to be melted into magma that rises to create new igneous rock and metamorphoses the lithosphere through which it rises, subjecting rocks to the stress of collision—assisting in the making of new rocks from old.

Key Terms and Concepts

asthenosphere 45
brittle 44
compressive stress 44
continental drift 43
convection cell 58
convergent plate boundary 53
Curie temperature 48
divergent plate boundary 52
ductile 44
elastic deformation 44
elastic limit 44
hot spot 57
island arc 55
lithosphere 45
paleomagnetism 48
plastic deformation 44
plate tectonics 43
polar-wander curve 51
rupture 44
seafloor spreading 49
shearing stress 44
strain 44
stress 44
subduction zone 54
tectonics 43
tensile stress 44
transform fault 56

Exercises

Questions for Review

1. What causes strain in rocks? How do elastic and plastic materials differ in their behavior?
2. What is plate tectonics, and how are continental drift and seafloor spreading related to it?
3. Define the terms *lithosphere* and *asthenosphere.* Where are the lithosphere and asthenosphere found?
4. Describe two kinds of paleomagnetic evidence supporting the theory of plate tectonics.
5. Cite at least three kinds of evidence, other than paleomagnetic evidence, for plate tectonics.

6. Explain how a subduction zone forms and what occurs at such a plate boundary.
7. What brings about continent-continent convergence, and what happens then?
8. What are hot spots, and how do they help to determine rates and directions of plate movements?
9. Describe how convection in the asthenosphere may drive the motion of lithospheric plates. Alternatively, how might plate motions churn the plastic asthenosphere?
10. Describe the rock cycle in terms of plate tectonics, making specific reference to the creation of new igneous, metamorphic, and sedimentary rocks.

For Further Thought

1. The moon's structure is very different from that of earth. For one thing, the moon has an approximately 1000-kilometer-thick lithosphere. Would you expect plate-tectonic activity similar to that on earth to occur on the moon? Why or why not?
2. The Atlantic Ocean is approximately 5000 km wide. If each plate has moved away from the Mid-Atlantic Ridge at an average speed of 1.5 cm per year, how long did the Atlantic Ocean basin take to form? What would the spreading rate be in miles per hour?

Suggested Readings/References

Bonatti, E. 1987. The rifting of continents. *Scientific American* 256 (March): 96–103.

Conrad, C. P., and Lithgow-Bertelloni, C. 2002. How mantle slabs drive plate tectonics. *Science* 298 (4 October): 207–9.

DePaolo, D., and Manga, M. 2003. Deep origin of hotspots—the mantle plume model. *Science* 300 (9 May) 920–21.

Gardner, J. V., Dartnell, P., Gibbons, H., and MacMillan, D. 2000. Exposing the sea floor: High-resolution multibeam mapping along the U.S. Pacific coast. USGS Fact Sheet 013–00.

Gramling, C. 2006. Uncharted territory: Utraslow ridges hold new clues to crust's formation. *Science News* 169 (1 April): 202–4.

Hassler, D. R., Goehring, L., and Fisher, C. 2002. The complex world along mid-ocean ridges. *Geotimes* (December): 14–18.

Hoffman, K. A. 1988. Ancient magnetic reversals: Clues to the geodynamo. *Scientific American* 258 (May): 76–83.

Lutz, R. A., Shank, T. M., and Evans, R. 2001. Life after death in the deep sea. *American Scientist* 89 (September–October): 422–31.

Murphy, J. B., and Nance, R. D. 2004. How do supercontinents assemble? *American Scientist* 92 (July–August): 324–33.

Park, J., and Levin, V. 2002. Seismic anisotropy: Tracing plate dynamics in the mantle. *Science* 296 (19 April): 485–89.

Tarduno, J. A., et al. 2003. The Emperor seamounts: Southward motion of the Hawaiian hotspot plume in Earth's mantle. *Science* 301 (22 August): 1064–69.

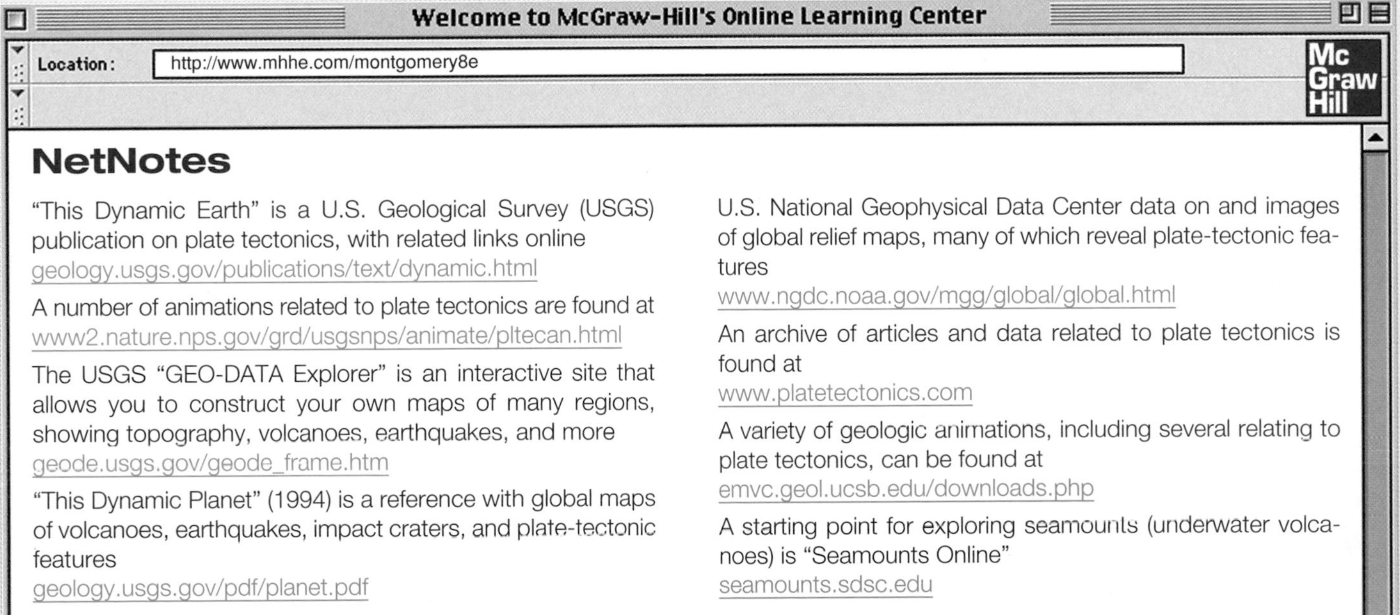

Many interactive geology exercises on plate-tectonic and other processes can be found at the "Discover Our Earth" site
www.discoverourearth.org

You can investigate the complex and sometimes bizarre communities of organisms around the vents at seafloor spreading ridges
oceanexplorer.noaa.gov

The Indian and Australian plates were once regarded as one; the story of how they were found to be breaking apart is at
www.columbia.edu/cu/pr/95/18688.html

4 Earthquakes

"The Alaska earthquake on March 27, 1964, commonly referred to as the 'Good Friday Earthquake,' struck at 5:36 P.M. A few members and staff of the Anchorage Petroleum Club were preparing for an early dinner. I was one of those members. The Petroleum Club was located on the top of the Anchorage-Westward Hotel, a fifteen-story building, which, at the time, was the tallest building in Anchorage. The outside walls of the dining room were made up entirely of windows to allow a full view of the mountains to the east.

"With some preliminary tremors, the earthquake gave warning to those in the dining room to move away from the windows just in case it became worse—and it did. There was barely time to move to the center of the dining room when the first strong shock hit the building, throwing almost everyone to the floor. For the next three minutes, due to the extreme swaying, it was impossible to do anything but remain on the floor. The swaying motion was so great that from the floor, it seemed that the building was about to go over. The noise generated by shock waves hitting the building sounded like the pounding of a giant sledge hammer.

After the earthquake, looking down at the Fourth Avenue area of downtown Anchorage, it was difficult to visualize what that area had looked like before. For several blocks east of the hotel the north side of Fourth Avenue had dropped down due to slumping and sliding during the earthquake.

"The city seemed so quiet. In addition to the visible destruction, Anchorage was without water, electric power, and natural gas. At least it was still daylight, and the temperature just below freezing.

"Those who could retrieve their belongings from the hotel proceeded to the YMCA, where they served as volunteers, setting up

The west side of this mountain along the Neelum River near Makhri village collapsed in the magnitude-7.6 Pakistan earthquake of 2005 (landslide at upper right). The village was virtually destroyed, the course of the river changed, and the water muddied with sediment from the slide.
Satellite image by GeoEye. © 2007. All Rights Reserved

dormitories for transients and serving hot coffee and what food there was on hand. It was the beginning of a long night. The magnitude-8.4 main shock was followed, over the next 24 hours, by ten aftershocks that were over magnitude 6.0 on the Richter scale, and by many smaller ones." (Contributed by Dr. Kenneth G. Smith, Adjunct Professor, Dallas Baptist University)

Terrifying as the 1964 Alaskan earthquake was, it caused only 115 deaths and $540 million in property damage. With the 1989 Loma Prieta, California, earthquake, despite the more than $5 billion in property damage (figure 4.1A), only sixty-three lives were lost. The 1995 Kobe, Japan, earthquake caused over 5200 deaths, and damages estimated at $95–140 billion (figure 4.1B). The 2004 Sumatran earthquake claimed nearly 300,000 lives. A sampling of historic earthquakes is presented in table 4.1 The casualties and the property damage from earthquakes result from a variety of causes. This chapter reviews the causes and effects of earthquakes and explores how their devastation can be minimized.

A

B

FIGURE 4.1 (A) Damage from the 1989 Loma Prieta earthquake in California: collapse of I-880, Oakland, California. (B) Overturned section of Hanshin Expressway, eastern Kobe, Japan. This expressway was built in the 1960s to then-existing seismic-design standards. Elevated freeways are common space-saving devices in densely populated areas.
(A) Photograph by D. Keefer, from U.S. Geological Survey Open-FIle Report 89-687. (B) Photograph by Christopher Rojahn, Applied Technology Council

TABLE 4.1

Selected Major Historic Earthquakes

Year	Location	Magnitude*	Deaths	Damages†
342	Turkey		40,000	
365	Crete: Knossos		50,000	
1290	China	6.75	100,000	
1456	Italy	(XI)	40,000 +	
1556	China	(XI)	830,000	
1667	USSR: Shemakh	6.9	80,000	
1693	Italy, Sicily		100,000	$5–25 million
1703	Japan	8.2	200,000	$5–25 million
1737	India: Calcutta		300,000	$1–5 million
1755	Portugal: Lisbon	(XI)	62,000	
1780	Iran: Tabriz		100,000	
1850	China	(X)	20,600	
1854	Japan	8.4	3200	
1857	Italy: Campania	6.5	12,000	

Continued

TABLE 4.1

Continued

Year	Location	Magnitude*	Deaths	Damages†
1868	Peru, Chile	8.5	52,000	$300 million
1906	United States: San Francisco	8.3	700	
1906	Ecuador	8.8	1000	
1908	Italy: Calabria	7.5	58,000	
1920	China: Kansu	8.5	200,000	
1923	Japan: Tokyo	8.3	143,000	$2.8 billion
1933	Japan	8.9	3000 +	
1939	Chile	8.3	28,000	$100 million
1939	Turkey	7.9	30,000	
1960	Chile	9.5	5700	$675 million
1964	Alaska	9.2	115	$540 million
1967	Venezuela	7.5	1100	over $140 million
1970	N. Peru	7.8	66,800	$250 million
1971	United States: San Fernando	6.4	60	$500 million
1972	Nicaragua	6.2	5000	$800 million
1976	Guatemala	7.5	23,000	$1.1 billion
1976	NE Italy	6.5	1000	$8 billion
1976	China: Tangshan	8.0	655,000	
1978	Iran	7.4	20,000	
1983	Turkey	6.9	2700	
1985	Chile	7.8	177	$1.8 billion
1985	SE Mexico	8.1	5600 +	$5–25 million
1988	USSR: Afghanistan	6.8	25,000	
1989	United States: Loma Prieta, California	7.1	63	$5.6–7.1 billion
1990	Iran	6.4	40,000 +	over $7 million
1994	United States: Northridge, California	6.8	57	$13–20 billion
1995	Japan: Kobe	7.2	5200 +	$95–140 billion
1998	Afghanistan	6.9	4000	
1999	Turkey: Izmit	7.6	17,000	$3–6.5 billion
1999	Taiwan	7.7	2300	$14 billion
2001	Northwestern India (Gujarat)	7.7	20,000	$1.3–5 billion
2001	El Salvador and off coast (2 events)	6.6, 7.6	1160	$2.8 billion
2001	United States: Washington State	6.8	0	$1–4 billion
2002	Afghanistan (2 events)	7.4, 6.1	1150	
2002	United States: Central Alaska	7.9	0	
2004	Northern Sumatra	9.1	283,000†	
2005	Northern Sumatra	8.6	1300†	
2005	Pakistan	7.6	86,000†	

Note: This table is not intended to be a complete or systematic compilation but to give examples of major damaging earthquakes in history.

Primary source (pre-1988 entries): *Catalog of Significant Earthquakes 2000 B.C.–1979* and supplement to 1985, World Data Center A, Report SE-27. Later data from National Earthquake Information Center.

*Where actual magnitude is not available, maximum Mercalli intensity is sometimes given in parentheses (see table 4.3). Earlier magnitudes are Richter magnitudes; for most recent events, and 1960 Chilean and 1964 Alaskan quakes, moment magnitudes are given.

†Estimated in 1979 dollars for pre-1980 entries. In many cases, precise damages are unknown.

EARTHQUAKES—TERMS AND PRINCIPLES

Basic Terms

Major earthquakes dramatically demonstrate that the earth is a dynamic, changing system. Earthquakes, in general, represent a release of built-up stress in the lithosphere. They occur along **faults,** planar breaks in rock along which there is displacement of one side relative to the other. Sometimes, the stress produces new faults or breaks; sometimes, it causes slipping along old, existing faults. When movement along faults occurs gradually and relatively smoothly, it is called **creep** (figure 4.2). Creep—sometimes termed *aseismic slip,* meaning fault displacement without significant earthquake activity—can be inconvenient but rarely causes serious damage.

When friction between rocks on either side of a fault prevents the rocks from slipping easily or when the rock under stress is not already fractured, some elastic deformation will occur before failure. When the stress at last exceeds the rupture strength of the rock (or the friction along a preexisting fault), a sudden movement occurs to release the stress. This is an **earthquake,** or *seismic slip.* With the sudden displacement and associated stress release, the rocks snap back elastically to their previous dimensions; this behavior is called **elastic rebound** (figure 4.3; recall the discussion of elastic behavior from chapter 3).

Faults come in all sizes, from microscopically small to thousands of kilometers long. Likewise, earthquakes come in all sizes, from tremors so small that even sensitive instruments can barely detect them, to massive shocks that can level cities. (Indeed, the "aseismic" movement of creep is actually characterized by many microearthquakes, so small that they are typically not felt at all.) The amount of damage associated with an earthquake is partly a function of the amount of accumulated energy released as the earthquake occurs.

The point on a fault at which the first movement or break occurs during an earthquake is called the earthquake's **focus,** or hypocenter (figure 4.4). In the case of a large earthquake, for example, a section of fault many kilometers long may slip, but there is always a point at which the first movement occurred, and this point is the focus. *Deep-focus earthquakes* are those with focal depths over 100 km, which distinguishes them from *shallow-focus earthquakes.* The point on the earth's surface directly above the focus is called the **epicenter.** When news accounts tell where an earthquake occurred, they report the location of the epicenter.

FIGURE 4.2 Curbstone offset by creep along the Hayward Fault in Hayward, California, from 1974 (top) to 1993 (bottom).
Photograph by Sue Hirschfeld, courtesy NOAA/National Geophysical Data Center

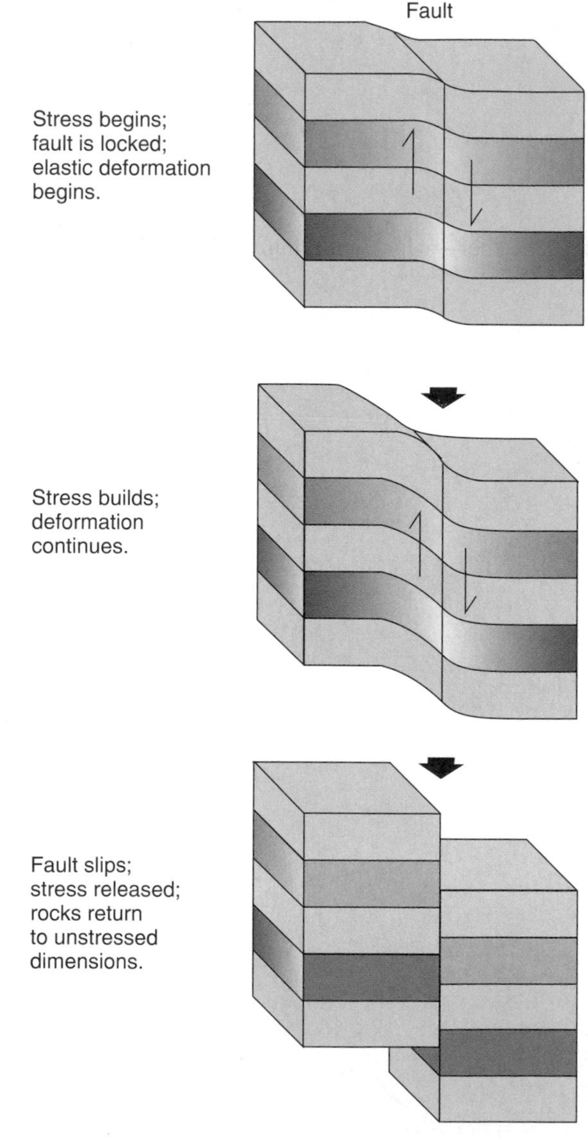

FIGURE 4.3 The phenomenon of elastic rebound: After the fault slips, the rocks spring back elastically to an undeformed condition.

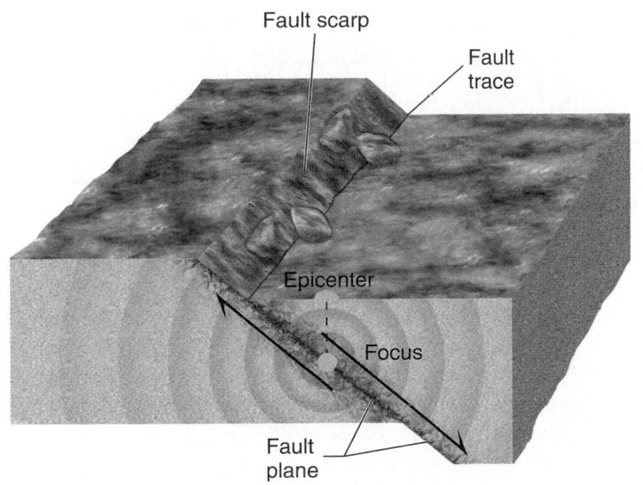

FIGURE 4.4 Simplified diagram of a fault, showing the focus, or hypocenter (point of first break along the fault), and the epicenter (point on the surface directly above the focus). Seismic waves dissipate the energy released by the earthquake as they travel away from the fault zone. Fault scarp is cliff formed along the fault plane at ground surface; fault trace is the line along which the fault plane intersects the ground surface.

Faults are sometimes described in terms of the nature of the displacement—how the rocks on either side move relative to each other—which commonly reflects the nature of the stresses involved. Certain types of faults are characteristic of particular plate-tectonic settings. In describing the orientation of a fault (or other planar feature, such as a layer of sedimentary rock), geologists use two measures. The *strike* is the compass orientation of the line of intersection of the fault plane with the earth's surface (e.g., the fault trace in figure 4.4). The *dip* of the fault is the angle the plane makes with the horizontal, a measure of the steepness of slope of the fault plane. A **strike-slip fault,** then, is one along which the displacement is parallel to the strike (horizontal). A transform fault is a type of strike-slip fault and reflects stresses acting horizontally. A **dip-slip fault** is one in which the displacement is vertical, as with the fault shown in figure 4.4. Rift valleys, whether along seafloor spreading ridges or on continents, are commonly bounded by steeply sloping dip-slip faults resulting from the tensional stress of rifting (recall figures 3.14 and 3.17A, or the opening image of chapter 3). Convergent boundaries are characterized by **thrust faults,** more shallowly sloping dip-slip faults along which compressional stresses force rocks on one side of the fault over the rocks below (figure 4.5).

Earthquake Locations

As shown in figure 4.6, the locations of major earthquake epicenters are concentrated in linear belts. These belts correspond to plate boundaries; recall figure 3.18. Not all earthquakes occur at plate boundaries, but most do. These areas are where plates jostle, collide with, or slide past each other, where relative plate movements may build up very large stresses, where

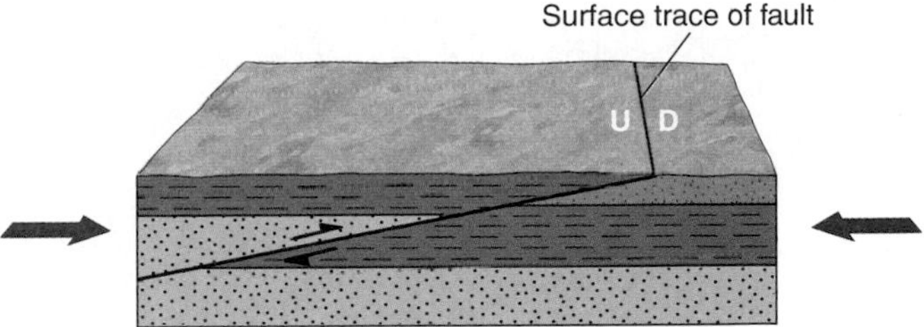

FIGURE 4.5 Along a thrust fault, compressional stress pushes one block up and over the other; "U" and "D" indicate relative motion up or down dip along the fault.

major faults or breaks may already exist on which further movement may occur. (On the other hand, intraplate earthquakes certainly occur and may be quite severe, as explored later in the chapter.)

A map showing only deep-focus earthquakes would look somewhat different: the spreading ridges would have disappeared, while subduction zones would still show by their frequent deeper earthquakes. The explanation is that earthquakes occur in the lithosphere, where rocks are elastic, capable of storing energy as they deform and then rupturing or slipping suddenly. In the plastic asthenosphere, material simply flows under stress. Therefore, the deep-focus earthquakes are concentrated in subduction zones, where elastic lithosphere is pushed deep into the mantle.

It was, in fact, the distribution of earthquake foci at subduction boundaries that helped in the recognition of the subduction process. Look, for example, at the northern and western margins of the Pacific Ocean in figure 4.6. Adjacent to each seafloor trench is a region in which earthquake foci are progressively deeper with increasing distance from the trench. A similar distribution is seen along the west side of Central and South America. This pattern is called a *Benioff zone,* named for the scientist who first mapped extensively these dipping planes of earthquake foci that we now realize reveal subducting plates.

SEISMIC WAVES AND EARTHQUAKE SEVERITY

Seismic Waves

When an earthquake occurs, it releases the stored-up energy in **seismic waves** that travel away from the focus. There are several types of seismic waves. ***Body waves*** (P waves and S waves) travel through the interior of the earth. **P waves** are compressional waves. As P waves travel through matter, the matter is alternately compressed and expanded. P waves travel through the earth, then, much as sound waves travel through air. A compressional sort of wave can be illustrated with a Slinky® toy by stretching the coil to a length of ten feet or so along a smooth surface, holding one end in place, pushing in the other

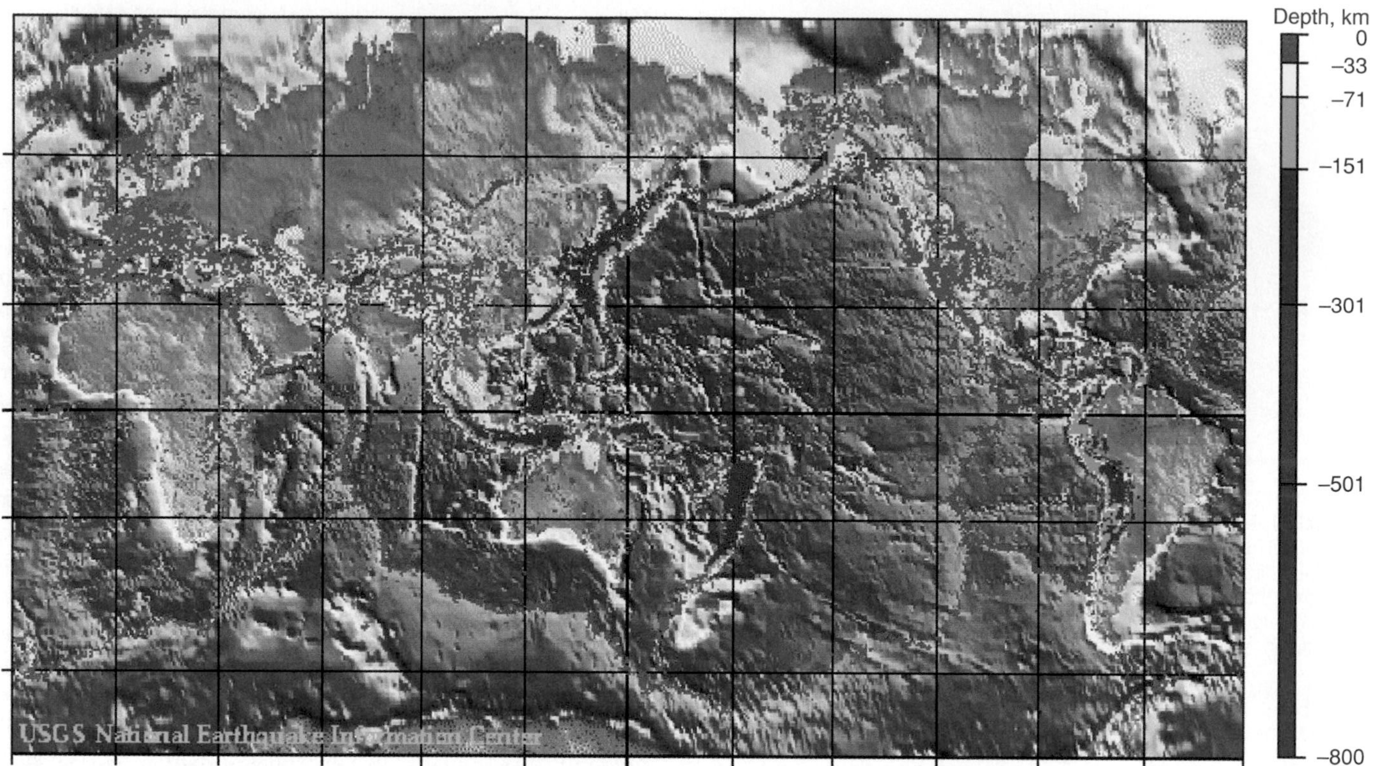

FIGURE 4.6 World seismicity, 1979–1995. Color of symbol indicates depth of focus.
Image courtesy of U.S. Geological Survey

end suddenly, and then holding that end still also. A pulse of compressed coil travels away from the end moved. **S waves** are shear waves, involving a side-to-side motion of molecules. Shear-type waves can also be demonstrated with a Slinky® by stretching the coil out along the ground as before, but this time twitching one end of the coil sideways (perpendicular to its length). As the wave moves along the length of the Slinky®, the loops of coil move sideways relative to each other, not closer together and farther apart as with the compressional wave. See also figure 4.7.

Seismic ***surface waves*** are somewhat similar to surface waves on water, discussed in chapter 7. That is, they cause rocks and soil to be displaced in such a way that the ground surface ripples or undulates. Surface waves also come in two types: Some cause vertical ground motions, like ripples on a pond; others cause horizontal shearing motions. The surface waves are larger in amplitude—amount of ground displacement—than the body waves from the same earthquake. Therefore, most of the shaking and resultant structural damage from earthquakes is caused by the surface waves.

Locating the Epicenter

Earthquake epicenters can be located using seismic body waves. Both types of body waves cause ground motions that are detectable using a **seismograph.** P waves travel faster through rocks than do S waves. Therefore, at points some distance from the scene of an earthquake, the first P waves arrive somewhat before the first S waves; indeed, the two types of body waves are actually known as *primary* and *secondary* waves, reflecting these arrival-time differences.

The difference in arrival times of the first P and S waves is a function of distance to the earthquake's epicenter. The effect can be illustrated by considering a pedestrian and a bicyclist traveling the same route, starting at the same time. If the cyclist can travel faster, he or she will arrive at the destination first. The longer the route to be traveled, the greater the difference in time between the arrival of the cyclist and the later arrival of the pedestrian. Likewise, the farther the receiving seismograph is from the earthquake epicenter, the greater the time lag between the first arrivals of P waves and S waves. This is illustrated graphically in figure 4.8A.

Once several recording stations have determined their distances from the epicenter in this way, the epicenter can be located on a map (figure 4.8B). If the epicenter is 200 kilometers from station *X,* it is located somewhere on a circle with a 200-kilometer radius around point *X.* If it is also found to be 150 kilometers from station *Y,* the epicenter must fall at either of the two points that are both 200 kilometers from *X* and 150 kilometers from *Y.* If the epicenter is also determined to be 100 kilometers from point *Z,* its position is uniquely identified.

In practice, complicating factors, such as inhomogeneities in the crust, make epicenter location somewhat more difficult. Results from more than three stations are usually required, and computers assist in reconciling all the data. The general principle, however, is as described.

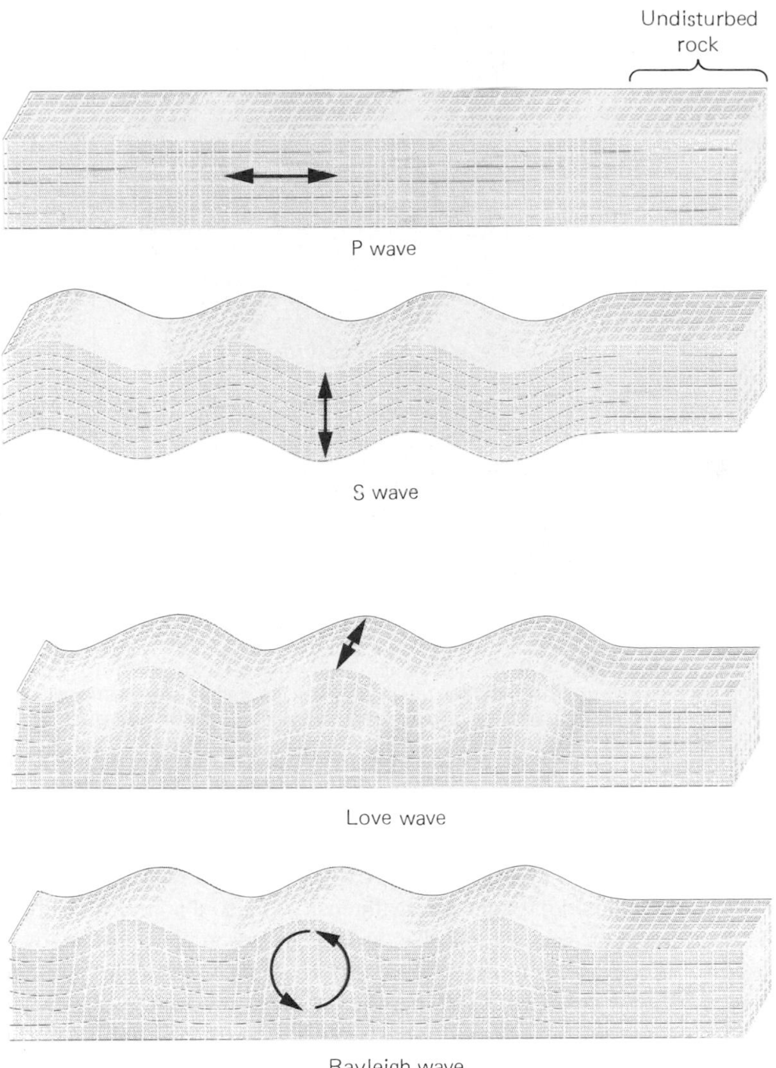

FIGURE 4.7 P waves and S waves are types of seismic body waves; Love and Rayleigh waves are surface waves that differ in the type of ground shaking they cause.

Magnitude and Intensity

All of the seismic waves represent energy release and transmission; they cause the ground shaking that people associate with earthquakes. The amount of ground motion is related to the **magnitude** of the earthquake. Historically, earthquake magnitude has most commonly been reported in this country using the *Richter magnitude scale,* named after geophysicist Charles F. Richter, who developed it.

A Richter magnitude number is assigned to an earthquake on the basis of the amount of ground displacement or shaking that it produces near the epicenter. The amount of ground motion is measured by a seismograph, and the size of the largest (highest-amplitude) seismic wave on the seismogram is determined. The value is adjusted for the particular type of instrument and the distance of the station from the earthquake epicenter (because ground motion naturally decreases with increasing distance from the site of the earthquake) so that different measuring stations in different places will arrive at approximately the same estimate of the ground displacement as it would have been measured close to the epicenter. From this adjusted ground displacement, the Richter magnitude value is assigned. The Richter scale is logarithmic, which means that an earthquake of magnitude 4 causes ten times as much ground movement as one of magnitude 3, one hundred times as much as one of magnitude 2,

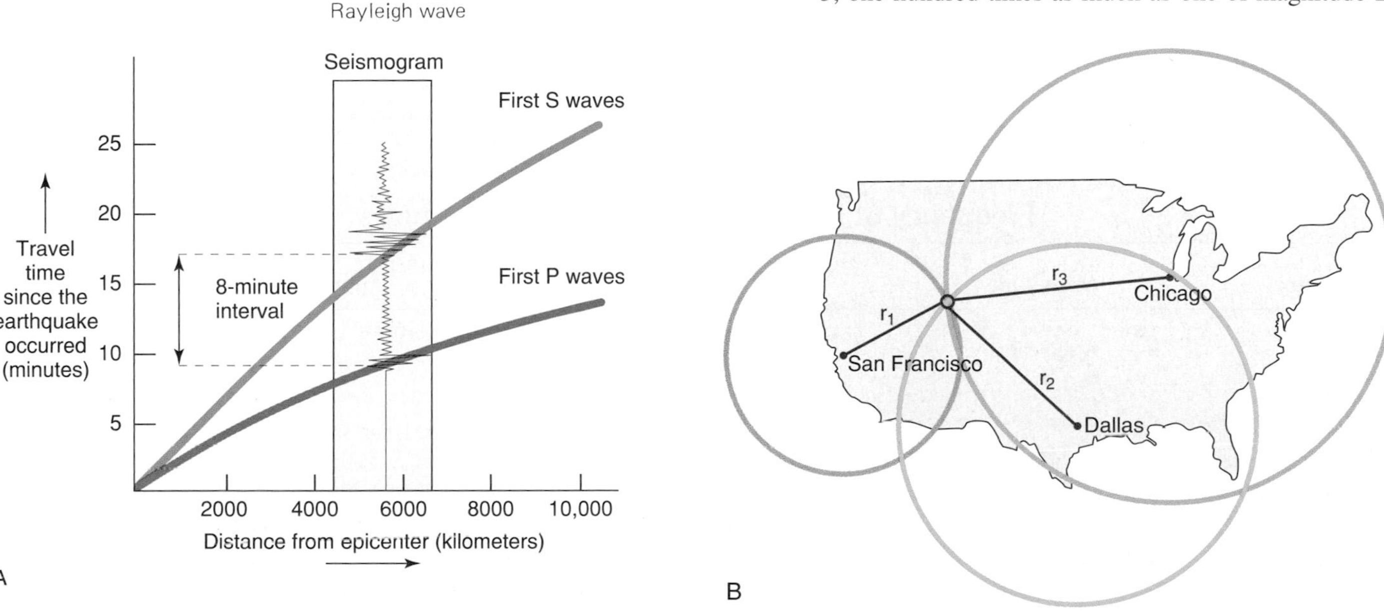

FIGURE 4.8 Use of seismic waves in locating earthquakes. (A) Difference in times of first arrivals of P waves and S waves is a function of the distance from the earthquake focus. This seismogram has a difference in arrival times of 8 minutes; the earthquake occurred about 5500 km away. (B) Triangulation using data from several seismograph stations allows location of an earthquake's epicenter: If the epicenter is 1350 km from San Francisco, it is on a circle of that radius around the city. If it is also found to be 1900 km from Dallas, it must fall at one of only two points. If it is further determined to be 2400 km from Chicago, its location is uniquely determined: northeastern Utah, near Salt Lake City.

and so on (see figure 4.9). The amount of energy released rises even faster with increased magnitude, by a factor of about 30 for each unit of magnitude: An earthquake of magnitude 4 releases approximately thirty times as much energy as one of magnitude 3, and nine hundred times as much as one of magnitude 2. There is no upper limit to the Richter scale. The largest recorded earthquakes have had Richter magnitudes of about 8.9. Although we only hear of the very severe, damaging earthquakes, there are, in fact, hundreds of thousands of earthquakes of all sizes each year. Table 4.2 summarizes the frequency and energy release of earthquakes in different Richter magnitude ranges.

Measurement of earthquake size in terms of Richter magnitude is something of an oversimplification. The scale was developed in California; earthquakes in other areas, with different rock types that respond differently to seismic waves, may not be readily compared on the basis of Richter magnitudes. For very large earthquakes especially, a better measure of relative energy release is *moment magnitude* (denoted M_W). Moment magnitude takes into account the area of break on the fault surface, the displacement along the fault during the earthquake, and the strength of the rock. For comparison, the largest recorded earthquake (in Chile in 1960) had a Richter magnitude of 8.9, an estimated moment magnitude of 9.5. There are other magnitude scales also. Still, the media persistently report Richter magnitudes (M_L) for newsworthy quakes. What all the magnitude scales have in common is a logarithmic character.

An alternative way of describing the size of an earthquake is by the earthquake's **intensity.** Intensity is a measure of the earthquake's effects on humans and on surface features. It is not a unique characteristic of an earthquake, nor is it defined on a precise quantitative basis. The surface effects produced by an earthquake of a given magnitude vary considerably as a result of such factors as local geologic conditions, quality of

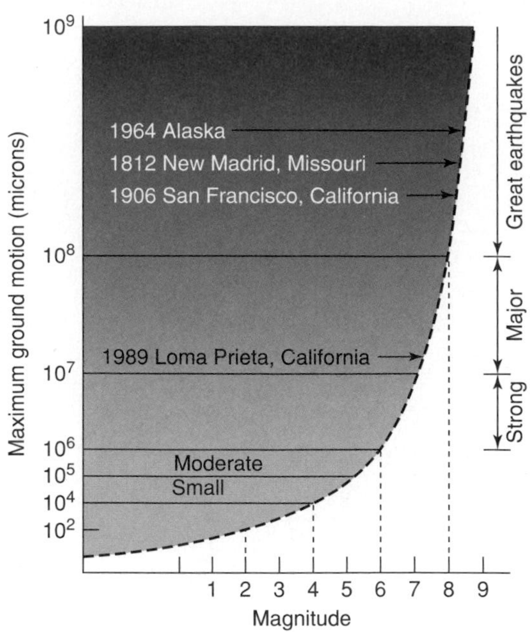

FIGURE 4.9 Ground motion increases by a factor of 10 for each unit increase in Richter magnitude; so, an earthquake of magnitude 7 causes 10 times the displacement (shaking) of one of magnitude 6. (10^9 microns = 1 meter)

construction, and distance from the epicenter. A single earthquake, then, can produce effects of many different intensities in different places (figure 4.10), although it will have only one magnitude assigned to it. Intensity is also a somewhat subjective measure in that it is usually based on observations by or impressions of individuals; different observers in the same spot may assign different intensity values to a single earthquake. Nevertheless, intensity is a more direct indication of the impact of a particular seismic event on humans in a given place than is

TABLE 4.2

Frequency of Earthquakes of Various Magnitudes

Descriptor	Magnitude	Number per Year	Approximate Energy Released (ergs)
great	8 and over	1 to 2	over 5.8×10^{23}
major	7–7.9	18	$2–42 \times 10^{22}$
strong	6–6.9	120	$8–150 \times 10^{20}$
moderate	5–5.9	800	$3–55 \times 10^{19}$
light	4–4.9	6200	$1–20 \times 10^{18}$
minor	3–3.9	49,000	$4–72 \times 10^{16}$
very minor	<3	[mag. 2–3 about 1000/day] [mag. 1–2 about 8000/day]	below 4×10^{16}

Source: Frequency data and descriptors from National Earthquake Information Center.

Note: For every unit increase in Richter magnitude, ground displacement increases by a factor of 10, while energy release increases by a factor of 30. Therefore, most of the energy released by earthquakes each year is released not by the hundreds of thousands of small tremors, but by the handful of earthquakes of magnitude 7 or larger.

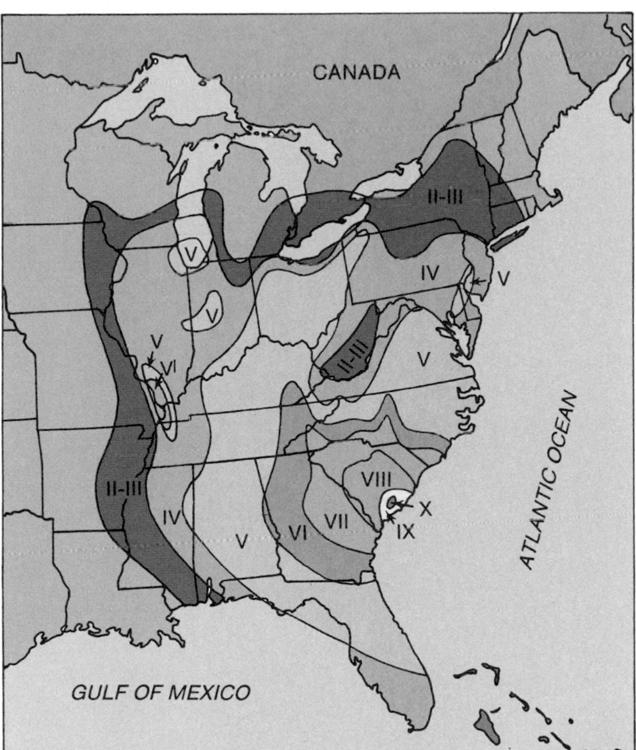

FIGURE 4.10 Zones of different Mercalli intensity for the Charleston, South Carolina earthquake of 1886. (Note how far away that earthquake could be felt!)
Data from U.S. Geological Survey

magnitude. The extent of damage at each intensity level is, in turn, related to the maximum ground velocity and acceleration experienced. Thus, even in uninhabited areas, intensities can be estimated if the latter data have been measured.

Several dozen intensity scales are in use worldwide. The most widely applied intensity scale in the United States is the Modified Mercalli Scale, summarized in table 4.3.

TABLE 4.3

Modified Mercalli Intensity Scale (abridged)

Intensity	Description
I	Not felt.
II	Felt by persons at rest on upper floors.
III	Felt indoors—hanging objects swing. Vibration like passing of light trucks.
IV	Vibration like passing of heavy trucks. Standing automobiles rock. Windows, dishes, and doors rattle; wooden walls or frame may creak.
V	Felt outdoors. Sleepers wakened. Liquids disturbed, some spilled; small objects may be moved or upset; doors swing; shutters and pictures move.
VI	Felt by all; many frightened. People walk unsteadily; windows and dishes broken; objects knocked off shelves, pictures off walls. Furniture moved or overturned; weak plaster cracked. Small bells ring. Trees and bushes shaken.
VII	Difficult to stand. Furniture broken. Damage to weak materials, such as adobe; some cracking of ordinary masonry. Fall of plaster, loose bricks, and tile. Waves on ponds; water muddy; small slides along sand or gravel banks. Large bells ring.
VIII	Steering of automobiles affected. Damage to and partial collapse of ordinary masonry. Fall of chimneys, towers. Frame houses moved on foundations if not bolted down. Changes in flow of springs and wells.
IX	General panic. Frame structures shifted off foundations if not bolted down; frames cracked. Serious damage even to partially reinforced masonry. Underground pipes broken; reservoirs damaged. Conspicuous cracks in ground.
X	Most masonry and frame structures destroyed with their foundations. Serious damage to dams and dikes; large landslides. Rails bent slightly.
XI	Rails bent greatly. Underground pipelines out of service.
XII	Damage nearly total. Large rock masses shifted; objects thrown into the air.

EARTHQUAKE-RELATED HAZARDS AND THEIR REDUCTION

Earthquakes can have a variety of harmful effects, some obvious, some more subtle. As noted earlier, earthquakes of the same magnitude occurring in two different places can cause very different amounts of damage, depending on such variables as the nature of the local geology, whether the area affected is near the coast, and whether the terrain is steep or flat.

Ground Motion

Movement along the fault is an obvious hazard. The offset between rocks on opposite sides of the fault can break power lines, pipelines, buildings, roads, bridges, and other structures that actually cross the fault. Such offset can be large: In the 1906 San Francisco earthquake, maximum strike-slip displacement across the San Andreas fault was more than 6 meters. Planning and thoughtful design can reduce the risks. For example, power lines and pipelines can be built with extra slack where they cross a fault zone, or they can be designed with other features to allow some "give" as the fault slips and stretches them. Such considerations had to be taken into account when the Trans-Alaska Pipeline was built, for it crosses several known, major faults along its route. The engineering proved its value in the magnitude-7.9 2002 Denali Fault earthquake (figure 4.11).

FIGURE 4.11 Bends in the Trans-Alaska Pipeline allow some flex, and the fact that it sits loosely on slider beams where it crosses the Denali Fault also helped prevent rupture of the pipeline in 2002.
Photo courtesy Alyeska Pipeline Service Company and U.S. Geological Survey

Fault displacement aside, the *ground shaking* produced as accumulated energy is released through seismic waves causes damage to and sometimes complete failure of buildings, with the surface waves—especially shear surface waves—responsible for most of this damage. Shifts of even a few tens of centimeters can be devastating, especially to structures made of weak materials such as adobe or inadequately reinforced concrete (figure 4.12), and as shaking continues, damage may become

A

B

FIGURE 4.12 The nature of building materials, as well as design, influences how well buildings survive earthquakes. (A) In the 1999 Düzce, Turkey, earthquake, traditional timber-frame buildings (foreground) were more resilient than concrete structures (rear) in which the reinforcing pillars pulled out of their foundations. (B) One person was killed in the collapse of a five-story tower at St. Joseph's Seminary during the Loma Prieta earthquake; the rest of the complex stood intact.
(A) Photograph by Roger Bilham, (B) photograph by H. G. Wilshire, both courtesy U.S. Geological Survey

progressively worse. Such effects, of course, are most severe on or very close to the fault, so the simplest strategy would be not to build near fault zones. However, many cities have already developed near major faults. Sometimes, cities are rebuilt many times in such places. The ancient city of Constantinople—now Istanbul—has been leveled by earthquakes repeatedly throughout history. Yet there it sits. This raises the question of designing "earthquake-resistant" buildings to minimize damage.

Engineers have studied how well different types of building have withstood real earthquakes. Scientists can conduct laboratory experiments on scale models of skyscrapers and other buildings, subjecting them to small-scale shaking designed to simulate the kinds of ground movement to be expected during an earthquake. On the basis of their findings, special building codes for earthquake-prone regions can be developed.

This approach, however, presents many challenges. There are a limited number of records of just how the ground does move in a severe earthquake. To obtain the best such records, sensitive instruments must be in place near the fault zone beforehand, and those instruments must survive the earthquake. (Sometimes alternative methods will provide useful information: after the 1995 Kobe earthquake, geoscientists studied videotapes from security cameras, using the recorded motions of structures and furnishings to deduce the ground motions during the earthquake.) Even with good records from an actual earthquake, laboratory experiments may not precisely simulate real earthquake conditions, given the number of variables involved. Such uncertainties raise major concerns about the safety of dams and nuclear power plants near active faults. In an attempt to circumvent the limitations of scale modeling, the United States and Japan set up in 1979 a cooperative program to test earthquake-resistant building designs. The tests included experiments on a full-sized, seven-story, reinforced-concrete structure, in which ground shaking was simulated by using hydraulic jacks. Results were not entirely as anticipated on the basis of earlier modeling studies, which indicates that full-scale experiments and observations may be critical to designing optimum building codes in earthquake-prone areas.

A further complication is that the same building codes cannot be applied everywhere. For example, not all earthquakes produce the same patterns of ground motion. The 1994 Northridge earthquake underscored the dependence of reliable design on accurately anticipating ground motion. The epicenter was close to that of the 1971 San Fernando earthquake that destroyed many freeways (figure 4.13). They were rebuilt to be earthquake-resistant, assuming the predominantly strike-slip displacement that is characteristic of the San Andreas. The Northridge quake, however, had a strong dip-slip component, for it occurred along a previously unmapped, buried thrust fault—and many of the rebuilt freeways collapsed again.

It is also important to consider not only how structures are built, but what they are built *on*. Buildings built on solid rock (bedrock) seem to suffer far less damage than those built on deep soil. In the 1906 San Francisco earthquake, buildings erected on filled land reclaimed from San Francisco Bay suffered up to

A

B

FIGURE 4.13 Freeway damage from the 1971 San Fernando earthquake (A) was repeated in the 1994 Northridge quake (B) because the design of the rebuilt overpasses did not account for the atypical ground motion experienced in 1994. This interchange of I-5 and C-14 was under construction in 1971; it was not subsequently retrofitted for enhanced earthquake resistance.

(A) Photograph courtesy National Information Service for Earthquake Engineering, University of California, Berkeley, Karl V. Steinbrugge, photographer; (B) Photograph by M. Celebi, U.S. Geological Survey

four times more damage from ground shaking than those built on bedrock; the same general pattern was repeated in the 1989 Loma Prieta earthquake (figure 4.14). The extent of damage in Mexico City resulting from the 1985 Mexican earthquake (figure 4.15) was partly a consequence of the fact that the city is underlain by weak layers of volcanic ash and clay. Most smaller and older buildings lacked the deep foundations necessary to reach more stable sand layers at depth. Many of these buildings collapsed completely. Acapulco suffered far less damage, although it was much closer to the earthquake's epicenter, because it stands firmly on bedrock. These observations are directly linked to the fact that ground shaking is amplified in unconsolidated materials (figure 4.16). The fact that underlying geology influences surface damage is demonstrated by the fact that intensity is not directly correlated with proximity to the epicenter (figures 4.17, 4.10).

FIGURE 4.14 Car crushed under third story of a failed building when the first two floors collapsed and sank. Some of the most severe damage in San Francisco from the 1989 Loma Prieta earthquake was here in the Marina district, built on filled land subject to liquefaction.
Photograph by J. K. Nakata, U.S. Geological Survey

FIGURE 4.15 Building failure in Mexico City from the 1985 earthquake: pancake-style collapse of fifteen-story, reinforced-concrete structure.
Photograph by M. Celebi, USGS Photo Library, Denver, CO.

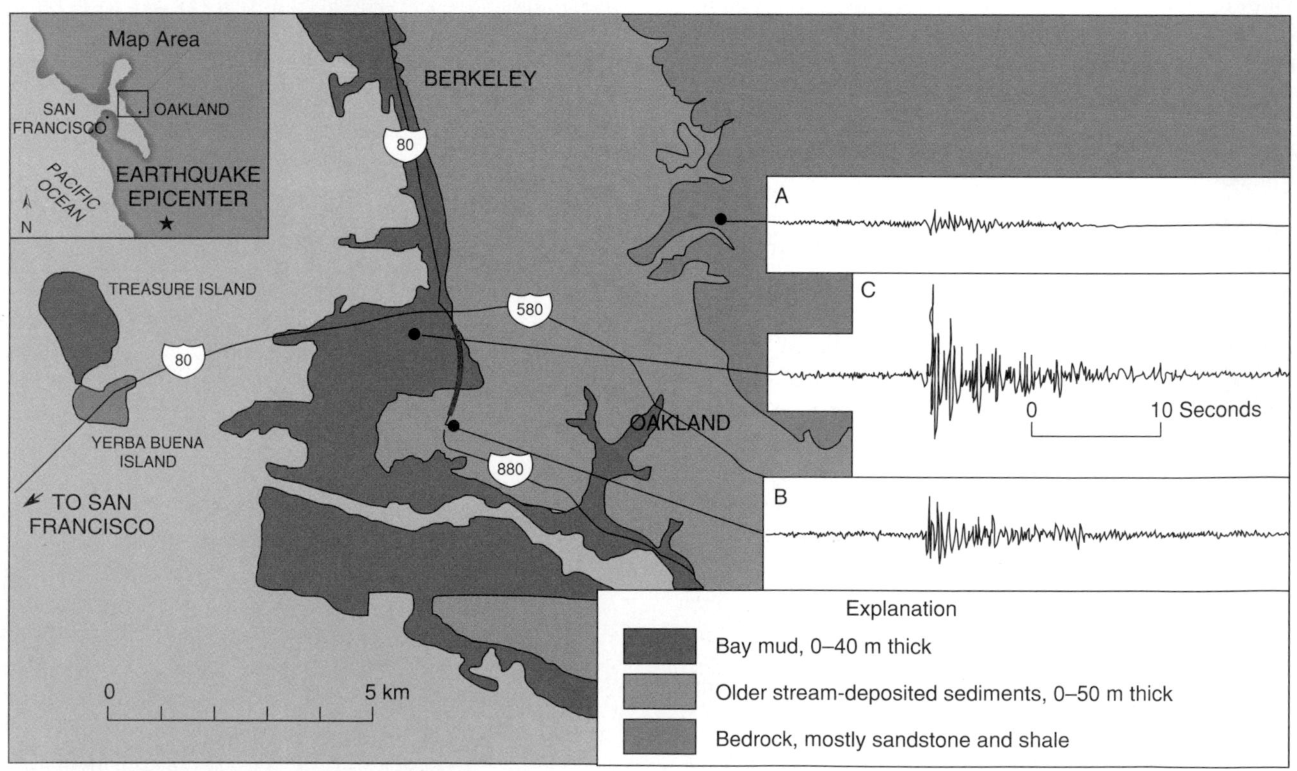

FIGURE 4.16 Ground shaking from the Loma Prieta earthquake, only moderate in bedrock (trace A), was amplified in younger river sediments (B) and especially in more recent, poorly consolidated mud (C).
Source: R. A. Page et al., Goals, Opportunities, and Priorities for the USGS Earthquake Hazards Reduction Program, *U.S. Geological Survey Circular 1079*

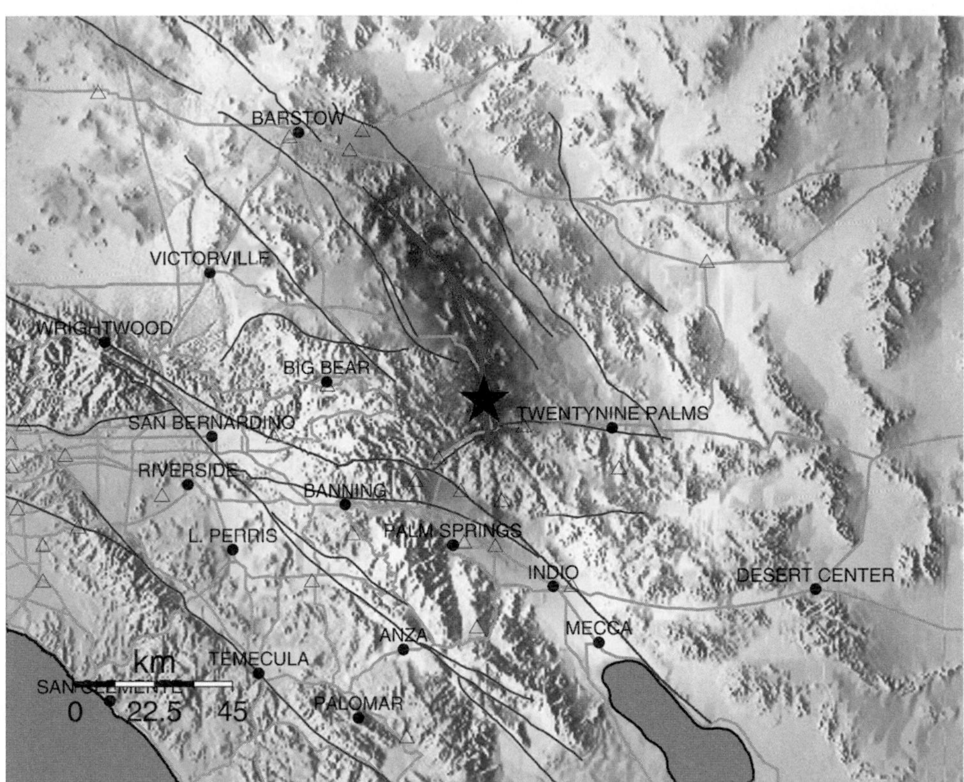

PERCEIVED SHAKING	Not felt	Weak	Light	Moderate	Strong	Very strong	Severe	Violent	Extreme
POTENTIAL DAMAGE	none	none	none	Very light	Light	Moderate	Moderate/Heavy	Heavy	Very Heavy
PEAK ACC.(%g)	*<.17*	*.17-1.4*	*1.4-3.9*	*3.9-9.2*	*9.2-18*	18-34	34-65	65-124	>124
INSTRUMENTAL INTENSITY	I	II-III	IV	V	VI	VII	VIII	IX	X+

FIGURE 4.17 Distribution of intensities for 1992 Landers earthquake (epicenter marked by star). These intensities are determined instrumentally from the maximum measured ground acceleration, reported as a percentage of *g*, earth's gravitational acceleration constant (the acceleration experienced by an object dropped near the surface).

Image courtesy ShakeMap Working Group, U.S. Geological Survey

The characteristics of the earthquakes in a particular region also must be taken into account. For example, severe earthquakes are generally followed by many **aftershocks,** earthquakes that are weaker than the principal tremor. The main shock usually causes the most damage, but when aftershocks are many and are nearly as strong as the main shock, they may also cause serious destruction.

The duration of an earthquake also affects how well a building survives it. In reinforced concrete, ground shaking leads to the formation of hairline cracks, which then widen and develop further as long as the shaking continues. A concrete building that can withstand a one-minute main shock might collapse in an earthquake in which the main shock lasts three minutes. Many of the California building codes, used as models around the world, are designed for a 25-second main shock, but earthquake main shocks can last ten times that long.

A final point to keep in mind is that even the best building codes are typically applied only to new construction. When a major city is located near a fault zone, thousands of vulnerable older buildings may already have been built in high-risk areas. The costs to redesign, rebuild, or even modify all of these buildings would be staggering. Most legislative bodies are reluctant to require such efforts; indeed, many do nothing even about municipal buildings built in fault zones.

Ground Failure

Landslides (figure 4.18) can be a serious secondary earthquake hazard in hilly areas. As will be seen in chapter 8, earthquakes are one of the major events that trigger slides on unstable slopes. The best solution is not to build in such areas. Even if a whole region is hilly, detailed engineering studies of rock and soil properties and slope stability may make it possible to avoid the most dangerous sites. Visible evidence of past landslides is another indication of especially dangerous areas.

Ground shaking may cause a further problem in areas where the ground is very wet—in filled land near the coast or in places with a high water table. This problem is **liquefaction.** When wet soil is shaken by an earthquake, the soil particles

A

B

FIGURE 4.18 (A) Landslide from the magnitude-6.5 earthquake in Seattle, Washington in 1965 made Union Pacific Railway tracks unuseable. (B) Landslide from an unnamed peak in the Alaska Range, less than ten miles west of the Trans-Alaska Pipeline, triggered by 2002 Denali Fault earthquake.
(A) Photograph courtesy University of California at Berkeley/NOAA, (B) courtesy U.S. Geological Survey Earthquake Hazards program

may be jarred apart, allowing water to seep in between them, greatly reducing the friction between soil particles that gives the soil strength, and causing the ground to become somewhat like quicksand. When this happens, buildings can just topple over or partially sink into the liquefied soil; the soil has no strength to support them. The effects of liquefaction were dramatically illustrated in Niigata, Japan, in 1964. One multistory apartment building tipped over to settle at an angle of 30 degrees to the ground while the structure remained intact! (See figure 4.19.) Liquefaction was likewise a major cause of damage from the Loma Prieta earthquake and the Kobe earthquake. Telltale signs of liquefaction include "sand boils," formed as liquefied soil bubbles to the surface during the quake (figure 4.20). In some areas prone to liquefaction, improved underground drainage systems may be installed to try to keep the soil drier, but little else can be done about this hazard, beyond avoiding the areas at risk. Not all areas with wet soils are subject to liquefaction; the nature of the soil or fill plays a large role in the extent of the danger.

Tsunamis and Coastal Effects

Coastal areas, especially around the Pacific Ocean basin where so many large earthquakes occur, may also be vulnerable to **tsunamis.** These are seismic sea waves, sometimes improperly called "tidal waves," although they have nothing to do with tides. The name derives from the Japanese for "harbor wave," which is descriptive of their behavior. When an undersea or near-shore earthquake occurs, sudden movement of the sea floor may set up waves traveling away from that spot, like ripples in a pond

FIGURE 4.19 Effects of soil liquefaction during an earthquake in Niigata, Japan, 1964. The buildings, which were designed to be earthquake-resistant, simply tipped over intact.
Photograph courtesy of National Geophysical Data Center

FIGURE 4.20 Sand boils in field near Watsonville, California.
Photograph by J. Tinsley, from U.S. Geological Survey Open-File Report 89-687

Box 4.1 "Earthquake-Proof" Buildings?

When the shaking started, Maria did all the right things. She gathered Juan Carlos, Pepe, and baby Conchita together and shepherded them quickly into the doorway leading to the apartment's back hall, where they huddled tightly together under a stout table that Maria had put beneath the sturdy arch "just in case." There they stayed while the building swayed and creaked and vases and dishes toppled and crashed for what seemed forever, but was really only about two minutes. When all was still and quiet again, they crept cautiously out of the apartment and the building, joining the rest of the residents in the street. Everyone was frightened, but safe. A house on the corner had collapsed (fortunately, the owners were away on vacation), but the apartment building was standing.

Still, it wasn't in perfect shape. There were large, gaping cracks in the walls and on one side of the foundation. Some of the bricks on the facade had fallen off. The building seemed to be a little out of plumb, too: doors were jamming, and they couldn't all be closed properly after the quake. It was becoming obvious that the tenants couldn't just go back in, start picking up the pieces (literally), and go on with business as usual. In fact, it looked as if they would be moving out, for some time at least, while repairs were made.

The building superintendent, Mr. Hernandez, was fuming. Maria heard him muttering about architects and building codes. It seemed that he'd been assured that this apartment building was "up to code" for earthquakes. So why, he wanted to know, was it such a mess? And he wasn't the only person who was staring glumly at a damaged building and asking those questions.

As it turned out, the real problem was one of communication and expectations, not faulty building codes or construction. It is common in areas with building codes or design guidelines that address earthquake resistance in detail to identify levels of expected structural soundness. For most residential and commercial buildings, "substantial life safety" may be the objective: the building may be extensively damaged, but casualties should be few; see figure 1. (The structure, in fact, may have to be razed and rebuilt.) Designing for "damage control" means that in addition to protecting those within, the structure should be repairable after the earthquake. This will add to the up-front costs, though it should save both repair costs and "down time" after a major quake. The highest (and most expensive) standard is design for "continued operation," meaning little or no structural damage at all; this would be important for hospitals, fire stations, and other buildings housing essential services, and also for facilities that might present hazards if the buildings failed (nuclear power plants, manufacturing facilities using or making highly toxic chemicals). This is the sort of standard to which the Trans-Alaska pipeline was (fortunately) designed: it was built with a possible magnitude-8 earthquake on the Denali fault in mind, so it came through the magnitude-7.9 earthquake in 2002, with its 14 feet of horizontal fault displacement, without a break.

In fact, Maria's apartment building did exactly what it was designed to do. "Substantial life safety" was the goal, and it was achieved. While the residents weren't pleased to find they'd be moving out for a while, and the owner certainly wasn't pleased at the loss of income during the repairs, nobody was seriously hurt.

Anyone living, or buying property, in an area where earthquakes are a significant concern would be well-advised to pin down just what level of earthquake resistance is promised in a building's design!

FIGURE 1 The concept of "substantial life safety"—and its limitations—are illustrated by this classic scene from the Mission district following the 1906 San Francisco earthquake. Nobody died here, but neither can business go on as usual! *Photograph by G. K. Gilbert, courtesy U.S. Geological Survey*

caused by a dropped pebble. Contrary to modern movie fiction, tsunamis are not seen as huge breakers in the open ocean that topple ocean liners in one sweep. In the open sea, tsunamis are only unusually broad swells on the water surface. As tsunamis approach land, the water piles up and they can develop into large breaking waves, just as ordinary ocean waves become breakers as the undulating waters "touch bottom" near shore. Tsunamis, however, can easily be over 15 meters high in the case of larger earthquakes. Several such waves may wash over the coast in succession; between waves, the water may be pulled swiftly seaward, emptying a harbor or bay, and perhaps pulling unwary onlookers along. Tsunamis can travel very quickly—speeds of

1000 kilometers/hour (600 miles/hour) in the open ocean are not uncommon—and tsunamis set off on one side of the Pacific may still cause noticeable effects on the other side of the ocean. A tsunami set off by a 1960 earthquake in Chile was still vigorous enough to cause 7-meter-high breakers when it reached Hawaii some fifteen hours later, and twenty-five hours after the earthquake, the tsunami was detected in Japan.

Given the speeds at which tsunamis travel, little can be done to warn those near the earthquake epicenter, but people living some distance away can be warned in time to evacuate, saving lives, if not property. In 1948, two years after a devastating tsunami hit Hawaii, the U.S. Coast and Geodetic Survey established a Tsunami Early Warning System, based in Hawaii. Whenever a major earthquake occurs in the Pacific region, sea-level data are collected from a series of monitoring stations around the Pacific. If a tsunami is detected, data on its source, speed, and estimated time of arrival can be relayed to areas in danger, and people can be evacuated as necessary. The system does not, admittedly, always work perfectly. In the 1964 Alaskan earthquake, disrupted communications prevented warnings from reaching some areas quickly enough. Also, residents' responses to the warnings were variable: Some ignored the warnings, and some even went closer to shore to watch the waves, often with tragic consequences. Two tsunami warnings were issued in early 1986 following earthquakes in the Aleutians. They were largely disregarded. Fortunately, in that instance, the feared tsunamis did not materialize.

There is no tsunami warning system in either the Caribbean sea, where the potential for tsunamis exists, or in the Indian Ocean, devastated by the tsunami from the 2004 Sumatran earthquake (Box 4.2). In the latter case, seismologists outside the area alerted local governments to the possibility that such an enormous earthquake might produce a sizeable tsunami. The word did not reach all affected coastal areas in time. Even where it did, some officials evidently chose not to issue public warnings, in part for fear of frightening tourists. Such behavior likely contributed to the death toll.

Even in the absence of tsunamis, there is the possibility of coastal flooding from sudden subsidence as plates shift during an earthquake. Areas that were formerly dry land may be permanently submerged and become uninhabitable (figure 4.21). Conversely, uplift of sea floor may make docks and other coastal structures useless, though this rarely results in casualties.

FIGURE 4.21 Flooding in Portage, Alaska, due to tectonic subsidence during 1964 earthquake.
Photograph by G. Plafker, courtesy USGS Photo Library, Denver, CO.

Fire

A secondary hazard of earthquakes in cities is fire, which may be more devastating than ground movement. In the 1906 San Francisco earthquake, 70% of the damage was due to fire, not simple building failure. As it was, the flames were confined to a 10-square-kilometer area only by dynamiting rows of buildings around the burning section. Fires occur because fuel lines and tanks and power lines are broken, touching off flames and fueling them. At the same time, water lines also are broken, leaving no way to fight the fires effectively, and streets fill with rubble, blocking fire-fighting equipment. Putting numerous valves in all water and fuel pipeline systems helps to combat these problems because breaks in pipes can then be isolated before too much pressure or liquid is lost.

EARTHQUAKE PREDICTION AND FORECASTING

Millions of people already live near major fault zones. For that reason, prediction of major earthquakes could result in many saved lives. Some progress has been made in this direction, but as the complexity of seismic activity is more fully recognized, initial optimism has been tempered somewhat.

Seismic Gaps

Maps of the locations of earthquake epicenters along major faults show that there are stretches with little or no seismic activity, while small earthquakes continue along other sections of the same fault zone. Such quiescent, or dormant, sections of otherwise-active fault zones are called **seismic gaps.** They apparently represent "locked" sections of faults along which friction is preventing slip. These areas may be sites of future

Box 4.2 Megathrust Makes Mega-Disaster

The tectonics of southern and southeast Asia are complex. East of the India and Australia plates lies the Sunda Trench, marking a long subduction zone. The plate-boundary fault has been called a "megathrust"—a very, very large thrust fault.

For centuries, the fault at the contact between the small Burma Plate and the India Plate had been locked. The edge of the Burma Plate had been warped downward, stuck to the subducting India Plate. The fault let go on 26 December 2004, in an earthquake so large that its precise moment magnitude—somewhere between 9.1 and 9.3—is still being debated. Even at 9.1, it would be the third-largest earthquake in the world since 1900, behind the 1960 Chilean quake and 1964 Alaskan quake. During the Sunda Trench earthquake, significant slip occurred over most of the southern half of the Burma Plate. Aftershocks covered the whole plate (figure 1).

When the fault ruptured, the freed leading edge of the Burma Plate snapped upward by as much as 5 meters. The resulting abrupt shove on the water column above set off the deadly tsunami. It reached the Sumatran shore in minutes, but it took hours to reach India and, still later, Africa (figure 2).

How high onshore the tsunami surged varied not only with distance from the epicenter but with coastal geometry. The highest runup was observed on gently sloping shorelines and where water was funneled into a narrowing bay. Detailed studies after the earthquake showed that in some places the tsunami waves reached heights of 31 meters (nearly 100 feet). Destruction was profound and widespread (figure 3). An irony of tsunami behavior is that fishermen well out at sea were unaffected by, and even unaware of, the tsunami until they returned to port.

Most of the estimated 283,000 deaths from this earthquake were in fact due to the tsunami, not building collapse or other consequences of ground shaking. Thousands of people never found or accounted for were presumed to have been swept out to sea and drowned. Tsunami deaths occurred as far away as Africa.

And the Sunda Trench had more in store. When one section of a fault zone slips in a major earthquake, regional stresses are shifted, which can bring other sections closer to rupture. On 28 March 2005, a magnitude 8.6 earthquake (itself the seventh-largest worldwide since 1900) ruptured the section of the fault zone just south of the 2004 break (figure 4). Though no major tsunami resulted this time, more than 1300 people died. Many more, already traumatized by the 2004 disaster, were badly frightened again. Given the region's tectonics, more significant earthquakes can certainly be expected. Development of an Indian Ocean tsunami warning system, now under discussion, could help reduce casualties.

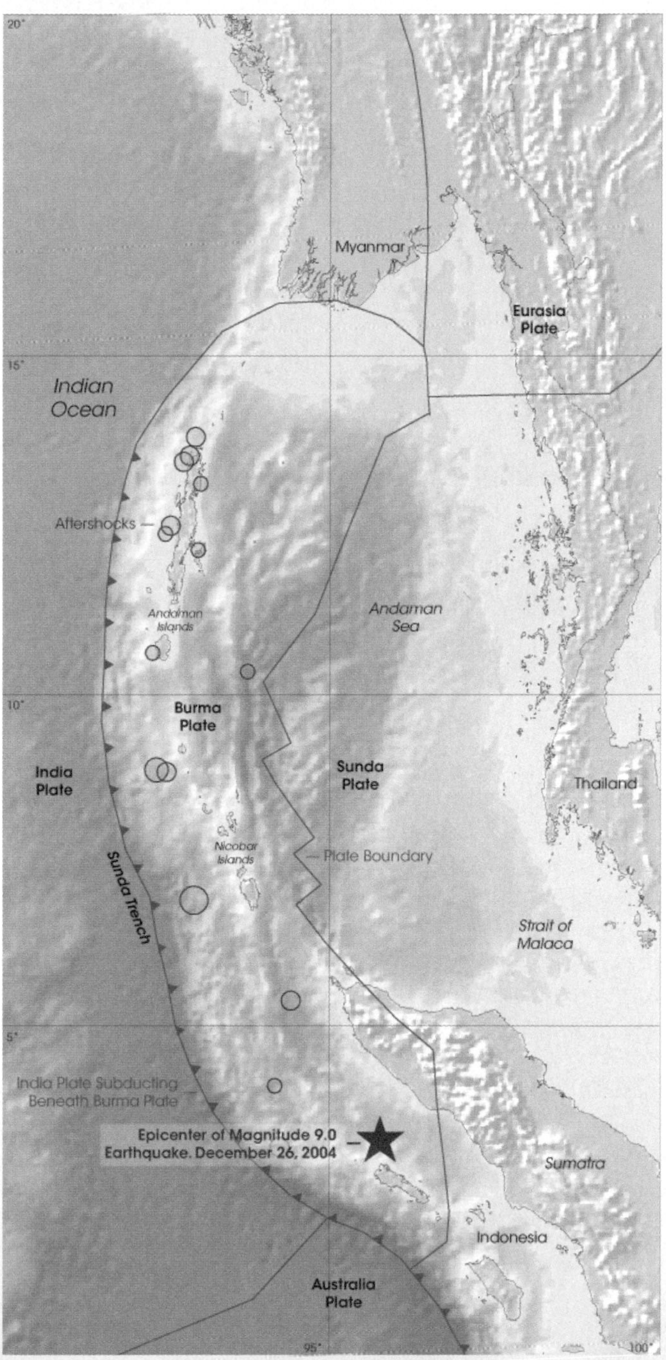

FIGURE 1 The December 2004 Sunda Trench earthquake ruptured a 400 km length of the southern Burma Plate, but aftershocks—only a few epicenters of which are shown here—affected the whole plate.
Image courtesy USGS/NASA

(Box 4.2 Continued)

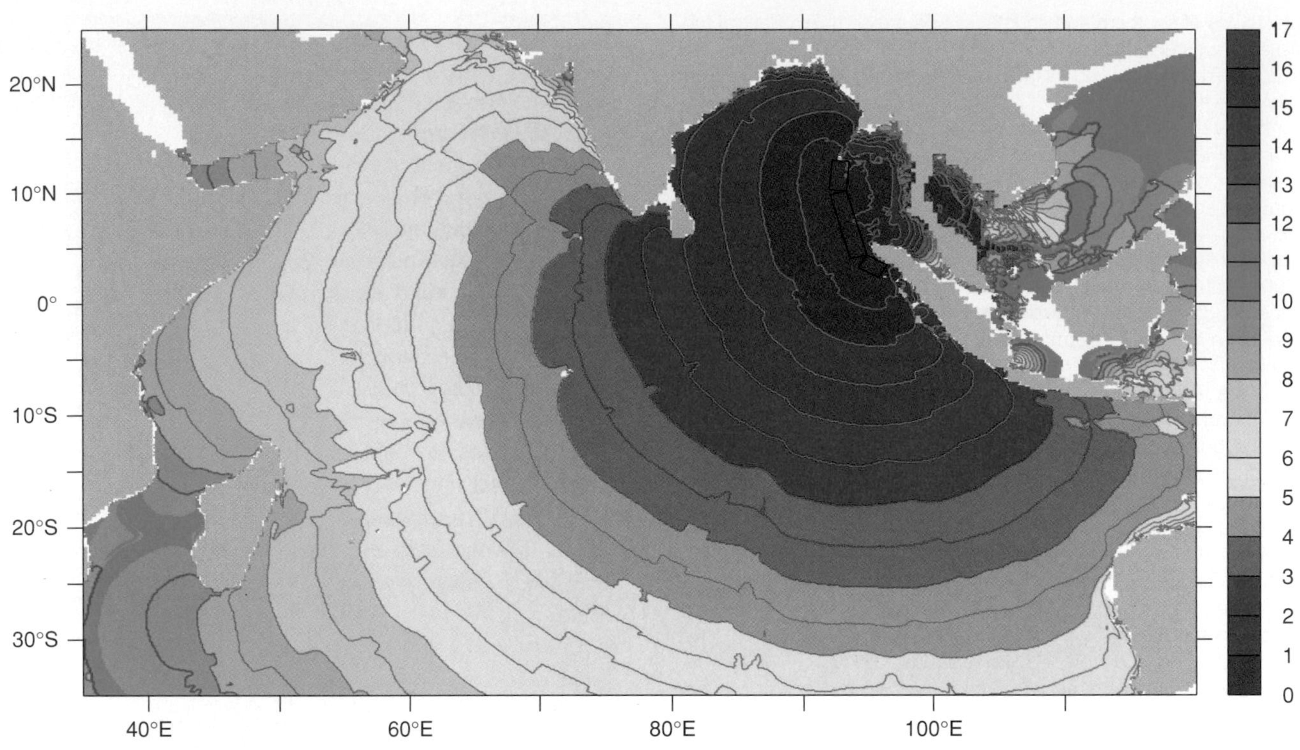

FIGURE 2 Travel time, in hours, for the tsunami generated by the 26 December 2004 earthquake.
Image courtesy NOAA Pacific Marine Environmental Lab Tsunami Research Program

FIGURE 3 Gleebruk, Indonesia before (A) and after (B) the tsunami. Photographs were taken 12 April 2004 and 2 January 2005, respectively.
Photographs © Digital Globe.

(Box 4.2 Continued)

FIGURE 4 Colored areas show ruptures associated with earthquakes in 1833, 1861, and 2004 along the Sunda Trench. Stress shifts following the December 2004 earthquake (epicenter at red star) may well have contributed to the March 2005 earthquake just to the south (yellow star). Aftershocks from the latter quake (yellow dots) suggest rupture of both the gap between the 1861 and 2004 rupture zones, and a large part of the former.
Map courtesy U.S. Geological Survey Earthquake Hazards Program

serious earthquakes. On either side of a locked section, stresses are being released by earthquakes. In the seismically quiet locked sections, friction is apparently sufficient to prevent the fault from slipping, so the stresses are simply building up. The concern, of course, is that the accumulated energy will become so great that, when that locked section of fault finally does slip again, a very large earthquake will result.

Recognition of these seismic gaps makes it possible to identify areas in which large earthquakes may be expected in the future. The 1989 Loma Prieta, California, earthquake occurred in what had been a seismic gap along the San Andreas fault. The subduction zone bordering the Banda Plate had been rather quiet seismically for many decades before the December 2004 quake; areas further south along the Sunda Trench still are. A major seismic gap along the subduction zone along the western side of Central America is likely to be the site of the next major earthquake to cause serious damage to Mexico City. Earthquake-cycle theory, discussed later in this chapter, may be a tool for anticipating major earthquakes that will fill such gaps abruptly.

Earthquake Precursors and Prediction

In its early stages, earthquake prediction was based particularly on the study of earthquake **precursor phenomena,** things that happen or rock properties that change prior to an earthquake. Many different possibilities have been examined. For example, the ground surface may be uplifted and tilted prior to an earthquake. Seismic-wave velocities in rocks near the fault may change before an earthquake; so may electrical resistivity (the resistance of rocks to electric current flowing through them). Changes have been observed in the levels of water in wells and/or in the content of radon (a radioactive gas that occurs naturally in rocks as a result of decay of trace amounts of uranium within them). Chinese investigators have cited examples of anomalous behavior of animals prior to an earthquake.

The hope, with the study of precursor phenomena, has been that one could identify patterns of precursory changes that could be used with confidence to issue earthquake predictions precise enough as to time to allow precautionary evacuations or other preparations. Unfortunately, the precursors have not proven to be very reliable. Not only does the length of time over which precursory changes are seen before an earthquake vary, but the pattern of precursors—which parameters show changes, and of what sorts—also varies; and, most problematic, some earthquakes seem quite unheralded by recognizable precursors at all. Loma Prieta was one of these; so was Northridge, and so was Kobe. After decades of precursor studies, no consistently useful precursors have been recognized, though the search continues.

Current Status of Earthquake Prediction

Since 1976, the director of the U.S. Geological Survey has had the authority to issue warnings of impending earthquakes and other potentially hazardous geologic events (volcanic eruptions, landslides, and so forth). An Earthquake Prediction Panel reviews scientific evidence that might indicate an earthquake threat and makes recommendations to the director regarding the issuance of appropriate public statements. These statements could range in detail and immediacy from a general notice to residents in a fault zone of the existence and nature of earthquake hazards there, to a specific warning of the anticipated time, location, and severity of an imminent earthquake.

In early 1985, the panel made its first endorsement of an earthquake prediction. They agreed that an earthquake of Richter magnitude about 6 could be expected on the San Andreas fault near Parkfield, California, in January 1988 $\pm$ 5 years. The prediction was based, in large part, on cyclic patterns of seismicity in the area: On average, major earthquakes occur there every twenty-two years, and the last had been in 1966 (figure 4.22A). Both the degree of confidence and the estimated probability were very high. Yet the critical time period came and went—and no Parkfield earthquake, even 25 years after the targeted date. Meanwhile, local residents were angered by perceived damage to property values and tourism resulting from the prediction.

The long-awaited quake finally happened in late September 2004. Despite the fact that this fault segment has been heavily instrumented and closely watched since the mid-1980s in anticipation of this event, there were no clear immediate precursors to prompt a short-term warning of it. Seismologists are now combing their records in retrospect to see what subtle precursory changes they may have missed, and they are launching a program of drilling into the fault to improve their understanding of its behavior.

In the People's Republic of China, tens of thousands of amateur observers and scientists work on earthquake prediction. In February 1975, after months of smaller earthquakes, radon anomalies, and increases in ground tilt followed by a rapid increase in both tilt and microearthquake frequency, the scientists predicted an imminent earthquake near Haicheng in northeastern China. The government ordered several million people out of their homes into the open. Nine and one-half hours later, a major earthquake struck, and many lives were saved because people were not crushed by collapsing buildings. The next year, they concluded that a major earthquake could be expected near Tangshan, about 150 kilometers southeast of Beijing. In this case, however, they could only say that the event was likely to occur sometime during the following two months. When the earthquake—magnitude over 8.0 with aftershocks up to magnitude 7.9—did occur, there was no immediate warning, no sudden change in precursor phenomena, and hundreds of thousands of people died.

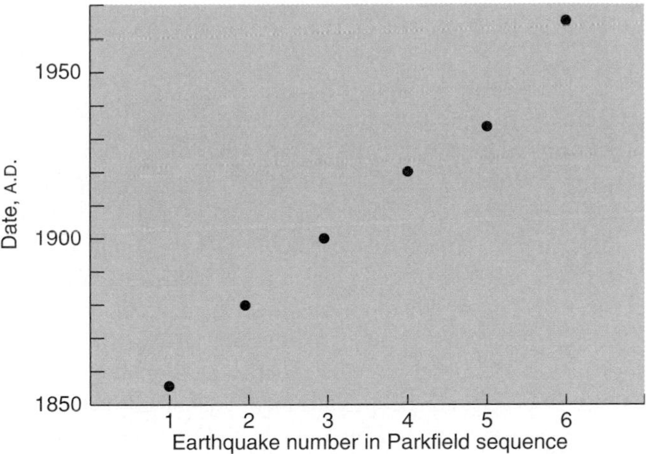

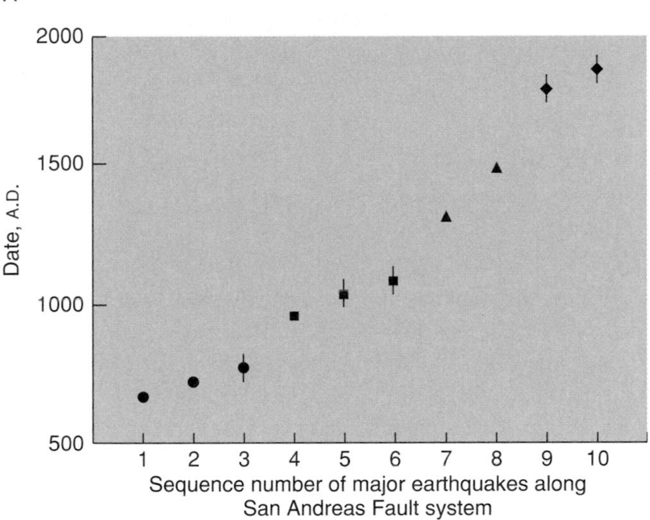

FIGURE 4.22 Periodicity of earthquakes assists in prediction/forecasting efforts. (A) The Parkfield prediction was based on an average interval of 22 years between earthquakes of magnitude 5.5 or greater. (B) Long-term records of major earthquakes on the San Andreas show both broad periodicity and some tendency toward clustering of 2 to 3 earthquakes in each active period. Prehistoric dates are based on carbon-14 dating of faulted peat deposits.

Source: (B) Data from U.S. Geological Survey Circular 1079

Only four nations—Japan, Russia, the People's Republic of China, and the United States—have had government-sponsored earthquake prediction programs. Such programs typically involve intensified monitoring of active fault zones to expand the observational data base, coupled with laboratory experiments designed to increase understanding of precursor phenomena and the behavior of rocks under stress. Even with these active research programs, scientists cannot monitor every area at once, and, as already noted, earthquake precursors are inconsistent and often absent altogether.

In the United States, earthquake predictions have not been regarded as reliable or precise enough to justify such actions as large-scale evacuations. In the near term, it seems that the more feasible approach is earthquake *forecasting,* identifying levels of earthquake probability in fault zones within relatively broad time windows, as described below. This at least allows for long-term preparations, such as structural improvements. Ultimately, short-term, precise predictions that can save more lives are still likely to require better recognition and understanding of precursory changes.

The Earthquake Cycle and Forecasting

Studies of the dates of large historic and prehistoric earthquakes along major fault zones have suggested that they may be broadly periodic, occurring at more-or-less regular intervals (figure 4.22). This is interpreted in terms of an **earthquake cycle** that would occur along a (non-creeping) fault segment: a period of stress buildup, sudden fault rupture in a major earthquake, followed by a brief interval of aftershocks reflecting minor lithospheric adjustments, then another extended period of stress buildup (figure 4.23). The fact that major faults tend to break in segments is well documented: The Anatolian fault zone, site of the 1999 Turkish earthquakes, is one example; the San Andreas is another; so is the thrust fault along the Sunda Trench (box 4.2 figure 4). The rough periodicity can be understood in terms of two considerations. First, assuming that the stress buildup is primarily associated with the slow, ponderous, inexorable movements of lithospheric plates, which at least over decades or centuries will move at fairly constant rates, one might reasonably expect an approximately constant rate of buildup of stress (or accumulated elastic strain). Second, the rocks along a given fault zone will have particular physical properties, which allow them to accumulate a certain amount of energy before failure, or fault rupture, and that amount would be approximately constant from earthquake to earthquake. Those two factors together suggest that periodicity is a reasonable expectation. Once the pattern for a particular fault zone is established, one may use that pattern, together with measurements of strain accumulation in rocks along fault zones, to project the time window during which the next major earthquake is to be expected along that fault zone and to estimate the likelihood of the earthquake's occurrence in any particular time period. Also, if a given fault zone or segment can store only a certain amount of accumulated energy before failure, the maximum size of earthquake to be expected can be estimated.

This is, of course, simplification of what can be a very complex reality. For example, though we speak of "The San Andreas fault," it is not a single simple, continuous slice through the lithosphere, nor are the rocks it divides either identical or homogeneous on either side. Both stress buildup and displacement

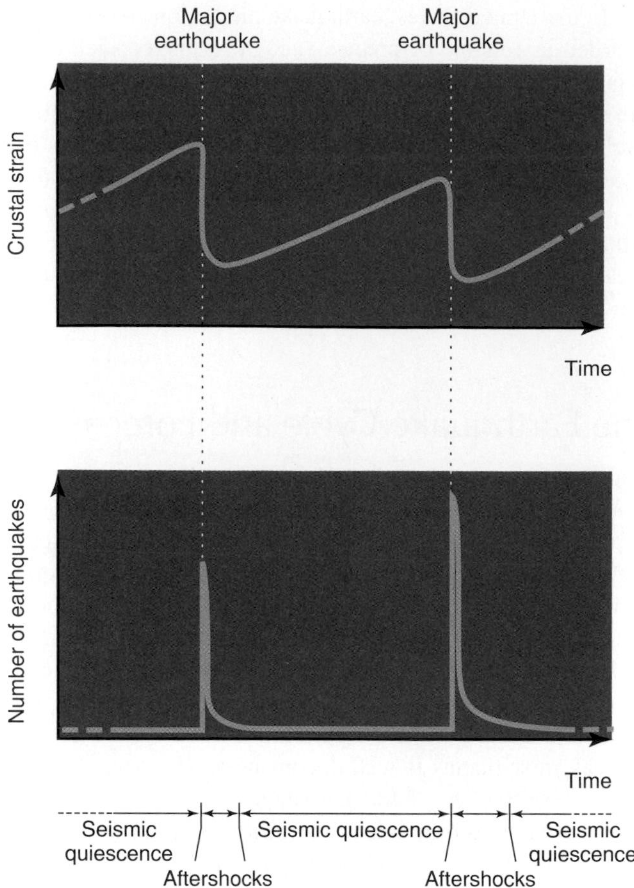

FIGURE 4.23 The "earthquake cycle" concept.
Source: After U.S. Geological Survey Circular 1079, modified after C. DeMets (pers. comm. 2006)

are distributed across a number of faults and small blocks of lithosphere. Earthquake forecasting is correspondingly more difficult. Nevertheless, the U.S. Geological Survey's Working Group on California Earthquake Probabilities publishes forecasts for different segments of the San Andreas (figure 4.24). Note that in this example, the time window in question is the 30-year period 1988–2018. Such projections, longer-term analogues of the meteorologist's forecasting of probabilities of precipitation in coming days, are revised as small (and large) earthquakes occur and more data on slip in creeping sections are collected. For example, after the 1989 Loma Prieta earthquake, which substantially shifted stresses in the region, the combined probability of a quake of magnitude 7 or greater either on the peninsular segment of the San Andreas or on the nearby Hayward fault (locked since 1868) was increased from 50 to 67%. Forecasts are also being refined through recent research on fault interactions and the ways in which one earthquake can increase or decrease the likelihood of slip on another segment, or another fault entirely.

Concerns Related to Earthquake Prediction

Assuming that routine earthquake prediction becomes reality, some planners have begun to look beyond the scientific questions to possible social or legal complications of such predictions. For example, the logistics of evacuating a large urban area on short notice if a near-term earthquake prediction is made are a real concern. Many urban areas have hopelessly snarled traffic every rush hour; what happens if everyone in the city wants to leave at once? Some critics say that a rapid evacuation might involve more casualties than the eventual earthquake. And what if a city is evacuated, people are inconvenienced or hurt, perhaps property is damaged by vandals and/or looters, and then the predicted earthquake never comes? Will the issuers of the warning be sued? Will future warnings be ignored? If and when earthquake prediction accuracy improves, someone will have to wrestle with difficult questions such as these.

Public Response to Earthquake Hazards

A step toward making the geologic hazard warning system work would be increasing public awareness of earthquakes as a hazard. Several other nations are far ahead of the United States in this regard. In the People's Republic of China, vigorous public-education programs (and several major modern earthquakes) have made earthquakes a well-recognized hazard. "Earthquake drills," which stress orderly response to an earthquake warning, are held in Japan on the anniversary of the 1923 Tokyo earthquake, in which nearly 150,000 people died.

By contrast, in the United States, surveys have repeatedly shown that even people living in such high-risk areas as along the San Andreas fault are often unaware of the earthquake hazard. Many of those who *are* aware believe that the risks are not very great and indicate that they would not take any special action even if a specific earthquake warning were issued. Past mixed responses to tsunami warnings underscore doubts about public reaction to earthquake predictions.

Many public officials have a corresponding lack of understanding or concern. Earthquake prediction aside, changes in land use, construction practices, and siting could substantially reduce the risk of property damage from earthquakes. Earthquake-prone areas also need comprehensive disaster-response plans. Within the United States, California is a leader in taking necessary actions in these areas. However, even there, most laws aimed at earthquake-hazard mitigation have been passed in spurts of activity following sizeable earthquakes, and many do not provide for retrofitting existing structures.

An important first step to improving public response is education to improve public understanding of earthquake phenomena and hazards. The National Earthquake Hazards

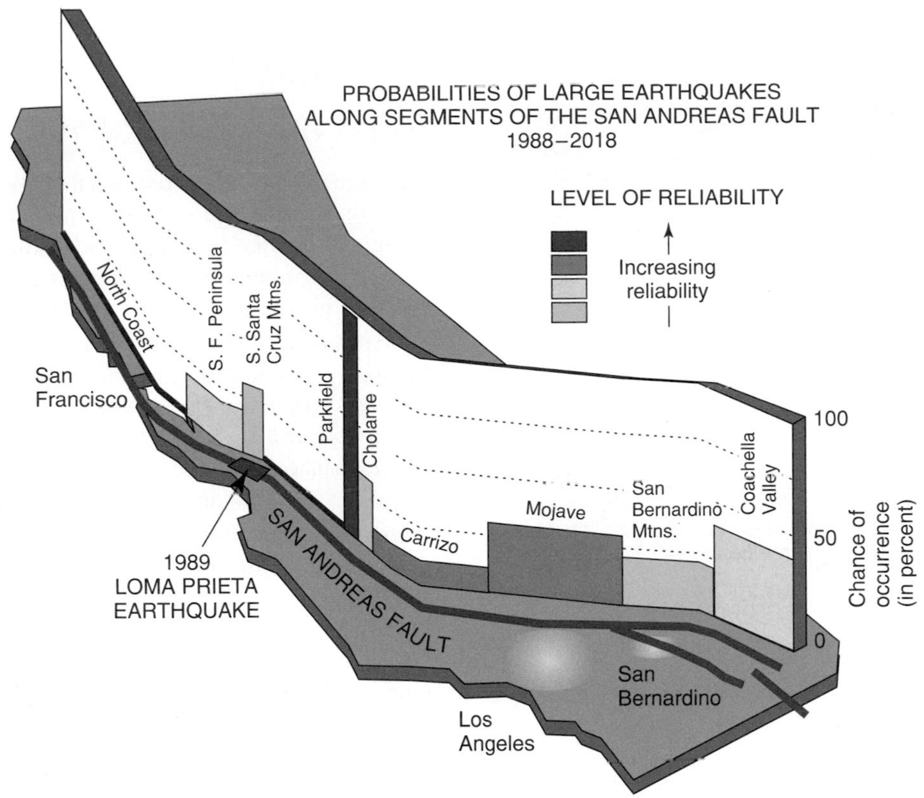

FIGURE 4.24 Earthquake forecast map issued in 1988 by the Working Group on California Earthquake Probabilities. Height of bar indicates estimated probability of a magnitude 6.5 to 7 earthquake in the period 1988 to 2018; color of bar indicates confidence in forecast.
Source: After U.S. Geological Survey Circular 1079

Reduction Program includes such an educational component. The need for it was reinforced in 1990 by public reaction to an unofficial earthquake "prediction." In that year, a biologist-turned-climatological-consultant predicted a major earthquake for the New Madrid fault zone (see section on Other Potential Problem Areas). The "prediction" was based on the fact that on 3 December 1990, the alignment of sun, moon, and earth would result in maximum tidal (gravitational) stress on the solid earth—which, presumably, would tend to trigger earthquakes on long-quiet faults. Geologists had previously investigated this concept and found no correlation between tidal stresses and major earthquakes; the "prediction" was not endorsed by the scientific community.

Unfortunately, geologists could not declare absolutely that an earthquake would not occur, only that the likelihood was extremely low. Meanwhile, in the New Madrid area, there was widespread panic: stores were stripped of emergency supplies; people scheduled vacations for the week in question (public-service personnel were forbidden to do so); schools were closed for that week. In the end, as some reporters described it, the biggest news was the traffic jam as all the curious left town after the earthquake didn't happen. The geologic community was very frustrated at indiscriminate media coverage, which certainly contributed to public concern, and is worried about public response to future, scientific prediction efforts.

EARTHQUAKE CONTROL?

Since earthquakes are ultimately caused by forces strong enough to move continents, human efforts to stop earthquakes from occurring would seem to be futile. However, there has been speculation about the possibility of moderating of some of earthquakes' most severe effects. If locked faults, or seismic gaps, represent areas in which energy is accumulating, perhaps enough to cause a major earthquake, then releasing that energy before too much has built up might prevent a major catastrophe.

In the mid-1960s, the city of Denver began to experience small earthquakes. They were not particularly damaging, but they were puzzling. In time, geologist David Evans suggested a connection with an Army liquid-waste-disposal well at the nearby Rocky Mountain Arsenal. The Army denied any possible link, but a comparison of the timing of the earthquakes with the quantities of liquid pumped into the well at different times showed a very strong correlation (figure 4.25). The earthquake

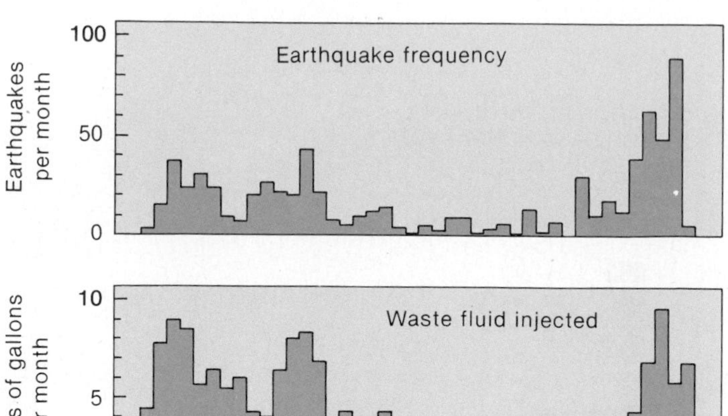

FIGURE 4.25 Correlation between waste disposal at the Rocky Mountain Arsenal (bottom diagram) and the frequency of earthquakes in the Denver area (top).

Adapted from David M. Evans, "Man-made Earthquakes in Denver," Geotimes, *10 (9):11–18, May/June 1966. Used by permission*

foci were also concentrated near the arsenal. It seemed likely that the liquid was "lubricating" old faults, allowing them to slip. In other words, the increased pore pressure resulting from the pumping-in of fluid decreased the shear strength, or resistance to shearing stress, along the fault zone. Other observations and experiments around the world have supported the concept that fluids in fault zones may facilitate movement along a fault.

Such observations at one time prompted speculation that *fluid injection* might be used along locked sections of major faults to allow the release of built-up stress. Unfortunately, geologists are far from sure of the results to be expected from injecting fluid (probably water) along large, locked faults. There is certainly no guarantee that only small earthquakes would be produced. Indeed, in an area where a fault had been locked for a long time, injecting fluid along that fault could lead to the release of all the stress at once, in a major, damaging earthquake. If a great deal of energy had accumulated, it is not even clear that all that energy *could* be released entirely through small earthquakes (recall table 4.2). The possible casualties and damage are a tremendous concern, as are the legal and political consequences of a serious, human-induced earthquake. The technique might be more safely used in major fault zones that have not been seismically quiet for long, where less stress has built up, to prevent that stress from continuing to build. The first attempts are also likely to be made in areas of low population density to minimize the potential danger. Certainly, much more research would be needed before this method could be applied safely and reliably on a large scale.

FUTURE EARTHQUAKES IN NORTH AMERICA?

Areas of Widely Recognized Risk

Southern Alaska sits above a subduction zone. On 27 March 1964, it was the site of the second-largest earthquake of the twentieth century in the world (estimated at approximately Richter magnitude 8.4, moment magnitude 9.2). The main shock, which lasted three to four minutes, was felt more than 1200 kilometers from the epicenter. About 12,000 aftershocks, many over magnitude 6, occurred during the next two months, and more continued intermittently through the following year. The associated uplifts—up to 12 meters on land and over 15 meters in places on the sea floor—left some harbors high and dry and destroyed habitats of marine organisms. Downwarping of 2 meters or more flooded other coastal communities. Tsunamis destroyed four villages and seriously damaged many coastal cities; they were responsible for about 90% of the 115 deaths. The tsunamis traveled as far as Antarctica and were responsible for sixteen deaths in Oregon and California. Landslides—both submarine and aboveground—were widespread and accounted for most of the remaining casualties. Total damage was estimated at \$300 million (see figure 4.26). The death toll might have been far higher had the earthquake not occurred in the early evening, on the Good Friday holiday, before peak tourist or fishing season. The area can certainly expect more earthquakes—including severe ones.

When the possibility of another large earthquake in San Francisco is raised, geologists debate not "whether" but "when?" At present, that section of the San Andreas fault has been locked for some time. The last major earthquake there was in 1906. At that time, movement occurred along at least 300 kilometers, and perhaps 450 kilometers, of the fault.

If another earthquake of the size of the 1906 event were to strike the San Francisco area during a busy workday, some estimates put the expected number of casualties at 50,000 or more, including the number of people at risk of drowning in case of dam failures in the area, with tens of billions of dollars in property damage from ground shaking alone. Scientists working on earthquake prediction hope that they will at least be able to give fairly precise warning of the next major Californian tremor by the time it occurs. At present, the precursor phenomena, insofar as they are understood, do not indicate that a great earthquake will occur near San Francisco within the next few years. However, the unheralded Loma Prieta and Parkfield earthquakes were none too reassuring in this regard.

Events of the past few years have suggested that there may, in fact, be a greater near-term risk along the southern San Andreas, near Los Angeles, than previously thought. In the spring and summer of 1992, a set of three significant earthquakes, including the Landers quake, occurred east of the southern San Andreas. The 1994 Northridge earthquake then occurred on a previously unrecognized buried thrust fault. The

A

B

C

FIGURE 4.26 Examples of property damage from the 1964 earthquake in Anchorage, Alaska. (A) Fourth Avenue landslide: Note the difference in elevation between the shops and street on the right and left sides. (B) Wreckage of Government Hill School, Anchorage. (C) Landslide in the Turnagain Heights area, Anchorage. Close inspection reveals a number of houses amid the jumble of down-dropped blocks in the foreground.
Photographs courtesy USGS Photo Library, Denver, CO.

southern San Andreas in this region, which last broke in 1857, has remained locked through all of this, so the accumulated strain has not been released by this cluster of earthquakes. Some believe that their net effect has been to increase the stress and the likelihood of failure along this section of the San Andreas. Moreover, in the late 1800s, there was a significant increase in numbers of moderate earthquakes around the northern San Andreas, followed by the 1906 San Francisco earthquake. The 1992–1994 activity may represent a similar pattern of increased activity building up to a major earthquake along the southern San Andreas. Time will tell.

Other Potential Problem Areas

Perhaps an even more dangerous situation is that in which people are not conditioned to regard their area as hazardous or earthquake-prone. The central United States is an example. Though most Americans think "California" when they hear "earthquake," the strongest and probably the most devastating series of earthquakes ever in the United States occurred in the vicinity of New Madrid, Missouri, during 1811–1812. The three strongest shocks are each estimated to have had Richter magnitudes over 8; they were spaced over almost two months, from 16 December 1811 to 7 February 1812. Towns were leveled, or drowned by flooding rivers. Tremors were felt from Quebec to New Orleans and along the east coast from New England to Georgia. Lakes were uplifted and emptied, while elsewhere the ground sank to make new lakes; boats were flung out of rivers. Aftershocks continued for more than *ten years.* Total damage and casualties have never been accurately estimated. It is the severity of those earthquakes that makes the central U.S. a high-risk area on the U.S. seismic-risk map (figure 4.27).

The potential damage from even one earthquake as severe as the worst of the New Madrid shocks is enormous: 12 million people now live in the area. Yet very few of those 12 million are probably fully aware of the nature and extent of the risk, and recent events may have caused them to take such risks less seriously. The big earthquakes there were a long time ago, beyond the memory of anyone now living. Minor earthquakes along the New Madrid fault zone in the summer of 1987 only briefly brought the danger to public attention. More attention was focused on the area by the (unfulfilled) amateur earthquake prediction made in late 1990, mentioned earlier. Beneath the midcontinent is a failed rift, a place where the continent began to rift apart, then stopped. Long and deep

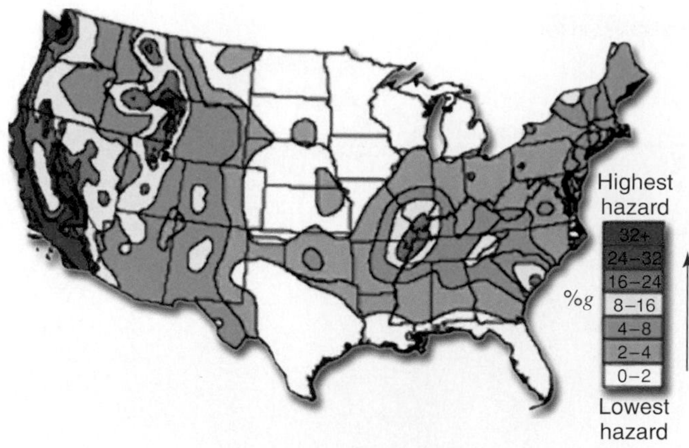

FIGURE 4.27 U.S. Seismic Hazard Map. The map reflects potential intensity of possible earthquakes rather than just frequency. Scale shows maximum acceleration (as % of *g;* see figure 4.17) that might be experienced in the next 50 years.

From U.S. Geological Survey National Seismic Hazard Mapping Project

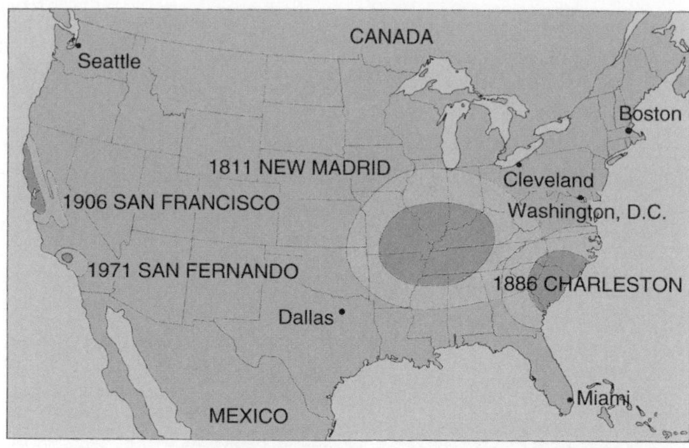

FIGURE 4.28 Part of the danger that midcontinent, intraplate earthquakes represent is related to the much broader areas likely to be affected, given the much more efficient transmission of seismic energy in the compact rocks that underlie these areas. Inner (dark tan) zone in each case experienced Mercalli intensity VII or above; outer (light tan), intensity VI or above.

Source: After U.S. Geological Survey Professional Paper 1240-B

faults in the lithosphere there remain a zone of weakness in the continent and a probable site of more major tremors. It is just a matter of time. Unfortunately, the press gave far more coverage to the 1990 "earthquake prediction" and to public reaction to it than to the scientific basis for prediction efforts generally or to the underlying geology and the nature of the hazard there in particular. As a result, and because there then was no major New Madrid earthquake in 1990, there is a real danger that any future official, scientific earthquake predictions for this area will not be taken seriously enough. The effects may be far-reaching, too, when the next major earthquake comes. Seismic waves propagate much more efficiently in the solid, compact rocks of the continental interior than in, for example, the complexly faulted, jumbled rocks around the San Andreas fault (figure 4.28). Therefore, when the next major earthquake occurs in the New Madrid fault zone, the consequences will be widespread.

There are still other hazards. The 1886 Charleston, South Carolina, earthquake was a clear indication that severe earthquakes can occur in the east also, as is reflected in the seismic-hazard map and in figure 4.28. Much of the southeastern United States, however, is blanketed in sediment, which somewhat complicates assessment of past activity and future hazards.

In the Pacific Northwest, there is seismicity associated with subduction (figure 4.29). Recent measurements among the mountains of Olympic National Park reveal crustal shortening and uplift that may indicate locking along the subduction zone to the west. An earthquake there could be as large as the 1964 Alaskan earthquake—also subduction-related—but with far more potential for damage and lost lives among the more than 10 million people in the Seattle/Portland area. A bit of a wakeup call in this regard was the magnitude-6.8 earthquake that shook the region in 2001. Though no lives were lost, over $1 billion in damage resulted.

Even in regions where the broad tectonic framework is well recognized, complexities in plate-tectonic details can produce seismic surprises for the public. A good example is the

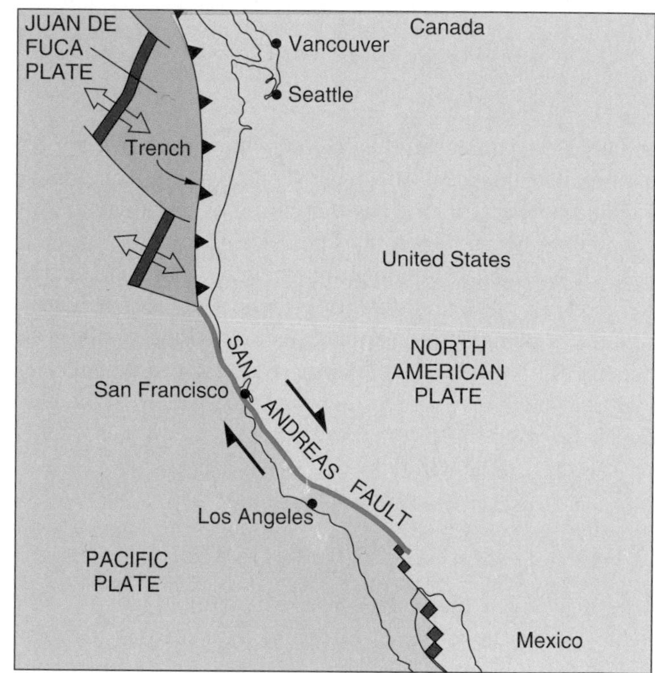

FIGURE 4.29 At the northern end of the San Andreas Fault lies a subduction zone under the Pacific Northwest.

Modified from U.S. Geological Survey.

2002 Denali Fault earthquake, which was *not* directly related to the subduction zone that underlies Anchorage (figure 4.30). Fortunately, Trans-Alaska Pipeline designers were well aware of the Denali Fault's potential. The 2002 earthquake there was, in fact, the fourteenth largest recorded in the United States.

Canada is, on the whole, much less seismically active than the United States. Still, each year, about 300 earthquakes (only

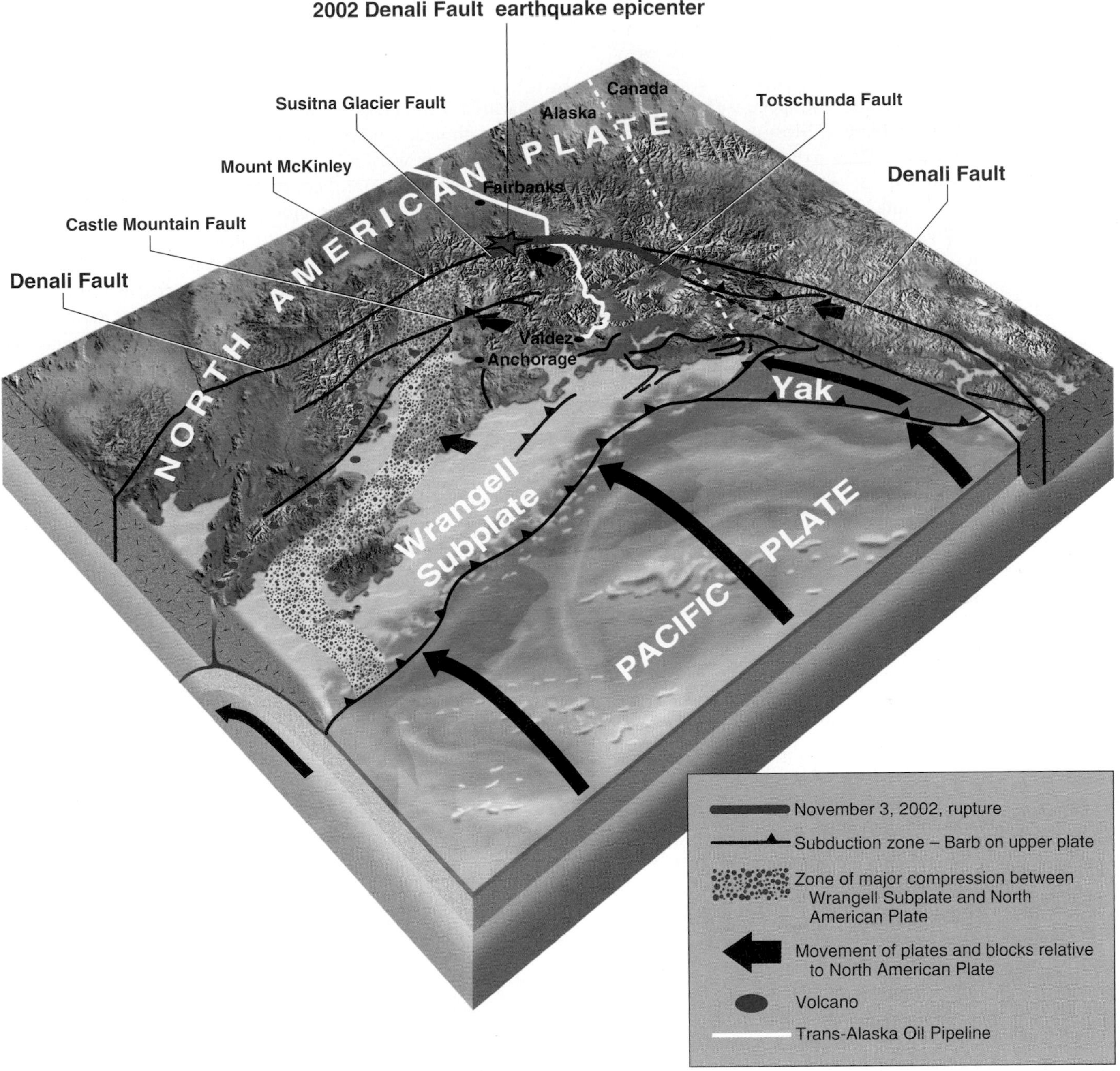

FIGURE 4.30 While the subduction zone is the dominant tectonic feature under southern Alaska, the Denali Fault is a strike-slip fault related to motions of small subplates, the Wrangell and the Yakutat ("Yak").
After USGS Fact Sheet 014-03

four of magnitude greater than 4) occur in eastern Canada, where an apparent zone of crustal weakness exists in the continental interior. Numerous other earthquakes—again, mostly small ones—occur in western Canada, primarily associated with the subduction zone there. Fortunately, severe Canadian earthquakes have been rare: a magnitude-7.2 event, complete with tsunami, off the Grand Banks of Newfoundland in 1929; an earthquake swarm with largest event magnitude 5.7 in central New Brunswick in 1982; a magnitude-7.3 quake under Vancouver Island in 1946; and a total of seven quakes of magnitude 6 or greater in the area of southwestern Canada in the last century. Still, the potential for significant earthquakes clearly continues to exist, while public awareness, particularly in eastern Canada, is likely to be low.

SUMMARY

Earthquakes result from sudden slippage along fault zones in response to accumulated stress. Most earthquakes occur at plate boundaries and are related to plate-tectonic processes. Earthquake hazards include ground rupture and shaking, liquefaction, landslides, tsunamis, coastal flooding, and fires. The severity of damage is determined not only by the size of an earthquake but also by underlying geology and the design of affected structures. While earthquakes cannot be stopped, their negative effects can be limited by (1) designing structures in active fault zones to be more resistant to earthquake damage; (2) identifying and, wherever possible, avoiding development in areas at particular risk from earthquake-related hazards; (3) increasing public awareness of and preparedness for earthquakes in threatened areas; (4) refining and expanding tsunami warning systems and public understanding of appropriate response; and (5) learning enough about patterns of seismicity over time along fault zones, and about earthquake precursor phenomena, to make accurate and timely predictions of earthquakes and thereby save lives. Until precise predictions on a short timescale become feasible, longer-term earthquake forecasts may serve to alert the public to general levels of risk and permit structural improvements and planning for earthquake response.

Key Terms and Concepts

aftershocks 77
body waves 69
creep 68
dip-slip fault 69
earthquake 68
earthquake cycle 85
elastic rebound 68
epicenter 68
fault 68
focus 68
intensity 72
liquefaction 77
magnitude 71
precursor phenomena 84
P waves 68
seismic gap 80
seismic waves 69
seismograph 70
strike-slip fault 69
surface waves 70
S waves 70
thrust fault 69
tsunami 78

Exercises

Questions for Review

1. Explain the concept of fault creep and its relationship to the occurrence of damaging earthquakes.
2. Why must rocks behave elastically in order for earthquakes to occur?
3. How do strike-slip and dip-slip faults differ?
4. In what kind of plate-tectonic setting might you find (a) a strike-slip fault? (b) a thrust fault?
5. What is an earthquake's focus? Its epicenter? Why are deep-focus earthquakes concentrated in subduction zones?
6. Name the two kinds of seismic body waves, and explain how they differ.
7. On what is the assignment of an earthquake's magnitude based? Is magnitude the same as intensity? Explain.
8. List at least three kinds of earthquake-related hazards, and describe what, if anything, can be done to minimize the danger that each poses.
9. What is a seismic gap, and why is it a cause for concern?
10. Describe the concept of earthquake cycles and their usefulness in forecasting earthquakes.
11. Explain the distinction between earthquake prediction and earthquake forecasting, and describe the kinds of precautions that can be taken on the basis of each.
12. Evaluate fluid injection as a possible means of minimizing the risks of large earthquakes.
13. What mechanism, if any, exists in the United States for warning people of earthquake hazards?
14. Areas identified as high risk on the seismic-risk map of the United States may not have had significant earthquake activity for a century or more. Why, then, are they mapped as high-risk regions?

For Further Thought

1. Research several earthquake predictions made within the last decade in the United States or elsewhere. How specific were the predictions? How accurate? What was the public response, if any?
2. Research the post-1988 history of earthquakes along the San Andreas fault, compare with forecast data of figure 4.24, and comment.
3. Investigate the history of any modern earthquake activity in your own area or in any other region of interest. What geologic reasons are given for this activity? How probable is significant future activity, and how severe might it be? (The U.S. Geological Survey or state geological surveys might be good sources of such information.)
4. Consider the issue of earthquake prediction from a public-policy standpoint. How precise would you wish a prediction to be before

making a public announcement? Ordering an evacuation? What kinds of logistical problems might you anticipate?

5. Check out current/recent seismicity (a) for the United States and Canada or (b) for the world, over a period of several weeks. (See NetNotes for data source locations.) Tabulate by magnitude and compare with table 4.2. Are the recent data consistent? Comment.

6. Look up your birthday, or other date(s) of interest, in "Today in Earthquake History" (http://earthquake.usgs.gov/learning/) and see what you find! Investigate the tectonic setting of any significant earthquake listed.

Suggested Readings/References

Achenbach, J. 2006. The new big one: Where will it strike? *National Geographic* (April): 120–147.

Bleier, T., and Freund, F. 2005. Earthquake alarm. *IEEE Spectrum* (December): 22–27.

Briggs, R. W., et al. 2006. Deformation and slip along the Sunda megathrust in the Great 2005 Nias-Simeulue Earthquake. *Science* 311 (31 March): 1897–1901.

Christenson, G. E. 2003. Life along the fault: Reducing risk in Utah. *Geotimes* (April): 14–17.

Earthquakes. 1993. From *The Citizens' Guide to Geologic Hazards,* Amer. Inst. of Professional Geologists, pp. 38–53.

Ellis, M., Gomberg, J., and Schweig, E. 2001. Indian earthquake may serve as analog for New Madrid earthquakes. *EOS 82* (7 August): 345–50.

Hildenbrand, T. G., et al. 1995. Crustal geophysics gives insight into New Madrid seismic zone. *EOS 76* (14 February): 65–69.

Hough, S. 2005. Earthquakes: Predicting the unpredictable? *Geotimes* (March): 18–23.

Johnston, A. C., and Kantner, L. R. 1990. Earthquakes in stable continental crust. *Scientific American* 262 (March): 68–75.

Johnston, M. J. S. 1993. *The Loma Prieta, California, earthquake of October 17, 1989—preseismic observations.* U.S. Geological Survey Professional Paper 1550-C.

Kanamori, H., and Brodsky, E. E. 2001. The physics of earthquakes. *Physics Today* (June): 34–40.

Miller, G. 2005. The tsunami's psychological aftermath. *Science* 309 (12 August): 1030–33.

Petroski, H. 1994. Broken bridges. *American Scientist* 82 (July–August): 318–21.

Reid, T. R. 1995. Kobe wakes to a nightmare. *National Geographic* (July): 116–36.

Somerville, P. 1995. Kobe earthquake: An urban disaster. *EOS 76* (7 February): 49–51.

Spence, W., Herrmann, R. B., Johnston, A. C., and Reagor, G. 1993. *Responses to Iben Browning's prediction of a 1990 New Madrid, Missouri, earthquake.* U.S. Geological Survey Circular 1083.

Stein, R. S. 2003. Earthquake conversations. *Scientific American* 275 (January): 72–79.

U.S. Geological Survey. 1996. *USGS response to an urban earthquake: Northridge '94.* USGS Open-File Report 96–263.

———. 2000. *Implications for earthquake risk reduction in the United States from the Kocaeli, Turkey, earthquake of August 17, 1999.* U.S. Geological Survey Circular 1193.

U.S. Geological Survey Staff. 1990. The Loma Prieta, California, earthquake: An anticipated event. *Science* 247 (January): 286–93.

Zeizel, A. J. 1998. Links across the Pacific: U.S.–Japan earthquake loss prevention strategies. *Geotimes* (April): 20–23.

Zoback, M. L. 2006. The 1906 earthquake and a century of progress in understanding earthquakes and their hazards. *GSA Today* 16 (April/May): 4–11.

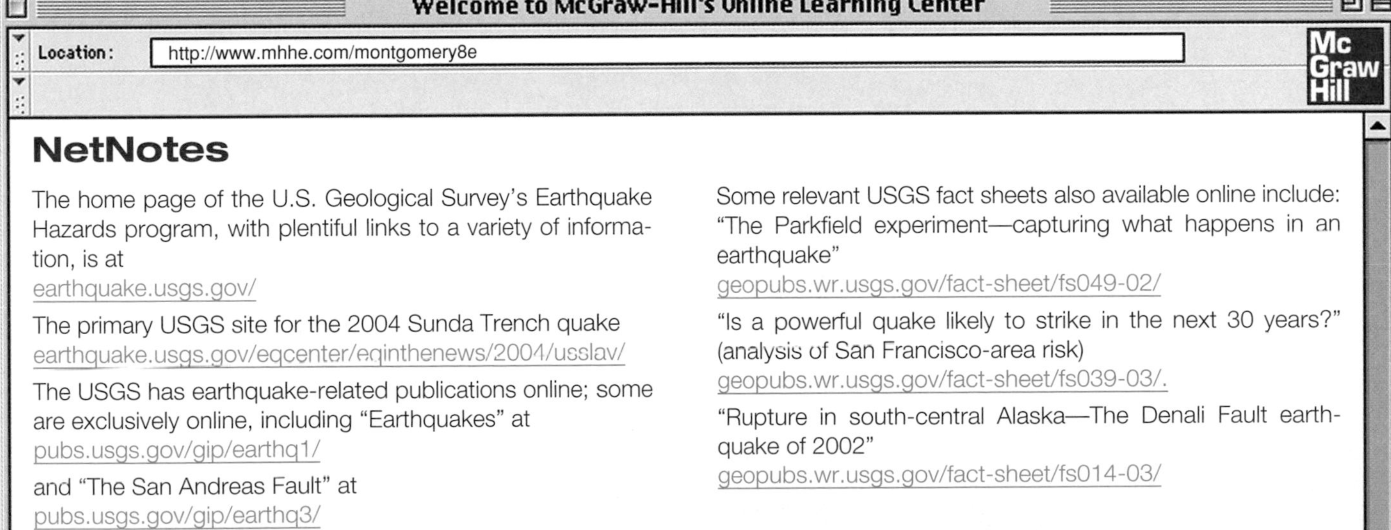

NetNotes

The home page of the U.S. Geological Survey's Earthquake Hazards program, with plentiful links to a variety of information, is at
earthquake.usgs.gov/

The primary USGS site for the 2004 Sunda Trench quake
earthquake.usgs.gov/eqcenter/eqinthenews/2004/usslav/

The USGS has earthquake-related publications online; some are exclusively online, including "Earthquakes" at
pubs.usgs.gov/gip/earthq1/

and "The San Andreas Fault" at
pubs.usgs.gov/gip/earthq3/

Some relevant USGS fact sheets also available online include:
"The Parkfield experiment—capturing what happens in an earthquake"
geopubs.wr.usgs.gov/fact-sheet/fs049-02/

"Is a powerful quake likely to strike in the next 30 years?" (analysis of San Francisco-area risk)
geopubs.wr.usgs.gov/fact-sheet/fs039-03/.

"Rupture in south-central Alaska—The Denali Fault earthquake of 2002"
geopubs.wr.usgs.gov/fact-sheet/fs014-03/

"Monitoring earthquake shaking in buildings to reduce loss of life and property"
pubs.usgs.gov/fs/2003/fs068-03/

Tsunami simulations and links to many tsunami-related sites
walrus.wr.usgs.gov/tsunami/

The International Tsunami Information Center
ioc3.unesco.org/itic/

Information on the Canadian National Earthquake Hazards Program—including a link to information on the use of seismology in monitoring compliance with the nuclear test ban treaty—can be found at "Earthquakes Canada"
www.seismo.nrcan.gc.ca/

The Global Seismic Hazard Assessment Program has unveiled its final report (in 1999) and global seismic-risk maps at
www.seismo.ethz.ch/GSHAP/index.html

Seismological Society of America home page
www.seismosoc.org/

The National Information Service for Earthquake Engineering maintains an online database of over 10,000 images of photos and art works (some hundreds of years old) relating to earthquakes.
nisee.berkeley.edu/

The Federal Emergency Management Agency has funded EQNET, a link to many related sites, at
www.eqnet.org

The Advanced National Seismic System is designed to improve U.S. ability to respond to earthquake, volcanic, and tsunami disasters
www.anss.org/

The information-rich sites inspired by the centennial of the 1906 San Francisco earthquake:
bancroft.berkeley.edu/collections/earthquake and fire www.
exploratorium.edu/faultline/index.html

A joint venture of the USGS, California Institute of Technology, and California Division of Mines and Geology to create ShakeMaps of earthquake intensity
www.trinet.org/shake

Have you experienced an earthquake? Help create a real-time intensity map:
pasadena.wr.usgs.gov/shake/

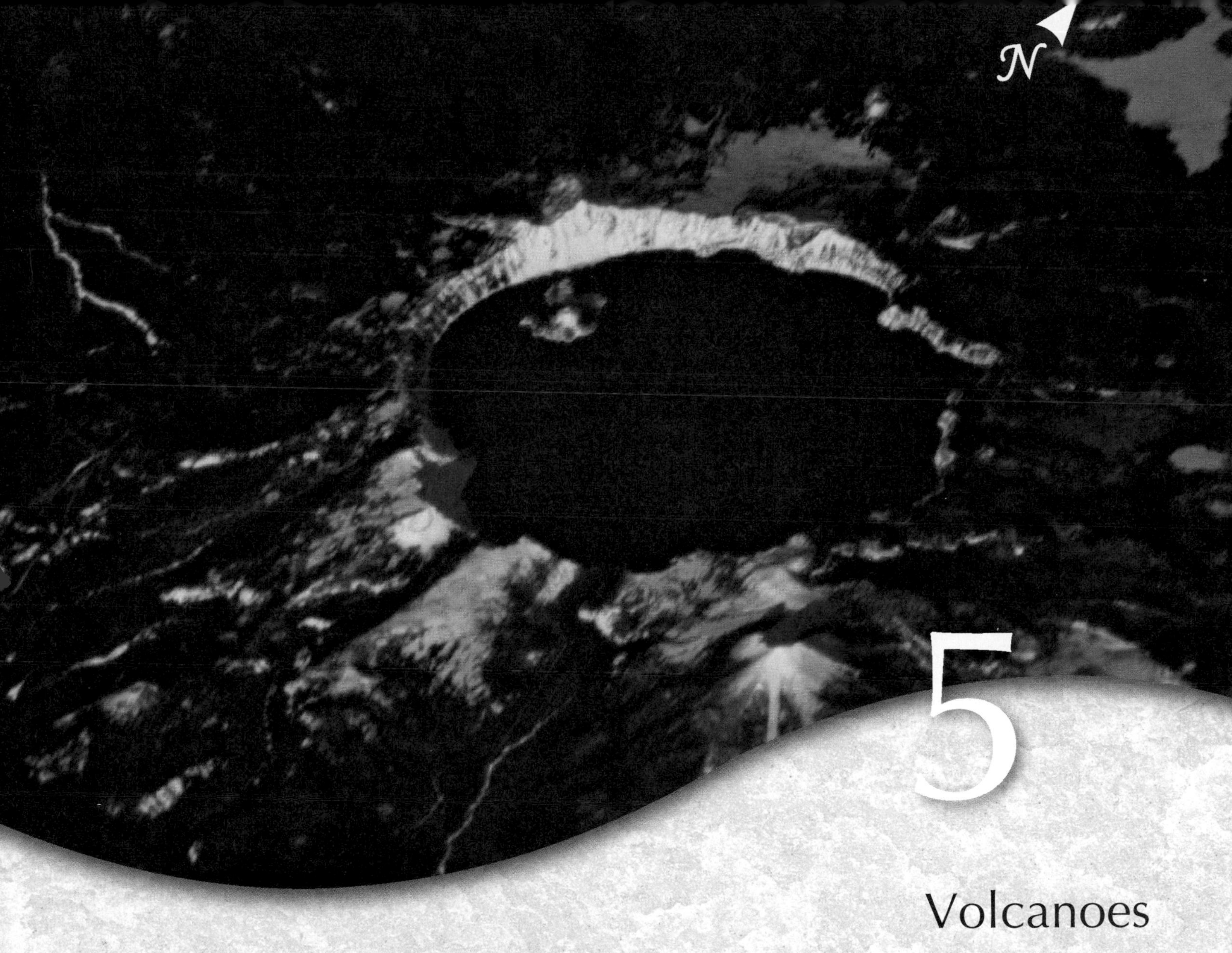

5 Volcanoes

On the morning of 18 May 1980, thirty-year-old David Johnston, a volcanologist with the U.S. Geological Survey, was watching the instruments monitoring Mount St. Helens in Washington State. He was one of many scientists keeping a wary eye on the volcano, expecting some kind of eruption, perhaps a violent one, but uncertain of its probable size. His observation post was more than 9 kilometers from the mountain's peak. Suddenly, he radioed to the control center, "Vancouver, Vancouver, this is it!"

It was indeed. Seconds later, the north side of the mountain blew out in a massive blast that would ultimately cost nearly $1 billion in damages and twenty-five lives, with another thirty-seven persons missing and presumed dead. The mountain's elevation was reduced by more than 400 meters. David Johnston was among the casualties. His death illustrates that even the experts still have much to learn about the ways of volcanoes. That fact has been reinforced by a dozen more volcanologists' deaths since 1990—in Japan, Indonesia, Colombia, and Ecuador—as scientists studying active volcanoes were unexpectedly overcome by sudden eruptive activity.

Mount St. Helens is not the only source of volcanic danger in the United States. A dozen or more peaks in the western United States could potentially become active, and other areas also may present some threat. For example, the spectacular geysers, hot springs, and other thermal features of Yellowstone National Park reflect its fiery volcanic past and may foretell an equally lively future. Another area of concern is the Long Valley–Mammoth Lakes area of California, where deformation of the ground surface and increased seismic activity may

Crater Lake, five miles wide and 2000 feet deep, formed from the explosion/collapse of Mount Mazama about 7700 years ago. The volume of debris ejected was about twice that from Krakatoa's 1883 eruption. Wizard Island, within the lake, is a cinder cone.
Image courtesy NASA

A

B

FIGURE 5.1 (A) Million-dollar Wahalua visitors' center in Hawaii Volcanoes National Park, aflame as it is engulfed by lava in 1989. (B) Ruins of the visitors' center after the lavas cooled.
Photograph (A) by J. D. Griggs, USGS Photo Library, Denver, CO.

be evidence of rising magma and impending eruption. And since 1983, Kilauea has been intermittently active, swallowing roads and homes as it builds Hawaii ever larger (figure 5.1).

Like earthquakes, volcanoes are associated with a variety of local hazards; in some cases, volcanoes even have global impacts. The dangers from any particular volcano, which are a function of what may be called its eruptive style, depend on the kind of magma it erupts and on its geologic and geographic settings. This chapter examines different kinds of volcanic phenomena and ways to minimize the dangers.

MAGMA SOURCES AND TYPES

Temperatures at the earth's surface are too low to melt rock, but temperature, like pressure, increases with depth. Some of this internal heat is left over from the earth's formation; more is constantly generated by the decay of naturally occurring radioactive elements in the earth. The majority of magmas originate in the upper mantle, at depths between 50 and 250 kilometers (30 and 150 miles), where the temperature is high enough and the pressure is low enough that the rock can melt, wholly or partially. They are typically generated in one of three plate-tectonic settings: (1) at divergent plate boundaries, both ocean ridges and continental rift zones; (2) over subduction zones; and (3) at "hot spots," isolated areas of volcanic activity that are not associated with current plate boundaries.

The tectonic setting in which magma is produced (which determines the available source material) and the extent of melting control the composition of the resulting melt. The major compositional variables among magmas are the proportions of silica (SiO_2), iron, and magnesium. In general, the more silica-rich magmas are poorer in iron and magnesium, and vice versa. A magma's composition, in turn, influences its physical properties, which determine the way it erupts—quietly or violently—and therefore the particular types of hazards it represents. A magma (or rock) relatively rich in iron and magnesium (and ferromagnesian silicates) is sometimes described as *mafic;* relatively silica-rich rocks tend to be rich in feldspar as well, and such rocks and magmas are therefore sometimes called *felsic.*

Figure 5.2 relates the various common volcanic rock types and their plutonic equivalents. Note that in this context, "silica-poor" still means 45 to 50% SiO_2; all of the principal minerals in most volcanic rocks are silicates. The most silica-rich magmas may be up to 75% SiO_2.

The asthenosphere is extremely rich in ferromagnesians ("ultramafic"). It is also very close to its melting temperature, which accounts for its plastic behavior. Aside from an increase in temperature, two things can trigger melting in the asthenosphere. One is a reduction in pressure. This promotes melting at a spreading ridge or rift zone, as noted in chapter 3: pressure from the overlying plates is reduced as they split and part and asthenosphere wells up from below. It also accounts for melting in the rising plume of mantle at a hot spot, for as plume material rises to shallower depths, the pressure acting on it is reduced, and again some melting may occur. The other factor that can promote melting of the asthenosphere is the addition of fluids, such as water, somewhat as salt sprinkled on an icy sidewalk promotes melting of the ice. A subducting oceanic plate carries water down with it, in the seafloor rocks themselves, which interacted with seawater as they formed, and in the sediments subducted along with the lithospheric plate. As

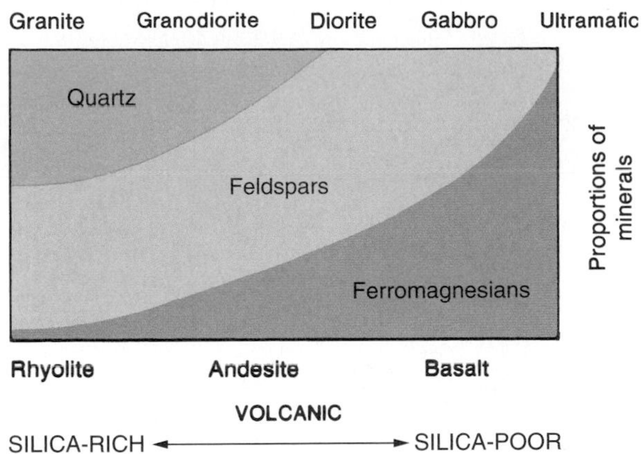

FIGURE 5.2 Common volcanic rock types (bottom labels) and their plutonic equivalents (top). The rock names reflect varying proportions of silica, iron, and magnesium, and thus of common silicate minerals. Rhyolite is the fine-grained, volcanic compositional equivalent of granite, and so on.

the subducted rocks and sediments go down, they are heated by the surrounding warm mantle, and eventually they become hot enough that the fluids are driven off, rising into the asthenosphere above and starting it melting.

The partial melts of asthenosphere that result from these processes are typically mafic. This accounts for the **basalt** (mafic volcanic rock) that forms new sea floor at spreading ridges. At a continental rift zone, the asthenosphere-derived melt is also mafic, but the rising hot basaltic magma may warm the more granitic continental crust enough to produce some silica-rich melt also—making **rhyolite,** the volcanic equivalent of granite—or the basaltic magma may incorporate some crustal material to make **andesite,** intermediate in composition between the mafic basalt and felsic rhyolite.

The magma typical of a hot-spot volcano in an ocean basin is similar to that of a seafloor spreading ridge: basaltic melt derived from partial melting of the very mafic/ultramafic mantle, which will remain basaltic even if it should interact with the basaltic sea floor on its way to the surface. A hot spot under a continent may produce more felsic magmas, just as can occur in a continental rift zone.

Magmatic activity in a subduction zone can be complex (figure 5.3). The bulk of melt generated is likely to be mafic, derived from the asthenosphere as water is released from the subducted sediments and slab of lithosphere. There may be contributions to the melt from melting of the sediments and, as subduction and slab warming proceed, even from some melting of the slab itself. Partial melting of basaltic slab material would add a more andesitic component. Melting of the basaltic crust and mafic upper mantle of the subducted slab, or of the asthenosphere above it, would tend to produce basaltic or perhaps andesitic magma. Sediments derived from continental crust would be relatively silica-rich, so the addition of melted sediments to the magma could tend to make it more felsic.

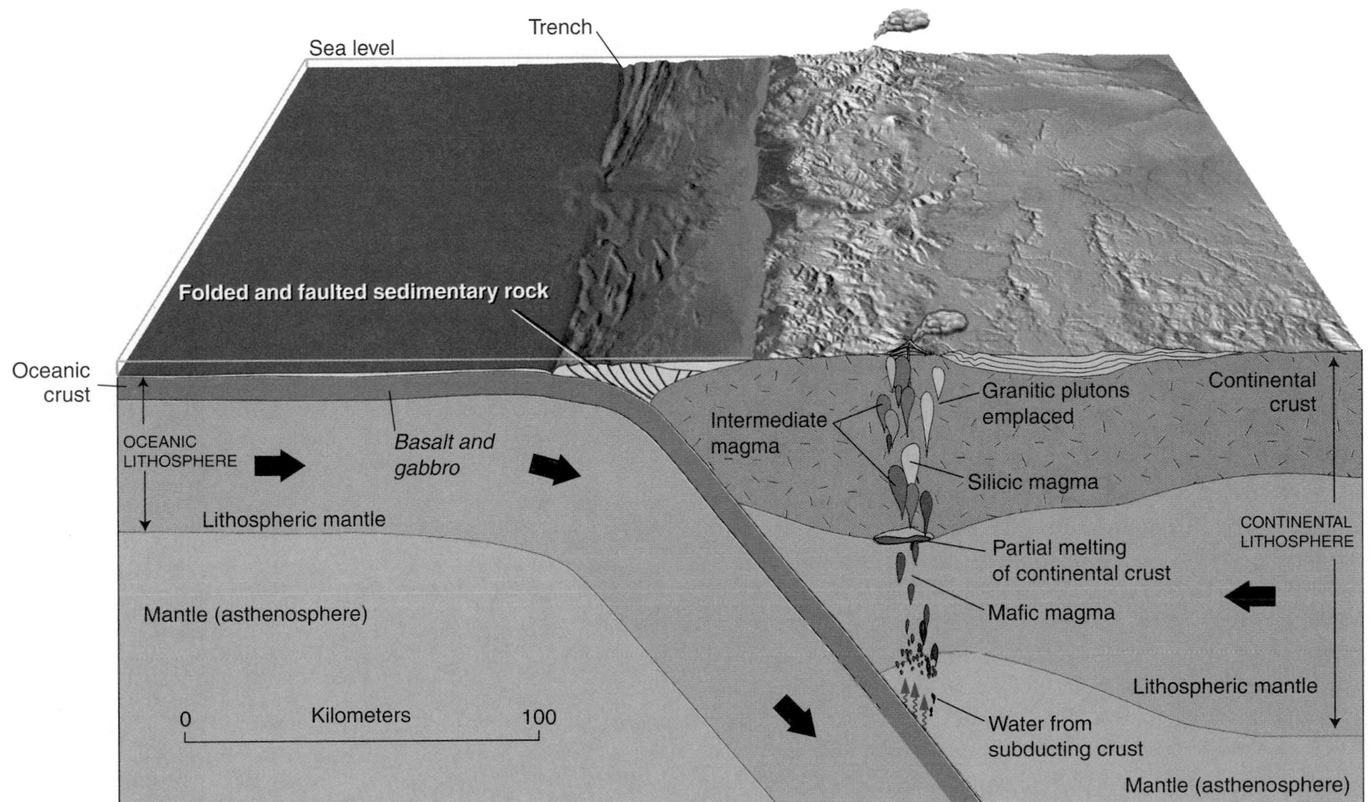

FIGURE 5.3 In a subduction zone, water and other fluids from the downgoing slab and sediments promote melting of asthenosphere; the hot, rising magma, in turn, may assimilate or simply melt overlying lithosphere, producing a variety of melt compositions.

Interaction with the overriding plate adds further complexity. Incorporation of crust from an overriding continent as the magma rises toward the surface adds more silica, and ultimately, an andesitic-to-rhyolitic melt may result. Heat from the rising magma can melt adjacent continental rocks to produce a separate granitic melt. In an island-arc setting, the overriding plate is not silica-rich, so the lavas erupted at the surface are basaltic to andesitic in composition.

The silica-poor mafic lavas that produce basalt are characteristically "thin," or low in viscosity, so they flow very easily. The more felsic (silica-rich) andesitic and rhyolitic lavas are more viscous, thicker and stiffer, and flow very sluggishly. Magmas also contain dissolved water and gases. As magma rises, pressure on it is reduced, and these gases begin to escape. From the more fluid, silica-poor magmas, gas escapes relatively easily. The viscous, silica-rich magmas tend to trap the gases, which may lead, eventually, to an explosive eruption. The physical properties of magmas and lavas influence the kinds of volcanic structures they build, as well as their eruptive style.

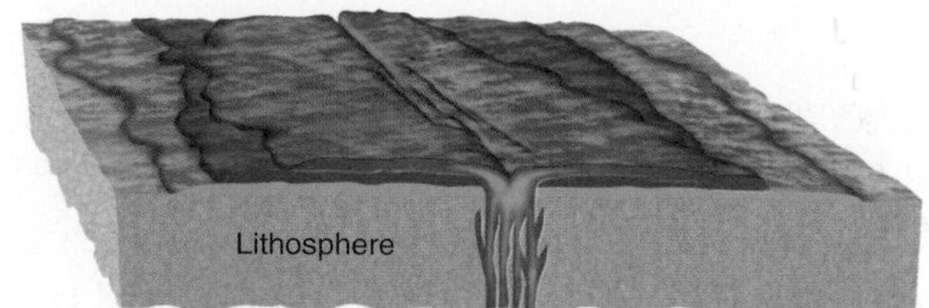

FIGURE 5.4 Schematic diagram of a continental fissure eruption. (At a spreading ridge, the magma has generally solidified before it can spread very far sideways at the surface, quenched by cold seawater.)

KINDS AND LOCATIONS OF VOLCANIC ACTIVITY

Given the definition of volcanic rock—igneous rock formed at or near the earth's surface—most volcanic rock is actually created at the seafloor spreading ridges, where magma fills cracks in the lithosphere and crystallizes close to the surface. Spreading ridges spread at the rate of only a few centimeters per year, but there are some 50,000 kilometers (about 30,000 miles) of these ridges presently active in the world. All in all, that adds up to an immense volume of volcanic rock. However, most of this activity is out of sight under the oceans, where it is largely unnoticed, and it involves quietly erupting mafic magma, so it presents no dangers to people.

Continental Fissure Eruptions

The outpouring of magma at spreading ridges is an example of **fissure eruption,** the eruption of magma out of a crack in the lithosphere, rather than from a single pipe or vent (see figure 5.4). (The structure of any fissure that transects the whole lithosphere is much more complex than shown in that figure, with multiple cracks and magma pathways.) There are also examples of fissure eruptions on the continents.

One example in the United States is the Columbia Plateau, an area of over 150,000 square kilometers (60,000 square miles) in Washington, Oregon, and Idaho, covered by layer upon layer of basalt, piled up nearly a mile deep in places (figure 5.5). This area may represent the ancient beginning of a continental rift that ultimately stopped spreading apart. It is no longer active but serves as a reminder of how large a volume of magma can come welling up from the asthenosphere when zones of weakness in the lithosphere provide suitable openings. Even larger examples of such "flood basalts" on continents, covering more than a million square kilometers, are found in India and in Brazil.

FIGURE 5.5 Flood basalts. (A) Areal extent of Columbia River flood basalts. (B) Multiple lava flows, one atop another, can be seen in an outcrop of these flows in Washington state.
(B) Photograph by P. Weis, U.S. Geological Survey

Individual Volcanoes—Locations

Most people think of volcanoes as eruptions from the central vent of some sort of mountainlike object. Figure 5.6 is a map of such volcanoes presently or recently active in the world, and it shows a particularly close association between volcanoes and subduction zones. Proximity to an active subduction zone, then, is an indication of possible volcanic, as well as earthquake, hazards. The so-called Ring of Fire, the collection of volcanoes rimming the Pacific Ocean, is really a ring of subduction zones. A few volcanoes are associated with continental rift zones—for instance, Kilimanjaro and other volcanoes of the East African Rift system.

As with earthquakes, a few volcanic anomalies are not associated with plate boundaries. Hawaii is one: an isolated hot spot in the mantle apparently accounts for the Hawaiian Islands, as discussed in chapter 3. Other hot spots may lie under the Galápagos Islands, Iceland, and Yellowstone National Park, among other examples (figure 5.7). What, in turn, accounts for the hot spots is not clear. Some have suggested that hot spots represent regions within the mantle that have slightly higher concentrations of heat-producing radioactive elements and,

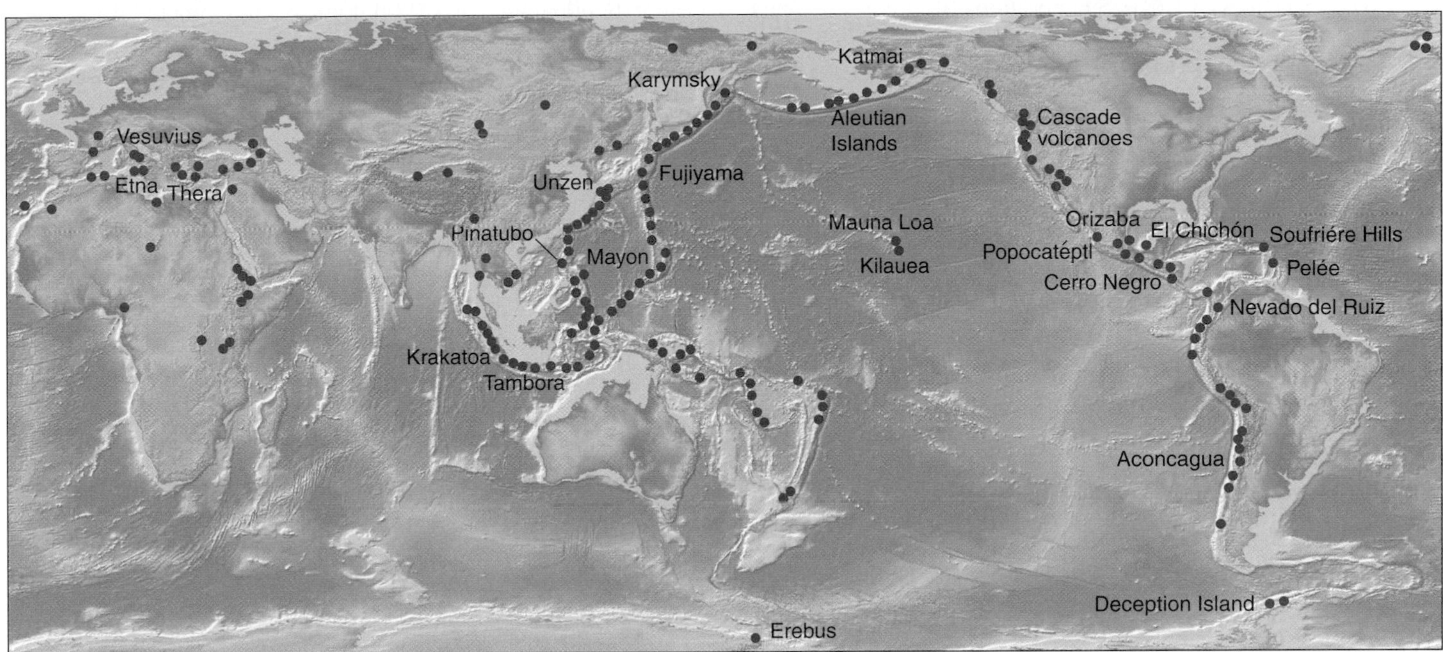

FIGURE 5.6 Map showing locations of recently active volcanoes of the world. (Submarine volcanoes not shown.)

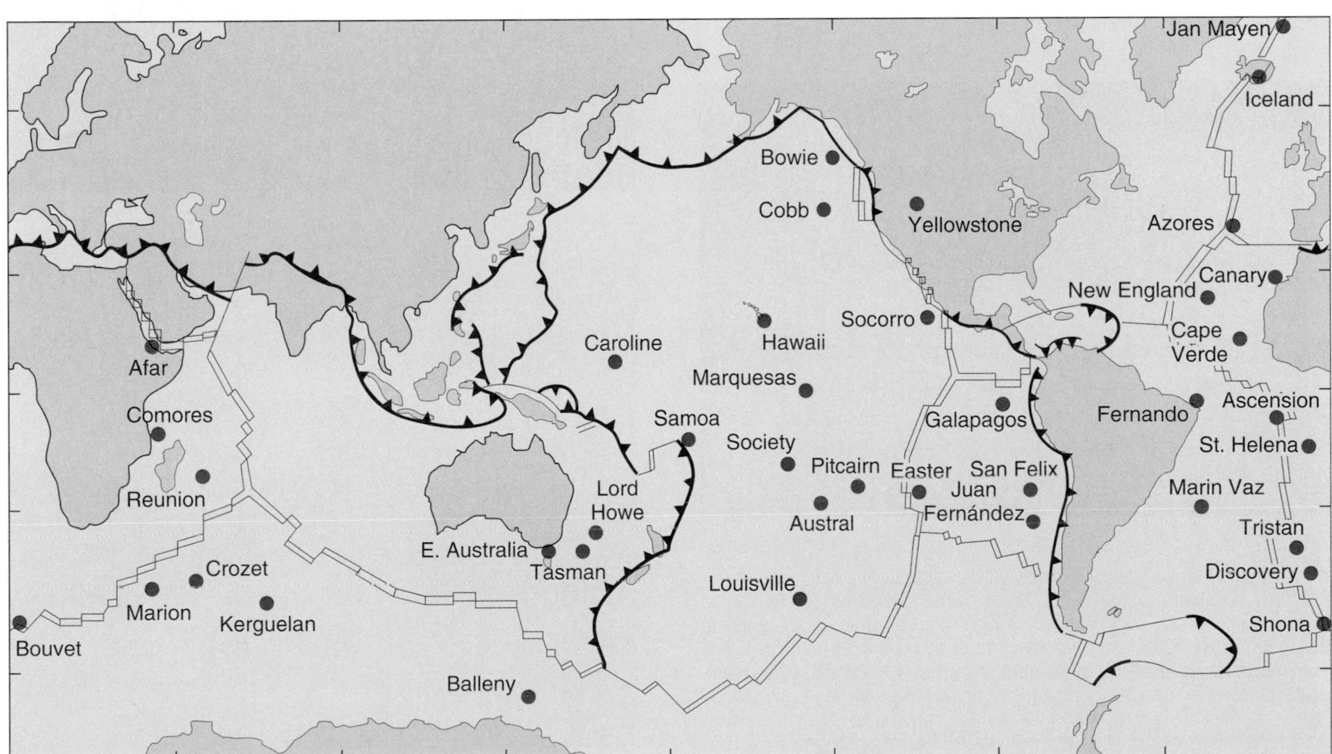

FIGURE 5.7 Selected prominent "hot spots" around the world. Some coincide with plate boundaries; most do not.

therefore, more melting and more magma. Others have suggested that the cause of hot spots lies in similar concentrations of radioactive elements in the outer core and that plumes of warm material rise from the core/mantle boundary to create them. Whatever the cause of hot spots, they are long-lived features; recall from chapter 3 that the volcanoes formed as the Pacific plate moved over the now-Hawaiian hot spot span some 70 million years.

The volcanoes of figure 5.6, eruptions from pipes or vents rather than from fissures, can be categorized by the kinds of structures they build. The structure, in turn, reflects the kind of volcanic material erupted and, often, the tectonic setting. It also provides some indication of what to expect in the event of future eruptions.

Shield Volcanoes

As noted, the mafic basaltic lavas are comparatively fluid, so they flow very freely and far when erupted. The kind of volcano they build, consequently, is very flat and low in relation to its diameter and large in areal extent. This low, shieldlike shape has led to the term **shield volcano** for such a structure (figure 5.8). Though the individual lava flows may be thin—a few meters or less in thickness—the buildup of hundreds or thousands of flows through time can produce quite large objects. The Hawaiian Islands are all shield volcanoes. Mauna Loa, the largest peak on the island of Hawaii, rises 3.9 kilometers (about 2.5 miles) above sea level. If measured properly from its true base, the sea floor, it is much more impressive: about 10 kilometers (6 miles) high and 100 kilometers in diameter at its base. With such broad, gently sloping shapes, the islands do not necessarily even look much like volcanoes from sea level.

Cinder Cones

When magma wells up toward the surface, the pressure on it is reduced, and dissolved gases try to bubble out of it and escape. The effect is much like popping the cap off a soda bottle: The soda contains carbon dioxide gas under pressure, and when the pressure is released by removing the cap, the gas comes bubbling out.

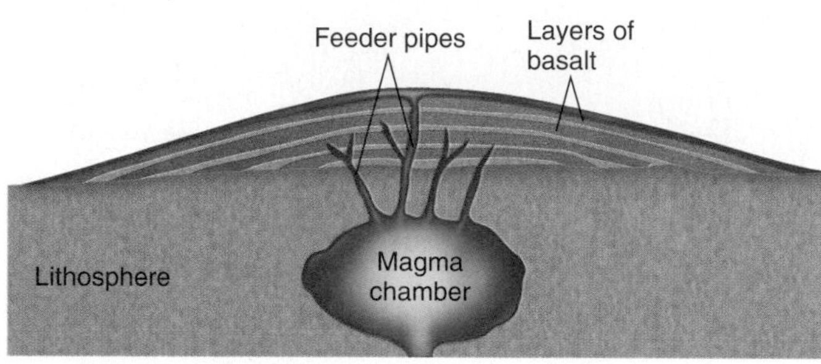

A

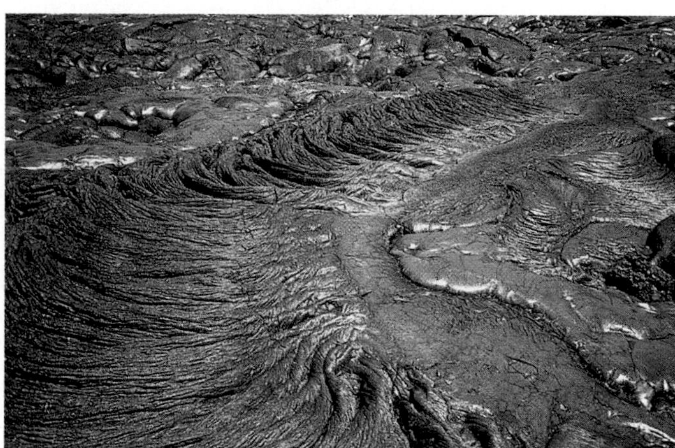

C

B

D

FIGURE 5.8 Shield volcanoes and their characteristics. (A) Simplified diagram of a shield volcano in cross section. (B) Very thin lava flows, like these on Kilauea in Hawaii, are characteristic of shield volcanoes. (C) Fluidity of Hawaiian lavas is evident even after they have solidified. This ropy-textured surface is termed *pahoehoe* (pronounced "pa-hoy-hoy"). (D) Mauna Loa, an example of a shield volcano, seen from low altitude. Note the gently sloping shape. The peak of Mauna Kea rises at rear of photograph.
Photographs (B) and (D) courtesy of USGS Photo Library, Denver, CO.

Sometimes, the built-up gas pressure in a rising magma is released more suddenly and forcefully by an explosion that flings bits of magma and rock out of the volcano. The magma may freeze into solid pieces before falling to earth. The bits of violently erupted volcanic material are described collectively as **pyroclastics,** from the Greek words for fire (*pyros*) and broken (*klastos*). The most energetic pyroclastic eruptions are more typical of volcanoes with thicker, more viscous silicic lavas because the thicker lavas tend to trap more gases. Gas usually escapes more readily and quietly from the thinner basaltic lavas, though even basaltic volcanoes may sometimes put out quantities of ash and small fragments.

The fragments of pyroclastic material can vary considerably in size from very fine, flourlike, gritty volcanic ash through cinders ranging up to golf-ball-size pieces, to large volcanic blocks that may be the size of a house (figure 5.9). Blobs of liquid lava may also be thrown from a volcano; these are volcanic "bombs."

While basaltic magmas generally erupt as fluid lava flows, they sometimes produce small volumes of chunky volcanic cinders that fall close to the vent from which they are thrown. The cinders may pile up into a very symmetric cone-shaped heap known as a **cinder cone** (figure 5.10).

Composite Volcanoes

Many volcanoes erupt somewhat different materials at different times. They may emit some pyroclastics, then some lava, then more pyroclastics, and so on. Volcanoes built up in this layercake fashion are called **stratovolcanoes,** or, alternatively, **composite volcanoes,** because they are built up of layers of more than one kind of material (figure 5.11). The mix of lava

A

B

C

D

FIGURE 5.9 Volcanic ash from Mount St. Helens (A), bombs from Mauna Kea (B), and cinders from Sunset Crater, Arizona (C) are all types of pyroclastics (which sometimes are produced even by the placid shield volcanoes). Bombs are molten, or at least hot enough to be plastic, when erupted, and may assume a streamlined shape in the air. (D) is volcanic breccia (at Mt. Lassen) formed of welded hot pyroclastics.

Photograph (A) by D. F. Wieprecht, (B) by J. P. Lockwood, both courtesy U.S. Geological Survey

A

B

FIGURE 5.10 Paricutín (Mexico), a classic cinder cone. (A) Night view shows formation by accumulation of pyroclastics flung out of the vent. (B) Shape of the structure revealed by day is typical symmetric form of cinder cones.

(B) Photograph by K. Segerstrom; both courtesy of USGS Photo Library, Denver, CO.

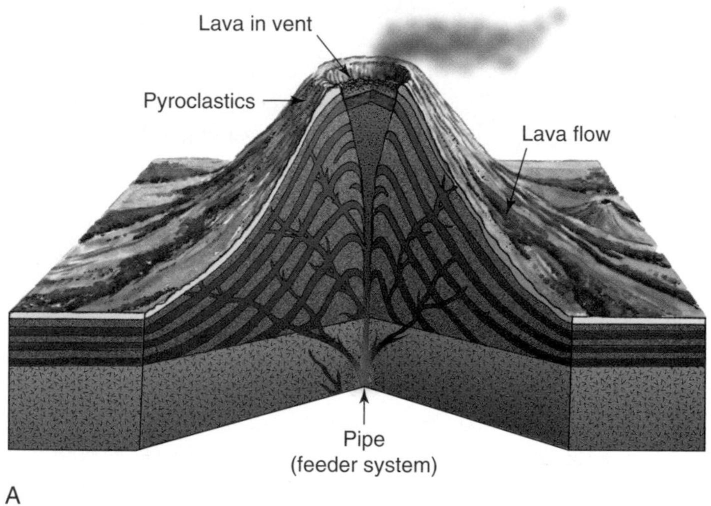

A

B

FIGURE 5.11 (A) Schematic cutaway view of a composite volcano (stratovolcano), formed of alternating layers of lava and pyroclastics. (B) Two composite volcanoes of the Cascade Range: Mount St. Helens (foreground) and Mt. Rainier (rear); photograph predates 1980 explosion of Mount St. Helens.
Photograph (B) by D. R. Mullineux, USGS Photo Library, Denver, CO.

and pyroclastics allows them to grow larger than either cinder cones or volcanic domes. Most of the potentially dangerous volcanoes worldwide are composite volcanoes. All tend to have fairly stiff, gas-charged andesitic lavas that sometimes flow and sometimes trap enough gases to erupt explosively with a massive blast and a rain of pyroclastic material. We will look at these volcanoes in more detail later.

When not erupting explosively, the slow-flowing rhyolitic and andesitic lavas tend to ooze out at the surface like thick toothpaste from a tube, piling up close to the volcanic vent, rather than spreading freely. The resulting structure is a compact and steep-sided **lava dome** (figure 5.12). Often, a lava dome will build in the crater of a composite volcano after an explosive eruption, as with Mount St. Helens since 1980.

FIGURE 5.12 Multiple lava domes have formed in the summit crater of Mount St. Helens since the 1980 eruption.
Photograph courtesy of U.S. Geological Survey

The relative sizes of these three types of volcanoes are shown in figure 5.13.

HAZARDS RELATED TO VOLCANOES

Lava

Many people instinctively assume that lava is the principal hazard presented by a volcano. Actually, lava generally is not life-threatening. Most lava flows advance at speeds of only a few kilometers an hour or less, so one can evade the advancing lava readily even on foot. The lava will, of course, destroy or bury any property over which it flows. Lava temperatures are typically over 850°C (over 1550°F), and basaltic lavas can be over 1100°C (2000°F). Combustible materials—houses and forests, for example—burn at such temperatures. Other property may simply be engulfed in lava, which then solidifies into solid rock.

Lavas, like all liquids, flow downhill, so one way to protect property is simply not to build close to the slopes of the volcano. Throughout history, however, people *have* built on or near volcanoes, for many reasons. They may simply not expect the volcano to erupt again (a common mistake even today). Also, soil formed from the weathering of volcanic rock may be very fertile. The Romans cultivated the slopes of Vesuvius and other volcanoes for that reason. Sometimes, too, a volcano is the only land available, as in the Hawaiian Islands or Iceland.

Iceland sits astride the Mid-Atlantic Ridge, rising above the sea surface because it is on a hot spot as well as a spreading ridge, and thus a particularly active volcanic area. Off the southwestern coast of the main island are several smaller volcanic islands. One of these tiny islands, Heimaey, accounts for the processing of about 20% of Iceland's major export, fish.

Heimaey's economic importance stems from an excellent harbor. In 1973, the island was split by rifting, followed by several months of eruptions. The residents were evacuated within hours of the start of the event. In the ensuing months, homes, businesses, and farms were set afire by falling hot pyroclastics or buried under pyroclastic material or lava. When the harbor itself, and therefore the island's livelihood, was threatened by encroaching lava, officials took the unusual step of deciding to fight back.

The key to the harbor's defense was the fact that the colder lava becomes, the thicker, more viscous, and slower-flowing it gets. When cooled enough, it may solidify altogether. When a large mass of partly solidified lava has accumulated at the advancing edge of a flow, it begins to act as a natural dam to stop or slow the progress of the flow. How could one cool a lava flow more quickly? The answer surrounded the island—water. Using pumps on boats and barges in the harbor, the people of Heimaey directed streams of water at the lava's edge. They also ran pipes (prevented from melting, in part, by the cool water running through them) across the solid crust that had started to form on the flow, and positioned them to bring water to places that could not be reached directly from the harbor. And it worked: part of the harbor was filled in, but much of it was saved, along with the fishing industry on Heimaey (see figure 5.14). In fact, the remaining harbor, partially closed off and thus sheltered by the newly formed rock, was in some respects improved by the increased protection.

This bold scheme succeeded, in part, because the needed cooling water was abundantly available and, in part, because the lava was moving slowly enough already that there was time to undertake flow-quenching operations before the harbor was filled. Also, the economic importance of the harbor justified the expense of the effort, estimated at about $1.5 million U.S. dollars. Similar efforts in other areas have not always been equally

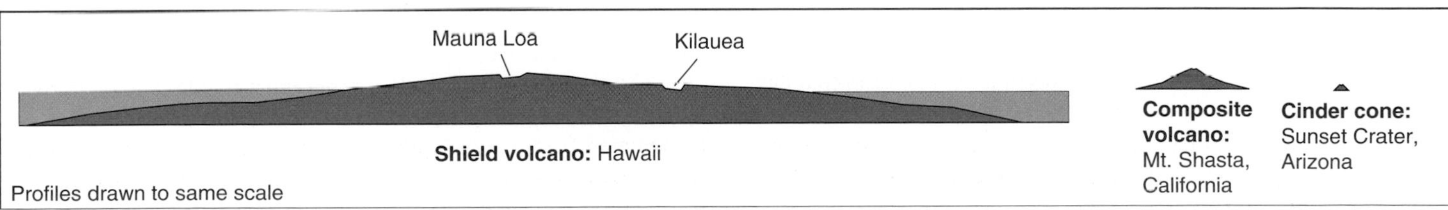

FIGURE 5.13 The voluminous, free-flowing lavas of shield volcanoes can build huge edifices; composite volcanoes and cinder cones are much smaller, as well as steeper-sided.

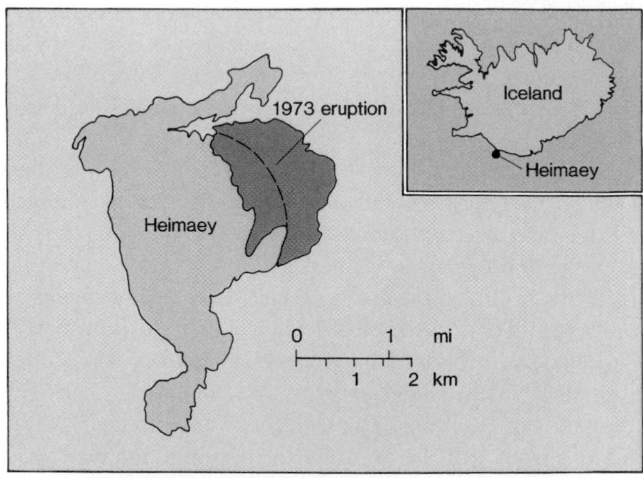

A

B

FIGURE 5.14 Impact of lava flows on Heimaey, Iceland. (A) Map showing extent of lava filling the harbor of Heimaey after 1973 eruption. (B) Lava-flow control efforts on Heimaey.

Source: (A) Data from R. Decker and B. Decker, Volcanoes, *copyright © 1981 by W.H. Freeman and Company. Photograph (B) courtesy of USGS Photo Library, Denver, CO.*

successful. (Still, the idea did form the basis for a dramatic sequence in the fictional movie *Volcano*!)

Where it is not practical to arrest the lava flow altogether, it may be possible to divert it from an area in which a great deal of damage may be done to an area where less valuable property is at risk. Sometimes, the motion of a lava flow is slowed or halted temporarily during an eruption because the volcano's output has lessened or because the flow has encountered a natural or artificial barrier. The magma contained within the solid crust of the flow remains molten for days, weeks, or sometimes months. If a hole is then punched in this crust by explosives, the remaining fluid magma inside can flow out and away. Careful placement of the explosives can divert the flow in a chosen direction.

A procedure of this kind was tried in Italy, where, in early 1983, Mount Etna began another in an intermittent series of eruptions (ongoing as this is written). By punching a hole in a natural dam of old volcanic rock, officials hoped to divert the latest lavas to a broad, shallow, uninhabited area. Unfortunately, the effort was only briefly successful. Part of the flow was diverted, but within four days, the lava had abandoned the planned alternate channel and resumed its original flow path. Later, new flows threatened further destruction of inhabited areas.

Lava flows may be hazardous to property, but in one sense, they are at least predictable: Like other fluids, they flow downhill. Their possible flow paths can be anticipated, and once they have flowed into a relatively flat area, they tend to stop. Other kinds of volcanic hazards can be trickier to deal with, and they affect much broader areas.

Pyroclastics

Pyroclastics—fragments of hot rock and spattering lava—are often more dangerous than lava flows. Pyroclastics erupt more suddenly, explosively, and spread faster and farther. The largest blocks and volcanic bombs present an obvious danger because of their size and weight. For the same reasons, however, they usually fall quite close to the volcanic vent, so they affect a relatively small area.

The sheer volume of the finer ash and dust particles can make them as severe a problem, and they can be carried over a much larger area. Also, ashfalls are not confined to valleys and low places. Instead, like snow, they can blanket the countryside. The 18 May 1980 eruption of Mount St. Helens (see aftermath, figure 5.15) was by no means the largest such eruption ever recorded, but the ash from it blackened the midday skies more than 150 kilometers away, and measurable ashfall was detected halfway across the United States. Even in areas where only a few millimeters of ash fell, transportation ground to a halt as drivers skidded on slippery roads and engines choked on airborne dust. Volcanic ash is also a health hazard, both uncomfortable and dangerous to breathe. The cleanup effort required to clear the debris strewn about by Mount St. Helens was enormous. As just one example, it has been estimated that 600,000 tons of ash landed on the city of Yakima, Washington, more than 100 kilometers away.

Past explosive eruptions of other violent volcanoes have been equally devastating, or more so, in terms of pyroclastics. When the city of Pompeii was destroyed by Mount Vesuvius in A.D. 79, it was buried not by lava, but by ash. (This is why extensive excavation of the ruins has been possible.) Contemporary accounts suggest that most residents had ample time to escape as the ash fell, though many chose to stay, and died. They and the town were ultimately buried under tons of ash. Pompeii's very existence was forgotten until some ruins were discovered during the late 1600s. The 1815 explosion of Tambora in Indonesia blew out an estimated 100 cubic kilometers of debris, and the prehistoric explosion of Mount Mazama,

FIGURE 5.15 The 1980 eruption of Mount St. Helens blew off the top 400 meters (over 1300 feet) of mountain. Twenty years later, the landscape was still blanketed in thick ash, gullied by runoff water.

in Oregon, which created the basin for the modern Crater Lake (shown in the chapter-opener photo) was larger still.

Lahars

Volcanic ash and water can combine to create a fast-moving volcanic mudflow called a **lahar.** Victims are engulfed and trapped in dense mud, which may also be hot from the heat of the ash. Such mudflows, like lava flows, flow downhill. They may tend to follow stream channels, choking them with mud and causing floods of stream waters. Flooding produced in this way was a major source of damage near Mount St. Helens. In A.D. 79, Herculaneum, closer to Vesuvius than was Pompeii, was partially invaded by volcanic mudflows, which preserved the bodies of some of the casualties in mud casts. Following the 1991 eruption of Mount Pinatubo in the Philippines, drenching typhoon rains soaked the ash blanketing the mountain's slopes, triggering devastating lahars (figure 5.16). Nor is rain an essential ingredient; with a snow-capped volcano, the heat of the falling ash melts snow and ice on the mountain, producing a lahar. The 1985 eruption of Nevado del Ruiz, in Colombia, was a deadly example (figure 5.17). The swift and sudden mudflows that swept down its steep slopes after its snowy cap was partially melted by hot ash were the principal cause of deaths in towns below the volcano.

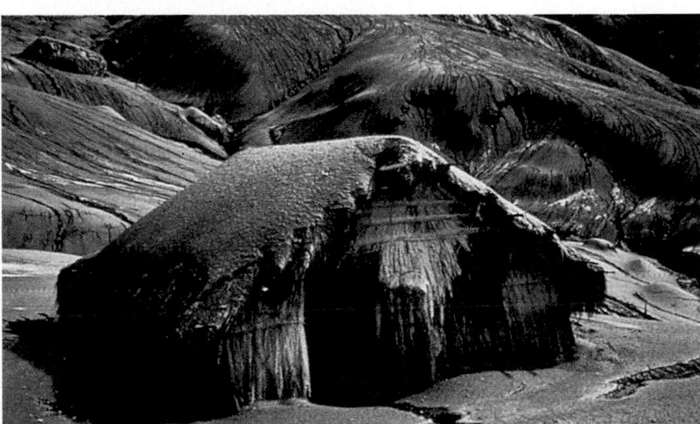

FIGURE 5.16 Some houses were swamped by lahars. Typhoon rains dropping soaking ash on roofs also contributed to casualties when the overburdened roofs collapsed.

Photograph by R. P. Hoblitt, both courtesy U.S. Geological Survey

A

B

FIGURE 5.17 (A) Aerial view of Abacan River channel in Angeles City near Clark Air Force Base, Philippines. A mudflow has taken out the main bridge; makeshift bridges allow pedestrians to cross the debris. (B) The town of Armero was destroyed by lahars from Nevado del Ruiz in November 1985; more than 23,000 people died. These ruins emerged as the flows moved on and water levels subsided.

(A) Photograph by T. J. Casadevall, (B) Photograph by R. Janda, USGS Cascade Volcano Observatory

Even after the immediate danger is past, when the lahars have flowed, long-term impacts remain. They often leave behind stream channels partially filled with mud, their water-carrying capacity permanently reduced, which aggravates ongoing flood hazards in the area. To dredge this mud to restore channel capacity can be very costly, and areas far from the volcano can be thus affected. This was a significant problem in the case of Mount St. Helens.

Pyroclastic Flows—Nuées Ardentes

Another special kind of pyroclastic outburst is a deadly, denser-than-air mixture of hot gases and fine ash that forms a hot **pyroclastic flow,** quite different in character from a rain of ash and cinders. A pyroclastic flow is sometimes known as a *nuée ardente* (from the French for "glowing cloud"). A nuée ardente is very hot—temperatures can be over 1000°C in the interior—and it can rush down the slopes of the volcano at more than 100 kilometers per hour (60 miles per hour), charring everything in its path. Such pyroclastic flows accompanied the major eruption of Mount St. Helens in 1980 (figure 5.18).

FIGURE 5.18 Pyroclastic flow from Mount St. Helens.

Photograph by P. W. Lipman, USGS Photo Library, Denver, CO.

Perhaps the most famous such event in recent history occurred during the 1902 eruption of Mont Pelée on the Caribbean island of Martinique. The volcano had begun erupting weeks before, emitting both ash and lava, but was believed by many to pose no imminent threat to surrounding towns. Then, on the morning of May 8, with no immediate advance warning, a nuée ardente emerged from that volcano and swept through the nearby town of St. Pierre and its harbor. In a period of about three minutes, an estimated 25,000 to 40,000 people died or were fatally injured, burned or suffocated. The single reported survivor in the town was a convicted murderer who had been imprisoned in the town dungeon, underground, where he was shielded from the intense heat. He spent four terrifying days buried alive without food or water before rescue workers dug him out. Figure 5.19 gives some idea of the devastation.

Just as some volcanoes have a history of explosive eruptions, so, too, many volcanoes have a history of eruption of pyroclastic flows. Again, the composition of the lava is linked to the likelihood of such an eruption. While the emergence of a nuée ardente may be sudden and unheralded by special warning signs, it is not generally the first activity shown by a volcano during an eruptive stage. Steam had issued from Mont Pelée for several weeks before the day St. Pierre was destroyed, and lava had been flowing out for over a week. This suggests one possible strategy for avoiding the dangers of these hot pyroclastic/gas flows: When a volcano known or believed to be capable of such eruptions shows signs of activity, leave.

Sometimes, however, human curiosity overcomes both fear and common sense, even when danger is recognized. A few days before the destruction of St. Pierre, the wife of the U.S. consul there wrote to her sister:

> This morning the whole population of the city is on the alert, and every eye is directed toward Mont Pelée, an extinct volcano. Everybody is afraid that the volcano has taken into its heart to burst forth and destroy the whole island. Fifty years ago, Mont Pelée burst forth with terrific force and destroyed everything within a radius of several miles. For several days, the mountain has been bursting forth in flame and immense quantities of lava are flowing down its sides. *All the inhabitants are going up to see it. ...* (Garesche 1902; *emphasis added*)

FIGURE 5.19 St. Pierre, Martinique, West Indies, was destroyed by a nuée ardente from Mont Pelée, 1902.
Photograph by Underwood and Underwood, courtesy Library of Congress

Evacuations can themselves be disruptive. If, in retrospect, the danger turns out to have been a false alarm, people may be less willing to heed a later call to clear the area. There are other volcanoes in the Caribbean similar in character to Mont Pelée. In 1976, La Soufrière on Guadeloupe began its most recent eruption. Some 70,000 people were evacuated and remained displaced for months, but only a few small explosions occurred. Government officials and volcanologists were blamed for disruption of people's lives, needless as it turned out to be. When Soufrière Hills volcano on Montserrat in the West Indies began to erupt in 1995, it soon became apparent that evacuation was necessary. Since then, casualties have been few, but property damage is extensive and continuing (figure 5.20).

Fortunately, officials took the 1980 threat of Mount St. Helens seriously after its first few signs of reawakening. The immediate area near the volcano was cleared of all but a few essential scientists and law enforcement personnel. Access to hundreds of square kilometers of terrain around the volcano was severely restricted for both residents and workers (mainly loggers). Others with no particular business in the area were banned altogether. Many people grumbled at being forced from their homes. A few flatly refused to go. Numerous tourists and reporters were frustrated in their efforts to get a close look at the action. When the major explosive eruption came, there were casualties, but far fewer than there might have been. On a normal spring weekend, 2000 or more people would have been on the mountain, with more living or camping nearby. Considering the influx of the curious once eruptions began and the numbers of people turned away at roadblocks leading into the restricted zone, it is possible that, with free access to the area, tens of thousands could have died. Many lives were likewise saved by timely evacuations near Mount Pinatubo in 1991, made possible by careful monitoring and supported by intensive educational efforts by the government.

Toxic Gases

In addition to lava and pyroclastics, volcanoes emit a variety of gases (figure 5.21). Water vapor, the primary component, is not a threat, but several other gases are. Carbon dioxide is nontoxic, yet can nevertheless be dangerous at high concentrations, as described below. Other gases, including carbon monoxide, various sulfur gases, and hydrochloric acid, are actively

A

B

FIGURE 5.20 (A) By the time this ash cloud loomed over Plymouth on 27 July 1996, the town had been evacuated; the potential for pyroclastic eruptions of Soufrière Hills volcano was well recognized. (B) By March 2003, pyroclastic flows had covered much of the mountain.

(A) Photograph © Gary Sego, (B) photograph © Montserrat Volcano Observatory Government of Montserrat and British Geological Survey; used by permission of the Director, MVO

poisonous. In the 1783 eruption of Laki in Iceland, toxic fluorine gas emissions apparently accounted for many of the 10,000 casualties. Many people have been killed by volcanic gases even before they realized the danger. During the A.D. 79 eruption of Vesuvius, fumes overcame and killed some unwary observers. Again, the best defense against harmful volcanic gases is common sense: Get well away from the erupting volcano as quickly as possible.

A bizarre disaster, indirectly attributable to volcanic gases, occurred in Cameroon, Africa, in 1986. Lake Nyos is one of a series of lakes that lie along a rift zone that is the site of intermittent volcanic activity. On 21 August 1986, a massive cloud of carbon dioxide (CO_2) gas from the lake flowed out over nearby towns and claimed 1700 lives. The victims were not poisoned; they were suffocated. Carbon dioxide is about 50% denser than air, so such a cloud tends to settle near the

FIGURE 5.21 "Vog" (volcanic fog) from the Pu'u O'o vent on Kilauea's east rift drifts across the evening sky. A blend of water vapor, carbon dioxide, sulfur gases, and acids such as hydrochloric, it can be toxic, especially to those with respiratory problems.

ground and to disperse only slowly. Anyone engulfed in that cloud would suffocate from lack of oxygen, and without warning, for CO_2 is both odorless and colorless. Furthermore, this is an ongoing threat distinct from obvious eruptions. The source of the gas in this case is believed to be near-surface magma below. The CO_2 apparently seeps up into the lake, where it accumulates in lake-bottom waters. Seasonal overturn of the lake waters as air temperatures change can lead to its release. Currently, both Lake Nyos and nearby Lake Monoun (where 37 people dropped dead on a road in 1984, in what is now believed to have been a similar CO_2 release) are being studied and monitored closely, and efforts to release CO_2 from Lake Nyos harmlessly have been initiated.

Steam Explosions

Some volcanoes are deadly not so much because of any characteristic inherent in the particular volcano but because of where they are located. In the case of a volcanic island, large quantities of seawater may seep down into the rock, come close to the hot magma below, turn to steam, and blow up the volcano like an overheated steam boiler. This is called a **phreatic eruption** or explosion.

The classic example is Krakatoa (Krakatau) in Indonesia, which exploded in this fashion in 1883. The force of its explosion was estimated to be comparable to that of 100 million tons of dynamite, and the sound was heard 3000 kilometers away in Australia. Some of the dust shot 80 kilometers into the air, causing red sunsets for years afterward, and ash was detected over an area of 750,000 square kilometers. Furthermore, the shock of the explosion generated a tsunami that built to over 40 meters high as it came onshore! Krakatoa itself was an uninhabited island, yet its 1883 eruption killed an estimated 36,000 people, mostly in low-lying coastal regions inundated by tsunamis.

In earlier times, when civilization was concentrated in fewer parts of the world, a single such eruption could be even more destructive. During the fourteenth century B.C., the volcano Santorini on the island of Thera exploded and caused a tsunami that wiped out many Mediterranean coastal cities. Some scholars have suggested that this event contributed to the fall of the Minoan civilization, as well as to the Atlantis legend.

Landslides and Collapse

The very structure of a volcano can become unstable and become a direct or indirect threat. As rocks weather and weaken, they may come tumbling down the steep slopes of a composite volcano as a landslide that may bury a valley below, or dam a stream to cause flooding. Just removing that weight from the side of the volcano may allow gases trapped under pressure to blast forth through the weakened area. This, in fact, is what happened at Mount St. Helens in 1980. With an island or coastal volcano, a large landslide crashing into the sea could trigger a tsunami. Such an event in 1792 at Mt. Mayuyama, Japan killed 15,000 people on the opposite shore of the Ariaka Sea. Scientists studying Kilauea are concerned that a combination of weathering, the weight of new lava, and undercutting by erosion along the coast may cause breakoff of a large slab from the island of Hawaii, which could likewise cause a tsunami; underwater imaging has confirmed huge coastal slides in the past.

Secondary Effects: Climate and Atmospheric Chemistry

A single volcanic eruption can have a global impact on climate, although the effect may be only brief. Intense explosive eruptions put large quantities of volcanic dust high into the atmosphere. The dust can take several months or years to settle, and in the interim, it partially blocks out incoming sunlight, thus causing measurable cooling. After Krakatoa's 1883 eruption, worldwide temperatures dropped nearly half a degree Celsius, and the cooling effects persisted for almost ten years. The larger 1815 eruption of Tambora in Indonesia caused still more dramatic cooling: 1816 became known around the world as the "year without a summer." Such experiences support fears of a "nuclear winter" in the event of a nuclear war because modern nuclear weapons are powerful enough to cast many tons of fine dust into the air, as well as to trigger widespread forest fires that would add ash to the dust.

The meterological impacts of volcanoes are not confined to the effects of volcanic dust. The 1982 eruption of the Mexican volcano El Chichón did not produce a particularly large quantity of dust, but it did shoot volumes of unusually sulfur-rich gases into the atmosphere. These gases produced clouds of sulfuric acid droplets (*aerosols*) that spread around the earth. Not only do the acid droplets block some sunlight, like dust, but in time, they also fall back to earth as acid rain, the consequences of which are explored in later chapters.

The 1991 eruptions of Mount Pinatubo in the Philippines (figure 5.22) had long-lasting, global effects on climate that were not only measurable by scientists but noticeable to a great many other people around the world. In addition to quantities of visible dust, Pinatubo pumped an estimated 20 million tons of sulfur dioxide (SO_2) into the atmosphere. While ash falls out of the atmosphere relatively quickly, the mist of sulfuric-acid aerosols formed from stratospheric SO_2 stays airborne for several years. This acid-mist cloud slowly encircled the earth, most concentrated close to the equator near Pinatubo's latitude (figure 5.23). At its densest, the mist scattered up to 30% of incident sunlight and decreased by 2 to 4% the amount of solar radiation reaching earth's surface to warm it. Average global temperatures decreased by about 0.5°C (1°F), but the decline was not uniform; the greatest cooling was observed in the intermediate latitudes of the Northern Hemisphere (where the United States is located). Mount Pinatubo thus received the credit—or the blame—for the unusually cool summer of 1992 in the United States. The gradual natural removal of Pinatubo-generated ash and later SO_2 from the atmosphere resulted in a rebound of temperatures in the lower atmosphere by 1994 (figure 5.24).

Another, more subtle impact of this SO_2 on atmospheric chemistry may present a different danger. The acid seems to aggravate ozone depletion (see discussion of ozone and the "ozone hole" in chapter 18), which means increased risk of skin cancer to those exposed to strong sunlight.

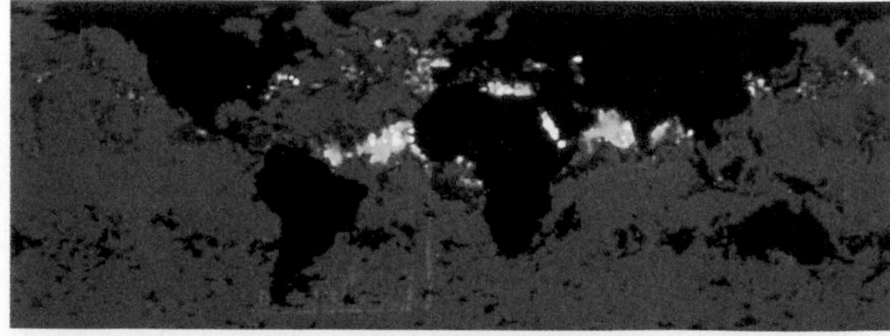

5/28 – 6/6/91 (before eruptions)

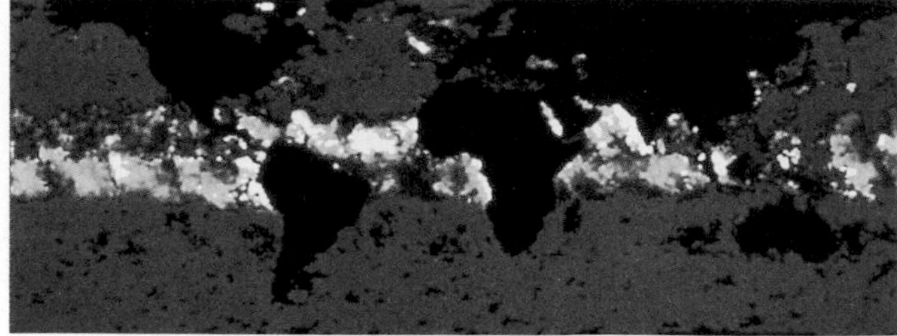

7/4 – 7/10/91 (after circling the earth)

FIGURE 5.23 Satellites tracked the path of the airborne sulfuric-acid mist formed by SO_2 from Mount Pinatubo; winds slowly spread it into a belt encircling the earth (bright yellow in figure).

Image by G. J. Orme, Department of the Army, Ft. Bragg, NC.

FIGURE 5.22 Ash and gas from Mount Pinatubo was shot into the stratosphere, and had an impact on climate and atmospheric chemistry worldwide. Eruption of 12 June 1991.

Photograph by K. Jackson, U.S. Air Force

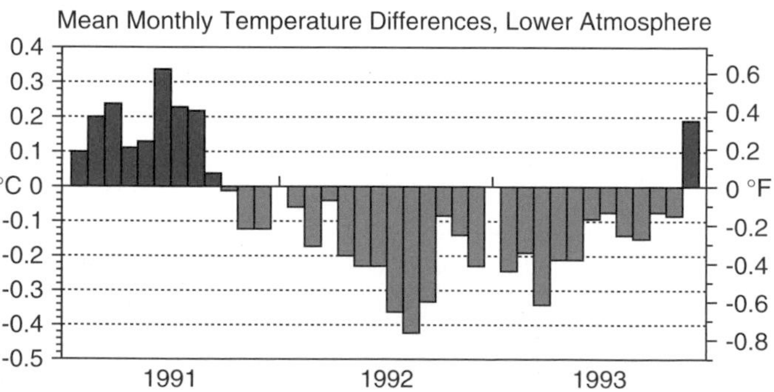

FIGURE 5.24 Effect of 1991 eruption of Pinatubo on near-surface (lower-atmosphere) air temperatures. Removal of ash and dust from the air was relatively rapid; sulfate aerosols persisted longer. The major explosive eruption occurred in mid-June 1991.

Source: Data from National Geophysical Data Center

Box 5.1 Degrees of Risk: Life on a Volcano's Flanks

The island of Hawaii is really a complex structure built of five overlapping shield volcanoes. Even where all the terrain is volcanic, however, degrees of future risk can be determined from the types, distribution, and ages of past eruptions (figure 1A).

The so-called East Rift of Kilauea in Hawaii has been particularly active in recent years (figure 1B). Less than ten years after a series of major eruptions in the 1970s, a new subdivision—Royal Gardens—was begun below the East Rift. In 1982–1983, lava flows from a renewed round of eruptions

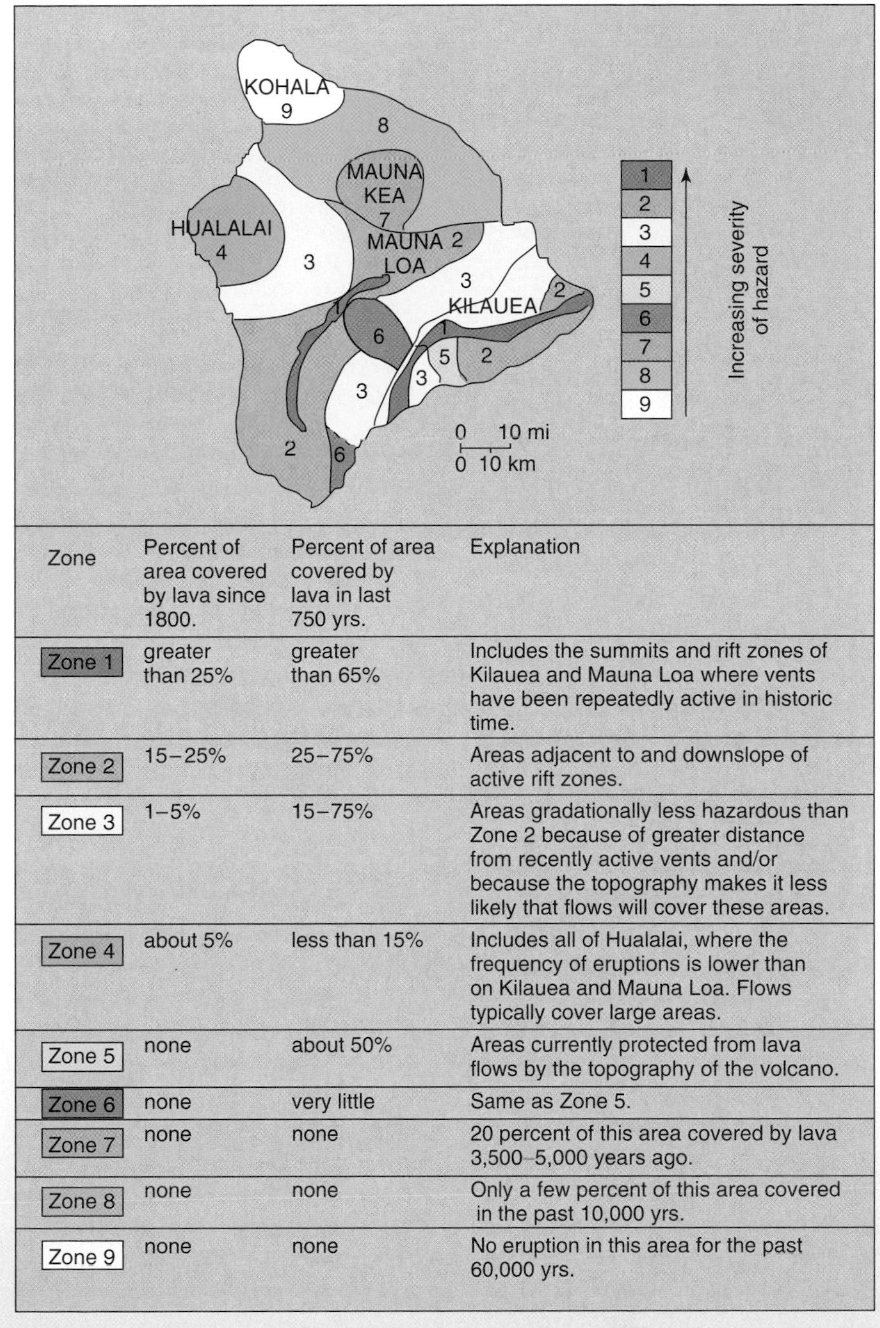

Zone	Percent of area covered by lava since 1800.	Percent of area covered by lava in last 750 yrs.	Explanation
Zone 1	greater than 25%	greater than 65%	Includes the summits and rift zones of Kilauea and Mauna Loa where vents have been repeatedly active in historic time.
Zone 2	15–25%	25–75%	Areas adjacent to and downslope of active rift zones.
Zone 3	1–5%	15–75%	Areas gradationally less hazardous than Zone 2 because of greater distance from recently active vents and/or because the topography makes it less likely that flows will cover these areas.
Zone 4	about 5%	less than 15%	Includes all of Hualalai, where the frequency of eruptions is lower than on Kilauea and Mauna Loa. Flows typically cover large areas.
Zone 5	none	about 50%	Areas currently protected from lava flows by the topography of the volcano.
Zone 6	none	very little	Same as Zone 5.
Zone 7	none	none	20 percent of this area covered by lava 3,500–5,000 years ago.
Zone 8	none	none	Only a few percent of this area covered in the past 10,000 yrs.
Zone 9	none	none	No eruption in this area for the past 60,000 yrs.

FIGURE 1 (A) Volcanic-hazard map of Hawaii showing five volcanoes and major associated rift zones. Degree of risk is related to topography (being downslope from known vents is worse), and extent of recent flows in the area. Zones 6–9 have experienced little or no volcanism for over 750 years; in Zone 1, 65% or more of the area has been covered by lava in that time, and more than 25% since 1800. (Continued)
(A) After C. Heliker, Volcanic and Seismic Hazards of Hawaii, *U.S. Geological Survey, 1991*

(Box 5.1 Continued)

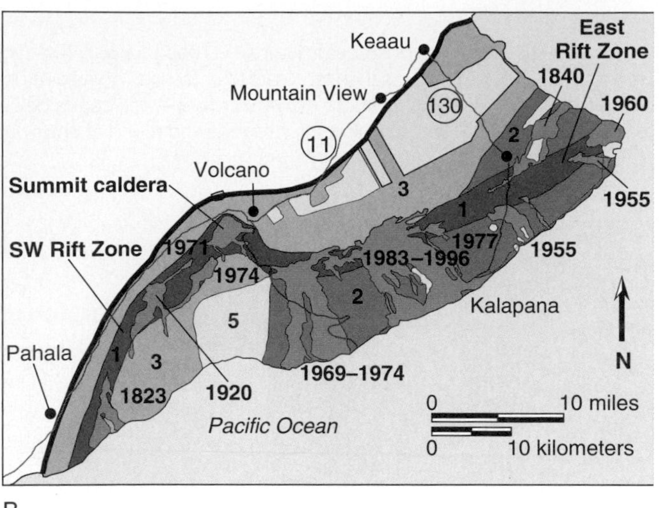

B

FIGURE 1 Continued. (B) Hazard map for Kilauea's East Rift Zone; note the number of relatively recent flows.
(B) Courtesy U.S. Geological Survey/Hawaii Volcanoes Observatory

reached down the slopes of the volcano and quietly obliterated new roads and houses. The eruptions have continued intermittently over more than two decades since then (figure 2). The photographs in figure 3 illustrate some of the results. Even those whose homes have survived are hardly unaffected. They may be isolated by a mile of lava from the nearest road to a grocery store, have no water or electricity, and be unable to insure what they have left—never mind sell the property in which they invested.

When asked why anyone would have built in such a spot, where the risks were so obvious, one native shrugged and replied, "Well, the land was cheap." It will probably remain so, too—a continuing temptation to unwise development—in the absence of zoning restrictions.

Over centuries, Hawaiians and other residents of active volcanic areas have commonly become philosophical about such activity. Anyone living in Hawaii is, after all, living on a volcano, and as noted, the nonviolent eruptive style of these shield volcanoes means that they rarely take lives. Staying out of the riskiest areas greatly reduces even the potential for loss of property. Still, as populations grow and migrate, more people may be tempted to move into areas that might better be avoided altogether! As outsiders unfamiliar with volcanoes move into an area like Hawaii, they may unknowingly settle in higher-risk areas. Sometimes, too, a particularly far-reaching flow will obliterate a town untouched for a century or more, where residents might have become just a little bit complacent, hazard maps or no.

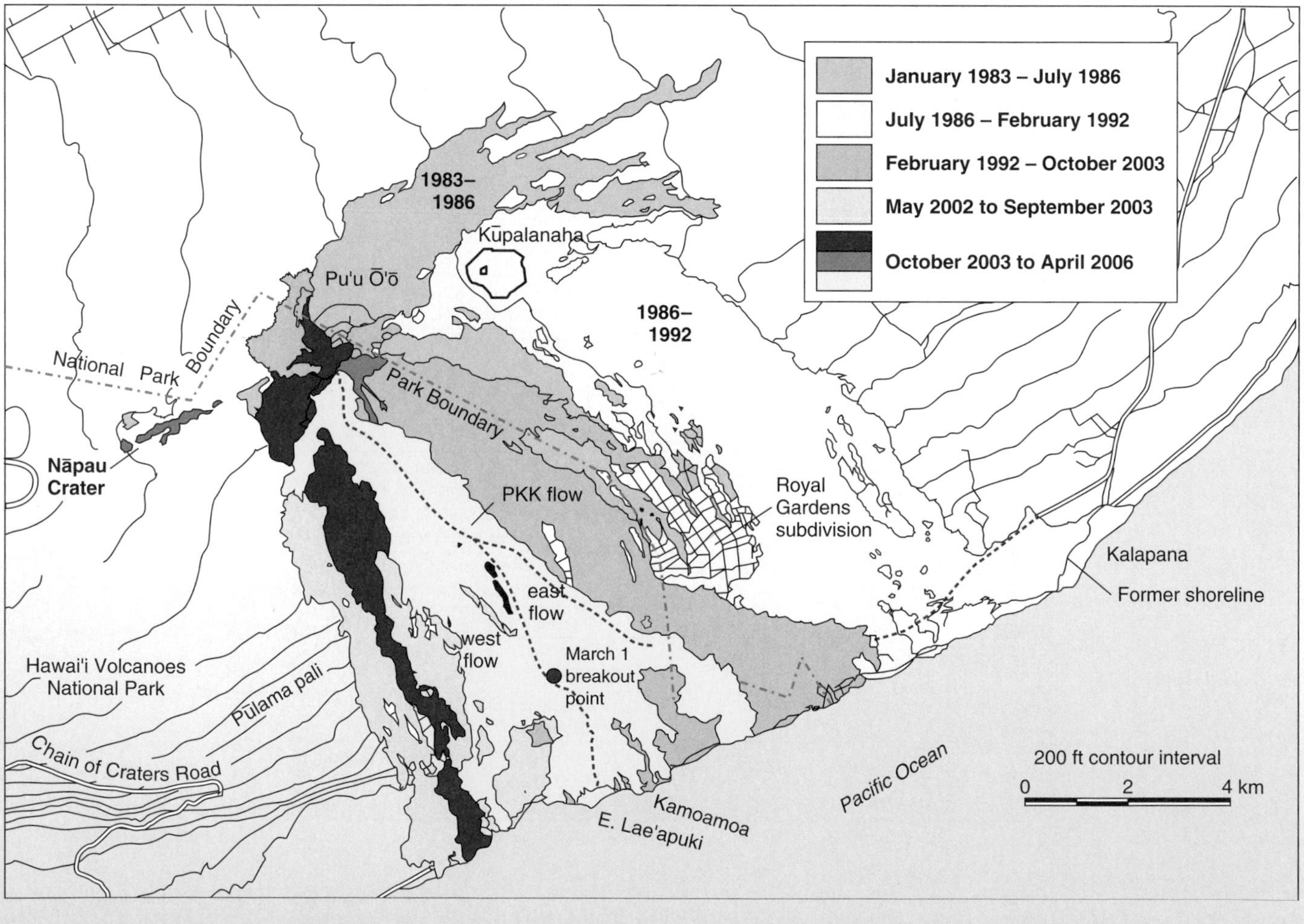

FIGURE 2 An eruptive phase can continue for decades, with activity shifting geographically over time. Note location of subdivision featured in box figure 3.
Map courtesy U.S. Geological Survey/Hawaii Volcanoes Observatory

(Box 5.1 Continued)

A

B

C

FIGURE 3 (A) Lava flowing over a nearly new road in the Royal Gardens subdivision, Kilauea, Hawaii, 1983. Hot lava can even ignite asphalt. (B) By 1986, destruction in Royal Gardens and Kalapana Gardens subdivisions was widespread. Here, vegetation has died from heat; note engulfed truck at rear of photograph. (C) By 1989, advancing tongues of lava had stretched to the coast and threatened more property. Water was sprayed both on flow fronts (to attempt to quench and halt them) and on structures (to cool and dampen them to suppress fire). But volumes of lava kept coming, and eventually structures and whole towns were overrun; this same building is the one featured in figure 5.1. Even the ruin in figure 5.1B is now entombed under younger lava flows, and vanished.

(B) and (C) Photographs by J. D. Griggs, all courtesy USGS Photo Library, Denver, CO.

ISSUES IN PREDICTING VOLCANIC ERUPTIONS

Classification of Volcanoes by Activity

In terms of their activity, volcanoes are divided into three categories: active; dormant, or "sleeping"; and extinct, or "dead." A volcano is generally considered **active** if it has erupted within recent history. When the volcano has not erupted recently but is fresh-looking and not too eroded or worn down, it is regarded as **dormant:** inactive for the present but with the potential to become active again. Historically, a volcano that has no recent eruptive history and appears very much eroded has been considered **extinct,** or very unlikely to erupt again.

Unfortunately, precise rules for assigning a particular volcano to one category or another do not exist. As volcanologists learn more about the frequency with which volcanoes erupt, it has become clear that volcanoes differ widely in their normal patterns of activity. Statistically, a "typical" volcano erupts once every 220 years, but 20% of all volcanoes erupt less than once every 1000 years, and 2% erupt less than once in 10,000 years. Long quiescence, then, is no guarantee of extinction. Just as knowledge of past eruptive style allows anticipation of the kinds of hazards possible in future eruptions, so knowledge of the eruptive history of any given volcano is critical to knowing how likely future activity is, and on what scales of time and space.

The first step in predicting volcanic eruptions is monitoring, keeping an instrumental eye on the volcano. However, there are not nearly enough personnel or instruments available to monitor every volcano all the time. There are an estimated 300 to 500 active volcanoes in the world (the inexact number arises from not knowing whether some are truly active, or only dormant). Monitoring those alone would be a large task. Commonly, intensive monitoring is undertaken only after a volcano shows some signs of near-term activity (see the discussion of volcanic precursors that follows). Dormant volcanoes might become active at any time, so, in principle, they also should be monitored; they can sometimes become very active very quickly. Theoretically, extinct volcanoes can be safely ignored, but that assumes that extinct volcanoes can be distinguished from the long-dormant. Vesuvius had been regarded as extinct until it destroyed Pompeii and Herculaneum in A.D. 79. Mount Pinatubo had been classed as dormant prior to 1991—it had not erupted in more than 400 years—then, within a matter of two months, it progressed from a few steam emissions to a major explosive eruption. The detailed eruptive history of many volcanoes is imperfectly known over the longer term.

The Volcanic Explosivity Index

The **Volcanic Explosivity Index** (VEI) has been developed as a way to characterize the relative sizes of explosive eruptions. It takes into account the volume of pyroclastics produced, how high into the atmosphere they rose, and the length of the eruption. The VEI scale, like earthquake magnitude scales, is exponential, meaning that each unit increase on the VEI scale represents about a tenfold increase in size/severity of eruption. Also, not surprisingly, there are far more small eruptions than large.

Several eruptions discussed in this chapter are plotted on the VEI diagram in figure 5.25. Knowing the size of eruption of which a given volcano is capable is helpful in anticipating future hazards. At one time, it was thought that the VEI could also be used to project the probable climatic impacts of volcanic eruptions. However, it seems that the volume of sulfur dioxide emitted is a key factor in an eruption's effect on climate, and that is neither included in the determination of VEI nor directly related to the size of the eruption as measured by VEI.

Volcanic Eruption Precursors

What do scientists look for when monitoring a volcano? One common advance warning of volcanic activity is seismic activity (earthquakes). The rising of a volume of magma and gas up through the lithosphere beneath a volcano puts stress on the rocks of the lithosphere, and the process may produce months of small (and occasionally large) earthquakes. Increased seismicity at Pinatubo was a major factor prompting evacuations. In fall 2004, when earthquakes at Mount St. Helens became both more frequent and shallower (figure 5.26A), volcanologists also took notice. Monitoring intensified; aircraft were

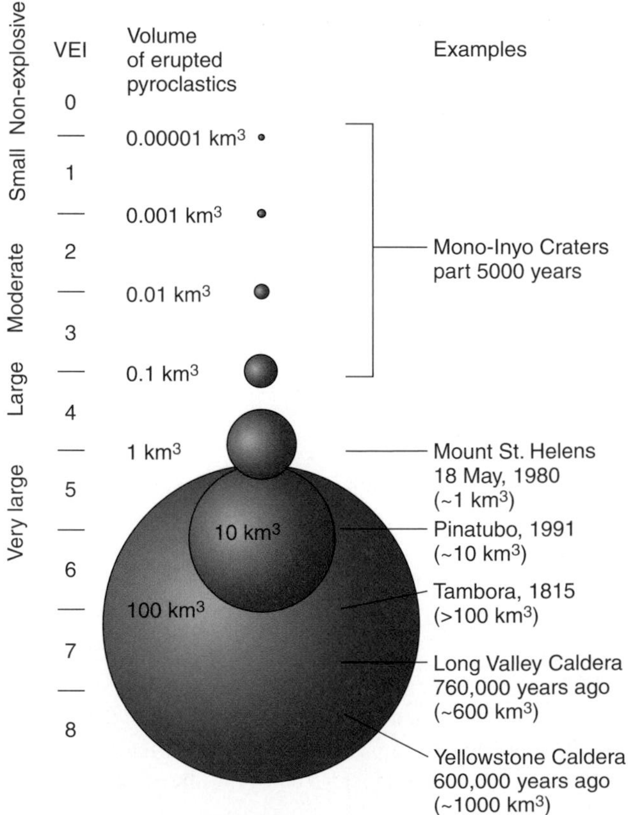

FIGURE 5.25 The Volcanic Explosivity Index indicates relative sizes of explosive eruptions.
After USGS Volcano Hazards Program.

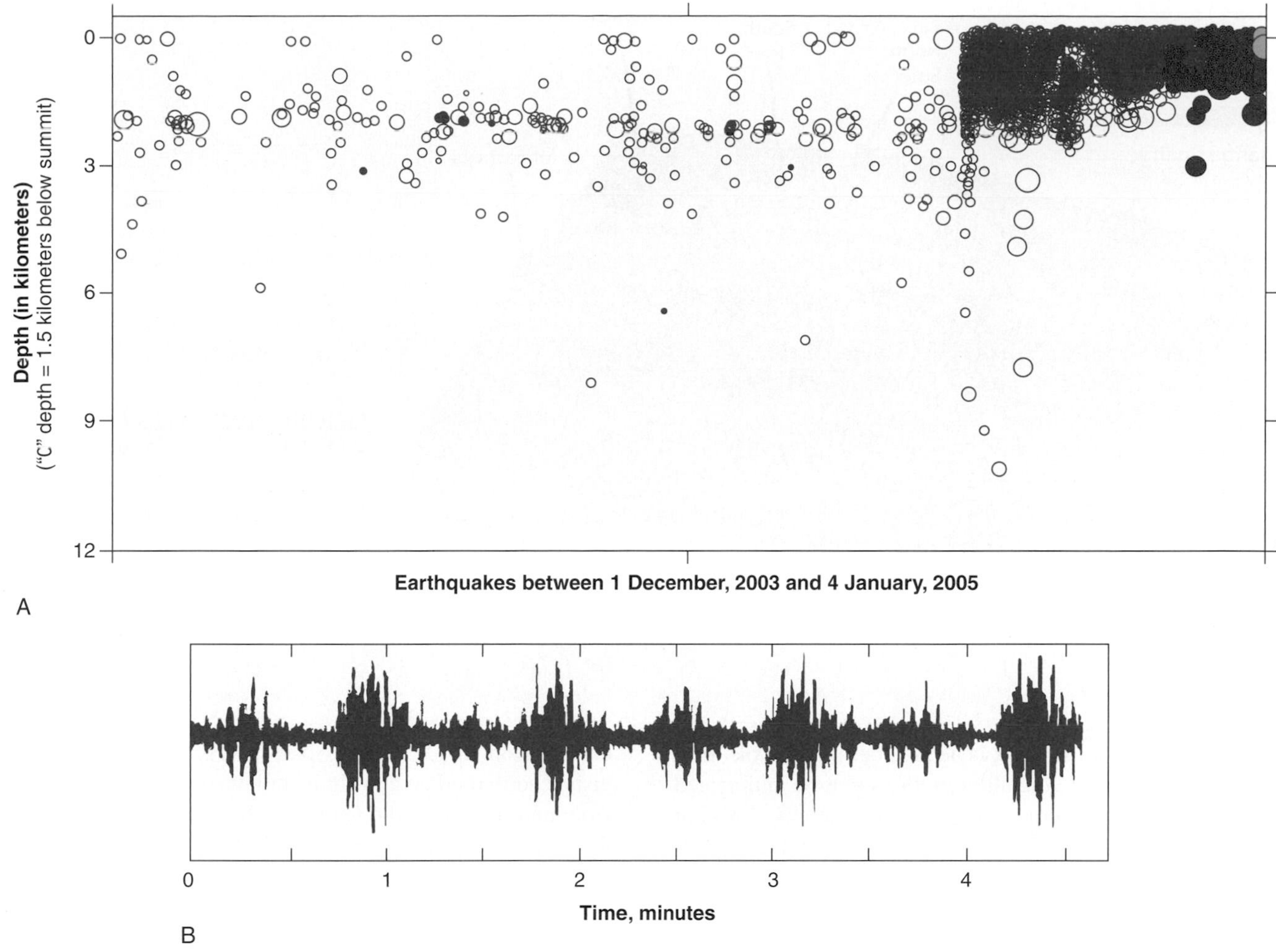

FIGURE 5.26 (A) Depths of earthquake foci below Mount St. Helens as a function of time. Size of circle represents size of quake; virtually all were below magnitude 4. (B) Harmonic-tremor signal.

(A) Modified after USGS Cascades Volcano Observatory; data from Pacific Northwest Seismograph Network. (B) After U.S. Geological Survey

diverted around the mountain in case of ash eruptions. Some small ash emissions have occurred intermittently since then, but the current eruptive phase seems mainly to involve renewed lava-dome building in the summit crater, the results of which were shown in figure 5.12.

Mount St. Helens is also a place where *harmonic tremors* have been recognized as indicating a possible impending eruption. These continuous, rhythmic tremors (figure 5.26B) are very different in character from earthquakes' seismic waves discussed in chapter 4. Their precise origin and significance are still being investigated.

Bulging, tilt, or uplift of the volcano's surface is also a warning sign. It often indicates the presence of a rising magma mass, the buildup of gas pressure, or both. On Kilauea, eruptions are preceded by an inflating of the volcano as magma rises up from the asthenosphere to fill the shallow magma chamber. Unfortunately, it is not possible to specify exactly when the swollen volcano will crack to release its contents. That varies from eruption to eruption with the pressures and stresses involved and the strength of the overlying rocks. This limitation is not unique to Kilauea. Uplift, tilt, and seismic activity may indicate that an eruption is approaching, but geologists do not yet have the ability to predict its exact timing. However, their monitoring tools are becoming more sophisticated. Historically, tilt was monitored via individual meters placed at a limited number of points on the volcano. Now, satellite radar interferometry allows a richer look at the regional picture (figure 5.27).

Other possible predictors of volcanic eruptions are being evaluated. Changes in the mix of gases coming out of a volcano may give clues to impending eruptions; SO_2 content of the escaping gas shows promise as a precursor, perhaps because more SO_2 can escape as magma nears the surface. Surveys of ground surface temperatures may reveal especially warm areas where magma is particularly close to the surface and about to break through. As with earthquakes, there have been reports that volcanic eruptions have been sensed by animals, which have behaved strangely for some hours or days before the event. Perhaps animals are sensitive to some changes in the earth that scientists have not thought to measure.

FIGURE 5.27 "Interferogram" showing uplift, 1997–2001, in Three Sisters volcanic area. Uplift is attributed to magma emplacement several miles deep in the crust. Each full color cycle represents close to 1″ of uplift; total uplift shown is nearly 6″ (14 cm). This method is now being applied at Yellowstone and elsewhere.

Interferogram by C. Wicks, USGS Volcano Hazards Program

Certainly, much more work is needed before the exact timing and nature of major volcanic eruptions can be anticipated consistently. The recognition of impending eruptions of Mount St. Helens, Pinatubo, and Soufrière Hills was obviously successful in saving many lives through evacuations and restrictions on access to the danger zones. Those danger zones, in turn, were effectively identified by a combination of mapping of debris from past eruptions and monitoring the geometry of the current bulging. Mount Pinatubo displayed increasing evidence of forthcoming activity for two months before its major eruption of 15 June 1991—local earthquakes, increasing in number, and becoming concentrated below a bulge on the volcano's summit a week before that eruption; preliminary ash emissions; a sudden drop in gas output that suggested a plugging of the volcanic "plumbing" that would tend to increase internal gas pressure and explosive potential. Those warnings, combined with aggressive and comprehensive public warning procedures and, eventually, the evacuation of over 80,000 people, greatly reduced the loss of life. In 1994, geoscientists monitoring Rabaul volcano in Papua New Guinea took a magnitude-5 earthquake followed by increased seismic activity generally as the signal to evacuate a town of 30,000 people. The evacuation, rehearsed before, was orderly—and hours later, a five-day explosive eruption began. The town was covered by a meter or more of wet volcanic ash, and a quarter of the buildings were destroyed, but no one died.

These successes clearly demonstrate the potential for reduction in fatalities when precursory changes occur and local citizens are educated to the dangers. On the other hand, the exact moment of the 1980 Mount St. Helens explosion was not known until seconds beforehand, and it is not currently possible to tell whether or just when a pyroclastic flow might emerge from a volcano like Soufrière Hills. Nor can volcanologists readily predict the volume of lava or pyroclastics to expect from an eruption or the length of the eruptive phase, although they can anticipate the likelihood of an explosive eruption or other eruption of a particular style based on tectonic setting and past behavior.

Response to Eruption Predictions

When data indicate an impending eruption that might threaten populated areas, the safest course is evacuation until the activity subsides. However, a given volcano may remain more or less dangerous for a long time. An active phase, consisting of a series of intermittent eruptions, may continue for months or years. In these instances, either property must be abandoned for prolonged periods, or inhabitants must be prepared to move out, not once but many times. The Soufrière Hills volcano on Montserrat began its latest eruptive phase in July 1995. By the end of 2000, dozens of eruptive events—ash venting, pyroclastic flows, ejection of rocks, building or collapse of a dome in the summit—had occurred, and more than half the island remained off-limits to inhabitants (figure 5.28). Given the uncertain nature of eruption prediction at present, too, some precautionary evacuations will continue to be shown, in retrospect, to have been unnecessary, or unnecessarily early.

Accurate prediction and assessment of the hazard is particularly difficult with volcanoes reawakening after a long dormancy because historical records for comparison with current data are sketchy or nonexistent. Beside the Bay of Naples is an old volcanic area known as the Phlegraean Fields. In the early 1500s, the town of Tripergole rose up some 7 meters as a chamber of gas and magma bulged below, and the region was

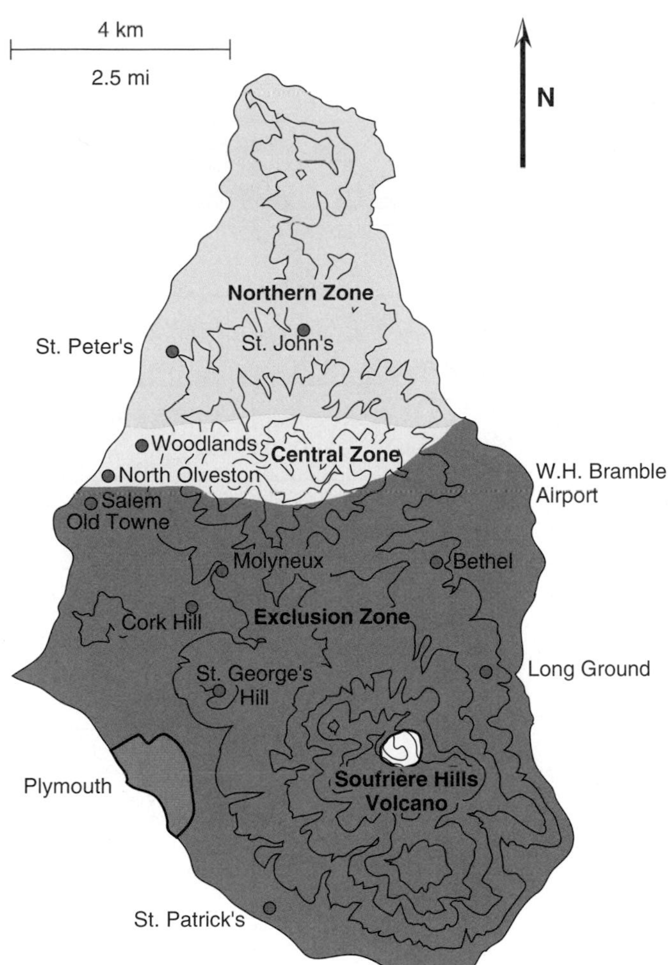

FIGURE 5.28 Hazard zones on Montserrat (1997). *Exclusion Zone* was off-limits to all but scientific monitoring and National Security personnel (note that it included the capital city, Plymouth, and the island's one airport). *Central Zone* was open to residents but as a hard-hat area, and all occupants had to be prepared to evacuate quickly, 24 hours a day. Only *Northern Zone* was unrestricted.

Source: Montserrat Volcano Observatory

shaken by earthquakes. Then, suddenly, a vent opened, and in three days, the town was buried under a new 140-meter-high cinder cone. For the next four centuries, the only thermal activity had been steaming fumaroles and hot springs. But in 1970–1972 and again in 1982–1984, the nearby town of Pozzuoli was uplifted nearly 3 meters. During the last episode of bulging, it was shaken by over 4000 earthquakes in a single year, leaving the town so badly damaged that many of its 80,000 residents had to sleep in "tent cities" outside the town for months. Business could be conducted in town by day, but the town had to be evacuated at night. The threat and the disruption in the people's lives persisted for years. These episodes are believed to be related to the presence of a shallow magma chamber. Volcanologists are still unsure of what sort of volcanic activity can be expected here, how large in scale, or how soon.

And while some volcanoes, like Mount St. Helens or Mount Pinatubo, may produce weeks or months of warning signals as they emerge from dormancy to full-blown eruption, allowing time for public education and reasoned response, not all volcanoes' awakenings may be so leisurely.

PRESENT AND FUTURE VOLCANIC HAZARDS IN THE UNITED STATES

Hawaii

Several of the Hawaiian volcanoes are active or dormant, and eruptions are frequent in some places. It is, however, not practical to evacuate the islands indefinitely. Fortunately, it is possible to identify areas of relatively higher and lower risk (recall Box 5.1), though few areas can be considered risk-free, and identifying high-risk areas does not prevent people from trying to live there.

Cascade Range

Mount St. Helens is only one of a set of volcanoes that threaten the western United States and southwestern Canada. There is subduction beneath the Pacific Northwest, which is the underlying cause of continuing volcanism there (figure 5.29). Several more of the Cascade Range volcanoes have shown signs of reawakening. Lassen Peak was last active between 1914 and 1921, not so very long ago geologically. Its products are very similar to those of Mount St. Helens; violent eruptions are possible. Mount Baker (last eruption, 1870), Mount Hood (1865), and Mount Shasta (active sometime within the last 200 years) have shown seismic activity, and steam is escaping from these and from Mount Rainier, which last erupted in 1882. In fact, at least nine of the Cascade peaks are presently showing thermal activity of some kind (steam emission or hot springs). The eruption of Mount St. Helens has in one sense been useful: It has focused attention on this threat. Scientists are watching many of the Cascade Range volcanoes very closely now; there is particular concern about possible mudflow dangers from Mount Rainier (figure 5.30). With skill, and perhaps some luck, they may be able to recognize when others of these volcanoes may be close to eruption in time to avert disaster.

The Aleutians

South-central Alaska and the Aleutian island chain sit above a subduction zone. This makes the region vulnerable not only to earthquakes, as noted in chapter 4, but also to volcanic eruptions (figure 5.31A). The volcanoes are typically andesitic with composite cones, and pyroclastic eruptions are common (figure 5.31B). In 1986, Augustine began a period of eruption; in late 1989, Redoubt volcano followed suit, following twenty-three years of quiescence, with the eruptive phase heralded by a period of increased seismicity and steam emission. Neither volcano lies close to a densely populated area. Nevertheless, they have presented special hazards. Jet aircraft flying through

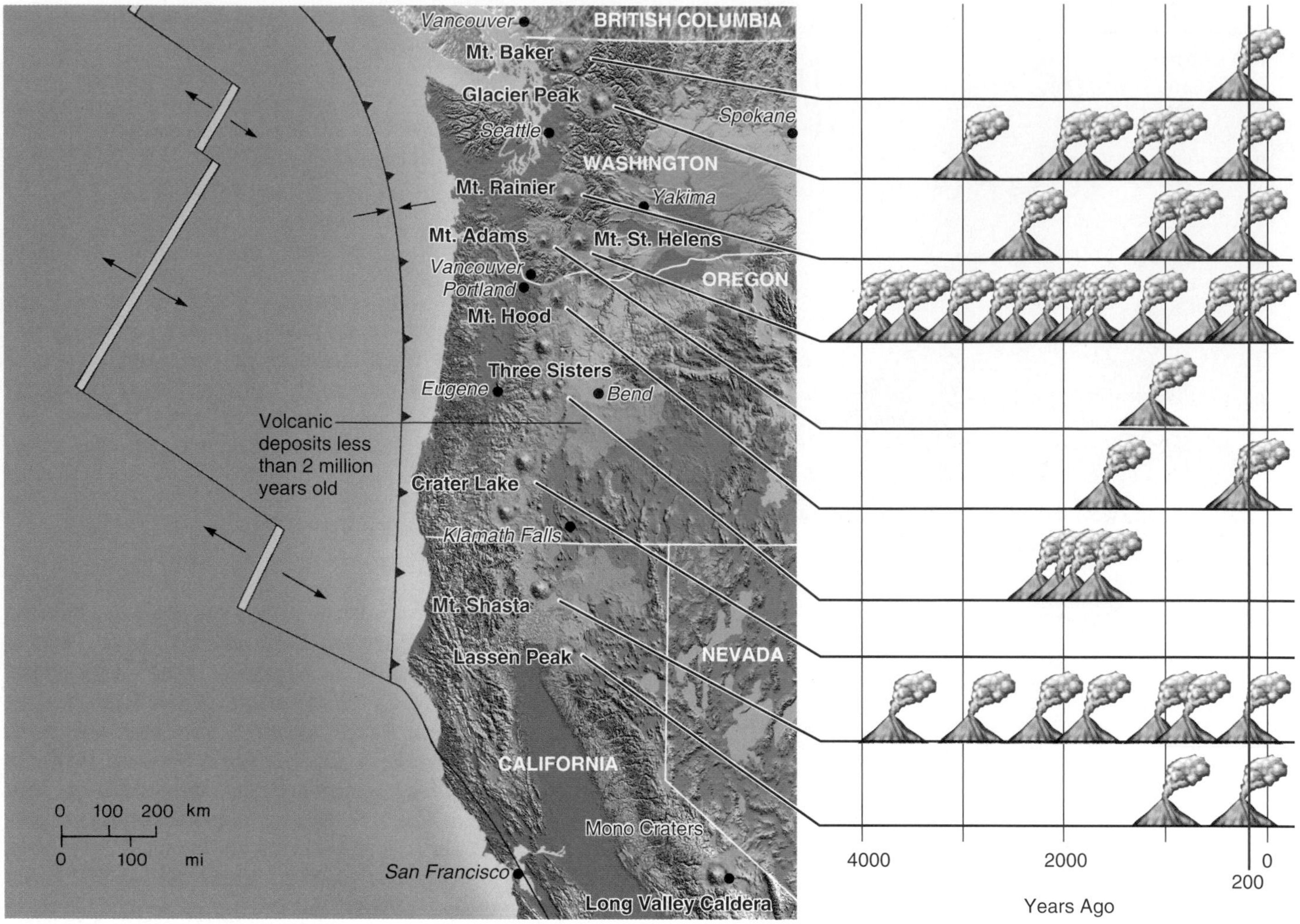

FIGURE 5.29 The Cascade Range volcanoes and their spatial relationship to the subduction zone and to major cities (plate-boundary symbols as in figure 3.6). Shaded area is covered by young volcanic deposits less than 2 million years old.
Eruptive histories from USGS Open-File Report 94–585.

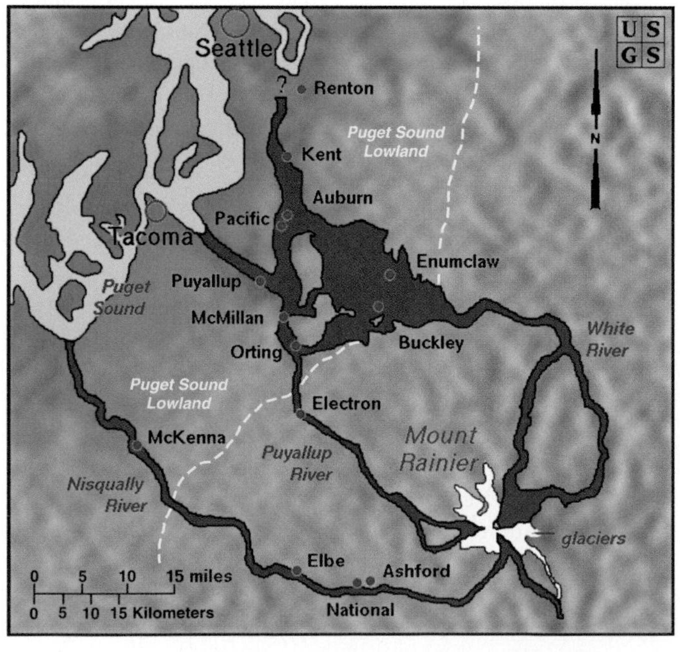

FIGURE 5.30 Over the last 5000 to 6000 years, huge mudflows have poured down stream valleys and into low-lying areas, tens of miles from Mount Rainier's summit. Even Tacoma or Seattle might be threatened by renewed activity on a similar scale.
Source: Data from T. W. Sisson, USGS Open-File Report 95–642

clouds of ash from the erupting Augustine experienced some engine trouble. It was subsequently discovered that volcanoes' andesitic ash melts at temperatures close to the operating temperatures of jet engines. Thereafter, pilots were advised to fly clear of such ash clouds, and a warning system was put in place by the National Oceanic and Atmospheric Administration. Despite this, in 1989 a KLM 747 flew through an ash cloud from Redoubt at 37,000 feet, lost power in all four engines, and plunged over 13,000 feet before pilots were able to restart the engines. This near-disaster, and other incidents worldwide, perhaps ensured that future ash-cloud warnings will be taken more seriously. A remaining problem is that much monitoring of the ash clouds is done by satellite, and ash and ordinary clouds cannot always be distinguished in the images. However,

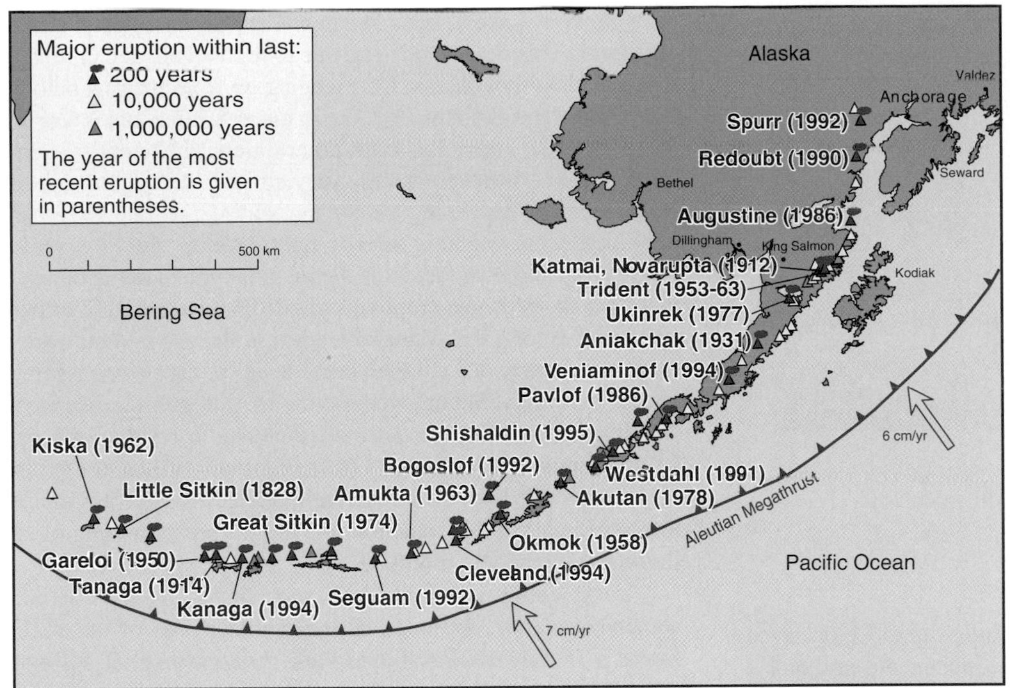

A

B

FIGURE 5.31 The Aleutians are a region of active volcanism—fortunately, a rather sparsely populated one. (A) Map of southwestern Alaska showing major volcanic peaks. (B) Augustine volcano, 2006 eruption; note grey and black lahars and pyroclastic flows, and how far they traveled from the summit.

(A) After map by C. J. Nye, courtesy Alaska Volcano Observatory; (B) Photograph by K. Bull, AVO/Alaska Division of Geology and Geophysics Surveys

volcanologists now coordinate with the National Weather Service to issue forecast maps of airborne ash plumes based on volcanic activity and wind velocity at different altitudes. This is especially important as Anchorage is a key stopover on long-distance flights in the region.

Long Valley and Yellowstone Calderas

Finally, there are areas that do not now constitute an obvious hazard but that are causing some geologists concern because of increases in seismicity or thermal activity, and their past histories. One of these is the Mammoth Lakes area of California. In 1980, the area suddenly began experiencing earthquakes. Within one forty-eight-hour period in May 1980, four earthquakes of magnitude 6 rattled the region, interspersed with hundreds of lesser shocks. Mammoth Lakes lies within the Long Valley Caldera, a 13-kilometer-long oval depression formed during violent pyroclastic eruptions 700,000 years ago. Smaller eruptions occurred around the area as recently as 50,000 years ago.

A **caldera** is an enlarged volcanic crater, which may be formed either by an explosion enlarging an existing crater or by collapse of a volcano after a magma chamber within has emptied. The summit caldera of Mauna Loa (figure 5.8D) has formed predominantly by collapse as the magma chamber has become depleted, and enlarged as the rocks on the rim have weathered and crumbled. The caldera at Crater Lake, shown in the chapter-opener photo, formed by a combination of collapse and explosion (figure 5.32). In the case of a massive explosive eruption such as the one that produced the Long Valley caldera, the original volcanic cone is obliterated altogether. The fear is that any future eruptions may also be on a grand scale.

Geophysical studies have shown that a partly molten mass of magma close to 4 kilometers across currently lies below the Long Valley caldera. Furthermore, since 1975, the center of the caldera has bulged upward more than 25 centimeters (10 inches); at least a portion of the magma appears to be rising in the crust. In 1982, seismologists realized that the patterns of earthquake activity in the caldera were disturbingly similar to those associated with the eruptions of Mount St. Helens. In May 1982, the director of the U.S. Geological Survey issued a formal notice of potential volcanic hazard for the Mammoth Lakes/Long Valley

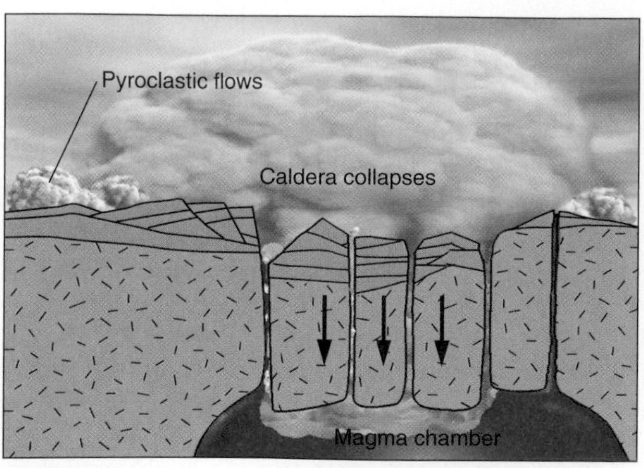

FIGURE 5.32 Formation of a caldera by collapse of rock over the partially emptied magma chamber, with ejection of pyroclastics.
After C. Bacon, U.S. Geological Survey

area. A swarm of earthquakes of magnitudes up to 5.6 occurred there in January of 1983. In 1989, following an earthquake swarm under Mammoth Mountain (just southwest of the town of Mammoth Lakes), trees began to die over an area of about 100 acres. This was found to be due to high levels of CO_2 in the soil, presumably volcanic CO_2 seeping up from magma below. Normal soil gas is about 1% CO_2; in the tree-kill zone, it was up to 95%! So far, there has been no eruption, but scientists continue to monitor developments very closely, seeking to understand what is happening below the surface so that they can anticipate what volcanic activity may develop and how soon. One of the unknown factors is the magma type in question, and therefore the probable eruptive style: Both basalts and rhyolites are found among the young volcanics in the area. Meanwhile, the area has provided a lesson in the need for care when issuing hazard warnings. Media overreaction to, and public misunderstanding of, the 1982 hazard notice created a residue of local hard feelings, many residents believing that tourism and property values had both been hurt, quite unnecessarily. It remains to be seen what public reaction to a more-urgent warning of an imminent threat here might be.

Long Valley is not the only such feature being watched somewhat warily. As noted earlier, another area of uncertain future is Yellowstone National Park. Yellowstone, at present, is notable for its geothermal features—geysers, hot springs, and fumaroles (figure 5.33A). All of these features reflect the presence of hot rocks at shallow depths in the crust below the park. Until recently, it was believed that the heat was just left over from the last cycle of volcanic activity in the area, which began 600,000 years ago. Recent studies, however, suggest that in fact, considerable fluid magma remains beneath the park, which is also still a seismically active area. Yellowstone is actually an intracontinental hot-spot volcano (figure 5.33B). One reason

A

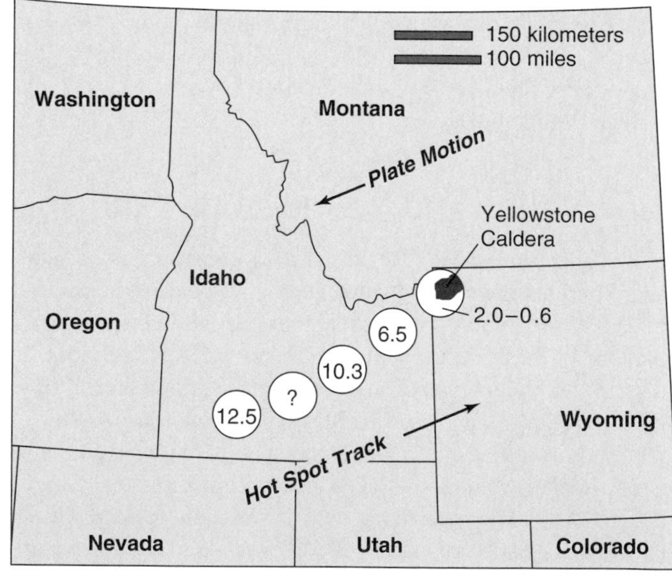

B

FIGURE 5.33 (A) Continuing thermal activity at Yellowstone National Park is extensive, with many boiling-water pools, steaming springs, and geysers. (B) The underlying cause is a hot spot. The track of North America across the Yellowstone hot spot is traced by volcanics; ages (circled) in millions of years.
(B) Data and base map from U.S. Geological Survey

for concern is the scale of past eruptions: In the last cycle, 1000 cubic kilometers (over 200 cubic miles) of pyroclastics were ejected! Indeed, the size of the caldera itself (figure 5.34) suggests that the potential scale of future eruptions is large, and this is confirmed by the volume of pyroclastics ejected in three huge explosive eruptions over the last 2 million years or so (figure 5.35). Eruptions of this scale, with a VEI of 8, lead to the term "supervolcano," which has caught public attention.

Whether eruption on that scale is likely in the foreseeable future, no one knows. Scientists continue to monitor uplift and changes in seismic and thermal activity at Yellowstone. The research goes on.

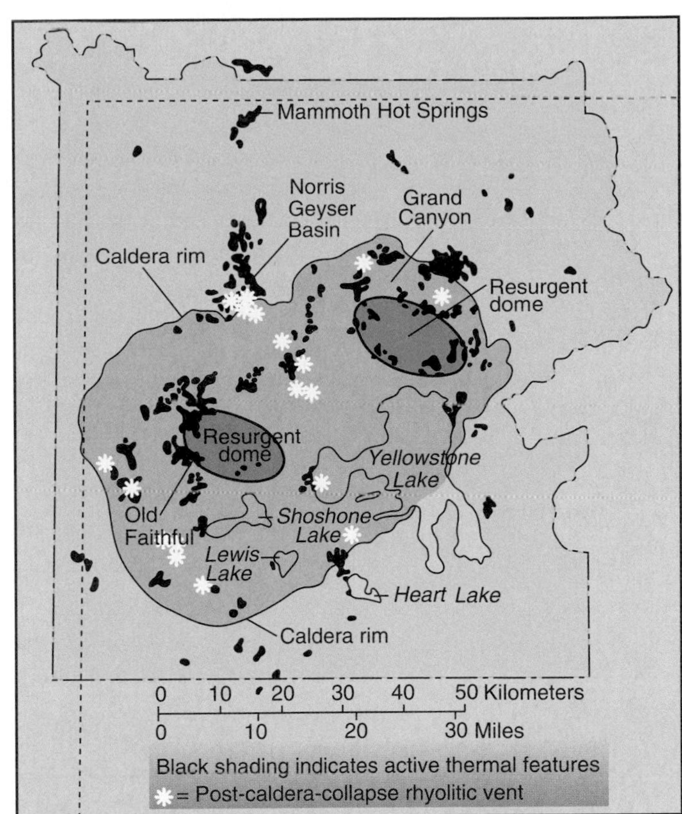

FIGURE 5.34 The size of the caldera (pink area), continuing presence of magma under the "resurgent domes," and scale of past eruptions causes concern about the future of the region.

Used with permission from Yellowstone Magmatic Evolution, *copyright © 1984 by the National Academy of Sciences. Courtesy of the National Academy Press, Washington, D.C.*

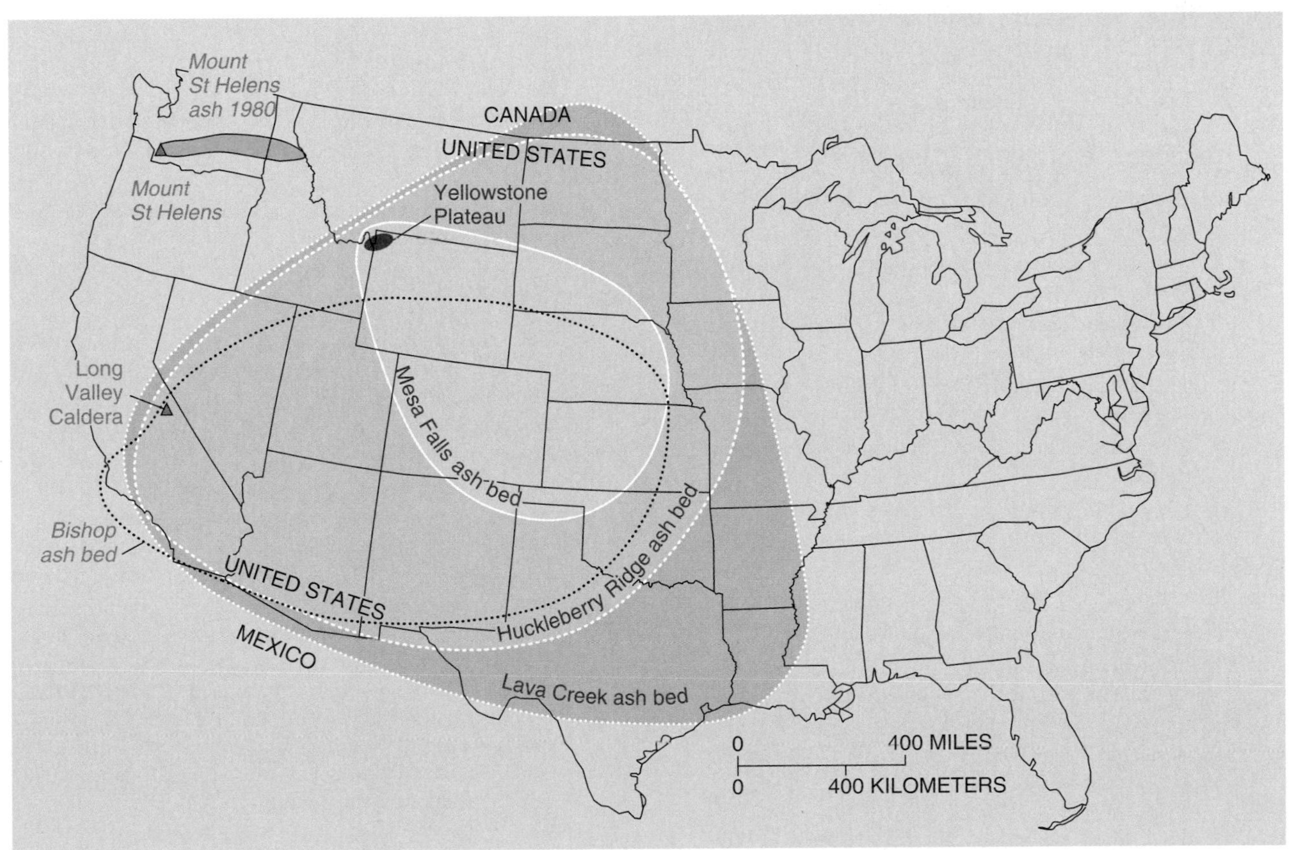

FIGURE 5.35 Major explosive eruptions of Yellowstone caldera produced the Mesa Falls, Huckleberry Ridge, and Lava Creek ash falls. The Lava Creek event was shown at VEI 8 in figure 5.25. Dashed black outline is Bishop ash fall from the Long Valley Caldera in eastern California.

After USGS Fact Sheet 2005–3024

SUMMARY

Most volcanic activity is concentrated near plate boundaries. Volcanoes differ in eruptive style and in the kinds of dangers they present. Those along spreading ridges and at oceanic hot spots tend to be more placid, usually erupting a fluid, basaltic lava. Subduction-zone volcanoes are supplied by more-viscous, silica-rich, gas-charged andesitic or rhyolitic magma, so, in addition to lava, they may emit large quantities of pyroclastics and pyroclastic flows, and they may also erupt explosively. The type of volcanic structure built is directly related to the composition/physical properties of the magma. At present, volcanologists can detect the early signs that a volcano may erupt in the near future (such as bulging, increased seismicity, or increased thermal activity), but they cannot predict the exact timing or type of eruption. Individual volcanoes, however, show characteristic eruptive styles (as a function of magma type) and patterns of activity. Therefore, knowledge of a volcano's eruptive history allows anticipation of the general nature of eruptions and of the likelihood of renewed activity in the near future.

Key Terms and Concepts

active volcano 114
andesite 97
basalt 97
caldera 119
cinder cone 101
composite volcano 101
dormant volcano 114
extinct volcano 114
fissure eruption 98
lahar 105
lava dome 102
phreatic eruption 109
pyroclastic flow 106
pyroclastics 101
rhyolite 97
shield volcano 100
stratovolcano 101
Volcanic Explosivity Index 114

Exercises

Questions for Review

1. Define a fissure eruption, and give an example.
2. The Hawaiian Islands are all shield volcanoes. What are shield volcanoes, and why are they not especially hazardous to life?
3. The eruptive style of Mount St. Helens is quite different from that of Kilauea in Hawaii. Why?
4. What are pyroclastics? Identify a kind of volcanic structure that pyroclastics may build.
5. Describe two strategies for protecting an inhabited area from an advancing lava flow.
6. Describe a way in which a lahar may develop, and a way to avoid its most likely path.
7. What is a nuée ardente, and why is a volcano known for producing these hot pyroclastic flows a special threat during periods of activity?
8. Explain the nature of a phreatic eruption, and give an example.
9. How may volcanic eruptions influence global climate?
10. Discuss the distinctions among active, dormant, and extinct volcanoes, and comment on the limitations of this classification scheme.
11. What is the Volcanic Explosivity Index?
12. Describe two precursor phenomena that may precede volcanic eruptions.
13. What is the underlying cause of present and potential future volcanic activity in the Cascade Range of the western United States?
14. What is the origin of volcanic activity at Yellowstone, and why is it sometimes described as a "supervolcano"?

For Further Thought

1. Before the explosion of Mount St. Helens in May 1980, scientists made predictions about the probable extent of and kinds of damage to be expected from a violent eruption. Investigate those predictions, and compare them with the actual effects of the blast.
2. How near to you is the closest active volcano? The closest dormant one? Are there any ancient lava flows or pyroclastic deposits in your area, and if so, how old are the youngest of these?
3. As this is written, Kilauea continues an active phase that has lasted over two decades. Check out the current activity report of the Hawaii Volcanoes Observatory (see NetNotes). Or, check the latest information of the Cascades Volcano Observatory for the current status of Mount St. Helens, or the Alaska Volcano Observatory for information on Aleutian volcanoes. Or, go exploring other sites for information on other active volcanoes worldwide. Identify the tectonic setting of the volcano(es) and relate to the nature of any current activity.

Suggested Readings/References

Bindeman, I. N. 2006. The secrets of supervolcanoes. *Scientific American* (June): 36–43.

Bruce, V. 2001. *No apparent danger.* New York: HarperCollins. (Describes two deadly volcanic eruptions, of Galeras and Nevado del Ruiz.)

Crockett, J.S., et al. 2005. Mount St. Helens reawakens. *EOS* 86 (18 January): 25, 29.

Decker, R., and Decker, B. 1998. *Volcanoes.* 3d ed. New York: W. H. Freeman.

Francis, P., and Self, S. 1987. Collapsing volcanoes. *Scientific American* 256 (June): 90–97.

Garesche, W. A. 1902. *Complete story of the Martinique and St. Vincent horrors.* L. G. Stahl.

Gore, R. 1984. A prayer for Pozzuoli. *National Geographic* 165: 614–25.

Halbwachs, M., et al. 2004. Degassing the "killer lakes" Nyos and Monoun, Cameroon. *EOS* 85 (27 July) 281, 285.

Hill, D. P., Pollitz, F., and Newhall, C. 2002. Earthquake-volcano interactions. *Physics Today* (November): 41–47.

Krakauer, J. 1996. Mount Rainier: Washington's tranquil threat. *Smithsonian* (July): 32–41.

Neal, C. A., et al. 1997. Volcanic ash—Danger to aircraft in the North Pacific. USGS Fact Sheet 030–97.

Perkins, S. 2003. Danger in the air. *Science News* 164 (13 September): 168–70.

Pfeiffer, T. 2003. Mount Etna's ferocious future. *Scientific American* 288 (April): 58–65.

Pinatubo Volcano Observatory Team. 1991. Lessons from a major eruption: Mt. Pinatubo, Philippines. *EOS, Trans. American Geophysical Union* 72 (3 December): 545–55.

Pinna, M. 2002. Mt. Etna ignites. *National Geographic* (February): 68–87.

Rampino, M. R., and Self, S. 1984. The atmospheric effects of El Chichón. *Scientific American* 250 (January): 48–57.

Sever, M. 2006. Italy's hidden hazard. *Geotimes* (April): 36–37.

Smith, E. I., Keenan, D. L., and Plank, T. 2002. Episodic volcanism and hot mantle: Implications for volcanic hazard studies at the proposed nuclear waste repository at Yucca Mountain, Nevada. *GSA Today* (April): 4–10.

"Stalking nature's most dangerous beasts," a special section of a half-dozen articles on volcanic hazards. 2003. *Science* 299 (28 March): 2015–30.

Stewart, D. 2006. Resurrecting Pompeii. *Smithsonian* (February): 60–68.

Stone, R. 2004. Iceland's doomsday scenario? *Science* 306 (19 November): 1278–81.

Williams, A. R. 1997. Under the volcano: Montserrat. *National Geographic* (July): 59–75.

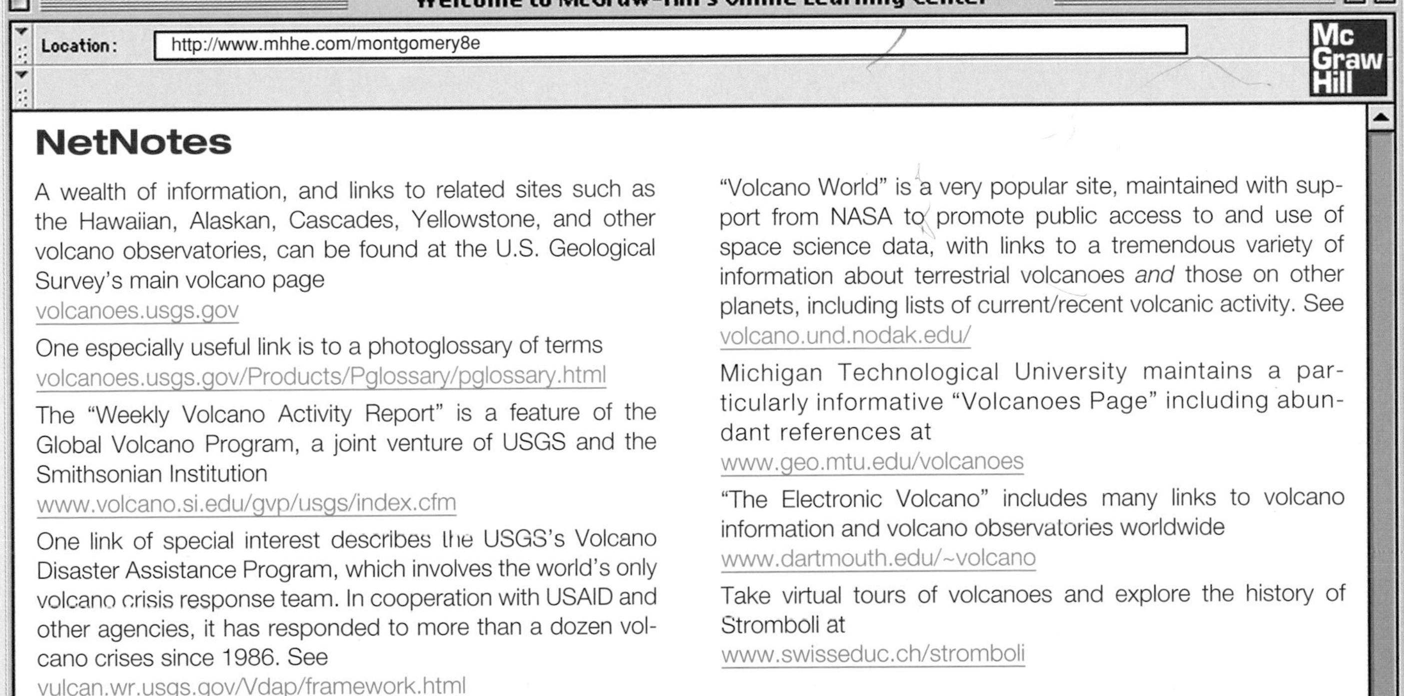

NetNotes

A wealth of information, and links to related sites such as the Hawaiian, Alaskan, Cascades, Yellowstone, and other volcano observatories, can be found at the U.S. Geological Survey's main volcano page
volcanoes.usgs.gov

One especially useful link is to a photoglossary of terms
volcanoes.usgs.gov/Products/Pglossary/pglossary.html

The "Weekly Volcano Activity Report" is a feature of the Global Volcano Program, a joint venture of USGS and the Smithsonian Institution
www.volcano.si.edu/gvp/usgs/index.cfm

One link of special interest describes the USGS's Volcano Disaster Assistance Program, which involves the world's only volcano crisis response team. In cooperation with USAID and other agencies, it has responded to more than a dozen volcano crises since 1986. See
vulcan.wr.usgs.gov/Vdap/framework.html

"Volcano World" is a very popular site, maintained with support from NASA to promote public access to and use of space science data, with links to a tremendous variety of information about terrestrial volcanoes *and* those on other planets, including lists of current/recent volcanic activity. See
volcano.und.nodak.edu/

Michigan Technological University maintains a particularly informative "Volcanoes Page" including abundant references at
www.geo.mtu.edu/volcanoes

"The Electronic Volcano" includes many links to volcano information and volcano observatories worldwide
www.dartmouth.edu/~volcano

Take virtual tours of volcanoes and explore the history of Stromboli at
www.swisseduc.ch/stromboli

A wealth of volcano information, imagery, and eruption accounts, including information on extraterrestrial volcanoes, is at
www.geology.sdsu.edu/how_volcanoes_work

The U.S. Geological Survey now makes many reports and fact sheets available online, including:

"Monitoring ground deformation from space" (2005)
pubs.usgs.gov/fs/2005/3025/2005–3025.pdf

"Mount Mazama and Crater Lake: Growth and destruction of a Cascade volcano" (2002)
pubs.usgs.gov/fs/2002/fs092-02/

"Mount Rainier — Learning to live with volcanic risk" (2002)
pubs.usgs.gov/fs/2002/fs034-02/

"Mount St. Helens erupts again" (2005)
vulcan.wr.usgs.gov/Volcanoes/MSH/Publications/FS2005-3036/FS2005-3036.html

"Steam explosions, earthquakes, and volcanic eruptions—What's next in Yellowstone's future?" (2005)
pubs.usgs.gov/fs/2005/3024/

"Tracking changes in Yellowstone's restless volcanic system" (2004)
pubs.usgs.gov/fs/fs100-03/

SECTION THREE

SURFACE PROCESSES

Many areas are threatened, or at least affected, by the hazards associated with the internal geologic processes discussed in section 2. Every place on earth is subject to one or more of the various kinds of *surface processes* to be examined in this section. These are the geologic processes with causes and effects at or near earth's surface. They involve principally the actions of water, ice, wind, and gravity. Local climate plays a major role in determining the relative importance of these processes. This is one reason for interest in possible future changes in global and local climate.

The earth's surface is, geologically, a very active place. It is here that we see the interplay of the internal heat of the earth (which builds mountains and shifts the land), the external heat from the sun (which drives the wind and provides the energy behind the hydrologic cycle), and the inexorable force of gravity (which constantly tries to pull everything down to the same level).

Chapters 6–9 explore several kinds of surface processes: stream-related processes, including flooding; coastal processes of erosion and mass transport; mass movements, the downhill march of material; and the effects of ice and wind in sculpturing the land. Chapter 10 examines some relationships among these various processes and climate, and considers possible long-term climate trends and consequences.

Wind and water shape topography at coastline of Africa's Namib Desert. Some dunes reach 300 meters (nearly 1000 feet) in height.
Image courtesy U.S. Geological Survey EROS Data Center.

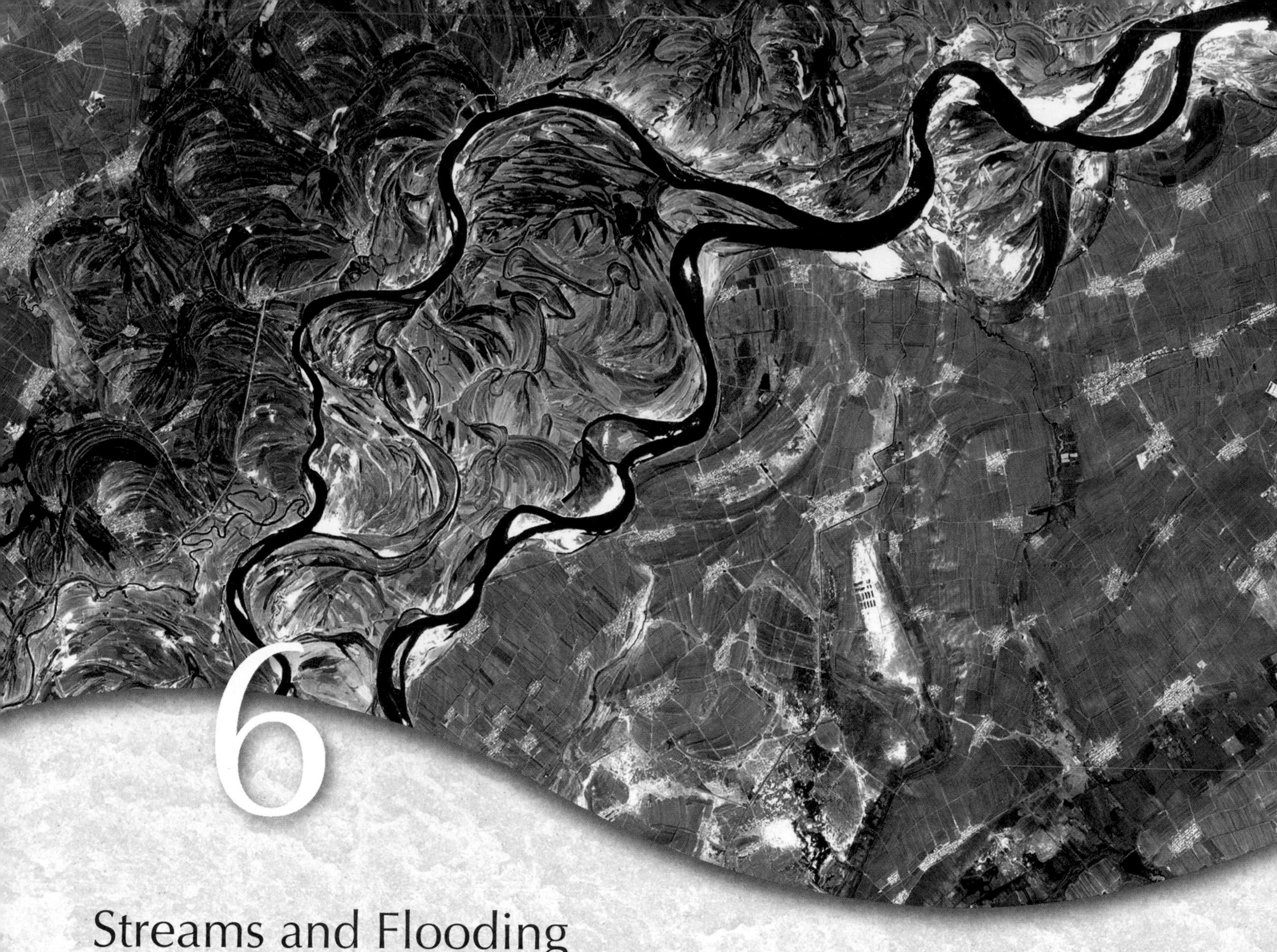

6 Streams and Flooding

Water is the single most important agent sculpturing the earth's surface. Mountains may be raised by the action of plate tectonics and volcanism, but they are shaped primarily by water. Streams carve valleys, level plains, and move tremendous amounts of sediment from place to place.

They have also long played a role in human affairs. Throughout human history, streams have served as a vital source of fresh water, and often of fish for food. Before there were airplanes, trucks, or trains, there were boats to carry people and goods over water. Before the widespread use of fossil fuels, flowing water pushing paddlewheels powered mills and factories. For these and other reasons, many towns sprang up and grew along streams. This, in turn, has put people in the path of floods when the streams overflow.

Floods are probably the most widely experienced catastrophic geologic hazards. On the average, in the United States alone, floods annually take nearly 100 lives and cause well over $1 billion in property damage. The 1993 flooding in the Mississippi River basin took 48 lives and caused an estimated $15–20 billion in damages.

Most floods do not make national headlines, but they may be no less devastating to those affected by them. Some floods are the result of unusual events, such as the collapse of a dam, but the vast majority are a perfectly normal, and to some extent predictable, part of the natural functioning of streams. Before discussing flood hazards, therefore, we will examine how water moves through the hydrologic cycle and also look at the basic characteristics and behavior of streams.

China's Songhua River meanders across the Manchurian plain. Abandoned meanders, some water-filled, make oxbows and oxbow lakes.
Photograph courtesy NASA/GSFC/METI/ERSDAC/JAROS and the U.S./Japan ASTER science team

THE HYDROLOGIC CYCLE

The **hydrosphere** includes all the water at and near the surface of the earth. Most of it is believed to have been outgassed from the earth's interior early in its history, when the earth's temperature was higher; icy planetesimals added late to the accreting earth's surface may have contributed some, too. Now, except for occasional minor additions from volcanoes bringing up "new" water from the mantle, and small amounts of water returned to the mantle with subducted lithosphere, the quantity of water in the hydrosphere remains essentially constant.

All of the water in the hydrosphere is caught up in the **hydrologic cycle,** illustrated in figure 6.1. The largest single reservoir in the hydrologic cycle, by far, consists of the world's oceans, which contain over 97% of the water in the hydrosphere; lakes and streams together contain only 0.016% of the water. The main processes of the hydrologic cycle involve evaporation into and precipitation out of the atmosphere. Precipitation onto land can reevaporate (directly from the ground surface or indirectly through plants by evapotranspiration), infiltrate into the ground, or run off over the ground surface. Surface runoff may occur in streams or by unchanneled overland flow. Water that percolates into the ground may also flow (see chapter 11) and commonly returns, in time, to the oceans. The oceans are the principal source of evaporated water because of their vast areas of exposed water surface.

The total amount of water moving through the hydrologic cycle is large, more than 100 million billion gallons per year. A portion of this water is temporarily diverted for human use, but it ultimately makes its way back into the natural global water cycle by a variety of routes, including release of municipal sewage, evaporation from irrigated fields, or discharge of

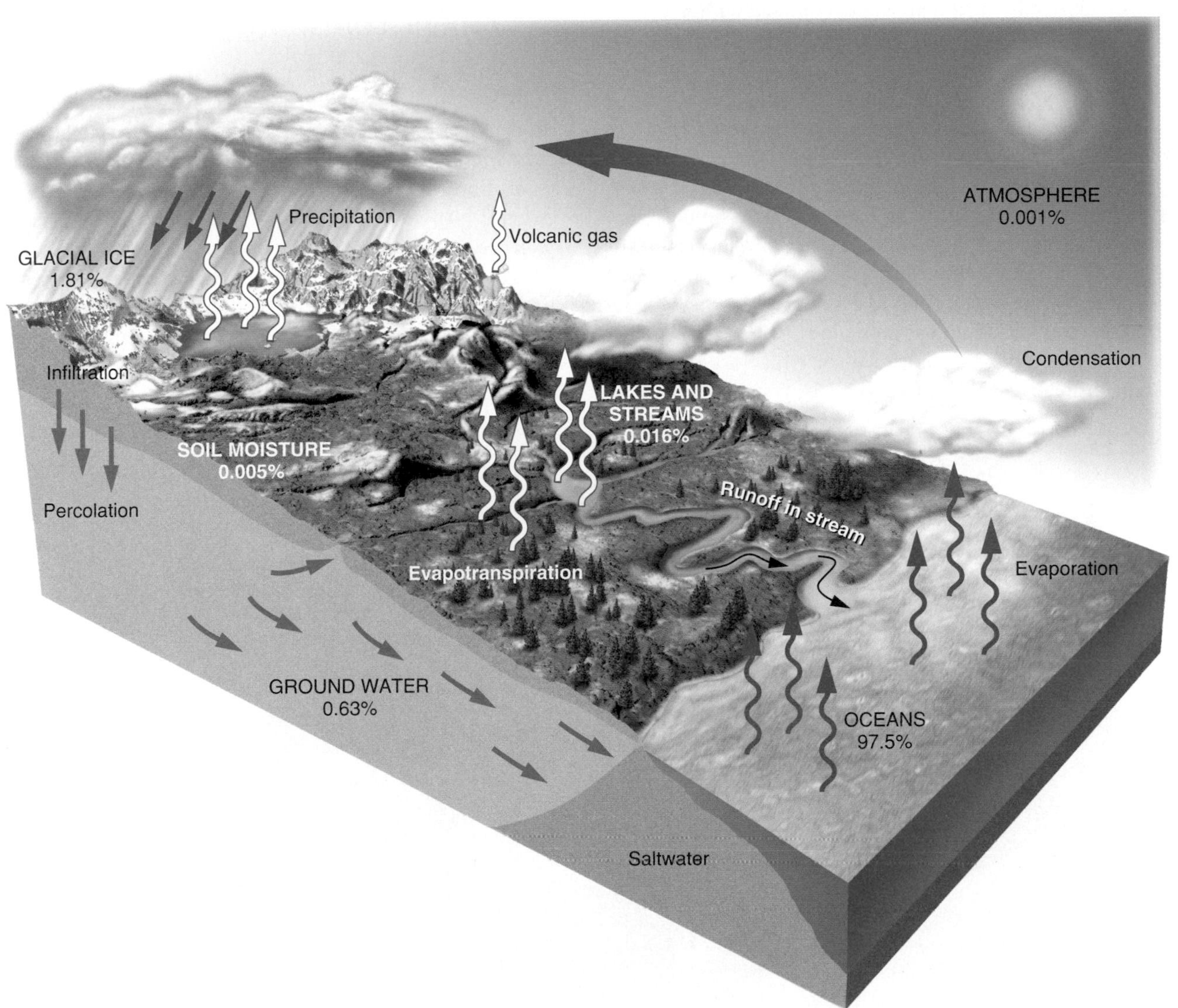

FIGURE 6.1 Principal processes and reservoirs of the hydrologic cycle (reservoirs in capital letters). Straight lines show movement of liquid water; wavy lines show water vapor. (Limited space prevents showing all processes everywhere they occur.)

industrial wastewater into streams. Water in the hydrosphere may spend extended periods of time—even tens of thousands of years—in storage in one or another of the water reservoirs shown in figure 6.1—a glacier, an ocean, ground water—but from the longer perspective of geologic history, it is still regarded as moving continually through the hydrologic cycle. In the process, it may participate in other geologic cycles and processes: for example, streams eroding rock and moving sediment, and subsurface waters dissolving rock and transporting dissolved chemicals, are also contributing to the rock cycle. We revisit the hydrologic cycle in the context of water resources in chapter 11.

STREAMS AND THEIR FEATURES

Streams—General

A **stream** is any body of flowing water confined within a channel, regardless of size, although people tend, informally, to use the term *river* for a relatively large stream. It flows downhill through local topographic lows, carrying away water over the earth's surface. The region from which a stream draws water is its **drainage basin** (figure 6.2), or *watershed.* A *divide* separates drainage basins. The size of a stream at any point is related in part to the size (area) of the drainage basin upstream from that point, which determines how much water from falling snow or rain can be collected into the stream. Stream size is also influenced by several other factors, including climate (amount of precipitation and evaporation), vegetation or lack of it, and the underlying geology, all of which are explained in more detail later in the chapter. That is, a stream carves its channel (in the absence of human intervention), and the channel carved is broadly proportional to the volume of water that must typically be accommodated. Long-term, sustained changes in precipitation, land use, or other factors that change the volume of water customarily flowing in the stream will be reflected in corresponding changes in channel geometry.

The size of a stream may be described by its **discharge,** the volume of water flowing past a given point (or, more precisely, through a given cross section) in a specified length of time. Discharge is the product of channel cross section (area) times average stream velocity (figure 6.3). Historically, the conventional unit used to express discharge is cubic feet per second; the analogous metric unit is cubic meters per second. Discharge may range from less than one cubic foot per second on a small creek to millions of cubic feet per second in a major river. For a given stream, discharge will vary with season and weather, and may be influenced by human activities as well.

Sediment Transport

Water is a powerful agent for transporting material. Streams can move material in several ways. Heavier debris may be rolled, dragged, or pushed along the bottom of the stream bed as its *traction load.* Material of intermediate size may be carried in short hops along the stream bed by a process called *saltation* (figure 6.4). All this material is collectively described as the *bed load* of the stream. The *suspended load* consists of material that is light or fine enough to be moved along suspended in the stream, supported by the flowing water. Suspended sediment clouds a stream and gives the water a muddy appearance. Finally, some substances may be completely dissolved in the water, to make up the stream's *dissolved load.*

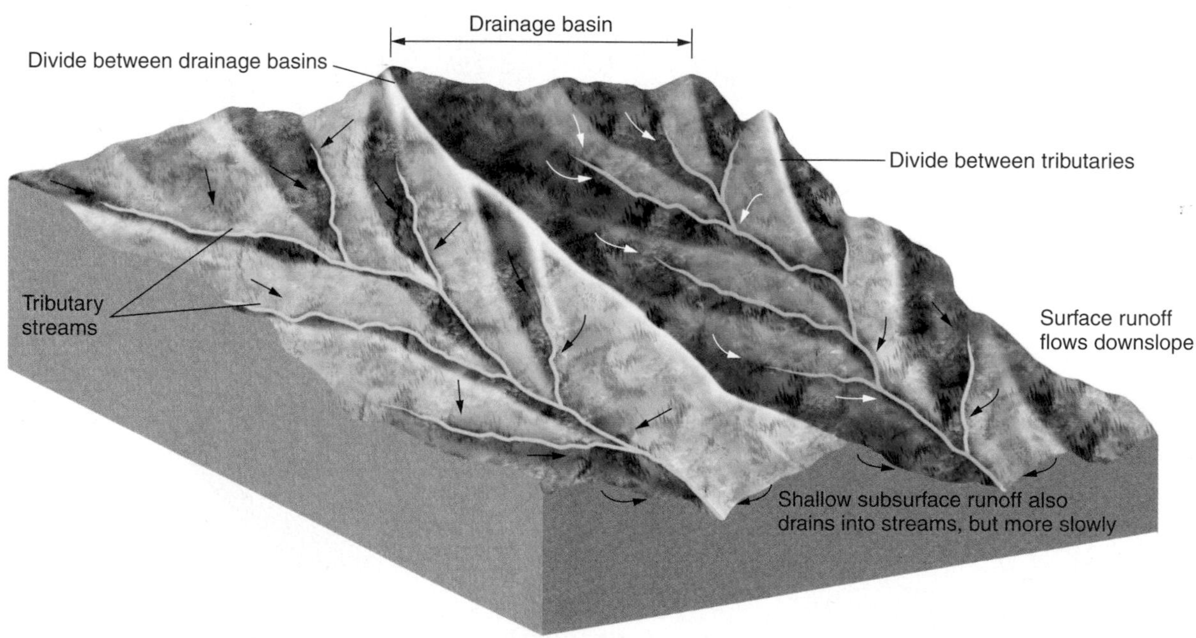

FIGURE 6.2 Streams and their drainage basins.

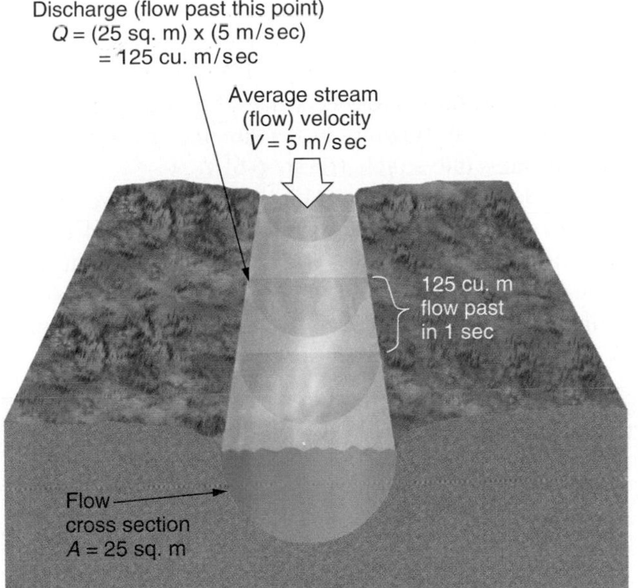

FIGURE 6.3 Discharge (Q) equals channel cross section (area) times average velocity: $Q = A \times V$. In a real stream, both channel shape and flow velocity vary, so reported discharge is necessarily an average. In any cross section, velocity near the center of the channel will be greater than along bottom and sides of the channel.

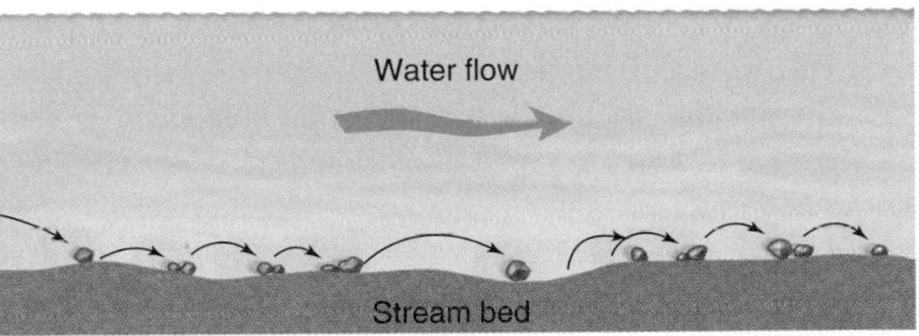

FIGURE 6.4 Schematic of saltation, one process of sediment transport. Particles move in short jumps, jarred loose by water or other particles and briefly carried in the stream flow.

The total quantity of material that a stream transports by all these methods is called, simply, its **load.** Stream **capacity** is a measure of the total load of material a stream can move. Capacity is closely related to discharge: the faster the water flows, and the more water is present, the more material can be moved. How much of a load is actually transported also depends on the availability of sediments or soluble material: a stream flowing over solid bedrock will not be able to dislodge much material, while a similar stream flowing through sand or soil may move considerable material. Vegetation can also influence sediment transport, by preventing the sediment from reaching the stream at all, or by blocking its movement within the stream channel.

Velocity, Gradient, and Base Level

Stream velocity is related partly to discharge and partly to the steepness (pitch or angle) of the slope down which the stream flows. The steepness of the stream channel is called its **gradient.** It is calculated as the difference in elevation between two points along a stream, divided by the horizontal distance between them along the stream channel (the water's flow path). All else being equal, the higher the gradient, the steeper the channel (by definition), and the faster the stream flows. Gradient and velocity commonly vary along the length of a stream (figure 6.5). Nearer its source, where the first perceptible trickle of water signals the

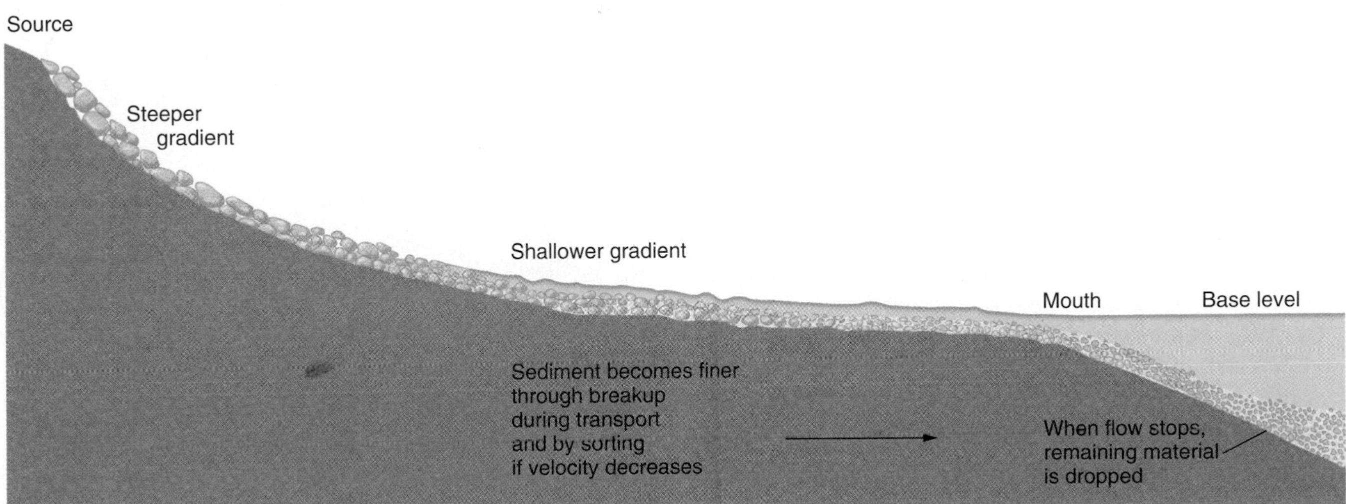

FIGURE 6.5 Typical longitudinal profile of a stream in a temperate climate. Note that the gradient decreases from source to mouth. Particle sizes may decrease with breakup downstream. Total discharge, capacity, and velocity may all increase downstream despite the decrease in gradient (though commonly, velocity decreases with decreasing gradient). Conversely, in a dry climate, water loss by evaporation and infiltration into the ground may *decrease* discharge downstream.

stream's existence, the gradient is usually steeper, and it tends to decrease downstream. Velocity may or may not decrease correspondingly; the effect of decreasing gradient may be counteracted by other factors, including increased water volume as additional tributaries enter the stream, friction between water and stream bed, and changes in the channel's width and depth (increasing cross section), but overall, discharge does tend to increase downstream, at least in moist climates.

By the time the stream reaches its end, or mouth, which is usually where it flows into another body of water, the gradient is typically quite low. Near its mouth, the stream is approaching its **base level,** which is the lowest elevation to which the stream can erode downward. For most streams, base level is the water (surface) level of the body of water into which they flow. For streams flowing into the ocean, for instance, base level is sea level. The closer a stream is to its base level, the lower the stream's gradient. It may flow more slowly in consequence, although the slowing effect of decreased gradient may be compensated by an increase in discharge due to widening of the channel and a corresponding increase in channel cross section. The downward pull of gravity causes a stream to cut down toward its base level. Counteracting this erosion is the influx of fresh sediment into the stream from the drainage basin. Over time, natural streams tend toward a balance, or equilibrium, between erosion and deposition of sediments. The **longitudinal profile** of such a stream (a sketch of the stream's elevation from source to mouth) assumes a characteristic concave-upward shape (see figure 6.5).

Velocity and Sediment Sorting and Deposition

Variations in a stream's velocity along its length are reflected in the sediments deposited at different points. The more rapidly a stream flows, the larger and denser are the particles moved. The sediments found motionless in a stream bed at any point are those too big or heavy for that stream to move at that point. Where the stream flows most quickly, it carries gravel and even boulders along with the finer sediments. As the stream slows down, it starts dropping the heaviest, largest particles—the boulders and gravel—and continues to move the lighter, finer materials along. If stream velocity continues to decrease, successively smaller particles are dropped: the sand-sized particles next, then the clay-sized ones. In a very slowly flowing stream, only the finest sediments and dissolved materials are still being carried. If a stream flows into a body of standing water, like a lake or ocean, the stream's flow velocity drops to zero, and all the remaining suspended sediment is dropped.

The relationship between the velocity of water flow and the size of particles moved accounts for one characteristic of stream-deposited sediments: They are commonly **well sorted** by size or density, with materials deposited at a given point tending to be similar in size or weight. If the stream is still carrying a substantial load as it reaches its mouth, and it then flows into still waters, a large, fan-shaped pile of sediment, a **delta,** may be built up (figure 6.6A). A similarly shaped feature, an **alluvial fan,** is formed when a tributary stream flows into a more slowly flowing, larger stream, or a stream flows from mountains into a plain (figure 6.6B).

An additional factor controlling the particle size of stream sediments is physical breakup and/or dissolution of the sediments during transport. That is, the farther the sediments travel, the longer they are subjected to collision and dissolution, and the finer they tend to become. Stream-transported sediments

A

B

FIGURE 6.6 (A) The Mississippi River delta (satellite photo). The river carries a million tons of sediment each day. Here, the suspended sediment shows as a milky, clouded area in the Gulf of Mexico. The sediment deposition gradually built Louisiana outward into the gulf. Now, upstream dams have cut off sediment influx, and the delta is eroding in many places. (B) Overlapping alluvial fans formed as mountain streams flow out into a dry plain near Albuquerque, New Mexico.

(A) Photo courtesy NASA

may thus tend, overall, to become finer downstream, whether or not the stream's velocity changes along its length.

Channel and Floodplain Evolution

When a stream is flowing rapidly and its gradient is steep, downcutting is rapid. The result is typically a steep-sided valley, V-shaped in cross section, and relatively straight (figure 6.7). But streams do not ordinarily flow in straight lines for very long. Small irregularities in the channel cause local fluctuations in velocity, which result in a little erosion where the water flows strongly against the side of the channel and some deposition of sediment where it slows down a bit. Bends, or **meanders,** thus begin to form in the stream. Once a meander forms, it tends to enlarge and also to shift downstream. It is eroded on the outside and downstream side of the meander, the **cut bank,** where the water flows somewhat faster (and the channel, too, tends to be a little deeper there as a result); **point bars,** consisting of sediment deposited on the insides of meanders, build out the banks in those parts of the channel (figure 6.8). The rates of lateral movement of meanders can range up to tens or even hundreds of meters per year, although rates below 10 meters/year (35 feet/year) are more common on smaller streams.

Obstacles or irregularities in the channel may slow flow enough to cause localized sediment deposition there. If the sediment load of the stream is large, these channel islands can build up until they reach the surface, effectively dividing the channel in a process called *braiding.* If the sediment load is very large in relation to water volume, the **braided stream** may develop a complex pattern of many channels that divide and rejoin, shifting across a broad expanse of sediment (figure 6.9).

FIGURE 6.7 Classic V-shaped valley carved by a rapidly downcutting stream. Grand Canyon of the Yellowstone River.

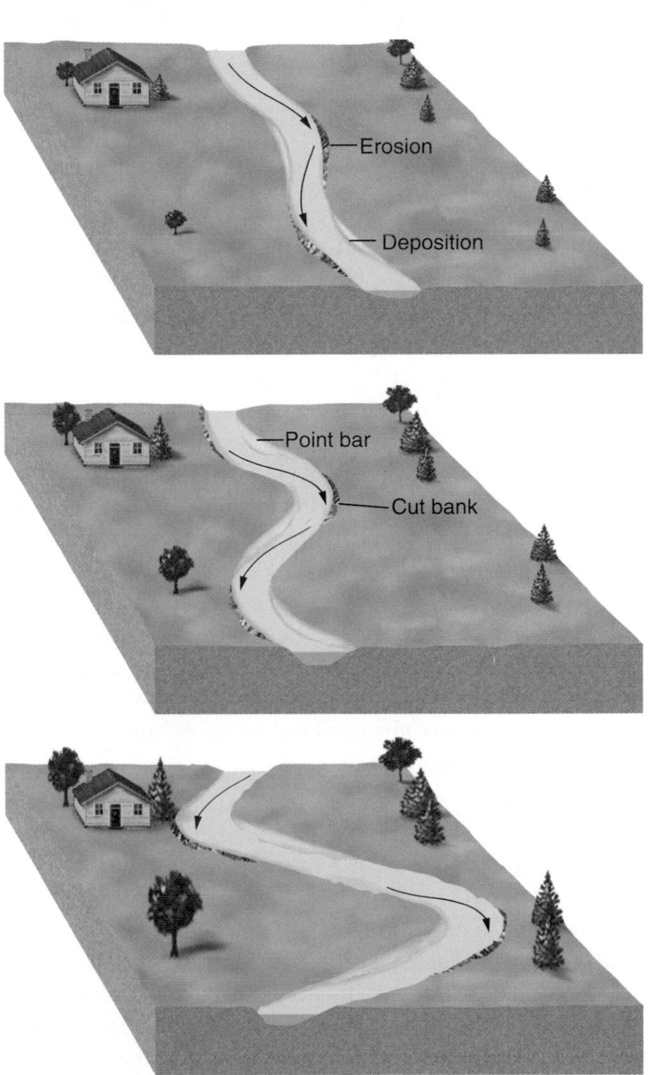

FIGURE 6.8 Development of meanders. Channel erosion is greatest at the outside and on the downstream side of curves, at the cut bank; deposition of point bars occurs in sheltered areas on the inside of curves. In time, meanders migrate both laterally and downstream and can enlarge as well, as shown here. See also figure 6.10.

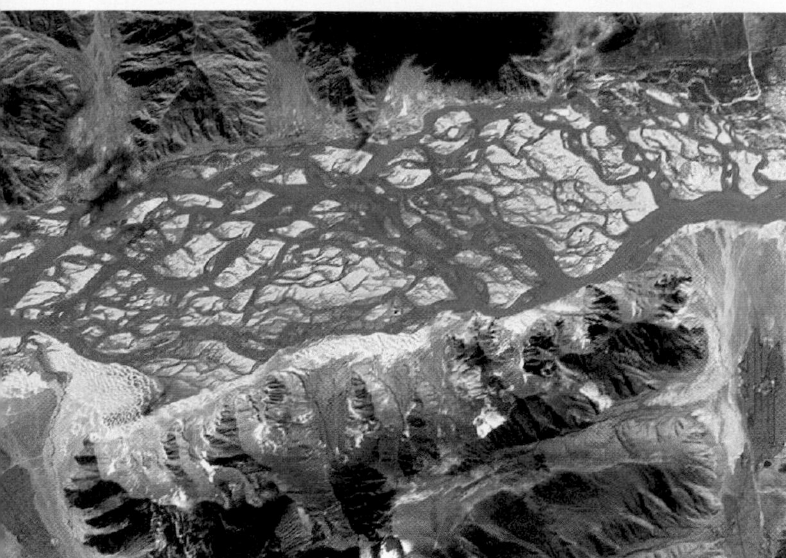

FIGURE 6.9 Braided streams have many channels dividing and rejoining. Brahmaputra River, Tibet.

Image courtesy of Earth Science and Image Analysis Laboratory, NASA Johnson Space Center

Over a period of time, the combined effects of erosion on the cut banks and deposition of point bars on the inside banks of meanders, downstream meander migration, and sediment deposition when the stream overflows its banks together produce a broad, fairly flat expanse of land covered with sediment around the stream channel proper. This is the stream's **floodplain,** the area into which the stream spills over during floods. The floodplain, then, is a normal product of stream evolution over time (see figures 6.10 and 6.11).

Meanders do not broaden or enlarge indefinitely. Very large meanders represent major detours for the flowing stream. At times of higher discharge, especially during floods, the stream may make a shortcut, or cut off a meander, abandoning the old, twisted channel for a more direct downstream route. The cutoff meanders are called **oxbows.** These abandoned channels may be left dry, or they may be filled with standing water, making *oxbow lakes.* Some oxbows and oxbow lakes are shown in the chapter-opener photo.

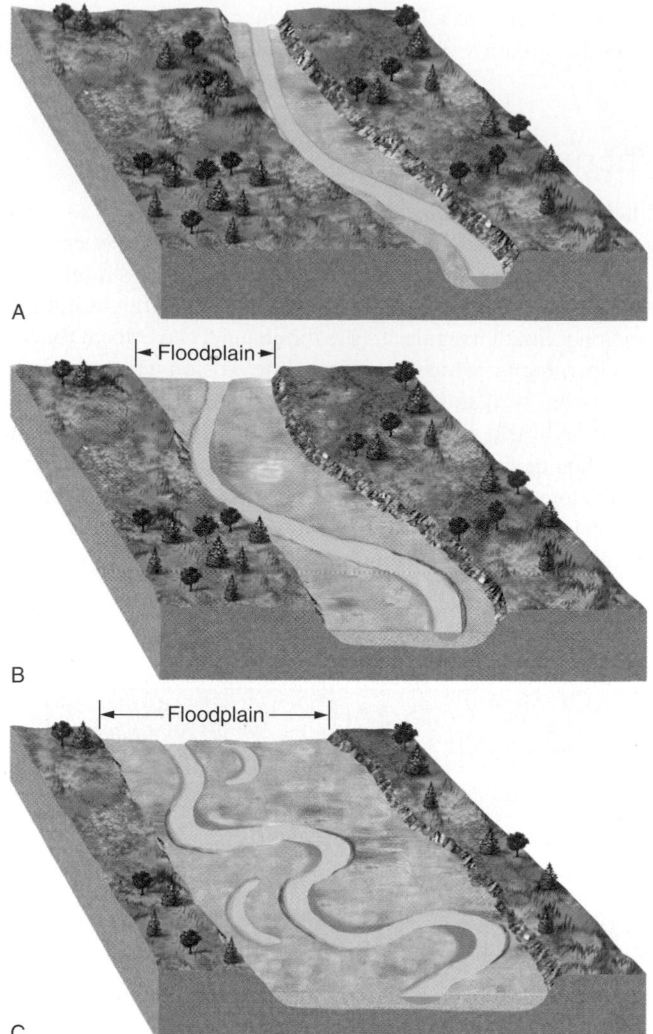

FIGURE 6.10 Meandering and sediment deposition during floods contribute to development of a floodplain. (A) Initially, the stream channel is relatively straight. (B) Small bends enlarge and migrate over time. (C) Broad, flat floodplain is developed around stream channel.

FLOODING

The size of an unmodified stream channel is directly related to the quantity of water that it usually carries. That is, the volume of the channel is approximately sufficient to accommodate the average maximum discharge reached each year. Much of the year, the surface of the water is well below the level of the stream banks. In times of higher discharge, the stream may overflow its banks, or **flood.** This may happen, to a limited extent, as frequently as every two to three years with streams in humid regions. More-severe floods occur correspondingly less often.

The vast majority of stream floods are linked to precipitation (rain or snow). When rain falls or snow melts, some of the water sinks into the ground (**infiltration**). Below ground, **percolation** moves it through soil and rock. Some of the water left on the surface evaporates directly back into the atmosphere. The rest of the water becomes surface runoff, flowing downhill over the surface under the influence of gravity into a puddle, lake, ocean, or stream.

Factors Governing Flood Severity

Many factors together determine whether a flood will occur. The quantity of water involved and the rate at which it enters the stream system are the major factors. When the water input

FIGURE 6.11 Meandering stream in the Alaskan interior.

The McGraw-Hill Companies, Inc./John A. Karachewski, photographer

exceeds the capacity of the stream to carry that water away downstream within its channel, the water overflows the banks. Worldwide, the most intense rainfall events occur in Southeast Asia, where storms have drenched the region with up to 200 centimeters (80 inches) of rain in less than three days. (To put such numbers in perspective, that amount of rain is more than double the average *annual* rainfall for the United States!) In the United States, several regions are especially prone to heavy rainfall events: the southern states, vulnerable to storms from the Gulf of Mexico; the western coastal states, subject to prolonged storms from the Pacific Ocean; and the midcontinent states, where hot, moist air from the Gulf of Mexico can collide with cold air sweeping down from Canada. Streams that drain the Rocky Mountains are likely to flood during snowmelt, especially when rapid spring thawing follows a winter of unusually heavy snow. Even areas that usually have only moderate precipitation can occasionally experience intense storms. A period of prolonged heavy rains in December 1996 and January 1997 was primarily responsible for severe flooding in the Pacific Northwest, California, and Nevada. Rapid snow melt in the upper Midwest in 1997 choked the Red River, flooding Grand Forks, North Dakota, and other cities in the United States and Canada.

The rate of surface runoff is influenced by the extent of infiltration, which, in turn, is controlled by the soil type and how much soil is exposed. Soils, like rocks, vary in porosity and permeability. A very porous and permeable soil allows a great deal of water to sink in relatively fast. If the soil is less permeable or is covered by artificial structures, the proportion of water that runs off over the surface increases. Once even permeable soil is saturated with water, any additional moisture is necessarily forced to become part of the surface runoff.

Topography also influences the extent or rate of surface runoff: The steeper the terrain, the more readily water runs off over the surface and the less it tends to sink into the soil. Water that infiltrates the soil, like surface runoff, tends to flow downgradient, and may, in time, also reach the stream. However, the subsurface runoff water, flowing through soil or rock, generally moves much more slowly than the surface runoff. The more gradually the water reaches the stream, the better the chances that the stream discharge will be adequate to carry the water away without flooding. Therefore, the relative amounts of surface runoff and subsurface (groundwater) flow, which are strongly influenced by the near-surface geology of the drainage basin, are fundamental factors affecting the severity of stream flooding.

Vegetation may reduce flood hazards in several ways. The plants may simply provide a physical barrier to surface runoff, decreasing its velocity and thus slowing the rate at which water reaches a stream. Plant roots working into the soil loosen it, which tends to maintain or increase the soil's permeability, and hence infiltration, thus reducing the proportion of surface runoff. Plants also absorb water, using some of it to grow and releasing some slowly by evapotranspiration from foliage. All of these factors reduce the volume of water introduced directly into a stream system.

All other factors being equal, local flood hazard may vary seasonally or as a result of meteorologic fluctuations. Just as soil already saturated from previous storms cannot readily absorb more water, so the solidly frozen ground of cold regions prevents infiltration; a midwinter rainstorm in such a region may produce flooding with a quantity of rain that could be readily absorbed by the soil in summer. The extent and vigor of growth of vegetation varies seasonally also, as does atmospheric humidity and thus evapotranspiration.

Flood Characteristics

During a flood, the water level of a stream is higher than usual, and its velocity and discharge also increase as the greater mass of water is pulled downstream by gravity. The higher volume (mass) and velocity together produce the increased force that gives floodwaters their destructive power (figure 6.12). The elevation of the water surface at any point is termed the **stage** of the stream. A stream is at *flood stage* when stream stage exceeds bank height. The magnitude of a flood can be described by either the maximum discharge or maximum stage reached. The stream is said to **crest** when the maximum stage is reached. This may occur within minutes of the influx of water, as with flooding just below a failed dam. However, in places far downstream from the water input or where surface runoff has been slowed, the stream may not crest for several days after the flood episode begins. In other words, just because the rain has stopped does not mean that the worst is over, as was made abundantly evident in the 1993 Midwestern flooding.

FIGURE 6.12 The heightened erosion by fast-flowing floodwaters can threaten structures. Bridge scour on Rio Puerco, New Mexico.
Photograph by Steven D. Craigg, U.S. Geological Survey

Flooding may afflict only a few kilometers along a small stream, or a region the size of the Mississippi River drainage basin, depending on how widely distributed the excess water is. Floods that affect only small, localized areas (or streams draining small basins) are sometimes called **upstream floods.** These are most often caused by sudden, locally intense rainstorms and by events like dam failure. Even if the total amount of water involved is moderate, the rapidity with which it enters the stream can cause it temporarily to exceed the stream channel capacity. The resultant flood is typically brief, though it can also be severe. Because of the localized nature of the excess water during an upstream flood, there is unfilled channel capacity farther downstream to accommodate the extra water. Thus, the water can be rapidly carried down from the upstream area, quickly reducing the stream's stage.

Flash floods are a variety of upstream flood, characterized by especially rapid rise of stream stage. They can occur anywhere that surface runoff is rapid, large in volume, and funneled into a relatively restricted area. Canyons are common natural sites of flash floods (figure 6.13). Urban areas may experience flash floods in a highway underpass or just a dip in the road that becomes a temporary stream as runoff washes over it. Flash floods can even occur in deserts after a cloudburst, though in this case, the water level is likely to subside quickly as water sinks into parched ground rather than draining away in a stream.

Floods that affect large stream systems and large drainage basins are called **downstream floods.** These floods more often result from prolonged heavy rains over a broad area or from extensive regional snowmelt. Downstream floods usually last longer than upstream floods because the whole stream system is choked with excess water. The spring 1990 flooding in Texas, Oklahoma, and throughout the south-central United States or the summer 1993 flooding in the Mississippi River basin (figure 6.14) are excellent examples of downstream flooding, brought on by weeks of above-average rainfall.

Stream Hydrographs

Fluctuations in stream stage or discharge over time can be plotted on a **hydrograph** (figure 6.15). Hydrographs spanning long periods of time are very useful in constructing a picture of the "normal" behavior of a stream and of that stream's response to flood-causing events. Logically enough, a flood shows up as a peak on the hydrograph. The height and width of that peak and its position in time relative to the water-input event(s) depend, in part, on where the measurements are taken relative to where the excess water is entering the system. Upstream, where the drainage basin is smaller and the water need not travel so far to reach the stream, the peak is likely to be sharper—with a higher crest and more rapid rise and fall of the water level—and to occur sooner after the influx of water. By the time that water pulse has moved downstream to a lower point in the drainage basin, it will have dispersed somewhat, so that the arrival of the water spans a longer time. The peak will be spread out, so that

A

FIGURE 6.13 (A) This canyon in Zion National Park is an example of a setting in which dangerous flash floods are possible, and steep canyon walls offer no route of quick retreat to higher ground. (B) Damage from Big Thompson Canyon (Colorado) flash flooding. The greater force associated with deeper, swifter floodwaters causes severe damage.
(B) Photograph courtesy USGS Photo Library, Denver, CO.

B

the hydrograph shows both a later and a broader, gentler peak (see figure 6.16A). A short event like a severe cloudburst will tend to produce a sharper peak than a more prolonged event like several days of steady rain or snowmelt, even if the same amount of water is involved. Sometimes, the flood-causing event also causes permanent changes in the drainage system (figure 6.16B).

Hydrographs can be plotted with either stage or discharge on the vertical axis. Which parameter is used may depend on the purpose of the hydrograph. For example, stage may be the more relevant to knowing how much of a given floodplain area will be affected by the floodwaters. As a practical matter, too, measurement of stream stage is much simpler, so accurate stage data may be more readily available.

Flood-Frequency Curves

Another way of looking at flooding is in terms of the frequency of flood events of differing severity. Long-term records make it possible to construct a curve showing discharge as a function of recurrence interval for a particular stream or section of one. The sample **flood-frequency curve** shown in figure 6.17 indicates that for this stream, on average, an event producing a discharge of 675 cubic feet/second occurs once every ten years, an event

FIGURE 6.14 Mississippi River Basin flooding, Summer 1993. (A) The Missouri River (entering from top of image) and Kansas River (left) have spread broadly across their floodplains. Within the flooded areas, the rivers' normal, meandering channels are outlined by vegetation (red in image). Kansas City, Missouri, municipal airport is just east of the junction of the rivers. (B) The flooding turned rural areas into lakes dotted with trees and scattered houses; in places, it was hard to pick the river channel out of the vast expanse of water.

(A) Image courtesy NASA. (B) Photograph courtesy of C. S. Melcher, U.S. Geological Survey

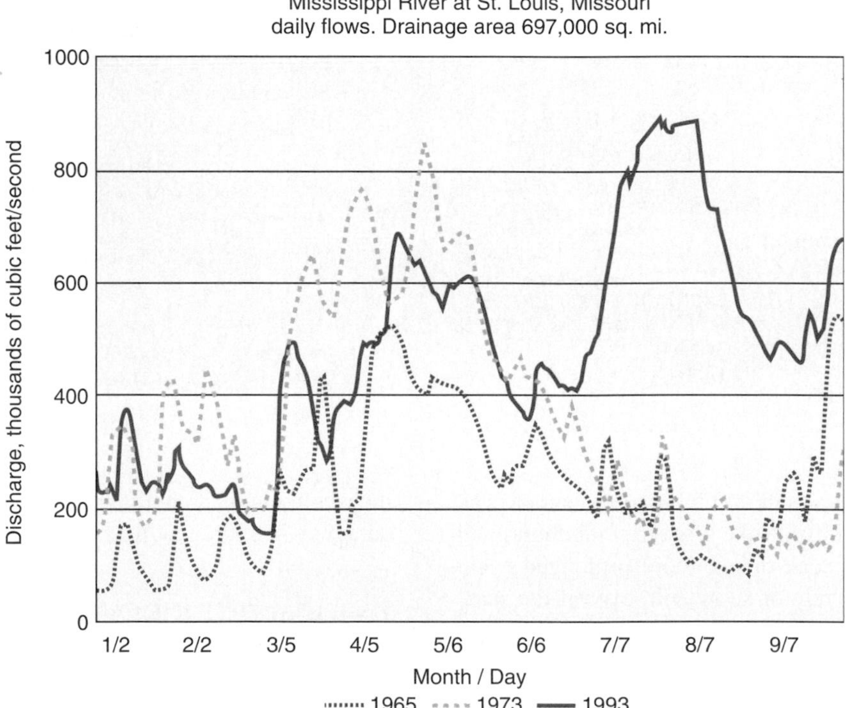

FIGURE 6.15 The summer 1993 flooding in the upper Mississippi River basin was not only unusual in its timing (compare with 1965 and 1973 patterns) but, in many places, broke flood-stage and/or discharge records set in 1973. In a typical year, snowmelt in the north plus spring rains lead to a rise in discharge in late spring. Summer is generally dry. Heavy 1993 summer rains were exceptional and produced exceptional flooding.

Hydrograph data courtesy U.S. Geological Survey

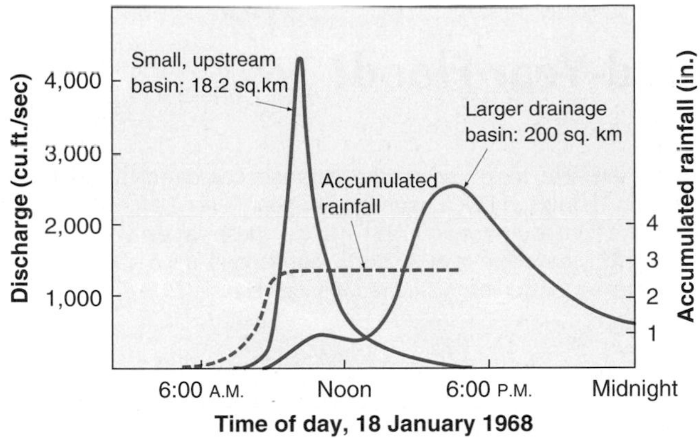

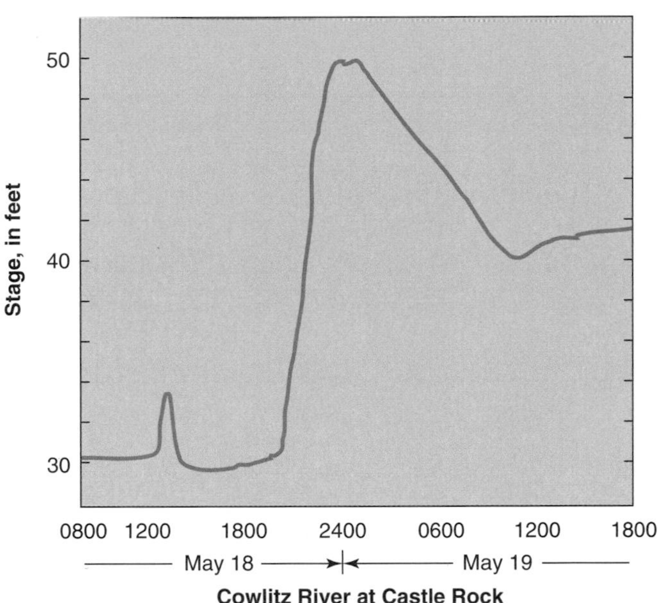

FIGURE 6.16 (A) Flood hydrographs for two different points along Calaveras Creek near Elmendorf, Texas. The flood has been caused by heavy rainfall in the upper reaches of the drainage basin. Here, flooding quickly follows the rain. The larger stream lower in the drainage basin responds more sluggishly to the input. (B) Mudflow into the Cowlitz River from the 18 May 1980 eruption of Mount St. Helens caused a sharp rise in stream stage (height), and flooding; infilling of the channel by sediment resulted in a long-term increase in stream stage—higher water level—for given discharge and net water depth.

(A) Source Water Resources Division, U.S. Geological Survey; (B) after U.S. Geological Survey

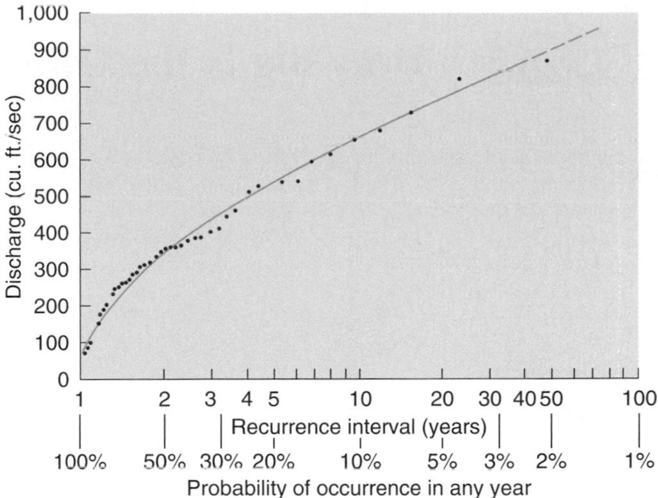

FIGURE 6.17 Flood-frequency curve for Eagle River at Red Cliff, Colorado. Records span 46 years, so estimates of the sizes of moderately frequent floods are likely to be fairly accurate, though the rare larger floods are hard to project.

USGS Open-File Report 79–1060

with discharge of 350 cubic feet/second occurs once every two years, and so on. A flood event can then be described by its **recurrence interval:** how frequently a flood of that severity occurs, *on average,* for that stream.

Alternatively (and preferably), one can refer to the *probability* that a flood of given size will occur in any one year; this probability is the inverse of the recurrence interval. For the stream of figure 6.17, a flood with a discharge of 675 cubic feet/second is called a "ten year flood," meaning that a flood of that size occurs about once every ten years, or has a 10% (1/10) probability of occurrence in any year; a discharge of 900 cubic feet/second is called a "forty-year flood" and has a 2.5% (1/40) probability of occurrence in any year, and so on.

Flood-frequency (or flood-probability) curves can be extremely useful in assessing regional flood hazards. If the severity of the one-hundred- or two-hundred-year flood can be estimated, then, even if such a rare and serious event has not occurred within memory, scientists and planners can, with the aid of topographic maps, project how much of the region would be flooded in such events. They can speak, for example, of the "one-hundred-year floodplain" (the area that would be water-covered in a one-hundred-year flood). Such information is useful in preparing flood-hazard maps, an exercise that is also aided by the use of aerial or satellite photography, as was shown in figure 6.14. Such maps, in turn, are helpful in siting new construction projects to minimize the risk of flood damage or in making property owners in threatened areas aware of dangers.

Unfortunately, the best-constrained part of the curve is, by definition, that for the lower-discharge, more frequent, higher probability, less-serious floods. Often, considerable guesswork is involved in projecting the shape of the curve at the high-discharge end. Reliable records of stream stages, discharges, and the extent of past floods typically extend back only a few decades. Many areas, then, may never have recorded a fifty- or one-hundred-year flood. Moreover, when the odd severe flood does occur, how does one know whether it was a sixty-year flood, a one-hundred-year flood, or whatever? The high-discharge events are rare, and even when they happen, their recurrence interval can only be estimated. Considerable uncertainty can exist, then, about just how bad a particular stream's one-hundred- or two-hundred-year flood might be. This problem is explored further in box 6.1, "How Big Is the One-Hundred-Year Flood?"

Box 6.1 How Big Is the One-Hundred-Year Flood?

A common way to estimate the recurrence interval of a flood of given size is as follows: Suppose the records of maximum discharge (or maximum stage) reached by a particular stream each year have been kept for N years. Each of these yearly maxima can be given a rank M, ranging from 1 to n, 1 being the largest, n the smallest. Then the recurrence interval R of a given annual maximum is defined as

$$R = \frac{(N + 1)}{M}$$

This approach assumes that the N years of record are representative of "typical" stream behavior over such a period; the larger N, the more likely this is to be the case.

For example, table 1 shows the maximum one-day mean discharges of the Big Thompson River, as measured near Estes Park, Colorado, for twenty-five consecutive years, 1951–1975. If these values are ranked 1 to 25, the 1971 maximum of 1030 cubic feet/second is the seventh largest and, therefore, has an estimated recurrence interval of

$$\frac{(25 + 1)}{7} = 3.71 \text{ years},$$

or a 27% probability of occurrence in any year.

Suppose, however, that only ten years of records are available, for 1966–1975. The 1971 maximum discharge happens to be the largest in that period of record. On the basis of the shorter record, its estimated

TABLE 1

Calculated Recurrence Intervals for Discharges of Big Thompson River at Estes Park, Colorado

		For Twenty-Five-Year Record		For Ten-Year Record	
Year	Maximum Mean One-Day Discharge (cu. ft./sec.)	M (rank)	R (years)	M (rank)	R (years)
1951	1220	4	6.50	3	3.67
1952	1310	3	8.67	2	5.50
1953	1150	5	5.20	4	2.75
1954	346	25	1.04	10	1.10
1955	470	23	1.13	9	1.22
1956	830	13	2.00	6	1.83
1957	1440	2	13.00	1	11.00
1958	1040	6	4.33	5	2.20
1959	816	14	1.86	7	1.57
1960	769	17	1.53	8	1.38
1961	836	12	2.17		
1962	709	19	1.37		
1963	692	21	1.23		
1964	481	22	1.18		
1965	1520	1	26.00		
1966	368	24	1.08	10	1.10
1967	698	20	1.30	9	1.22
1968	764	18	1.44	8	1.38
1969	878	10	2.60	4	2.75
1970	950	9	2.89	3	3.67
1971	1030	7	3.71	1	11.00
1972	857	11	2.36	5	2.20
1973	1020	8	3.25	2	5.50
1974	796	15	1.73	6	1.83
1975	793	16	1.62	7	1.57

Source: Data from U.S. Geological Survey Open-File Report 79–681.

(Box 6.1 Continued)

recurrence interval is (10 + 1)/1 = 11 years, corresponding to a 9% probability of occurrence in any year. Alternatively, if we look at only the first ten years of record, 1951–1960, the recurrence interval for the 1958 maximum discharge of 1040 cubic feet/second (a maximum discharge of nearly the same size as that in 1971) can be estimated at 2.2 years, meaning that it would have a 45% probability of occurrence in any one year.

Which estimate is "right"? Perhaps none of them, but their differences illustrate the need for long-term records to smooth out short-term anomalies in streamflow patterns, which are not uncommon, given that climatic conditions can fluctuate significantly over a period of a decade or two.

The point is further illustrated in figure 1. It is rare to have one hundred years or more of records for a given stream, so the magnitude of fifty-year, one-hundred-year, or larger floods is commonly estimated from a flood-frequency curve. Curves *A* and *B* in figure 1 are based, respectively, on the first and last ten years of data from the last two columns of table 1, the data points represented by X symbols for *A* and open circles for *B*. These two data sets give estimates for the size of the one-hundred-year flood that differ by more than 50%. These results, in turn, both differ somewhat from an estimate based on the full 25 years of data (curve *C*, solid circles). This graphically illustrates how important long-term records can be in the projection of recurrence intervals of larger flood events. The flood-frequency curve in figure 6.17, based on nearly fifty years of record, inspires more confidence—if recent land-use changes have been minimal, at least.

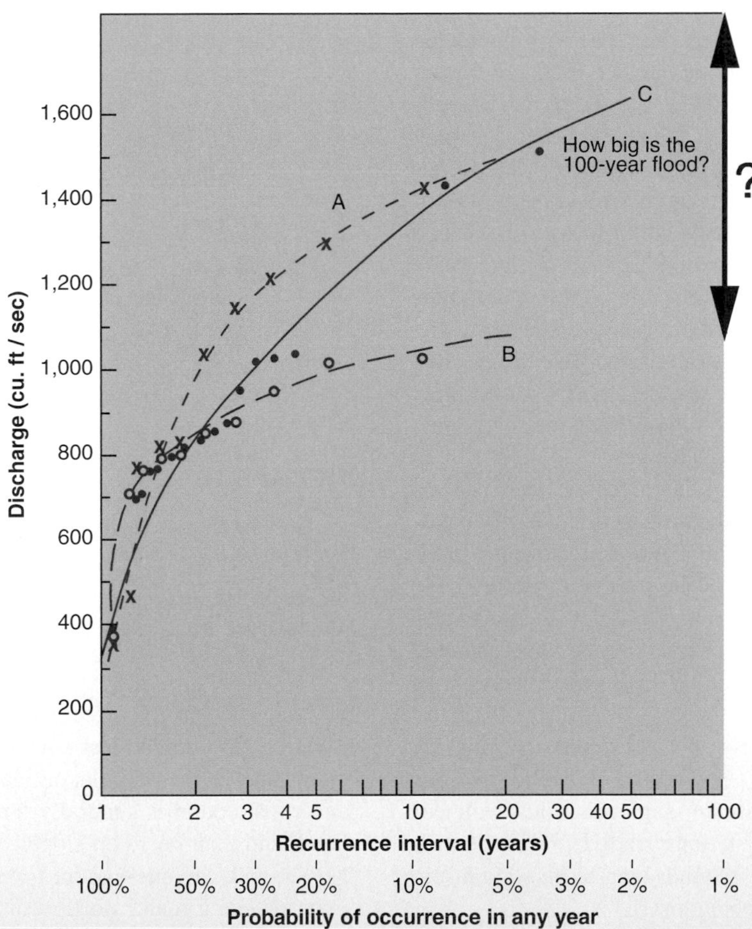

FIGURE 1 Flood-frequency curves for a given stream can look very different, depending on the length of record available and the particular period studied.

It is important to remember, too, that the recurrence intervals or probabilities assigned to floods are *averages.* The recurrence-interval terminology may be misleading in this regard, especially to the general public. Over many centuries, a fifty-year flood should occur an average of once every fifty years; or, in other words, there is a 2% chance that that discharge will be exceeded in any one year. However, that does not mean that two fifty-year floods could not occur in successive years, or even in the same year. The probability of two fifty-year flood events in one year is very low (2% × 2%, or .04%), but it is possible. The Chicago area, in 1986–1987, experienced two one-hundred-year floods within a ten-month period. Therefore, it is foolish to assume that, just because a severe flood has recently occurred in an area, it is in any sense "safe" for a while. Statistically, another such severe flood probably will not happen for some time, but there is no guarantee of that. Misleading though it may be, the recurrence-interval terminology is still in common use by many agencies involved with flood-hazard mapping and flood-control efforts, such as the U.S. Army Corps of Engineers.

Another complication is that streams in heavily populated areas are affected by human activities, as we shall see further in the next section. The way a stream responded to 10 centimeters of rain from a thunderstorm a hundred years ago may be quite different from the way it responds today. The flood-frequency curves are therefore changing with time. A flood-control dam that can moderate spikes in discharge may reduce the magnitude of the hundred-year flood. Except for measures specifically designed for flood control, most human activities have tended to aggravate flood hazards and to decrease the recurrence intervals of high-discharge events. In other words, what may have been a one-hundred-year flood two centuries ago might be a fifty-year flood, or a more frequent one, today. Figure 6.18 illustrates both types of changes.

The challenge of determining flood recurrence intervals from historical data is not unlike the forecasting of earthquakes by applying the earthquake-cycle model to historic and prehistoric seismicity along a fault zone. The results inevitably have associated uncertainties, and the more complex the particular geologic setting, the more land use and climate have changed over time, and the sparser the data, the larger the uncertainties will be.

Economics plays a role, too. A 1998 report to Congress by the U.S. Geological Survey showed that over the previous 25 years, the number of active stream-gaging stations declined: Stations abandoned exceeded stations started or reactivated. The ability to accumulate or extend long-term streamflow records in many places is thus being lost.

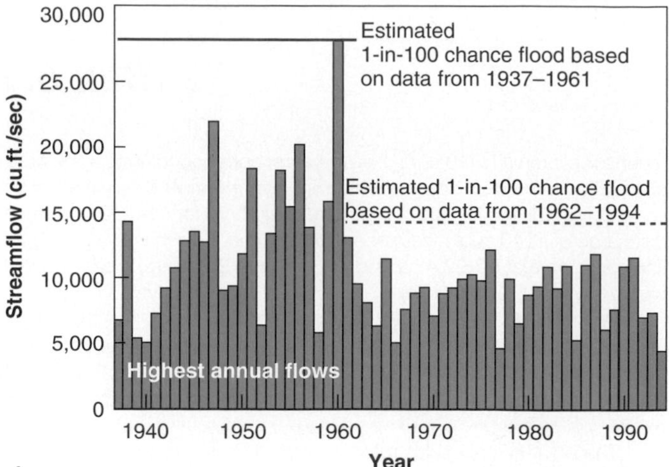

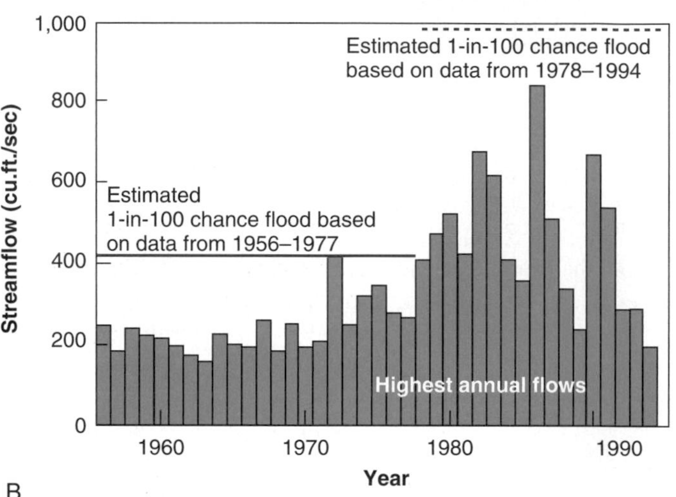

FIGURE 6.18 The size of the "1-in-100 chance flood" (the "hundred-year flood") can change dramatically as a result of human activities. (A) For the Green River at Auburn, Washington, the projected size of this flood has been halved by the Howard Hanson Dam. (B) Rapid urban development in the Mercer Creek Basin since 1977 has more than doubled the discharge of the projected hundred-year flood.
Modified after USGS Fact Sheet 229–96

CONSEQUENCES OF DEVELOPMENT IN FLOODPLAINS

Why would anyone live in a floodplain? One reason might be ignorance of the extent of the flood hazard. Someone living in a stream's one-hundred- or two-hundred-year floodplain may be unaware that the stream could rise that high, if historical flooding has all been less severe. In mountainous areas, floodplains may be the only flat or nearly flat land on which to build, and construction is generally far easier and cheaper on nearly level land than on steep slopes. Around a major river like the Mississippi, the one-hundred- or two-hundred-year floodplain may include a major portion of the land for miles around, as became painfully evident in 1993, and it may be impractical to leave that much real estate entirely vacant. Farmers have settled in floodplains since ancient times because flooding streams deposit fine sediment over the lands flooded, replenishing nutrients in the soil and thus making the soil especially fertile. Where rivers are used for transportation, cities may have been built deliberately as close to the water as possible. And, of course, many streams are very scenic features to live near. Obviously, the more people settle and build in floodplains, the

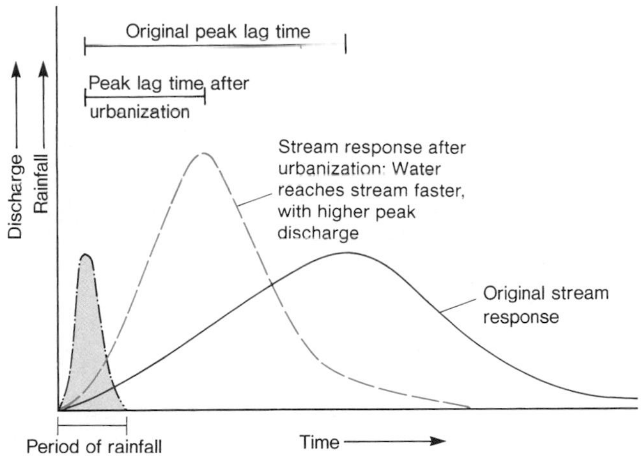

FIGURE 6.19 Hydrograph reflecting modification of stream response to precipitation following urbanization: Peak discharge increases, lag time to peak decreases.

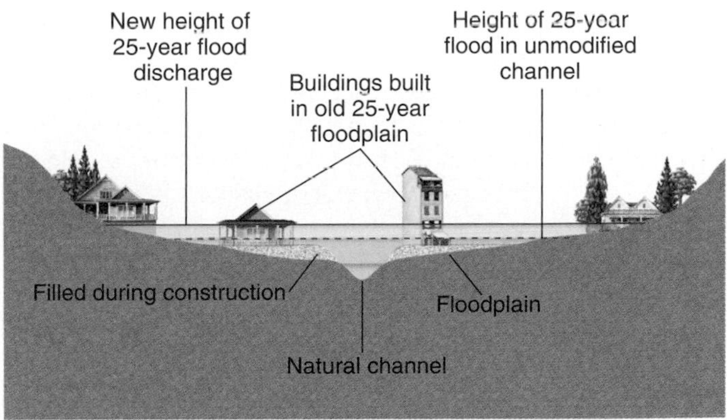

FIGURE 6.20 How floodplain development increases flood stage for given discharge.

more damage flooding will do. What people often fail to realize is that floodplain development can actually increase the likelihood or severity of flooding.

As mentioned earlier, two factors affecting flood severity are the proportion and rate of surface runoff. The materials extensively used to cover the ground when cities are built, such as asphalt and concrete, are relatively impermeable and greatly reduce infiltration (and groundwater replenishment, as will be seen in chapter 11). Therefore, when considerable area is covered by these materials, surface runoff tends to be much more concentrated and rapid than before, increasing the risk of flooding. This is illustrated by figure 6.19. If *peak lag time* is defined as the time lag between a precipitation event and the peak flood discharge (or stage), then, typically, that lag time decreases with increased urbanization, where the latter involves covering land with impermeable materials and/or installing storm sewers. The peak discharge (stage) also increases, as was evident in figure 6.18B.

Buildings in a floodplain also can increase flood heights (see figure 6.20). The buildings occupy volume that water formerly could fill, and a given discharge then corresponds to a higher stage (water level). Floods that occur are more serious. Filling in floodplain land for construction similarly decreases the volume available to stream water and further aggravates the situation.

Measures taken to drain water from low areas can likewise aggravate flooding along a stream. In cities, storm sewers are installed to keep water from flooding streets during heavy rains, and, often, the storm water is channeled straight into a nearby stream. This works fine if the total flow is moderate enough, but by decreasing the time normally taken by the water to reach the stream channel, such measures increase the probability of flooding. The same is true of the use of tile drainage systems in farmland. Water that previously stood in low spots in the field and only slowly reached the stream after infiltration instead flows swiftly and directly into the stream, increasing the flood hazard for those along the banks.

Both farming and urbanization also disturb the land by removing natural vegetation and, thus, leaving the soil exposed. The consequences are twofold. As noted earlier, vegetation can decrease flood hazards somewhat by providing a physical barrier to surface runoff, by soaking up some of the water, and through plants' root action, which keeps the soil looser and more permeable. Vegetation also can be critical to preventing soil erosion. When vegetation is removed and erosion increased, much more soil can be washed into streams. There, it can fill in, or "silt up," the channel, decreasing the channel's volume and thus reducing the stream's capacity to carry water away quickly.

It is not obvious whether land-use or climatic changes, or both, are responsible, but it is interesting to note that long-term records for the Red River near Grand Forks, North Dakota, show an increase in peak discharge over time leading up to the catastrophic 1997 floods (figure 6.21).

STRATEGIES FOR REDUCING FLOOD HAZARDS

Restrictive Zoning and "Floodproofing"

Short of avoiding floodplains altogether, various approaches can reduce the risk of flood damage. Many of these approaches—restrictive zoning, special engineering practices—are similar to strategies applicable to reducing damage from seismic and other geologic hazards. And commonly, existing structures present particular challenges.

A first step is to identify as accurately as possible the area at risk. Careful mapping coupled with accurate stream discharge data should allow identification of those areas threatened by floods of different recurrence intervals. Land that could be inundated often—by twenty-five-year floods, perhaps—might best be restricted to land uses not involving much building. The

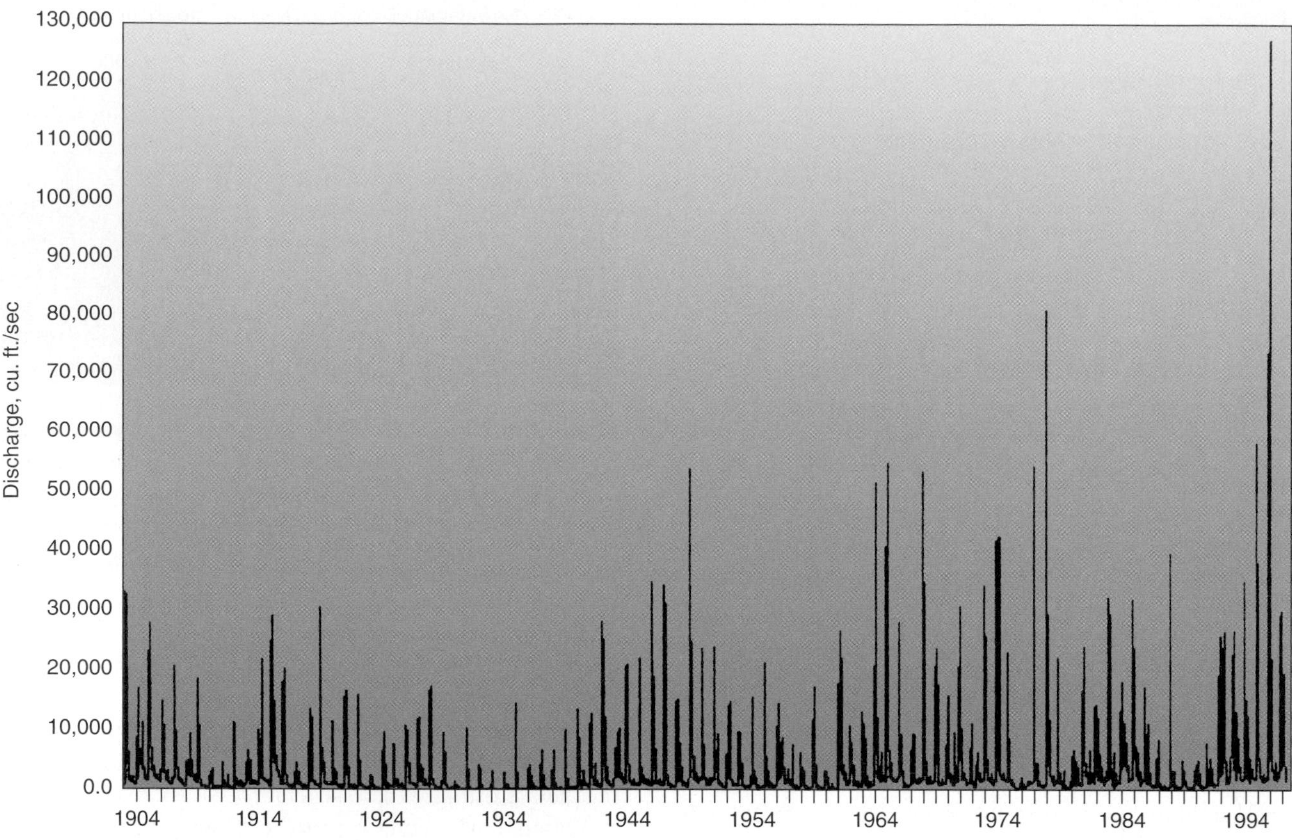

FIGURE 6.21 Nearly a century of peak-discharge records for the Red River of the North at Grand Forks, North Dakota, shows both decade-scale fluctuations (see 1914–1944) and a generally increasing long-term trend.
Data from U.S. Geological Survey

land could be used, for example, for livestock grazing-pasture or for parks or other recreational purposes.

Unfortunately, there is often economic pressure to develop and build on floodplain land, especially in urban areas. Local governments may feel compelled to allow building at least in the outer fringes of the floodplain, perhaps the parts threatened only by one-hundred- or two-hundred-year floods. In these areas, new construction can be designed with the flood hazard in mind. Buildings can be raised on stilts so that the lowest floor is above the expected two-hundred-year flood stage, for example. While it might be better not to build there at all, such a design at least minimizes interference by the structure with flood flow and also the danger of costly flood damage to the building. A major limitation of any floodplain zoning plan, whether it prescribes design features of buildings or bans buildings entirely, is that it almost always applies only to new construction. In many places, scores of older structures are already at risk. Programs to move threatened structures off floodplains are few and also costly when many buildings are involved. Usually, any moving occurs *after* a flood disaster, when individual structures or, rarely, whole towns are rebuilt out of the floodplain, as was done with Valmeyer, Illinois, after the 1993 Mississippi River floods. Strategies to modify the stream or the runoff patterns may help to protect existing structures without moving them, and are much more common.

Retention Ponds, Diversion Channels

If open land is available, flood hazards along a stream may be greatly reduced by the use of **retention ponds** (figure 6.22). These ponds are large basins that trap some of the surface runoff, keeping it from flowing immediately into the stream. They may be elaborate artificial structures; old, abandoned quarries; or, in the simplest cases, fields dammed by dikes of piled-up soil. The latter option also allows the land to be used for other purposes, such as farming, except on those rare occasions of heavy runoff when it is needed as a retention pond. Retention ponds are frequently a relatively inexpensive option, provided that ample undeveloped land is available, and they have the added advantage of not altering the character of the stream.

A similar strategy is the use of *diversion channels* that come into play as stream stage rises; they redirect some of the water flow into areas adjacent to the stream where flooding will cause minimal damage. The diversion of water might be into farmland or recreational land and away from built-up areas, to reduce loss of life and property damage.

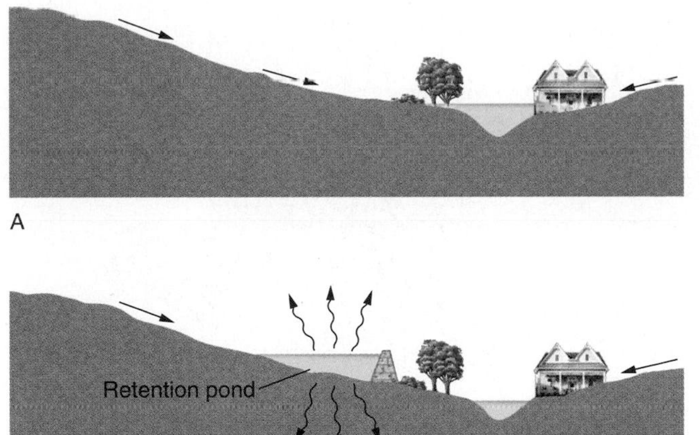

FIGURE 6.22 Use of retention pond to moderate flood hazard. (A) Before the retention pond, fast surface runoff caused flooding along the stream. (B) The retention pond traps some surface runoff, prevents the water from reaching the stream quickly, and allows for slow infiltration or evaporation instead.

Channelization

Channelization is a general term for various modifications of the stream channel itself that are usually intended to increase the velocity of water flow, the volume of the channel, or both. These modifications thus increase the discharge of the stream and hence the rate at which surplus water is carried away.

The channel can be widened or deepened, especially where soil erosion and subsequent sediment deposition in the stream have partially filled in the channel. Care must be taken, however, that channelization does not alter the stream dynamics too greatly elsewhere. Dredging of Missouri's Blackwater River in 1910 enlarged the channel and somewhat increased the gradient, thereby increasing stream velocities. As a result, discharges upstream from the modifications increased. This, in turn, led to more stream bank erosion and channel enlargement upstream and to increased flooding below.

Alternatively, a stream channel might be rerouted—for example, by deliberately cutting off meanders to provide a more direct path for the water flow. Such measures do tend to decrease the flood hazard upstream from where they are carried out. In the 1930s, the Mississippi River below Cairo, Illinois, was shortened by a total of 185 kilometers (115 miles) in a series of channel-straightening projects, with perceptible reduction in flooding as a result.

A meandering stream, however, often tends to keep meandering or to revert to old meanders. Channelization is not a one-time effort. Constant maintenance is required to limit erosion in the straightened channel sections and to keep the river in the cutoffs. The erosion problem along channelized or modified streams has in some urban areas led to the construction of wholly artificial stream channels made of concrete or other resistant materials. While this may be effective, it is also costly and frequently unsightly as well.

The Mississippi also offers a grand-scale example of the difficulty of controlling a stream's choice of channel: the Atchafalaya. This small branch diverges from the Mississippi several hundred miles above its mouth. Until a logjam blocking it was removed in the nineteenth century, it diverted little of the river's flow. Once that blockage was cleared, the water began to flow increasingly down the Atchafalaya, rather than the main branch of the Mississippi. The increased flow, in turn, scoured a larger channel, allowing more of the water to flow via the Atchafalaya. The water available for navigation on the lower Mississippi, and the sediment it carried to build the Mississippi delta, diminished. In the late 1950s, the U.S. Army Corps of Engineers built the first of a series of structures designed to limit the diversion of water into the Atchafalaya. The 1973 flooding taxed the engineering, and more substantial structures were built, but they didn't exactly represent a permanent solution: 2003 saw the start of another massive flood-control/water-management Corps of Engineers project in the Atchafalaya basin, with a projected cost of over $250 million.

Channelization has ecological impacts as well. Wetlands may be drained as water is shifted more efficiently downstream. Streambank habitat is reduced as channels are straightened and shortened, and fish may adapt poorly to new streamflow patterns and streambed configuration.

Finally, a drawback common to many channelization efforts (and one not always anticipated beforehand) is that by causing more water to flow downstream faster, channelization often increases the likelihood of flooding downstream from the alterations. Meander cutoff, for instance, increases the stream's gradient by shortening the channel length over which a given vertical drop occurs.

Levees

The problem that local modifications may increase flood hazards elsewhere is sometimes also a complication with the building of **levees,** raised banks along a stream channel. Some streams form low, natural levees along the channel through sediment deposition during flood events (figure 6.23A). These levees may purposely be enlarged, or created where none exist naturally, for flood control (figure 6.23B). Because levees raise the height of the stream banks close to the channel, the water can rise higher without flooding the surrounding country. This ancient technique was practiced thousands of years ago on the Nile by the Egyptian pharaohs. It may be carried out alone or in conjunction with channelization efforts, as on the Mississippi.

Confining the water to the channel, rather than allowing it to flow out into the floodplain, however, effectively shunts the water downstream faster during high-discharge events, increasing flood risks downstream. It artificially raises the stage of the stream for a given discharge, which can increase the risks upstream, too, as the increase in stage is propagated above the confined section. Another large problem is that levees may make people feel so safe about living in the floodplain that,

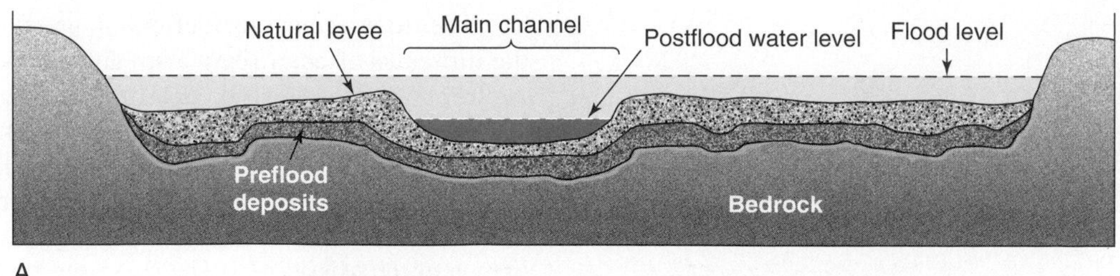

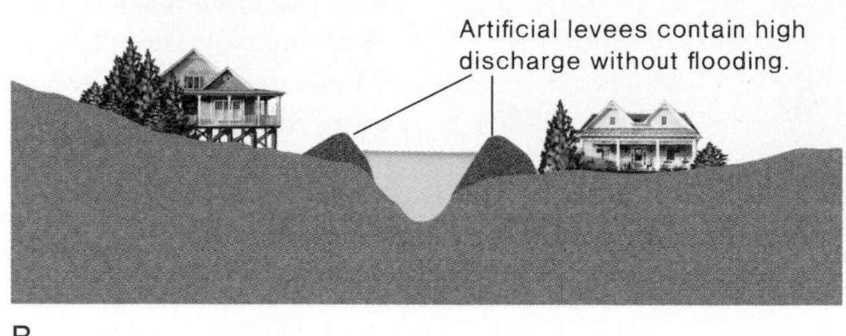

FIGURE 6.23 (A) When streams flood, waters slow quickly as they flow onto the floodplain, and therefore tend to deposit sediment, especially close to the channel where velocity first drops. (B) Artificial levees are designed to protect floodplain land from flooding by raising the height of the stream bank.

in the absence of restrictive zoning, development will be far more extensive than if the stream were allowed to flood naturally from time to time. If the levees have not, in fact, been built high enough and an unanticipated, severe flood overtops them, or if they simply fail and are breached during a high-discharge event, far more lives and property may be lost as a result. Problems with levees were abundantly demonstrated in the 1993 Mississippi River flooding, when an estimated 80% of privately built levees failed, as in figure 6.24A (though most federally built levees held). Also, if levees are overtopped, water is then trapped outside the stream channel, where it may stand for some time after the flood subsides until infiltration and percolation return it to the stream (figure 6.24B).

Levees alter sedimentation patterns, too. During flooding, sediment is deposited in the floodplain outside the channel. If the stream and its load are confined by levees, increased sedimentation may occur in the channel. This will raise stream stage for a given discharge—the channel bottom is raised, so stage rises also—and either the levees must be continually raised to compensate, or the channel must be dredged, an additional expense. Jackson Square in New Orleans stands at the elevation of the wharf in Civil War days; behind the levee beside the square, the Mississippi flows 10–15 feet higher, largely due to sediment accumulation in the channel. This is part of why the city was so vulnerable during Hurricane Katrina. See also box 6.2 "Life on the Mississippi."

FIGURE 6.24 The negative sides of levees as flood-control devices. (A) If they are breached, as here in the 1993 Mississippi River floods, those living behind the levees may suddenly and unexpectedly find themselves inundated. (B) Breached or overtopped levees dam water behind them. The stream stage may drop rapidly after the flood crests, but the water levels behind the levees remain high. Sandbagging was ineffective in preventing flooding by the Kishwaukee River, DeKalb, Illinois, in July 1983. By the time this picture was taken from the top of the levee, the water level in the river (right) had dropped several meters, but water trapped behind the levee (left) remained high, prolonging the damage to property behind the levee.

(A) Photograph courtesy C. S. Melcher, U.S. Geological Survey

Flood-Control Dams and Reservoirs

Yet another approach to moderating streamiflow to prevent or minimize flooding is through the construction of flood-control dams at one or more points along the stream. Excess water is held behind a dam in the reservoir formed upstream and may then be released at a controlled rate that does not overwhelm the capacity of the channel beyond. Additional benefits of constructing flood-control dams and their associated reservoirs (artificial lakes) may include availability of the water for irrigation, generation of hydroelectric power at the dam sites, and development of recreational facilities for swimming, boating, and fishing at the reservoir.

The practice also has its drawbacks, however. Navigation on the river—both by people and by aquatic animals—may be restricted by the presence of the dams. Also, the creation of a reservoir necessarily floods much of the stream valley behind the dam and may destroy wildlife habitats or displace people and their works. (China's massive Three Gorges Dam project is displacing over a million people and flooding numerous towns.) Fish migration can be severely disrupted. As this is being written, the issue of whether to breach (dismantle) a number of hydropower dams in the Pacific Northwest in efforts to restore populations of endangered salmon is being hotly debated as the Army Corps of Engineers weighs the alternatives; the final disposition of the issue may lie with Congress.

If the stream normally carries a high sediment load, further complications arise. The reservoir represents a new base level for the stream above the dam (figure 6.25). When the stream flows into that reservoir, its velocity drops to zero, and it dumps its load of sediment. Silting-up of the reservoir, in turn, decreases its volume, so it becomes less effective as a flood-control device. Some reservoirs have filled completely in a matter of decades, becoming useless. Others have been kept clear only by repeated dredging, which can be expensive and presents the problem of where to dump the dredged sediment. At the same time, the water released below the dam is free of sediment and thus may cause more active erosion of the channel there. Or, if a large fraction of water in the stream system is impounded by the dam, the greatly reduced water volume downstream may change the nature of vegetation and of habitats there. Recognition of some negative long-term impacts of the Glen Canyon Dam system led, in 1996, to the first of three experiments in controlled flooding of the Grand Canyon to restore streambank habitat and shift sediment in the canyon, scouring silted-up sections of channel and depositing sediment at higher elevations to restore beach habitats for wildlife. The long-term consequences of these actions are still being evaluated.

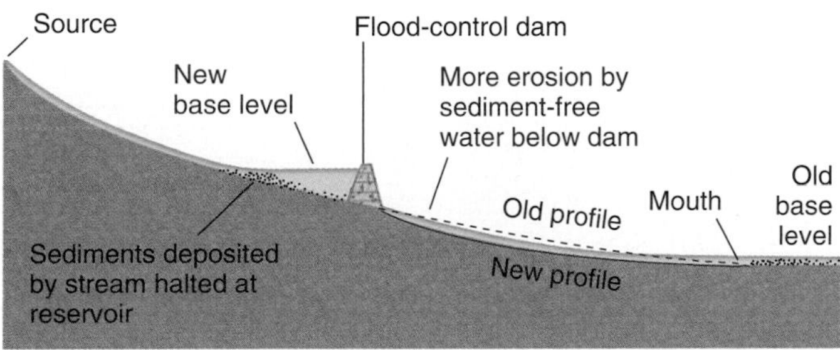

FIGURE 6.25 Effects of dam construction on a stream: change in base level, sediment deposition in reservoir, reshaping of the stream channel.

Some large reservoirs, such as Lake Mead behind Hoover Dam, have even been found to cause earthquakes. The reservoir water represents an added load on the rocks, increasing the stresses on them, while infiltration of water into the ground under the reservoir increases pore pressure sufficiently to reactivate old faults. Since Hoover Dam was built, at least 10,000 small earthquakes have occurred in the vicinity. Filling of the reservoir behind the Konya Dam in India began in 1963, and earthquakes were detected within months thereafter. In December 1967, an earthquake of magnitude 6.4 resulted in 177 deaths and considerable damage there. At least ninety other cases of reservoir-induced seismicity are known. Earthquakes caused in this way are usually of low magnitude, but their foci, naturally, are close to the dam. This raises concerns about the possibility of catastrophic dam failure caused, in effect, by the dam itself.

If levees increase residents' sense of security, the presence of a flood-control dam may do so to an even greater extent. People may feel that the flood hazard has been eliminated and thus neglect even basic floodplain-zoning and other practical precautions. But dams may fail; or, the dam may hold, but the reservoir fill abruptly with a sudden landslide (as in the Vaiont disaster described in chapter 8) and flooding occur anyway from the suddenly displaced water. Floods may result from intense rainfall or runoff *below* the dam. Also, a dam/reservoir complex may not necessarily be managed so as to maximize its flood-control capabilities. For obvious reasons, dams and reservoirs often serve multiple uses (e.g., flood control plus hydroelectric power generation plus water supply). But for maximally effective flood control, one wants maximum water-storage *potential,* while for other uses, it is generally desirable to maximize actual water *storage*—leaving little extra volume for flood control. The several uses of the dam/reservoir system may thus tend to compete with, rather than to complement, each other.

Flood Warnings?

Many areas of the United States have experienced serious floods (figure 6.26), and their causes are varied (Table 6.2). Depending on the cause, different kinds of precautions are possible. The areas that would be affected by flooding associated with the 1980 explosion of Mount St. Helens could be anticipated from topography and an understanding of lahars, but the actual eruption was too sudden to allow last-minute evacuation. The coastal flooding from a storm surge—discussed further in chapter 7—typically allows time for evacuation as the storm approaches, but damage to coastal property is inevitable. Dam failures are sudden; rising water from spring snowmelt, much slower. As long as people inhabit areas prone to flooding, property will be destroyed when the floods happen, but with better warning of impending flash floods, more lives can be saved.

Box 6.2 Life on the Mississippi: A Case for More Levees?

The Mississippi River is a long-term case study in levee-building for flood control. It is the highest-discharge stream in the United States and the third largest in the world (figure 1). The first levees were built in 1717, after New Orleans was flooded. Further construction of a patchwork of local levees continued intermittently over the next century. Following a series of floods in the mid-1800s, there was great public outcry for greater flood-control measures. In 1879, the Mississippi River Commission was formed to oversee and coordinate flood-control efforts over the whole lower Mississippi area. The commission urged the building of more levees. In 1882, when over $10 million (a huge sum for the time) had already been spent on levees, severe floods caused several hundred breaks in the system and prompted renewed efforts. By the turn of the century, planners were confident that the levees were high enough and strong enough to withstand any flood. Flooding in 1912–1913, accompanied by twenty failures of the levees, proved them wrong. By 1926, there were over 2900 kilometers (1800 miles) of levees along the Mississippi, standing an average of 6 meters high; nearly half a billion cubic meters of earth had been used to construct them. Funds totalling more than twenty times the original cost estimates had been spent.

In 1927, the worst flooding recorded to that date occurred along the Mississippi. More than 75,000 people were forced from their homes; the levees were breached in 225 places; 183 people died; nearly 50,000 square kilometers of land were flooded; and there was an estimated $500 million worth of damage. This was regarded as a crisis of national importance, and thereafter, the federal government took over primary management of the flood-control measures. In the upper parts of the drainage basin, the government undertook more varied efforts—for example, five major flood-control dams were built along the Missouri River in the 1940s in response to flooding there—but the immense volume of the lower Mississippi cannot be contained in a few artificial reservoirs. The building up and reinforcing of levees continued.

The fall and winter of 1972–1973 were mild and wet over much of the Mississippi's drainage basin. By early spring, the levels in many tributary streams were already high. Prolonged rain in March and April contributed to record-setting flooding in the spring of 1973. The river remained above flood stage for more than three months in some places. Over 12 million acres of land were flooded, 50,000 people were evacuated, and over $400 million in damage was done, not counting secondary effects, such

FIGURE 1 The Mississippi River's drainage basin covers a huge area. Note also the continental divide (red line), east of which streams drain into the Atlantic Ocean or Gulf of Mexico, and west of which they flow into the Pacific.

(Box 6.2 Continued)

as loss of wildlife and the estimated 240 million tons of sediment that were washed down the Mississippi, much of it fertile soil irrecoverably eroded from farmland. All this devastation occurred despite more than 250 years of ever-more-extensive (and expensive) flood-control efforts.

Still worse flooding was yet to come. A sustained period of heavy rains and thunderstorms in the summer of 1993 led to flooding that caused 50 deaths, over $10 billion in property and crop damage, and, locally, record-breaking flood stages and/or levee breaches at various places in the Mississippi basin. Some places remained swamped for weeks before the water finally receded (figure 2; recall also figure 6.14). We haven't stopped the flooding yet—and we may now be facing possible increased risks from long-term climate change.

Inevitably, the question of what level of flood control is adequate or desirable arises. The greater the flood against which we protect ourselves, the more costly the efforts. We could build still higher levees, bigger dams, and larger reservoirs and spillways, and (barring dam or levee failure) be safe even from a projected five-hundred-year flood or worse, but the costs would be astronomical and could exceed the property damage expected from such events. By definition, the probability of a five-hundred-year or larger flood is extremely low. Precautions against high-frequency flood events seem to make sense. But at what point, for what probability of flood, does it make more economic or practical sense simply to accept the real but small risk of the rare, disastrous, high-discharge flood events?

The same questions are now being asked about New Orleans in the wake of Hurricane Katrina, discussed in chapter 7.

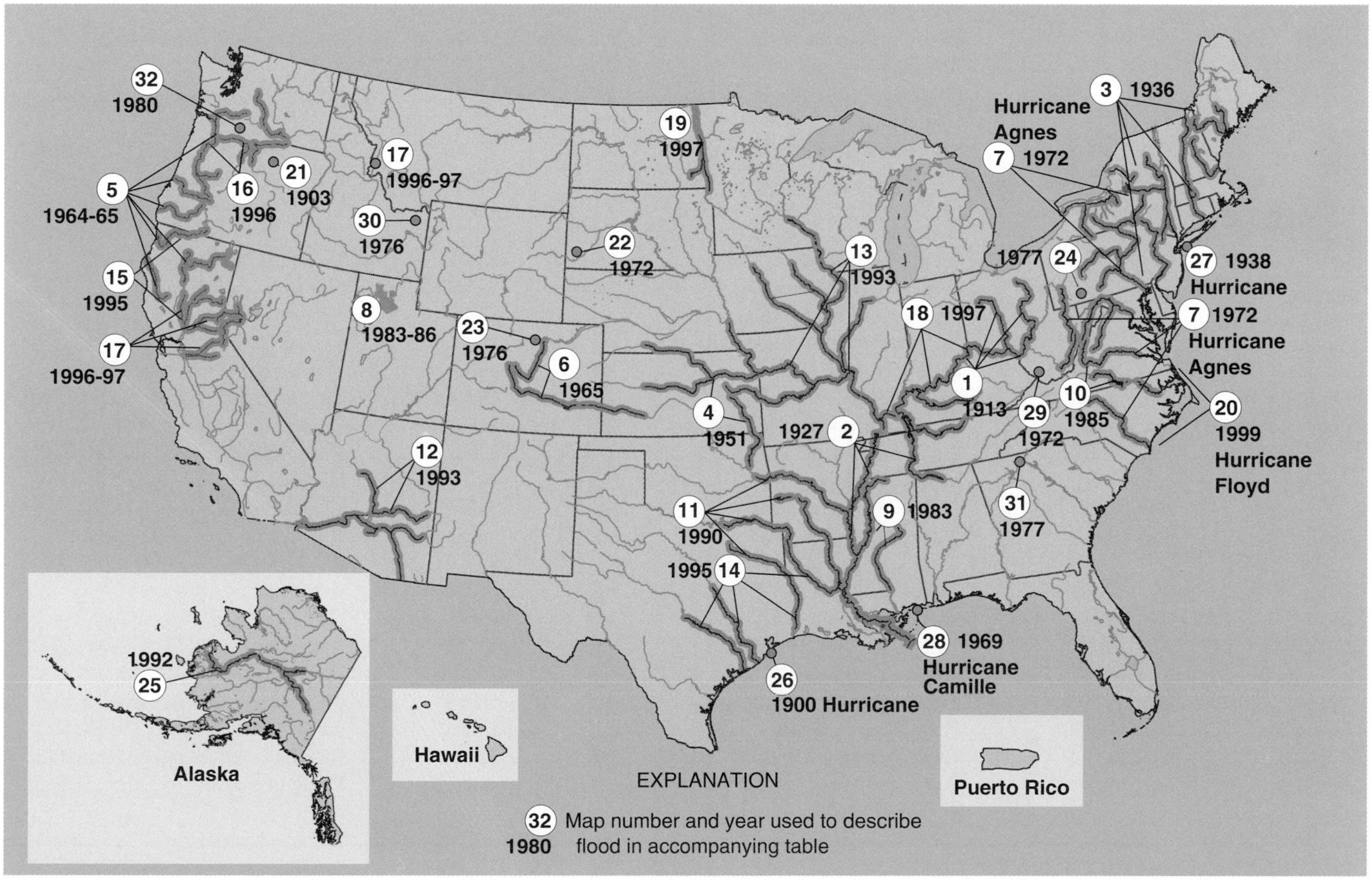

FIGURE 2 Flooding in Davenport, Iowa, Summer 1993.
© Doug Sherman/Geofile

FIGURE 6.26 Significant U.S. floods of the twentieth century. Map number key is in table 6.2. Hurricane Katrina, in 2005, affected much the same area as Camille.
Modified after USGS Fact Sheet 024-00

TABLE 6.2

Significant Floods of the Twentieth Century

Flood Type	Map No.	Date	Area or Stream with Flooding	Reported Deaths	Approximate Cost (uninflated)*	Comments
Regional flood	1	Mar.–Apr. 1913	Ohio, statewide	467	$143M	Excessive regional rain.
	2	Apr.–May 1927	Mississippi River from Missouri to Louisiana	unknown	$230M	Record discharge downstream from Cairo, Illinois.
	3	Mar. 1936	New England	150+	$300M	Excessive rainfall on snow.
	4	July 1951	Kansas and Neosho River Basins in Kansas	15	$800M	Excessive regional rain.
	5	Dec. 1964–Jan. 1965	Pacific Northwest	47	$430M	Excessive rainfall on snow.
	6	June 1965	South Platte and Arkansas Rivers in Colorado	24	$570M	14 inches of rain in a few hours in eastern Colorado.
	7	June 1972	Northeastern United States	117	$3.2B	Extratropical remnants of Hurricane Agnes.
	8	Apr.–June 1983 June 1983–1986	Shoreline of Great Salt Lake, Utah	unknown	$621M	In June 1986, the Great Salt Lake reached its highest elevation and caused $268M more in property damage.
	9	May 1983	Central and northeast Mississippi	1	$500M	Excessive regional rain.
	10	Nov. 1985	Shenandoah, James, and Roanoke Rivers in Virginia and West Virginia	69	$1.25B	Excessive regional rain.
	11	Apr. 1990	Trinity, Arkansas, and Red Rivers in Texas, Arkansas, and Oklahoma	17	$1B	Recurring intense thunderstorms.
	12	Jan. 1993	Gila, Salt, and Santa Cruz Rivers in Arizona	unknown	$400M	Persistent winter precipitation.
	13	May–Sept. 1993	Mississippi River Basin in central United States	48	$20B	Long period of excessive rainfall.
	14	May 1995	South-central United States	32	$5–6B	Rain from recurring thunderstorms.
	15	Jan.–Mar. 1995	California	27	$3B	Frequent winter storms.
	16	Feb. 1996	Pacific Northwest and western Montana	9	$1B	Torrential rains and snowmelt.
	17	Dec. 1996–Jan. 1997	Pacific Northwest and Montana	36	$2–3B	Torrential rains and snowmelt.
	18	Mar. 1997	Ohio River and tributaries	50+	$500M	Slow moving frontal system.
	19	Apr.–May 1997	Red River of the North in North Dakota and Minnesota	8	$2B	Very rapid snowmelt.
	20	Sept. 1999	Eastern North Carolina	42	$6B	Slow-moving Hurricane Floyd.
Flash flood	21	June 14, 1903	Willow Creek in Oregon	225	unknown	City of Heppner, Oregon, destroyed.
	22	June 9–10, 1972	Rapid City, South Dakota	237	$160M	15 inches of rain in 5 hours.
	23	July 31, 1976	Big Thompson and Cache la Poudre Rivers in Colorado	144	$39M	Flash flood in canyon after excessive rainfall.
	24	July 19–20, 1977	Conemaugh River in Pennsylvania	78	$300M	12 inches of rain in 6–8 hours.
Ice-jam flood	25	May 1992	Yukon River in Alaska	0	unknown	100-year flood on Yukon River.
Storm-surge flood	26	Sept. 1900	Galveston, Texas	6,000+	unknown	Hurricane.
	27	Sept. 1938	Northeast United States	494	$306M	Hurricane.
	28	Aug. 1969	Gulf Coast, Mississippi and Louisiana	259	$1.4B	Hurricane Camille.
Dam-failure flood	29	Feb. 2, 1972	Buffalo Creek in West Virginia	125	$60M	Dam failure after excessive rainfall.
	30	June 5, 1976	Teton River in Idaho	11	$400M	Earthen dam breached.
	31	Nov. 8, 1977	Toccoa Creek in Georgia	39	$2.8M	Dam failure after excessive rainfall.
Mudflow flood	32	May 18, 1980	Toutle and lower Cowlitz Rivers in Washington	60	unknown	Result of eruption of Mt. St. Helens.

After U.S. Geological Survey Fact Sheet 024-00.

*[M, million; B, billion]

With that in view, the U.S. Geological Survey (USGS) and National Weather Service (NWS) have established a partnership to provide flash-flood warnings. The USGS contributes real-time stream-stage measurements, and data on the relationship between stage and discharge for various streams. To this information, NWS adds historical and forecast precipitation data and applies models that relate stream response to water input. The result is a stream-stage projection (a projected hydrograph) from which NWS can identify areas at risk of near-term flooding and issue warnings accordingly. Later comparison of the model projections with actual water input and stream response (figure 6.27) allows refinement of the models for better, more precise future warnings.

Recently, attention has been given to the problem of international flood forecasting when the affected nation occupies only a small part of the upstream drainage basin and cannot itself do all the necessary monitoring. As an extreme example, Bangladesh occupies only the lowest 7% of the Ganges-Brahmaputra-Meghna drainage basin. NASA is proposing a Global Precipitation Measurement system, to begin in 2010, to help address this problem, giving such nations real-time access to critical precipitation data throughout the relevant drainage basin.

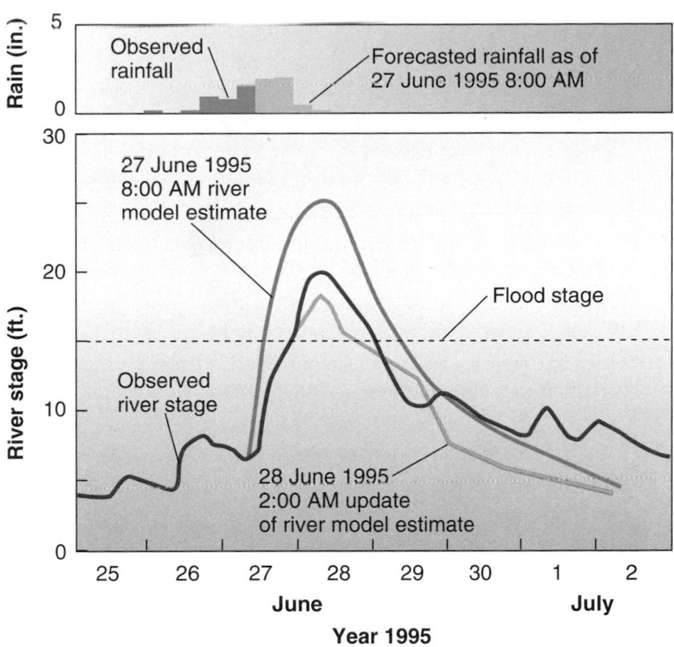

FIGURE 6.27 Comparison of projected and measured stream stage, Rappahannock River at Remington, Virginia, during 1995 flood.
After USGS Fact Sheet 209–95

SUMMARY

Streams are active agents of sediment transport. They are also powerful forces in sculpturing the landscape, constantly engaged in erosion and deposition, their channels naturally shifting over the earth's surface.

Stream velocity is a strong control on both erosion and sediment transport. The size of a stream is most commonly measured by its discharge. Flooding is the normal response of a stream to an unusually high input of water in a short time. Regions at risk from flooding, and the degree of risk, can be identified if accurate maps and records about past floods and their severity are available. However, records may not extend over a long enough period to permit precise predictions about the rare, severe floods. Moreover, human activities may have changed regional runoff patterns or stream characteristics over time, making historical records less useful in forecasting future problems, and long-term climate change adds uncertainty as well. Strategies designed to minimize flood damage include restricting or prohibiting development in floodplains, controlling the kinds of floodplain development, channelization, and the use of retention ponds, levees, and flood-control dams. Unfortunately, many flood-control procedures have drawbacks, one of which may be increased flood hazards elsewhere along the stream. Flash-flood warnings can be an important tool for saving lives, if not property, by making timely evacuations possible.

Key Terms and Concepts

Exercises

Questions for Review

1. Define stream load, and explain what factors control it.
2. Why do stream sediments tend to be well sorted? (Relate sediment transport to variations in water velocity.)
3. Explain how enlargement and migration of meanders contribute to floodplain development.
4. Discuss the relationship between flooding and (a) precipitation, (b) soil characteristics, (c) vegetation, and (d) season.
5. How do upstream and downstream floods differ? Give an example of an event that might cause each type of flood.
6. What is a flood-frequency curve? What is a recurrence interval? In general, what is a major limitation of these measures?
7. Describe two ways in which urbanization may increase local flood hazards. Sketch the change in stream response as it might appear on a hydrograph.
8. Channelization and levee construction may reduce local flood hazards, but they may worsen the flood hazards elsewhere along a stream. Explain.
9. Outline two potential problems with flood-control dams.
10. List several appropriate land uses for floodplains that minimize risks to lives and property.
11. What kinds of information are used to develop flash-flood warnings?

For Further Thought

1. Investigate the availability of flood-hazard maps for a nearby stream. On what sorts of data are the maps based? How complete are the data, and how up-to-date are the maps?
2. For a river that has undergone channel modification, investigate the history of flooding before and after that modification, and compare the costs of channelization to the costs of any flood damages. (The Army Corps of Engineers often handles such projects and may be able to supply the information.)
3. Consider the implications of various kinds of floodplain restrictions and channel modifications that might be considered by the government of a city through which a river flows. List criteria on which decisions about appropriate actions might be based.
4. Check on the latest research results on the Grand Canyon flooding and report/comment on how those results relate to the stated flood objectives.
5. Select a major flood event, and search the Internet for (a) data on that flooding, and (b) relevant precipitation data for the same region. Relate the two, commenting on probable factors contributing to the flooding.
6. If a flash-flood warning is issued by NWS in your area, note the specifics; later, examine the actual stream hydrographs for the area (see NetNotes for the USGS real-time data site). How close was the forecast? What might account for any differences?

Suggested Readings/References

Carter, J. M., Williamson, J. E., and Teller, R.W. 2002. *The 1972 Black Hills–Rapid City Flood revisited.* U.S. Geological Survey.

Collier, M. P., Webb, R. H., and Andrews, E. D. 1997. Experimental flooding in Grand Canyon. *Scientific American* (January): 82–89.

Hossain, F., and Katiyar, N. 2006. Improving flood forecasting in international river basins. *EOS* 87 (31 January): 49, 54.

Konrad, C.P. 2003. Effects of urban development on floods. USGS Fact Sheet FS-076-03.

Leopold, L. B. 1994. *A view of the river.* Cambridge, Mass.: Harvard University Press.

Long, M. E. 1997. The Grand managed Canyon. *National Geographic* (July): 117–35.

Lucchitta, I., and Leopold, L. B. 1999. Floods and sandbars in the Grand Canyon. *GSA Today* 9 (April): 1–7.

Mairson, A. 1994. The great flood of '93. *National Geographic* 185 (January): 44–87.

Mason, R. R. Jr., and Weiger, B. A. 1995. Stream gaging and flood forecasting. USGS Fact Sheet FS 209–95.

Pennisi, E. 2004. The Grand (Canyon) experiment. *Science* 306 (10 December): 1884–86.

Perry, C. A. 2000. Significant floods in The United States during the 20th century—USGS measures a century of floods. USGS Fact Sheet 024-00.

Petroski, H. 2006. Levees and other raised ground. *American Scientist* 94 (January–February): 7–11.

Pinter, N., Thomas, R., and Wlosinski, J.H. 2001. Assessing flood hazard on dynamic rivers. *EOS* 82(31 July): 333, 338–39.

Rigby, J. G., et al. 1998. *The 1997 New Year's floods in western Nevada.* Nevada Bureau of Mines and Geology Special Publication 23.

Theroux, P. 1997. The imperiled Nile delta. *National Geographic* (January): 2–35.

Watson, B. 1996. A town makes history by rising to new heights. *Smithsonian* (June): 110–20.

Welcome to McGraw-Hill's Online Learning Center

Location: http://www.mhhe.com/montgomery8e

NetNotes

The U.S. Geological Survey provides a variety of hydrologic and water-resource information. Their official site for maps and graphs on stream flow, with links to NWS and other related data, is "Waterwatch":
water.usgs.gov/waterwatch/

For real-time hydrologic (stream-gaging) data for many states and links to historical streamflow data see
water.usgs.gov/public/realtime.html

The 1998 USGS report to Congress on the decline of stream-gaging stations is at
water.usgs.gov/streamgaging

Discussion of the deliberate flooding of the Grand Canyon is found at
water.usgs.gov/public/wid/FS_089-96/FS_089-96.html

with a related followup study, "Effects of flooding on plant production downstream from Glen Canyon Dam", at
pubs.usgs.gov/fs/FS-060-99/

Many flood-related USGS Fact Sheets, including those in the reference list, are available online, often in both PDF and HTML format; see
water.usgs.gov/wid/indexlist.html

Some of special note include
"Large floods in the United States: Where they happen and why" (USGS Circular 1245, 2003)
pubs.usgs.gov/circ/2003/circ1245/pdf/circ1245.pdf

"The world's largest floods, past and present: Their causes and magnitudes" (USGS Circular 1254, 2004)
pubs.usgs.gov/circ/2004/circ1254/pdf/circ1254.pdf

"Flood hazards—A national threat" (USGS Fact Sheet 2006–3026, January 2006)
Pubs.usgs.gov/fs/2006/3026

"*StreamStats:* A U.S. Geological Survey Web application for stream information" (USGS Fact Sheet FS 2004–3115, 2005)
md.water.usgs.gov/publications/fs-2004-3115/

The National Water Conditions site offers streamflow data (current) over the United States and southern Canada; it is a joint effort of USGS and Environment Canada's Climate Information Branch, at
water.usgs.gov/nwc

Images and information on the Scientific Assessment and Strategy Team (SAST), formed to assess the 1993 Mississippi River basin flooding and to develop a long-term floodplain management plan, are found in USGS Fact Sheet 10399 at
erg.usgs.gov/isb/pubs/factsheets/fs10399.html

and links to a variety of other information/sites on that flood are found at
ghrc.msfc.nasa.gov/

The U.S. Army Corps of Engineers maintains a page with links to much information about the 1993 Mississippi flooding
pubs.usgs.gov/gip/ocoee2/

For flash-flood safety information:
www.nws.noaa.gov/om/brochures/ffbro.htm

Information about the 1997 Red River flooding is available in USGS Open-File Report 97–575 at
nd.water.usgs.gov/pubs/ofr/ofr97575/index.html

and photographs at
nd.water.usgs.gov/index/flood.html

The Great Salt Lake, which has no outlet streams, flooded badly in the 1980s; see
edcwww.cr.usgs.gov/earthshots/slow/GreatSaltLake/GreatSaltLake

The Dartmouth Flood Observatory at Dartmouth College has created a site through which information on and high resolution images of floods since 1 January 1994 can be accessed at
www.dartmouth.edu/artsci/geog/floods/index.html

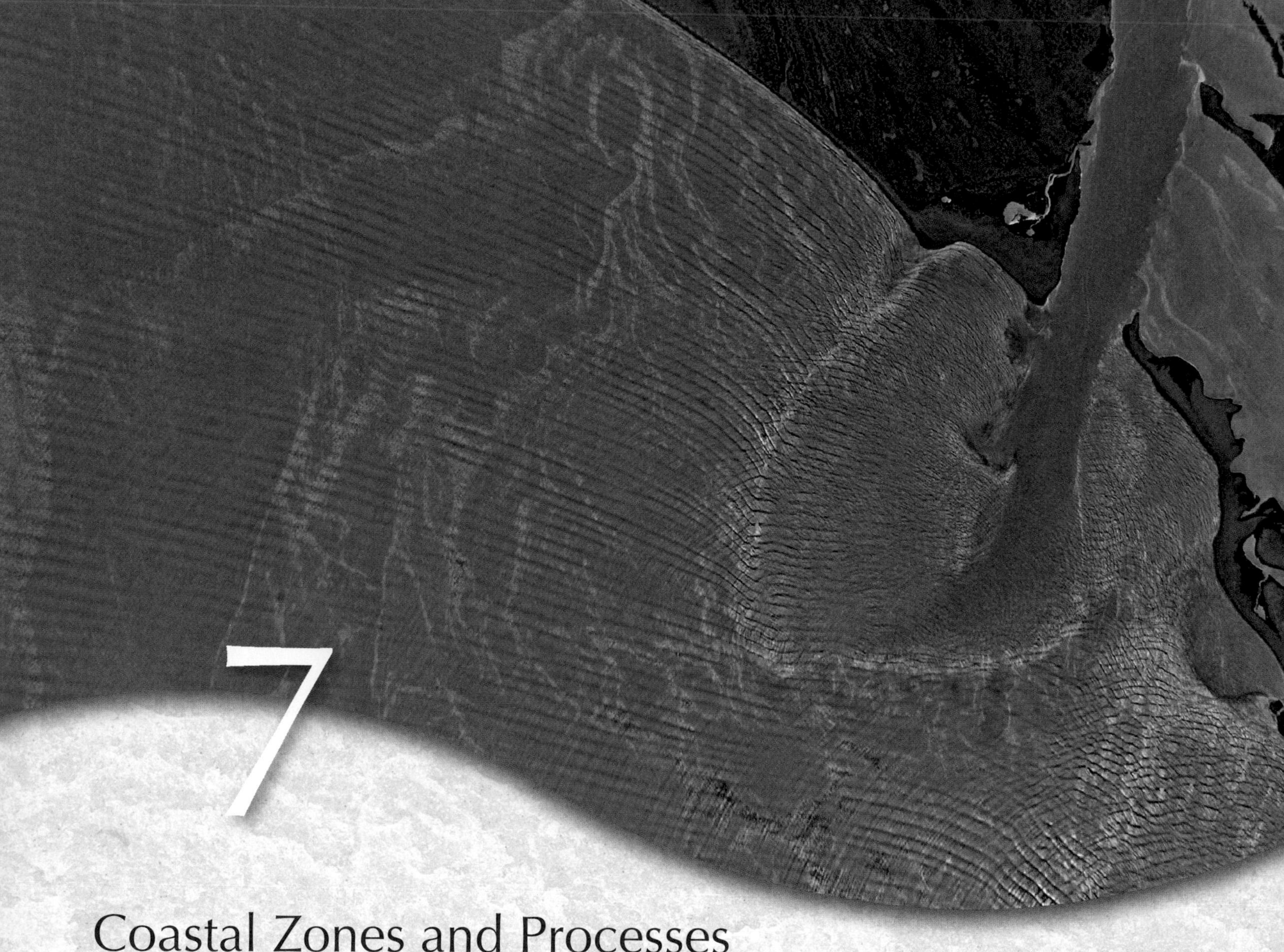

7

Coastal Zones and Processes

Coastal areas vary greatly in character and in the kinds and intensities of geologic processes that occur along them. They may be dynamic and rapidly changing under the interaction of land and water, or they may be comparatively stable. What happens in coastal zones is of direct concern to a large fraction of the people in the United States: 30 of the 50 states abut a major body of water (Atlantic or Pacific Ocean, Gulf of Mexico, one of the Great Lakes), and those 30 states are home to approximately 85% of the nation's population. To the extent that states become involved in coastal problems, they then affect all those people—in fact, about half of that 85% actually live in the coastal regions, not just in coastal states, and that population is growing rapidly (figure 7.1). When European settlers first colonized North America, many settled on the coast for easy access to shipping (for both imports and exports). Now, not only do the resultant coastal cities draw people to their employment, educational, and cultural opportunities; scenery and recreational activities in less-developed areas attract still more to the shore. Coastal areas also provide temporary or permanent homes for 45% of U.S. threatened or endangered species, 30% of waterfowl, and 50% of migratory songbirds. In this chapter, we will briefly review some of the processes that occur at shorelines, look at several different types of shorelines, and consider the impacts that various human activities have on them.

Sunlight reflecting off the water surface reveals complex wave patterns as the tide goes out in the Gulf of California. The freshwater stream entering from top of image adds a lobe of fresh surface water (outlined by whitish boundary) and further complexity.
Image courtesy NASA

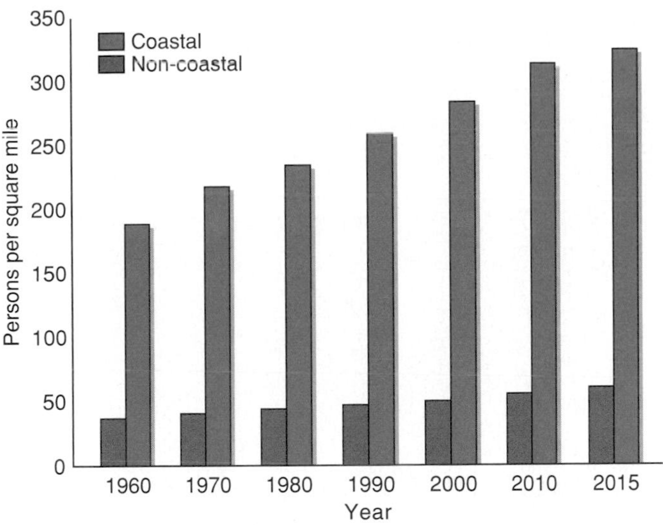

FIGURE 7.1 The density of population in coastal counties in the United States is far greater than in interior counties, growing rapidly, and is projected to continue to do so.

Data from National Oceanic and Atmospheric Administration, 2003

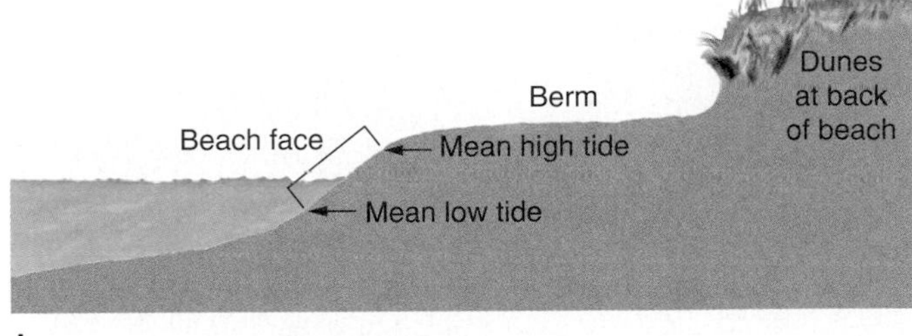

A

B

C

FIGURE 7.2 A beach. (A) Schematic beach profile. The beach may be backed by a cliff rather than by sand dunes. (B) The various components of the profile are clearly developed on this Hawaiian beach. (C) Where the sediment is coarser, as on this cobbled Alaskan shore, the slope of the beach face may be much steeper.

NATURE OF THE COASTLINE

The nature of a coastal region is determined by a variety of factors, including tectonic setting, the materials present at the shore, and the energy with which water strikes the coast.

One factor that influences the geometry of a coastline is plate tectonics. Continental margins can be described as *active* or *passive* in a tectonic sense, depending upon whether there is or is not an active plate boundary at the continent's edge. The western margin of North America is an **active margin,** one at which there is an active plate boundary, with various sections of it involved in subduction or transform faulting. An active margin is often characterized by cliffs above the waterline, a narrow continental shelf, and a relatively steep drop to oceanic depths offshore. The eastern margin of North America is a **passive margin,** far removed from the active plate boundary at the Mid-Atlantic Ridge. It is characterized by a broad continental shelf and extensive development of broad beaches and sandy offshore islands.

A **beach** is a gently sloping surface washed over by the waves and covered by sediment (figure 7.2). The sand or sediment of a beach may have been produced locally by wave erosion, transported overland by wind from behind the beach, or delivered to the coast and deposited there by streams or coastal currents. The **beach face** is that portion regularly washed by the waves as tides rise and fall. The *berm* is the flatter part of the beach landward of the beach face.

Waves and Tides

Waves and currents are the principal forces behind natural shoreline modification. Waves are induced by the flow of wind across the water surface, which sets up small undulations in that surface. The shape and apparent motion of waves reflect the changing geometry of the water surface; the actual motion of water molecules is quite different. While a wave may seem to travel long distances across the water, the water molecules are actually rising and falling locally in circular orbits that grow smaller with depth (figure 7.3A). (Think about a cork bobbing on rippling water: the cork mainly rises and falls as the ripples move across the surface.) As waves approach the shore and "feel bottom," or in other words, when water shallows to the point that the orbits are disrupted, the waves develop into breakers (figure 7.3B,C).

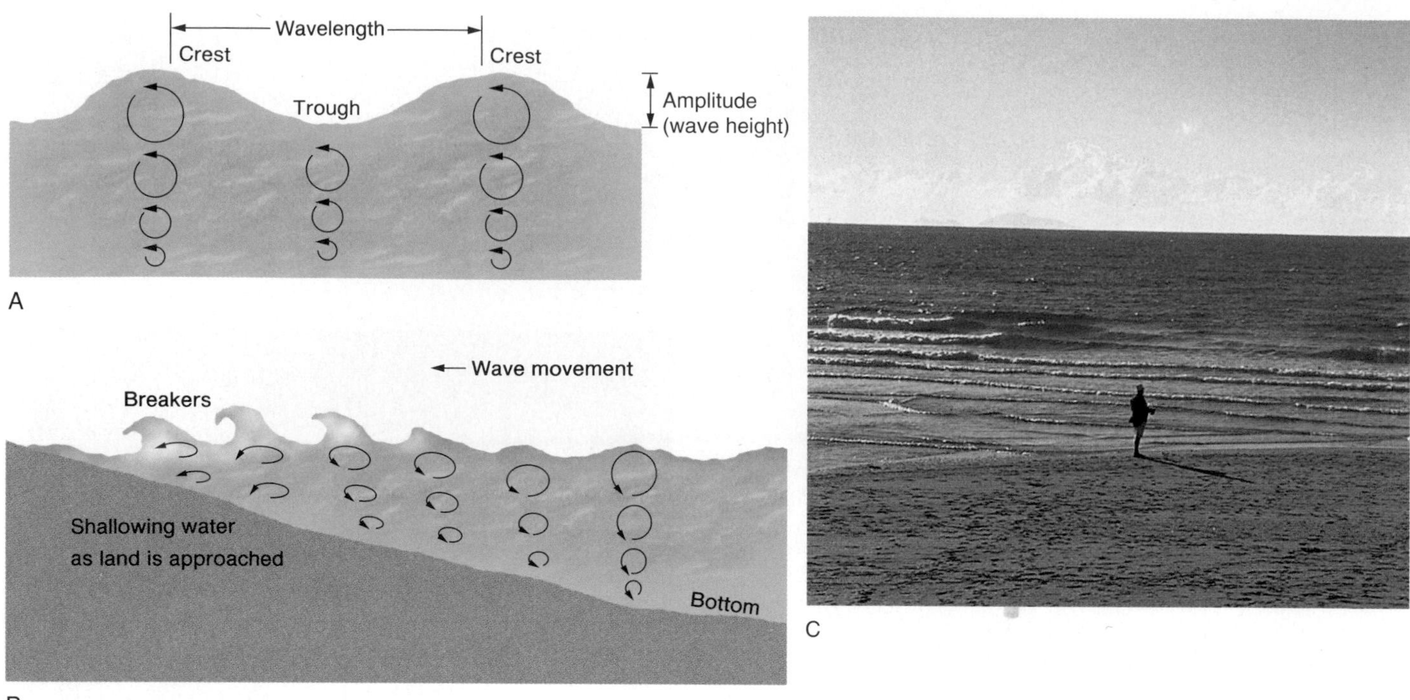

FIGURE 7.3 (A) Waves are undulations in the water surface, beneath which water moves in circular orbits. (B) Breakers develop as waves approach shore and orbits are disrupted. (C) Evenly spaced natural breakers developing on a gently sloping shore off the Australian coast. Note also the regular spacing of wave crests in the chapter-opening image.

The most effective erosion of solid rock along the shore is through wave action—either the direct pounding by breakers or the grinding effect of sand, pebbles, and cobbles propelled by waves, which is called **milling,** or *abrasion.* Logically, the erosive effects are concentrated at the waterline, where breaker action is most vigorous (figure 7.4). If the water is salty, it may cause even more rapid erosion through solution and chemical weathering of the rock. Erosion and transport of sediment are more complex, and will be discussed in some detail later in the chapter.

Water levels vary relative to the elevation of coastal land, over the short term, as a result of tides and storms. *Tides* are the periodic regional rise and fall of water levels as a consequence of the effects of the gravitational pull of the sun and moon on the watery envelope of oceans surrounding the earth. The earth's rotation tends to cause a bulge in this water mass, greatest near the equator where the velocity of the surface is most rapid, and decreasing near the poles. Superimposed on this effect of the earth's rotation, and acting at angles to it, are the effects of the gravitational pull of the sun and especially the moon. The closer an object, the stronger its gravitational attraction. The moon therefore pulls most strongly on matter on the side of the earth facing it, least strongly on the opposite side. The combined effects of gravity and rotation cause bulges in the water envelope, two that are especially pronounced: one facing the moon, where the moon's gravitational pull on the water is greatest, and one on the opposite side of the earth, where the gravitational pull is weakest and rotational forces dominate. As the rotating earth spins through these bulges of water each day, overall water level at a given point on the surface rises and falls twice a day. This is the phenomenon recognized as tides. At high tide, waves reach higher up the beach face; more of the beach face is exposed at low tide. Tidal extremes are greatest when sun, moon, and earth are all aligned, and the sun and moon are thus pulling together. The resultant tides are *spring tides* (figure 7.5A). (They have nothing to do with the spring season of the year.) When the sun and moon are pulling at right angles to each other, the difference between high and low tides is minimized. These are *neap tides* (figure 7.5B). The magnitude of water-level fluctuations in any one spot is also controlled, in part, by the local underwater topography. Storms can create further variations in water height and in intensity of erosion, as discussed later in the chapter.

Over longer periods, water levels may shift systematically up or down as a result of tectonic processes or from changes related to glaciation or human activities such as extraction of ground water or oil. These relative changes in land and water height may produce distinctive coastal features and may also aggravate coastal problems, as will be seen later in the chapter.

Sediment Transport and Deposition

Just as water flowing in streams moves sediment, so do waves and currents. When waves approach a beach at an angle, as the water washes up onto and down off the beach, it also moves laterally along the shoreline, creating a **longshore current**

A

B

FIGURE 7.4 (A) Sea arch formed by wave action along the California coast. Erosion is most intense at the waterline, undercutting the rock above. (B) Continued erosion causes the collapse of an arch, leaving monument-like *sea stacks.*

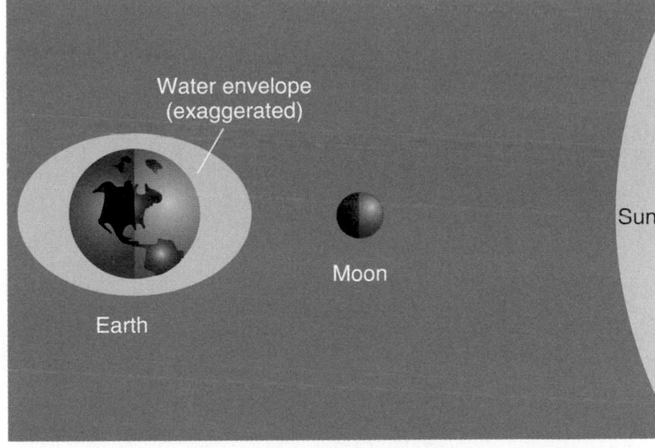

A

B

FIGURE 7.5 Tides. (A) Spring tides: Sun, moon, and earth are all aligned (near times of full and new moons). (B) Neap tides: When moon and sun are at right angles, tidal extremes are reduced. Note that these sketches are *not* to scale, and the sun, relatively, is *much* farther away.

(figure 7.6). Likewise, any sand caught up by and moved along with the flowing water is not carried straight up the beach but is transported at an angle to the shoreline. The net result is **littoral drift,** sand movement along the beach in the same general direction as the motion of the longshore current. Currents tend to move consistently in certain preferred directions on any given beach, which means that, over time, they continually transport sand from one end of the beach to the other. On many beaches where this occurs, the continued existence of the beach is assured by a fresh supply of sediment produced locally by wave erosion or delivered by streams.

The more energetic the waves and currents, the more and farther the sediment is moved. Higher water levels during storms (discussed below) also help the surf reach farther inland, perhaps all the way across the berm to dunes at the back of the beach. But a ridge of sand or gravel may survive the storm's fury, and may even be built higher as storm winds sweep sediment onto it. This is one way in which a *beach ridge* forms at the back of a beach. Where sediment supply is plentiful, as near the mouth of a sediment-laden river, or where the collapse of sandy cliffs adds sand to the beach, the beach may build outward into the water, new beach ridges forming seaward of the

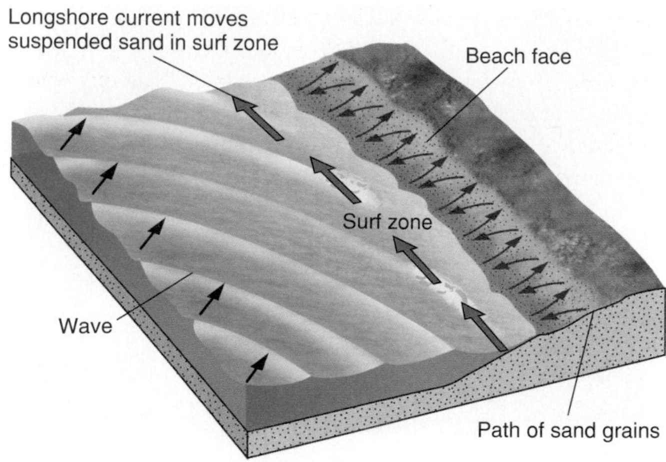

FIGURE 7.6 Longshore currents and their effect on sand movement. Waves push water and sand ashore at an angle; gravity drains water downslope again, leaving some sand behind.

old, producing a series of roughly parallel ridges that may have marshy areas between them.

Storms and Coastal Dynamics

Unconsolidated materials, such as beach sand, are especially readily moved and rapidly eroded during storms. The low air pressure associated with a storm causes a bulge in the water surface. This, coupled with strong onshore winds piling water against the land, can result in unusually high water levels during the storm, a storm **surge.** The temporarily elevated water level associated with the surge, together with unusually energetic wave action and greater wave height (figure 7.7), combine to attack the coast with exceptional force. The gently sloping expanse of beach (berm) along the outer shore above the usual high-tide line may suffer complete overwash, with storm waves reaching and beginning to erode dunes or cliffs beyond (figure 7.8). The post-storm beach profile typically shows landward recession of dune crests, as well as removal of considerable material from the zone between dunes and water (figure 7.8D).

Hurricanes Hugo in 1989, Andrew in 1992, and Katrina in 2005 illustrated the impact of storm surges especially

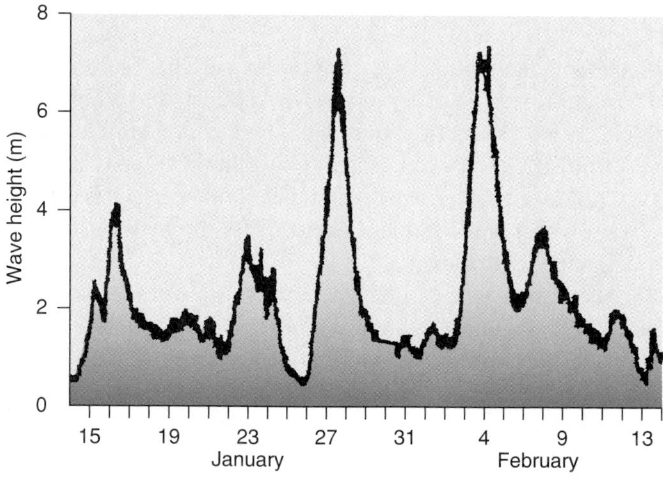

FIGURE 7.7 Storms in January and February 1998 pounded the east coast of the United States. Wave heights at Ocean City, Maryland, normally less than 1 meter (about 3 feet), were sometimes as high as 7 meters during the storms.

U.S. Geological Survey Center for Coastal Geology

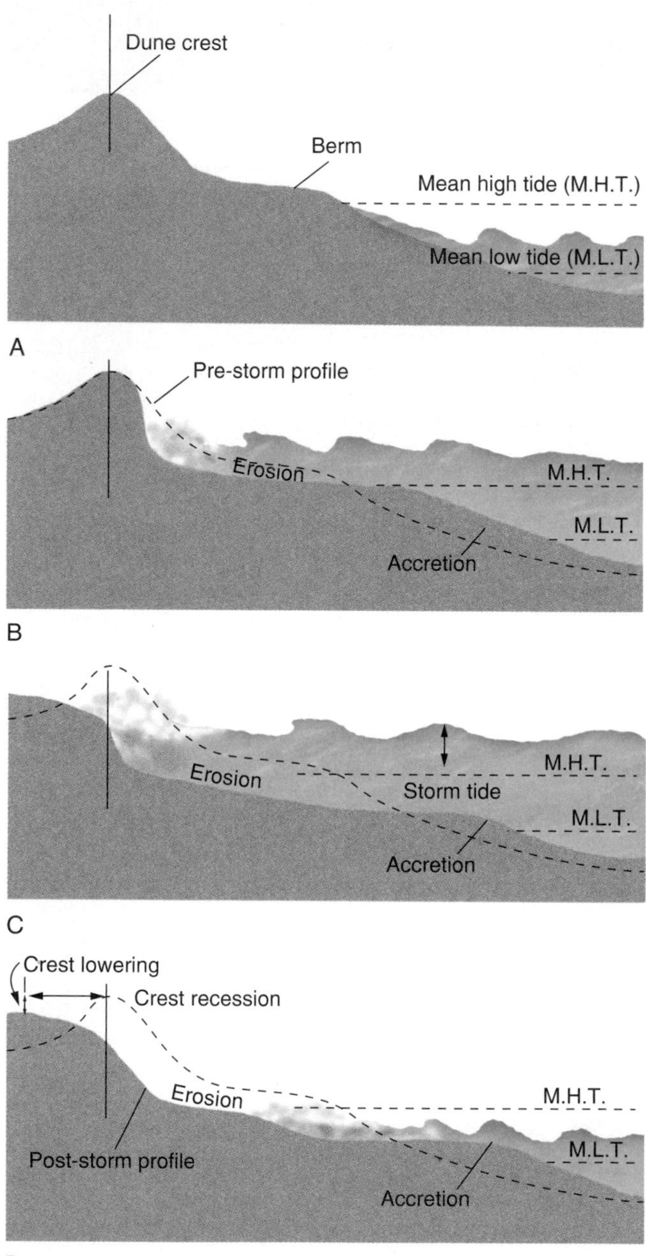

FIGURE 7.8 Alteration of shoreline profile due to accelerated erosion by unusually high storm tides. (A) Normal wave action. (B) Initial surge of storm waves. (C) Storm wave attack at front of dune. (D) After storm subsides, normal wave action resumes.

Source: Shore Protection Guidelines, *U.S. Army Corps of Engineers*

dramatically. To appreciate the magnitude of the effects, see table 7.1, and note that (considering the gentle slopes of many coastal regions) evacuations miles inland may be necessary in response to an anticipated surge of "only" 10 to 20 feet. (The highest storm surge ever recorded was 13 meters—about 42 feet!—from the Bathurst Bay, Australia, hurricane of 1899.) At least hurricanes are now tracked well enough that there is warning to make evacuation possible. Still, the damage to immobile property can be impressive (figure 7.9; table 7.2). Furthermore, as was shown with Hurricane Katrina, not everyone can or will heed a call to evacuate, and where a large city is involved, evacuation can involve large logistic problems.

TABLE 7.1

Saffir-Simpson Hurricane Scale

Category	Wind Speed (in miles per hour)	Storm Surge (feet above normal tides)	Evacuation
No. 1	74–94	4.5	No.
No. 2	96–110	6–8	Some shoreline residences and low-lying areas, evacuation required.
No. 3	111–130	9–12	Low-lying residences within several blocks of shoreline, evacuation possibly required.
No. 4	131–155	13–18	Massive evacuation of all residences on low ground within 2 miles of shore.
No. 5	155	18	Massive evacuation of residential areas on low ground within 5–10 miles of shore possibly required.

Source: National Weather Service, NOAA.

TABLE 7.2

A Sampling of Major U.S. Hurricanes

Hurricane	Date	Category at Landfall	Deaths*	Damages (Billions)†
Galveston, TX	1900	4	8000+	$31
Southeast FL/MS/AL	1926	4	243	$84
New England	1938	3	600	$19
Hazel (SC/NC)	1954	4	95	$ 8
Carla (TX)	1961	4	46	$ 8
Betsy (southeast FL/LA)	1965	3	75	$14
Camille (MS/southeast U.S.)	1969	5	256	$13
Agnes (northwest FL/northeast U.S.)	1972	1	122	$12
Hugo (SC)	1989	4	<30	$11
Andrew (southeast FL/LA)	1992	4	<30	$38
Opal (northwest FL/AL)	1995	3	<30	$ 3
Floyd (NC)	1999	2	57	$ 4.5
Isabel (Southeast U.S.)	2003	4	50	$ 1.7
Katrina (LA, MS, AL)	2005	4	~1850	$75

*Precise data not available where "<30" is indicated.

†Normalized to 1998 dollars and taking into account increases in local population and wealth, as well as inflation, since the hurricane occurred. In other words, for older hurricanes, this figure estimates the damage had they occurred in 1998.

Data for pre-1998 events from C. W. Lundsea and Pielke and Lundsea (1998).

FIGURE 7.9 Damage in Pass Christian, Mississippi, from Hurricane Katrina. Though the storm occurred in late August, wind-stripped trees are leafless; most houses have been obliterated entirely; seawall has failed and road is damaged.
Photograph courtesy USGS Coastal and Marine Geology Program

EMERGENT AND SUBMERGENT COASTLINES

Causes of Long-Term Sea-Level Change

Over the long stretches of geologic time, tectonics plays a subtle role. At times of more rapid seafloor spreading, there is more warm, young sea floor. In general, heat causes material to expand, and warm lithosphere takes up more volume than cold, so the volume of the ocean basins is reduced at such times, and water rises higher on the continents. Short-term tectonic effects can also have long-term consequences: as plates move, crumple, and shift, continental margins may be uplifted or dropped down. Such movements may shift the land by several meters in a matter of seconds to minutes, and then cease, abruptly changing the geometry of the land/water interface and the patterns of erosion and deposition.

The plastic quality of the asthenosphere means that the lithosphere sits somewhat buoyantly on it and can sink or rise in response to loading or unloading. For example, in regions overlain and weighted down by massive ice sheets during the last ice age, the lithosphere was downwarped by the ice load. Lithosphere is so rigid, and the drag of the viscous mantle so great, that tens of thousands of years later, the lithosphere is still slowly springing back to its pre-ice elevation. Where the thick ice extended to the sea, a consequence is that the coastline is slowly rising relative to the sea. This is well documented, for instance, in Scandinavia, where the rebound still proceeds at rates of up to 2 centimeters per year (close to 1 inch per year). Conversely, in basins being filled by sediment, like the Gulf of Mexico, the sediment loading can cause slow sinking of the lithosphere.

Glaciers also represent an immense reserve of water. As this ice melts, sea levels rise worldwide. This is perhaps the most widely experienced ongoing factor in changing coastal water levels; its consequences will be explored further later in the chapter. Global warming aggravates the sea-level rise in another way, too: warmed water expands, and this effect, multiplied over the vast volume of the oceans, could cause more sea-level rise than the melting of the ice itself, at least in the near future. There is also some concern that, if a large chunk of the Antarctic ice sheet, currently perched like a shelf over water, were suddenly to break off and plunge into the ocean, like a giant ice cube dropped into a full glass, a much more sudden sea-level rise would result.

Locally, pumping large volumes of liquid from underground has caused measurable—and problematic—subsidence of the land. In Venice, Italy, groundwater extraction that began in the 1930s had caused subsidence of up to 150 centimeters (close to 6 feet) by 1970, when the pumping was stopped—by which time a number of structures had been flooded. Oil extraction in Long Beach, California, caused similar subsidence problems earlier in the twentieth century.

Signs of Changing Relative Sea Level

The elevation at which water surface meets land depends on the relative heights of land and water, either of which can change. From a practical standpoint, it tends to matter less to people whether *global* sea level is rising or falling than whether the water is getting higher relative to the local coast (*submergent* coastline), or the reverse (*emergent* coastline). Often, geologic evidence will reveal the local trend.

A distinctive coastal feature that develops where the land is rising and/or the water level is falling is a set of **wave-cut platforms** (figure 7.10). Given sufficient time, wave action tends to erode the land down to the level of the water surface. If the land rises in a series of tectonic shifts and stays at each new elevation for some time before the next movement, each rise results in the erosion of a portion of the coastal land down to the new water level. The eventual product is a series of steplike terraces. These develop most visibly on rocky coasts, rather than coasts consisting of soft, unconsolidated material. The surface of each such step—each platform—represents an old water-level marker on the continent's edge.

When, on the other hand, global sea level rises or the land drops, one result is that streams that once flowed out to sea now have the sea rising partway up the stream valley from the mouth. A portion of the floodplain may be filled by encroaching seawater, forming a **drowned valley** (figure 7.11). A variation on the same theme is produced by the advance and retreat of coastal glaciers during major ice ages. Submergence of beach features, too, can signal a change in relative sea level (figure 7.12).

Glaciers flowing off land and into water do not just keep eroding deeper underwater until they melt. Ice floats; freshwater glacier ice floats especially readily on denser salt water. The carving of glacial valleys by ice, therefore, does not extend long distances offshore. During the last ice age, when sea level worldwide was as much as 100 meters lower than it now is,

Wave erosion tends to level land to sea level.

Land uplift and/or drop in sea level leads to a new "step" cut at new sea level.

A

B

FIGURE 7.10 Wave-cut platforms. (A) Wave-cut platforms form when land is elevated or sea level falls. (B) Wave-cut platforms at Mikhail Point, Alaska.

Photograph (B) by J. P. Schafer, USGS Photo Library, Denver, CO.

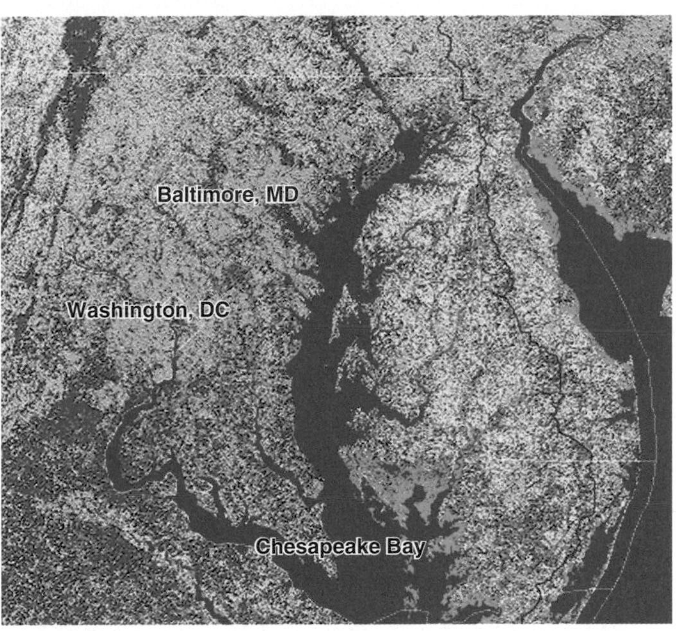

FIGURE 7.11 Chesapeake Bay is a drowned valley, as seen in this enhanced satellite image.

Image courtesy of Scott Goetz, The Woods Hole Research Center

FIGURE 7.12 Global sea-level change is swamping beach ridges at Cape Krusenstern National Monument, Alaska.

Photograph courtesy National Park Service

alpine glaciers at the edges of continental landmasses cut valleys into what are now submerged offshore areas. With the retreat of the glaciers and the concurrent rise in sea level, these old glacial valleys were emptied of ice and partially filled with water. That is the origin of the steep-walled fjords common in Scandinavian countries (especially Norway) and Iceland.

Present and Future Sea-Level Trends

Much of the coastal erosion presently plaguing the United States is the result of gradual but sustained global sea-level rise, from the combination of melting of glacial ice and expansion of the water itself as global temperatures rise, discussed earlier. The sea-level rise is currently estimated at about ⅓ meter (1 foot) per century. While this does not sound particularly threatening, two additional factors should be considered. First, many coastal areas slope very gently or are nearly flat, so that a small rise in sea level results in a far larger inland retreat of the shoreline (figure 7.13). Rates of shoreline retreat from rising sea level have, in fact, been measured at several meters per year in some low-lying coastal areas. A sea-level rise of ⅓ meter (about 1 foot) would erode beaches by 15 to 30 meters in the northeast, 65 to 130 meters in California, and up to 300 meters in Florida—and the average commercial beach is only about 30 meters wide now. Second, the documented rise of atmospheric carbon-dioxide levels, discussed in chapter 10, suggests that increased greenhouse-effect heating will melt the ice caps and glaciers more rapidly, as well as warming the oceans more, accelerating the sea-level rise; some estimates put the anticipated rise in sea level at 1 meter by the year 2100. Aside from the havoc this would create on beaches, it would flood 30 to 70% of U.S. coastal wetlands (figure 7.14).

Consistently rising sea level is a major reason why shoreline-stabilization efforts repeatedly fail. It is not just that the high-energy coastal environment presents difficult engineering

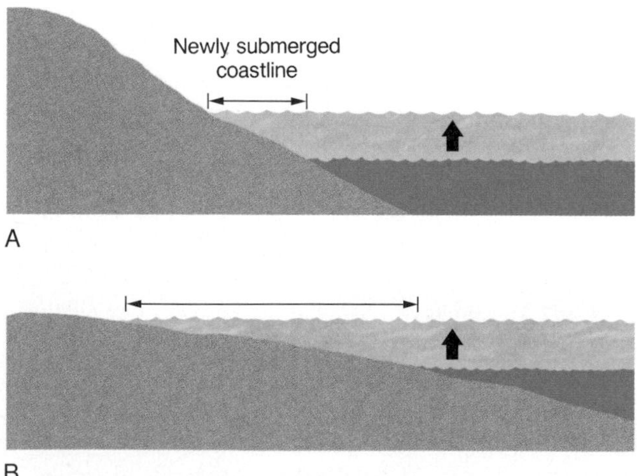

FIGURE 7.13 Effects of small rises in sea level on steeply sloping shoreline (A) and on gently sloping shore (B). Dashed lines indicate new, higher sea level. The same rise inundates a much larger expanse of land in case (B).

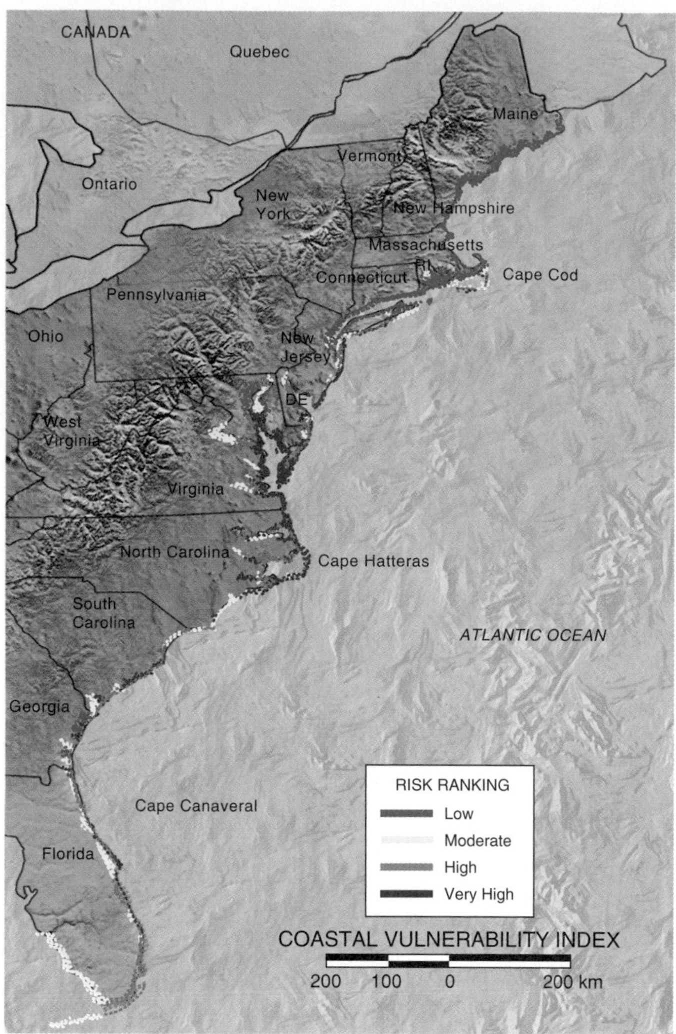

FIGURE 7.14 (A) Relative vulnerability of coastal regions to change resulting from rising sea level depends on factors such as topography and materials present. (B) Rising sea level causes flooding of coastal wetlands, increases salinity of estuaries; cypress in this flooded swamp died as a result.

(A) After USGS Fact Sheet FS-076-00 (B) Photograph from Jim Allen, USGS National Wetlands Research Center

A

B

C

FIGURE 7.15 The Cape Hatteras Lighthouse is a historic landmark. (A) It had been threatened for many years (note attempted shoreline stabilization structure in the 1996 photo). (B) Sandbagging had also been tried. (C) Finally the lighthouse was moved, at a cost of $11.8 million.

(A) and (C) Photographs courtesy USGS Center for Coastal Geology; (B) Photograph by R. Dolan, U.S. Geological Survey

problems. The problems themselves are intensifying as a rising water level brings waves farther and farther inland, pressing ever closer to and more forcefully against more and more structures developed along the coast.

COASTAL EROSION AND "STABILIZATION"

Coasts are dynamic, beaches especially so. Natural, undeveloped beaches achieve an equilibrium between sediment addition and sediment removal (at least in the absence of sustained shifts in sea level or climate). In fair weather, waves washing up onto the beach carry sand from offshore to build up the beach; storms attack the beach and dunes and carry sediment back offshore. Where longshore currents are present, the beach reflects the balance between sediment added from the up-current end—perhaps supplied by a stream delta—and sediment carried away down-current. Where sandy cliffs are eroding, they supply fresh sediment to the beach to replace what is carried away by waves and currents. Pressure to try to control or influence the natural flux of sediment arises when human development appears threatened and/or economic losses loom.

The approaches used fall into three broad categories. We will look at examples of each. *Hard structural stabilization,* as the term suggests, involves the construction of solid structures. *Soft structural stabilization* methods include sand replenishment and dune rebuilding or stabilization. These methods may be preferred for their more naturalistic look, but their effects can be fleeting. Nonstructural strategies, the third type, involve land-use restrictions, prohibiting development or mandating minimum setback from the most unstable or dynamic shorelines; this is analogous to restrictive zoning in fault zones or floodplains, and may meet with similar resistance. The move of the Cape Hatteras lighthouse (figure 7.15) is also an example of a nonstructural approach, sometimes described as "strategic retreat."

Beach Erosion, Protection, and Restoration

Beachfront property owners, concerned that their beaches (and perhaps their houses and businesses also) will wash away, erect structures to try to "stabilize" the beach, which generally end up further altering the beach's geometry. One commonly used method is the construction of one or more groins or jetties (figure 7.16)—long, narrow obstacles set more or less perpendicular to the shoreline. By disrupting the usual flow and velocity patterns of the currents, these structures change the shoreline. Currents slowed by such a barrier tend to drop their load of sand up-current from it. Below (down-current from) the barrier, the water picks up more sediment to replace the lost load, and the beach is eroded. The common result is that a formerly stable, straight shoreline develops an unnatural scalloped shape. The beach is built out in the area up-current of the groins and eroded landward below. Beachfront properties in the eroded zone may be more severely threatened after construction of the "stabilization" structures than they were before.

Interference with sediment-laden waters can cause redistribution of sand along beachfronts. A marina built out into such waters may cause some deposition of sand around it and, perhaps, in the protected harbor. Farther along the beach, the now unburdened waters can more readily take on a new sediment load of beach sand. Breakwaters, too, though constructed primarily to moderate wave action, may cause sediment redistribution (figure 7.17).

Even modifications far from the coast can affect the beach. One notable example is the practice, increasingly common over the last century or so, of damming large rivers for flood control, power generation, or other purposes. As indicated in chapter 6, one consequence of the construction of artificial reservoirs is that the sediment load carried by the stream is trapped behind the dam. Farther downstream, the cutoff of sediment supply to coastal beaches near the mouth of the stream can lead to erosion of the sand-starved beaches. It may be a difficult problem to solve, too, because no readily available alternate supply of sediment might exist near the problem beach areas. Flood-control dams, especially those constructed on the Missouri River, have reduced the sediment load delivered to the Gulf of Mexico by the Mississippi River by more than half over the last forty years, initiating rapid coastal erosion in parts of the Mississippi delta.

Where beach erosion is rapid and development (especially tourism) is widespread, efforts have sometimes been made to import replacement sand to maintain sizeable beaches. The initial cost of such an effort can be very high. Moreover, if no steps are, or can be, taken to reduce the erosion rates, the sand will have to be replenished over and over, at ever-rising cost, unless the beach is simply to be abandoned. Typically, the reason for beach "nourishment" or "replenishment" efforts is, after all, that natural erosion has removed much of the sand; it is then perhaps not surprising that continued erosion may make short work of the restored beach. Still, it was very likely a shock to those who paid $2.5 million for a 1982 beach-replenishment project in Ocean City, New Jersey, that the new beach lasted only 2½ months.

In many cases, too, it has not been possible—or has not been thought necessary—to duplicate the mineralogy or grain size of the sand originally lost. The result has sometimes been further environmental deterioration. When coarse sands are replaced by finer ones, softer and muddier than the original sand, the finer material more readily stays suspended in the water, clouding it. This is not only unsightly, but can also be deadly to organisms. Off Waikiki Beach in Hawaii and off Miami Beach, delicate coral reef communities have been damaged or killed by this extra water turbidity (cloudiness). Replacing finer sand

FIGURE 7.16 When groins or jetties perpendicular to the shoreline disrupt longshore currents, deposition occurs up-current, erosion below. Ocean City, New Jersey.
Photograph courtesy of S. Jeffress Williams

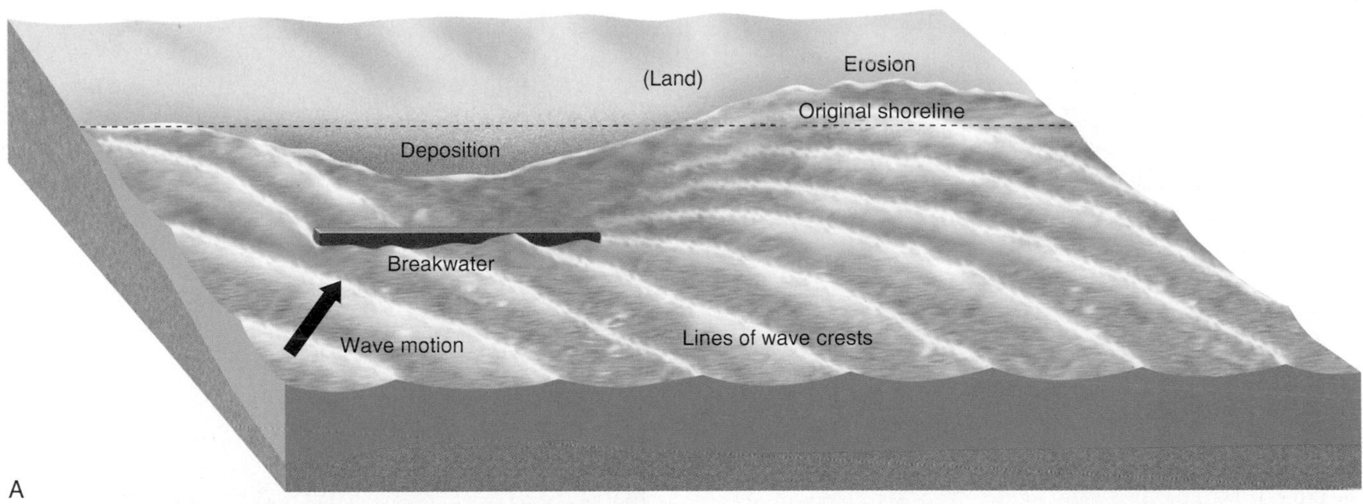

FIGURE 7.17 (A) Sediment is deposited behind a breakwater; erosion occurs down-current through the action of water that has lost its original sediment load. (B) Breakwaters and groins together combat beach erosion at Presque Isle State Park, Pennsylvania.
Photo by Ken Winters, U.S. Army Corps of Engineers

with coarser, conversely, may steepen the beach face (recall figures 7.2B and C), which may make the beach less safe for small children or poor swimmers.

Cliff Erosion

Both waves and currents can vigorously erode and transport loose sediments, such as beach sand, much more readily than solid rock. Erosion of sandy cliffs may be especially rapid. Removal of material at or below the waterline undercuts the cliff, leading, in turn, to slumping and sliding of sandy sediments and the swift landward retreat of the shoreline.

Not all places along a shoreline are equally vulnerable. Jutting points of land, or headlands, are more actively under attack than recessed bays because wave energy is concentrated on these headlands by **wave refraction,** deflection of the waves around irregularities in the coastline (figure 7.18). Wave refraction occurs because waves "touch bottom" first as they approach the projecting points, which slows the waves down; wave motion continues more rapidly elsewhere. The net result is deflection of the waves around and toward the headlands,

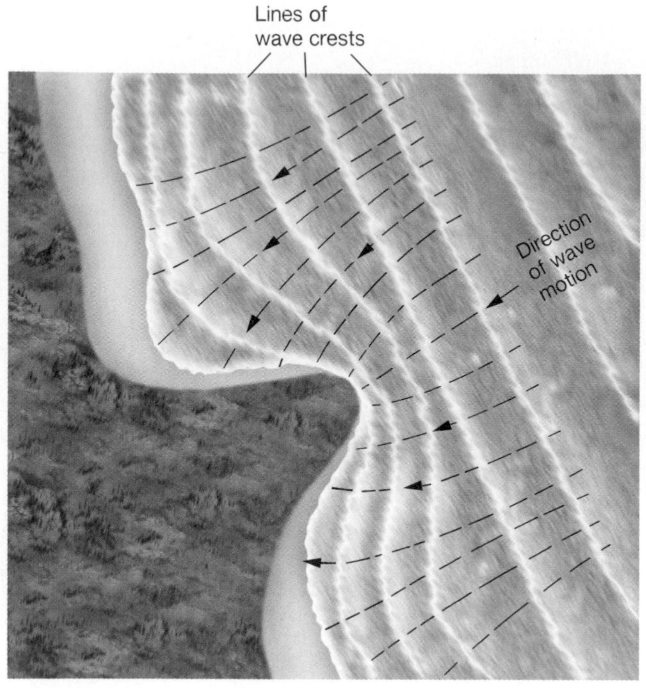

FIGURE 7.18 Wave refraction. (A) This sketch-map illustrates how the energy of waves approaching a jutting point of land is focused on it by refraction. (B) Wave refraction caused by nearshore coral reefs, Oahu, Hawaii. See also refraction around breakwater in figure 7.17B, and where waves meet freshwater lens in chapter-opening image.

where the wave energy is thus focused. (The effect is not unlike the refraction and focusing of light by a lens.) The long-term tendency is toward a rounding-out of angular coastline features, as headlands are eroded and sediment is deposited in the lower energy environments of the bays (figure 7.19).

Various measures, often as unsuccessful as they are expensive, may be tried to combat cliff erosion. A fairly common practice is to place some kind of barrier at the base of the cliff to break the force of wave impact (figure 7.20). The protection may take the form of a solid wall *(seawall)* of concrete or other material, or a pile of large boulders or other blocky debris *(riprap).* If the obstruction is placed only along a short length of cliff directly below a threatened structure (as in figure 7.20A), the water is likely to wash in beneath, around, and behind it, rendering it largely ineffective. Erosion continues on either side of the barrier; loss of the protected structure will be delayed but probably not prevented.

Wave energy may also bounce off, or be reflected from, a length of smooth seawall, to erode the beach more quickly, or attack a nearby unprotected cliff with greater force. The Galveston seawall (figure 7.21) was built to protect the city after the deadly 1900 hurricane. One unintended result was the total loss of the beach in front of the seawall.

FIGURE 7.19 As waves approach this section of the Oregon coast, projecting rocks are subject to highest-energy wave action; sand accumulates in recessed bays.

A

B

FIGURE 7.20 Some shore-protection structures. (A) Riprap near El Granada, California. Erosion of the sandy cliff will continue unabated on either side of riprap. (B) More substantial (and expensive) seawall protects estate at Newport, Rhode Island.
(A) Photograph courtesy USGS Photo Library, Denver, CO.

FIGURE 7.21 The Galveston seawall—note absence of beach!
Photograph © Laurence Parent

ESPECIALLY DIFFICULT COASTAL ENVIRONMENTS

Many coastal environments are unstable, but some, including barrier islands and estuaries, are particularly vulnerable either to natural forces or to human interference.

Barrier Islands

Barrier islands are long, low, narrow islands paralleling a coastline somewhat offshore from it (figure 7.22). Exactly how or why they form is not known. Some models suggest that they have formed through the action of longshore currents on delta

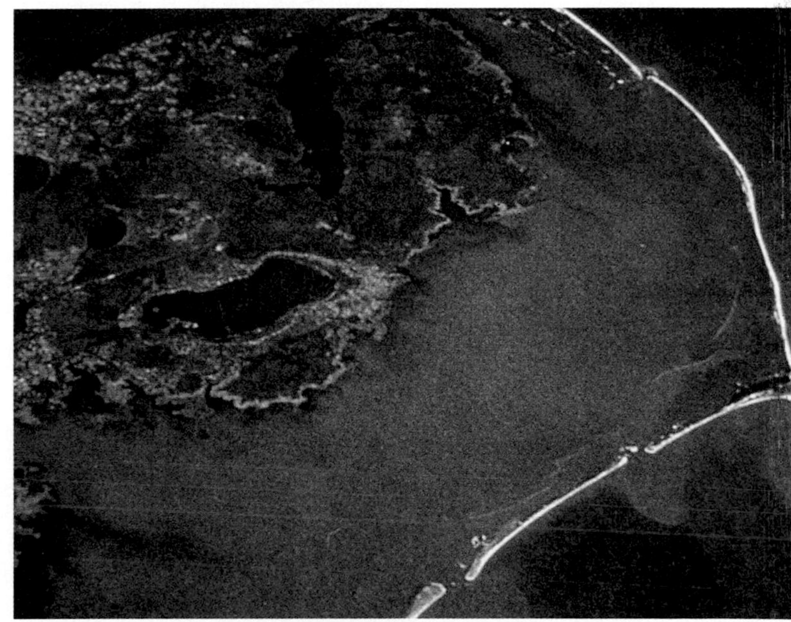

FIGURE 7.22 Barrier islands: North Carolina's Outer Banks.
Photograph by R. Dolan, USGS Photo Library, Denver, CO.

Box 7.1 The Great Lakes—A Somewhat Controllable Coastline?

The Great Lakes are freshwater lakes, but they are large enough to share many of the coastal dynamics of the seacoast, including significant shoreline erosion.

The water levels in the Great Lakes fluctuate for a variety of reasons. An obvious factor is variation in the input from precipitation and associated runoff to the lakes. Winds and storm surges temporarily raise local lake levels. Evaporation, water consumption in areas adjacent to the lakes, and outflow reduce water levels. Outflow from one lake to the next is through dams, locks, and power-generation facilities. Lake levels can, therefore, be regulated artificially, and the International Joint Commission makes policy decisions concerning lake-level adjustments. Still, there is a limit to how much control they can exercise, given the huge volume of water involved. Over the last century, the level of Lake Michigan has fluctuated over a range of about 2 meters, which may not sound like much, but can make a big difference in coastal erosion.

Periods of high water peaked in 1929, the early 1950s, 1968–69 (on Lakes Superior and Erie), 1973–74, and the mid-1980s. A 1971 study showed more than one-third of the Great Lakes coastline to be eroding significantly despite the presence in some places of shoreline protection structures. Beach-replenishment and shoreline-protection efforts are active in many places (figure 1). The low-lying beaches may expand during periods of low water and shrink as water levels rise, but the march of the cliffs is in one direction only: inland (figure 2). In those places most severely under attack, cliff edge retreat has occurred at rates averaging more than 2 meters (6 feet) per year for over a decade, and obviously, cliffs don't rebuild themselves when water levels drop. The costs and impacts of high water levels, especially when areas of extensive development are threatened, are enormous (figure 3). A new and growing concern now is the possible impact of long-term climate change: what if the region receives consistently increased precipitation as a result, and the water levels just keep rising?

A

B

FIGURE 1 (A) Beach nourishment project at St. Joseph, Michigan. (B) Looking at the adjacent shoreline, it is clear why this beach restoration at Sims Park, Ohio, on Lake Erie, requires substantial protection structures.

(A) Photograph courtesy U.S. Army Corps of Engineers; (B) Photograph by Ken Winters, U.S. Army Corps of Engineers

(Box 7.1 Continued)

A B

FIGURE 2 Shoreline erosion at Great Lakes, Michigan, during a storm (A) and the result (B).
(A) Photograph @ Jim Rider/AP/Wide World Photos; (B) @ The Times of Northwest Indiana, Sam Riche/AP/Wide World Photos

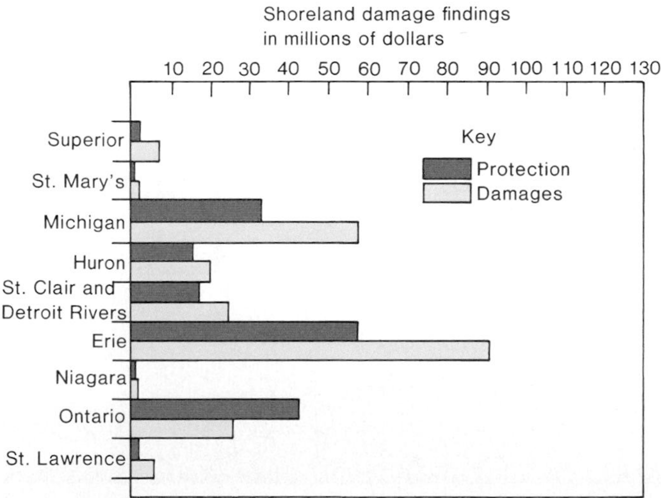

FIGURE 3 Cost of Great Lakes shore damage during 1972–74, 1975, and 1976, by lake area. Total damages were $231 million, total cost of protective structures was $170 million.

sands deposited at the mouths of streams. Their formation may also require changes in relative sea level. However they have formed, they provide important protection for the water and shore inland from them because they constitute the first line of defense against the fury of high surf from storms at sea. The barrier islands themselves are extremely vulnerable, partly as a result of their low relief. Water several meters deep may wash right over these low-lying islands during unusually high storm tides, such as occur during hurricanes (figure 7.23). Strong storms may even slice right through narrow barrier islands, separating people from bridges to the mainland, as in Hurricane Isabel in 2003 and Katrina in 2005.

Because they are usually subject to higher-energy waters on their seaward sides than on their landward sides, most barrier islands are retreating landward with time. Typical rates of retreat on the Atlantic coast of the United States are 2 meters (6 feet) per year, but rates in excess of 20 meters per year have been noted. Clearly, such settings represent particularly unstable locations in which to build, yet the aesthetic appeal of long beaches has led to extensive development on privately owned stretches of barrier islands (figure 7.24). About 1.4 million acres of barrier islands exist in the United States, and approximately 20% of this total area has been developed. Thousands of structures, including homes, businesses, roads, and bridges, are at risk.

On barrier islands, shoreline-stabilization efforts—building groins and breakwaters and replenishing sand—tend to be especially expensive and, frequently, futile. At best, the benefits are temporary. Construction of artificial stabilization structures may easily cost tens of millions of dollars and, at the same time, destroy the natural character of the shoreline and even the beach, which was the whole attraction for developers in the first place. At least as costly an alternative is to keep moving buildings or rebuilding roads ever farther landward as the beach before them erodes (recall the Cape Hatteras lighthouse). Expense aside, this is clearly a "solution" for the short term only, and in many cases, there is no place left to retreat anyway (consider figure 7.23). Other human activities exacerbate the plight of barrier islands: dams on coastal rivers in Texas have trapped much sediment, starving fragile barrier islands there of vital sand supply. Considering the inexorable rise of global sea level and that more barrier-island land is being submerged more frequently, the problems will only get worse.

FIGURE 7.23 Photographs of Dauphin Island, Alabama, immediately before Hurricane Lili (top) and after Hurricanes Ivan (middle) and Katrina (bottom). Note extreme erosion, landward shift of sand, and loss of houses due to Katrina. Structure at lower left of bottom photo is an oil rig that broke loose, drifted here, and ran aground.
Photographs courtesy USGS Coastal and Marine Geology Program

Estuaries

An **estuary** is a body of water along a coastline, open to the sea, in which the tide rises and falls and in which fresh and salt water meet and mix to create brackish water. San Francisco Bay, Chesapeake Bay, Puget Sound, and Long Island Sound are examples. Some estuaries form at the lower ends of stream valleys, especially in drowned valleys. Others may be tidal flats in which the water is more salty than not.

Over time, the complex communities of organisms in estuaries have adjusted to the salinity of that particular water. The salinity itself reflects the balance between freshwater input, usually river flow, and salt water. Any modifications that alter this balance change the salinity and can have a catastrophic impact on the organisms. Also, water circulation in estuaries is often very limited. This makes them especially vulnerable to pollution; because they are not freely flushed out by vigorous water flow, pollutants can accumulate. Unfortunately, many

FIGURE 7.24 Many barrier-island areas are heavily developed and highly vulnerable. South Dade County, Florida.
Photograph © Alan Schein Photography/Corbis

FIGURE 7.25 The Zuider Zee, a partially filled estuary in the Netherlands (1973 Landsat satellite photo). Patches of reclaimed land (polders), with date on which each was drained. Markerwaard was diked but not drained, for use as a freshwater reserve.
Image courtesy U.S. Geological Survey, EROS Data Center

vital wetlands areas are estuaries under pressure from environmental changes. One factor is that many of the world's large coastal cities—San Francisco and New York, for example—are located beside estuaries that may receive polluted effluent water. Another is that demand for the fresh water that would otherwise flow into estuaries is diminishing that flow. The salinity of San Francisco Bay has been rising markedly as fresh water from rivers flowing into the bay is consumed en route.

The problem with estuaries is not so much that natural changes in estuaries affect us but that we may be changing the estuaries for the worse. For example, where land is at a premium, estuaries are frequently called on to supply more. They may be isolated from the sea by dikes and pumped dry or, where the land is lower, wholly or partially filled in. Naturally, this further restricts water flow and also generally changes the water's chemistry. In addition, development may be accompanied by pollution. All of this greatly stresses the organisms of the estuary. Depending on the local geology, additional problems may arise. It was already noted in chapter 4 that buildings erected on filled land rather than on bedrock suffer much greater damage from ground shaking during earthquakes. This was abundantly demonstrated in San Francisco in 1906 and again in 1989. Yet the pressure to create dry land for development where no land existed before is often great.

One of the most ambitious projects involving land reclamation from the sea in an estuary is that of the Zuider Zee in the Netherlands (figure 7.25). The estuary itself did not form until a sandbar was breached in a storm during the thirteenth century. Two centuries later, with a growing population wanting more farmland, the reclamation was begun. Initially, the North Sea's access to the estuary was dammed up, and the flow of fresh water into the Zuider Zee from rivers gradually changed the water in it from brackish to relatively fresh. (This was necessary if the reclaimed land was to be used for farming; most crops grow poorly, if at all, in salty ground.) Portions were filled in to create dry land, while other areas were isolated by dikes and then pumped dry. More than half a million acres of new land have been created here.

Certainly, this has been a costly and prolonged effort. The marine organisms that had lived in the Zuider Zee estuary were necessarily destroyed when it was converted to a freshwater system. Continual vigilance over and maintenance of the dikes is required because if the farmlands were flooded with salt water, they would be unsuitable for growing crops for years.

In the Netherlands, where land is at a premium, the benefits of filling in the Zuider Zee may outweigh the costs, environmental and economic. Destruction of *all* estuaries would, however, result in the elimination of many life forms uniquely adapted to their brackish-water environment. Even organisms that do not live their whole lives in estuaries may use the estuary's quiet waters as breeding grounds, and so these populations also would suffer from habitat destruction. Where other land exists for development, it may be argued that the estuaries are too delicate and valuable a system to tamper with.

COSTS OF CONSTRUCTION—AND RECONSTRUCTION—IN HIGH-ENERGY ENVIRONMENTS

Historically, the most concentrated damage has resulted from major storms, and the number of people and value of property at risk only increases with the shift in population toward the coast, illustrated in figure 7.1. (Recall also table 7.2.) U.S. federal disaster relief funds, provided in barrier-island and other coastal areas, can easily run into hundreds of millions

Box 7.2 Hurricane Katrina and Coastal Vulnerability

Even in a year notable for the number of major hurricanes that occurred, Katrina stood out. Though it was only category 4 at landfall and weakened quickly thereafter, it was unusually large (figure 1), with hurricane-force winds extending 200 km (125 miles) outward from its center. Hurricane Camille, a category-5 storm in 1969, was less than half Katrina's size.

The height of storm surge depends not only on the intensity (category) of a hurricane, but on its size, speed, air pressure, and other factors, including local coastal topography. A big, slow-moving storm like Katrina can build up a tremendous surge, locally reported at up to 10 meters (over 30 feet)! Barrier islands, predictably, were hard hit, as illustrated in figure 7.23. On the mainland, things weren't necessarily any better (figure 2; recall also figure 7.9).

And then there was New Orleans. Much of the city actually lies below sea level, by as much as 5 meters, the result of decades of slow sinking. It has been protected by levees, not just along the Mississippi, but along the shore of Lake Pontchartrain, which is really a large estuary (visible in figure 1). The levees along Lake Pontchartrain had been built with a category 3 hurricane in mind. Unfortunately, Katrina was worse. Once some poorly constructed levees were overtopped, too, erosion behind them caused them to fail altogether. The result was inundation of a huge area of the city (figure 3) and standing water for weeks thereafter, as levees were patched and the water then pumped out.

The remaining question for many is: Now what? Some advocate abandoning the lowest, most flood-prone sections of New Orleans. Some suggest adding fill to raise rebuilt structures in such areas. Others advocate higher, stronger levees, and/or systems of gates to stem the influx of storm surge into Lake Pontchartrain. The costs of such enhanced flood-control measures are in the tens of billions of dollars. Meanwhile, New Orleans continues to sink, at rates recently measured at up to 28 mm (over 1 inch) per year. Likely causes include subsidence related to groundwater extraction, draining of wetlands, loss of sediments from the Mississippi, gradual displacement along a local fault, and settling of sediments; the latter may be exacerbated by the weight of additional fill or levees. Global sea-level rise will only worsen the problems. It will be interesting to see how the situation is resolved.

FIGURE 1 Hurricane Katrina, 29 August 2005. Eye is north of Lake Pontchartrain, the large embayment above the Mississippi delta along the southeastern edge of Louisiana. *Image by Jeff Schmaltz, MODIS Land Rapid Response Team*

(Box 7.2 Continued)

FIGURE 2 Section of shoreline, Biloxi, Mississippi, after Katrina. Even large buildings have been damaged—or in this case, moved entirely off their foundations; smaller ones are simply gone.
Photograph courtesy USGS Coastal and Marine Geology Program

FIGURE 3 Flooding in New Orleans from Katrina: Flooded areas in the city appear as dark brown or purplish, between Lake Pontchartrain at the top and the Mississippi River winding through the city across center of image.
Image courtesy Jesse Allen, Earth Observatory, using data obtained courtesy of the NASA/GSFC/METI/JAROS, and U.S./Japan ASTER Science Team

of dollars after each such storm. Increasingly, the question is asked, does it make sense to continue subsidizing and maintaining development in such very risky areas? More and more often, the answer being given is "No!" From 1978 to 1982, $43 million in federal flood insurance (see chapter 19) was paid in damage claims to barrier-island residents, which far exceeded the premiums they had paid. So, in 1982, Congress decided to remove federal development subsidies from about 650 miles of beachfront property and to eliminate in the following year the availability of federal flood insurance for these areas. It simply did not make good economic sense to encourage unwise development through such subsidies and protection. Still, more than a million flood-insurance policies representing over $1 billion in coverage were still outstanding half a dozen years later in vulnerable coastal areas.

A continuing (and growing) concern is one of fairness in the distribution of costs of coastal protection. While some stabilization and beach-replenishment projects are locally funded, many are undertaken by the U.S. Army Corps of Engineers. Thus, millions of federal taxpayer dollars may be spent—over and over—to protect a handful of private properties of individuals choosing to locate in high-risk areas. A national study done in March 2000 reported that beach renourishment must typically be redone every 3 to 7 years. For example, from 1962–1995, the beach at Cape May, New Jersey, had to be renourished ten times, at a total cost of $25 million; from 1952–1995, the Ocean City, New Jersey beach was supplied with fresh sand 22 times, costing over $83 million altogether. Reported costs of beach maintenance and renourishment in developed areas along the Florida and Carolina coasts ranged up to $17.5 million per mile. When Hurricane Katrina hit, about $150 million in beach-replenishment projects were already underway in Florida; Katrina undid some of that work and created a need for more.

Most barrier-island areas are owned by federal, state, or local governments. Even where they are undeveloped, much money has been spent maintaining access roads and various other structures, often including beach-protection structures. It is becoming increasingly obvious that the most sensible thing to do with many such structures is to abandon the costly and ultimately doomed efforts to protect or maintain them and simply to leave these areas in their dynamic, natural, rapidly changing state, most often as undeveloped recreation areas. Certainly there are many safer, more stable, and more practical areas in which to build. However, considerable political pressure may be brought to bear in favor of continuing, and even expanding, shoreline-protection efforts.

Recognition of Coastal Hazards

Conceding the attraction of beaches and coastlines and the likelihood that people will want to continue living along them, can at least the most unstable or threatened areas be identified so that the problems can be minimized? That is often possible, but it depends both on observations of present conditions and on some knowledge of the area's history.

The best setting for building near a beach or on an island, for instance, is at a relatively high elevation (5 meters or more above normal high tide, to be above the reach of most storm tides) and in a spot with many high dunes between the proposed building site and the water, to supply added protection. Thick vegetation, if present, will help to stabilize the beach sand. Also, information about what has happened in major storms in the past is very useful. Was the site flooded? Did the overwash cover the whole island? Were protective dunes destroyed? A key factor in determining a "safe" elevation, too, is overall water level. On a lake, not only the short-term range of storm tide heights but also the long-term range in lake levels must be considered. On a seacoast, it would be important to know if that particular stretch of coastline was emergent or submergent over the long term. Around the Pacific Ocean basin, possible danger from tsunamis should not be overlooked.

On either beach or cliff sites, one very important factor is the rate of coastline erosion. Information might be obtained from people who have lived in the area for some time. Better and more reliable guides are old aerial photographs, if available from the U.S. or state geological survey, county planning office, or other sources, or detailed maps made some years earlier that will show how the coastline looked in past times. Comparison with the present configuration and knowledge of when the photos were taken or maps made allows estimation of the rate of erosion (figure 7.26). It also should be kept in mind that the shoreline retreat may be more rapid in the future than it has been in the past as a consequence of a faster-rising sea level, as discussed earlier. On cliff sites, too, there is landslide potential to consider, and in a seismically active area, the dangers are greatly magnified.

Storms, of course, commonly accelerate change. Aerial photographs and newer tools such as scanning airborne laser altimetry allow monitoring of changes and suggest the possible magnitude of future storm damage. The U.S. Geological Survey and other agencies maintain archives of images that can be useful here.

It is advisable to find out what shoreline modifications are in place or are planned, not only close to the site of interest but elsewhere along the coast. These can have impacts far away from where they are actually built. Sometimes, aerial or even satellite photographs make it possible to examine the patterns of sediment distribution and movement along the coast, which should help in the assessment of the likely impact of any shoreline modifications. The fact that such structures exist or are being contemplated is itself a warning sign! A history of repairs to or rebuilding of structures not only suggests a very dynamic coastline, but also the possibility that protection efforts might have to be abandoned in the future for economic reasons.

FIGURE 7.26 Building about to be lost to cliff erosion. This cliff at Moss Beach in San Mateo County, California, has retreated more than 50 meters in a century. Past cliff positions are known from old maps and photographs; arrows indicate position of cliff at corresponding dates.
Photograph by K. R. LaJoie, USGS Photo Library, Denver, CO.

SUMMARY

Many coastal areas are rapidly changing. Accelerated by rising sea levels worldwide, erosion is causing shorelines to retreat landward in most areas, often at rates of more than a meter a year. Sandy cliffs and barrier islands are especially vulnerable to erosion. Storms, with their associated surges and higher waves, cause especially rapid coastal change. Efforts to stabilize beaches are generally expensive, often ineffective over the long (or even short) term, and frequently change the character of the shoreline. They may also cause unforeseen negative consequences to the coastal zone and its organisms. Demand for development has led not only to construction on unstable coastal lands, but also to the reclamation of estuaries to create more land, to the detriment of plant and animal populations.

Key Terms and Concepts

Exercises

Questions for Review

1. High storm tides may cause landward recession of dunes. Explain this phenomenon, using a sketch if you wish.
2. Evaluate the use of riprap and seawalls as cliff protection structures.
3. Explain longshore currents and how they cause littoral drift.
4. Sketch a shoreline on which a jetty has been placed to restrict littoral drift; indicate where sand erosion and deposition will subsequently occur and how this will reshape the shoreline.
5. Discuss the pros and cons of sand replenishment as a strategy for stabilizing an eroding beach.
6. Describe three ways in which the relative elevation of land and sea may be altered. What is the present trend in global sea level?
7. Briefly explain the formation of (a) wave-cut platforms and (b) drowned valleys.
8. What are barrier islands, and why have they proven to be particularly unstable environments for construction?
9. What is an estuary? Why do estuaries constitute such distinctive coastal environments?
10. Briefly describe at least two ways in which the dynamics of a coastline over a period of years can be investigated.

For Further Thought

1. Choose any major coastal city, find out how far above sea level it lies, and determine what proportion of it would be inundated by a relative rise in sea level of (a) 1 meter and (b) 5 meters. Consider what defensive actions might be taken to protect threatened structures in each case. (It may be instructive to investigate what has happened to Venice, Italy, under similar circumstances.)
2. Consider the vulnerability of barrier-island real estate, and think about what kinds of protections and controls you might advocate: How much should a prospective purchaser of the property be told about beach stability? Should local property owners pay for shoreline-protection structures, or should state or federal taxes be used? Should property owners be prevented from developing the land, for their own protection? What would you consider the most important factors in your decisions? What are the main unknowns?
3. If you live in a near-coastal area, calculate how soon water would reach your home at a rate of sea-level rise of about 1 meter per century.
4. The fictional film *The Day After Tomorrow* depicts a supposed storm surge that immerses the Statue of Liberty up to her neck. In round numbers, that indicates a surge of about 250 feet. Use the data from table 7.1 to make a graph of surge height versus wind velocity, and extrapolate to estimate the wind speed required to create a surge that large. Compare your estimate with the highest measured natural wind speed, 231 mph, and consider the plausibility of the movie's depiction.

Suggested Readings/References

Ackerman, J. 1997. Islands at the edge. *National Geographic* (August): 2–29.

Bohannon, J. 2005. A sinking city yields some secrets. *Science* 309 (23 September): 1978–80.

Dolan, R., and Goodell, H. G. 1986. Sinking cities. *American Scientist* 74 (January–February): 38–47.

Douglas, B. C., and Peltier, W. R. 2002. The puzzle of global sea-level rise. *Physics Today* (March): 35–40.

Fischetti, M. 2001. Drowning New Orleans. *Scientific American* 285 (October): 1–11.

——— 2006. Protecting New Orleans. *Scientific American* (February): 64–71.

Hayes, B. 2005. Natural and unnatural disasters. *American Scientist* 93 (November–December): 496–99.

Horton, T. 1993. Hanging in the balance: Chesapeake Bay. *National Geographic* 183 (June): 3–35.

Kantha, L. 2006. Time to replace the Saffir-Simpson hurricane scale? *EOS* 87 (3 January): 3, 6.

Phillips, A. 2000. Tall order: Cape Hatteras lighthouse makes tracks. *National Geographic* (May): 98–105.

Pielke, R. A. Jr., and Landsea, C. W. 1998. Normalized Atlantic hurricane damage 1925–1995. *Weather Forecasting* 13: 621–31.

Pilkey, O. H. 2005. Beaches awash with politics. *Geotimes* (July): 38–39, 50.

Thompson, D. A., and Yocom, T. G. 1993. Uncertain ground. *Technology Review* (August–September): 20–29.

Urban, M. J. 2006. Setting the stage for disaster: A look at the conditions surrounding Katrina's landfall in New Orleans. *The Professional Geologist* (January/February): 6–9.

U.S. Geological Survey. 1997. Coastal wetlands and global change: Overview. USGS Fact Sheet FS-089-97.

Walden, D. 1990. Raising Galveston. *American Heritage of Invention and Technology* (Winter): 8–18.

Williams, T. 2006. The last line of defense. *Audubon* (May–June): 56–71.

Welcome to McGraw-Hill's Online Learning Center

Location: http://www.mhhe.com/montgomery8e

Mc Graw Hill

NetNotes

The National Oceanic and Atmospheric Administration (NOAA), alone or in collaboration with other agencies, conducts a great deal of oceanographic and coastal research. For example, see the following:

Marine geology and geophysics
www.ngdc.noaa.gov/mgg/

Great Lakes Environmental Research Labs
www.glerl.noaa.gov/data/data.html

NOAA/National Geophysical Data Center coastal programs
www.ngdc.noaa.gov/mgg/coastal/coastal.html

Excellent data on hurricanes and other major storms are available through Frequently Asked Questions at
www.aoml.noaa.gov/hrd/tcfaq/

For online images of hurricanes and other major storms:
www.ncdc.noaa.gov/ol/satellite/olimages.html

The U.S. Fish and Wildlife Service Coastal Program deals more with coastal biological environments than coastal erosion; see
www.fws.gov/cep/cepcode.html

Federal Emergency Management Agency Fact Sheet on Hurricanes:
www.fema.gov/library

USGS Western Region Marine and Coastal Surveys includes links to case studies and illustrations of a variety of coastal and nearshore areas at
walrus.wr.usgs.gov/hazards/erosion.html

The USGS Center for Coastal and Marine Geology offers links to a variety of case histories and images as well as basic information:
coastal.er.usgs.gov/

USGS programs relating to the Great Lakes are described at
water.usgs.gov/wid/html/gl.html

Among online USGS publications relevant to this chapter are:

Collins, B. Minasian, D., Reiss, T., and Gibbons, H. 2006. USGS scientists revisit New Orleans levee breaks to collect high-accuracy survey data.
soundwaves.usgs.gov/2006/05/

Williams, S. J., Dodd, K., and Gohn, K. K. 1997. Coasts in crisis (USGS Circular 1075)
pubs.usgs.gov/circ/c1075/

See also an online publication of changes to the San Francisco Bay estuary at
sfbay.wr.usgs.gov/

The California Coastal Records Project is developing an extensive archive of photographs to document change along the California coast:
californiacoastline.org

8

Mass Movements

While the internal heat of the earth drives mountain-building processes, just as inevitably, the force of gravity acts to tear the mountains down. Gravity is the great leveler. It tugs constantly downward on every mass of material everywhere on earth, causing a variety of phenomena collectively called **mass wasting,** or **mass movements,** whereby geological materials are moved downward, commonly downslope, from one place to another. The movement can be slow, subtle, almost undetectable on a day-to-day basis but cumulatively large over days or years. Or the movement can be sudden, swift, and devastating, as in a rockslide or avalanche. Mass movements need not even involve slipping on a slope: vertical movements include surface subsidence, as from extraction of oil or ground water, or collapse into sinkholes as described further in chapter 11.

Landslide is a general term for the results of rapid mass movements. Overall, landslide damage is greater than one might imagine from the usually brief coverage of landslides in the news media. In the United States alone, landslides and other mass movements cause over $3.5 billion in property damage every year, and 25 to 50 deaths. Many landslides occur quite independently of human activities. In some areas, active steps have been taken to control downslope movement or to limit its damage. On the other hand, certain human activities aggravate local landslide dangers, usually as a result of failure to take those hazards into account.

Large areas of this country—and not necessarily only mountainous regions—are potentially at risk from landslides (figure 8.1). It is important to realize, too, that given the area represented in that figure, it is necessarily a generalization, representing *regional* landslide hazard levels associated with the characteristic regional topography, soil type, etc. One can live in a high-risk spot within an area generally classified as low-risk on this map!

Laguna Beach, CA landslide, 1 June 2005.
Photograph by Gerald Bawden, U.S. Geological Survey

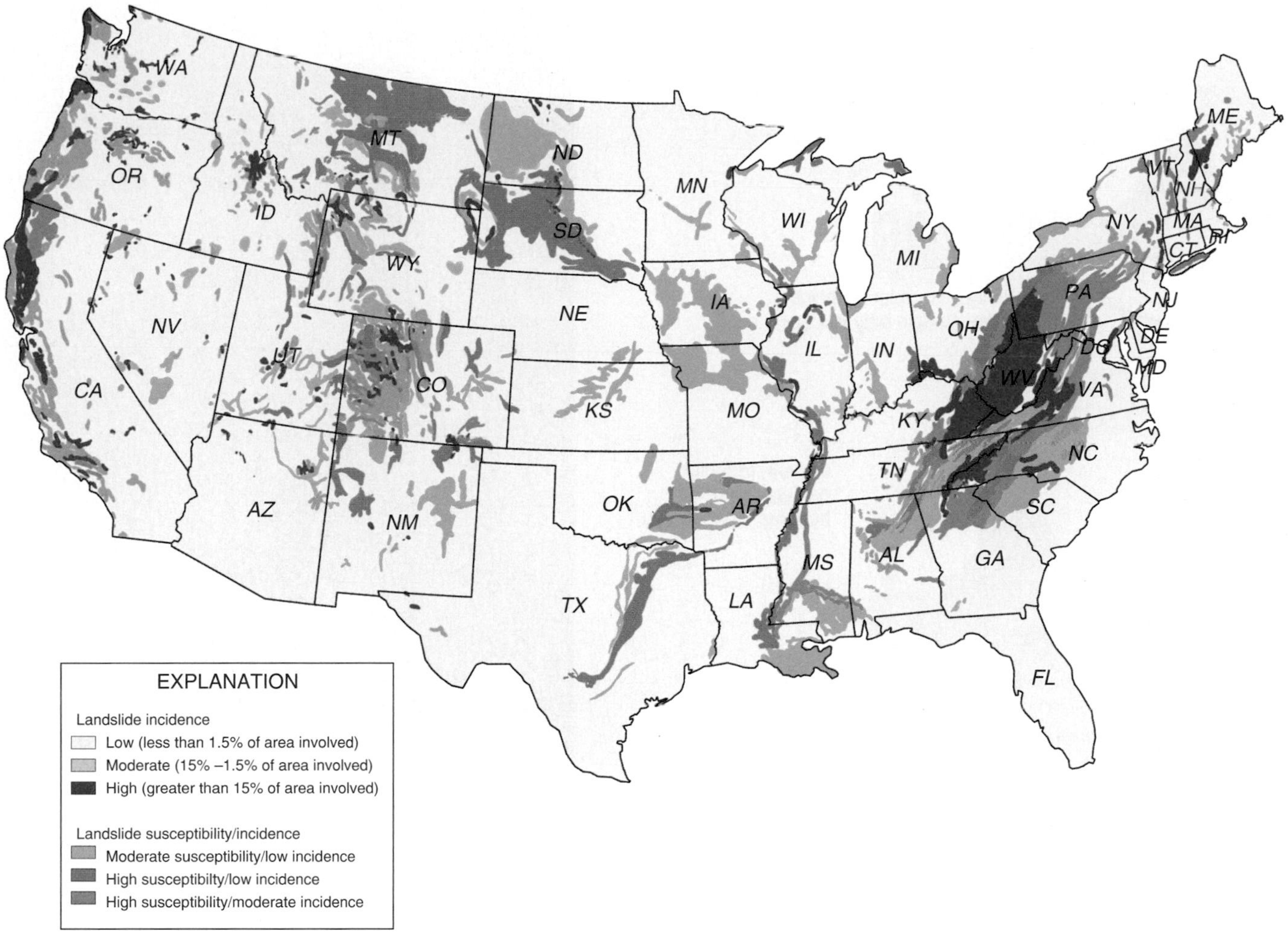

WA
OR
CA
NV
ID
MT
WY
UT
AZ
NM
CO
ND
SD
NE
KS
OK
TX
MN
IA
MO
AR
LA
WI
IL
MI
IN
OH
KY
TN
MS
AL
GA
FL
SC
NC
VA
WV
PA
NY
ME
VT
NH
MA
RI
CT
NJ
DE
MD
DC
EXPLANATION
Landslide incidence
Low (less than 1.5% of area involved)
Moderate (15% –1.5% of area involved)
High (greater than 15% of area involved)
Landslide susceptibility/incidence
Moderate susceptibility/low incidence
High susceptibilty/low incidence
High susceptibility/moderate incidence

FIGURE 8.1 Landslide potential of regions in the United States.
After U.S. Geological Survey Open-File Report 97-289

FACTORS INFLUENCING SLOPE STABILITY

Basically, mass movements occur whenever the downward pull of gravity overcomes the forces—usually frictional—resisting it. The downslope pull tending to cause mass movements, called the **shearing stress,** is related to the mass of material and the slope angle. Counteracting the shearing stress is *friction,* or, in a coherent solid, **shear strength.** When shearing stress exceeds frictional resistance or the shear strength of the material, sliding occurs. Therefore, factors that increase shearing stress, decrease friction, or decrease shear strength tend to increase the likelihood of sliding, and vice versa. Application of these simple principles is key to reducing landslide hazards.

Friction
Shearing stress
Gravity
Friction
Shearing stress
Gravity

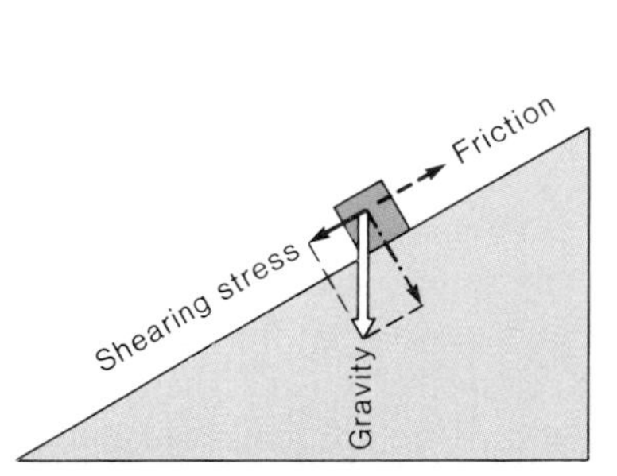

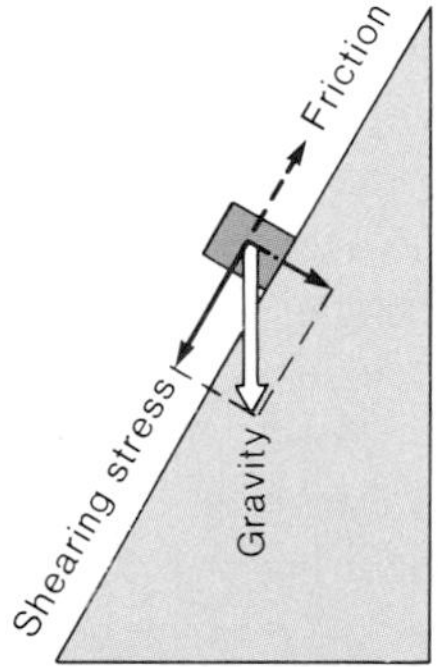

FIGURE 8.2 Effects of Slope Geometry on Slide Potential. The mass of the block and thus the total downward pull of gravity are the same in both cases, but the steeper the slope, the greater the shearing stress component. An increase in pore pressure can decrease frictional resistance to shearing stress.

Effects of Slope and Materials

All else being equal, the steeper the slope, the greater the shearing stress (figure 8.2) and therefore the greater the likelihood of slope failure. For dry, unconsolidated material, the **angle of repose** is the maximum slope angle at which the material is stable (figure 8.3). This angle varies with the material. Smooth, rounded particles tend to support only very low-angle slopes

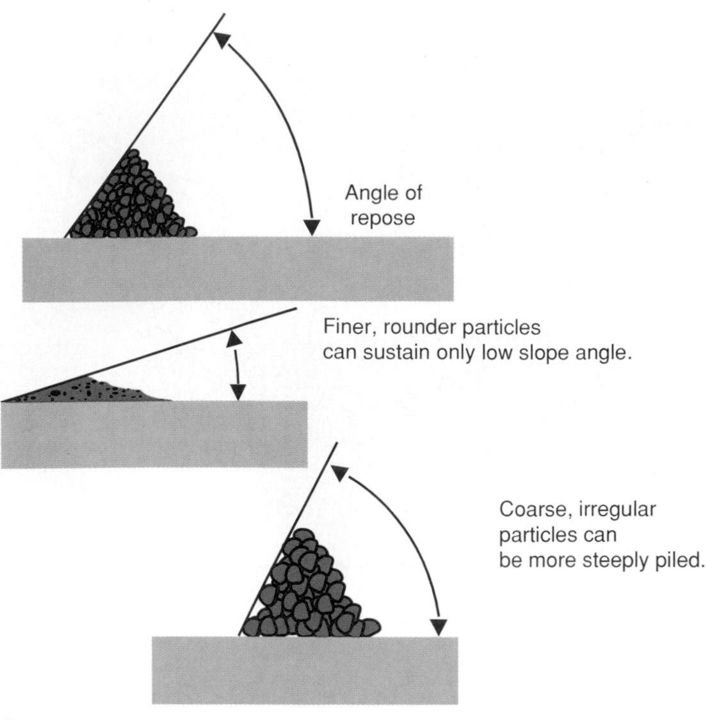

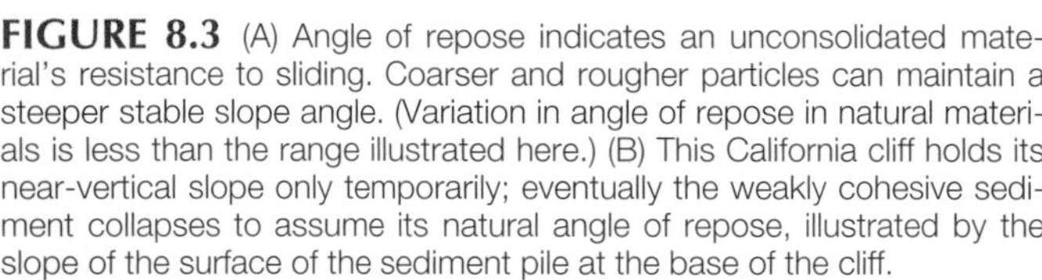

FIGURE 8.3 (A) Angle of repose indicates an unconsolidated material's resistance to sliding. Coarser and rougher particles can maintain a steeper stable slope angle. (Variation in angle of repose in natural materials is less than the range illustrated here.) (B) This California cliff holds its near-vertical slope only temporarily; eventually the weakly cohesive sediment collapses to assume its natural angle of repose, illustrated by the slope of the surface of the sediment pile at the base of the cliff.
(B) Photograph by J. T. McGill, USGS Photo Library, Denver, CO.

(imagine trying to make a heap of marbles or ball bearings), while rough, sticky, or irregular particles can be piled more steeply without becoming unstable. Other properties being equal, coarse fragments can usually maintain a steeper slope angle than fine ones. The tendency of a given material to assume a constant characteristic slope can be seen in such diverse geologic forms as cinder cones (figure 8.4A) and sand dunes (figure 8.4B).

Solid rock can be perfectly stable even at a vertical slope but may lose its strength if it is broken up by weathering or fracturing. Also, in layered sedimentary rocks, there may be weakness along bedding planes, where different rock units are imperfectly held together; some units may themselves be weak or even slippery (clay-rich layers, for example). Such planes of weakness are potential slide or failure planes.

Slopes may be steepened to unstable angles by natural erosion by water or ice. Erosion also can undercut rock or soil, removing the support beneath a mass of material and thus leaving it susceptible to falling or sliding. This is a common contributing factor to landslides in coastal areas and along stream valleys (figure 8.5).

Over long periods of time, slow tectonic deformation also can alter the angles of slopes and bedding planes, making them steeper or shallower. This is most often a significant factor in young, active mountain ranges, such as the Alps or the coast ranges of California. Steepening of slopes by tectonic movements has been suggested as the cause of a large rockslide in Switzerland in 1806 that buried the township of Goldan under a block of rock nearly 2 kilometers long, 300 meters across, and 60 to 100 meters thick. In such a case, past evidence of landslides may be limited in the immediate area because the dangerously steep slopes have not existed long in a geologic sense.

Effects of Fluid

The role of fluid in mass movements is variable. Addition of some moisture to dry soils may increase adhesion, the tendency of the particles to stick together (it takes damp sand to make a sand castle). However, saturation of unconsolidated materials reduces the friction between particles that otherwise provides cohesion and strength, and the reduced friction can destabilize

A

B

FIGURE 8.4 (A) Angle of repose of volcanic cinders is shown not only by the smoothly sloping sides of each individual cinder cone in the summit crater of Haleakala volcano in Hawaii, but also by the consistent slope and shape from cone to cone. (B) The uniform slope on the face of this sand dune is maintained by slumping to the sand's angle of repose. Oregon Dunes National Recreation Area.

FIGURE 8.5 Undercutting of this sea cliff (despite protective structures), coupled with heavy rains, caused slope failure in Bodega Bay, CA.

FIGURE 8.6 Instability in expansive clay seriously damaged this road near Boulder, Colorado.
Photograph courtesy USGS Photo Library, Denver, CO.

a slope. As was described in chapter 4, elevated pore-fluid pressure can trigger earthquakes by facilitating slip in faulted rocks under tectonic stress. It is also effective in promoting sliding in rocks under stress due to gravity, reducing the force holding the rock to the plane below (vector perpendicular to slope in figure 8.2). The very mass of water in saturated soil adds extra weight, and thus extra downward pull.

Aside from its role in erosion, water can greatly increase the likelihood of mass movements in other ways. It can seep along bedding planes in layered rock, reducing friction and making sliding more likely. The expansion and contraction of water freezing and thawing in cracks in rocks or in soil can act as a wedge to drive chunks of material apart *(frost wedging). Frost heaving,* the expansion of wet soil as it freezes and the ice expands, loosens and displaces the soil, which may then be more susceptible to sliding when it next thaws.

Some soils rich in clays absorb water readily; one type of clay, montmorillonite, may take up twenty times its weight in water and form a very weak gel. Such material fails easily under stress. Other clays expand when wet, contract when dry, and can destabilize a slope in the process (figure 8.6; see also chapter 20).

Sudden, rapid landslides commonly involve a triggering mechanism. Heavy rainfall or rapid melting of abundant snow can be a very effective trigger, quickly adding weight, decreasing friction, and increasing pore pressure. As with flooding, the key is not simply how much water is involved, but how rapidly it is added and how close to saturated the system was already. Tropical regions subject to intense cloudbursts may also be subject to frequent slides in thick tropical soils—which may tend to be clay-rich—and in severely weathered rock (figure 8.7).

Effects of Vegetation

Plant roots, especially those of trees and shrubs, can provide a strong interlocking network to hold unconsolidated materials together and prevent flow (figure 8.8). In addition, vegetation takes up moisture from the upper layers of soil and can thus reduce the overall moisture content of the mass, increasing its shear strength. Moisture loss through the vegetation by transpiration also helps to dry out sodden soil more quickly. Commonly, then, vegetation tends to increase slope stability. However, the plants also add weight to the slope. If the added weight is large and the root network of limited extent, the vegetation may have a destabilizing effect instead. Also, plants that take up so much water that they dry out soil until it loses its cohesion can aggravate the slide hazard.

A relatively recently recognized hazard is the fact that when previously forested slopes are bared by fires or by logging, they may become much more prone to sliding than before. Previous illegal logging on slopes above the town of Guinsaugon in the Philippines was blamed for the February 2006 mudslide, triggered by heavy rains, that engulfed the town and killed an estimated 1800 of its 1857 residents.

Earthquakes

Landslides are a common consequence of earthquakes in hilly terrain, as noted in chapter 4. Seismic waves passing through rock stress and fracture it. The added stress may be as much as half that already present due to gravity. Ground shaking also jars apart soil particles and rock masses, reducing the friction that holds them in place. In the 1964 Alaskan earthquake, the Turnagain Heights section of Anchorage was heavily damaged by landslides (recall figure 4.26C). California contains not only many fault zones but also many sea cliffs and hillsides prone to landslides during earthquakes.

One of the most lethal earthquake-induced landslides occurred in Peru in 1970. An earlier landslide had already occurred below the steep, snowy slopes of Nevados Huascarán, the highest peak in the Peruvian Andes, in 1962, without the help of an earthquake. It had killed approximately 3500 people. In 1970, a magnitude-7.7 earthquake centered 130 kilometers to the west shook loose a much larger debris avalanche that buried most of the towns of Yungay and Ranrahirca and more than 18,000 people (figure 8.9). Once sliding had begun, escape was impossible: Some of the debris was estimated to have moved at 1000 kilometers per hour (about 600 miles per hour). The steep mountains of Central and South America, sitting above subduction zones, continue to suffer earthquake-induced slides (figure 8.10).

Earthquakes in and near ocean basins may also trigger submarine landslides. Sediment from these may be churned up into a denser-than-water suspension that flows downslope to the sea floor, much as pyroclastic flows flow down from a

FIGURE 8.7 Landslide (mudflow) following heavy rains in Brazil.
Photograph by F. O. Jones, USGS Photo Library, Denver, CO.

FIGURE 8.8 Examination of this Alaskan roadcut shows surface soil held in place by grass roots, erosion and slope failure below.

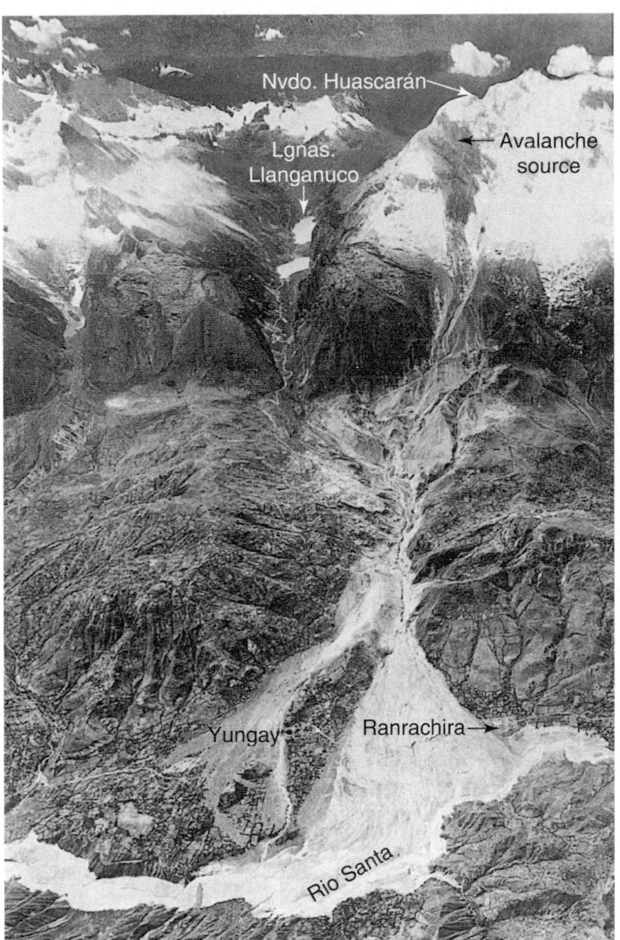

FIGURE 8.9 The Nevados Huascarán debris avalanche, Peru, 1970.
Photograph courtesy USGS Photo Library, Denver, CO.

volcano. These *turbidity currents* not only carry sediment to the deep-sea floor; they are forceful enough to break cables, and they could therefore disrupt transatlantic communications in the days when all transatlantic telephone calls were carried by undersea cables!

Quick Clays

A geologic factor that contributed to the 1964 Anchorage landslides and that continues to add to the landslide hazards in Alaska, California, parts of northern Europe, and elsewhere is a material known as "quick" or "sensitive" clay. True **quick clays** (figure 8.11) are most common in northern polar latitudes. The grinding and pulverizing action of massive glaciers can produce a **rock flour** of clay-sized particles, less than 0.02 millimeter (0.0008 inch) in diameter. When this extremely fine material is deposited in a marine environment, and the sediment is later uplifted above sea level by tectonic movements, it contains salty pore water. The sodium chloride in the pore water acts as a glue, holding the clay particles together. Fresh water subsequently infiltrating the clay washes out the salts, leaving a delicate, honeycomblike structure of particles. Seismic-wave vibrations break the structure apart, reducing the strength of the quick clay by as much as twenty to thirty times, creating a finer-grained equivalent of quicksand that is highly prone to sliding. Failure of the Bootlegger Clay, a quick clay underlying Anchorage, Alaska, was responsible for the extent of damage from the 1964 earthquake. Nor is a large earthquake necessary to trigger failure; vibrations from passing vehicles can also do it. So-called **sensitive clays** are somewhat similar in behavior to quick clays but may be formed from different materials. Weathering of volcanic ash, for example, can produce a sensitive-clay sediment. Such deposits are not uncommon in the western United States.

TYPES OF MASS WASTING

In the broadest sense, even subsidence of the ground surface is a form of mass wasting because it is gravity-driven. Subsidence includes slow downward movement as ground water or oil is extracted, and the rapid drop of sinkhole formation, described in chapter 11. Here we focus on downslope movements and displacement of distinct masses of rock and soil.

When downslope movement is quite slow, even particle by particle, the motion is described as **creep.** Soil creep, which is often triggered by frost heaving, occurs more commonly than rock creep. Though gradual, creep may nevertheless leave telltale signs of its occurrence so that areas of particular risk can be avoided. For example, building foundations may gradually weaken and fail as soil shifts; roads and railway lines may be disrupted. Often, soil creep causes serious property damage, though lives are rarely threatened. A related phenomenon, *solifluction,* describes slow movement of wet soil over impermeable material. It will be considered further in connection with permafrost, discussed in chapter 10.

Mass movements may be subdivided on the basis of the type of material moved and the characteristic type or form of movement. The material moved can range from unconsolidated, fairly fine material (for example, soil or snow) to large, solid masses of rock. *Landslide* is a nonspecific term for rapid mass movements in rock or soil. There are several types of landslides, and a particular event may involve more than one type of motion. The rate of motion is commonly related to the proportion of moisture: frequently, the wetter the material, the faster the movement. In general, the more rapid the movement, the greater the likelihood of casualties. A few examples may

FIGURE 8.10 Slide on Pan-American Highway east of Ropango, El Salvador, triggered by magnitude-7.6 earthquake in January 2001. *Photograph by E. L. Harp, U.S. Geological Survey*

FIGURE 8.11 Example of quick clay. Turnagain Heights, Anchorage, Alaska.

clarify the different types of mass movements, which are summarized in figure 8.12.

Falls

A **fall** is a free-falling action in which the moving material is not always in contact with the ground below. Falls are most often **rockfalls** (figure 8.13). They frequently occur on very steep slopes when rocks high on the slope, weakened and broken up by weathering, lose support as materials under them erode away. Falls are common along rocky coastlines where cliffs are undercut by wave action and also at roadcuts through solid rock. The coarse rubble that accumulates at the foot of a slope prone to rockfalls is **talus.** As the pounding water undercuts the resistant rock layer that has created Niagara Falls (figure 8.14), repeated rockfalls cause the falls to move slowly upstream, while talus accumulates at the base of the falls.

Slumps and Slides

In a **slide,** a fairly cohesive unit of rock or soil slips downward along a clearly defined surface or plane. Rockslides most often involve movement along a bedding plane between successive

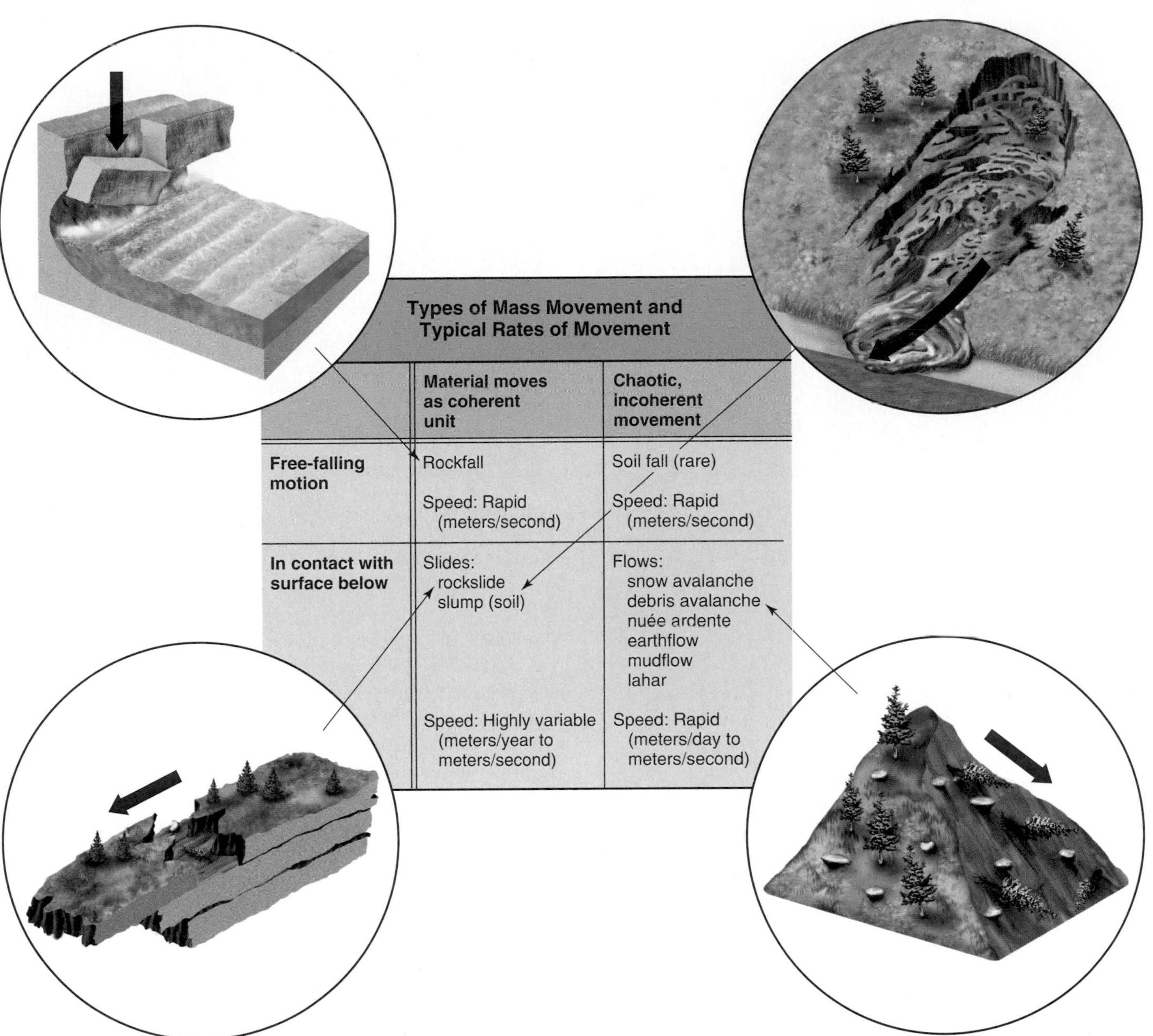

Types of Mass Movement and Typical Rates of Movement

	Material moves as coherent unit	Chaotic, incoherent movement
Free-falling motion	Rockfall Speed: Rapid (meters/second)	Soil fall (rare) Speed: Rapid (meters/second)
In contact with surface below	Slides: rockslide slump (soil) Speed: Highly variable (meters/year to meters/second)	Flows: snow avalanche debris avalanche nuée ardente earthflow mudflow lahar Speed: Rapid (meters/day to meters/second)

FIGURE 8.12 Comparison of nature of various types of mass movements.

layers of sedimentary rocks, or slippage where differences in original composition of the sedimentary layers results in a weakened layer or a surface with little cohesion (figure 8.15). Rockslides may be complex, with multiple rock blocks slipping along the same plane.

Slumps can occur in rock or soil. In a soil slump, a rotational movement of the soil mass typically accompanies the downslope movement. A *scarp* may be formed at the top of the slide. Usually, the surface at the top of the slide is relatively undisturbed. The lower part of the slump block may end in a flow (figure 8.16). A rockslide that does not move very far may also be a slump.

Flows and Avalanches

In a **flow,** the material moved is not coherent but moves in a more chaotic, disorganized fashion, with mixing of particles within the flowing mass, as a fluid flows. Flows of unconsolidated material are extremely common. They need not involve only soils. Snow avalanches are one kind of flow; nuées

FIGURE 8.13 Rockfall in Colorado National Monument. Note accumulation of talus at base of cliff, from past falls.

FIGURE 8.14 Niagara Falls are moving slowly upstream as a consequence of rockfalls. This is also a site at which the International Joint Commission regulates Great Lakes outflow.
Photograph by Ken Winters, courtesy U.S. Army Corps of Engineers

FIGURE 8.15 A rockslide on Karakoram Highway in Hunza Valley, Pakistan. Slip occurred along sloping joints in the rock.
© Robert Holmes/Corbis

ardentes, or pyroclastic flows (see chapter 5), are another, associated only with volcanic activity. Where soil is the flowing material, these phenomena may be described as *earthflows* (fairly dry soil; figure 8.17) or *mudflows* (if saturated with water), which include the volcanic lahars. A flow involving a wide variety of materials—soil, rocks, trees, and so on—together in a single flow is a **debris avalanche,** or *debris flow* (figure 8.18). Regardless of the nature of the materials moved, all flows have in common the chaotic, incoherent movement of the particles or objects in the flow.

CONSEQUENCES OF MASS MOVEMENTS

Just as landslides can be a result of earthquakes, other unfortunate events, notably floods, can be produced by landslides. A stream in the process of cutting a valley may create unstable slopes. Subsequent landslides into the valley can dam up the stream flowing through it, creating a natural reservoir. The filling of the reservoir makes the area behind the earth dam uninhabitable, though it usually happens slowly enough that lives are not lost.

Heavy rains and snows in Utah in 1982–1983 caused landslides in unprecedented numbers. In one case, an old 3-million-cubic-meter slide near Thistle, Utah, was reactivated. It blocked Spanish Fork Canyon, creating a large lake and cutting off one transcontinental railway line and a major highway route into central Utah (figure 8.19). Shipments of coal and other supplies were disrupted. Debris flows north of Salt Lake City also forced hundreds of residents to evacuate and destroyed some homes, and water levels in reservoirs in the Wasatch Plateau

A

B

FIGURE 8.16 (A) Slump and flow at La Conchita, California, in spring 1995. Houses at the toe were destroyed by the debris flow, but no one was killed. (B) Ten years later, part of the 1995 slide remobilized after a period of heavy rain, flowing out over four blocks of the town, and ten people died. What is perhaps more disturbing is that recent studies suggest that the La Conchita slides are just a small part of a much larger complex slide; see scarp at upper left.
(A) Photograph by R. L. Schuster, (B) photograph by Jonathan Godt, both courtesy U.S. Geological Survey

FIGURE 8.17 Recent earthflows in grassland, Gilroy, California.
Photograph © The McGraw-Hill Companies, Inc./Doug Sherman, photographer

FIGURE 8.18 Puget Peak debris avalanche following 1964 earthquake, Alaska.
Photograph by G. Plafker, USGS Photo Library, Denver, CO.

FIGURE 8.19 An old, 3-million-cubic-meter slide near Thistle, Utah, was reactivated in the wet spring of 1983. It blocked Spanish Fork Canyon and cut off highway and rail routes. Residents were driven from their homes as the lake behind the slide grew.
Photograph courtesy USGS Photo Library, Denver, CO.

were lowered to reduce dangers of landslides into the reservoirs or breaching of the dams.

A further danger is that the unplanned dam formed by a landslide will later fail. In 1925, an enormous rockslide a kilometer wide and over 3 kilometers (2 miles) long in the valley of the Gros Ventre River in Wyoming blocked that valley, and the resulting lake stretched to 9 kilometers long. After spring rains and snowmelt, the natural dam failed. Floodwaters swept down the valley below. Six people died, but the toll might have been far worse in a more populous area.

Impact of Human Activities

Given the factors that influence slope stability, it is possible to see many ways in which human activities increase the risk of mass movements. One way is to clear away stabilizing vegetation. As noted earlier, where clear-cutting logging operations have exposed sloping soil, landslides of soil and mud may occur far more frequently or may be more severe than before.

Many types of construction lead to oversteepening of slopes. Highway roadcuts, quarrying or open-pit mining operations, and construction of stepped home-building sites on hillsides are among the activities that can cause problems (figure 8.20). Where dipping layers of rock are present, removal of material at the bottom ends of the layers may leave large masses of rock unsupported, held in place only by friction between layers. Slopes cut in unconsolidated materials at angles higher than the angle of repose of those materials are by nature unstable, especially if there is no attempt to plant stabilizing vegetation (figure 8.21). In addition, putting a house above a naturally unstable or artificially steepened slope adds weight to the slope, thereby increasing the shear stress acting on the slope. Other activities connected with the presence of housing developments on hillsides can increase the risk of landslides in more subtle ways. Watering the lawn, using a septic tank for sewage disposal, and even putting in an in-ground swimming pool from which water can seep slowly out are all activities that increase the moisture content of the soil and render the slope more susceptible to slides. On the other hand, the planting of suitable vegetation can reduce the risk of slides.

Recent building-code restrictions have limited development in a few unstable hilly areas, and some measures have been taken to correct past practices that contributed to landslides. (Chapter 19 includes an illustration of the effectiveness of legislation to restrict building practices in Los Angeles County.) However, landslides do not necessarily cease just because ill-advised practices that caused or accelerated them have been stopped. Once activated or reactivated, slides may continue to slip for decades. The area of Portuguese Bend in the Palos Verdes Peninsula in Los Angeles County is a case in point.

In the 1950s, housing developments were begun in Portuguese Bend. In 1956, a 1-kilometer-square area began to slip, though the slope in the vicinity was less than 7 degrees, and within months, there had been 20 meters of movement. What activated this particular slide is unclear. Most of the homes used cesspools for sewage disposal, which added fluid to the ground, potentially increasing fluid pressure. On the other hand, the county highway department, in building a road across what became the top of this slide, added a lot of fill, and thus weight. In any event, the resultant damage was extensive: cracks developed within the sliding mass; houses on and near the slide were damaged or destroyed; a road built across the base of the slide had to be rebuilt repeatedly. Over $10 million in property damage resulted from the slippage (figure 8.22).

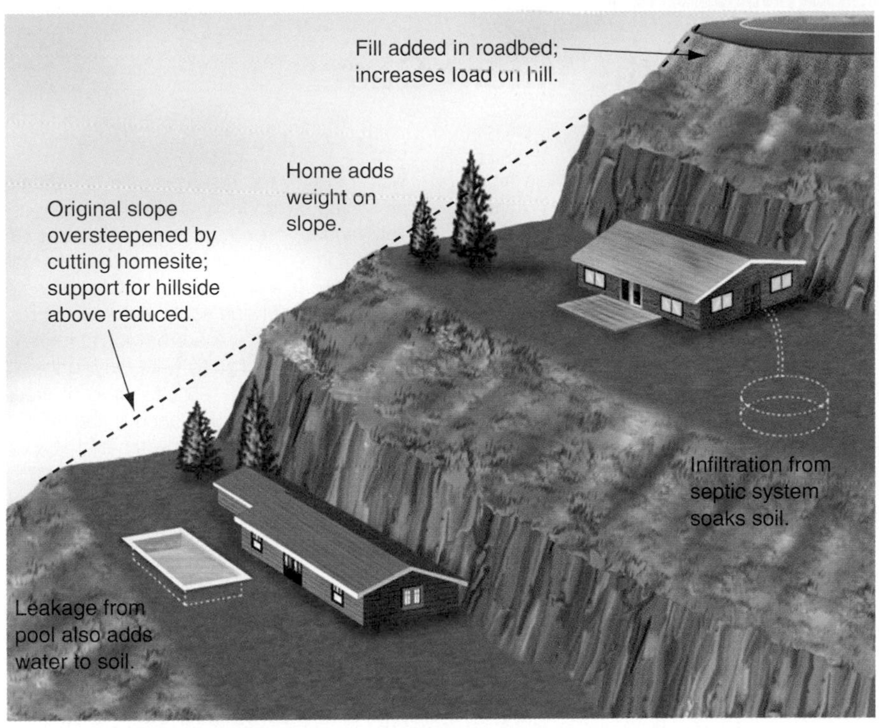

FIGURE 8.20 Effects of construction and human habitation on slope stability.

FIGURE 8.21 Failure of steep slope along a roadcut now threatens the cabin above.

FIGURE 8.22 Damage from the Portuguese Bend Landslide. Logs have been inserted underneath this house to prop it up and prevent collapse.

Photograph courtesy of Martin Reiter

Box 8.1 The Case of the Vaiont Dam

The Vaiont River flows through an old glacial valley in the Italian Alps. The valley is underlain by a thick sequence of sedimentary rocks, principally limestones with some clay-rich layers, that were folded and fractured during the building of the Alps. The sedimentary units of the Vaiont Valley are in the form of a syncline, or trough-shaped fold, so the beds on either side of the valley dip down toward the valley (figure 1). The rocks themselves are relatively weak and particularly prone to sliding along the clay-rich layers. Evidence of old rockslides can be seen in the valley. Extensive solution of the carbonates by ground water further weakens the rocks.

The Vaiont Dam, built for power generation, is the highest "thin-arch" dam in the world. It is made of concrete and stands over 265 meters (875 feet) high at its highest point. Modern engineering methods were used to stabilize the rocks in which it is based. The capacity of the reservoir behind the dam was 150 million cubic meters—initially.

The obvious history of landslides in the area originally led some to object to the dam site. Later, the very building of the dam and subsequent filling of the reservoir aggravated an already-precarious situation. As the water level in the reservoir rose following completion of the dam, pore pressures of ground water in the rocks of the reservoir walls rose also. This tended to buoy up the rocks and to swell the clays of the clay-rich layers, further decreasing their strength, making sliding easier.

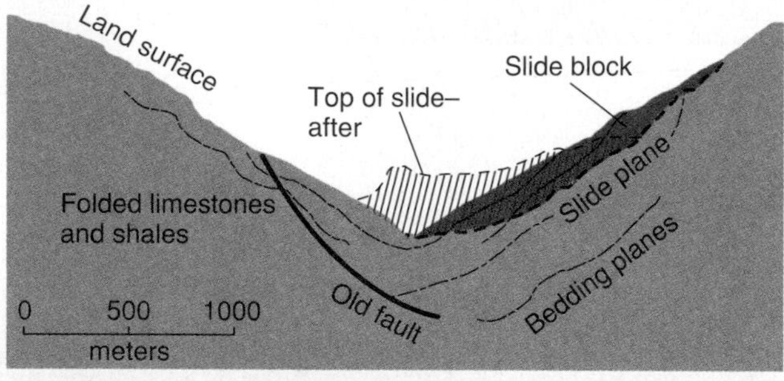

FIGURE 1 Geologic cross section of the Vaiont valley.
G. A. Kiersch, "The Vaiont Reservoir Disaster," Mineral Information Service *18(7):129–138, 1965. State of California, Division of Mines and Geology.*

In 1960, a block of 700,000 cubic meters of rock slid from the slopes of Monte Toc on the south wall into the reservoir. Creep was noted over a still greater area. A set of monitoring stations was established on the slopes of Monte Toc to track any further movement. Measurable creep continued. In 1960–1961, the rate of creep occasionally reached 25 to 30 centimeters (10 to 12 inches) per *week,* and the total volume of rock affected by the creep was estimated at about 200 million cubic meters.

Late summer and early fall of 1963 were times of heavy rainfall in the Vaiont Valley. The rocks and soils were saturated. The reservoir level rose by over 20 meters, and pore pressures in the rocks increased. By mid-September 1963, measured creep rates were consistently about 1 centimeter per day. What was still not realized was that the rocks were slipping not in a lot of small blocks, but as a single, coherent mass.

Animals sensed the danger before people did. On October 1, animals that had been grazing on those slopes of Monte Toc moved off the hillside and reportedly refused to return.

The rains continued, and so did the creep. Creep rates grew to 20 to 30 centimeters per day. Finally, on October 8, the engineers realized that all the creep-monitoring stations were moving together. They also discovered that a far larger area of hillside was moving than they had thought. They tried to reduce the water level in the reservoir by opening the gates of two outlet tunnels. But it was too late, and the water level continued to rise: The silently creeping mass had begun to encroach on the reservoir, decreasing its capacity significantly. On the night of October 9, in the midst of another downpour, the disaster struck.

A resident of the town of Casso, on the slopes above the reservoir, later reported that, at first, there was just a sound of rolling rocks, ever louder. Then a gust of wind hit his house, smashing windows and raising the roof so that the rain poured in. The wind suddenly died; the roof collapsed.

Worse, the movement has continued for decades since, slow but unstoppable, at a rate of about 3 meters per year. Some portions of the slide have moved 70 meters.

Human activities can increase the hazard of landslides in still other ways. Irrigation and the use of septic tanks increase the flushing of water through soils and sediments. In areas underlain by quick or sensitive clays, these practices may hasten the washing-out of salty pore waters and the destabilization of the clays, which may fail after the soils are drained. Even cleanup after one slide may reduce stability and contribute to the next slide: Often, the cleanup involves removing the toe of a slump or flow—for example, where it crosses a road. This removes support for the land above and may oversteepen the local slope, too. Artificial reservoirs may cause not only earthquakes, but landslides, too. As the reservoirs fill, pore pressures in rocks along the sides of the reservoir increase, and the strength of the rocks to resist shearing stress can be correspondingly decreased. The case of the Vaiont reservoir disaster (see Box 8.1, "The Case of the Vaiont Dam") is a classic in this respect. Chapter 20 describes other examples of ways in which engineering activities may influence land stability.

A Compounding of Problems: the Venezuelan Coast

On the north coast of Venezuela, the terrain is mountainous, with steep slopes and narrow stream channels. A few people settled on the slopes, risking landslides; larger towns developed

(Box 8.1 Continued)

A chunk of hillside, over 240 million cubic meters in volume, had slid into the reservoir (figure 2). The shock of the slide was detected by seismometers in Rome, in Brussels, and elsewhere around Europe. That immense movement, occurring in less than a minute, set off the shock wave of wind that rattled Casso and drew the water 240 meters *upslope* out of the reservoir after it. Displaced water crashed over the dam in a wall 100 meters (over 325 feet) above the dam crest and rushed down the valley below. The water wave was still over 70 meters high some 1.5 kilometers downstream at the mouth of the Vaiont Valley, where it flowed into the Piave River. The energy of the rushing water was such that some of it flowed *upstream* in the Piave for more than 2 kilometers. Within about five minutes, nearly 3000 people were drowned and whole towns wiped out.

It is a great tribute to the designer and builder of the Vaiont Dam that the dam itself held through all of this, resisting forces far beyond the design specifications; it still stands today. It is also very much to the discredit of those who chose the site that the dam and reservoir were ever put in such a spot. Ample evidence of slope instability dated back long before the conception of the dam. With more thorough study of the rocks' properties and structure and the possible effects of changing water levels, the Vaiont tragedy could have been avoided.

History often seems destined to repeat itself. In the late 1970s, the Tablacha Dam on Rio Mantaro, Peru, was threatened by a similar failure (figure 3). Foliated metamorphic rocks dip toward the dam and reservoir. A portion of an ancient slide was reactivated, most likely by the increasing pore pressure due to reservoir filling. The situation is much like that of the Vaiont Dam. In this case, however, the Peruvian government spent $40 million on stabilization efforts—building a 460,000-cubic-meter earth buttress at the toe of the slide block, and adding drainage, rock bolts, and other measures. So far, they seem to be succeeding.

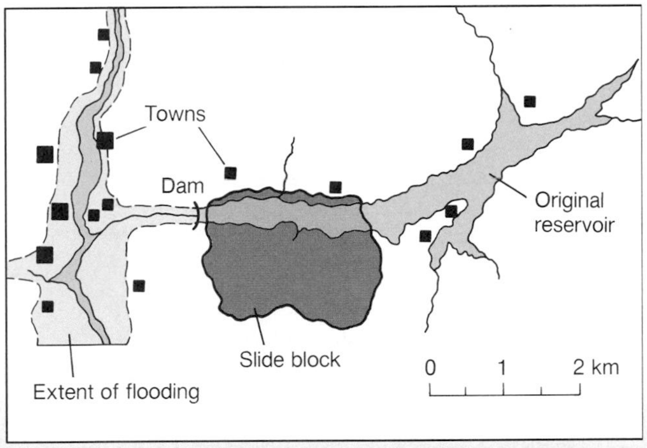

FIGURE 2 The landslide at the Vaiont reservoir and resultant flood damage.
G. A. Kiersch, "The Vaiont Reservoir Disaster," Mineral Information Service *18(7):129–138, 1965. State of California, Division of Mines and Geology.*

FIGURE 3 Reactivated slide block (identified by red arrows) above Tablacha Dam and reservoir, Peru.
Photograph by J. T. McGill, USGS Photo Library, Denver, CO.

in the only flat areas available, beside the streams or in alluvial plains at the mouths of the valleys (figure 8.23).

The first two weeks of December 1999 were hardly dry, with 293 millimeters (nearly 12 inches) of rain, but the real problem was the deluge of another 911 millimeters (close to *3 feet*) of rain that fell December 14–16. Runoff from the steep slopes was swift, and flash floods quickly rose in the valleys. Worse yet, thousands of debris avalanches from the hillsides cascaded into the same valleys, mingling with floodwaters to make huge debris flows that swept out into the alluvial plains and through the towns (figure 8.24). An estimated 30,000 people died from the consequences of flooding and landslides combined.

Nor is the danger now past, for much rock and soil remains in those stream channels, to be remobilized by future runoff. And there are few safe sites in the region for those who choose to stay and rebuild.

POSSIBLE PREVENTIVE MEASURES

Avoiding the most landslide-prone areas altogether would greatly limit damages, of course, but as is true with fault zones, floodplains, and other hazardous settings, developments may already exist in areas at risk, and economic pressure for more development can be strong. In areas such as the Venezuelan coast, there may be little land *not* at risk. Population density, too, coupled with lack of "safe" locations, may push development into unsafe areas.

Sometimes, local officials just resign themselves to the existence of a mass-movement problem and perhaps take some steps to limit the resulting damage (figure 8.25). In places where the structures to be protected are few or small, and the slide zone is narrow, it may be economically feasible to bridge structures and simply let the slides flow over them. For example, this

FIGURE 8.23 Quebrada San Julián (foreground) and Caraballeda (rear) were built, respectively, in a floodplain and on an alluvial plain; but the terrain leaves little choice.
Photograph by M. C. Larsen and H. T. Sierra, U.S. Geological Survey

might be done to protect a railway line or road running along a valley from avalanches—either snow or debris—from a particularly steep slope. This solution would be far too expensive on a large scale, however, and no use at all if the base on which the structure was built were sliding also.

There are some other steps that can be taken to reduce actual landslide hazards.

Slope Stabilization

If a slope is too steep to be stable under the load it carries, any of the following steps will reduce slide potential: (1) reduce the slope angle, (2) place additional supporting material at the foot of the slope to prevent a slide or flow at the base of the slope, or (3) reduce the load (weight, shearing stress) on the slope by removing some of the rock or soil (or artificial structures) high on the slope (figure 8.26). These measures may be used in combination. Depending on just how unstable the slope, they may need to be executed cautiously. If earthmoving equipment is being used to remove soil at the top of a slope, for example, the added weight of the equipment and vibrations from it could possibly trigger a landslide.

A

C

B

FIGURE 8.24 Damage in Caraballeda: (A) Overview of Los Corales sector. (B) A close look at this apartment building shows that at their peak, debris flows flowed through the lowest three floors. (C) Debris-flow deposits in channel at San José de Galipán. Sizes of largest boulders moved—estimated to weigh up to 400 tons—confirm the force of the flow.
Photographs by M. C. Larsen and H. T. Sierra, U.S. Geological Survey

A

B

C

FIGURE 8.25 Measures to Reduce the Consequences of Slope Instability. (A) Chain-link fencing draped over roadcut to protect road from rockfall. (B) Shelter built over railroad track or road in snow avalanche area diverts snow flow. (C) Example of avalanche-protection structure in the Canadian Rockies, in Canada's Glacier National Park.

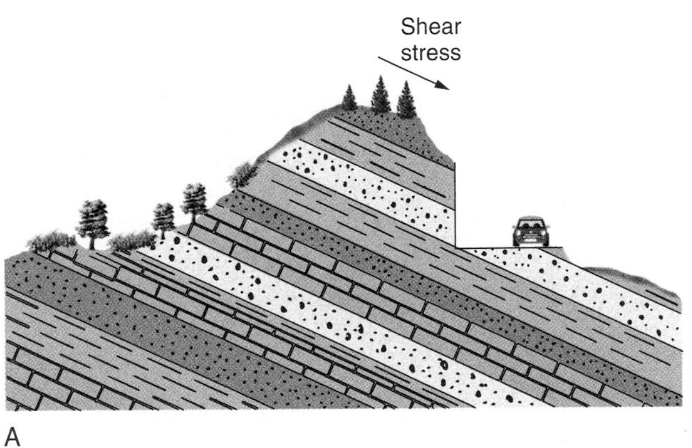

A

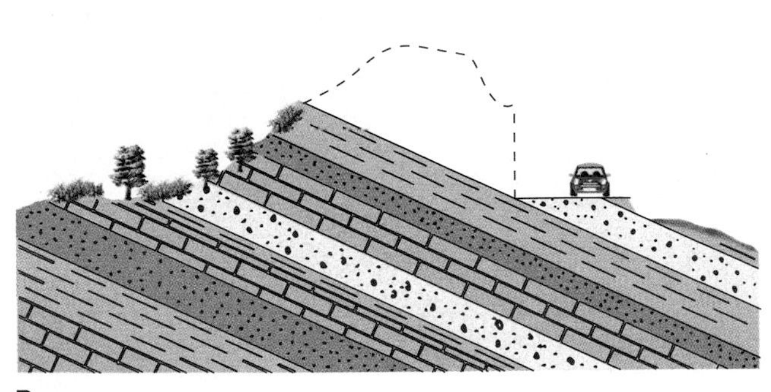

B

C

FIGURE 8.26 Slope Stabilization by Slope Reduction and Removal of Unstable Material along Roadcut. (A) Before: Roadcut leaves steep, unsupported slope. If the shale layers (grey) are relatively impermeable, fluid may accumulate along bedding planes, promoting sliding. (B) After: Material removed to reduce slope angle and load. (C) This roadcut near Denver, CO has been "terraced," cut into steps to break up the long slope while reducing the overall slope angle. Roadbed for one direction of traffic appears at lower right; retaining wall for the other is two steps up.

To stabilize exposed near-surface soil, ground covers or other vegetation may be planted (preferably fast-growing materials with sturdy root systems). But sometimes plants are insufficient, and the other preventive measures already described impractical, in a particular situation. Then, retaining walls or other stabilization structures can be built against the slope to try to hold it in place (figure 8.27). Given the distribution of stresses acting on retaining walls, the greatest successes of this kind have generally been low, thick walls placed at the toe of a fairly coherent slide to stop its movement. High, thin walls have been less successful (figure 8.28).

Since water can play such a major role in mass movements, the other principal strategy for reducing landslide hazards is to decrease the water content or pore pressure of the rock or soil. This might be done by covering the surface completely with an impermeable material and diverting surface runoff above the slope. Alternatively, subsurface drainage might be undertaken. Systems of underground boreholes can be drilled to increase drainage, and pipelines installed to carry the water out of the slide area (figure 8.29). All such moisture-reducing techniques naturally have the greatest impact where rocks or soils are relatively permeable. Where the rock or soil is fine-grained and drains only slowly, hot air may be blown through boreholes to help dry out the ground. Such moisture reduction reduces pore pressure and increases frictional resistance to sliding.

Other slope-stabilization techniques that have been tried include the driving of vertical piles into the foot of a shallow slide to hold the sliding block in place. The procedure works only where the slide is comparatively solid (loose soils may simply flow between piles), with thin slides (so that piles can be driven deep into stable material below the sliding mass), and on low-angle slopes (otherwise, the shearing stresses may simply snap the piles). So far, this technique has not been very effective.

The use of rock bolts to stabilize rocky slopes and, occasionally, rockslides has had greater success (figure 8.30). Rock bolts have long been used in tunneling and mining to stabilize rock walls. It is sometimes also possible to anchor a rockslide with giant steel bolts driven into stable rocks below the slip plane. Again, this works best on thin slide blocks of very coherent rocks on low-angle slopes.

Procedures occasionally used on unconsolidated materials include hardening unstable soil by drying and baking it with heat (this procedure works well with clay-rich soils) or by treating with portland cement. By far the most common strategies, however, are modification of slope geometry and load, dewatering, or a combination of these techniques. The more ambitious engineering efforts are correspondingly expensive and usually reserved for large construction projects, not individual homesites.

A

B

FIGURE 8.27 (A) Retaining wall of concrete blocks on slope cut in old slide debris, Pacific Palisades area, California. (B) Side view of retaining structure formed by rock rubble encased in wire mesh, Lake Tahoe, Nevada. Note that this is also stepped rather like the roadcut in figure 8.26.

(A) Photograph by J. T. McGill, USGS Photo Library, Denver, CO.

A

B

FIGURE 8.28 Slapping concrete sheets on a steep, rubbly hillside in Rock Creek Valley, Montana, is not altogether effective; (B) photograph, taken several years after (A), shows slipping and buckling, and telltale road repair below.

A

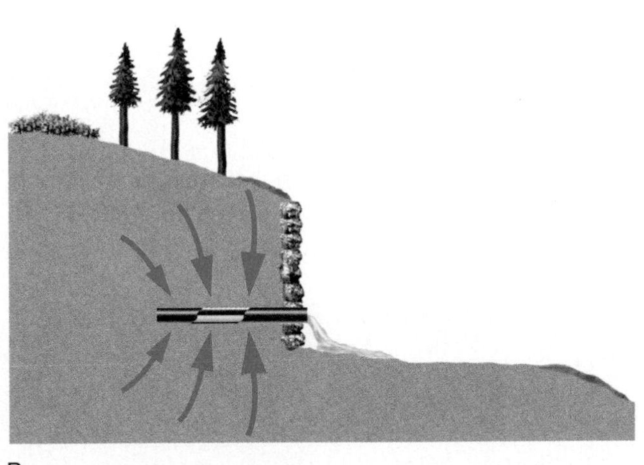

B

FIGURE 8.29 Improved drainage enhances slope stability by reducing pore pressure. (A) Before: Water trapped in wet soil causes movement, pushing down retaining wall. (B) After: Water drains through pipe, allowing wall to keep slope stable.

Determination of the absolute limits of slope stability is an imprecise science. This was illustrated in Japan in 1969, when a test hillside being deliberately saturated with water unexpectedly slid prematurely, killing several of the researchers. The same point is made less dramatically every time a slope stabilization effort fails.

Recognizing the Hazards

Mass movements occur as a consequence of the basic characteristics of a region's climate, topography, and geology, independent of human activities. Where mass movements are naturally common, they tend to recur repeatedly in the same places. Recognition of past mass movements in a region indicates a need for caution and serious consideration of the hazard in any future development. Such recognition can also save lives.

Even when recent activity is limited, the potential can often be seen. The granite domes of Yosemite have been shaped by the flaking-off of massive curving slabs of rock (figure 8.31A), but these rockfalls are infrequent. A 1996 fall that killed one park visitor and injured 14 (figure 8.31B) has focused the attention of the National Park Service on the rockfall hazards and how to identify the highest-hazard regions.

Recognizing past rockfalls can be quite simple in vegetated areas. Large chunks of rock are inhospitable to most vegetation, so rockfalls tend to remain barren of trees and plants. The few trees that do take hold may only contribute to further breakup by root action. Lack of vegetation may also mark the paths of past debris avalanches or other soil flows or slides

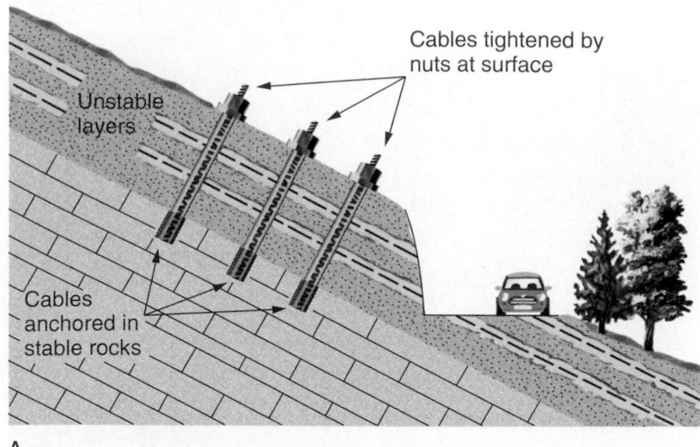

A

B

FIGURE 8.30 (A) Installation of rock bolts to stabilize a slope. The bolts are steel cables anchored in cement. Tightening nuts at the surface pulls unstable layers together and anchors them to the bedrock below. (B) Rock bolts used to stabilize a slope in the Bighorn Mountains, Wyoming.

Photograph © Doug Sherman/Geofile

A

B

FIGURE 8.31 (A) A close look at the granite domes of Yosemite shows curved slabs of rock at the surface; some above valleys are poised to fall. (B) In 1996, a 200-ton block crashed down, shattering on impact with a blast that leveled thousands of trees.

(B) Photograph by Gerry Wieczorek, U.S. Geological Survey

(figure 8.32). These scars on the landscape point plainly to slope instability.

Landslides are not the only kinds of mass movements that recur in the same places. Snow avalanches disappear when the snow melts, of course, but historical records of past avalanches can pinpoint particularly risky areas. Records of the character of past volcanic activity and an examination of a volcano's typical products can similarly be used to recognize a particular volcano's tendency to produce pyroclastic flows.

Very large slumps and slides may be less obvious, especially when viewed from the ground. The coherent nature of rock and soil movement in many slides means that vegetation growing atop the slide may not be greatly disturbed by the movement. Aerial photography or high-quality topographic maps can be helpful here. In a regional overview, the mass movement often shows up very clearly, revealed by a scarp at the head of a slump or an area of hummocky, disrupted topography relative to surrounding, more-stable areas (recall figures 8.5 and 8.16).

With creep or gradual soil flow, individual movements are short-distance and the whole process is slow, so vegetation may continue to grow in spite of the slippage. More detailed observation, however, can reveal the movement. For example, trees are biochemically "programmed" to grow vertically upward. If the soil in which trees are growing begins to creep downslope, tree trunks may be tilted, and this indicates the soil movement. Further tree growth will continue to be vertical. If slow creep is prolonged over a considerable period of time, curved tree trunks may result (figure 8.33A). (Trees growing in slanted fractures in rock, however, may develop curved trunks through similar growth tendencies in the absence of creep; this is a better

FIGURE 8.32 Areas prone to landslides may be recognized by the failure of vegetation to establish itself on unstable slopes. Rock Creek Valley, Montana.

A

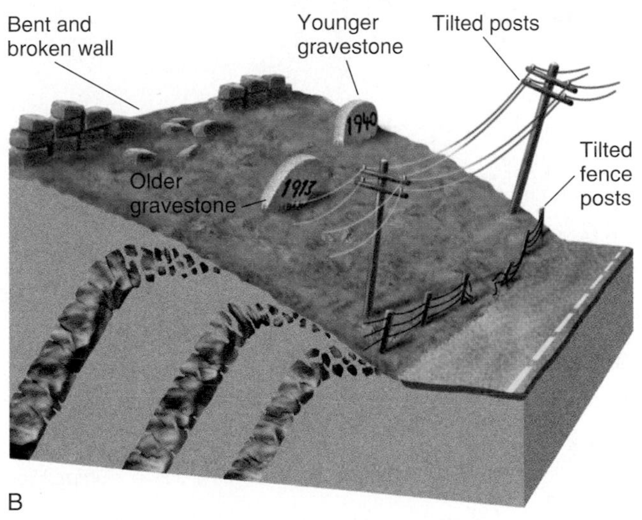

B

FIGURE 8.33 (A) Tilted tree trunks are a consequence of creep, as seen here on a weathered slope in Bryce Canyon National Park. (B) Other signs of creep: tilted monuments, fences, utility poles.

indicator of movement in unconsolidated material.) Inanimate objects can reflect soil creep, too. Slanted utility poles and fences and the tilting-over of once-vertical gravestones or other monuments also indicate that the soil is moving (figure 8.33B). The ground surface itself may show cracks parallel to (across) the slope.

A prospective home buyer can look for additional signs that might indicate unstable land underneath (figure 8.34). Ground slippage may have caused cracks in driveways, garage floors, freestanding brick or concrete walls, or buildings; cracks in walls or ceilings are especially suspicious in newer buildings that would not yet normally show the settling cracks common in old structures. Doors and windows that jam or do not close properly may reflect a warped frame due to differential movement in the soil and foundation. Sliding may have caused leaky swimming or decorative pools, or broken utility lines or pipes. If movement has already been sufficient to cause obvious structural damage, it is probable that the slope cannot be stabilized adequately, except perhaps at very great expense. Possible sliding should be investigated particularly on a site with more than 15% slope (15 meters of rise over 100 meters of horizontal distance), on a site with much steeper slopes above or below it, or in any area where landslides are a recognized problem. (Given appropriate conditions, slides may occur on quite shallow slopes.)

A

B

FIGURE 8.34 Evidence of slope instability in the Pacific Palisades of California includes broken walls (A) and, in extreme cases, slumped yards (B).

Photographs by J. T. McGill, USGS Photo Library, Denver, CO.

Landslide Warnings?

In early 1982, severe rain-triggered landslides in the San Francisco Bay area killed 25 people and caused over $66 million in damages. In response to this event, the U.S. Geological Survey began to develop a landslide warning system. The basis of the warning system was the development of quantitative relationships among rainfall intensity (quantity of water per unit time), storm duration, and a variety of slope and soil characteristics relating to slope stability—slope angle, pore fluid pressure, shear strength, and so on. These relationships were formulated using statistical analyses of observational data on past landslides. For a given slope, it was possible to approximate threshold values of storm intensity and duration above which landsliding would become likely, given the recent precipitation history at the time (how saturated the ground was prior to the particular storm of concern).

The system, though incomplete, was tested in February of 1986. Warnings were broadcast as special weather advisories on local radio and television stations. These warnings were rather specific as to time, more general as to area at risk (necessarily so, given the complexity of the geology and terrain and lack of highly detailed data on individual small areas). Some local government agencies recommended evacuations, and many were able to plan emergency responses to the landslides before they occurred. Many landslides did occur; total estimated landslide damage was $10 million, with one death. Of ten landslides for which the times of occurrence are known precisely, eight took place when forecast.

The models need refinement. The ultimate goal would be *predictions* that are precise both as to time and as to location. To achieve this, too, more extensive data on local geology and topography are needed. Certainly the public is not yet accustomed to landslide warnings, so response is uneven. But the successes of the 1986 efforts do suggest that landslide prediction has considerable potential to reduce casualties and enhance the efficiency of agencies' responses to such events. Since then, the U.S. Geological Survey and other agencies have expanded monitoring of potentially unstable slopes in key areas.

One such area is in Puget Sound, where a wet winter in 1996–1997 produced numerous landslides (figure 8.35). Thereafter, monitoring stations were established on particularly unstable slopes, to track pore pressure in relation to rainfall (figure 8.36). Data will need to be collected for some time before the relationship between pore pressure and sliding becomes clear enough that landslide warnings can be based on these data.

FIGURE 8.35 Landslide in Puget Sound, 1996.
Photograph © Teresa Tamura/The Seattle Times

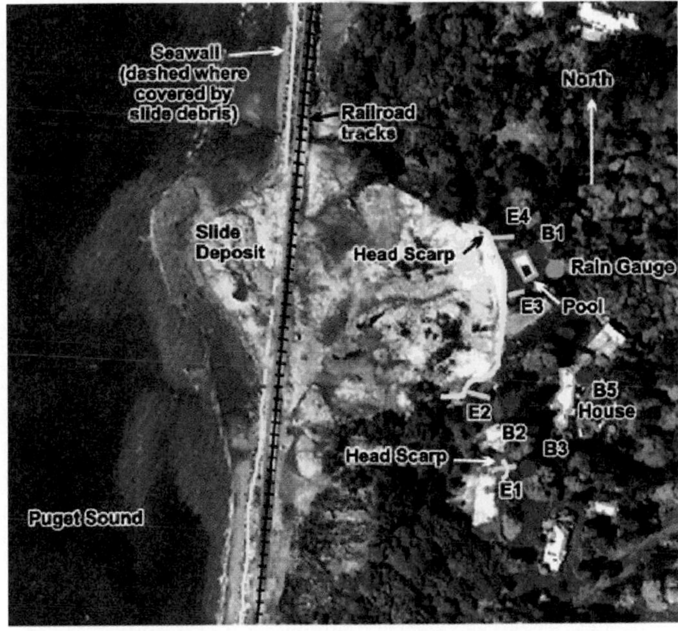

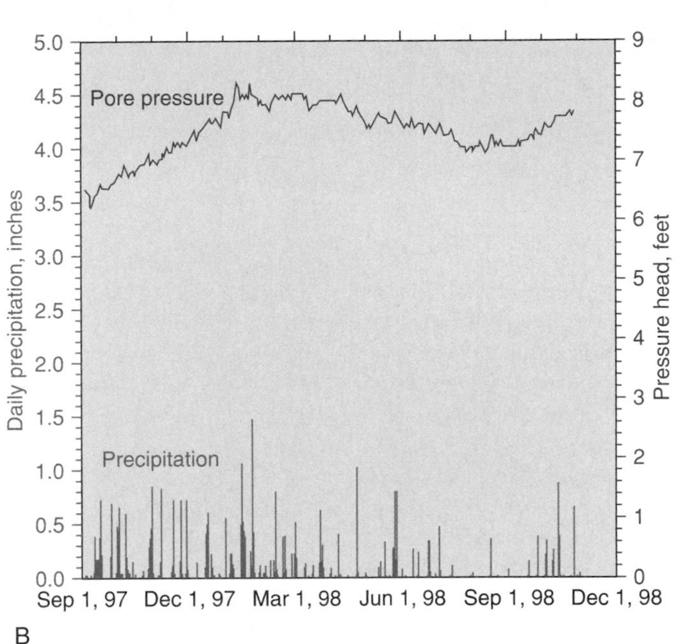

FIGURE 8.36 (A) Landslide-monitoring site at Woodway, Washington. Note proximity of house and pool to slide scarps. (B) Seasonal fluctuations in rainfall are reflected in seasonal fluctuations in pore pressure. Such observations over time help to distinguish dangerously elevated pore pressures likely to trigger sliding.
Courtesy U.S. Geological Survey

SUMMARY

Mass movements occur when the shearing stress acting on rocks or soil exceeds the shear strength of the material to resist it. Gravity provides the main component of shearing stress. The basic susceptibility of a slope to failure is determined by a combination of factors, including geology (nature and strength of materials), geometry (slope steepness), and moisture content. Sudden failure usually also involves a triggering mechanism, such as vibration from an earthquake, addition of moisture, steepening of slopes (naturally or through human activities), or removal of stabilizing vegetation. Landslide hazards may be reduced by such procedures as modifying slope geometry, reducing the weight acting on the slope, planting vegetation, or dewatering the rocks or soil. Damage and loss of life can also be limited by restricting or imposing controls on development in areas at risk, and where adequate data are available, through landslide warning systems.

Key Terms and Concepts

Exercises

Questions for Review

1. Explain in general terms why landslides occur. What two factors particularly influence slope stability?
2. Earthquakes are one landslide-triggering mechanism; water can be another. Evaluate the role of water in mass movements.
3. What are quick clays and sensitive clays?
4. Describe at least three ways in which development on hillsides may aggravate landslide hazards.
5. Give two ways to recognize soil creep and one way to identify sites of past landslides.
6. Briefly explain the distinctions among falls, slides, and flows.
7. Common slope-stabilization measures include physical modifications to the slope itself or the addition of stabilizing features. Choose any three such strategies and briefly explain how they work.

For Further Thought

1. Experiment with the stability of a variety of natural unconsolidated materials (soil, sand, gravel, and so forth). Pour the dry materials into separate piles; compare angles of repose. Add water, or put an object (load) on top, and observe the results.
2. Find aerial or other photographs of your area of the country (a local library might be able to provide them). Or, search the Internet for sources—you might start with the USGS photo archive at libraryphoto.er.usgs.gov

   Examine them for evidence of past mass wasting.
3. Consider how you might respond to a known local landslide hazard or a warning. How definite a warning would you look for before evacuating? Would you take any precautions prior to the issuance of an official warning if conditions seemed to be increasing the risk?
4. Check the latest monitoring data for the Woodway, Washington, site (see NetNotes). Consider whether it would make you feel more or less secure to live where such monitoring had been undertaken.

Suggested Readings/References

Chadwick, J., Thackray, G., Dorsch, S., and Glenn, N. 2005. Landslide surveillance: New tools for an old problem. *EOS* (15 March): 109, 114.

El Niño Response Group (USGS). 1998. Slope failure and shoreline retreat during northern California's latest El Niño. *GSA Today* 8 (August): 1–6.

Harp, E. L., et al. 2003. Landslides and liquefaction triggered by the M 7.9 Denali Fault earthquake of 3 November 2002. *GSA Today* (August): 4–10.

Highland, L. M., and Brown, W. L., III. 1996. Landslides—The natural hazard sleepers. *Geotimes* (January): 16–19.

Keefer, D. K., et al. 1987. Real-time landslide warning during heavy rainfall. *Science* 238: 921–25.

Kiersch, G. A. 1965. The Vaiont reservoir disaster. *Mineral Information Service* 18 (7): 129–38.

Larsen, M. K., Wieczorek, G. F., Eaton, L. S., Morgan, B. A., and Torres-Sierra, H. 2001. Natural hazards on alluvial fans: The Venezuela debris flow and flash flood disaster. USGS Fact Sheet FS 103-01.

Nuhfer, E. B., Proctor, R. J., and Moser, P. H., et al. 1993. *The Citizens' Guide to Geologic Hazards.* Arrada, Colo. American Institute of Professional Geologists. See chapters on "Landslides and avalanches" and "Subsidence."

Radbruch-Hall, D. H., Colton, R. B., Davies, W. E., Lucchitta, I., Varnes, D. J., and Skipp, B. A. 1997. Landslide overview map of the conterminous United States. U.S. Geological Survey Professional Paper 1183.

Schuster, R. L., and Highland, L. M. 2001. Socioeconomic and environmental impacts of landslides in the Western Hemisphere. USGS Open-File Report 01-0276.

Slossan, J. E., Keene, A. G., and Johnson, J. A., eds. 1992. *Landslides/landslide mitigation.* Geological Society of America Reviews in Engineering Geology v. IX.

U.S. Geological Survey. 1997. Debris-flow hazards in the United States. USGS Fact Sheet 176–97.

———. 1999. Land subsidence in the United States. USGS Circular 1182.

———. 2000. Landslide hazards. USGS Fact Sheet FS-071-00.

———. 2001. *Natural hazards on alluvial fans:* The Venezuela debris flow and flash flood disaster. USGS Fact Sheet FS-103-01.

Varnes, D. J. 1984. *Landslide hazard zonation: A review of principles and practice.* Paris: United Nations Educational, Scientific, and Cultural Organization (UNESCO).

Wilson, R., and Pike, R. 2005. Watching for landslide weather in California. *Geotimes* (November): 28–31.

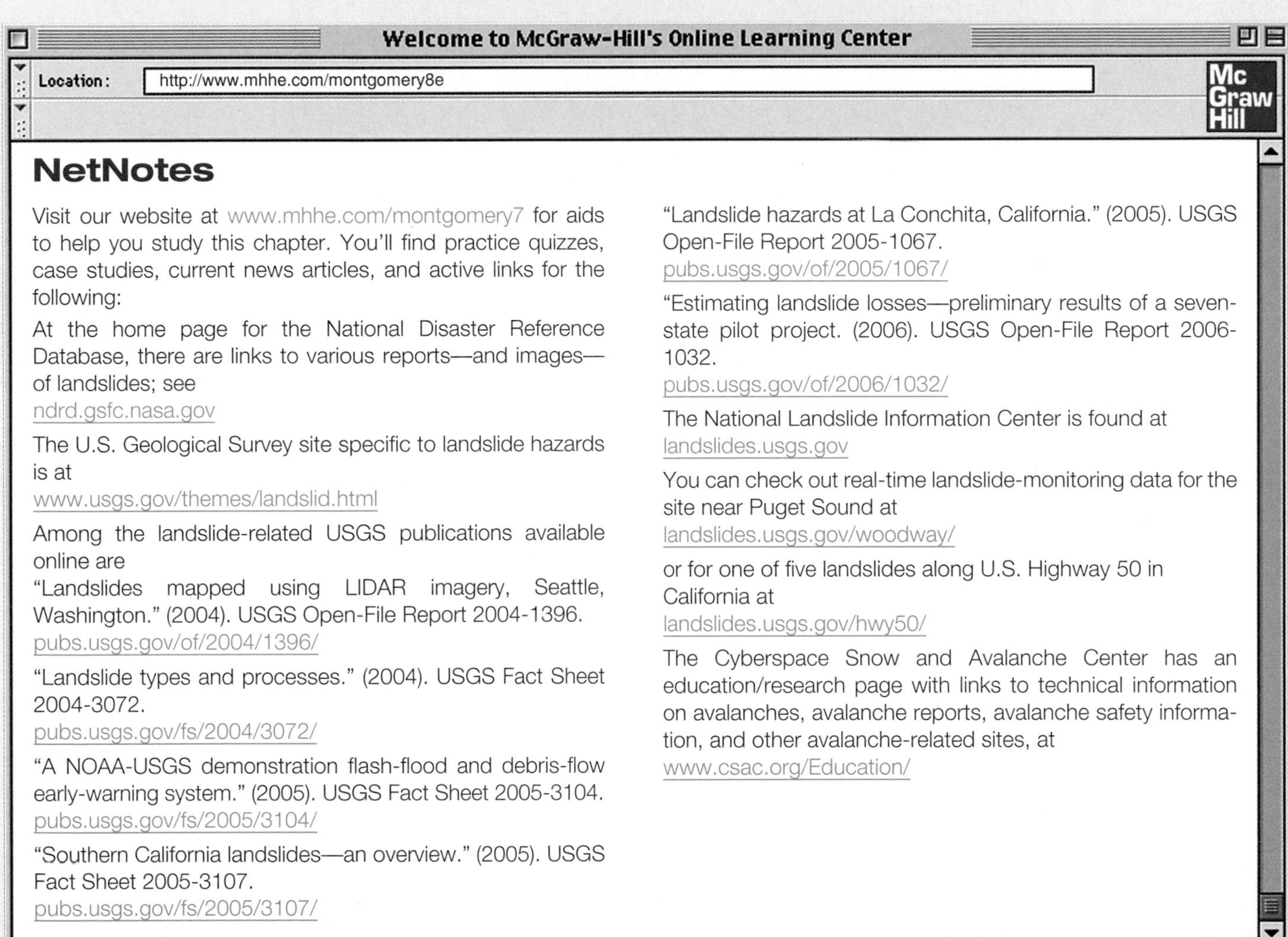

NetNotes

Visit our website at www.mhhe.com/montgomery7 for aids to help you study this chapter. You'll find practice quizzes, case studies, current news articles, and active links for the following:

At the home page for the National Disaster Reference Database, there are links to various reports—and images—of landslides; see
ndrd.gsfc.nasa.gov

The U.S. Geological Survey site specific to landslide hazards is at
www.usgs.gov/themes/landslid.html

Among the landslide-related USGS publications available online are

"Landslides mapped using LIDAR imagery, Seattle, Washington." (2004). USGS Open-File Report 2004-1396.
pubs.usgs.gov/of/2004/1396/

"Landslide types and processes." (2004). USGS Fact Sheet 2004-3072.
pubs.usgs.gov/fs/2004/3072/

"A NOAA-USGS demonstration flash-flood and debris-flow early-warning system." (2005). USGS Fact Sheet 2005-3104.
pubs.usgs.gov/fs/2005/3104/

"Southern California landslides—an overview." (2005). USGS Fact Sheet 2005-3107.
pubs.usgs.gov/fs/2005/3107/

"Landslide hazards at La Conchita, California." (2005). USGS Open-File Report 2005-1067.
pubs.usgs.gov/of/2005/1067/

"Estimating landslide losses—preliminary results of a seven-state pilot project. (2006). USGS Open-File Report 2006-1032.
pubs.usgs.gov/of/2006/1032/

The National Landslide Information Center is found at
landslides.usgs.gov

You can check out real-time landslide-monitoring data for the site near Puget Sound at
landslides.usgs.gov/woodway/

or for one of five landslides along U.S. Highway 50 in California at
landslides.usgs.gov/hwy50/

The Cyberspace Snow and Avalanche Center has an education/research page with links to technical information on avalanches, avalanche reports, avalanche safety information, and other avalanche-related sites, at
www.csac.org/Education/

9

Ice and Glaciers, Wind and Deserts

Fundamentally, the topics of this chapter and the next are related to heat—solar heat. Without solar heat to drive evaporation, there would be no precipitation to make glaciers. Differential solar heating of land and water, and thus of the air above them, drives the winds. Seasonal variations in temperature in a particular region are related to the changing intensity of incident sunlight as the earth, tilted on its axis, revolves around the sun over the course of a year (figure 9.1). Over longer periods, a variety of factors may lead to warming or cooling trends worldwide.

Glaciers are more than just another agent by which our landscape is shaped. They represent a very large reserve of (temporarily solid) water. Past glacial activity has also left behind sediments that provide gravel for construction, store subsurface water for future consumption, and contribute to farmland fertility.

Except during severe storms, wind is rarely a serious hazard. As an agent of change of the earth's surface, wind is considerably less efficient than water or ice. Nevertheless, it can locally play a significant role, particularly with regard to erosion. The effects of wind are most often visible in deserts, where little vegetation or development is present to mask or counter its effects.

Over a human lifetime, the extent of glaciation, the expanse of deserts, and the overall global climate generally vary little (though

Alaska is the most extensively glaciated state in the United States. Complex of alpine glaciers in College Fjord, Alaska.
Photograph by W. A. Montgomery

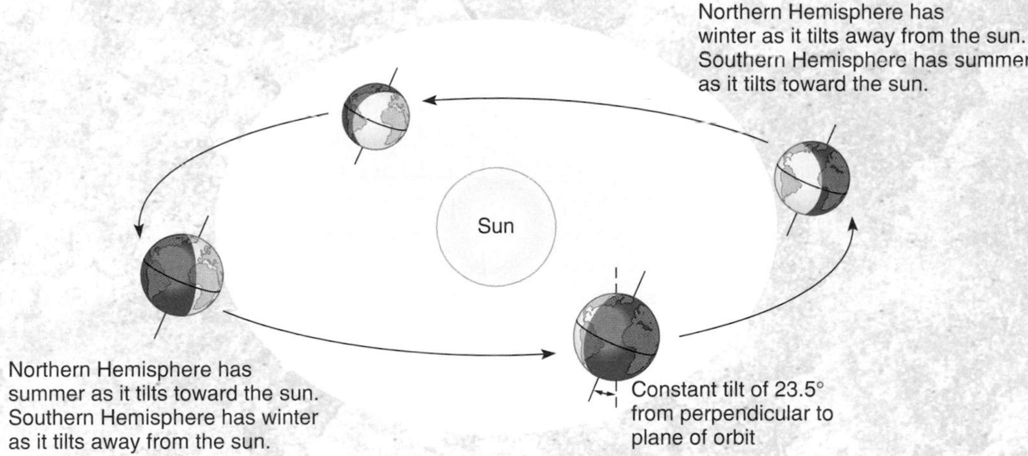

FIGURE 9.1 Seasonal warming and cooling results depending on whether your hemisphere is tilted toward or away from the sun at that time; the effect is most pronounced near the poles.

there may be local exceptions). Over the longer term, significant changes can occur. We have evidence in the geologic record for a number of episodes during which vast ice masses covered major portions of the continents. Today, human activities are demonstrably changing the chemistry of our atmosphere, which may lead to permanent changes in climate, through global warming. We have noted two consequences: melting of glaciers worldwide, and thermal expansion of oceans, with resultant rise in net sea level, inundating low-lying coastal regions. There is also concern that on a warmer globe, arid lands will be more extensive. Desertification caused by human activities is already a serious concern of those who wonder how the earth's still-growing population will be fed; local or regional drought aggravated by global warming may compound the problem. Chapter 10 will explore the subject of climate more broadly.

GLACIERS AND GLACIAL FEATURES

A **glacier** is a mass of ice that moves over land under its own weight, through the action of gravity. A quantity of snow sufficient to form a glacier does not fall in a single winter, so for glaciers to develop, the climate must be cold enough that some snow and ice persist year-round. This, in turn, requires an appropriate combination of altitude and latitude. Glaciers tend to be associated with the extreme cold of polar regions. However, temperatures also generally decrease at high altitudes (figure 9.2), so glaciers can exist in mountainous areas even in tropical or subtropical regions. Three mountains in Mexico have glaciers, but these glaciers are all at altitudes above 5000 meters (15,000 feet). Similarly, there are glaciers on Mount Kenya and Mount Kilimanjaro in east Africa, within a few degrees north or south of the equator, but only at high altitudes.

Glacier Formation

For glaciers to form, there must be sufficient moisture in the air to provide the necessary precipitation. Glaciers are, after all, large reservoirs of frozen water. In addition, the amount of winter snowfall must exceed summer melting so that the snow accumulates year by year. Most glaciers start in the mountains as snow patches that survive the summer. Slopes that face the poles (north-facing in the Northern Hemisphere, southfacing in the Southern Hemisphere), protected from the strongest sunlight, favor this survival. So do gentle slopes, on which snow can pile up thickly instead of plummeting down periodically in avalanches.

As the snow accumulates, it is gradually transformed into ice. The weight of overlying snow packs it down, drives out

FIGURE 9.2 The snow line on a mountain range reflects decreasing temperature at higher elevation. Olympic Range, Washington.

much of the air, and causes it to recrystallize into a coarser, denser form (*firn*), an intermediate texture between snow and ice, and finally to a compact solid mass of interlocking ice crystals (figure 9.3). The conversion of snow into solid glacial ice may take only a few seasons or several thousand years, depending on such factors as climate and the rate of snow accumulation at the top of the pile, which hastens compaction. Eventually, the mass of ice becomes large enough that, if there is any slope to the terrain, it begins to flow downhill—or, on more level ground, the ice begins to flow outward from the thickest accumulations. The moving ice is then a true glacier.

Types of Glaciers

Glaciers are divided into two types on the basis of size and occurrence. The most numerous today are the **alpine glaciers,** also known as mountain or valley glaciers. As these names suggest, they are glaciers that occupy valleys in mountainous terrain, most often at relatively high altitudes. Most of the estimated 70,000 to 200,000 glaciers in the world today are alpine glaciers. A number of these can be seen in the chapter-opener photograph.

The larger and rarer **continental glaciers** are also known as *ice caps* (generally less than 50,000 square kilometers in area) or *ice sheets* (larger). They can cover whole continents and reach thicknesses of a kilometer or more. Though they are far fewer in number than the alpine glaciers, the continental glaciers comprise far more ice. At present, the two principal continental glaciers are the Greenland and the Antarctic ice sheets. The Antarctic ice sheet is so large that it could easily cover the forty-eight contiguous United States. The geologic record indicates that, at several times in the past, even more extensive ice sheets existed on earth.

Movement and Change of Glaciers

Glaciers do not behave as rigid, unchanging lumps of ice. Their flow is plastic, like that of asphalt on a warm day (or of the convecting upper mantle), and different parts of glaciers move at different rates. At the base of a glacier, or where it scrapes against valley walls, the ice moves more slowly. In the higher, more central portions, it flows more freely and rapidly (figure 9.4). Where it is cold enough for snow to fall, fresh material accumulates, adding to the weight of ice and snow that pushes the glacier downhill. Parts of a glacier may locally flow up an incline for a short distance. However, the net flow of the glacier must be downslope, under the influence of gravity. The downslope movement may be either down the slope of the underlying land surface, or in the downslope direction of the *glacier's* surface, from areas of thicker ice to thinner, regardless of the underlying topography (also known as flow "down gradient"), as is typical for ice sheets (figure 9.5).

That flow is typically slow, a few tens of meters per year. However, alpine glaciers may sometimes *surge* at rates of several tens of meters per day for a few days or weeks. In addition to flowing plastically under its own weight, a glacier may slide on meltwater accumulated at the base of the ice. It may be that some surges occur when that meltwater is dammed up and prevented from draining away until an unusually large quantity has accumulated to lubricate the glacier's slide downhill.

At some point, the advancing edge of the glacier terminates, either because it flows out over water and breaks up, creating icebergs by a process known as **calving,** or, more commonly, because it has flowed to a place that is warm enough that ice loss by melting or evaporation is at least as rapid as the rate at which new ice flows in (figure 9.6). The set of processes by which ice is lost from a glacier is collectively called **ablation.** The **equilibrium line** is the line on the glacial surface where there is no net gain or loss of material.

FIGURE 9.3 Thin slice of glacier ice viewed under a microscope in polarized light. The interlocking ice crystals resemble the compact texture of an igneous or metamorphic rock.

Photograph by A. Gow, U.S. Army Corps of Engineers, courtesy NOAA/National Geophysical Data Center

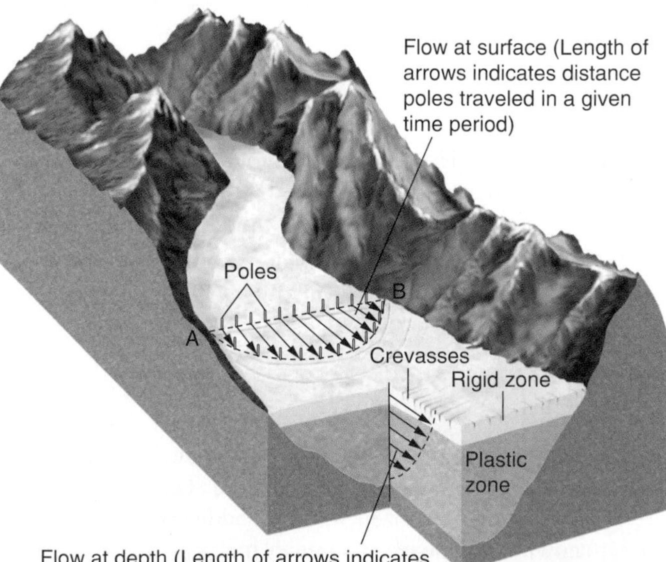

FIGURE 9.4 Glacier flow is mainly plastic; flow is fastest at center, surface. Crevasses form as the brittle surface ice cracks.

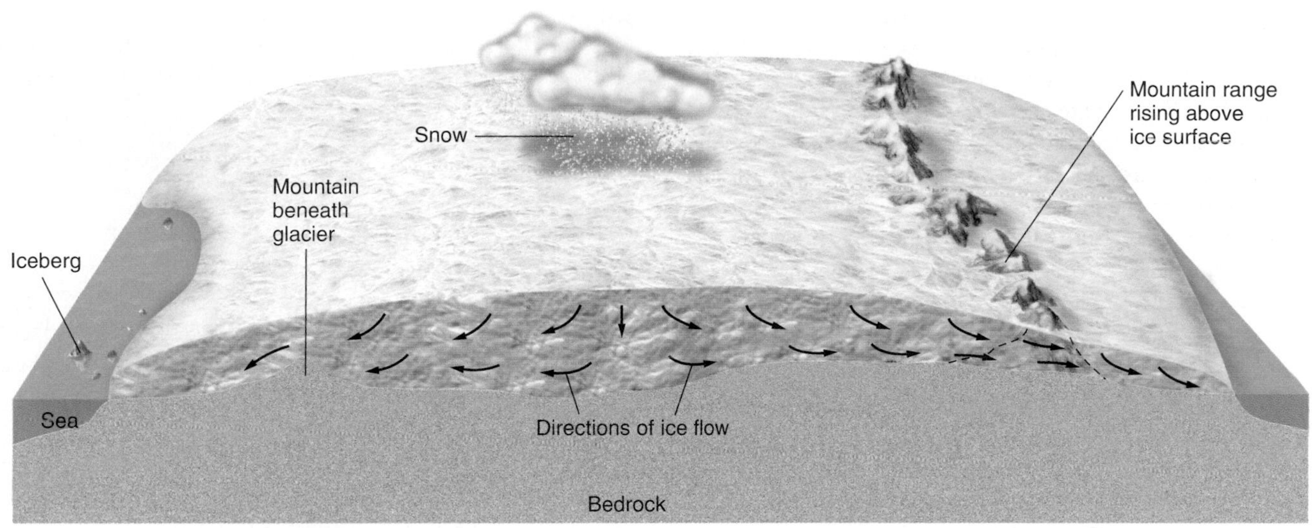

FIGURE 9.5 An ice sheet flows outward from its thickest point.

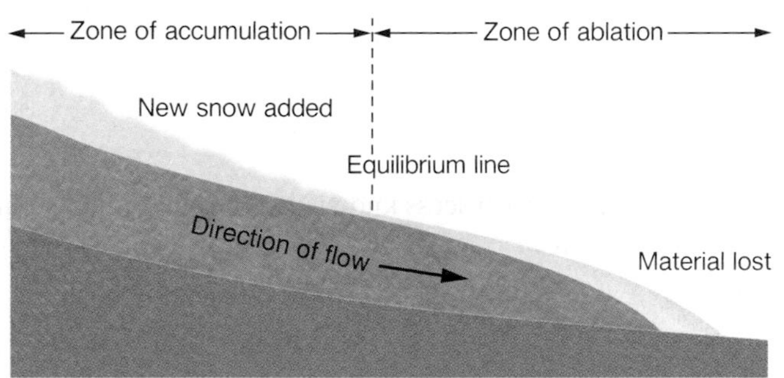

FIGURE 9.6 Schematic of longitudinal cross section of glacier. In the zone of accumulation, the addition of new material exceeds loss by melting or evaporation; the reverse is true in the zone of ablation, where there is a net loss of ice.

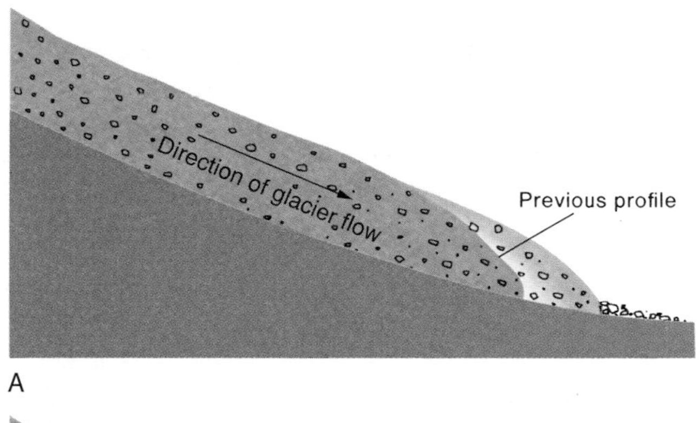

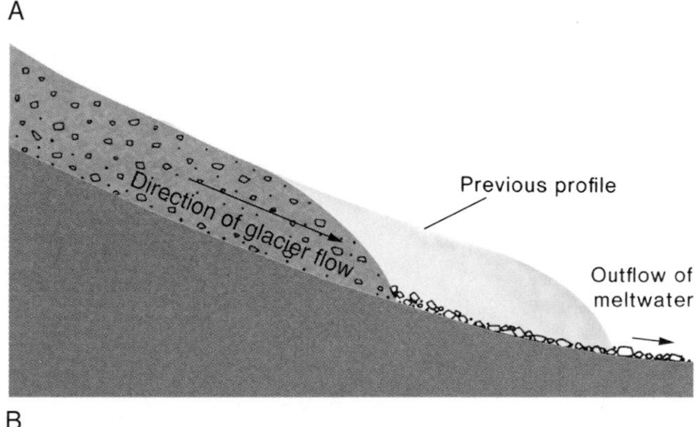

FIGURE 9.7 Glacial advance and retreat. (A) Advance of the glacier moves both ice and sediment downslope. (B) Ablation, especially melting, causes apparent retreat of glacier, accompanied by deposition of sediment from the lost ice volume.

Over the course of a year, the size of a glacier varies somewhat (figure 9.7). In winter, the rate of accumulation increases, and melting and evaporation decrease. The glacier becomes thicker and longer; it is then said to be advancing. In summer, snowfall is reduced or halted and melting accelerates; ablation exceeds accumulation. The glacier is then described as "retreating," although, of course, it does not flow backward, uphill. The leading edge simply melts back faster than the glacier moves forward. Over many years, if the climate remains stable, the glacier achieves a dynamic equilibrium, in which the winter advance just balances the summer retreat, and the average extent of the glacier remains constant from year to year.

More significant are the sustained increases or decreases in glacier size that result from consistent cooling or warming trends. An unusually cold or snowy period spanning several years would be reflected in a net advance of the glacier over that period, while a warm period would produce a net retreat. Such phenomena have been observed in recent history. From the mid-1800s to the early 1900s, a worldwide retreat of alpine glaciers far up into their valleys was observed. Beginning about 1940, the trend then seemed to reverse, and glaciers advanced (which may have been due to cooling associated with air pollution, as will be explored in later chapters). More recently, a clear global warming trend has resumed, with corresponding shrinkage of many glaciers. It is important to bear in mind in all discussions of "global" climate changes, however, that considerable local climate variations can exist, as a result of local topography, geography, precipitation, winds and water currents, and many other factors. So, for example, though a majority of alpine glaciers worldwide are currently retreating, and over several decades that trend has strengthened so that the net mass of ice in these glaciers is decreasing, some glaciers may still be stable or advancing (table 9.1). Similarly, while the edges of the Greenland ice sheet are melting back rapidly, the center is thickening as a result of increased precipitation (figure 9.8).

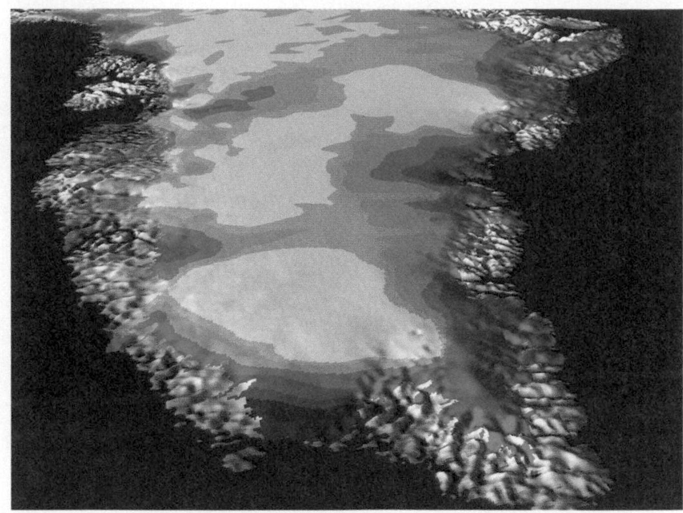

FIGURE 9.8 Satellite topographic measurements show that while the center of the Greenland ice sheet is thickening, the edges are melting away. Net loss of ice is currently estimated to be over 50 cubic kilometers (more than 12 cubic miles) per year.
Image courtesy NASA Goddard Space Flight Center Scientific Visualization Studio

Glacial Erosion and Deposition

A glacier's great mass and solidity make it especially effective in erosion. Only because streams far outnumber glaciers is liquid water, overall, a more significant agent in shaping the earth's surface than is ice. Glaciers do, however, leave distinctive features by which their past extent and movements can be tracked, long after the ice is gone.

Like a stream, an alpine glacier carves its own valley. The solid ice carves a somewhat different shape, however: A glacial valley is typically U-shaped (figure 9.9). On a grand scale, thick continental glaciers may gouge out a series of large troughs in the rock over which they move; the Finger Lakes of upstate New York occupy valleys deepened by an ice sheet. On a very small scale, bits of rock that have become frozen into the ice at the base of a glacier act like sandpaper on solid rock below, making fine, parallel scratches, called **striations,** on the underlying surface (figure 9.10). Such signs not only indicate

TABLE 9.1

Response of Cascade (U.S.), Swiss, and Italian Glaciers to Global Climate Change

	Advancing	Stationary	Retreating	Gone
North Cascades, Washington				
1967	19	8	20	0
1974	20	5	22	0
1985	5	10	32	0
1995	0	0	47	2
2005	0	0	43	4
Switzerland				
1967	31	14	55	
1975	70	3	28	
1985	49	13	42	
1995	9	13	72	
2005	0	7	86	
Italy				
1981	25	10	10	
1985	25	6	14	
1995	8	9	104	
2005	1	5	126	

Source: Data from M. Pelto, personal communication (2007).

Because observations were not always made of the same glaciers each time, total number of glacier observations reported at each location varies.

FIGURE 9.9 Typical U-shaped glacial valley. Rocky Mountain National Park.

the past presence of glaciers but also the direction in which they flowed. Erosion by the scraping of ice or included sediment on the surface underneath is **abrasion.** Water can also seep into cracks in rocks at the base of the glacier and refreeze, attaching rock fragments to the glacier. As the ice moves on, small bits of the rock are torn away in a process known as *plucking*. Plucking at the head of an alpine glacier contributes to the formation of a *cirque,* a bowl-shaped depression (figure 9.11A). Where alpine glaciers flow side by side, the wall of rock between them may be thinned to a sharp ridge, an *arête* (figure 9.11B); erosion by several glaciers around a single peak produces a *horn* (figure 9.11C), such as the Matterhorn.

Glaciers also transport and deposit sediment very efficiently. Debris that has been eroded from the sides of an alpine glacier's valley or that has fallen onto the glacier from peaks above tends to be carried along on or in the ice, regardless of the size, shape, or density of the sediment. It is not likely to be deposited until the ice carrying it melts or evaporates, and then all the debris, from large boulders to fine dust, is dropped together. Sediment deposited directly from the melting ice is called **till.** Till is characteristically angular and poorly sorted material (figure 9.12), as contrasted with stream-deposited sediment. Some of the till may be transported and redeposited by the meltwater, whereupon the sediment is called **outwash.** Till and outwash together are two varieties of glacial **drift.**

A landform made of till is called a **moraine.** There are many different types of moraine. One of the most distinctive is a curving ridge of till piled up at the nose of a glacier. Some of it may have been pushed along ahead of the glacier as it advanced. Much of it, however, is accumulated at the glacier's end by repeated annual advances and retreats; if the net extent of the glacier remains fairly constant for some years, the effect is like that of a conveyor belt carrying sediment to the end of

FIGURE 9.10 Striations on a rock surface show direction of glacial flow. Notice that the grooves extend continuously across the fractures breaking up the outcrop but are absent where the scratched surface rock has weathered away. Lassen Park, California.

A

B

C

FIGURE 9.11 Erosional features of glaciers: (A) cirque, (B) arête, and (C) horn, with arêtes spreading out from it.

FIGURE 9.12 Till deposited by an Alaskan glacier shows characteristically poor sorting; fragments are also angular and blocky, not rounded. This pile of till is about two meters high.

the glacier. This is an *end moraine.* When the glacier begins a long-term retreat, a *terminal moraine* is left to mark the farthest advance of the ice (figure 9.13). Moraines are useful in mapping the extent of past glaciation on both a local and a continental scale.

Glaciers are responsible for much of the topography and near-surface geology of northern North America. The Rocky Mountains and other western mountain ranges show abundant evidence of carving by alpine glaciers. The Great Lakes occupy basins carved and deepened by the most recent ice sheets and were once filled with glacial meltwater as the ice sheets retreated. Much of the basic drainage pattern of the Mississippi River was established during the surface runoff of the enormous volumes of meltwater; the sediments carried along with that same meltwater are partially responsible for the fertility of midwestern farmland (see "Loess" later in the chapter). Infiltration of a portion of the meltwater resulted in groundwater supplies tapped today for drinking water. The importance of the legacy of past glaciers should not be overlooked just because their extent at present is more limited. Better understanding of past widespread glaciation may also help in modeling future climate changes and consequences.

Ice Ages and Their Possible Causes

The term *Ice Age* is used by most people to refer to a time beginning in the relatively recent geologic past when extensive continental glaciers covered vastly more area than they now do. There have been many cycles of advance and retreat of the ice sheets during this epoch, known to geologists as the Pleistocene, which spanned the time from about 2 million to 10,000 years ago. At its greatest extent, Pleistocene glaciation in North America at one time or another covered essentially all of Canada and much of the northern United States (figure 9.14).

A

B

FIGURE 9.13 (A) Formation of end moraine occurs primarily through repeated annual cycles of deposition of till by melting ice. An advancing glacier deposits till in the zone of ablation and pushes material ahead of ice front; then the glacier retreats, leaving a ridge of till behind. (B) Moraine left by retreating glacier in Denali National Park, Alaska. The terminal moraine is overgrown with vegetation.

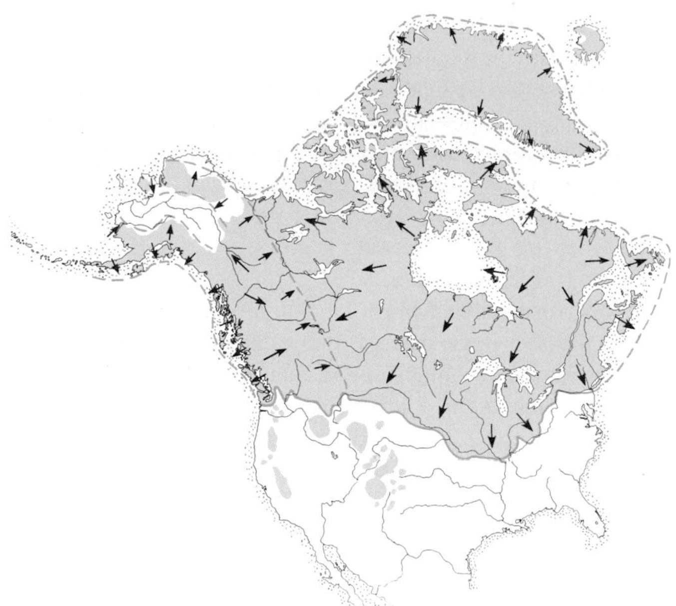

FIGURE 9.14 Extent of glaciation in North America during the Pleistocene epoch. Arrows show direction of ice movement. Note that not every bit of land within the area colored in blue was necessarily ice-covered, and the maximum ice extent shown is for the whole Ice Age, not each glacial advance.

After C. S. Denny, National Atlas of the United States, *U.S. Geological Survey*

The ice was well over a kilometer thick over most of that area. Climatic effects extended well beyond the limits of the ice, with cooler temperatures and more precipitation common in surrounding lands, and worldwide sea levels lowered as vast volumes of water were locked up as ice in ice sheets. During the "interglacials," the ice sheets retreated; we are likely now in just such an interglacial time.

Pleistocene glaciation can be studied in some detail because it was so recent, and the evidence left behind is thus fairly well preserved. It is not, however, the only large-scale continental glaciation in earth's past. There have been at least half a dozen ice ages over the earth's history, going back a billion years or more. As previously noted, the predrift reassembly of the pieces of the continental jigsaw puzzle is aided by matching up features produced by past ice sheets on what are now separate continents.

An underlying question is what kinds of factors might be involved in creating conditions conducive to the formation of such immense ice masses, bearing in mind that ice a kilometer thick represents far more than a few winters' snow. The proposed causes of ice ages fall into two groups: those that involve events external to the earth and those for which the changes arise entirely on earth. They differ in time scale, as well as in magnitude.

One possible external cause, for example, would be a significant change in the sun's energy output. Present cycles of sunspot activity cause variations in sunlight intensity, which should logically result in temperature fluctuations worldwide. However, the variations in solar-energy output would have to be about ten times as large as they are in the modern sunspot cycle to account even for short-term temperature fluctuations observed on earth. To cause an ice age of thousands of years' duration, any cooling trend would have to last much longer than the eleven years of the present sunspot cycle. Another problem with linking solar activity fluctuations with past ice ages is simply lack of evidence. Although means exist for estimating temperatures on earth in the past, scientists have yet to conceive of a way to determine the pattern of solar activity in ancient times. Thus, there is no way to test the hypothesis, to prove or disprove it.

Since the advent of plate-tectonic theory, a novel suggestion for a cause of ice ages has invoked continental drift. Prior to the breakup of Pangaea, all the continental landmasses were close together. This concentration of continents meant much freer circulation of ocean currents elsewhere on the globe and more circulation of warm equatorial waters to the poles. The breakup of Pangaea disrupted the oceanic circulation patterns. The poles could have become substantially colder as their waters became more isolated. This, in turn, could have made sustained development of a thick ice sheet possible. A limitation of this mechanism is that it accounts for only the Pleistocene glaciation, while considerable evidence exists for extensive glaciation over much of the Southern Hemisphere—India, Australia, southern Africa, and South America—some 200 million years ago, just *prior* to the breakup of Pangaea. Beyond that, not enough is known about continental positions prior to the formation of Pangaea to determine whether continental drift and resulting changes in oceanic circulation patterns could be used to explain earlier ice ages.

Another possible earth-based cause of ice ages might involve blocking of incoming solar radiation by something in the atmosphere. We have observed that volcanic dust and sulfuric-acid droplets in the atmosphere can cause measurable cooling, but with the modern examples at least, the effects were significant only for a year or two. In none of these cases was the cooling serious enough or prolonged enough to cause an ice age. Even some of the larger prehistoric eruptions, such as those that occurred in the vicinity of Yellowstone National Park or the explosion that produced the crater in Mount Mazama that is now Crater Lake, would have been inadequate individually to produce the required environmental change, and again, the dust and acid droplets from any one of these events would have settled out of the atmosphere within a few years. However, there have been periods of more intensive or frequent volcanic activity in the geologic past. The cumulative effects of multiple eruptions during such an episode might have included a sustained cooling trend and, ultimately, ice-sheet formation, but at present, the data are inadequate to demonstrate a clear link to actual ice ages.

A better-supported proposed cause of ice ages is related to the observed variation in the tilt of the earth's axis in space and the eccentricity of the earth's orbit around the sun. That orbit is not perfectly circular, so the distance between sun and earth is not constant, and the shape of the orbit varies. Axial tilt varies somewhat over time relative to the earth's orbital plane. Such changes do not significantly change the total amount of sunlight reaching earth, but they do affect its distribution, the amount of sunlight falling at different latitudes (and therefore the relative surface heating expected), and the time of year at which a region receives its maximum sunlight (figure 9.15). Together, they result in long-term cyclic variations, over time scales of thousands of years, in local insolation (incident sunlight), named *Milankovitch cycles* after the scientist who recognized them in the early twentieth century. Could they account for ice advances and retreats by changing the relative heating of poles, equator, and midlatitudes?

Some evidence supports this hypothesis. When glaciers are extensive and large volumes of water are locked up in ice, sea level is lowered. The reverse is true as ice sheets melt. It is possible to use ages of Caribbean coral reefs to date high stands of sea level over the last few hundred thousand years, and these periods of high sea level correlate well with periods during which maximum solar radiation fell on the mid-Northern Hemisphere. Conversely, then, periods of reduced sun exposure are apparently correlated with ice advance (lower sea level). However, this precise a record does not go back very far in time, nor is the correlation perfect even for the period of relatively complete data. And while Milankovitch cycles may account for ice advances and retreats during the Ice Age, it is less clear that they can account for its initiation in the first place. It may also be that changes in insolation lead to changes in precipitation patterns over the oceans, and thus to changes in seawater salinity and mixing patterns, which, in turn, would alter the distribution of organisms in the oceans and the oceans' effectiveness as a "sink" for atmospheric carbon. (To understand what this has to do with heating, see chapter 10 on the greenhouse effect.)

The processes and phenomena that may have caused climatic upheavals in the past are many, and, as yet, there is insufficient evidence to demonstrate that any single model is correct. The answer may lie in a combination of several factors. As we will see in chapter 10, global climate involves complex interactions among several earth systems.

WIND AND ITS GEOLOGIC IMPACTS

Air moves from place to place over the earth's surface mainly in response to differences in pressure, which are often related to differences in surface temperature. Solar radiation falls most

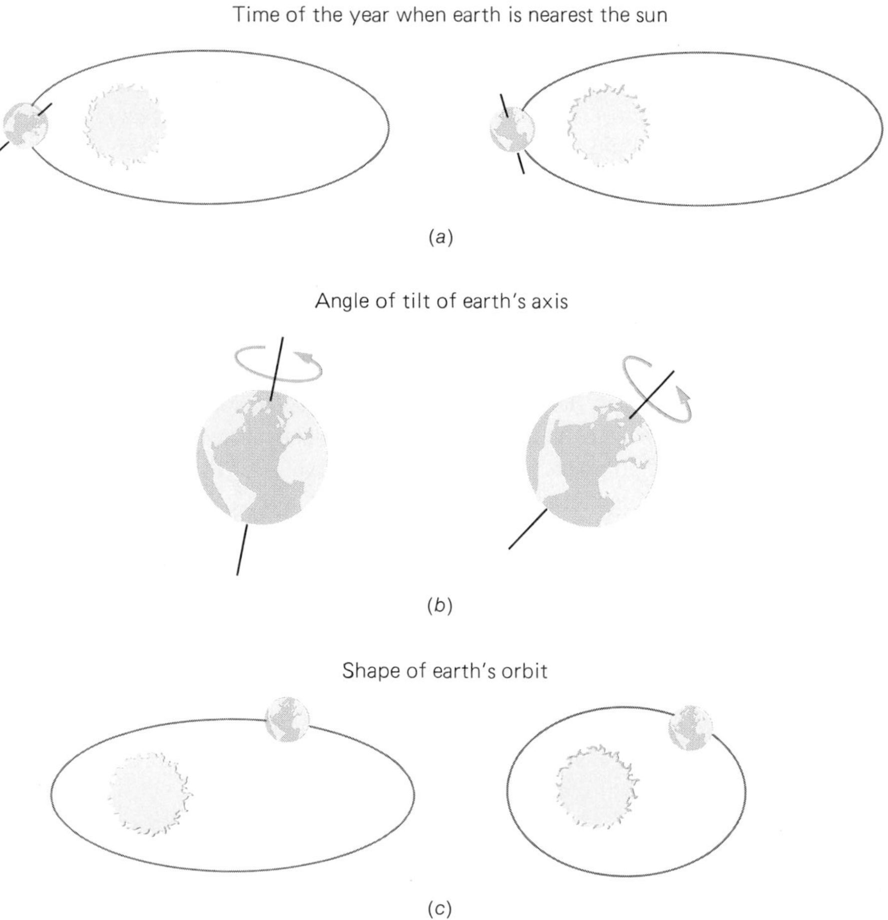

FIGURE 9.15 Variations in earth's orbit and axial tilt lead to cyclic variations, over thousands of years, in the amount of solar radiation reaching the surface in different parts of the globe and at different seasons.

Box 9.1 Glaciers as a Water Source

Glaciers are not only large masses of ice; they can also be viewed as an important freshwater reserve. Approximately 75% of the fresh water on earth is stored as ice in glaciers. The world's supply of glacial ice is the equivalent of sixty years of precipitation over the whole earth.

In regions where glaciers are large or numerous, glacial meltwater may be the principal source of summer streamflow. By far the most extensive glaciers in the United States are those of Alaska, which cover about 3% of the state's area. Summer streamflow from these glaciers is estimated at nearly 50 trillion gallons of water! Even in less-glaciated states like Montana, Wyoming, California, and Washington, streamflow from summer meltwater amounts to tens or hundreds of billions of gallons.

Anything that modifies glacial melting patterns can profoundly affect regional water supplies. Dusting the glacial ice surface with something dark, like coal dust, increases the heating of the surface, and hence the rate of melting and water flow. Conversely, cloud seeding over glaciated areas could be used to increase precipitation and the amount of water stored in the glaciers. Increased meltwater flow can be useful not only to increase water supplies but also to achieve higher levels of hydroelectric power production. Techniques for modifying glacial meltwater flow are not now being used in the United States, in part because the majority of U.S. glaciers are in national parks, primitive areas, or other protected lands. However, some of these techniques have been practiced in parts of the former Soviet Union and China. Global warming could be expected to increase the release of meltwater, as it is already decreasing water storage in alpine glaciers. In the short term, this may produce more abundant surface water, but at the expense of the ice reserve for the future.

Decreased glacier volume may have more subtle water-supply effects. Early in the melt season in alpine regions, some of the meltwater is stored in pore space in glacial snow and ice. As glacial melting proceeds through spring and summer, this stored water adds to glacial meltwater volume, as the glacial snow/ice containing it melts, enhancing water supplies in many regions. If glacier size decreases, so does this water storage capacity. Recent studies of the North Cascade Glacier Climate Project have documented two resultant problems: increased likelihood of spring flooding, and decreased meltwater flow later in the season. In the Andes and Himalayas, where agriculture depends heavily on glacial meltwater runoff, glacial retreat may signal future reductions in crop productivity. The vanishing glaciers of Mount Kilimanjaro (figure 1) have historically provided vital water supplies to local people in dry seasons and when monsoon rains fail to arrive; what happens when those glaciers are gone? Such issues are one aspect of broader concerns about global climate change.

FIGURE 1 The dormant stratovolcano Kilimanjaro was given its Swahili name, which translates as "shining mountain," for its extensive glacier cap. As seen in this April 2003 image, the glaciers are now nearly gone; they are expected to vanish altogether by 2020.
Courtesy Earth Observatory Lab, Johnson Space Center

directly on the earth near the equator, and, consequently, solar heating of the atmosphere and surface is more intense there. The sun's rays are more dispersed near the poles. On a nonrotating earth of uniform surface, that surface would be heated more near the equator and less near the poles; correspondingly, the air over the equatorial regions would be warmer than the air over the poles. Because warm, less-dense (lower-pressure) air rises, warm near-surface equatorial air would rise, while cooler, denser polar air would move into the low-pressure region. The rising warm air would spread out, cool, and sink. Large circulating air cells would develop, cycling air from equator to poles and back. These would be atmospheric convection cells, analogous to the mantle convection cells associated with plate motions.

The actual picture is considerably more complicated. The earth rotates on its axis, which adds a net east-west component to air flow as viewed from the surface. Land and water are heated differentially by sunlight, with surface temperatures on the continents generally fluctuating much more than temperatures of adjacent oceans. Thus, the pattern of distribution of land and water influences the distribution of high- and low-pressure regions and further modifies air flow. In addition, the earth's surface is not flat; terrain irregularities introduce further complexities in air circulation. Friction between moving air masses and land surfaces can also alter wind direction and speed; tall, dense vegetation may reduce near-surface wind speeds by 30 to 40% over vegetated areas. A generalized view of actual global air circulation patterns is shown in figure 9.16. Different regions of the earth are characterized by different prevailing wind directions. Most of the United States is in a zone of westerlies, in which winds generally blow from west/southwest to east/northeast. Local weather conditions and the details of local geography produce regional deviations from this pattern on a day-to-day basis.

Air and water have much in common as agents shaping the land. Both can erode and deposit material; both move material

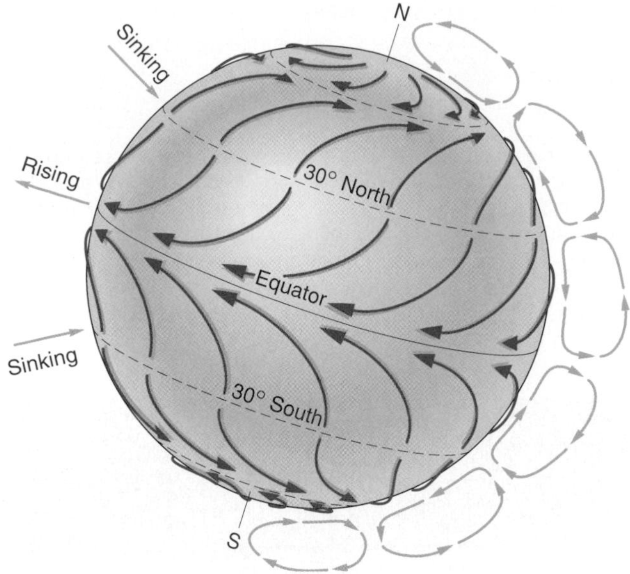

FIGURE 9.16 Principal present atmospheric circulation patterns.

more effectively the faster they flow; both can move particles by rolling them along, by saltation, or in suspension (see chapter 6). Because water is far denser than air, it is much more efficient at eroding rocks and moving sediments and has the added abilities to dissolve geologic materials and to attack them physically by freezing and thawing. On average worldwide, wind erosion moves only a small percentage of the amount of material moved by stream erosion. Historically, the cumulative significance of wind erosion is approximately comparable to that of glaciers. Nevertheless, like glaciers, winds can be very important in individual locations particularly subject to their effects.

Wind Erosion

Like water, wind erosion acts more effectively on sediment than on solid rock, and wind-related processes are especially significant where the sediment is exposed, not covered by structures or vegetation, in such areas as deserts, beaches, and unplanted (or incorrectly planted) farmland. In dry areas like deserts, wind may be the major or even the sole agent of sediment transport.

Wind erosion consists of either abrasion or deflation. *Wind abrasion* is the wearing-away of a solid object by the impact of particles carried by wind. It is a sort of natural sandblasting, analogous to milling by sand-laden waves. Where winds blow consistently from certain directions, exposed boulders may be planed off in the direction(s) from which they have been abraded, becoming **ventifacts** ("wind-made" rocks; figure 9.17). If wind velocity is too low to lift the largest transported particles very high above the ground, tall rocks may show undercutting close

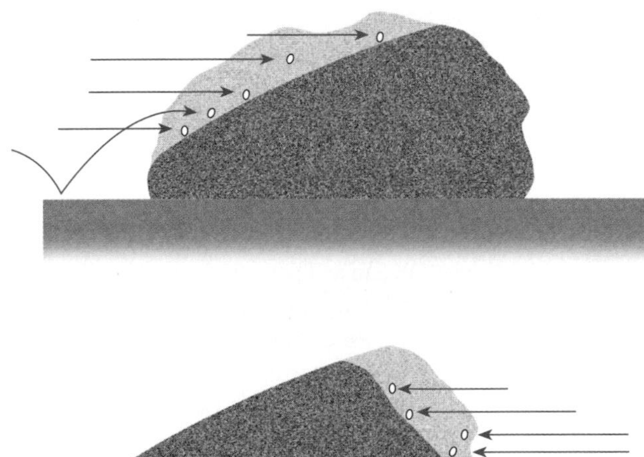

FIGURE 9.17 Wind abrasion on low-lying rocks results in the planing of rock surfaces. (A) If the wind is predominantly from one direction, rocks are planed or flattened on the upwind side. With a persistent shift in wind direction, additional facets are cut in the rock. (B) Example of a ventifact: rock polished and faceted by windblown sand at San Gorgonio Pass near Palm Springs, California.
Photograph © Doug Sherman/Geofile

to ground level (figure 9.18A). Abrasion can also cause serious property damage. Desert travelers caught in windstorms have been left with cars stripped of paint and windshields so pitted and frosted that they could no longer be seen through. Abrasion likewise scrapes paint from buildings and can erode construction materials such as wood or soft stone.

Deflation is the wholesale removal of loose sediment, usually fine-grained sediment, by the wind (figure 9.18B). In barren areas, a combination of deflation and erosion by surface runoff may proceed down to some level at which larger rocks are exposed, and these larger rocks protect underlying fine material from further erosion. Such an effect may contribute to the maintenance of a surface called *desert pavement* (figure 9.19). A desert pavement surface, once established, can be quite stable. Finer sediment added to the surface may be swept between coarser chunks, later settling below them, while the pavement continues to protect the fine material against loss. However, if the protective coarse-rock layer is disturbed—for example, by construction or resource-exploration activities—newly exposed fine sediment may be subject to rapid wind erosion.

The importance of vegetation in retarding erosion was demonstrated especially dramatically in the United States

A

B

FIGURE 9.18 (A) Undercutting of granite boulder by near-surface wind abrasion; Enchanted Rock, TX. (B) Deflation at The Pinnacles, Namburg National Park, Australia.

(A) Photograph © Ken Wagner/Visuals Unlimited, (B) Photograph © Michael Fogden/Animals Animals/Earth Scenes

FIGURE 9.19 Desert pavement developed near Death Valley National Monument, CA.

Photograph © Gordon Whitten/Corbis Images

during the early twentieth century. After the Civil War, there was a major westward migration of farmers to the flat or gently rolling land of the Great Plains. Native prairie grasses and wildflowers were removed and the land plowed and planted to crops. While adequate rainfall continued, all was well. Then, in the 1930s, several years of drought killed the crops. The native prairie vegetation had been suitably adapted to the climate, while many of the crops were not. Once the crops died, there was nothing left to hold down the soil and protect it from the west winds sweeping across the plains. This was the Dust Bowl period. We look further at the Dust Bowl and the broader problem of minimizing soil erosion on cropland in chapter 12.

Wind Deposition

Where sediment is transported and deposited by wind, the principal depositional feature is a **dune,** a low mound or ridge, usually made of sand. Dunes start to form when sediment-laden winds slow down. The lower the velocity, the less the wind can carry. The coarser and heavier particles are dropped first. The deposition, in turn, creates an obstacle that constitutes more of a windbreak, causing more deposition. Once started, a dune can grow very large. A typical dune is 3 to 100 meters high, but dunes as high as 200 meters are found. What ultimately limits a dune's size is not known, but it is probably some aspect of the nature of the local winds.

When the word *dune* is mentioned, most people's reaction is *sand dune.* However, the particles can be any size, ranging from sand down to fine dust, and they can even be snow or ice crystals, although sand dunes are by far the most common. The particles in a given set of dunes tend to be similar in size. Eolian, or wind-deposited, sediments are well sorted like many stream deposits, and for the same reason: The velocity of flow controls the size and weight of particles moved. The coarser the particles, the stronger was the wind forming the dunes. The orientation of the dunes reflects the prevailing wind direction (if winds show a preferred orientation), with the more shallowly sloping side facing upwind (figure 9.20).

A

B

FIGURE 9.20 The formation of dunes. (A) Huge sand dunes at Great Sand Dunes National Monument formed as windswept sand piled up against the Sangre de Cristo mountains. (B) Snow ripples, mimicking dunes in form and formation, on barren ground in winter. Note ground bared by deflation between ripples.

Dune Migration

Dunes move if the wind blows predominantly from a single direction. As noted previously, a dune assumes a characteristic profile in cross section, gently sloping on the windward side, steeper on the downwind side. With continued wind action, particles are rolled or moved by saltation up the shallower slope. They pile up a bit at the peak, like windswept snow cornices on a mountain, then tumble down the steeper face, or **slip face,** which tends to assume a slope at the angle of repose of sand (or whatever size of particle is involved; recall figure 8.4B). The net effect of these individual particle shifts is that the dune moves slowly downwind (figure 9.21A). As layer upon layer of sediment slides down the slip face, slanted *crossbeds* develop in the dune (figure 9.22).

Migrating dunes, especially large ones, can be a real menace as they march across roads and even through forests and over buildings (figure 9.21B, C). During the Dust Bowl era, farmland and buildings were buried under shifting, windblown soil. The costs to clear and maintain roads along sandy beaches and through deserts can be high because dunes can move several meters or more in a year. The usual approach to dune stabilization is to plant vegetation. However, since many dunes exist, in part, because the terrain is too dry to support vegetation, such efforts are likely to be futile. Aside from any water limitations, young plants may be difficult to establish in

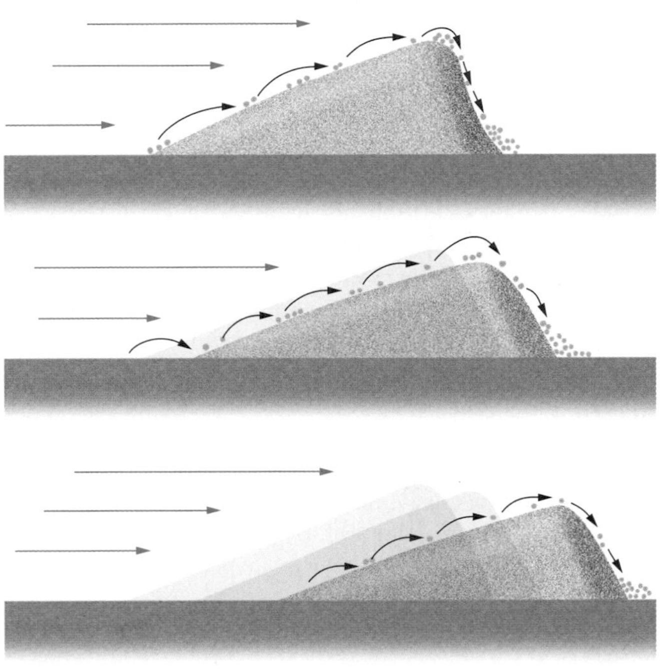

A

C

B

FIGURE 9.21 Dune migration and its consequences. (A) A schematic of dune migration. (B) Marching sand dunes encroaching on full-grown trees, and (C) slowly burying a trail. Oregon Dunes National Recreation Area.

FIGURE 9.22 Crossbeds in a sand dune. Here, several sets of beds meeting at oblique angles show the effects of shifting wind directions, changing sediment deposition patterns. Note also wind-produced ripples in foreground. Oregon Dunes National Recreation Area.

FIGURE 9.23 Example of loess. Norton County, KS
Photograph by L. B. Buck, USGS Photo Library, Denver, CO.

shifting dune sands because their tiny roots may not be able to secure a hold.

Loess

Rarely is the wind strong enough to move sand-sized or larger particles very far or very rapidly. Fine dust, on the other hand, is more easily suspended in the wind and can be carried many kilometers before it is dropped. A deposit of windblown silt is known as **loess** (figure 9.23). The fine mineral fragments in loess are in the range of 0.01 to 0.06 millimeter (0.0004 to 0.0024 inch) in diameter.

The principal loess deposits in the United States are in the central part of the country, and their spatial distribution provides a clue to their source (figure 9.24). They are concentrated around the Mississippi River drainage basin, particularly on the east sides of major rivers of that basin. Those same rivers drained away much of the meltwater from retreating ice sheets in the last ice age. Glaciers grinding across the continent, then, were apparently the original producers of this sediment. The sediment was subsequently washed down the river valleys, and the lightest material was further blown eastward by the prevailing west winds.

As noted previously, the great mass of ice sheets enables them to produce quantities of finely pulverized *rock flour*. Because dry glacial erosion does not involve as much chemical

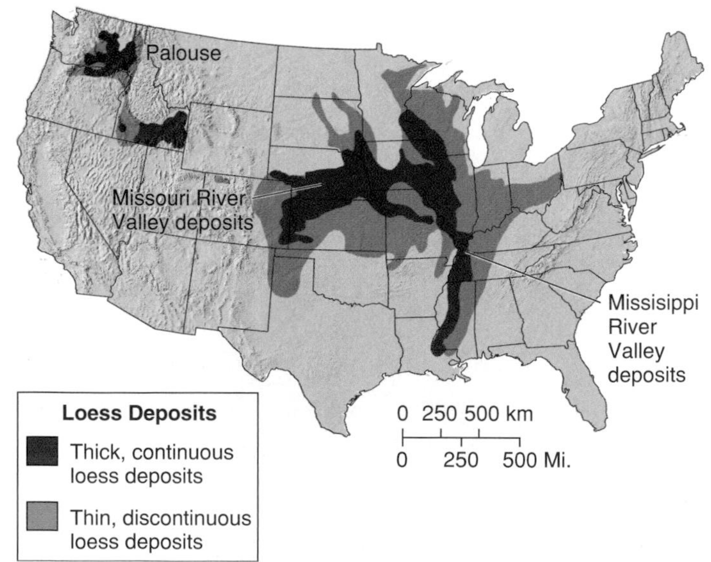

FIGURE 9.24 Much of the loess distribution in the central United States is close to principal stream valleys supplied by glacial meltwater. Loess to the southwest may have been derived from western deserts.

weathering as stream erosion, many soluble minerals are preserved in glacial rock flour. These minerals provide some plant nutrients to the farmland soils now developed on the loess. Because newly deposited loess is also quite porous and open in structure, it has good moisture-holding capacity. These two characteristics together contribute to making the farmlands developed on midwestern loess particularly productive.

Not all loess deposits are of glacial origin. Loess deposits form wherever there is an abundant supply of very fine sediment. Loess derived from the Gobi Desert covers large areas of China, for example, and additional loess deposits are found close to other major deserts.

Loess does have drawbacks with respect to applications other than farming. While its light, open structure is reasonably strong when dry and not heavily loaded, it may not make suitable foundation material. Loess is subject to hydrocompaction, a process by which it settles, cracks, and becomes denser and more consolidated when wetted, to the detriment of structures built on top of it. The very weight of a large structure can also cause settling and collapse.

DESERTS AND DESERTIFICATION

Many of the features of wind erosion and deposition are most readily observed in **deserts.** Deserts can be defined in a variety of ways. A desert may be defined as a region with so little vegetation that only a limited population (human or animal) can be supported on that land. It need not be hot or even, technically, dry. Ice sheets are a kind of desert with plenty of water—but not liquid. In more temperate climates, deserts are characterized by very little precipitation, commonly less than 10 centimeters (about 4 inches) a year, but they may be consistently hot, cold, or variable in temperature depending on the season or time of day. The distribution of the arid regions of the world (exclusive of polar deserts) is shown in figure 9.25.

Causes of Natural Deserts

A variety of factors contribute to the formation of a desert, so deserts come in several different types, depending on their origins.

One factor is moderately high surface temperatures. Most vegetation, under such conditions, requires abundant rainfall and/or slow evaporation of what precipitation does fall. The availability of precipitation is governed, in part, by the global air-circulation patterns shown in figure 9.16. Warm air holds more moisture than cold. Similarly, when the pressure on a mass of air is increased, the air can hold more moisture. Air spreading outward from the equator at high altitudes is chilled and at low pressure, since air pressure and temperature decrease with increasing altitude. Thus, the air holds little moisture. When that air circulates downward, at about 30 degrees north and south

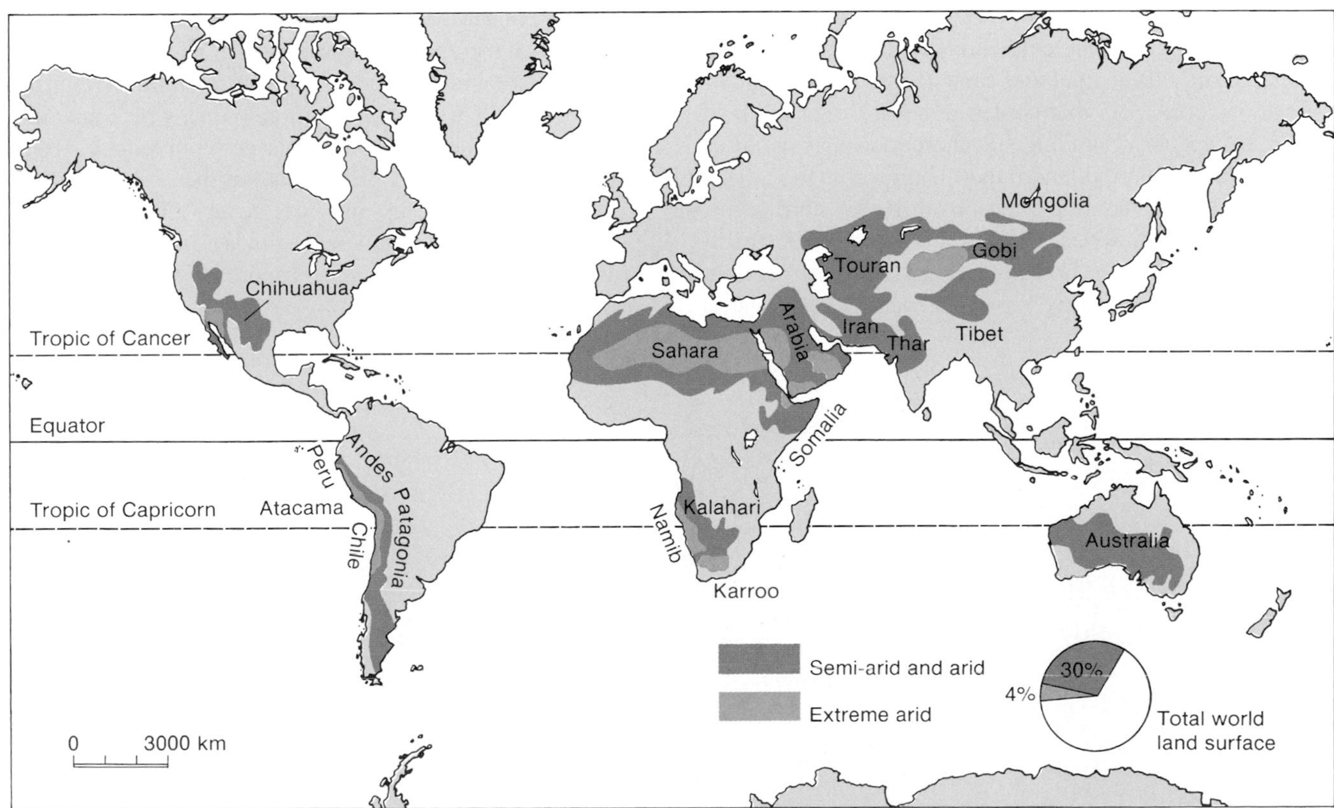

FIGURE 9.25 Distribution of the world's arid lands.
Data from A. Goudie and J. Wilkinson, The Warm Desert Environment. *Copyright © 1977 by Cambridge University Press*

latitudes, it is warmed as it approaches the surface, and also subjected to increasing pressure from the deepening column of air above it. It can then hold considerably more water, so when it reaches the earth's surface, it causes rapid evaporation. Note in figure 9.25 that many of the world's major deserts fall in belts close to these zones of sinking air at 30 degrees north and south of the equator. These are the subtropical-latitude deserts.

Topography also plays a role in controlling the distribution of precipitation. A high mountain range along the path of principal air currents between the ocean and a desert area may be the cause of the latter's dryness. As moisture-laden air from over the ocean moves inland across the mountains, it is forced to higher altitudes, where the temperatures are colder and the air thinner (lower pressure). Under these conditions, much of the moisture originally in the air mass is forced out as precipitation, and the air is much drier when it moves farther inland and down out of the mountains. In effect, the mountains cast a **rain shadow** on the land beyond (figure 9.26). Rain shadows cast by the Sierra Nevada of California and, to a lesser extent, by the southern Rockies contribute to the dryness and desert regions of the western United States.

Because the oceans are the major source of the moisture in the air, distance from the ocean (in the direction of air movement) can by itself be a factor contributing to the formation of a desert. The longer an air mass is in transit over dry land, the greater chance it has of losing some of its moisture through precipitation. This contributes to the development of deserts in continental interiors. On the other hand, even coastal areas can have deserts under special circumstances. If the land is hot and the adjacent ocean cooled by cold currents, the moist air coming off the ocean will be cool and carry less moisture than air over a warmer ocean. As that cooler air warms over the land and becomes capable of holding still more moisture, it causes rapid evaporation from the land rather than precipitation. This phenomenon is observed along portions of the western coasts of Africa and South America. Polar deserts can also be attributed to the differences in moisture-holding capacity between warm and cold air: Air traveling from warmer latitudes to colder near-polar ones will tend to lose moisture by precipitation, so less remains to fall as snow near the poles, and the limited evaporation from cold high-latitude oceans contributes little additional moisture to enhance local precipitation. Thick polar ice caps, then, reflect effective preservation of what snow does fall, rather than heavy precipitation.

Desertification

Climatic zones shift over time. In addition, topography changes, global temperatures change, and plate motions move landmasses to different latitudes. Amid these changes, new deserts develop in areas that previously had more extensive vegetative cover. The term **desertification,** however, is generally restricted to apply only to the relatively rapid development of deserts caused by the impact of human activities.

The exact definition of the lands at risk is difficult. *Arid* lands are commonly defined as those with annual rainfall of less than 250 millimeters (about 10 inches); *semiarid* lands, 250 to 500 millimeters (10–20 inches); and extremely arid lands, those areas that may have at least 12 consecutive months without rainfall. The extent to which vegetation will thrive in low-precipitation areas also depends on such additional factors as temperature and local evaporation rates. Many of the arid lands border true desert regions. Desertification does not involve the advance or expansion of desert regions as a result of forces originating within the desert. Rather, desertification is a patchy conversion of dry-but-habitable land to uninhabitable desert as a consequence of land-use practices (perhaps accelerated by such natural factors as drought).

Vegetation in dry lands is, by nature, limited (figure 9.27). At the same time, it is a precious resource, which may in various cases provide food for people or for livestock, wood for shelter or energy, and protection for the soil from erosion. Desertification typically involves severe disturbance of that vegetation. The environment is not a resilient one to begin with, and its deterioration, once begun, may be irreversible and even self-accelerating.

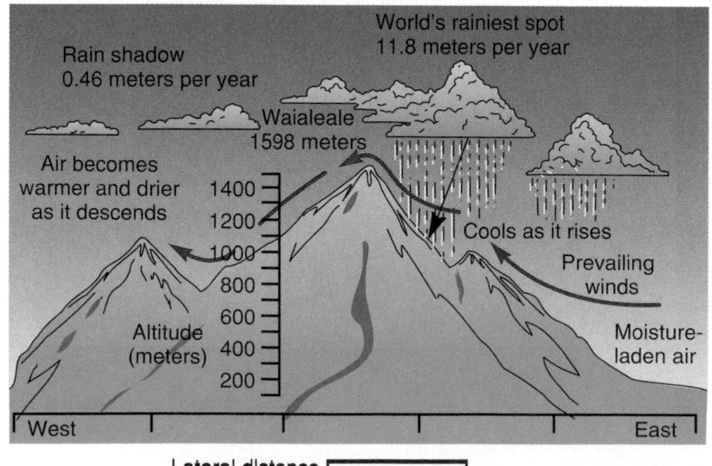

FIGURE 9.26 Rain shadows can occur even on an island in the ocean! Mt. Waialeale, on the Hawaiian island of Kauai, is described as the "world's rainiest spot," receiving over 450 inches of rain a year. But downwind from it, in its rain shadow, annual rainfall is just 18 inches. With air that is drier to start with — as in most parts of the world — the dryness in the rain shadow is correspondingly more extreme.

FIGURE 9.27 The inability of deserts to sustain large numbers of people or larger animals is obvious from scarce vegetation. Monument Valley, Arizona.

Photograph © The McGraw-Hill Companies, Inc./Doug Sherman, photographer

On land used for farming, native vegetation is routinely cleared to make way for crops. While the crops thrive, all may be well. If the crops fail, or if the land is left unplanted for a time, several consequences follow. One, as in the Dust Bowl, is erosion. A second, linked to the first, is loss of soil fertility. The topmost soil layer, richest in organic matter, is most nutrient-rich and also is the first lost to erosion. A third result may be loss of soil structural quality. Under the baking sun typical of many dry lands, and with no plant roots to break it up, the soil may crust over, becoming less permeable. This increases surface runoff, correspondingly decreasing infiltration by what limited precipitation does fall. That, in turn, decreases reserves of soil moisture and ground water on which future crops may depend. All of these changes together make it that much harder for future crops to succeed, and the problems intensify.

Similar results follow from the raising of numerous livestock on the dry lands. In drier periods, vegetation may be reduced or stunted. Yet it is precisely during those periods that livestock, needing the vegetation not only for food but also for the moisture it contains, put the greatest grazing pressure on the land. The soil may again be stripped bare, with the resultant deterioration and reduced future growth of vegetation as previously described for cropland.

Natural drought cycles thus play a role in desertification. However, in the absence of intensive human land use, the degradation of the land during drought is typically less severe, and the natural systems in the arid lands can recover when the drought ends. On a human time scale, desertification—permanent conversion of marginal dry lands to desert—is generally observed only where human activities are also significant.

Desertification is cause for concern because it effectively reduces the amount of arable (cultivatable) land on which the world depends for food. An estimated 600 million people worldwide now live on the arid lands; by recent UN definition, close to 40% of the world's land surface is "drylands," and that *excludes* the true deserts. All of those lands, in some measure, are potentially vulnerable to desertification. More than 10% of those 600 million people live in areas identified as actively undergoing desertification now; overall, up to 70% of the drylands are undergoing significant degradation (figure 9.28). Some projections suggest that, early in the twenty-first century, one-third of the world's once-arable land will be rendered useless for the culture of food crops as a consequence of desertification and attendant soil deterioration. The recent famine in Ethiopia may have been precipitated by a drought, but it has been prolonged by desertification brought on by overuse of land incapable of supporting concentrated human or animal populations.

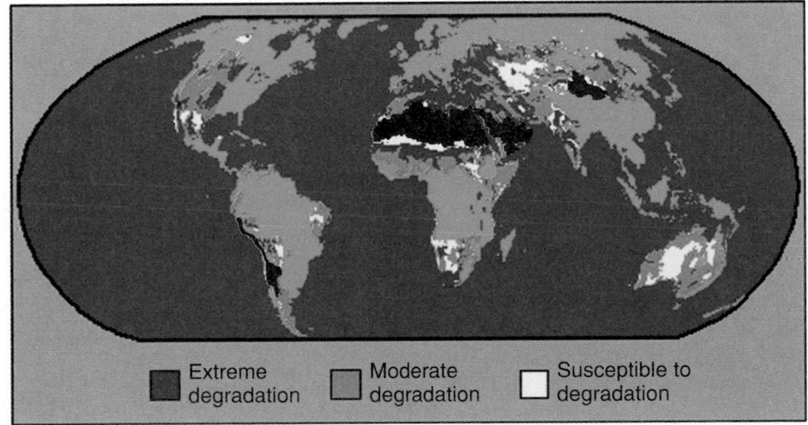

FIGURE 9.28 Degradation—loss of soil, water resources, or biodiversity—has reduced productivity over about 70% of the world's "drylands," and the rest are vulnerable. Over 100 nations are potentially affected. (Black areas are deserts already.)

Base map from UNEP Dryland Degradation Map, 1993. Reprinted by permission

Desertification in the United States?

Though desertification makes news most often in the context of Third World countries, the process is ongoing in many other places. In the United States, a large portion of the country is potentially vulnerable: Much of the western half of the country can be classified as semiarid on the basis of its low precipitation. An estimated one-million-plus square miles of land—more than a third of this low-rainfall area—has undergone severe desertification, characterized by loss of desirable native vegetation, seriously increased erosion, and reduced crop yields. The problems are caused or aggravated by intensive use of surface water and overgrazing, and the population of these areas—especially the Sun Belt—is growing.

Why, then, so little apparent concern? The main reason is that so far, the impact on humans has been modest. Other parts of the country can supply needed food. The erosion and associated sediment-redistribution problems are not yet highly visible to most people. Groundwater supplies are being tapped both for water supplies and for irrigation, creating the illusion of adequate water; but as will be seen in chapter 11, groundwater users may be living on borrowed time, or at least borrowed water that is destined to run out. As that happens, it will become painfully obvious that too much pressure from human use was brought to bear on dry lands.

SUMMARY

Glaciers past and present have sculptured the landscape not only in mountainous regions but over wide areas of the continents. They leave behind U-shaped valleys, striated rocks, piles of poorly sorted sediment (till) in a variety of landforms (moraines), and outwash. Meltwater from glaciers supplies surface and ground water. Most present glaciers are alpine glaciers. The two major ice sheets remaining are in Greenland and Antarctica.

Wind moves material much as flowing water does, but less forcefully. As an agent of erosion, wind is less effective than water but may have significant effects in dry, exposed areas, such as beaches, deserts, or farmland. Wind action creates well-sorted sediment deposits, as dunes or in blankets of fine loess. The latter can improve the quality of soil for agriculture. However, loess typically makes a poor base for construction. Features created by wind are most obvious in deserts, which are sparsely vegetated and support little life. The extent of unproductive desert and arid lands may be increasing through desertification brought on by intensive human use of these fragile lands.

Key Terms and Concepts

ablation 202
abrasion 205
alpine glacier 202
calving 202
continental glacier 202
deflation 211
desert 215
desertification 216
drift 205
dune 212
equilibrium line 202
glacier 201
loess 214
moraine 205
outwash 205
rain shadow 216
slip face 213
striations 204
till 205
ventifact 210

Exercises

Questions for Review

1. Discuss ways in which glaciers might be manipulated for use as a source of water.
2. Briefly describe the formation and annual cycle of an alpine glacier.
3. What is a moraine? How can moraines be used to reconstruct past glacial extent and movements?
4. What is an ice age? Choose any two proposed causes of past ice ages and evaluate the plausibility of each. (Is the effect on global climate likely to have been large enough? long enough? Is there any geologic evidence to support the proposal?)
5. How is sunlight falling on the earth's surface a factor in wind circulation?
6. By what two principal processes does wind erosion occur?
7. Briefly describe the process by which dunes form and migrate.
8. What is loess? Must the sediment invariably be of glacial derivation, as much U.S. loess appears to be?
9. Assess the significance of loess to (a) farming and (b) construction.

10. What is desertification? Describe two ways in which human activities contribute to the process.

For Further Thought

1. During the Pleistocene glaciation, a large fraction of the earth's land surface may have been covered by ice. Assume that 10% of the 149 million square kilometers was covered by ice averaging 1 kilometer thick. How much would sea level have been depressed over the 361 million square kilometers of oceans? You may find it interesting to examine a bathymetric (depth) chart of the oceans to see how much new land this would have exposed.
2. It is estimated that since 1960, sea level has risen about 18 millimeters due to added water from melting glaciers. Using ocean-area data from problem 1, estimate the volume of ice melted.
3. After the next windstorm, observe the patterns of dust distribution (or, if in a snowy climate and season, snow distribution), and try to relate them to the distribution of obstacles that may have altered wind velocity.

Suggested Readings/References

Ackerman, J. 2000. New eyes on the oceans. *National Geographic* (October): 86–115.

Berner, E. K., and Berner, R. A. 1996. *Global environment: Water, air, and geochemical cycles.* Upper Saddle River, N.J.: Prentice-Hall.

Bindschadler, R.A., *et al.* 1998. What is happening to the West Antarctic Ice Sheet? *EOS* (2 June): 257, 264–65.

Easterling, D. R., Meehl, G. A., Parmesan, C., Changnon, S. A., Karl, T. R., and Mearns, L. O. 2000. Climate extremes: Observations, modeling, and impacts. *Science* 289 (22 September): 2068–74.

Gasse, F. 2002. Kilimanjaro's secrets revealed. *Science* 298 (18 October): 548–49.

Global deserts outlook. 2006. United Nations Environment Programme.

Issar, A. S. 1995. Climatic change and the history of the Middle East. *American Scientist* (July–August): 350–55.

Kerr, R. A. 2003. Warming Indian Ocean wringing moisture from the Sahel. *Science* 302 (10 October): 210–11.

Montanez, I., and Soreghan, G. S. 2006. Lessons learned from deep-time ice ages. *Geotimes* (March): 24–27.

Pelto, M. S. 1993. Changes in water supply in alpine regions due to glacier retreat. Amer. Inst. of Physics.

Perkins, S. 2004. Thin skin. *Science News* 165 (3 January): 11–13. (Relates to desert pavement.)

Ray, L. L. 1992. *The great Ice Age.* U.S. Geological Survey.

Sturm, M., Peruvich, D. K., and Serreze, M. C. 2003. Meltdown in the North. *Scientific American* (October): 60–67.

Thomas, R. H. 2001. Remote sensing reveals shrinking Greenland ice sheet. *EOS* 82 (21 August): 369–73.

Walker, A. S. 1992. *Deserts: Geology and resources.* U.S. Geological Survey.

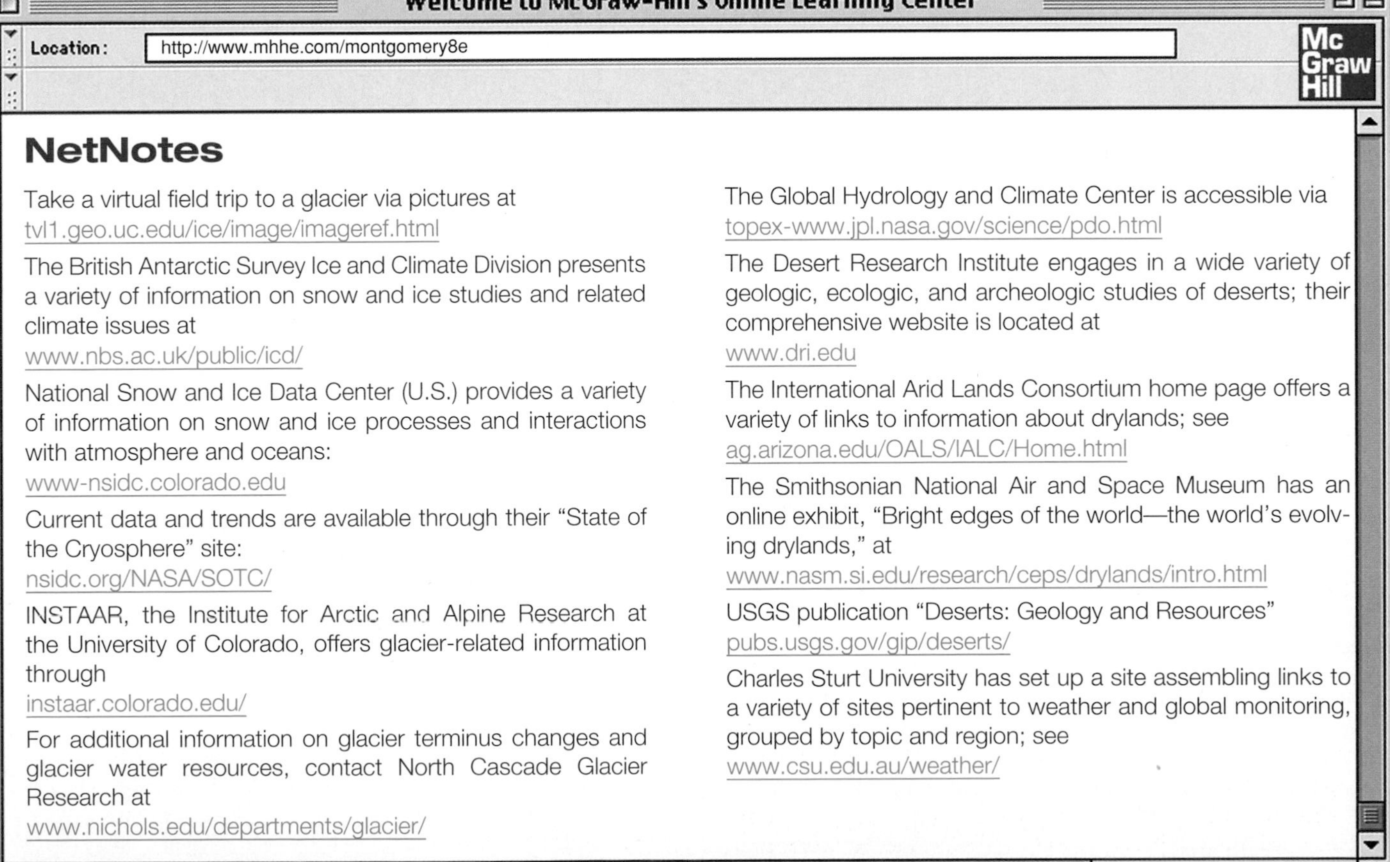

NetNotes

Take a virtual field trip to a glacier via pictures at tvl1.geo.uc.edu/ice/image/imageref.html

The British Antarctic Survey Ice and Climate Division presents a variety of information on snow and ice studies and related climate issues at www.nbs.ac.uk/public/icd/

National Snow and Ice Data Center (U.S.) provides a variety of information on snow and ice processes and interactions with atmosphere and oceans: www-nsidc.colorado.edu

Current data and trends are available through their "State of the Cryosphere" site: nsidc.org/NASA/SOTC/

INSTAAR, the Institute for Arctic and Alpine Research at the University of Colorado, offers glacier-related information through instaar.colorado.edu/

For additional information on glacier terminus changes and glacier water resources, contact North Cascade Glacier Research at www.nichols.edu/departments/glacier/

The Global Hydrology and Climate Center is accessible via topex-www.jpl.nasa.gov/science/pdo.html

The Desert Research Institute engages in a wide variety of geologic, ecologic, and archeologic studies of deserts; their comprehensive website is located at www.dri.edu

The International Arid Lands Consortium home page offers a variety of links to information about drylands; see ag.arizona.edu/OALS/IALC/Home.html

The Smithsonian National Air and Space Museum has an online exhibit, "Bright edges of the world—the world's evolving drylands," at www.nasm.si.edu/research/ceps/drylands/intro.html

USGS publication "Deserts: Geology and Resources" pubs.usgs.gov/gip/deserts/

Charles Sturt University has set up a site assembling links to a variety of sites pertinent to weather and global monitoring, grouped by topic and region; see www.csu.edu.au/weather/

10

Climate—Past, Present, and Future

We all experience and deal with weather on a daily basis—sun and clouds, wind, rain, snow—and that weather can vary considerably from day to day or even hour to hour. *Climate* is, in a sense, weather averaged over long periods of time; as such, it is less variable on a human timescale. Both weather fluctuations and regional climate can affect the operation or intensity of the surface processes examined in the previous four chapters, and thus the impact they have on human activities. While we can't do much about a day's weather, it has become increasingly evident that humans *can* influence climate on a regional or global scale. This chapter explores various influences on and effects of climate, evidence for climate change, and possible consequences of those changes. Such information will be relevant to discussions of such diverse topics as energy sources, air pollution, and environmental law and policy in future chapters.

The warm Gulf Stream shows complex eddies as it departs from North America and flows across the North Atlantic. Image from April 2005. (5 °C = 41 °F;23 °C = 73 °F)
Image by Norman Kuring, MODIS Ocean Team

MAJOR CONTROLS ON GLOBAL CLIMATE; THE GREENHOUSE EFFECT

Climate is the result of the interplay of a number of factors. The main source of energy input to the earth is sunlight, which warms the land surface, which, in turn, radiates heat into the atmosphere. Globally, how much surface heating occurs is related to the sun's energy output and how much of the sunlight falling on earth actually reaches the surface. Incoming sunlight may be blocked by cloud cover or, as we have previously seen, by dust and sulfuric acid droplets from volcanic eruptions. In turn, heat (infrared rays) radiating outward from earth's surface may or may not be trapped by certain atmospheric gases, a phenomenon known as the **greenhouse effect.**

On a sunny day, it is much warmer inside a greenhouse than outside it. Light enters through the glass and is absorbed by the ground, plants, and pots inside. They, in turn, radiate heat: infrared radiation, not visible light. Infrared rays, with longer wavelengths than visible light, cannot readily escape through the glass panes; the rays are trapped, and the air inside the greenhouse warms up. The same effect can be observed in a closed car on a bright day.

In the atmosphere, molecules of various gases, especially water vapor and carbon dioxide, act similarly to the greenhouse's glass. Light reaches the earth's surface, warming it, and the earth radiates infrared rays back. But the longer-wavelength infrared rays are trapped by these gas molecules, and a portion of the radiated heat is thus trapped in the atmosphere. Hence the term "greenhouse effect." (See figure 10.1.) As a result of the greenhouse effect, the atmosphere stays warmer than it would if that heat radiated freely back out into space. In moderation, the greenhouse effect makes life as we know it possible: without it, average global temperature would be closer to $-17°C$ (about 1°F) than the roughly 15°C (59°F) it now is. However, one can have too much of a good thing.

The evolution of a technological society has meant rapidly increasing energy consumption. Historically, we have relied most heavily on carbon-rich fuels—wood, coal, oil, and natural gas—to supply that energy. These probably will continue to be important energy sources for several decades at least. One combustion by-product that all of these fuels have in common is carbon dioxide gas (CO_2).

A "greenhouse gas" in general is any gas that traps infrared rays and thus promotes atmospheric warming. Water vapor is, in fact, the most abundant greenhouse gas in the earth's atmosphere, but human activities do not substantially affect its abundance, and it is in equilibrium with surface water and oceans. Excess water in the atmosphere readily falls out as rain or snow. Some of the excess carbon dioxide is removed by geologic processes (see chapter 18), but since the start of the so-called Industrial Age in the mid-nineteenth century, the amount of carbon dioxide in the air has increased by an estimated 30%—and its concentration continues to climb (figure 10.2). If the heat trapped by carbon dioxide were proportional to the concentration of carbon dioxide in the air, the increased carbon dioxide would by now have caused sharply increased greenhouse-effect heating of the earth's atmosphere.

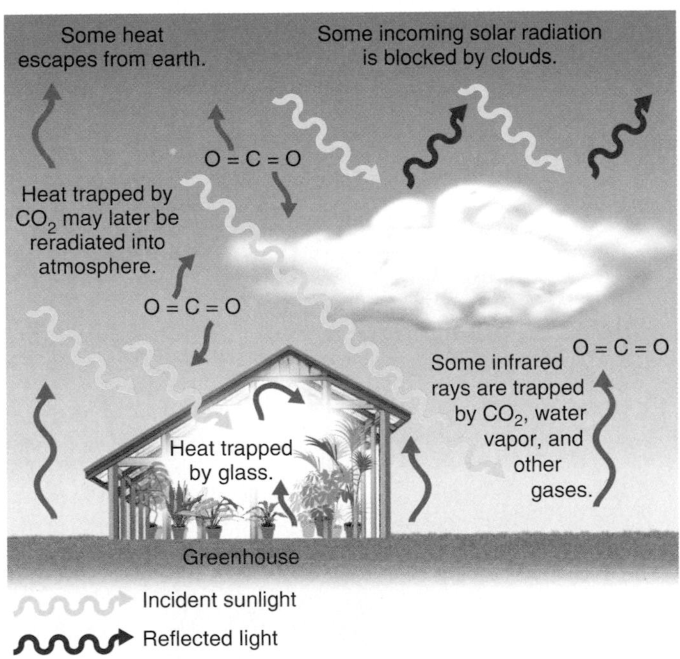

FIGURE 10.1 The "greenhouse effect" (schematic). Both glass and air are transparent to visible light. Like greenhouse glass, CO_2 and other "greenhouse gases" in the atmosphere trap infrared rays radiating back from the sun-warmed surface of the earth.

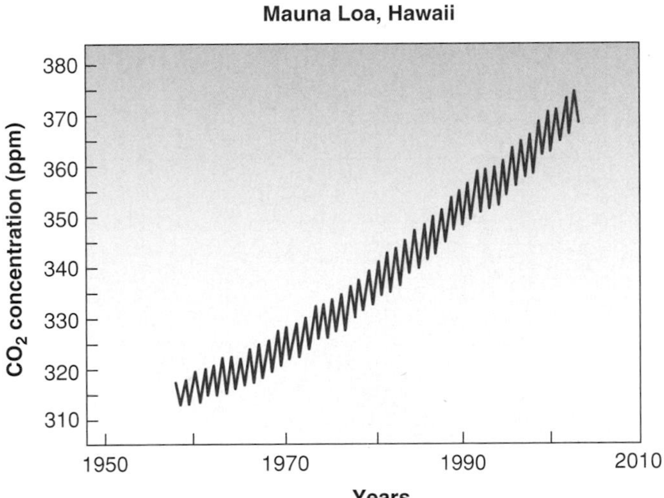

FIGURE 10.2 Rise in atmospheric CO_2 over past several decades is clear. (Zigzag pattern reflects seasonal variations in local uptake by plants.) Preindustrial levels in ice cores were about 280 ppm (parts per million). One ppm is 0.0001%.
Data from Dave Keeling and Tim Whorf, Scripps Institution of Oceanography

So far, the actual temperature rise has been much more moderate; the earth's climate system is more complex. Indeed, measuring that increase directly, and separating it from the "background noise" of local weather variations, has been difficult. Locally, at any given time in earth's history, a wide variety of local climates can obviously exist simultaneously—icy glaciers, hot deserts, steamy rain forests, temperate regions—and a given region may also be subject to wide seasonal variations in temperature and rainfall. One of the challenges in determining global temperature trends, then, is deciding just how to measure global temperature at any given time. Commonly, several different kinds of data are used, including air temperatures, over land and sea, at various altitudes, and sea surface temperatures. Satellites help greatly by making it possible to survey large areas quickly.

Altogether, since the start of the Industrial Age, global surface temperature has risen about 0.8°C (1.4°F) (figure 10.3). Trivial as this may sound, it is already having obvious and profound impacts in many parts of the world. An even larger concern is what the future may hold.

CLIMATE AND ICE REVISITED

Early discussions of climate change related to increasing greenhouse-gas concentrations in the atmosphere tended to focus heavily on the prospect of global warming and the resultant melting of earth's reserves of ice, especially the remaining ice sheets. If *all* the ice melted, sea level could rise by close to 75 meters (about 250 feet) from the added water alone, leaving aside thermal expansion of that water. About 20% of the world's land area would be submerged. Many millions, perhaps billions, of people now living in coastal or low-lying areas would be displaced, since a large fraction of major population centers grew up along coastlines and rivers. The Statue of Liberty would be up to her neck in water. The consequences to the continental United States and southern Canada are illustrated in figure 10.4. Also, raising of the base levels of streams draining into the ocean would alter the stream channels and could cause significant flooding along major rivers, and this would not require nearly such drastic sea-level rise.

Such large-scale melting of ice sheets would take time, perhaps several thousand years. On a shorter time scale, however, the problem could still be significant. Continued intensive fossil-fuel use could double the level of carbon dioxide in the atmosphere before the middle of this century. This would produce a projected temperature rise of 4° to 8°C (7°–14°F), which would be sufficient to melt at least the West Antarctic ice sheet completely in a few hundred years. The resulting 3- to 6-meter rise in sea level, though it sounds small, would nevertheless be enough to flood many coastal cities and ports, along with most of the world's beaches; see, for example, figure 10.5. This would be both inconvenient and extremely expensive. (The only consolation is that the displaced inhabitants would have decades over which to adjust to the changes.) Meanwhile, the smaller alpine glaciers are clearly dwindling (figure 10.6).

Arctic polar ice is also demonstrably shrinking, in both area (figure 10.7) and thickness (figure 10.8). It is particularly vulnerable because, instead of sitting on long-chilled rock, it floats on the circulating, warming ocean. The good news is that, just as melting ice in a glass does not cause the liquid to overflow, melting sea ice will not cause a rise of sea level. However, concerns about Arctic ecosystems are being raised; for example, polar bears that hunt seals on the ice in Hudson Bay suffer when a shorter ice season reduces their food intake.

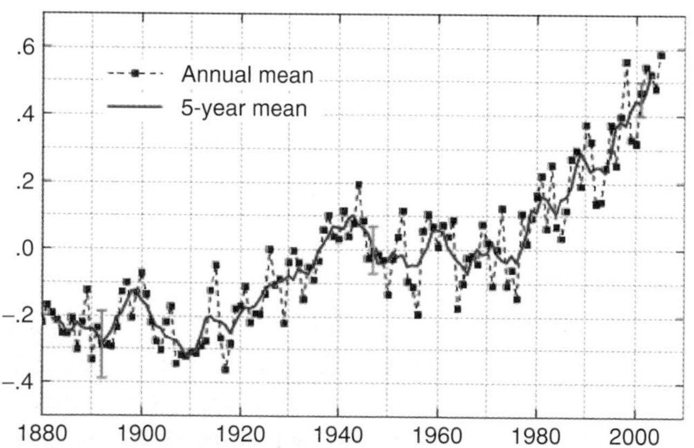

FIGURE 10.3 Global temperature rise emerges from background noise; blue bars indicate uncertainties.

Image by J. Hansen, R. Ruedy, M. Sato, and K. Lo, NASA Goddard Institute for Space Studies

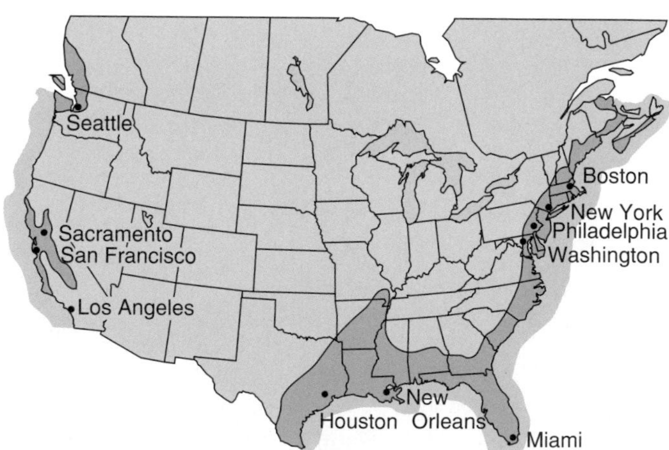

FIGURE 10.4 The flooding (blue-shaded area) that would result from melting all the world's ice sheets. The resultant 75-meter rise in sea level would flood a large fraction of major cities in the United States and elsewhere. However, such extensive melting will not occur in the near future, at least.

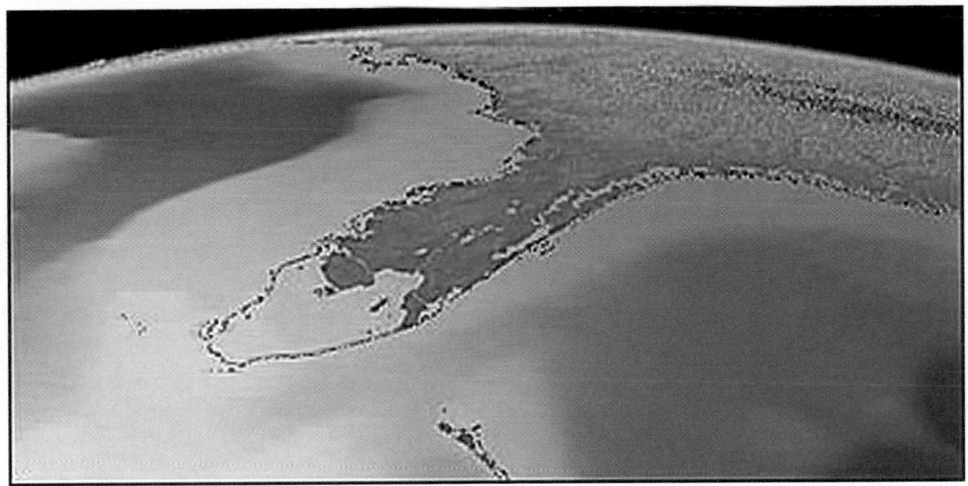

FIGURE 10.5 Impact on Florida of melting either the West Antarctic or the Greenland ice sheet, which would result in a sea-level rise of about 5 meters (16 feet).
Image courtesy William Haxby/NASA

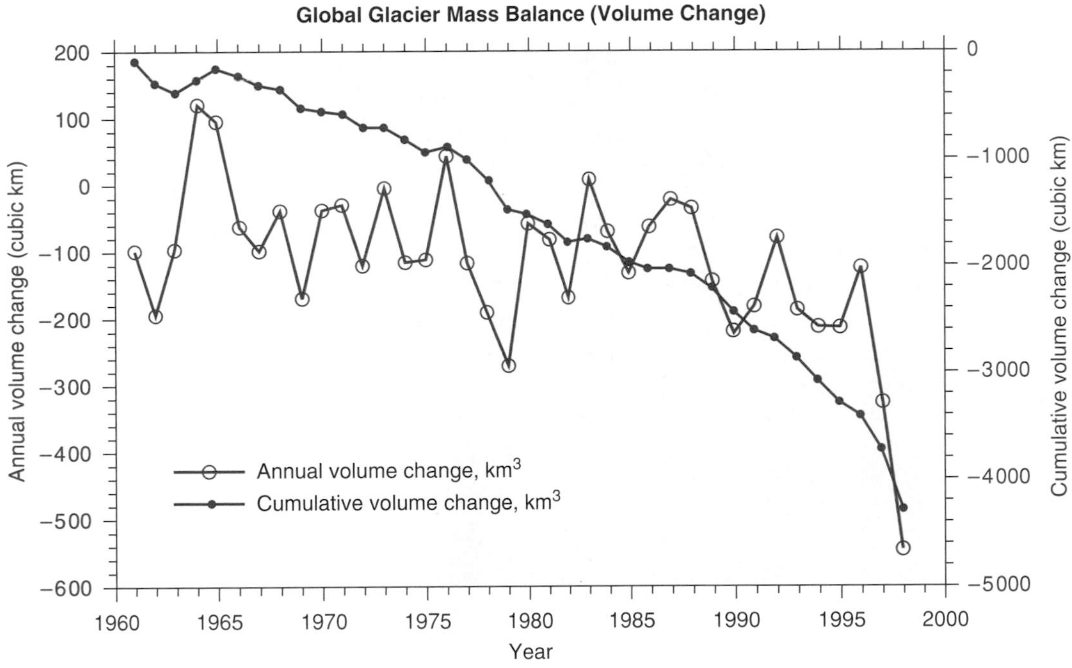

FIGURE 10.6 The volume of ice in alpine glaciers fluctuates, but the trend is clear: downward.
NASA National Snow and Ice Data Center

The Hidden Ice: Permafrost

Warming is affecting more than the visible ice of cold regions. In alpine climates, winters are so cold, and the ground freezes to such a great depth, that the soil does not completely thaw even in the summer. The permanently frozen zone is **permafrost** (figure 10.9). The meltwater from the thawed layer cannot infiltrate the frozen ground below, so the terrain is often marshy (figure 10.10). Structures built on or in permafrost may be very stable while it stays frozen but sink into the muck when it thaws; vehicles become mired in the sodden ground, so travel may be impossible when the upper soil layers thaw.

Studies have shown that permafrost is also being lost as the world warms. Where some permafrost remains, the thaw penetrates deeper, and the frozen-ground season is shorter. Some areas are losing their permafrost altogether, and lakes drain

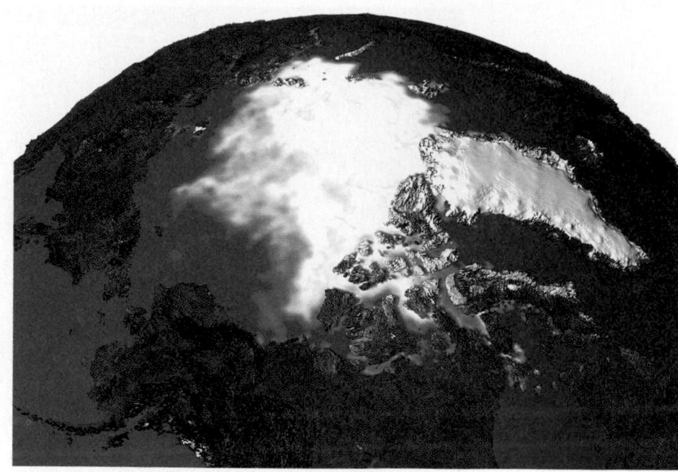

FIGURE 10.7 Illustrations of minimum (summer) Arctic sea ice extent in 1979 (left image) and 2003 (right).
Image courtesy Scientific Visualizations Studio, NASA GSFC

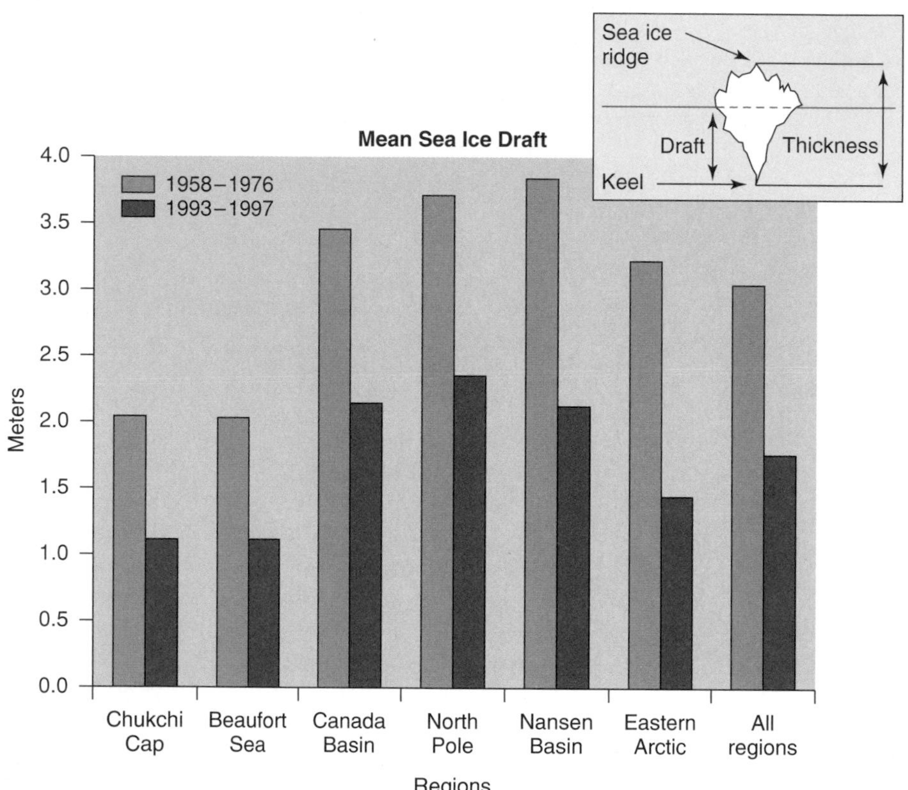

FIGURE 10.8 Thickness of Arctic sea ice is decreasing.
Figure from NASA National Snow and Ice Data Center

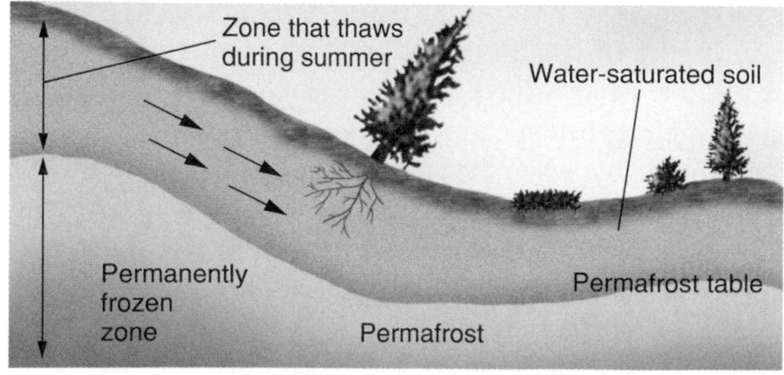

FIGURE 10.9 In cold climates, a permanently frozen soil zone, permafrost, may be found beneath the land surface.

FIGURE 10.10 A portion of the north coast of the National Petroleum Reserve in Alaska. Note elongated topographic features, a result of ice-sheet flow during the last Ice Age. Differences in suspended sediment content cause differences in water color in the many lakes.
Image courtesy NASA/GSFC/METI/ERSDAC/JAROS, and the U.S./Japan ASTER Science Team

abruptly when the last of the ice below melts away. All these changes are having effects on ecosystems, animal migrations, and the traditional lifestyles of some native peoples. They are even affecting oil exploration and the Trans-Alaska Pipeline, as we will see in later chapters.

OCEANS AND CLIMATE

One reason that there is no simple correlation between atmospheric greenhouse-gas concentrations and land surface temperatures is the oceans. That huge volume of water represents a much larger thermal reservoir than the tenuous atmosphere. Coastal regions may have relatively mild climates for this reason: As air temperatures drop in winter, the ocean is there to release some heat to the atmosphere; as air warms in summer, the ocean water absorbs some of the heat. The oceans store and transport a tremendous amount of heat around the globe. A thorough discussion of the role of the oceans in global and local climate is beyond the scope of this chapter, but it is important to appreciate the magnitude of that role.

We have already seen that the oceans supply most of the water vapor to make rain and snow. Water absorbs heat when vaporizing, and more water will evaporate from warmer surface water. Heat from sea surface water supplies the energy to spawn hurricanes and other tropical storms. While meteorologists have not yet seen a systematic increase in the *number* of U.S. hurricanes each year, recent research does suggest a trend toward increasing *intensity* of hurricanes—not good news for residents of the Gulf Coast.

The Thermohaline Circulation

Broadly, oceanic circulation is driven by a combination of winds, which push surface currents, and differences in density within the oceans' waters, related to temperature and salinity. Cold water is denser than warm, and at a given temperature, density increases with increasing salinity. The roles of temperature and salinity are reflected in the name **thermohaline circulation** (remember, "halite" is sodium chloride, salt, the most abundant dissolved material in the oceans) given to the large-scale circulation of the oceans (figure 10.11). Among the notable features are the Gulf Stream, carrying warmed equatorial water to the North Atlantic, and the sinking of cold water in the polar regions. The informal name of "ocean conveyor" is sometimes given to this circulation system in recognition of its heat-transport role.

In the climate records described later in the chapter, scientists have found evidence of episodes of "abrupt climate change" in some locations. One proposed mechanism for sudden cooling of the North Atlantic is disruption of the thermohaline circulation to this region, reducing the influx of warm water. For instance, extensive melting of the Greenland ice sheet could create a pool of cool, fresh surface water that would block the incoming warm flow; or, increased water evaporation around a warmer equator could increase the salinity of the surface water so much that it would sink rather than flowing northward. Much research is needed to test such hypotheses. In any case, to a geologist, "abrupt" climate change means change on a time scale of decades to centuries—not days to weeks as in recent movie fiction!

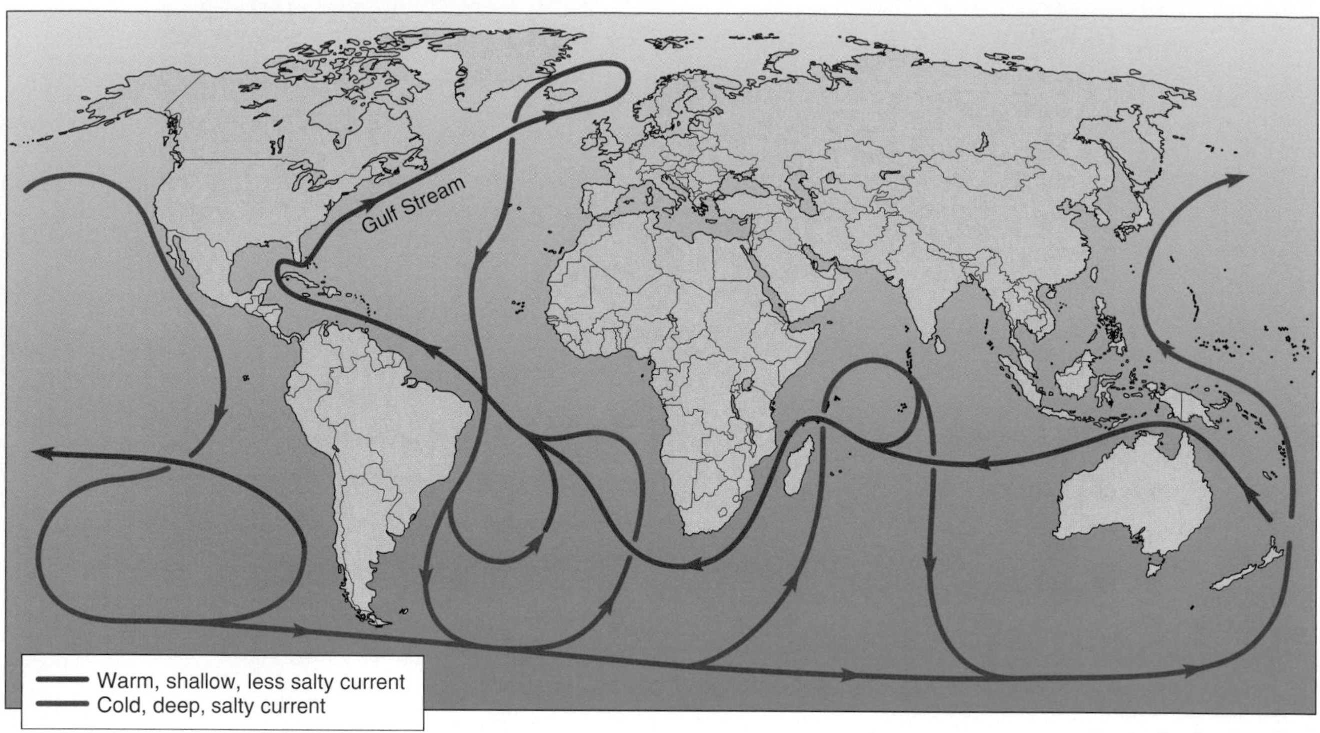

FIGURE 10.11 The thermohaline circulation, moving water—and thus heat—around the globe.

El Niño

Most of the vigorous circulation of the oceans is confined to the near-surface waters. Only the shallowest waters, within 100 to 200 meters of the surface, are well mixed by waves, currents, and winds, and warmed and lighted by the sun. The average temperature of this layer is about 15°C (60°F).

Below the surface layer, temperatures decrease rapidly to about 5°C (40°F) at 500 to 1000 meters below the surface. Below this is the so-called *deep layer* of cold, slow-moving, rather isolated water. The temperature of this bottommost water is close to freezing and may even be slightly below freezing (the water is prevented from freezing solid by its dissolved salt content and high pressure). This cold, deep layer originates largely in the polar regions and flows very slowly toward the equator.

When winds blow offshore, they push the warm surface waters away from the coastline also. This, in turn, creates a region of low pressure and may result in *upwelling* of deep waters to replace the displaced surface waters (figure 10.12). The deeper waters are relatively enriched in dissolved nutrients, in part because few organisms live in the cold, dark depths to consume those nutrients. When the nutrient-laden waters rise into the warm, sunlit zone near the surface, they can support abundant plant life and, in turn, animal life that feeds on the plants. Many rich fishing grounds are located in zones of coastal upwelling. The west coasts of North and South America and of Africa are subject to especially frequent upwelling events.

From time to time, however, for reasons not precisely known, the upwelling is suppressed for a period of weeks or longer, as warm waters from the western South Pacific extend eastward to South America (figure 10.13A). Abatement of coastal winds, for one, reduces the pressure gradient driving the upwelling. The reduction in upwelling of the fertile cold waters has a catastrophic effect on the Peruvian anchoveta industry. Such an event is called *El Niño* ("the [Christ] Child") by the fishermen because it commonly occurs in winter, near the Christmas season.

At one time, El Niño events were believed to be rather sporadic and isolated, local events. Now it is recognized that they are cyclic in nature, occurring every four to seven years as part of the El Niño–Southern Oscillation (referring to the southern oceans). The opposite of an El Niño, a situation with unusually cold surface waters off western South America, has been named *La Niña* (figure 10.13B). Extensive shifts in seawater surface temperatures, in turn, cause large-scale changes in evaporation, and thus precipitation, as well as wind-circulation

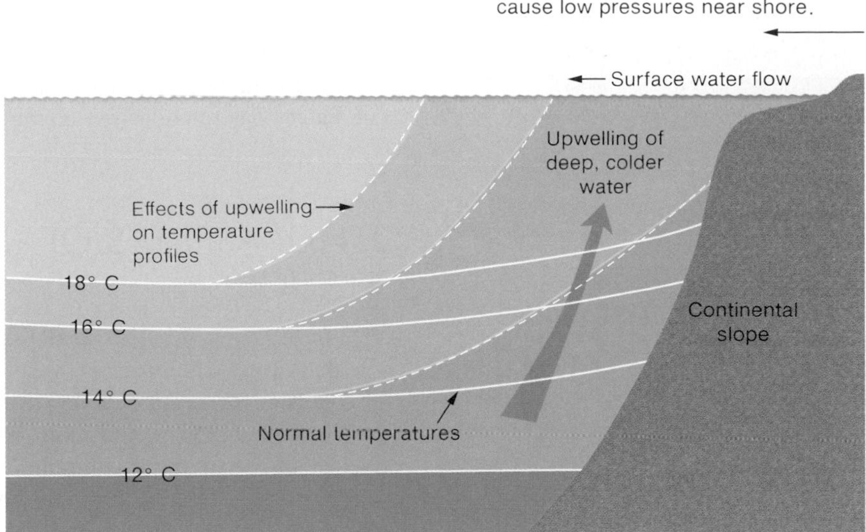

FIGURE 10.12 Warm surface water normally overlies colder seawater at depth. When offshore winds blow warm waters away from shore and create local low pressure, upwelling of colder deep waters may occur.

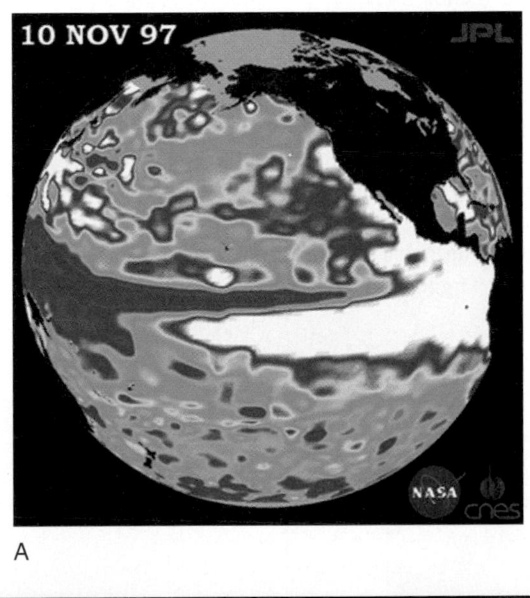

A

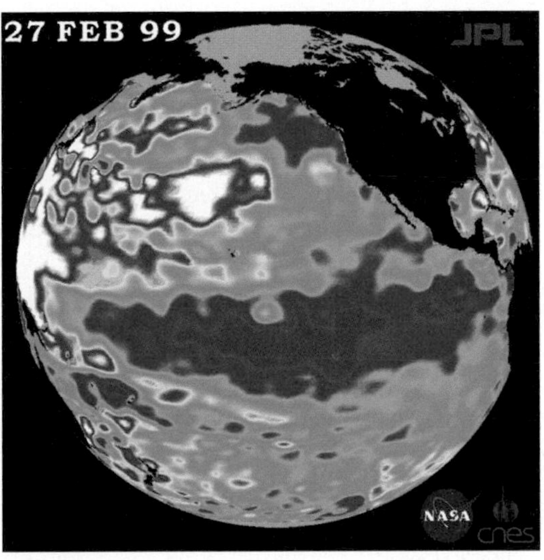

B

FIGURE 10.13 (A) El Niño: Unusually warm water appears as higher sea-level elevation; warm water expands. (B) La Niña: Lower sea-level elevations show areas of cooler surface waters.
Images courtesy of JPL/NASA

patterns (figure 10.14). Climatic changes associated with El Niño/La Niña events generally include changes in frequency, intensity, and paths of Pacific storms, short-term droughts and floods in various regions of the world, and changes in the timing and intensity of the monsoon season in India. Heavy rains associated with El Niño events have been blamed for major landslides along the Pacific coast of the United States. With respect to global climate change, one concern is that warmer oceans may mean more frequent, intense, and/or prolonged El Niño episodes, with especially devastating consequences

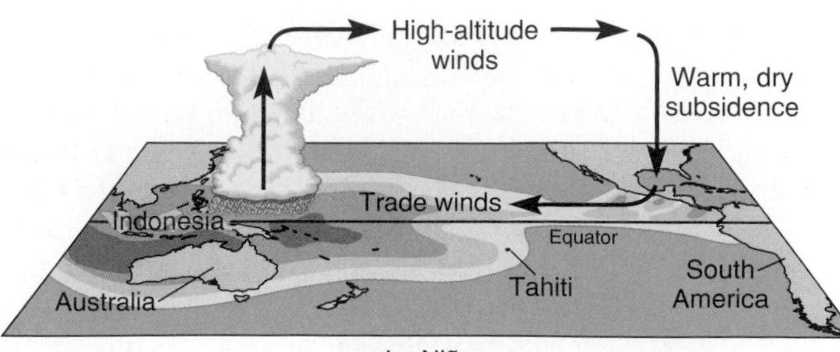

La Niña

Tahiti

El Niño

FIGURE 10.14 Under La Niña conditions, warm, wet air over the western Pacific dumps monsoon rains on southeast Asia as it rises and cools; when the air sinks and warms again, over the eastern Pacific, it is drying. During El Niño, surface waters in the eastern Pacific are warmer, which shifts the patterns of evaporation and precipitation; the monsoon season is much dryer, while more rain falls on the western United States.

to those nations whose much-needed rainy season will be disrupted.

About 1996, investigators first recognized that underlying the El Niño/La Niña is a longer-term phenomenon, dubbed the Pacific Decadal Oscillation (PDO), with a cycle of fluctuating Pacific Ocean surface temperatures over a period of 20–30 years. The PDO, too, affects patterns of evaporation, precipitation, and wind. Still more recently, an Atlantic Multidecadal Oscillation (AMO) has been identified and named. It is entirely possible that more such cycles important to our understanding of climate/ocean relationships remain to be recognized.

OTHER ASPECTS OF GLOBAL CHANGE

As we have seen in previous sections, "global warming" means more than just a little warmer weather everywhere. There are consequences for ice amount and distribution, oceanic circulation, patterns of storms, and more, as will be explored below. It is very important to bear in mind, too, that "global" climate changes do not affect all parts of the world equally or in the same ways.

Changes in wind-flow patterns and amounts and distribution of precipitation will cause differential impacts in different areas, not all of which will be equally resilient. A region of temperate climate and ample rainfall may not be seriously harmed by a temperature change of a few degrees, one way or the other, or several inches more or less rainfall. A modest temperature rise in colder areas may actually increase plant vigor and the length of the growing season (figure 10.15), and some dry areas may enjoy increased productivity with increased rainfall.

Warming and moisture redistribution also have a potential downside, of course. In many parts of the world, agriculture is already only marginally possible because of hot climate and little rain. A temperature rise of only a few degrees could easily make living and farming in these areas impossible. Recent projections suggest that summer soil-moisture levels in the

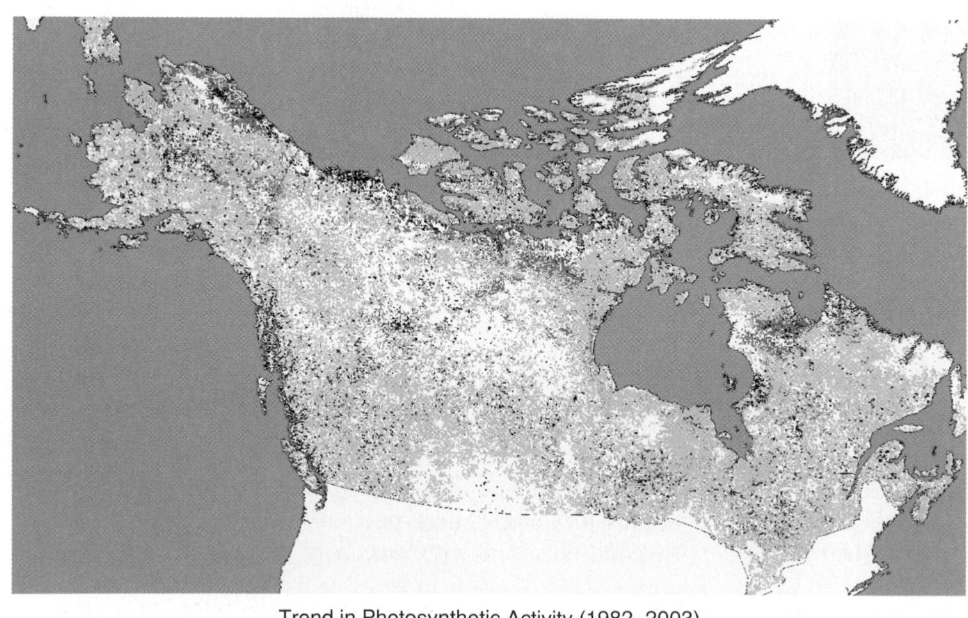

FIGURE 10.15 Plant activity in the northernmost reaches of Alaska and Canada is picking up with global warming, but it is declining over a much larger area farther south.
Image courtesy NASA, based on data from Scott Goetz, Woods Hole Research Center

Northern Hemisphere could drop by up to 40% with a doubling in atmospheric CO_2. This is a sobering prospect for farmers in areas where rain is barely adequate now, for in many places, irrigation is not an option.

It has been argued that rising CO_2 levels should be beneficial in promoting plant growth. After all, photosynthesis involves using CO_2 and water and solar energy to manufacture more complex compounds and build the plant's structure. However, controlled experiments have shown that not all types of plants grow significantly more vigorously given higher CO_2 concentrations in their air, and even those that do may show enhanced growth only for a limited period of time, not indefinitely.

Agriculture aside, places where temperature or moisture conditions already make living marginal are especially vulnerable to small increases in temperature or decreases in rainfall and soil moisture. There is reason to believe that global warming may produce more extremes—catastrophic flooding from torrential storms, devastating droughts, "killer heat waves" like that costing hundreds of lives in Chicago in 1995. Because of the many variables involved and the immense complexity of the calculations, global-climate computer models must also limit the resolution at which effects are calculated. Typical models operate on areal units, or cells, that are hundreds of kilometers on a side, meaning that small local climatic anomalies may be missed. It is therefore difficult to anticipate the effects of global climate change at the local level.

Public-health professionals have identified yet another threat related to global change. Climate influences the occurrence and distribution of a number of diseases. Mosquito-borne diseases, such as malaria and dengue, will be influenced by conditions that broaden or reduce the range of their particular carrier species; the incidence of both has increased in recent years and is currently expected to expand further (figure 10.16). Waterborne parasites and bacteria can thrive in areas of increased rainfall. Efforts to anticipate, and plan responses to, the climate-enhanced spread of disease are growing.

United Nations studies further suggest that less-developed nations, with fewer resources, may be much less able to cope with the stresses of climate change. Yet many of these nations are suffering the most immediate impacts: The very existence of some low-lying Pacific island nations is threatened by sea-level rise; reduced soil moisture and higher temperature quickly affect the world's "drylands", where many poorer countries are located; the tropical diseases are spreading most obviously in the Third World where medical care is limited. This differential impact may have implications for global political stability as well.

EVIDENCE OF CLIMATES PAST

An understanding of past climatic fluctuations would be helpful in developing models of possible future climate change, but the evidence available is somewhat limited. Aspects of local climate may often be deduced from the geologic record, especially the sedimentary-rock record. For example, the now-vegetated dunes of the Nebraska Sand Hills (figure 10.17) are

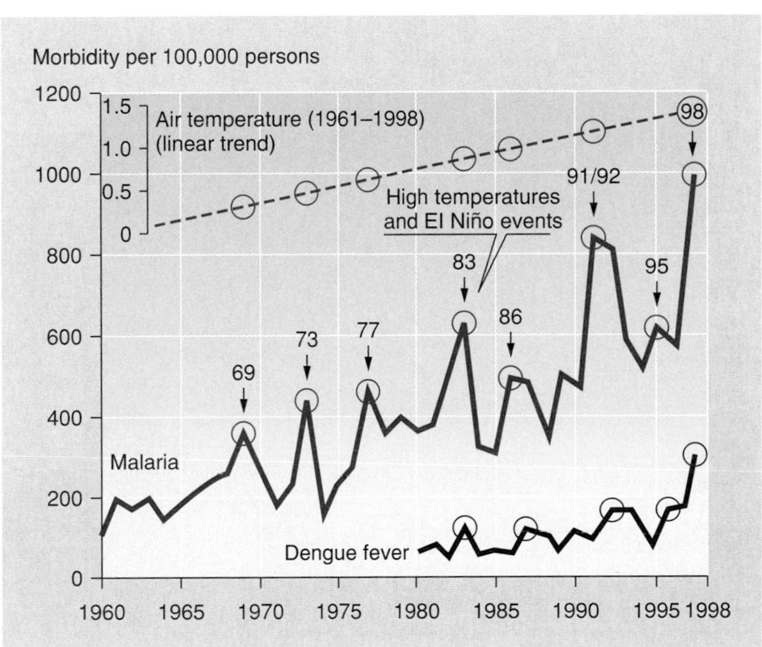

FIGURE 10.16 Occurrence of malaria and dengue fever in Colombia is correlated with temperature, spiking in warmer El Niño years, but also trending upward overall as temperatures rise.

Graph from United Nations Environment Programme/GRID-Arendal

FIGURE 10.17 Aerial photograph of the Nebraska Sand Hills area, formerly a desert, shows well-preserved dune shapes, now covered by vegetation under moister modern conditions.
Photograph © Tom Bean/Corbis Images

remnants of an arid "sand sea" in the region 18,000 years ago, when conditions must have been much drier than they are now. Much of North Africa was fertile, wheat-growing land in the time of the Roman Empire; small reductions in rainfall, and the resultant crop failures, may have contributed to the Empire's fall. We have noted that glacial deposits may be recognized in now-tropical regions. Coal deposits indicate warm, wet conditions, conducive to lush plant growth, perhaps in a swampy setting. Some idea of global climate at a given time may be gained from compiling such data from a variety of localities (taking into account changes in latitude related to continental drift); but those records are far from complete.

A special challenge in reconstructing past climate is that we do not have direct measurements of ancient temperatures. These must be determined indirectly using various methods, sometimes described as "proxies" for direct temperature records. For example, from marine sediments comes evidence of water-temperature variations. The proportion of calcium carbonate ($CaCO_3$) in Pacific Ocean sediments in a general way reflects water temperature, because the solubility of calcium carbonate is strongly temperature-related: it is more soluble in cold water than in warm.

A particularly powerful tool involves variations in the proportions of oxygen isotopes in geological materials. By far the most abundant oxygen isotope in nature is oxygen-16 (^{16}O); the heaviest is oxygen-18 (^{18}O). Because they differ in mass, so do molecules containing them, and the effect is especially pronounced when oxygen makes up most of the mass, as in H_2O. Certain natural processes, including evaporation and precipitation, produce fractionation between ^{16}O and ^{18}O, meaning that the relative abundances of the two isotopes will differ between two substances, such as ice and water, or water and water vapor. See figure 10.18. As water evaporates, the ligher $H_2{}^{16}O$ evaporates preferentially, and the water vapor will then be isotopically "lighter" (richer in ^{16}O, poorer in ^{18}O) than the residual water. Conversely, as rain or snow condenses and falls, the precipitation will be relatively enriched in the heavier $H_2{}^{18}O$. As water vapor evaporated from equatorial oceans drifts toward the poles, depositing $H_2{}^{18}O$-enriched precipitation, the remaining water vapor becomes progressively lighter isotopically, and so does subsequent precipitation; snow falling near the poles has a much lower $^{18}O/^{16}O$ ratio than tropical rain, or seawater. The fractionation between ^{18}O and ^{16}O in coexisting water vapor and rain or snow is temperature-dependent, more pronounced at lower temperatures.

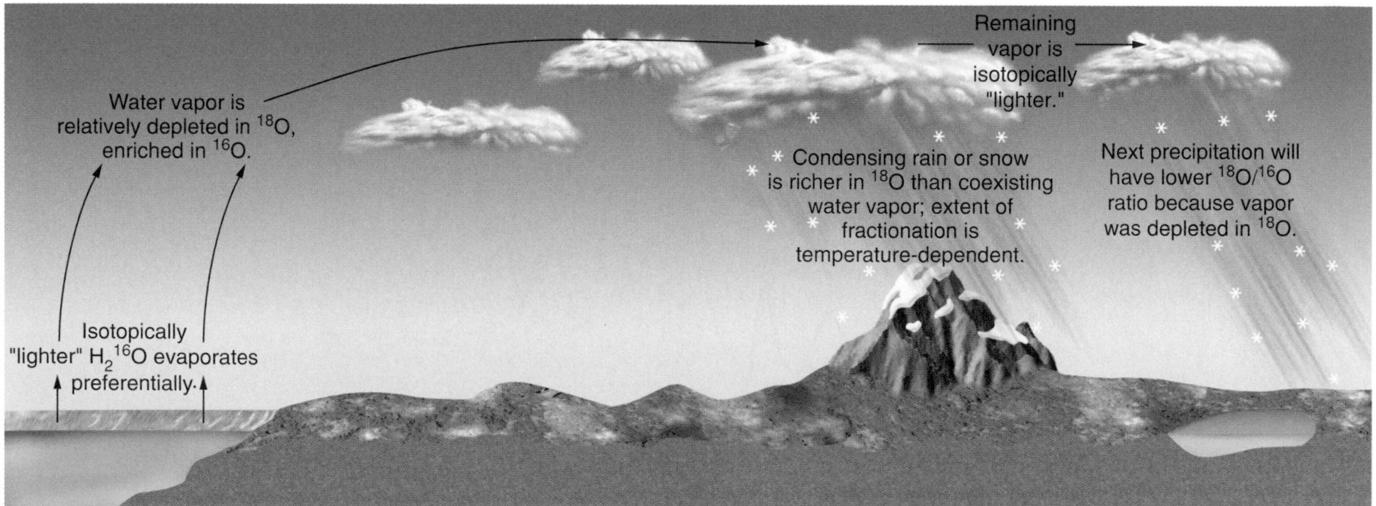

FIGURE 10.18 Mass differences between ^{18}O and ^{16}O lead to different $^{18}O/^{16}O$ ratios in coexisting water and water vapor; the size of the difference is a function of temperature.

Therefore, at a given latitude, variations in the $^{18}O/^{16}O$ ratio of precipitation reflect variations in temperature. (Similarly, variations in the $^{18}O/^{16}O$ ratio in the oxygen-rich carbonate [$CaCO_3$] or silica [SiO_2] skeletons of marine microorganisms, resulting from mineral-water fractionation of oxygen isotopes, give evidence of variations in the temperature of the near-surface seawater from which these skeletons precipitated and have been used to track past El Niño cycles.)

The longevity of the massive continental glaciers—sometimes hundreds of thousands of years—makes them useful in preserving evidence of both air temperature and atmospheric composition in the past. Tiny bubbles of contemporary air may be trapped as the snow is converted to dense glacial ice. The oxygen-isotope composition of the ice reflects seawater O-isotope composition and the air temperature at the time the parent snow fell. If those temperatures can be correlated with other evidence, such as the carbon-dioxide content of the air bubbles or the presence of volcanic ash layers in the ice from prehistoric explosive eruptions, much information can be gained about possible causes of past climatic variability (figure 10.19), and the climate-prediction models may be refined accordingly.

The longest ice-core records come from Antarctica (figure 10.20); they span about 400,000 years. Records from the Greenland ice sheet go back about 160,000 years, and over that period, show a pattern very similar to the Antarctic results. The data show substantial temperature variations, spanning a range of about 10°C (20°F), some very rapid warming trends, and obvious correlations between temperature and the concentrations of carbon dioxide (CO_2) and methane (CH_4). Correlation does not show a cause-and-effect relationship, however. The warming might be a result of sharp increases in greenhouse gases in the atmosphere, or the cause (see discussion of climate feedbacks in the next section). One fact is clear: The *highest* greenhouse-gas concentrations seen in the ice cores are about the levels in our atmosphere just prior to the start of the Industrial Age. Since then, we have pushed CO_2 concentrations about 100 ppm higher still, and the numbers continue to climb. We have no records to show the climatic results of such high CO_2 levels.

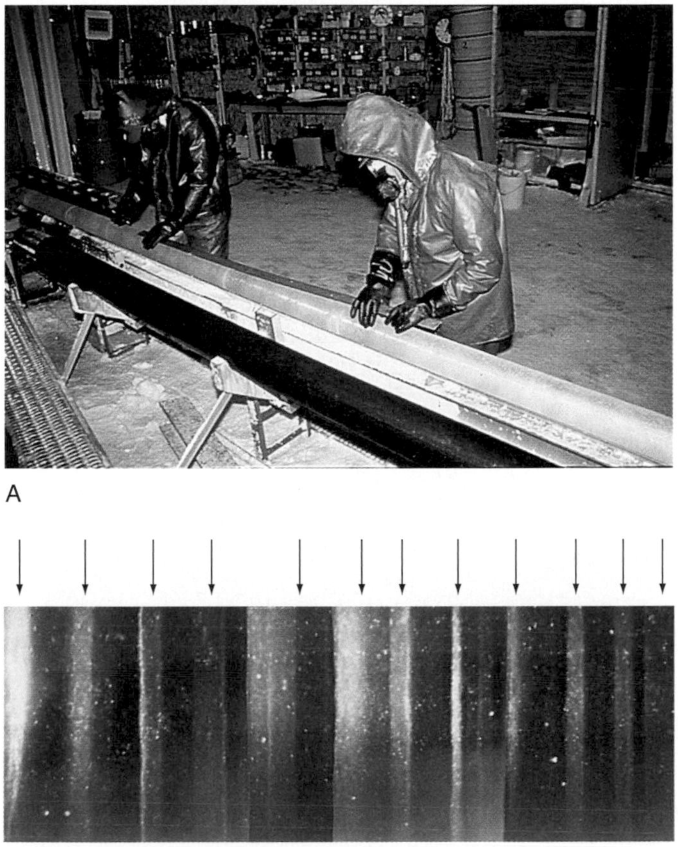

A

B

FIGURE 10.19 Ice coring in Greenland and Antarctica supports climate-trend investigations. (A) Cores must be kept frozen to preserve structure and trapped gases, so researchers must work in the cold. (B) Section of Greenland core shows annual layers defined by summer dust blown onto ice. Core sections are dated by counting these layers.

(A) Photograph by Kendrick Taylor, DRI, University of Nevada–Reno; (B) by A. Gow, U.S. Army Corps of Engineers; both courtesy NOAA/National Geophysical Data Center

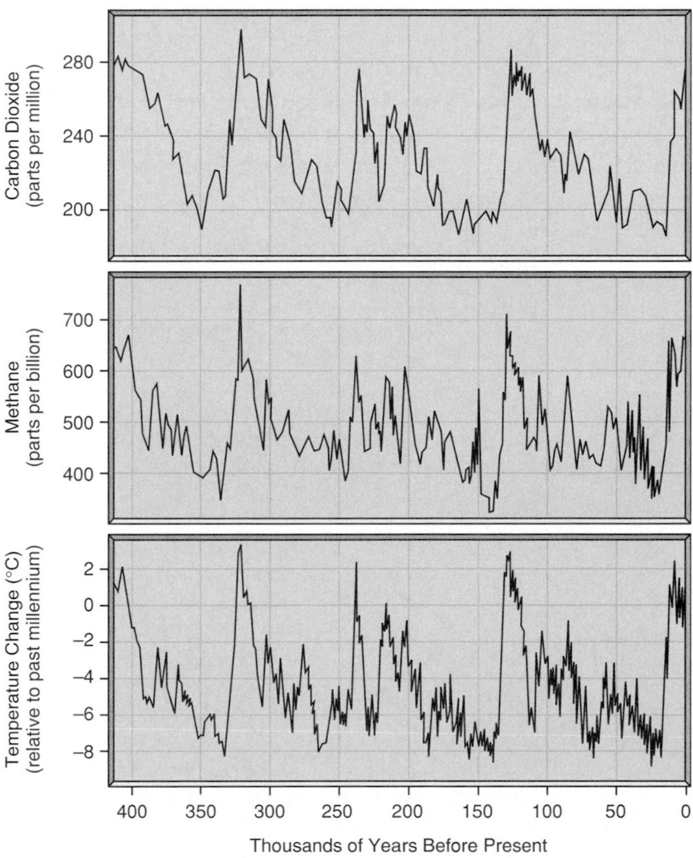

FIGURE 10.20 Ice-core data from Greenland and Antarctica reveal strong correlations between concentrations of the greenhouse gases carbon dioxide and methane, and temperature changes as indicated by oxygen isotopes.

Box 10.1 Climate—Blowing Hot and Cold

"I've *had* it with this snow!" fumed Chad, hanging up the shovel with a clatter. "Three feet so far and it isn't even January. 'Global warming,' my foot! Coldest December on record, the weather report said."

His sister Diana didn't try to argue, but when he came inside and yanked off his boots, he noticed her working on the computer, looking at maps. "Here, check this out," she called out. "Looks like you're right about December, but look at the whole year." She showed him one map revealing that indeed, most of North America had been colder in December 2000 than was typical for Decembers over the last 50 years (figure 1). However, comparing whole years, 2000 was clearly relatively *warmer,* there and over most of the rest of the world, too (figure 2). "I've found this great website where you can make maps to compare temperatures between different time periods that you pick. You can change some of the

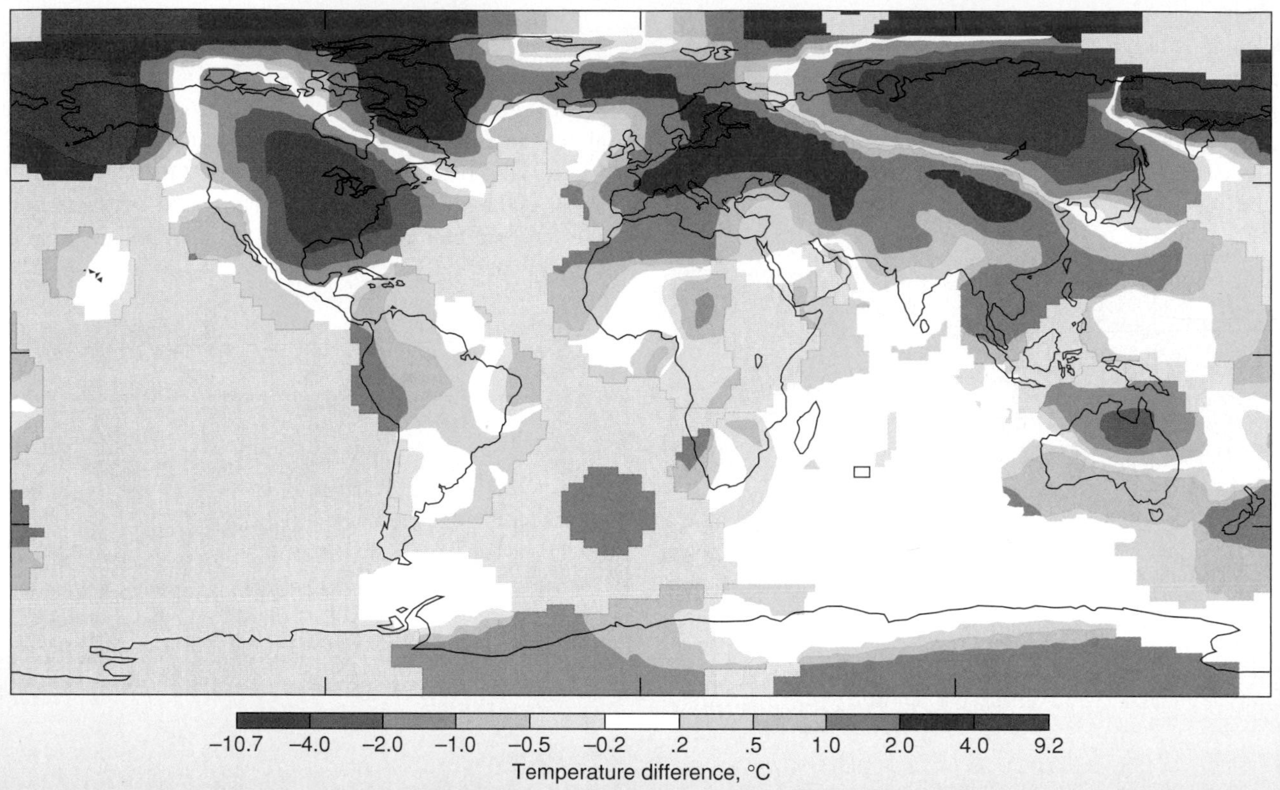

FIGURE 1 December 2000 temperatures compared with December temperatures for 1951–1999.
Image courtesy Goddard Institute for Space Studies/NASA

WHITHER FOR THE FUTURE?—CLIMATE FEEDBACKS, PREDICTIVE UNCERTAINTY

We have already seen some of the many short-term responses of earth systems to global warming associated with increasing greenhouse gases. Trying to project the longer-term trends is complicated, in part, by the fact that some changes tend to reinforce a particular trend, while others tend to counter it and reverse climatic direction. A few examples will illustrate the challenge.

We see that as temperatures rise, one consequence is loss of glacial cover. This decreases the average *albedo* (reflectivity) of the earth's surface, which should mean more heating of the land by sunlight, and more infrared radiation back out into the atmosphere to be captured by greenhouse gases, and thus more heating, more ice melting, and so on. This is an example of positive feedback, a warming trend reinforced.

But wait—if air temperatures rise, evaporation of water from the oceans would tend to increase. More water in the air means more clouds. More clouds, in turn, would reflect more sunlight back into space before it could reach and heat the

(Box 10.1 Continued)

other features, too—"and shortly they were both trying differing combinations of parameters and watching the resultant maps plotted up.

The site Diana had found was one maintained by NASA's Goddard Institute for Space Studies; the mapmaking program, at http://data.giss.nasa.gov/gistemp/maps/

Modern temperature data let us look at global and regional trends over decades or centuries. But geologic processes and changes often occur on much longer time scales, and local, short-term weather may not reflect global or long-term trends. Having unusual weather? Check the longer-term trends before you jump to conclusions. Then consider figure 10.20 and the extent to which 50 or 100 years of climate data may indicate future trends and over what time scale. See also the "For Further Thought" exercise at the end of the chapter to examine regional differences in warming over the last century.

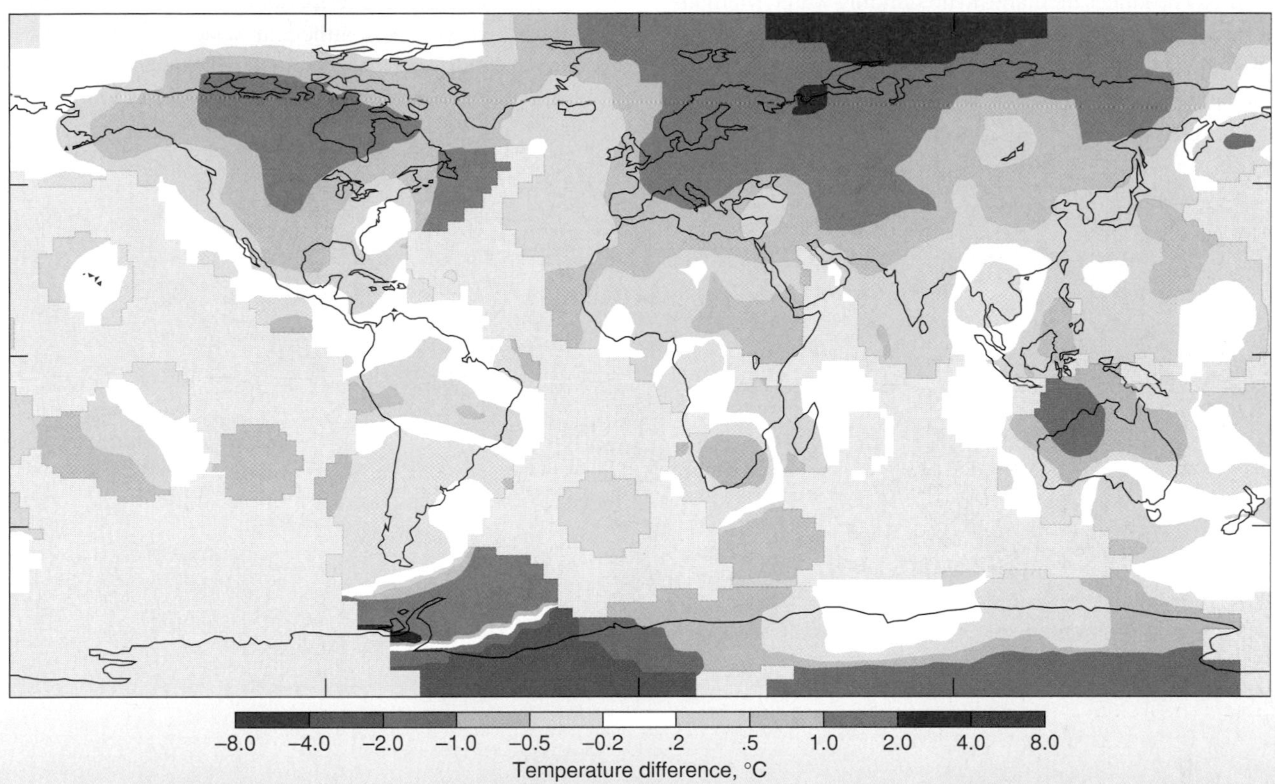

FIGURE 2 Temperatures for the year 2000 (December 1999–November 2000) relative to annual averages for 1951 through 1999. Note that here and in figure 1, the temperature scale is not linear.
Image courtesy Goddard Institute for Space Studies/NASA

surface. This is an example of negative feedback, in which warming induces changes that tend to cause cooling. Some speculate that over centuries, the world might then cool off, perhaps even enough to initiate another ice advance or ice age far in the future. In fact, NASA has identified cloud cover as the single largest source of uncertainty in projections of greenhouse-effect heating.

Now consider the oceans. Atmospheric carbon dioxide can dissolve in seawater, and more dissolves in colder water. So if global warming warms sea surface water, it may release some dissolved CO_2 back into the atmosphere, where it can trap more heat to further warm the surface waters (positive feedback). Or, will melting of polar and glacier ice put so much cool, fresh water on the sea surface that there will be more solution of atmospheric CO_2, removing it from the atmosphere and reducing the heating (negative feedback)?

Still more complexities exist. For instance, as permafrost regions thaw, large quantities of now-frozen organic matter in the soil can begin to decompose, adding CO_2 and/or CH_4 to the atmosphere—another positive-feedback scenario. If overall, the positive feedbacks dominate, one could eventually have a "runaway greenhouse" effect such as exists on the surface of Venus, where the CO_2-rich atmosphere keeps the planet's surface hot enough to melt lead.

Beyond water vapor, CO_2, and CH_4, there are still more greenhouse gases produced by human activity, including nitrous oxide (N_2O), a product of internal-combustion engines, and the synthetic chlorofluorocarbons. Many of these are even *more* effective at greenhouse-effect heating of the atmosphere than is CO_2. So is methane. Methane is produced naturally during some kinds of decay of organic matter and during the digestive processes of many animals. Human activities that increase atmospheric methane include extraction and use of fossil fuels, raising of domestic livestock (cattle, dairy cows, goats, camels, pigs, and others), and the growing of rice in rice paddies, where bacterial decay produces methane in the standing water. Methane concentrations have also risen, but currently, CH_4 is still far less abundant than CO_2; its concentration in the atmosphere is below 2 parts per million (ppm). If the amounts of minor but potent greenhouse gases continue to rise, they will complicate climate predictions tied primarily to CO_2 projections.

In any case, the scenarios explored here are likely to be of practical concern mainly to future generations. The more immediate concern for coming decades is continued warming as CO_2 levels continue to rise. A 2001 report by the Intergovernmental Panel on Climate Change projected further global temperature increases of 0.8 to 2.6°C (1.4 to 4.7°F) by 2050 and 1.4 to 5.8°C (2.5 to 10.4°F) by 2100, depending on model assumptions. Considering the effects of the much smaller temperature rise of the last century or so, the implications are substantial. This spurs interest in energy sources that do not produce CO_2 (chapter 15), research aimed at better understanding of the global carbon cycle (chapter 18), and international agreements to limit greenhouse-gas emissions (chapter 19).

SUMMARY

Modern burning of fossil fuels has been increasing the amount of carbon dioxide in the air. The resultant greenhouse-effect heating is melting alpine glaciers, edges of the Greenland ice cap, Arctic sea ice, and permafrost and contributing to the thermal expansion of seawater, causing a rise in global sea level. If this continues, flooding of many coastal areas could result. Global warming may influence the oceans' thermohaline circulation and the occurrence of El Niño events. Other probable consequences include changes in global weather patterns, including amount and distribution of precipitation, which may have serious implications for agriculture, and more-intense storms. Warming appears to be expanding the occurrence of certain diseases, thus affecting human health. Cores taken from ice sheets provide data on past temperature and greenhouse-gas fluctuations over several hundred thousand years.

Climatic projections for the future are complicated by the number and variety of factors to be considered, including other greenhouse gases, a variety of positive and negative climate feedback mechanisms, and complex interactions between oceans and atmosphere.

Key Terms and Concepts

greenhouse effect 221

permafrost 223

thermohaline circulation 225

Exercises

Questions for Review

1. Explain the greenhouse effect and its relationship to modern industrialized society.
2. Why is Arctic sea ice more vulnerable to melting than many land-based glaciers?
3. What is permafrost? Why is traveling in permafrost regions difficult during the summer thaw?
4. What drives the thermohaline circulation in the oceans, and how does this "conveyor" affect climate around the North Atlantic?
5. What is an El Niño event? What is its apparent significance to climate and to the productivity of coastal fishing grounds? Explain.
6. How is global warming related to the spread of certain diseases?
7. How are ice cores used to investigate relationships between atmospheric greenhouse gases and temperature? What is the "proxy" for temperature?
8. Briefly describe one positive and one negative feedback process relating to global warming.
9. Water vapor and methane are both greenhouse gases. Why is the focus so often on carbon dioxide instead?

For Further Thought

1. NASA's Goddard Institute for Space Studies identified 2005 as the warmest, or tied for warmest, ever recorded. (See earthobservatory.nasa.gov/Newsroom/NewImages/images.php3?img_id=17166.) Examine the map, noting where the most marked warming occurred. In light of the material in this chapter, consider what the cause(s) and consequences of such pronounced warming in that region might be. Also, see if you can determine from the map if 2005 was or was not an El Niño year.

Suggested Readings/References

Ackerman, J. 2000. New eyes on the oceans. *National Geographic* (October): 86–115.

Analysis links Pacific decadal variability to drought and streamflow in United States. 1999. *EOS* 80 (21 December): 621–26.

Bard, E. 2002. Climate shock: Abrupt changes over millennial time scales. *Physics Today* (December): 32–38.

Carslaw, K. S., Harrison, R. G., and Kirkby, J. 2002. Cosmic rays, clouds, and climate. *Science* 298 (29 November): 1732–37.

Crowley, T. J. 2000. Causes of climate change over the past 1000 years. *Science* 289 (14 July): 270–77.

Curtis, S. 2005. El Niño: Predicting causes and consequences. *Geotimes* (March): 30–33.

Earth's variable climatic past, 2001. (A series of articles on climate change.) *Science* 292 (27 April): 657–93.

Easterling, D. R., Meehl, G. A., Parmesan, C., Changnon, S. A., Karl, T. R., and Mearns, L. O. 2000. Climate extremes: Observations, modeling, and impacts. *Science* 289 (22 September): 2068–74.

El Niño: Nature's vicious cycle. 1999. *National Geographic* (March): 72–95.

Gonzalez, J.E., *et al.* 2005. Urban heat islands developing in coastal tropical cities. *EOS* 86 (18 October): 397, 403.

Gutzler, D. S. 2000. Evaluating global warming: A post-1990s perspective. *GSA Today* 10 (October): 1–7.

Hansen, J. 2004. Defusing the global warming time bomb. *Scientific American* (March): 68–77.

Issar, A. S. 1995. Climatic change and the history of the Middle East. *American Scientist* (July–August): 350–55.

Levitus, S., Antonov, J. I., Boyer, T. P., and Stephens, C. 2000. Warming of the world ocean. *Science* 287 (24 March): 2225–29.

Martens, P. 1999. How will climate change affect human health? *American Scientist* 87 (November–December): 534–41.

Perry, T. S. 1993. Modeling the world's climate. *IEEE Spectrum* 30 (July): 33–42.

Rogers, D. J., and Randolph, S. E. 2000. The future spread of malaria in a future, warmer world. *Science* 289 (8 September): 1763–69.

Ruddiman, W.F. 2005. How did humans first alter global climate? *Scientific American* (March): 46–53.

Seager, R. 2006. The source of Europe's mild climate. *American Scientist* 94 (July-August): 334–41.

Sturm, M., Peruvich, D. K., and Serreze, M. C. 2003. Meltdown in the North. *Scientific American* (October): 60–67.

Taylor, K. 1999. Rapid climate change. *American Scientist* 87 (July–August): 320–27.

Trouble in polar paradise. 2002. (A series of articles relating to climate change.) *Science* 297 (30 August): 1489–1510.

Weart, S. 2003. The discovery of rapid climate change. *Physics Today* (August): 30–36.

Zorpetle, G. 1993. Sensing climate change. *IEEE Spectrum* 30 (July): 20–27.

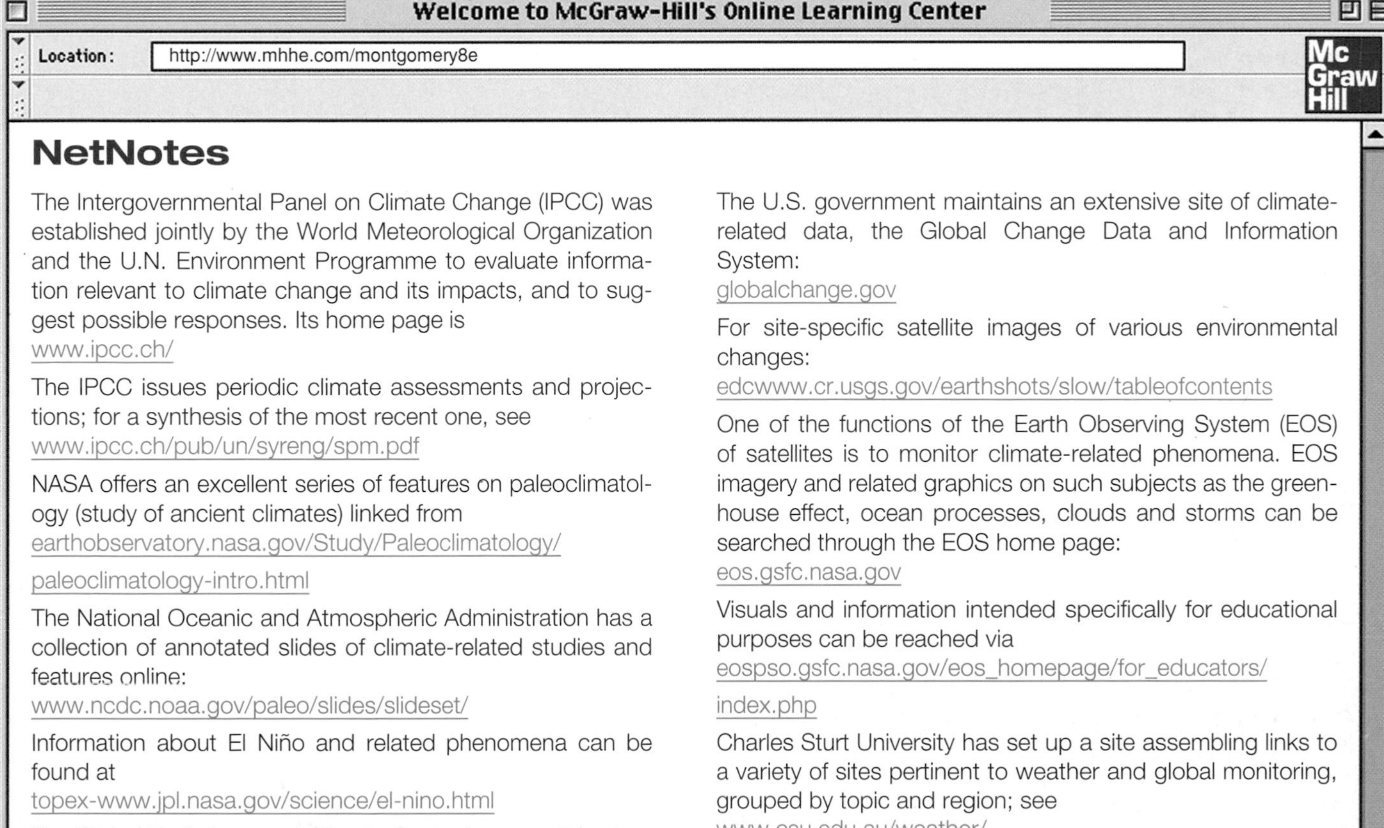

NetNotes

The Intergovernmental Panel on Climate Change (IPCC) was established jointly by the World Meteorological Organization and the U.N. Environment Programme to evaluate information relevant to climate change and its impacts, and to suggest possible responses. Its home page is
www.ipcc.ch/

The IPCC issues periodic climate assessments and projections; for a synthesis of the most recent one, see
www.ipcc.ch/pub/un/syreng/spm.pdf

NASA offers an excellent series of features on paleoclimatology (study of ancient climates) linked from
earthobservatory.nasa.gov/Study/Paleoclimatology/paleoclimatology-intro.html

The National Oceanic and Atmospheric Administration has a collection of annotated slides of climate-related studies and features online:
www.ncdc.noaa.gov/paleo/slides/slideset/

Information about El Niño and related phenomena can be found at
topex-www.jpl.nasa.gov/science/el-nino.html

The Global Hydrology and Climate Center is accessible via
topex-www.jpl.nasa.gov/science/pdo.html

The U.S. government maintains an extensive site of climate-related data, the Global Change Data and Information System:
globalchange.gov

For site-specific satellite images of various environmental changes:
edcwww.cr.usgs.gov/earthshots/slow/tableofcontents

One of the functions of the Earth Observing System (EOS) of satellites is to monitor climate-related phenomena. EOS imagery and related graphics on such subjects as the greenhouse effect, ocean processes, clouds and storms can be searched through the EOS home page:
eos.gsfc.nasa.gov

Visuals and information intended specifically for educational purposes can be reached via
eospso.gsfc.nasa.gov/eos_homepage/for_educators/index.php

Charles Sturt University has set up a site assembling links to a variety of sites pertinent to weather and global monitoring, grouped by topic and region; see
www.csu.edu.au/weather/

Human impacts of global change are evaluated by the Consortium for International Earth Science Information Network; their Socioeconomic Data and Applications Center is accessible at
sedac.ciesin.org *or* www.ciesin.org

Indicators of Arctic climatic change are found at
www.arctic.noaa.gov/detect

Another major site for environmental-change data, the Center for Data on Ice, Atmosphere and Climate of Oak Ridge National Laboratory, is at
cdiac.esd.ornl.gov/

In 2000, the National Assessment of the Potential Consequences of Climate Variability and Change was released; it can be found at
www.gcrio.org/NationalAssessment *or*
sedac.ciesin.org/NationalAssessment

The British Antarctic Survey Ice and Climate Division presents a variety of information on snow and ice studies and related climate issues at
www.nbs.ac.uk/public/icd/

National Snow and Ice Data Center (U.S.) provides a variety of information on snow and ice processes and interactions with atmosphere and oceans:
www-nsidc.colorado.edu

Current data and trends are available through their "State of the Cryosphere" site:
nsidc.org/NASA/SOTC/

instaar.colorado.edu/

For additional information on glacier terminus changes and glacier water resources, contact North Cascade Glacier Research at
www.nichols.edu/departments/glacier/

Graphical information on climate past is available through the publication "Global temperature in past centuries"
www.ngdc.noaa.gov/paleo/ei/ei_cover.html

A tutorial on climate from Manchester Metropolitan University, U.K., is found at
www.ace.mmu.ac.uk/Resources/gcc/contents.html

Trends in soil moisture from 1948 to 2002 are illustrated at
www.nsf.gov/news/mmg/media/images/pr05003pdsi_h.jpg

SECTION FOUR

RESOURCES

In a general sense, resources are all those things that are necessary or important to human life and civilization, that have some value to individuals and/or to society. Implicit in the term is the availability of a supply that can be drawn on as needed in the future. The requirement that resources have some value to people means that what constitutes a resource may change with time or social context. Many of the fuels, building materials, metals, and other substances important to modern technological civilization (which are thus resources in a modern context) were of no use to the earliest cave dwellers. In chapters 11 through 15, we will survey earth resources, that subset of all resources involved in or formed by geologic processes.

In the early 1970s, the Arab oil embargo precipitated sharp increases in the prices of gasoline and heating oil, long lines at gas stations, and a lot of talk about an "energy crisis." Ten years later, news reports were full of the "oil glut" and of widespread layoffs among petroleum industry employees. Now, at the start of the twenty-first century, some are speaking of the "new energy crisis." What explains these wildly fluctuating views?

There is more and more discussion nowadays of "mining" water, and some warn of impending water shortages. Yet over 70% of the earth's surface is covered by water. Why, then, is there concern about water resources?

The world hardly seems to be running out of the surface accumulations of rock and mineral fragments and other debris that make up soil. However, in many places, soil is eroding far faster than new soil is being produced by natural processes, and the fertility of the remaining soil is decreasing. This may have serious implications for the world's food supply.

The chapters that follow describe the various processes by which many mineral and fuel deposits and soils form and the pathways by which water moves through the environment. Certain resources are **renewable** (replaceable on a human timescale), while others are not. We will look at the rates at which resources

Even ground water can be "mined" and locally used up. The threat of depletion is perhaps more obvious in the Sahara (A) than in Kansas (B), but in both cases, center-pivot irrigation systems are tapping dwindling groundwater supplies. (Larger circles in (A) and smaller ones in (B) are about the same size, approximately ½ mile in diameter.)

(A) Image by Robert Simmon, based on data acquired by the Landsat 7 Science Team. Landsat data is archived and distributed by the USGS EROS Data Center (B) Image courtesy NASA/GSFC/METI/ERSDAC/JAROS, and U.S./Japan ASTER Science Team.

are being consumed, and for those in finite supply, how long they may be expected to last. In some cases, it may become necessary to look to different resources in the future from those that have been used in the past. We will also consider some of the potential adverse environmental impacts of the use of our current and possible future resources.

RESOURCES, PEOPLE, AND STANDARDS OF LIVING

Resource supplies have become a pressing problem, in part because of the surge in the world's population discussed in chapter 1. The more people on the earth, the more water consumed, the more fuel burned, the more minerals used, and so on. Yet population growth alone is not the whole story. If the rate of population growth is compared with the rate of energy consumption or use of many mineral resources as a function of time (figure R.1), it is apparent that the rates of resource use are increasing even faster than the population! Not only are there more and more people, but *per-capita* resource consumption is rising.

This is mainly a result of elevated standards of living, both for people in underdeveloped, or developing, countries and for those in developed ones. Most societies continue to develop increasingly sophisticated technologies and greater use of machinery, and to have lifestyles that involve increasing quantities of manufactured goods (home furnishings, clothes, cars, and so on). Therefore, each person tends to account for the consumption of more and more energy and mineral resources. The connection between per-capita energy consumption and standard of living can be seen in figure R.2. In it, energy consumption is plotted against gross domestic product (GDP), a measure of the value of all goods and services produced. In a general way, per-capita GNP reflects the level of technological development and standard of living. Increased per-capita consumption, then, together with a fast-growing world population, makes questions of resource supply even more pressing.

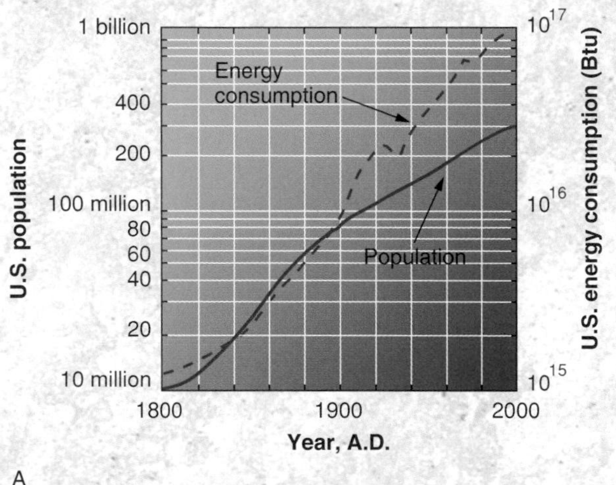

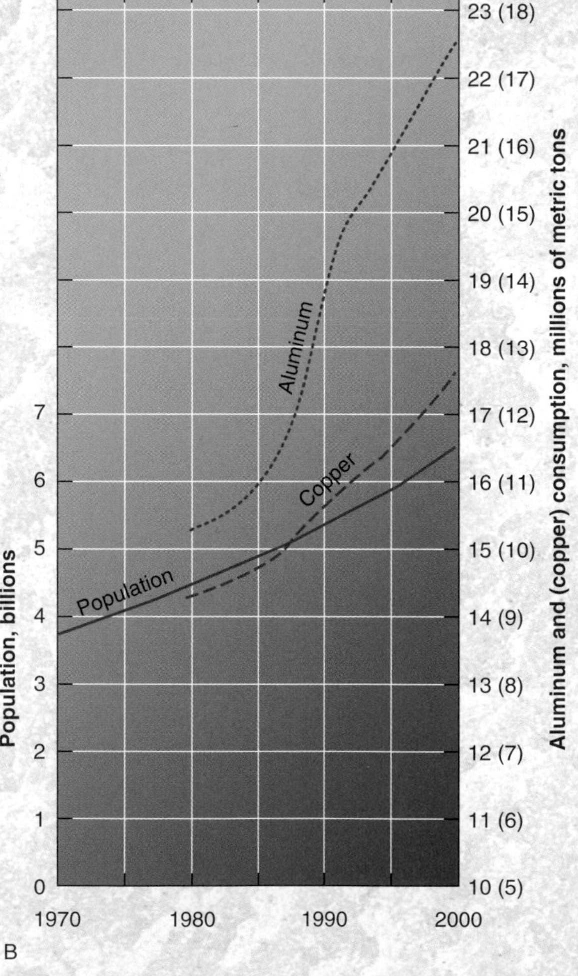

FIGURE R.1 Comparison of rates of population growth and of energy and mineral-resource consumption. (A) U.S. energy consumption has been rising faster than the population, except during the Great Depression and the Arab oil embargo of the early 1970s. (B) Worldwide consumption of copper (dashed line) and aluminum (dotted line) is also increasing faster than population (solid line), and this pattern is true for many industrial materials.

Source: (A) Data from Earl Cook, "The Flow of Energy in an Industrialized Society," Scientific American, *September 1971, updated through 2000, and U.S. Geological Survey. (B) Data from World Resources Institute and U.S. Geological Survey*

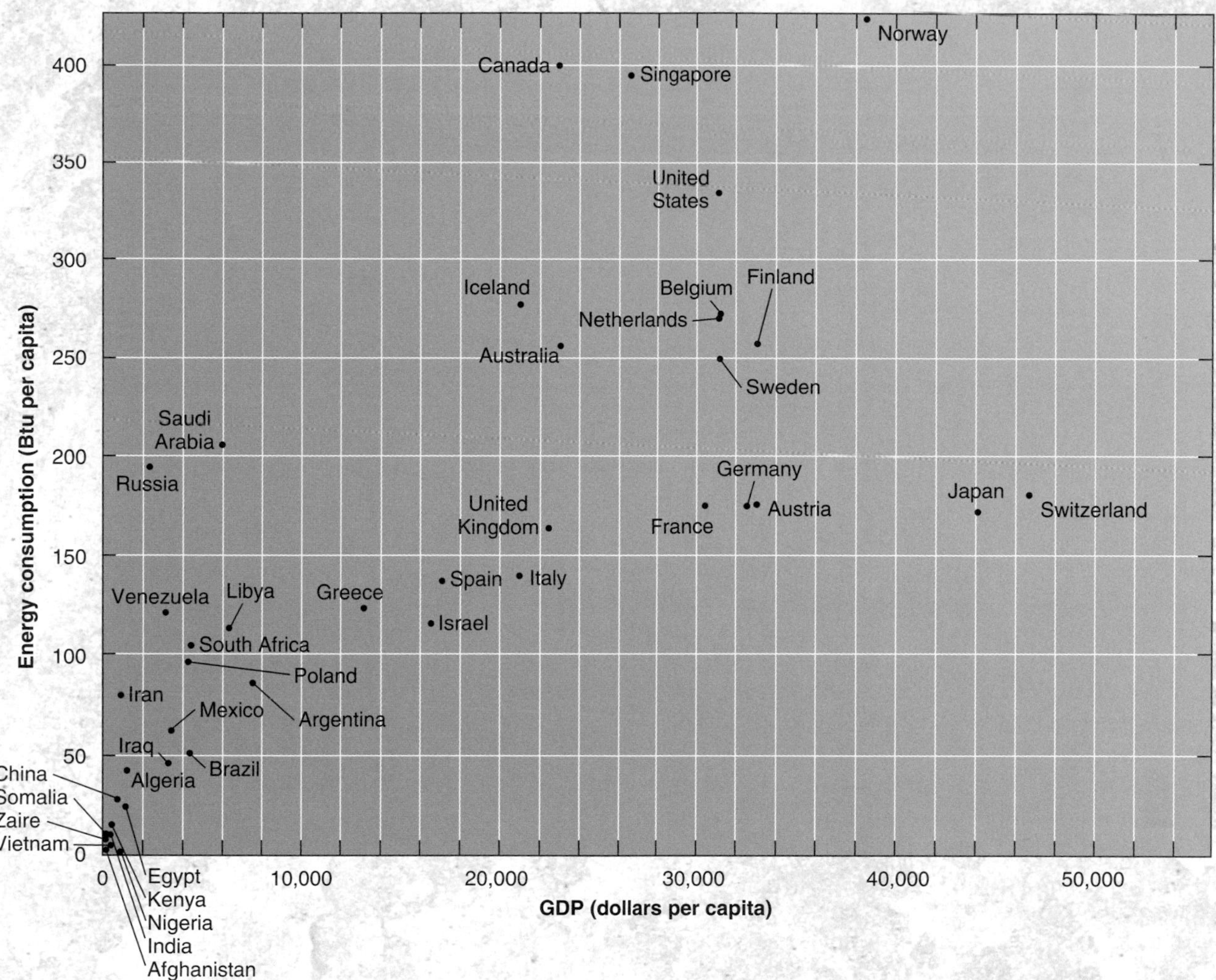

FIGURE R.2 There is a variable but generally positive correlation between GDP and energy consumption: The more energy consumed, the higher the value of goods and services produced, and generally, the higher the level of technological development as well.
Source: Data from International Energy Annual 2001, *U.S. Energy Information Administration, and* 2002 World Population Data Sheet, *Population Reference Bureau*

This is an ongoing concern internationally. The question of *sustainable development* is a major focus of the United Nations: How can developing nations continue to advance in technology and standard of living, and how can development continue elsewhere, without causing serious environmental damage or excessive consumption and eventual exhaustion of finite resources? Resource limitations are an important part of discussions of sustainable development, along with such issues as safe waste disposal, threats from pollution, wildlife habitat destruction, and loss of biodiversity.

PROJECTION OF RESOURCE SUPPLY AND DEMAND

Projecting the lengths of time that existing supplies of various resources are likely to last is a complex task, requiring accurate assessment of the available supplies and an estimate of how fast they will be used. With regard to the first part of the task, certain terms need to be defined.

Up to this point, the phrase *earth resources* has been used informally to describe any useful or valuable geologic materials. In discussing the supply/demand question, however, distinctions are made between *reserves* and various kinds of *resources,* as summarized in figure R.3. The **reserves** are that quantity of a given material that has been found and that can be recovered economically with existing technology. Usually, the term is used only for material not already consumed, as distinguished from **cumulative reserves,** which include the quantity of the material already used up in addition to the remaining (unused) reserves. The reserves represent the most conservative estimate of how much of a given metal, mineral, or fuel remains unused. Beyond that, several additional categories of resources can be identified.

Those deposits that have already been found but that cannot presently be profitably exploited are the **subeconomic resources** (also known as **conditional resources**). Some may

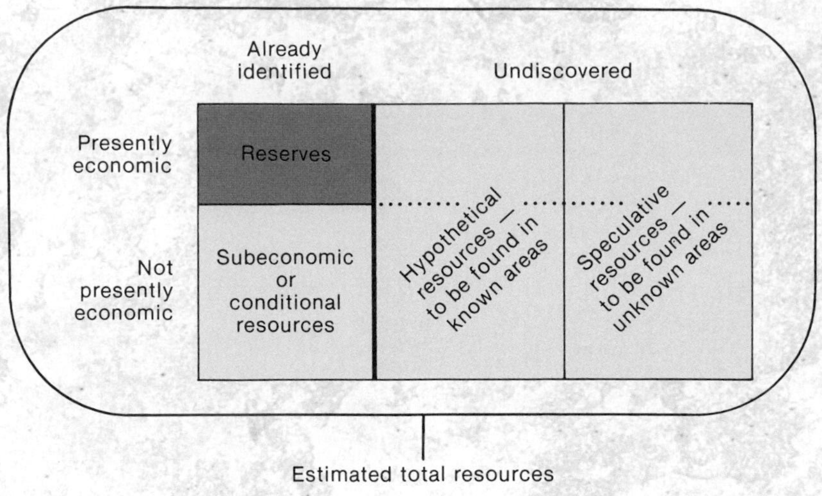

FIGURE R.3 Different categories of reserves and resources.

be exploitable with existing technology but contain ore that is too low-grade or fuel that is too dispersed to produce a profit at current prices. Others may require further advances in technology before they can be exploited. Still, these deposits have at least been located, and the quantity of material they represent can be estimated fairly well.

Then there are the undiscovered resources. These are sometimes subdivided into **hypothetical resources,** additional deposits expected to be found in areas in which some deposits of the material of interest have already been found, and the **speculative resources,** those deposits that might be found in explored or unexplored regions where deposits of the material are not already known to occur. Estimates of undiscovered resources are extremely rough by nature, and it would be unwise to count too heavily on deposits that have not even been found yet, especially for the near future.

Consequently, the reserves are the amounts of materials normally used to make supply projections. Even those figures may be somewhat imprecise, especially on a worldwide basis. Many mining and energy companies and some nations prefer to reveal as little about their unexploited assets as possible. In some countries, firms are taxed, in part, on the size of their remaining reserves, so their public estimates may be on the low side. Still, over past decades, scientists have combined published information on reserves and mineral production data with basic geologic knowledge to make what are believed to be reasonably good estimates of worldwide reserves. Estimates of U.S. reserves are presumably more accurate.

Predictions of how long supplies of various resources will last depend very strongly on projections of future demand. The subject of exponential growth was explored in chapter 1. For many mineral and energy resources, demand has been growing exponentially over decades, which in some cases leads to startlingly short projections for the length of time reserves of certain materials will last. Moreover, any such calculations done on a global scale necessarily ignore significant political realities arising from the fact that geologic resources are hardly uniformly distributed around the world.

Economics influence both sides of the problem, for costs of materials affect both consumer demand for, and profitability of exploitation or extraction of, particular minerals or fuels. The various aspects of supply/demand problems are explored more fully with respect to specific materials in the chapters that follow, particularly chapters 13 through 15.

11 Water as a Resource

Some aspects of surface-water flow, including the hydrologic cycle, were already considered in chapter 6. We now take a larger view of water on earth, to consider water as a resource. The importance of water availability for domestic use, agriculture, and industry is apparent. What may be less obvious is that water, or the lack of it, may control the extent to which we can develop certain other resources, such as fossil fuels. To begin, we will look at the earth's water supply and its distribution in nature. This involves considering not only surface water but also ground water and ice.

Table 11.1 shows how the water in the hydrosphere is distributed. Several points emerge immediately from the data. One is that there is, relatively speaking, little fresh liquid water on the earth. Most of the fresh water is locked up as ice, mainly in the large polar ice caps. Even the ground water beneath continental surfaces is not all fresh. These facts underscore the need for restraint in our use of fresh water. From the long-term geologic perspective, water is a renewable resource, but local supplies may be inadequate in the short term. Given the relative volumes of fresh surface water and of ground water, it is clear why one might consider ground water as a source of supply. The geologic setting in which it occurs influences not only how much is locally available, but how accessible it is.

The ability of ground water to dissolve and transport material can be revealed by the minerals deposited when the water reaches the surface. Mammoth Terraces, Yellowstone National Park.

TABLE 11.1

The Water in the Hydrosphere

Reservoir	Percentage of Total Water*	Percentage of Fresh Water†	Percentage of Unfrozen Fresh Water
oceans	97.54	—	—
ice	1.81	73.9	—
ground water	0.63	25.7	98.4
lakes and streams			
salt	0.007	—	—
fresh	0.009	0.36	1.4
atmosphere	0.001	0.04	0.2

Source: Data from J. R. Mather, *Water Resources,* 1984, John Wiley & Sons, Inc., New York.

*These figures account for over 99.9% of the water. Some water is also held in organisms (the biosphere).

†This assumes that all ground water is more or less fresh, since it is not all readily accessible to be tested and classified.

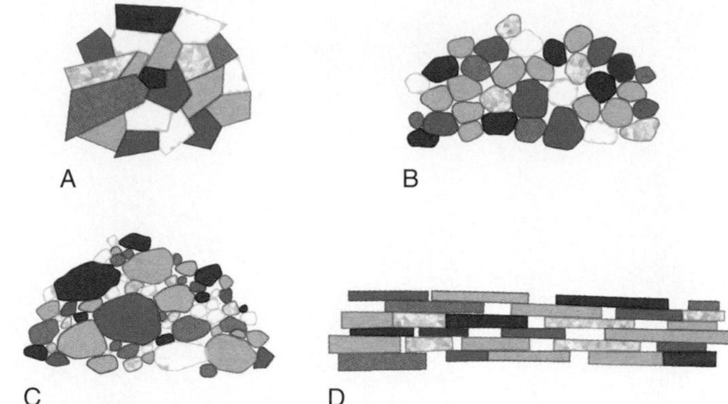

FIGURE 11.1 Porosity and permeability vary with grain shapes and the way grains fit together. (A) Low porosity in an igneous rock. (B) A sandstone made of rounded grains similar in size has more pore space. (C) Poorly sorted sediment: fine grains fill pores between coarse ones. (D) Packing of plates of clay in a shale may result in high porosity but low permeability. In all cases, fracturing or weathering may increase porosity and permeability beyond that resulting from grain shapes and packing.

FLUID STORAGE AND MOBILITY: POROSITY AND PERMEABILITY

Porosity and permeability involve the ability of rocks or other mineral materials (sediments, soils) to contain fluids and to allow fluids to pass through them. **Porosity** is the proportion of void space in the material—holes or cracks, unfilled by solid material, within or between individual mineral grains—and is a measure of how much fluid the material can store. Porosity may be expressed either as a percentage (for example, 1.5%) or as an equivalent decimal fraction (0.015). The pore spaces may be occupied by gas, liquid, or a combination of the two. **Permeability** is a measure of how readily fluids pass through the material. It is related to the extent to which pores or cracks are interconnected, and to their size—larger pores have a lower surface-to-volume ratio so there is less frictional drag to slow the fluids down. Porosity and permeability of geologic materials are both influenced by the shapes of mineral grains or rock fragments in the material, the range of grain sizes present, and the way in which the grains fit together (figure 11.1).

Igneous and metamorphic rocks consist of tightly interlocking crystals, and they usually have low porosity and permeability unless they have been broken up by fracturing or weathering. The same is true of many of the chemical sedimentary rocks, unless cavities have been produced in them by slow dissolution. Clastic sediments, however, accumulate as layers of loosely piled particles. Even after compaction or cementation into rock, they may contain more open pore space. Well-rounded equidimensional grains of similar size can produce sediments of quite high porosity and permeability. (This can be illustrated by pouring water through a pile of marbles.) Many sandstones have these characteristics; consider, for example, how rapidly water drains into beach sands. In materials containing a wide range of grain sizes, finer materials can fill the gaps between coarser grains. Porosity can thereby be reduced, though permeability may remain high. Clastic sediments consisting predominantly of flat, platelike grains, such as clay minerals or micas, may be porous, but because the flat grains can be packed closely together parallel to the plates, these sediments may not be very permeable, especially in the direction perpendicular to the plates. Shale is a rock commonly made from such sediments. The low permeability of clays can readily be demonstrated by pouring water onto a slab of artists' clay.

Figure 11.2 gives some representative values of porosity and permeability. These properties are relevant not only to discussions of groundwater availability and use, but also to issues such as stream flooding, petroleum resources, water pollution, and waste disposal.

SUBSURFACE WATERS

If soil on which precipitation falls is sufficiently permeable, infiltration occurs. Gravity continues to draw the water downward until an impermeable rock or soil layer is reached, and the water begins to accumulate above it. (The impermeable material itself may or may not contain significant pore water.) Immediately above the impermeable material is a volume of rock or soil that is water-saturated, in which water fills all the accessible pore space, the **saturated zone,** or **phreatic zone. Ground water** is the water in the saturated zone. Usually, ground water is found, at most, a few kilometers into the crust. In the deep crust and below, pressures on rocks are so great that compression closes up any pores that water might fill. Above the saturated zone is rock or soil in which the pore spaces are filled partly with water, partly with air: the **unsaturated zone,**

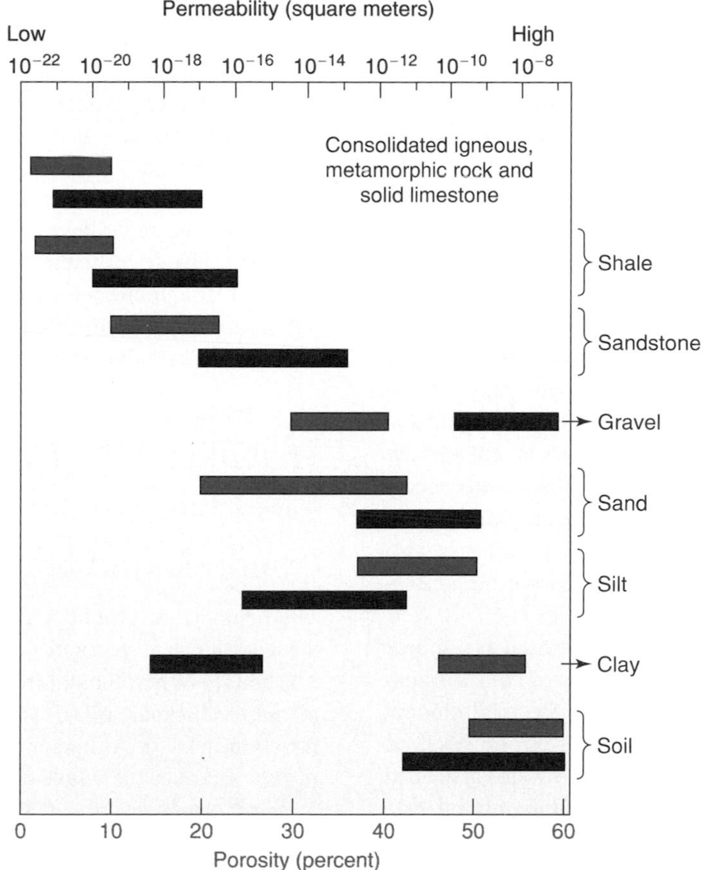

FIGURE 11.2 Typical ranges of porosities (blue bars) and permeabilities (brown) of various geological materials.

or **vadose zone.** The water in unsaturated soil is **soil moisture,** and is often an important factor in agricultural productivity. All of the water occupying pore space below the ground surface is, logically, called *subsurface water,* a term that thus includes ground water, soil moisture, and water in unsaturated rocks. The **water table** is the top surface of the saturated zone, where the saturated zone is not confined by overlying impermeable rocks. (See also the discussion of aquifer geometry in the next section.) These relationships are illustrated in figure 11.3.

The water table is not always below the ground surface. Where the water table locally intersects the ground surface, the result may be a lake, stream, spring, or wetland; the water's surface is the water table. The water table below ground is not flat like a tabletop, either. It may undulate with the surface

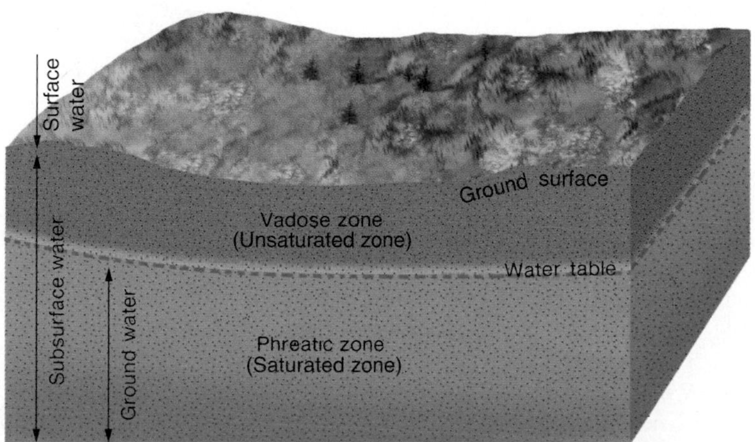

FIGURE 11.3 Nomenclature of surface and subsurface waters. Ground water is water in the saturated zone, below the water table.

topography and with the changing distribution of permeable and impermeable rocks underground. The height of the water table varies, too. It is highest when the ratio of input water to water removed is greatest, typically in the spring, when rain is heavy or snow and ice accumulations melt. In dry seasons, or when local human use of ground water is intensive, the water table drops, and the amount of available ground water remaining decreases. Ground water can flow laterally through permeable soil and rock, from higher elevations to lower, from areas of higher pressure or potential (see below) to lower, from areas of abundant infiltration to drier ones, or from areas of little groundwater use toward areas of heavy use. Ground water may contribute to streamflow, or water from a stream may replenish ground water (figure 11.4). The processes of infiltration and migration or percolation by which ground water is replaced are collectively called **recharge.** Groundwater **discharge** occurs where ground water flows into a stream, escapes at the surface in a spring, or otherwise exits the aquifer. The term can also apply to the water moving through an aquifer; see the discussion of Darcy's Law later in the chapter.

If ground water drawn from a well is to be used as a source of water supply, the porosity and permeability of the surrounding rocks are critical. The porosity controls the total amount of water available. In most places, there are not vast pools of open water underground to be tapped, just the water dispersed in the rocks' pore spaces. This pore water amounts, in most cases, to a few percent of the rocks' volume at best, although some sands and gravels may occasionally have porosities of 50% or more. The permeability, in turn, governs both the rate at which water can be withdrawn and the rate at which recharge can occur. All the pore water in the world would be of no practical use if it were so tightly locked in the rocks that it could not be pumped out. A rock that holds enough water and transmits it rapidly enough to be useful as a source of water is an **aquifer.** Many of the best aquifers are sandstones or other coarse clastic sedimentary rocks, but any other type of rock may serve if it is sufficiently porous and permeable—a porous or fractured limestone, fractured basalt, or weathered granite, for instance. One can describe rocks of very low and low permeability, respectively, as *aquicludes* or *aquitards.* An **aquitard** is a rock that may store a considerable quantity of water, but in which water flow is slowed, or retarded; that is, its permeability is low, regardless of its porosity. Shales are common aquitards. Historically, the term *aquiclude* was used to describe an extreme aquitard, a rock that is essentially impermeable on a human timescale; but virtually no rock would be impermeable indefinitely, and the term aquiclude has fallen into disuse.

AQUIFER GEOMETRY AND GROUNDWATER FLOW

Confined and Unconfined Aquifers

The behavior of ground water is controlled to some extent by the geology and geometry of the particular aquifer in which it is found. When the aquifer is directly overlain only by permeable rocks and soil, it is described as an **unconfined aquifer** (figure 11.5). An unconfined aquifer may be recharged by infiltration over the whole area underlain by that aquifer, if no impermable layers above stop the downward flow of water from surface to aquifer. If a well is drilled into an unconfined aquifer, the water will rise in the well to the same height as the water table in the aquifer. The water must be actively pumped up to the ground surface.

A **confined aquifer** is bounded above and below by low-permeability rocks (aquitards). Water in a confined aquifer may be under considerable pressure from the adjacent rocks, or as a consequence of lateral differences in elevation within the aquifer. The confining layers prevent the free flow of water

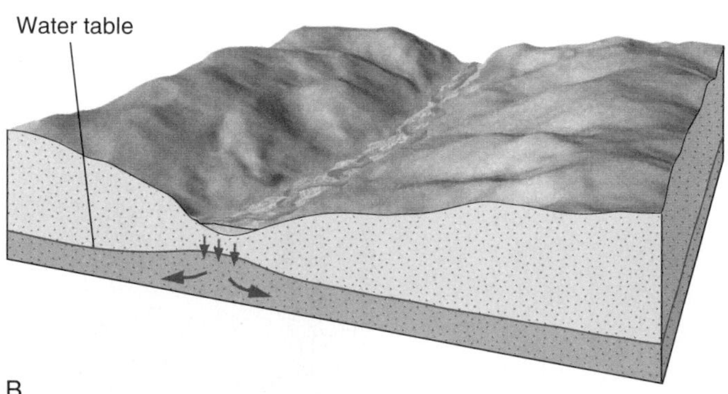

FIGURE 11.4 (A) In this natural system, the stream drains water from the region—not only surface runoff, but ground water from adjacent saturated rocks and soil that slowly drains into the stream. Such slow groundwater influx explains why many streams can keep flowing even when there has been no rain for weeks. The stream represents a site of groundwater discharge. (B) Here, water from the stream channel percolates downward to add to the groundwater supply.

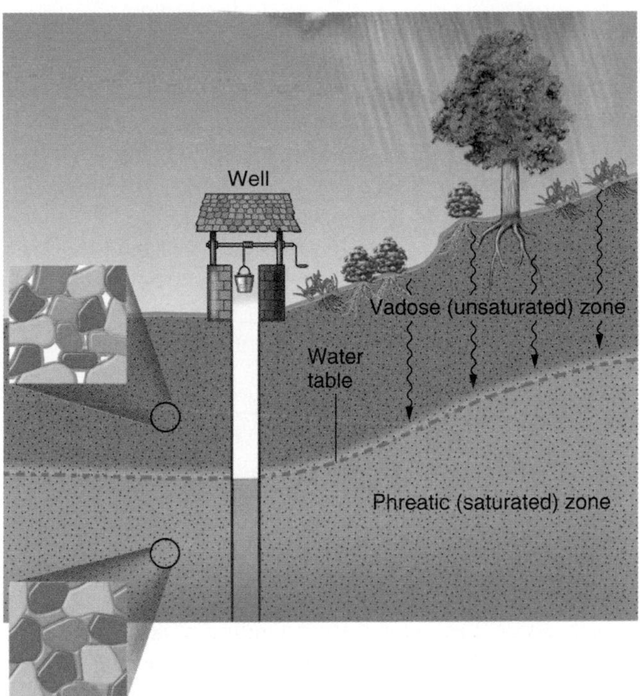

FIGURE 11.5 An unconfined aquifer. Water rises in an unpumped well just to the height of the water table.

to relieve this pressure. At some places within the aquifer, the water level in the saturated zone may be held above or below the level it would assume if the aquifer were unconfined and the water allowed to spread freely (figure 11.6).

If a well is drilled into a confined aquifer, the water can rise above its level in the aquifer because of this extra hydrostatic (fluid) pressure. This is called an **artesian system.** The water in an artesian system may or may not rise all the way to the ground surface; some pumping may still be necessary to bring it to the surface for use. In such a system, rather than describing the height of the water table, geologists refer to the height of the **potentiometric surface,** which represents the height to which the water's pressure would raise the water if the water were unconfined. This level will be somewhat higher than the top of the confined aquifer where its rocks are saturated, and it may be above the ground surface, as shown in figure 11.6.

Advertising claims notwithstanding, this is all that "artesian water" means—artesian water is no different chemically from, and no purer, better-tasting, or more wholesome than, any other ground water; it is just under natural pressure. Water towers create the same effect artificially: After water has been pumped into a high tower, gravity increases the fluid pressure in the water-delivery system so that water rises up in the pipes of individual users' homes and businesses without the need for many pumping stations. The same phenomenon can be demonstrated very simply by holding up a water-filled length of rubber tubing or hose at a steep angle. The water is confined by the hose, and the water in the low end is under pressure from the weight of the water above it. Poke a small hole in the top side of the lower end of the hose, and a stream of water will shoot up to the level of the water in the hose.

Darcy's Law and Groundwater Flow

How readily ground water can move through rocks and soil is governed by permeability, but where and how rapidly it actually does flow is also influenced by differences in **hydraulic head** (potential energy) from place to place. Ground water flows from areas of higher hydraulic head to lower. The height of the water table (in an unconfined aquifer) or of the potentiometric surface (confined aquifer) reflects the hydraulic head at each point in the aquifer. All else being equal, the greater the difference in hydraulic head between two points, the faster

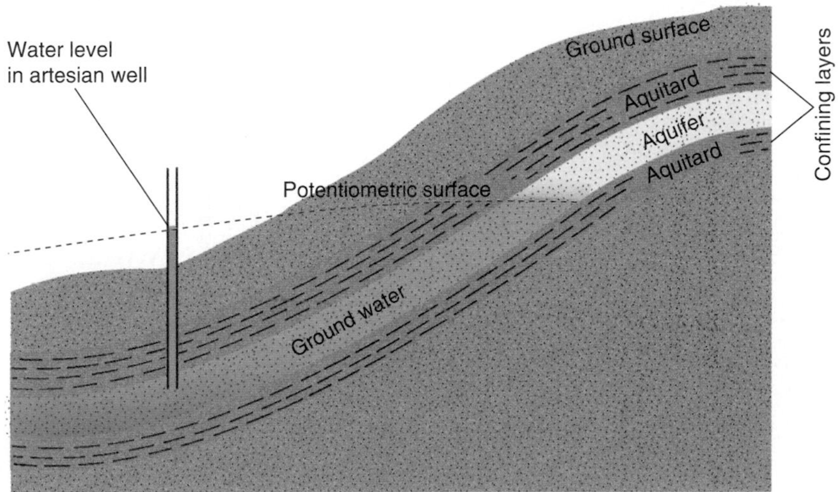

FIGURE 11.6 Natural internal pressure in a confined aquifer system creates artesian conditions, in which water may rise above the apparent (confined) local water table. The aquifer here is a sandstone; the confining layers, shale.

ground water will flow between them. This is a key result of **Darcy's Law,** which can be expressed this way

$$Q = K \cdot A \cdot \frac{\Delta h}{\Delta l}$$

where Q = discharge, A = cross-sectional area, K is a parameter known as *hydraulic conductivity* that takes into account both the permeability of the rock or soil and the viscosity and density of the flowing fluid, and ($\Delta h/\Delta l$) is the *hydraulic gradient,* the difference in hydraulic head between two points (Δh) divided by the distance between them (Δl). (The hydraulic gradient of ground water is analogous to a stream's gradient.) See figure 11.7.

Note that this relationship works not only for water, but for other fluids, such as oil; the value of K will change with changes in viscosity, but the underlying behavior is the same. Also, Darcy's Law somewhat resembles the expression for stream discharge, with flow velocity represented by $\left(K\frac{\Delta h}{\Delta l}\right)$.

Other Factors in Water Availability

More complex local geologic conditions can make it difficult to determine the availability of ground water without thorough study. For example, locally occurring lenses or patches of relatively impermeable rocks within otherwise permeable ones may result in a *perched water table* (figure 11.8). Immediately above the aquitard is a local saturated zone, far above the true regional water table. Someone drilling a well above this area could be deceived about the apparent depth to the water table and might also find that the perched saturated zone contains very little total water. The quantity of water available would be especially sensitive to local precipitation levels and fluctuations. Using ground water over the longer term might well require drilling down to the regional water table, at a much greater cost.

As noted earlier, ground water flows (albeit often slowly). Both flow paths (directions) and locations of recharge zones may have to be identified in assessing water availability. For example, if water is being consumed very close to the recharge area and consumption rate exceeds recharge rate, the stored water may be exhausted rather quickly. If the point of extraction is far down the flow path from the recharge zone, a larger reserve of stored water may be available to draw upon. Like stream systems, aquifer systems also have divides from which ground water flows in different directions, which determine how large a section of the aquifer system can be tapped at any point. Groundwater divides may also separate polluted from unpolluted waters within the same aquifer system.

CONSEQUENCES OF GROUNDWATER WITHDRAWAL

Lowering the Water Table

When ground water must be pumped from an aquifer, the rate at which water flows in from surrounding rock to replace that which is extracted is generally slower than the rate at which water is taken out. In an unconfined aquifer, the result is a circular lowering of the water table immediately around the well, which is called a **cone of depression** (figure 11.9A). A similar feature is produced on the surface of a liquid in a drinking glass when it is sipped hard through a straw. When there are many closely spaced wells, the cones of depression of adjacent wells may overlap, further lowering the water table between wells (figure 11.9B). If over a period of time, groundwater withdrawal rates consistently exceed recharge rates, the regional water table may drop. A clue that this is happening is the need for wells throughout a region to be drilled deeper periodically to keep the water flowing. Cones of depression can likewise develop in potentiometric surfaces. When artesian ground water is withdrawn at a rate exceeding the recharge rate, the potentiometric surface can be lowered (figure 11.10).

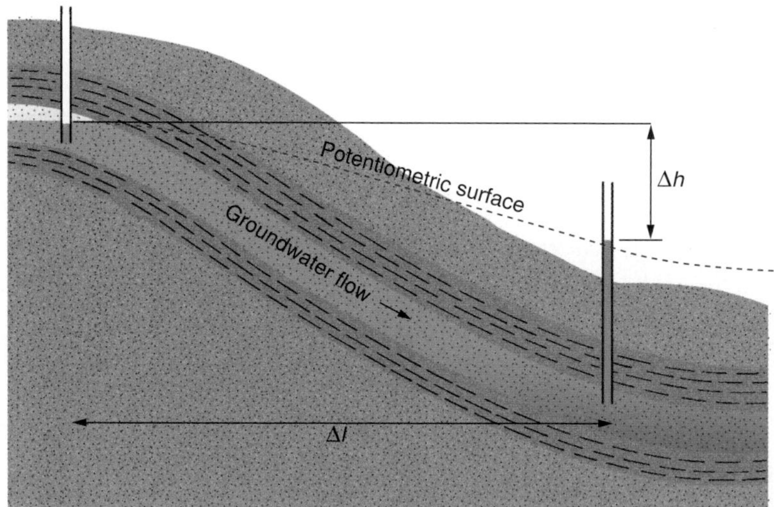

FIGURE 11.7 Darcy's Law relates groundwater flow rate, and thus discharge, to hydraulic gradient. Height of water in wells reflects relative hydraulic head.

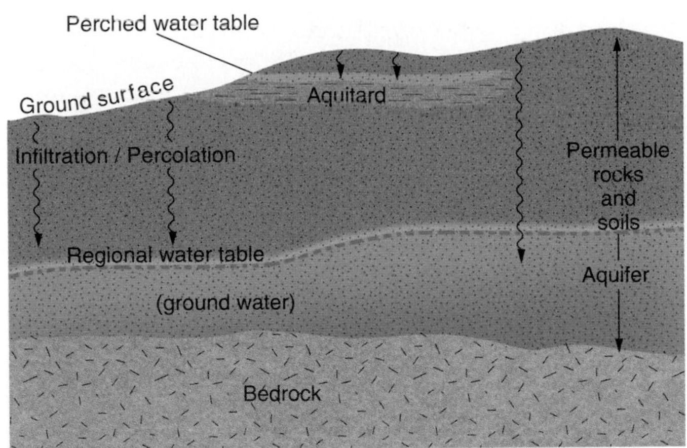

FIGURE 11.8 A perched water table may create the illusion of a shallow regional water table.

These should be warning signs, for the process of deepening wells to reach water cannot continue indefinitely, nor can a potentiometric surface be lowered without limit. In many areas, impermeable rocks are reached at very shallow depths. In an individual aquifer system, the lower confining layer may be far above the bedrock depth. Therefore, there is a "bottom" to the groundwater supply, though without drilling, it is not always possible to know just where it is. Groundwater flow rates are highly variable, but in many aquifers, they are of the order of only meters or tens of meters per year. Recharge of significant amounts of ground water, especially to confined aquifers with limited recharge areas, can thus require decades or centuries.

Where ground water is being depleted by too much withdrawal too fast, one can speak of "mining" ground water (figure 11.11). The idea is not necessarily that the water will never be recharged but that the rate is so slow on the human timescale as to be insignificant. From the human point of view, we may indeed use up ground water in some heavy-use areas. Also, as we will see in the next section, human activities may themselves reduce natural recharge, so ground water consumed may not be replaced, even slowly.

Compaction and Surface Subsidence

Lowering of the water table may have secondary consequences. The aquifer rocks, no longer saturated with water, may become compacted from the weight of overlying rocks. This decreases

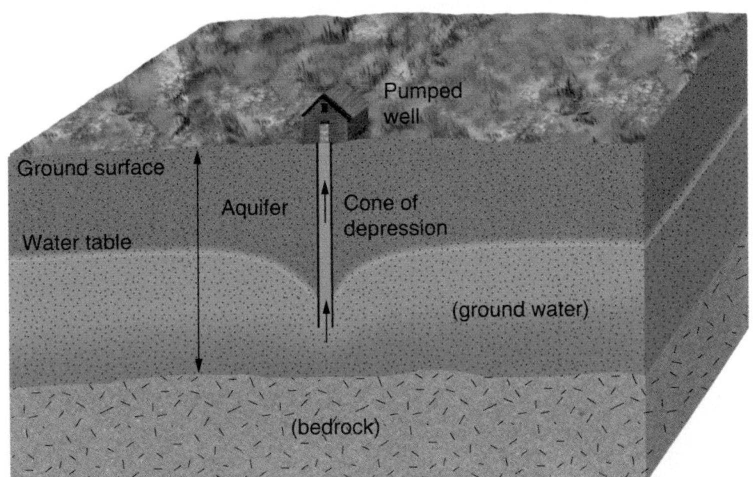

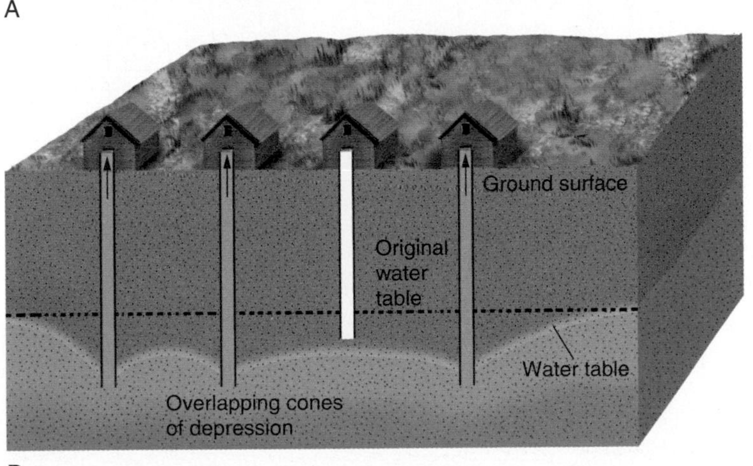

FIGURE 11.9 The formation and effect of cones of depression. (A) Lowering of the water table in a cone of depression around a pumped well in an unconfined aquifer. (B) Overlapping cones of depression lead to net lowering of the water table; shallower wells may run dry.

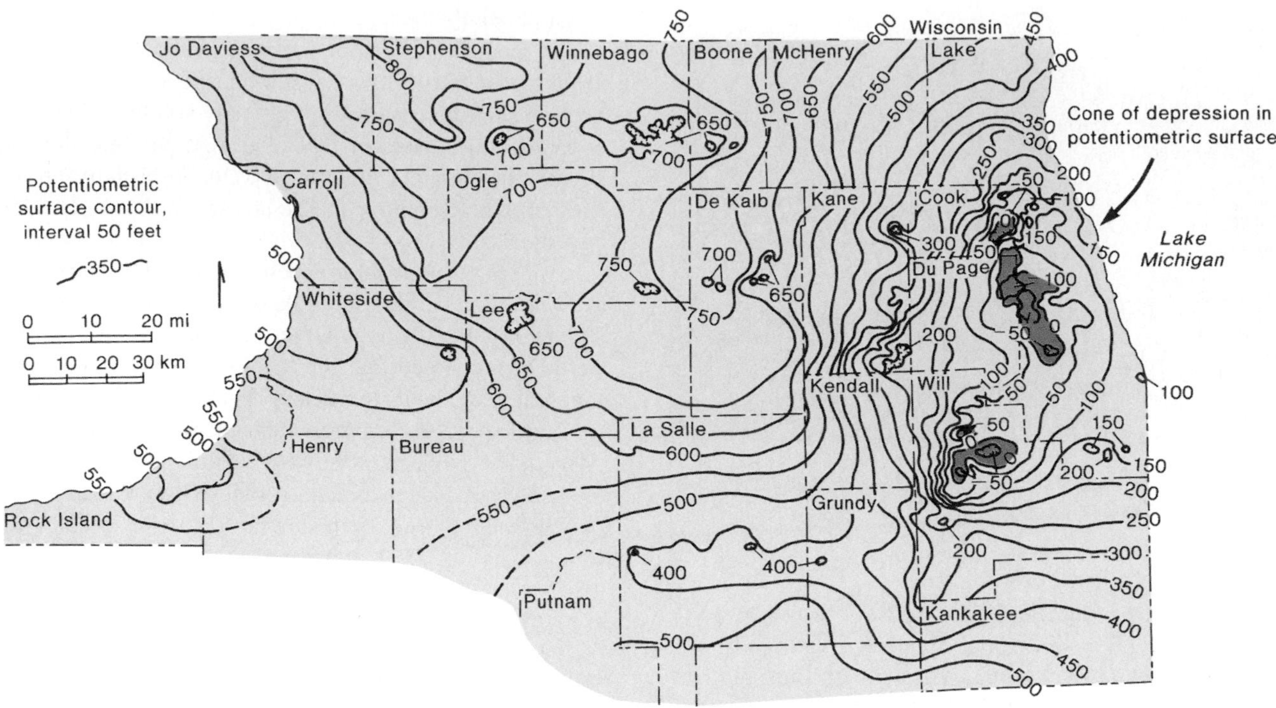

FIGURE 11.10 Potentiometric surface of the principal aquifer system in northern Illinois. Contours are in feet above sea level. The shaded area denotes where the potentiometric surface is below sea level. The city of Chicago is located in Cook County.

From R. T. Sasman, et al., *"Water Level Decline and Pumpage in Deep Wells in the Chicago Region, 1966–1971,"* Illinois State Water Survey Circular 113, *1973. Reprinted by permission*

their porosity, permanently reducing their water-holding capacity, and may also decrease their permeability. At the same time, as the rocks below compact and settle, the ground surface itself may subside. Where water depletion is extreme, the surface subsidence may be several meters. Lowering of the water table/potentiometric surface also may contribute to *sinkhole* formation, as described in a later section.

At high elevations or in inland areas, this subsidence causes only structural problems as building foundations are disrupted. In low-elevation coastal regions, the subsidence may lead to extensive flooding, as well as to increased rates of coastal erosion. The city of Venice, Italy, is in one such slowly drowning coastal area. Many of its historical, architectural, and artistic treasures are threatened by the combined effects of the gradual rise in worldwide sea levels, the tectonic sinking of the Adriatic coast, and surface subsidence from extensive groundwater withdrawal. Drastic and expensive engineering efforts will be needed to save them. The groundwater withdrawal began about 1950. Since 1969, no new wells have been drilled, groundwater use has declined, and subsidence from this cause has apparently stopped, but the city remains awash. Closer to home, the Houston/Galveston Bay area has suffered subsidence of up to several meters from a combination of groundwater withdrawal and, locally, petroleum extraction. Some 80 sq. km (over 30 sq. mi.) are permanently flooded, and important coastal wetlands have been lost. In the San Joaquin Valley in California,

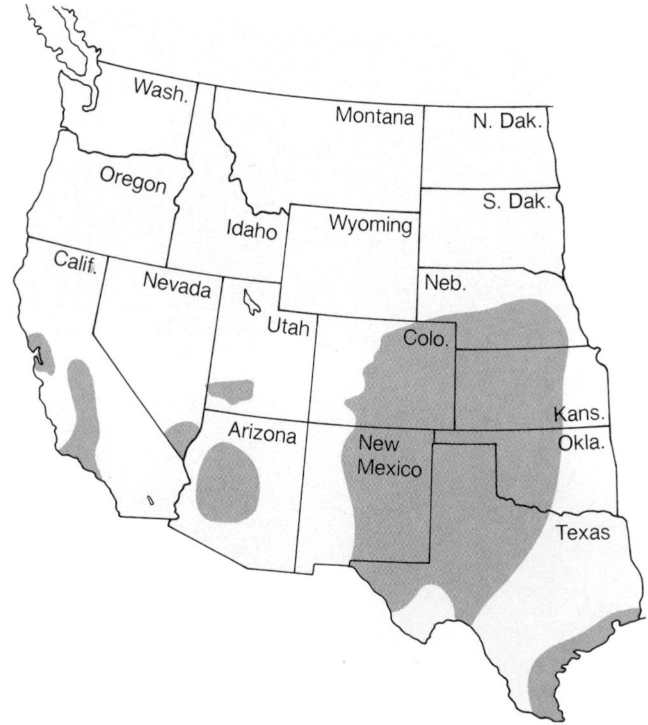

FIGURE 11.11 Areas of the arid western United States in which ground water is clearly being "mined," with withdrawal rates exceeding recharge.

Source: U.S. Water Resources Council

half a century of groundwater extraction, much of it for irrigation, produced as much as 9 meters of surface subsidence (figure 11.12).

In such areas, simple solutions, such as pumping water back underground, are unlikely to work. The rocks may have been permanently compacted: This is the case in Venice, where subsidence may have been halted but rebound is not expected because clayey rocks in the aquifer system are irreversibly compacted, and compaction has occurred in the San Joaquin Valley, too. Also, what water is to be employed? If groundwater use is heavy, it may well be because there simply is no great supply of fresh surface water. Pumping in salt water, in a coastal area, will in time make the rest of the water in the aquifer salty, too. And, presumably, a supply of fresh water is still needed for local use, so groundwater withdrawal often cannot be easily or conveniently curtailed.

FIGURE 11.12 Subsidence of as much as 9 meters (almost 30 feet) occurred between 1925 and 1977 in the San Joaquin Valley, California, as a consequence of groundwater withdrawal. Signs indicate former ground surface elevation. Photograph was taken in December 1976.
Photograph courtesy Richard O. Ireland, U.S. Geological Survey

Saltwater Intrusion

A further problem arising from groundwater use in coastal regions, aside from the possibility of surface subsidence, is **saltwater intrusion** (figure 11.13). When rain falls into the ocean, the fresh water promptly mixes with the salt water. However, fresh water falling on land does not mix so readily with saline ground water at depth because water in the pore spaces in rock or soil is not vigorously churned by currents or wave action. Fresh water is also less dense than salt water. So the fresh water accumulates in a lens, which floats above the denser salt water. If water use approximately equals the rate of recharge, the freshwater lens stays about the same thickness.

However, if consumption of fresh ground water is more rapid, the freshwater lens thins, and the denser saline ground water, laden with dissolved sodium chloride, moves up to fill in pores emptied by removal of fresh water. *Upconing* of salt water below cones of depression in the freshwater lens may also occur. Wells that had been tapping the freshwater lens may begin pumping unwanted salt water instead, as the limited freshwater supply gradually decreases. Moreover, once a section of an aquifer becomes tainted with salt, it cannot readily be "made fresh" again. Saltwater intrusion destroyed useful aquifers beneath Brooklyn, New York, in the 1930s and is a significant problem in many coastal areas nationwide: the southeastern and Gulf coastal states (including the vicinity of Cape Canaveral, Florida), some densely populated parts of California, and Cape Cod in Massachusetts.

IMPACTS OF URBANIZATION ON GROUNDWATER RECHARGE

Obviously, an increasing concentration of people means an increased demand for water. We saw in chapter 6 that urbanization may involve extensive modification of surface-water runoff patterns and stream channels. Insofar as it modifies surface runoff and the ratio of runoff to infiltration, urbanization also influences groundwater hydrology.

Impermeable cover—buildings, asphalt and concrete roads, sidewalks, parking lots, airport runways—over one part of a broad area underlain by an unconfined aquifer has relatively little impact on that aquifer's recharge. Infiltration will continue over most of the region. In the case of a confined aquifer, however, the available recharge area may be very limited, since the overlying confining layer prevents direct downward infiltration in most places (figure 11.14). If impermeable cover is built over the recharge area of a confined aquifer, then, recharge can be considerably reduced, thus aggravating the water-supply situation.

Filling in wetlands is a common way to provide more land for construction. This practice, too, can interfere with recharge, especially if surface runoff is rapid elsewhere in the area. A marsh or swamp in a recharge area, holding water for long periods, can be a major source of infiltration and recharge. Filling it in so water no longer accumulates there, and worse yet, topping the

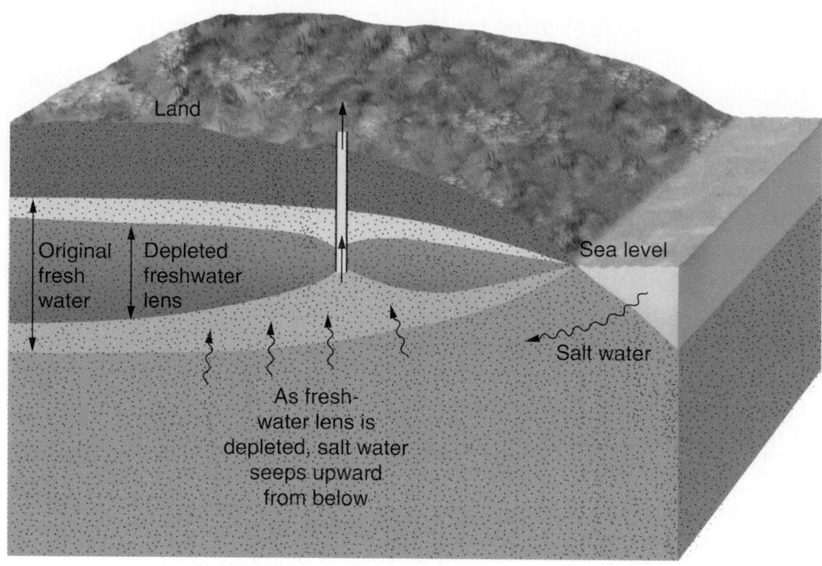

FIGURE 11.13 Saltwater intrusion in a coastal zone. If groundwater withdrawal exceeds recharge, the lens of fresh water thins, and salt water flows into more of the aquifer system from below. "Upconing" of saline water also occurs below a cone of depression. Similar effects may occur in an inland setting where water is drawn from a fresh groundwater zone underlain by more saline waters.

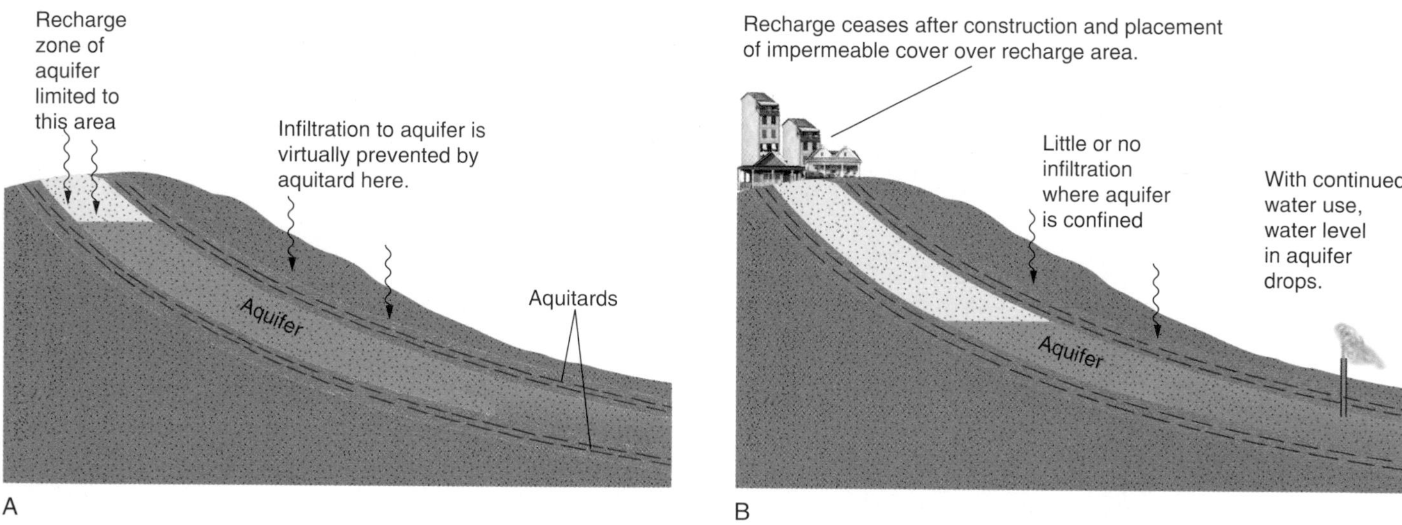

FIGURE 11.14 Recharge to a confined aquifer. (A) The recharge area of this confined aquifer is limited to the area where permeable rocks intersect the surface. (B) Recharge to the confined aquifer may be reduced by placement of impermable cover over the limited recharge area. (Vertical scale exaggerated.)

fill with impermeable cover, again may greatly reduce local groundwater recharge.

In steeply sloping areas or those with low-permeability soils, well-planned construction that includes artificial recharge basins can aid in increasing groundwater recharge (figure 11.15A). The basin acts similarly to a flood-control retention pond in that it is designed to catch some of the surface runoff during high-runoff events (heavy rain or snowmelt). Trapping the water allows more time for infiltration and thus more recharge in an area from which the fresh water might otherwise be quickly lost to streams and carried away.

Recharge basins are a partial solution to the problem of areas where groundwater use exceeds natural recharge rate, but, of course, they are effective only where there is surface runoff to catch, and they rely on precipitation, an intermittent water source. Increasingly, artificial recharge involves diverting streams, as shown in figure 11.15B.

KARST AND SINKHOLES

Broad areas of the contiguous United States are underlain by carbonate rocks (limestone and dolomite) or beds of rock salt or gypsum, chemical sediments deposited in shallow seas (figure 11.16). One characteristic that these rock types have in

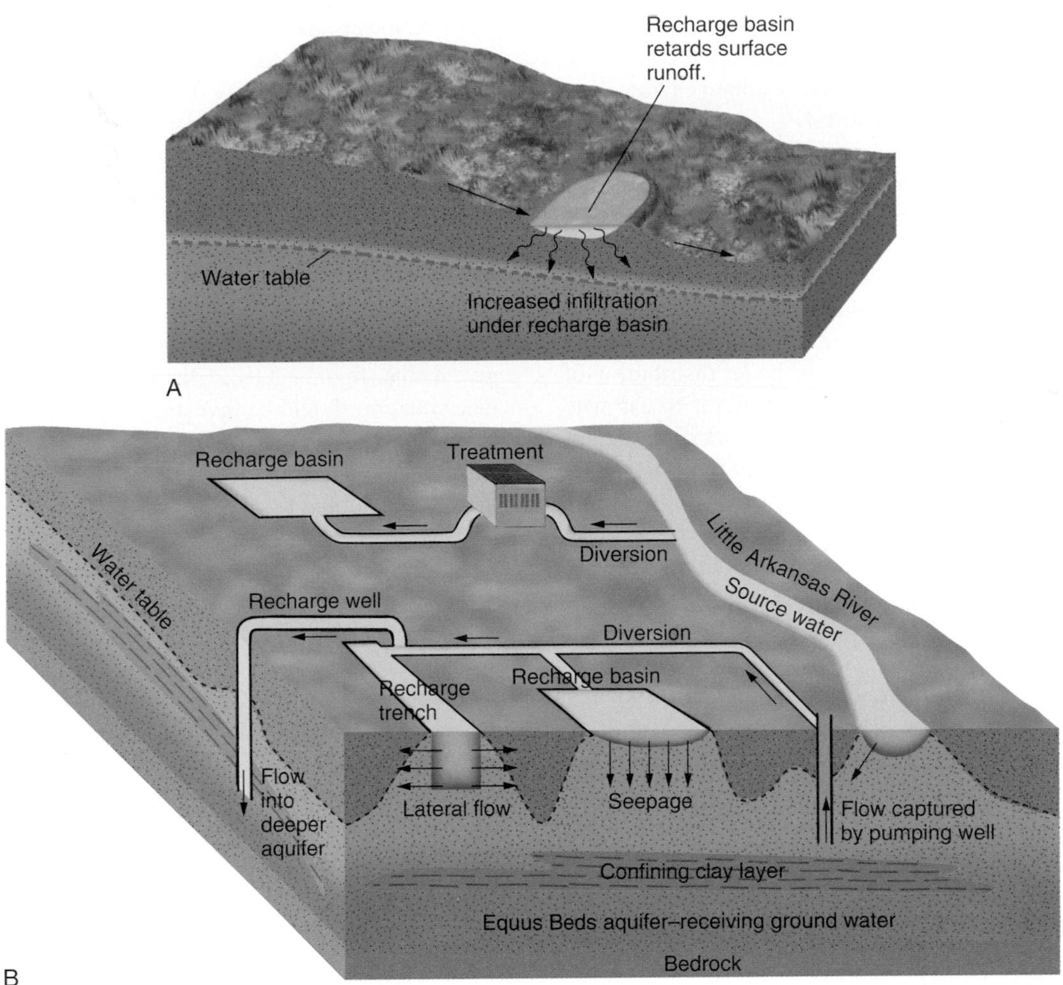

FIGURE 11.15 (A) Artificial recharge basins can aid recharge by slowing surface runoff. (B) USGS artificial-recharge demonstration project near Wichita, Kansas, is designed to stockpile water to meet future needs.
(B) after U.S. Geological Survey

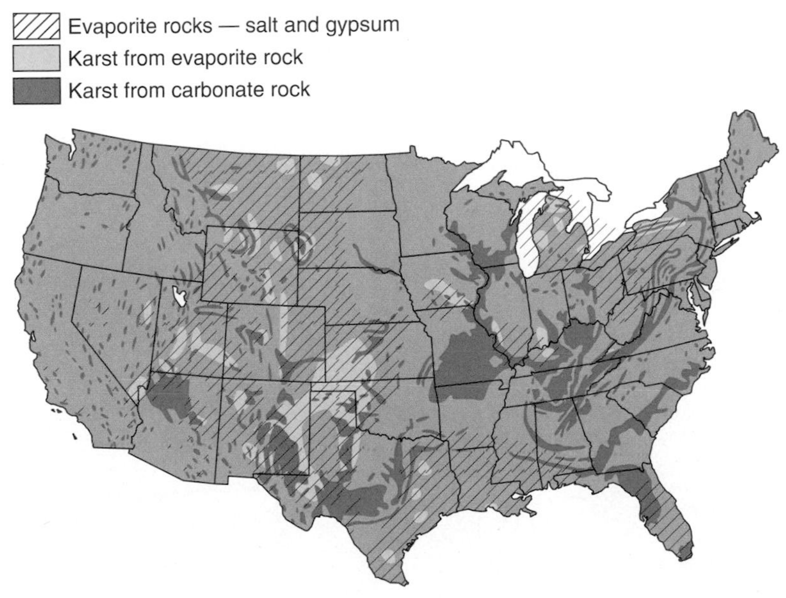

FIGURE 11.16 Highly soluble rocks underlie more than half of the contiguous United States.
From USGS Fact Sheet 165-00

common is that they are extremely soluble in water. Dissolution of these rocks by subsurface water, and occasional collapse or subsidence of the ground surface into the resultant cavities, creates a distinctive terrain known as **karst.**

Underground, the solution process forms extensive channels, voids, and even large caverns (figure 11.17). This creates large volumes of space for water storage and, where the voids are well interconnected, makes rapid groundwater movement possible. Karst aquifers, then, can be sources of plentiful water; in fact, 40% of the ground water used in the United States for drinking water comes from karst aquifers. However, the news in this regard is not all good. First, the irregular distribution of large voids and channels makes it more difficult to estimate available water volume or to determine groundwater flow rates and paths. Second, because the water flow is typically much more rapid than in non-karst aquifers, recharge and discharge occur on shorter timescales, and water supply may be less predictable. Contaminants can spread rapidly to pollute the water in karst aquifer systems, and the natural filtering that occurs in many non-karst aquifers with finer pores and passages may not occur in karst systems.

Over long periods of time, underground water dissolving large volumes of soluble rocks, slowly enlarging underground caverns, can also erode support for the land above. There may be no obvious evidence at the surface of what is taking place until the ground collapses abruptly into the void, producing a **sinkhole** (figures 11.17, 11.18).

The collapse of a sinkhole may be triggered by a drop in the water table as a result of drought or water use that leaves rocks and soil that were previously buoyed up by water pressure unsupported. Failure may also be caused by the rapid input of large quantities of water from heavy rains that wash overlying soil down into the cavern, as happened in late 2004 when hurricanes drenched Florida, or by an increase in subsurface water flow rates.

Sinkholes come in many sizes. The larger ones are quite capable of swallowing many houses at a time: They may be over 50 meters deep and cover several tens of acres. If one occurs in a developed area, a single sinkhole can cause millions of dollars in property damage. Sudden sinkhole collapses beneath bridges, roads, and railways have occasionally caused accidents and even deaths.

Sinkholes are rarely isolated phenomena. As noted, limestone, gypsum, and salt beds deposited in shallow seas can extend over broad areas. Therefore, where there is one sinkhole in a region underlain by such soluble rocks, there are likely to be others. An abundance of sinkholes in an area is a strong hint that more can be expected. Often, the regional topography will give a clue (figure 11.19). Clearly, in such an area, the subsurface situation should be investigated before buying or building a home or business. Circular patterns of cracks on the ground or conical depressions in the ground surface may be early signs of trouble developing below.

WATER QUALITY

As noted earlier, most of the water in the hydrosphere is in the very salty oceans, and almost all of the remainder is tied up in ice. That leaves relatively little surface or subsurface water for potential freshwater sources. Moreover, much of the water on and in the continents is not strictly fresh. Even rainwater, long the standard for "pure" water, contains dissolved chemicals of various kinds, especially in industrialized areas with substantial air pollution. Once precipitation reaches the ground, it reacts with soil, rock, and organic debris, dissolving still more chemicals naturally, aside from any pollution generated by human activities. Water quality thus must be a consideration when evaluating water supplies.

Pollution of both surface water and ground water, and some additional measures of water quality used particularly in describing pollution, will be discussed in chapter 17. Here, we focus on aspects of natural water quality.

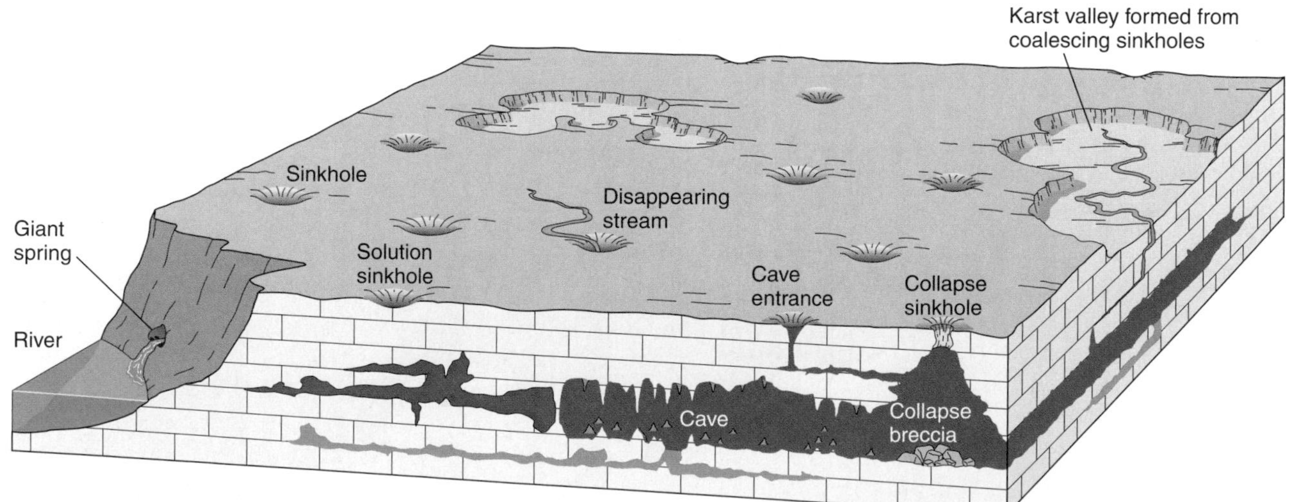

FIGURE 11.17 Dissolution of soluble rocks underground creates large voids below the surface and, often, leads to surface subsidence above.

FIGURE 11.18 Sinkhole at Winter Park, Florida. Collapse on May 8–9, 1981 was caused in part by drought. The sinkhole was over 100 meters across and 30 meters deep; estimated damage, over $2 million. Note cars for scale.

Photograph courtesy USGS Water Resources Division

A

B

FIGURE 11.19 (A) Sinkholes developed over limestone, near Timaru, New Zealand. (B) Satellite image of karst topography in eastern Florida, showing many round lakes formed in sinkholes. (Red is false color, and corresponds to green vegetation.)

(A) © G. R. "Dick" Roberts/Natural Sciences Image Library. (B) Photograph courtesy NASA

Measures of Water Quality

Water quality may be described in a variety of ways. A common approach is to express the amount of a dissolved chemical substance present as a concentration in parts per million (ppm) or, for very dilute substances, parts per billion (ppb). These units are analogous to weight percentages (which are really "parts per hundred") but are used for lower (more dilute) concentrations. For example, if water contains 1 weight percent salt, it contains one gram of salt per hundred grams of water, or one ton of salt per hundred tons of water, or whatever unit one wants to use. Likewise, if the water contains only 1 ppm salt, it contains one gram of salt per million grams of water (about 250 gallons), and so on. For comparison, the most abundant dissolved constituents in seawater can be measured in parts per thousand (magnesium, sulfate) or even percent (sodium, chloride).

Another way to express overall water quality is in terms of *total dissolved solids* (TDS), the sum of the concentrations of all dissolved solid chemicals in the water. How low a level of TDS is required or acceptable varies with the application. Standards might specify a maximum of 500 or 1000 ppm TDS for drinking water; 2000 ppm TDS might be acceptable for watering livestock; industrial applications where water chemistry is important (in pharmaceuticals or textiles, for instance) might need water even purer than normal drinking water.

Yet describing water in terms of total content of dissolved solids does not present the whole picture: At least as important as the quantities of impurities present is *what* those impurities are. If the main dissolved component is calcite (calcium carbonate) from a limestone aquifer, the water may taste fine and be perfectly wholesome with well over 1000 ppm TDS in it. If iron or sulfur is the dissolved substance, even a few parts per million may be enough to make the water taste bad, though it may not be actually unhealthful. Many synthetic chemicals that have leaked into water through improper waste disposal are toxic even at concentrations of 1 ppb or less.

Other parameters also may be relevant in describing water quality. One is pH, which is a measure of the acidity or alkalinity of the water. The pH of water is inversely related to acidity: the lower the pH, the more acid the water. (Water that is neither acid nor alkaline has a pH of 7.) For health reasons, concentrations of certain bacteria may also be monitored in drinking-water supplies.

A water-quality concern that has only recently drawn close attention is the presence of naturally occurring radioactive elements that may present a radiation hazard to the water consumer. Uranium, which can be found in most rocks, including those serving commonly as aquifers, decays through a series of steps. Several of the intermediate decay products pose special hazards. One—radium—behaves chemically much like calcium and therefore tends to be concentrated in the body in bones and teeth; see Box 11.1, "What's in the Water?" Another—radon—is a chemically inert gas but is radioactive itself and decays to other radioactive elements in turn. Radon leaking into indoor air from water supplies contributes to indoor air pollution (see chapter 18). High concentrations of radium and/or radon in ground water may result from decay of uranium in the aquifer itself or, in the case of radon, from seepage out of adjacent uranium-rich aquitards, especially shales. Other elements emerging as water-quality concerns—arsenic and mercury—are discussed in chapter 17.

Hard Water

Aside from the issue of health, water quality may be of concern because of the particular ways certain dissolved substances alter water properties. In areas where water supplies have passed through soluble carbonate rocks, like limestone, the water may be described as "hard." **Hard water** simply contains substantial amounts of dissolved calcium and magnesium. When calcium and magnesium concentrations reach or exceed the range of 80 to 100 ppm, the hardness may become objectionable.

Perhaps the most irritating routine problem with hard water is the way it reacts with soap, preventing the soap from lathering properly, causing bathtubs to develop rings and laundered clothes to retain a gray soap scum. Hard water or water otherwise high in dissolved minerals may also leave mineral deposits in plumbing and in appliances, such as coffeepots and steam irons. Primarily for these reasons, many people in hard-water areas use water softeners, which remove calcium, magnesium, and certain other ions from water in exchange for added sodium ions. The sodium ions are replenished from the salt (sodium chloride) supply in the water softener. (While softened water containing sodium ions in moderate concentration is unobjectionable in taste or household use, it may be of concern to those on diets involving restricted sodium intake.) The "active ingredient" in water softeners is a group of hydrous silicate minerals known as *zeolites.* The zeolites have an unusual capacity for *ion exchange,* a process in which ions loosely bound in the crystal structure can be exchanged for other ions in solution.

Overall, groundwater quality is highly variable. It may be nearly as pure as rainwater or saltier than the oceans. Some representative analyses of different waters in the hydrosphere are shown in table 11.2 for reference.

TABLE 11.2

Concentrations of Some Dissolved Constituents in Rain, River Water, and Seawater

	Concentration (ppm)						
		Rainwater		River Water			
Constituent	1-year avg., Mid-Atlantic States	Single storm, California	1-year avg., Inland U.S.	Amazon River	Mississippi River	World avg. (est.)	Average Seawater
silica (SiO_2)	—	0.3	—	7.0	6.7	13	6.4
calcium (Ca)	0.65	0.8	1.41	4.3	42	15	400
sodium (Na)	0.56	9.4	0.42	1.8	25	6.3	10,500
potassium (K)	0.11	0.0	—	—	2.9	2.3	380
magnesium (Mg)	0.14	1.2	—	1.1	12	4.1	1350
chloride (Cl)	0.57	17	0.22	1.9	30	7.8	19,000
fluoride (F)	—	—	—	0.2	0.2	—	1.3
sulfate (SO_4)	2.18	7.6	2.14	3.0	56	11	2700
bicarbonate (HCO_3)	—	4	—	19	132	58	142
nitrate (NO_3)	0.62	0	—	0.1	2.4	1	0.5

Source: J. D. Hem, *Study and Interpretation of the Chemical Characteristics of Natural Water,* U.S. Geological Survey Water-Supply Paper 1473, 1970, pp. 11, 12, and 50. (Ground water composition is so variable that representative analyses cannot be determined.)

Box 11.1 What's in the Water?

David, fraternity president, was scanning the house's mail. The item from the water department he assumed was going to tell them that water rates were going up—but that wasn't it. Startled, he began to read aloud to the brothers lounging around the house living room.

"Water Update. The City Public Water Supply wishes to advise its customers that the maximum allowable concentration for radium in drinking water has been exceeded during the past year. The maximum allowable concentration for radium, as determined by the state Pollution Control Board, is 5 picocuries per liter of water. Samples taken over the past year indicate the average level to be 7.2 picocuries per liter. A portion of the radium, which is ingested, remains in the bone. The radiation given off by the radium, because of its high energy, causes damage to surrounding tissue. A dose of 5 picocuries per liter may result in the development of bone cancer in a very small portion of the population. At the levels detected, the short-term risk is minimal and no special precautions need to be taken by the consumer at this time—"

Before he could read any further, David was interrupted by a stream of questions. "Radium? In our water?" "Bone cancer? And we're not supposed to worry?" "What's a picocurie??" "So what are they *doing* about it?"

The last one David could answer, at least partially. The notice went on to discuss how the city was going to drill some new wells into different aquifers with water having lower radium concentrations and how it was also going to do some kind of water treatment—"ion exchange softening," according to the notice.

"Well, that makes sense," observed Marcus, the chemistry major, who had walked in on the tail end of the discussion.

"Marcus! Just the man we need. Does this really make sense to you?"

By the time Marcus had finished explaining, the brothers were reassured. It was only after he'd turned in for the night that Adam, who hadn't been worried because he preferred bottled water anyhow, suddenly started to wonder, "Yes, but I *shower* in that city water every day! Should I give that up, too?"

Marcus had started out with some of the basic terms. A *curie* is a unit of radioactivity (named after Marie Curie, who first isolated radium from uranium ore). *Pico-* is a Greek prefix, like "milli-" or "micro-," but for even smaller fractions; a picocurie equals one millionth of one millionth of a curie (or in scientific notation, 10^{-2} curies). Translated to radioactive decay rate, 5 picocuries per liter means that the amount of radium in the water is enough that about 10 atoms of radium decay every minute per liter (about a quart) of water.

Pulling out an introductory chemistry book, Marcus had then explained the bone connection. In the periodic table (described in chapter 2), elements are arranged in a way that reflects similarities of chemical behavior, which vary in a regular way with atomic number. In particular, elements in the same column tend to behave similarly, because they have similar electronic structures. The elements in the second column (figure 1), known collectively as the alkaline-earth elements, have two electrons in the outermost shell of the neutral atom. Thus, they all tend to form cations of +2 charge by losing those two electrons, and they tend to substitute for each other in minerals and other chemical compounds. In this column, with calcium (Ca), are strontium (Sr) and radium (Ra). Bones and teeth are built of calcium-rich compounds. In the mid-1900s, fallout from atmospheric nuclear-weapons tests included radioactive isotopes of Sr; grazing cows concentrated some of that Sr, along with Ca, in their milk, which then could contribute Sr to the bones and teeth of children drinking lots of milk. Radium in drinking water may likewise be concentrated in the body in bones and teeth, where there is some potential for damage from radiation as the radium decays. As Marcus pointed out, though, not only are we all surrounded by radiation—cosmic rays from space, natural radioactivity in rocks and soils around us—we, and all our foods and the air we breathe, are also radioactive, for all contain carbon-14 (and humans contain a number of other radioactive isotopes, as well). So it's hardly that we live in a radiation-free world in the first place.

Marcus had gone on to point out that the fact that radium behaves like calcium can help in addressing this particular issue, too. Water softeners remove from water not only much of the calcium and magnesium (which is also in the second column of the periodic table), but of radium as well. Therefore, the tap water actually drunk in the house, which was softened, would contain far less radium than the city's figures.

And the next day, Marcus was also able to reassure Adam that yes, he *definitely* ought to keep taking showers. The body absorbs nutrients through the digestive system, not the skin. Just as taking a milk bath wouldn't do a thing for adding calcium to his bones, so the water in the shower streaming over him wouldn't put radium in his bones, even if the water hadn't been softened.

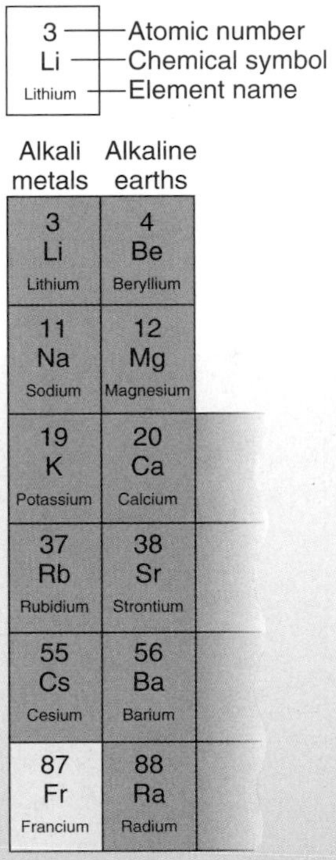

3 — Atomic number
Li — Chemical symbol
Lithium — Element name

Alkali metals	Alkaline earths
3 Li Lithium	4 Be Beryllium
11 Na Sodium	12 Mg Magnesium
19 K Potassium	20 Ca Calcium
37 Rb Rubidium	38 Sr Strontium
55 Cs Cesium	56 Ba Barium
87 Fr Francium	88 Ra Radium

FIGURE 1 Expanded view of left side of the periodic table.

WATER USE, WATER SUPPLY

General U.S. Water Use

Inspection of the U.S. water budget overall would suggest that ample water is available for use (figure 11.20). Some 4200 billion gallons of precipitation fall on this country each day; subtracting 2750 billion gallons per day lost to evapotranspiration still leaves a net of 1450 billion gallons per day for streamflow and groundwater recharge. Water-supply problems arise, in part, because the areas of greatest water availability do not always coincide with the areas of concentrated population or greatest demand, and also because a portion of the added fresh water quickly becomes polluted by mixing with impure or contaminated water.

People in the United States use a large amount of water. Biologically, humans require about a gallon of water a day per person, or, in the United States, about 300 million gallons per day for the country. Yet, Americans divert, or "withdraw," about 400 *billion* gallons of water each day—about 1800 gallons per person—for cooking, washing, and other household uses, for industrial processes and power generation, and for livestock and irrigation, a wide range of *"offstream"* water uses. Another several trillion gallons of water are used each day to power hydroelectric plants (*"instream"* use). Of the total water withdrawn, more than 100 billion gallons per day are *consumed,* meaning that the water is not returned as wastewater. Most of the consumed water is lost to evaporation; some is lost in transport (for example, through piping systems).

It might seem easier to use surface waters rather than subsurface waters for water supplies. Why, then, worry about using ground water at all? One basic reason is that, in many dry areas, there is little or no surface water available, while there may be a substantial supply of water deep underground. Tapping the latter supply allows us to live and farm in otherwise uninhabitable areas.

Then, too, streamflow varies seasonally. During dry seasons, the water supply may be inadequate. Dams and reservoirs allow a reserve of water to be accumulated during wet seasons for use in dry times, but we have already seen some of the negative consequences of reservoir construction (chapter 6). Furthermore, if a region is so dry at some times that dams and reservoirs are necessary, then the rate of water evaporation from the broad, still surface of the reservoir may itself represent a considerable water loss and aggravate the water-supply problem.

Precipitation (the prime source of abundant surface runoff) varies widely geographically (figure 11.21), as does population density. In many areas, the concentration of people far exceeds what can be supported by available local surface waters, even during the wettest season.

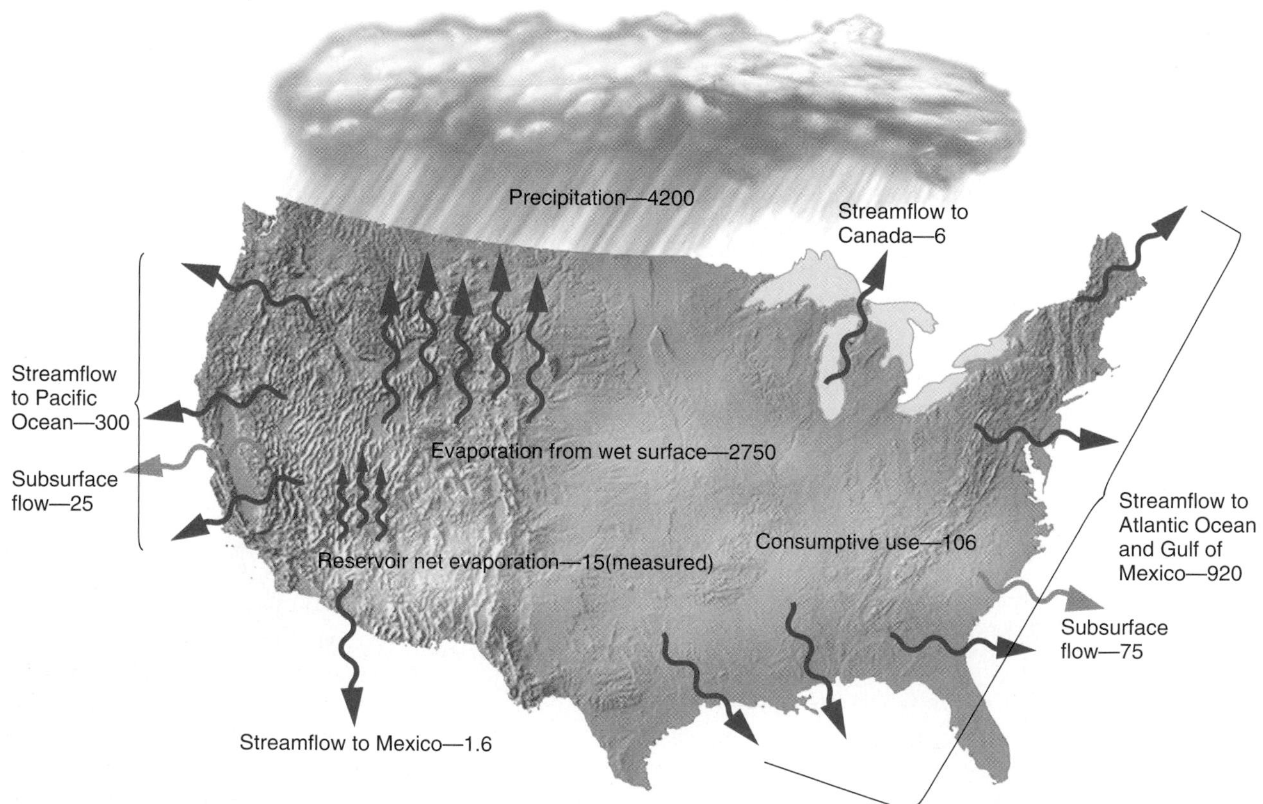

FIGURE 11.20 In terms of gross water flow, U.S. water budget seems ample. Figures are in billions of gallons per day.
Source: Data from The Nation's Water Resources 1975–2000, *U.S. Water Resources Council*

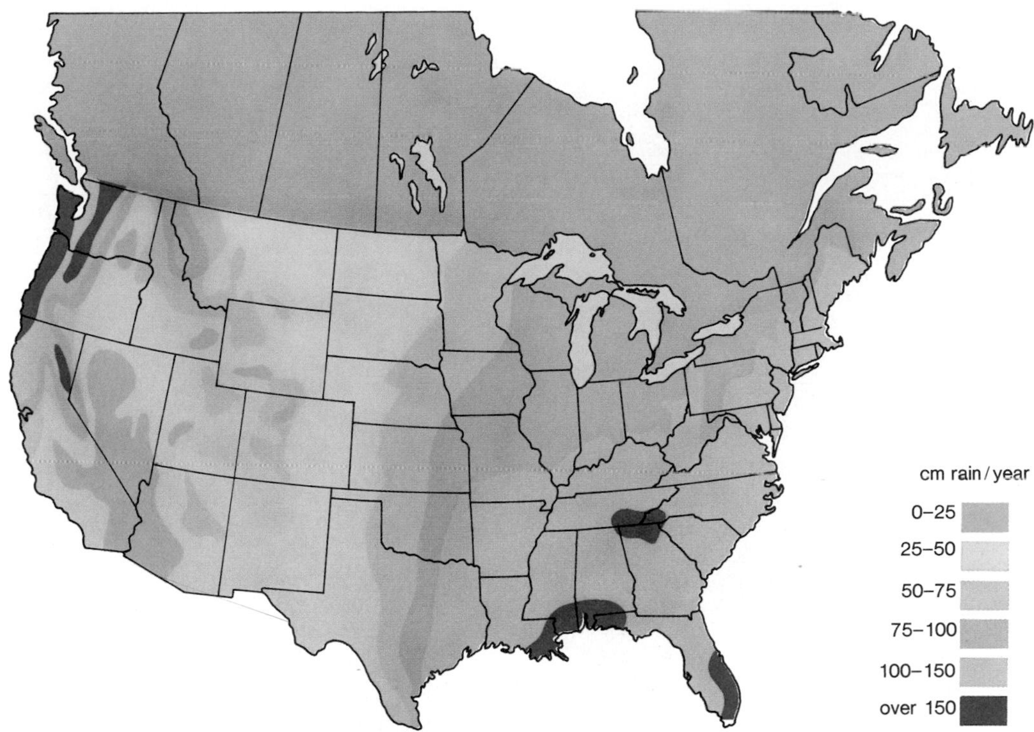

FIGURE 11.21 Average annual precipitation in the contiguous United States. (One inch equals 2.5 cm.)
Source: U.S. Water Resources Council

Also, streams and large lakes have historically been used as disposal sites for untreated wastewater and sewage, which makes the surface waters decidedly less appealing as drinking waters. A lake, in particular, may remain polluted for decades after the input of pollutants has stopped if there is no place for those pollutants to go or if there is a limited input of fresh water to flush them out. Ground water from the saturated zone, on the other hand, has passed through the rock of an aquifer and has been naturally filtered to remove some impurities—soil or sediment particles and even the larger bacteria—although it can still contain many dissolved chemicals (and in fact, once polluted, can remain so for some time, a problem explored further in chapter 17).

Finally, ground water is by far the largest reservoir of unfrozen fresh water. For a variety of reasons, then, underground waters may be preferred as a supplementary or even sole water source.

Regional Variations in Water Use

Water withdrawal varies greatly by region (figure 11.22), as does water consumption. These two quantities are not necessarily directly related, for the fraction of water withdrawn that is actually consumed depends, in part, on the principal purposes for which it is withdrawn. This, in turn, influences the extent to which local water use may deplete groundwater supplies. Of the fresh water withdrawn, only about 25% is ground water; but many of the areas of heaviest groundwater use rely on that water for irrigation, which consumes a large fraction of the water withdrawn.

Aside from hydropower generation, several principal categories of water use can be identified: municipal (public) supplies (home use and some industrial use in urban and suburban areas), rural use (supplying domestic needs for rural homes and watering livestock), irrigation (a form of rural use, too, but worthy of special consideration for reasons to be noted shortly), self-supplied industrial use (use other than for electricity generation, by industries for which water supplies are separate from municipal sources), and thermoelectric power generation (using fuel—coal, nuclear power, etc.—to generate heat). The quantities withdrawn for and consumed by each of these categories of users are summarized in figure 11.23. The sources and disposition of water for the four major water uses are shown in table 11.3.

A point that quickly becomes apparent from table 11.3 is that while power generation may be the major water *user,* agriculture is the big water *consumer.* Power generators and industrial users account for more than half the water withdrawn, but nearly all their wastewater is returned as liquid water at or near the point in the hydrologic cycle from which it was taken. Most of these users are diverting surface waters and dumping wastewaters (suitably treated, one hopes!) back into the same lake or stream. Together, industrial users and power generation *consume* only about 10 billion gallons per day, or 10% of the total.

A much higher fraction of irrigation water is consumed: lost to evaporation, lost through transpiration from plants, or

FIGURE 11.22 U.S. variations in water withdrawals by state. Compare regional patterns with figure 11.21.
Source: U.S. Geological Survey Water Resources Division, 1995 Estimated Water Use

TABLE 11.3

Source and Disposition of Water by Use, 1990

	Thermoelectric Power Generation	Irrigation	Industrial	Domestic
Withdrawal, millions of gallons per day	195,000	137,000	27,800	25,300
Source				
Surface water	100%	63%	67%	1%
Ground water		37%	14%	13%
Public supply			19%	86%
Disposition				
Consumption	2%	56%	15%	23%
Conveyance loss		20%		
Return flow	98%	24%	85%	77%

Data from U.S. Geological Survey.

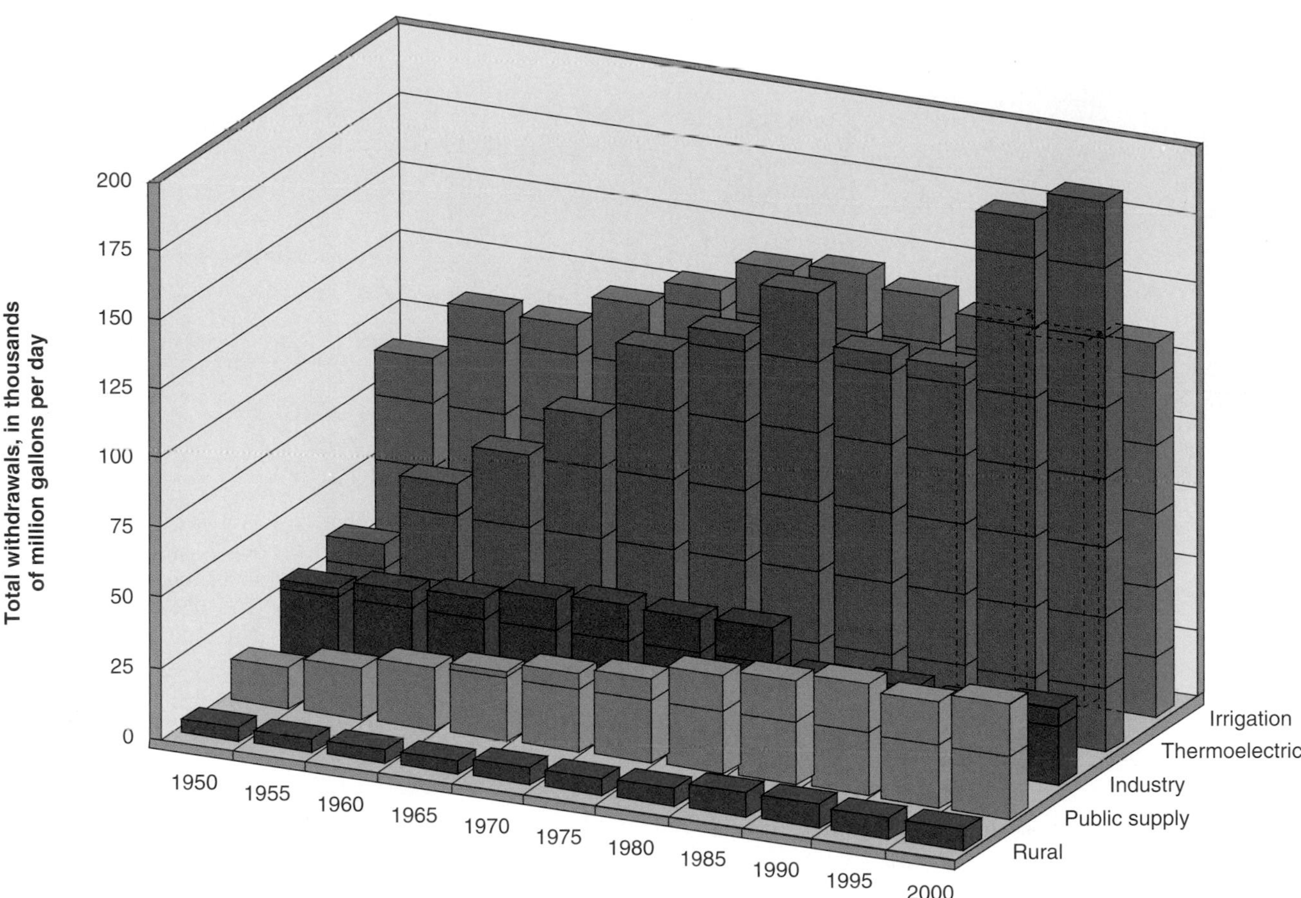

FIGURE 11.23 Water withdrawal by sector, 1950–2000.
Source: U.S. Geological Survey National Circular 1081 and 1995 Estimated Water Use

lost because of leakage from ditches and pipes. Moreover, close to 40% of the water used for irrigation is ground water. Most of the water lost to evaporation drifts out of the area to come down as rain or snow somewhere far removed from the irrigation site. It then does not contribute to the recharge of aquifers or to runoff to the streams from which the water was drawn.

The regional implications of all this can be examined through figures 11.22 and 11.24. The former indicates the relative volumes of water withdrawn, by state. The latter figure shows that many states with high overall water withdrawal use their water extensively for irrigation. Where irrigation use of water is heavy, water tables have dropped by tens of meters and streams have been drained nearly dry, while we have become increasingly dependent on the crops.

Nor is water supply an issue only in the western United States, by any means. It is an international political matter as well (table 11.4). In most of the cases cited, the central issues relate either to water availability or to the quality of the available water. One country's water-use preferences and practices can severely disadvantage another, especially if they share a common water source (figure 11.25).

CASE STUDIES IN WATER CONSUMPTION

There is no lack of examples of water-supply problems; they may involve lakes, streams, or ground water. Problems of surface-water supply may be either aggravated or alleviated (depending on location) as precipitation patterns shift in reflection of normal climate cycles and/or global change induced by human activities. Where ground water is at issue, however, it is often true that relief through recharge is not in sight, as modern recharge may be insignificant. Many areas are drawing on "fossil" groundwater recharged long ago—for example, the upper Midwest uses ground water recharged over 10,000 years ago as continental ice sheets melted. See also the areas

FIGURE 11.24 Water use for irrigation, by state.
Source: U.S. Geological Survey Water Resources Division

TABLE 11.4

International Water Disputes

River	Countries in Dispute	Issues
Nile	Egypt, Ethiopia, Sudan	Siltation, flooding, water flow/diversion
Euphrates, Tigris	Iraq, Syria, Turkey	Reduced water flow, salinization
Jordan, Yarmuk, Litani, West Bank aquifer	Israel, Jordan, Syria, Lebanon	Water flow/diversion
Indus, Sutlei	India, Pakistan	Irrigation
Ganges-Brahmaputra	Bangladesh, India	Siltation, flooding, water flow
Salween	Myanmar, China	Siltation, flooding
Mekong	Cambodia, Laos, Thailand, Vietnam	Water flow, flooding
Paraná	Argentina, Brazil	Dam, land inundation
Lauca	Bolivia, Chile	Dam, salinization
Rio Grande, Colorado	Mexico, United States	Salinization, water flow, agrochemical pollution
Rhine	France, Netherlands, Switzerland, Germany	Industrial pollution
Maas, Schelde	Belgium, Netherlands	Salinization, industrial pollution
Elbe	Czechoslovakia, Germany	Industrial pollution
Szamos	Hungary, Romania	Industrial pollution

Note that in most cases, central issues to the disputes involve adequacy and/or quality of water supply.

Source: Michael Renner, *National Security: The Economic and Environmental Dimensions,* Worldwatch Paper 89, p. 32. Copyright © 1989 Worldwatch Institute, Washington, D.C. www.worldwatch.org. Reprinted by permission.

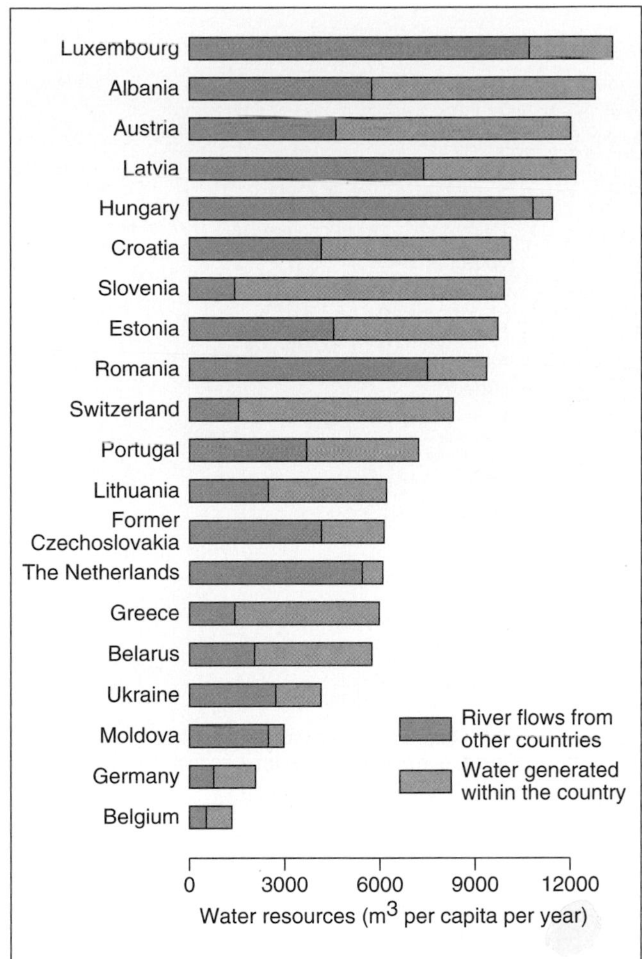

FIGURE 11.25 "Internal" and "external" contributions to renewable water resources in selected European countries. While some (e.g., Switzerland) control most of their own water resources, others (e.g., Hungary) do not.

Source: The Dobris Report, U.N. Environment Programme, figure 5.4

highlighted in the opening image for this section. Water availability and population distribution and growth together paint a sobering picture of coming decades in many parts of the world (figure 11.26).

The Colorado River Basin

The Colorado River's drainage basin drains portions of seven western states (figure 11.27). Many of these states have extremely dry climates, and it was recognized decades ago that some agreement would have to be reached about which region was entitled to how much of that water. Intense negotiations during the early 1900s led in 1922 to the adoption of the Colorado River Compact, which apportioned 7.5 million acre-feet of water per year each to the Upper Basin (Colorado, New Mexico, Utah, and Wyoming) and the Lower Basin (Arizona, California, and Nevada). (One acre-foot is the amount of water required to cover an area of 1 acre to a depth of 1 foot; it is more than 300,000 gallons.) No provision was made for Mexico, into which the river ultimately flows.

Rapid development occurred throughout the region. Huge dams impounded enormous reservoirs of water for irrigation or for transport out of the basin. In 1944, Mexico was awarded by treaty some 1.5 million acre-feet of water per year, but with no stipulation concerning water quality. Mexico's share was to come from the surplus above the allocations within the United States or, if that was inadequate, equally from the Upper and Lower Basins.

Heavy water use has led to a reduction in both water flow and water quality in the Colorado River. The reduced flow results not only from diversion of the water for use but also from large evaporation losses from the numerous reservoirs in the system. The reduced water quality is partly a consequence of that same evaporation, concentrating dissolved minerals, and of selective removal of fresh water. Also, many of the streams flow through soluble rocks, which then are partially dissolved, increasing the dissolved mineral load. By 1961, the water delivered to Mexico contained up to 2700 ppm TDS, and partial crop failures resulted from the use of such saline water for irrigation. In response to protests from the Mexican government, the United States agreed in 1974 to build a desalination plant at the U.S.-Mexican border to reduce the salinity of Mexico's small share of the Colorado.

It has now been shown that sufficient water flow simply does not exist in the Colorado River basin even to supply the full allocations to the U.S. users. The streamflow measurements made in the early 1900s were grossly inaccurate and apparently also were made during an unusually wet period. Estimates put the total flow in 1990 at only 15.6 million acre-feet, and of that, an estimated 2 million acre-feet are lost to evaporation, especially from reservoirs. The 1922 agreement and Mexican treaties still work tolerably well only because the Upper Basin states are using little more than half their allotted water. The dry Lower Basin continues to need ever more water as its population grows. Arizona has already gone to court against California over the issue of which state is entitled to how much water; the consequence is that California has been compelled to *reduce* its use of Colorado River water. And California has grown very dependent on that water: The Colorado River Aqueduct, completed in 1941, alone carries over 1.2 million acre-feet of water per year from Lake Havasu to supply Los Angeles and San Diego, and California's 4.4 million acre-feet is already the lion's share of the Lower Basin's allocation.

Moreover, the possibility of greatly increased future water needs looms for another reason: energy. The western states contain a variety of energy resources, conventional and new. Chief among these are coal and oil shale. Extraction of fuel from oil shale is, as we will see in chapter 14, a water-intensive process. In addition, both the coal and oil shale are very likely to be strip-mined, and current laws require reclamation of strip-mined land. Abundant water will be needed to reestablish vegetation on the reclaimed land. Already, energy interests have water-rights claims to more than 1 million acre-feet of water in the Colorado oil shale area alone. The streamflow there, as

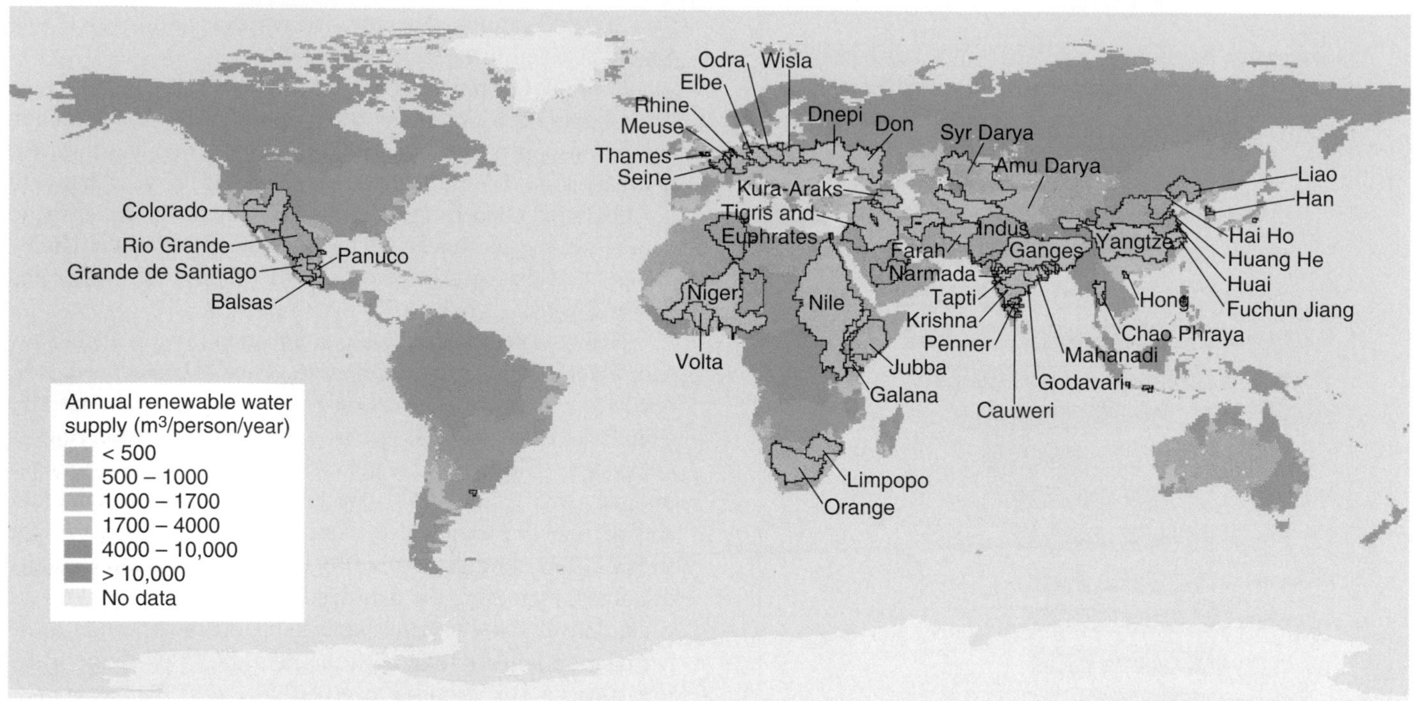

FIGURE 11.26 World Resources Institute projections suggest inadequate water for many by 2025. Outlined river basins have populations over 10 million. An adequate supply of water is considered to be at least 1700 cubic meters per person per year. Where renewable supplies are less than this, the population may be draining nonrenewable supplies—or suffering without.
Source: Earth Trends 2001, *World Resources Institute*

FIGURE 11.27 The Colorado River Basin.
Source: U.S. Geological Survey Water Supply Circular 1001

in many other parts of the Colorado River basin, is already overappropriated. The crunch will get worse if more claims are fully utilized and additional claims filed with the development of these energy resources. The unconventional fuels have so far remained uneconomic to exploit; but the potential (and associated potential water demand) is still there.

The High Plains (Ogallala) Aquifer System

The Ogallala Formation, a sedimentary aquifer, underlies most of Nebraska and sizeable portions of Colorado, Kansas, and the Texas and Oklahoma panhandles (figure 11.28). The most productive units of the aquifer are sandstones and gravels. Farming in the region accounts for about 25% of U.S. feed-grain exports and 40% of wheat, flour, and cotton exports. More than 14 million acres of land are irrigated with water pumped from the Ogallala, and with good reason: yields on irrigated land may be triple the yields on similar land cultivated by dry farming (no irrigation).

The Ogallala's water was, for the most part, stored during the retreat of the Pleistocene continental ice sheets. Present recharge is negligible over most of the region. The original groundwater reserve in the Ogallala is estimated to have been approximately 2 billion acre-feet. But each year, farmers draw from the Ogallala more water than the entire flow of the Colorado River.

Soon after development in the plains began, around 1940, the water table began to decline noticeably. In areas of heavy use—Texas, southwestern Kansas, the Oklahoma panhandle—it had dropped over 100 feet by 1980. That observation, together with simple arithmetic comparing the total water reserve with current withdrawal rates (and considering that the saturated thickness is less than 200 feet over two-thirds of the aquifer's

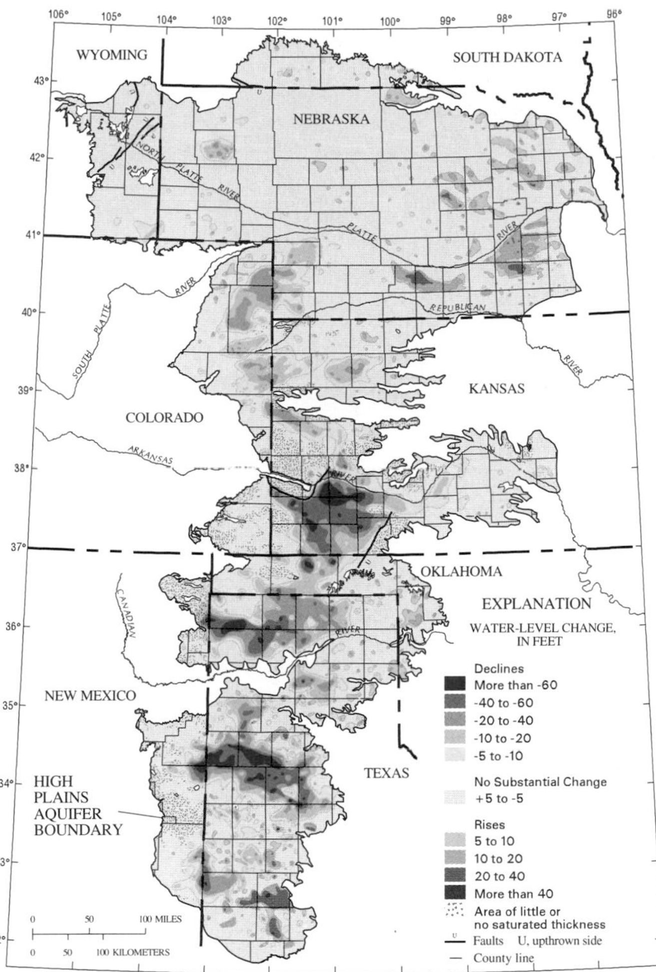

FIGURE 11.28 Changes in water levels in the Ogallala Formation, 1980–1997. Only in a few places, such as along the Platte River, has the water table risen significantly, in response to short-term precipitation increases; the overall long-term trend is definitely down.

Image from V. L. McGuire, 2004, Water-level changes in the High Plains aquifer, pre-development to 2002, 1980 to 2002, and 2001 to 2002: U.S. Geological Survey Fact Sheet 2004-3026, p. 6

area), suggested that depletion within a matter of decades was possible. What then?

Reversion to dry farming, where possible at all, would greatly diminish yields. Reduced vigor of vegetation could lead to a partial return to preirrigation, Dust-Bowl-type conditions. Alternative local sources of municipal water are in many places not at all apparent. Planners in Texas and Oklahoma have advanced ambitious water-transport schemes as solutions. Each such scheme would cost billions of dollars, perhaps tens of billions. Water-transport systems of this scale could take a decade or longer to complete from the time the public commits itself to the projects. Even if and when the transport networks are finished, the cost of the water might be ten times what farmers in the region can comfortably afford to pay if their products are to remain fully competitive in the marketplace. Meanwhile, the draining of the Ogallala Formation continues day by day.

It is noteworthy that there has been little incentive in the short term for the exercise of restraint and that this is due in large measure to federal policies. Price supports encourage the growing of crops like cotton that require irrigation in the southern part of the region. The government cost-shares in soil-conservation programs and provides for crop-disaster payments, thus giving compensation for the natural consequences of water depletion. And, in fact, federal tax policy provides for groundwater depletion allowances (tax breaks) for High Plains area farmers using pumped ground water, with larger breaks for heavier groundwater use. The results of all these policies aren't surprising.

Still, water-resource professionals are convincing more and more of those dependent on the Ogallala to conserve. Federal, state, and local water-resource specialists are working with farmers and communities to slow water depletion. More-efficient irrigation technology and timing have somewhat reduced water use. Some Ogallala users are trying more creative measures: For example, some farmers in Kansas and Nebraska pump fresh ground water from their land for local towns, which return their treated wastewater to irrigate the fields. Already, depletion has greatly slowed in the northern part of the region, aided, in part, by unusually heavy precipitation in the late 1990s. Over the longer term, further emphasis on conservation, plus modifications of existing laws that promote groundwater use, could substantially extend the useful life of the Ogallala aquifer.

The Aral Sea

The Aral Sea lies on the border of Kazakhstan and Uzbekistan. For decades, water from rivers draining into the Aral Sea has been diverted to irrigate land for growing rice and cotton. Clearly, the depletion cannot continue at comparable rates. From 1973 to 1987, the Aral Sea dropped from fourth to sixth largest lake in the world. Its area has shrunk by more than half, its water volume by over 60% (figure 11.29). Its salinity has increased from 10% to over 23%, and a local fishing industry has failed; the fish could not adapt to the rapid rise in salt content. As the lake has shrunk, former lake-bed sediments—some containing residues of pesticides washed off of the farmland—and salt have been exposed. Meanwhile, the climate has become dryer, and dust storms now regularly scatter the salt and sediment over the region. Ironically, the salt deposited on the fields by the winds is reducing crop yields. Respiratory ailments are also widespread as the residents breathe the pesticide-laden dust. Here, then, there are serious issues even beyond those of water supply, which are acute enough.

Lake Chad

Lake Chad, on the edge of the Sahara Desert in West Africa, is another example of a disappearing lake. It was once one of the largest bodies of fresh water in Africa, close in area to Lake Erie. But like Lake Erie, it was relatively shallow—Lake Chad, indeed, was only 5 to 8 meters deep before the depletion of recent decades—so the volume of water was modest from the outset. A combination of skyrocketing demand for irrigation

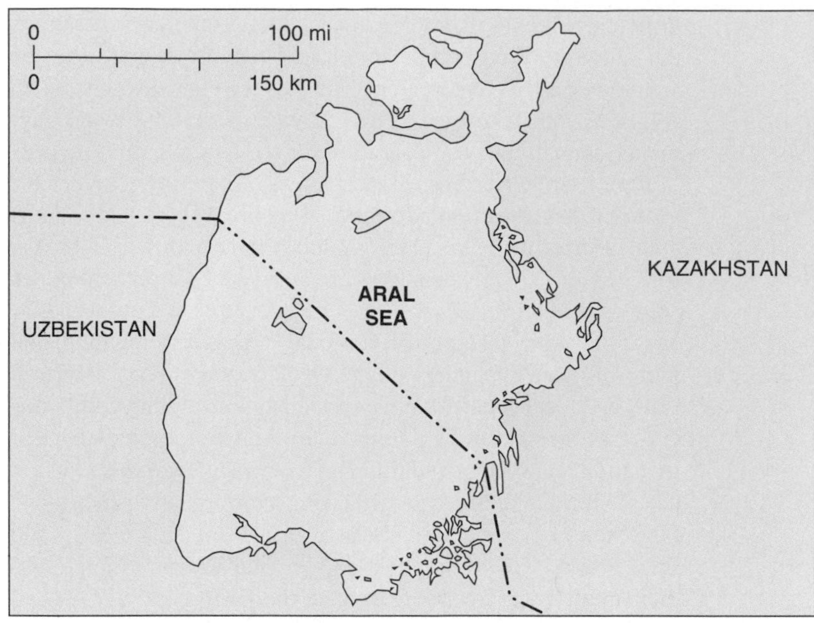

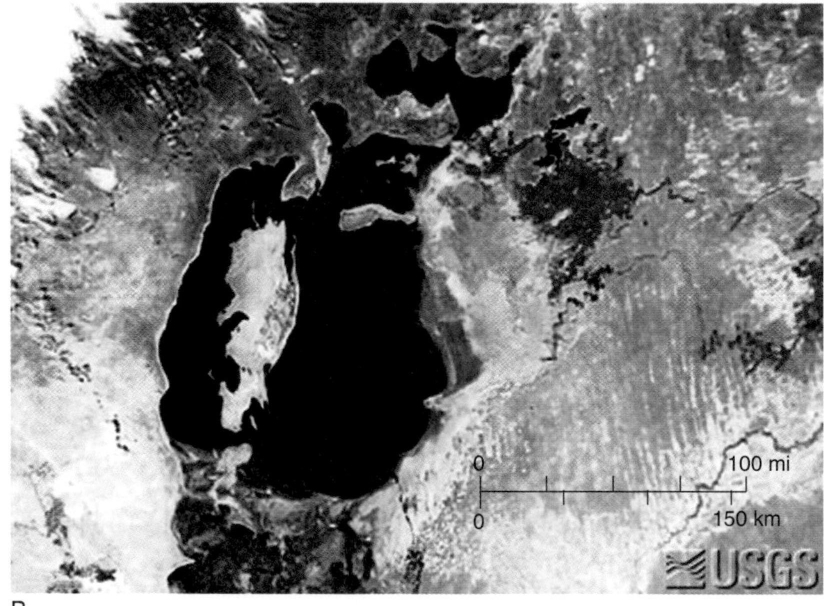

FIGURE 11.29 The shrinking of the Aral Sea. (A) Map shows the extent of the lake in 1973. (B) By 1997, shrinkage is obvious in this satellite image, by comparison with (A). *(B) Image courtesy U.S. Geological Survey*

water from the four countries bordering the lake (Chad, Niger, Nigeria, and Cameroon), plus several decades of declining rainfall, have shrunk Lake Chad to about one-twentieth its former size (figure 11.30). While it still supports vegetation in its former lake bed, the volume of useable water remaining in Lake Chad is severely limited.

EXTENDING THE WATER SUPPLY

Conservation

The most basic approach to improving the U.S. water-supply situation is conservation. Water is wasted in home use every day—by long showers; inefficient plumbing; insistence on lush, green lawns even in the heat of summer; and in dozens of other ways. Raising livestock for meat requires far more water per pound of protein than growing vegetables for protein. Still, municipal and rural water uses (excluding irrigation) together account for only about 10% of total U.S. water consumption.

The big water drain is plainly irrigation, and that use must be moderated if the depletion rate of water supplies is to be reduced appreciably. For example, the raising of crops that require a great deal of water could be shifted, in some cases at least, to areas where natural rainfall is adequate to support them. Irrigation methods can also be made more efficient so that far less water is lost by evaporation. This can be done, for instance,

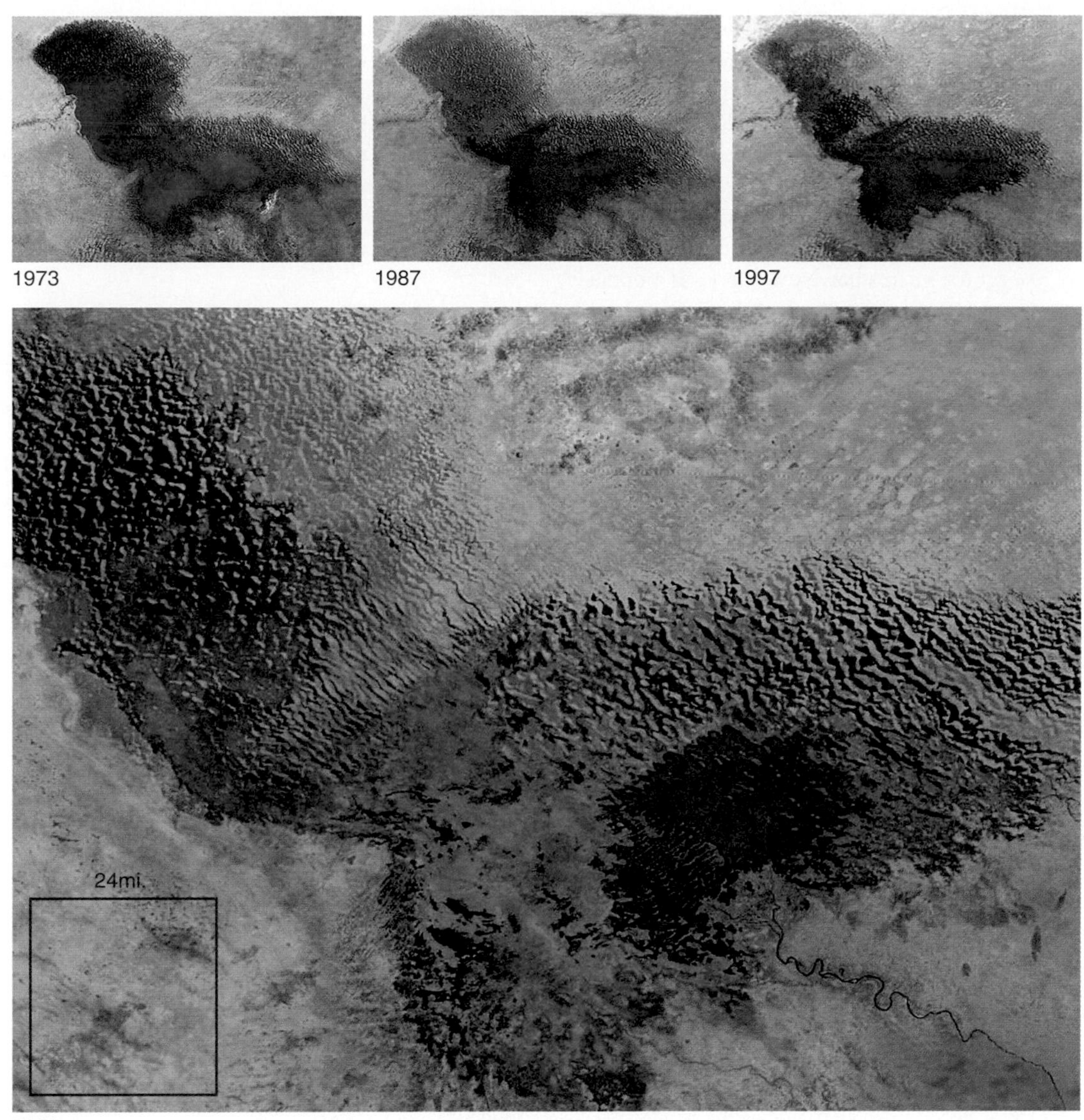

FIGURE 11.30 Africa's Lake Chad, vanishing right before our eyes. Top three images are false-color composites, in which the red area around the (blue) lake is vegetation; vegetation is green in bottom (2001) image.
Images courtesy NASA/Goddard Space Flight Center Scientific Visualization Studio

by drip irrigation. Instead of running irrigation water in open ditches from which evaporation loss is high, the water can be distributed via pipes with tiny holes from which water seeps slowly into the ground at a rate more closely approaching that at which plants use it. The more efficient methods are often considerably more expensive, too. However, they have become more attractive as water prices have been driven up by shortages.

Domestic use can be reduced in a variety of ways. For example, lawns can be watered morning or evening when evaporation is less rapid than at midday; or one can forgo traditional lawns altogether in favor of ground covers that don't need watering. Stormwater can be directed into recharge basins rather than dumped into sewer systems. Increasingly, municipalities in dry areas are looking to recycle their wastewater (discussed further in chapter 16).

Interbasin Water Transfer

In the short term, conservation alone will not resolve the imbalance between demand and supply. New sources of supply are needed. Part of the supply problem, of course, is purely local. For example, people persist in settling and farming in areas that may not be especially well supplied with fresh water, while other areas with abundant water go undeveloped. If the people cannot be persuaded to be more practical, perhaps the water can be redirected. This is the idea behind interbasin transfers—moving surface waters from one stream system's drainage basin to another's where demand is higher.

California pioneered the idea with the Los Angeles Aqueduct. The aqueduct was completed in 1913 and carried nearly 150 million gallons of water per day from the eastern slopes of the Sierra Nevada to Los Angeles. In 1958, the system was expanded to bring water from northern California to the southern part of the state. More and larger projects have been undertaken since. Bringing water from the Colorado River to southern coastal California, for example, required the construction of over 300 kilometers (200 miles) of tunnels and canals. Other water projects have transported water over whole mountain ranges. It should be emphasized, too, that such projects are not confined to the drier west: for example, New York City draws on several reservoirs in upstate New York. If population density or other sources of water demand are high, a local supply shortfall can occur even in an area thought of as quite moist.

Dozens of interbasin transfers of surface water have been proposed. Political problems are common even when the transfer involves diverting water from one part of a single state to another. In 1982, it was proposed to expand the aqueduct system to carry water from northern California to the south; 60% of the voters in the southern part of the state were in favor, but 90% of those in the north were opposed, and the proposition lost. The opposition often increases when transfers among several states are considered. In the 1990s, officials in states around the Great Lakes objected to a suggestion to divert some lake water to states in the southern and southwestern United States. The problems may be far greater when transfers between nations are involved. Various proposals have been made to transfer water from little-developed areas of Canada to high-demand areas in the United States and Mexico. Such proposals, which could involve transporting water over distances of thousands of kilometers, are not only expensive (one such scheme, the North American Water and Power Alliance, had a projected price of $100 billion), they also presume a continued willingness on the part of other nations to share their water. Sometimes, too, the diversion, in turn, causes problems in the region from which the water is drawn (figure 11.31).

Desalination

Another alternative for extending the water supply is to improve the quality of waters not now used, purifying them sufficiently to make them usable. Desalination of seawater, in particular, would allow parched coastal regions to tap the vast ocean reservoirs. Also, some ground waters are not presently used for water supplies because they contain excessive concentrations of dissolved materials. There are two basic methods used to purify water of dissolved minerals: filtration and distillation (figure 11.32).

In a filtration system, the water is passed through fine filters or membranes to screen out dissolved impurities. An advantage of this method is that it can rapidly filter great quantities of water. A large municipal filtration operation may produce several billion gallons of purified water per day. A disadvantage is that the method works best on water not containing very

FIGURE 11.31 Diversion of surface water that formerly flowed into Mono Lake, California, has caused lake level to drop, exposing sedimentary formations originally built underwater by reaction of lake water with spring water seeping up from below. The good news is that steps have recently been taken to reduce diversion, and the lake level is beginning to rise again, slowly.

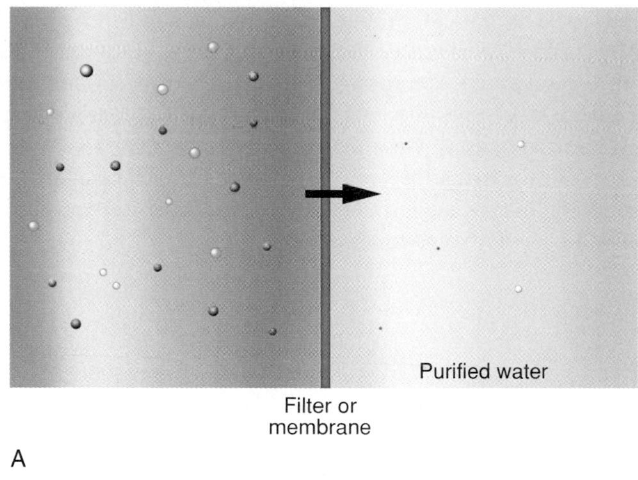

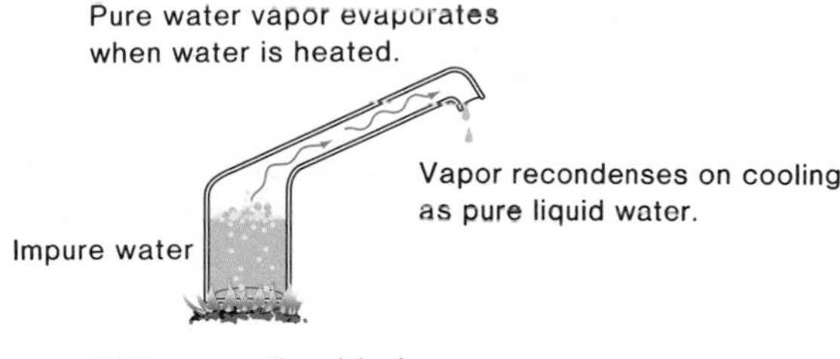

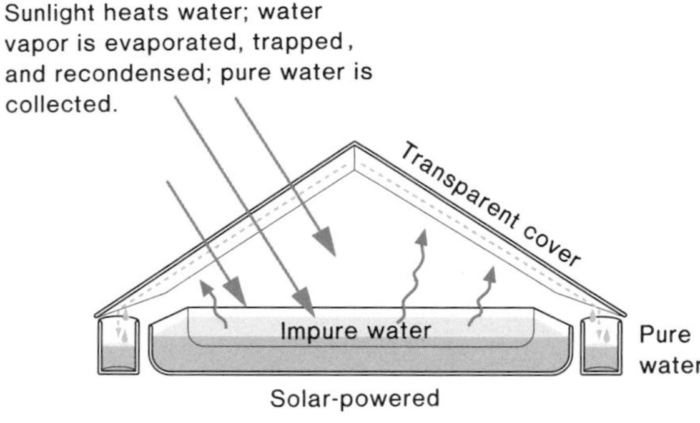

FIGURE 11.32 Methods of Desalination. (A) Filtration (simplified schematic): Dissolved and suspended material (represented by dots at left) is screened out by very fine filters. (B) Distillation: As water is heated, pure water is evaporated, then recondensed for use. Dissolved and suspended materials stay behind.

high levels of dissolved minerals. Pumping anything as salty as seawater through the system quickly clogs the filters. This method, then, is most useful for cleaning up only moderately saline ground waters or lake or stream water.

Distillation involves heating or boiling water full of dissolved minerals. The water vapor driven off is pure water, while the minerals stay behind in what remains of the liquid. Because this is true regardless of how concentrated the dissolved minerals are, the method works fine on seawater as well as on less saline waters.

A difficulty, however, is the nature of the necessary heat source. Furnaces fired by coal, gas, or other fuels can be used, but any fuel may be costly in large quantity, and many conventional fuels are becoming scarce. The sun is an alternative possible heat source. Sunlight is free and inexhaustible, and some solar desalination facilities already exist. Their efficiency is limited by the fact that solar heat is low-intensity heat. If a large quantity of desalinated water is required rapidly, the water to be heated must be spread out shallowly over a large area, or the rate of water output will be slow. A large city might need a solar desalination facility covering thousands of square kilometers to provide adequate water, and construction on such a scale would be prohibitively expensive even if the space were available.

Desalinated water may be five to ten times more costly to deliver than water pumped straight from a stream or aquifer. For most homeowners, the water bill is a relatively minor expense, so a jump in water costs, if necessitated by the use of desalinated water, would not be a great hardship. Water for irrigation, however, must be both plentiful and cheap if the farmer is to compete with others here and abroad who need not irrigate and if the cost of food production is to be held down. Desalinated water in most areas is prohibitively expensive for irrigation use. Unless ways can be found to reduce drastically the cost of desalinated water, agriculture will continue to drain limited and dwindling surface- and groundwater supplies. Water from the ocean will not be a viable alternative for large-scale irrigation for some time.

SUMMARY

Most of the water in the hydrosphere at any given time is in the oceans; most of the remaining fresh water is stored in ice sheets. Relatively little water is found in lakes and streams. A portion of the fresher waters on the continents—both surface water and ground water—is diverted for human use. The availability of ground water is influenced by such factors as the presence of suitable aquifers, water quality, and rate of recharge relative to rate of water use. Where ground water is plentiful and subsurface rocks are soluble, dissolution may create caves below ground and sinkholes at the surface, producing a distinctive landscape (karst). Karst aquifers may be productive, with rapid water flow, but the supply may be erratic and rapidly polluted.

In the United States, more than half of offstream water use (withdrawal) is for industrial purposes and thermoelectric power generation. However, these industrial applications generally consume very little water. Of the water actually consumed in the United States, approximately 80% is as water loss associated with irrigation. While three-fourths of the water *withdrawn* for use is surface water, close to half of the water *consumed* is ground water, which, in many cases, is being consumed faster than recharge can replace it. Adverse consequences of such rapid consumption of ground water include lowering water tables, surface subsidence, and, in coastal areas, saltwater intrusion. Conservation, interbasin transfers of surface water, and desalination are possible ways to extend the water supplies of high-demand regions. Desalinated water is presently too expensive to use for irrigation on a large scale, which means that it is unlikely to have a significant impact on the largest consumptive water use for some time to come.

Key Terms and Concepts

aquifer 244
aquitard 244
artesian system 245
cone of depression 246
confined aquifer 244
Darcy's Law 246
discharge 244
ground water 242
hard water 254
hydraulic head 245
karst 252
permeability 242
phreatic zone 242
porosity 242
potentiometric surface 245
recharge 244
saltwater intrusion 249
saturated zone 242
sinkhole 252
soil moisture 243
unconfined aquifer 244
unsaturated zone 242
vadose zone 243
water table 243

Exercises

Questions for Review

1. Explain the importance of porosity and of permeability to groundwater availability for use.
2. Define the following terms: *ground water, water table,* and *potentiometric surface.*
3. How is an artesian aquifer system formed?
4. What is Darcy's Law, and what does it tell us about groundwater flow?
5. Explain how sinkholes develop. What name is given to a landscape in which sinkholes are common?
6. In what ways do karst aquifers differ from non-karst aquifers, in the context of water supply?
7. Explain three characteristics used to describe water quality.
8. What is *hard water,* and why is it often considered undesirable?
9. Describe two possible consequences of groundwater withdrawal exceeding recharge.
10. Explain the process of saltwater intrusion.
11. In what ways may urbanization affect groundwater recharge?
12. Industry is the big water *user,* but agriculture is the big water *consumer.* Explain.
13. Compare and contrast filtration and distillation as desalination methods, noting advantages and drawbacks of each.
14. What factor presently limits the potential of desalination to alleviate agricultural water shortages?

For Further Thought

1. Where does your water come from? What is its quality? How is it treated, if at all, before it is used? Is a long-term shortage likely in your area? If so, what plans are being made to avert it?
2. Compare figures 11.21 and 11.22. Speculate on the reasons for some of the extremes of water withdrawal shown in figure 11.22, and consider the likelihood that this withdrawal is predominantly *consumption* in the highest-use states.
3. Some seemingly minor activities are commonly cited as greatwater-wasters. One is letting the tap run while brushing one's teeth. Try plugging the drain while doing this; next, measure the volume of water accumulated during that single toothbrushing (a measuring cup and bucket may be helpful). Consider the implications for water use, bearing in mind the 300 million persons in the United States. Read your home water meter before and after watering the lawn or washing the car, and make similar projections.
4. Select an area, or country outside the United States, and investigate current precipitation and water use. See NetNotes for potential Internet sites at which to start.

Suggested Readings/References

American Meteorological Society. 2001. *Water in the earth system.* Boston, Mass.: AMS.

Bartolino, J. R., and Cunningham, W. L. 2003. *Ground-water depletion across the nation.* USGS Fact Sheet 103–03.

Bourne, J. K., Jr. 2005. Salton Sea. *National Geographic* (February): 88–107.

Cobb, J., and Currens, J. 2001. Karst: The stealthy hazard. *Geotimes* (May): 14–17.

Fetter, C. W. 2001. *Applied hydrogeology,* 4th ed. New York: Prentice Hall.

Galloway, D., Jones, D. R., and Ingebreitsen, S. E. 1999. *Land subsidence in the United States.* U.S. Geological Survey Circular 1182.

Gleick, P. H. 2001. Making every drop count. *Scientific American* (February): 40–45.

Heppenheimer, T. A. 1991. The man who made Los Angeles possible. *Invention and Technology* (Spring/Summer): 11–18.

Howe, J. 2004. The Great Southwest salt saga. *Wired* (November): 156–65.

Martinez, J. D., Johnson, K. S., and Neal, J. T. 1998. Sinkholes in evaporite rocks. *American Scientist* 86 (January–February): 38–51.

McGuire, V. L., *et al.* 2003. *Water in storage and approaches to ground-water management, High Plains Aquifer, 2000.* U.S. Geological Survey Circular 1243.

Popkin, B. P. 2004. Water resource challenges in Egypt and India with an Iraqi sidebar. *The Professional Geologist* (May/June): 41–43.

Robert, L. 2004. Hijacking the Rio Grande. *Geotimes* (May): 26–29.

Scott, T. M. 2002. Florida's springs in jeopardy. *Geotimes* (May): 16–20.

U.S. Geological Survey. 2004. *Estimated use of water in the United States in 2000. USGS Circular 1268.*
[online] (downloadable in PDF format)
http://water.usgs.gov/circ1268/

Vandas, S. J., Winter, T. C., and Battaglin, W. A. 2002. *Water and the environment.* Alexandria, VA: Amer. Geological Inst.

Vörösmarty, C., *et al.* 2004. Humans transforming the global water system. *EOS* (30 November): 509, 513–14.

Waller, R. M. 1989. *Ground water and the rural homeowner.* U.S. Geological Survey publication.

White, K., and Mattingly, D. J. 2006. Ancient lakes of the Sahara. *American Scientist* 94 (January–February): 58–65.

Winter, T. C., Harvey, J. W., Franke, O. L., and Alley, W. M. 1999. *Ground water and surface water: A single resource.* U.S. Geological Survey Circular 1139.

Zwingle, E. 1993. Ogallala aquifer: Wellspring of the High Plains. *National Geographic* 184 (March): 80–109.

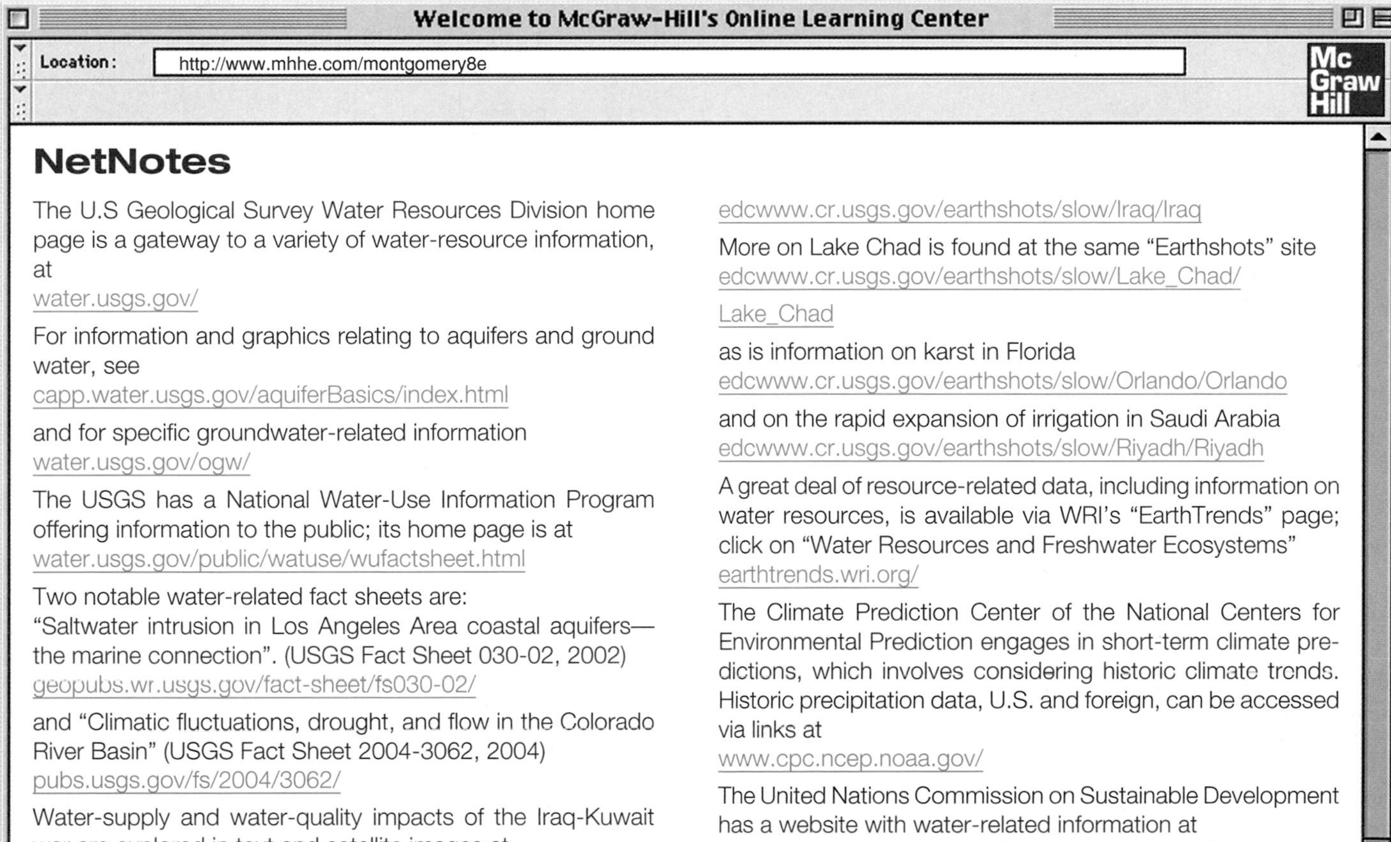

NetNotes

The U.S Geological Survey Water Resources Division home page is a gateway to a variety of water-resource information, at
water.usgs.gov/

For information and graphics relating to aquifers and ground water, see
capp.water.usgs.gov/aquiferBasics/index.html

and for specific groundwater-related information
water.usgs.gov/ogw/

The USGS has a National Water-Use Information Program offering information to the public; its home page is at
water.usgs.gov/public/watuse/wufactsheet.html

Two notable water-related fact sheets are:
"Saltwater intrusion in Los Angeles Area coastal aquifers—the marine connection". (USGS Fact Sheet 030-02, 2002)
geopubs.wr.usgs.gov/fact-sheet/fs030-02/

and "Climatic fluctuations, drought, and flow in the Colorado River Basin" (USGS Fact Sheet 2004-3062, 2004)
pubs.usgs.gov/fs/2004/3062/

Water-supply and water-quality impacts of the Iraq-Kuwait war are explored in text and satellite images at
edcwww.cr.usgs.gov/earthshots/slow/Iraq/Iraq

More on Lake Chad is found at the same "Earthshots" site
edcwww.cr.usgs.gov/earthshots/slow/Lake_Chad/
Lake_Chad

as is information on karst in Florida
edcwww.cr.usgs.gov/earthshots/slow/Orlando/Orlando

and on the rapid expansion of irrigation in Saudi Arabia
edcwww.cr.usgs.gov/earthshots/slow/Riyadh/Riyadh

A great deal of resource-related data, including information on water resources, is available via WRI's "EarthTrends" page; click on "Water Resources and Freshwater Ecosystems"
earthtrends.wri.org/

The Climate Prediction Center of the National Centers for Environmental Prediction engages in short-term climate predictions, which involves considering historic climate trends. Historic precipitation data, U.S. and foreign, can be accessed via links at
www.cpc.ncep.noaa.gov/

The United Nations Commission on Sustainable Development has a website with water-related information at
www.un.org/esa/sustdev/sdissues/water/water.htm

12

Soil as a Resource

Soil does not, at first glance, strike many people as a resource requiring special care for its preservation. In most places, even where soil erosion is active, a substantial quantity seems to remain underfoot. Associated problems, such as loss of soil fertility and sediment pollution of surface waters, are even less obvious to the untutored eye and may be too subtle to be noticed readily. Nevertheless, soil is an essential resource on which we depend for the production of the major portion of our food. Soils vary in their suitability not only for agriculture but also for construction and other purposes. Unfortunately, soil erosion is a significant and expensive problem in an increasing number of places as human activities disturb more and more land. In this chapter, we will examine the nature of soil, its formation and properties, aspects of the soil erosion problem, and some strategies for reducing erosion.

Soil formation involves weathering in place; if erosion is too rapid, it cannot accumulate. Here in the Badlands, soil develops and supports plant growth only where the surface is relatively flat and erosion is therefore slow.

SOIL FORMATION

Soil is defined in different ways for different purposes. Engineering geologists define soil very broadly to include all unconsolidated material overlying bedrock. Soil scientists restrict the term *soil* to those materials capable of supporting plant growth and distinguish it from *regolith,* which encompasses all unconsolidated material at the surface, fertile or not. (The loose material on the lunar surface is described as regolith.) Conventionally, the term *soil* implies little transportation away from the site at which the soil formed, while the term *sediment* indicates matter that has been transported and redeposited by wind, water, or ice.

Soil is produced by *weathering,* a term that encompasses a variety of chemical, physical, and biological processes acting to break down rocks and minerals. It may be formed directly from bedrock, or from further breakdown of transported sediment such as glacial till. The relative importance of the different kinds of weathering processes is largely determined by climate. Climate, topography, the composition of the material from which the soil is formed, the activity of organisms, and time govern a soil's final composition.

Soil-Forming Processes: Weathering

Mechanical weathering is the physical breakup of rocks without changes in the rocks' composition. In a cold climate, with temperatures that fluctuate above and below freezing, water in cracks repeatedly freezes and expands, forcing rocks apart, then thaws and contracts or flows away. Crystallizing salts in cracks may have the same wedging effect. Whatever the cause, the principal effect of mechanical weathering is the breakup of large chunks of rock into smaller ones. In the process, the total exposed surface area of the particles is increased (figure 12.1).

Chemical weathering involves the breakdown of minerals by chemical reaction with water, with other chemicals dissolved in water, or with gases in the air. Minerals differ in the kinds of chemical reactions they undergo. Calcite (calcium carbonate) tends to dissolve completely, leaving no other minerals behind in its place. Calcite dissolves rather slowly in pure water but more rapidly in acidic water. Many natural waters are slightly acidic; acid rainfall or acid runoff from coal strip mines (see chapter 14) is more so and causes more rapid dissolution. This is, in fact, becoming a serious problem where limestone and its metamorphic equivalent—marble—are widely used for outdoor sculptures and building stone. Calcite dissolution is gradually destroying delicate sculptural features and eating away at the very fabric of many buildings in urban areas and where acid rain is common.

Silicates tend to be somewhat less susceptible to chemical weathering and leave other minerals behind when they are attacked. Feldspars principally weather into clay minerals. Ferromagnesian silicates leave behind insoluble iron oxides and hydroxides and some clays, with other chemical components being dissolved away. Those residual iron compounds are responsible for the reddish or yellowish colors of many soils. In most climates, quartz is extremely resistant to chemical weathering, dissolving only slightly. Representative weathering reactions are shown in table 12.1.

The susceptibility of many silicates to chemical weathering can be inferred from the conditions under which the silicates formed. Given several silicates that have crystallized from the same magma, those that formed at the highest temperatures tend to be the least stable, or most easily weathered, and vice versa, probably because the lower-temperature silicates have structures with more interlocking of silica tetrahedra in two and three dimensions. A rock's tendency to weather chemically is determined by its mineral composition. For example, a gabbro (the coarsely crystalline equivalent of basalt), formed at high temperatures and rich in ferromagnesian minerals, generally weathers more readily than a granite rich in quartz and low-temperature feldspars.

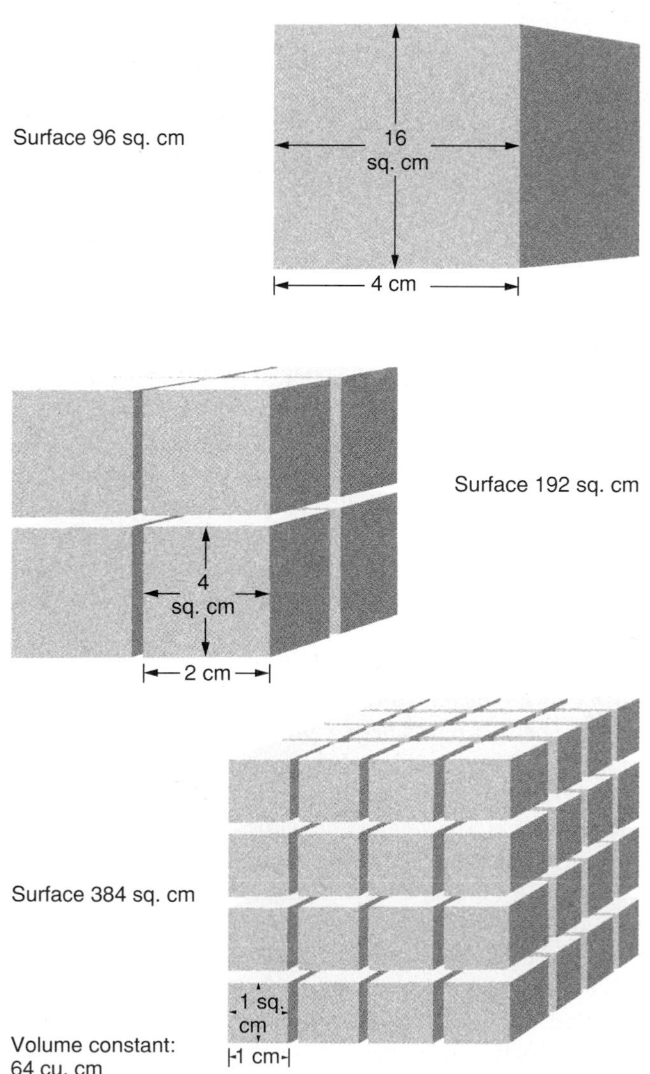

FIGURE 12.1 Mechanical breakup increases surface area and surface-to-volume ratio.

TABLE 12.1

Some Chemical Weathering Reactions

Solution of calcite (no solid residue)

$CaCO_3 + 2\ H^+ = Ca^{2+} + H_2O + CO_2$ (gas)

Breakdown of ferromagnesians (possible mineral residues include iron compounds and clays)

$FeMgSiO_4$ (olivine) $+ 2\ H^+ = Mg^{2+} + Fe(OH)_2 + SiO_2$*

$2\ KMg_2FeAlSi_3O_{10}(OH)_2$ (biotite) $+ 10\ H^+ + \frac{1}{2}\ O_2$ (gas) $=$
$2\ Fe(OH)_3 + Al_2Si_2O_5(OH)_4$ (kaolinite, a clay) $+ 4\ SiO_2$* $+$
$2\ K^+ + 4\ Mg^{2+} + 2\ H_2O$

Breakdown of feldspar (clay is the common residue)

$2\ NaAlSi_3O_8$ (sodium feldspar) $+ 2\ H^+ + H_2O =$
$Al_2Si_2O_5(OH)_4 + 4\ SiO_2$* $+ 2\ Na^+$

Solution of pyrite (making dissolved sulfuric acid, H_2SO_4)

$2\ FeS_2 + 5\ H_2O + \frac{15}{2}\ O_2$ (gas) $= 4\ H_2SO_4 + Fe_2O_3 \cdot H_2O$

Notes: Hundreds of possible reactions could be written; the above are only examples of the common kinds of processes involved.

All ions (charged species) are dissolved in solution; all other substances, except water, are solid unless specified otherwise.

Commonly, the source of the H^+ ions for solution of calcite and weathering of silicates is carbonic acid, H_2CO_3, formed by solution of atmospheric CO_2.

*Silica is commonly removed in solution.

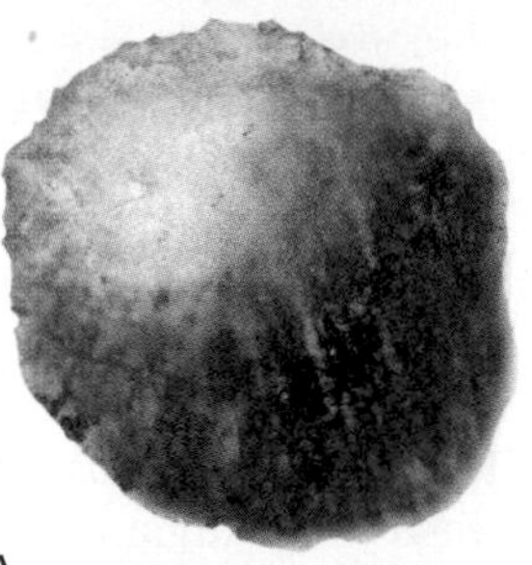

A

B

FIGURE 12.2 (A) Angular fragments are rounded by weathering. Corners are attacked on three surfaces, edges on two. In time, fragments become rounded as a result. (B) Fractured Hawaiian basalt weathers into rounded chunks.

Climate plays a major role in the intensity of chemical weathering. Most of the relevant chemical reactions involve water. All else being equal, then, the more water, the more chemical weathering. Also, most chemical reactions proceed more rapidly at high temperatures than at low ones. Therefore, warm climates are more conducive to chemical weathering than cold ones.

The rates of chemical and mechanical weathering are interrelated. Chemical weathering may speed up the mechanical breakup of rocks if the minerals being dissolved are holding the rock together by cementing the mineral grains, as in some sedimentary rocks. Increased mechanical weathering may, in turn, accelerate chemical weathering through the increase in exposed surface area, because it is only at grain surfaces that minerals, air, and water interact. The higher the ratio of surface area to volume—that is, the smaller the particles—the more rapid the chemical weathering. The vulnerability of surfaces to attack is also shown by the tendency of angular fragments to become rounded by weathering (figure 12.2).

The organic component of soil is particularly important to its fertility and influences other physical properties as well. Biological processes are also important to the formation of soil. Biological weathering effects can be either mechanical or chemical. Among the mechanical effects is the action of tree roots in working into cracks to split rocks apart (figure 12.3). Chemically, many organisms produce compounds that may react with and dissolve or break down minerals. Plants, animals, and microorganisms develop more abundantly and in greater variety in warm, wet climates. Mechanical weathering is generally the dominant process only in areas where climatic conditions have limited the impact of chemical weathering and biological effects—that is, in cold or dry areas.

Finally, airborne chemicals and sediments may *add* components to soil—for example, sulfate from acid rain, salts from sea spray in coastal areas, clay minerals and other fine-grained minerals blown on the wind.

FIGURE 12.3 Tree roots working into cracks in this granite break up the rock mass. Granite in Yosemite National Park.

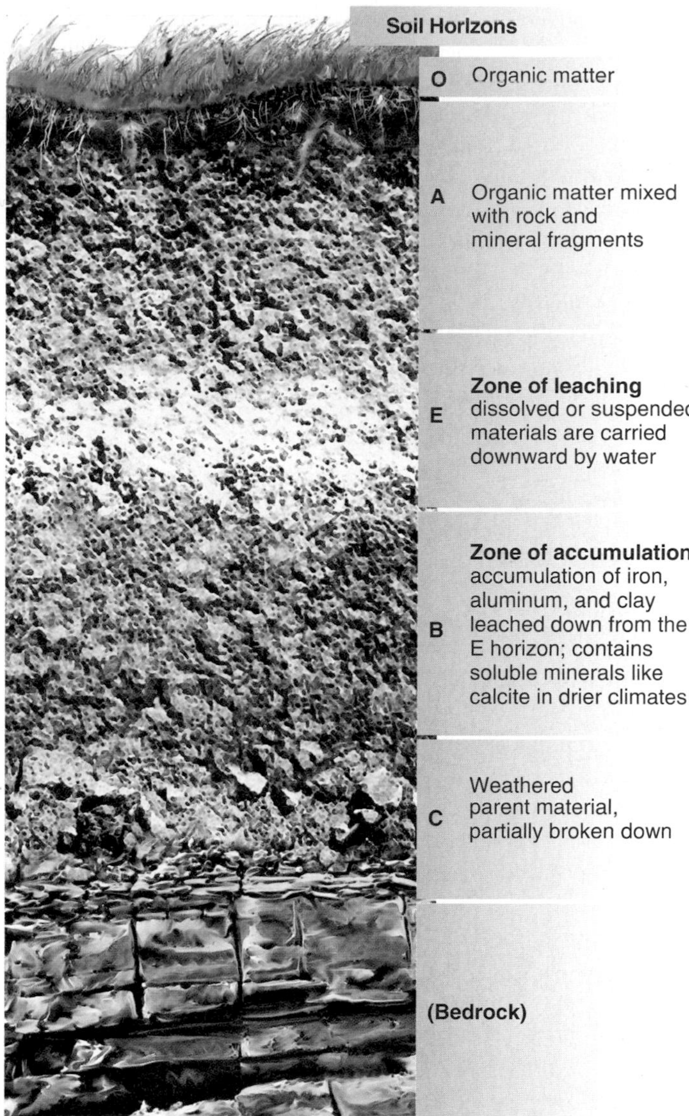

FIGURE 12.4 A generalized soil profile. Individual horizons can vary in thickness. Some may be locally absent, or additional horizons or subhorizons may be identifiable.

Soil Profiles, Soil Horizons

The result of mechanical, chemical, and biological weathering, together with the accumulation of decaying remains from organisms living on the land and any input from the atmosphere, is the formation of a blanket of soil between bedrock and atmosphere. A cross section of this soil blanket usually reveals a series of zones of different colors, compositions, and physical properties. The number of recognizable zones and the thickness of each vary. A basic, generalized soil profile as developed directly over bedrock is shown in figure 12.4.

At the very top is the **O horizon,** consisting wholly of organic matter, whether living or decomposed—growing plants, decaying leaves, and so on. Below that is the **A horizon.** It consists of the most intensively weathered rock material, being the zone most exposed to surface processes, mixed with organic debris from above. Unless the local water table is exceptionally high, precipitation infiltrates down through the A horizon and below. In so doing, the water may dissolve soluble minerals and carry them away with it. This process is known as **leaching,** and it may be especially intense just below the A horizon, as acids produced by the decay of organic matter seep downward with percolating water. The **E horizon,** below the A horizon, is therefore also known as the **zone of leaching.** Fine-grained minerals, such as clays, may also be washed downward through this zone.

Many of the minerals leached or extracted from the E horizon accumulate in the layer below, the **B horizon,** also known as the **zone of accumulation.** Soil in the B horizon has been somewhat protected from surface processes. Organic matter from the surface is largely absent from the B horizon. It may contain relatively high concentrations of iron and aluminum oxides, clay minerals, and, in drier climates, even soluble minerals such as calcite. Below the B horizon is a zone consisting principally of very coarsely broken-up bedrock and little else. This is the **C horizon,** which does not resemble our usual idea of soil at all. Similar zonation—though without bedrock at the base of the soil profile—is found in soils developed on transported sediment.

The boundaries between adjacent soil horizons may be sharp or indistinct. In some instances, one horizon may be divided into several recognizable subhorizons. Subhorizons also may exist that are gradational between A and B or B and C. It is also possible for one or more horizons locally to be absent from the soil profile. Some of the diversity of soil types is illustrated in the next section. All variations in the soil profile arise from the different mix of soil-forming processes and starting materials found from place to place. The overall total thickness of soil is partly a function of the local rate of soil formation and partly a function of the rate of soil erosion. The latter reflects the work of wind and water, the topography, and often the extent and kinds of human activities.

CHEMICAL AND PHYSICAL PROPERTIES OF SOILS

Mechanical weathering merely breaks up rock without changing its composition. In the absence of pollution, wind and rainwater rarely add many chemicals, and runoff water may carry away some leached chemicals in solution. Thus, chemical weathering tends to involve a net subtraction of elements from rock or soil. One control on a soil's composition, then, is the composition of the material from which it is formed. If the bedrock or parent sediment is low in certain critical plant nutrients, the soil produced from it will also be low in those nutrients, and chemical fertilizers may be needed to grow particular crops on that soil even if the soil has never been farmed before. The extent and balance of weathering processes involved in soil formation then determine the extent of further depletion of the soil in various elements. The weathering processes also influence the mineralogy of the soil, the compounds in which its elements occur. The physical properties of the soil are affected by its mineralogy, the texture of the mineral grains (coarse or fine, well or poorly sorted, rounded or angular, and so on), and any organic matter present.

FIGURE 12.5 Extensive weathering of iron-rich Hawaiian basalt, enriched with decaying vegetation, produces red-brown soil.

Color, Texture, and Structure of Soils

Soil *color* tends to reflect compositional characteristics. Soils rich in organic matter tend to be black or brown, while those poor in organic matter are paler in color, often white or gray. When iron is present and has been oxidized by reaction with oxygen in air or water, it adds a yellow or red color (figure 12.5; recall also figure 1.20, the Martian surface). Rust on iron or steel is also produced by oxidation of iron.

Soil *texture* is related to the sizes of fragments in the soil. The U.S. Department of Agriculture recognizes three size components: sand (grain diameters 2–0.05 mm), silt (0.05–0.002 mm), and clay (less than .002 mm). Soils are named on the basis of the dominant grain size(s) present (figure 12.6). An additional term, *loam,* describes a soil that is a mixture of all three particle sizes in similar proportions (10 to 30% clay, the balance nearly equal amounts of sand and silt). Soils of

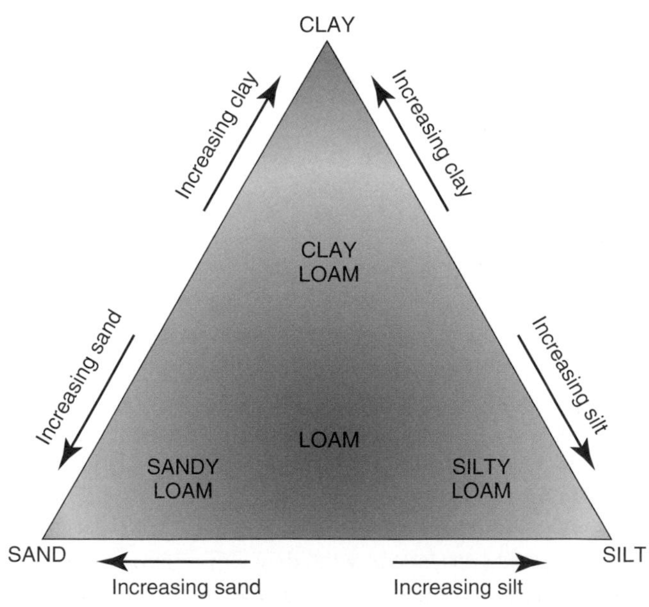

FIGURE 12.6 Soil-texture terminology reflects particle size, not mineralogy. Diagram simplified from U.S. Department of Agriculture classification scheme.

intermediate particle-size distribution have mixed names: for example, "silty clay" would be a soil consisting of about half silt-sized particles, half clay-sized particles. The significance of soil texture is primarily the way it influences drainage. Sandy soils, with high permeability (like many sandstone aquifers), drain quickly. In an agricultural setting, this may mean a need for frequent watering or irrigation. In a flood-control context, it means relatively rapid infiltration. Clay-rich soils, by contrast, may hold a great deal of water but be relatively slow to drain, by virtue of their much lower permeability, a characteristic shared by their lithified counterpart, shale.

Soil *structure* relates to the soil's tendency to form lumps or clods of soil particles. These clumps are technically called *peds,* their name deriving from the Latin root *pedo-,* meaning "soil." A soil that clumps readily may be more resistant to erosion; some of the fine-grained soils developed on fine loess in the United States and in China may erode very readily. On the other hand, soil consisting of very large peds with large cracks between them may be a poor growing medium for small plants with fine roots. Abundant organic matter may promote the aggregation of soil particles into crumblike peds especially conducive to good plant growth; mechanical weathering can break up larger clumps into smaller, just as it breaks up rock fragments.

Soil Classification

Classifying minerals offers a convenient way to discuss groups of minerals of similar composition or structure, and specifying a mineral's classification immediately provides some information about it. Likewise, soil classifications can indicate something of a soil's composition and perhaps its origins, which in turn may have implications for its suitability for agriculture or construction, or its vulnerabity to degradation.

Early soil classification schemes emphasized compositional differences among soils and thus principally reflected the effects of chemical weathering. The resultant classification into two broad categories of soil was basically a climatic one. The **pedalfer** soils were seen as characteristic of more humid regions. Where the climate is wetter, there is naturally more extensive leaching of the soil, and what remains is enriched in the less-soluble oxides and hydroxides of aluminum and iron, with clays accumulated in the B horizon. The term *pedalfer* comes from the prefix *pedo-* and the Latin words for aluminum (*alumium*) and iron (*ferrum*). In North America, pedalfer-type soils are found in higher-rainfall areas like the eastern United States and most of Canada. Pedalfer soils are typically acidic. Where the climate is drier, such as in the western and especially the southwestern United States, leaching is much less extensive. Even quite soluble compounds like calcium carbonate remain in the soil, especially in the B horizon. From this observation came the term **pedocal** for the soil of a dry climate. The presence of calcium carbonate makes pedocal soils more alkaline.

One problem with this simple classification scheme is that, to be strictly applied, the soils it describes must have formed over suitable bedrock. For example, a rock poor in iron and aluminum, such as a pure limestone or quartz sandstone, does not leave an iron- and aluminum-rich residue, no matter how extensively leached or intensely weathered. Still, the terms *pedalfer* and *pedocal* can be used generally to indicate, respectively, more and less extensively leached soils.

Modern soil classification has become considerably more sophisticated and complex. Various schemes may take into account characteristics of present soil composition and texture, the type of bedrock on which the soil was formed, the present climate of the region, the degree of "maturity" of the soil, and the extent of development of the different soil horizons. Different countries have adopted different schemes. The United Nations Education, Scientific, and Cultural Organization (UNESCO) world map uses 110 different soil map units. This is, in fact, a small number compared to the number of distinctions made under other classification schemes.

The U.S. comprehensive soil classification, known as the Seventh Approximation, has twelve major categories (orders), which are subdivided through five more levels of classification into a total of some 12,000 soil series. (Even soil scientists may need reference books to keep all of the distinctions straight!) A short summary of the twelve orders is presented for reference in table 12.2. Some of the orders are characterized by a particular environment of formation and the distinctive soil properties resulting from it—for example, the histosols, which are wetland soils. Others are characterized principally by physical properties—for example, the entisols, which lack horizon zonation, and the vertisols, in which the upper layers are mixed because these soils contain expansive clays. The practical significance of the vertisols lies in the considerable engineering problems posed by expansive clays (see chapter 20). Most of the oxisols and some ultisols, on the other hand, are soils of a type that has serious implications for agriculture, especially in much of the Third World: lateritic soils, which will be discussed below. Figure 12.7 illustrates some of the variety possible among soils. Figure 12.8 shows broad distribution of soil types worldwide.

SOILS AND HUMAN ACTIVITIES

Soils and their characteristics are not only critical to agricultural activities; their properties influence the stability of construction projects. In the area of soil erosion, human activities may aggravate or moderate the problems—and sometimes may be influenced by the erosion in turn. This section samples several relationships between soil and human activities.

Lateritic Soil

Lateritic soil is common to many less-developed nations and poses special agricultural challenges. A **laterite** may be regarded as an extreme kind of pedalfer. Lateritic soils develop in tropical climates with high temperatures and heavy rainfall, so they are severely leached. Even quartz may have been dissolved out of the soil under these conditions. Lateritic soil may

TABLE 12.2

The Twelve Soil Orders, with Descriptions

Order	Description
Alfisols	Formed in semiarid to moist areas; deeper layers enriched in clay
Andisols	Soils of cool volcanic areas with moderate to heavy precipitation
Aridosols	Common desert soils, dry and rich in soluble minerals such as calcite, halite, or gypsum
Entisols	Soils with little development of horizons, formed on recently deposited or rapidly eroding parent materials
Gelisols	Soils with near-surface permafrost or that show evidence of mixing by frost action
Histosols	Wetland soils rich in organic matter
Inceptisols	Soils of semiarid to humid climates with limited evidence of weathering or horizon development
Mollisols	The common soils of relatively dry grasslands, with a high content of organic matter near the surface
Oxisols	Highly leached soils of tropical and subtropical climates, which often have indistinct horizons
Spodosols	Soils common under conifer forests in humid regions, with aluminum oxides and organic matter leached from the surface and deposited below
Ultisols	Moderately leached soils of temperate, humid climates
Vertisols	Soils rich in expansive clays (clays that expand when wet, shrink and cause cracking when dry)

Source: After U.S. Department of Agriculture

contain very little besides the insoluble aluminum and iron compounds. (Indeed, lateritic weathering has produced useful mineral deposits, as described in Chapter 13.) Soils of the lush tropical rain forests are commonly lateritic, which seems to suggest that lateritic soils have great farmland potential. Surprisingly, however, the opposite is true, for two reasons.

The highly leached character of lateritic soils is one reason. Even where the vegetation is dense, the soil itself has few soluble nutrients left in it. The forest holds a huge reserve of nutrients, but there is no corresponding reserve in the soil. Further growth is supported by the decay of earlier vegetation. As one plant dies and decomposes, the nutrients it contained are quickly taken up by other plants or leached away. If the forest is cleared to plant crops, most of the nutrients are cleared away with it, leaving little in the soil to nourish the crops. Many natives of tropical climates practice a slash-and-burn agriculture, cutting and burning the jungle to clear the land. Some of the nutrients in the burned vegetation settle into the topsoil temporarily, but relentless leaching by the warm rains makes the soil nutrient-poor and infertile within a few growing seasons. Nutrients could, in principle, be added through synthetic chemical fertilizers. However, many of the nations in regions of lateritic soil are among the poorer developing countries, and vast expenditures for fertilizer are simply impractical.

Even with fertilizers available, a second problem with lateritic soils remains. A clue to the problem is found in the term *laterite* itself, which is derived from the Latin for "brick." A lush rain forest shields lateritic soil from the drying and baking effects of the sun, while vigorous root action helps to keep the soil well broken up. When its vegetative cover is cleared and it is exposed to the baking tropical sun, lateritic soil can quickly harden to a solid, bricklike consistency that resists infiltration by water or penetration by crops' roots. What crops can be grown provide little protection for the soil and do not slow the hardening process very much. Within five years or less, a freshly cleared field may become completely uncultivatable. The tendency of laterite to bake into brick has been used to advantage in some tropical regions, where blocks of hardened laterite have served as building materials. Still, the problem of how to maintain crops remains. Often, the only recourse is to abandon each field after a few years and clear a replacement. This results, over time, in the destruction of vast tracts of rain forest where only a moderate expanse of farmland is needed. Also, once the soil has hardened and efforts to farm it have been abandoned, it may revegetate only very slowly, if at all.

In Indochina, in what is now Cambodian jungle, are the remains of the Khmer civilization that flourished from the ninth to the sixteenth centuries. There is no clear evidence of why that civilization vanished. A major reason may have been the difficulty of agriculture in the region's lateritic soil. (It is clear that the phenomenon of laterite solidification was occurring at that time: Chunks of hardened laterite were used to construct temples at Angkor Wat and elsewhere.) Perhaps the Mayas, too, moved north into Mexico to escape the problems of lateritic soils. In Sierra Leone in west Africa, increased clearing of forests for firewood, followed by deterioration of the exposed soil, has reduced the estimated carrying capacity of the land to 25 persons per square kilometer; the actual population is already over 40 persons per square kilometer. Part of the concern over deforestation associated with harvesting

FIGURE 12.7 A spectrum of soils. Though we may not notice it when looking down from above, soils in cross section vary widely in color, texture, structure, and mineralogy. Soil (A) is an entisol, a low-fertility, sandy soil developed on glacial till in northern Michigan; (B) an inceptisol, forming on layered volcanic ash in Japan; (C) an aridosol, topped with desert pavement, Arizona; (D) a mollisol with visible white carbonate nodules in the B and C horizons; (E) a spodosol with pale, leached A horizon, in upstate New York; and (F) a histosol of decomposed mud over peat, southern Michigan. Figure 12.5 showed an aridosol.

Photographs from Marlbut Memorial Slides, courtesy Soil Science Society of America

of hardwoods in South America relates to loss of habitat and possible extinction of unique species (loss of biodiversity), but in part it relates to the fact that what is left behind is barren lateritic soil. Today, some countries with lateritic soil achieve agricultural success only because frequent floods deposit fresh, nutrient-rich soil over the depleted laterite. The floods, however, cause enough problems that flood-control efforts are in progress or under serious consideration in many of these areas of Africa and Asia. Unfortunately, successful flood control could create an agricultural disaster for such regions.

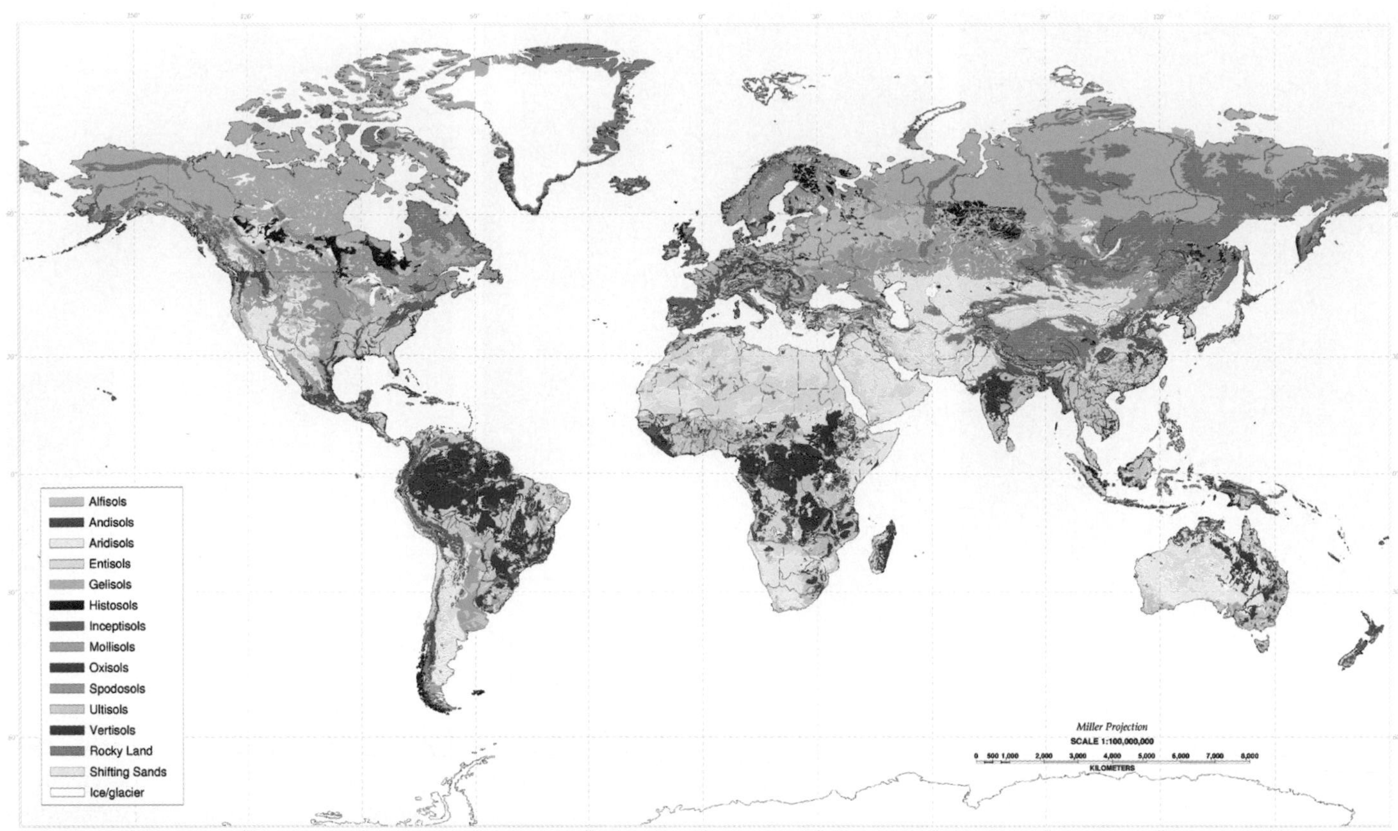

FIGURE 12.8 Distribution of major soil types worldwide.
Map from World Soil Resources Office of USDA Natural Resources Conservation Service

Wetland Soils

The lateritic soils are leached, oxidized, and inclined to bake hard; the soils of wetlands, by contrast, tend to be rich in accumulated organic matter, reduced because the decaying organic matter consumes dissolved oxygen, and soft. (Recall figure 12.7F) Wetlands provide vital habitat for waterbirds and distinctive ecological niches for other organisms and can act as natural retention ponds for floodwaters. Some act as settling ponds, reducing sediment pollution and also water pollution from contaminants carried on the sediment. Unfortunately, wetlands have not always been properly appreciated; many of these swampy areas have been drained for farmland or land for development, or simply to provide water wanted elsewhere for irrigation. In drier climates, reduced precipitation can exacerbate wetland loss, as in the case of Lake Chad, described in chapter 11. As the value of wetlands has become more widely recognized, concerted regulatory efforts to preserve, protect, or restore wetlands have expanded in the United States (figure 12.9) and elsewhere. Direct human impacts aside, however, there is an added threat to many coastal wetlands: rising sea level, as noted in chapter 8.

FIGURE 12.9 Even where undeveloped land is plentiful, wetlands can serve special roles. This wetland, near Anchorage, Alaska, is protected as a wildlife refuge. A raised wooden boardwalk (at right) allows visitors to view and photograph from the perimeter without disturbing birds or animals.

Soil Erosion

Weathering is the breakdown of rock or mineral materials in place; *erosion* involves physical removal of material from one place to another. Soil erosion is caused by the action of water and wind. Rain striking the ground helps to break soil particles loose (figure 12.10). Surface runoff and wind together carry away loosened soil. The faster the wind and water travel, the larger the particles and the greater the load they move. Therefore, high winds cause more erosion than calmer ones, and fast-flowing surface runoff moves more soil than slow runoff. This, in turn, suggests that, all else being equal, steep and unobstructed slopes

FIGURE 12.10 The impact of a single raindrop loosens many soil particles.
Photograph courtesy of USDA Soil Conservation Service

are more susceptible to erosion by water, for surface runoff flows more rapidly over them. Flat, exposed land is correspondingly more vulnerable to wind erosion. The physical properties of the soil also influence its vulnerability to erosion.

Rates of soil erosion can be estimated in a variety of ways. Over a large area, erosion due to surface runoff may be judged by estimating the sediment load of streams draining the area. On small plots, runoff can be collected and its sediment load measured. Controlled laboratory experiments can be used to simulate wind and water erosion and measure their impact. Because wind is harder to monitor comprehensively, especially over a range of altitudes, the extent of natural wind erosion is more difficult to estimate. Generally, it is less significant than erosion through surface-water runoff, except under drought conditions (for example, during the Dust Bowl era).

The Dust Bowl area proper, although never exactly defined, comprised close to 100 million acres of southeastern Colorado, northeastern New Mexico, western Kansas, and the Texas and Oklahoma panhandles. The farming crisis there during the 1930s resulted from an unfortunate combination of factors: clearing or close grazing of natural vegetation, which had been better adapted to the local climate than were many of the crops; drought that then destroyed those crops; sustained winds; and poor farming practices.

Once the crops had died, there was nothing to hold down the soil. The action of the wind was most dramatically illustrated during the fierce dust storms that began in 1932. The storms were described as "black blizzards" that blotted out the sun (figure 12.11A). Black rain fell in New York, black snow in Vermont, as windblown dust moved eastward. People choked

A

B

FIGURE 12.11 (A) A 1930s dust storm in the Dust Bowl. (B) Barren fields buried under windblown soil.
Photographs courtesy of USDA Soil Conservation Service

on the dust, some dying of suffocation or of a "dust pneumonia" similar to the silicosis miners develop from breathing rock dust.

As the dust from Kansas and Oklahoma settled on their desks, politicians in Washington, D.C., realized that something had to be done. In April 1935, a permanent Soil Conservation Service was established to advise farmers on land use, drainage, erosion control, and other matters. By the late 1930s, concerted efforts by individuals and state and federal government agencies to improve farming practices to reduce wind erosion—together with a return to more normal rainfall—had considerably reduced the problems. However, renewed episodes of drought and/or high winds in the 1950s, the winter of 1965, and the mid-1970s led to similar, if less extreme, dust storms and erosion. In recent years, the extent of wind damage has actually been limited somewhat by the widespread use of irrigation to maintain crops in dry areas and dry times. But, as sources of that irrigation water are depleted, future spells of combined drought and wind may yet produce more scenes like those in figure 12.11.

If all the soil in an area is lost, farming clearly becomes impossible. Long before that point, however, erosion becomes a cause for concern. The topsoil of the A horizon, with its higher content of organic matter and nutrients, is especially fertile and suitable for agriculture, and it is the topsoil that is lost first as the soil erodes. Estimates indicate that conserving the nutrients now being lost could save U.S. farmers $20 billion annually in fertilizer costs. In addition, the organic-matter-rich topsoil usually has the best structure for agriculture—it is more permeable, more readily infiltrated by water, and retains moisture better than deeper soil layers.

Soil erosion from cropland leads to reduced crop quality and reduced agricultural income. Six inches of topsoil loss in western Tennessee has reduced corn yields by over 40%, and such relationships are not uncommon. Even when the nutrients required for adequate crop growth are added through fertilizers, other chemicals that contribute to the nutritional quality of the food grown may be lacking; in other words, the food itself may be less healthful. Also, the soil eroded from one place is deposited, sooner or later, somewhere else. If a large quantity is moved to other farmland, the crops on which it is deposited may be stunted or destroyed (recall figure 12.11B), although small additions of fresh topsoil may enrich cropland and make it more productive.

A subtle consequence of soil erosion in some places has been increased persistence of toxic residues of herbicides and pesticides in the soil. The loss of nutrients and organic matter through topsoil erosion may decrease the activity of soil microorganisms that normally speed the breakdown of many toxic agricultural chemicals. Many of these chemicals, which contribute significantly to water pollution, also pollute soils.

Another major soil-erosion problem is sediment pollution. In the United States, about 750 million tons per year of eroded sediment end up in lakes and streams. This decreases the water quality and may harm wildlife. The problem is still more acute when those sediments contain toxic chemical residues, as from agricultural herbicides and pesticides: The sediment is then both a physical and a potential chemical pollutant. A secondary consequence of this sediment load is the infilling of stream channels and reservoirs, restricting navigation and decreasing the volume of reservoirs and thus their usefulness for their intended purposes, whether for water supply, hydropower, or flood control. Tens of millions of dollars are spent each year to dredge and remove unwanted sediment deposits. The problems are widespread; see figure 12.12.

In recent years, there has been an increase in several kinds of activities that may increase soil erosion in places other than farms and cities. One is strip mining, which leaves behind readily erodable piles of soil and broken rock, at least until the land is reclaimed. Another is the use of off-road recreational vehicles (ORVs). This last is a special problem in dry areas where vegetation is not very vigorous or easily reestablished. Fragile plants can easily be destroyed by a passing ORV, leaving the land bare and vulnerable to intensified erosion (figure 12.13).

Figure 12.14 shows that erosion during active urbanization (highway and building construction and so forth) is considerably more severe than erosion on any sort of undeveloped land. Erosion during highway construction in the Washington, D.C., area, for example, was measured at rates up to 75 tons/acre/year. However, the total U.S. area undergoing active development is only about 1.5 million acres each year, as compared to nearly 400 million devoted to cropland. Also, construction disturbance is of relatively short duration. Once the land is covered by buildings, pavement, or established landscaping, erosion rates typically drop to below even natural, predevelopment levels. (Whether this represents an optimum land use is another issue.) The problems of erosion during construction, therefore, may be very intensive but are typically localized and short-lived.

Soil Erosion versus Soil Formation

Estimates of the total amount of soil erosion in the United States vary widely, but estimates by the U.S. Department of Agriculture Natural Resources Conservation Service put the figure at several billion tons per year. To a great extent, human activities, such as construction and especially farming (figure 12.15), account for the magnitude of soil loss. On cropland, average erosion losses from wind and water together have been estimated at 5.6 tons per acre per year. In very round numbers, that is about a 0.04 centimeter (0.02 inch) average thickness of soil removed each year. The rate of erosion from land under construction might be triple that (figure 12.15A). Erosion rates from other intensely disturbed lands, such as unreclaimed strip mines, might be at least as high as those for construction sites.

How this compares with the rate of soil *formation* is difficult to generalize because the rate of soil formation is so sensitive to climate, the nature of the parent rock material, and other factors. Upper limits on soil formation rates, however, can be determined by looking at soils in glaciated areas of the northern

A

Explanation

Area in which erosion or sedimentation is a significant problem

Water resources region boundary

Nature of erosion or sedimentation problems

- Cropland or rangeland erosion or sedimentation
- Urbanization, mining, or industrial and highway construction
- Natural erosion of stream channels
- Shoreline, streambank, or gully erosion
- Sedimentation of farm ponds, lakes, water supply, and flood-control channels

FIGURE 12.12 (A) Storm runoff washes soil off this Tennessee field. (B) Erosion and associated sediment deposition in stream channels is a common problem in many parts of the United States.
(A) photograph by Tim McCabe, courtesy USDA Natural Resources Conservation Service; (B) After The Nation's Water Resources 1975–2000, *U.S. Water Resources Council*

A

B

FIGURE 12.13 Erosion by off-road vehicles is not necessarily a new problem. (A) Widespread destruction of vegetation, and soil erosion, result when ORV users refuse to stay on marked trails, and often, the drier the land, the more severe the impact. ORV-track damage to Bureau of Land Management lands at Indian Creek, Utah, near Livermore, California. (B) Oregon Trail Ruts were carved in soft sedimentary rocks by the passage of many westward-bound wagon trains.

(A) Photo © David Wasserman/Brand X Pictures/Alamy; (B) Photograph © Lowell Georgia/Corbis Images

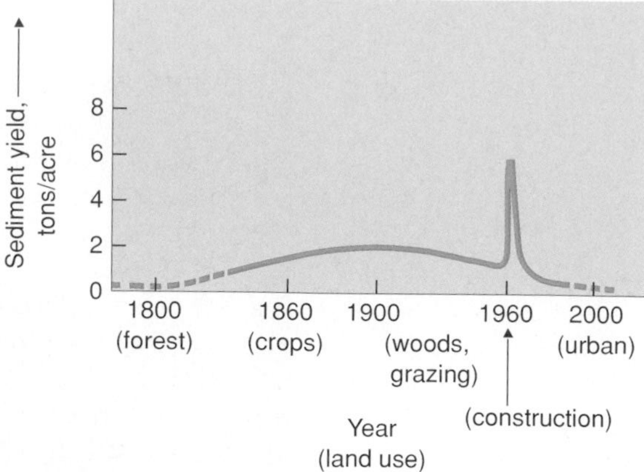

FIGURE 12.14 The impact of urbanization and other human activities on soil erosion rates. Construction may cause even more intensive erosion than farming, but far more land is affected by farming, and for a longer time. In this study of the Washington, D.C. area, erosion rates rose sharply during a highway construction boom around 1960; once construction was completed, erosion became negligible.

Reprinted from "A Cycle of Sedimentation and Erosion in Urban River Channels," M. G. Wolman from Geografiska Annaler, *Vol. 49, series A, figure 1, by permission of Scandinavian University Press*

A

B

FIGURE 12.15 Examples of water erosion of soil. (A) Gully erosion of soil bank exposed during construction, Des Moines, Iowa. (B) Gullying caused by surface runoff over farmland.

Photograph (A) by Lynn Betts, both courtesy of USDA Natural Resources Conservation Service

United States last scraped clean by the glaciers tens of thousands of years ago. Where the parent material was glacial till, which should weather more easily than solid rock, perhaps 1 meter (about 3 feet) of soil has formed in 15,000 years in the temperate, fairly humid climate of the upper Midwest. The corresponding average rate of 0.006 centimeters (0.002 inches) of soil formed per year is an order of magnitude less than the average cropland erosion rate. Furthermore, soil formation in many areas and over more resistant bedrocks is slower still. In some places in the Midwest, virtually no soil has formed on glaciated bedrocks, and in the drier southwest, soil formation is also likely to be very much slower. It seems clear that erosion is stripping away farmland far faster than the soil is being replaced. In other words, soil in general—and high-quality farmland—is also a nonrenewable resource in many populated areas.

Strategies for Reducing Erosion

The wide variety of approaches for reducing erosion on farmland basically involve either reducing the velocity of an eroding agent or protecting the soil from its effects. Under the latter heading come such practices as leaving stubble in the fields after a crop has been harvested and planting cover crops in the off-season between cash crops. In either case, the plants' roots help to hold the soil in place, and the plants themselves, to some extent, shield the soil from wind and rain (figure 12.16).

The lower the wind or water runoff velocity, the less material carried. Surface runoff may be slowed on moderate slopes by contour plowing (figure 12.17A). Plowing rows parallel to the contours of the hill, perpendicular to the direction of water flow, creates a ridged land surface so that water does not rush downhill as readily. Other slopes may require terracing (figure 12.17B), whereby a single slope is terraced into a series

FIGURE 12.16 On this Iowa cornfield, stubble has been left in the field over winter; soil is cultivated and planted in one step in the spring.
Photograph by Lynn Betts, courtesy of USDA Natural Resources Conservation Service

A

B

C

FIGURE 12.17 Decreasing slope of land, or breaking up the slope, decreases erosion by surface runoff. (A) Contour-plowed field in northern Iowa. (B) Terracing creates "steps" of shallower slope from one long, steeper slope. (C) Terraces become essential on very steep slopes, as with these terraces carved out of jungle.
(A) Courtesy of Tim McCabe/NRCS/USDA; (B) photograph courtesy of USDA Soil Conservation Service; (C) © The McGraw-Hill Companies, Inc./Barry Barker, photographer

A

C

B

FIGURE 12.18 Creating near-ground irregularities decreases wind velocity. (A) Windbreaks retard wind erosion by decreasing wind velocity. (B) Strip cropping also breaks up the surface topography, disrupting sweeping winds. Allamakee County, Iowa. (C) Contour strip cropping of corn and alfalfa in northeastern Iowa.

(A) Photograph © John D. Cunningham/Visuals Unlimited. (B) & (C) Courtesy of Tim McCabe/NRCS/USDA

of shallower slopes, or even steps that slant backward into the hill. Again, surface runoff makes its way down the shallower slope more slowly, if at all, and therefore carries far less sediment with it. Grass is typically left on the steeper slopes at the edges of the terraces to protect them from erosion. Terracing has, in fact, been practiced since ancient times. Both terracing and contour plowing, by slowing surface runoff, increase infiltration and enhance water conservation as well as soil conservation. Often, terracing and contouring are used in conjunction (figures 12.17B,C).

Wind can be slowed down by planting hedges or rows of trees as windbreaks along field borders or in rows perpendicular to the dominant wind direction (figure 12.18A) or by erecting low fences, like snow fences, similarly arrayed. This does not altogether stop soil movement, as shown by the ridges of soil that sometimes pile up along the windbreaks. However, it does reduce the distance over which soil is transported, and some of the soil caught along the windbreaks can be redistributed over the fields. Strip cropping (figure 12.18B), alternating crops of different heights, slows near-ground wind by making the land surface more irregular. Combining strip cropping and contouring (figure 12.18C) may help reduce both wind and water erosion.

A major obstacle to erosion control on farmland is cost. Many of the recommended measures—for example, terracing and planting cover crops—can be expensive, especially on a large scale. Even though the long-term benefits of reduced erosion are obvious and may involve long-term savings or increased income, the effort may not be made if short-term costs seem too high. The benefits of a particular soil conservation measure may be counterbalanced by substantial drawbacks, such as reduced crop yields. Also, if strict erosion-control standards are imposed selectively on farmers of one state (as by a state environmental agency, for instance), those farmers are at a competitive disadvantage with respect to farmers elsewhere who are not faced with equivalent expenses. Even meeting the same standards does not cost all farmers everywhere equally. Increasingly, the government is becoming involved in cost-sharing for erosion-reduction programs so that the financial burden does not fall too heavily on individual farmers. Since the Dust Bowl era, over $20 billion in federal funds have been spent on soil conservation efforts.

The USDA Natural Resources Conservation Service reports that a total of 2.1 billion tons of soil was lost from cropland in 1992 (figure 12.19A), 40% of it from just five states (Texas, Minnesota, Iowa, Montana, and Kansas); such losses are concentrated in states with significant agricultural activity, naturally enough, but local climate, soil types, and topography influence vulnerability to erosion, and which kind predominates (figure 12.19B). Clear progress has been made in erosion reduction over the past two decades (figure 12.20), though the rate of progress has slowed.

Other strategies can minimize erosion in nonfarm areas. For example, in areas where vegetation is sparse, rainfall limited, and erosion a potential problem, ORVs should be restricted to prescribed trails. In the case of urban construction projects, one reason for the severity of the resulting erosion is that it is common practice to clear a whole area, such as an entire housing-project site, at the beginning of the work, even though only a small portion of the site is actively worked at any given time. Clearing the land in stages, as needed, minimizes the length of time the soil is exposed and reduces urban erosion. Mulch or temporary ground covers can protect soil that must be left exposed for longer periods. Stricter mining regulations requiring reclamation of strip-mined land (see chapter 13) are already significantly reducing soil erosion and related problems in these areas.

Where erosion cannot be eliminated, sediment pollution may nevertheless be reduced. "Soil fences" that act somewhat like snow fences, and hay-lined soil traps through which runoff water must flow, may reduce soil runoff from construction sites. On either construction sites or farmland, surface runoff water may be trapped and held in ponds to allow suspended sediment to settle out in the still ponds before the clarified water is released; such settling ponds are illustrated in chapter 17.

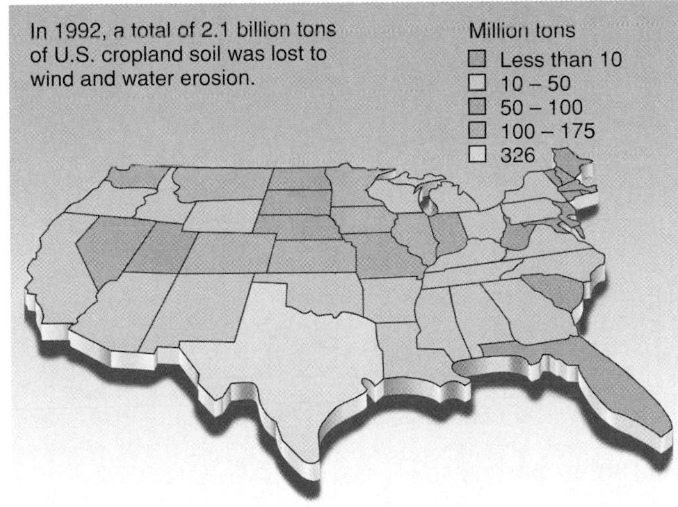

A

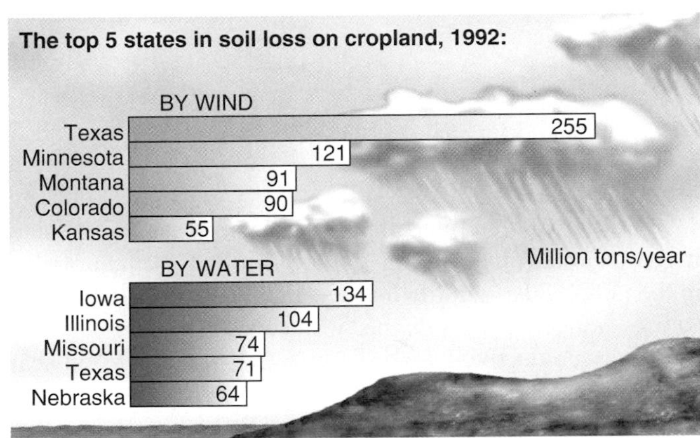

B

FIGURE 12.19 Soil erosion in the United States has been reduced substantially, but it remains a significant problem on cropland. It is particularly acute in a limited number of states (A); whether wind or water is the more serious threat varies by location (B), and total loss per state is also determined by land area under cultivation.

Source: National Resources Inventory April 1995, *USDA Natural Resources Conservation Service*

Irrigation and Soil Chemistry

We have noted leaching as one way in which soil chemistry is modified. Application of fertilizers, herbicides, and pesticides —whether on agricultural land or just one's own backyard—is another way that soil composition is changed, in this case by the addition of a variety of compounds through human activities. Less-obvious additions can occur when irrigation water redistributes soluble minerals.

In dry climates, for example, irrigation water may dissolve salts in pedocal soils; as the water evaporates near the surface, it can redeposit those salts in near-surface soil from which they had previously been leached away. As the process continues, enough salt may be deposited that plant growth suffers (figure 12.21A).

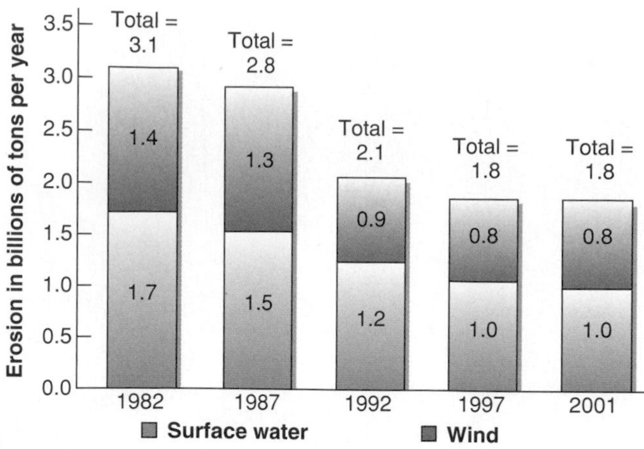

FIGURE 12.20 Erosion reduction on cropland and land removed from production due to erosion susceptibility under the CRP (Conservation Reserve Plan) of 1995.

Data from USDA Natural Resources Conservation Service, Natural Resources Inventory 2001

A

B

FIGURE 12.21 (A) Salt buildup on land surface in Colorado, due to irrigation: salty water wicks to the drier surface and evaporates, leaving salty crust. (B) Kesterson Wildlife Refuge, with selenium-contaminated ponds. These would be an attractive habitat for waterfowl, were it not for the toxicity of the waters.

(A) Photograph by Tim McCabe, courtesy of USDA Natural Resources Conservation Service; (B) courtesy of USGS Photo Library, Denver, CO.

A different kind of chemical-toxicity problem traceable to irrigation was recognized in California's San Joaquin Valley in the mid-1980s. Wildlife managers in the Kesterson Wildlife Refuge noticed sharply higher incidence of deaths, deformities, and reproductive failures in waterfowl nesting in and near the artificial wetlands created by holding ponds for irrigation runoff water (figure 12.21B). In the 1960s and 1970s, that drainage system had been built to carry excess runoff from the extensively irrigated cropland in the valley, precisely to avoid the kind of salinity buildup just described. The runoff was collected in ponds at Kesterson and seemed to create beneficial habitat. What had not initially been realized is that the irrigated soils were also rich in selenium.

Selenium is an element present in trace amounts in many soils. Small quantities are essential to the health of humans (note that a recommended daily allowance is shown on your multivitamin/mineral-supplement package!), livestock, and other organisms. As with many essential nutrients, however, large quantities can be toxic, and selenium tends to accumulate in organisms that ingest it and to increase in concentration up a food chain. From selenium-rich rocks in the area, selenium-rich soils were derived. This was no particular problem until the soluble selenium was dissolved in the irrigation water, drained into the ponds at Kesterson, and concentrated there by evaporation in the ponds, where it built up through the algae→insects→fish→birds food chain. In time, it reached levels obviously toxic to the birds.

The short-term solution at Kesterson was to fill in the ponds. The episode underscored the more general issue of the need to consider the effects of adding water to soils which, in dry climates, may contain quantities of soluble minerals that, when mobilized in solution, can present a new hazard.

The Soil Resource—The Global View

Soil degradation is a concern worldwide. Desertification, erosion, deterioration of lateritic soil, contamination from pollution, and other chemical modification by human activity all contribute to reduced soil quality, fertility, and productivity. The problems are widespread (figure 12.22A); the contributing causes vary regionally (figure 12.22B).

Land area is finite. Differences in population density as well as soil quality create great disparities in per-capita arable land by region of the globe, and the numbers decline as population grows and as arable land is covered up by development (figure 12.23). Soil degradation simply further diminishes the availability of the farmland needed to feed the world.

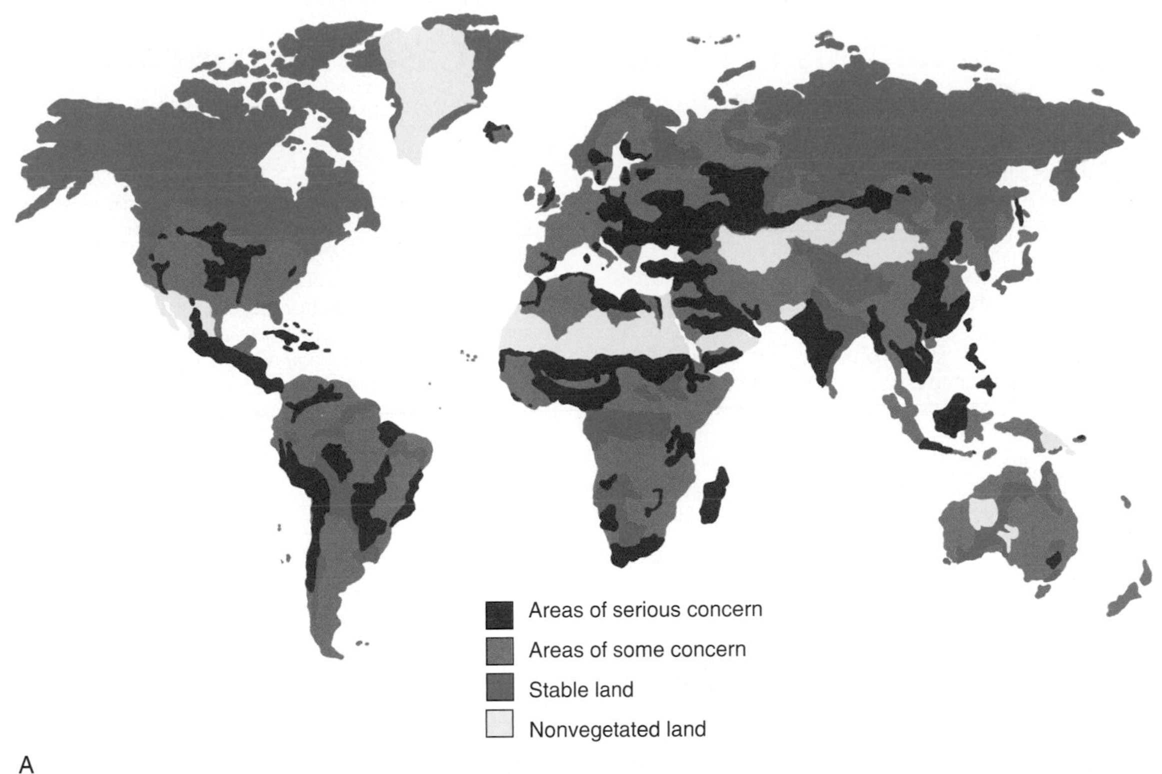

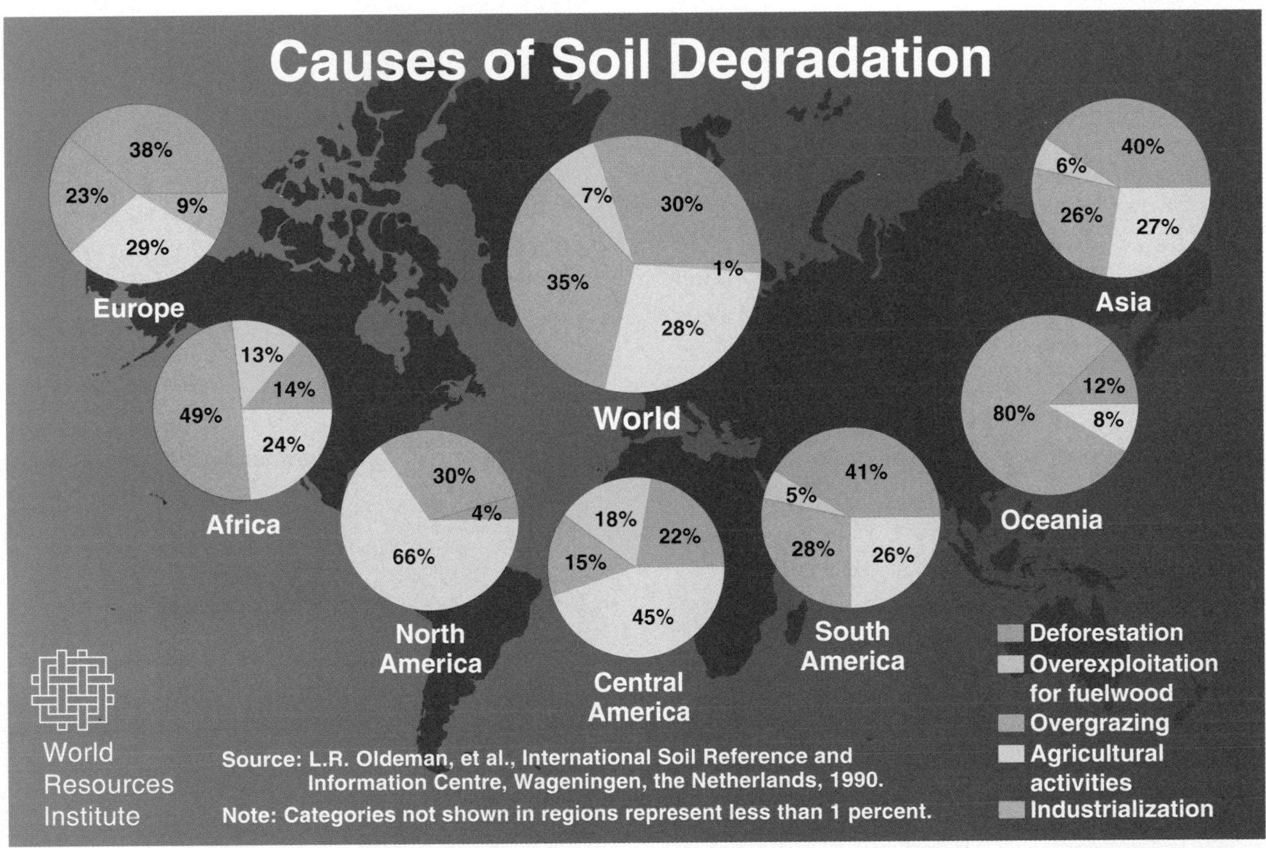

FIGURE 12.22 Soil degradation of various kinds is a global problem. (A) Six continents are significantly affected by soil degradation; on four (Africa, Australia, Europe, and South America), little land is considered stably vegetated. (B) The causes of soil degradation, however, vary around the world.
Sources: (A) data from Global Resource Information Database of U.N. Environment Programme; (B) from World Resources Institute

Box 12.1 Plantations in Paradise—A Case of Secondary Consequences

Earlier in the chapter, we focused on the characteristics of tropical soils and how they change with farming. Problems associated with agricultural activities in such areas are by no means limited to Third World countries, or to the soil itself. Broad areas of the Hawaiian Islands are devoted to pineapple and sugarcane crops. Practices associated with the latter are particularly controversial.

When the sugarcane is ready to be harvested, it is set afire. The leaves are consumed; only the charred canes remain. This serves several purposes: Like slash-and-burn methods of clearing land, it restores nutrients to the soil; it reduces the volume and weight of material that has to be hauled to the sugar mill, removing the unwanted leaves in the field; and the heat kills insects that might otherwise linger to attack the next planting. However, it also contributes to air pollution, producing an acrid-sweet brown smoke that is unsightly (figure 1) and occasionally disrupts traffic when visibility on nearby roads becomes too poor.

The cane fields are at least generally on fairly level ground, where the huge cane-hauling trucks can operate, so sediment runoff between crops is reduced. Some of the pineapple plantations occupy much steeper terrain, and runoff has created problems offshore, among coral-reef communities. Corals require clear (as well as warm) waters to thrive. A major reason that Australia's Great Barrier Reef could grow so spectacular is that relatively few rivers carry sediment off the east side of the continent. Increased sediment runoff makes the water murkier, and corals die (figure 2). The World Resources Institute estimates that between 20% and 25% of reefs around the world are threated by inland pollution and sediment erosion, and another 30% by coastal development, which can also increase sediment runoff.

A

B

FIGURE 1 The brown smoky haze from burning sugar cane is distinctive and very visible (A); the process clears the cane field and restores nutrients to the tropical soil for the next crop (B).

A

B

FIGURE 2 (A) In clear tropical waters, corals and associated reef communities can thrive. (B) Sediment runoff clouds the water, and the coral reef and ecosystem suffer.

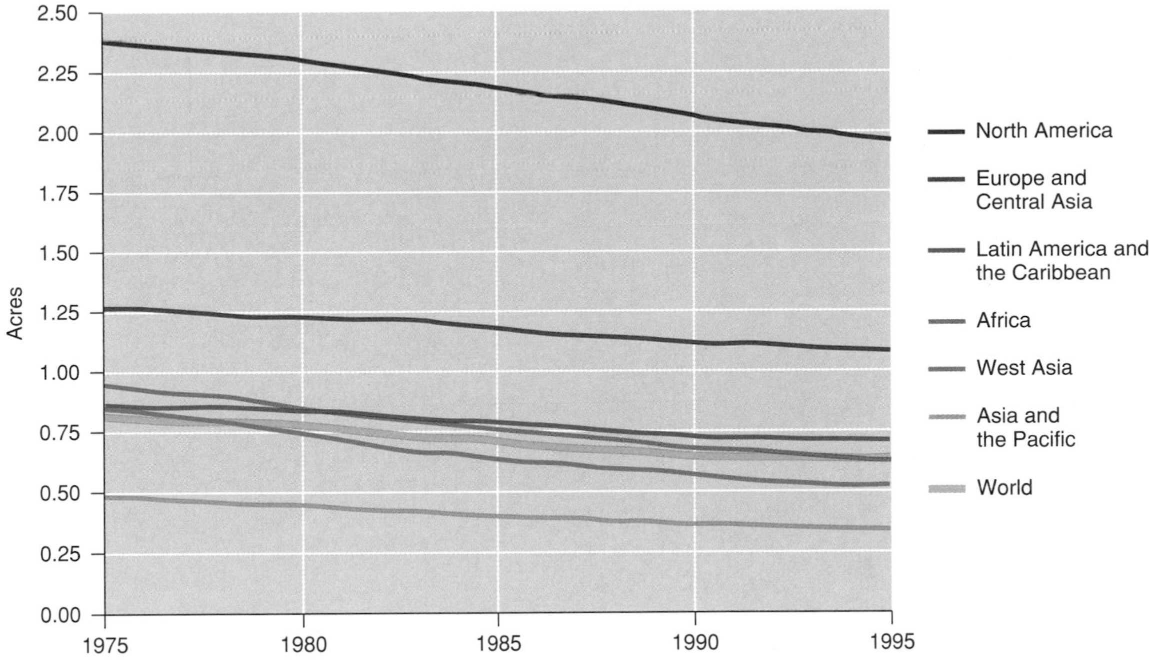

FIGURE 12.23 Population growth, the spread of cities and suburbs onto arable land, and loss of arable land through degradation all reduce available farmland per capita; the resource is especially scarce in Asia.
Data from the U.N. Environment Programme Global Resource Information Database

SUMMARY

Soil forms from the breakdown of rock materials through mechanical, chemical, and biological weathering. Weathering rates are closely related to climate. The character of the soil reflects the nature of the parent material and the kinds and intensities of weathering processes that formed it. The wetter the climate, the more leached the soil. The lateritic soil of tropical climates is particularly unsuitable for agriculture: Not only is it highly leached of nutrients, but when exposed, it may harden to a bricklike solidity. The texture of soil is a major determinant of its drainage characteristics; soil structure is related to its suitability for agriculture. Agricultural practices may also alter soil chemistry, whether through direct application of additional chemicals or through irrigation-induced changes in natural cycles.

Soil erosion by wind and water is a natural part of the rock cycle. Where accelerated by human activity, however, it can also be a serious problem, especially on farmland or, locally, in areas subject to construction or strip mining. Erosion rates far exceed inferred rates of soil formation in many places. A secondary problem is the resultant sediment pollution of lakes, streams, and nearshore ocean waters. Strategies to reduce soil erosion on farmland include terracing, contour plowing, planting or erecting windbreaks, the use of cover crops, strip cropping, and minimum-tillage farming. Elsewhere, restriction of ORVs, more selective clearing of land during construction, the use of sediment traps and settling ponds, and careful reclamation of strip-mined areas could all help to minimize soil erosion.

Key Terms and Concepts

A horizon 273
B horizon 273
chemical weathering 271
C horizon 273
E horizon 273
laterite 275
leaching 273
mechanical weathering 271
O horizon 273
pedalfer 275
pedocal 275
soil 271
zone of accumulation 273
zone of leaching 273

Exercises

Questions for Review

1. Briefly explain how the rate of chemical weathering is related to (a) the amount of precipitation, (b) the temperature, and (c) the amount of mechanical weathering.
2. Sketch a generalized soil profile and indicate the A horizon, B horizon, C horizon, E horizon, O horizon, zone of leaching, and zone of accumulation. Is such a profile always present?
3. How do pedalfer and pedocal types of soil differ, and in what kind of climate is each more common?
4. The lateritic soil of the tropical jungle is poor soil for cultivation. Explain.
5. Soil erosion during active urbanization is far more rapid than it is on cultivated farmland, and yet the majority of soil-conservation efforts are concentrated on farmland. Why?
6. Cite and briefly describe three strategies for reducing cropland erosion.
7. Irrigation can, over time, cause harmful changes in soil chemistry that reduce crop yields or create toxic conditions in runoff water. Explain briefly.

For Further Thought

1. Choose any major agricultural region of the United States and investigate the severity of any soil erosion problems there. What efforts, if any, are being made to control that erosion, and what is the cost of those efforts?
2. Visit the site of an active construction project; examine it for evidence of erosion, and erosion-control activities.
3. How does your state fare with respect to soil erosion? What factors can you identify that contribute to the extent—or the lack—of soil erosion? (The National Resources Inventory 2001 may be a place to start, the Natural Resources Conservation Service or your local geological survey may be additional resources.)
4. Assume that solid rock can be eroded at a rate comparable to that at which it can be converted to soil, and use the rate of 0.006 centimeter per year as representative. How long would it take to erode away a mountain 1500 meters (close to 5000 feet) high?

Suggested Readings/References

Bohn, H. L., McNeal, B. L., and O'Connor, G. A. 2001. *Soil chemistry.* 3d ed. New York: John Wiley & Sons, Inc.

Crosson, P. R., and Rosenberg, N. J. 1989. Strategies for agriculture. *Scientific American* 261 (September): 128–35.

El-Baz, F. 1993. Saving the Sphinx. *Geotimes* (May): 12–17.

Lal, R. 1987. Managing the soils of sub-Saharan Africa. *Science* 236: 1069–76.

Loynachan, T. E., *et al.* 2006. *Soils, society, and the environment.* Alexandria, VA: Amer. Geological Inst.

Nizeyimana, E. L., Petersen, G. W., and Warner, E. D. 2002. Tracking farmland loss. *Geotimes* (January): 18–20.

Martini, I. P., and Chesworth, W., eds. 1992. *Weathering, soils, and paleosols.* New York: Elsevier.

Ollier, C., and Pain, C. 1997. *Regolith, soils and landforms.* New York: John Wiley & Sons, Inc.

Pimentel, D., *et al.* 1995. Environmental and economic costs of soil erosion and conservation benefits. *Science* (24 February): 1117–23.

Schultz, H. D., and Hadeler, A., eds. 2003. *Geochemical processes in soil and ground water.* New York: John Wiley & Sons, Inc.

Trimble, S. W., and Crosson, P. 2000. U.S. soil erosion rates—myth and reality. *Science* 289 (14 July): 248–50.

U.S. Department of Agriculture Natural Resources Conservation Service. 2003. *National resources inventory 2001.* [online] http://www.nrcs.usda.gov/technical/land/nri01/

World Resources Institute. 1994. *World Resources 1994–95.* (Chapter 6, "Food and Agriculture.") New York: Oxford University Press.

World Resources Institute, United Nations Environment Programme, United Nations Development Programme, and The World Bank. 1996. *World Resources 1996–97.* (Chapter 9, "Forests and Land Cover"; Chapter 10, "Food and Agriculture.") New York: Oxford University Press.

Welcome to McGraw-Hill's Online Learning Center

Location: http://www.mhhe.com/montgomery8e

NetNotes

The U.S. Department of Agriculture Natural Resources Conservation Service (formerly USDA Soil Conservation Service) site, which includes on extensive photo gallery, is located at www.nrcs.usda.gov/

Their World Soil Resources Office site is soils.usda.gov/use/worldsoils

and the National Soil Information System at nasis.nrcs.usda.gov/

with a general "Soils" website at soils.usda.gov/

A set of online posters illustrating and describing the twelve soil orders is found at soils.usda.gov/technical/soil_orders

The National Park Service engages in soil inventory and mapping activities: www2.nature.nps.gov/geology/soils

The Smithsonian Institution and Soil Science Society of America have an online soils exhibit: www.soils.org/smithsonian/index.html

The National Wetlands Research Center is at www.nwrc.gov/

The Canadian Soil Information System is at sis.agr.gc.ca/cansis

The International Union of Soil Sciences Working Group on Land Degradation and Desertification's International Soil Reference and Information Center is www.isric.nl/

and the International Erosion Control Association at www.ieca.org

Soil-related information at WRI's EarthTrends site can be found by clicking on "Agriculture and Food" at http://earthtrends.wri.org/

A feature on the use of soil-chemistry mapping to address the selenium contamination at Kesterson is at minerals.cr.usgs.gov/gips/na/0elemap.htm#elemap

13 Mineral and Rock Resources

Though we may not often think about it, virtually everything we build or create or use in modern life involves rock, mineral, or fuel resources. In some cases, this is obvious: We can easily see the steel and aluminum used in vehicles and appliances; the marble or granite of a countertop or floor; the gold, silver, and gemstones in jewelry; the copper and nickel in coins. But our dependence on these resources is much more pervasive. Asphalt highways (which themselves use petroleum resources) are built on crushed-rock roadbeds. A wooden house is built on a slab or foundation of concrete (which requires sand and limestone to make), held together with metal nails, finished inside with wallboard (made with gypsum), wired with copper or aluminum, given a view by windows (glass made from quartz-rich sand); the bricks of a brick house are made from clay. Your toothpaste cleans your teeth using mineral abrasives (such as quartz, calcite, corundum, or phosphate minerals), reduces cavities with fluoride (from the mineral fluorite), is probably colored white with mineral-derived titanium dioxide (the same pigment as in white paint), and, if it sparkles, may contain mica flakes. The computers and other electronic devices on which we increasingly rely would not exist without metals for circuits and semiconductors.

The bulk of the earth's crust is composed of fewer than a dozen elements. In fact, as noted in table 1.2, eight chemical elements make up more than 98% of the crust. Many of the elements *not* found in abundance in the earth's crust, including industrial and precious metals, essential components of chemical fertilizers, and elements like uranium that serve as energy sources, are vitally important to society. Some of these are found in very minute amounts in the average rock of the continental crust: copper, 100 ppm; tin, 2 ppm; gold, 4 ppb. Clearly, then, many useful elements must be mined from very atypical rocks. This chapter begins with a look at the occurrences of a variety of rock and mineral resources and then examines the U.S. and world supply-and-demand picture.

Quarrying sand and gravel, Wasatch Front, Utah.
Photograph © The McGraw-Hill Companies/John A. Karachewski, photographer

ORE DEPOSITS

An **ore** is a rock in which a valuable or useful metal occurs at a concentration sufficiently high, relative to average rocks, to make it economically worth mining. A given ore deposit may be described in terms of the enrichment or **concentration factor** of a metal of interest:

$$\text{Concentration factor of a given metal in an ore} = \frac{\text{Concentration of the metal in the ore deposit}}{\text{Concentration of the metal in average continental crust}}$$

The higher the concentration factor, the richer the ore, and the less of it needs to be mined to extract a given amount of the metal.

In general, the minimum concentration factor required for profitable mining is inversely proportional to the average crustal concentration: If ordinary rocks are already fairly rich in the metal, it need not be concentrated much further to make mining it economic, and vice versa. Metals like iron or aluminum, which make up about 6% and 8% of average continental crust, respectively, need only be concentrated by a factor of 4 to 5 times for mining to be profitable. Copper must be enriched about 100 times relative to average rock, while mercury (average concentration 80 ppb) must be enriched to about 25,000 times its average concentration before the ore is rich enough to mine profitably. The exceptions to this rule are those few extremely valuable elements, such as gold, that are so high-priced that a small quantity justifies a considerable amount of mining. At several hundred dollars an ounce, gold need be found in concentrations only a few thousand times its average 4 ppb to be worth mining. This inverse relationship between average concentration and the concentration factor required for profitable mining further suggests that economic ore deposits of the relatively abundant metals (such as iron and aluminum) might be more plentiful than economic deposits of rarer metals, and this is indeed commonly the case.

The value of the mineral or metal extracted and its concentration in a particular deposit, then, are major factors determining the profitability of mining a specific deposit. The economics are naturally sensitive to world demand. If demand climbs and prices rise in response, additional, not-so-rich ore deposits may be opened up; a fall in price causes economically marginal mines to close. The discovery of a new, rich deposit of a given ore may make other mines with poorer ores noncompetitive and thus uneconomic in the changing market. The practicality of mining a specific ore body may also depend on the mineral(s) in which a metal of interest is found, because this affects the cost to extract the pure metal. Three iron deposits containing equal concentrations of iron are not equally economic if the iron is mainly in oxides in one, in silicates in another, and in sulfides in the third. The relative densities of the iron-bearing minerals, compared to other minerals in the rock, will affect how readily those iron-rich minerals can be separated from the rest; the nature and strengths of the chemical bonds in the iron-rich phases will influence the energy costs to break them down to extract the iron, and the complexity of chemical processes required; the concentration of iron in the iron-rich minerals (rather than in the rock as a whole) will determine the yield of iron from those minerals, after they are separated. These factors and others influence the economics of mining.

By definition, ores are somewhat unusual rocks. Therefore, it is not surprising that the known economic mineral deposits are very unevenly distributed around the world. (See, for example, figure 13.1.) The United States controls about 50% of the world's known molybdenum deposits and about 15% of the lead. But although the United States is the major world consumer of aluminum, using over 20% of the total produced, it has virtually no domestic aluminum ore deposits; Australia, Guinea, and Jamaica together control one-half of the world's aluminum ore. Congo (Kinshasa, formerly Zaire) accounts for nearly half the known economic cobalt deposits, Chile for almost 30% of the copper, and China, Malaysia, and Brazil for, respectively, 27%, 16%, and 16% of the tin. South Africa controls nearly half the world's known gold ore, over 75% of the chromium, and almost 90% of the platinum-group metals. These great disparities in mineral wealth among nations are relevant to later discussions of world supply/demand projections.

TYPES OF MINERAL DEPOSITS

Deposits of economically valuable rocks and minerals form in a variety of ways. This section does not attempt to review them all but describes some of the more important processes involved.

Igneous Rocks and Magmatic Deposits

Magmatic activity gives rise to several different kinds of deposits. Certain igneous rocks, just by virtue of their compositions, contain high concentrations of useful silicate or other minerals. The deposits may be especially valuable if the rocks are coarse-grained, so that the mineral(s) of interest occur as large, easily separated crystals. **Pegmatite** is the term given to unusually coarse-grained igneous intrusions (figure 13.2). In some

FIGURE 13.1 Proportions of world reserves of some nonfuel minerals controlled by various nations. Although the United States is a major consumer of most metals, it is a major producer of very few.

Source: Data from Mineral Commodity Summaries 2000, *U.S. Geological Survey*

FIGURE 13.2 Pegmatite, a very coarse-grained plutonic rock, may yield large, valuable crystals. This sample contains tourmaline crystals (green) up to about 15 cm (6″) long; crystals in pegmatites may be much larger still.

pegmatites, single crystals may be over 10 meters (30 feet) long. Feldspars, which provide raw materials for the ceramics industry, are common in pegmatites. Many pegmatites also are enriched in uncommon elements. Examples of rarer minerals mined from pegmatites include tourmaline (used both as a gemstone and for crystals in radio equipment) and beryl (mined for the metal beryllium when crystals are of poor quality, or used as the gemstones aquamarine and emerald when found as clear, well-colored crystals).

Other useful minerals may be concentrated within a cooling magma chamber by gravity. If they are more or less dense than the magma, they may sink or float as they crystallize, instead of remaining suspended in the solidifying silicate mush, and accumulate in thick layers that are easily mined (figure 13.3). Chromite, an oxide of the metal chromium, and magnetite, one of the iron oxides, are both quite dense. In a magma of suitable bulk composition, rich concentrations of these minerals may form in the lower part of the magma chamber during crystallization of the melt. The dense precious metals, such as gold and platinum, may also be concentrated during magmatic crystallization. Even where disseminated throughout an igneous rock body, a very valuable mineral commodity may be worth mining. This is true, for example, of diamonds. One reason for the rarity of diamonds is that they must be formed at extremely high pressures, such as those found within the mantle, and then brought rapidly up into the crust. They are mined primarily from igneous rocks called **kimberlites,** which occur as pipelike intrusive bodies that must have originated in the mantle. Even where only a few gem-quality diamonds are scattered within many tons of kimberlite, their high unit value makes the mining profitable.

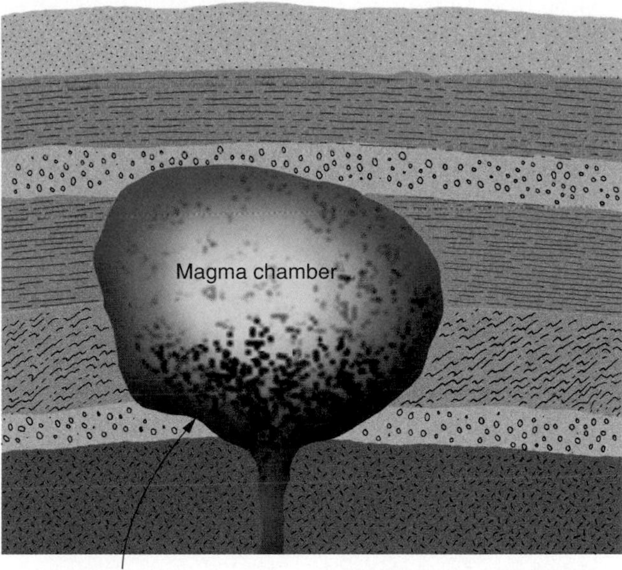

FIGURE 13.3 Formation of magmatic ore deposit by gravitational settling of a dense mineral during crystallization.

Hydrothermal Ores

Not all mineral deposits related to igneous activity form within igneous rock bodies. Magmas have water and other fluids dissolved in or associated with them. Particularly during the later stages of crystallization, the fluids may escape from the cooling magma, seeping through cracks and pores in the surrounding rocks, carrying with them dissolved salts, gases, and metals. These warm fluids can leach additional metals from the rocks through which they pass. In time, the fluids cool and deposit their dissolved minerals, creating a **hydrothermal** (literally, "hot water") ore deposit (figure 13.4).

The particular minerals deposited vary with the composition of the hydrothermal fluids, but worldwide, a great variety of metals occur in hydrothermal deposits: copper, lead, zinc, gold, silver, platinum, uranium, and others. Because sulfur is a common constituent of magmatic gases and fluids (figure 13.5), the ore minerals are frequently sulfides. For example, the lead in lead ore deposits is found in galena (PbS); zinc, in sphalerite (ZnS); and copper, in a variety of copper and copper-iron sulfides ($CuFeS_2$, CuS, Cu_2S, and others).

The hydrothermal fluids need not all originate within the magma. Sometimes, circulating subsurface waters are heated sufficiently by a nearby cooling magma to dissolve, concentrate, and redeposit valuable metals in a hydrothermal ore deposit. Or the fluid involved may be a mix of magmatic and nonmagmatic fluids. Indeed, some hydrothermal deposits are found far from any obvious magma source, and those fluids might have been of metamorphic origin.

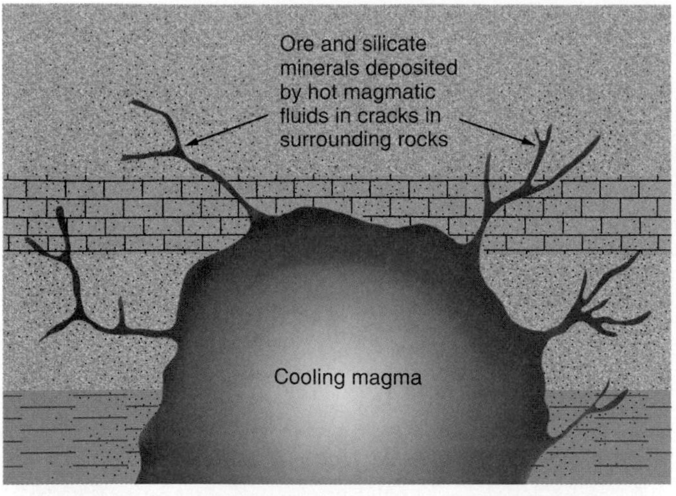

A

B

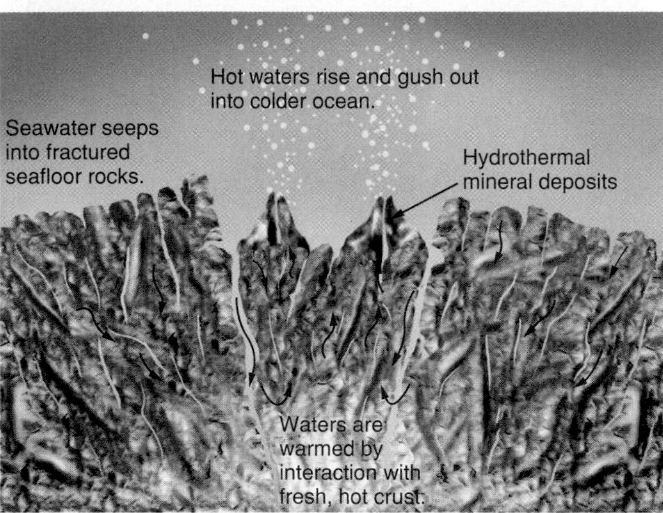

C

FIGURE 13.4 Hydrothermal ore deposits form in various ways. (A) Ore deposition in veins around a magma chamber. Less-concentrated ore is disseminated through the rock as fluids seep outward from magma. (B) Hydrothermal sulfides permeate this basaltic rock. The most visible are pyrite (brassy gold crystals). (C) Sulfides can be deposited by hydrothermal circulation around a spreading ridge.
Photograph (B) by W. R. Normark, USGS Photo Library, Denver, CO.

FIGURE 13.5 Sulfur deposition around a fumarole, Kilauea, Hawaii.
Photograph by J. C. Ratté, USGS Photo Library, Denver, CO.

Relationship to Plate Margins

The relationship between magmatic or metamorphic activity and the formation of many ore deposits suggests that hydrothermal and igneous-rock deposits should be especially common in regions of extensive magmatic activity—that is, plate boundaries. This is generally true. A striking example of this link can be seen in figure 13.6A, which shows the locations of a certain type of copper and molybdenum deposit of igneous origin in North and South America. Comparison with figure 3.18 shows that these locations correspond closely to present or recent subduction zones. The same plate-tectonic influence can be seen in figure 13.6B: The precious-metal deposits in the western United States are predominantly hydrothermal deposits related to magmatic activity associated with plate collision and subduction along the western margin of the continent.

Scientists using submersible research vessels to investigate underwater spreading ridges have discovered hydrothermal

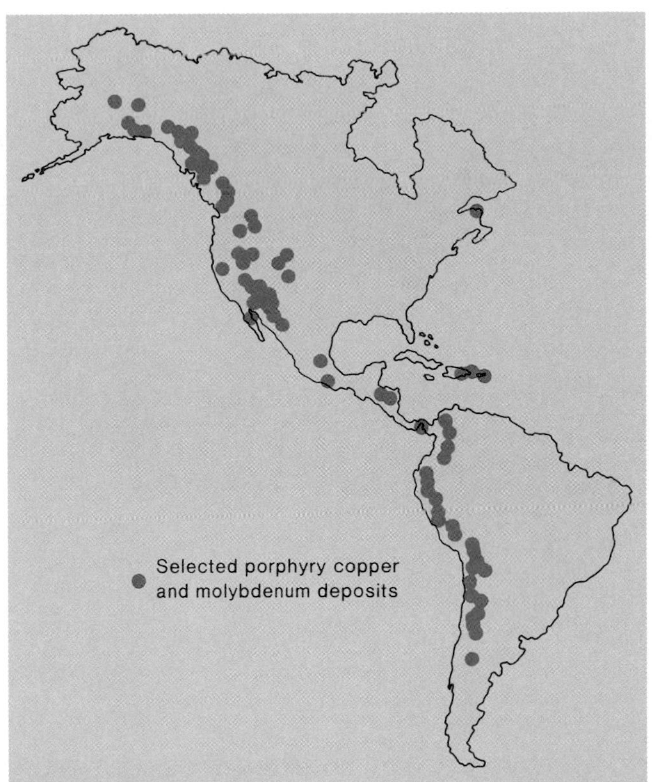

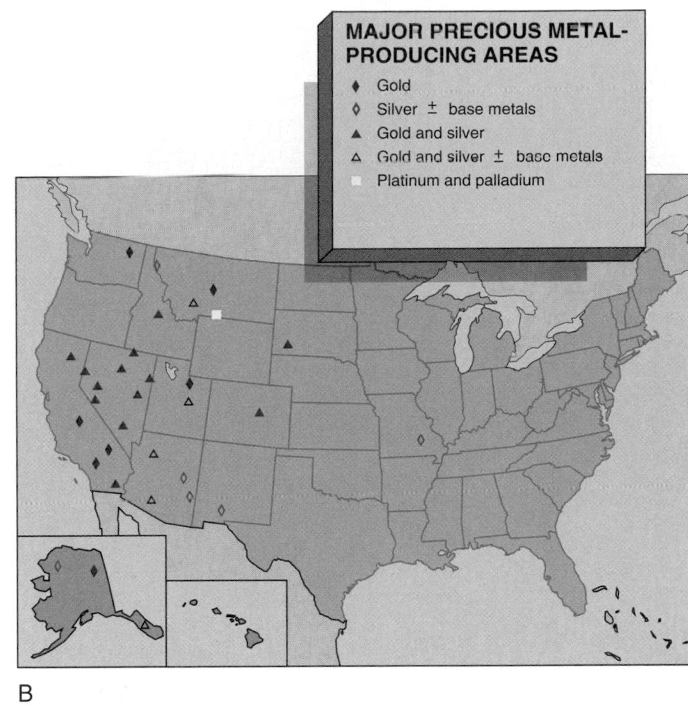

FIGURE 13.6 (A) Distribution of copper and molybdenum deposits. (Deposit in eastern Canada relates to an ancient subduction zone, rather than to modern subduction.) (B) Major precious-metal-producing areas in the United States. ("Base metals" are the more common metals, such as copper and lead, as distinguished from "precious metals" like gold and silver.)

(A) Source: Data from M. L. Jensen and A. M. Bateman, Economic Mineral Deposits, *3d ed. Copyright © 1981 John Wiley and Sons, Inc., New York. (B) Source: Data from* Mineral Commodity Summaries 2003, *U.S. Bureau of Mines*

fluids gushing from the sea floor at places along these ridges, depositing sulfides as they cool (figure 13.7A). Many of these hydrothermal vents have been nicknamed "black smokers" for the dark cloud of suspended sulfide minerals issuing from them. Similar activity probably accounts for the metal-rich muds at the bottom of the Red Sea, which is also a spreading rift zone and the site of a developing ocean basin (see figure 13.7B).

Sedimentary Deposits

Sedimentary processes can also produce economic mineral deposits. Some such ores have been deposited directly as chemical sedimentary rocks. Layered sedimentary iron ores, called **banded iron formation,** are an example (figure 13.8). Iron-rich layers (predominantly hematite or magnetite) alternate with silicate- or carbonate-rich layers. These large deposits, which may extend for tens of kilometers, are, for the most part, very ancient. Their formation is believed to be related to the development of the earth's atmosphere. As pointed out in chapter 1, the early atmosphere would have lacked free oxygen. Under those conditions, iron from the weathering of continental rocks would have been very soluble in the oceans. As photosynthetic organisms began producing oxygen, that oxygen would have reacted with the dissolved iron and caused it to precipitate. If the majority of large iron ore deposits formed in this way, during a time long past when the earth's surface chemistry was very different from what it is now, it follows that similar deposits are unlikely to form now or in the future.

Other sedimentary mineral deposits can form from seawater, which contains a variety of dissolved salts and other chemicals. When a body of seawater trapped in a shallow sea dries up, it deposits these minerals in **evaporite** deposits (figure 13.9). Some evaporites may be hundreds of meters thick. Ordinary table salt, known mineralogically as halite, is one mineral commonly mined from evaporite deposits. Others include gypsum and salts of the metals potassium and magnesium.

Other Low-Temperature Ore-Forming Processes

Streams also play a role in the formation of mineral deposits. They are rarely the sites of primary formation of ore minerals. However, as noted in chapter 6, streams often deposit sediments

A

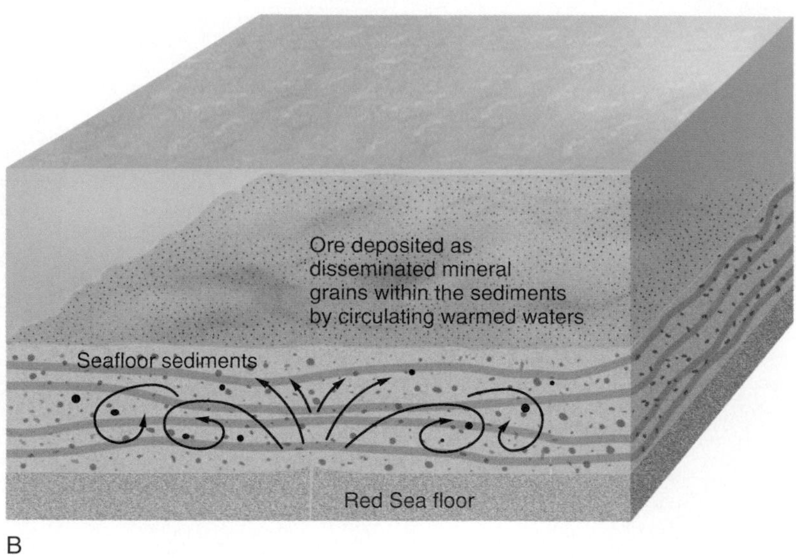

FIGURE 13.7 Effects of hydrothermal activity along the East Pacific Rise. (A) An active "black smoker chimney" vent. The black "smoke" is mainly fine-grained iron sulfide. (B) Sulfides are also being deposited in Red Sea muds.
(A) Photograph © B. Murton/Southhampton Oceanography Centre/Photo Researchers

FIGURE 13.8 Banded iron formation; this sample has been deformed during metamorphism.

FIGURE 13.9 Rock salt from an evaporite deposit.

well sorted by size and density. The sorting action can effectively concentrate certain weathering-resistant, dense minerals in places along the stream channel. The currents of a coastal environment can also cause sediment sorting and selective concentration of minerals. Such deposits, mechanically concentrated by water, are called **placers.** The minerals of interest are typically weathered out of local rocks, then transported, sorted, and concentrated while other minerals are dissolved or swept away (figure 13.10). Gold, diamonds, and tin oxide are examples of minerals that have been mined from the sands and gravels of placer deposits. (Indeed, the ability to separate

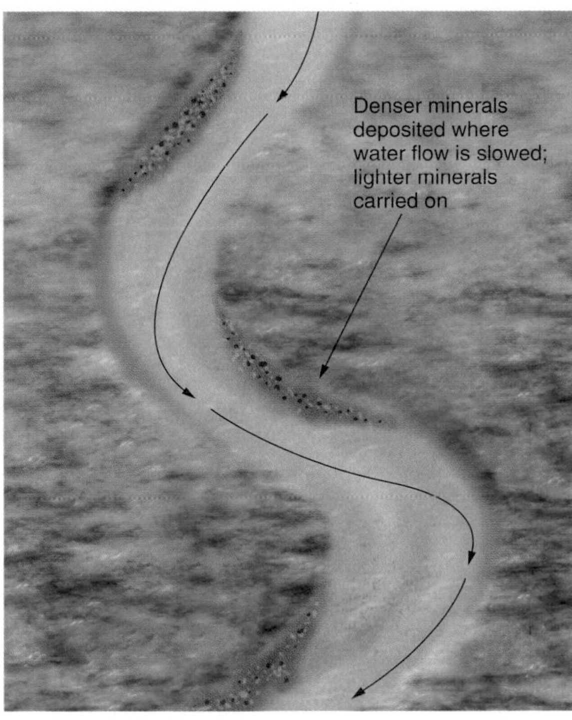

A

B

FIGURE 13.10 (A) Formation of a placer deposit. (B) Since placer deposits typically occur as loose gravels, mining them can be relatively easy; here, water flowing from a giant nozzle washes gold-bearing gravels to sluices, in the Yukon, Canada.

(B) Photograph © Steve McCutcheon/Visuals Unlimited

high-density gold from other minerals in water by gravity is the underlying principle of gold panning.)

Even weathering alone can produce useful ores by leaching away unwanted minerals, leaving a residue enriched in some valuable metal. As described in chapter 12, the extreme leaching of tropical climates gives rise to lateritic soils from which nearly everything has been leached except for aluminum and iron compounds. Many of the aluminum deposits presently being mined were enriched in aluminum to the point of being economic by lateritic weathering. (Note that the countries identified in figure 13.1 as having substantial aluminum-ore reserves are wholly or partially in tropical or subtropical climate zones.) The iron content of the laterites is generally not as high as that of the richest sedimentary iron ores described earlier. However, as the richest of those ore deposits are mined out, it may become profitable to mine laterites for iron as well as for aluminum.

Metamorphic Deposits

The mineralogical changes caused by the heat or pressure of metamorphism also can produce economic mineral deposits. Graphite, used in "lead" pencils, in batteries, as a lubricant, and for many applications where its very high melting point is vital, is usually mined from metamorphic deposits. Graphite consists of carbon, and one way in which it can be formed is by the metamorphism of coal, which, as will be seen in chapter 14, is also carbon-rich. Asbestos, as previously noted, is not a single mineral but a general term applied to a group of fibrous silicates that are formed by the metamorphism of igneous rocks rich in ferromagnesian minerals, with the addition of water. The asbestos minerals are used less in insulation now but are still valuable for their heat- and fire-resistant properties, especially because their unusual fibrous character allows them to be woven into cloth.

MINERAL AND ROCK RESOURCES—EXAMPLES

Dozens of minerals and rocks have some economic value. In this section, we survey a sampling of these resources, noting their occurrences and principal applications. Figure 13.11 represents per-capita consumption of a number of these materials in the United States.

Metals

The term *mineral resources* usually brings metals to mind first. Overwhelmingly, the most heavily used metal is iron. Fortunately, it is also one of the most common metals. Nearly all iron ore is used for the manufacture of iron and especially steel products. It is mined principally from the ancient sedimentary

Crushed stone
5600 kg
Sand and gravel
4250 kg
Salt
181 kg
Gypsum
128 kg
Phosphate
136 kg
Potash
20 kg
Iron ore
192 kg
Aluminum
23 kg
Zinc
4.6 kg
Lead
5.0 kg
Copper
7.7 kg

FIGURE 13.11 Per-capita consumption of mineral and rock resources in the United States, 2005. (1 kg = 2.2 lb).
Source: Data from Mineral Commodity Summaries 2006, *U.S. Geological Survey*

deposits described in the previous section but also from some laterites and from concentrations of magnetite in some igneous bodies.

Aluminum is another relatively common metal, and it is the second most widely used. Its light weight, coupled with its strength, make it particularly useful in the transportation and construction industries; it is also widely used in packaging, especially for beverage cans. Aluminum is the third most common element in the crust, but there it is most often found in silicates, from which it is extremely difficult to extract. Most of what is mined commercially is bauxite, an aluminum-rich laterite in which the aluminum is found as a hydroxide. Even in this form, the extraction of the aluminum is somewhat difficult and energy-intensive: 3 to 4% of the electricity consumed in the United States is used just in the production of aluminum metal from aluminum ore.

Many less-common but also important metals, including copper, lead, zinc, nickel, cobalt, and others, are found in sulfide ore deposits. Sulfides occur frequently in hydrothermal deposits and may also be concentrated in igneous rocks. Copper, lead, and zinc may also be found in sedimentary ores; some laterites are moderately rich in nickel and cobalt. Clearly, these metals may be concentrated into economically valuable deposits in a variety of ways. Copper is primarily used for electrical applications because it is an excellent conductor of electricity; it is also used in the construction and transportation industries. An important use of lead is in batteries; among its many other applications, it is a component of many solders and is used in paints and ceramics. The zinc coating on steel cans (misnamed "tin cans") keeps the cans from rusting, and zinc is also used in the manufacture of brass and other alloys.

The so-called precious metals—gold, silver, and platinum—seem to have a special romance about them that most metals lack, but they also have some unique practical uses. Gold is used not only for jewelry, in the arts, and in commerce, but also in the electronics industry and in dentistry. It is particularly valued for its resistance to tarnishing. Silver's principal use, accounting for nearly half the silver consumed in the United States, is for photographic materials (for example, film), and the next broadest applications are in electronics. Platinum is an excellent *catalyst,* a substance that promotes chemical reactions. Currently, close to half the platinum used in the United States goes into automobile emissions-control systems, with the rest finding important applications in the petroleum and chemical industries, in electronics, and in medicine, among other areas. All of the precious metals can be found as native metals, most frequently in igneous or hydrothermal ore deposits. Silver also commonly forms sulfide minerals. Most gold and silver production in this country is, in fact, a by-product of mining ores of more abundant metals like copper, lead, and zinc, with which small amounts of precious metals may be associated.

Nonmetallic Minerals

Another by-product of mining sulfides is the nonmetal sulfur. Sulfur may also be recovered from petroleum during refining, from volcanic deposits (sulfur is sometimes precipitated as pure native sulfur from fumes escaping from volcanic vents, as seen in figure 13.5), and from evaporites. The primary use of sulfur (90% of it in this country) is for the manufacture of sulfuric acid for industrial purposes.

Several important minerals are recovered from evaporite deposits. The most abundant is halite, or rock salt, used principally as a source of the sodium and chlorine of which it is composed, and secondarily for road salt, either directly or through the production of other salts from it. Halite has many lower-volume applications, including, of course, seasoning food as table salt. Gypsum, essential to the manufacture of plaster, portland cement, and wallboard for construction, is another evaporite mineral. Others include phosphate rock and potassium-rich potash, key ingredients of the synthetic fertilizers on which much of U.S. agriculture depends.

As noted in chapter 2, clay is not a single mineral, but a group of layered hydrous silicates that are formed at low temperature, commonly by weathering, and that are abundant in sedimentary deposits in the United States. The diversity of clay minerals leads to a variety of applications, from fine ceramics to the making of clay piping and other construction materials, the processing of iron ore, and drilling for oil. In the last case, the clay is mixed with water and sometimes other minerals to make "drilling mud," which lubricates the drill bit; the low permeability of clays also enables them to seal off porous rock layers encountered during drilling.

Rock Resources

When it comes to overall mass, the quantities used of the various metals and minerals mentioned previously shrink to insignificance beside the amount of rock, sand, and gravel used. In the United States in 2005, over 1.2 billion tons of sand and gravel were used in construction, especially in making cement and concrete; another 1.6 billion tons of crushed rock were consumed for fill and other applications (such as limestone for making cement). That corresponds to several tons of rock products for every person in the country. In addition, over 31 million tons of quartz-rich sand were used in industry, particularly for glassmaking (a pure quartz sand is nearly all silica, SiO_2, the main ingredient in glass). Also, nearly 1½ million tons of dimension and facing stone were used—slate for flagstones, and various other attractive and/or durable rocks, such as marble, granite, sandstone, and limestone, for monuments and building facings and interior surfaces.

MINERAL SUPPLY AND DEMAND

Assessments of the adequacy of known supplies of various rock and mineral resources naturally require information both on those supplies and on the rates at which they are, and will be, consumed. We have already examined some of the variables on the supply side of the calculation, including geologic and economic factors. Large uncertainties also exist on the demand side.

For example, for most metals, growth in worldwide demand ranged from just over 2% per year to nearly 10% per year from World War II to the mid-1970s. More recently, world economic conditions restrained the rising demand, at least temporarily. Over the period from 1976 to 1982, demand for most minerals grew much more slowly than it had in prior decades, and for a few materials, demand actually declined. As the global economy has rebounded, demand has tended to increase, with regional fluctuations such as a decline in demand in Asia during the difficult economic times of the 1990s. Improvements in the standard of living for billions of people living in less-developed countries also increase demand pressures, as is illustrated in figure 13.12.

U.S. Mineral Production and Consumption

All told, the United States consumes huge quantities of mineral and rock materials, relative to both its population and its production of most minerals. Per-capita consumption of many materials was shown in figure 13.11. Table 13.1 and figure 13.13 give a global perspective by showing 2005 U.S. primary production and consumption data and the percentages of world totals that these figures represent. Keep in mind when looking at the consumption figures that only about 4.5% of the people in the world live in the United States.

The relationship of domestic consumption to domestic production can be complex. "Primary production" is production from new ore, excluding production from recycling or from partially processed ore imported from elsewhere; total

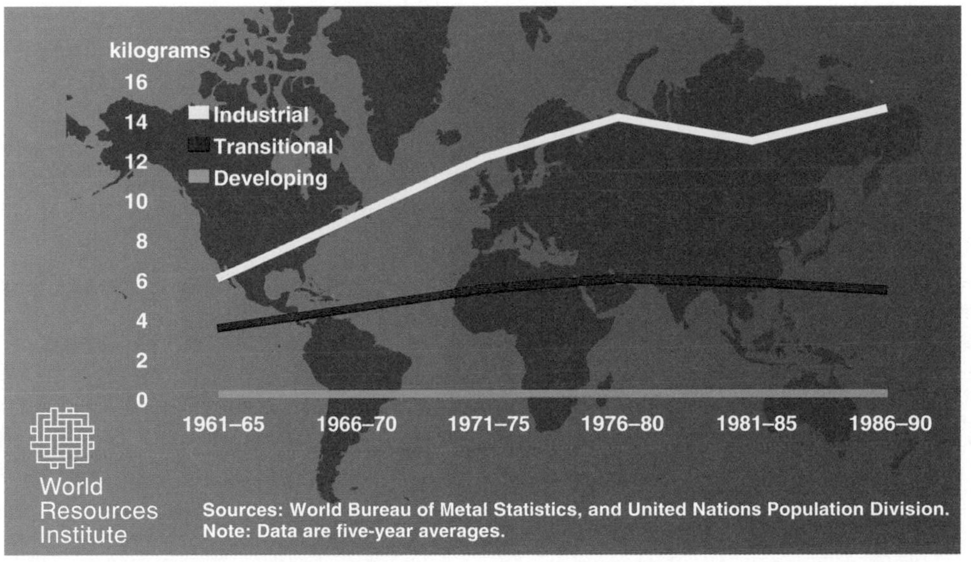

FIGURE 13.12 Per-capita aluminum consumption in more- and less-developed countries.
Source: World Resources Institute

TABLE 13.1

U.S. Production and Consumption of Rock and Mineral Resources, 2005[†]

Material	U.S. Primary Production	U.S. Primary Production as % of World	U.S. Consumption	U.S. Consumption as % of World[‡]	U.S. Production as % of U.S. Consumption[■] Primary	Total
Metals						
aluminum[x]	2500	8.0	6800	21.8	37	53
chromium	0	0	550	3.0	0	31
cobalt	0	0	11.0	21.0	0	22
copper	1180	7.9	2290	15.4	52	54
iron ore	55,000	3.6	57,000	3.8	96	96
lead	440	13.4	1490	45.4	30	113
manganese	0	0	760	7.8	0	0
nickel	0	0	221	14.7	0	0
tin	0	0	65.8	23.5	0	14
zinc	760	7.5	1370	13.6	55	85
gold	250	10.2	195	8.0	128	179
silver	1300	6.4	7720	38.0	17	40
platinum group	18.4	4.2	ø	ø	ø	ø
Nonmetals						
clays	41,900	*	36,900	*	114	114
gypsum	17,500	15.9	37,900	34.4	46	69
phosphate	38,300	25.6	40,500	27.4	94	94
potash	1200	3.9	6100	19.7	20	20
salt	45,900	21.8	53,600	25.5	86	86
sulfur	8840	13.8	12,400	19.4	71	71
Construction						
sand and gravel, construction	1,260,000	*	1,260,000	*	100	100
stone, crushed	1,650,000	*	1,660,000	*	99	99
stone, dimension and facing	1460	*	**	*	**	**

[†]All production and consumption figures in thousands of metric tons, except for gold, silver, and platinum-group metals, for which figures are in metric tons. One metric ton = 1000 kg = 2200 lb.

[‡]Assumes that overall, world production approximates world consumption. This may not be accurate if, for example, recycling is extensive.

[■]"Primary" and "Total" headings refer to U.S. production; total production includes production from recycled scrap.

*World data not available

øConsumption data not available

**Consumption not available in tonnage. However, $2.4 billion worth of dimension stone was imported (value of domestic production, $285 million, of which $146 million worth was exported).

[x]Partly from bauxite imports

Source: Mineral Commodity Summaries 2006, U.S. Geological Survey.

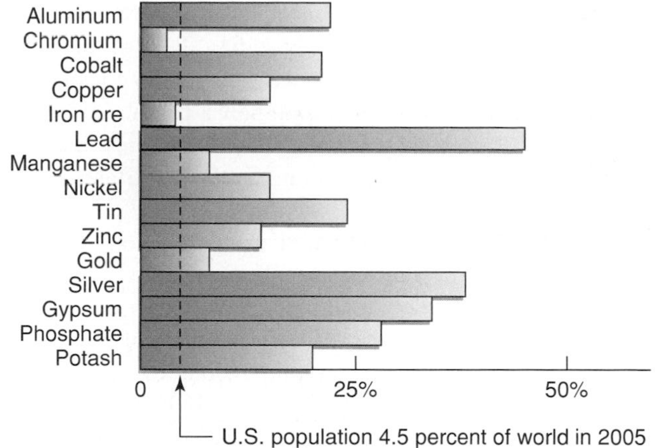

FIGURE 13.13 United States share of global consumption of selected materials. Assumes that world consumption of each commodity approximately equals world production.

Source: Data from Mineral Commodity Summaries 2006, *U.S. Geological Survey*

production includes the latter kinds of production as well. (This accounts for differences in figures for metals in the last two columns of table 13.1.) Sometimes, the United States even imports materials of which it has ample domestic supplies, perhaps to stay on good trade terms with countries having other, more critical commodities that the United States also wishes to import or because the cost of complying with U.S. environmental regulations makes it cheaper to import the commodity from another nation. For example, in the case of molybdenum, in 2005, U.S. primary production was 56,900 metric tons, against consumption of 34,200 tons—yet we imported 23,600 tons; in 2002, we produced 300 metric tons of gold and consumed 170 but imported 125 tons. Flow into and out of government stockpiles can also be significant: For mercury in 1992, total production was 160 metric tons, consumption about 500 tons; but we imported only 100 tons and exported 400, reflecting releases of about 500 tons from the National Defense Stockpile and U.S. Department of Energy.

Of course, where production falls short of consumption, importing is necessary to meet domestic demand. Moreover, low domestic production may not necessarily mean that the United States is conserving its own resources, but that it has none to produce or, at least, no economic deposits. This is true, for example, of chromium, an essential ingredient of stainless steel, and manganese, which is currently irreplaceable in the manufacture of steel. Currently, for materials such as these, the United States is almost wholly dependent on imports (figure 13.14).

Leaving aside the unanswerable political question of how long any country can count on an uninterrupted flow of necessary mineral imports, how well supplied is the world, overall, with minerals? How imminent are the shortages? We will address these questions below.

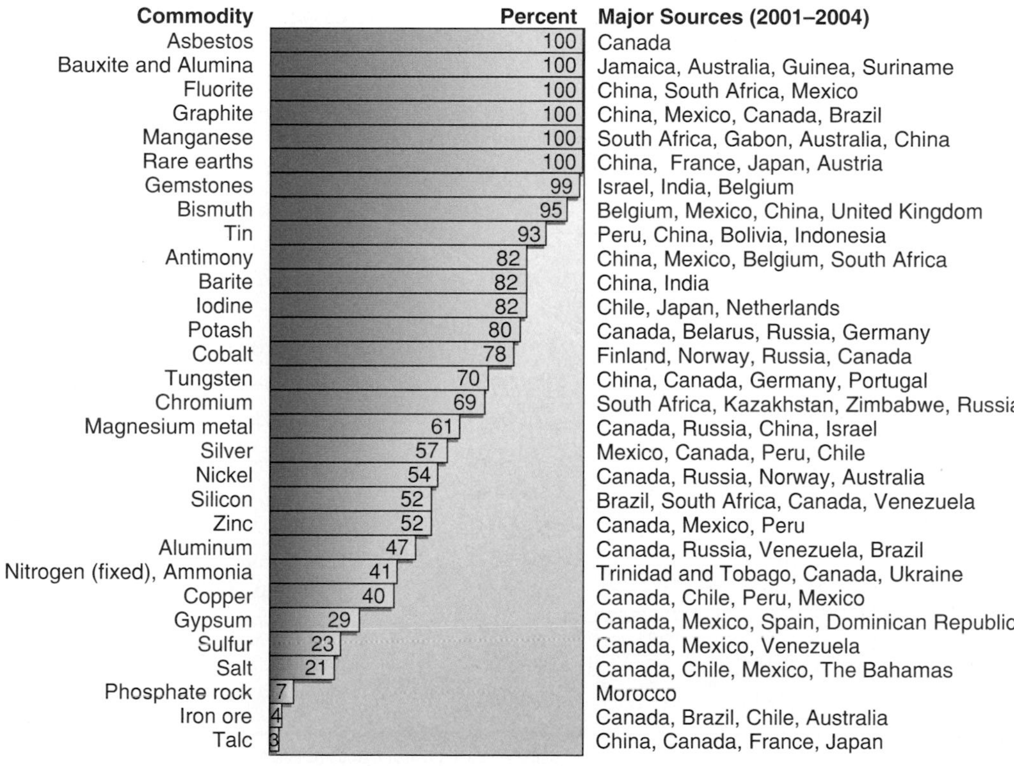

Commodity	Percent	Major Sources (2001–2004)
Asbestos	100	Canada
Bauxite and Alumina	100	Jamaica, Australia, Guinea, Suriname
Fluorite	100	China, South Africa, Mexico
Graphite	100	China, Mexico, Canada, Brazil
Manganese	100	South Africa, Gabon, Australia, China
Rare earths	100	China, France, Japan, Austria
Gemstones	99	Israel, India, Belgium
Bismuth	95	Belgium, Mexico, China, United Kingdom
Tin	93	Peru, China, Bolivia, Indonesia
Antimony	82	China, Mexico, Belgium, South Africa
Barite	82	China, India
Iodine	82	Chile, Japan, Netherlands
Potash	80	Canada, Belarus, Russia, Germany
Cobalt	78	Finland, Norway, Russia, Canada
Tungsten	70	China, Canada, Germany, Portugal
Chromium	69	South Africa, Kazakhstan, Zimbabwe, Russia
Magnesium metal	61	Canada, Russia, China, Israel
Silver	57	Mexico, Canada, Peru, Chile
Nickel	54	Canada, Russia, Norway, Australia
Silicon	52	Brazil, South Africa, Canada, Venezuela
Zinc	52	Canada, Mexico, Peru
Aluminum	47	Canada, Russia, Venezuela, Brazil
Nitrogen (fixed), Ammonia	41	Trinidad and Tobago, Canada, Ukraine
Copper	40	Canada, Chile, Peru, Mexico
Gypsum	29	Canada, Mexico, Spain, Dominican Republic
Sulfur	23	Canada, Mexico, Venezuela
Salt	21	Canada, Chile, Mexico, The Bahamas
Phosphate rock	7	Morocco
Iron ore	4	Canada, Brazil, Chile, Australia
Talc	3	China, Canada, France, Japan

FIGURE 13.14 Proportions of U.S. needs for selected minerals supplied by imports as a percentage of apparent material consumption, 2005. Principal sources of imports indicated at right; note the variety of nations represented.

Source: Data from Mineral Commodity Summaries 2006, *U.S. Geological Survey*

World Mineral Supply and Demand

Certain assumptions must underlie any projections of "lifetimes" of mineral reserves. One is future demand. For reasons already outlined, demand can fluctuate both up and down. Furthermore, not all commodities follow the same trends. Making the assumption that world production annually approximates demand, at least for "new" mine production, we find that just over one recent decade, annual demand for iron ore remained approximately constant; for copper and zinc, it increased by 10% and 20%, respectively; for lead, it declined by 7%. Rather than attempt to make precise demand estimates for the future, the following discussion assumes demand equal to primary production at constant 2005 levels.

Table 13.2 shows projections of the lifetimes of selected world mineral reserves, assuming that constant demand. Note that, for most of the metals, the present reserves are projected to last only a few decades, even at these constant consumption levels. Consider the implications if consumption resumes the rapid growth rates of the mid-twentieth century.

Such projections also presume unrestricted distribution of minerals so that the minerals can be used as needed and where needed, regardless of political boundaries or such economic factors as nations' purchasing power. It is interesting to compare global projections with corresponding data for the United States only (table 13.3). Even at constant 2005 consumption rates, the United States has less than half a century's worth of reserves of most of these materials. Of some metals, including chromium, manganese, and platinum, the United States has virtually no reserves at all.

Some relief can be anticipated from the economic component of the definition of reserves. For example, as currently identified reserves are depleted, the law of supply and demand will drive up minerals' prices. This, in turn, will make some of what are presently subeconomic resources profitable to mine; some of those deposits will effectively be reclassified as reserves. Improvements in mineral-processing technology may accomplish the same result. In addition, continued exploration can be expected to locate additional economic deposits, particularly as exploration methods become more sophisticated.

The technological advances necessary for the development of some of the subeconomic resources, however, may not be achieved rapidly enough to help substantially. Also, if the move is toward developing lower- and lower-grade ore, then by definition, more and more rock will have to be processed and more and more land disturbed to extract a given quantity of mineral

TABLE 13.2

World Production and Reserves Statistics, 2005*

Material	Production	Reserves	Projected Lifetime of Reserves (years)	Estimated Resources
bauxite	165,000	25,000,000	152	55,000,000–75,000,000
chromium	18,000	475,000[°]	26	12,000,000
cobalt	52.4	7000	133	15,000
copper	14,900	470,000	32	2,300,000[†]
iron ore	1,520,000	160,000,000	105	over 800,000,000
lead	3280	67,000	20	1,500,000
manganese	9790	430,000	44	**
nickel	1500	62,000	41	130,000
tin	280	6100	22	**
zinc	10,100	220,000	22	1,900,000
gold	2450	42,000	17	n.a.
silver	20,300	270,000	13	n.a.
platinum group	434	71,000	164	100,000
gypsum	110,000	**	**	**
phosphate	148,000	18,000,000	122	**
potash	31,000	8,300,000	268	250,000,000

*All production, reserve, and resource figures in thousands of metric tons, except for gold, silver, and platinum-group metals, for which figures are in metric tons.

**Resources and reserves estimated as "large" but not quantified.

[°]Some additional world reserve figures not available

[†]Includes 0.7 billion tons copper estimated to occur in manganese nodules; "extensive" resources of nickel are also projected to occur in these nodules.

Source: Mineral Commodity Summaries 2006, U.S. Geological Survey.

TABLE 13.3

Projected Lifetimes of U.S. Mineral Reserves (assuming complete reliance on domestic reserves)*

Material	Reserves	Projected Lifetime (years)
bauxite†	20,000	7.1
chromium	0	0
cobalt	0	0
copper	35,000	15
iron ore	6,900,000	121
lead	8100	5.4
manganese	0	0
nickel	0	0
tin	20	0.3
zinc	30,000	22
gold	2700	14
silver	25,000	3.2
platinum group	900	**
gypsum	700,000	18
phosphate	1,200,000	30
potash	90,000	15

*Reserves in thousands of metric tons, except for gold, silver, and platinum-group metals, for which figures are in metric tons.

**Accurate consumption data not available.

†Note that bauxite consumption is only a partial measure of total aluminum consumption; additional aluminum is consumed as refined aluminum metal, of which there are no reserves.

Source: Mineral Commodity Summaries 2006, U.S. Geological Survey.

or metal. Some of the consequences of mining activities, which will have increasing impact as more mines are developed, are discussed later in this chapter.

What other prospects exist for averting future mineral-resource shortages?

MINERALS FOR THE FUTURE: SOME OPTIONS CONSIDERED

One way to extend nonrenewable resources, in theory, would be to reduce consumption rates, or at least hold them steady. In practice, this seems unlikely to occur, even in the developed nations where resource consumption is already high.

The United States has the world's highest rate of car ownership: In 2000, over 55% of U.S. households owned two or more cars, and 17%, three or more, for total vehicle ownership of about 760 vehicles per 1000 people. (The global average at that time was estimated at about 120 cars per 1000 people.) Many factors—trends toward smaller households, necessity, more leisure for travel, desires for convenience or status—put upward pressure even on those U.S. car-ownership numbers. Or consider the impact of new technologies: for example, microwave ovens, personal computers, or cellular phones (increasing in number so fast that some telephone companies in urban areas are scrambling to create new area codes to accommodate them). In many cases, the new devices are additions to, rather than replacements for, other devices already owned (conventional ovens, typewriters, stationary in-home phones). All represent consumption of materials (as well as energy, to manufacture and often to use). While it is perhaps perfectly natural for people to want to acquire goods that increase comfort, convenience, efficiency, or perceived quality of life, the result may be increases in already high resource-consumption rates (figure 13.15): In the United States, from 1950 to 1990,

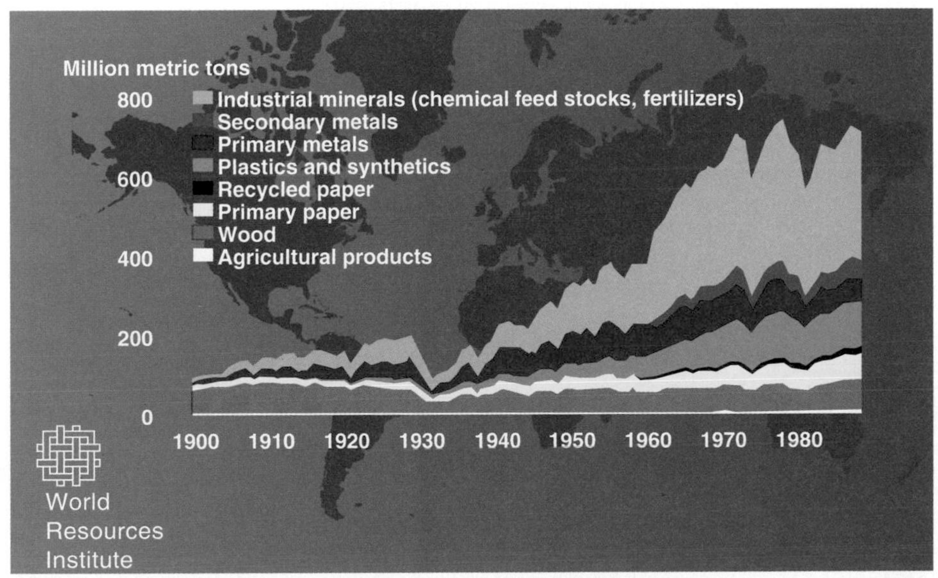

FIGURE 13.15 Trends in U.S. material consumption, 1900–1990.
Source: World Resources Institute

the population grew by about 65%, while materials consumption increased twice as fast, by more than 130%. Most of the increase was in plastics and industrial minerals. In 2006, the Mineral Information Institute estimated that every American born would, over a lifetime, account for the consumption of *3.7 million pounds* of mineral and fuel resources.

Furthermore, even if the technologically advanced nations were to restrain their appetite for minerals, they represent a minority of the world's people. If the less-developed nations are to achieve a standard of living for their citizens that is even remotely comparable to that in the most industrialized nations, vast quantities of minerals and energy will be needed. For example, 2000 data put total motor-vehicle ownership at only 10 per 1000 people in China and 7 per 1000 in India, only about 1% of the U.S. vehicle-ownership rate. Moreover, as pointed out in chapter 1, those same countries represent large and growing segments of the global population. Consider the resource implications if such nations increased their car ownership rates just tenfold (leaving them still below the world average!) and correspondingly increased their consumption of the consumer goods plentiful in the most developed countries.

If demand cannot realistically be cut, supplies must be increased or extended. We must develop new sources of minerals, in either traditional or nontraditional areas, or we must conserve our minerals better so that they will last longer.

New Methods in Mineral Exploration

Most of the "easy ores," or near-surface mineral deposits in readily accessible places, have probably already been discovered, at least in developed and/or well-explored regions. Fortunately, a variety of new methods are being applied to the search. Geophysics provides some assistance. Rocks and minerals vary in density, and in magnetic and electrical properties, so changes in rock types or distribution below the earth's surface can cause small variations in the gravitational or magnetic field measured at the surface, as well as in the electrical conductivity of the rocks. Some ore deposits may be detected in this way. As a simple example, because many iron deposits are strongly magnetic, large magnetic anomalies may indicate the presence and the extent of a subsurface body of iron ore. Radioactivity is another readily detected property; uranium deposits may be located with the aid of a Geiger counter.

Geochemical prospecting is another idea that is gaining applications. It may take a variety of forms. Some studies are based on the recognition that soils reflect, to a degree, the chemistry of the materials from which they formed. The soil over a copper-rich ore body, for instance, may itself be enriched in copper relative to surrounding soils. Surveys of soil chemistry over a wide area can help to pinpoint likely spots to dig in search of ores. Occasionally, plants can be sampled instead of soil: Certain plants tend to concentrate particular metals, making the plants very sensitive indicators of locally high concentrations of those metals. Ground water may contain traces of metals leached from ore deposits along its flow path (figure 13.16). Even soil gases can supply clues to ore deposits. Mercury is a very volatile metal, and high concentrations of mercury vapor have been found in gases filling soil pore spaces over mercury ore bodies. Geochemical prospecting has also been used in petroleum exploration.

Remote sensing methods are becoming increasingly sophisticated and valuable in mineral exploration. These methods rely on detection, recording, and analysis of wave-transmitted energy, such as visible light and infrared radiation, rather than on direct physical contact and sampling. Aerial photography is one example, satellite imagery another. Remote sensing, especially using satellites, is a quick and efficient way to scan broad areas, to examine regions having such rugged topography or

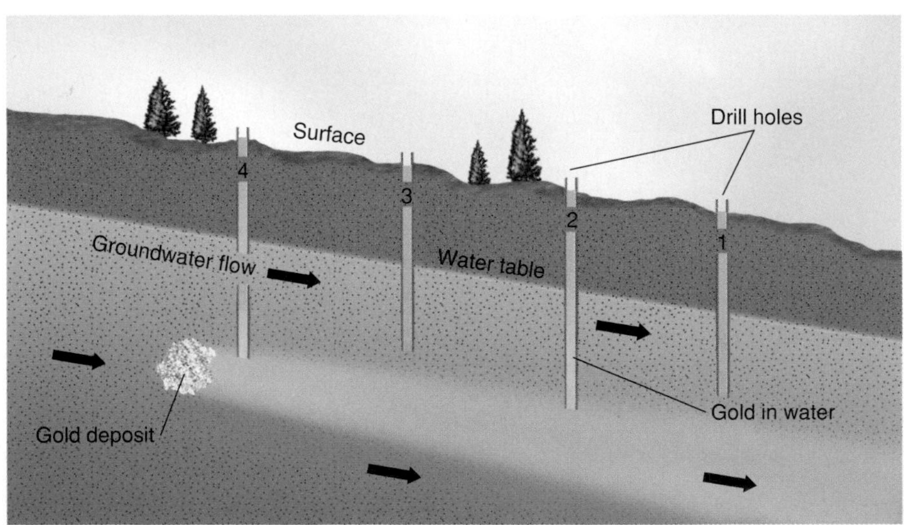

FIGURE 13.16 Groundwater sampling can detect trace metals in concentrations as low as a few parts per trillion; even low-solubility metals like gold can thus be detected down the flow path from an ore deposit, and mapping anomalies helps locate it.
After U.S. Geological Survey

hostile climate that they cannot easily be explored on foot or with surface-based vehicles, and to view areas to which ground access is limited for political reasons. Probably the best known and most comprehensive earth satellite imaging system is the one initiated in 1972, known as Landsat until sold by the government to private interests, and renamed Eosat.

Landsat satellites orbit the earth in such a way that images can be made of each part of the earth. Each orbit is slightly offset from the previous one, with the areas viewed on one orbit overlapping the scenes of the previous orbit. Each satellite makes fourteen orbits each day; complete coverage of the earth takes eighteen days.

The sensors in the Landsat satellites do not detect all wavelengths of energy reflected from the surface. They do not take photographs in the conventional sense. They are particularly sensitive to selected green and red wavelengths in the visible light spectrum and to a portion of the infrared (invisible heat radiation, with wavelengths somewhat longer than those of red light). These wavelengths were chosen because plants reflect light most strongly in the green and the infrared. Different plants, rocks, and soils reflect different proportions of radiation of different wavelengths. Even the same feature may produce a somewhat different image under different conditions. Wet soil differs from dry; sediment-laden water looks different from clear; a given plant variety may reflect a different radiation spectrum depending on what trace elements it has concentrated from the underlying soil or how vigorously it is growing. Landsat images can be powerful mapping tools. Multiple images can be joined into mosaics covering whole countries or continents. In recent decades, additional satellites and other spacecraft have offered still more opportunities for imaging earth's surface, and resolution has been increasing as well.

Basic geologic mapping, identification of geologic structures, and resource exploration are only some of the applications of such imagery (figure 13.17). It is especially useful when some *ground truth* can be obtained— information gathered by direct surface examination (best done at the time of imaging, if vegetation is involved). Alternatively, images of inaccessible regions can be compared with images of other regions that have been mapped and sampled directly, and the similarities in imagery characteristics used to infer the actual geology or vegetation. See also figure 13.18 and appendix B.

Finally, advances in geologic understanding also play a role in mineral exploration. Development of plate-tectonic theory has helped geologists to recognize the association between particular types of plate boundaries and the occurrence of certain kinds of mineral deposits. This recognition may direct the search for new ore deposits. For example, because geologists know that molybdenum deposits are often found over existing subduction zones, they can logically explore for more molybdenum deposits

A

B

FIGURE 13.17 Landsat satellite images may reveal details of the geology that will aid in mineral exploration. Vegetation, also sensitive to geology, may enhance the image. (A) View of South Africa, dry season. (B) Same view, rainy season. Recognizable geologic features include a granite pluton (round feature at top) and folded layers of sedimentary rock (below).

Images courtesy NASA

FIGURE 13.18 Data from the Airborne Visible and Infrared Imaging Spectrometer (AVIRIS) make prospecting much simpler. The lefthand image of Cuprite, Nevada, in true color, reveals little about mineral distribution. Central image reveals distribution of hydroxide, carbonate, and sulfate minerals, some associated with gold in the region. Right image distinguishes among different iron-bearing minerals, refining understanding of the geology.
Images courtesy United States Geological Survey Spectroscopy Lab

in other present or past subduction zones. Also, the realization that many of the continents were once united can suggest likely areas to look for more ores (figure 13.19). If mineral deposits of a particular kind are known to occur near the margin of one continent, similar deposits might be found at the corresponding edge of another continent once connected to it. If a mountain belt rich in some kind of ore seems to have a counterpart range on another continent once linked to it, the same kind of ore may occur in the matching mountain belt. Such reasoning is partly responsible for the supposition that economically valuable ore deposits may exist on Antarctica. (Some of the legal/political consequences of such speculations are explored in chapter 19; fundamental concerns about the wisdom of mineral exploration in the fragile Antarctic environment have motivated international agreements to restrict resource exploration there.)

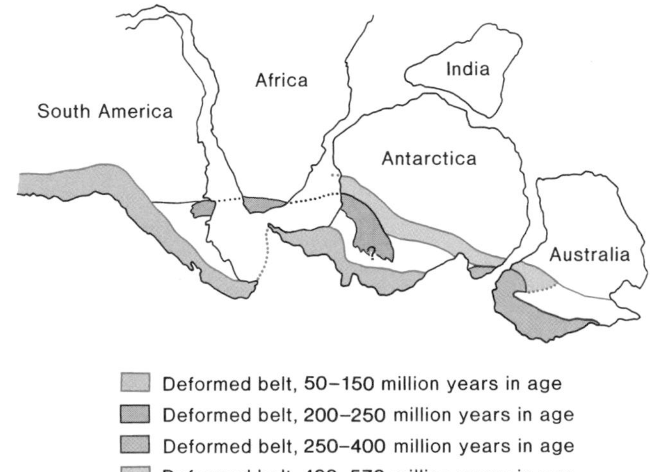

FIGURE 13.19 Pre-continental-drift reassembly of landmasses suggests locations of possible ore deposits in unexplored regions by extrapolation from known deposits on other continents.
Source: Data from C. Craddock et al., Geological Maps of Antarctica, *Antarctic Map Folio Series, folio 12, copyright 1970 by the American Geographical Society*

Marine Mineral Resources

Increased consideration is also being given to seeking mineral wealth in unconventional places. In particular, the sea may provide partial solutions to some mineral shortages. Seawater itself contains dissolved in it virtually every chemical element. However, most of what it contains is dissolved halite, or sodium chloride. Most needed metals occur in far lower concentrations. Vast volumes of seawater would have to be processed to extract small quantities of such metals as copper and gold. The costs would be very high, especially since existing technology is not adequate to extract, selectively and efficiently, a few specific metals of interest. Several types of underwater mineral deposits have greater potential.

During the last Ice Age, when a great deal of water was locked in ice sheets, worldwide sea levels were lower than at present. Much of the area of the now-submerged continental shelves was dry land, and streams running off the continents flowed out across the exposed shelves to reach the sea. Where the streams' drainage basins included appropriate source rocks, these streams might have produced valuable placer deposits. With the melting of the ice and rising sea levels, any placers on the continental shelves were submerged. Finding and mining these deposits may be costly and difficult, but as land deposits are exhausted, placer deposits on the continental shelves may well be worth seeking. Already, more than 10% of world tin production is from such marine placers; potentially significant amounts of titanium, zirconium, and chromium could also be mined from them if the economics were right.

The hydrothermal ore deposits forming along some sea-floor spreading ridges are another possible source of needed metals. In many places, the quantity of material being deposited is too small, and the depth of water above would make recovery prohibitively expensive. However, the metal-rich muds of the Red Sea contain sufficient concentrations of such metals as copper, lead, and zinc that some exploratory dredging is underway, and several companies are interested in the possibility of mining those sediments. Along a section of the Juan de Fuca ridge, off the coast of Oregon and Washington, hundreds of thousands of tons of zinc- and silver-rich sulfides may already have been deposited, and the hydrothermal activity continues.

Perhaps the most widespread undersea mineral resource, as well as the most frequently and seriously discussed as a practical resource for the near future, is **manganese nodules.** These are lumps up to about 10 centimeters in diameter, composed mostly of manganese minerals. They also contain lesser but potentially valuable amounts of copper, nickel, cobalt, platinum, and other metals. Indeed, the value of the minor metals may be a greater motive for mining manganese nodules than the manganese itself. The nodules are found over much of the deep-sea floor, in regions where sedimentation rates are slow enough not to bury them (figure 13.20A). At present, the costs of recovering these nodules would be high compared to the costs of mining the same metals on land, and the technical problems associated with recovering the nodules from beneath several kilometers of seawater remain to be worked out. Still, manganese nodules represent a sufficiently large metal resource that they have become a subject of International Law of the Sea conferences (see chapter 19). With any ocean resources outside territorial waters—manganese nodules, hydrothermal deposits at spreading ridges, fish, whales, or whatever—questions of ownership, or of who has the right to exploit these resources, inevitably lead to heated debate. Potential environmental impacts of recovering such marine resources have also not been fully analyzed.

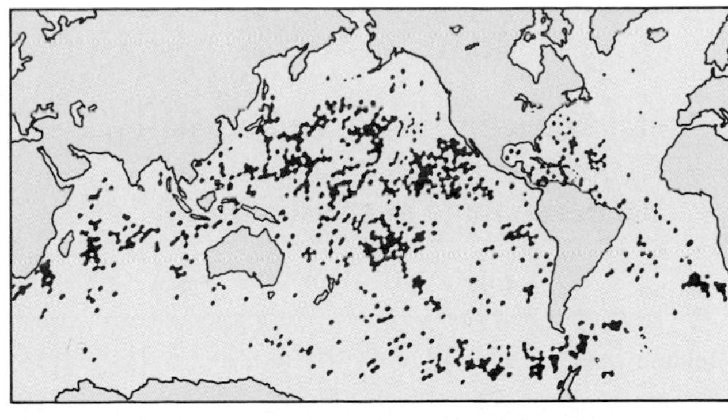

A

B

FIGURE 13.20 Manganese nodules. (A) Manganese-nodule distribution on the sea floor. (B) Manganese nodules on the floor of the northeast Pacific Ocean. A nodule grows by only about 1 centimeter in diameter over 1 million years, so they are exposed only where sedimentation rates are very low.

(A) Data from G. R. Heath, "Manganese Nodules: Unanswered Questions," Oceanus *Vol. 25, No. 3, pp. 37–41. 1982. Copyright © Woods Hole Oceanographic Institute. (B) Photograph © Institute of Oceanographic Sciences/NERC/Photo Researchers*

Conservation of Mineral Resources

Overall need for resources is likely to increase, but for selected individual materials that are particularly scarce, perhaps demand can be moderated. One way is by making substitutions. One might, for certain applications, replace a very rare metal with a more abundant one. The extent to which this is likely to succeed is limited, since reserves of most metals are quite limited, as was clear from tables 13.2 and 13.3. Substituting one metal for another reduces demand for the second while increasing demand for the first; the focus of the shortage problem shifts, but the problem persists. Nonmetals are replacing metals in other applications. Unfortunately, many of these nonmetals are plastics or other materials derived from petroleum, which has simply contributed to petroleum consumption. As we will see in chapter 14, supplies of that resource are limited, too. For some applications, ceramics or fiber products can be substituted; however, sometimes the physical, electrical, or other properties required for the particular application demand the use of metals.

The most effective way to extend mineral reserves, for some metals at least, may be through recycling. Overall use of

TABLE 13.4

Metal Recycling in the United States, 1986–2005 (recycled scrap as percentage of consumption)

Metal	1986	1992	1999	2005
aluminum	15	30	20	16
chromium	21	26	20	29
cobalt	15	25	34	27
copper	22	42	33	30
lead	50	62	68	72
manganese	0	0	0	0
nickel	25	30	38	39
zinc	7	29	26	26

Source: Mineral Commodity Summaries 1990, 1993, 2000 and 2006, U.S. Geological Survey.

the metals may increase, but if ways can be found to reuse a significant fraction of these metals repeatedly, proportionately less new metal will need to be extracted from mines, reducing—or at least moderating increases in—primary production. Some metals are already extensively recycled, at least in the United States (table 13.4 lists a few examples). Worldwide, recycling is less widely practiced, in part because the less-industrialized countries have accumulated fewer manufactured products from which materials can be retrieved. Still, given the limited U.S. reserves of many metals, recycling can be a very important means of extending those reserves. Additional benefits of recycling include a reduction in the volume of the waste-disposal problem and a decrease in the extent to which more land must be disturbed by new mining activities.

Unfortunately, not all materials lend themselves equally well to recycling. Among those that work out best are metals that are used in pure form in sizeable pieces—copper in pipes and wiring, lead in batteries, and aluminum in beverage cans. The individual metals are relatively easy to recover and require minimal effort to purify for reuse. Recycling can be more energy-efficient than producing new metal by mining and refining new ore, which results in the dual benefits of saving energy and cutting costs.

Where different materials are intermingled in complex manufactured objects, it is more difficult and costly to extract individual metals. Consider trying to separate the various metals from a refrigerator, a lawn mower, or a computer. The effort required to do that, even if it were technically possible, would make most of the materials recovered far too costly and thus noncompetitive with new production. Only in a few rare cases are metals valuable enough that the recycling effort may be worthwhile; for example, there is interest in recovering the platinum from catalytic converters in exhaust systems.

Alloys present special problems. The United States uses tens of millions of tons of steel each year, and some of it is indeed recycled scrap. However, steel is not a single chemical substance. It is iron alloyed with one or more other elements, and, often, the composition is very specific to the application. Chromium is added to make stainless steel; titanium, molybdenum, or tungsten steels are high-strength steels; other alloys are used when other properties are important. If the composition of the alloy is critical to a particular application, clearly one cannot just toss any old steel scrap into the recycling pot. Each type of steel would have to be recycled individually. Facilities for separate collection and recycling of many different compositions of steel are rare. Progress is nevertheless being made: The U.S. automotive industry in 2005 recycled over 17 million vehicles by means of 200^{+} car shredders, and contributed over 13 million tons of shredded scrap steel to the steel industry, which, in turn, produces 100 million tons of new steel each year. As with aluminum, it takes much less energy to recycle old steel scrap than to manufacture new steel.

Some materials are not used in discrete objects at all. The potash and phosphorus in fertilizers are strewn across the land and cannot be recovered. Road salt washes off streets and into the soil and storm sewers. Lead in gasoline, now essentially phased out, was emitted with exhaust fumes into the atmosphere. These things clearly cannot be recovered and reused.

For the foregoing and other reasons, it is unrealistic to expect that all minerals can ever be wholly recycled. However, more than half of the U.S. consumption of certain metals already is being recovered from old scrap. Where this is possible, the benefits are substantial.

IMPACTS OF MINING-RELATED ACTIVITIES

Mining and mineral-processing activities can modify the environment in a variety of ways, and few areas in the United States are untouched by mining activities (figure 13.21). Most obvious is the presence of the mine itself. Both underground mines and surface mines have their own sets of associated impacts. Though safety is improving, the hazards of mining activities to the miners should also not be overlooked: According to the National Safety Council, mining is the most dangerous occupation in the United States, with 22 deaths per 100,000 workers per year in 2004 (followed by agriculture at 21 and construction at 11 deaths per 100,000). Aspects of these hazards will be noted in the next chapter. Both underground and surface mines can also be sources of water pollution, as explored further in chapters 14 and 17.

Underground mines are generally much less apparent than surface mines. They disturb a relatively small area of the land's surface close to the principal shaft(s). Waste rock dug out of the mine may be piled close to the mine's entrance, but in most underground mines, the tunnels follow the ore body as closely as possible to minimize the amount of non-ore rock

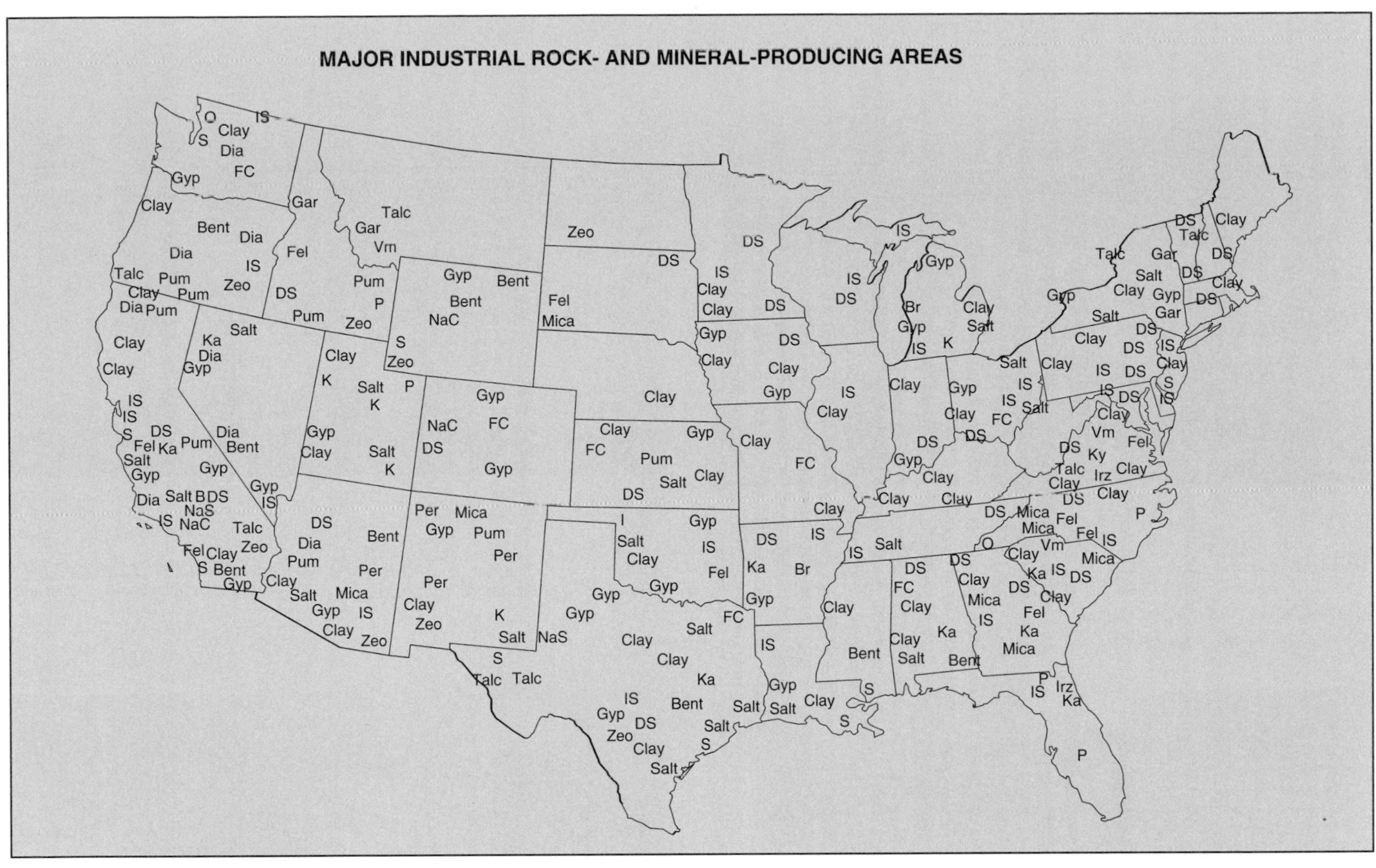

Mineral symbols

Bent Bentonite	Fel Feldspar	Ka Kaolin	Pum Pumice and pumicite
B Borates	Gar Garnet	Ky Kyanite	Salt Salt
Br Bromine	Gyp Gypsum	Mica Mica	NaC Soda ash
Clay Common clay	Irz Ilmenite, rutile, and zircon	O Olivine	NaS Sodium sulfate
Dia Diatomite	IS Industrial sand	Per Perlite	S Sulfur
DS Dimension stone	I Iodine	P Phosphate	Talc Talc
FC Fire clay		K Potash	Vm Vermiculite
			Zeo Zeolites

FIGURE 13.21 Even a partial map of sites of mining for industrial rock and mineral resources shows that these activities are widespread in the United States. Still more places are affected by mining for metal ores.
Source: Mineral Commodity Summaries 2003, *U.S. Bureau of Mines*

to be removed and, thus, mining costs. When mining activities are complete, the shafts can be sealed, and the area often returns very nearly to its pre-mining condition. However, near-surface underground mines occasionally have collapsed years after abandonment, when supporting timbers have rotted away or ground water has enlarged the underground cavities through solution (figure 13.22). In some cases, the collapse occurred so long after the mining had ended that the existence of the old mines had been entirely forgotten.

Surface-mining activities consist of either open-pit mining (including quarrying) or strip-mining. Quarrying extracts rock to be used either intact (for building or facing stone) or crushed (as for cement-making, road bed, etc.); see figure 13.23A and the chapter-opener photograph. Open-pit mining is practical when a large, three-dimensional ore body is located near the surface (figure 13.23B). Most of the material in the pit contains the valuable commodity and is extracted for processing. Both procedures permanently change the topography, leaving a large hole. The exposed rock may begin to weather and, depending on the nature of the ore body, may release pollutants into surface runoff water.

Strip-mining, more often used to extract coal than mineral resources, is practiced most commonly when the material of interest occurs in a layer near and approximately parallel to the surface (figure 13.23C). Overlying vegetation, soil, and rock are stripped off, the coal or other material is removed, and the waste rock and soil is dumped back as a series of **spoil banks** (figure 13.24A). Until fairly recently, that was all that was

A

B

FIGURE 13.22 Subsurface mining activities sometimes affect the earth's surface. (A) Collapse of land surface over old abandoned copper mine in Arizona. Note roads, trees, and houses for scale. (B) Subsidence pits and troughs over abandoned underground coal mines in Sheridan County, Wyoming.

Photograph (A) by H. E. Malde, (B) by C. R. Dunrud, both courtesy USGS Photo Library, Denver, CO.

A

B

C

FIGURE 13.23 Surface mining. (A) Granite quarrying at the Rock of Ages quarry, Barre, Vermont. (B) The world's largest open-pit mine: Bingham Canyon, Utah. Over $6 billion worth of minerals have been extracted from this one mine over the decades of its operation. (C) Strip-mining for manganese in South Africa.

Photograph (A) The McGraw-Hill Companies, Inc./Doug Sherman, photographer. (B) courtesy of Kennecott; (C) courtesy USGS Photo Library, Denver, CO.

Box 13.1 What's in a 12-Ounce Can?

Amie waltzed into the kitchen and, with her best basketball form, neatly sailed her empty cola can over her shoulder, across the kitchen, into the garbage. Two of her housemates reacted simultaneously.

"Nice shot," Jason observed. "Hey! That's recyclable!" was Mimi's startled comment.

"Okay, okay—but the big blue box is too easy a target," responded Amie, fishing out the can and dropping it in the recycling bin with a clatter. "And does it really make all that much difference? I mean, it's nice to cut down the trash, I guess, but what's one can in the scheme of things?"

Paul, who had been curled up with a book in the corner, looked up. "Two points. One: Think globally, act locally. Two: How much water in a 12-ounce can?" In the puzzled silence that followed, he explained, "To *you,* this is one can. Maybe an ounce or so of alumnimum. But if every person in the United States recycled one can, that's 300 million cans, which is—what?" Pulling out his calculator, he concluded, "—almost 9400 tons of aluminum. And if everybody in the *world* recycled one can, that's about 6½ billion people, and that's . . . wait a minute . . . it's over 200,000 tons! *Now* do you see it?"

"Babies don't recycle," objected Amie.

"No, but you use a lot more than one can a year, or even a week. Anyway, the exact number isn't the point. It's the principle," was Mimi's quick reply.

"So where do the 12 ounces of water come in?" asked Jason.

"Well, the point there is not the 12 ounces physically in the can; it's the "hidden" water, and energy, and all that—it's a discussion I remember from high school. What does it take to make a can?"

"Metal," the other three replied, practically in unison.

"And what else?"

"Well, energy to make the can," replied Mimi, "and I remember hearing that with aluminum, it takes about twenty times as much energy to make a new can from aluminum ore as it does to clean up and remanufacture the can from recycled cans, because it's hard to get pure aluminum from its ore."

"That's why it was such a big deal to put aluminum on top of the Washington Monument," chimed in Jason, the history major. "It was really expensive to make aluminum metal at the time."

"Okay, so you recycle the can, you save on trash disposal and energy. Also mining, I suppose, since you already have the aluminum," continued Amie.

"And if you're mining less, you're disturbing less land," observed geographer Mimi, "and putting fewer miners at risk, too. I suppose the mining and the metal refining must use water, too, or you wouldn't have brought up this 'hidden' water bit."

"So keep going. Where does your energy come from? Chances are the power generation uses water, too, one way or another," Paul pointed out. "So does mine reclamation, generally, which might be an issue depending on what you're mining and where."

The discussion went on for a while, but Paul had made his point about thinking beyond the obvious, to appreciate the full impact of recycling, largely inspired by a long-ago discussion of a little pamphlet called "How much water in a 12-ounce can?" (See reference list.)

done. The broken-up material of the spoil banks, with its high surface area, is very susceptible to both erosion and chemical weathering. Chemical and sediment pollution of runoff from spoil banks was common. Vegetation reappeared on the steep, unstable slopes only gradually, if at all (figure 13.24B). Now, much stricter laws govern strip-mining activities, and reclamation of new strip mines is required. Reclamation usually involves regrading the area to level the spoil banks and to provide a more gently sloping land surface (figure 13.24C); restoring the soil; replanting grass, shrubs, or other vegetation; and, where necessary, fertilizing and/or watering the area to help establish vegetation. The result, when successful, can be land on which evidence of mining activities has essentially been obliterated (figure 13.24D).

Naturally, reclamation is much more costly to the mining company than just leaving spoil banks behind. Costs can reach several thousand dollars per acre. Also, even when conscientious reclamation efforts are undertaken, the consequences may not be as anticipated. For example, the inevitable changes in topography after mining and reclamation may alter the drainage patterns of regional streams and change the proportion of runoff flowing to each. In dry parts of the western United States, where every drop of water in some streams is already spoken for, the consequences may be serious for landowners downstream. In such areas, the water needed to sustain the replanted vegetation may also be in very short supply. An additional problem is presented by old abandoned, unreclaimed mines that were operated by now-defunct companies: Who is responsible for reclamation, and who pays? (Superfund, discussed in chapter 16, was created in part to address a similar problem with toxic-waste dumps.)

Mineral processing to extract a specific metal from an ore can also cause serious environmental problems. Processing generally involves finely crushing or grinding the ore. The fine waste materials, or **tailings,** that are left over may end up heaped around the processing plant to weather and wash away, much like spoil banks (figure 13.25). Generally, traces of the ore are also left behind in the tailings. Depending on the nature of the ore, rapid weathering of the tailings may leach out harmful elements, such as mercury, arsenic, cadmium, and uranium, to contaminate surface and ground waters. Uncontrolled usage of uranium-bearing tailings as fill and in concrete in Grand Junction, Colorado, has resulted in radioactive buildings that must be extensively torn up and rebuilt before they are safe for occupation.

The chemicals used in processing are often hazardous also. For example, cyanide is commonly used to extract gold from its ore. Water pollution from poorly controlled runoff of cyanide solutions can be a significant problem. Additional

A

B

C

D

FIGURE 13.24 Strip-mining and land reclamation. (A) Spoil banks, Rainbow coal strip mine, Sweetwater County, Wyoming. (B) Grading of spoils at Indian Head coal strip mine, Mercer County, North Dakota. (C) Revegetation of ungraded spoils at Indian Head mine is slow. (D) Reclaimed portion of Indian Head mine one year after seeding.

Photographs by H. E. Malde, USGS Photo Library, Denver, CO.

A

B

FIGURE 13.25 As the Bingham Canyon mine (figure 13.23B) is approached, huge tailings piles stand out against the mountains (A). Closer inspection of the tailings (B) shows obvious erosion by surface runoff. Local residents report metallic-tasting water, attributed to copper leached from tailings.

concerns arise as attention turns to trying to recover additional gold from old tailings by "heap-leaching," using cyanide solutions percolating through the tailings piles. Gold is valuable, and heap-leaching old tailings is relatively inexpensive, especially in comparison to mining new ore. But unless the process is very carefully controlled, both metals and cyanide can contaminate local surface and ground waters. Smelting to extract metals from ores may, depending on the ores involved and on emission controls, release arsenic, lead, mercury, and other potentially toxic elements along with exhaust gases and ash. Sulfide-ore processing also releases the sulfur oxide gases that are implicated in the production of acid rain (chapter 18). These same gases, at high concentrations, can destroy nearby vegetation: During the twentieth century, one large smelter in British Columbia destroyed all conifers within 19 *kilometers* (about 12 miles) of the operation. More-stringent pollution controls are curbing such problems today. Still, mineral processing can be a huge potential source of pollutants if not carefully executed. Worldwide, too, not all nations are equally sensitive to the hazards; residents of areas with weak pollution controls may face significant health risks.

SUMMARY

Economically valuable mineral deposits occur in a variety of specialized geologic settings—igneous, sedimentary, and metamorphic. Both the occurrence of and the demand for minerals are very unevenly distributed worldwide. Projections for mineral use, even with conservative estimates of consumption levels, suggest that present reserves of most metals and other minerals could be exhausted within decades. As shortages arise, some currently subeconomic resources will be added to the reserves, but the quantities are unlikely to be sufficient to extend supplies greatly. Other strategies for averting further mineral shortages include applying new exploration methods to find more ores, looking to undersea mineral deposits not exploited in the past, and recycling metals to reduce the demand for newly mined material. Decreasing mining activity would minimize the negative impacts of mining, such as disturbance of the land surface and the release of harmful chemicals through accelerated weathering of pulverized rock or as a consequence of mineral-processing activities.

Key Terms and Concepts

banded iron formation 297
concentration factor 293
evaporite 297
hydrothermal 295
kimberlites 295
manganese nodules 309
ore 293
pegmatite 293
placer 298
remote sensing 306
spoil banks 311
tailings 313

Exercises

Questions for Review

1. Explain how economics and concentration factor relate to the definition of an ore.
2. Describe two examples of magmatic ore deposits.
3. Hydrothermal ore deposits tend to be associated with plate boundaries. Why?
4. What is an evaporite? Give an example of a common evaporite mineral.
5. Explain how stream action may lead to the formation of placer deposits. Why is there interest in exploring for placers on the continental shelves?
6. As mineral reserves are exhausted, some resources may be reclassified as reserves. Explain.
7. How might plate-tectonic theory contribute to the search for new ore deposits?
8. What metallic mineral resource is found over much of the deepsea floor, and what political problem arises in connection with it?
9. Why are aluminum and lead comparatively easy to recycle, while steel is less so?
10. Describe one hazard associated with underground mining.
11. What steps are involved in strip-mine reclamation? Can land always be fully restored to its pre-mining condition with sincere effort? Explain.
12. Why are tailings from mineral processing a potential environmental concern?

For Further Thought

1. Choose an area that has been subjected to extensive surface mining (examples include the coal country of eastern Montana or Appalachia and the iron ranges of Upper Michigan). Investigate

the history of mining activities in this area and of legislation relating to mine reclamation. Assess the impact of the legislation.

2. Select one metallic mineral resource and investigate its occurrence, distribution, consumption, and reserves. Evaluate the impact of its customary mining and extraction methods. (How and where is it mined? How much energy is used in processing it? What special pollution problems, if any, are associated with the processing?) Can this metal be recycled, and if so, from what sorts of material?
3. Choose a specific type and size of can, glass bottle, or other object made of one metal or mineral material, and carry out a more quantitative analysis of the kind discussed in "What's in a 12-ounce can?" of the energy, water, and material input involved in its manufacture and its recycling. Or, look around you, choose an everyday object, and investigate what mineral materials go into it and where they come from. (The Mineral Information Institute's website might be a place to start on the latter.)

Suggested Readings/References

Briskey, J. A., et al. 2001. It's time to know the planet's mineral resources. *Geotimes* (March): 14–19.

Craig, J. R., Vaughan, D. J., and Skinner, B. J. 2001. *Resources of the earth.* 3d ed. Englewood Cliffs, N.J.: Prentice-Hall.

Davis, L. L. 1992. Construction—the essential role of minerals. *Minerals Today* (April): 12–17.

Glasby, G. P. 2000. Lessons learned from deep-sea mining. *Science* 289 (28 July): 551–53.

Haxel, G. B., Hedrick, J. B., and Orris, G. J. 2002. *Rare earth elements—critical resources for high technology.* USGS Fact Sheet 087-02.

Hodges, C. A. 1995. Mineral resources, environmental issues, and land use. *Science* (2 June): 1305–12.

James, I. C. II, Kammerer, J. C., and Murray, C. R. 1976. *How much water in a 12-ounce can?* U.S. Geological Survey.

Langer, W. H., Drew, L. J., and Sachs, J. S. 2004 *Aggregate and the environment.* Amer. Geol. Institute.

Lumsden, G. I., ed. 1992. *Geology and the environment in western Europe.* Oxford: Clarendon Press.

Rona, P. A. 2003. Resources of the sea floor. *Science* 299 (31 January): 673–74.

Scott, S. 2002. Minerals on land, minerals in the sea. *Geotimes* (December): 19–23.

Souren, A. W. M. G. 2000. Living with cyanide. *The Geochemical News* 105 (October 16–26).

Sznopek, J. L. 2006. *Drivers of U.S. mineral demand.* USGS Open-File Report 2006-1025.

U.S. Geological Survey. 1999. *Fertilizers—sustaining global food supplies.* USGS Fact Sheet FS-155-99.

———. 2000. *Minerals in our environment.* [online] http://wrgis.wr.usgs.gov/open-file/of00-144/

———. 2001. *Obsolete computers, "gold mine" or high-tech trash? Resource recovery from recycling.* USGS Fact Sheet 060-01 [online] http://pubs.usgs.gov/fs/fs060-01/

U.S. Geological Survey 2006. *Mineral commodity summaries.* [online] http://minerals.usgs.gov/minerals/pubs/mcs/2006/mcs2006.pdf

———. *Minerals yearbook.* (published annually.) [online] http://minerals.usgs.gov/minerals/pubs/commodity/myb

Vanecek, M., ed. 1994. *Mineral deposits of the world.* New York: Clavier.

Virta, R. L. 2001. *Some facts about asbestos.* USGS Fact Sheet FS-012-01.

Zumberge, J. H. 1979. Mineral resources and geopolitics in Antarctica. *American Scientist 67* (January/February): 68–76.

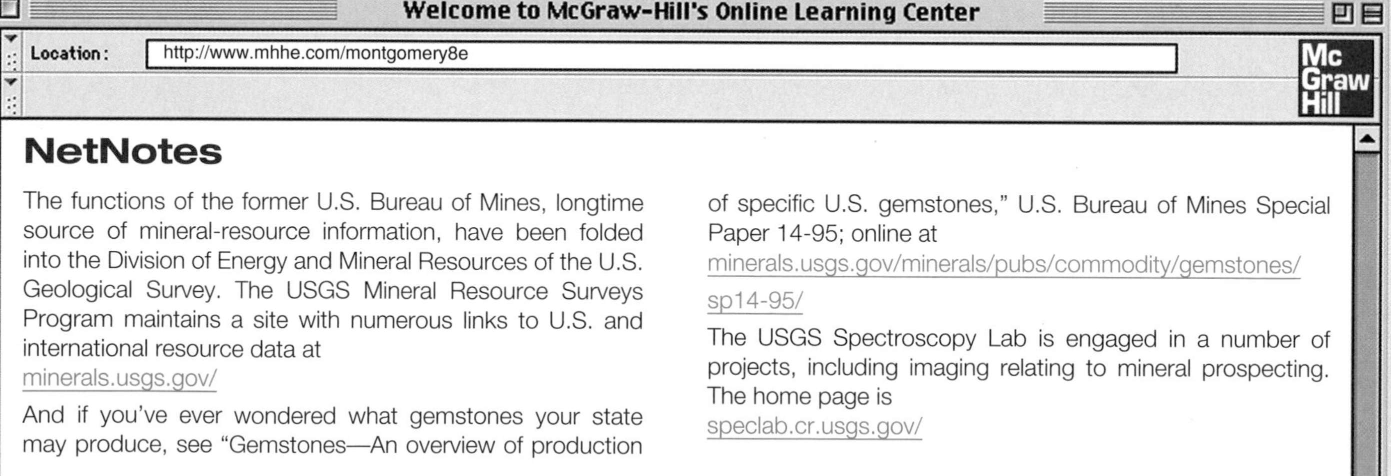

NetNotes

The functions of the former U.S. Bureau of Mines, longtime source of mineral-resource information, have been folded into the Division of Energy and Mineral Resources of the U.S. Geological Survey. The USGS Mineral Resource Surveys Program maintains a site with numerous links to U.S. and international resource data at

minerals.usgs.gov/

And if you've ever wondered what gemstones your state may produce, see "Gemstones—An overview of production of specific U.S. gemstones," U.S. Bureau of Mines Special Paper 14-95; online at

minerals.usgs.gov/minerals/pubs/commodity/gemstones/sp14-95/

The USGS Spectroscopy Lab is engaged in a number of projects, including imaging relating to mineral prospecting. The home page is

speclab.cr.usgs.gov/

and information specifically about the Cuprite, Nevada, study is at
speclab.cr.usgs.gov/map.intro.html

Maps of mineral-resource data can be created online
mrdata.usgs.gov/

The U.S. Minerals Management Service (MMS) oversees offshore mineral and fuel leasing arrangements, with special emphasis on the Gulf of Mexico, the Pacific Basin, and Alaska; see
www.mms.gov/

A variety of publications of the Environmental Studies Program of MMS can be found at
mmspub.mms.gov/eppd/sciences/esp/index.htm

The U.S. Office of Surface Mining is found at
www.osmre.gov/

The Mineral Information Institute is a nonprofit organization dedicated to educating the public about minerals and other geologic resources; they offer a wealth of educational materials at
www.mii.org/

Natural Resources Canada (NRC), which deals with a broad range of resources, including minerals, energy, and forestry, has its home page at
www.nrcan.gc.ca

The United Nations Commission for Sustainable Development site is found at
www.un.org/esa/sustdev/csd.htm

World Resources Institute's EarthTrends site includes resource-consumption tables; see
earthtrends.wri.org

14 Energy Resources—Fossil Fuels

The earliest and, at first, only energy source used by primitive people was the food they ate. Then, wood-fueled fires produced energy for cooking, heat, light, and protection from predators. As hunting societies became agricultural, people used energy provided by animals—horsepower and ox power. The world's total energy demands were quite small and could readily be met by these sources.

Gradually, animals' labor was replaced by that of machines, new and more sophisticated technologies were developed, and demand for manufactured goods rose. These factors all greatly increased the need for energy and spurred the search for new energy sources (figure 14.1). It was first wood, then primarily the fossil fuels that eventually met those needs, and they continue to dominate our energy consumption (figure 14.2). They also pose certain environmental problems. All contribute to the carbon-dioxide pollution in the atmosphere, and several can present other significant hazards such as toxic spills and land subsidence (oil), or sulfur pollution, acid runoff, and mine collapse (coal).

The term *fossil* refers to any remains or evidence of ancient life. Energy is stored in the chemical bonds of the organic compounds of living organisms. The ultimate source of that energy is the sun, which drives photosynthesis in plants. The **fossil fuels** are those energy sources that formed from the remains of once-living organisms. These include oil, natural gas, coal, and fuels derived from oil shale and tar sand. When we burn them, we are using that stored energy. The differences in the physical properties among the various fossil fuels arise from differences in the starting materials from which the fuels formed and in what happened to those materials after the organisms died and were buried within the earth.

A fully developed oil field is a network of pipelines connecting wells and drilling structures (white dots). In this image of the Permian Basin of west Texas and southeastern New Mexico, the edges of thick layers of sedimentary rock can be seen along the sides of the plateau at bottom of image.
Image courtesy of Earth Sciences and Image Analysis Laboratory, NASA Johnson Space Center

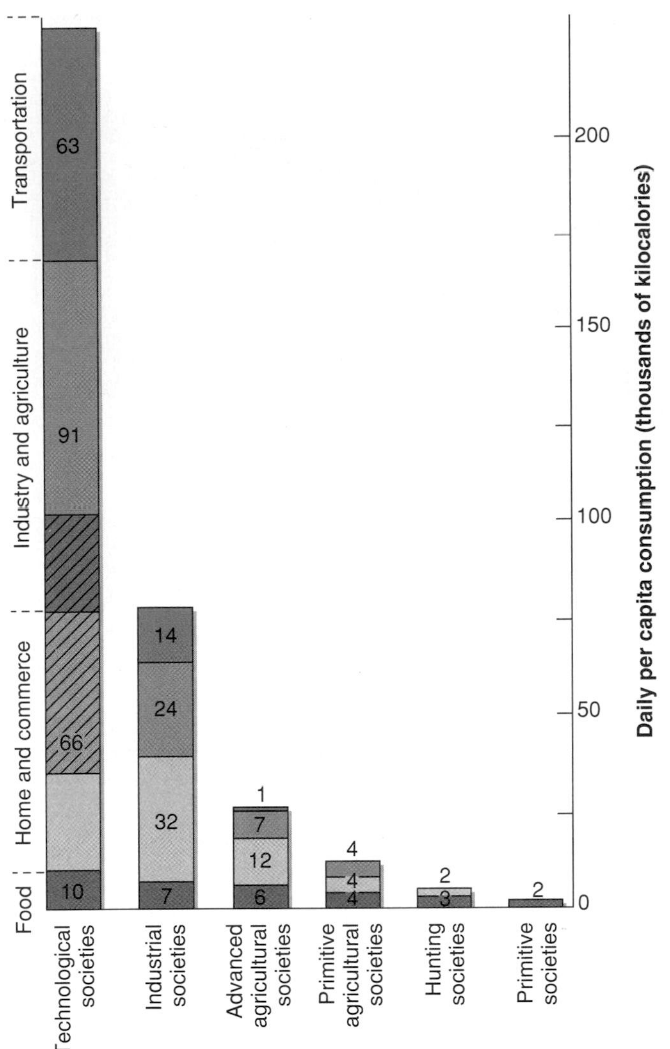

FIGURE 14.1 As societies evolve from primitive to agricultural to industrialized, the daily per capita consumption of energy increases sharply. The body needs, on average, about 2000 kilocalories per day to live (the "kilocalorie," a measure of energy, is the food calorie that dieters count). However, in the United States in 2001, per capita energy consumption was 237,000 kilocalories per day: 34% for residential/commercial uses, 39% for industry, 27% for transportation. Of the total, about 37% was consumed in the generation of electricity. Note that, as societies advance, even the energy associated with the food individuals eat increases. Members of modern societies do not consume 10,000 kilocalories of food energy directly, but behind the 2,000 or so actually consumed are large additional energy expenditures associated with agriculture, food processing, and preservation, and even the conversion of plants to meat by animals, which in turn are consumed as food. The shaded area in the far left bar of the graph is electricity usage.

Source: Graph after Earl Cook, "The Flow of Energy in an Industrialized Society," Scientific American, *© 1971.*

FORMATION OF OIL AND NATURAL GAS DEPOSITS

Petroleum is not a single chemical compound. Liquid petroleum, or **oil,** comprises a variety of liquid hydrocarbon compounds (compounds made up of different proportions of the elements carbon and hydrogen). There are also gaseous hydrocarbons (**natural gas**), of which the compound methane (CH_4) is the most common. How organic matter is transformed into liquid and gaseous hydrocarbons is not fully understood, but the main features of the process are outlined in the following section.

The production of a large deposit of any fossil fuel requires a large initial accumulation of organic matter, which is rich in carbon and hydrogen. Another requirement is that the organic debris be buried quickly to protect it from the air so that decay by biological means or reaction with oxygen will not destroy it.

Microscopic life is abundant over much of the oceans. When these organisms die, their remains can settle to the sea floor. There are also underwater areas near shore, such as on many continental shelves, where sediments derived from continental erosion accumulate rapidly. In such a setting, the starting requirements for the formation of oil are satisfied: There is an abundance of organic matter rapidly buried by sediment. Oil and much natural gas are believed to form from such accumulated marine microorganisms. Continental oil fields often reflect the presence of marine sedimentary rocks below the surface. Additional natural gas deposits not associated with oil may form from deposits of plant material in sediment on land.

As burial continues, the organic matter begins to change. Pressures increase with the weight of the overlying sediment or rock; temperatures increase with depth in the earth; and slowly, over long periods of time, chemical reactions take place. These reactions break down the large, complex organic molecules into simpler, smaller hydrocarbon molecules. The nature of the hydrocarbons changes with time and continued heat and pressure. In the early stages of petroleum formation in a marine deposit, the deposit may consist mainly of larger hydrocarbon molecules ("heavy" hydrocarbons), which have the thick, nearly solid consistency of asphalt. As the petroleum matures, and as the breakdown of large molecules continues, successively "lighter" hydrocarbons are produced. Thick liquids give way to thinner ones, from which are derived lubricating oils, heating oils, and gasoline. In the final stages, most or all of the petroleum is further broken down into very simple, light, gaseous molecules—natural gas. Most of the maturation process occurs in the temperature range of 50° to 100°C (approximately 120° to 210°F). Above these temperatures, the remaining hydrocarbon is almost wholly methane; with further temperature increases, methane can be broken down and destroyed in turn (see figure 14.3).

A given oil field yields crude oil containing a distinctive mix of hydrocarbon compounds, depending on the history of the material. The refining process separates the different types of hydrocarbons for different uses (see table 14.1). Some of the heavier hydrocarbons may also be broken up during refining into smaller, lighter molecules through a process called *cracking,* which allows some of the lighter compounds such as gasoline to be produced as needed from heavier components of crude oil.

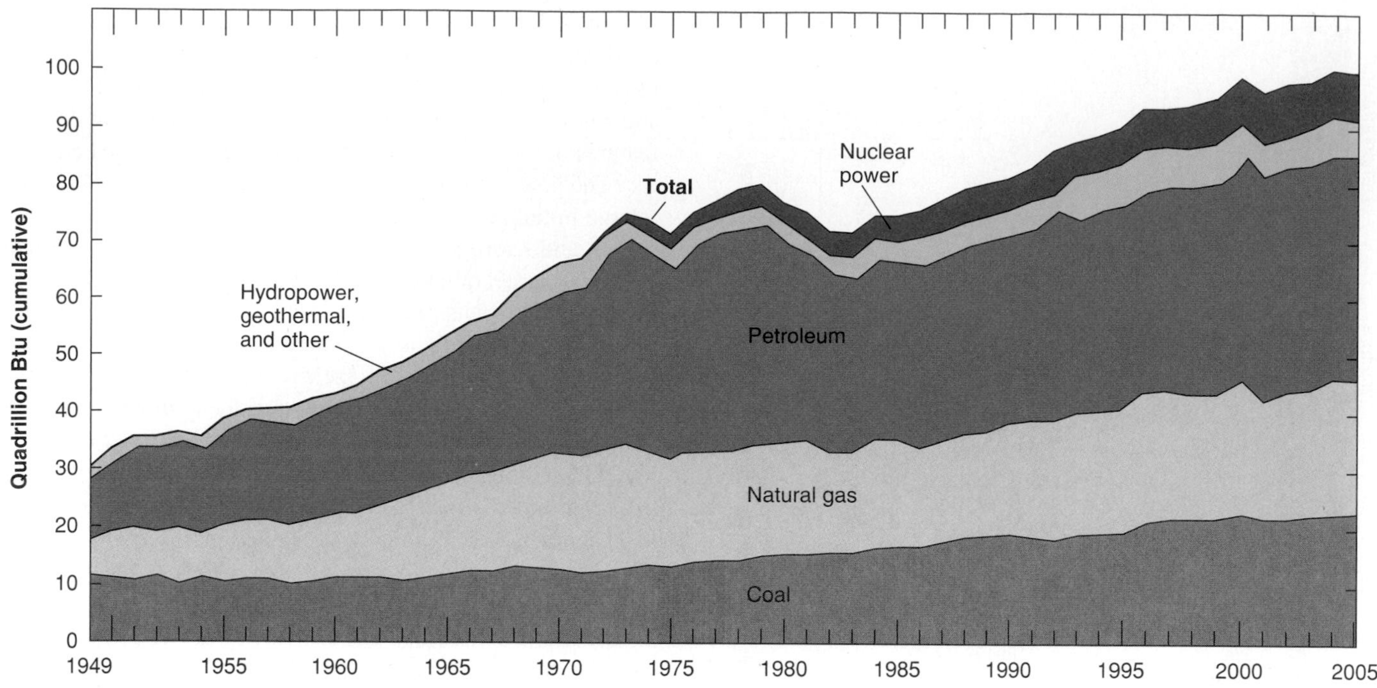

FIGURE 14.2 U.S. energy consumption, 1949–2005. The trend is clearly upward, except following the Arab oil embargo of the early 1970s, and in the early 1980s, when high prices and a sluggish economy combined to depress energy consumption. Note that figures are expressed in terms of energy derived from each source so that the contributions can be more readily compared. For energy equivalents of the several fossil fuels, see the table of units.
Source: Annual Energy Review 2005, *U.S. Energy Information Administration, Department of Energy.*

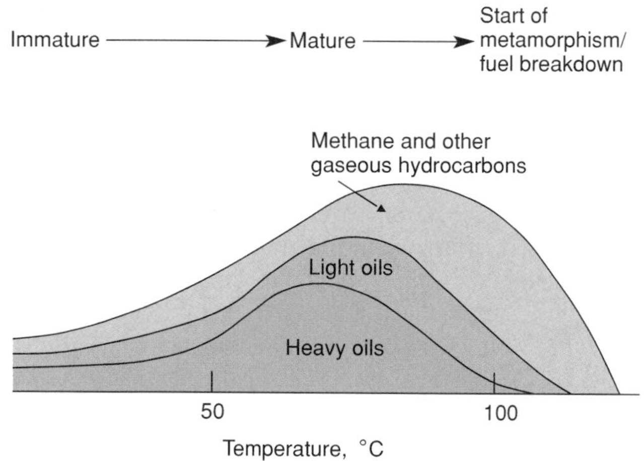

FIGURE 14.3 The process of petroleum maturation in a marine deposit, in simplified form.
Source: After J. Watson, Geology and Man, 1983, *Allen and Unwin, Inc.*

Once the solid organic matter is converted to liquids and/or gases, the hydrocarbons can migrate out of the rocks in which they formed. Such migration is necessary if the oil or gas is to be collected into an economically valuable and practically usable deposit. The majority of petroleum source rocks are fine-grained clastic sedimentary rocks of low permeability, from which it would be difficult to extract large quantities of oil or gas quickly. Despite the low permeabilities, oil and gas are able to migrate out of their source rocks and through more permeable rocks over the long spans of geologic time. The pores, holes, and cracks in rocks in which fluids can be trapped are commonly full of water. Most oils and all natural gases are less dense than water, so they tend to rise as well as to migrate laterally through the water-filled pores of permeable rocks.

Unless stopped by impermeable rocks, oil and gas may keep rising right up to the earth's surface. At many known oil and gas seeps, these substances escape into the air or the oceans or flow out onto the ground. These natural seeps, which are one of nature's own pollution sources, are not very efficient sources of hydrocarbons for fuel if compared with present drilling and extraction processes, although asphalt from seeps in the Middle East was used in construction five thousand years ago.

Commercially, the most valuable deposits are those in which a large quantity of oil and/or gas has been concentrated and confined (trapped) by impermeable rocks in geologic structures (see figure 14.4). The reservoir rocks in which the oil or gas has accumulated should be relatively porous if a large quantity of petroleum is to be found in a small volume of rock and should also be relatively permeable so that the oil or

TABLE 14.1

Fuels Derived from Liquid Petroleum and Gas

	Material	Principal Uses
Heavier hydrocarbons	waxes (for example, paraffin)	candles
↑	heavy (residual) oils	heavy fuel oils for ships, power plants, and industrial boilers
	medium oils	kerosene, diesel fuels, aviation (jet) fuels, power plants, and domestic and industrial boilers
	light oils	gasoline, benzene, and aviation fuels for propeller-driven aircraft
↓	"bottled gas" (mainly butane, C_4H_{10})	primarily domestic use
Lighter hydrocarbons	natural gas (mostly methane, CH_4)	domestic/industrial use and power plants

Source: Data from J. Watson, *Geology and Man,* copyright © 1983 Allen and Unwin, Inc.

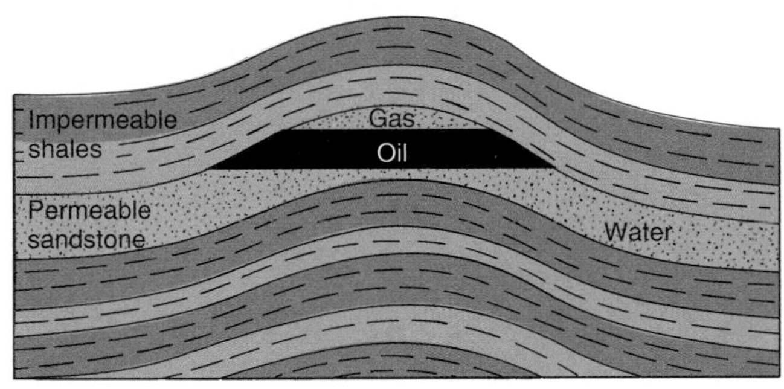

Impermeable shales
Gas
Oil
Permeable sandstone
Fault
C

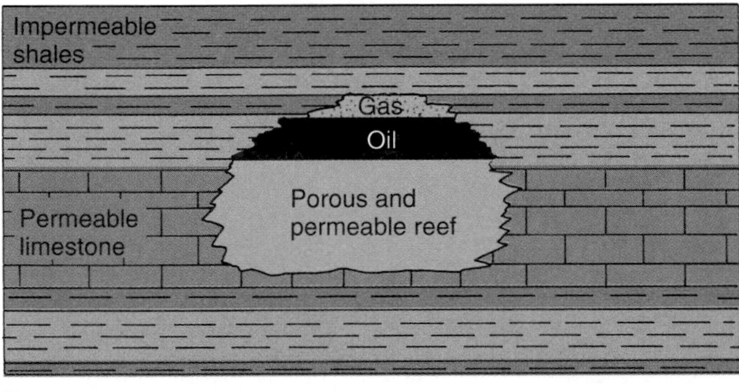

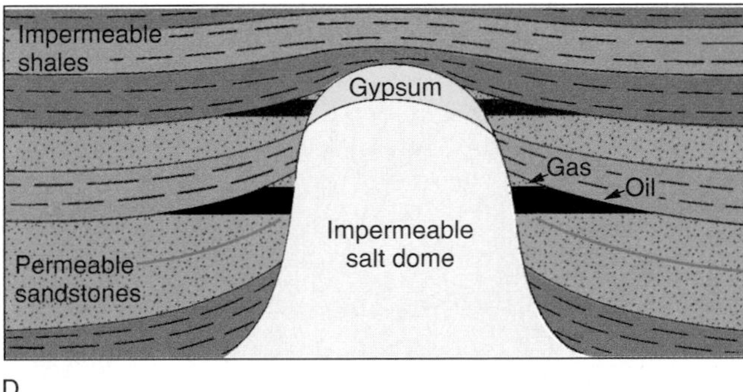

FIGURE 14.4 Types of petroleum traps. (A) A simple fold trap. (B) Petroleum accumulated in a fossilized ancient coral reef. (C) A fault trap. (D) Petroleum trapped against an impermeable salt dome, which has risen up from a buried evaporite deposit.

gas flows out readily once a well is drilled into the reservoir. If the reservoir rocks are not naturally very permeable, it may be possible to fracture them artificially with explosives or with water or gas under high pressure to increase the rate at which oil or gas flows through them.

The amount of time required for oil and gas to form is not known precisely. Since virtually no petroleum is found in rocks younger than 1 to 2 million years old, geologists infer that the process is comparatively slow. Even if it took only a few tens of thousands of years (a geologically short period), the world's oil and gas is being used up far faster than significant new supplies could be produced. Therefore, oil and natural gas are among the **nonrenewable** energy sources. We have an essentially finite supply with which to work.

SUPPLY AND DEMAND FOR OIL AND NATURAL GAS

As with minerals, the most conservative estimate of the supply of an energy source is the amount of known reserves, "proven" accumulations that can be produced economically with existing technology. A more optimistic estimate is total resources, which include reserves, plus known accumulations that are technologically impractical or too expensive to tap at present, plus some quantity of the substance that is expected to be found and extractable. As with minerals, estimates of energy reserves are thus sensitive both to price fluctuations and to technological advances.

Oil

Oil is commonly discussed in units of barrels (1 barrel = 42 gallons). Worldwide, over one trillion barrels of oil have been consumed, according to estimates by the U.S. Geological Survey. The estimated remaining reserves are about 1.2 trillion barrels (figure 14.5). That does not sound too ominous until one realizes that close to half of the consumption has occurred in the last quarter-century or so. Also, global demand continues to increase as more countries advance technologically.

Oil supply and demand are very unevenly distributed around the world. Some low-population, low-technology, oil-rich countries (Libya, for example) may be producing fifty or a hundred times as much oil as they themselves consume. At the other extreme are countries like Japan, highly industrialized and energy-hungry, with no petroleum reserves at all. The United States alone consumes about 25% of the oil used worldwide, more than all of Europe together, and about three times as much as China or Japan, the next-highest individual-country consumers. As with mineral resources, the size of petroleum resources of a given nation are not correlated with geographic size. Many have speculated that much of the international interest in the middle East is motivated as much by concern for its estimated 60% share of world oil reserves as by humanitarian or other motives.

The United States initially had perhaps 10% of all the world's oil resources. Already over 400 billion barrels have been produced and consumed; remaining U.S. proven *reserves* are under 22 billion barrels. For more than three decades, the United States has been using up about as much domestic oil each year as the amount of new domestic reserves discovered, or proven, and in many years somewhat more, so the net U.S. oil reserves have generally been decreasing year by year (see table 14.2). Domestic production has likewise been declining (figures 14.6 and 14.7).

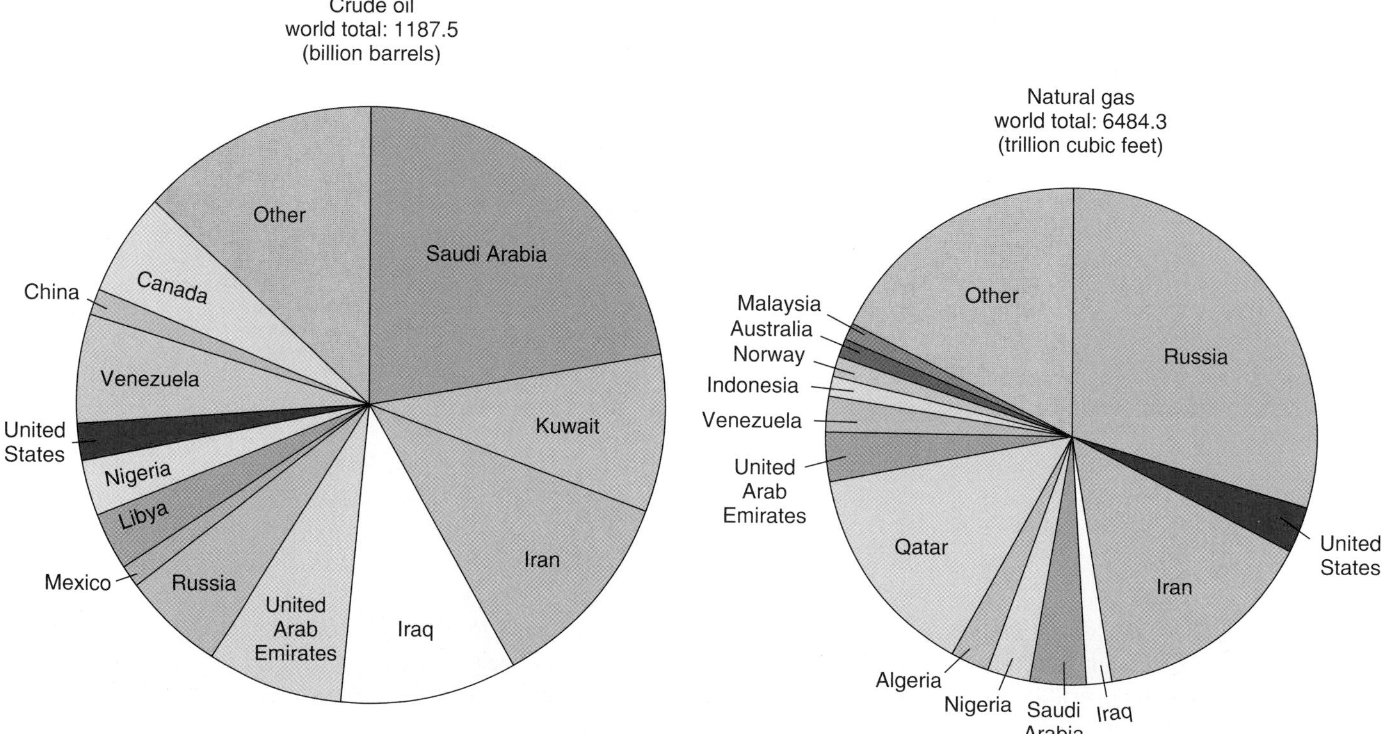

FIGURE 14.5 Estimated proven world reserves of crude oil and natural gas, January 2006.
Source: Averages of estimates of BP Statistical Review, Oil and Gas Journal *and* World Oil, *as summarized in* International Energy Annual 2006, *U.S. Energy Information Administration.*

TABLE 14.2

Proven U.S. Reserves Crude Oil and Natural Gas, 1982–2006

Year	Crude Oil (billions of barrels)	Natural Gas (trillions of cu ft)
1982	29.5	209.3
1986	28.3	201.1
1990	27.6	177.6
1994	23.6	171.9
1998	22.7	141.8
2002	22.4	183.5
2006	21.4	192.5

Source: Data from *Annual Energy Review 1999*, and *International Energy Annual 2001* and *2006*. U.S. Energy Information Administration, Department of Energy.

So, the United States has for many years relied heavily on imported oil to meet part of its oil demand. Without it, U.S. reserves would dwindle more quickly. The United States consumes nearly 7 billion barrels of oil per year, to supply about 40% of all the energy used. Currently, about 60% of the oil consumed is imported from other countries. That percentage has increased steadily since the early 1980s. Principal sources currently are Canada, Mexico, Saudi Arabia, Venezuela, and Nigeria, with other nations contributing as well (for example, in 2005, the United States was importing half a million barrels a day from Iraq). Simple arithmetic demonstrates that the rate of oil consumption in the United States is very high compared even to the estimated total U.S. resources available. Some of those resources have yet to be found, too, and will require time to discover and develop. Both U.S. and world oil supplies could be nearly exhausted within decades (figure 14.7), especially considering the probable acceleration of world energy demands. On occasion, explorationists do find the rare, very large concentrations of petroleum—the deposits on Alaska's North Slope and beneath Europe's North Sea are examples. Yet even these make only a modest difference in the long-term picture. The Prudhoe Bay oil field, for instance, represented reserves of only about 13 billion barrels, a great deal for a single region, but less than two years' worth of U.S. consumption. U.S. oil production peaked at 9.64 million barrels per day in 1970, and has declined steadily since then, dipping to 5.12 million barrels per day in 2005. Concern about dependence on imported oil led to the establishment of the Strategic Petroleum Reserve in 1977, but the amount of oil in the SPR stayed at about 550 million barrels from 1980 to 2000, while the number of days' worth of imports that it represents has declined from a high of 115 in 1985 to 50 in 2002 (figure 14.8). Only renewed addition of imported oil, to raise SPR stocks to about 685 million barrels in 2005, slowed the decline in the number of days of imports held in the SPR, but that figure was still only 55 in 2005. Such facts increase pressure to develop new domestic oil sources; see Box 14.1.

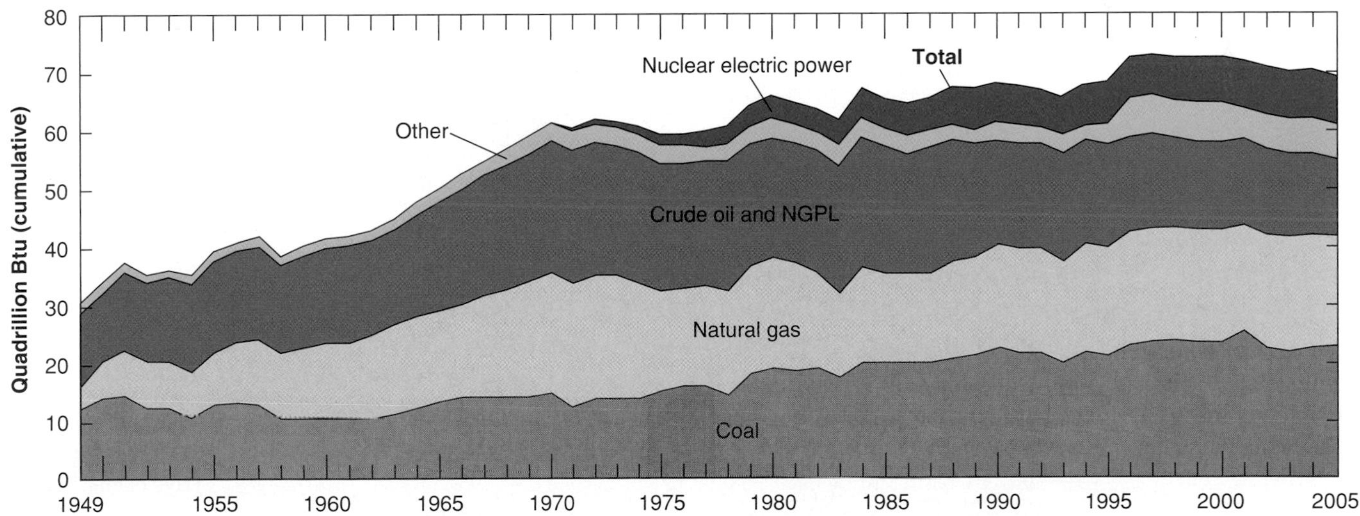

FIGURE 14.6 U.S. energy production since the mid-twentieth century, by energy source. Note that domestic production has leveled off despite sharp rises in fossil-fuel prices and rising consumption. (Natural gas plant liquids (NGPL) are liquid fuels produced during refining/processing of natural gas.)
Source: Annual Energy Review 2005, *U.S. Energy Information Administration.*

Box 14.1 A Case of Conflicting Priorities: The Arctic National Wildlife Refuge—To Drill or Not to Drill?

The Prudhoe Bay field on the North slope of Alaska is the largest U.S. oil field ever found. It has yielded 9.7 billion barrels of oil so far, with an estimated remaining reserve of 3.3 billion barrels. Its richness justified the $8 billion cost of building the 800-mile Trans-Alaska pipeline. West of Prudhoe Bay lies the 23-million-acre tract now known as the National Petroleum Reserve–Alaska (NPRA), first established as Naval Petroleum Reserve No. 4 by President Warren Harding in 1923. East of Prudhoe Bay is the 19-million-acre Arctic National Wildlife Refuge (ANWR), established in 1980 (figure 1).

Geologic characteristics of, and similarities among, the three areas suggested that oil and gas might very well be found in both NPRA and ANWR. Thus when establishing ANWR, Congress deferred a final decision on the future of the coastal portion, some 1.5 million acres known as the "1002 Area" after the section of the act that addressed it.

Little petroleum exploration occurred in NPRA until the late 1990s. In 1996, oil was found in the area between NPRA and Prudhoe Bay. Then, beginning in 2000, exploration wells began to find oil and gas in NPRA.

Estimating potential oil and gas reserves in a new region is a challenge. Drilling and seismic and other data constrain the types and distribution of subsurface rocks. Samples from drill cores, and comparisons with geologically similar areas, allow projection of quantities of oil and gas that may be present. Only a portion of the in-place resources may be technologically recoverable—perhaps 30–50% of the oil, 60–70% of the gas. And of course, the market price of each commodity will determine how much is economically recoverable. Combining all of these considerations results in a range of estimates of oil and gas reserves in NPRA and the ANWR 1002 Area (figure 2). At crude-oil prices of $40 per barrel, the analysis suggested a mean probability of close to 7 billion barrels of economically recoverable oil each in NPRA and in the ANWR 1002 Area. Moreover, it is believed that the oil accumulations in the ANWR 1002 Area are, on average, larger, meaning that they could be developed more efficiently. (Development of the gas reserves would require another pipeline.) At $60/barrel, the projected reserves would be somewhat higher, though note in figure 2 that they increase only moderately as oil price rises from $30 to $40 per barrel.

Those who favor drilling in the ANWR 1002 Area point to the current massive U.S. dependence on foreign oil, the financial benefits to Alaskans of additional oil production, the fact that there are tens of millions of acres of other wilderness. They also note that oil production technology has become more sophisticated so as to leave less of a "footprint"—for example, one can drill one well at the surface that splays out in multiple directions at depth rather than having to drill multiple wells from the surface to tap the same deposits. Those opposed to drilling note the extreme environmental fragility of alpine ecosystems in general, the biological richness of the 1002 Area in particular (the ANWR has been called the Alaskan Serengeti), and the fact that the United States consumes close to 7 billion barrels of crude oil in just a year (so, why risk the environment for so relatively little oil?). The debate continues.

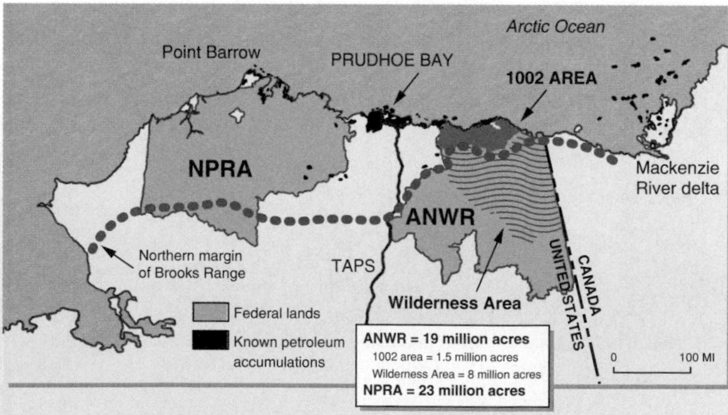

FIGURE 1 The NPRA and ANWR flank the petroleum-rich Prudhoe Bay area. The future of the "1002 Area" of the ANWR is currently being hotly debated.
After USGS Fact Sheet 040-98.

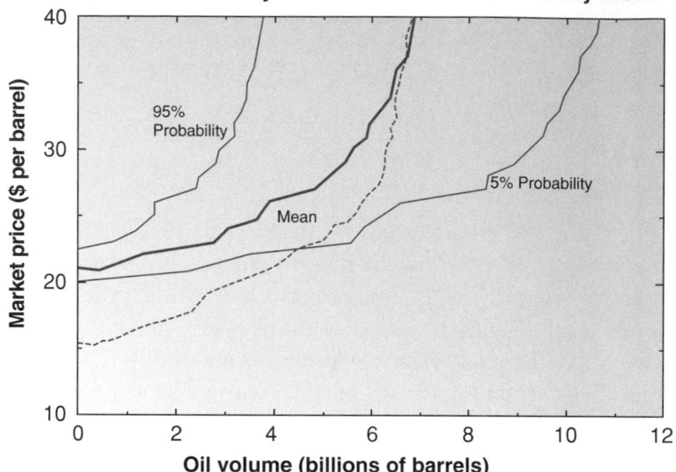

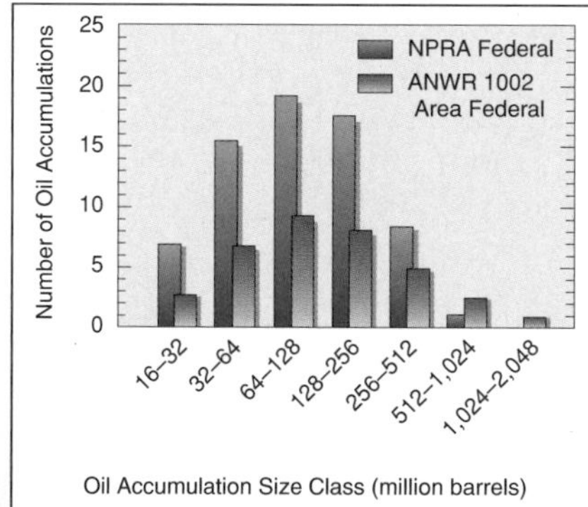

Market Price ($/barrel)	Economically Recoverable Oil (billion barrels) NPRA	ANWR
15	0	0
–	–	–
20	0	3.2
21	0.4	4.0
22	1.3	4.4
23	2.8	5.0
24	3.1	5.2
25	3.7	5.6
26	4.0	5.8
27	4.8	6.0
28	5.1	6.2
29	5.4	6.3
30	5.6	6.3
–	–	–
35	6.4	6.6
40	6.9	6.8

Table of mean values

FIGURE 2 U.S. Geological Survey estimates of economically recoverable oil in the federally owned portions of NPRA and the ANWR 1002 Area (other areas are Native lands), and size distribution of individual oil accumulations (inset graph). Red lines are for NPRA; dashed blue line is mean for ANWR 1002 Area.
After USGS Fact Sheet 045-02.

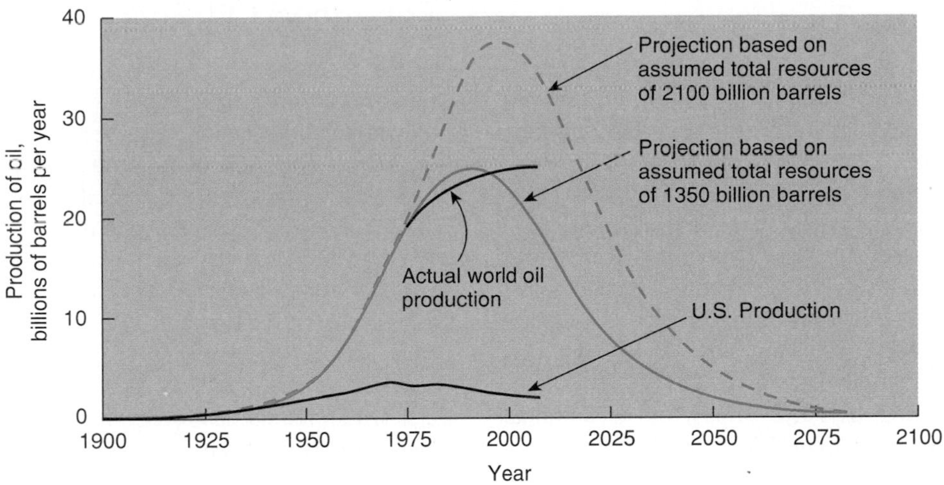

FIGURE 14.7 When this classic projection by M. King Hubbert was first widely published in the early 1970s, many people were shocked to discover that the total supply of oil appeared so limited. In fact, the total resource estimates had long existed in the geologic literature. The supply problem was made acute by the rapid rise in consumption starting in the 1950s. Hubbert accurately predicted the slower rise in world production of oil after 1975, which to a great extent is the result of depletion of reserves, especially easily accessible ones. U.S. production peaked in 1970 for that reason. Current projections of the peak in world oil production, using revised USGS estimates of total recoverable world oil resources of about 3000 billion barrels, place the peak within about five years of 2020, depending upon model assumptions.

Source: Projections from M. King Hubbert, "The Energy Resources of the Earth," Scientific American, *Sept. 1971, p. 69. Production data from* Annual Energy Review 2005 *and* International Energy Annual 2006, *U.S. Energy Information Administration.*

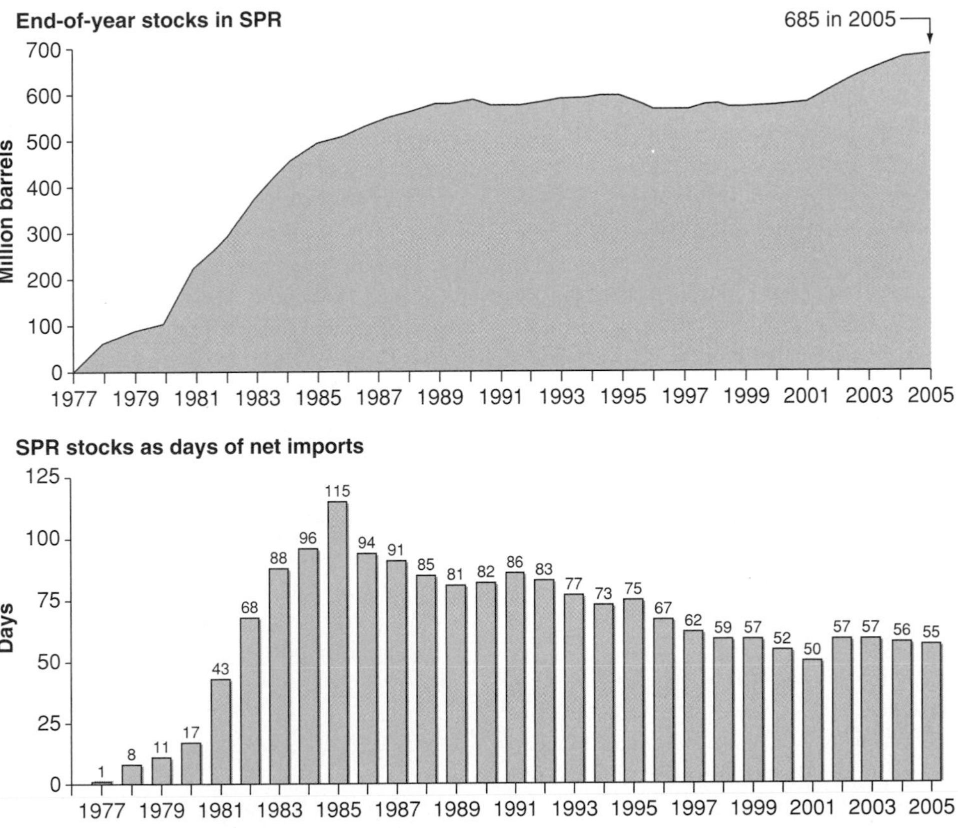

FIGURE 14.8 Strategic Petroleum Reserve (SPR) was established to provide a "cushion" in the event of disruption of imports of oil. The length of time the SPR will last declines as consumption of imported oil rises.

Source: Annual Energy Review 2005, *U.S. Energy Information Administration, Department of Energy.*

Natural Gas

The supply/demand picture for conventional natural gas is similar to that for oil (see figures 14.2 and 14.6). Natural gas presently supplies 20 to 25% of the energy used in the United States. The United States has proven natural gas reserves of under 200 trillion cubic feet. However, roughly 22 trillion cubic feet are consumed per year, and most years, less is found in new domestic reserves than the quantity consumed. The net result is, again, generally declining reserves (table 14.2). It is noteworthy that until the mid-1980s, U.S. natural-gas production roughly equaled consumption. Since then, we have been importing increasing amounts of natural gas; imports now account for about 20% of consumption. The U.S. supply of conventional natural gas, too, could be used up in a matter of decades.

Future Prospects

Many people tend to assume that, as oil and natural gas supplies dwindle and prices rise, there will be increased exploration and discovery of new reserves so that we will not really run out of these fuels. But the supplies are ultimately finite, after all.

Moreover, while decreasing reserves have prompted exploration of more areas, most regions not yet explored have been neglected precisely because they are unlikely to yield appreciable amounts of petroleum. The high temperatures involved in the formation of igneous and most metamorphic rocks would destroy organic matter, so oil would not have formed or been preserved in these rocks. Nor do these rock types tend to be very porous or permeable, so they generally make poor reservoir rocks as well, unless fractured. The large portions of the continents underlain predominantly by igneous and metamorphic rocks are, for the most part, very unpromising places to look for oil. Some such areas are now being explored, but with limited success. Other, more promising areas may be protected or environmentally sensitive.

The costs of exploration have gone up, and not simply due to inflation. In the United States, drilling for oil or gas costs an average of nearly $300 per *foot;* the average oil well drilled is about 6000 feet deep (it was about 4000 feet in 1949); even with increasingly sophisticated methods of exploration to target areas most likely to yield economic hydrocarbon deposits, more than half of exploratory wells come up dry. Costs for drilling offshore are substantially higher than for land-based drilling, too, and many remaining sites where hydrocarbons might be sought are offshore. Yield from producing wells is also declining, from a peak average of 18.6 barrels of oil per well per day in 1972 to 10.1 barrels in 2005.

Despite a quadrupling in oil prices between 1970 and 1980 (*after* adjustment for inflation), U.S. proven reserves continued to decline. They are declining still, despite oil prices at near-record levels as this is written. This is further evidence that higher prices do not automatically lead to proportionate, or even appreciable, increases in fuel supplies. Also, many major oil companies have been branching out into other energy sources beyond oil and natural gas, suggesting an expectation of a shift away from petroleum in the future.

Enhanced Oil Recovery

A few techniques are being developed to increase petroleum production from known deposits. An oil well initially yields its oil with minimal pumping, or even "gushes" on its own, because the oil and any associated gas are under pressure from overlying rocks, or perhaps because the oil is confined like water in an artesian system. Recovery using no techniques beyond pumping is *primary recovery*. When flow falls off, water may be pumped into the reservoir, filling empty pores and buoying up more oil to the well (*secondary recovery*). Primary and secondary recovery together extract an average of one-third of the oil in a given trap, though the figure varies greatly with the physical properties of the oil and host rocks in a given oil field. On average, then, two-thirds of the oil in each deposit has historically been left in the ground. Thus, additional *enhanced recovery* methods have attracted much interest.

Enhanced recovery comprises a variety of methods beyond conventional secondary recovery. Permeability of rocks may be increased by deliberate fracturing, using explosives or even water under very high pressure. Carbon dioxide gas under pressure can be used to force out more oil. Hot water or steam may be pumped underground to warm thick, viscous oils so that they flow more easily and can be extracted more completely. There have also been experiments using detergents or other substances to break up the oil.

All of these methods add to the cost of oil extraction. That they are now being used at all is largely a consequence of higher oil prices in the last two decades and increased difficulty of locating new deposits. Researchers in the petroleum industry believe that, from a technological standpoint, up to an additional 40% of the oil initially in a reservoir might be extractable by enhanced-recovery methods. This would substantially increase oil reserves. A further positive feature of enhanced-recovery methods is that they can be applied to old oil fields that have already been discovered, developed by conventional methods, and abandoned, as well as to new discoveries. In other words, numerous areas in which these techniques could be used (if the economics were right) are already known, not waiting to be found. It should, however, be kept in mind that enhanced recovery may involve increases in problems such as ground subsidence or ground-water pollution that arise also with conventional recovery methods. Even conventional drilling uses large volumes of drilling mud to cool and lubricate the drill bits. The drilling muds may become contaminated with oil and must be handled carefully to avoid polluting surface or ground water. When water is more extensively used in fracturing rock or warming the oil, the potential for pollution is correspondingly greater.

Alternate Natural Gas Sources

During coal formation, quantities of methane-rich gas are also produced. Most coal deposits contain significant amounts of **coal-bed methane.** Historically, this methane has been treated as a hazardous nuisance, potentially explosive, and its incidental release during coal mining has contributed to rising methane concentrations is the atmosphere. However, an estimated 100 trillion cubic feet of methane that would be economically recoverable with current technology exists in-place in U.S. coal beds. If it could be extracted for use rather than wasted, it could contribute significantly to U.S. gas reserves. And we know where it is, because U.S. coal beds are already well mapped, as noted in the later discussion of coal. The major technological obstacle to recovery of coal-bed methane is the need to extract and dispose of the water that typically pervades coal beds, water that is often saline and requires careful disposal to avoid polluting local surface or ground water. However, coal-bed methane is already being produced in the Wasatch Plateau area of Utah and the Powder River Basin in Wyoming. As extraction technology is improved and the geology of coal-bed methane further studied, increased development of this resource is expected.

Some evidence suggests that additional natural gas reserves exist at very great depths in the earth. Thousands of meters below the surface, conditions are sufficiently hot that any oil would have been broken down to natural gas (recall figure 14.3). This natural gas, under the tremendous pressures exerted by the overlying rock, may be dissolved in the water filling the pore spaces in the rock, much as carbon dioxide is dissolved in a bottle of soda. Pumping this water to the surface is like taking off the bottle cap: The pressure is released, and the gas bubbles out. Enormous quantities of natural gas might exist in such **geopressurized zones;** recent estimates of the amount of potentially recoverable gas in this form range from 150 to 2000 trillion cubic feet.

Special technological considerations are involved in developing these deposits. It is difficult and very expensive to drill to these deep, high-pressure accumulations. Also, many of the fluids in which the gas is dissolved are very saline brines that cannot be casually released at the surface without damage to the quality of surface water; the most effective means of disposal is to pump the brines back into the ground, but this is costly. On the plus side, however, the hot fluids themselves may represent an additional geothermal energy source (see the discussion of geothermal energy in chapter 15). Geopressurized natural gas may well become an important supplementary fossil fuel in the future, though its total potential and the economics of its development are poorly known. It is unlikely that enough can be found and produced soon enough to solve near-future energy problems.

Still more recently, the significance of **methane hydrates** has been increasingly discussed. Gas (usually methane) hydrates are crystalline solids of gas and water molecules (figure 14.9). These hydrates have been found to be abundant in arctic regions and in marine sediments. The U.S. Geological Survey has estimated that the amount of carbon found in gas hydrates may

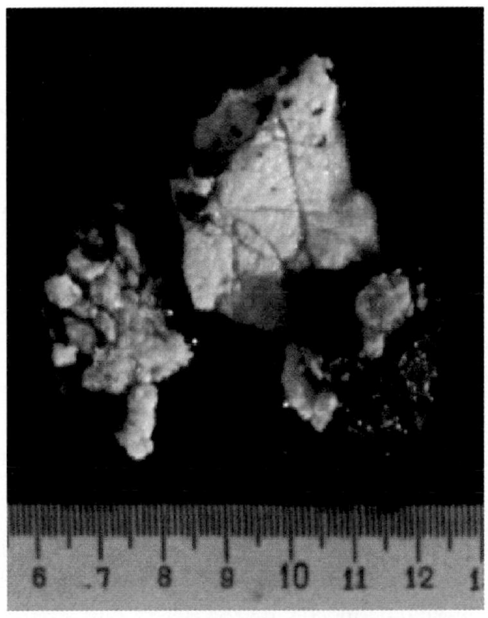

A

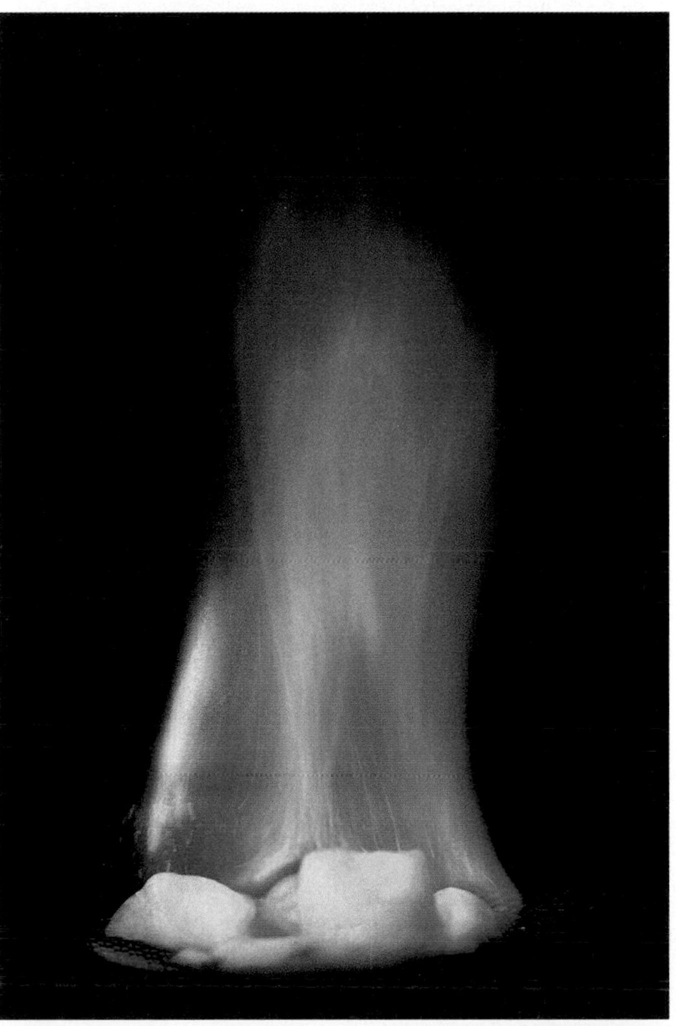

B

FIGURE 14.9 (A) Irregular chunks of gas hydrate from sediments in the Sea of Okhotsk. Scale is in centimeters. (B) Methane released from a piece of hydrate burning.

(A) Photograph courtesy U.S. Geological Survey. (B) Photograph by J. Pinkston and L. Stern , U. S. Geological Survey

be twice that in all known fossil fuels on earth (figure 14.10). Methane in methane hydrate thus represents a huge potential natural gas resource. Just how to tap methane hydrates for their energy is not yet clear.

Beyond that, new questions are being raised about methane hydrates and the greenhouse effect. In arctic regions, the hydrates are stabilized by the low temperature. If these arctic sediments were warmed enough to break down some of the hydrates, methane gas would be released into the atmosphere, which would further increase greenhouse-effect heating. Considering that methane is a far more efficient greenhouse gas than is CO_2 and that the amount of methane locked in hydrates now is estimated at about 3000 times the methane currently in the atmosphere, the potential impact is large and represents yet another source of uncertainty in global-climate modeling.

In the meantime, scientists are studying the feasibility of using fracturing techniques to initiate enhanced gas recovery from "tight" (low-permeability) sandstones in the Rocky Mountains and gas-bearing shales in the Appalachians. These projects are still in the experimental stages. As with geopressurized gas, the quantity of recoverable gas in these rocks is uncertain. Recent estimates, however, range from 60 to 840 trillion cubic feet. A factor that limits their current attractiveness is lack of price competitiveness, a problem they share with fuels derived from oil shale and tar sand, as will be seen later in this chapter.

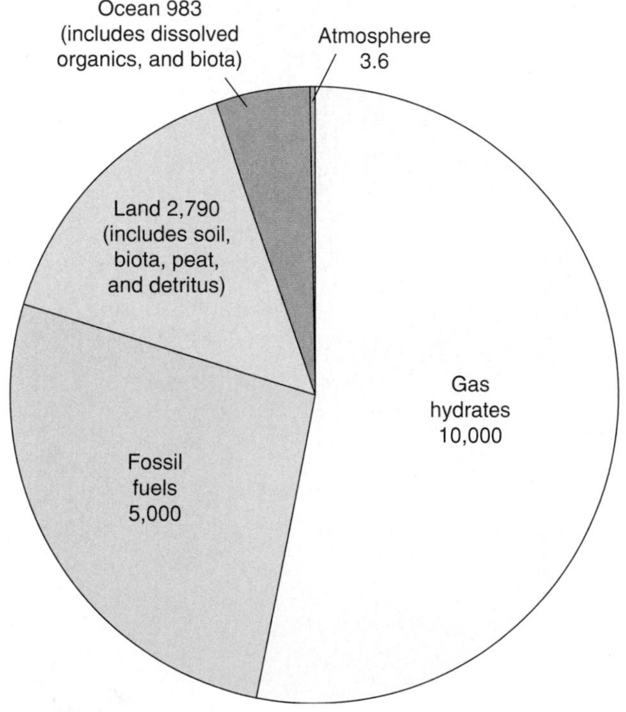

FIGURE 14.10 The amount of carbon now believed to exist in gas hydrates exceeds that in all known fossil fuels and other organic-carbon reservoirs combined. Units are billions of tons of carbon.
Data from U.S. Geological Survey.

Conservation

Conservation of oil and gas is, potentially, a very important way to stretch remaining supplies. More and more individuals, businesses, and governments in industrialized societies are practicing energy conservation out of personal concern for the environment, from fear of running out of nonrenewable fuels, or simply for basic economic reasons when energy costs have soared. The combined effects of conservation and economic recession can be seen in the flattening of the U.S. energy-consumption curve in figure 14.2 during the mid-1970s to mid-1980s. However, beginning about 1990, energy consumption resumed rising steadily, reaching an all-time high in 2004 before slackening slightly. Some of the factors in rising energy consumption can be climatic—unusual heat in July and August, unusual cold in November and December—but other consumption increases are discretionary. Low gasoline prices, for example, make fuel-efficient cars and other fuel-conservation methods such as carpooling less urgent priorities for some consumers. Worldwide, too, the United States has long had relatively inexpensive gasoline and consumed it freely as a result (figure 14.11). See also Box 14.2, later in this chapter.

Conservation can buy some much-needed time to develop alternative energy sources. However, world energy consumption will probably not decrease significantly in the near future. Even if industrialized countries consistently adopt more energy-efficient practices, demand in the many nonindustrialized countries is expected to continue to rise. Many technologically less-developed countries view industry and technology as keys to better, more prosperous lives for their people. They resent being told to conserve energy by nations whose past voracious energy consumption has largely led to the present squeeze.

From time to time, poor global economic conditions may depress energy use in the industrialized countries, as has happened late in the twentieth century worldwide. Such events may lead to talk of an "oil glut" on world markets and to temporarily falling prices, but this surplus really represents just a short-term, temporary excess of current production over current demand. Assuming that world demand for energy stays at least as high as it is now, we face a crisis of dwindling petroleum and natural-gas supplies over the next few decades. In the meantime, heavy reliance on oil and gas continues to have some serious environmental consequences, notably oil spills.

OIL SPILLS

Natural oil seeps are not unknown. The LaBrea Tar Pits of prehistoric times are an example. In fact, it is estimated that oil rising up through permeable rocks escapes into the ocean at the rate of 600,000 tons per year, though many natural seeps seal themselves as leakage decreases remaining pressure. Tankers that flush out their holds at sea continually add to the oil pollution of the oceans and, collectively, are a significant source

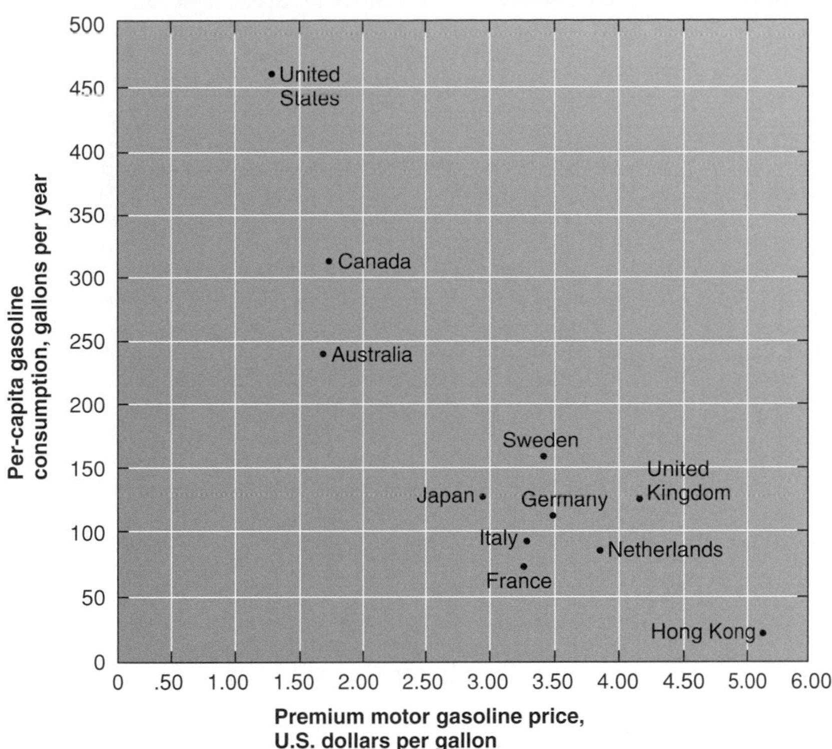

FIGURE 14.11 Among industrialized nations worldwide, there is an apparent inverse correlation between gasoline price and gasoline consumption. It remains to be seen if recent sharp rises in U.S. gasoline prices will change long-established patterns of high gasoline use.

Data from International Energy Annual 2006, *U.S. Energy Information Administration.*

of such pollution. The oil spills about which most people have been concerned, however, are the large, sudden, catastrophic spills that occur in two principal ways: from accidents during drilling of offshore oil wells and from wrecks of oil tankers at sea. Oil spills represent the largest negative impacts from the extraction and transportation of petroleum, although as a source of water pollution, they are less significant volumetrically than petroleum pollution from careless disposal of used oil, amounting to only about 5% of petroleum released into the environment. And although we tend to hear only about the occasional massive, disastrous spill, there are many more that go unreported in the media: The U.S. Coast Guard reports about 10,000 spills in and around U.S. waters each year, totaling 15 to 25 million gallons of oil annually.

Drilling accidents have been a growing concern as more areas of the continental shelves are opened to drilling. Normally, the drill hole is lined with a steel casing to prevent lateral leakage of the oil, but on occasion, the oil finds an escape route before the casing is complete. This is what happened in Santa Barbara in 1979, when a spill produced a 200-square-kilometer (about 80-square-mile) oil slick. Alternatively, drillers may unexpectedly hit a high-pressure pocket that causes a blowout. This was the cause of a spill in the Gulf of Mexico in 1979. And in 2005, Hurricane Katrina damaged oil drilling, processing, and storage facilities in the Gulf of Mexico, causing more than 100 spills totaling over 8 million gallons of oil!

Tanker disasters are potentially becoming larger all the time. The largest supertankers now being built are longer than the Empire State Building is high (over 335 meters) and can carry nearly 2 million barrels of oil (about 80 million gallons). The largest single marine spill ever resulted from the wreck of the *Amoco Cadiz* near Portsall, France, in 1978; the bill for cleaning up what could be recovered of the 1.6 million barrels spilled was more than $50 million, and at that, negative environmental impacts were still detectable seven years later.

In 1991, the war in the Persian Gulf demonstrated yet another possible cause of major oil spills: destruction of major pipelines and refinery facilities (figure 14.12).

FIGURE 14.12 Oil slick along the Persian Gulf after the 1991 Gulf War.

Photograph courtesy Hazardous Materials Response Branch, Jeremy Gault, NOAA.

Box 14.2 Energy Prices, Energy Uses

"Thank goodness gas prices are coming back down!" sighed Fran to her husband Karl. "The van's great for hauling Chris and the soccer team around, but it's nice to get some change back when I fill up."

"Great. So gasoline goes down, and the heating bill triples, thanks to the oil," Karl replied. "You can't tell me somebody isn't playing games with the prices!"

"Well," piped up Chris from the sofa, "maybe if gas stayed expensive for a while, people would drive less or conserve more, and then they wouldn't pollute so much, either."

"If you really want not to pollute," contributed Chris's sister Denise, "get an electric car! That won't pollute at all. It'd be great. The freeway is horribly smelly now when the traffic backs up."

"Right," said Chris, "and your electricity is going to come from where? Oh, sure, the *freeway* would be less smelly, and the air around it would be better for the people living there. But that electricity has to be generated someplace, using some kind of fuel. Besides, what happens when your next rolling blackout hits, the way it happened in California? No wheels either!"

The discussion went on and touched on several factors that influence both energy prices and energy consumption. On the price side, the underlying cost of the commodity may vary little except when political events constrain availability (figure 1); as this is written, ongoing fighting in Iraq and elsewhere in the Middle East, and reduced production from hurricane-damaged drilling facilities in the Gulf of Mexico, are together helping to drive oil prices to record highs. Short-term supply-and-demand issues are often much more significant influences on retail price. As noted in the chapter, crude oil must be refined into a range of petroleum products, and refiners make projections about the relative amounts of gasoline, heating oil, etc. that will be needed at different times of year. Unexpected developments (such as a fierce, prolonged cold spell in heating-oil country) can drive up prices until more of the needed fuel can be produced. Likewise with natural gas: Although it's readily delivered by pipeline, it is only extracted and transported so fast, and if use is unexpectedly heavy for some time, prices will again rise until demand slackens or supply catches up.

Consumers can moderate their own costs through conservation but often feel little incentive to do so if financial pressures are low. Gasoline prices generally fell in the 1980s and 1990s, and consumption tended to rise in complementary fashion (figure 2). In part, this can be related to the rise in popularity of vans, sport-utility vehicles, and other vehicles with notably lower mileage than is mandated for passenger cars by the Environmental Protection Agency (figure 3). Whether a sharp recent price rise will depress consumption remains to be seen.

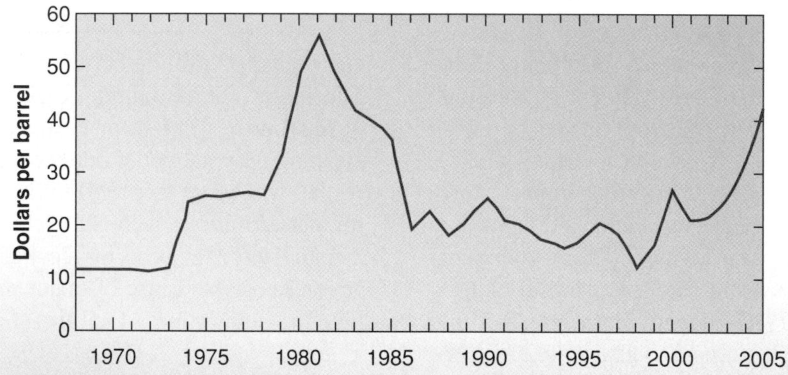

FIGURE 1 The Arab oil embargo and other events in the Middle East led to a big spike in crude-oil prices in the 1980s. Prices then subsided for nearly two decades, but damage to U.S. production by Hurricane Katrina, and wars in the Middle East, have recently driven prices sharply higher. Normalized to year 2000 dollars.
Source: U.S. Energy Information Administration.

When an oil spill occurs, the oil, being less dense than water, floats. The lightest, most volatile hydrocarbons start to evaporate initially, decreasing the volume of the spill somewhat but polluting the air. Then a slow decomposition process sets in, due to sunlight and bacterial action. After several months, the mass may be reduced to about 15% of the starting quantity, and what is left is mainly thick asphalt lumps. These can persist for many months more. Oil is toxic to marine life, causes water birds to drown when it soaks their feathers, may decimate fish and shellfish populations, and can severely damage the economies of beach resort areas.

A variety of approaches to damage control following an oil spill have been tried. In calm seas, if a spill is small, it may be contained by floating barriers and picked up by specially designed "skimmer ships" that can skim up to fifty barrels of oil per hour off the water surface. Some attempts have been made to soak up oil spills with peat moss, wood shavings, and even chicken feathers. Large spills or spills in rough seas are a greater problem. When the tanker *Torrey Canyon* broke up off Land's End, England, in 1967, several strategies were tried, none very successfully. First, an attempt was made to burn off the spill. By the time this was tried, the more flammable, volatile compounds had evaporated, so aviation fuel was poured over the spill, and bombs were dropped to ignite it! This did not really work very well, and in any event, it would have produced a lot of air pollution. Some French workers used ground chalk

(Box 14.2 Continued)

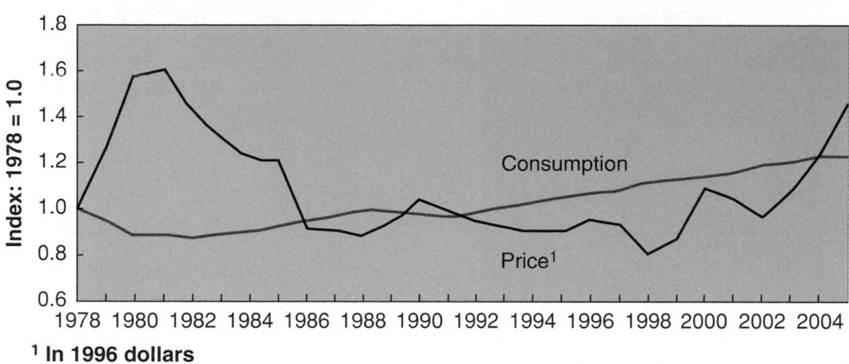

FIGURE 2 In the 1980s and 1990s, U.S. gasoline prices generally declined, and consumption rose. As this is written, consumption has not yet shown a response to recent price increases.
Source: U.S. Energy Information Administration.

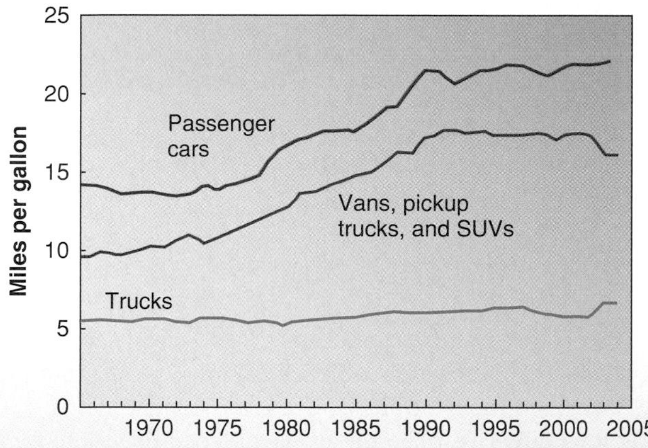

FIGURE 3 Mileage of passenger cars has risen steadily—largely due to EPA mandates; mileage of vans, SUVs, and pickup trucks has lagged consistently behind, and the gap has widened in recent years.
Source: U.S. Energy Information Administration.

And what about the electric car? At present, only about 10,000 electric cars are in use in the United States; they make up only about 2% of the so-called alternative-fueled vehicles. Chris's observation was correct, too: Their main application at present is in easing *local* pollution problems. Electricity still has to be generated somehow (and a great deal of energy is lost in the process, as will be explored in chapter 15). The electric car may operate cleanly and efficiently, but making that electricity still consumes energy and creates pollution somewhere. More-efficient hybrid-technology cars have also been developed, but it will be a long time before their numbers on the highways are significant. Still, improving the fuel economy of motor vehicles, of whatever sort, could substantially reduce oil consumption, considering that motor gasoline accounts for nearly half the petroleum consumption in the United States.

to absorb and sink the oil. Sinking agents like chalk, sand, clay, and ash can be effective in removing an oil spill from the sea surface, but the oil is no healthier for marine life on the ocean bottom. Furthermore, the oil may separate out again later and resurface. The British mixed some 2 million gallons of detergent with part of the spill, hoping to break up the spill so that decomposition would work more rapidly. The detergents, in turn, turned out to be toxic to some organisms, too.

It is the potential for concentrated negative local impact that raises particular concerns about oil spills, such as the spill off the Galápagos Islands in early 2001. Similar concerns arose with the 1989 grounding of the *Exxon Valdez* in Prince William Sound.

The port of Valdez lies at the southern end of the Trans-Alaska Pipeline (figure 14.13). Through that pipeline flow 1.6–1.7 million barrels of oil a day. From Valdez, the oil is transported by tanker to refinery facilities elsewhere. Prince William Sound, in which Valdez is located, is home to a great variety of wildlife, from large mammals (including whales, dolphins, sea otters, seals, and sea lions) to birds, shellfish, and fish (of which salmon and herring are particularly important economically).

Early in the morning of 23 March 1989, the tanker *Exxon Valdez,* loaded with 1.2 million barrels (50 million gallons) of crude oil from Alaska's North Slope, ran aground on Bligh Island in Prince William Sound. The result was the worst oil spill yet in U.S. waters. Response to the spill was delayed for

FIGURE 14.13 The oil tanker port at Valdez, Prince William Sound, Alaska, southern end of the Trans-Alaska Pipeline. (Note classic glacial valley in background.)

ten to twelve hours after the accident, in part because an oil-containment barge happened to be disabled. More than 10 million gallons of oil escaped, eventually to spread over more than 900 square miles of water. The various cleanup efforts cost Exxon an estimated $2.5 billion.

Skimmers recovered relatively little of the oil, as is typical: Three weeks after the accident, only about 5% of the oil had been recovered. Chemical dispersants were not very successful (perhaps because their use was delayed by concerns over *their* impact on the environment), nor were attempts to burn the spill. Four days after the accident, the oil had emulsified by interaction with the water into a thick, mousse-like substance, difficult to handle, too thick to skim, impossible to break up or burn effectively. Oil that had washed up on shore had coated many miles of beaches and rocky coast (figure 14.14).

Thousands of birds and marine mammals died or were sickened by the oil. The annual herring season in Valdez was cancelled, and salmon hatcheries were threatened. When other methods failed, it seemed that only slow degradation over time would get rid of the oil. Given the size of the spill and the cold Alaskan temperatures, prospects seemed grim; microorganisms that might assist in the breakdown grow and multiply slowly in the cold. It appeared highly likely that the spill would be very persistent.

So, in a $10-million experiment, Exxon scientists sprayed miles of beaches in the Sound with a fertilizer solution designed to stimulate the growth of naturally occurring microorganisms that are known to "eat" oil, thereby contributing to its decomposition. Within two weeks, treated beaches were markedly cleaner than untreated ones. Five months later, the treated beaches still showed higher levels of those microorganisms than did untreated

FIGURE 14.14 Beach cleanup workers in oil-splattered raingear mop oil from beach on LaTouche Island, Prince William Sound.
Photograph from Alaska State Archive slide, courtesy Oil Spill Public Information Center.

beaches. This type of treatment isn't universally successful. The fertilizer solution runs off rocky shores, so this treatment won't work on rocky stretches of the oil-contaminated coast; nor is it as effective once the oil has seeped deep into beach sands. But microorganisms may indeed be the future treatment of choice.

Analysis of the success of the *Valdez* cleanup is ongoing. Examination of the long-term effects of another strategy, hot-water washing of beaches, has suggested that it may actually do more harm than good. The heat can kill organisms that had survived the oil, the pressure of the hot-water blast can drive oil deeper into the sediments where degradation is slower, and the water treatment may wash oil onto beaches and into tidepools that were unaffected by the original spill.

Such experiences, in which many cleanup methods are tried but none is notably effective, are typical. Perhaps the best, most environmentally benign prospect for future oil spills is the development of specialized, "oil-hungry" microorganisms that will eat the spill for food and thus get rid of it. Scientists are currently developing suitable bacterial strains. For the time being, no good short-term solution exists to the problems posed by a major oil spill. Preventive measures can be increased—for instance, double-hulled tankers and tugboat escorts are now mandatory for oil tankers in Prince William

Sound, Alaska—but as long as demand for petroleum remains high, some accidents are, perhaps, inevitable.

The search for solutions continues. In November 2002, the tanker *Prestige* went down off the Iberian coast. The Spanish government's solution was to tow the tanker further offshore and sink it, on the presumption that the estimated 60,000 tons of oil still in the tanker would solidify in the deeper, colder water. Alas, not so. For months thereafter, oil continued to leak to the surface—about 125 tons a day, which ultimately blanketed more than 900 km of coastline from Portugal to France, costing an estimated $1 billion in cleanup costs and prompting a spate of lawsuits that are ongoing as this is written.

It may be some consolation to keep in mind the long perspective of geologic time. The short-term local impact may be severe both economically and ecologically. Over the long term, such intense environmental stresses may lead to increases in numbers and/or diversity of local organisms.

COAL

Prior to the discovery and widespread exploitation of oil and natural gas, wood was the most commonly used fossil fuel, followed by coal as the industrial age began in earnest in the nineteenth century. However, coal was bulky, cumbersome, and dirty to handle and to burn, so it fell somewhat out of favor, particularly for home use, when the liquid and gaseous fossil fuels became available. Currently, though it supplies only about 23% of total U.S. energy needs, coal plays a key role in the energy picture, for coal-fired power plants account for half of U.S. electric power generation. With oil and gas supplies dwindling, many energy users are looking again at the potential of coal for other applications also.

Formation of Coal Deposits

Coal is formed not from marine organisms, but from the remains of land plants. A swampy setting, in which plant growth is lush and where there is water to cover fallen trees, dead leaves, and other debris, is especially favorable to the initial stages of coal formation. The process requires **anaerobic** conditions, in which oxygen is absent or nearly so, since reaction with oxygen destroys the organic matter. (A pile of dead leaves in the street does not turn into fuel, and in fact, over time, leaves very little solid residue.)

The first combustible product formed under suitable conditions is *peat.* Peat can form at the earth's surface, and there are places on earth where peat can be seen forming today. Further burial, with more heat, pressure, and time, gradually dehydrates the organic matter and transforms the spongy peat into soft brown coal (**lignite**) and then to the harder coals (**bituminous** and **anthracite**); see figure 14.15. As the coals become harder, their carbon content increases, and so does the amount of heat released by burning a given weight of coal. The hardest, high-carbon coals (especially anthracite), then, are the most desirable as fuels because of their potential energy yield. As with

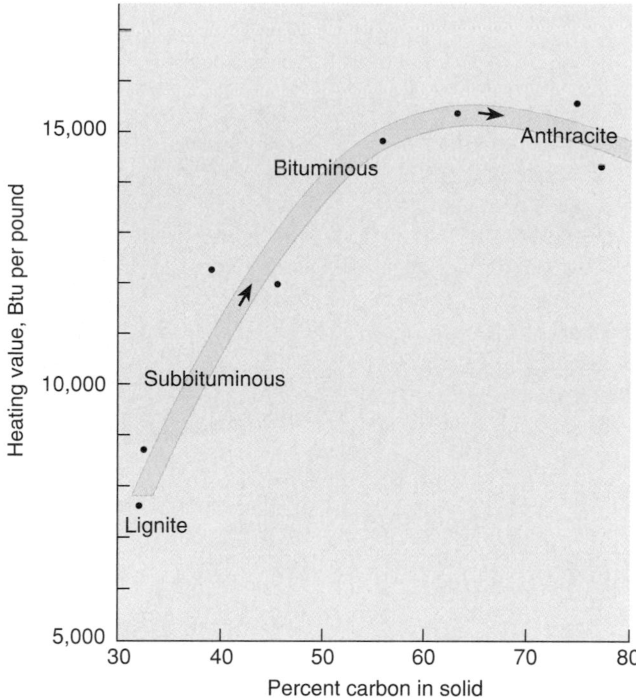

FIGURE 14.15 Change in character of coal with increasing application of heat and pressure. There is a general trend toward higher carbon content and higher heat value with increasing grade, though heat value declines somewhat as coal tends toward graphite.

oil, however, the heat to which coals can be subjected is limited: overly high temperatures lead to metamorphism of coal into graphite.

The higher-grade coals, like oil, apparently require formation periods that are long compared to the rate at which coal is being used. Consequently, coal too can be regarded as a nonrenewable resource. However, the world supply of coal represents an energy resource far larger than that of petroleum, and this is also true with regard to the U.S. coal supply. In terms of energy equivalence, U.S. coal reserves represent more than 50 times the energy in the remaining oil reserves and 40 times the energy of remaining U.S. conventional natural-gas reserves.

Coal Reserves and Resources

Coal resource estimates are subject to fewer uncertainties than corresponding estimates for oil and gas. Coal is a solid, so it does not migrate. It is found, therefore, in the sedimentary rocks in which it formed; one need not seek it in igneous or metamorphic rocks. It occurs in well-defined beds that are easier to map than underground oil and gas concentrations. And because it formed from land plants, which did not become widespread until 400 million years ago, one need not look for coal in more ancient rocks.

The estimated world reserve of coal is about one trillion tons; total resources are estimated at over 10 trillion tons. The United States is particularly well supplied with coal, possessing about 25% of the world's reserves, over 270 billion tons

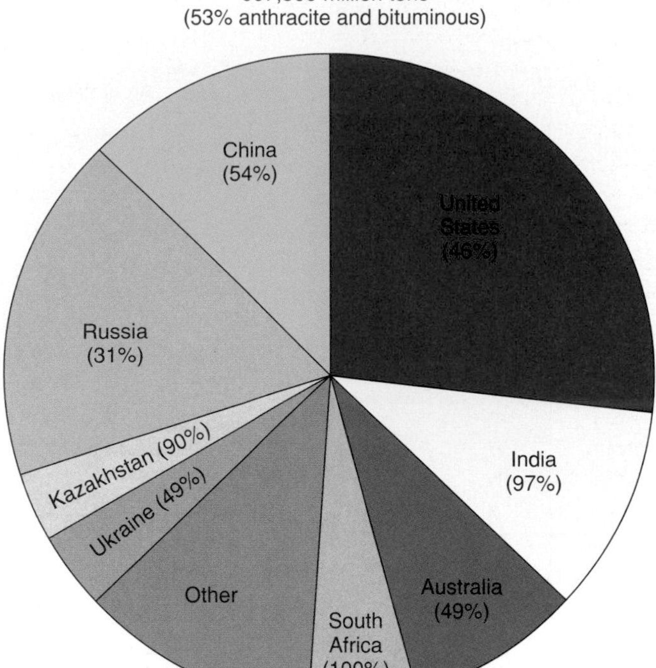

FIGURE 14.16 World coal reserves, in millions of tons. Sum of anthracite and bituminous for each country in parentheses.

Source: Data from International Energy Annual 2006, *U.S. Energy Information Administration, Department of Energy.*

of recoverable coal (figure 14.16). Total U.S. coal resources may approach ten times that amount, and most of that coal is yet unused and unmined. At present, coal provides between 20 and 25% of the energy used in the United States. While the United States has consumed probably half of its total petroleum resources, it has used up only a few percent of its coal. Even if only the reserves are counted, the U.S. coal supply could satisfy U.S. energy needs for more than 200 years at current levels of energy use, if coal could be used for all purposes. As a supplement to other energy sources, coal could last many centuries (figure 14.17). Unfortunately, heavy reliance on coal has drawbacks.

Limitations on Coal Use

A primary limitation of coal use is that solid coal is simply not as versatile a substance as petroleum or natural gas. A major shortcoming is that it cannot be used directly in most forms of modern transportation, such as automobiles and airplanes. (It was, of course, used to power trains in the past, but coal-fired locomotives quickly fell out of favor when cleaner fuels became available.) Coal is also a dirty and inconvenient fuel for home heating, which is why it was abandoned in favor of oil or natural gas. Therefore, given present technology, coal simply cannot be adopted as a substitute for oil in all applications.

Coal can be converted to liquid or gaseous hydrocarbon fuels—gasoline or natural gas—by causing the coal to react with steam or with hydrogen gas at high temperatures. The conversion processes are called **gasification** (when the product is gaseous) and **liquefaction** (when the product is liquid fuel). Both processes are intended to transform coal into a cleaner-burning, more versatile fuel, thus expanding its range of possible applications.

Commercial coal gasification has existed on some scale for 150 years. Many U.S. cities used the method before inexpensive natural gas became widely available following World War II. Europeans continued to develop the technologies into the 1950s. At present, only experimental coal-gasification plants are operating in the United States.

Current gasification processes yield a gas that is a mixture of carbon monoxide and hydrogen with a little methane. The heat derived from burning this mix is only 15 to 30% of what can be obtained from burning an equivalent volume of natural gas.

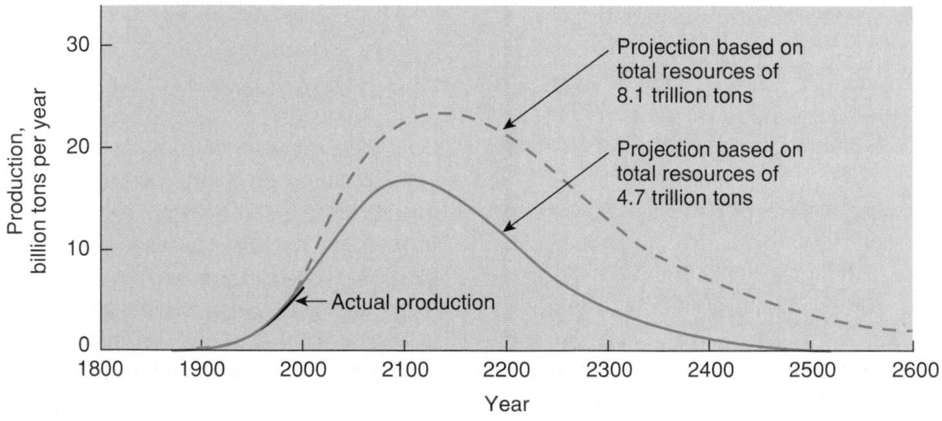

FIGURE 14.17 Projected world coal production, assuming coal as the principal substitute for petroleum. Note what a small fraction of coal has been consumed, and compare that with oil (figure 14.7).

Source: Projections from M. King Hubbert, "The Energy Resources of the Earth," Scientific American, *Sept. 1971, p. 69. Actual production data to 2001 from* Annual Energy Review 2001, *U.S. Energy Information Administration, Department of Energy.*

Because the low heat value makes it uneconomic to transport this gas over long distances, it is typically burned only where produced. Technology exists to produce high-quality gas from coal, equivalent in heat content to natural gas, but it is presently uneconomical when compared to natural gas. Research into improved technologies continues.

Pilot projects are also underway to study *in situ* underground coal gasification, through which coal would be gasified in place and the gas extracted directly. The expected environmental benefits of not actually mining the coal include reduced land disturbance, water use, air pollution, and solid waste produced at the surface. Potential drawbacks include possible groundwater pollution and surface subsidence over gasified coal beds. Underground gasification may provide a means of using coals that are too thin to mine economically or that would require excessive land disruption to extract. However, in the near future, simple extraction of coal-bed methane is likely to yield economical gas more readily.

Like gasification, the practice of generating liquid fuel from coal has a longer history than many people realize. The Germans used it to produce gasoline from their then-abundant coal during World War II; South Africa has a liquefaction plant in operation now, producing gasoline and fuel oil. A variety of technologies exist, and several noncommercial pilot plants have operated in the United States. Technological advances made during the 1980s slashed costs by 60% and greatly improved yields (to about 70% of original coal converted to liquid fuel). Even so, their liquid fuel products have not historically been economically competitive with conventional petroleum, particularly given declining oil prices in the 1990s. However, petroleum prices have recently risen sharply. The higher gasoline prices rise, the more economically viable coal liquefaction becomes, and the sooner commercial development is likely.

ENVIRONMENTAL IMPACTS OF COAL USE

Gases

A major problem posed by coal is the pollution associated with its mining and use. Like all fossil fuels, coal produces carbon dioxide (CO_2) when burned. In fact, it produces significantly more carbon dioxide per unit energy released than oil or natural gas. (It is the burning of the carbon component—making CO_2—that is the energy-producing reaction when coal is burned; when hydrocarbons are burned, the reactions produce partly CO_2, partly H_2O as their principal by-products, the proportions varying with the ratio of carbon to hydrogen in the particular hydrocarbon(s).) The relationship of carbon dioxide to greenhouse-effect heating was discussed in chapter 10. The additional pollutant that is of special concern with coal is sulfur.

The sulfur content of coal can be more than 3%, some in the form of the iron sulfide mineral pyrite (FeS_2), some bound in the organic matter of the coal itself. When the sulfur is burned along with the coal, sulfur gases, notably sulfur dioxide (SO_2), are produced. These gases are poisonous and are extremely irritating to eyes and lungs. The gases also react with water in the atmosphere to produce sulfuric acid, a very strong acid. This acid then falls to earth as acid rainfall. Acid rain falling into streams and lakes can kill fish and other aquatic life. It can acidify soil, stunting plant growth. It can dissolve rock; in areas where acid rainfall is a severe problem, buildings and monuments are visibly corroding away because of the acid. The geology of acid rain is explored further in chapter 18.

Oil can contain appreciable sulfur derived from organic matter, too, but most of that sulfur can be removed during the refining process, so that burning oil releases only about one-tenth the sulfur gases of burning coal. Some of the sulfur can be removed from coal prior to burning, but the process is expensive and only partially effective, especially with organic sulfur. Alternatively, sulfur gases can be trapped by special devices ("scrubbers") in exhaust stacks, but again, the process is expensive (in terms of both money and energy) and not totally efficient. From the standpoint of environmental quality, then, low-sulfur coal (1% sulfur or less) is more desirable than high-sulfur coal for two reasons: first, it poses less of a threat to air quality, and second, if the sulfur must be removed before or after burning, there is less of it to remove, so stricter emissions standards can be met more cheaply. On the other hand, much of the low-sulfur coal in the United States, especially western coal, is also lower-grade coal (see figure 14.18), which means that more of it must be burned to yield the same amount of energy. This dilemma is presently unresolved.

Another toxic substance of growing concern with increasing coal use is mercury. A common trace element in coal, mercury is also a very volatile (easily-vaporized) metal, so it can be released with the waste gases. The consequences are examined in chapter 17.

Ash

Coal use also produces a great deal of solid waste. The ash residue left after coal is burned typically ranges from 5 to 20% of the original volume. The ash, which consists mostly of noncombustible silicate minerals, also contains toxic metals. If released with waste gases, the ash fouls the air. If captured by scrubbers or otherwise confined within the combustion chamber, this ash still must be disposed of. If exposed at the surface, the fine ash, with its proportionately high surface area, may weather very rapidly, and the toxic metals such as selenium and uranium can be leached from it, thus posing a water-pollution threat. Uncontrolled erosion of the ash could likewise cause sediment pollution. The magnitude of this waste-disposal problem should not be underestimated. A single coal-fired electric power plant can produce over a million tons of solid waste a year. There is no obvious safe place to dump all that waste. Some is used to make concrete, and it has been suggested that the ash could actually serve as a useful source of some of the rarer metals concentrated in it. At present, however, most of the ash from coal combustion is deposited in landfills.

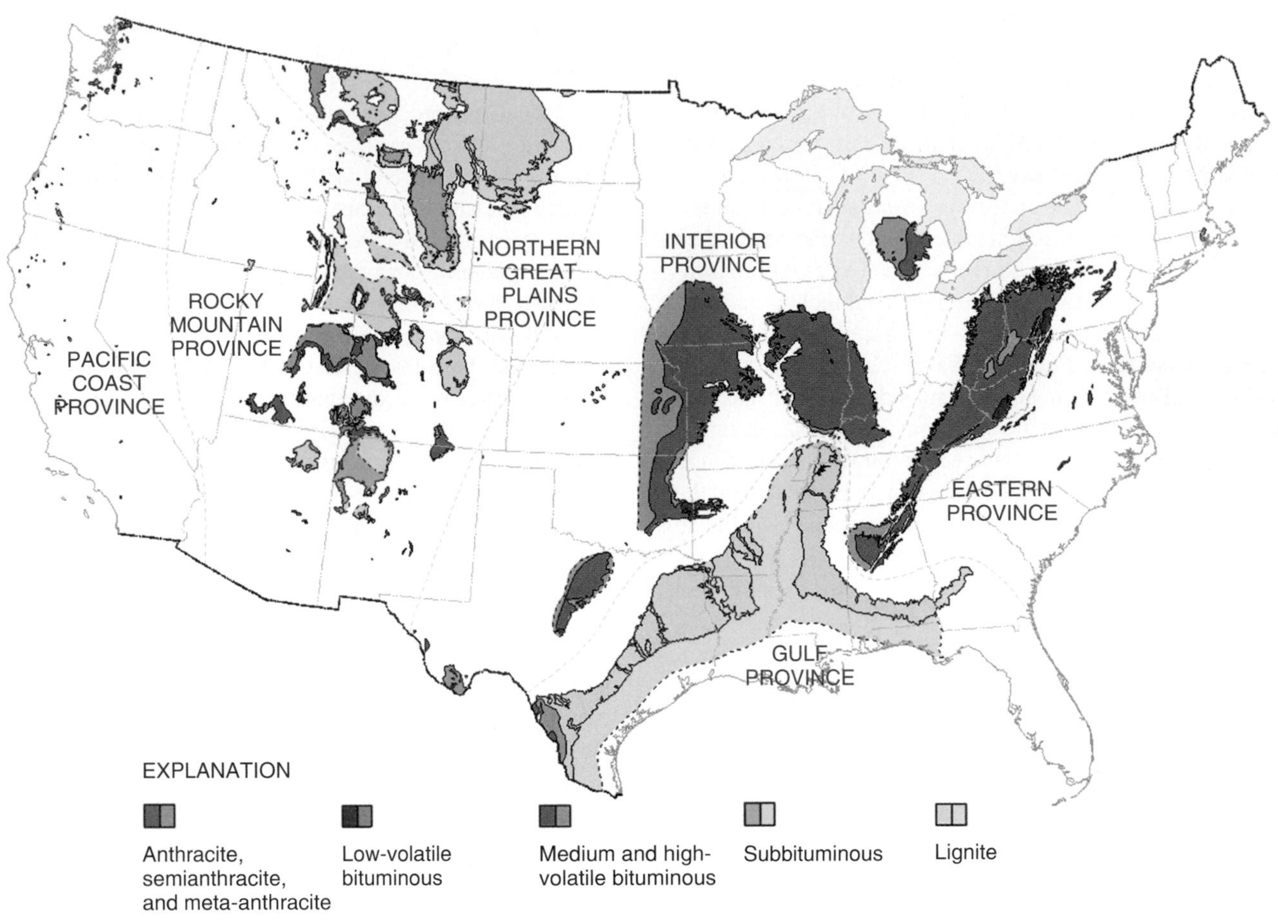

FIGURE 14.18 Distribution of U.S. coal fields. Within each color, darker shading represents current or likely future reserves; lighter shades are deposits of unknown or lower quality.
Source: U.S. Geological Survey.

Coal-Mining Hazards and Environmental Impacts

Coal mining poses further problems. Underground mining of coal is notoriously dangerous, as well as expensive. Mines can collapse; miners may contract black lung disease from breathing the dust; there is always danger of explosion from pockets of natural gas that occur in many coal seams. A tragic coal-mine fire in Utah in December 1984 and a methane explosion in a Chinese coal mine in 2004 that killed over 100 people were reminders of the seriousness of the hazards. Recent evidence further suggests that coal miners are exposed to increased cancer risks from breathing the radioactive gas radon, which is produced by natural decay of uranium in rocks surrounding the coal seam.

Even when mining has ceased, fires may start in underground mines (figure 14.19). An underground fire that may have been started by a trash fire in an adjacent dump has been burning under Centralia, Pennsylvania, since 1962; over 1000 residents have been relocated, at a cost of more than $40 million, and collapse and toxic fumes threaten the handful of remaining residents resisting evacuation. And while Centralia's fire may be among the most famous, it is by no means the only such fire, and they are a problem in other nations as well. China relies heavily on coal for its energy; uncontrolled coal-mine fires there have cost over $100 million in lost coal, the carbon-dioxide emissions from them are estimated to account for several percent of global CO_2 emissions, and the associated pollution is believed to be aggravating respiratory problems in China. Mining of India's largest coal field began in 1894; the first fires there started in 1916, and fifty years later, the fires could be found across the whole coal field. They burn on today.

The rising costs of providing a safer working environment for underground coal miners, together with the greater technical difficulty (and thus lower productivity) of underground mines, are largely responsible for a steady shift in coal-mining methods in the United States, from about 20% surface mining around 1950 to over 65% surface mining by 2000. Surface

FIGURE 14.19 Fire out of control in abandoned underground coal mine in Sheridan County, Wyoming.
Photograph by C. R. Dunrud, USGS Photo Library, Denver, CO.

mining continues to dominate, as demand for low-sulfur western coal increases (figure 14.20). The positive side of this shift in methods is that it has been accompanied by a sharp decline in U.S. coal-mine fatalities (figure 14.21); strip-mining of coal is far safer for the miners.

A significant portion of U.S. coal, particularly in the west, occurs in thick beds very close to the surface. It is relatively cheap to strip off the vegetation and soil and dig out the coal at the surface, thus making strip-minable coal very attractive economically in spite of land reclamation costs. Still, strip-mining presents its own problems. Strip-mining in general was discussed in chapter 13 (see also figure 13.24). A particular problem with strip-mining coal involves, again, the sulfur in the coal. Not every bit of coal is extracted from the surrounding rock. Some coal and its associated sulfur are left behind in the waste rock in spoil banks. Pyrite is also common in the shales with which coal often is interlayered, and these pyritic shales form part of the spoil banks. The sulfur in the spoils can react with water and air to produce runoff water containing sulfuric acid:

$$\underset{\text{pyrite}}{2\,FeS_2} + \underset{\text{water}}{5\,H_2O} + \underset{\text{oxygen}}{\frac{15}{2}O_2} = \underset{\text{sulfuric acid}}{4\,H_2SO_4} + \underset{\text{hydrated iron oxide}}{Fe_2O_3 \cdot H_2O}$$

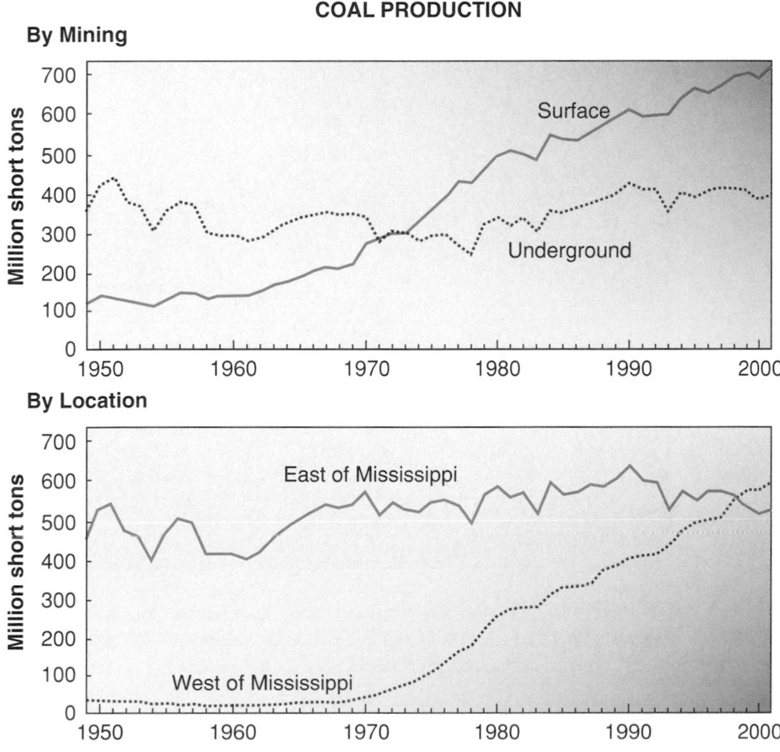

FIGURE 14.20 Rise in proportion of surface versus underground mining parallels relative increase in production of western coal.
Source: Data from Annual Energy Review 2001, *U.S. Energy Information Administration, Department of Energy.*

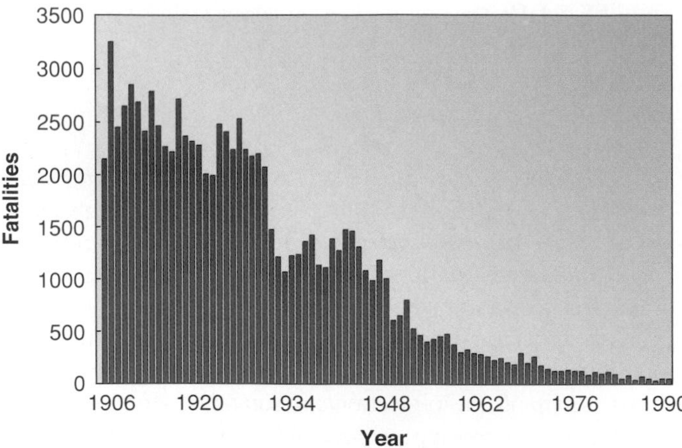

FIGURE 14.21 Drop in coal-mine fatalities reflects partly safer mining practices, partly the shift from underground mining to strip-mining.
Source: U.S. Bureau of Mines, Minerals Today, *Oct. 1991, p. 19.*

FIGURE 14.22 Abandoned coal strip mine, Fulton County, Illinois. There is no lack of moisture, but plants will not thrive in acid soil.
Photograph © Charles Rotkin/Corbis Images.

Because plants grow poorly in very acid conditions, this acid slows revegetation of the area by stunting plant growth or even preventing it altogether (figure 14.22). The acid runoff can also pollute area ground and surface waters, killing aquatic plants and animals in lakes and streams and contaminating the water supply. Very acidic water also can be particularly effective at leaching or dissolving some toxic elements from soils, further contributing to water pollution. Coal strip mines, like others, can be reclaimed, but in addition to regrading and replanting the land, it is frequently necessary to replace the original topsoil (if it was saved) or to bring in fresh soil so that sulfur-rich rocks are not left exposed at the surface, where weathering is especially rapid. (Even underground coal mines may have associated acid drainage, but water circulation there is generally more restricted.) The question of water availability to support plant regrowth is of special concern in the western United States: 50% of U.S. coal reserves are found there, 40% of these reserves would be surface-mined, and most regions underlain by western coal are very dry (figure 14.23).

When coal surface mines are reclaimed, efforts are commonly made to restore original topography—although if a thick coal seam has been extracted, this may not be easy. The artificial slopes of the restored landscape may not be altogether stable, and as natural erosional processes begin to sculpture the land surface, natural slope adjustments in the form of slumps and slides may occur. In dry areas, thought must be given to how the reclamation will modify drainage patterns (surface and subsurface). In wetter areas, the runoff may itself be a problem, gullying slopes and contributing to sediment pollution of surface waters until new vegetation is established. In short, reclamation can be a multifaceted challenge.

FIGURE 14.23 Rainbow coal strip mine, Sweetwater County, Wyoming. Note the very thin coal beds (black) relative to waste rock (brown).
Photograph by H.E. Malde, USGS Photo Library, Denver, CO.

OIL SHALE

Oil shale is very poorly named. The rock, while always sedimentary, need not be a shale, and the hydrocarbon in it is not oil! The potential fuel in oil shale is a sometimes-waxy solid called **kerogen,** which is formed from the remains of plants, algae, and bacteria. Kerogen is not a single compound; the term is a general one describing organic matter that is not readily dissolved either in water or in organic solvents, and the term can also be applied to solid precursor organic compounds of oil and natural gas in other types of rocks. The physical properties of kerogen dictate that the oil shale must be crushed and heated to distill out the hydrocarbon as "shale oil," which then is refined somewhat as crude oil is to produce various liquid petroleum products.

The United States has about two-thirds of the world's known supply of oil shale (see locations, figure 14.24), for a total estimated resource of 2 to 5 trillion barrels of shale oil. For a number of reasons, the United States is not yet using this apparently vast resource to any significant extent, and it may not be able to do so in the near future.

One reason for this is that much of the kerogen is so widely dispersed through the oil shale that huge volumes of rock must be processed to obtain moderate amounts of shale oil. Even the richest oil shale yields only about three barrels of shale oil per ton of rock processed. The cost is not presently competitive with that of conventional petroleum, and was not competitive even at the peak oil prices of the 1980s. Nor have large-scale processing facilities yet been built—present plants are strictly experimental. More-efficient and cheaper processing technologies are needed.

Another problem is that a large part of the oil shale is located at or near the surface. At present, the economical way to mine it, therefore, appears to be surface- or strip-mining with its attendant land disturbance. Most of the richest oil shale is located in areas of the western United States that are already chronically short of water (figure 14.25). The dry conditions would make revegetation of the land after strip-mining especially difficult. Some consideration has been given to *in situ* extraction of the kerogen, warming the rock in place and using pressure to force out the resulting fluid hydrocarbon, without mining the rock, to minimize land disruption. But the necessary technology is poorly developed, though a pilot project does exist, and the process leaves much of the fuel behind.

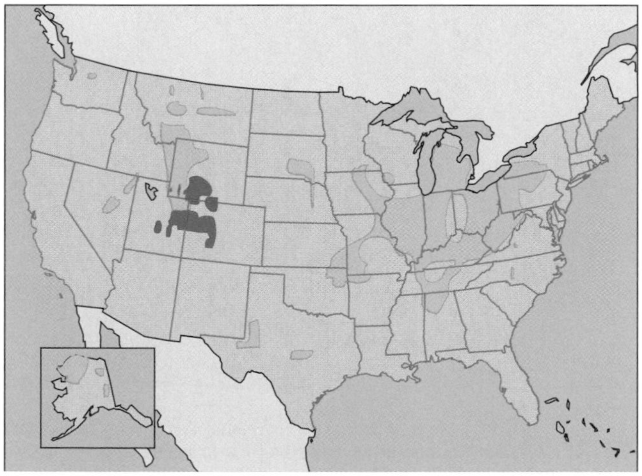

FIGURE 14.24 Distribution of U.S. oil shale deposits. The richest of these deposits is the Green River formation (dark green); most of the rest are noneconomic.

Source: Data from J. W. Smith, "Synfuels: Oil Shale and Tar Sands," in Perspectives on Energy, *3d edited by Lon C. Ruedisili and Morris W. Firebaugh, 1982, Oxford University Press, NY.*

FIGURE 14.25 Oil shale of the Green River Formation.

The water shortage presents a further problem. Current processing technologies require large amounts of water—on the order of three barrels of water per barrel of shale oil produced. Just as water for reclamation is in short supply in the west, so, too, is the water to process the oil shale.

Finally, since the volume of the rock actually increases during processing, it is possible to end up with a 20 to 30% larger volume of waste rock to dispose of than the original volume of rock mined. Aside from any problems of accelerated weathering of the crushed material, there remains the basic question of where to put it. Because it will not all fit back into the space mined, the topography will inevitably be altered.

Scientists and economists differ widely in the extent to which they see oil shale as a promising alternative to conventional oil and gas. Certainly, the water-shortage, waste-disposal, and land-reclamation problems will have to be solved before shale oil can be used on a large scale. It is unlikely that those problems can be solved within the next few decades, although over the longer term, oil shale may become an important resource, and the higher conventional oil prices rise, the more competitive "shale oil" becomes.

TAR SAND

Tar sands are sedimentary rocks containing a very thick, semisolid, tarlike petroleum. The heavy petroleum in tar sands is believed to be formed in the same way and from the same materials as lighter oils. Tar-sand deposits may represent very immature petroleum deposits, in which the breakdown of large molecules has not progressed to the production of the lighter liquid and gaseous hydrocarbons. Alternatively, the lighter compounds may have migrated away, leaving this dense material behind. Either way, the tar is too thick to flow out of the rock. Like oil shale, tar sand presently must be mined, crushed, and heated to extract the petroleum, which can then be refined into various fuels.

Many of the environmental problems associated with oil shale likewise apply to tar sand. Because the tar is disseminated through the rock, large volumes of rock must be mined and processed to extract significant amounts of petroleum. Many tar sands are near-surface deposits, so the mining method used is commonly strip-mining. The processing requires a great deal of water, and the amount of waste rock after processing may be larger than the original volume of tar sand. The negative impact of tar-sand production will naturally increase as production increases.

In Canada, the extensive Athabasca tar sand (figure 14.26) is estimated to contain about a trillion barrels of petroleum, of which 25 to 35% would be recoverable. Indeed, Canadian supplies of "unconventional" petroleum resources far exceed that nation's "conventional" oil deposits. Actual development has lagged somewhat behind aspirations, in part because of environmental concerns. Two open-pit mining operations and associated refinery facilities are functioning in the province of Alberta, yielding over 200,000 barrels of oil a day. To reduce the impact of mining, pilot projects involving *in situ* extraction of the "tar" are underway. One, nicknamed the "huff-and-puff" procedure, involves injection of hot steam for several weeks to warm the oil, after which the less-viscous fluid can be pumped out for a month or so before it chills and stiffens and reheating is necessary. The United States, however, has virtually no tar-sand resources, so it cannot look to tar sand to solve its domestic energy problems even if the environmental difficulties could be overcome.

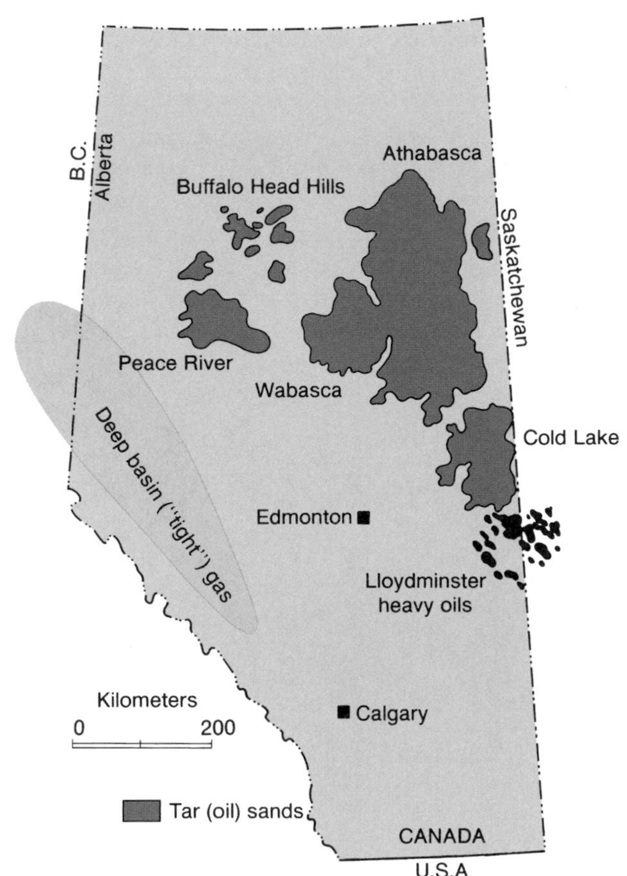

FIGURE 14.26 Locations of unconventional petroleum deposits in western Canada.

After R.M. Procter, et al., *1984,* Oil and Natural Gas Resources of Canada, *Energy, Mines and Resources Canada. Reproduced with permission of the Minister of Supply and Services Canada, 1994.*

SUMMARY

The United States now relies on fossil fuels for over 85% of its energy: About 40% comes from oil, and 20 to 25% each from natural gas and coal. All the fossil fuels are nonrenewable energy sources. Known oil and gas supplies have been seriously depleted and may be exhausted within decades. Remaining coal supplies are much larger, but many hazards and unsolved environmental problems are associated with heavy reliance on coal: acid rain, acid runoff, mine accidents and fires, strip-mine reclamation issues, and more. Technical, environmental, and economic problems also limit the potential of other fossil fuels—petroleum derived from oil shale or tar sand—to become major energy sources. It seems likely that other, non-fossil energy alternatives will be essential in the future.

Key Terms and Concepts

anaerobic 333
anthracite 333
bituminous 333
coal-bed methane 326
fossil fuel 318
gasification 334
geopressurized zones 327
kerogen 339
lignite 333
liquefaction 334
methane hydrate 327
natural gas 319
nonrenewable 321
oil 319
oil shale 339
petroleum 319
tar sand 340

Exercises

Questions for Review

1. A society's level of technological development strongly influences its per-capita energy consumption. Explain.
2. What are fossil fuels?
3. Briefly describe how oil and gas deposits form and mature.
4. Compare and contrast past and projected U.S. consumption of petroleum and coal.
5. What is enhanced oil recovery, and why is it of interest? Give two examples of the method.
6. Explain the nature of geopressurized natural gas resources, and note at least one obstacle to their exploitation.
7. What are methane hydrates, and where are they found? Why are they both a substantial potential energy resource, and a concern with respect to global climate change?
8. What is coal? Why is it a less versatile fuel than oil, and how might its versatility be increased?
9. What advantages does coal-bed methane have over other "unconventional" gas sources? Describe one problem associated with its extraction.
10. What air-pollution problems are associated particularly with coal, relative to other fossil fuels?
11. List and describe at least three potential negative environmental impacts of coal mining.
12. What are oil shales and tar sands?
13. Oil shale and tar sand share several drawbacks. Cite and explain several of these.

For Further Thought

1. Select a particular region or major city and investigate its energy consumption. Identify the principal energy sources and the proportion that each contributes; see whether these proportions are significantly different now from what they were ten and twenty-five years ago. To what extent have the types and quantities of energy used been sensitive to economic factors?
2. If possible, visit an area that has been or is being drilled for oil or mined for coal. Observe any visible signs of negative effects on the environment, and note any efforts being made to minimize such impacts.
3. Compare the distribution of precipitation given in figure 11.21 with the locations of principal U.S. oil-shale deposits (figure 14.24). Consider the implications for the economics of developing these fuel sources. Where *would* you get the water, economics aside?
4. Check the current status of the *Exxon Valdez* cleanup analysis. How many oil spills of similar size have occurred since 1989, where have they been, and what cleanup strategies have been tried? (See NetNotes for some sources.)

Suggested Readings/References

Ahlbrandt, T. S., and McCabe, P. J. 2002. Global petroleum resources: A view to the future. *Geotimes* (November): 14–18.

Appenzeller, T. 2006. The high cost of cheap coal. *National Geographic* (March): 96–123.

Bird, K. J., and Housekenecht, D. W. 2002. *USGS 2002 petroleum resources assessment of the National Petroleum Reserve in Alaska.* USGS Fact Sheet 045-02.

Chiles, J. R. 2001. Getting in deep. *Smithsonian:* 68–74.

Collett, T. S. 2004. Gas hydrates as a future energy source. *Geotimes* (November): 24–27.

Crawley, J. 2004. Coal-bed methane for today and tomorrow. *The Professional Geologist* (July/August): 50–53.

Dickens, G. R. 2004. Methane hydrate and abrupt climate change. *Geotimes* (November): 18–22.

Greb, S. F., Eble, C. F., Peters, D. C., and Papp, A. R. 2006. *Coal and the environment.* Alexandria, VA: Amer. Geol. Institute.

Harrison, W. F., and Testa, S. M. 2004. *Petroleum and the environment.* Alexandria, VA: Amer. Geol. Inst.

Kleinberg, R. L., and Brewer, P. G. 2001. Probing gas hydrate deposits. *American Scientist* 89 (May–June): 244–51.

Krajick, K. 2005. Fire in the hole. *Smithsonian* (May): 52–62.

Mitchell, J. G. 1999. In the wake of the spill: Ten years after the *Exxon Valdez. National Geographic* 193 (March): 96–117.

Peterson, C. H., et al. 2003. Long-term ecosystem response to the Exxon Valdez oil spill. *Science* 302 (19 December): 2082–86.

Sever, Megan. 2006. Oil spill: first responders. *Geotimes* (February): 33–35.

Stracher, G. B. 2002. Coal fires. *Geotimes* (October): 36–39.

———. 2003. *International energy outlook 2003.* Washington, D.C.: U.S. Department of Energy.

U.S. Geological Survey. 2000. *USGS World petroleum assessment 2000.* USGS Fact Sheet FS-070-00.

———. 2000. *Coal-bed methane: Potential and concerns.* USGS Fact Sheet FS-123-00.

———. 2001. *USGS national coal resource assessment.* USGS Fact Sheet FS-020-01.

vander Veen, C. J. 2006. Reevaluating Hubbert's prediction of U.S. peak oil. *EOS* 87(16 May): 199, 202.

Wallace, S. 2005. ANWR: The great divide. *Smithsonian* (October): 48–56.

Welcome to McGraw-Hill's Online Learning Center

Location: http://www.mhhe.com/montgomery8e

Mc Graw Hill

NetNotes

The U.S. Department of Energy's Energy Information Administration publishes online a wealth of energy-related data, including the volumes *Annual Energy Review, International Energy Annual,* and *International Energy Outlook,* all downloadable in PDF. The general Energy Information Administration site is located at
www.eia.doe.gov/

An excellent summary of the history and current trends of energy use in the United States is at
www.eia.gov/emeu/aer/eh/frame.html

Through the EIA "bookshelf" and other links, shorter items can be found, including "Where does my gasoline come from?"
www.eia.doe.gov/neic/brochure/gaso6/gasoline.htm

A recent EIA analysis concerning "peak oil" is "Long-term world oil supply scenarios: The future is neither as bleak nor as rosy as some assert," by J.H. Wood, G.R. Long, and D.F. Morehouse (2004)
www.eia.doe.gov/pub/oil_gas/petroleum/
feature_articles/2004/worldoilsupply/pdf/itwos.pdf

General information about energy is accessible from USGS via
www.usgs.gov/themes/energy.html

The USGS Energy Resource Surveys Program home page is at
energy.usgs.gov/

and USGS World Energy Project home page is
energy.cr.usgs.gov/energy/WorldEnergy/WEnergy.html

The USGS National Assessment of Oil and Gas Online (NOGA Online) is at
energy.cr.usgs.gov/oilgas/noga/

and the USGS World Petroleum Assessment 2000 is at
pubs.usgs.gov/dds/dd_060/

Several pertinent USGS publications online:
"Arctic National Wildlife Refuge, 1002 Area, Petroleum assessment, 1998, including economic analysis" (Fact Sheet 0028-01, 2001)
pubs.usgs.gov/fs/fs-0028-01/fs-0028-01.pdf

"U.S. Geological Survey 2002 petroleum resource assessment of the National Petroleum Reserve in Alaska (NPRA)" (Fact Sheet 045-02, 2002)
pubs.usgs.gov/fs/2002/fs045-02/fs045-02.pdf

"Oil and gas assessment of Central North Slope, Alaska, 2005" (Fact Sheet 2005-3043)
pubs.usgs.gov/fs/2005/3043/

"Economic analysis of undiscovered oil and gas of the Central North Slope of Alaska, 2005" (Fact Sheet 2005-3120)
pubs.usgs.gov/fs/2005/3120/

"Geology and resources of some world oil-shale deposits" (Scientific Investigations Report 2005-5294, 2005)
pubs.usgs.gov/sir/2005/5294/pdf/sir5294_508.pdf

Information on world energy trends from World Resources Institute can be found at their EarthTrends site
earthtrends.wri.org/

The Office of Surface Mining, which provides information specifically about coal mining, can be reached at
www.osmre.gov/osm.htm

The Oil Spill Public Information Center in Alaska was established by the U.S. Department of Justice to provide information about the *Valdez* spill. It is now housed at the Alaska Resource Library and Information Services
www.arlis.org

15

Energy Resources—Alternative Sources

In the discussion of the (nonrenewable) fossil fuels in chapter 14, the point was made that, with the conspicuous exception of coal, most of the U.S. supply of recoverable fossil fuels could be exhausted within decades. It was further pointed out that coal is not the most environmentally benign of energy sources. Alternative energy sources for the future are thus needed, both to supply essential energy and to spare the environment as much disruption as possible. And this is true worldwide, for globally, too, fossil fuels are the primary energy source (figure 15.1).

The extent to which alternative energy sources are required and how soon they will be needed is directly related to future world energy demand, which is difficult to predict precisely. In general, consumption can be expected to rise as population increases and as standards of living improve. However, the correlation between energy consumption and standard of living is perhaps even less direct than the correlation between mineral consumption and living standards.

By way of a simplified example, consider the addition of central heating to a home. The furnace is likely to contain about the same quantity of material, regardless of its type or efficiency; the drain on mineral reserves is approximately fixed. The very addition of that furnace represents a certain jump in both mineral and energy consumption. However, depending on the efficiency of the unit, the "tightness" of the home's insulation, and the climate, the amount of fuel the furnace must consume to maintain a certain level of heat in the home will vary enormously. The best-insulated modern homes may require as little as one-tenth the energy for heating as the average U.S. home.

One point to bear in mind with hydropower is the potential effect of sustained drought. The rich green vegetation here is growing on what was the bottom of Lake Powell before a six-year drought set in, sharply curtailing power generation from the Glen Canyon Dam; what is left of the flow of the Colorado River appears at the right side of the valley.

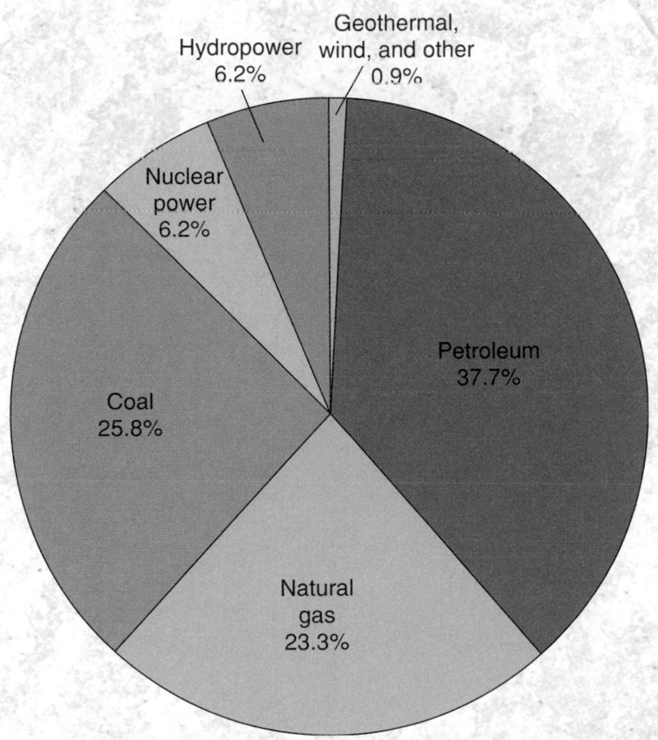

FIGURE 15.1 World energy production by source, 2004.
Source: Data from International Energy Annual 2004, *U.S. Energy Information Administration, Department of Energy.*

Also, the quantity of material used to manufacture the furnace will not change whether the homeowner maintains the temperature at 60°F or 80°F, but the amount of energy consumed plainly will. And, of course, the climate in which the home is built influences its heating costs. Both cold and heat, taken to extremes, cause increased energy demand—for heating and air conditioning, respectively. Thus there are considerable regional variations in energy consumption in the United States because of varying population density, degree of industrialization, and climatic variations.

In the area of transportation, similar variability arises. As standards of living rise to the point that motorized transportation becomes commonplace, consumption of both materials and energy rises. However, the amount of energy used depends heavily on the mode of transportation chosen and its fuel efficiency. It is more fuel-efficient to transport fifty people in a bus than in a dozen automobiles, but the distribution of people and their travel patterns must be clustered appropriately in order for mass transportation to work in a practical sense. Just among passenger cars, fuel efficiency varies greatly (which has substantial implications for possible energy conservation; see table 15.1). When projecting future energy use in developing countries, the underlying population base is important, too: An increase of 50 vehicles per thousand persons in China, with population over 1 billion, means far more added automobiles than an increase of 200 vehicles per thousand in South Korea, with a population of only about 50 million, less than one-twentieth of China's.

Thus, in deciding just how much energy various alternative sources must supply, assumptions must be made not only about the rates of increase in standards of living (growth of GNP) but also about the degree of energy efficiency with which the growth is achieved.

TABLE 15.1

Fuel Efficiency, Fuel Consumption

	Passenger Cars	Vans, Pickup Trucks, and SUVs
Number of vehicles	136,430,000	91,850,000
Average miles per gallon (est.)	22.4	16.2
Average annual miles driven per vehicle	12,500	11,040
Total annual miles driven by vehicle type	1,705 billion	1,014 billion
Total gasoline consumption, gallons	76.1 billion	62.6 billion

If the vans/pickups/SUVs achieved the same average mpg as the passenger cars (22.4), total gasoline consumed by this group would be only 45.3 billion gallons, for a saving of 17.3 billion gallons (412 million barrels) of petroleum per year.

If the passenger cars achieved mileage equal to the EPA city rating for the most efficient gasoline-powered car of the 2006 model year (32), passenger cars would consume only 53.3 billion gallons of gasoline, saving 22.8 billion gallons (543 million barrels) per year.

If the passenger cars achieved the 60 mpg city mileage rating of the most efficient gas-electric hybrid of 2006, their total gasoline consumption would be just 28.4 billion gallons, saving 47.7 billion gallons (1.14 billion barrels) per year.

Data from U.S. EPA and *Highway Statistics 2004,* Federal Highway Safety Administration, U.S. Department of Transportation.

Other factors that influence both demand and projections include prices of various forms of energy (which recent history suggests is difficult to assess years, let alone decades, into the future) and public policy decisions (to pursue or not to pursue nuclear power, for example) in those nations that are major energy consumers.

Figure 15.2 shows one recent projection of world energy demand to 2030. Whether or not it is perfectly accurate, it clearly indicates substantial anticipated growth in energy use. If fossil fuels are inadequate in quantity and/or environmentally problematic, one or more alternatives must be far more extensively developed. This chapter surveys some possibilities. Evaluation of each involves issues of technical practicality, environmental consequences, and economic competitiveness.

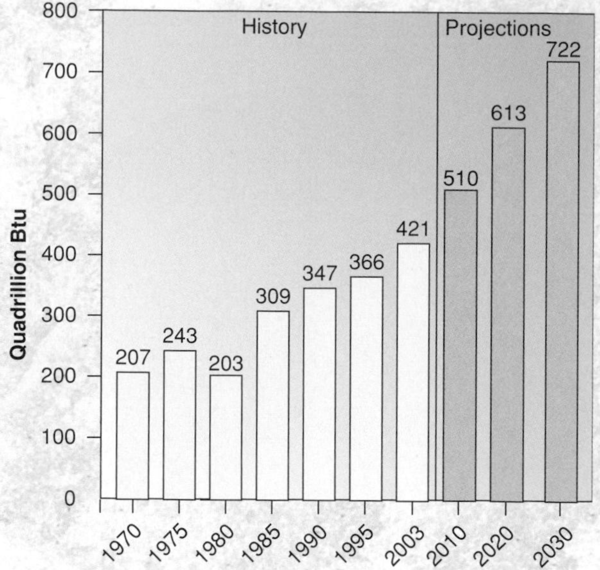

FIGURE 15.2 World energy consumption, historic and projected to 2030. Projected increase from 2003 to 2030 is about 70%.

Source: International Energy Outlook 2002 *and* International Energy Outlook 2006, *U.S. Energy Information Administration.*

NUCLEAR POWER—FISSION

Fission—Basic Principles

The phrase *nuclear power* actually comprises two different types of processes with different advantages and limitations. **Fission** is the splitting apart of atomic nuclei into smaller ones, with the release of energy. **Fusion** is the combining of smaller nuclei into larger ones, also releasing energy. Currently, only one of these processes is commercially feasible: fission. Fission basics are outlined in figure 15.3. (For a review of the terminology related to atoms, see chapter 2.) Very few isotopes—some 20 out of more than 250 naturally occurring isotopes—can undergo fission spontaneously, and do so in nature. Some additional nuclei can be induced to split apart, and some naturally fissionable nuclei can be made to split up more rapidly, thus increasing the rate of energy release. The fissionable nucleus of most interest in modern nuclear power reactors is the isotope of uranium with 92 protons and 143 neutrons, uranium-235.

A uranium-235 nucleus can be induced to undergo fission by firing another neutron into the nucleus. The nucleus splits into two lighter nuclei (not always the same two) and releases additional neutrons as well as energy. Some of the newly released neutrons can induce fission in other nearby uranium-235 nuclei, which, in turn, release more neutrons and more energy in breaking up, and so the process continues in a **chain reaction.** A controlled chain reaction, with a continuous, moderate release of energy, is the basis for fission-powered reactors (figure 15.4). The energy released heats cooling water that circulates through the reactor's core. The heat removed from the core is transferred through a heat exchanger to a second water loop in which steam is produced. The steam, in turn, is used to run turbines to produce electricity.

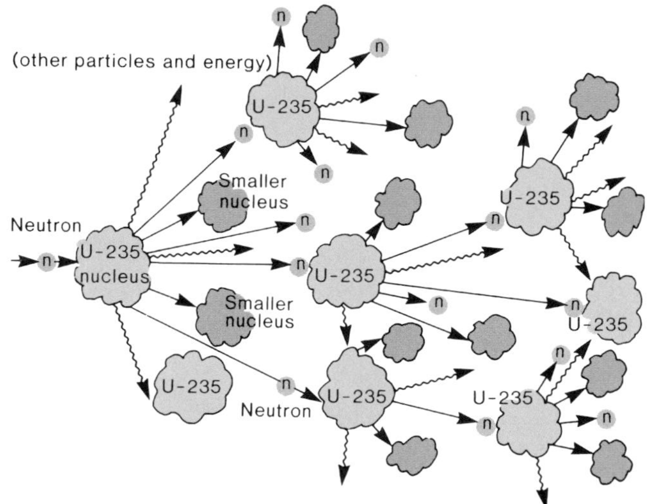

FIGURE 15.3 Nuclear fission and chain reaction involving uranium-235 (schematic). Neutron capture by uranium-235 causes fission into two smaller nuclei plus additional neutrons, other subatomic particles, and energy. Released neutrons, in turn, cause fission in other uranium-235 nuclei. As U-235 nuclei are used up, reaction rate slows; eventually fresh fuel must replace "spent" fuel.

This scheme is somewhat complicated by the fact that a chain reaction is not sustained by ordinary uranium. Only 0.7% of natural uranium is uranium-235. The material must be processed to increase the concentration of this isotope to several percent of the total to produce reactor-grade uranium. As the reactor operates, the uranium-235 atoms are split and destroyed so that, in time, the fuel is so depleted (spent) in this isotope that it must be replaced with fresh enriched uranium.

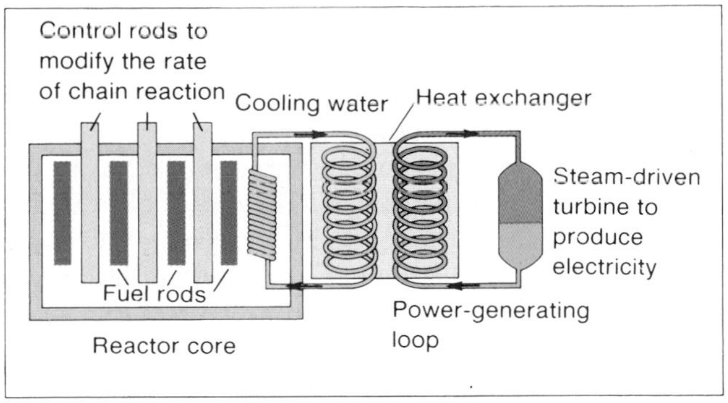

FIGURE 15.4 Schematic diagram of conventional nuclear fission reactor. Heat is generated by chain reaction; withdrawing or inserting control rods between fuel elements varies rate of reaction, and thus rate of release of heat energy. Cooling water also serves to extract heat for use. Heat is transferred to power loop via heat exchanger, so the cooling water, which may contain radioactive contaminants, is isolated from the power-generating equipment.

TABLE 15.2

U.S. Uranium Reserve and Resource Estimates

Recoverable Costs	Reserves (million pounds of U_3O_8)	Resources, Including Reserves (million pounds)
\$30/lb U_3O_8	300	3800
\$50/lb U_3O_8	900	6400
\$100/lb U_3O_8	1400	9800

Source: Data from Energy Information Administration, U.S. Department of Energy.

Actual uranium prices (in constant 1983 dollars) rose from under \$20/lb U_3O_8 in the early 1970s, to over \$70/lb on the spot market in 1976, then declined sharply to under \$15/lb in 1990, and have remained depressed since, largely as a result of reduced demand for power-plant fuel. Domestic production virtually ceased in the early 1990s. Peak production of recent years has been below 5 million pounds U_3O_8, while over 60 million pounds were imported in 2005.

The Geology of Uranium Deposits

Worldwide, 95% of known uranium reserves are found in sedimentary or metasedimentary rocks. In the United States, the great majority of deposits are found in sandstone. They were formed by weathering of uranium source rocks, followed by uranium migration in and deposition by ground water.

Minor amounts of uranium are present in many crustal rocks. Granitic rocks and carbonates may be particularly rich in uranium (meaning that its concentration in these may be in the range of ppm to tens of ppm; in most rocks, it is even lower). In granites, the uranium is concentrated in the late stages of magma crystallization; its large atoms fit relatively poorly in most early-crystallizing silicate minerals. Uranium is concentrated in carbonate rocks during precipitation of the carbonates from water, including seawater. As uranium-bearing rocks weather under near-surface conditions, that uranium goes into solution: uranium is particularly soluble in an oxygen-rich environment. The uranium-bearing solutions then infiltrate and join the groundwater system. As they percolate through permeable rocks, such as sandstone, they may encounter chemically reducing conditions, created by some factor such as an abundance of carbon-rich organic matter, or sulfide minerals, in shales bounding the sandstone. Under reducing conditions, the solubility of uranium is much lower. The dissolved uranium is then precipitated and concentrated in these reducing zones. Over time, as great quantities of uranium-bearing ground water percolate slowly through such a zone, a large deposit of precipitated uranium ore may form.

World estimates of available uranium are somewhat difficult to obtain, partly because the strategic importance of uranium leads to some secrecy. Even for the United States, the reserve estimates are strongly sensitive to price, as one would expect. Table 15.2 summarizes present reserve and resource estimates at different price levels. With the type of nuclear reactor currently in commercial operation in the United States, nuclear electricity-generating capacity probably could not be increased to much more than four times present levels by the year 2020 without serious fuel shortages. This increase in production would supply less than 15% of total U.S. energy needs. In other words, the rare isotope uranium-235 is in such short supply that the United States could use up our reserves within several decades, assuming no improvements in reactor technology and no significant uranium imports. Given the declining interest in nuclear fission power in the United States, this may not be a significant concern; however, worldwide, new fission plants continue to be added, so means of increasing the fuel supply for fission reactors are an issue. And if public opinion in the United States should also shift toward expanded use of fission power, fuel supply will be an issue here, too.

Extending the Nuclear Fuel Supply

Uranium-235 is not the only possible fission-reactor fuel, although it is the most plentiful naturally occurring one. When an atom of the far more abundant uranium-238 absorbs a neutron, it is converted into plutonium-239, which is, in turn, fissionable. Uranium-238 makes up 99.3% of natural uranium and over 90% of reactor-grade enriched uranium. During the chain reaction inside the reactor, as freed neutrons move about, some are captured by uranium-238 atoms, making plutonium. Spent fuel could be *reprocessed* to extract this plutonium, which could be purified into fuel for future reactors, as well as to re-enrich the remaining uranium in uranium-235. How reprocessing would alter the nuclear fuel cycle is shown in figure 15.5. Fuel reprocessing with recovery of both plutonium and uranium could reduce the demand for "new" enriched uranium by an estimated 15%.

A **breeder reactor** can maximize the production of new fuel. Breeder reactors produce useful energy during operation,

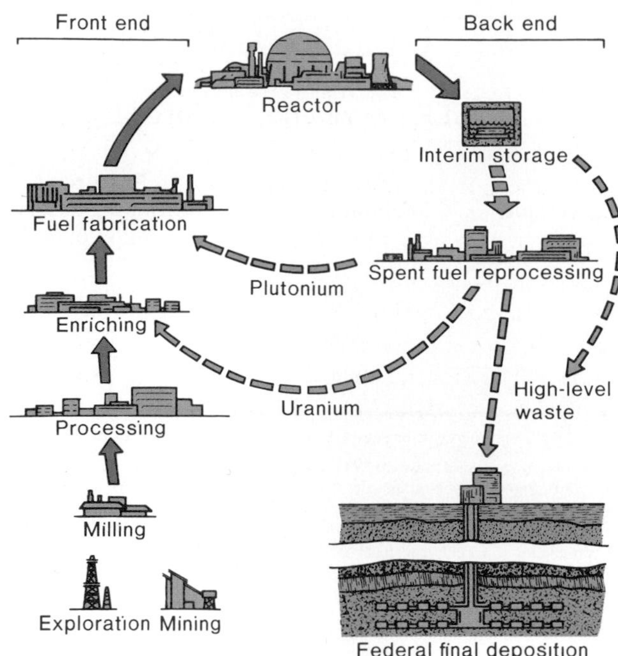

FIGURE 15.5 The nuclear fuel cycle, as it currently operates and as it would function with fuel reprocessing.
Source: U.S. Energy Information Administration, Department of Energy.

just as conventional "burners" using up uranium-235 do, by fission in a sustained chain reaction within the reactor core. In addition, they are designed so that surplus neutrons not required to sustain the chain reaction are used to produce more fissionable fuels from suitable materials, like plutonium-239 from uranium-238, or thorium-233, which is derived from the common isotope thorium-232. Breeder reactors could synthesize more nuclear fuel for future use than they would actively consume while generating power.

Breeder-reactor technology is more complex than that of conventional water-cooled burners. The core coolant is liquid metallic sodium; the reactor operates at much higher temperatures. The costs would be substantially higher than for burner reactors, perhaps close to $10 billion per reactor by the time any breeders could be completed in the United States. The breeding process is slow, too: The break-even point after which fuel produced would exceed fuel consumed might be several decades after initial operation. If the nuclear-fission option is to be pursued vigorously in the twenty-first century, the reprocessing of spent fuels and the use of breeder reactors are essential. Yet at present, reprocessing is minimal in the United States, and in early 1985, Congress cancelled funding for the experimental Clinch River breeder reactor, in part because of high estimated costs. No commercial breeders currently exist in the United States, and only a handful of breeder reactors are operating in the world, though some—notably the Superphénix reactor in France—are functioning successfully. Even conventional reactors have fallen out of favor in the United States. There are several reasons for this declining enthusiasm.

FIGURE 15.6 Three Mile Island near Harrisburg, Pennsylvania; damaged reactor remains shut down, while others are still operative.
© Doug Sherman/Geofile.

Concerns Related to Nuclear Reactor Safety

A major concern regarding the use of fission power is reactor safety. In normal operation, nuclear power plants release very minor amounts of radiation, which are believed to be harmless. (A general discussion of radiation and its hazards is presented in chapter 16.) The small but finite risk of damage to nuclear reactors through accident or deliberate sabotage is more worrisome to many.

One of the most serious possibilities is a so-called loss-of-coolant event, in which the flow of cooling water to the reactor core would be interrupted. Resultant overheating of the core might lead to **core meltdown,** in which the fuel and core materials would deteriorate into a molten mass that might or might not melt its way out of the containment building and thus release high levels of radiation into the environment, depending upon the design of the reactor and containment building. A partial loss of coolant, with 35 to 45% meltdown, occurred at Three Mile Island in 1979 (figure 15.6).

No matter how far awry the operation of a commercial power plant might go, and even if there were a complete loss of coolant, the reactor could *not* explode like an atomic bomb. Bomb-grade fuels must be much more highly enriched in the fissionable isotope uranium-235 for the reaction to be that intensive and rapid. Also, the newest reactors have additional safety features designed to reduce the risk of accident. However, an ordinary explosion originating within the reactor (or by saboteurs' use of conventional explosives) could, if large enough, rupture both the containment building and reactor core and thus release large amounts of radioactive material. The serious accident at Chernobyl reinforced reservations about reactor safety in many people's minds; see Box 15.1.

Plant siting is another problem. Siting nuclear plants close to urban areas puts more people potentially at risk in case of accident; placing the plants far from population centers where energy is needed means more transmission loss of electricity (which already claims nearly 10% of electricity generated). Proximity to water is often important for cooling purposes but makes water pollution in case of mishap more likely. There are also concerns about the structural integrity of nuclear plants located close to fault zones—the Diablo Canyon station in California, the Humboldt Bay nuclear plant, and others; several proposed reactor sites have been rejected when investigation revealed nearby faults.

Concerns Related to Fuel and Waste Handling

The mining and processing of uranium ore are operations that affect relatively few places in the United States (figure 15.7). Nevertheless, they pose hazards because of uranium's natural radioactivity. Miners exposed to the higher radiation levels in uranium mines have experienced higher occurrence rates of some types of cancer. Carelessly handled tailings from processing plants have exposed others to radiation hazards; recall the case of Grand Junction, Colorado, mentioned in chapter 13, where radioactive tailings were unknowingly mixed into concrete used in construction.

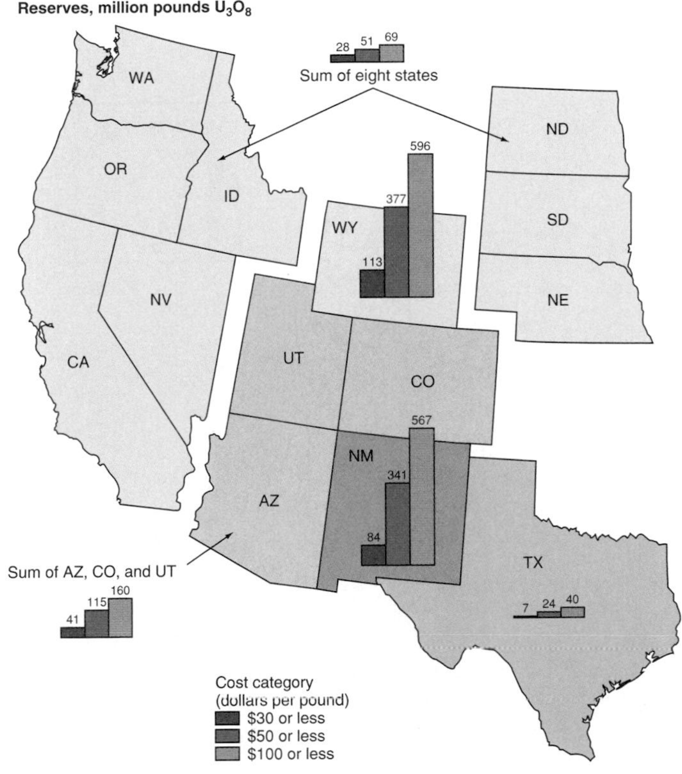

FIGURE 15.7 Locations of U.S. uranium reserves.
Source: Annual Energy Review 1999, *U.S. Energy Information Administration, Department of Energy.*

The use of reprocessing or breeder reactors to produce and recover plutonium to extend the supply of fissionable fuel poses special problems. Plutonium itself is both radioactive and chemically toxic. Of greater concern to many people is that, as a readily fissionable material, it can also be used to make nuclear weapons. Extensive handling, transport, and use of plutonium would pose a significant security problem and would therefore require very tight inventory control to prevent the material from falling into hostile hands.

The radioactive wastes from the production of fission power are another concern. Radiation hazards and radioactive-waste disposal are considered within the broader general context of waste disposal in chapter 16. Here, two aspects of the problem are highlighted. First, radioactive materials cannot be treated by chemical reaction, heating, and so on to make them nonradioactive. In this respect, they differ from many toxic chemical wastes that can be broken down by appropriate treatment. Second, there has been sufficient indecision about the best method of radioactive-waste disposal and the appropriate site(s) for it that none of the radioactive wastes generated anywhere in the world have been disposed of permanently to date. Currently, the wastes are in temporary storage while various disposal methods are being explored, and many of the temporary waste-holding sites are filled almost to capacity. Clearly, acceptable waste-disposal methods must be identified and adopted, if only to dispose of wastes already accumulated.

Nuclear plants have another unique waste problem. The bombardment of the reactor core and structure by neutrons and other atomic debris from the fission process converts some of the structural materials to radioactive ones and changes the physical properties of others, weakening the structure. At some point, then, the plant must be **decommissioned**—taken out of operation, broken down, and the most radioactive parts delivered to radioactive-waste disposal sites. In late 1982, an electricity-generating plant at Shippingport, Pennsylvania, became the first commercial U.S. fission power plant to face decommissioning, after twenty-five years of operation. Typical costs of demolition and disposal are over $100 million per reactor in the United States, and the process may take a decade or more. In other countries, cost estimates run as high as $500 million per reactor, and dozens of reactors are being, or have been, decommissioned worldwide. A total of 28 U.S. reactors, including Shippingport, had been retired by the close of 2005. Historically, the nominal lifetime allowed by regulatory agencies was about 30 years; currently, it is about 40 years; with rigorous plant monitoring and replacement of some key components, it may be possible to operate a fission-powered electricity-generating plant for 70 years. Still, sooner or later, decommissioning will take its fiscal

Box 15.1 Crisis at Chernobyl

On 26 April 1986, an accident occurring at the nuclear power plant at Chernobyl, Ukraine, in the former USSR, dwarfed any U.S. reactor accident in scale and regional impact and reinforced many fears about the safety of the nuclear industry. The reactor core overheated; an explosion and core meltdown occurred. The initial overheating of the core resulted from a still-unexplained power surge to 50% of reactor capacity from standby status. A runaway chain reaction ensued. What then caused the explosion is not clear. Possibly, leaks allowed water and steam to hit the hot graphite of the reactor core, producing hydrogen gas, which is highly explosive. The explosion, in turn, may have disrupted emergency cooling systems and also prevented the insertion of control rods to slow the reaction.

Quantities of radioactive material escaped from the reactor building, drifting with atmospheric circulation over Scandinavia and eastern Europe (figure 1). The official report was that thirty-one deaths were directly attributable to the accident; 203 persons were hospitalized for acute radiation sickness; 135,000 people were evacuated. Downwind, preventive measures to minimize possible health impacts were suggested: For example, some people consumed large doses of natural iodine to try to saturate the thyroid gland (which concentrates iodine) and thus prevent its absorption

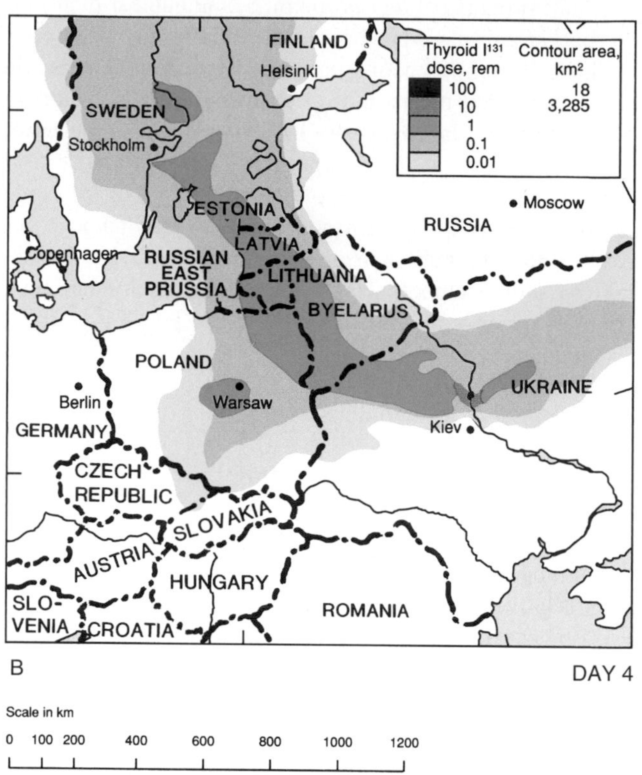

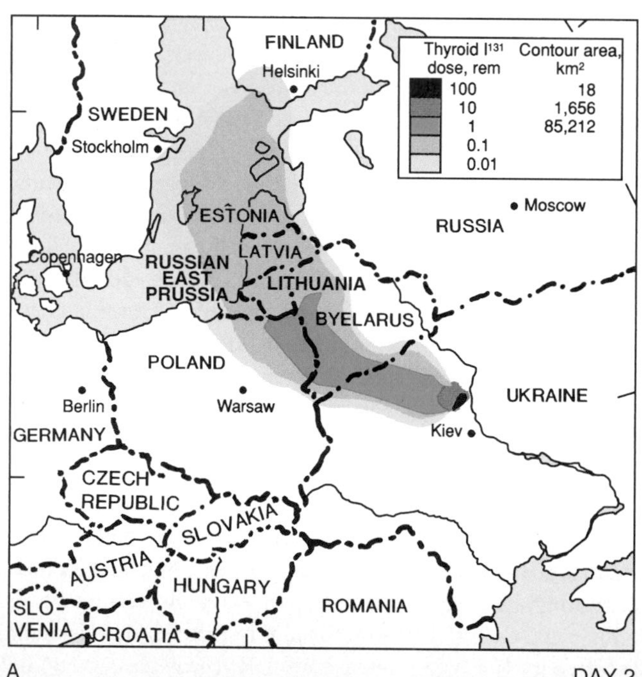

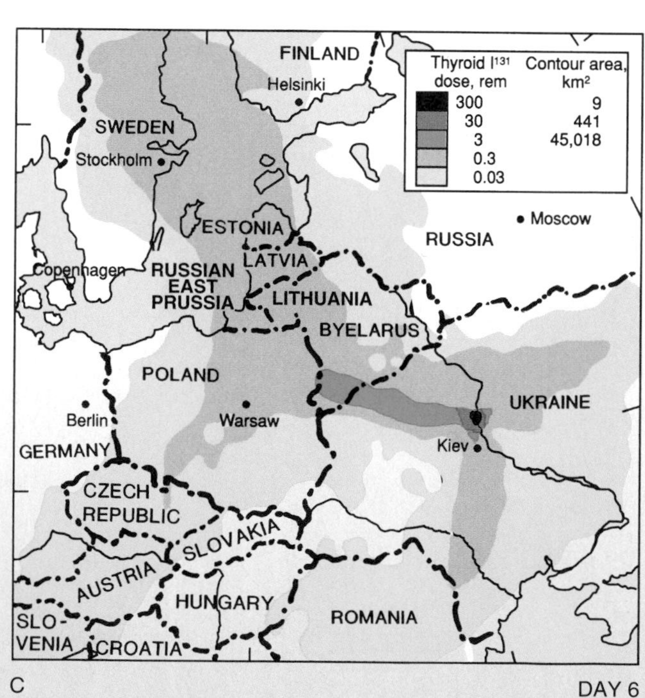

FIGURE 1 A cloud of radioactive iodine (iodine-131) drifts and spreads out from the Chernobyl reactor site in the week after the accident. For comparison, maximum *annual* radiation emission permitted at the boundary of a nuclear facility in the United States is 25 rem. (A *rem* is a unit of radiation exposure that takes into account both the energy and nature of the radiation, so that different types of radiation can be compared in terms of their effects on the human body.) All maps are as of 0000 Greenwich Mean Time (GMT); the accident began on 26 April 1986 at 1:23 GMT. (A) Day 2 (28 April); (B) Day 4; (C) Day 6.

Source: Data from Mark A. Fischetti, "The Puzzle of Chernobyl" in IEEE Spectrum *23, July 1986, pp. 38–39. Copyright © 1986 IEEE.*

(Box 15.1 Continued)

of the radioactive iodine in fallout from the reactor. The republics most severely affected were Ukraine and Byelarus.

Some of the long-range health effects of the accident will not be known for decades, partly because some of these consequences develop long after radiation exposure and partly because the effects of radiation are not always distinguishable from the effects of other agents (see chapter 16).

Later studies have begun to document increases in medical problems attributable to radiation—thyroid cancer in children, chromosome damage—near the reactor site. Estimates vary widely, but put the increased lifetime risk of contracting radiogenic cancer at 0 to 0.02% in Europe, 0.003% in the Northern Hemisphere overall. Over the long term, total cancer cases resulting from the Chernobyl incident are projected at 5000 to 100,000.

Can a Chernobyl-style accident happen at a commercial nuclear reactor in the United States? Basically, no, because the design of these reactors is fundamentally different from the design of the Chernobyl reactor (though a few research reactors in this country have a core of Chernobyl type). In any reactor, some material must serve as a *moderator,* to slow the neutrons streaming through the core enough that they can interact with nuclei to sustain the chain reaction. In the Chernobyl reactor, the moderator is graphite; the core is built largely of graphite blocks. In commercial U.S. reactors (and most reactors in western nations), the moderator is water. Should the core of a water-moderated reactor overheat, the water would vaporize. Steam is a poor moderator, so the rate of chain reaction would slow down. (Emergency cooling of the core would still be needed, but as the reaction slowed, so would the rate of further heat production.) However, graphite remains solid up to extremely high temperatures, and hot graphite and cold graphite are both effective moderators. In a graphite reactor, therefore, a runaway chain reaction is not stopped by changes in the moderator. At Chernobyl, as the graphite continued to overheat, it began to burn, turning the reactor core into the equivalent of a giant block of radioactive charcoal. (To put it out, workers eventually had to smother it in concrete.) The fuel rods in the burning core ruptured and melted from the extreme heat. The core meltdown, together with the explosion, led to the release of high levels of radioactive materials into the environment. The extent of that release was also far greater for the Chernobyl reactor than it would have been for a U.S. reactor suffering a similar accident because the containment structures of U.S. reactors are much more substantial and effective. The accident focused international attention sharply on the issue of reactor safety and also on the reactor-operator interface (operator error may have contributed to the accident). Out of this disaster have come a number of constructive suggestions for the improvement of reactor safety, both in design and in the operation of existing reactors. The incident also emphasized that, like air pollution, radioactive fallout ignores international boundaries.

and waste-disposal toll. In the United States and worldwide, reactors are aging (figure 15.8).

Risk Assessment, Risk Projection

Much of the debate about energy sources focuses on hazards: what kinds, how many, how large. Many of the risks associated with various energy sources may not be immediately obvious. No energy source is risk-free, but, then, neither is living. The question is, what constitutes "acceptable" risk (and, acceptable to whom)? Underground miners of both coal and uranium face increased health risks, but the power consumer does not see those risks directly. Those living downstream from a dam are at risk; so are those downwind from a nuclear or fossil-fuel plant.

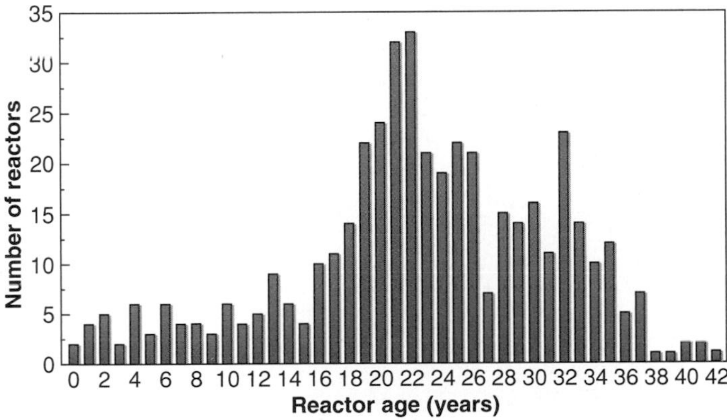

FIGURE 15.8 Age distribution of nuclear reactors worldwide, June 2006. Few face decommissioning yet, but many may in the next decade or two.

Source: International Atomic Energy Agency.

A power plant of 1-billion-watt capacity serves the electricity needs of about 1 million people. One study in the mid-1970s projected accidental deaths at 0.2 per year for a uranium-fueled plant this size (virtually all associated with mining), compared to an estimated 2.6 to 4 deaths per year for a comparable coal-fired plant, depending upon mining method, including significant risks associated with processing and transporting the very much larger volume of coal involved. By way of comparison, in the U.S. in 2003, per million people, there were 154 deaths from motor-vehicle accidents, 67 from accidental poisoning, 11 from drowning, 59 from falls, and 3 from accidents involving firearms.

A further consideration when comparing risks may be how well-defined the risks are. That is, estimates of risks from fires, automobile accidents, and so on are actuarial, based on considerable actual past experience and hard data. By contrast, projections of risks associated with extensive use of fission power are based on limited past experience, which does not even include the worst possible events, plus educated guesses about future events based, in part, on hard-to-quantify parameters like probability of human errors. In 1975, a report prepared for the U.S. Nuclear Regulatory Commission estimated the risk of core meltdown at one per million reactor-years of operation. (To date, total global fission-reactor operations add up to less than 10,000 reactor-years.) Since the accidents at Three Mile Island and particularly Chernobyl, and given the numerous other less serious "events" at other commercial reactors, many have begun to speculate that the risk estimate is too low. There is, however, no way to confirm or refute this suspicion directly in the absence of more data and experience.

In the early 1970s, it was predicted that 25% of U.S. energy would be supplied by nuclear power by the year 2000. At the end of 2005, there were 104 nuclear power plants (figure 15.9) operating in the United States (down from a 1990 peak of 112), and they accounted for only 8% of U.S. energy production. Moreover, during the last twenty years, nuclear plant cancellations have far exceeded new orders; since 1978, there have been no new orders at all, and several utilities have decided to power their new generating plants with coal, not nuclear fission.

Nuclear plants do have lower fueling and operating costs than coal-fired plants. The small quantity of fuel required for their operation can also be more easily stockpiled against interruption by strikes or transportation delays. However, nuclear plants have become more costly to build than coal-fired plants and now require longer to plan, construct, and license (nine to twelve years for U.S. nuclear plants; six to ten for coal). Especially in times of high interest rates, these economic and time factors loom large. Those, together with increasingly vocal public opposition to nuclear power plants (itself a contributing factor in many licensing delays), have made the coal option more appealing to many utilities.

Worldwide, reliance on nuclear power varies widely, accounting for only 2 to 5% of electricity production in such countries as Mexico, India, Pakistan, and China, but over 30% in much of Europe, and in some countries (Slovakia, Belgium, France, Lithuania), over 50% (figure 15.10). Some nations have turned to nuclear fission for lack of fossil-fuel resources. With new international emphasis on limiting CO_2 emissions, too, the clean operation of nuclear fission plants enhances their appeal to some. In mid-2006, there were 442 operable nuclear fission plants in the world, and 27 more under construction, including 7 in India and 4 each in China and Russia. Still, many projections anticipate a declining role for fission-power generation globally over coming decades. If so, it is not clear what nonfossil alternatives could replace it.

Different people weigh the pros and cons of nuclear fission power in different ways. For many, the uncertainties and potential risks outweigh the benefits. For others, the problems associated with using coal (the most readily available and likely alternative) appear at least as great.

NUCLEAR POWER—FUSION

Nuclear *fusion* is the opposite of fission. As noted earlier, fusion is the process by which two or more smaller atomic nuclei combine to form a larger one, with an accompanying release of energy. It is the process by which the sun generates its vast amounts of energy. In the sun, simple hydrogen nuclei containing one proton are fused to produce helium. For technical reasons, fusion of the heavier hydrogen isotopes deuterium (nucleus containing one proton and one neutron) and tritium

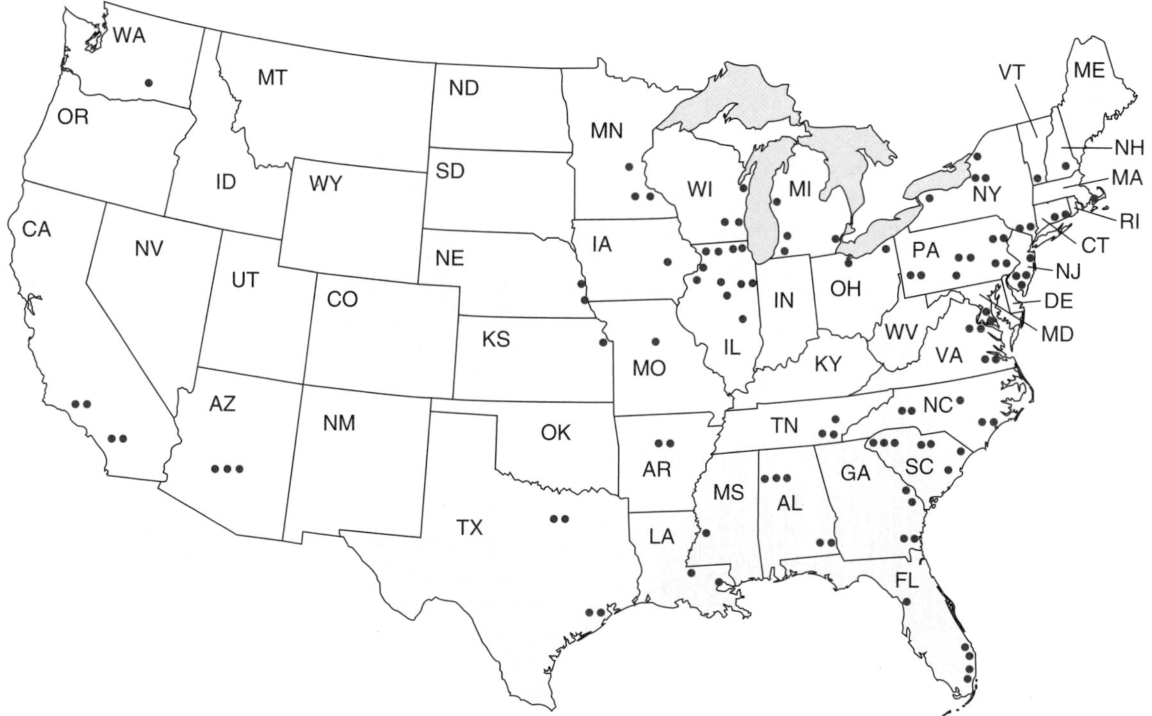

FIGURE 15.9 Distribution of U.S. nuclear power plants. In December 2005, there were 104 operable plants, and three for which construction permits had been granted. (Due to space limitations, symbols do not represent actual locations; the number of plants in each state *is* accurate.)

Source: Data from Annual Energy Review 2005, *U.S. Energy Information Administration, and International Atomic Energy Agency.*

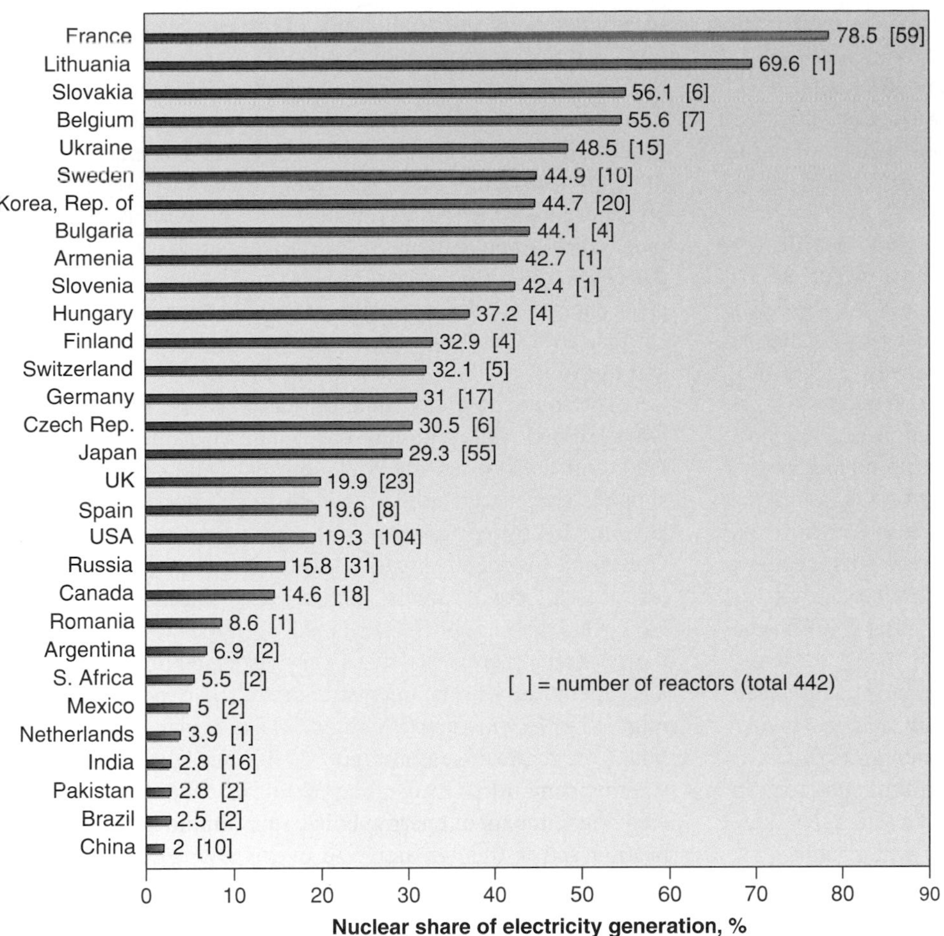

FIGURE 15.10 Percentage of electricity generated by nuclear fission varies greatly by country, and not simply in proportion to numbers of reactors; where electricity consumption is moderate, a few reactors can account for a large share.

Source: Data for 2005; from International Atomic Energy Agency.

(one proton and two neutrons) would be easier to achieve on earth. That fusion reaction is diagrammed in figure 15.11. (More recently, it has been suggested that fusion of helium-3 may be more technically feasible. However, helium-3 is extremely rare on earth. It has been further suggested that helium-3 be harvested from the moon's surface, but the cost and difficulty of this means that it is not a near-term prospect.)

Hydrogen is plentiful because it is a component of water; the oceans contain, in effect, a huge reserve of hydrogen, an essentially inexhaustible supply of the necessary fuel for fusion. This is true even considering the scarcity of deuterium, which is only 0.015% of natural hydrogen. (Tritium is rarer still and would probably have to be produced from the comparatively rare metal lithium. However, the magnitude of energy release from fusion reactions is such that this is not a serious constraint.) The principal product of the projected fusion reactions—helium—is a nontoxic, chemically inert, harmless gas. There could be some mildly radioactive light-isotope by-products of fusion reactors, but they would be much less hazardous than many of the products of fission reactors.

Since fusion is a far "cleaner" form of nuclear power than fission, why not use it? The principal reason is technology, or lack of it. To bring about a fusion reaction, the reacting nuclei must be brought very close together at extremely high temperatures (millions of degrees at least). The natural tendency of hot gases is to expand, not come together, and no known physical material could withstand such temperatures to contain the reacting nuclei. The techniques being tested in laboratory fusion experiments are elaborate and complex, either involving containment of the fusing materials with strong magnetic fields or using lasers to heat frozen pellets of the

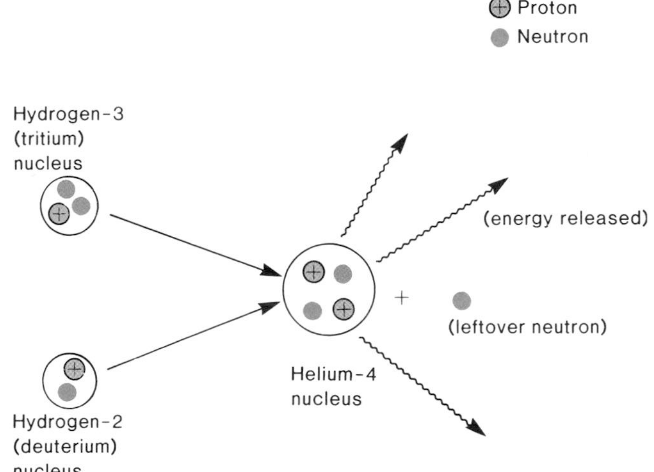

FIGURE 15.11 Schematic diagram of one nuclear fusion reaction. Other variants are possible.

reactants very rapidly. At best, the experimenters have been able to achieve the necessary conditions for fractions of a second, and the energy required to bring about the fusion reactions has exceeded the energy released thereby. Certainly, the present technology does not permit the construction of commercial fusion reactors in which controlled, sustained fusion could be used as a source of energy.

Scientists in the field estimate that several decades of intensive, expensive research will be needed before fusion can become a commercial reality. (Late in the year 2000, a speaker at the annual meeting of the Geological Society of America dubbed a talk on fusion "Twenty years away for twenty years.") Even then, it will be costly. Some projections suggest that a commercial fusion electricity-generating plant could cost tens of billions of dollars. Nevertheless, the abundance of the fuel supply and the relative cleanness of fusion (as compared to both fission and fossil-fuel power generation) make it an attractive prospect, at least for later in the twenty-first century. Fusion, however, is a means of generating electricity in stationary power plants only. This fact limits its potential contribution toward satisfying total energy needs.

It should be noted that in 1989, two scientists announced what would have been an immensely important technical breakthrough: "cold fusion," or fusion at room temperature in a beaker. The experimental setup involved palladium electrodes in a beaker full of "heavy" (deuterium-enriched) water. The idea was that the electric current would split the water molecules, and deuterium ions from the water would be driven into one of the electrodes, where some would fuse. Evidence cited to support the occurrence of cold fusion included heat in excess of what could be accounted for by chemical processes and current flow in the apparatus, and the presence of tritium and helium (expected products of deuterium fusion) afterwards. If fusion could be achieved under such simple, inexpensive conditions, the principal obstacle to the use of fusion power would be removed, and this vast resource would become readily and cheaply available. Hundreds of scientists rushed to replicate the experiments. Unfortunately, the overwhelming majority failed to reproduce any of the anomalies reported; some proposed that any helium found was simply a contaminant from the atmosphere or introduced in the electrodes. The consensus is that whatever is going on in such apparatus is not "cold fusion," and commercial fusion power is still a long way off.

SOLAR ENERGY

The earth intercepts only a small fraction of the energy radiated by the sun. Much of that energy is reflected or dissipated in the atmosphere. Even so, the total solar energy reaching the earth's surface far exceeds the energy needs of the world at present and for the foreseeable future. The sun can be expected to go on shining for approximately 5 billion years—in other words, the resource is inexhaustible, which contrasts sharply with nonrenewable sources like uranium or fossil fuels. Sunlight falls on the earth without any mining, drilling, pumping, or disruption of the land. Sunshine is free; it is not under the control of any company or cartel, and it is not subject to embargo or other political disruption of supply. The *use* of solar energy is essentially pollution-free, at least in the sense that the absorption of sunlight for heat or the operation of a solar cell for electricity are very "clean" processes (though solar electricity has other environmental costs; see "Solar Electricity"). It produces no hazardous solid wastes, air or water pollution, or noise. For most current applications, solar energy is used where it falls, thereby avoiding transmission losses. All these features make solar energy an attractive option for the future. Several practical limitations on its use also exist, however, particularly in the short term.

The solar energy reaching the earth is dissipated in various ways. Some is reflected into space; some heats the atmosphere, land, and oceans, driving ocean currents and winds. The sun supplies the energy needed to cause evaporation and thus to keep the hydrologic cycle going; it allows green plants to generate food by photosynthesis. More than ample energy is still left over to provide for all human energy needs, in principle. However, the energy is distributed over the whole surface of the earth. In other words, it is a very dispersed resource: Where large quantities of solar energy are used, solar energy collectors must cover a wide area. Sunlight is also variable in intensity, both from region to region (figure 15.12) and from day to day as weather conditions change. The two areas in which solar energy can make the greatest immediate contribution are in space heating and in the generation of electricity, uses that together account for about two-thirds of U.S. energy consumption.

Solar Heating

Solar space heating typically combines direct use of sunlight for warmth with some provision for collecting and storing additional heat to draw on when the sun is not shining. It is typically employed on the scale of an individual building. The simplest approach is *passive-solar heating,* which does not require

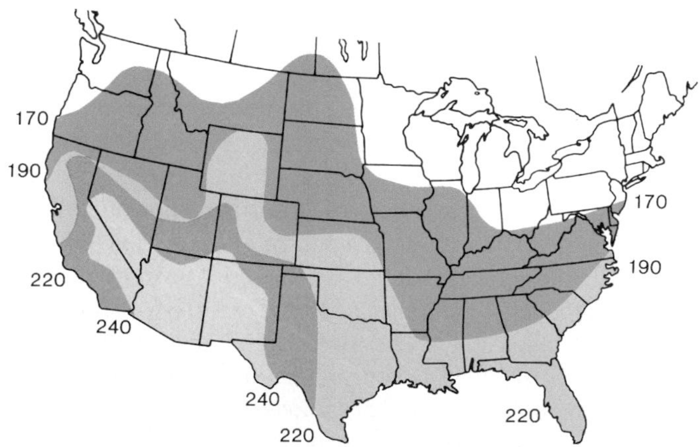

FIGURE 15.12 Distribution of solar energy over the continental United States. Maximum insolation occurs over the southwestern region. Numbers are watts per square meter; boundaries between zones connect points of equal insolation.

Source: Data from Energy Strategies: Toward a Solar Future, *ed. by S. W. Kendall and S. J. Nadis, Union of Concerned Scientists, 1980, Ballinger Publishing Co.*

mechanical assistance. The building design should allow the maximum amount of light to stream in through south and west windows during the cooler months. This heats the materials inside the house, including the structure itself, and the radiating heat warms indoor air. Media used specifically for storing heat include water—in barrels, tanks, even indoor swimming pools—and the rock, brick, concrete, or other dense solids used in the building's construction (figure 15.13A). These supply a thermal mass that radiates heat back when needed, in times of less (or no) sunshine. Additional common features of passive-solar design (figure 15.13B) include broad eaves to block sunshine during hotter months (feasible because the sun is higher in the sky during summer than during winter) and drapes or shutters to help insulate window areas during long winter nights. Other possible variations on passive-solar heating systems are too numerous to describe here.

Active-solar heating systems usually involve the mechanical circulation of solar-heated water (figure 15.13C). The flat solar collectors are water-filled, shallow boxes with a glass surface to admit sunlight and a dark lining to absorb sunlight and help heat the water. The warmed water is circulated either directly into a storage tank or into a heat exchanger through which a tank of water is heated. The solar-heated water can provide both space heat and a hot-water supply. If a building already uses conventional hot-water heat, incorporating solar collectors is not necessarily extremely expensive (especially considering the free "fuel" to be used). With the solar collectors mounted on the roof, an active-solar system does not require the commitment of any additional land to the heating system—another positive feature. The method can be as practical for urban row houses or office buildings as for widely spaced country homes.

While solar heating may be adequate by itself in mild, sunny climates, in areas subject to prolonged spells of cloudiness or extreme cold, a conventional backup heating system is almost always needed. In the latter areas, then, solar energy can greatly reduce but not wholly eliminate the need for some consumption of conventional fuels. It has been estimated that, in the United States, 40 to 90% of most homes' heating requirements could be supplied by passive-solar heating systems, depending on location. It is usually more economical to design and build passive-solar technology features into a new structure initially than to incorporate them into an existing building later (retrofit). (A few 100%-solar homes have been built, even in such severe climates as northern Canada. However, the variety of solar-powered devices, extra insulation, and other features required can add tens of thousands of dollars to the cost of such homes, especially when an existing home is retrofitted, pricing this option out of the reach of many home buyers. Moreover, superinsulation can aggravate indoor air pollution; see chapter 18.)

Solar Electricity

Direct production of electricity using sunlight is accomplished through **photovoltaic cells,** also called simply "solar cells" (see figure 15.14). In simplest form, they consist of two layers

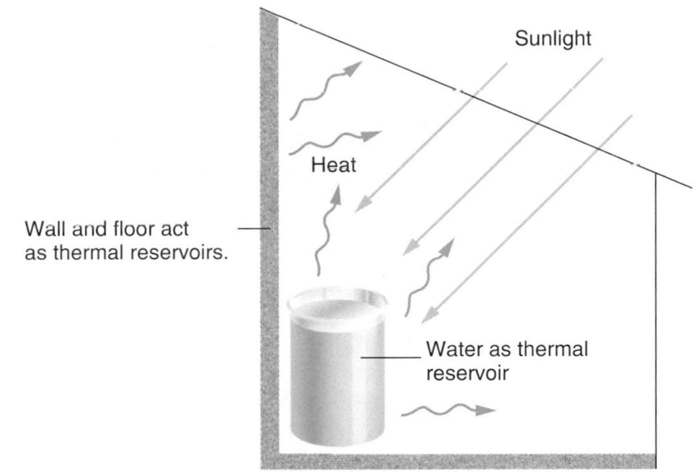

A

B

C

FIGURE 15.13 (A) Basics of passive-solar heating with water or structural materials as thermal reservoir: Sunlight streams into greenhouse with glass roof and walls, heat is stored for nights and cloudy days. (B) Design features of home and landscaping can optimize use of sun in colder weather, provide protection from it in summer. (C) A common type of active-solar heating system with a pump to circulate the water between the collector and the heat exchanger/storage tank.

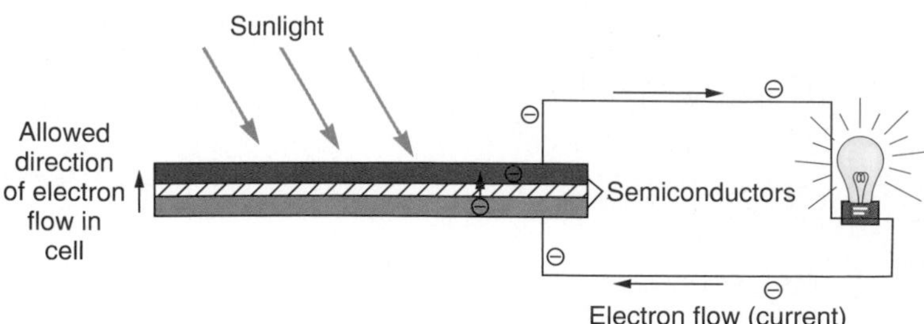

FIGURE 15.14 In a photovoltaic cell, incident sunlight dislodges electrons in top layer, which flow through wire as electric current, to return to the other side of the cell. Accumulated electrons move back to upper layer of cell and the cycle continues.

of semiconductor material sandwiched together, with a barrier between that allows electrons to flow predominantly in one direction only. Sunlight striking the exposed side can dislodge some electrons, which flow as electric current through a circuit, to return to the cell and continue the cycle.

Solar cells have no moving parts and, like solar heating systems, do not emit pollutants during operation. For many years, they have been the principal power source for satellites and for remote areas difficult to reach with power lines (figure 15.15). A major limitation on solar-cell use has historically been cost, which is several times higher per unit of power-generating capacity than for either fossil-fuel or nuclear-powered generating plants. This has restricted the appeal of home-generated solar electricity. The high cost is partly a matter of technology (present solar cells are not very efficient, though they are being improved) and partly one of scale (the industry is not large enough to enjoy the economies of mass production).

Currently, low solar-cell efficiency and the diffuse character of sunlight continue to make photovoltaic conversion an inadequate option for energy-intensive applications, such as many industrial and manufacturing operations. Even in the areas of strongest sunlight in the United States, incident radiation is of the order of 250 watts per square meter. Operating with commercial solar cells of about 20% efficiency means power generation of only 50 watts per square meter. In other words, to keep one 100-watt light bulb burning would require at least 2 square meters of collectors (with the sun always shining). A 100- megawatt power plant would require 2 square kilometers of collectors (nearly one square mile!). This represents a large commitment of both land and the mineral resources from which the collectors are made.

FIGURE 15.15 Solar electricity is very useful in remote areas. High in the mountains of Denali National Park, at the base camp that is the takeoff point for expeditions to climb Denali (Mount McKinley), solar cells (right) power vital communications equipment.

The most efficient photovoltaic cells use potentially toxic materials, such as gallium and arsenic. These toxic materials already present health hazards in both the mining and manufacturing phases. For the collector array alone, a 100-megawatt solar electric plant operating with basic solar cells would use at least an estimated 30,000 to 40,000 tons of steel, 5000 tons of glass, and 200,000 tons of concrete. A nuclear plant, by contrast, would require about 5000 tons of steel and 50,000 tons of concrete; a coal-fired plant, still less.

Siting such a sizable array requires a substantial commitment and disturbance of land. Its presence could alter patterns of evaporation and surface runoff. These considerations could be especially critical in desert areas, which are the most favorable sites for such facilities from the standpoint of intensity and constancy of incident sunlight. Construction could also disturb desert-pavement surfaces and accelerate erosion.

Clearly, then, for large-scale applications, it would be important to concentrate the solar energy being used to generate electricity. This can be done using arrays of parabolic mirrors, which can focus sunlight from a broader area onto a small expanse of solar cells. Or, the solar energy can be concentrated to heat water or another medium to run turbines to generate electricity, just as other fuels heat water to make steam for power generation. This approach uses no photovoltaic cells at all (figure 15.16).

Storing solar electricity is also a more complex matter than storing heat. For individual homeowners, batteries may suffice, but no wholly practical scheme for large-scale storage has been devised, despite advances in battery technology. Some of the proposals are shown in figure 15.17.

For the time being, it appears that solar-generated electricity could eventually supply perhaps 10 to 15% of U.S. electricity needs, but currently, it accounts for less than half a percent.

FIGURE 15.16 To make solar electricity without photovoltaic cells, solar heat can be used to make steam to power turbines, but it must first be concentrated, as here in the California desert, where parabolic mirrors focus sunlight on tubes of water.

Photograph © The McGraw-Hill Companies, Inc./Doug Sherman, photographer.

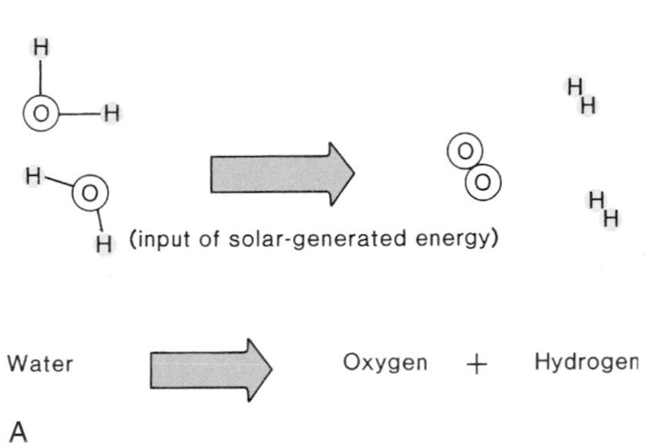

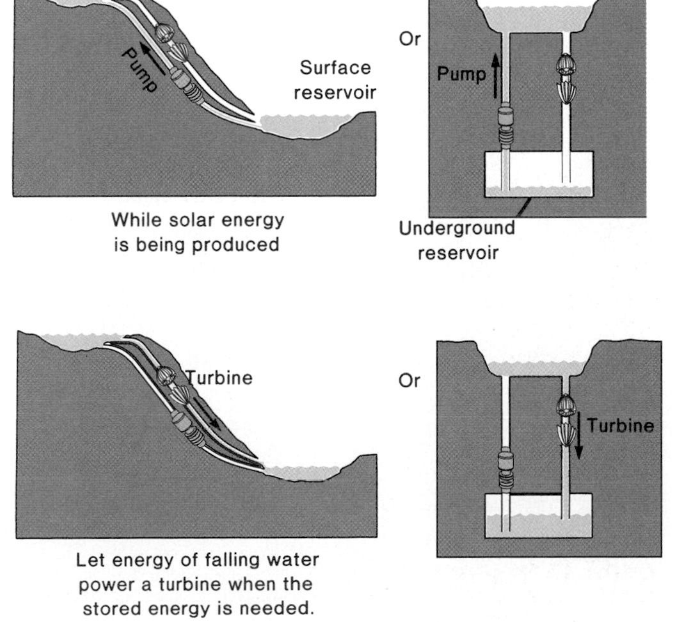

FIGURE 15.17 Some possible schemes for storing the energy of solar-generated electricity. (A) Use solar electricity to break up water molecules into hydrogen and oxygen; recombine them later (burn the hydrogen) to release energy. (B) Use solar energy to pump water up in elevation; when the energy is needed, let the water fall back and use it to generate hydropower.

Major improvements in efficiency of generation and in storage technology are needed before the sun can be the principal source of electricity. It may be still longer before solar electricity can contribute to the transportation energy budget.

GEOTHERMAL POWER

The earth contains a great deal of heat, most of it left over from its early history, some continually generated by decay of radioactive elements in the earth. Slowly, this heat is radiating away and the earth is cooling down, but under normal circumstances, the rate of heat escape at the earth's surface is so slow that we do not even notice it and certainly cannot use it. If the heat escaping at the earth's surface were collected over an average square meter for a year, it would be sufficient only to heat about 2 gallons of water to the boiling point, though local heat flow can be substantially higher, for example in young volcanic areas.

Magma rising into the crust from the mantle brings unusually hot material nearer the surface. Heat from the cooling magma heats any ground water circulating nearby (figure 15.18). This is the basis for generating **geothermal energy.** The magma-warmed waters may escape at the surface in geysers and hot springs, signalling the existence of the shallow heat source below (figure 15.19). More subtle evidence of the presence of hot rock at depth comes from sensitive measurements of heat flow at the surface, the rate at which heat is being conducted out of the ever-cooling earth: High heat flow signals unusually high temperatures at shallow depths. High heat flow and recent (or even current) magmatic activity go together and, in turn, are most often associated with plate boundaries. Therefore, most areas in which geothermal energy is being tapped extensively are along or near plate boundaries (figure 15.20).

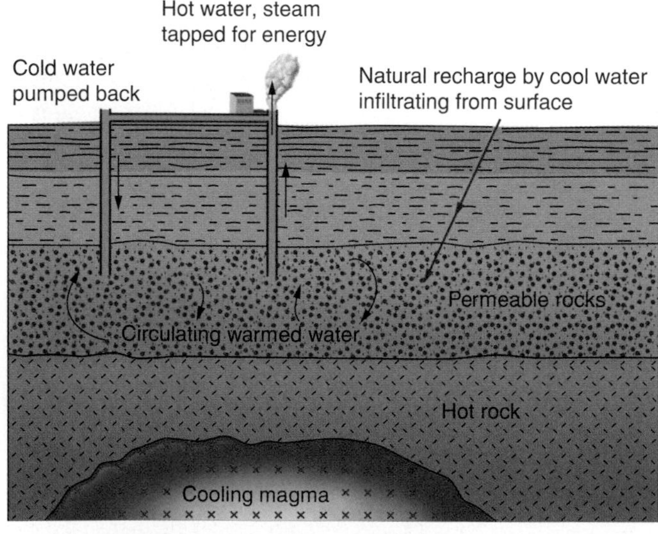

FIGURE 15.18 Geothermal energy is utilized by tapping circulating warmed ground water.

Conventional Geothermal Energy

Exactly how the geothermal energy is used depends largely on how hot the system is. In some places, the ground water is warmed, but not enough to turn to steam. Still, the water temperatures may be comparable to what a home heating unit produces (50 to 90°C, or about 120 to 180°F). Such warm waters can be circulated directly through homes to heat them. This is being done in Iceland and in parts of the former Soviet Union.

Other geothermal areas may be so hot that the water is turned to steam. The steam can be used, like any boiler-generated steam produced using conventional fuels, to run electric generators. The largest U.S. geothermal-electricity operation is The Geysers, in California (figure 15.21), which has operated since 1960 and now has a generating capacity of close to 2 billion watts. In 1989, The Geysers and six smaller geothermal areas together generated close to 10 billion kilowatt-hours of electricity (0.3% of total energy consumed) in this country, though its output has since declined, as described below. Other steam systems are being used in Larderello, Italy, and in Japan, Mexico, the Philippines, and elsewhere. Altogether, there are about forty sites worldwide where geothermal electricity is actively being developed.

Where most feasible, geothermal power is quite competitive economically with conventional methods of generating

FIGURE 15.19 One of the many thermal features in Yellowstone National Park: Lone Star Geyser. Structure is built by deposition of dissolved minerals.

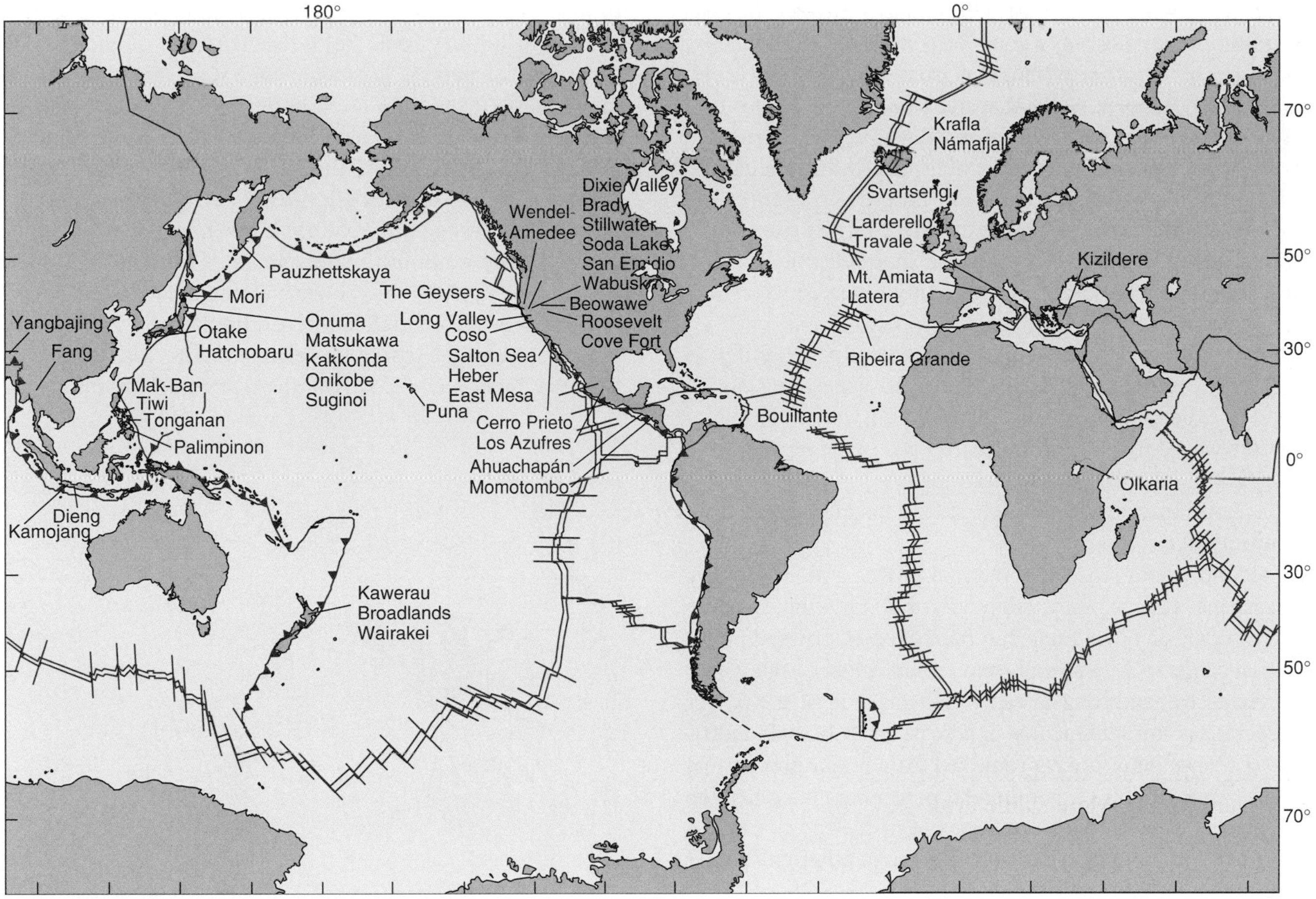

FIGURE 15.20 Geothermal power plants worldwide.
Source: Figure prepared by L. J. Patrick Muffler and Ellen Lougee, U.S. Geological Survey; plate boundaries supplied by Charles DeMets, University of Wisconsin at Madison.

FIGURE 15.21 The Geysers geothermal power complex, California, is the largest such facility in the world.
Photograph © The McGraw-Hill Companies, Inc./John A. Karachewski, photographer.

electricity. The use of geothermal steam is also largely pollution-free. Some sulfur gases derived from the magmatic heat source may be mixed with the steam, but these certainly pose no more serious a pollution problem than sulfur from coal burning. Moreover, there are no ash, radioactive-waste, or carbon-dioxide problems as with other fuels. Warm geothermal *waters* may be a somewhat larger problem. They frequently contain large quantities of dissolved chemicals (recall the chapter-opener photo for chapter 11) that can clog or corrode pipes (a potentially significant problem that will increase operational costs) or may pollute local ground or surface waters if allowed to run off freely. Sometimes, there are surface subsidence problems, as at Wairakei (New Zealand), where subsidence of up to 0.4 meters per year has been measured. Subsidence problems may be addressed by reinjecting water, but the long-term success of this strategy is unproven.

While the environmental difficulties associated with geothermal power are relatively small, three other limitations severely restrict its potential. First, each geothermal field can only be used for a period of time—a few decades, on average—before the rate of heat extraction is seriously reduced. This is because rocks conduct heat very poorly. That characteristic can easily be demonstrated by turning over a flat, sunbaked rock on a bright but cool day. The surface exposed to the sun may be hot to the touch, but the heat will not have traveled far into the rock; the underside stays cool for some time. Conversely, as hot water or steam is withdrawn from a geothermal field, it is replaced by cooler water that must be heated before use. Initially, the heating can be rapid, but in time, the permeable rocks become chilled to such an extent that water circulating through them heats too slowly or too little to be useful. The heat of the magma has not been exhausted, but its transmittal into the permeable rocks is slow. Some time must then elapse before the permeable rocks are sufficiently reheated to resume normal operations.

Steam pressure at The Geysers has declined rapidly in recent years, forcing the idling of some of its generating capacity: By 1991, despite a capacity of 2 billion watts, electricity production was only 1½ billion watts; power generation had declined to half its 1987 peak within a decade, and by 2001 was around 1.2 billion watts, with the first four installed generating units entirely idle.

A second limitation of geothermal power is that not only are geothermal power plants stationary, but so is the resource itself. Oil, coal, or other fuels can be moved to power-hungry population centers. Geothermal power plants must be put where the hot rocks are, and long-distance transmission of the power they generate is not technically practical, or, at best, is inefficient. Most large cities are far removed from major geothermal resources. Also, of course, geothermal power cannot contribute to such energy uses as transportation.

The total number of sites suitable for geothermal power generation is the third limitation. Clearly, plate boundaries cover only a small part of the earth's surface, and many of them are inaccessible (seafloor spreading ridges, for instance). Not all have abundant circulating subsurface water in the area, either. Even accessible regions that do have adequate subsurface water may not be exploited. Yellowstone National Park has the highest concentration of thermal features of any single geothermal area in the world, but because of its scenic value and uniqueness, the decision was made years ago not to build geothermal power plants there.

Alternative Geothermal Sources

Many areas away from plate boundaries have heat flow somewhat above the normal level, and rocks in which temperatures increase with depth more rapidly than in the average continental crust. The **geothermal gradient** is the rate of increase of temperature with increasing depth in the earth. Even in ordinary crust, the geothermal gradient is about 30° C/kilometer (about 85° F/mile). Where geothermal gradients are at least 40° C/kilometer, even in the absence of much subsurface water, the region can be regarded as a potential geothermal resource of the **hot-dry-rock** type. Deep drilling to reach usefully high temperatures must be combined with induced circulation of water pumped in from the surface (and perhaps artificial fracturing to increase permeability, as in some oil and gas fields) to make use of these hot rocks. The amount of heat extractable from hot-dry-rock fields is estimated at more than ten times that of natural hot-water and steam geothermal fields, just because the former are much more extensive. Most of the regions identified as possible hot-dry-rock geothermal fields in the United States are in thinly populated western states with restricted water supplies (figure 15.22). There is, therefore, a large degree of uncertainty about how much of an energy contribution they could ultimately make. Certainly, hot-dry-rock geothermal energy will be less economical than that of the hot-water or steam fields where circulating water is already present. Although some experimentation with hot-dry-rock fields is underway—Los Alamos National Laboratory has an

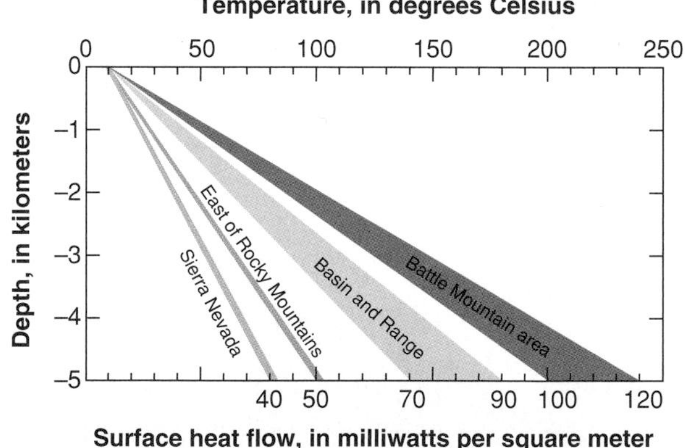

FIGURE 15.22 The area east of the Rocky Mountains has a geothermal gradient and surface heat flow typical of world average continental crust; selected areas west of the Rockies have more energy potential. The faster temperature increases with depth, the closer to the surface are usefully warm rocks, and—all else being equal—the greater the geothermal-energy potential.
After USGS Circular 1249.

experimental project in operation in the Jemez Mountains of New Mexico—commercial development in such areas is not expected in the immediate future.

The geopressurized natural gas zones described in chapter 14 represent a third possible type of geothermal resource. As hot, gas-filled fluids are extracted for the gas, the heat from the hot water might, in theory, also be used to generate electricity. Substantial technical problems probably will be associated with developing this resource, as was noted. Moreover, it probably would be economically infeasible to pump the spent water back to such depths for recycling. No geopressurized zones are presently being developed for geothermal power, even on an experimental basis.

Finally, at the scale of individual buildings, there is growing interest in a very different geothermal technology: geothermal heat pumps. Such a device takes advantage of the fact that rocks and soils conduct heat poorly, so seasonal surface temperature variations do not penetrate very far underground: Over much of the U.S., at a depth of 20 feet, temperatures fluctuate only a few degrees over the year, and average 50 to 60°F (10–15°C). Pipes laid in the deep soil can circulate water between this zone of fairly constant temperature and a heat exchanger at the surface, drawing on earth's heat to warm the building in winter when the air is colder, and/or carrying away heat from the building into the soil when surface air is hotter. No emissions are generated in the process. Supplementary heating or cooling may also be required, but the EPA reports that typically, homeowners save 30 to 70% on heating and 20 to 50% on cooling using such "geoexchange" systems, and over 1 million such systems are currently installed in the United States.

HYDROPOWER

The energy of falling or flowing water has been used for centuries. It is now used primarily to generate electricity. Hydroelectric power has consistently supplied a small percentage of U.S. energy needs for several decades; it currently provides about 3% of U.S. energy (about 6% of U.S. electricity). The principal requirements for the generation of substantial amounts of hydroelectric power are a large volume of water and the rapid movement of that water. Nowadays, commercial generation of hydropower typically involves damming up a high-discharge stream, impounding a large volume of water, and releasing it as desired, rather than operating subject to great seasonal variations in discharge. The requirement of plentiful surface water is reflected in large regional variations in hydropower generation (figure 15.23).

Hydropower is a very clean energy source in that the water is not polluted as it flows through the generating equipment. No chemicals are added to it, nor are any dissolved or airborne pollutants produced. The water itself is not consumed during power generation; it merely passes through the generating equipment. In fact, water use for hydropower in the United States is estimated to be more than 2½ times the average annual surface-water runoff of the nation, which is possible because the same water can pass through multiple hydropower dams along a stream. Hydropower is renewable as long as the streams continue to flow. Its economic competitiveness with other sources is demonstrated by the fact that nearly one-third of U.S. electricity-generating plants are hydropower plants; worldwide, about 7% of all energy consumed is hydropower.

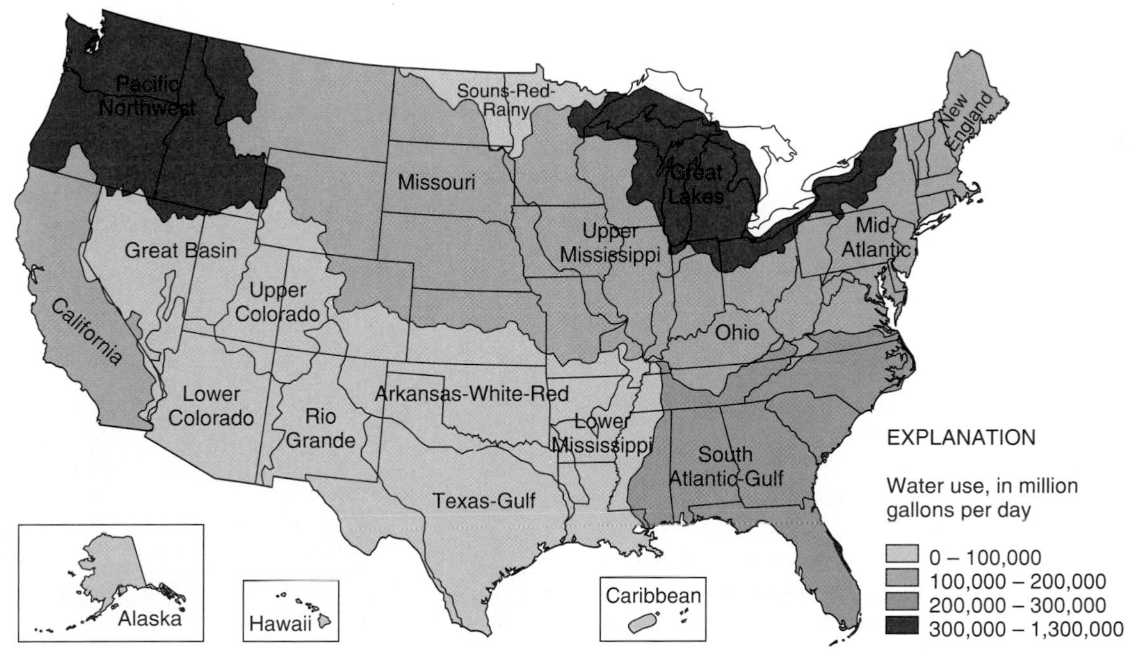

FIGURE 15.23 Hydropower generation in the United States is concentrated where streamflow is plentiful.
Source: 1995 Estimated Water Use in the United States, *U.S. Geological Survey.*

The Federal Power Commission has estimated that the potential energy to be derived from hydropower in the United States, if it were tapped in every possible location, is about triple current hydropower use. In principle, then, hydropower could supply one-fifth of U.S. electricity if consumption remained near present levels. However, development is unlikely to occur on such a scale.

Limitations on Hydropower Development

We have already considered, in chapters 6 and 11, some of the problems posed by dam construction, including silting-up of reservoirs, habitat destruction, water loss by evaporation, and even, sometimes, earthquakes. Evaluation of the risks of various energy sources must also consider the possibility of dam failure. There are over a thousand dams in the United States (not all constructed for hydropower generation). Several dozen have failed within the twentieth century. Aside from age and poor design or construction, the reasons for these failures may include geology itself. Fault zones often occur as topographic lows, and streams thus frequently flow along fault zones. It follows that a dam built across such a stream is built across a fault zone, which may be active or may be reactivated by filling the reservoir. Not all otherwise-suitable sites, then, are safe for hydropower dams.

Indeed, the late 1990s saw some reversals in U.S. hydropower development. For the first time, the Federal Energy Regulatory Commission ordered the dismantling of a dam, over the owner's objections: the Edwards Dam in Maine, the removal of which began in July of 1999. The move was a response to concerns over the cumulative environmental impact of 162 years of the dam's existence, and as it was a small facility (accounting for less than 1% of Maine's electric power), opposition to the dismantling was limited. More-vigorous objections are being made to the proposed removal of four dams on the Snake River in Washington state. Concern for local salmon populations is among the reasons for the removal, but as these dams and reservoirs are more extensively used both for power generation and to supply irrigation water, their destruction would have more negative impact on local people than did the Edwards Dam removal. As of this writing, these dams' ultimate fate is uncertain.

Other sites with considerable power potential may not be available or appropriate for development. Construction might destroy a unique wildlife habitat or threaten an endangered species. It might deface a scenic area or alter its natural character; suggestions for additional power dams along the Colorado River that would have involved backup of reservoir water into the Grand Canyon were met by vigorous protests. Many potential sites—in Alaska, for instance—are just too remote from population centers to be practical unless power transmission efficiency is improved.

An alternative to development of many new hydropower sites would be to add hydroelectric-generating facilities to dams already in place for flood control, recreational purposes, and so on. Although the release of impounded water for power generation alters streamflow patterns, it is likely to have far less negative impact than either the original dam construction or the creation of new dam/reservoir complexes. Still, it is clear that flood control and power generation are somewhat conflicting aims: the former requires considerable reserve storage capacity, while the latter is enhanced by impounding the maximum volume of water.

Like geothermal power, conventional hydropower is also limited by the stationary nature of the resource. In addition, hydropower is more susceptible to natural disruptions than other sources considered so far. Just as torrential precipitation and 100-year floods are rare, so, too, are prolonged droughts—but they do happen, as residents of the Midwest were reminded in 1988. In fact, U.S. hydropower generation declined by about 25% from 1986 to 1988, largely as a consequence of drought. A western drought that began in 1999 dropped water levels in Lake Powell, the reservoir behind Glen Canyon Dam (figure 15.24), by more than 100 feet, reducing water storage by more than 50% and hydropower generation at the dam from a 1997 high of 6.7 billion kilowatt-hours (kwh) to just 3.2 billion kwh in 2005. Heavier reliance on hydropower would thus leave many energy consumers vulnerable to interruption of service in times of extreme weather.

For various reasons, then, it is unlikely that numerous additional hydroelectric power plants will be developed in the United States. This clean, cheap, renewable energy source can continue indefinitely to make a modest contribution to energy use, but it cannot be expected to supply much more energy in the future than it does now. Still, hydropower is an important renewable energy source in the United States (figure 15.25).

Worldwide, future hydropower development is expected to be most rapid in Asia, but opposition exists there, too: The Three Gorges Dam project in China, slated for completion in 2009, continues despite international protests from environmental groups. Serious droughts in 1999 in Latin and South America sharply reduced generating capacity at existing hydroelectric

FIGURE 15.24 Glen Canyon Dam hydroelectric project, Arizona. In dry surroundings such as these, evaporation losses from reservoirs are high, and can exacerbate regional water-supply problems. Photo taken in 2005 shows drought-lowered level of Lake Powell; note the "bathtub ring" behind the dam that indicates the normal level of the reservoir when full. See also the chapter-opening photo.

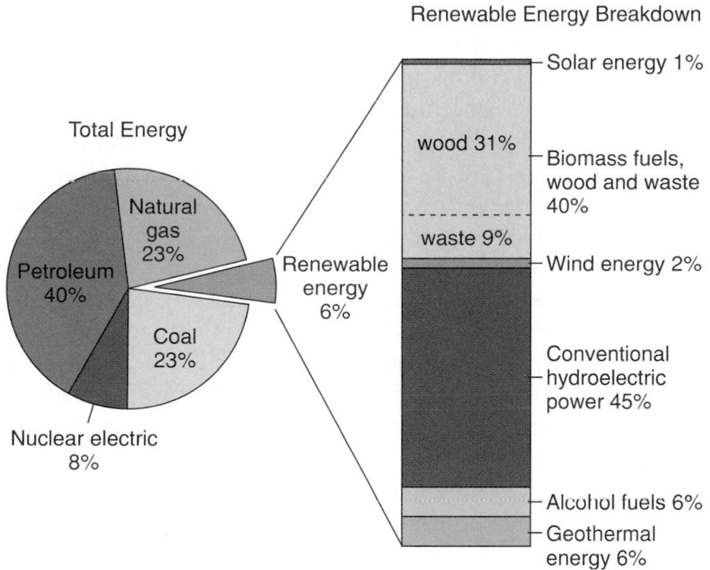

FIGURE 15.25 Hydropower is currently the dominant renewable energy source in the United States, though renewable sources still provide a relatively small proportion of energy used.

Source: Annual Energy Review 2005, *U.S. Energy Information Administration.*

facilities, forcing Mexico to buy electricity from the United States and causing Chile to look toward developing more fossil-fuel power-generating capacity as an alternative. (See further discussion of dams in chapter 20.)

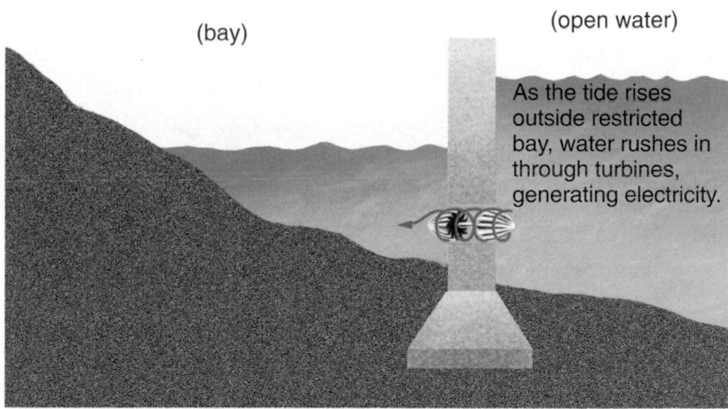

FIGURE 15.26 Tidal-power generation uses flowing water to generate electricity, as with conventional hydropower.

TIDAL POWER

All large bodies of standing water on the earth, including the oceans and large lakes like the Great Lakes, show tides. Why not also harness this moving water as an energy source? Unfortunately, the energy represented by tides is too dispersed in most places to be useful. Average beach tides reflect a difference between high-tide and low-tide water levels of about 1 meter. A commercial tidal-power electricity-generating plant requires at least 5 meters difference between high and low tides for efficient generation of electricity and a bay or inlet with a narrow opening that could be dammed to regulate the water flow in and out (figure 15.26). The proper conditions exist in very few places in the world. Tidal power is being used in a small way at several locations—at the Bay of Fundy in Nova Scotia, near St. Malo, France, in the Netherlands, and in the former Soviet Union. Worldwide, the total potential of tidal power is estimated at only 2% of the energy potential of conventional hydropower.

OCEAN THERMAL ENERGY CONVERSION

Ocean thermal energy conversion (OTEC) is another clean, renewable technology that is currently in the developmental stages. It exploits the temperature difference between warm surface water and the cold water at depth. Either the warm water is vaporized and used directly to run a turbine, or its heat is used to vaporize a "working fluid" to do so; the vapor is recondensed by chilling with the cold water. No fuel is burned, no emissions released, and the vast scale of the oceans assures a long life for such facilities (in contrast, for instance, to geothermal fields). Where the water is vaporized, the vapor is pure water, which, when recondensed, can be used for water supply—this is essentially a distillation process. The cold seawater can be used for other purposes, such as air conditioning or aquaculture.

However, specialized conditions are needed for OTEC to be a productive energy source. The deep cold water must be accessible near shore; coastlines with broad shelves are unsuitable. The temperature difference between warm and cold seawater must be at least 40°F (22°C) year-round, which is true only near the equator (figure 15.27). Thus, this technology is most suitable for select tropical islands. (An experimental OTEC facility exists on Hawaii, but has yet to produce electricity.) And there is concern about possible damage to reefs and coastal ecosystems from the pipes and other equipment. The near-term potential of OTEC is consequently very limited.

WIND ENERGY

Because the winds are ultimately powered by the sun, wind energy can be regarded as a variant of solar energy. It is clean and, like sunshine, renewable indefinitely (at least for 5 billion

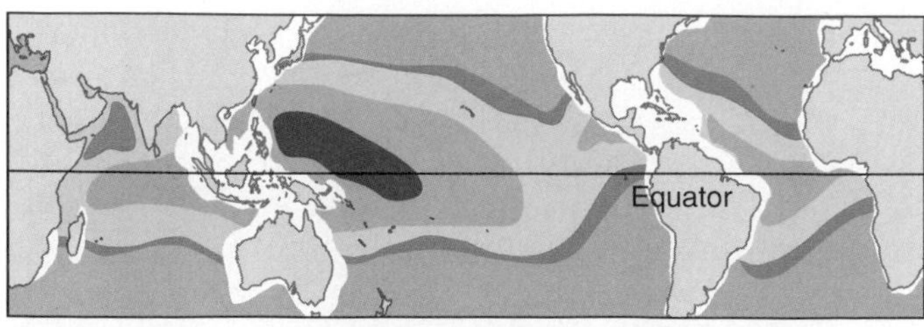

FIGURE 15.27 Given the thermal and other requirements of OTEC, tropical islands are likely to be the first sites for its development.
After National Renewable Energy Lab.

years or so). Wind power has been utilized to some extent for more than two thousand years; the windmills of the Netherlands are probably the best-known historic example. Today, there is considerable interest in making more extensive use of wind power. Windmills currently are most widely used for pumping ground water and for generating electricity for individual homes and farms.

In most places, the cost of commercial wind power on a large scale is not now competitive with conventionally generated electricity, but that may simply be because windmills are not now mass-produced and are, therefore, relatively expensive. Technological design improvements to increase efficiency are also possible.

Wind energy shares certain limitations with solar energy. It is dispersed, not only in two dimensions but in three: It is spread out through the atmosphere. Wind is also erratic, highly variable in speed both regionally and locally. The regional variations in potential power supply are even more significant than they may appear from average wind velocities because windmill power generation increases as the cube of wind speed. So if wind velocity doubles, power output increases by a factor of 8 ($2 \times 2 \times 2$).

Figure 15.28 shows that most of the windiest places in the United States are rather far removed physically from most of the heavily populated and heavily industrialized areas. As with other physically localized power sources, the technological difficulty of long-distance transmission of electricity will limit wind power's near-term contribution in areas of high electricity consumption until transmission efficiencies improve.

Even where average wind velocities are great, strong winds do not always blow. This presents the same storage problem as does solar electricity, and it has likewise not yet been solved

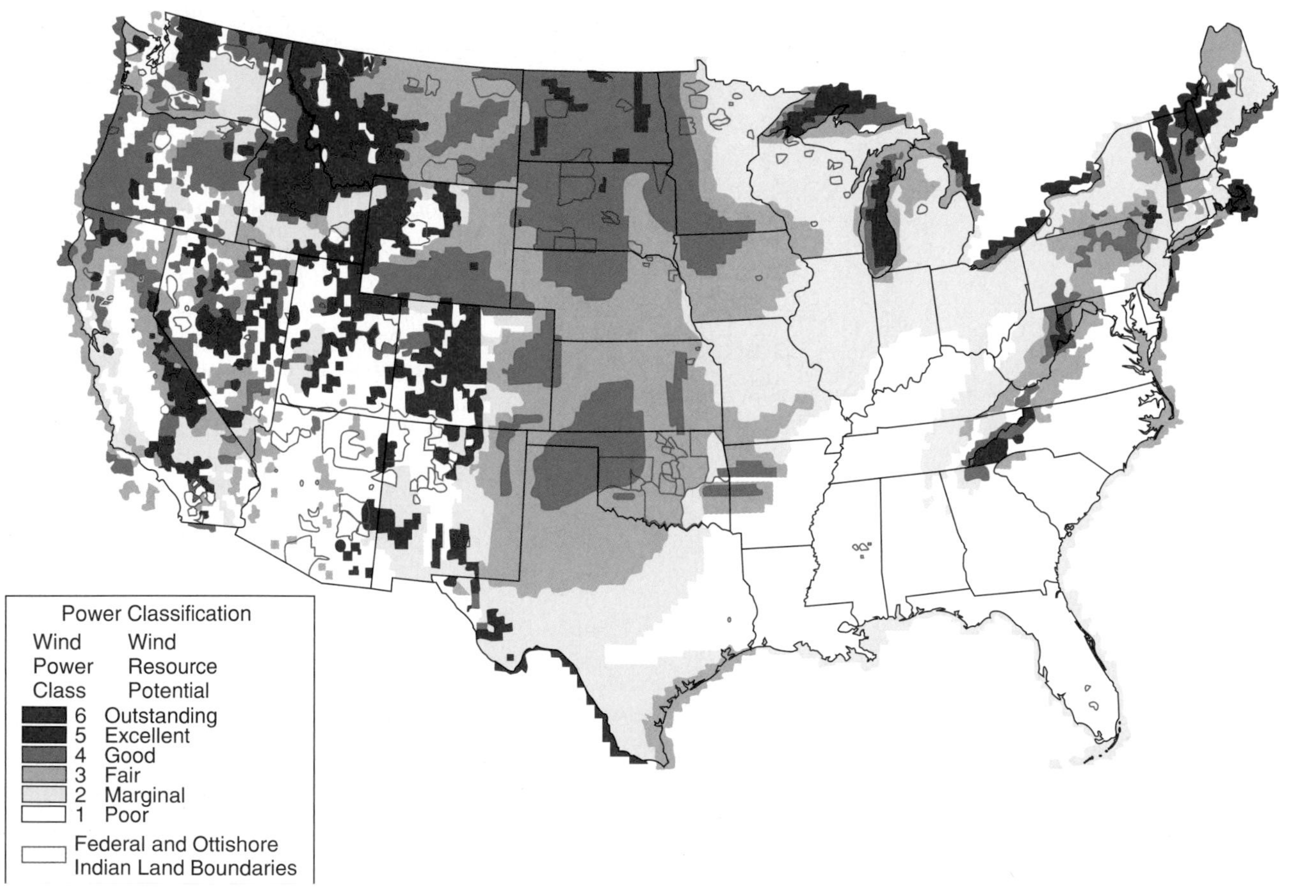

FIGURE 15.28 Wind resource potential in the 48 contiguous United States.
Source: National Renewable Energy Lab, U.S. Department of Energy.

satisfactorily. In the near future, wind-generated electricity might most effectively be used to supplement conventionally generated power when wind conditions are favorable, with conventional plants bearing most or all of the demand load at other times. Still, design improvements have increased the reliability of windmills; the newest wind turbines are available for power generation at least 95% of the time.

Winds blow both more strongly and more consistently at high altitudes. Tall windmills, perhaps a mile high, may provide more energy with less of a storage problem. However, such tall windmills are technically more difficult to build and require substantial quantities of material.

The ultimate potential of wind energy is unclear. Certainly, blowing wind represents far more total energy than we can use, but most of it cannot be harnessed. Most schemes for commercial wind-power generation of electric power involve "wind farms," concentrations of many wind generators in a few especially favorable windy sites (figure 15.29). The limits to wind-power use would then include the area that can be committed to wind-generator arrays, as well as the distance the electricity can be transmitted without excessive loss in the power grid. About 1000 1-megawatt windmills are required to generate as much power as a sizable conventional coal- or nuclear-powered electric generating plant. The windmills have to be spread out, or they block each other's wind flow. Spacing them at four windmills per square kilometer requires about 250 square kilometers (100 square miles) of land to produce energy equivalent to a 1000-megawatt power plant. However, the land need not be devoted exclusively to the wind-powered generators. Farming and the grazing of livestock, common activities in the Great Plains area, could continue on the same land concurrently with power generation. The storage problem, however, would still remain. Other concerns relating to wind farms include the aesthetic impact (the windmill arrays must naturally be very exposed); interference with and death to migrating birds, which already occurs on existing wind farms; the noise associated with a large number of windmills operating; and disruption of communications—TV, radio, cellular telephones, perhaps even aircraft communications.

At one time, it was projected that the United States might produce 25 to 50% of its electricity from wind power by the year 2000, but the necessary major national program for wind-energy development did not materialize. This percentage would represent a significant fraction of anticipated total energy needs, and it also would contribute to conserving nonrenewable fuels. However, no governmental program of any such scale is presently active or contemplated. In fact, in 2005, wind-powered electric generating facilities in the United States accounted for only about 1% of U.S. generating capacity, and of actual electricity generation, despite a sharp recent rise in capacity (figure 15.30). At that, the United States is a world leader in generation of electricity from wind power, although plans for wind-power development in Europe are more ambitious than those of the United States over the next decade.

BIOMASS

The term **biomass** technically refers to the total mass of all the organisms living on earth. In an energy context, the term has become a catchall for various ways of deriving energy from organisms or from their remains. Indirectly, biomass-derived energy is ultimately solar energy, since most biomass energy sources are plant materials, and plants need sunlight to grow. Biomass fuels could also be thought of as "unfossilized fuels" because they represent fuels derived from living or recent organisms rather than ancient ones and hence have not been modified extensively by geologic processes acting over long periods of time.

The possibilities for biomass-derived energy are many. These fuels fall into three broad categories: wood, waste, and alcohol fuels. Some are used alone, others "cofired" (burned in combination with conventional fuels; see figure 15.31. All biomass fuels are burned to release their energy, so they share the carbon-dioxide-pollution problems of fossil fuels. Some, like wood, also contribute particulate air pollutants. However, unlike the fossil fuels, biomass fuels are renewable. We can, in principle, produce and replenish them on the same time scale as

FIGURE 15.29 Wind-turbine array near Palm Springs, California. Different elements of the array can take advantage of various velocities of wind. Some of the smaller turbines turn even in light wind, while the larger turbines require stronger winds to drive them.

© The McGraw-Hill Companies, Inc./Doug Sherman, photographer.

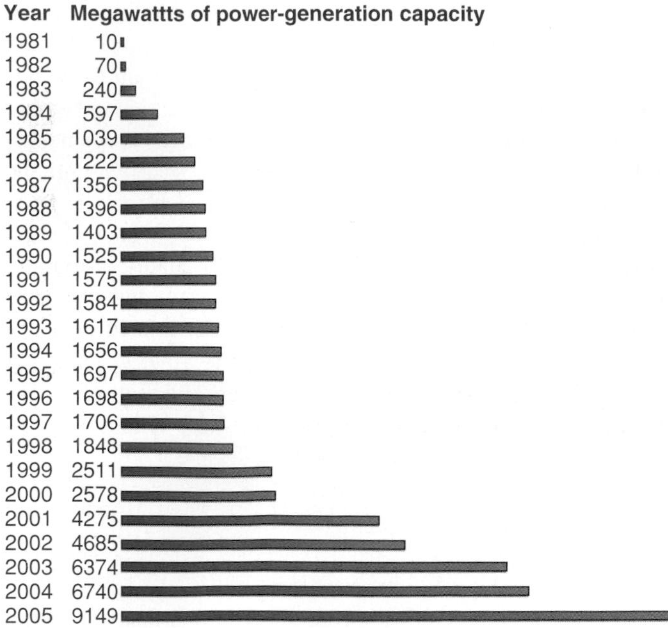

FIGURE 15.30 U.S. wind power installed capacity, 1981–2005.
Sources: U.S. Department of Energy Wind Energy Program, National Renewable Energy Lab, and American Wind Energy Association.

that on which we consume them. This is their appeal, together with the fact that we can produce them domestically, thus reducing dependence on energy imports.

Waste-Derived Fuels

In some sense, much of the wood burned as biomass fuel is also waste, especially from logging, lumber-milling, and similar operations. However, most of the truly waste-related biomass fuels involve agricultural or other wastes that would historically have been burned in a field or dumped in a landfill. For example, there is increasing interest in burning waste plant materials after a crop is harvested. Some such crop wastes may require processing beforehand; sugar-cane waste is an example. The combustible portion of urban refuse can be burned to provide heat for electric generating plants. More-extensive use of incineration as a means of solid-waste disposal, both in the United States and abroad, would further increase the use of biomass for energy worldwide.

Some waste-derived fuels are liquids. Research is ongoing on ways to derive inoffensive liquid fuels from animal manures, which are rich in organic matter. Experimental vehicles have been designed to run on used vegetable oil from food-frying operations. Some diesel-powered vehicles can run on *biodiesel,* a general term for fuels derived from vegetable oil or animal fats (including wastes), or a blend of such oil with petroleum diesel fuel. The alcohol fuels discussed below are now mainly produced from corn, but processes to derive alcohol from plant wastes rather than potential food are being developed.

Another waste-derived biomass fuel growing in use, *biogas,* could be called "gas from garbage." When broken down in the absence of oxygen, organic wastes yield a variety of gaseous products. Some of these are useless, or smelly, or toxic, but among them is methane (CH_4), the same compound that predominates in natural gas. Sanitary-landfill operations, described in chapter 16, are suitable sites for methane production as organic wastes in the refuse decay. Straight landfill gas is too full of impurities to use alone, but landfill-derived gas can be blended with purer, pipelined natural gas (another example of cofiring) to extend the gas supply. Methane can also be produced from decaying manures. Where large quantities of animal waste are available, as for example on feedlots, it may also be possible to create biogas by microbial digestion

A

B

FIGURE 15.31 (A) Truck dumping wood chips to be burned at the Tracy Biomass Plant, and electricity-generating plant in Tracy, CA. (B) Biomass is cofired with coal at the Northern Indiana Public Service Company generating station in Bailey, IN; the biomass contributes about 5% of the energy.
(A) Photograph by Andrew Carlin, Tracy Operations, (B) photograph by Kevin Craig; both courtesy National Renewable Energy Lab.

Box 15.2 Electricity's Hidden Energy Costs

Sheila breezed in the door, dropped her books on the kitchen table, and glanced at the mail. "Hey, cool!" she exclaimed, picking up a piece of paper.

"Since when is an electric bill 'cool'?" demanded her brother Frank, rolling his eyes.

"No, but look—they're telling us how all the power was generated, and the pollution it created!" Sheila waved the sheet under Frank's nose. (See figure 1.)

"Wow. Most of the power's nuclear, but hardly any of the waste is."

Such considerations are very much part of current discussions about energy sources. At one time, the main concern was availability of the fossil fuels. Now global focus has shifted to emission of greenhouse gases like CO_2, and the relative pollutant outputs of different energy sources are a significant part of the debate. With nuclear power, however, the issue of waste disposal is also a concern.

A contributing factor in rising U.S. energy use (and associated pollution) is what has been described as the "electrification" of U.S. households, as we acquire more and more appliances that run on electricity. By 2001, 99% of U.S. households had a color television, and two-thirds had two or more, while 90% had a VCR or DVD player, and 34% a large-screen TV; 96% had an oven, the majority electric; and 50% had central air conditioning. From 1978 to 2001, the households having a microwave oven went from 8 to 86%; from 1990 to 2001 the percentage with personal computers increased from 16 to 60, and this percentage continues to rise.

A hidden issue with electricity is efficiency, or lack of it. Only *one-third* of the energy consumed in generating electricity is delivered to the end user as power, so as electricity consumption grows, the corresponding energy consumption grows three times as fast (figure 2). The other two-thirds is lost, most as waste heat during the process of conversion from the energy of the power sources (for example, the chemical energy of fossil fuels) to electricity, and the rest in transmission and distribution. This would be something to keep in mind in choosing among appliances, heating systems, and so on, powered by different forms of energy, such as electricity versus natural gas.

The inefficiency of the ways we now generate our electricity also has a bearing, indirectly, on the utility of hydrogen fuel cells. In a hydrogen fuel cell, which operates somewhat like a battery, hydrogen and oxygen are combined to generate electricity, typically to power a vehicle, and this process is actually quite efficient. The only "waste" product is water, so fuel-cell-powered cars are very clean, and the technology already exists. So far, so good. But if we are to run millions of vehicles on fuel cells, where will the hydrogen come from? The obvious answer is water, of which we have a plentiful global supply. However, the usual method for separating water into hydrogen and oxygen rapidly and in quantity uses electricity. Thus, before we can embrace fuel cells on a large scale, we need either to figure out how to generate still *more* electricity cleanly and sustainably, or to develop alternate ways to produce the necessary large quantities of hydrogen.

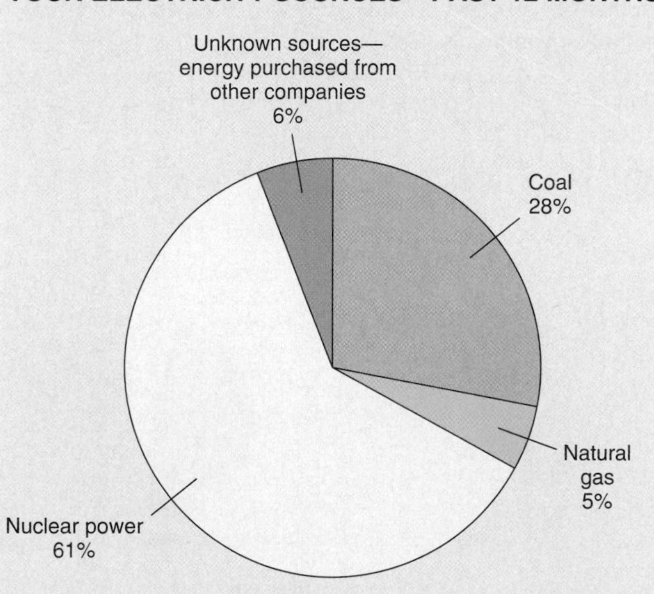

AVERAGE EMISSIONS and NUCLEAR WASTE per 1000 kilowatt-hours (kWhs) PRODUCED from known sources for the past 12 months	
Carbon dioxide	659.84 lb.
Nitrogen oxides	1.69 lb.
Sulfur dioxide	2.97 lb.
High-level nuclear waste	0.0037 lb.
Low-level nuclear waste	0.0003 cu. ft.

FIGURE 1 Typical profile of municipal electricity supply, a mix of sources with varying waste and pollution consequences. Note that amounts of coal ash generated are not tabulated; it is not a waste product monitored and regulated in the same way as the gaseous air pollutants or radioactive waste.

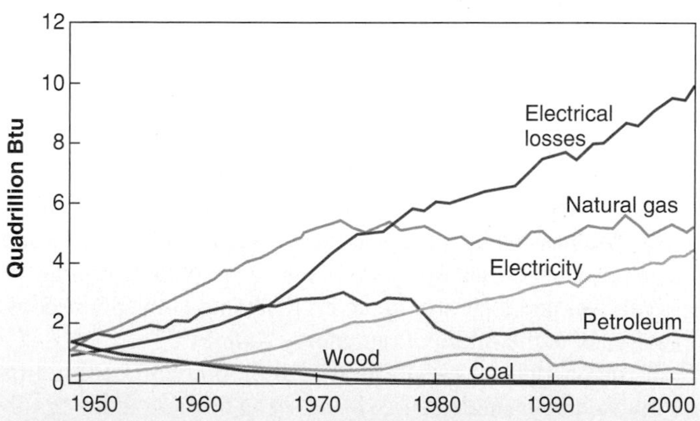

FIGURE 2 Residential and commercial energy consumption in the United States is increasingly in the form of electricity, with correspondingly larger losses. This diagram is for residential energy consumption.
Source: Annual Energy Review 2002, *U.S. Energy Information Administration.*

in enclosed tanks, again reducing waste-disposal and pollution problems while generating useful fuel.

Alcohol Fuels

One biomass fuel that has received special attention is alcohol. Initially, it was extensively developed mainly for incorporation into *gasohol.* Gasohol as originally created was a blend of 90% gasoline and 10% alcohol, the proportions reflecting a mix on which conventional gasoline engines could run. Gasohol came into vogue following the OPEC oil embargo of the 1970s. Its popularity waned when oil prices subsequently declined, though most gasoline still contains some ethanol (the most common type of fuel alcohol).

The higher the proportion of alcohol in the mix, of course, the further the gasoline can be stretched. There is, in principle, no reason why engines cannot be designed to run on 20%, 50%, or 100% alcohol, and, in fact, vehicles have been made to run solely on alcohol. Recently, increasing numbers of vehicles have been designed to run on *E85,* a blend of 85% ethanol, 15% gasoline. Such vehicles, most common in the Midwest and South, are now the majority of alcohol-fueled vehicles. Some can run either on E85 or on regular gasoline, though they must be specifically designed to do so.

One important consideration is that to grow, harvest, and derive ethanol from corn or other starting plant material takes energy. Depending on the process, indeed, it can take *more* energy than is released burning the ethanol. Thus the alcohol fuels as currently developed are not so much a major new source of energy as a way to reduce dependence on imported liquid petroleum—if the energy to make the ethanol comes from something else.

Government policy can encourage expanded use of alcohol and alcohol/gasoline blends. The Clean Air Act Amendments of 1990 included provisions for reduced vehicle emissions, and alcohol is somewhat cleaner-burning. The Energy Policy Act of 1992 required gradual replacement of some government, utility-company, and other "fleet" vehicles with "alternative-fuel" vehicles, which would include vehicles powered by straight alcohol. (This is more immediately feasible for fleet vehicles because they can be refueled at central locations stocking their special fuels.) The same act extends an excise-tax break on ethanol-blend fuels. Large gains in ethanol consumption remain to be realized, however. In 2005, only about 548,000 vehicles in the United States were designed to run on alternative fuels, and of these, only 27% used an alcohol-based fuel (most commonly, E85).

SUMMARY

As we contemplate the limits on supplies of petroleum and conventional natural gas and the environmental impacts of coal, we find an almost bewildering variety of alternatives available. None is as versatile as liquid and gaseous petroleum fuels, or as immediately and abundantly available as coal. Some of the already-viable, clean alternatives are "placebound" and, in any case, have limited ultimate potential (e.g., hydropower, geothermal power). Nuclear fission produces minimal emissions but entails waste-disposal problems, and concerns about reactor safety; the fuel reprocessing necessary if fission-power use is greatly expanded raises security concerns. Solar and wind energy, though free and clean to use, are so diffuse, and so variable over time and space, that without substantial technological advances they are impractical for energy-intensive applications. Biomass fuels, like fossil fuels, yield carbon dioxide (and perhaps other pollutants also). This incomplete sampling of alternatives and their pros and cons illustrates some of the complexity about the future of our energy sources.

The energy-use picture in the future probably will not be dominated by a single source as has been the case in the fossil-fuel era. Rather, a blend of sources with different strengths, weaknesses, and specialized applications is likely. Different nations, too, will make different choices, for reasons of geology or geography, economics, or differing environmental priorities (figure 15.32).

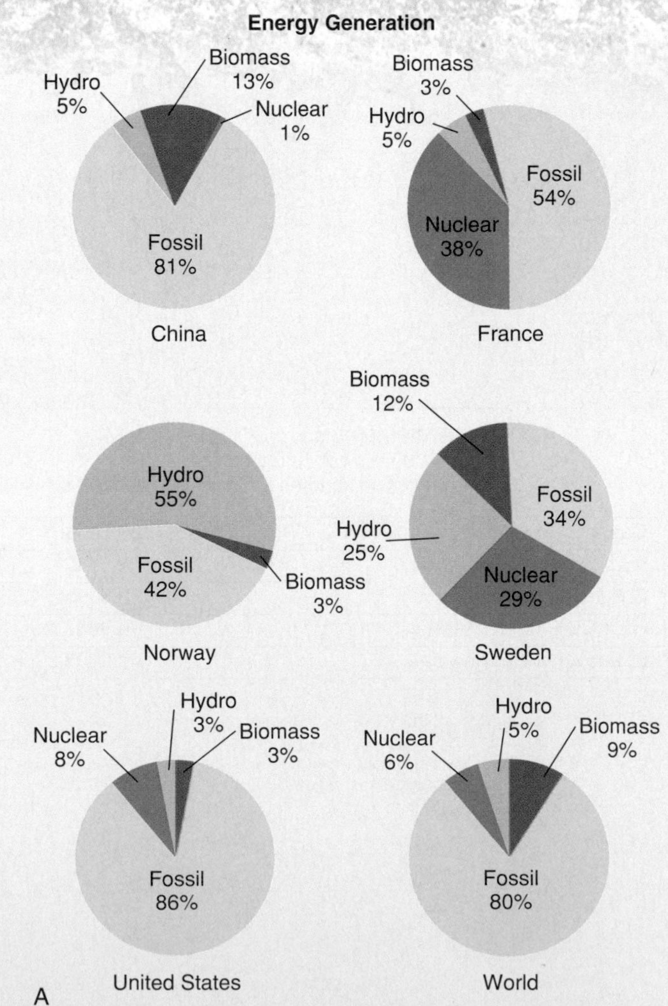

FIGURE 15.32 (A) Shares of fossil and non-fossil fuels in energy consumption for selected countries, 2004. (B) In part, national choices in renewable energy are dictated by geography and geology.

(A) Source: International Energy Agency and World Resources Institute. (B) Map courtesy United Nations Environment Programme/GRIDA.

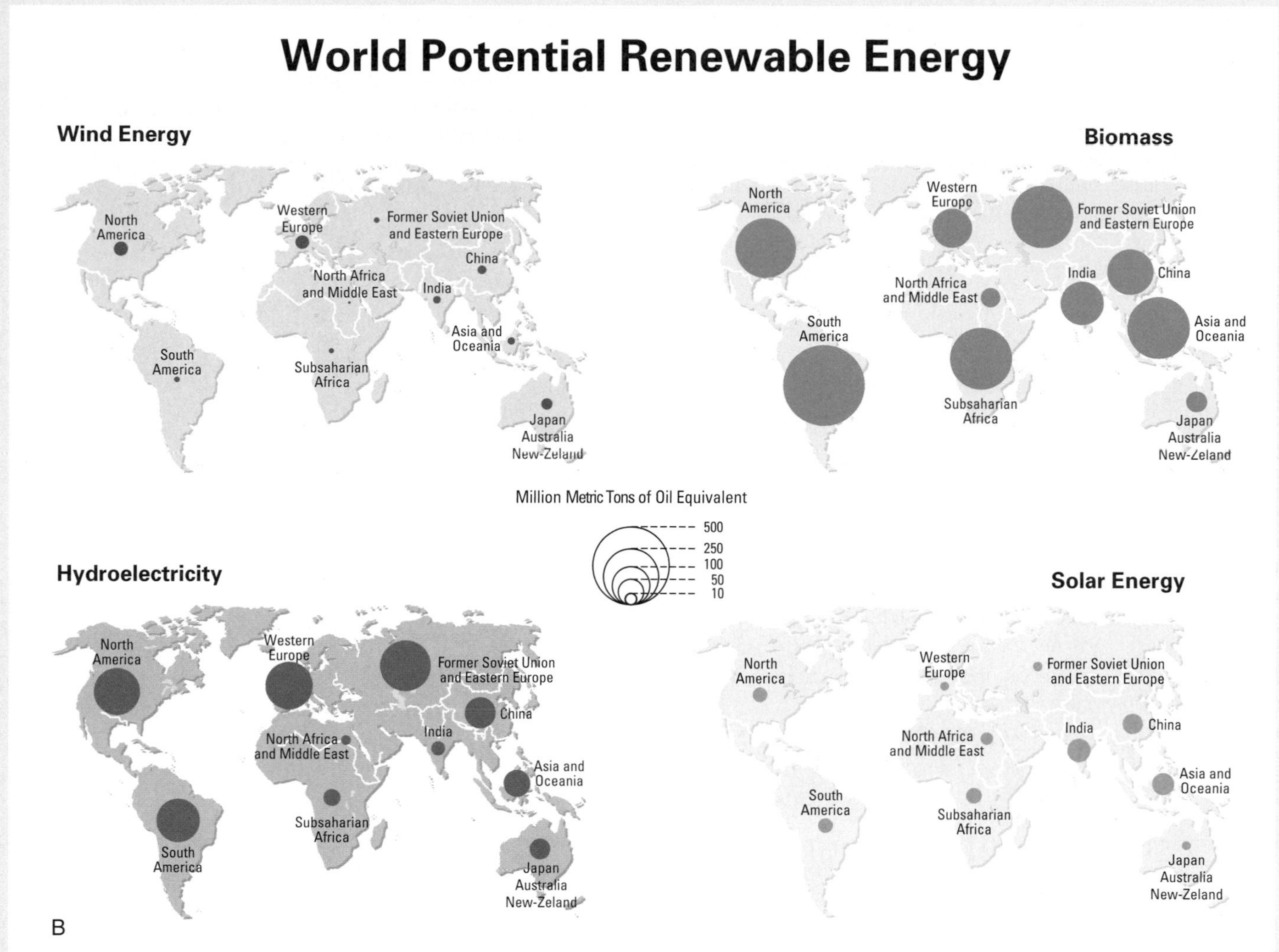

FIGURE 15.32 *Continued*

In the meantime, as the alternatives are explored and developed, vigorous efforts to conserve energy would yield much-needed time to make a smooth transition from present to future energy sources. However, considerable pressures toward increasing energy consumption exist in countries at all levels of technological development.

Key Terms and Concepts

biomass 365
breeder reactor 347
chain reaction 346
core meltdown 348
decommissioning 349
fission 346
fusion 346
geothermal energy 358
geothermal gradient 360
hot-dry-rock 360
ocean thermal energy conversion 363
photovoltaic cells 355

Exercises

Questions for Review

1. Briefly describe the nature of the fission chain reaction used to generate power in commercial nuclear power plants. How is the energy released utilized?
2. If the nuclear power option is pursued, breeder reactors and fuel reprocessing will be necessary. Why? What additional safety and security concerns are then involved?
3. What is "decommissioning" in a nuclear-power context?
4. Describe the fusion process, and evaluate its advantages and present limitations.
5. In what areas might solar energy potentially make the greatest contributions toward our energy needs? Explain.
6. What technological limitations do solar and wind energy presently share?
7. Explain the nature of geothermal energy and how it is extracted.
8. What factors restrict the use of geothermal energy in time and in space? How do hot-dry-rock geothermal areas expand its potential?
9. Assess the potential of (a) conventional hydropower and (b) tidal power to help solve impending energy shortages.
10. What is the basis of ocean thermal energy conversion; in what areas is it most viable and why?
11. What are biomass fuels? Describe two examples.
12. Choose any two energy sources from this chapter and compare/contrast them in terms of the negative environmental impacts associated with each.

For Further Thought

1. While most alternative energy sources must be developed on a large scale, solar energy (at least for space heating) can be installed building by building. For your region, explore the feasibility and costs of conversion to solar heating.
2. Investigate the history of a completed commercial nuclear-power plant project. How long a history does the project have? What were the cost projections of the plant and the energy consumption projections on which those costs were justified? Have the projections held up? Did the plant suffer regulatory or other construction delays? How long has it been operating, and when is it projected for decommissioning?
3. Consider having a home energy audit (often available free or at nominal cost through a local utility company) to identify ways to conserve energy.
4. Choose any nonfossil fuel and investigate its development over recent decades, in the United States, in another nation, or globally. How significant a factor is it in worldwide energy production currently? How do its economics compare with fossil fuels? What are its future prospects? (The Internet may provide particularly current data.)

Suggested Readings/References

American Association for the Advancement of Science. 2004. Not so simple. (Special section with emphasis on climate, emissions, and hydrogen.) *Science* 305 (13 August): 957–76.

Ashley, S. 2005. On the road to fuel-cell cars. *Scientific American* (March): 62–69.

Bezdek, R. H., and Wendling, R.M. 2005. Fuel efficiency and the economy. *American Scientist* 93 (March–April): 132–39.

Duffield, W. A., and Sass, J. H. 2003. *Geothermal Energy—Clean power from the earth's heat.* U.S. Geological Survey Circular 1249.

Gellings, C. W., and Yeager, K.E. 2004. Transforming the electric infrastructure. *Physics Today* (December): 45–51.

Golay, M. W. 1990. Longer life for nuclear plants. *Technology Review* (May/June): 25–30.

Kazimi, M. S. 2003. Thorium fuel for nuclear energy. *American Scientist* 91 (September–October): 408–15.

Loewen, E. P. 2004. Heavy-metal nuclear power. *American Scientist* 92 (November–December): 522–31.

Massachusetts Institute of Technology. 2002. Energy. *Technology Review* 105 (January–February). (Whole special issue on energy.)

Moreira, N. 2005. Growing expectations. *Science News* 168 (1 October): 218–20.

Parfit, M. 2005. Future power. *National Geographic* (August): 2–31.

Petit, C. 2006. It's scary. It's expensive. It could save the earth. *National Geographic* (April): 54–63.

Petroski, H. 2003. Fuel cells. *American Scientist* 91 (September–October): 398–402.

Pool, R. 1990. Cold fusion: Only the grin remains. *Science* 250 (9 November): 754–55.

Scientific American, Inc. 2006. Energy's future: Beyond carbon. (Special issue devoted to energy-related material.) *Scientific American* (September).

Stone, R. 2006. The long shadow of Chernobyl. *National Geographic* (April): 32–53.

Sweet, W., ed. 1997. A nuclear reconnaissance. *IEEE Spectrum* (November) 23–48. (A series of shorter articles.)

Sweet, W., and Bretz, E. A., eds. 1999. Toward carbon-free energy. *IEEE Spectrum* (November): 28–56. (A series of short articles, focusing in part on growing third-world energy demands.)

Tenenbaum, D. 1995. Tapping the fire down below. *Technology Review* (January): 38–47. (Pertains to geothermal power generation, at Los Alamos and in general.)

U.S. Department of Energy, Energy Information Administration. 1983. *Commercial nuclear power: Prospects for the U.S. and the world.* Washington, D.C.: U.S. Government Printing Office.

Wald, M. L. 2003. Dismantling nuclear reactors. *Scientific American* (March): 60–69.

Walker, J. S., 2004. *Three Mile Island: A nuclear crisis in perspective.* Berkeley, CA: Univ. of Calif. Press.

World Resources Institute. 2003. *World Resources 2001–02.* New York: Oxford University Press.

Worldwatch Institute. 2001. *State of the world 2001.* New York: W. W. Norton and Company. See especially Ch. 5, "Decarbonizing the energy economy," and Ch. 6, "Making better transportation choices."

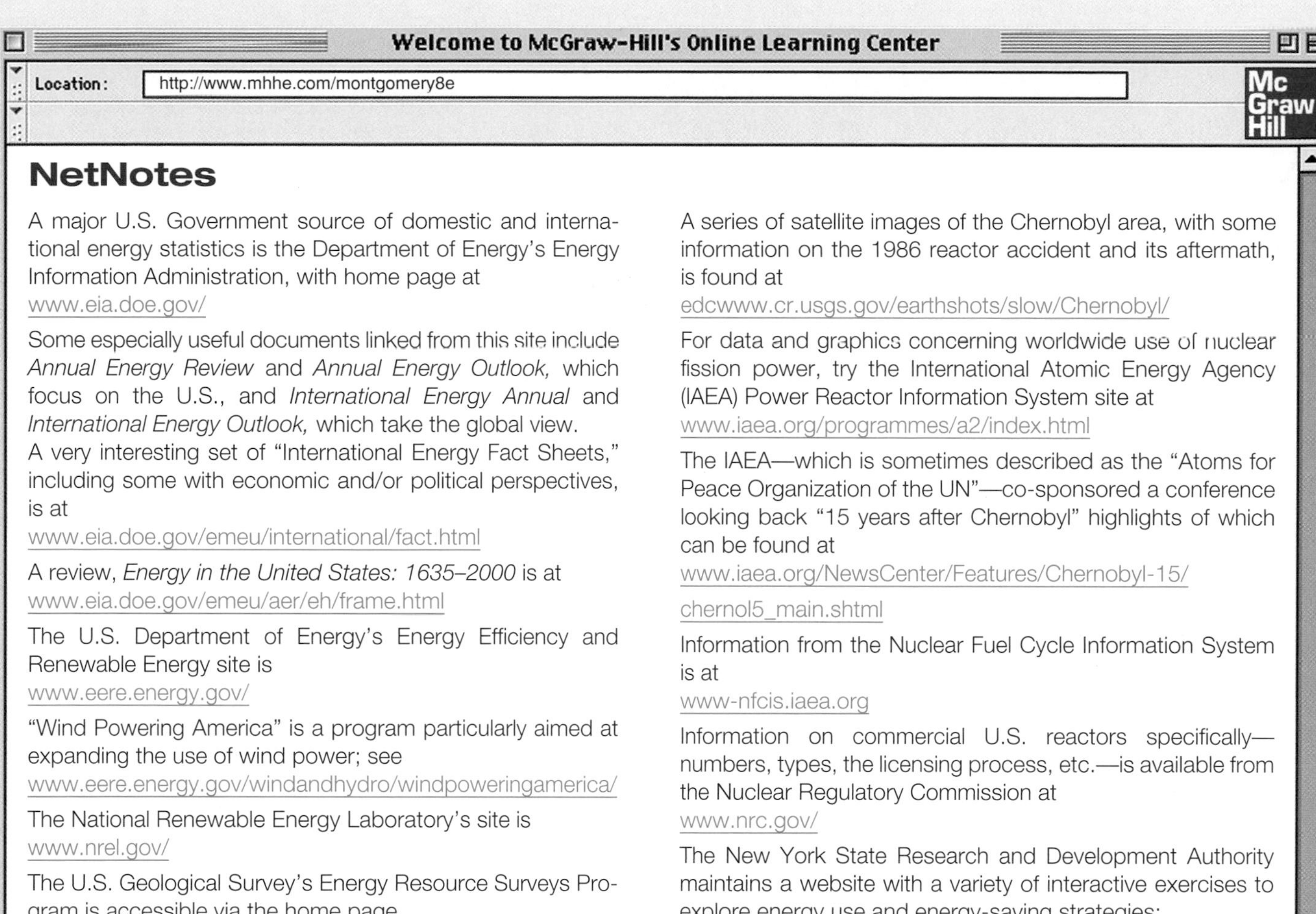

NetNotes

A major U.S. Government source of domestic and international energy statistics is the Department of Energy's Energy Information Administration, with home page at
www.eia.doe.gov/

Some especially useful documents linked from this site include *Annual Energy Review* and *Annual Energy Outlook,* which focus on the U.S., and *International Energy Annual* and *International Energy Outlook,* which take the global view.

A very interesting set of "International Energy Fact Sheets," including some with economic and/or political perspectives, is at
www.eia.doe.gov/emeu/international/fact.html

A review, *Energy in the United States: 1635–2000* is at
www.eia.doe.gov/emeu/aer/eh/frame.html

The U.S. Department of Energy's Energy Efficiency and Renewable Energy site is
www.eere.energy.gov/

"Wind Powering America" is a program particularly aimed at expanding the use of wind power; see
www.eere.energy.gov/windandhydro/windpoweringamerica/

The National Renewable Energy Laboratory's site is
www.nrel.gov/

The U.S. Geological Survey's Energy Resource Surveys Program is accessible via the home page
energy.usgs.gov/

A series of satellite images of the Chernobyl area, with some information on the 1986 reactor accident and its aftermath, is found at
edcwww.cr.usgs.gov/earthshots/slow/Chernobyl/

For data and graphics concerning worldwide use of nuclear fission power, try the International Atomic Energy Agency (IAEA) Power Reactor Information System site at
www.iaea.org/programmes/a2/index.html

The IAEA—which is sometimes described as the "Atoms for Peace Organization of the UN"—co-sponsored a conference looking back "15 years after Chernobyl" highlights of which can be found at
www.iaea.org/NewsCenter/Features/Chernobyl-15/
chernol5_main.shtml

Information from the Nuclear Fuel Cycle Information System is at
www-nfcis.iaea.org

Information on commercial U.S. reactors specifically—numbers, types, the licensing process, etc.—is available from the Nuclear Regulatory Commission at
www.nrc.gov/

The New York State Research and Development Authority maintains a website with a variety of interactive exercises to explore energy use and energy-saving strategies:
www.getenergysmart.org

SECTION FIVE

WASTE DISPOSAL, POLLUTION, AND HEALTH

Waste disposal and pollution are intimately related. Both concern the presence and handling of certain chemicals in our physical environment. There is, of course, nothing ominous about chemicals *per se,* despite the fact that the word prompts sinister associations in many people's minds. Water is a chemical; rocks are collections of chemicals; indeed, the human body is a very complex mass of chemicals. A *pollutant* sometimes is defined simply as a "chemical out of place," a substance found in sufficient concentration in some setting that it creates a nuisance or a hazard. Recently the concept of pollution has been broadened to include thermal pollution, the input of excess heat into a system (usually water). What might be called *biological pollution,* such as excessive growth of harmful microorganisms or algae, is often a further result of chemical or thermal pollution.

Some chemical deterioration of the environment is natural and occurs with or without human activities; natural oil seeps are just one example. Much of it, however, is the result of careless waste disposal or of the lack of deliberate waste-disposal efforts. A basic principle to keep in mind in dealing with these subjects, credited to biologist Dr. Barry Commoner, may be expressed as, "Nothing ever goes away." Sulfur gases and ash from a coal-fired power plant can be trapped by smokestack scrubbers rather than released into the air, but they must still be *put* somewhere afterward. A factory may not release toxic chemicals into a lake, but the chemicals have to go *someplace,* or else be thoroughly destroyed (and the products then disposed of). Out of sight may be out of mind, but not out of environmental circulation. In a real sense, as many pollution problems are reduced, waste-disposal problems are increased.

Much of the concern about pollution, in turn, reflects concern about possible negative effects on health, whether of humans or other organisms. This is an aspect of the broad field of *geomedicine,* the study of the relationships between geology and health. Centuries ago, scientists began to observe geographic patterns in the occurrence of certain diseases. For example, until the twentieth century, the northern half of the United States was known as the "goiter belt" for the high frequency of the iodine-deficiency disease goiter. Subsequent analysis revealed that the soils in that area were relatively low in iodine; consequently, the crops grown and the animals grazing on those lands constituted an iodine-deficient diet

Low-hanging smog over San Gabriel, CA, is both unsightly and unhealthy.
Photograph by Gene Daniels, courtesy U.S. Environmental Protection Agency. Photo by National Archives

for people living there. When their diets were supplemented with iodine, the occurrence of goiter decreased.

Such observations provided early evidence of the importance to health of *trace elements,* those elements that occur in very low concentrations in a system (rock, soil, plant, animal, etc.). Trace-elements' concentrations are typically a few hundred ppm or less. Their effects may be beneficial or toxic, depending on the element and the organism. To complicate matters further, the same element may be both beneficial and harmful, depending upon the dosage; fluorine is an example. Teeth are composed of a calcium phosphate mineral—apatite. Incorporation of some fluoride into its crystal structure makes the apatite harder, more resistant to decay. This, too, was recognized initially on the basis of regional differences in the occurrence of tooth decay. Persons whose drinking water contains fluoride concentrations even as low as 1 ppm show a reduction in tooth-decay incidence. This corresponds to a fluoride intake of 1 milligram per liter of water consumed; a milligram is one thousandth of a gram, and a gram is about the mass of one raisin. Fluoride may also be important in reducing the incidence of osteoporosis, a degenerative loss of bone mass commonly associated with old age. Normal total daily fluoride intake from food and water ranges from 0.3 to 5 milligrams.

Fluoride is not an unmixed blessing. Where natural water supplies are unusually high in fluoride, so that two to eight times the normal dose of fluoride is consumed, teeth may become mottled with dark spots. The teeth are still quite decay-resistant, however; the spots are only a cosmetic problem. Of more concern is the effect of very high fluoride consumption (twenty to forty times the normal dose), which may trigger abnormal, excess bone development (bone sclerosis) and calcification of ligaments. A *dose-response curve* is a graph illustrating the relative benefit or harm of a trace element or other substance as a function of the dosage. Fluoride in humans would be characterized by the dose-response curve shown in figure W.1.

The prospect of reduced tooth decay is the principal motive behind fluoridation of public water supplies where natural fluoride concentrations are low. Because extremely high doses of fluoride might be toxic to humans (fluorine is an ingredient of some rat poisons), there has been some localized opposition to fluoridation programs. However, the concern is a misplaced one, for the difference between a beneficial and toxic dose is enormous. The usual concentration achieved through purposeful fluoridation is 1 ppm (1 milligram per liter). A lethal dose of fluoride would be about 4 grams (4,000 milligrams). To take in that much fluoride through fluoridated drinking water, one would not only have to consume 4,000 liters (about 1,000 gallons) of water, but would have to do so within a day or so, because fluoride is not accumulative in the body; it is readily excreted.

As a practical matter, there are limits on the extent to which detailed dose-response curves can be constructed for humans. The limits are related to the difficulties in isolating the effects of individual trace elements or compounds and in having representative human populations exposed to the full range of doses of interest. Controlled experiments on plant or animal populations are possible, but there is then the difficulty of trying to extrapolate the results to humans with any confidence. It is also true that different individuals, and persons of different ages, may respond somewhat differently to any chemical, so any dose-response curve must either be specific to a particular population, or it must be only an approximation. Such considerations make it difficult to establish safe exposure limits for hazardous materials and permissible levels of pollutant emissions.

The challenge of isolating the effects on health of any one substance is also illustrated, in part, by the complex paths by which elements move from rocks or air into the body (figure W.2). A variety of natural processes and human activities modify their concentrations along the way. In addition, we are exposed to other geological materials (for example, airborne dust) that can affect

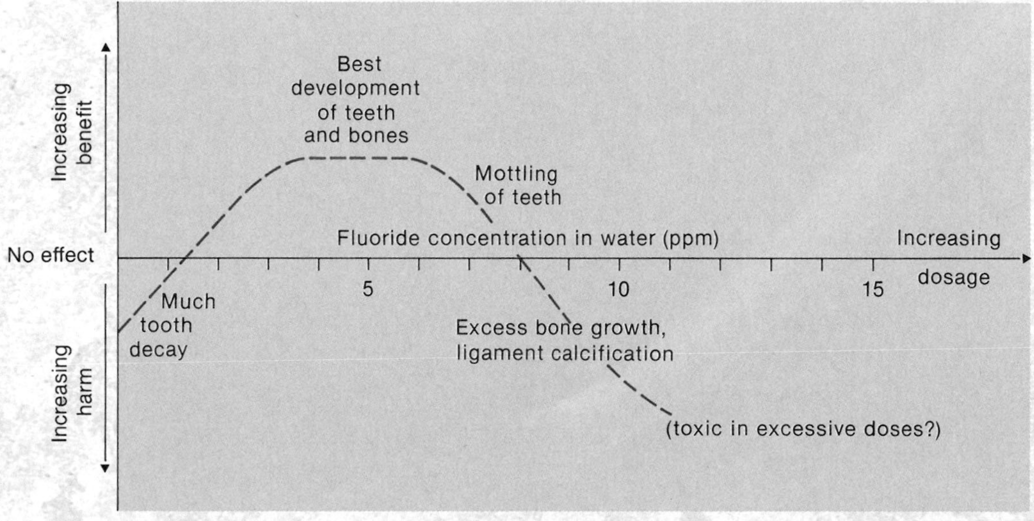

FIGURE W.1 Generalized dose-response curve for fluoride in humans.

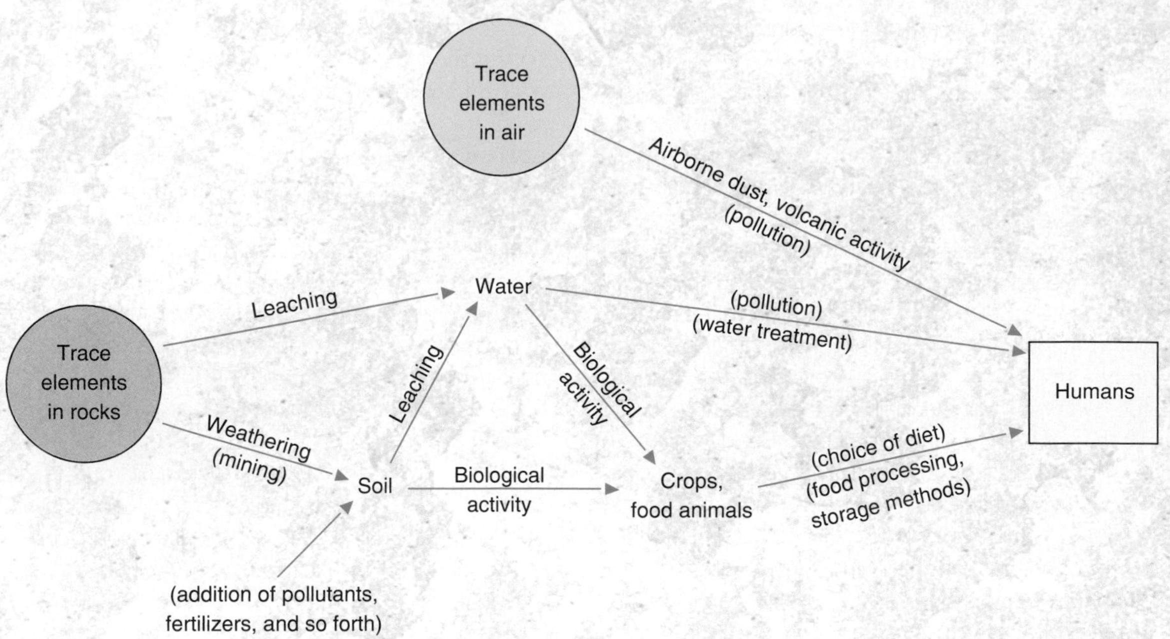

FIGURE W.2 Pathways through which trace elements enter the body, including processes through which trace-element concentrations may be modified. Human influences are in parentheses.

our health, and a huge and growing variety of synthetic chemicals as well as naturally occurring chemical compounds. Increasingly, geomedicine has expanded its scope to include all of the various substances, natural and synthetic, that make their way into the geological environment which can influence the health of humans and ecosystems.

In chapters 16 through 18, we survey various waste-disposal strategies and problems, and then consider some of the causes and consequences of water and air pollution. Geoscientists are increasingly involved in evaluating and limiting the associated hazards.

16 Waste Disposal

High-consumption technological societies tend to generate copious quantities of wastes. In the United States alone, each person generates, on average, more than 2 kilograms (about 4.4 pounds) of what is broadly called "garbage" every day. That represents an increase of more than 60% over 1960 per-capita waste production. Every five years, each average American generates a mass of waste greater than the mass of the Statue of Liberty! That, plus industrial, agricultural, and mineral wastes, together amount to an estimated total of 4 billion tons of solid waste produced in the United States each year. Most of the 40 billion gallons of water withdrawn daily by public water departments end up as sewage-tainted wastewater, and more concentrated liquid wastes are generated by industry. Each day, the question of where to put the growing accumulations of radioactive waste materials becomes more pressing. Proper, secure disposal of all these varied wastes is critical to minimizing environmental pollution. According to data from the Recycling Council of Ontario, North America has 8% of the world's population, consumes one-third of the world's resources—and produces almost half the world's nonorganic trash. In this chapter, we will survey various waste-disposal strategies and examine their pros and cons.

View to the south of Yucca Mountain showing coring activities. Extensive geologic studies have been undertaken to evaluate the likely long-term safety of the proposed radioactive-waste repository.
Photograph courtesy Office of Civilian Radioactive Waste Management/DOE

SOLID WASTES—GENERAL

The principal sources of solid waste in the United States are shown in figure 16.1. More than 50% of the wastes are linked to agricultural activities, with the dominant component being waste from livestock. Most of this waste is not highly toxic except when contaminated with agricultural chemicals, nor is it collected for systematic disposal. The volume of the problem, therefore, is seldom realized.

The other major waste source is the mineral industry, which generates immense quantities of spoils, tailings, slag, and other rock and mineral wastes. Materials such as tailings and spoils are generally handled on-site, as for example when surface mines are reclaimed. The amount of waste involved makes long-distance transportation or sophisticated treatment of the wastes uneconomical. The weathering of mining wastes can be a significant water-pollution hazard, depending on the nature of the rocks, with metals and sulfuric acid among the principal pollutants. Shielding the pulverized rocks from rapid weathering with a soil cover is a common control/disposal strategy. In addition, certain chemicals used to extract metals during processing are toxic and require special handling in disposal like other industrial wastes.

Much of the attention in solid-waste disposal is devoted to the comparatively small amount of municipal waste. It is concentrated in cities, is highly visible, and must be collected, transported, and disposed of at some cost. The following section explores historical and present strategies for dealing with municipal wastes.

Nonmining industrial wastes likewise command a relatively large share of attention because many industrial wastes are highly toxic. Most of the highly publicized unsafe hazardous-waste disposal sites involve improper disposal of industrial chemical wastes. Some of the principal industrial solid-waste sources are shown in figure 16.2.

While industrial wastes may supply the largest volumes of toxic materials, municipal waste is far from harmless. Aside from the organic materials like food waste and paper, a wide variety of poisons is used in every household: corrosive cleaning agents, disinfectants, solvents such as paint thinner and dry-cleaning fluids, insecticides and insect repellents, and so on. These toxic chemicals together represent a substantial, if more dilute, potential source of pollution if carelessly handled.

MUNICIPAL WASTE DISPOSAL

A great variety of materials collectively make up the solid-waste disposal problem that costs municipalities several billion dollars each year (figure 16.3). The complexity of the waste-disposal problem is thus compounded by the mix of different materials to be dealt with. The best disposal method for one kind of waste may not be appropriate for another.

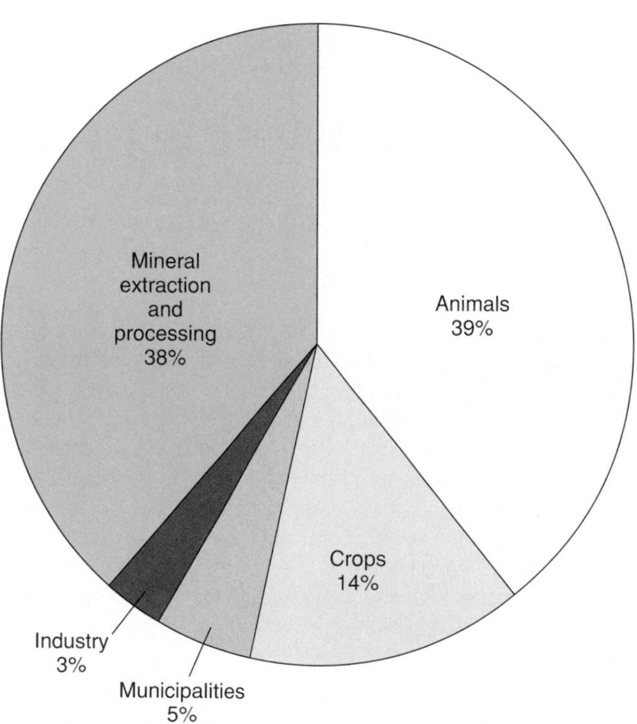

FIGURE 16.1 Principal sources of solid wastes. Note that industry and municipalities together account for less than 10% of the total.

Data from J. E. Fergusson, 1982. Inorganic Chemistry and the Earth. *New York: Pergamon Press.*

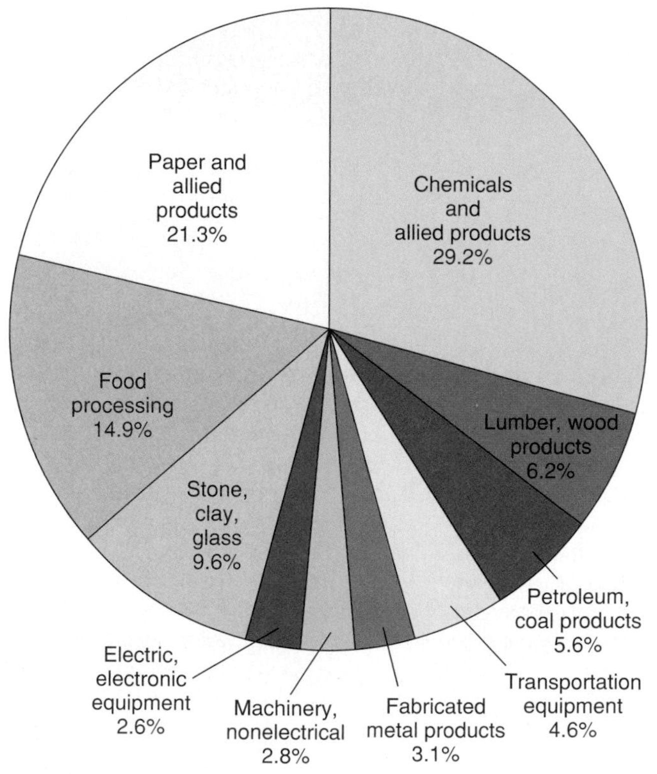

FIGURE 16.2 Principal industrial solid-waste sources; this is a breakdown of the 3% from figure 16.1.

Data from U.S. Environmental Protection Agency.

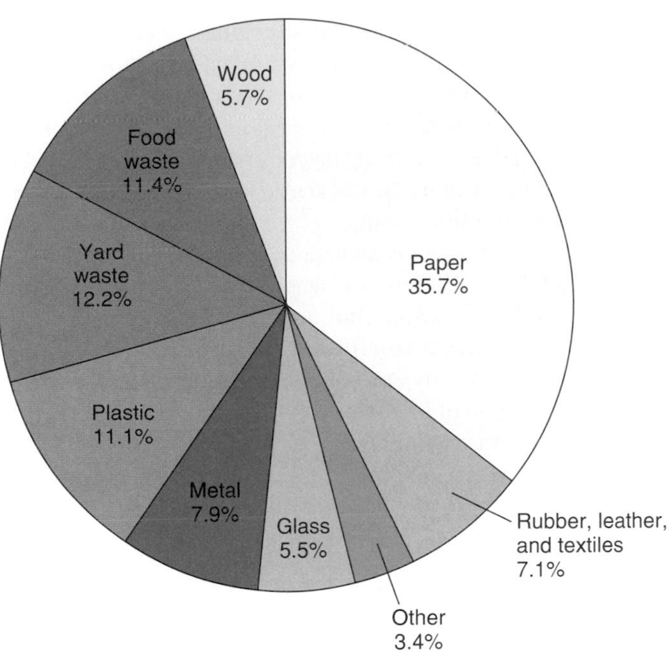

FIGURE 16.3 Typical composition of municipal solid waste (the 5% in figure 16.1) in 2001, before recycling. Individual municipalities' waste composition may vary considerably.
Data from U.S. Environmental Protection Agency.

A long-established method for solid-waste disposal that demands a minimum of effort and expense has been the open dump site. Drawbacks to such facilities are fairly obvious, especially to those having the misfortune to live nearby. Open dumps are unsightly, unsanitary, and generally smelly; they attract rats, insects, and other pests; they are fire hazards. Surface water percolating through the trash can dissolve out, or leach, harmful chemicals that are then carried away from the dump site in surface runoff or through percolation into ground water. Trash may be scattered by wind or water. Some of the gases rising from the dump may be toxic.

All in all, open dumps are an unsatisfactory means of solid-waste disposal. Yet even now, after decades of concerted efforts to close all open dump sites, the Environmental Protection Agency has estimated that there may still be several hundred thousand illicit open dumps in the United States.

Sanitary Landfills

Where open dumps have been phased out, they have most often been supplanted by **sanitary landfills** (figure 16.4). The basic method has been in use since the early twentieth century. In a basic sanitary-landfill operation, a layer of compacted trash is covered with a layer of earth at least once a day. The earth cover keeps out vermin and helps to confine the refuse. Landfills have generally been sited in low places—natural valleys, old abandoned gravel pits, or surface mines. When the site is full, a thicker layer of earth is placed on top, and the land can be used for other purposes, provided that the nature of the wastes and the design of the landfill are such that leakage of noxious gases or toxins is minimal (figures 16.4B). The most suitable uses include parks, pastureland, parking lots, and other facilities not requiring much excavation. The city of Evanston, Illinois, built a landfill up into a hill, and the now-complete "Mount Trashmore" is a ski area. Golf courses built over old landfill sites are increasingly common. Attempts to construct buildings on old landfill sites may be complicated by the limited excavation possible because of refuse near the land surface, by the settling of trash as it later decomposes (which can put stress on the structures), and by the possibility of pollutants escaping from the landfill site.

Pollutants can escape from improperly designed landfills in a variety of ways. Gases are produced in decomposing refuse in a landfill just as in an open dump site, though the particular gases differ somewhat as a result of the exclusion of air from landfill trash. Initially, decomposition in a landfill, as in an open dump, proceeds aerobically, producing such products as carbon

A

B

FIGURE 16.4 Sanitary landfills. (A) Basic principle of a sanitary landfill (partial cutaway view). The "sanitary" part is the covering of the wastes with soil each day. New landfills must meet still more-stringent standards. (B) New Lyme (Ohio) Landfill site after restoration.
(B) Photograph by Harry Weddington, courtesy U.S. Army Corps of Engineers.

dioxide (CO_2) and sulfur dioxide (SO_2). When oxygen in the covered landfill is used up, anaerobic decomposition yields such gases as methane (CH_4) and hydrogen sulfide (H_2S). If the surface soil is permeable, these gases may escape through it. Sealing the landfill to the free escape of gases can serve a twofold purpose: reduction of pollution, and retention of useful methane as described in chapter 15. The potential buildup of excess gas pressure requires some venting of gas, whether deliberate or otherwise. Where the quantity of usable methane is insufficient to be recoverable economically, the vented gases may be burned directly at the vent over the landfill site to break down noxious compounds.

If soil above or below a landfill is permeable, **leachate** (infiltrating water containing dissolved chemicals from the refuse) can escape to contaminate surface or ground waters (figure 16.5). This is a particular problem with landfills so poorly sited that the regional water table reaches the base of the landfill during part or all of the year. It is also a problem with older landfills that were constructed without impermeable liners beneath. Increasing awareness particularly of the danger of groundwater pollution has led to improvements in location and design of sanitary landfills. They are now ideally placed over rock or soil of limited permeability (commonly clay-rich soils, sediments, or sedimentary rocks, or unfractured bedrock) well above the water table. Where the soils are too permeable, liners of plastic or other waterproof material may be used to contain percolating leachate, or thick layers of low-permeability clay (several meters or more in thickness) may be placed beneath the site before infilling begins.

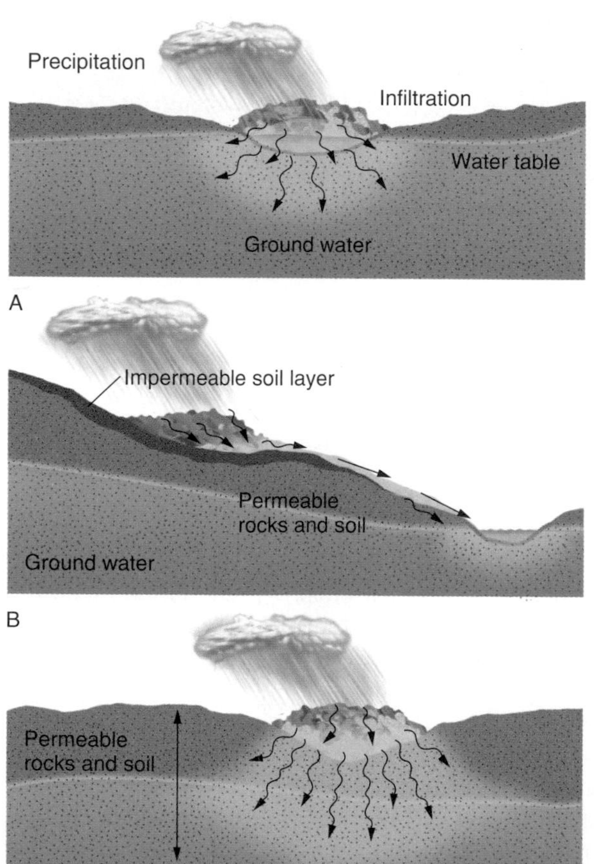

FIGURE 16.5 Leachate can escape from a poorly designed or poorly sited landfill to contaminate surface or ground waters. (A) Direct contamination of ground water; water table intersects landfill. (B) Leachate runs off over sloping land to pollute lake or stream below, despite impermeable material directly under landfill site. (c) Lack of impermeable liner below alows leachate to infiltrate to ground water.

Unfortunately, there is another potential problem with landfills sealed below with low-permeability materials, if the layers above are fairly permeable. Infiltrating water from the surface will accumulate in the landfill as in a giant bathtub, and the leachate may eventually spill out and pollute the surroundings (figure 16.6A). Use of low-permeability materials above as well as below the refuse minimizes this possibility (figure 16.6B). Alternatively, any accumulating leachate can be pumped out, but it then constitutes a further waste-disposal problem. Leachate problems are lessened in arid climates where precipitation is light, but unfortunately, most large U.S. cities are not located in such places. Thus, modern landfills are carefully monitored to detect developing leachate-buildup problems, and in some cases, leachate is pumped out to prevent leakage (though this does, in turn, create a liquid-waste-disposal problem).

A subtle pathway for the escape of toxic chemicals may be provided by plants growing on a finished landfill site that is not covered by an impermeable layer. As plant roots take up water, they also take up chemicals dissolved in the water, some of which, depending on the nature of the refuse, may be toxic. This possibility indicates the need for caution in using an old landfill site for cropland or pastureland.

Thus, a well-designed modern municipal landfill is a complex creation, not just a dump topped with dirt (figure 16.7). Still, however carefully crafted, a landfill involves a commitment of land. A rule of thumb for a sanitary landfill for municipal wastes is that one acre is needed for each 10,000 people each year, if the landfill is filled to a depth of about 3 meters per year. For a large city, that represents considerable real estate. True, land used for a sanitary landfill can later be used for something else when the site is full, but then another site must be found, and another, and another. All the time, the population grows and spreads and also competes for land. More than half the cities in the United States are pressed for landfill facilities, with municipal solid waste piling up at over 150 million tons per year.

Local residents typically resist when new landfill sites are proposed, a phenomenon sometimes described by the acronym NIMBY—"Not In My Back Yard." (Other variants that have appeared on wall posters at the Environmental Protection Agency include NIMFYE (Not In My Front Yard Either), NIMEY (Not In My Election Year), and NOPE (Not On Planet Earth)—but landfills always have to be put *somewhere.*) Between the political/social constraints and the increased geological constraints on landfill siting that have evolved as understanding of pollutant escape from disposal sites has improved, the number of potential sites for new landfill operations is limited, and as existing landfills fill, the total number of landfills has declined, from nearly 8000 in 1988 to fewer than 1800 in 2002. In states with high population density and limited landfill sites, the ratio of people to landfills may be 80,000 to 1 or

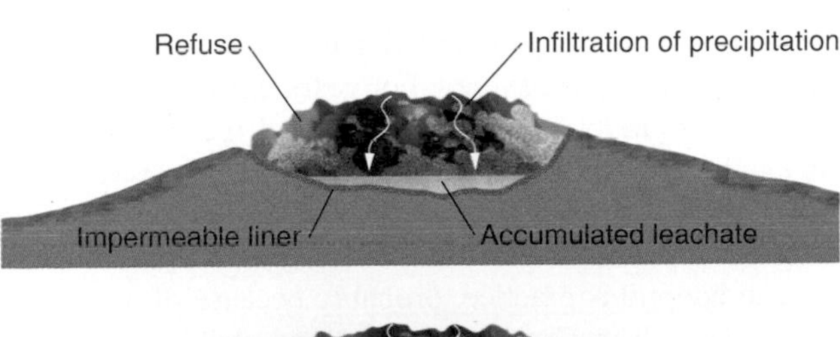

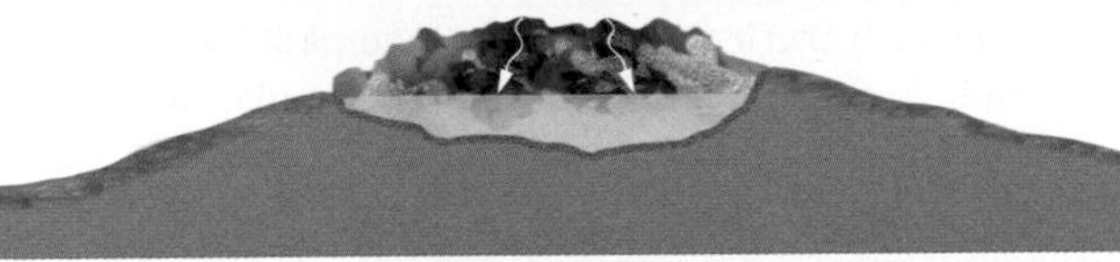

FIGURE 16.6 (A) The "bathtub effect" caused by the accumulation of infiltrated leachate above impermeable liner in unmonitored landfill. Overflow may not occur for months or years. (B) Installation of impermeable plastic membrane and gas vents, K. I. Sawyer Air Force Base, Michigan.
(B) Photograph by Harry Weddington, courtesy U.S. Army Corps of Engineers.

higher, and obviously, the greater this ratio, the more rapidly the landfills fill. One state in this situation, New Jersey, has to ship half its 11 million tons/year of municipal waste to other states' landfills. In efforts to conserve dwindling landfill volume, about half of the 50 states have decreed that residents can no longer include landscape waste (grass clippings, leaves, brush, and so on) with trash collected for landfill disposal; the landscape waste is collected separately for composting. The landfill squeeze is especially acute in the northeastern United States (figure 16.8).

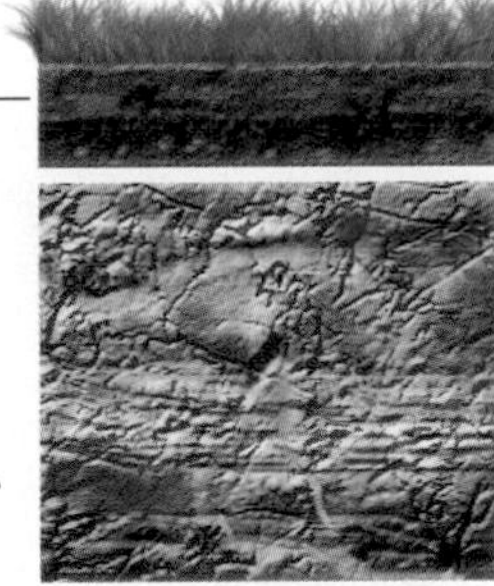

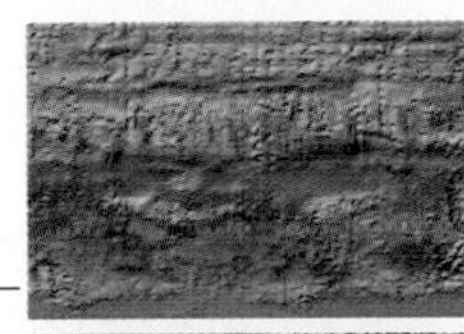

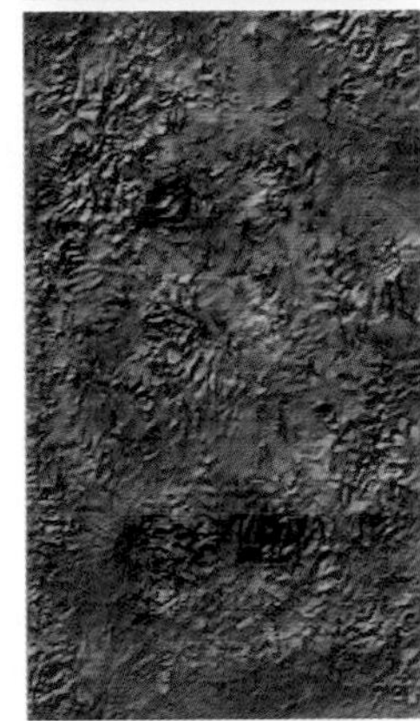
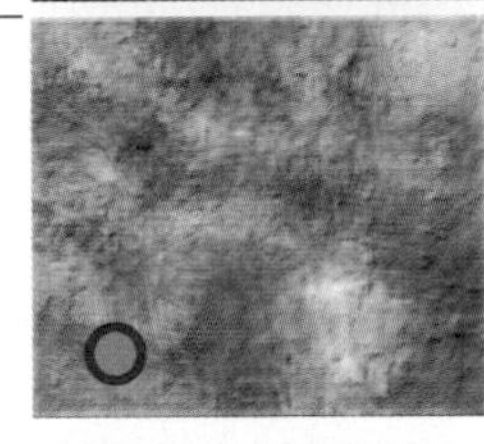
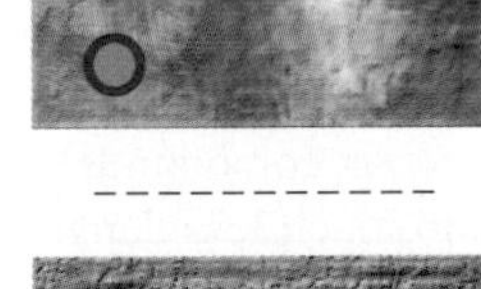

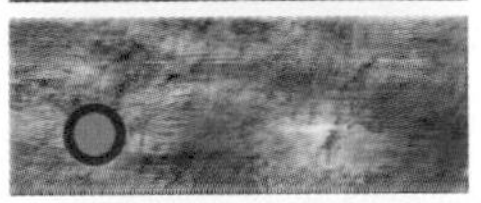
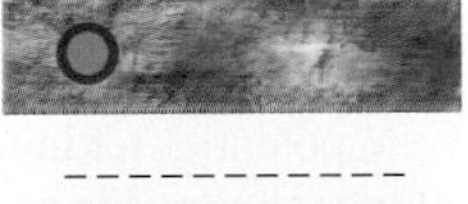
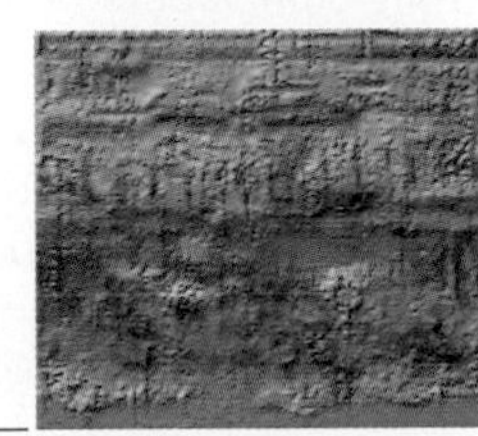

FIGURE 16.7 A properly designed modern municipal landfill provides for sealing below and (when completed) above, venting of gas, and collection of leachate. Monitoring wells may also be placed around the site.
After New York State Department of Environmental Conservation.

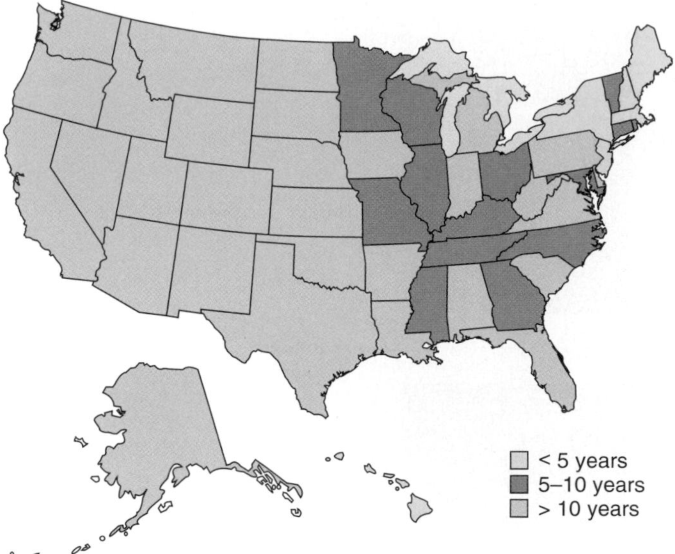

FIGURE 16.8 A 1996 assessment of remaining landfill capacity suggested imminent shortages, especially in the northeastern U.S. While overall national capacity remains adequate, local shortages can be acute. Since this study, the number of landfills has declined, but newer landfills are larger, maintaining essential capacity.

Source: Data from Directory and Atlas of Solid Waste Disposal Facilities, *map from U.S. Environmental Protection Agency.*

Incineration

Incineration as a means of waste disposal provides a partial solution to the space requirements of landfills. However, it is an imperfect solution, since burning wastes contributes to air pollution, adding considerable carbon dioxide (CO_2) if nothing else. At moderate temperatures, incineration may also produce a variety of toxic gases, depending on what is burned. For instance, plastics when burned can release chlorine gas and hydrochloric acid, both of which are toxic and corrosive, or deadly hydrogen cyanide; combustion of sulfur-bearing organic matter releases sulfur dioxide (SO_2); and so on.

The technology of incineration has been improved in recent years. Modern very-high-temperature (up to 1700°C, or 3000°F) incinerators break down hazardous compounds, such as the complex organic compounds, into much less dangerous ones, especially carbon dioxide (CO_2) and water (H_2O). Still, individual chemical *elements* are not destroyed. Volatile toxic elements, mercury and lead among others, may escape with the waste gases. Much of the volume of urban waste is noncombustible, even at high temperatures. Refuse that will not burn must be separated before incineration or removed afterward and disposed of in some other way—often in a landfill site. Harmful nonvolatile, noncombustible substances in the wastes will be concentrated in the residue. If the potential toxicity of the residue is high, it may require handling comparable to that for toxic industrial wastes, raising the net cost of waste disposal considerably. Still, general-purpose municipal incinerators can be operated quite cheaply and are useful in reducing the total volume of solid wastes by burning the paper, wood, and other combustibles before ultimate (landfill) disposal of the rest.

A further benefit of incineration can be realized if the heat generated thereby is recovered. For years, European cities have generated electricity using waste-disposal incinerators as the source of heat. The combined benefits of land conservation and energy production have led to extensive adoption of incineration in other nations (figure 16.9). The United States has been slower to adopt this practice, probably because of more abundant supplies of other energy sources (and perhaps land also).

Several U.S. cities, however, have put the considerable quantity of heat energy released by an incinerator to good use. The city of North Little Rock, Arkansas, turned to an incinerator when landfill space grew scarce. The steam-generating incinerator is saving the city an estimated \$50,000/year, while reducing required additional landfill volume by 95%. In Kansas City, Kansas, the American Walnut Company is burning waste sawdust and wood chips in a steam-generating boiler, saving itself natural gas previously used to help dispose of the wastes and even generating surplus steam, which it sells to other local users. Such fringe benefits of incineration contribute to increasing acceptance of incineration as a waste-disposal strategy, though as a heat source its efficiency is modest (table 16.1). There is

TABLE 16.1

Comparative Heat Values of Fuels and Wastes

Fuel	Btu/lb
Coal (anthracite)	13,500
Coal (bituminous)	14,000
Peat	3600
#2 fuel oil	18,000
Natural gas (Btu/cu ft)	1116
Mixed municipal solid waste	**4800**
Mixed paper	6800
Newsprint	7950
Corrugated	7043
Junk mail	6088
Magazines	5250
Mixed food waste	2370
Wax milk cartons	11,325
Polyethylene	18,687
Polystyrene	16,419
Mixed plastic	14,100
Tires	13,800
Leaves (10% moist.)	7984
Cured lumber	7300

While average munipal waste yields much less energy per pound than most fuels, incineration also reduces remaining waste volume.

Source: U.S. Environmental Protection Agency

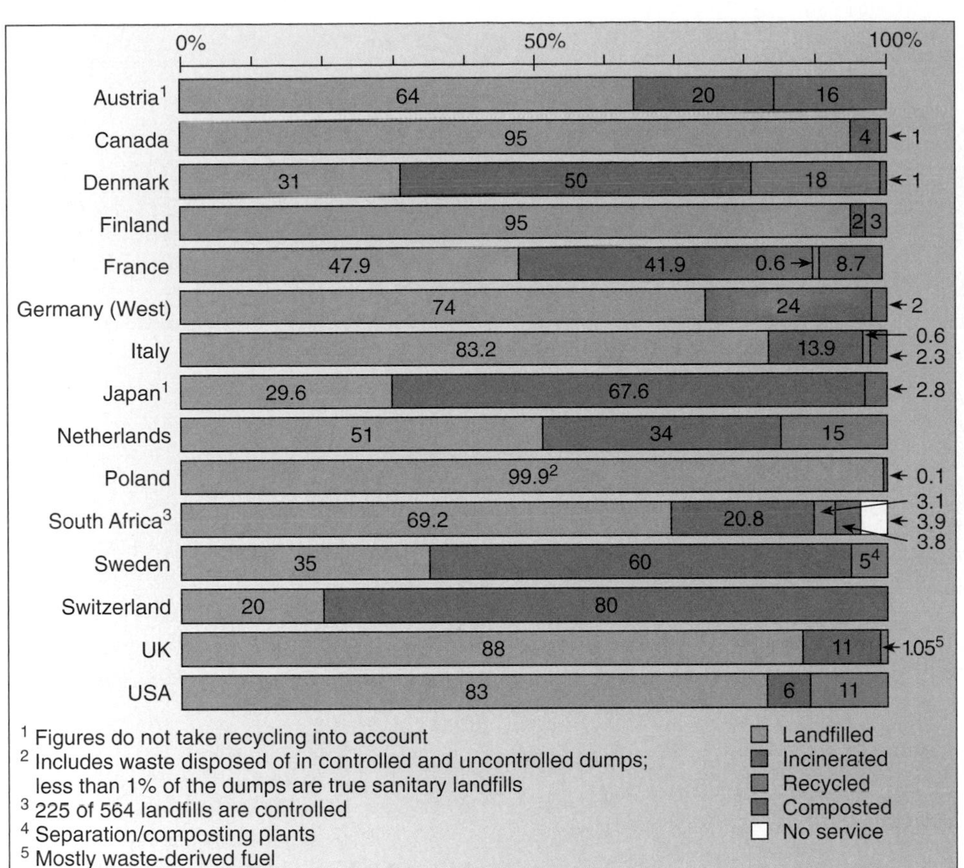

FIGURE 16.9 Proportions of municipal waste handled by various disposal methods in selected countries.

Data from Geology and the Environment in Western Europe, *ed. by G.I. Lumsden. Copyright © 1992 Clarendon Press, Oxford, England.*

still considerable public resistance to new incinerators, however, which is part of the reason that it may take three to five years from inception to opening of a new incinerator. Still, as landfills fill, *something* has to be done to reduce waste volume.

Ocean Dumping

A variant of land-based incineration, developed over the past several decades, is shipboard incineration in the open ocean. Following combustion, unburned materials are simply dumped at sea. This method has been applied to stockpiles of particularly hazardous chemical wastes. A 1981 report of the Environmental Protection Agency described the technique as "promising" for a variety of reasons, making the statement that "it has a minimal impact on the environment by removing the destruction site far from populated areas so that emissions are absorbed by the oceans," and noting that offshore incinerators "not handicapped by emission control requirements that apply to land-based units" could be very cost-effective. The desirability of this method plainly depends on one's point of view. It does not much matter if carbon dioxide is added to the air over land or over water; it still contributes to the rising carbon dioxide levels in the atmosphere. True, dumping the solid residues at sea puts them out of the sight of people, but if toxic materials are present and left unburned, they contribute to the pollution of the oceans to which the world turns increasingly for food. As noted in chapter 17, scientists really do not yet know the ultimate fate of many of the chemicals involved in water pollution. Perhaps this waste-disposal scheme should be embraced with some caution.

Ocean dumping *without* prior incineration has also been used for chemical wastes, municipal garbage, and other refuse. The potential for water pollution is apparent. In some cases, too, shifting currents have brought the waste back to shore rather than dispersing it in the oceans as intended. Increasing recognition of the dangers of dumping untreated wastes in the sea led to drastic curtailment of the practice by the Environmental Protection Agency in most areas. In the late 1980s, public outcry over incidents of wastes (including needles and other medical wastes) washing up onto beaches, high bacterial counts in nearshore waters, and deaths of marine mammals combined to add pressure on Congress to do something. The Ocean Dumping Ban Act of 1988 decreed that ocean dumping of U.S. sewage sludge and industrial waste would cease after 1991. Growing numbers of nations are recognizing ocean dumping as a pollution problem and moving to end it. Britain, the other nation dumping large quantities of sewage sludge, phased out the practice by 1998; it had ended ocean dumping of its industrial waste by 1993. However, the oceans remain a dumping site for one very high-volume waste product: *dredge spoils.*

Dredge spoils are sediments dredged from reservoirs and waterways to enlarge capacity or improve navigation. They are dumped in the oceans at the rate of about 200 million tons a year. At first, this may seem harmless enough; it is "just dirt." However, dumping fine sediments may harm or destroy marine organisms that cannot survive in sediment-clouded waters.

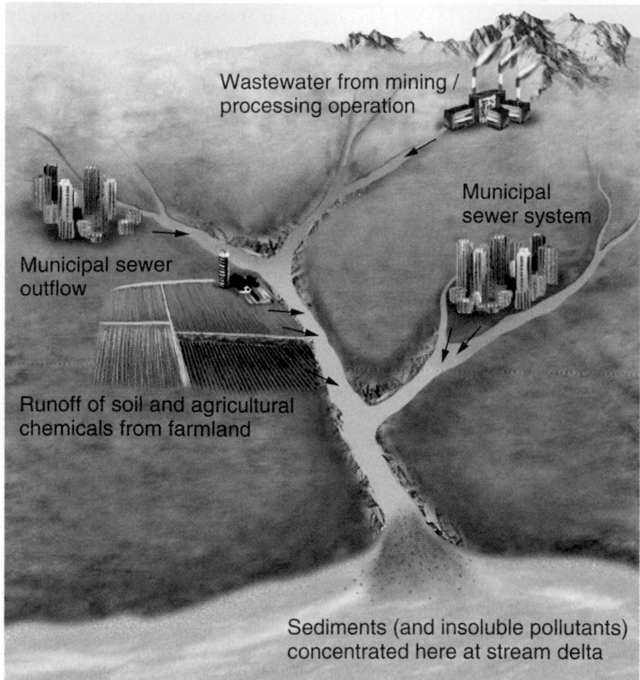

FIGURE 16.10 Concentration of pollutants in stream delta sediments. Pollutants that are solid or that attach themselves to solid particles accumulate along with the sediments and may become highly concentrated in dredge spoils. The ultimate sink for many such pollutants could be any one of several other depositional environments—along river systems, in estuaries, or on the continental shelf, for example.

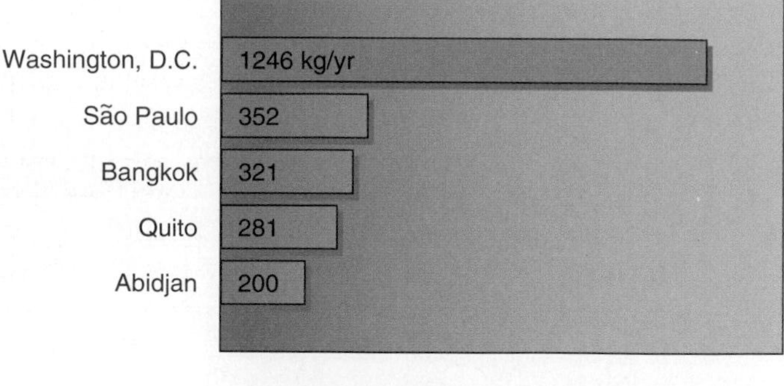

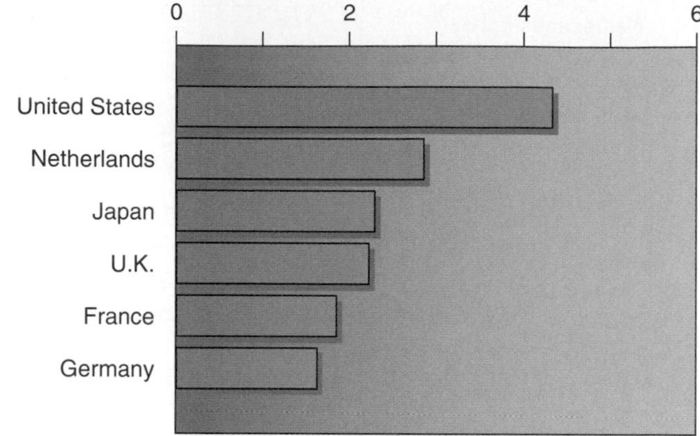

FIGURE 16.11 Relative amounts of solid waste generated per capita in selected cities and countries.

Data from World Resources 1995–96, *World Resources Institute, and U.S. Environmental Protection Agency.*

Also, many toxic chemicals dumped into water do not go into or stay in solution, but stick to the surfaces of sediment particles. Dumping sediments dredged from the mouth of a polluted river means dumping a load of such pollutants concentrated from the whole drainage basin (figure 16.10). When sediments dredged from fresher waters are dumped into saline waters, some adsorbed chemicals may be redissolved as a consequence of the change in water chemistry, posing a pollution threat even if the sediments themselves settle harmlessly to the bottom.

REDUCING WASTE VOLUME

The sheer volume of the solid-waste-disposal problem has led to a variety of attempts to reduce it. In industrialized countries, the problem can be especially acute (figure 16.11). As was already noted, incineration disposes of some waste, leaving less to be dealt with by other means. Another volume-reduction strategy is compaction, either in individual homes with trash compactors or at large municipal compaction facilities. Less volume means less landfill space used or the slower filling of available sites. However, it also means no reuse of any potentially recoverable material, and the slower decay of organic material (a concern if methane production is desired).

In that connection, it is perhaps worth noting that studies of landfill sites over the last two decades have revealed that decomposition in landfills takes place much more slowly than once believed, especially if the landfill is relatively dry. The University of Arizona's Garbage Project has found well-preserved food wastes two decades old, and readable newspapers and telephone books three or four decades old. Decomposition rates seem to be faster in wet waste—such as in the deeper layers of the world's largest landfill, New York City's Fresh Kills, which are below the water table—but modern landfills are most often designed to minimize water inflow. In the future, careful manipulation of water content and, perhaps, judicious addition of suitable microorganisms can maximize biodegradation (and production of useful methane) while minimizing the risks from toxic leachate.

Handling (Nontoxic) Organic Matter

On-site disposal—for example, with a home in-sink garbage disposal unit—is not really disposal at all. The practice merely diverts some organic matter to become part of the water-pollution problem. The organic-matter content of the water is increased (see chapter 17 for the consequences of this), and more of a load is placed on municipal sewage-treatment plants.

Organic matter can be turned to good use through composting, a practice long familiar to gardeners and farmers. Many kinds of plant wastes and animal manures can be handled this way. Partial decomposition of the organic matter by microorganisms produces a crumbly, brown material rich in plant nutrients. Finished compost is a very useful soil additive, improving soil structure and water-holding capacity, as well as adding nutrients. In the United States, composting is more often practiced by individuals. In parts of Europe and Asia, demand for finished compost has long made it economically practical to establish municipal composting facilities. The city of Auckland, New Zealand, began such a facility in 1960 after it became evident that remaining landfill sites were severely limited. The process is not without complications, however. Organic matter must be separated from glass, metal, and other noncompostables. Occasionally, chemical contamination of the refuse with weed-killers or other toxic substances makes some batches of compost unusable for agriculture. Still, sale of the finished compost helps to pay the cost of the composting operation in Auckland, and the volume of waste going to the landfill is reduced by a factor of about fifteen in the process.

As landfills fill, the United States is increasingly turning to composting too. By 1997, nearly half of the states had banned yard waste from landfills, and yard-waste composting programs had been established in nearly every state.

Recycling

In chapter 13, we considered recycling metals in the context of mineral resources and noted that an energy saving can often be realized in the process. Recycling and reuse are also waste-reduction strategies. Glass is not made from scarce commodities, but just as quartz is a weathering-resistant mineral, silica-rich glass is virtually indestructible in dumps and landfills—and along roadsides. It can be broken up, but it does not readily dissolve or break down. Reuse of returnable glass bottles requires only about one-third the energy needed to make new bottles. Recycling all glass beverage containers would reduce by over 5 million tons per year the amount of glass contributing to the solid-waste disposal problem. (Glasses vary in composition, so, like alloys, scrap glasses cannot be mixed indiscriminately and reprocessed—but once a glass bottle is manufactured, it can be reused many times, if carefully handled.)

Imposing deposits on beverage containers also provides a financial incentive not to litter. In Oregon, passage of a mandatory-deposit law for beverage [illegible] side litter by u[illegible] Oregon pioneered the idea in 1971, nine other states have followed suit. (However, voter resistance to the perceived inconvenience of having to return cans and bottles to recover deposits, and lobbying efforts by bottlers, have led to the defeat of "bottle bills" in some states, too.) Currently, the United States recycles about 19% of its glass; in countries such as Switzerland and the Netherlands, the proportion recycled approaches 75%.

Paper might also be recycled more extensively. In the United States, about 48% of the paper and paperboard we discard is recycled, and about 35% of paper used has some recycled content. Paper recycling is easiest and most effective when a single type of paper—newspaper, for example—is collected in quantity. That limits the variety of inks and other chemicals that must be handled during reprocessing. Mixtures of printed, waxed, and plasticized papers are somewhat harder to handle economically but can nevertheless be recycled. The Recycling Council of Ontario estimates that every ton of newspapers recycled means about 18 trees, and 3 cubic meters of landfill space, saved. Moreover, making paper from the recycled fibers requires 60% less energy than does manufacturing paper from newly cut trees. Still, for all the recycling (and talk of a "paperless society"), the United States continues to discard large amounts of paper (figure 16.12).

Plastics continue to be something of a disposal problem, though much less of one than they were believed to be a decade or more ago. The same durability that makes them useful also makes them difficult to break down when no longer needed, except by high-temperature combustion. Some degradable plastics have been developed to break down in the environment after a period of exposure to sunlight, weather, and microbial activity, but these plastics are suitable only for applications where they need only hold together for a short time—for example, fast-food containers. (And if a long useful life for such a product is not actually required, perhaps it did not need to be made of plastic, anyway.) Another difficulty in recycling plastics is similar to the problem with different steels: A mix of plastics, when reprocessed, is unlikely to have quite the right properties for any of the applications from which the various scrap plastics were derived. Still, the blend may be suitable for other uses, such as plastic piping, plastic lumber, or a shredded-plastic stuffing for upholstery.

One approach to facilitating plastic recycling is to mark those plastics that can be more easily recycled with the triangular symbol of three arrows head-to-tail that is widely used to represent recycling, and identify the basic type of plastic by a number (1, 2, 3, etc.) within the triangle. (Look on the bottom of a two-liter soda bottle, gallon plastic milk jug, or other plastic food container; see figure 16.13 for examples and key.) A remaining obstacle is that there must exist an identifiable market or demand for a *particular* plastic in order for its recycling to be economically feasible; so a given municipal waste hauler might be able to collect soda bottles and milk jugs for recycling, but not foam packing materials. Markets will vary by region. It [illegible] be realized that plastic is typically not recycled into the same form—old soda bottle to new soda bottle, for instance. In this respect, plastic recycling differs from that of other materials. A recycled aluminum can is likely to come back as a new aluminum can, a recycled glass bottle as a refilled or remanufactured bottle. A plastic's properties tend to change during recycling, however, and recycled plastic may be less strong than new plastic. So a soda bottle may be recycled, but it will be transformed into something else—such as fiber for carpeting, plastic trash bags, or "plastic lumber" for park benches.

Recent research has explored converting plastics into other compounds, such as waxes, for reuse, or developing novel applications for chopped or shredded recycled plastic—for

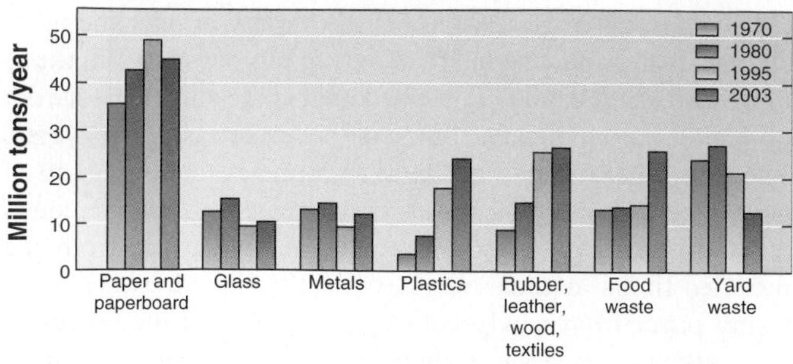

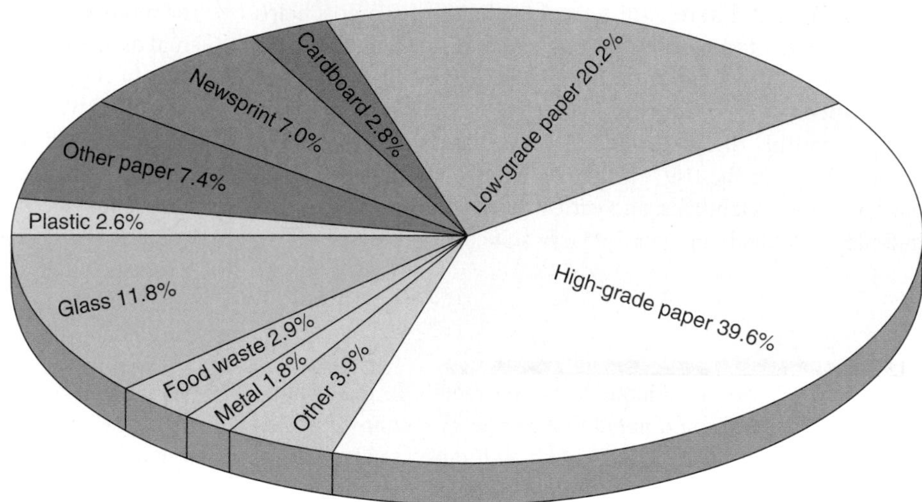

FIGURE 16.12 (A) Despite recycling, municipal solid waste in the United States includes large quantities of paper and, increasingly, plastics, among other materials. Note sharp drop in yard waste, largely due to mandatory composting. (B) The primary waste product of a large federal office building is paper, not all of which is recycled. *Source: U.S. Environmental Protection Agency.*

example, using it as fill for pillows, as insulation in clothing, or as an additive to fiberglass. There is particular interest in finding uses for the plastic-rich residue from the shredders used to chew up junked cars. Once the metals are extracted from the shredded product, most of the remaining 150 to 200 kilograms (300 to 400 pounds) of material per car is plastic. The challenge is finding the uses for the material that will make the whole process more economically practical.

There are additional general obstacles to recycling, some of which were mentioned earlier. Where recovery of materials from municipal refuse is desired, **source separation** is generally necessary. This means that individual homeowners, businesses, and other trash generators must sort that trash into categories—paper, plastic, glass, metal, and so on—prior to collection. While this has been successful in some communities, it is not now widely practiced, although where legally mandated, it can work well once residents adjust to the new realities. Increasingly, communities are focusing on ways to recycle household trash that do *not* require extensive source separation.

Recycling may conflict in some measure with other waste-disposal objectives. For example, recycling combustible materials reduces the energy output of municipal incinerators used to generate power. Paper recyclers are already encountering this problem, and as uses are found for recycled plastics, the same difficulty may arise with those materials.

In the United States, waste recovery for recycling or composting is increasing—from less than 10% of municipal wastes in 1985 to about 30% in 2003 (figure 16.14A). The recycling rates of individual materials vary considerably, in part reflecting differences in economics, or technical or practical constraints described previously (figure 16.14B).

Other Options

New and creative approaches to solid-waste problems are continually being developed. When Chicago's Edens Expressway was rebuilt in 1979–1980, 335,000 tons of old pavement were crushed and reused in the new roadbed, thereby saving that much asphalt (petroleum) and crushed stone, while reducing

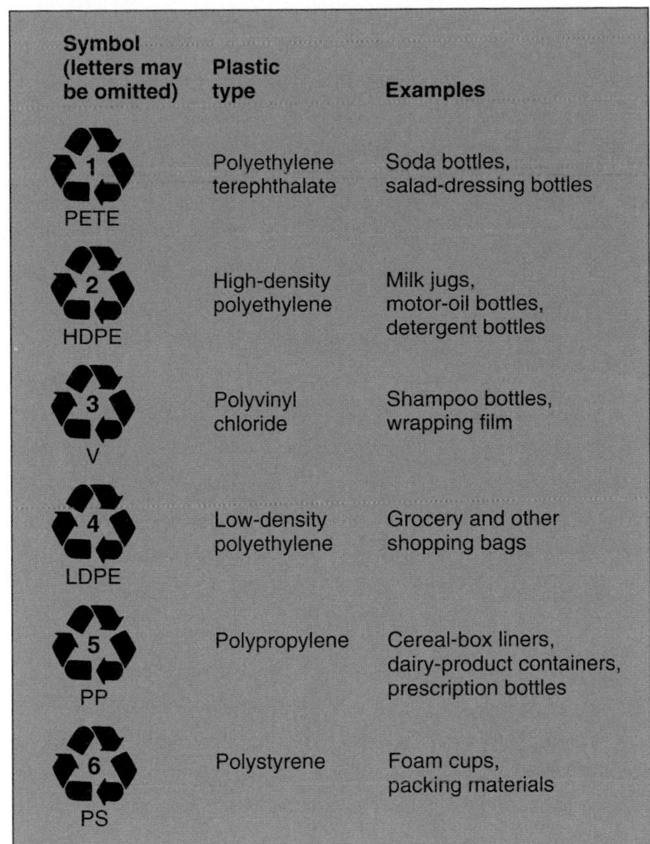

Symbol (letters may be omitted)	Plastic type	Examples
1 PETE	Polyethylene terephthalate	Soda bottles, salad-dressing bottles
2 HDPE	High-density polyethylene	Milk jugs, motor-oil bottles, detergent bottles
3 V	Polyvinyl chloride	Shampoo bottles, wrapping film
4 LDPE	Low-density polyethylene	Grocery and other shopping bags
5 PP	Polypropylene	Cereal-box liners, dairy-product containers, prescription bottles
6 PS	Polystyrene	Foam cups, packing materials

FIGURE 16.13 Use of standard symbols facilitates separation of distinct types of plastics for recycling. Sometimes the same symbol with a numeral "7" is used for "other." Certain products (e.g., milk jugs and soda bottles) are consistently made with one type of plastic; other (e.g., containers for detergents or dairy products) may be made of different types depending upon the desired properties—clear container versus opaque, flexible versus rigid, and so on.

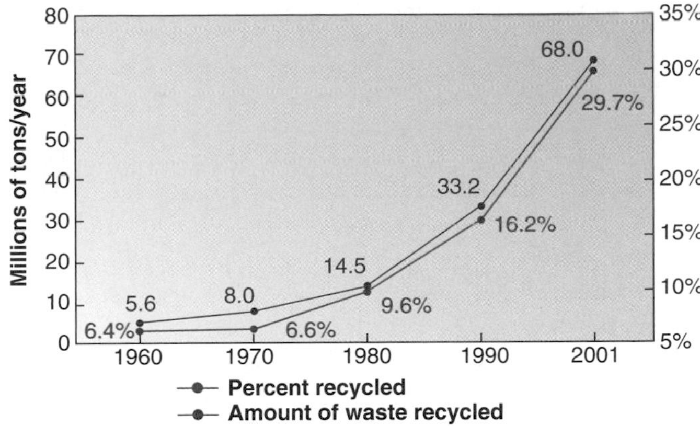

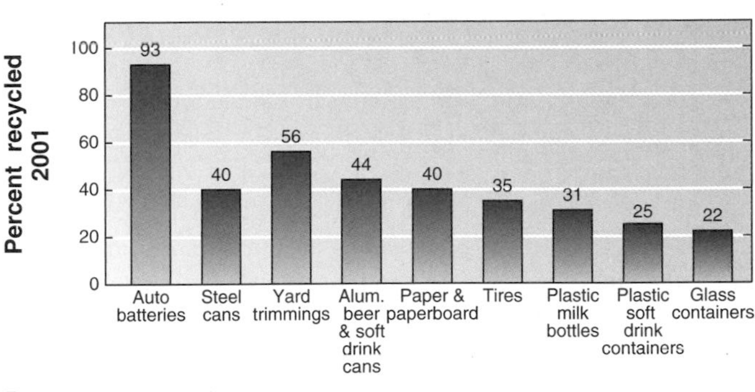

FIGURE 16.14 (A) Both the amount of municipal solid waste recycled, and the percentage, have risen sharply in the last two decades. (B) The recycling rate varies greatly by material; 2003 data.
Source: U.S. Environmental Protection Agency.

Chicago's solid-waste disposal problem by 335,000 tons. When the University of North Carolina at Asheville razed and rebuilt its three oldest dormitories in 2001, virtually all the scrap was recycled—steel beams converted to bike racks, bricks salvaged for use as paving for campus paths, and so on. The United States and other countries are developing "waste exchanges," whereby one industry's discarded waste becomes another's raw materials. Ideally, this both reduces waste disposal and saves the waste generator money: For example, an aluminum smelting firm in Maryland that had been paying $70 a ton in landfill fees to dispose of its leftover powdered fluorite found a taker in Pittsburgh who would pay $20 a ton; one company's waste isopropyl alcohol became another's cleaning solvent. The Canadian Waste Materials Exchange has dealt in such diverse materials as glove-leather scrap, oat hulls, and fish-processing waste.

Gradually, handling of municipal solid waste in the United States has shifted in the direction of recovery (for composting or recycling) and away from "disposal" (figure 16.15). Still, some materials are too difficult to reuse efficiently, for reasons already outlined. Others are toxic by-products of industrial processes and are not themselves useful or are too toxic for safe handling during extensive reprocessing. These will continue to require disposal. Also, many of the highly toxic industrial wastes are liquids rather than solids, and these wastes may require somewhat different handling from solid wastes.

LIQUID-WASTE DISPOSAL

Liquid-waste problems come in many forms. A notable example is the problem of used oil. Presently, over 1 billion gallons of used lubricants derived from petroleum are generated in the United States each year; 40% of this waste is poured into the ground or into storm drains, and the fate of another 20% is unknown. An increasing proportion of this "waste" is being reclaimed and recycled, but many individuals still simply discard it. Another high-volume problem, but a localized one, is liquid animal waste from stockyards, pig farms, and other facilities where large numbers of animals are concentrated. However, the two major types of liquid wastes are sewage, which is discussed in a subsequent section, and the more-concentrated, highly toxic, liquid waste by-products of industrial processes—acids, bases, organic solvents, and so on. This section focuses

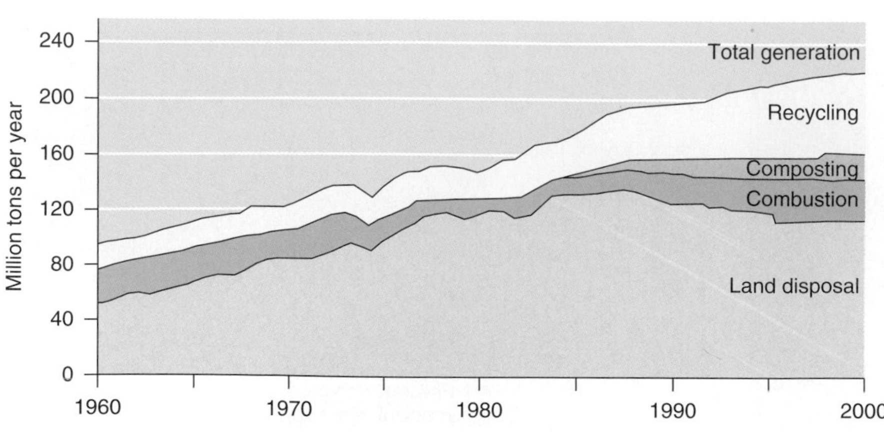

FIGURE 16.15 The trend in municipal waste disposal is away from traditional "disposal."
Source: U.S. Environmental Protection Agency.

on disposal strategies for the latter, major sources of which are shown in figure 16.16.

Handling of toxic liquid wastes has tended to follow one of two divergent paths. The *dilute-and-disperse* approach, based on the assumption that, if toxic substances are sufficiently diluted, they will be rendered harmless, has been the rationale behind much dumping into oceans and large lakes and rivers. With the increasing recognition of substances that are toxic even at levels below 1 ppb in water, including many of the complex organic solvents, agricultural chemicals, and others, the basic premise has been brought into question. Also, certain pollutants can accumulate in organisms and become more concentrated up a food chain (see chapter 17), which means that the chemicals can return to hazardous concentrations.

The opposite approach is the *concentrate-and-contain* alternative. Thoughtless disposal of concentrated wastes followed by inadequate containment has led to disasters like Love Canal, New York, or Woburn, Massachusetts (see also figure 16.17). In the past, some concentrated liquid industrial wastes have been dumped in trenches or pits directly and buried, while other wastes have been placed in metal or plastic containers and consigned to dumps or landfills. The disposal sites frequently were not evaluated with respect to their suitability as toxic-waste disposal sites and, over the longer term, the wastes were *not* contained. Metal drums rusted, plastic cracked and leaked, and wastes seeped out to contaminate ground water and soil. Superfund, discussed in a later section, was created in large part to address such issues.

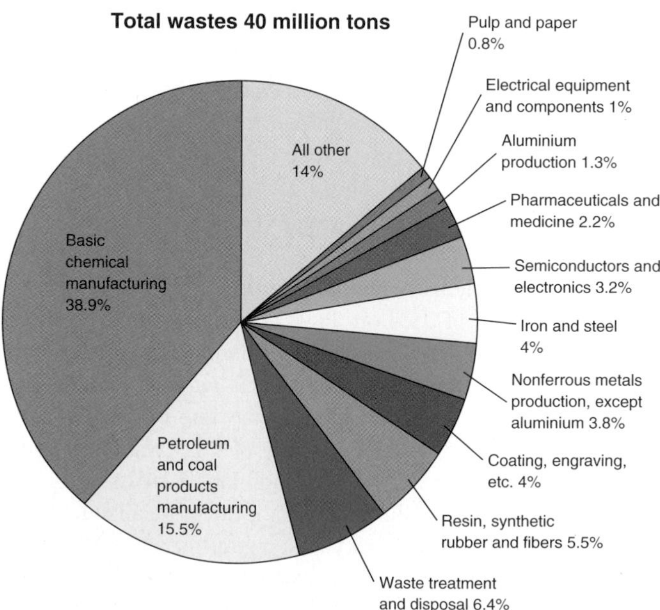

FIGURE 16.16 Selected industries among the 50 largest generators of hazardous waste (by weight) as identified by the U.S. Environmental Protection Agency, 2001.

Secure Landfills

For some years, waste-disposal specialists have believed that, in principle, it is possible to design a **secure landfill** site for toxic liquid wastes. An example of a recommended design is shown in figure 16.18. The wastes are put in sealed drums before disposal. Beneath the drums are layers of plastic and/or compacted clay to contain any unexpected leaks. Wells and piping are installed so that the ground water below and around the site can be checked periodically for any sign of leakage of the waste chemicals. Excess accumulating leachate can be pumped out before it leaks out. Such a system provides multiple safeguards against accidental environmental contamination, and the monitoring wells allow prompt detection of any leaks. The design shares many features with modern municipal landfills, but with still more provisions for waste containment and site monitoring.

Unfortunately, a growing body of evidence indicates that no site is truly secure, even if conscientiously designed. Carefully compacted clay may be very low in permeability but is probably never completely impermeable, especially over long time intervals. Chemical and biological reactions in the wastes and leachate can rupture or decompose plastic, and the stress caused by the weight of wastes and cover can fracture a clay liner. Even relatively innocuous municipal waste can prove hard to contain. When built, the "Mount Trashmore" landfill in Evanston, Illinois, was hailed for its state-of-the-art design. But monitoring wells later revealed detectable leakage of at least a

A

B

FIGURE 16.17 Careless toxic-waste disposal leads to pollution. (A) TNT-contaminated water seeps to the surface during excavation at Weldon Springs Ordnance Works, St. Louis, Missouri. (B) Once soils are contaminated, cleanup usually means digging out that soil, then disposing of it elsewhere. Thorium-contaminated soil, Wayne, New Jersey.

Photographs by Bill Empson, courtesy U.S. Army Corps of Engineers.

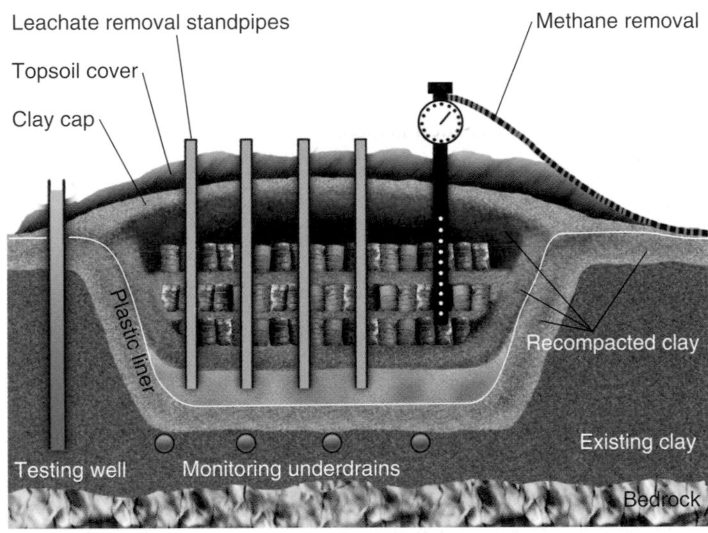

FIGURE 16.18 A secure landfill design for toxic-waste disposal, including provisions for leachate containment and for monitoring the chemistry of subsurface water nearby.

dozen volatile organic compounds deemed high-priority toxic pollutants by the Environmental Protection Agency, including benzene, toluene, vinyl chloride, and chloroform. Leakage from "secure" toxic-waste dumps, in which hundreds or thousands of barrels of concentrated toxic liquid chemicals are stored, has far more potential for harm. The historical response to detection of leakage from one toxic-waste dump has been to dig up as much of the hazardous material and contaminated soil as possible and transfer it to a more secure landfill. There is now some question as to whether a wholly different disposal method might be preferable.

Deep-Well Disposal

Another alternative for disposal of liquid industrial waste is injection into deep wells (figure 16.19). This method has been practiced since World War II. The rock unit selected to receive the wastes must be relatively porous and permeable (commonly, sandstone or fractured limestone), and it must be isolated by low-permeability layers (for example, shale) above and below. The subsurface geology must be known in sufficient detail that there is reasonable confidence that the disposal stratum remains isolated for some distance from the well site in all directions. Information about that geology may be derived from many sources: direct drilling to obtain core samples that provide a vertical section of the rock units present; geophysical studies that provide data on depths to and thicknesses of different rock layers, and on the distribution of ground water; geologic mapping on the basis of cores, surface outcrops, and geophysical data to interpolate between points sampled directly.

These disposal wells are hundreds to thousands of meters deep, far removed from the surface, and below the regional water table. The pore water in the disposal stratum should be brackish or saline water not suitable for a water supply. Where the well intersects any shallower aquifers that are or might be used for water supply, it must be snugly lined (cased) to prevent leakage of the wastes into those aquifers. Local well water is monitored to detect any accidental leaks promptly.

Movement of deep ground water is generally slow, and the assumption is that by the time the toxic chemicals have migrated far enough laterally to reach a usable aquifer or another body of water, they will have become sufficiently diluted not to pose a threat. This presumes knowledge of the toxicity of the chemicals in low concentrations. When the wastes are more or less dense than the ground water they displace and not miscible

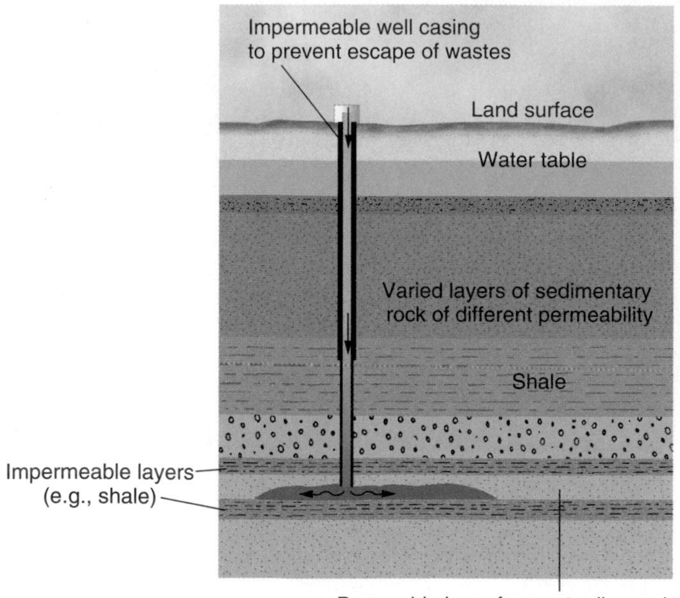

FIGURE 16.19 Deep-well disposal for liquid wastes. (A) Basic design: Wastes are placed in a deep permeable layer that is geographically and geologically isolated by low-permeability strata. (B) Containment of wastes assisted by geologic structures, much as petroleum is trapped.

with it, folds or other geologic structures may help to contain them and slow their spread, as oil traps contain petroleum (figure 16.19B). The behavior of chemicals that dissolve in the pore water is much less well understood. They can diffuse through the water more rapidly than the water itself moves so that even if deep groundwater transport is slow, contaminant migration may not be.

Costs for deep-well disposal are comparable to or somewhat less than those associated with "secure" landfill sites. The rate of waste disposal in a deep well is limited by the permeability of the rocks of the disposal stratum, while landfills have no equivalent limitation. The region's geology must be such that suitable strata exist for disposal by injection, while landfills can be constructed in a much greater variety of settings. Like landfills, deep injection wells may leak. Finally, as noted in chapter 4, deep-well waste injection may trigger earthquakes in faulted rocks. The various geologic constraints on possible deep-well disposal sites are thus more restrictive than those for landfill sites.

Other Strategies

There are now furnaces designed to incinerate fine streams of liquid. This permits the destruction of toxic liquid organic chemicals, often with no worse by-product than carbon dioxide, and may ultimately prove the best way to deal with them. Some liquid wastes can be neutralized or broken down by chemical treatment, which may avoid the necessity for ultrasecure disposal altogether. As with solid wastes, it may even be possible to use certain liquid "wastes" via waste exchanges. The nitric acid used in quantity by the electronics industry to etch silicon wafers can be neutralized to produce calcium nitrate and then incorporated in high-grade fertilizers. Spent acid used in the steel industry is rich in dissolved iron and can be used at geothermal power plants to control hydrogen sulfide gas emissions, which react with the iron in solution to precipitate iron sulfides.

These and other means of handling toxic liquid industrial wastes will continue to be developed. What these methods generally have in common is that they are designed to deal either with specialized types of waste or with limited quantities. Volumetrically, at least, a far larger liquid-waste disposal problem is posed by that very commonplace material, sewage.

SEWAGE TREATMENT

Problems arising from organic matter in water include oxygen depletion and "algal bloom," as described in chapter 17. These problems, and concern about the spread of disease through biological contamination of drinking-water supplies by *pathogenic* (disease-causing) organisms, provide the motivation for effective sewage disposal. Much of the approximately 40 billion gallons of water withdrawn for public water supplies in the United States each day winds up mixed with sewage. So does urban surface runoff water collected in storm drains, along with a portion of the water used in rural areas. Appropriate treatment strategy varies with population density and local geology.

Septic Systems

On an individual-user level, modern sewage treatment typically involves a septic system of some kind (figure 16.20). Wastes are first transferred to a settling tank in which solids settle out, to be broken down slowly through bacterial action. The remaining liquid carries a load of dissolved organic matter and of microorganisms—some pathogenic—whose metabolism requires little or no oxygen. The dissolved organic matter represents food for those microorganisms. The liquid is allowed to seep out through porous pipes into the soil of the **leaching field** or **adsorption field.** There, oxygen is available, in the pore spaces, and aerobic soil microorganisms that can use that oxygen in metabolizing the organic matter compete for the nutrients with the microorganisms in the sewage, breaking down the organic matter and destroying some pathogens.

Box 16.1 Decisions, Decisions . . .

"Well, *I'm* not wasting a foam cup for my coffee!" declared Bonnie firmly. "Paper makes a lot less waste. Besides, these cups are partly recycled paper. I get them from the same place I get my recycled-paper coffee filters."

"But when you've used up that paper cup, it's all trash," retorted Rod. "You can't recycle paper that's dirty with food waste. I, on the other hand, simply rinse my plastic foam cup, and voilà—I'm ready to recycle, having carefully observed the '6–PS' on the bottom, meaning that it's polystyrene, which our garbage-disposal company does take. I checked."

"Of course, you've just used—and dirtied—some water rinsing out the cup, Mr. Greener-than-thou. And my paper cup will biodegrade in the landfill and make methane! Your plastic cup is just going to take more energy to recycle—probably to make more trash-can liners like the one in this wastebasket."

"Well, it beats just making more trash we didn't have room for. And speaking of trash-can liners, your paper cup isn't going to do much of anything for years and years, neatly sealed in plastic in the landfill. Don't hold your breath waiting for that methane."

"Okay, but how about where the cups come from? Yours took oil to make; mine just took trees, and they're renewable."

Bonnie and Rod had progressed to the point of arguing the relative energy and materials efficiency of paper versus polystyrene cups when Jason sauntered in, waving his mug.

"I do hate to interrupt a good argument," he commented, "but may I suggest a superior alternative? What we have here is the basic china mug—made from abundant natural material, clay, and infinitely reuseable, unless I get clumsy."

There was a pause, and then Bonnie muttered, "It still needs washing, using water and detergent. So it's not perfect. At least I *hope* you wash that thing once in a while!"

This is a first look at the kind of analysis that can go into trying to choose the most environmentally benign of two or more alternatives. A complete comparison of paper versus polystyrene cups would require a good deal of research on the resource demands of, and pollution and waste produced by, the cutting of logs and milling and production of the paper cup, the drilling for oil to make and the process for the manufacturing of the polystyrene, and so on. Another common debate of a similar kind involves disposable paper versus reuseable cloth diapers. Both alternatives, in the latter case, use renewable raw materials, but the environmental consequences of their manufacture will differ; the disposables commonly end up in landfills, while the reuseables require water, detergents, and energy to clean. Again, a thorough, "cradle-to-grave" analysis of each alternative may be a good deal more complex, and the choices not as straightforward as they may initially appear.

Consider other everyday products that offer a choice of materials. That choice probably has waste-disposal and pollution, as well as resource-use, consequences. Think about what those might be and how you might decide which choice to make.

FIGURE 1 Paper, plastic, ceramic? The choice with the least negative environmental impact is not always obvious.

Passage through the soil, especially if it is fine-grained, also filters the liquid, removing remaining fine suspended solids and even the larger pathogenic organisms. Inorganic reactions in the soil can also break down undesirable compounds in the sewage. Ideally, by the time any of the liquid reaches either the surface or groundwater supplies, it has been purified of biological and many chemical contaminants. Some compounds do remain in solution, however; nitrate is ordinarily the most significant potential pollutant among them. (Should the settling tank require pumping out, disposal of the septage can be a further problem, for few sites will accept it, and municipal sewage-treatment facilities are reluctant to take on the added load.)

There are several geologic requirements for a properly functioning septic system. The soil must be sufficiently permeable that the fluids will flow through it rather than merely backing up in the septic tank, but not so permeable that the flow into water supplies or out at the surface occurs before the wastes have been sufficiently purified. Ordinarily, the water table should be well below the level of the septic system: first, to avoid immediate groundwater contamination by raw sewage; second, because oxygen levels in saturated soil are often too low to permit rapid aerobic breakdown of the organic matter. There must be sufficient soil depth so that the wastewater will be adequately filtered by the time it reaches either surface or bedrock; this filtration usually requires at least 60 centimeters of soil above the pipes and 150 centimeters below. The leaching field should not extend to within about 15 meters of any body of surface water, for similar reasons. (The exact dimensions required are controlled by soil characteristics.)

If a well is used to obtain drinking water on the same site, it must be far enough removed from the septic system that partially decomposed sewage does not reach the well, or else it should tap an aquifer that is isolated from the septic system by low-permeability material. Where many houses in a single area rely on septic systems, they must be spaced far enough apart so that they do not collectively saturate the soil with raw sewage and overwhelm the natural capacity of the soil and the microorganisms within it to handle the waste. The necessary spacing depends, in part, on the sizes of leaching fields to be accommodated. The required size of each leaching field is controlled, in turn, by soil permeability (table 16.2) and the number of persons to be served. Considerations such as these and

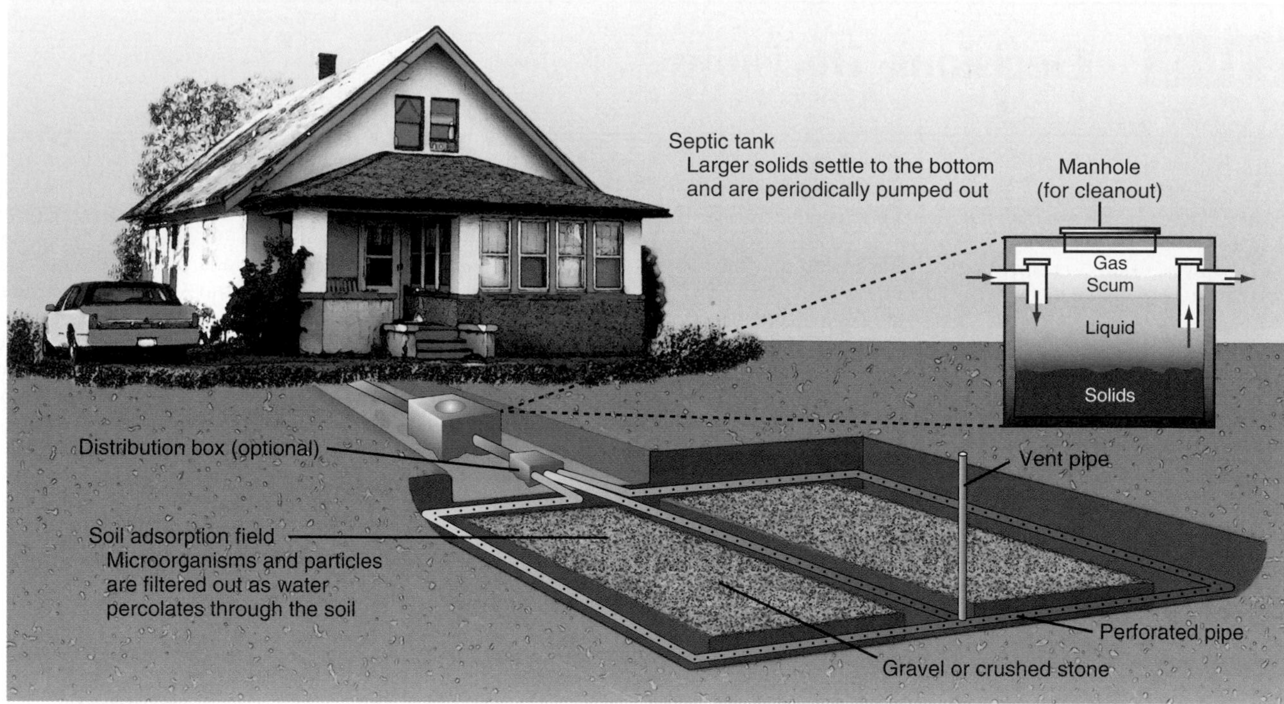

FIGURE 16.20 Basics of septic tank system: A settling tank for solids and the slow release of liquids into the soil of the leaching field for natural decomposition.

TABLE 16.2

Relationship between Soil Character and Required Extent of Leaching Field

Soil Type*	Adsorption Area Needed (square meters per bedroom)
gravel or coarse sand	6.5
fine sand	8.3
sandy loam	10.6
clay loam	13.9
sandy clay	16.2
clay with minor sand or gravel	23.1
heavy clay	unsuitable

*Arranged broadly in order of decreasing permeability from top to bottom of table.

Source: Data from *Water in Environmental Planning,* by T. Dunne and L. B. Leopold, p. 187. Copyright © 1978 W. H. Freeman and Company.

assessments of the potential impact of the nitrate released from these systems lead local authorities to stipulate minimum lot sizes where septic tanks are to be used. Typical lot sizes might be one-half acre to one acre per dwelling, but again, appropriate limits are controlled largely by local geology.

As is evident from the foregoing discussion, the sewage treatment associated with a septic system is entirely natural, and its thoroughness thus highly variable from site to site. One can enhance the suitability of the site—for example, by embedding the plumbing of the leaching field in "imported" sand if the soil is naturally too clay-rich to be suitable—but the natural soil chemistry, biology, and physical properties largely control the effectiveness of the system. Also, if toxic household wastes are dumped in the system, many will be untouched by the chemical processes and microbial activity that attack and decompose the organic wastes (and some might even kill those helpful microorganisms).

Just as natural wetlands can serve a water-cleaning role, so artificial wetlands and ponds can enhance a septic system and increase its capacity, with the aid of plants. The conventional subsurface leaching field can be replaced by one or more gravel-filled basins planted with wetland plants. While the effluent stays below the surface, the plants' roots reach down into the wet gravel. The plants consume some of the nutrients in the water, and their roots host quantities of the bacteria that decompose the sewage. The cleaned water may then be channeled into a pond supporting additional plant life and waterfowl. Most such schemes at present are small, experimental operations, but their aesthetic appeal can make them an attractive option where the wastewater load is not too great.

Municipal Sewage Treatment

In urbanized areas, population density is far too high to permit effective sewage treatment by septic systems. More than 75% of the U.S. population is now served instead by sewer systems.

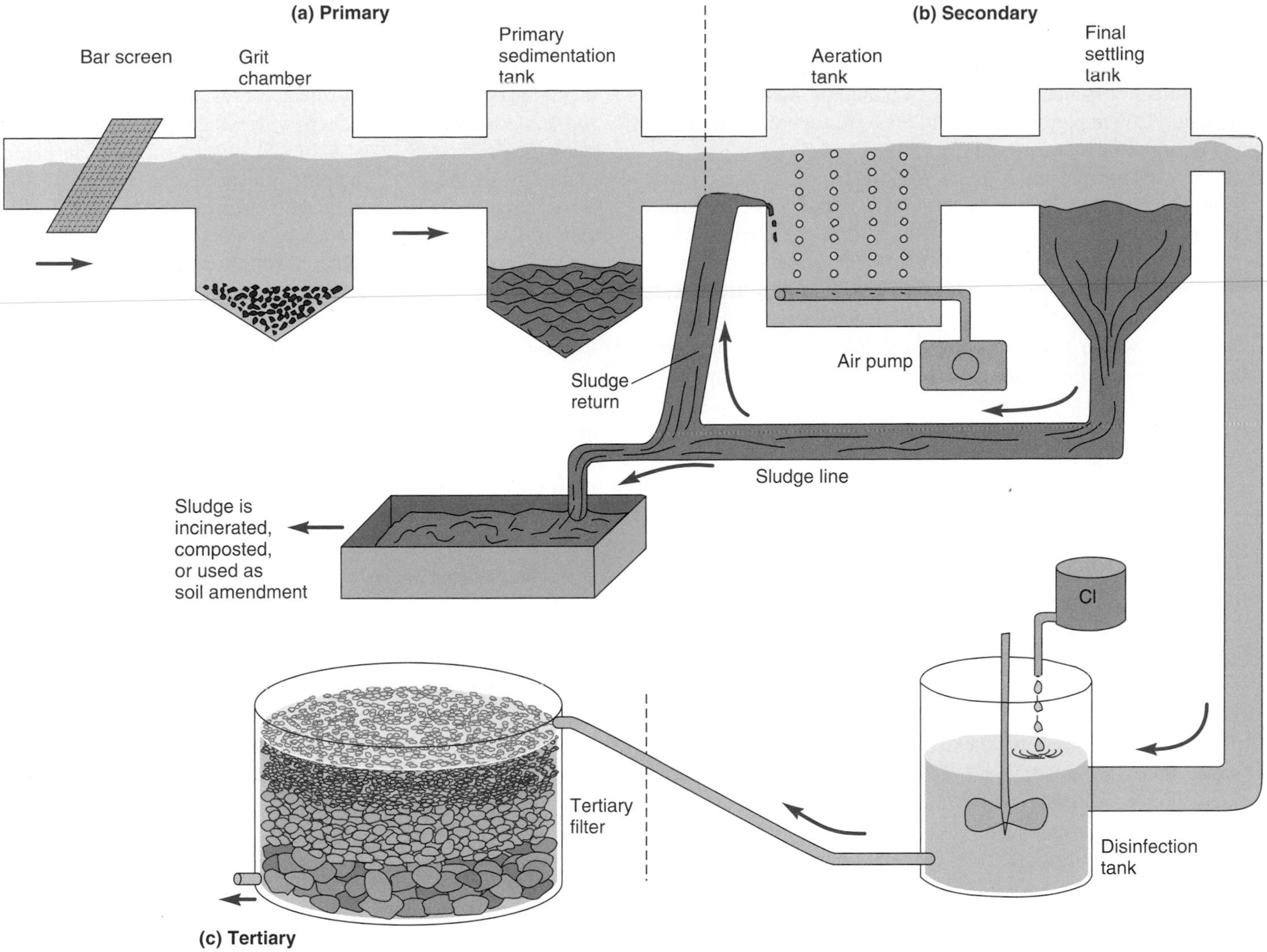

(a) Primary

Bar screen

Grit chamber

Primary sedimentation tank

(b) Secondary

Aeration tank

Final settling tank

Air pump

Sludge return

Sludge line

Sludge is incinerated, composted, or used as soil amendment

Cl

Tertiary filter

Disinfection tank

(c) Tertiary

A

In fewer than 5% of the cases is completely untreated sewage released from such systems, but the completeness of municipal sewage treatment does vary. The basic steps involved are summarized in figure 16.21A.

Primary treatment is usually physical processing only, involving removal of solids, including inorganic trash, fine sediment, and so on, that has been picked up by the wastewater. The screens remove large objects like floating trash. Sand and gravel then settle in the grit chamber. The flow is slowed, and suspended organic and inorganic matter settles into the bottom of the sedimentation tank. Oil or grease may have to be skimmed off the surface. Where industrial wastes are also present, some chemical treatment, such as neutralizing excess acidity or alkalinity, may be necessary.

Secondary treatment of the remaining liquid is mainly biological. The effluent from primary treatment is aerated, and bacteria and fungi act on the dissolved and suspended organic matter to break it down. The treated sewage passes through another settling tank to remove solids, which now contain the useful microorganisms, too. This sludge can be cycled back to

B

FIGURE 16.21 (A) Schematic diagram of primary, secondary, and tertiary stages of municipal sewage treatment. (B) Dallas Central Wastewater Treatment Plant.

(A) modified after U.S. Environmental Protection Agency; (B) photograph by Bill Empson, U.S. Army Corps of Engineers.

mix with fresh input water and air to continue the breakdown process. Sometimes, the water is also chlorinated to disinfect it when biological activity is completed. Primary plus secondary treatment together can reduce suspended solids and oxygen demand (see chapter 17) by about 90%, nitrogen by one-half, and phosphorus by one-third.

A weakness in many municipal treatment systems, even those that practice secondary treatment, is that large amounts of rain or meltwater runoff in a short time can overload the capacity of treatment plants. At such times, some raw sewage may be released untreated. Also, even secondary treatment has little impact on dissolved minerals and potentially toxic chemicals, which remain in the water. Recently, the safety of chlorination has been questioned because, in some communities where hydrocarbon impurities were already present in the water supply, chlorination was found to lead to the formation of chlorinated hydrocarbons, such as chloroform. Some of these compounds may be carcinogenic. However, the chlorination procedure often can be adjusted to minimize the production of chlorinated hydrocarbons. The value of chlorination in the destruction of pathogenic organisms probably far outweighs the danger from this largely controllable problem.

After secondary treatment, wastewater can also be subjected to various kinds of tertiary, or advanced, treatment and purified sufficiently to be recycled as drinking water. Fine filtration, passage through activated charcoal, distillation, further chlorination, and various chemical treatments to remove dissolved minerals are some possible forms of tertiary treatment. Such thorough tertiary wastewater treatment may cost up to five times what primary and secondary treatment together do, which is why tertiary treatment is uncommon. Only 2% of the U.S. population is served by systems undertaking tertiary sewage treatment. It is now practiced mainly in a few communities where water shortages compel water recycling. On the other hand, tertiary treatment of all U.S. municipal wastewater would cost only a few tens of dollars per person per year. As a means of extending dwindling drinking-water supplies, tertiary treatment is likely to find increasing use in the future.

Wastewater treatment can yield other benefits besides drinking water. The city of Arcata, California, uses sewage that has passed through primary treatment to supply water to marshes and ponds, where natural microbial action and plant growth purify the water further and remove nutrients from it. Ultimately, the water purity is such that the output water can be used at a fish hatchery. The Tapia Water Reclamation Facility in northwestern Los Angeles County takes 10 million gallons of sewage per day and transforms it into water of high enough quality to use for irrigation.

A by-product of sewage treatment is quantities of sludge, which can present a solid-waste disposal problem. Chicago's municipal sewage treatment facilities process 1.4 billion gallons of sewage and generate approximately 600 tons of sludge (on a dry weight basis) *every day*. What can be done with such wastes? One option, because sludge is rich in organic matter and nutrients, is to use it as fertilizer. In the early 1970s, for example, Chicago developed a plan for using some of its sludge to fertilize 11,000 acres of formerly strip-mined land during reclamation—thereby neatly solving several environmental problems at once. Austin, Texas, composts and transforms sludge into "Dillo Dirt" with which parks and athletic fields are fertilized; it is also sold to gardeners. (The name comes from the city's mascot, the armadillo.) The city of Milwaukee has been selling the sludge as the organic fertilizer "milorganite" for decades. However, much urban sludge contains unacceptably high concentrations of heavy metals or other toxic industrial chemicals, which have to be removed with the organic solids, and which are too dispersed through the sludge for efficient extraction. This may limit the use of such sludge fertilizer to land not to be used for food crops or pastureland (parks, golf courses, and so on) and may even necessitate its disposal in landfill sites like other solid wastes.

In speaking of sewage treatment, it is worth keeping in mind the disparities between the industrialized and other nations. According to the World Resources Institute, it has been estimated that over 90% of the sewage in developing countries is discharged into surface waters with no treatment whatever; for example, of India's 3100-plus cities and towns, only a few hundred have even partial sewage treatment. It should also be noted that while the focus here has been on U.S.-style sewage transport and treatment, which involve a great deal of water, treatment of human wastes does not necessarily require those water-intensive approaches. In this country, composting toilets are perhaps the most familiar example of a non-aqueous system. Other nations, where water is in shorter supply, have made larger-scale use of such approaches, and they become increasingly appealing as water shortages become more acute.

RADIOACTIVE WASTES

Radioactive wastes differ somewhat in character from chemical wastes. Thus, proposals for the disposal of solid and liquid radioactive wastes usually differ somewhat from the methods used for other wastes, although some radioactive materials are chemical toxins as well.

Radioactive Decay

Some isotopes, some of the possible combinations of protons and neutrons in atomic nuclei, are basically unstable. Sooner or later, unstable nuclei decay. In doing so, they release radiation: alpha particles (nuclei of helium-4, with two protons, two neutrons, and a +2 charge), beta particles (electrons, with a −1 charge), or gamma rays (electromagnetic radiation, similar to X rays or microwaves but shorter in wavelength and more penetrating). A given decay may emit more than one type of radiation.

Radioactive decay is a statistical phenomenon. It is impossible to predict the instant at which an atom of a radioactive element will decay. However, over a period of time, a fixed percentage of a group of atoms of a given radioactive isotope (radioisotope) will decay. The idea is somewhat like flipping a coin: One cannot know beforehand whether a given flip will come up heads or tails, but over a great many flips, one can expect that half will come up heads, half tails.

Each different radioisotope has its own characteristic rate of decay, often described in terms of a parameter called **half-life.** The half-life of a radioisotope is the length of time required for half the atoms of that isotope initially present in a system to decay. The concept is illustrated in figure 16.22. Suppose, for example, that the half-life of a particular radioactive isotope is ten years and that, initially, there are 2000 atoms of it present. After ten years, there will be about 1000 atoms left; after ten more years, 500 atoms; after ten more years, 250; and so on. After five half-lives, only a few percent of the original amount is left; after ten half-lives or more, the fraction left is vanishingly small.

Half-lives of different radioactive elements vary from fractions of a second to billions of years, but for any specific radioactive isotope, the half-life is constant. Uranium-238 has a half-life of about 4.5 billion years; carbon-14, 5730 years; radon-222, 3.82 days. The half-life of a given isotope is constant regardless of the physical or chemical state in which it exists, whether it is in a mineral or dissolved in water, in the atmosphere or deep in the crust at high temperature and pressure. Spontaneous radioactive decay cannot be accelerated or slowed down by chemical or physical means. An important consequence of this is that radioactive wastes will continue to be radioactive and each isotope will decay at its own rate regardless of how the wastes are handled. The wastes cannot be made to decay faster to get rid of the radioactive material more quickly, and they cannot be treated to make them nonradioactive. This is a key difference between radioactive wastes and many chemical toxins; many of the latter can, through proper treatment, be broken down or neutralized, which increases the disposal options available for dealing with them.

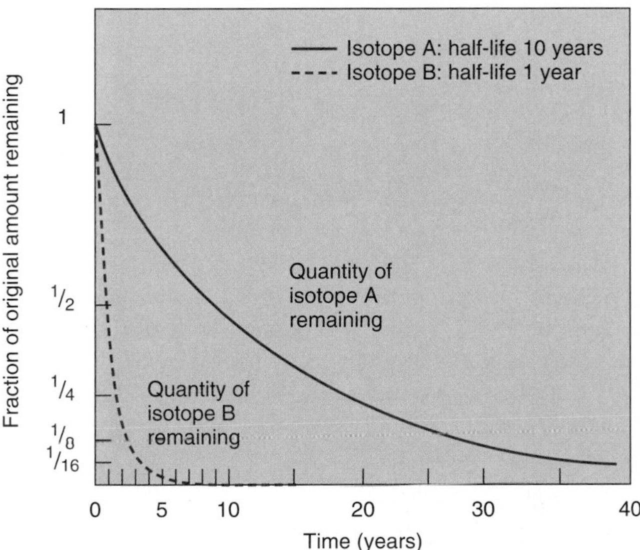

FIGURE 16.22 The phenomenon of radioactive decay: Each radioisotope decays at its own characteristic rate, defined by its half-life.

Effects of Radiation

Alpha and beta particles and gamma rays all are types of ionizing radiation. That is, they can strip off electrons from atoms or split molecules into pieces. Depending on which particular atoms or molecules in an organism are affected and the intensity of the radiation dose, the results could include genetic mutations, cancer, tissue burns—or nothing significant at all. Alpha particles are more massive but not very energetic. They can be very damaging but cannot travel far through matter; they can be stopped by a sheet of paper. Alpha-emitting radioisotopes are thus dangerous only when inhaled or ingested. Beta particles are lighter and less damaging but travel farther; they can be stopped by wood. Gamma rays are the most energetic and penetrating, but can be stopped by concrete.

How much damage is done by a given dose of radiation? How small a dose is harmless? Scientists do not know precisely. Most controlled radiation studies have been done on animals, not humans. The most definitive data on radiation effects on people come from accidental exposures of scientists or technicians to sizable doses of radiation or from observing the after-effects of the explosion of atomic weapons in Japan during World War II. The reactor accident at Chernobyl is further expanding the database. In those cases, the doses received by the victims are generally very imprecisely known, and their relevance to concerns about exposure to low levels of radiation is not clear in any case.

The effects of low doses of radiation on humans are particularly hard to quantify for several reasons. One is that some of the consequences, such as slowly developing cancers, do not appear for many years after exposure. By that time, it is much harder to link cause and effect clearly than it is, for example, in the case of severe radiation burns resulting immediately from a massive dose. A related problem is that many of the results of radiation exposure—like cancer or mutations—can have other causes, too. There are hosts of known or suspected *chemical* carcinogens and mutagens; how does one uniquely identify the cause of each particular cancer or mutation, years after exposure to whatever caused it?

Another uncertainty arises from the question of linearity in cause and effect. If a population is exposed to a certain dose of a certain type of radiation and one hundred cases of cancer result, does that mean that one-tenth the exposure would induce ten cases of cancer, and so on? Or is there a minimum harmful level of radiation exposure, below which we are not at risk?

We are, after all, surrounded by radiation. Cosmic rays bombard us from space; naturally occurring radioactive elements decay in the soil beneath our feet and in the wood and concrete in our homes. Perhaps, in consequence, humans have developed a tolerance of radiation doses that are not significantly above natural background levels. Nobody can avoid

Box 16.2 The Ghost of Toxins Past: Superfund

Disposal of identifiably toxic wastes in the United States is carefully controlled today. It has not always been so. In some cases, disposal companies hired by other firms to dispose of toxic wastes have done so illicitly and inappropriately. In other instances, the waste-generating companies dumped or buried their own wastes on-site. Decades later, when the firm responsible is long gone, pollutants may ooze from the land surface or may suddenly be detected in ground water. Sometimes, the waste disposer can be found but denies that that problem is serious and/or declines to spend the money to clean up the site. Sometimes, the responsible company is no longer in business.

In response to these and other similar situations, in late 1980, Congress passed the Comprehensive Environmental Response, Compensation, and Liability Act, which, among other provisions, created what is popularly known as Superfund. The original $1.6 billion trust fund was generated primarily through taxes on oil and on producers of forty-two specific hazardous chemicals. Superfund monies are intended to pay for immediate emergency cleanup of abandoned toxic-waste sites and of sites whose owners refuse to clean them up. This represents a major step toward prompt rectification of past waste-disposal abuses following recognition of the problem. Superfund was reauthorized several times, its trust fund increasing to $13.6 billion before its taxing authority expired in 1995.

Meanwhile, there has been increasing realization that the size of the problem is far greater than originally believed. At the time Superfund was created, Congress envisioned cleanup of about 400 toxic-waste sites. By 1987, over 950 "priority" sites had been identified (figure 1); that number later rose to over 1500. Many other sites posing less-acute risks were also identified. By 1995, the Environmental Protection Agency was working with state agencies and tribal councils to assess or clean up over 40,000 sites. Overall, EPA estimates that 1 in 4 Americans lives within 3 miles of one or more Superfund sites..

The EPA's attention is not restricted to conventional toxic-waste sites, either. Mining of vermiculite (a silicate used in insulation and to improve soil structure for growing plants) began at Libby, Montana in the early 1920s. The operation was highly productive, at its peak producing an estimated 80% of the vermiculite used worldwide. Unfortunately, in the 1960s it was realized that the Libby vermiculite was associated

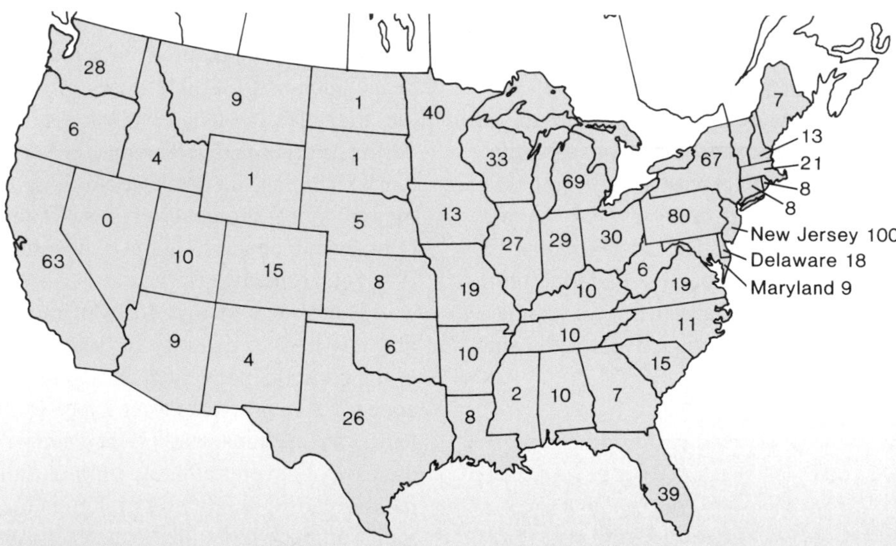

FIGURE 1 Distribution of the first 951 toxic-waste dump sites placed, or proposed for placement, on the EPA National Priorities List (NPL) for Superfund cleanup as of July 1987. *Source: U.S. Environmental Protection Agency.*

some natural exposure to radiation (table 16.3). We increase our exposure with every medical X ray, every jet airplane ride, every journey to a high mountaintop, and every time we wear a watch with a luminous dial. Indeed, each of us is radioactive, a consequence of naturally occurring carbon-14 and other radioisotopes. Therefore, any additional exposure from radioactive waste (or uranium in mine tailings, or other nonroutine sources) is just that, an incremental increase over our inevitable, constant exposure to radiation. Does such a small exposure to low-level radiation represent a significantly increased risk? How small is small? Frustratingly, perhaps, there are no simple, direct answers to questions such as these. In part because individuals' background exposure to radiation is universal, yet quite variable, it is virtually impossible to construct a definitive dose-response curve for low radiation doses. That uncertainty, together with a lack of public understanding of natural radiation exposure, contributes to the controversy concerning radioactive-waste disposal.

Nature of Radioactive Wastes

As noted in chapter 15, nuclear fission reactor wastes include many radioisotopes. The isotopes of most concern from the standpoint of radiation hazards are not those of very short half-life

(Box 16.2 Continued)

with significant asbestos, and not the relatively benign variety, chrysotile, discussed in chapter 2. Tests in the 1980s showed 4–7% asbestos in the processed vermiculite. The mine was closed in 1990. Cleanup of asbestos-contaminated soil around the processing plant is underway, along with further studies of possible risks to area residents who were not directly involved with the vermiculite mining and processing.

Remedial action may be in progress at hundreds of sites each year. By the end of fiscal year 2000, over 6400 removal actions—over 1000 at NPL sites—had been conducted. These "removals" are short-term, emergency actions where "imminent and substantial" health risks are found to exist. Many such sites then require more comprehensive long-term cleanup action.

More problem areas continue to be identified. The *average* cost of cleaning up a Superfund site is over $30 million, not counting any associated litigation, and some projections now put eventual total cleanup costs of Superfund sites at around *$1 trillion.* Where responsible parties can be identified, they do contribute to the cleanup costs—over $24 billion through fiscal year 2005—but massive cleanup bills remain. In part, federal appropriations have compensated for the loss of the original tax revenues. It has been suggested that additional funding might be provided by a tax on *disposers* of toxic waste, rather than the generators of it. There is concern, however, that a tax added at the disposal end might tend to encourage more illicit dumping of toxic waste, ultimately compounding an already large problem.

In the meantime, every bit of hazardous waste cleaned up and removed from one spot still has to be disposed of in another. . . . And Superfund deals only with abandoned or inoperative sites in need of cleanup. In the first decade of its operations, Superfund addressed cleanup operations at hundreds of sites on and off the priority list (figure 2); by the end of fiscal year 2005, 946 NPL Superfund sites had cleanup completed, with work on over 400 more in progress. Thousands of non-NPL abandoned toxic-waste sites remain to be addressed, and the EPA has also identified an estimated 2500 *active* disposal sites that may require corrective action, too. Waste minimization becomes increasingly appealing as these problematic sites multiply.

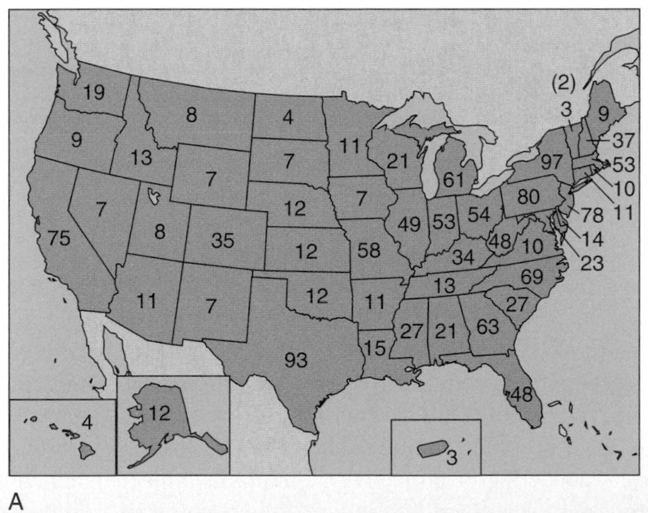

FIGURE 2 (A) Completed removals during the first decade of Superfund, 1980–1990. (B) The Berkeley Pit, Butte, Montana, part of the largest toxic-mine-waste site on the NPL. The 140-acre Pit is one of several former open-pit mines at the site, now filled hundreds of feet deep with acidic water containing copper, arsenic, zinc, lead, and other potentially toxic metals. To keep the water levels in these old open-pit mines low enough that they do not contaminate local ground and surface waters, pumping and water treatment may need to be continued indefinitely, so cleanup here may never be truly completed.

(A) Data from Superfund Emergency Response Action, Sixth Annual Report—Fiscal Year 1991, U.S. Environmental Protection Agency. (B) Photograph © Calvin Larsen/Photo Researchers, Inc.

(if an isotope has a half-life of 10 seconds, it is virtually all gone in a few minutes) or very long half-life (if an isotope's half-life is 10 billion years, very little of it will be decaying over a human lifetime). It is the elements with half-lives of the order of years to hundreds of years that pose the greatest problems: They are radioactive enough to be significant hazards, yet will persist in the environment for some time.

Some of the waste radioactive isotopes are also toxic chemical poisons, so they are dangerous independently of their radioactivity, even if their half-lives are long. Plutonium-239, with a half-life of 24,000 years, is one example of such an isotope. Other isotopes pose special hazards because the corresponding elements are biologically concentrated, often in one particular organ, leading to the possibility of a concentrated radiation source in the body. Included in this category are such products as iodine-131 (half-life, 8 days), which would be concentrated, like any iodine, in the human thyroid gland; iron-59 (45 days), which would be concentrated in the iron-rich hemoglobin in blood; cesium-137 (30 years), which is concentrated in the nervous system; and strontium-90 (29 years) which, like all strontium, tends to be concentrated with calcium—in bones, teeth, and milk.

Radioactive wastes are often classified collectively as **low level** or **high-level.** The division is an imprecise one. The

TABLE 16.3

Typical Natural Radiation Exposure

Source	Dosage, mrem/yr*
radon	200
external (cosmic rays, rocks, soil, masonry building materials)	55
medical (mainly X rays)	53
internal (in the body)—includes doses from food, water, air	39
consumer products	10
other, including normal operation of nuclear power plants	3
Where and how you live can affect your exposure	
worldwide radioactive fallout	4 mrem/yr
chest X ray	10–39 mrem
lower-gastrointestinal-tract X ray	500 mrem
jet travel—for each 2500 miles, add	1 mrem
TV viewing, 3 hours/day	1 mrem/yr
cooking with natural gas, add	6 mrem/yr

*Radiation dosages are expressed in units of *rems* or, for very small doses, *millirems* (mrem). A *rem* ("radiation equivalent man") is a unit that takes into account the type of radiation involved, the energy of that radiation, the amount of it to which one has been exposed, and the effect of that type of radiation on the body. The U.S. average annual dosage is 360 mrem.

Source: U.S. Department of Energy.

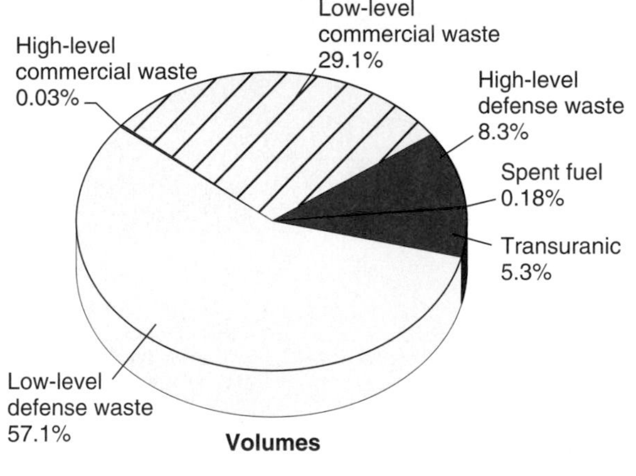

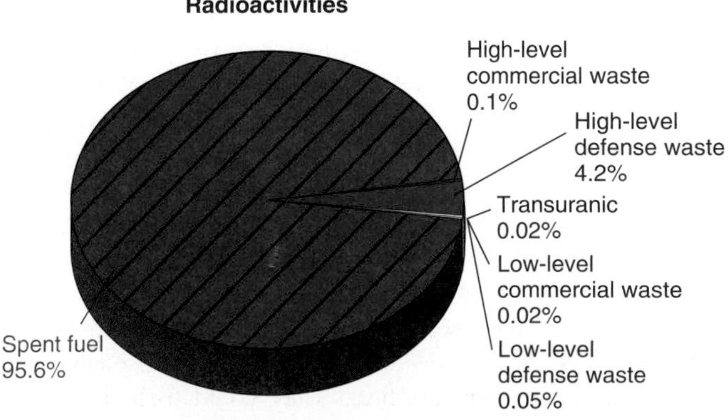

low-level wastes are relatively low-radioactivity wastes, not requiring extraordinary disposal precautions. Volumetrically, they account for over 90% of radioactive wastes generated (figure 16.23). Some low-level wastes are believed to be harmless even if released directly into the environment. In this category, for example, fall the routine small releases of radioactive gas from operating fission reactors. Liquid low-level wastes—from laundering of protective clothing, from decontamination processes, from floor drains—are often released also, diluted as necessary so that the concentration of radioactivity is suitably low. Solid low-level wastes, such as filters, protective clothing, and laboratory materials from medical and research laboratories, are commonly disposed of in landfills or held in temporary storage until the radioactivity has decreased enough that they can be consigned to a landfill. Currently, there are twenty temporary repositories and six commercial disposal sites for low-level wastes. For environmental and capacity reasons, including problems of waste leakage, some of the commercial disposal sites have been closed. Federal legislation required states to take the responsibility for their own low-level wastes in the early 1990s (though they may enter into multi-state compacts, and several of these exist). States that benefit most from nuclear power and research and other facilities using nuclear materials thus bear more of the burden of the disposal of the corresponding

FIGURE 16.23 By volume, high-level waste is a small fraction of the radioactive-waste-disposal problem—but it accounts for nearly 100% of the radioactivity, and thus the radiation hazard.
Data from U.S. Department of Energy report DOE/RW-0006.

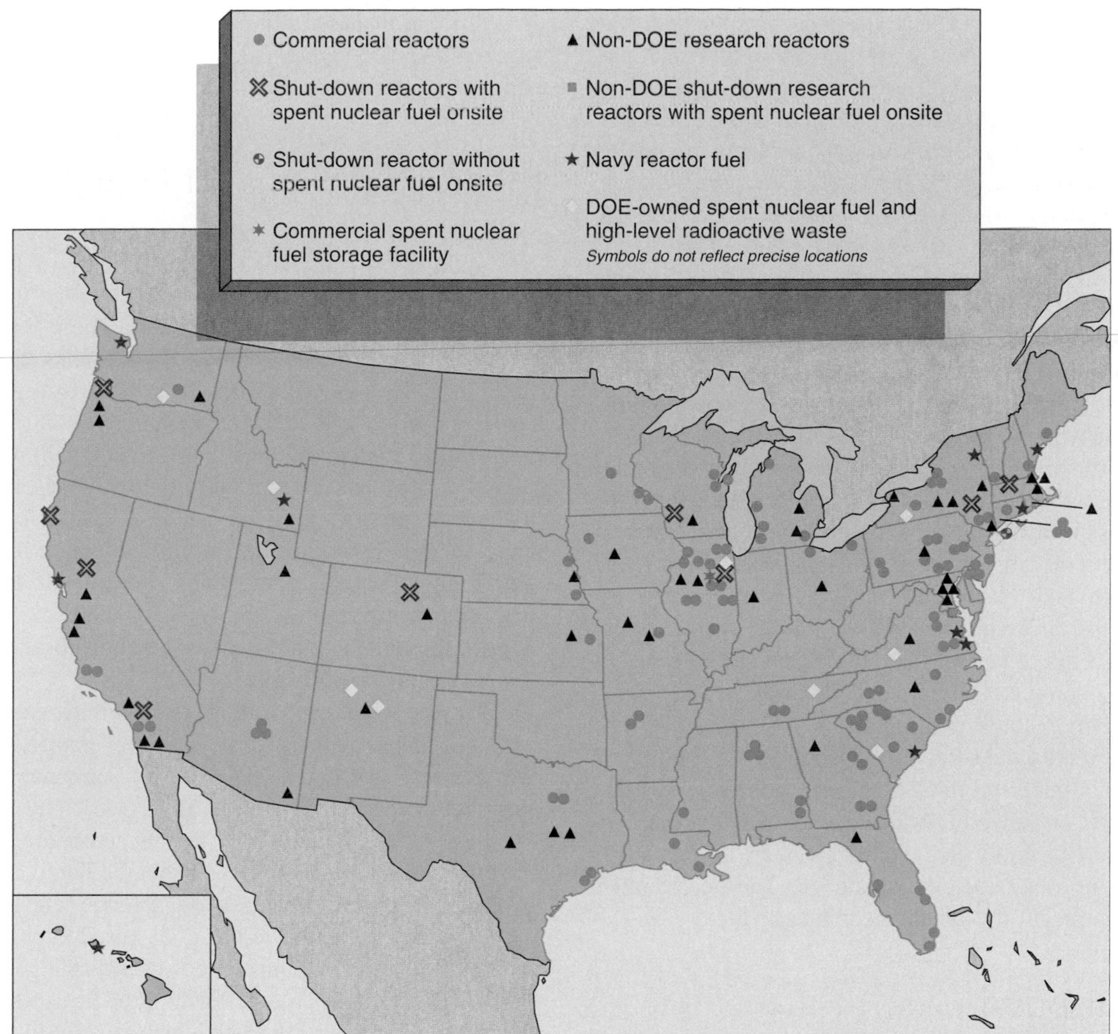

FIGURE 16.24 Locations (approximate) of spent nuclear fuel elements and high-level radioactive wastes in the U.S. awaiting disposal.
Source: From U.S. Department of Energy, Office of Civilian Radioactive Waste Management.

low-level wastes; and the transportation of those wastes is reduced. There is often concern over local disposal of low-level wastes (the NIMBY phenomenon again). As a result, no states have yet developed the mandated disposal sites to the point of actual disposal, and in some cases, planning is not far advanced. However, these issues are minimal compared to concerns over the much more intensely radioactive high-level wastes.

Spent reactor fuel rods and by-products from their fabrication and reprocessing are examples of high-level wastes, for which various more-or-less elaborate disposal schemes have been proposed. Currently, they sit in temporary storage—often on-site at nuclear reactors and other nuclear facilities—awaiting permanent disposal (figure 16.24). The basic disposal problem with the chemically varied high-level wastes is how to isolate them from the biosphere with some confidence that they will stay isolated for thousands of years or longer. In 1985, the Environmental Protection Agency specified that high-level wastes be isolated in such a way as to cause fewer than 1000 deaths in 10,000 years (the expected fatality rate associated with unmined uranium ore). The sections that follow survey some of the many strategies proposed for dealing with these high-level and other concentrated radioactive wastes.

Historical Suggestions: Space, Ice, and Plate Tectonics

Some of the more exotic schemes proposed in the past for disposal of high-level radioactive wastes involved putting wastes in space, perhaps rocketing them into the sun. Certainly, that would get rid of the wastes permanently—if the rocket launch were successful. However, many launches are not successful (as the shuttle *Challenger* explosion dramatically demonstrated) or are aborted, and the payloads return to earth, sometimes in an uncontrolled fashion. Furthermore, since many spacecraft would be required, the costs would be extremely high, even if successful launches could somehow be guaranteed.

Others have proposed putting encapsulated wastes under thick ice sheets, with Antarctica a suggested site. High-level radioactive wastes generate heat. They could melt their way down through the ice sheet, to be well isolated eventually by the overlying blanket of ice, which would refreeze above. But glacier ice flows, and it flows at different rates at the top and bottom of a glacier. How would one keep track of waste movement through time? What if future climatic changes were to melt the ice sheet? For various reasons, disposal of radioactive wastes in Antarctica is presently prohibited by international treaty, and no other, thinner ice sheets are seriously under consideration.

Subduction zones also have been considered as possible repositories for high-level wastes. The seafloor trenches above subduction zones are often sites of fairly rapid accumulation of sediments eroded from a continent. In theory, these sediments would cover the wastes initially, and the wastes would, in time, be carried down with the subducting plate into the mantle, well out of the way of surface-dwelling life. However, subduction is a slow process; the wastes would sit on the sea floor for some time before disappearing into the mantle. Even at sedimentation rates of several tens of centimeters per year, large waste containers would take years to cover. Seawater, especially seawater warmed by decaying wastes, is highly corrosive. Might the wastes not begin to leak out into the ocean before burial or subduction was complete? How would the wastes be attached to the down-going plate to ensure that they would actually be subducted? Currently, the behavior of sediments in subduction zones is not fully understood. Geologic evidence suggests that some sediments are indeed subducted with down-going oceanic lithosphere, but some, on the other hand, are "scraped off" by the overriding plate and not subducted. The uncertainties appear to be too great for this scheme to be satisfactory.

Seabed Disposal

The deep-sea floor is a relatively isolated place, one with which humans are unlikely to make accidental contact. Much of this area is covered by layers of clay-rich sediment that may reach hundreds of meters in thickness, and studies have shown that the clays have remained stable for millions of years. Interest is growing in radioactive-waste disposal schemes that involve sinking waste canisters into these deep, stable clays (in areas well removed from both active plate boundaries and any seafloor resources, notably manganese nodules). The low permeability of clays means restricted water circulation in the sediments. Many of the elements in the wastes, even if they did escape from the containers, would be readily adsorbed by the fine clay particles. Any dissolved elements escaping into the ocean would be entering the dense, cold (1 to 4°C, or 34 to 39°F), saline, deep waters that mix only very slowly with near-surface water. By the time any leaked material reached inhabited waters, it would be greatly diluted, and many of the shorter-half-life isotopes would have decayed. The long-term geologic stability of the deep-sea floor is appealing, especially to nations having radioactive wastes to dispose of but not located in particularly stable areas of the continents, or having such high population densities that they lack remote areas for disposal sites. An international Seabed Working Group currently has this option under active study. It has already concluded that seabed disposal is technically feasible but that more research on the long-term safety of the method is still needed.

Bedrock Caverns for Liquid Waste

Many of the high-level radioactive wastes are presently held in liquid form rather than solid, in part to keep the heat-producing radioactive elements dilute enough that they do not begin to melt their storage containers. These liquid wastes are currently stored in cooled underground tanks, holding up to 1 million gallons apiece, which from time to time have leaked spectacularly: At the federal radioactive-waste repository in Hanford, Washington, more than 300,000 gallons of high-level liquid wastes leaked from their huge storage tanks in the late twentieth century. (Currently, remediation efforts are in process, including bioremediation through which microorganisms break down organic compounds and convert toxic metals to less-mobile compounds; but these methods can do nothing to reduce the radioactivity. Secure initial disposal is clearly preferable to later remediation.)

Caverns hollowed out of low-permeability, unfractured rocks—igneous rocks such as basalt or granite, for example—have been suggested as more secure, permanent disposal sites for high-level liquid wastes. The rocks' physical properties would have to be thoroughly investigated beforehand to ensure that the caverns could indeed contain the liquid effectively. The wells leading down to the caverns, through which wastes would be pumped, would have to be tightly cased to prevent shallower leaks. The long-term concern, however, is the geologic stability of the bedrock unit chosen. Faulting or fracturing of the rocks would create conduits through which the highly radioactive liquid could escape rapidly, perhaps to the surface or into nearby aquifers. While it is possible to identify with a high degree of probability the areas that are likely to remain stable for thousands or millions of years, stability cannot be absolutely guaranteed.

Serious consideration has been given to solidifying liquid high-level wastes prior to disposal because of the much lower mobility of solids. Certain minerals, ceramics, and glasses have proven to be relatively resistant to leaching over centuries or even millennia in the natural environment. Conversion of the wastes into a relatively stable solid form would be a first step in many high-level waste disposal schemes. The waste would then be sealed in canisters and the canisters placed in some kind of bedrock cavern, old mine, or similar space. Solidification would not eliminate the uncertainty regarding bedrock stability, but it would render the wastes less susceptible to rapid escape if fracturing breached the storage volume.

Bedrock Disposal of Solid High-Level Wastes

A variety of bedrock types are being investigated worldwide as host rocks for solid high-level-waste disposal. The general design of a bedrock disposal site or repository for such wastes is shown in very simplified form in figure 16.25. It involves the **multiple barrier** (or *multibarrier*) **concept:** surrounding solid waste with several different types of materials to create multiple obstructions to waste leakage or invasion by ground water. A major variable is the nature of the host rock. Each of the rock types highlighted below has some particular characteristics that make it promising as a potential host rock, as well as some potential drawbacks.

Granite is a plentiful constituent of the continental crust. Granites are strong and very stable structurally following excavation. They have low porosity; the common minerals of granite (quartz, feldspars) are quite insoluble in a temperate climate. A concern with respect to granite is the possible presence of fractures (natural or created during excavation) that would allow solutions to permeate and wastes, perhaps, to escape. One might expect that, below a depth of about 1 kilometer, the pressures would close any fractures. However, field measurements suggest that this is not necessarily the case. Tests on granite are underway at the Nevada Test Site; the United States and Sweden also have a joint testing program in progress at Stripa, Sweden. Half a dozen nations worldwide are investigating granite as a potential host rock for radioactive-waste disposal.

Thick *basalts,* such as the Columbia River plateau basalts described in chapter 5, are another option. Fresh, unfractured basalt is a strong rock. Basalt consists of high-temperature minerals and sometimes glass, so it can withstand the elevated temperatures of high-level wastes, and it has fairly high thermal conductivity, to allow that heat to be dispersed. However, basalts commonly contain porous zones of gas bubbles; they may also weather easily. Both phenomena contribute to zones of weakness. Fractures in the basalt and even the planes between successive flows may provide water conduits.

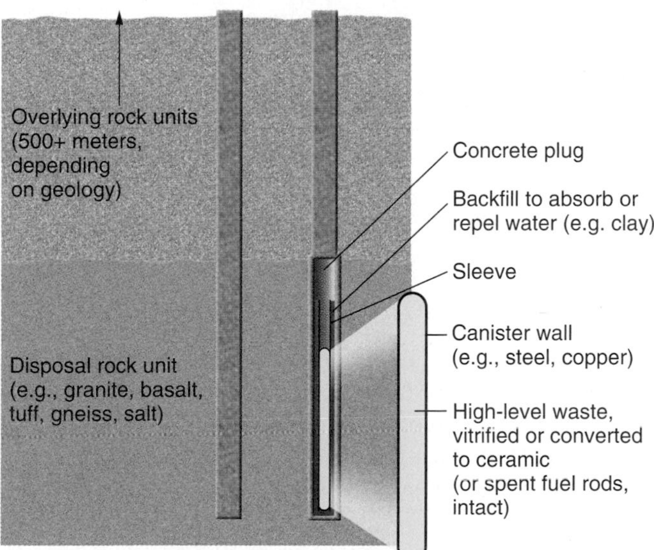

FIGURE 16.25 Schematic of a high-level-waste disposal site in bedrock: the multibarrier concept.

Massive deposits of *tuff,* pyroclastic materials deposited during past large-scale explosive volcanic eruptions, are another possible host rock for solid high-level wastes. *Welded tuffs* are those that were so hot at the time of deposition that they fused; these deposits are not unlike basalts in many of their physical properties. But they are very brittle and easily fractured, which raises questions about their integrity with respect to water flow. Other tuffs have been extensively altered, producing abundant *zeolites.* These hydrous silicate minerals—the same group used in water softeners—are potentially valuable in a waste-disposal site for their ion-exchange capacity. They should tend to capture and hold wastes that might begin to migrate from a repository. However, zeolitic tuffs are physically weaker, more porous, and more permeable than welded tuffs; furthermore, zeolites begin to break down and dehydrate at relatively low temperatures (100 to 200°C, or about 200 to 400°F), which reduces their ion-exchange potential and also releases water that might facilitate waste migration.

Shale and other clay-rich sedimentary rocks are another option. As noted earlier in the seabed-disposal discussion, clays may adsorb migrating elements, providing a measure of containment for the waste. Shale is also low in permeability, and somewhat plastic under stress, though less so than salt (see below). On the other hand, shaly rocks are weak, fracture readily, and may be interlayered with much more permeable sedimentary units. At elevated temperatures, clays, like zeolites, may decompose and dehydrate. Some clays are stable to higher temperatures than are most zeolites, but those that would most actively adsorb waste components tend to lose their adsorptive efficiency at those higher temperatures. So, overall, clays may not have clear advantages over zeolites in disposal sites. Belgium and Italy have investigated shaly rocks as disposal units.

In the United States and Germany, much consideration has been given to high-level waste disposal in thickly bedded *salt* deposits or salt domes. Salt has several special properties that may make it a particularly suitable repository. It has a high melting point, higher in fact than those of most rocks. Consequently, salt can withstand considerable heating by the radioactive wastes without melting. Although soluble under wet conditions, salt typically has low porosity and permeability. Under dry conditions, then, it can provide very "tight" storage with minimal leakage potential. Salt's low porosity and permeability result, in part, from its ability to flow plastically under pressure while remaining solid. It is therefore somewhat self-sealing. If it were fractured by seismic activity or by accumulated stress in the rocks, flow in the salt could seal the cracks. Most other rock types are more brittle under crustal conditions and do not so readily self-seal. Finally, evaporites are not particularly scarce resources. Using an old salt mine for radioactive-waste disposal would not be a hardship in that context.

Waste Isolation Pilot Plant (WIPP)

While commercial high-level wastes await permanent disposal, the U.S. government *has* inaugurated a disposal site in

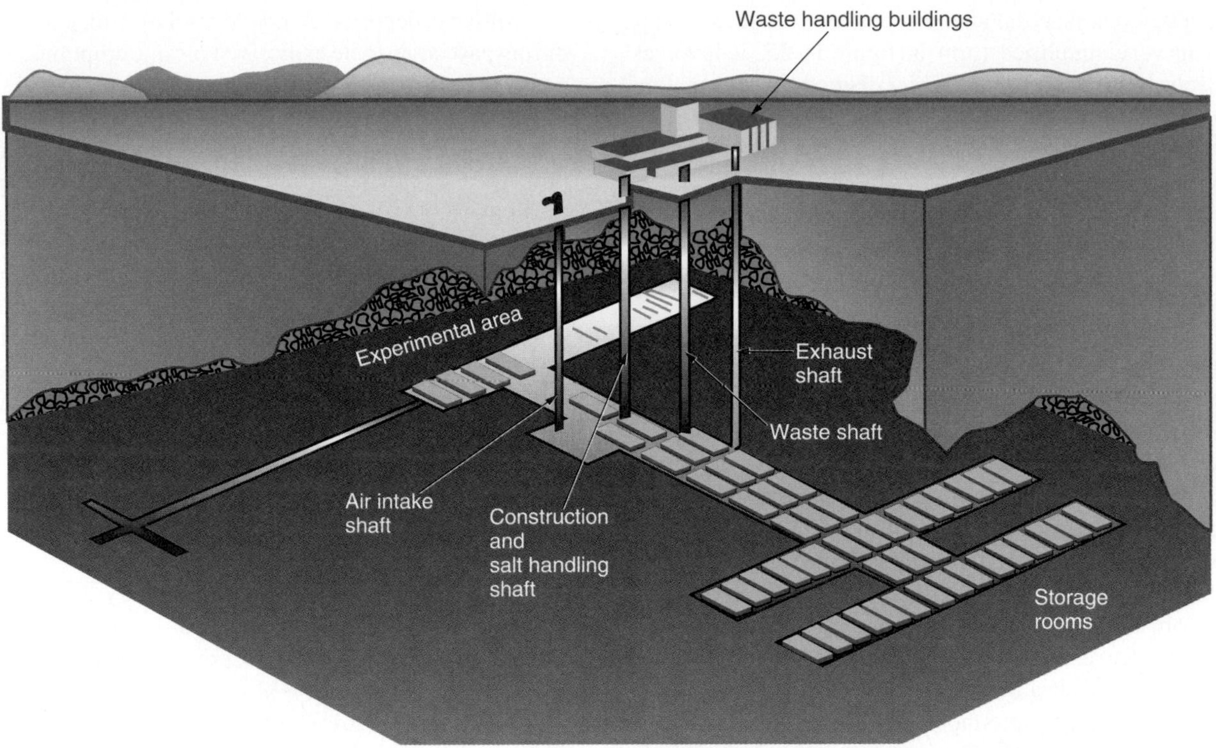

FIGURE 16.26 Waste Isolation Pilot Plant (WIPP) in New Mexico. (A) Plan of storage facility. (B) Waste canisters are emplaced, by remotely controlled machinery, into holes bored into salt walls.

Sources: (A) After Office of Environmental Restoration and Waste Management, Environmental Restoration and Waste Management: An Introduction, *U.S. Department of Energy. (B) Photograph courtesy of U.S. Department of Energy.*

southeastern New Mexico for some of its own high-level wastes, **transuranic wastes** (wastes containing radioactive elements heavier than uranium—including plutonium) created during the production of nuclear weapons. The Waste Isolation Pilot Plant, or WIPP, officially became the first U.S. underground repository for such wastes on 26 March 1999, when it received its first waste shipment. This event was the culmination of twenty years of site evaluation, planning, and construction following congressional authorization of the site in 1979.

The disposal unit at WIPP is bedded salt, underlain by more evaporites and overlain by mudstones. The disposal layer is 2150 feet below the surface. The region is now dry, so the water table is low, and examination of the geology indicates that it has been stable for over 200 million years, offering reassurance with respect to future stability and waste containment. The facility is shown in figure 16.26.

WIPP is expected to receive transuranic wastes from some two dozen sites across the nation, the wastes hauled in by truck in special, highly secure containers (figure 16.27). However, WIPP's authorization extends only to defense-related transuranic waste, not to commercial or other high-level radioactive wastes. For these, site analysis and development are still in progress.

The Long Road to Yucca Mountain

Much of the delay in the selection of a site for disposal of high-level radioactive waste has resulted from the need for thorough investigation of the geology of any proposed site. However, in the United States at least, there is also a political component to the delay. Simply put, nobody wants to host a high-level-waste disposal site.

The Nuclear Waste Policy Act of 1982 provided for the establishment of two high-level-waste disposal sites, one in the western United States, one in the East, both to be identified by the Department of Energy (DOE). The same Act created

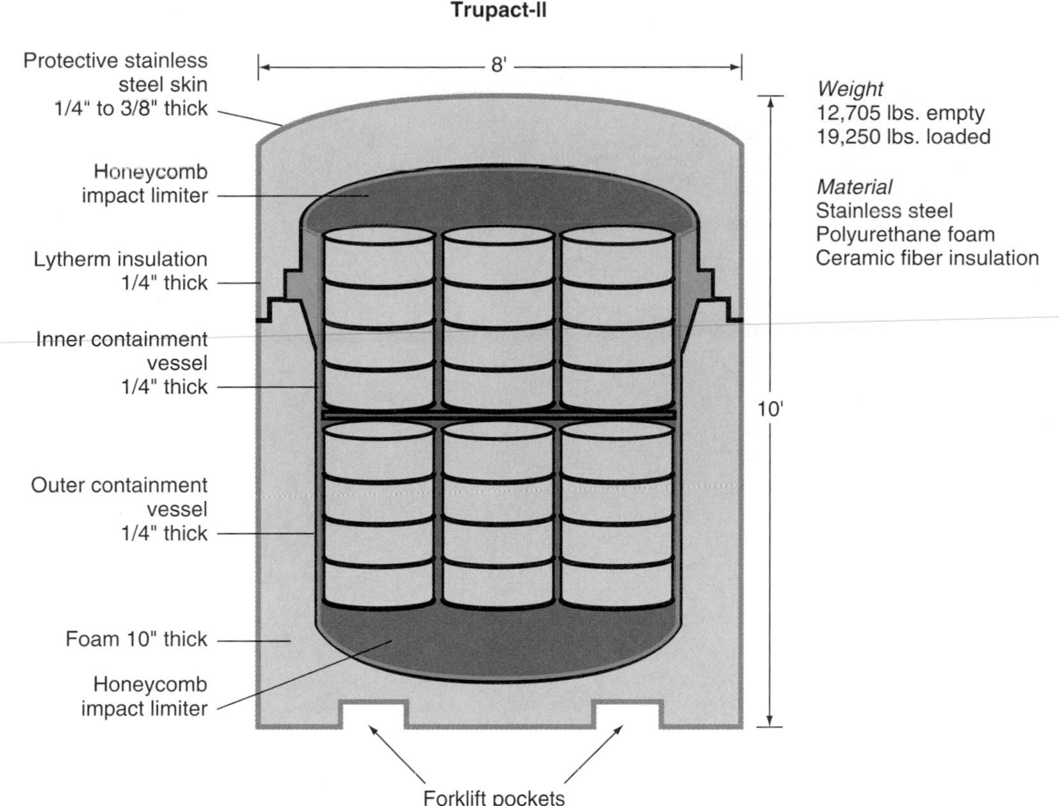

FIGURE 16.27 Wastes destined for WIPP are contained in specially designed vessels even before transport to the site. Similar vessels would be used for waste headed for Yucca Mountain.
Source: U.S. Department of Energy.

the Office of Civilian Radioactive Waste Management and established the Nuclear Waste Fund (NWF), supported by spent-fuel fee charges on waste generators and a one-mill-per-kilowatt-hour fee on electricity. This sounds small, but to date, utilities have contributed billions to this fund; NWF income for 1991 was over $900 million. The NWF was to finance disposal-site selection and investigation.

In 1986, DOE announced three candidates for the western site: Hanford, Washington (where disposal would be in basalt); Yucca Mountain, Nevada (ash-flow tuff); and Deaf Smith County, Texas (salt). All three states protested. So did Tennessee, where—at Oak Ridge—the waste-packaging site was to be built to prepare the wastes for disposal. Original plans would next have called for a billion-dollar testing program to be undertaken at each of the three proposed disposal sites, to be followed by selection of one to be *the* western site. Congress subsequently called a year-long halt to further action.

Then, in late 1987, Congress suddenly moved decisively to end the uncertainties—at least those of choice of site. It was decided that the Yucca Mountain site (figure 16.28) would be developed and investigation of the other sites halted, at a cost savings estimated at $4 billion. The state of Nevada would receive $20 million per year and "special consideration" of certain research grants by way of compensation.

Properties of the site that contribute to its attractiveness include (1) the tuff host rock; (2) generally arid climate, with no perennial streams and limited subsurface water percolation in the area; (3) low population density; (4) low regional water table, so the repository would be 200 to 400 meters above the water table; and (5) apparent geologic stability. The site is also located on lands already owned by the federal government, and overlaps the Nevada Test Site previously used for nuclear-weapons testing. Site-investigation activities include much more detailed structural, hydrologic, and geochemical studies, and more detailed analysis of geologic-stability issues. Seismicity is an obvious concern: there are a number of faults in the area, some related to the collapse of ancient volcanic calderas, some related to the formation of the mountains, some related to plate motions (figure 16.29). However, current seismicity is low, and evidence suggests limited local fault displacement for the last 10 million years; still, repository design will allow for earthquakes up to magnitude 6.8 or so, the largest plausible on the known faults. The most recent volcanism in the region is a cinder cone that may be as young as 20,000 years old, and basaltic volcanism is possible within the next 10,000 years or so. There are also questions about how likely the region is to stay as dry as it currently is, particularly in light of anticipated global climate change.

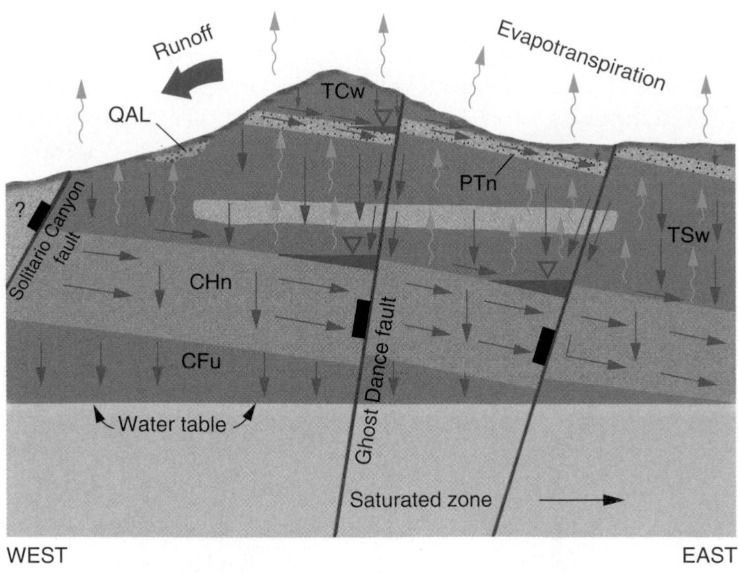

FIGURE 16.28 Proposed waste-disposal site at Yucca Mountain, Nevada. Generalized east-west section through Yucca Mountain area, showing water and water-vapor flow paths; the "welded" and "unwelded" units are tuffs. Repository level highlighted. See chapter-opening photograph for regional view.

After Site Characterization Plan Overview, *U.S. Department of Energy, 1988.*

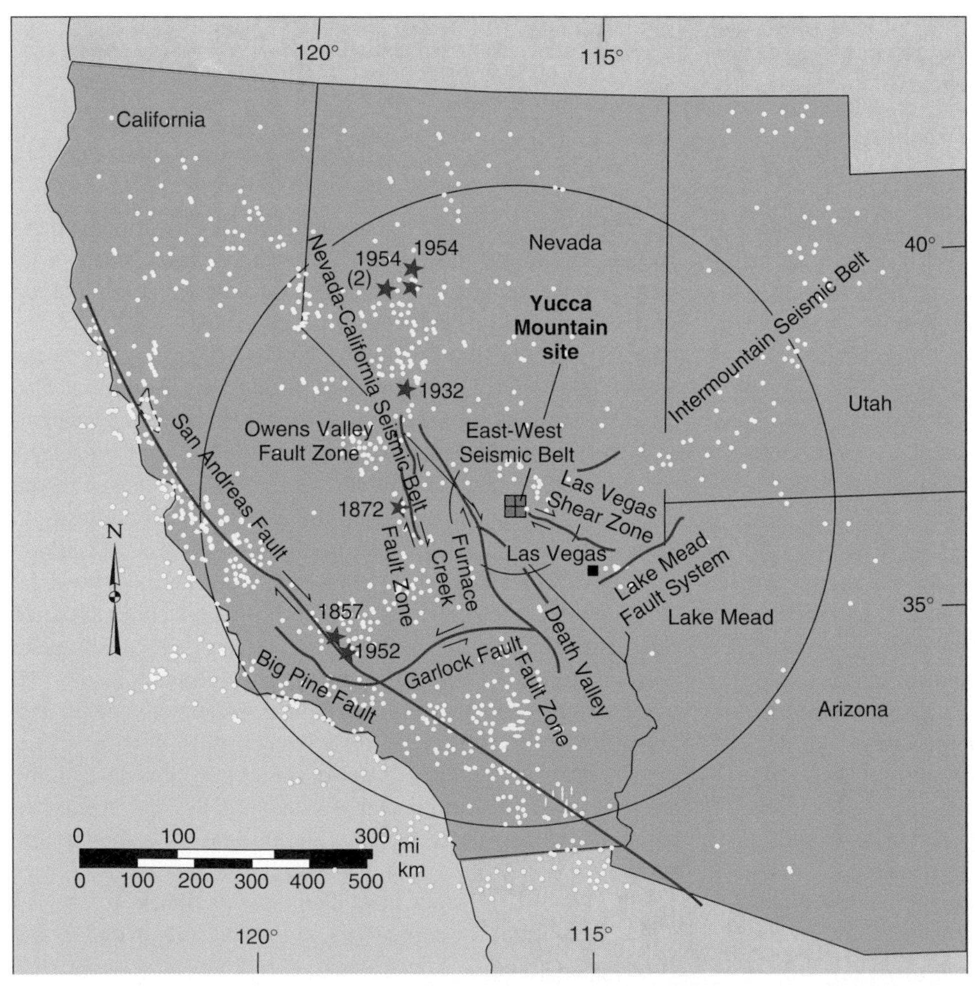

FIGURE 16.29 Seismicity in the south-western United States, 1969–1978, showing earthquakes of magnitude 4 or higher. Stars denote a magnitude of 6.5 or higher. Major faults in the region are also shown.

Source: After Site Characterization Plan Overview, *U.S. Department of Energy, 1988.*

The evaluation has been comprehensive. The site characterization *plan* was a nine-volume, 6000⁺-page document, the draft and final environmental impact statements much longer. Meanwhile, the high-level wastes remain on hold, in temporary storage, as the Yucca Mountain timetable slips. Site characterization began in 1986. At one time, the projection was for the start of construction in 1998 and first waste disposal in 2003. In fact, the final site analysis was not completed until late in fiscal year 2001, the final environmental impact statement was issued by the Department of Energy (DOE) in February 2002, and the site was officially designated by Congress and the president for a high-level waste repository in July 2003. (Objections are still being raised by the state of Nevada and others as this is written.) Nuclear Regulatory Commission review of DOE's 2008 license application is expected to take several years; currently, officials hope for construction to begin in time for waste disposal to begin in 2017, their estimate of the "best achievable" timetable.

No High-Level Radioactive Waste Disposal Yet

At least two dozen nations are making plans to dispose of high-level nuclear wastes; but so far, *no* high-level radioactive wastes have been consigned to permanent disposal sites elsewhere in the world by any nation. All this waste—some solid, some liquid—is in temporary storage awaiting selection and construction of disposal sites. Geologists can identify sites that have a very high probability of remaining geologically quiet and secure for many generations to come, but absolute guarantees are not possible. Uncertainties about the long-term integrity and geologic stability of any given site, coupled with increasing public resistance in regions with sites under consideration, have held up the disposal efforts. Worldwide, the accumulated wastes amount to millions of gallons of liquids, tens of thousands of tons of solids.

Radioactive-waste disposal remains an issue even if the nuclear-fission power option is not pursued. A growing number of nations in the industrialized world have accumulated high-level wastes, from past power generation, nuclear weapons production, and radiochemical research, including medical applications. All of that waste must ultimately be put somewhere in permanent disposal. (The only alternative is to keep the wastes at the surface, closely monitored and guarded to protect against accident or sabotage, assuming that long-term political stability is more likely than geologic stability—a debatable assumption, surely.) But it remains an open question where the wastes will be put, and how soon.

SUMMARY

Methods of solid-waste disposal include open dumping (being phased out for health reasons), use of landfills, incineration on land or at sea, and disposal in the oceans. Incineration contributes to air pollution but can also provide energy. Landfills require a commitment of space, although the land may be used for other purposes when filled. "Secure" landfills for toxic solid or liquid industrial wastes should isolate the hazardous wastes with low-permeability materials and include provisions for monitoring groundwater chemistry to detect any waste leakage. Increasing evidence suggests that construction of wholly secure landfill sites is not possible. Deep-well disposal and incineration are alternative methods for handling toxic liquid wastes.

The principal liquid-waste problem in terms of volume is sewage. Low-population areas can rely on the natural sewage treatment associated with septic tanks, provided that soil and site characteristics are suitable. Municipal sewage treatment plants in densely populated areas can be constructed to purify water so well that it can even be recycled as drinking water, although most municipalities do not go to such expensive lengths.

A complicating feature of radioactive wastes is that they cannot be made nonradioactive; each radioisotope decays at its own characteristic rate regardless of how the waste is handled. Given the long half-lives of many isotopes in high-level wastes and the fact that some are chemical poisons as well as radiation hazards, long-term waste isolation is required. Disposal in bedrock, tuff, salt deposits, or on the deep-sea floor are among the methods under particularly serious consideration worldwide. In the United States, the Waste Isolation Pilot Plant has begun to receive the government's transuranic waste; Yucca Mountain, Nevada, has been designated for development of a permanent disposal site for commercial high-level wastes, but actual disposal is still years away.

Key Terms and Concepts

adsorption field 388
half-life 393
high-level waste 395
leachate 378
leaching field 388
low-level waste 395
multiple barrier concept 399
sanitary landfill 377
secure landfill 386
source separation 384
transuranic wastes 400

Exercises

Questions for Review

1. What two kinds of activities generate the most solid wastes?
2. What advantages does a sanitary landfill have over an open dump? Describe three pathways through which pollutants may escape from an improperly designed landfill site.
3. Describe any three features of a modern municipal landfill that reduce the risk of pollution, and explain their function.
4. Landfills and incinerators have in common the fact that both can serve as energy sources. Explain.
5. Use of in-sink garbage disposal units in the home is sometimes described as "on-site disposal." Is this phrase accurate? Why or why not?
6. Compare the relative ease of recycling of (a) glass bottles, (b) paper, (c) plastics, (d) copper, and (e) steel, noting what factors make the practice more or less feasible in each case.
7. Compare the dilute-and-disperse and concentrate-and-contain philosophies of liquid-waste disposal.
8. Outline the relative merits and drawbacks of deep-well disposal and of incineration as disposal strategies for toxic liquid wastes.
9. What kinds of limitations restrict the use of septic tanks?
10. Municipal sewage treatment produces large volumes of sludge as a by-product. Suggest possible uses for this material, and note any factors that might restrict its use.
11. Describe two reasons that it is difficult to assess the hazards of low-level exposure to radiation.
12. Evaluate the advantages, disadvantages, and possible concerns relating to disposal of high-level radioactive wastes in (a) subduction zones, (b) sediments on the deep-sea floor, (c) tuff, and (d) bedded salt.
13. What characteristics of the Yucca Mountain site make it attractive as a possible disposal site for high-level wastes? Outline three geologic concerns relative to the long-term security of waste containment.

For Further Thought

1. Where does your garbage go? How is it disposed of? Is any portion of it utilized in some useful way? If your community has a recycling program, find out what is recycled, in what quantities, and what changes have occurred over the last five or ten years in materials recycled and levels of participation.
2. For a period of one or several weeks, weigh all the trash you discard. If every one of the 300 million people in the United States or the 30 million in Canada discarded the same amount of trash, how much would be generated in a year? Suppose that each of the more than 6 billion people on earth did the same. How much garbage would that make? Is this likely to be a realistic estimate of actual world trash generation? Why or why not?
3. Investigate your community's sewage treatment. What level is it? What is the quality of the outflow? How does the storm-sewer system connect with the sewage-treatment system?
4. Choose a single Superfund NPL site, or several sites within a state or region, and check the history and current status of cleanup activities there. A website including all the NPL (National Priorities List) sites by region, which would be a place to start, is at http://www.epa.gov/superfund/oerr/impm/products/nplsites/usmap.htm

Suggested Readings/References

Ackland, L. 1993. Radiation risks revisited. *Technology Review* (February/March): 56–61.

American Geological Institute. 2006. Turning trash into energy (a series of short articles by various authors). *Geotimes* (February): 18–24.

American Institute of Physics. 1997. *Physics Today* (June): 22–62. (Special issue on radioactive waste.)

Apted, M. 2004. Locating a radioactive waste repository in the Ring of Fire. *EOS* 85(9 November): 465, 467.

Christenson, S. C., and Cozzarelli, I. M. 2003. *The Norman Landfill environmental research site: What happens to the waste in landfills?* USGS Fact Sheet FS-040-03.

Hanks, T. C., Winograd, I. J., Anderson, R. E., Reilly, T. E., and Weeks, E. P. 1999. *Yucca Mountain as a radioactive-waste repository.* U.S. Geological Survey Circular 1184.

Jewell, W. J. 1994. Resource-recovery wastewater treatment. *American Scientist* 82 (July–August): 366–75.

Krajick, K. 1997. Mining the scrap heap for treasure. *Smithsonian* (May): 34–44.

Rathje, W. L. 1991. Once and future landfills. *National Geographic* (May): 117–34.

Royte, E. 2006. Corn plastic to the rescue? *Smithsonian* (August): 84–88.

Stone, R. F., Sagar, A. D., and Ashford, N. A. 1992. Recycling the plastic package. *Technology Review* (July): 49–56.

U.S. Department of Energy. 2000. *Site characterization progress report, Yucca Mountain, Nevada.* Washington, D.C.: U.S. Government Printing Office.

———. 2000. *The Waste Isolation Pilot Plant: Pioneering nuclear waste disposal.* Washington, D.C.: U.S. Government Printing Office.

———. 2003. *Why Yucca Mountain? Frequently asked questions.* Washington, D.C.: U.S. Government Printing Office.

U.S. Environmental Protection Agency. 1998. *How wastewater treatment works . . . the basics.* Report EPA 833-F-98-002.

———. 2003. *Municipal solid waste generation, recycling, and disposal in the United States: 2001 Facts and figures.* Report EPA530-S-03-011. (Also accessible online through the office of Solid Waste Management: http://www.epa.gov/OSW)

U.S. Geological Survey. 2001. *Obsolete computers, "gold mine" or high-tech trash?* USGS Fact Sheet 060-01; also online at pubs.usgs.gov/fs/fs060-01/

Zhou, W., Beck, B. F., and Josefczyk, R. C. 2005. Disposal of wastes in sinkholes: Hydrogeological significance, environmental implications, and appropriate application of dye tracing. *The Professional Geologist* (November/December): 46–51.

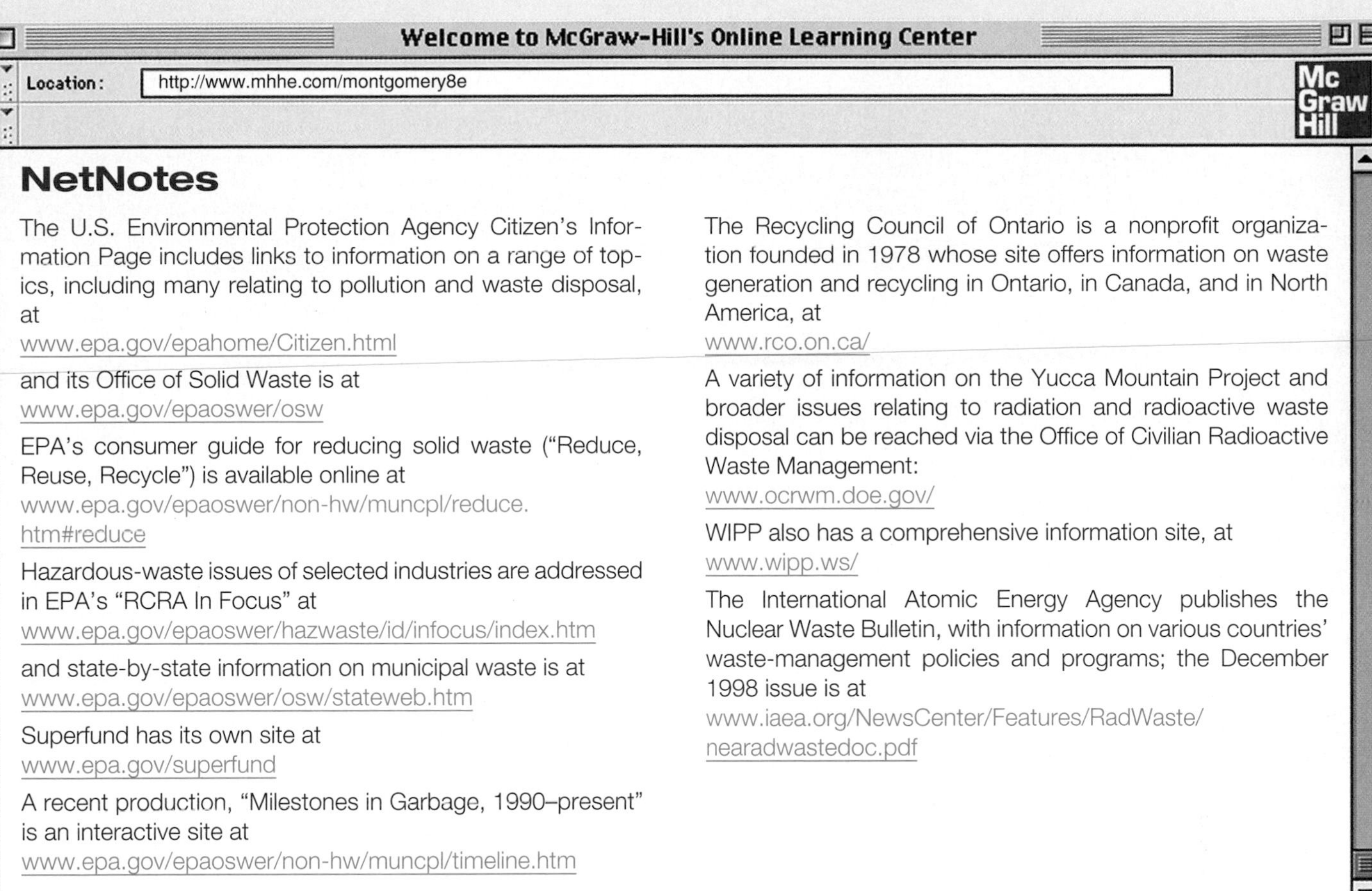

Welcome to McGraw-Hill's Online Learning Center

Location: http://www.mhhe.com/montgomery8e

NetNotes

The U.S. Environmental Protection Agency Citizen's Information Page includes links to information on a range of topics, including many relating to pollution and waste disposal, at
www.epa.gov/epahome/Citizen.html

and its Office of Solid Waste is at
www.epa.gov/epaoswer/osw

EPA's consumer guide for reducing solid waste ("Reduce, Reuse, Recycle") is available online at
www.epa.gov/epaoswer/non-hw/muncpl/reduce.htm#reduce

Hazardous-waste issues of selected industries are addressed in EPA's "RCRA In Focus" at
www.epa.gov/epaoswer/hazwaste/id/infocus/index.htm

and state-by-state information on municipal waste is at
www.epa.gov/epaoswer/osw/stateweb.htm

Superfund has its own site at
www.epa.gov/superfund

A recent production, "Milestones in Garbage, 1990–present" is an interactive site at
www.epa.gov/epaoswer/non-hw/muncpl/timeline.htm

The Recycling Council of Ontario is a nonprofit organization founded in 1978 whose site offers information on waste generation and recycling in Ontario, in Canada, and in North America, at
www.rco.on.ca/

A variety of information on the Yucca Mountain Project and broader issues relating to radiation and radioactive waste disposal can be reached via the Office of Civilian Radioactive Waste Management:
www.ocrwm.doe.gov/

WIPP also has a comprehensive information site, at
www.wipp.ws/

The International Atomic Energy Agency publishes the Nuclear Waste Bulletin, with information on various countries' waste-management policies and programs; the December 1998 issue is at
www.iaea.org/NewsCenter/Features/RadWaste/nearadwastedoc.pdf

17 Water Pollution

As noted in chapter 11, the more-or-less fresh ground water and surface water on the continents constitutes less than 1% of the water in the hydrosphere. Pollution has tainted much of the most accessible of this reserve: lakes, streams, water in shallow aquifers near major population centers. The number and variety of potentially harmful water pollutants are staggering.

In this chapter, we review some important categories of pollutants and the nature of the problems each presents. We also look at some possible solutions to existing water-pollution problems. While it is true that substantial quantities of surface water and ground water are unpolluted by human activities, much of this water is far removed from population centers that require it, and the quality of much of the most accessible water has indeed been degraded.

Some water pollution is obvious, as at this industrial site on the Calumet River. Not all toxins are visible, unfortunately.
Photo courtesy of the Evironmental Protection Agency (Region V)

GENERAL PRINCIPLES

Any natural water—rainwater, surface water, or ground water—contains dissolved chemicals. Some of the substances that find their way naturally into water are unhealthy to us or to other life forms, as, unfortunately, are some of the materials produced by modern industry, agriculture, and just people themselves.

Geochemical Cycles

All of the chemicals in the environment participate in geochemical cycles of some kind, similar to the rock cycle. Not all environmental cycles are equally well understood.

For example, a very simplified natural cycle for calcium is shown in figure 17.1. Calcium is a major constituent of most common rock types. When weathered out of rocks, usually by solution of calcium carbonate minerals or by the breakdown of calcium-bearing silicates, calcium goes into solution in surface water or ground water. From there it may be carried into the oceans. Some is taken up in the calcite of marine organisms' shells, in which form it may later be deposited on the sea floor after the organisms die. Some is precipitated directly out of solution into limestone deposits, especially in warm, shallow water; other calcium-bearing sediments may also be deposited. In time, the sediments are lithified, then metamorphosed, or subducted and mixed with magma from the mantle, or otherwise reincorporated into the continental crust as the cycle continues. Subcycles also exist. For example, some mantle-derived calcium is leached directly into the oceans by the warmed seawater or other hydrothermal fluids interacting with the fresh seafloor basalts at spreading ridges; after precipitation, it may be reintroduced into the mantle without cycling through the crust. Some dissolved weathering products are carried into continental waters—such as lakes—from which they may be directly redeposited in continental sediments, and so on.

Residence Time

Within the calcium cycle outlined, there are several important calcium reservoirs, including the mantle, continental crustal rocks, submarine limestone deposits, and the ocean, in which large quantities of calcium are dissolved. A measure of how rapidly the calcium cycles through each of these reservoirs is a parameter known as **residence time.** This can be defined in a variety of ways. For a substance that is at saturation level in a reservoir, so that the reservoir can hold no more of it, residence time is given by the relationship:

$$\text{Residence time (of a substance in a reservoir)} = \frac{\text{Capacity (of the reservoir to hold that substance)}}{\text{Rate of influx (of the substance into the reservoir)}}$$

This can be illustrated by a nongeologic example. Consider a hotel with 100 single rooms. Some lodgers stay only one night, others stay for weeks. On the average, though, ten people check out each day, so (assuming this is a popular hotel that can

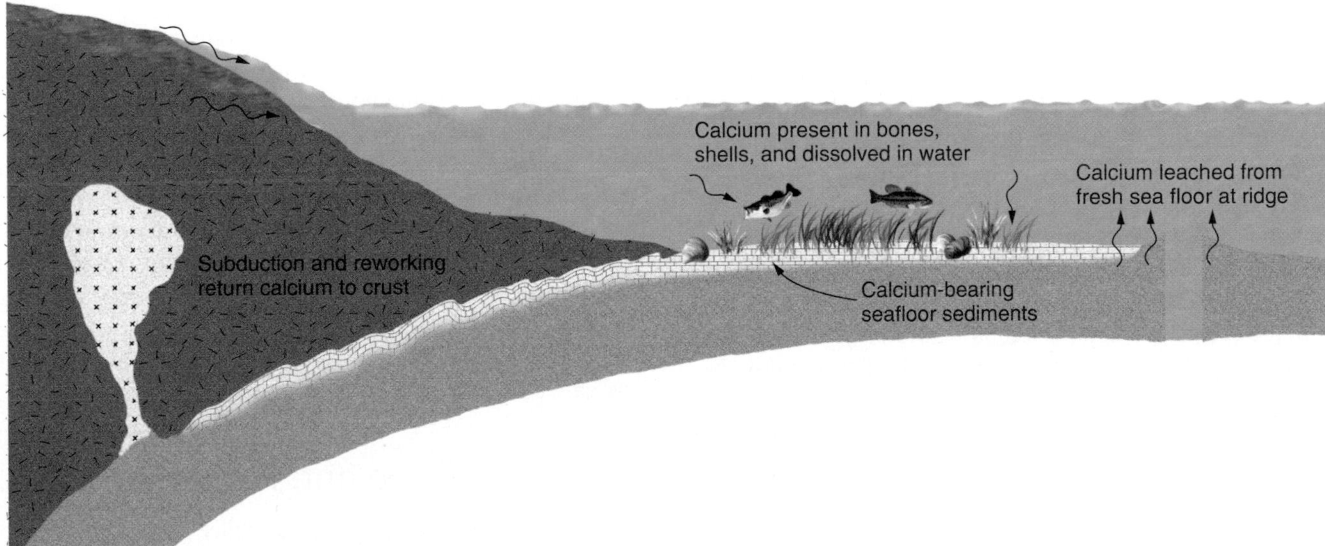

FIGURE 17.1 Simplified calcium cycle: Calcium weathered out of rocks is washed into bodies of water. Some remains in solution, some is taken up by organisms, and some is deposited in sediments. The deposited calcium is, in time, reworked through the rock cycle into new rock.

always fill its rooms) ten new people check in each day. The residence time of lodgers in the hotel, then, is

$$\text{Residence time} = \frac{100 \text{ persons}}{10 \text{ persons/day}} = 10 \text{ days}$$

In other words, people stay an average of ten days each at this hotel.

There are many other situations in which the quantity of a substance in a reservoir stays approximately constant (a *steady-state* situation) though the reservoir is not at capacity. In such a case, residence time could be defined as

$$\frac{\text{Amount of the substance in the reservoir}}{\text{Rate of input (or outflow) of the substance}}$$

To pursue our hotel analogy, consider a competing hotel in the next block that has similar turnover of clientele, but a much lower occupancy rate—only 50 of its 100 rooms are typically occupied. Then if 10 people check in and out each day, the residence time of patrons in the second hotel is only (50 persons)/(10 persons/day) = 5 days; each patron stays only half as long, on average, in the second hotel. Or, assuming for the moment that world population is a constant 6 billion, then if 80 million people are born each year and 80 million die (the numbers would have to be equal in order for the population to stay constant), the residence time of people on earth would be (6 billion) ÷ (80 million) = 75 years. That would not mean that everyone would live on earth for 75 years, but that would be the *average* life span.

Which brings us to a more qualitative way of viewing residence time: as the average length of time a substance remains in a system (or reservoir). This may be a more useful working definition for substances whose capacity is not limited in a given reservoir (the continental crust, for instance) or for substances that do not achieve saturation but are removed from the reservoir by decomposition after a time, as is the case with many synthetic chemicals that cycle through water or air.

Of various reservoirs in figure 17.1, the one in which the residence time of calcium can be calculated most directly is the ocean. The capacity of the ocean to hold calcium is basically controlled by the solubility of calcite. (If too much calcium and carbonate have gone into solution, some calcite is removed by precipitation.) The rate of influx can be estimated by measuring the amount of dissolved calcium in all major rivers draining into the ocean. The resulting calculated residence time of calcium in the oceans works out to about 1 million years. In other words, the average calcium atom normally floats around in the ocean for 1 million years from the time it is flushed or leached into the ocean until it is removed by precipitation in sediments or in shells.

Oceanic residence times for different elements vary widely (see table 17.1). Sodium, for instance, is put into the oceans less rapidly (in smaller quantity) than calcium, and the capacity of the ocean for sodium is much larger. (Sodium chloride, the main dissolved salt in the ocean, is very soluble; what sodium is removed is taken up by clay-rich sediments.) The residence time of sodium in the oceans, then, is longer than that of calcium: 100 million years. By contrast, iron compounds, as noted earlier, are not very soluble in the modern ocean. The residence time of iron in the ocean is only 200 years; iron precipitates out rather quickly once it is introduced into the sea. Residence times will differ for any given element in different reservoirs with different capacities and rates of influx.

Human activities may alter the natural figures somewhat, chiefly by altering the rate of influx. In the case of calcium, for instance, extensive mining of limestone for use in agriculture (to neutralize acid soil) and construction (for building stone and the manufacture of cement) increases the amount of calcium-rich material exposed to weathering; calcium salts are also used for salting roads, and these salts gradually dissolve away into runoff water. These additions are probably insufficient to alter appreciably the residence times of materials in very large reservoirs like the ocean. Locally, however, as in the case of a single lake or stream, they may change the rate of cycling and/or increase a given material's concentration throughout a cycle.

TABLE 17.1

Residence Times of Selected Major and Minor Elements in Seawater with Respect to Modern Influx through Surface Runoff

Element	Concentration (ppm)	Residence Time (years)
chlorine	18,980 (1.9%)	68,000,000
sodium	10,540 (1.0%)	100,000,000
magnesium	1270	12,000,000
calcium	400	1,000,000
potassium	380	7,000,000
bromine	60	100,000,000
silicon	3.0	18,000
phosphorus	0.07	180,000
aluminum	0.01	100
iron	0.01	200
cadmium	0.00011	500,000
mercury	0.00003	80,000
lead	0.00003	400

Source: Data from J. E. Fergusson, *Inorganic Chemistry and the Earth.* Copyright © 1982 Pergamon Press, Oxford, England.

Residence Time and Pollution

In principle, one can speak of residence times of more complex chemicals, too, including synthetic ones, in natural systems. Here, there is the added complication that not enough is really known about the behavior of many synthetic chemicals in the environment. There is no long natural history to use for reference when estimating the capacities of different reservoirs for recently created compounds. Some compounds' residence times are limited by rapid breakdown into other chemicals, but breakdown rates for many are slow or unknown. So, too, is the ultimate fate of many synthetic chemicals. Uncertainties of this sort make evaluation of the seriousness of industrial pollution very difficult.

Where residence times of pollutants are known in particular systems, they indicate the length of time over which those pollutants will remain a problem in those reservoirs, and thus the rapidity with which the pollutants will be removed from that system. Suppose, for example, that a factory has been discharging a particular toxic compound into a lake. Some of the compound is drained out of the lake by an outflowing stream. The residence time of the water, and thus of the toxic chemical in the lake, has been determined to be 20 years. If that is the average length of time the water and synthetic chemical remain in the lake, then if input of the toxin ceases, but flushing of water through the lake continues, then after one residence time (20 years), half of the chemical should have been removed from the lake; after a second 20 years, another half; and so on. One can then project, for instance, how long it would take for a safely low concentration to be reached. The calculations are more complex if additional processes are involved (if, for instance, some of the toxin decomposes, and some is removed into lake-bottom sediment), but the principles remain the same. However, substances removed from one system, unless broken down, will only be moved into another, where they may continue to pose a threat. If our hypothetical toxic chemical is at saturation concentration in a lake and is removed only by precipitation into the lake-bottom sediments, then if the rate of input into the lake doubles, so will the rate of output, and the concentration in the sediments will be twice as high.

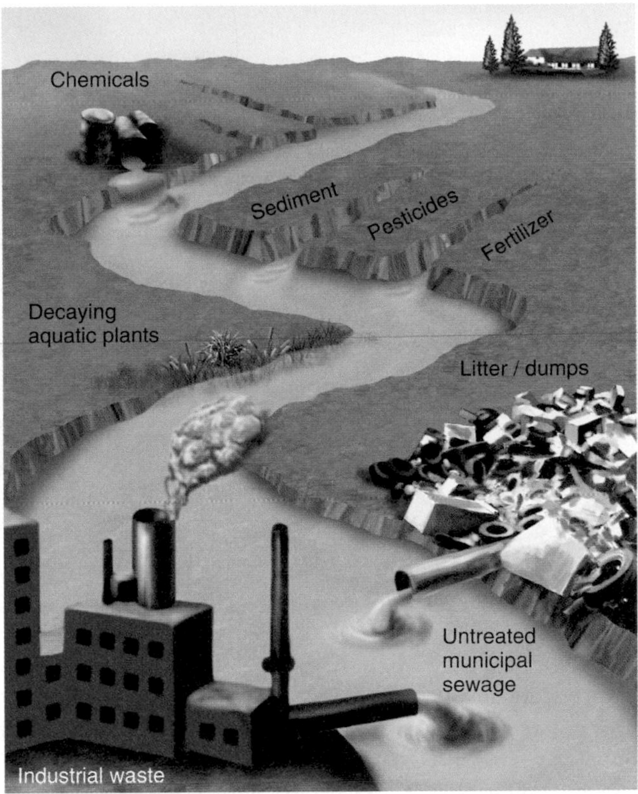

FIGURE 17.2 Point sources (such as industrial plants and storm-sewer outflow) and nonpoint sources (for example, fields and dispersed litter) of water pollution in a stream. Downward infiltration of contaminant-laden water from the nonpoint sources eventually also pollutes ground water.
Source: After USDA Soil Conservation Service.

Point and Nonpoint Pollution Sources

Sources of pollution may be subdivided into point sources and nonpoint sources (figure 17.2). **Point sources,** as their name suggests, are sources from which pollutants are released at one readily identifiable spot: a sewer outlet, a steel mill, a septic tank, and so forth. **Nonpoint sources** are more diffuse; examples would include fertilizer runoff from farmland, acid drainage from an abandoned strip mine, or runoff of sodium or calcium chloride from road salts, either directly off roadways or via storm drains. The point sources are often easier to identify as potential pollution problems. They are also easier to monitor systematically. It is a comparatively straightforward task to evaluate the quality of water from a single sewer outfall. Sampling runoff from a field in a representative way is more difficult.

The reverse problem—identifying the sources of water pollutants once a problem is recognized—can likewise be difficult. Sometimes, the source can be identified by the overall chemical "fingerprint" of the pollution, for rarely is a single substance involved. Different pollutant sources—municipal sewage, a factory's wastewater, a farmer's particular type of fertilizer—are characterized by a specific mix of compounds, which may, in some cases, be identified in similar proportions in polluted water.

ORGANIC MATTER

Organic matter in general (as distinguished from the smaller subset of toxic organic compounds, discussed in the next section) is the substance of living and dead organisms and their by-products. It includes a variety of materials, ranging from dead leaves settling in a stream to algae on a pond. Its most abundant and problematic form in the context of water pollution is human and animal wastes. Feedlots and other animal-husbandry activities create large concentrations of animal wastes (figure 17.3), and the trend in U.S. agriculture is toward ever-denser concentrations of cattle, or pigs, or poultry in such operations. Food-processing plants are other sources of large quantities of organic matter discharged in wastewater.

Sewage becomes mixed with a large quantity of wastewater not originally contaminated with it: Domestic wastewater is typically all channeled into one outlet, and in municipalities, domestic wastes, industrial wastes, and frequently even storm

FIGURE 17.3 Uncontrolled runoff from this livestock yard may be a major contributor of organic waste to local streams or shallow ground water.
Photo by Tim McCabe, courtesy USDA Natural Resources Conservation Service.

runoff collected in street drains all ultimately come together in the municipal sewer system. Moreover, sewage-treatment plants are commonly located beside a body of surface water, into which their treated water is discharged. To the extent that the wastewater is incompletely treated, the pollution of the surface water is increased.

Aside from its potential for spreading disease, organic matter creates another kind of water-pollution problem. In time, organic matter is broken down by microorganisms, especially bacteria. If there is ample oxygen in the water, **aerobic decomposition** occurs, consuming oxygen, facilitated by aerobic organisms. This depletes the dissolved oxygen supply in the water. Eventually, so much of the oxygen may be used up that further breakdown must proceed by **anaerobic decomposition** (without oxygen). Anaerobic decay produces a variety of noxious gases, including hydrogen sulfide (H_2S), the toxic gas that smells like rotten eggs, and methane (CH_4), which is unhealthy and certainly of no practical use in a body of water. More importantly, anaerobic decay signals oxygen depletion. Animal life, including fish, requires oxygen to survive—no dissolved oxygen, no fish. When the dissolved oxygen is used up, the water is sometimes described as "dead," though anaerobic bacteria and some plant life may continue to live and even to thrive.

Biochemical Oxygen Demand

The organic-matter load in a body of water is described by a parameter known as **biochemical oxygen demand,** or **BOD** for short. The BOD of a system is a measure of the amount of oxygen required to break down the organic matter aerobically; the more organic matter, the higher the BOD. The BOD may far exceed the actual amount of dissolved oxygen in the water. As oxygen is depleted, more tends to dissolve in the water to restore chemical equilibrium, but, initially, this gradual reoxygenation lags behind oxygen consumption. Measurements of water chemistry along a stream below a sizable source of organic waste matter characteristically define an **oxygen sag curve** (figure 17.4), a graph of dissolved-oxygen content as a function of distance from the waste source that shows sharp depletion near the source, recovering downstream. Persistent oxygen depletion is more likely to occur in a body of standing water, such as a lake or reservoir, than in a fast-flowing stream. Flowing water is better mixed and circulated, and more of it is

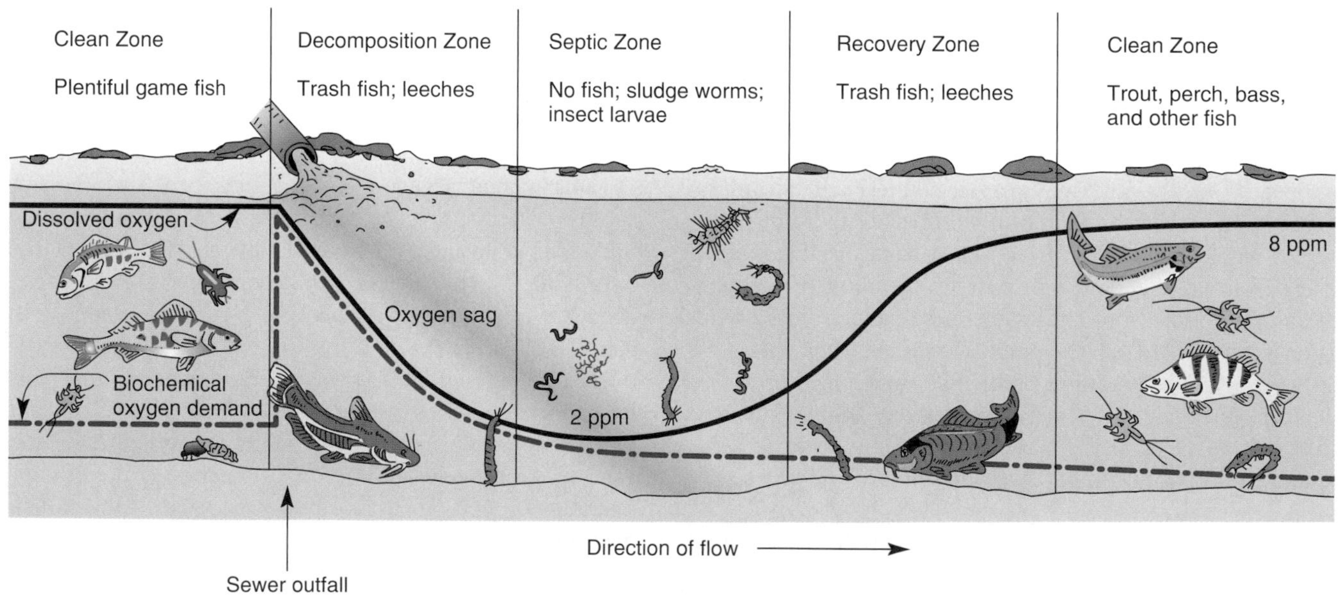

FIGURE 17.4 Typical oxygen sag curve downstream from an organic-matter input, and its effect on both dissolved oxygen and the fauna in the stream. The dissolved-oxygen content of this stream is normally 8 ppm; it drops to 2 ppm below the sewer outfall where BOD spikes abruptly.

regularly brought into contact with the air, from which it can take up more oxygen. Climate also plays a role (figure 17.5).

Eutrophication

The breakdown of excess organic matter not only consumes oxygen, but it also releases a variety of compounds into the water, among them nitrates, phosphates, and sulfates. The nitrates and phosphates, in particular, are critical plant nutrients, and an abundance of them in the water strongly encourages the growth of plants, including algae. Excess fertilizer runoff from farmland can also put excessive nutrients into water. This development is known as **eutrophication** of the water; the water body itself is then described as *eutrophic*. The exuberant algal growth that often accompanies or results from eutrophication has also been called "algal bloom" and may appear as slimy green scum on the water (see Box 17.1). Once such a condition develops, it acts to continue to worsen the water quality (figure 17.6). Algal growth proceeds vigorously in the photic (well-lighted) zone near the water surface until the plants are killed by cold or crowded out by other plants. The dead plants sink to the bottom, where they, in turn, become part of the organic-matter load on the water, increasing the BOD and, as they decay, re-releasing nutrients into the water. Again, the consequences are most acute in very still water, where bottom waters do not readily circulate to the surface to take up oxygen. It should be noted that eutrophication can be a normal process in the development of a natural lake. Pollution, however, may accelerate or enhance it to the point that it becomes problematic. Aside from the aesthetic impact of algal bloom, oxygen depletion in eutrophic waters can kill aquatic organisms such as fish that require oxygen to survive.

Thorough waste treatment, as described in chapter 16, reduces the impact of organic matter, at least in terms of oxygen demand. However, as already noted, breakdown of the complex organic molecules releases into the water simpler compounds that may serve as nutrients. Human or animal wastes are not the only agents in sewage that contribute to eutrophication, either. Fertilizer runoff from farmland is a factor, as noted later in the chapter. (See also figure 17.7.) Phosphates from detergents are potential nutrients, too. The phosphates are added to detergents to enhance their cleaning ability, in part by softening the water. Realization of the harmful environmental effects of phosphates in sewage has led to restrictions on the amounts used in commercial detergents, but some phosphate component is still allowed in most places. And most sewage treatment does not remove all of the dissolved nutrient (nitrate and phosphate) load.

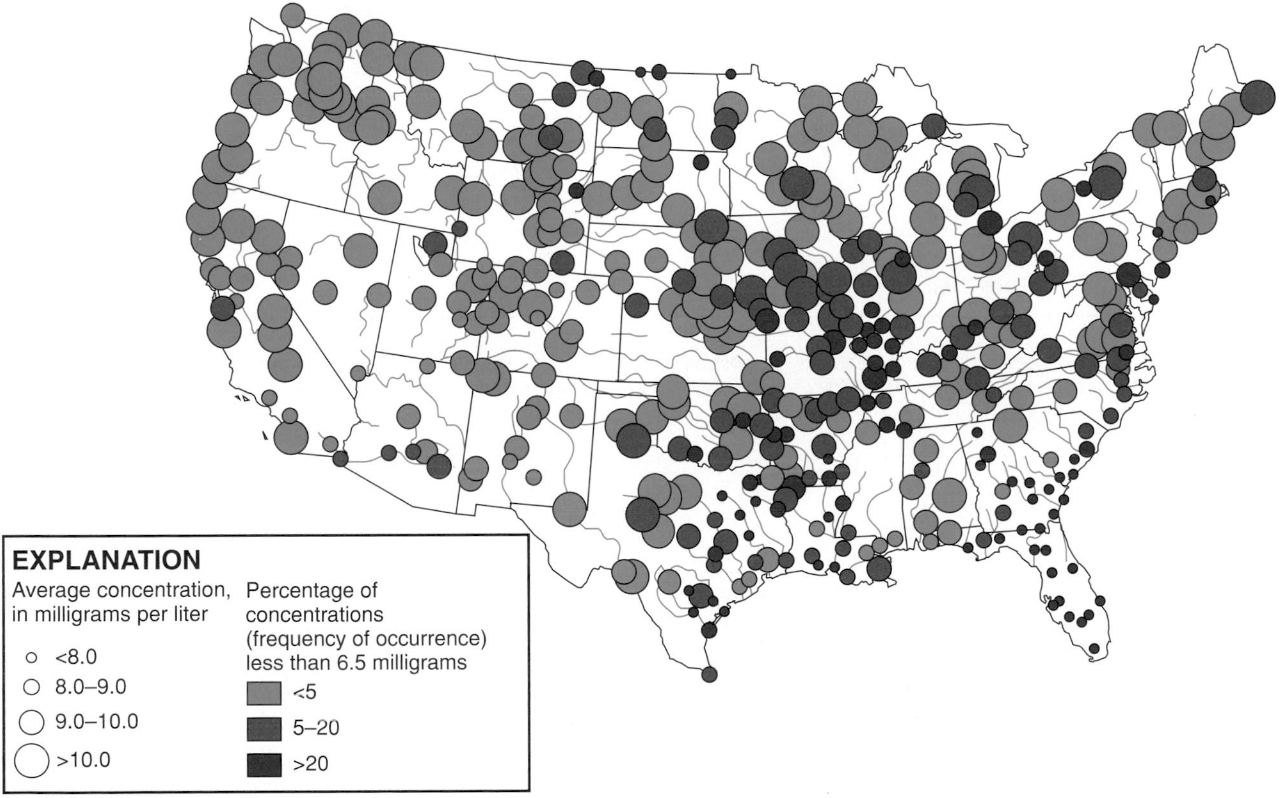

FIGURE 17.5 Dissolved-oxygen concentrations in U.S. surface waters in the 1980s. Low oxygen levels (small and/or dark circles) can be due to nutrient loading/eutrophication. Also, oxygen dissolves more readily in cold water than in warm, accounting for depressed oxygen concentrations in the southeastern United States.

Source: After U.S. Geological Survey Water Supply Paper 2400.

A

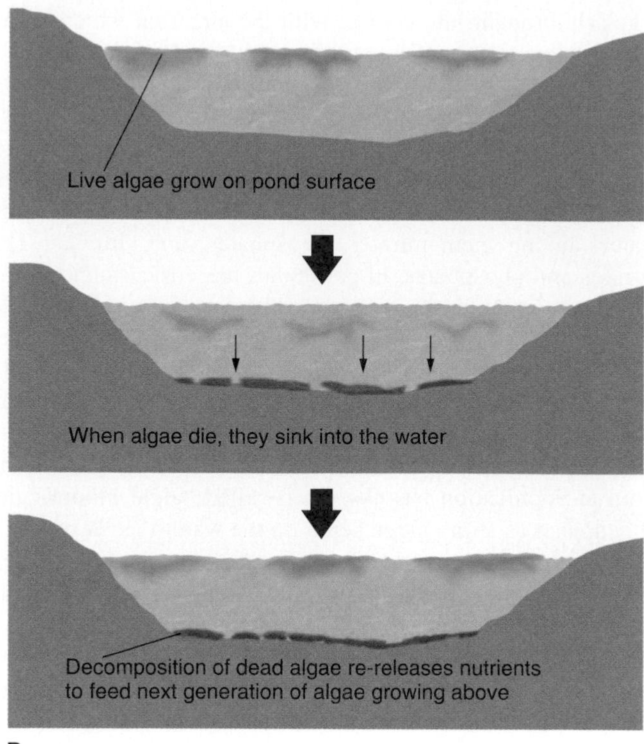

B

FIGURE 17.6 (A) In extreme cases, algal bloom may blanket the surface of a lake, as here in southern Wisconsin. (B) Schematic diagram showing effects of algal bloom on water quality: (top) Abundant growth of algae in sunlit shallow water when nutrients are abundant. (middle) In colder weather, algae die and sink to the lake bottom. (bottom) The next growing season, more algae thrive at the surface while older material decays at the lake bottom, increasing BOD and releasing more nutrients to fuel new growth.

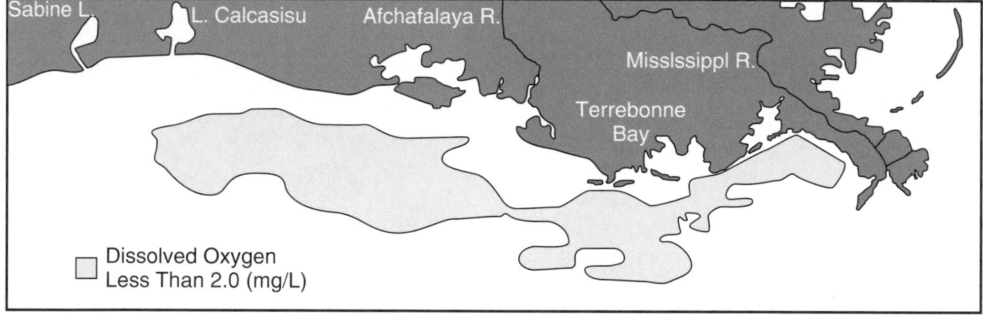

FIGURE 17.7 Nutrients supplied to the Gulf of Mexico by the Mississippi River support abundant growth of plankton (microscopic plants) near the water surface. When they die and settle toward the bottom, they increase BOD, deplete oxygen, and create a "dead zone" covering thousands of square miles. *Map courtesy U.S. Environmental Protection Agency Gulf of Mexico program.*

INDUSTRIAL POLLUTION

Hundreds of new chemicals are created by industrial scientists each year. The rate at which new chemicals are developed makes it impossible to demonstrate the safety of the new chemicals as fast as they are invented. To "prove" a chemical safe, it would be necessary to test a wide range of doses on every major category of organism—including people—at several stages in the life cycle. That simply is not possible, in terms of time, money, or laboratory space (not to mention the ethics of testing potential toxins on humans). Tests are therefore usually made on a few representative populations of laboratory animals to which the chemical is believed most likely to be harmful, or perhaps on tissue cultures. Whether other organisms will turn out to be unexpectedly sensitive to the chemical will not be known until after they have been exposed to it in the environment. The completely safe alternative would be to stop developing new chemicals, of course, but that would deprive society of many important medicinal drugs, as well as materials to make life safer, more comfortable, or more pleasant. The magnitude of the mystery about the safety of industrial chemicals is emphasized by a National Research Council report. A committee compiled a list of nearly 66,000 drugs, pesticides, and other industrial chemicals and discovered that *no* toxicity data at all were available for 70% of them; a complete health hazard evaluation was possible for *only 2%*. In 1990, the American

Box 17.1 Clogging with Kindness?

Samantha and Jon wandered along the path beside the lagoon. Near the footbridge, they came upon several groups of parents and children, feeding chunks of stale bread to a huge flock of noisy Canada geese.

"I wish they wouldn't do that," commented Jon (who happened to be a biology major).

"Why? They're cute! Look at that little boy running away from that enormous goose!" exclaimed Samantha.

"Oh, they're cute, I agree. But it's really no good for the geese—or for that matter, the lagoon," responded Jon.

"But the geese are obviously hungry," objected Samantha. "Look at the way they eat!"

"Yes, but they've come to depend on the human handouts," Jon pointed out. "And because they have such an easy food source, they hang around in the fall and winter when they should be flying south; and in the summer, their numbers are much larger than they'd be if they had to forage for themselves. If the winter weather gets bad and people stay inside, many will starve; and in the summer, the lagoon gets disgusting."

"How is that the geese's fault?"

"Well, they can't really help it, but let's just say there's an awful lot of organic fertilizer here in the summer," was Jon's reply.

Eutrophication isn't just a matter of farmland fertilizer runoff or sewage discharge. Some less obvious sources can be a problem, too. Jon's worry about the lagoon was well-founded (figure 1). Canada geese, ducks, and other migratory waterfowl are very adaptable and have made themselves at home on ponds in industrial parks, residential neighborhoods, and recreational sites. Indeed, they tend to prefer artificial ponds that lack dense vegetation nearby, as brush and shrubs might hide predators such as foxes, cats, and dogs. (Some golf courses have added border collies to their staffs to drive away flocks of geese that damage turf as well as fouling ponds.) The waterfowl contribute to algal bloom, which, in severe cases, can cover entire lake surfaces (recall figure 17.6A). When humans supply food, more birds congregate than would be supported by the natural food supplies in the area, and they also impose more of a pollutant load on local water bodies. Well-intentioned actions, like feeding the not-altogether-wild ducks and geese, can have undesirable consequences—as Jon noted—for the birds as well as the water quality.

A

B

FIGURE 1 Feeding flocks of birds at a pond can be an appealing activity, and on the face of it, may seem beneficial to the birds (A). However, the resultant excess nutrient loading of the water with organic fertilizer can be a problem, as here at the campus lagoon at Northern Illinois University (B).
(A) Photograph © J. A. Kraulis/Masterfile.

Chemical Society recorded the *ten millionth* entry in its registry of new chemicals created or identified since 1957. Complete toxicity assessment for that many substances would obviously be a formidable task.

Inorganic Pollutants—Metals

Many of the inorganic industrial pollutants of particular concern are potentially toxic metals. Manufacturing, mining, and mineral-processing activities can all increase the influxes of these naturally occurring substances into the environment and locally increase concentrations from harmless to toxic levels. Examples of metal pollutants commonly released in wastewaters from various industries are shown in table 17.2.

The element mercury is a case in point. In nature, the small amounts of mercury (1 to 2 ppb) found in rocks and soil are weathered out into waters (where mercury is typically found in concentrations of a few ppb) or into the air (less than 1 ppb), from which it is quickly removed with precipitation. Mercury in water is in time removed into sediments. Due to its low solubility, its residence time in the oceans is about 80,000 years.

Mercury is one of the **heavy metals,** a group that includes lead, cadmium, plutonium, and others. A feature that the heavy metals have in common is that they tend to accumulate in the bodies of organisms that ingest them. Therefore, their concentrations increase up a food chain (figure 17.8), a phenomenon known as **biomagnification.** Some marine algae may contain heavy metals at concentrations of up to 100 times that of the water in which they are living. Small fish eating the algae develop higher concentrations of heavy metals in their flesh, larger fish who eat the smaller fish concentrate the metals still further, and so on up to fish-eating birds or mammals.

TABLE 17.2

Principal Trace Metals in Industrial Wastewaters

Industry	Metals
mining and ore processing	arsenic, beryllium, cadmium, lead, mercury, manganese, uranium, zinc
metallurgy, alloys	arsenic, beryllium, bismuth, cadmium, chromium, copper, lead, mercury, nickel, vanadium, zinc
chemical industry	arsenic, barium, cadmium, chromium, copper, lead, mercury, tin, uranium, vanadium, zinc
glass	arsenic, barium, lead, nickel
pulp and paper mills	chromium, copper, lead, mercury, nickel
textiles	arsenic, barium, cadmium, copper, lead, mercury, nickel
fertilizers	arsenic, cadmium, chromium, copper, lead, mercury, manganese, nickel, zinc
petroleum refining	arsenic, cadmium, chromium, copper, lead, nickel, vanadium, zinc

Source: Data from J. E. Fergusson, *Inorganic Chemistry and the Earth.* Copyright © 1982 Pergamon Press, Oxford, England.

For those who are aware of the mercury in the silver fillings in teeth, a word of reassurance may first be in order. The "silver" that dentists put in teeth is really an amalgam of mercury and silver, but elemental mercury is not the highly toxic form, for it is not readily absorbed by the body; silver fillings pose no known health threat. The severely toxic form is *methylmercury,* an organic compound of mercury produced by microorganisms in water and taken up by plants and small organisms at the bottom of the food chain. It is as methylmercury that mercury accumulates up the food chain.

Even so, in most natural settings, heavy-metal accumulations in organisms are not very serious because the natural concentrations of these metals are low in waters and soils to begin with. The problems develop when human activities locally upset the natural cycle. Mining and processing heavy metals in particular can increase the rate at which the heavy metals weather out of rock into the environment, and these industries have also discharged concentrated doses of heavy metals into some bodies of water.

Methylmercury acts on the central nervous system and the brain. It can cause loss of sight, feeling, and hearing, as well as nervousness, shakiness, and death. By the time the symptoms become apparent, the damage is irreversible. The effects of mercury poisoning may, in fact, have been recognized long before their cause. In earlier times, one common use of mercury was in the making of felt. Some scholars have suggested that the idiosyncrasies of the Mad Hatter of *Alice in Wonderland* may actually have been typical of hat makers of the period, whose frequent handling of felt led to chronic mercury poisoning, commonly known as "hatter's shakes." Mercury has also been used as a fungicide. In Iraq, in 1971–1972, severe famine drove peasants to eat mercury-treated seed grain instead of saving it to plant. Five hundred died, and an estimated 7000 people were affected. See also Box 17.2.

Mercury is not the only toxic heavy metal. Cadmium poisoning was most dramatically demonstrated in Japan when cadmium-rich mine wastes were dumped into the upper Zintsu River. The people used the water for irrigation of the rice fields and for domestic use. Many developed *itai-itai* (literally, "ouch-ouch") disease, cadmium poisoning characterized by abdominal pain, vomiting, and other unpleasant symptoms. Fear of the toxic-metal effects of plutonium is one concern of those worried about radioactive-waste disposal.

For some industrial applications, substitutes for toxic metals can be found. Mercury is no longer used in the manufacture of felt, for instance. However, high concentrations of toxic metals may exist in natural geological materials. Indeed, the burning of coal is now the primary source of mercury pollution in the United States.

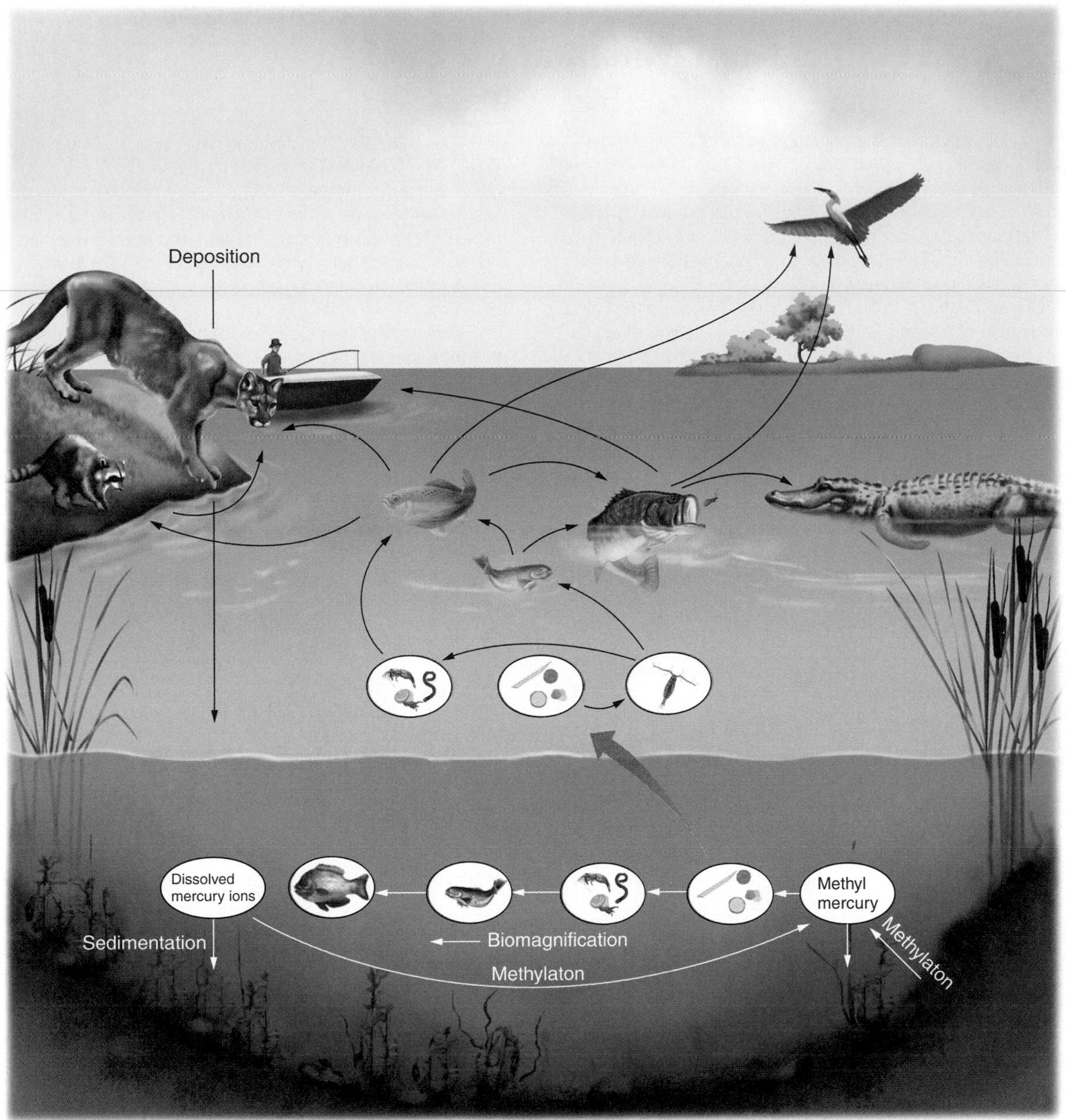

FIGURE 17.8 Simplified outline of mercury behavior in a body of water: Mercury deposited in the water may be converted to methylmercury, in which form it is concentrated by biomagnification. Microorganisms containing methylmercury are consumed by small fish, the larger fish eat the smaller fish, and so on up to birds and mammals higher up the chain. Accumulative toxins like the heavy metals may become very highly concentrated in creatures at the top of the chain.
Modified after USGS Fact Sheet 063–00.

Likewise, potentially toxic metals can occur naturally in concentrations above U.S. drinking-water standards. Arsenic is an example.

The toxic effects of large doses of arsenic are well documented. It can cause skin, bladder, and other cancers; in large doses, it is lethal. The effects of low doses are much less clear. In some cultures, arsenic is deliberately consumed (in moderation) to promote a clear complexion and glossy hair. It is accumulative in the body, and over time, some tolerance to its toxicity may be developed. Arsenic is not known to be an essential nutrient—there is no recommended minimum daily dose, as there is for iron, or zinc, or copper, or several other metals—but whether it is harmful even in the smallest quantities is not known. Drinking-water limits for such elements are established by extrapolation from high doses known to be toxic to a dose low enough to correspond to a perceived acceptable level of risk (e.g., one death per million persons, or serious illness per 100,000, or whatever) coupled with statistical studies of the

Box 17.2 Lessons from Minamata

In the early twentieth century, Minamata was a small coastal town in western Kyushu, Japan, with a long history as a village of farmers and fishermen. Industrial development was in its early stages. At the time, release of industrial wastewater into the ocean was common in many places worldwide, and so it was here. The major local industrial firm, Chisso Corporation, paid the local fishermen to compensate them for any possible damage to their livelihood as a result.

In 1932, Chisso began to manufacture acetaldehyde, a chemical used in the production of plastics. At that time, the production of acetaldehyde, in turn, required mercury. In the early 1950s, after the end of World War II, demand for plastics—and thus acetaldehyde—soared, as did production at Minamata.

It was the cats who soon provided the first clue to the developing tragedy, although its significance initially went unrecognized. The cats of Minamata began to behave oddly, "dancing," falling into the ocean and drowning. Then, strange symptoms appeared in people. They began to stumble, to slur their speech, to suffer convulsions and fits of trembling. Babies were born deformed. Some victims became paralyzed. Some died. The collection of symptoms was dubbed "Minamata disease," and researchers scrambled to identify its cause.

We know now that it was mercury poisoning. Mercury in Chisso's wastewater was flushed into Minamata Bay, the water of which only slowly mixes with the open ocean. As methylmercury, it was taken up in algae, then fish and shellfish, then the cats and the people who ate the fish, its concentration biomagnified along the way. Among the human population, the fishermen, who typically ate the most fish, were most acutely affected.

Fishing has long been prohibited in Minamata Bay, where even now the sediments are laden with mercury. A memorial garden has been planted to honor and commemorate the victims. The total number of victims may never be known precisely; milder cases of mercury poisoning have symptoms that cannot be tied uniquely to mercury. The toll of those acutely poisoned is grim enough: more than 1400 have died so far, and thousands more suffer lingering effects. A 2001 study suggested that tens of thousands of persons may have been affected altogether, including many living outside Minamata Bay.

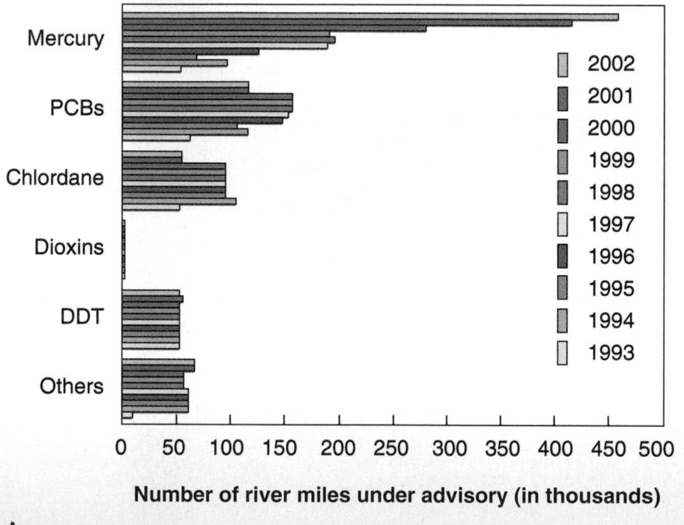

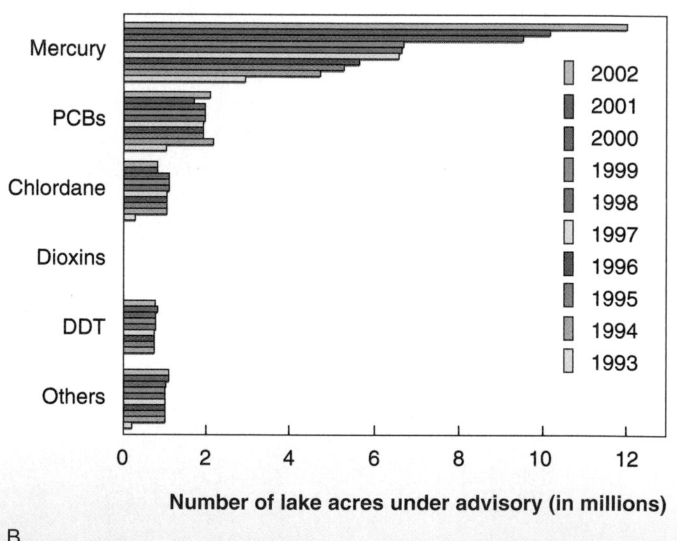

FIGURE 1 Number of river miles (A) and lake acres (B) under fish consumption advisories for various pollutants, 1993–2002. Note that the only fast-growing problem over this period is mercury.
Graphs after U.S. Environmental Protection Agency.

incidence of toxic effects in populations naturally exposed to various doses of the substance, often with an additional safety margin thrown in.

For many years, the drinking-water standard for arsenic was set at 50 ppb (or, 50 micrograms per liter of water; a microgram is one millionth of a gram, or about 1/30,000th of an ounce, and a liter is just over a quart). More-recent research led the National Research Council, in 1999, to call for a new, lower limit for arsenic in drinking water. In January 2001, the EPA published a new limit of 10 ppb, to take effect in January 2006 (to give municipalities time to adjust their processing of water accordingly). Objections were immediately raised, in part because natural arsenic levels in ground water in much of the country already exceed that level (figure 17.9), and in part because the evidence justifying the lower standard was not definitive. In March 2001, the administration withdrew the new standard, but it was later reimposed, and did indeed become effective in January 2006.

(Box 17.2 Continued)

As the toxicity and the accumulative quality of methylmercury were realized, testing of fish for mercury was initiated in many nations. The results were sufficiently alarming that during the 1960s and 1970s, people were advised to limit their intake of tuna and certain other fish with particularly high mercury contents. (Over 95% of the mercury in fish is in the form of methylmercury.) After some years of enhanced pollution control, particularly of point sources such as industrial wastewater, the mercury levels, and the concerns, abated.

However, that is far from the end of the story, for it turns out that wastewater is not the only contributor to mercury pollution problems.

Mercury is a very volatile (easily vaporized) metal. When mercury vapor is put into the atmosphere—whether by coal burning, waste incineration, smelting of mercury-bearing ores, battery manufacturing, or whatever—it disperses widely before being deposited on land and water. In recent decades, human activity has more than doubled the concentration of mercury in the atmosphere, correspondingly increasing that deposition. What begins as an air-pollution issue becomes a water-pollution problem.

The increasing use of coal for energy, together with expanded sampling, have led to sharply increasing numbers of mercury-related fish consumption advisories on lakes and rivers (figure 1). By 2002, some 18 states had 100% of their lakes and rivers under advisory for various pollutants, mercury chief among them (figure 2). The EPA has estimated that in some watersheds, the airborne mercury load to lakes and streams would have to be reduced by more than 75% to bring methylmercury concentrations in fish down to acceptable levels.

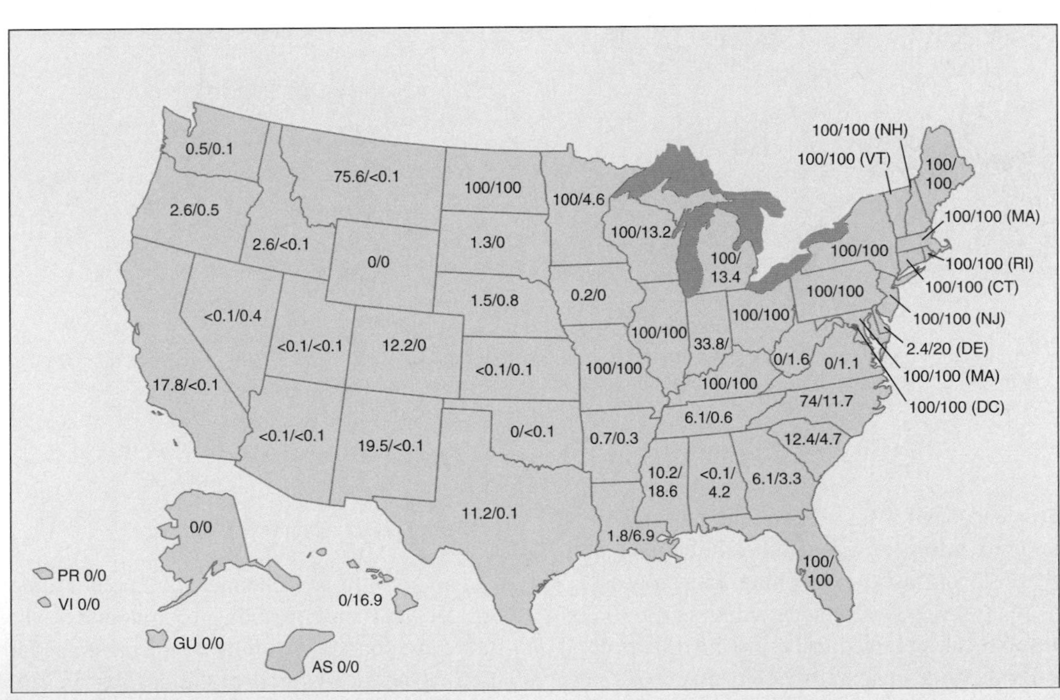

FIGURE 2 Percentage of lake acres/river miles under fish consumption advisories as of December 2002, by state. Note that lack of an advisory may reflect lack of sampling, not absence of problems.
After U.S. Environmental Protection Agency.

Other Inorganic Pollutants

Some nonmetallic elements commonly used in industry are also potentially toxic to aquatic life, if not to humans. For example, chlorine is widely used to kill bacteria in municipal water and sewage treatment plants and to destroy various microorganisms that might otherwise foul the plumbing in power stations. Released in wastewater, it can also kill algae and harm fish populations.

Acids from industrial operations used to be a considerably greater pollution problem before stringent controls on their release were imposed. Acid mine drainage, however, remains a serious source of surface- and groundwater pollution, especially in coal- and sulfide-mining areas. The acids, in turn, may leach additional toxic metals from rocks, tailings, or soil.

The toxic effects of certain asbestos minerals were not manifested or well defined until long after their initial release into the environment by human activities. Asbestos minerals

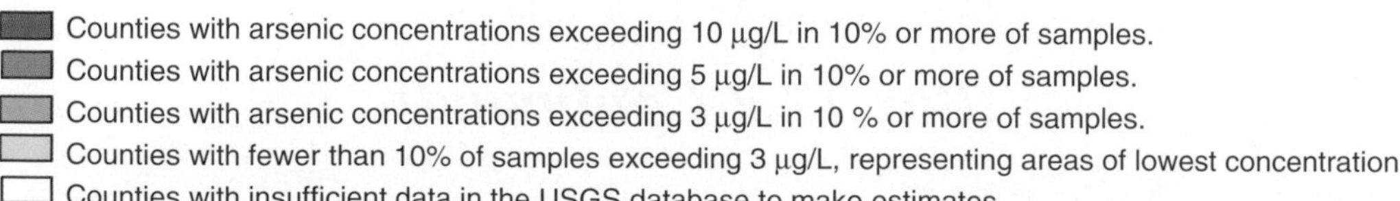

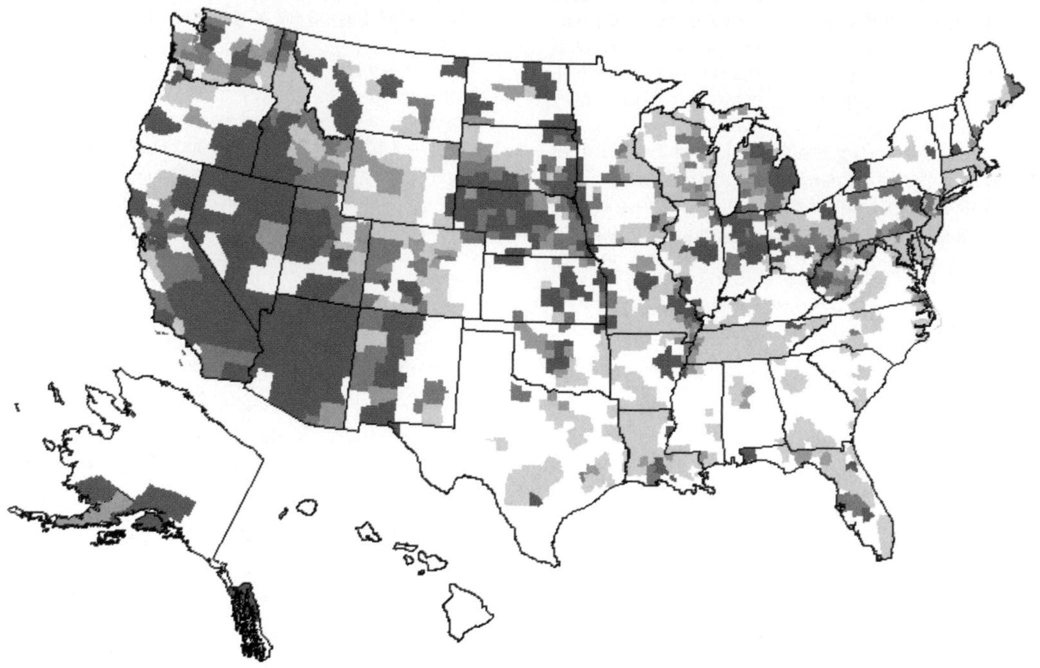

FIGURE 17.9 Arsenic in ground water in many parts of the United States naturally exceeds the new EPA drinking-water standard of a maximum of 10 micrograms per liter (μg/L), and new data suggest significant toxicity at still lower concentrations.
After USGS Fact Sheet 063-00.

have been prized for decades for their fire-retardant properties and have been used in ceiling tiles and other building materials for years. Wastes from asbestos mining and processing were dumped into many bodies of water, including the Great Lakes—they were, after all, believed to be just inert, harmless mineral materials. By the time the carcinogenic effects of some asbestos minerals were realized, asbestos workers had been exposed and segments of the public had been drinking asbestos-bearing waters for twenty years or more. The full impact of this sustained exposure is still being assessed. While the potential risk from inhalation is best documented, it is worth noting that ingestion of asbestos minerals has been implicated as a cause of cancer, at least in some cases: In Japan, many prefer powdered rice, and the powder is mainly pulverized talc that occurs geologically in association with some types of asbestos. It has been suggested that this may account for higher incidence of stomach cancer.

Unfortunately, the harmful effects of such inorganic pollutants are often slow to develop, which means that decades of testing might be required to reveal the dangers of such materials. The costs of such long-term testing are extremely high, generally prohibitively so.

Organic Compounds

The majority of new chemical compounds created each year are organic (carbon-containing) compounds. Many thousands of these compounds, naturally occurring and synthetic, are widely used as herbicides and pesticides, as well as in a variety of industrial processes. Examples include DDT and dioxin. Their negative effects in organisms vary with the particular type of compound: Some are carcinogenic, some are directly toxic to humans or other organisms, and others make water unpalatable. Some also accumulate in organisms as the heavy metals do.

Oil spills are another kind of organic-compound pollution. At least as much additional oil pollution occurs each year from the careless disposal of used crankcase oil, dumping of bilge from ships, and the runoff of oil from city streets during rainstorms. Underground tanks and pipelines may also leak, and drilling muds and waste brines (saline pore fluids) discarded in oil fields may be contaminated with petroleum. Oil spills into U.S. waters average over 10 million gallons per year.

Another type of organic-compound pollution is the result of the U.S. plastics industry's demand for production of nearly 7 billion pounds of vinyl chloride each year. Vinyl chloride vapors are carcinogenic, and it is not known how harmful traces of vinyl chloride in water may be.

Polychlorinated biphenyls (PCBs) were used for nearly twenty years as insulating fluid in electrical equipment and as plasticizers (compounds that help preserve flexibility in plastics). Laboratory tests revealed that PCBs in animals cause impaired reproduction, stomach and liver ailments, and other problems. PCB production in the United States was banned in 1977, but approximately 900 million pounds of PCBs had already been produced and some portion of that quantity released into the environment, where it remains.

A concern that arose in the 1990s is MTBE (methyl tert-butyl ether), a petroleum derivative added to gasoline. MTBE is used to promote more-complete burning of gasoline, to reduce carbon monoxide pollution in urban areas. However, MTBE has begun to appear in analyses of ground water in areas where it is widely used. Its health effects are not known precisely but it is regarded by the U.S. EPA as a potential carcinogen; it also has an objectionable taste and odor. A national study to assess the incidence of MTBE contamination in ground water is now in progress.

Problems of Control

Cost and efficiency are factors in pollution control. Those substances not to be released with effluent water (or air) must be retained. This becomes progressively more difficult as complete cleaning of the effluent is approached. First, there is no wholly effective way to remove pollutants from wastewater before discharge. No chemical process is 100% efficient. Second, as cleaner and cleaner output is approached, costs skyrocket. If it costs \$1 million to remove 90% of some pollutant from industrial wastewater, then removal of the next 9% (90% of the remaining 10% of the pollutant) costs an additional million dollars. To get the water 99.9% clean costs a total of \$3 million, and so on. Some toxin always remains, and the costs escalate rapidly as "pure" water is approached. At what point is it no longer cost-effective to keep cleaning up? That depends on the toxicity of the pollutant and on the importance (financial or otherwise) of the process of which it is the product. Furthermore, the toxins retained as a result of the cleaning process do not just disappear. They become part of the growing mass of industrial toxic waste requiring careful disposal.

The very nature of the ways in which hazardous compounds are used is an obvious problem. Herbicides and pesticides are commonly spread over broad areas, where it is typically impossible to confine them. The result, not surprisingly, is that they become nearly ubiquitous in the environment (figure 17.10).

Even when the threat posed by a toxic agent is recognized and its production actually ceases, it may prove long-lived in the environment. Once released and dispersed, it may be impossible to clean up. The only course is to wait for its natural destruction, which may take unexpectedly long. DDT is a case in point.

The compound DDT (dichlorodiphenyl trichloroethane) was discovered to act as an insecticide during the 1930s. Its first wide use during World War II, killing lice, ticks, and malaria-bearing mosquitoes, unquestionably prevented a great deal of suffering and many deaths. In fact, the scientist who first recognized DDT's insecticidal effects, Paul Muller, was subsequently awarded the Nobel Prize in medicine for his work. Farmers undertook wholesale sprayings with DDT to control pests in food crops. The chemical was regarded as a panacea for insect problems; cheery advertisements in the popular press promoted its use in the home.

Then the complications arose. One was that whole insect populations began to develop some resistance to DDT. Each time DDT was used, those individual insects with more natural resistance to its effects would survive in greater proportions than the population as a whole. The next generation would contain a

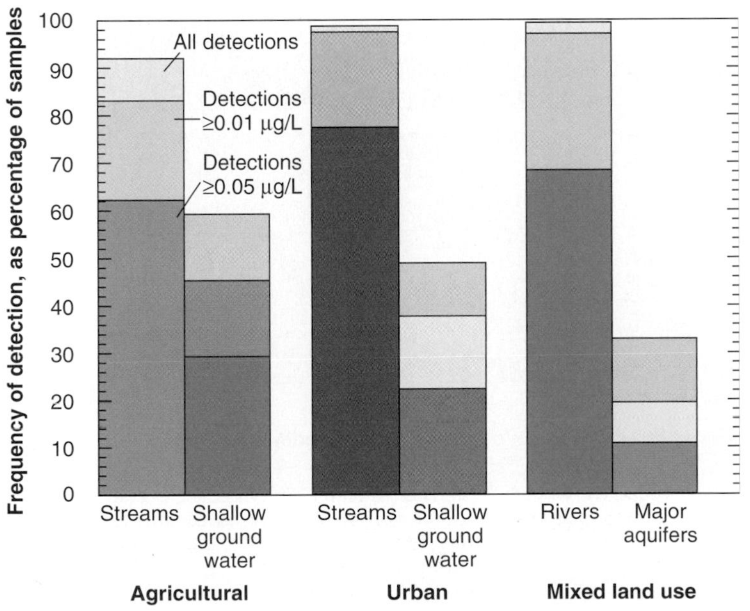

FIGURE 17.10 Herbicides and pesticides are everywhere in our water and, perhaps contrary to expectations, as common in urban as agricultural areas. Pesticides were detectable in nearly all stream samples in a national study.

Source: USGS Pesticides National Synthesis Project, 2000.

higher proportion of insects with some inherited resistance, who would, in turn, survive the next spraying in greater numbers, and so on. This development of resistance resulted in the need for stronger and stronger applications of DDT, which ultimately favored the development of ever-tougher insect populations that could withstand these higher doses. Insects' short breeding cycles made it possible for all of this to occur within a very few years.

Early on, DDT was also found to be quite toxic to tropical fish, which was not considered a big problem. Moreover, it was believed that DDT would break down fairly quickly in the environment. In fact, it did not. It is a persistent chemical in the natural environment. Also, like the heavy metals, DDT is an accumulative chemical. It is fat-soluble and builds up in the fatty tissues of humans and animals. Fish not killed by DDT nevertheless accumulated concentrated doses of it, which were passed on to fish-eating birds. Then another deadly effect of DDT was realized: It impairs calcium metabolism. In birds, this effect was manifested in the laying of eggs with very thin and fragile shells. Whole colonies of birds were wiped out, not because the adult birds died, but because not a single egg survived long enough to hatch. Robins picked up the toxin from DDT-bearing worms. Whole species were put at risk. All of this prompted Rachel Carson to write *Silent Spring* in 1962, warning of the insecticide's long-term threat.

Finally, the volume of data demonstrating the toxicity of DDT to fish, birds, and valuable insects such as bees became so large that, in 1972, the U.S. Environmental Protection Agency banned its use, except on a few minor crops and in medical emergencies (to fight infestations of disease-carrying insects). So persistent is it in the environment that twenty years later, fish often had detectable (though much lower) levels of DDT in their tissues. Encouragingly, some wildlife populations have recovered since DDT use in the United States was sharply curtailed. Still, DDT continues to be applied extensively in other countries.

DDT was undoubtedly a uniquely powerful tool for fighting insect pests transmitting serious diseases. It was neither the first nor the last new compound to be found—long after it came into general use—to have significantly toxic effects. This may be inevitable given the rate of synthesis of new compounds and the complexity of testing their safety, but one unfortunate result is the long-term accumulation of toxins in water, sediments, and organisms (Fig. 17.11).

Thermal Pollution

Thermal pollution, the release of excess or waste heat, is a by-product of the generation of power. The waste heat from automobile exhaust or heating systems contributes to thermal pollution of the atmosphere, but its magnitude is generally believed to be insignificant. Potentially more serious is the local thermal pollution of water by electric generating plants and other industries using cooling water.

Only part of the heat absorbed by cooling water can be extracted effectively and used constructively. The still-warm cooling water is returned to its source—most commonly a stream—and replaced by a fresh supply of cooler water. The resulting temperature increase usually exceeds normal seasonal fluctuations in the stream and results in a consistently higher temperature regime near the effluent source.

Fish and other cold-blooded organisms, including a variety of microorganisms, can survive only within certain temperature ranges. Excessively high temperatures may result in the wholesale destruction of organisms. More-moderate increases can change the balance of organisms present. For example, green algae grow best at 30 to 35° C (86 to 95° F); blue-green algae, at 35 to 40° C (95 to 104° F). Higher temperatures thus favor the blue-green algae, which are a poorer food source for fish and may be toxic to some. Many fish spawn in waters somewhat cooler than they can live in; a temperature rise of a few degrees might have no effect on existing adult fish populations but could seriously impair breeding and thus decrease future numbers. Furthermore, changes in water temperature change the rates of chemical reactions and critical chemical properties of the water, such as concentrations of dissolved gases, including the oxygen that is vital to fish.

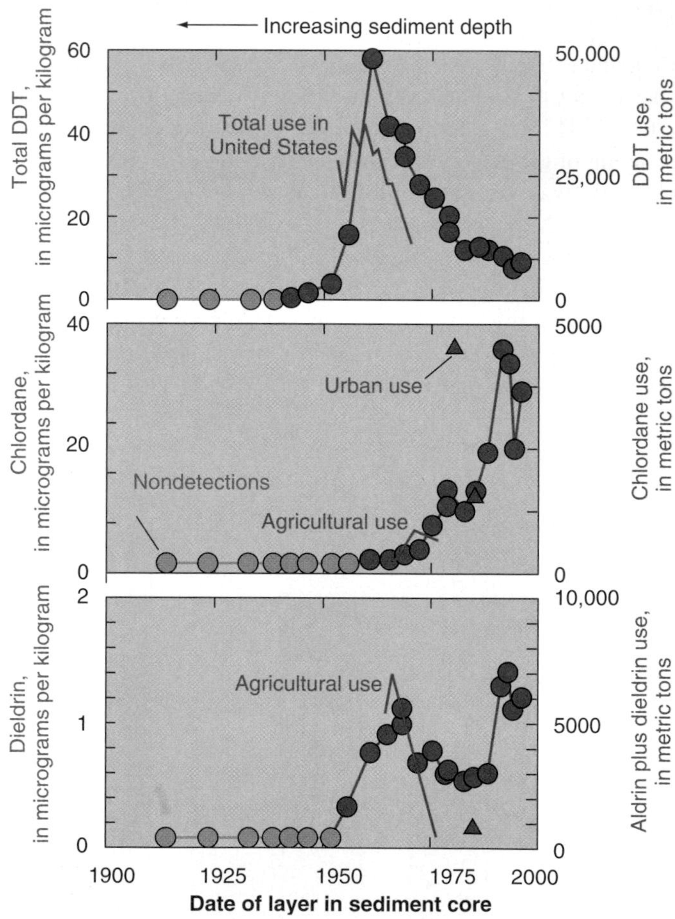

FIGURE 17.11 A study of sediments in White Rock Lake, Dallas, TX, showed continued accumulation of pesticides in sediments, even decades after the pesticides' last use: dieldrin was banned in 1974, while chlordane use continued into the late 1980s.
Source: U.S. Geological Survey Circular 1225.

Thermal pollution can be greatly reduced by holding water in cooling towers before release. (In practice, however, even where cooling towers are used, the released water is still at a somewhat elevated temperature.) The extent of thermal pollution is also restricted in both space and time. It adds no substances to the water that persist for long periods or that can be transported over long distances, and reduction in the output of excess waste heat results in an immediate reduction in the magnitude of the continuing pollution problems.

AGRICULTURAL POLLUTION

Some water-pollution problems resulting from agricultural activities are similar to ones already described. They are of special note because agricultural activities cover such broad areas.

Fertilizers and Organic Waste

The three principal constituents of commercial fertilizers are nitrates, phosphates, and potash. When applied to or incorporated in the soil, they are not immediately taken up by plants. The compounds must be soluble in order for plants to use them, but that means that they can also dissolve in surface-water and groundwater bodies. These plant foods then contribute to eutrophication problems. There is typically a clear correlation with applications of fertilizer (figure 17.12), and thus the concerns are greatest in such areas.

Reduction in fertilizer applications to the minimum needed, perhaps coupled with the use of slow-release fertilizers, would minimize the harmful effects. Nitrates could alternatively be supplied by the periodic planting of legumes (a plant group that includes peas, beans, and clovers), on the roots of which grow bacteria that fix nitrogen in the soil. This practice reduces the need to apply soluble synthetic fertilizers.

Unfortunately, the use of other natural fertilizers, like animal manures, fails to eliminate the water-quality problems associated with fertilizer runoff. Manures are, after all, organic wastes. Thus, they contribute to the BOD of runoff waters in addition to adding nutrients that lead to eutrophication.

Likewise, runoff of wastes from domestic animals, especially from commercial feedlots, is a potential problem. One concern that arises in connection with some such operations is not unlike the problem of municipal sewage-treatment plants overwhelmed by the added volume of runoff water from major storms. It is increasingly common for the wastes to be diverted into shallow basins where they may be partially decomposed and/or dried and concentrated for use as fertilizer. If poorly designed, the surrounding walls (often made of earth) may

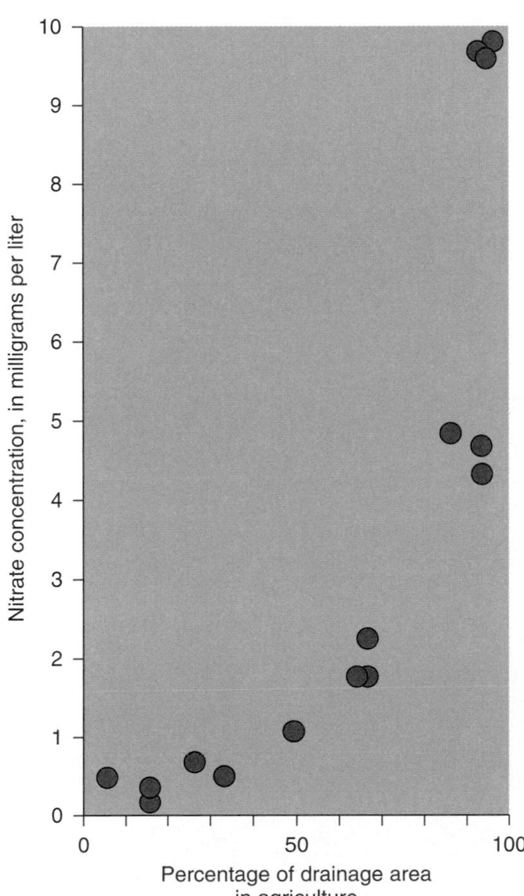

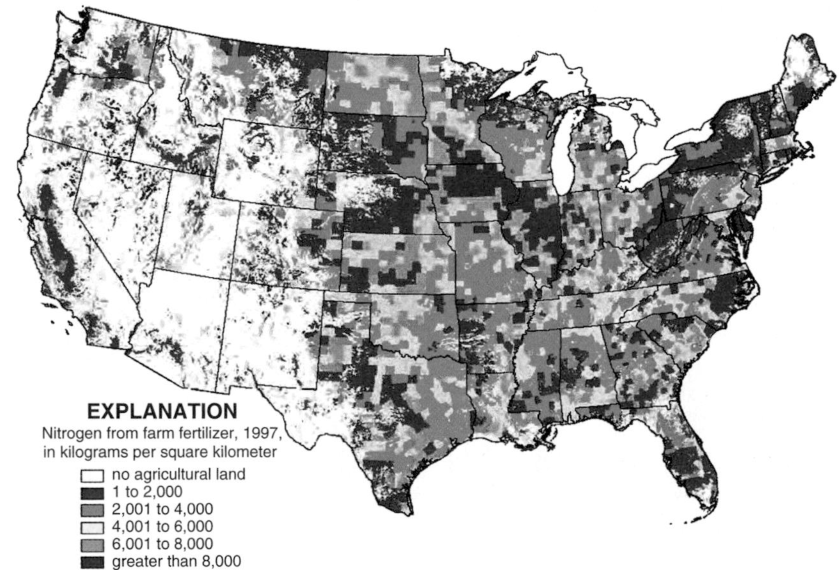

FIGURE 17.12 (A) Study in the Willamette Basin shows typical strong correlation of nitrate in spring stream runoff with proportion of drainage basin in agricultural use. (B) Potential nonpoint-source pollution from nitrogen in commercial fertilizer application, 1997.

Sources: (A) Data from U.S. Geological Survey Circular 1225, (B) From USGS Scientific Investigations Report 2006-5012.

fail, releasing the contents. Even if the structure holds, in an intense storm, the added water volume may cause overflowing of the wastes. One approach that has shown promise is to build broad channels downslope of the waste source, planted with grass or other vegetation, to act as filter strips for the runoff (figure 17.13). The filter strip catches suspended solids and slows fluid runoff, allowing time for decomposition of some dissolved constituents and nutrient uptake by the vegetation (depending on season), plus increased infiltration to reduce the volume of runoff leaving the site. Studies have shown reductions of up to 80% in dissolved phosphorous, nitrogen, and other constituents in the runoff water leaving the filter strip, though sometimes partly at the cost of increases in these substances in the ground water below the site.

An alternative, productive solution to the excess-manure problem would be more extensive use of these wastes for the production of methane for fuel. This has already been found to be economically practical where large numbers of animals are concentrated in one place (figure 17.14).

Sediment Pollution

Some cases of high suspended-sediment load in water occur naturally (figure 17.15). However, these are commonly localized situations, and may be temporary as well. In many agricultural areas, sediment pollution of lakes and streams is the most serious water-quality problem and is ongoing. Farmland and forest land together are believed to account for about 75% of the 3 billion tons of sediment supplied annually to U.S. waterways. Of this total, the bulk of sediment is derived from farmland; sediment yields from forest land are typically low.

Sediment pollution not only causes water to be murky and unpleasant to look at, swim in, or drink, it reduces the light available to underwater plants and blankets food supplies and

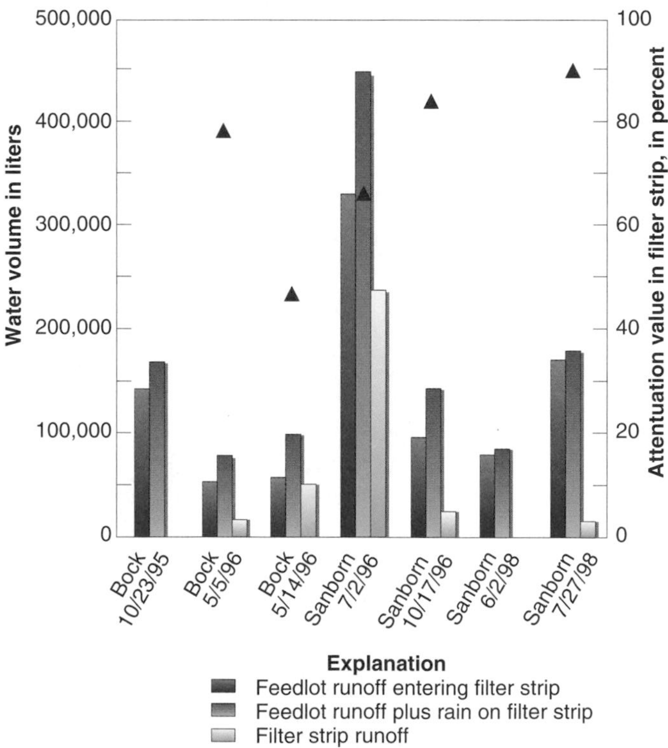

FIGURE 17.13 Studies show that a grassy filter strip below a feedlot or waste-holding lagoon can reduce runoff of nutrient-rich wastewater following storms (though it may increase those nutrients in local ground water). "Attenuation value" is the percentage reduction in surface-runoff volume due to the presence of the strip.

After U.S. Geological Survey Water-Resource Investigations Report 03-4036.

FIGURE 17.14 Anaerobic digester for pig manure generates biogas while containing waste.

Photograph courtesy California Polytechnic State University and National Renewable Energy Lab.

FIGURE 17.15 Sediment clouds stream waters. The stream at right is murky with suspended sediment, probably because of bank collapse. Note the contrast with the clear tributary at left, and the fact that the waters of the two streams do not immediately mix where they join. Junction of Soda Butte Creek and the Lamar River, Yellowstone National Park.

the nests of fish, thus reducing fish and shellfish populations. Channels and reservoirs may be filled in, as noted in earlier chapters. The sediment also clogs water filters and damages power-generating equipment. Some sediment transport is a perfectly natural consequence of stream erosion, but agricultural development typically increases erosion rates by four to nine times unless agricultural practices are chosen carefully.

Chapter 12 examined some strategies for limiting soil loss from farmland. From a water-quality standpoint, an additional approach would be to use settling ponds below fields subject to serious erosion by surface runoff (figure 17.16). These ponds do not stop the erosion from the fields, but by impounding the water, they cause the suspended soil to be dropped before the water is released to a lake or stream. Settling ponds can also be helpful below large construction projects, logging operations, or anywhere ground disturbance is accelerating erosion.

FIGURE 17.16 (A) Settling ponds enhance water quality and reduce soil loss, reducing sediment pollution and thereby also trapping pollutants adsorbed onto the sediment particles. Removal of sediment from the water, however, may increase erosion below the settling pond. (B) Heavy surface runoff may contribute to both flooding and soil erosion, muddying floodwaters with sediment pollution. This "watershed dam" in eastern Iowa traps the sediment-laden waters, reducing flooding downstream and containing the sediment, which will settle out of the basin's still water.

(B) Photograph by Lynn Betts, courtesy USDA Natural Resources Conservation Service.

Herbicides and Pesticides

Herbicides (chemical weed-killers) and pesticides (chemicals used to kill insect pests) are a significant source of pollution in agriculture. Most such compounds now in use are complex organic compounds of the kinds described in the "Industrial Pollution" section earlier in the chapter and are in some measure toxic to humans or other life-forms. U.S. farmers use more than a billion pounds of herbicides and pesticides each year, of which about 90% are synthetic organic compounds. The use of pesticides in agriculture doubled, and the use of herbicides more than quadrupled, from 1965 to 1980, though the use of both has leveled off since then.

The agricultural chemicals themselves are not the sole problem, either. Many of the chemicals used in producing the end-use products are themselves toxic and persistent (dioxin, for instance); so are some of the early products of herbicide and pesticide decomposition, when and if they do begin to break down.

Commonly, spring applications of agricultural chemicals are followed by concentration spikes in runoff water (figure 17.17);

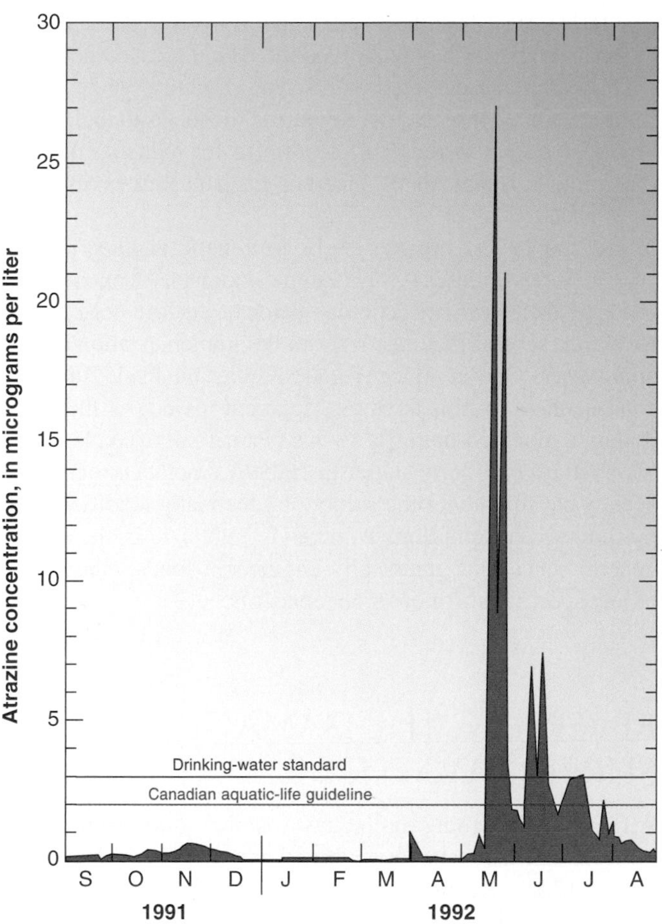

FIGURE 17.17 A study in the central Nebraska corn belt showed a sharp rise in pesticide runoff (here illustrated by atrazine) following early-spring application to fields. Spring and summer rains continued to wash off significant, though declining, amounts. While the average annual concentration in runoff did not exceed drinking-water standards, peak concentrations were up to ten times that standard.

Source: After USGS Circular 1225.

clearly, a great deal of what is applied is wasted. The potential risks could be reduced by eliminating routine applications of such chemicals as a sort of preventive medicine. Applying a remedy only when a problem is demonstrated to exist may initially reduce synthetic chemical use at the cost of crop yield. However, over the longer term, the results may be more economical, in part because the bugs or weeds will not as readily develop resistance to the substance being used.

A growing variety of nonchemical insect-reduction strategies are also finding increasing acceptance. Some are specific to a single pest. For instance, the larvae of a small wasp that is harmless to people and animals prey parasitically on tomato hornworms and kill them. A particular bacterium (*Bacillus thuringiensis*) attacks and destroys several types of borer insects but again produces no known ill effects in humans, mammals, birds, or fish, and leaves no persistent chemical residues on food.

Other remedies may be more broadly applicable. An example is the practice of sterilizing a large population of a particular insect pest by irradiation and then releasing the sterilized insects into the fields. Assuming the usual number of matings, the next generation of pests should be smaller in number because some of the matings will have involved infertile insects. Other growers bait simple insect traps with scents carefully synthesized to simulate female insects, thereby luring the males to their doom. Procedures such as these offer promise for reducing the need for synthetic chemicals of uncertain or damaging environmental impact.

In the 1990s, breakthroughs in genetic engineering created new possibilities—for example, corn plants modified to produce their own insecticidal defense against corn borers. However, several factors constrain the implementation of such approaches. One is concern for possible unintended negative consequences (in this example, apparent toxicity of the corn's pollen to monarch-butterfly larvae that eat milkweed leaves on which stray wind-borne pollen has fallen). Another is regulatory-agency concern about the safety of genetically modified crops for human consumption. Another is public resistance to the general concept of genetically engineered foods, independent of any specific health or safety concerns.

REVERSING THE DAMAGE—SURFACE WATER

All of the foregoing pollutants together have significantly affected surface waters over broad areas of the United States (figure 17.18). Reduction in pollutant input is one strategy for addressing such problems, but it requires time to succeed, especially with relatively stagnant waters such as lakes. A variety of more-aggressive treatment methods are available where indicated.

Many water pollutants, including phosphates, toxic organics, and heavy metals, can become attached to the surfaces of fine sediment particles on a lake bottom. After a long period of pollutant accumulation, the bottom sediments may contain a large reserve of undesirable or toxic chemicals, which could, in principle, continue to be re-released into the water long after input of new pollutants ceased. Wholesale removal of the contaminated sediments by dredging takes the pollutants permanently out of the system. However, they still remain to be disposed of, commonly in landfills.

Dredging operations must be done carefully to minimize the amount of very fine material churned back into suspension in the water. Fine resuspended sediment, with its high surface-to-volume ratio, tends to contain the highest concentrations of pollutants adsorbed onto grain surfaces. Also, if fine resuspended sediments remain suspended for some time, the increased water turbidity may be harmful to life in the lake. The process, moreover, can be expensive.

Dredging projects in the United States so far have tended to be small, usually involving less than 1 million cubic meters of sediment, and most are sufficiently recent that the long-term impacts are not known. However, the dredging of nutrient-laden sediments from Lake Trummen in Sweden during the early 1970s has been notably successful in reducing eutrophication by virtually eliminating nutrient recycling out of the sediments. In the near term, dredging may be used most in the United States for emergency cleanup of toxic wastes. When 250 gallons of PCBs spilled into the Dunwamish Waterway near Seattle in 1974, the EPA chose to dredge nearly 4 meters' thickness of contaminated sediment from the waterway and thereby succeeded in recovering 220 to 240 gallons of the spilled chemicals.

Another way to reduce the escape of contaminants from bottom sediments is to leave the sediments in place but to isolate them wholly or partially with a physical barrier. Over limited areas, like lagoons and small reservoirs, impermeable plastic liners have been placed on top of the sediments and held in place by an overlying layer of sand. The permanence of the treatment is questionable, and it has not been used on a large scale. Compacted clay layers of low permeability might, in principle, serve a similar purpose, but this treatment, to date, is a theoretical possibility only.

The addition of salts of aluminum, calcium, and iron to sediment changes the sediment chemistry, fixing phosphorous in the sediment and thereby reducing eutrophication. The technique has been used in small lakes in Wisconsin, Ohio, Minnesota, Washington, Oregon, and elsewhere. While not always successful, the method has often reduced phosphate levels in the water and corresponding algal growth. The cost compares favorably with the use of synthetic algicides to destroy the algae. However, careless overtreatment may prove toxic to fish, and, in any event, the treatment must be repeated every two to three years. An alternative, biological means of addressing eutrophication in ponds and lakes that has been developed recently involves introducing algae-eating (nonnative) fish that have been sterilized to prevent their overwhelming native species.

FIGURE 17.18 Surface-water pollution in the United States. (A) Point-source pollution. (B) Nonpoint-source pollution. In all cases, shaded areas represent sites of significant problems; symbols identify type(s).
Source: Modified from U.S. Water Resources Council, The Nation's Water Resources 1975–2000, *vol. 1, pp. 61–63.*

Decontamination is most often used in response to toxic-waste spills. The specific treatment methods depend on the toxin involved. In 1974, a toxic organic-chemical herbicide washed into Clarksburg Pond in New Jersey from an adjacent parking lot. Many fish were killed, and local wildlife dependent on the pond for water were endangered. The contamination also threatened to spread to ground water by infiltration and to the Delaware River to which the pond water flows overland. The EPA set up an emergency water-cleaning operation. The herbicide was removed principally by activated charcoal through which the water was filtered. Subsequent tests showed the ground water to be uncontaminated, and within two years, fish were again plentiful in the pond.

Artificial aeration addresses oxygen depletion in a lake. There are several possible procedures, including bubbling air or oxygen up through the waters, thereby providing more oxygen to oxygen-starved deep waters, or simply circulating the water mechanically, cycling up deep waters to the surface where they can dissolve oxygen directly from the atmosphere. Many small artificial ponds are now designed with fountains whose function is not purely decorative; spraying water into the air enhances aeration through the droplets' considerable surface area. When successful, aeration can transform water from an anaerobic to an aerobic condition, to the great benefit of fish populations (a common reason for the attempt). However, when air rather than pure oxygen is bubbled through the water

FIGURE 17.18 *(Continued)*

under pressure at depth, an excess of dissolved nitrogen may develop in the water, which is toxic to some fish.

GROUNDWATER POLLUTION

Groundwater pollution, whether from point or nonpoint sources, is especially insidious because it is not visible and often goes undetected for some time. Municipalities using well water must routinely test its quality. Homeowners relying on wells may be less anxious to go to the trouble or expense of testing, particularly if they are unaware of any potential danger. Also, in most instances, the passage of pollutants from their source into an aquifer used for drinking water is slow because that passage occurs by percolation of water through rock and soil, not by overland flow. There may therefore be a significant time lapse between the introduction of a pollutant into the system in one spot and its appearance in ground or surface water elsewhere. Conversely, groundwater pollution in karst areas, with their rapid drainage, may spread unexpectedly swiftly. The "out of sight, out of mind" aspect of ground water contributes to the problem; so does imperfect understanding of the interrelationships between surface and ground water.

The Surface–Ground Water Connection Explored

Given the nature of groundwater recharge, the process can readily introduce soluble pollutants along with recharge water. For example, agricultural pesticides are applied at the surface.

But those pesticides neither used up nor broken down may remain dissolved in infiltrating water as well as runoff, resulting in pesticide contamination of ground water, as previously described. Acid mine drainage, storm-sewer runoff, and other surface-water pollutants are likewise potential groundwater pollutants. The susceptibility of the ground water to pollution from such sources is a function of both inputs and local geology (figure 17.19).

Acid rain (discussed in chapter 18) can be a particular problem, because many toxic metals and other substances are more soluble in more acidic water. How severely groundwater chemistry is affected is related, in part, to the residence time of the ground water in its aquifer before it is extracted through a well or spring. That is, if water residence time is short, as with the shallow aquifers in figure 17.20, there is less opportunity for chemical reactions in the aquifer to buffer or moderate the

FIGURE 17.19 (A) Susceptibility of ground water to nitrate pollution is related not only to nitrogen applied in fertilizer (as shown in figure 17.12B) but to speed of drainage; "well-drained" soils are permeable soils through which water infiltrates readily. (B) In a national study of major aquifers, nitrate concentrations in ground water decreased as depth to aquifer increased; most samples exceeding the drinking-water standard of 10 milligrams per liter were from shallow aquifers. This may reflect more mixing of contaminated surface water with uncontaminated water in deep aquifers, or protection of deeper aquifers by thicker cover of low-permeability material.

Source: U.S. Geological Survey National Water Quality Assessment Program.

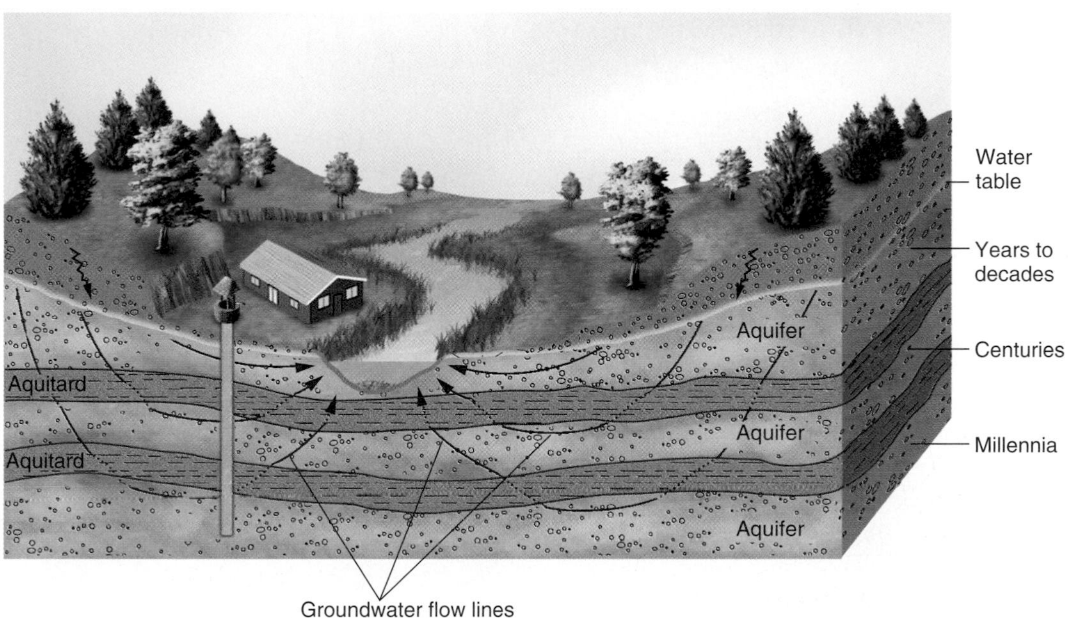

FIGURE 17.20 Groundwater flow rates may vary widely; in deep aquifers beneath multiple aquitards, centuries or more may elapse between infiltration in the recharge area and appearance in surface water, depending upon the rate of passage through aquitards.

impacts of acid rain. Longer residence times may allow more moderation of the chemistry of the acid waters. This is well illustrated by German studies of impacts of acid rain, which is locally a significant problem due to heavy reliance on the abundant German coal resources. Especially after major snowmelt events, there is rapid influx of acidified precipitation into groundwater systems. Springs fed by ground waters with short water residence times show not only high levels of sulfate and nitrate from sulfuric and nitric acid, but also elevated concentrations of a number of metals, due to solution in the acidic waters. Over time, the acidic waters can react, especially with aquifers containing dissolved carbonate; the acidity is reduced, along with the concentrations of many other dissolved constituents, while dissolved carbonate increases. The chemical impact on springs fed by ground waters with intermediate (months) or long (years) water residence times in the aquifers is correspondingly much less.

Altogether, substantial groundwater-quality problems exist in the United States (figure 17.21)—some natural (for example, high dissolved mineral content), some due to human activities (pollution). The nature of the problems and their causes can vary greatly from place to place. Not surprisingly, pesticides (or their breakdown products) are common in ground water beneath agricultural areas. Nitrate concentrations, too, tend to be higher in ground water below agricultural areas than in ground water under undeveloped areas. On the other hand, fertilizers, pesticides, and herbicides are not unique to agricultural regions; they may be used abundantly in residential areas, on

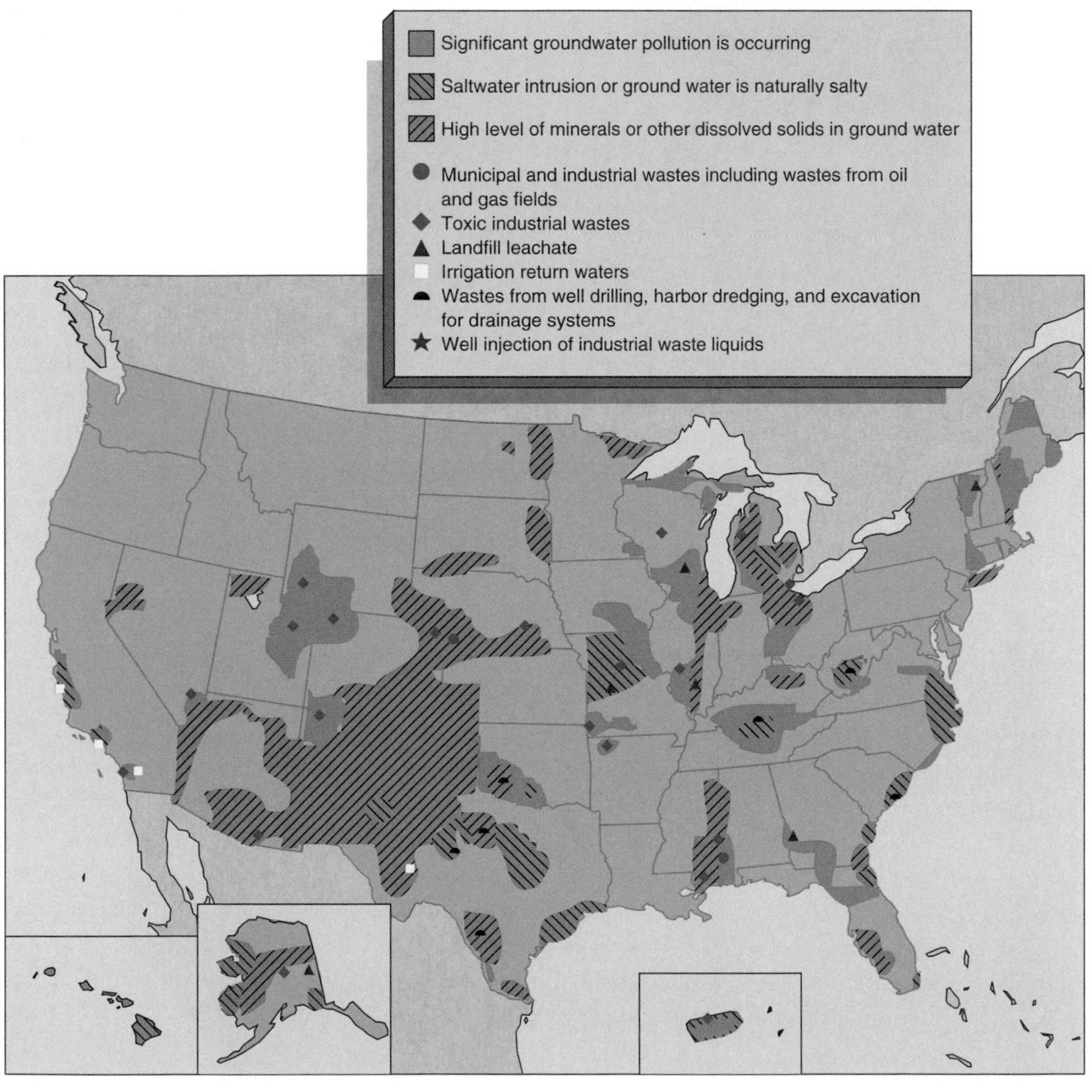

FIGURE 17.21 Groundwater pollution problems.
Source: Modified from U.S. Water Resources Council, The Nation's Water Resources 1975–2000, *vol. 1, pp. 61–63.*

golf courses and public gardens, and elsewhere. And where improvements in sewage treatment have been made to reduce ammonia concentrations, this has often been accomplished by converting the nitrogen in the ammonia to nitrate—thereby increasing nitrate concentrations in the water system.

Groundwater pollution has sometimes appeared decades after the industry or activity responsible for it has ceased to operate and has disappeared from sight and memory. For example, chemicals dumped or spilled into the soil long ago might not reach an aquifer for years. Even after the source has been realized, so large an area may have been contaminated that cleanup is impractical and/or prohibitively expensive. This is a problem with many old, abandoned toxic-waste dump sites. Groundwater pollution from nonpoint sources, like farmland, may also be so widespread that cleanup is not feasible. Some strategies for addressing groundwater contamination are explored below. With ground water, however, it may be especially important to avoid or limit pollution initially, because it is typically more difficult to clean up afterward than is polluted surface water. So, for example, there is pressure to replace underground fuel storage tanks of metal (which may corrode, and leak) with tanks of more corrosion-resistant materials, such as plastic or fiberglass. Pollution avoided is pollution that does not have to be treated. What is perhaps less obvious is the extent to which surface-water pollution may result in groundwater pollution too, given the interrelationships among water reservoirs in the hydrologic cycle.

Bearing in mind the potential of groundwater pollution from surface-water infiltration, one can also assess the vulnerability of a groundwater system to such pollution in terms of how well protected it is. Such an assessment may, in turn, suggest how closely the groundwater system should be monitored. For the sources of surface-water pollution are many. In addition to those already cited, there are a variety of casual pollution sources related to improper disposal: crankcase oil dumped into a storm sewer by a home mechanic, solvents and cleaning fluids poured down the drain by homeowners or dry cleaners or painters or furniture refinishers, oil that was deposited on roads by traffic flushed off and running into storm sewers in a heavy rain, even chemicals from a college science laboratory thoughtlessly flushed down a laboratory drain—all these, many simply reflecting individual carelessness, may pollute ground water.

Tracing Pollution's Path

Groundwater pollutants migrate from a point source in two ways. As ground water flows, it carries pollutants along. Dissolved constituents can also diffuse through the water, even when it is not moving, from water in which their concentrations are high toward areas of lower concentration. (If you drop a drop of dye or food coloring into a glass of water, the color will slowly disperse throughout the glass.) Depending on the nature of the pollutants, they may gradually break down as they move away from their source, as by reaction with aquifer rocks or through the action of microorganisms.

Where pollutant migration is primarily by groundwater flow, a **contaminant plume** develops down-gradient from the point source. This is a tongue of polluted ground water that may extend for hundreds of meters from the source (figure 17.22). Leachate plumes seep from leaky landfills, ill-designed toxic-waste sites, and other pollutant sources.

Their direction will be constrained by the prevailing direction of groundwater flow, but the extent of each plume and concentration of each contaminant must be determined by sampling via wells. Some pollutants float or sink if they are less or more dense than water and do not mix with it; some dissolve and diffuse very rapidly. Understanding the behavior of the plumes and the fate of the various pollutants is important to containing and cleaning up the contaminants; studies are underway at a number of current and former waste sites, such as the Norman Landfill from which the data of figure 17.22B come.

REVERSING THE DAMAGE—GROUND WATER

Earlier sections of this chapter noted various means of reducing the flow of pollutants into water. The approach of reducing or stopping the input of further pollutants, then waiting for natural processes to remove or destroy the pollutants already in the system, is often the *only* approach technically or economically possible in the case of contaminated ground water, which is relatively inaccessible. Systematic monitoring of ground water can be difficult without an extensive (and usually expensive) network of wells because, by the time the pollution problems are recognized, the contaminants have typically spread widely in the aquifer system. The slow migration and limited mixing of most ground waters complicates *in situ* treatment of the problems; bear in mind the way ground water occurs, in cracks and pores dispersed throughout a large volume of rock. Most commonly, then, polluted ground water is only treated after it is extracted for use.

Decontamination after Extraction

This approach may be called the **pump-and-treat** method. It is most often used when the water is needed for some purpose, such as a municipal water supply. The type of treatment depends on the intended water use and the particular pollutants involved.

Inorganic compounds can be removed from extracted ground water by adjusting its acidity (pH). Addition of alkalies may result in the precipitation of many heavy or toxic metals

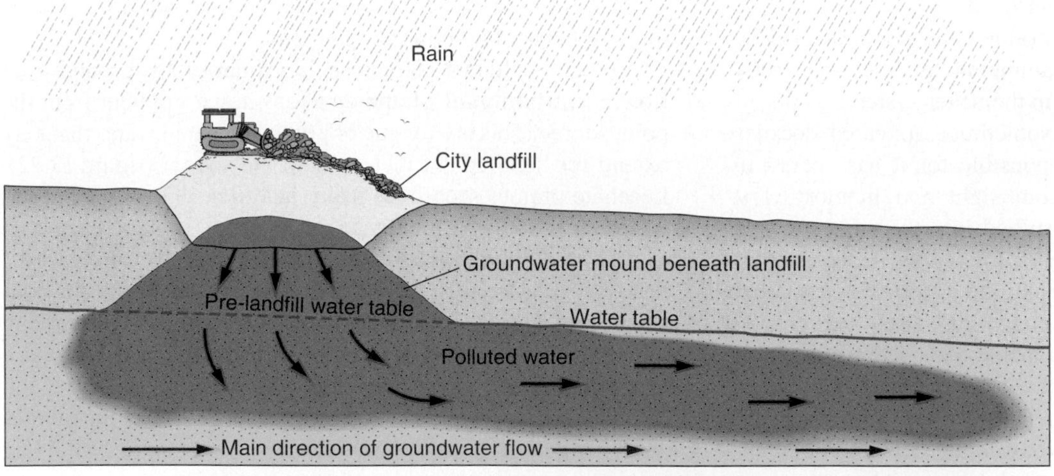

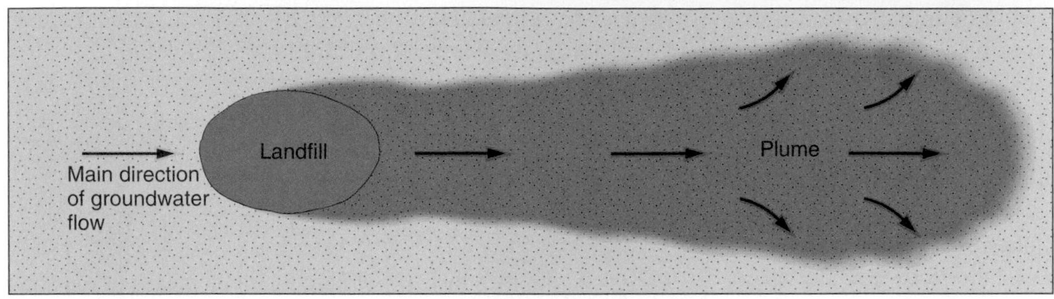

FIGURE 17.22 Groundwater contaminant plumes. (A) Schematic view, in cross section (top) and map view (bottom). Note that the plume tends to spread as it flows. When it reaches a well or body of surface water, serious pollution may appear very suddenly. (B) A groundwater contaminant plume from the Norman (Oklahoma) Landfill Environmental Research Site. Note that concentration distributions differ for different substances. (mg/L = milligrams per liter, approximately equal to ppm)

(B) After U.S. Geological Survey Fact Sheet 040-03.

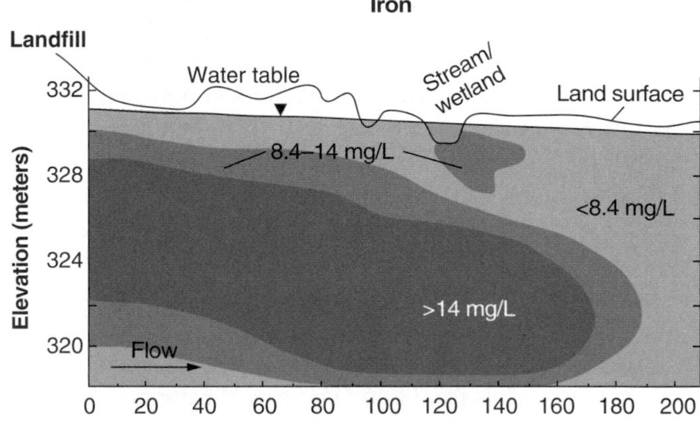

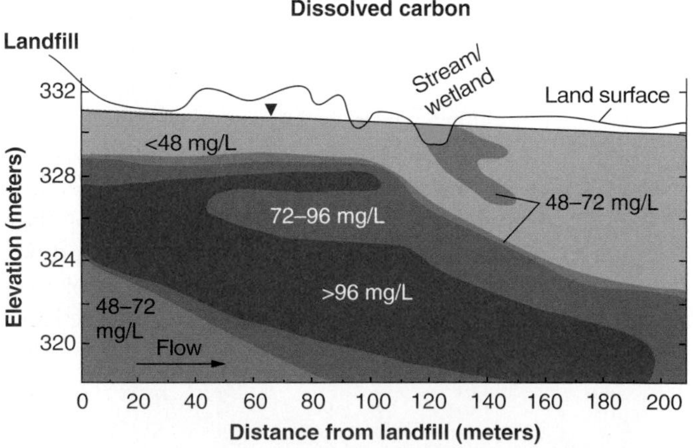

as hydroxides. Many of the same metals may be precipitated as sulfides, which are relatively insoluble, or as carbonates. All of these strategies yield a solid sludge (the precipitate) that contains the toxic metals and must still be disposed of.

Just as microorganisms can be used to decompose organic matter in sewage, microbial activity can break down a variety of organic compounds in ground water after it is extracted. The organisms can be mixed in bulk with the water; after treatment, the biomass must then be settled or filtered out. Alternatively, the organisms can be fixed on a solid substrate, and the contaminated water passed over them.

Air stripping encompasses a set of methods by which volatile organic pollutants are transferred from water into air and thus removed from the water. The mechanical details of the process vary, but each case involves some form of aeration of the water, followed by separation of the gas. The pollutants are still present in the gas phase and must be removed for alternate disposal.

Activated charcoal (activated carbon), used in a variety of filters, adsorbs organic compounds dissolved in ground water. Again, the compounds themselves remain intact and require disposal.

In short, treatment methods available *after* ground water is extracted are many and span the same broad range as methods of treating surface-water pollution or purifying a municipal water supply. But as long as the polluted ground water remains in the ground, unless it is contained (which is not common, because of costs), the contaminants can go right on spreading, carried with the water flow or diffusing through the fluid.

In Situ Decontamination

Treatment of contaminated ground water in place is possible only when the extent and nature of the pollution are well defined; the methods used are very site- and pollutant-specific. For inorganic pollutants, such as heavy metals, immobilization is one strategy. Injection wells placed within the contaminated zone are used to add chemicals that cause the toxic substances to precipitate and thus to become stationary. The uncontaminated water can then be extracted for use.

Biological decomposition (*bioremediation*) is effective on a broad range of organic compounds, many of which can be broken down by microorganisms. Organisms that will "eat" certain types of contaminants are injected into the contaminated ground water. The process can be stimulated, in individual cases, by the addition of oxygen or nutrients to accelerate growth of the microorganisms; additional microorganisms (indigenous or foreign) can be introduced to attack the particular compound(s). Not all organics can be eliminated in this way, and occasionally, objectionable residues affect water taste and odor. However, most of the twenty-four "priority organic pollutants" identified by the Environmental Protection Agency can be successfully attacked by biological means. Genetic engineering may also allow the development of strains of microorganisms targeted to specific pollutants, increasing the efficiency of bioremediation efforts.

With petroleum products that do not mix well with water, but tend to float on it (as with natural oil and gas deposits), wells that reach barely to water-table level may selectively pump out contaminant-rich fluid, leaving much purer water behind.

A more recently developed approach is the *permeable reactive barrier,* in which pollutants are broken down as the ground water flows through the barrier. Such an approach has successfully dealt with toxic organic compounds that previously escaped from the site of the Denver Federal Center (figure 17.23).

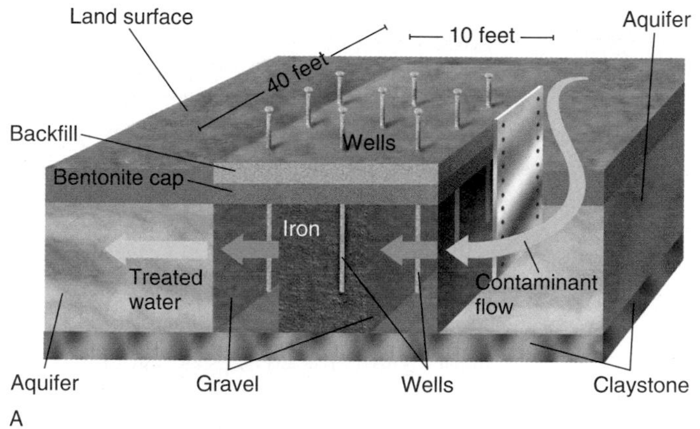

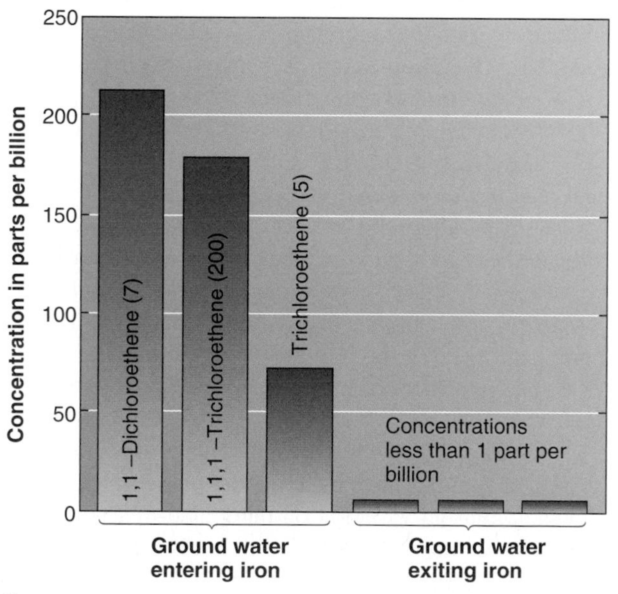

FIGURE 17.23 (A) Ground water flows through gravel into permeable barrier containing metallic iron, which reacts with and destroys the organic solvents, leaving clean water to flow on through the aquifer (B).
After U.S. Geological Survey, 2003.

Damage Control by Containment—The Rocky Mountain Arsenal

One human activity can have multiple environmental impacts. We noted in chapter 4 that deep-well disposal of liquid waste at the Rocky Mountain Arsenal, near Denver, Colorado, was identified as the cause of a series of small earthquakes in the area. Careless surface disposal of toxic wastes at the same site also resulted in contamination of local ground water. Beginning in 1943, wastewater containing a variety of organic and inorganic chemicals was discharged into unlined surface ponds. The principal contaminants were the organics diisomethylphosphonate and dichloropentadiene.

The water table in the shallow aquifer in the area is locally only 2 to 5 meters below the ground surface. Water from this aquifer is widely used for irrigation and for watering livestock. Incidents of severe crop damage were reported in the early 1950s. To limit further contamination, an asphalt-lined disposal pond was constructed in 1956, but, in time, the liner leaked. Contaminants already in the groundwater system continued to spread. When new complaints of damage to crops and livestock were made in the early 1970s, the Colorado Department of Health investigated. They detected a variety of toxic organic compounds in water and soil on the arsenal property; some of these compounds were present, though in much lower concentrations, in water drawn from municipal supply wells off the site. The Department of Health ordered a cessation of the waste discharges and cleanup of the existing pollution, with provision to prevent further discharge of pollutants from the arsenal property.

Because groundwater contamination was already so widespread, the only feasible approach to containing the further spread of pollutants was to halt them at the arsenal boundaries. This effort was combined with cleanup activity (figure 17.24). A physical barrier of clay-rich material was placed along the arsenal side of the boundary at several locations on the north and west sides, where the natural groundwater flow was outward from the arsenal. On the arsenal side of the barrier, wells extract the contaminated ground water. It is treated by a variety of processes, including filtration, chemical oxidation, air stripping, and passage through activated charcoal. The cleaned water is then reintroduced into the aquifer on the outward side of the barrier. The net groundwater flow regime outside of the arsenal has been only minimally disrupted, and pollutants in nearby water supplies have been greatly reduced. This, then, is a somewhat unusual variation on pump-and-treat: the water is not immediately used after treatment, but is returned, cleaned, to the aquifer system.

The efforts have not been inexpensive. Over $25 million has been spent on the control and decontamination activities, and the boundary barrier system will have to continue in operation indefinitely as the slowly migrating contaminants in the ground water continue to move toward the limits of the site. These activities illustrate both the decontamination and control success that can be achieved in selected cases and the value (practical and economic) of appropriate planning to prevent contamination in the first place.

NEW TECHNOLOGY MEETS PROBLEMS FROM THE PAST: CALIFORNIA GULCH SUPERFUND SITE, LEADVILLE, COLORADO

Mining near Leadville began in 1859; the sulfide-rich ores were mined for gold, silver, lead, and zinc. In the process, tailings piles were scattered over about 30 square kilometers (about 12 square miles) in and around Leadville. Sulfide-mineral residues in the tailings weather to produce sulfuric-acid-rich drainage. The acid drainage, in turn, leaches a number of metals, some of them toxic, including lead, cadmium, and arsenic (figure 17.25). The hazard is obvious where polluted surface drainage is visible. Where it is not, however, visual inspection does not help much in assessing degrees of risk from acid drainage. Recently, the use of high-resolution airborne spectrometers has provided a powerful tool for assessing such sites.

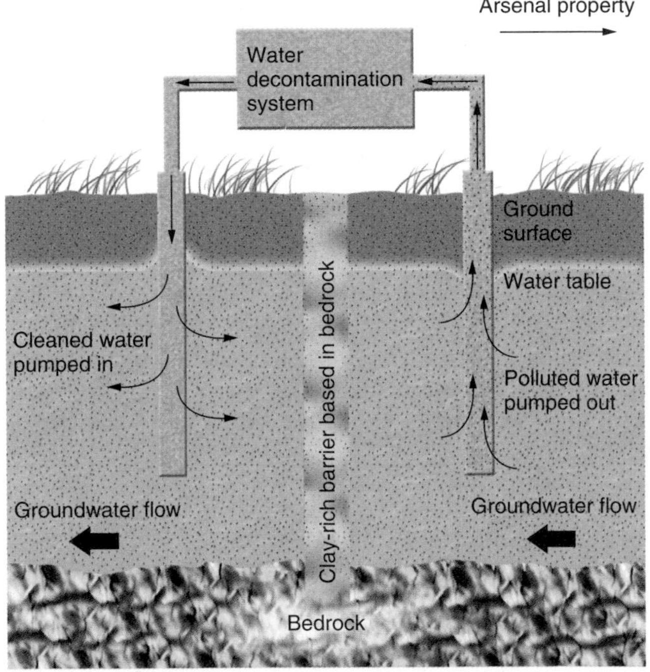

FIGURE 17.24 Sketch of boundary control system for contaminated ground water at Rocky Mountain Arsenal.

FIGURE 17.25 Obviously polluted runoff water from mine dump, Leadville, Colorado.

Photograph © David Hiser/Stone/Getty Images.

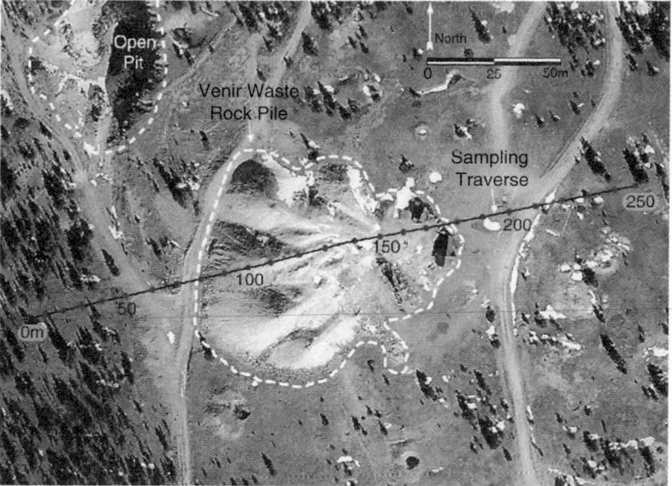

A

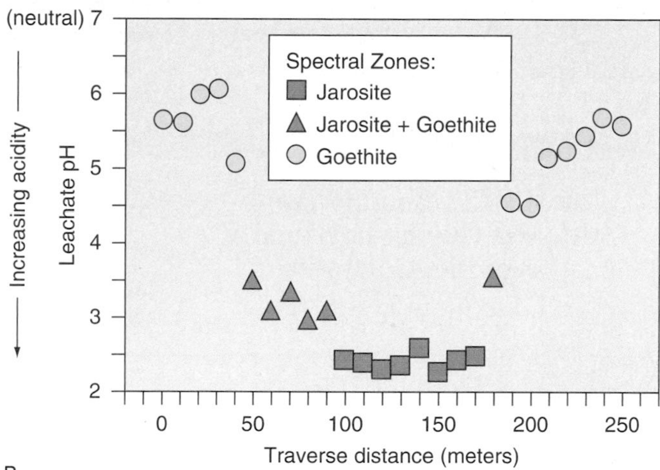

B

FIGURE 17.26 (A) Traverse track across a tailings pile at California Gulch. Red dots are sampling sites. (B) Different minerals found across the tailings pile correspond to different acidity (pH) of leachate; goethite is an iron hydroxide, jarosite a hydrous iron sulfate formed by weathering of sulfides in acid conditions. Many toxic metals at this site are much more leachable in acid waters.

Source: U.S. Geological Survey Spectroscopy Lab.

Investigators selected a single tailings pile and made an aerial traverse across it (figure 17.26A). Comparing spectral data to analysis of surface samples, they could identify zones of different mineralogy across the pile by the distinctive spectrographic signatures of the minerals: different minerals reflect different wavelengths of radiation, just as they reflect different wavelengths of visible light, resulting in different colors. These mineral zones corresponded to varying degrees of acidity of drainage (figure 17.26B). In general, increasing acidity (lower pH) correlated with higher solubility/leachability of metals. In short, the mineralogy—which could be determined by overflight—was an indicator of drainage geochemistry. This offers a much quicker way to map relative degrees of hazard from dissolved toxic metals over a broad area than does surface sampling and conventional chemical analysis (figure 17.27).

There are close to 50,000 sites in the United States alone, now inactive, where metals were once mined. Each of these represents a potential acid-drainage hazard. Having efficient ways to identify quickly those sites, or parts of sites, posing the greatest risk allows prioritization of cleanup efforts.

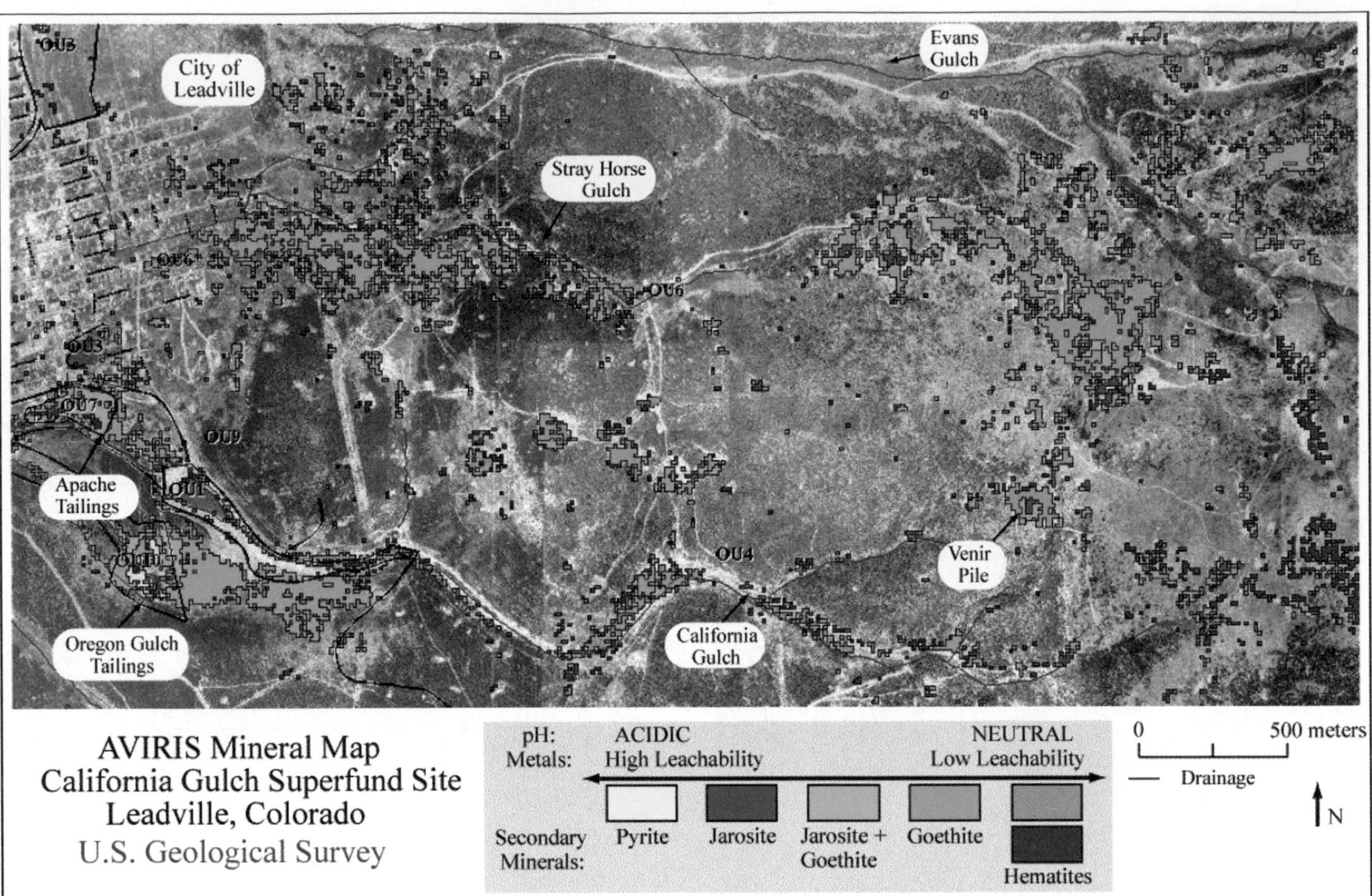

FIGURE 17.27 Airborne-spectroscopic scan shows distribution of minerals, which correspond to varying degrees of acidity of water and thus of leachability of metals; combined with drainage information, the image helps target areas of greatest water-pollution danger.
Source: U.S. Geological Survey Spectroscopy Lab.

SUMMARY

A variety of substances, some naturally occurring and some added by human activities, cycles through the hydrosphere. How long each substance spends in a particular reservoir is described by the residence time of that substance in the reservoir. The materials of greatest concern in the context of pollution are usually toxic ones with long residence times in the environment. Little is known about many synthetic chemicals—their movements, longevity in natural systems, and often even the extent of their toxicity.

In addition to adding many new and sometimes-toxic chemicals to the environment, industrial activities can unbalance the natural cycles of other harmful substances, such as the heavy metals. These elements share with some other chemicals the tendency to accumulate in organisms and increase in concentration up the food chain, becoming a greater hazard higher up in the chain. Power plants may present a thermal, rather than a chemical, pollution threat.

Sediment pollution is a problem particularly in agricultural areas, where better erosion control could greatly reduce it. Organic matter, whether contributed by domestic sewage, agriculture, or industry, increases the biochemical oxygen demand (BOD) of the water, making it inhospitable to oxygen-breathing organisms. Where the water is rich in plant nutrients from organic wastes, fertilizer runoff, or other sources, such as phosphate from detergents, a eutrophic condition develops that encourages undesirably vigorous growth of algae and other plant life.

Reduction in pollutant release into aqueous systems is the most common strategy for reducing pollution problems. As a general rule, pollutants from point sources are more easily controlled or confined than are pollutants from nonpoint sources. More-aggressive physical and/or chemical treatments are sometimes used in water-quality restoration efforts, particularly in the case of small, still water bodies, or to combat toxic-chemical spills. However, even if harmful materials are contained rather than released into the environment or are removed from water or underlying sediment, they still pose a waste-disposal problem. Groundwater pollution is a particular challenge, not only because it is often less readily detected, but because the nature of ground water makes it harder to treat *in situ*. Bioremediation and the use of permeable reactive barriers are among techniques that can successfully address groundwater pollution without disrupting flow patterns. New technological tools can assist in focusing efforts to remediate water-pollution problems on areas of most acute need.

Key Terms and Concepts

aerobic decomposition 410
anaerobic decomposition 410
biochemical oxygen demand (BOD) 410
biomagnification 414
contaminant plume 429
eutrophication 411
heavy metals 414
nonpoint source 409
oxygen sag curve 410
point source 409
pump-and-treat 429
residence time 407

Exercises

Questions for Review

1. Explain the concept of residence time; illustrate it with respect to some dissolved constituent in seawater or another water reservoir.
2. In what way do human activities most commonly alter the cycles of naturally occurring elements?
3. What is biomagnification, and why is it a particular concern of humans? Give an example of a chemical that is subject to biomagnification.
4. Why do potentially harmful health effects of organic compounds constitute a major area of concern?
5. What is BOD? How is it related to the oxygen sag curve often noted in streams below sources of organic-waste matter?
6. What is thermal pollution? From what activity does it principally originate? In what sense is it a less-worrisome kind of pollution than most types of chemical pollution?
7. Cite at least three possible sources of the nutrients that contribute to eutrophication of water; explain the concept. Why are eutrophic conditions generally considered undesirable?
8. What kinds of pollution can be reduced by the use of settling ponds? Explain.
9. Briefly describe two after-the-fact approaches to reducing water pollution, and note their limitations.

For Further Thought

1. As the largest freshwater lakes in the United States, the Great Lakes receive considerable attention and study. Find out the current status of pollution in Lake Erie, and compare it with any of the other Great Lakes.
2. Investigate what becomes of a local industry's wastewater. How is it treated before release? Where is it released? What pollutants remain and in what quantities?
3. Investigate your own water quality. Ask your local water department for information—or, if your water comes from a private source such as a well, contact the health department for information on how to have your water quality tested. See also NetNotes for relevant information from the U.S. EPA.

Suggested Readings/References

Endicott, K. 1998. The trembling edge of science. *Dartmouth Alumni Magazine* (April): 22–31. (Describes the mercury-poisoning death of chemist Karen Watterhahn and related issues.)

Environmental Protection Agency. 1990. Saving the nation's great water bodies. *EPA Journal* 16 (November/December). (This whole issue is devoted to pollution and other water-related problems.)

Komor, S. C., and Hansen, D. S. 2003. *Attenuation of runoff and chemical loads in grass filter strips at two cattle feedlots, Minnesota, 1995–98.* USGS Water-Resources Investigations Report 03-4036.

Laws, E. A. 2000. *Aquatic pollution: An introductory text.* 3rd ed. New York: John Wiley & Sons, Inc.

Pinsker, L. M. 2003. In search of the mercury solution. *Geotimes* (August): 16–21.

Raloff, J. 2004. Limiting dead zones. *Science News* 165 (12 June): 378–80.

Rinella, J. F., Hamilton, P. A., and McKenzie, S. W. 1993. Persistence of the DDT pesticide in the Yakima River Basin, Washington. U.S. Geological Survey Circular 1090.

Sampat, P. 2001. Uncovering groundwater pollution. Ch. 2 in *State of the World 2001.* New York: W.W. Norton & Co.

Sexton, K., Needham, L. L., and Pirkle, J. L. 2004. Human biomonitoring of environmental chemicals. *American Scientist* 92 (January–February): 38–45.

Sparks, D. L. (ed.). 2005. Toxic metals in the environment: The role of surfaces. *Elements* 1 (September): 193–226. (A set of six related articles by various authors.)

Spiro, T., and Stigliani, W. 2003. *Chemistry of the environment.* New York: Prentice-Hall.

Swayze, G. A., et al., 2000. Using imaging spectroscopy to map acidic mine waste. *Environmental Science and Technology* 34: 47–54.

U.S. Geological Survey. 1999. *The quality of our nation's waters—Nutrients and pesticides.* USGS Circular 1225; online at http://water.usgs.gov/pubs/circ/circ1225/

———. 2000. *Mercury in the environment.* USGS Fact Sheet 146-00; online at http://www.usgs.gov/themes/factsheet/146-00/

Vaughan, D. J. (ed.). 2006. Arsenic. *Elements* 2 (April): 71–107. (A series of related articles by various authors on the geology and hazards of arsenic in the environment.)

Vengosh, A. et al. 2004. Natural born contamination in Mediterranean groundwater. *Geotimes* (May): 20–25.

Welch, A. H., Watkins, S. A., Helsel, D. R., and Focazio, M. F. 2000. *Arsenic in groundwater resources of the United States.* U.S. Geological Survey Fact Sheet 063–00.

Zwillich, T. 2000. A tentative comeback for bioremediation. *Science* 289 (29 September): 2266–67.

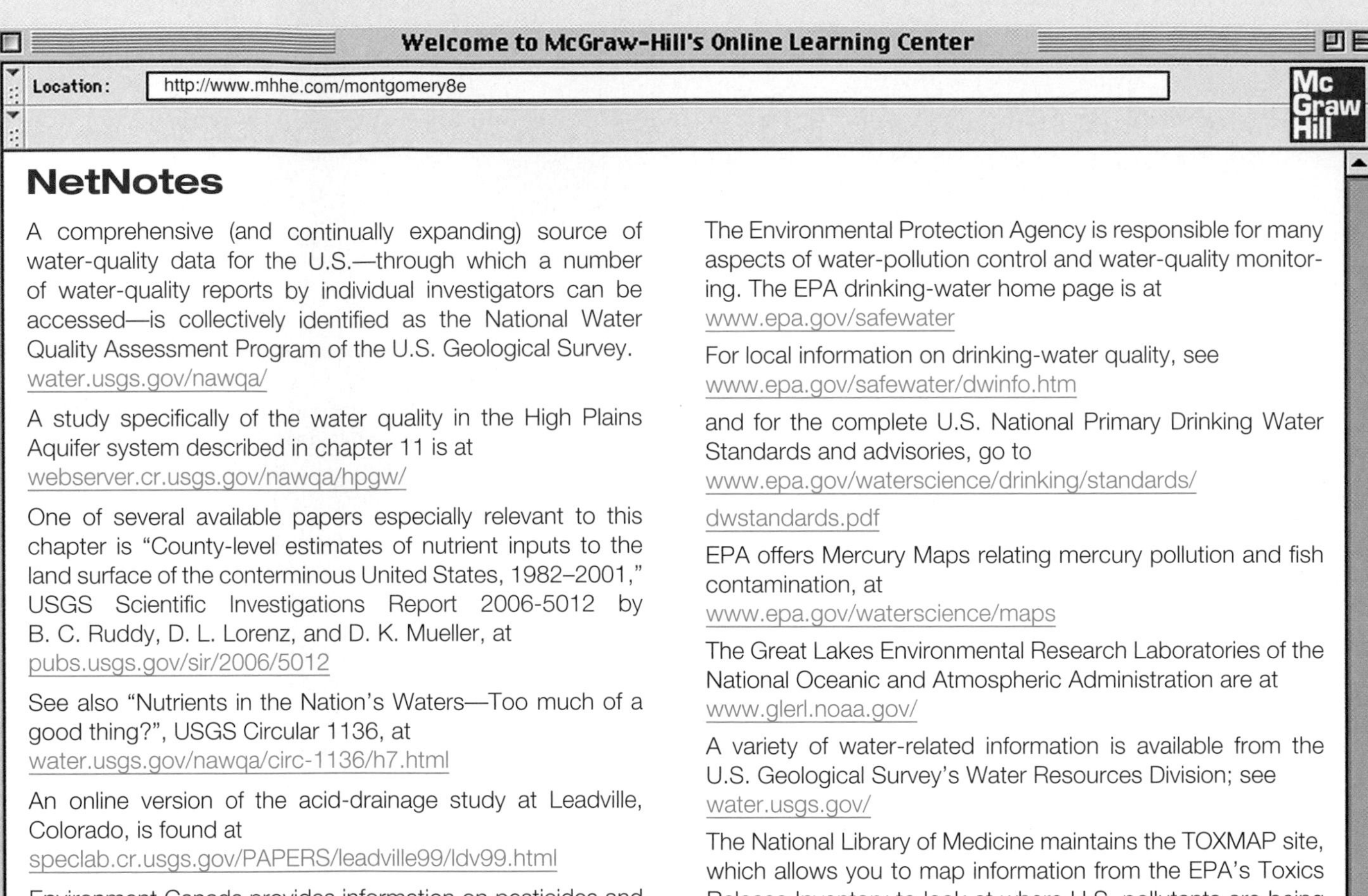

NetNotes

A comprehensive (and continually expanding) source of water-quality data for the U.S.—through which a number of water-quality reports by individual investigators can be accessed—is collectively identified as the National Water Quality Assessment Program of the U.S. Geological Survey.
water.usgs.gov/nawqa/

A study specifically of the water quality in the High Plains Aquifer system described in chapter 11 is at
webserver.cr.usgs.gov/nawqa/hpgw/

One of several available papers especially relevant to this chapter is "County-level estimates of nutrient inputs to the land surface of the conterminous United States, 1982–2001," USGS Scientific Investigations Report 2006-5012 by B. C. Ruddy, D. L. Lorenz, and D. K. Mueller, at
pubs.usgs.gov/sir/2006/5012

See also "Nutrients in the Nation's Waters—Too much of a good thing?", USGS Circular 1136, at
water.usgs.gov/nawqa/circ-1136/h7.html

An online version of the acid-drainage study at Leadville, Colorado, is found at
speclab.cr.usgs.gov/PAPERS/leadville99/ldv99.html

Environment Canada provides information on pesticides and alternatives to them at
www.ns.ec.gc.ca/epb/fiddle/alternat.html

The Environmental Protection Agency is responsible for many aspects of water-pollution control and water-quality monitoring. The EPA drinking-water home page is at
www.epa.gov/safewater

For local information on drinking-water quality, see
www.epa.gov/safewater/dwinfo.htm

and for the complete U.S. National Primary Drinking Water Standards and advisories, go to
www.epa.gov/waterscience/drinking/standards/dwstandards.pdf

EPA offers Mercury Maps relating mercury pollution and fish contamination, at
www.epa.gov/waterscience/maps

The Great Lakes Environmental Research Laboratories of the National Oceanic and Atmospheric Administration are at
www.glerl.noaa.gov/

A variety of water-related information is available from the U.S. Geological Survey's Water Resources Division; see
water.usgs.gov/

The National Library of Medicine maintains the TOXMAP site, which allows you to map information from the EPA's Toxics Release Inventory to look at where U.S. pollutants are being released and examine trends over time:
toxmap.nlm.nih.gov/toxmap/main/index.jsp

18 Air Pollution

Donora, Pennsylvania, 1948: "The fog closed over Donora on the morning of Tuesday, October 26th. . . . By Thursday, it had stiffened . . . into a motionless clot of smoke. That afternoon, it was just possible to see across the street . . . the air began to have a sickening smell, almost a taste . . ." (Roueché, 1953, p. 175). By Friday, residents with asthma and other lung disorders began to find it difficult to breathe. Then more of the residents took sick, becoming nauseous, coughing and choking, suffering from headaches and abdominal pains. The fog persisted for five days. Altogether, nearly 6000 people were stricken by the polluted fog; twenty of them died.

Donora was a mill town with a steel plant, a wire plant, and a zinc and sulfuric acid plant among its industries. The litany of toxic chemicals in its air identified during later investigations included fluoride, chloride, hydrogen sulfide, sulfur dioxide, and cadmium oxide, along with soot and ash. These materials were all routinely released into the air, but not generally with such devastating effects.

The events of late October 1948 in Donora might be called an acute air-pollution episode—sudden, obvious, and dramatic. Other still more serious single episodes are known. For four days in early December 1952, weather conditions trapped high concentrations of smoke and sulfur gases from coal burning in homes and factories over London, England; an estimated 3500 to 4000 people died in consequence, with many more made ill. Acute pollution events of this kind are, fortunately, rare. The health impact of moderately elevated levels of pollutants found widely over the earth, especially near urban areas, is more difficult to assess.

Air pollution, including soot and acid rain, takes a toll on building exteriors and outdoor sculptures, especially those made of limestone, like this medieval church in St. Emilion, France; note loss of sculptural detail, and solution pits in limestone bricks of wall at left.

Air pollution is costly, and not only in terms of health. The Council on Environmental Quality has estimated *direct* costs at $16 billion per year in the United States alone, including over $1 billion just for cleaning soiled items and $500 million in damage to crops and livestock. Worldwide, estimated costs of forest-product harvest reduction due to air pollution exceed $40 billion per year.

In addition, there are the sizable costs in illness, medical expenses, absenteeism, and loss of production, which are somewhat harder to quantify. One estimate suggests that reduction of air-pollution levels by 50% in major urban areas would save more than $2 billion per year in health costs. Implicit in the decreased health costs is an increase in longevity, decrease in illness, and improvement in quality of life among those now adversely affected by air pollution. The financial and human considerations together are powerful incentives for trying to limit air pollution.

ATMOSPHERIC CHEMISTRY—CYCLES AND RESIDENCE TIMES

The atmosphere consists of three principal elements. On average, nitrogen comprises nearly 77% of the total by weight; oxygen, about 23%; the inert gas argon, close to 1%. Locally, water vapor can be significant, up to 3% of the total. Everything else in the atmosphere together makes up much less than 1% of it.

Materials cycle through the atmosphere as they do through other natural reservoirs. One can speak of the residence times of gases or particles in the atmosphere just as one can discuss residence times of chemicals in the ocean (table 18.1). As with dissolved substances, residence times of gases are influenced by the amounts present in a given reservoir such as the atmosphere, and the rates of addition and removal. Oxygen, for example, is added to the atmosphere during photosynthesis by plants; it is removed by oxygen-breathing organisms, by solution in the oceans, by reaction with rocks during weathering, and by combustion. Its residence time in the atmosphere is estimated at 7 million years.

Carbon dioxide (CO_2) has an even more complex cycle (figure 18.1). It is added to the atmosphere by volcanic eruptions and as a product of respiration and combustion, and it is removed during photosynthesis and by solution in the oceans. It is further removed from the oceans by precipitation in carbonate sediments. The concentration of carbon dioxide in the atmosphere is

TABLE 18.1

Abundances and Residence Times of Some Elements and Gases in the Atmosphere

Substance	Average Concentration (by weight)	Estimated Residence Time
nitrogen		
molecular (N_2)	76.6%	44 million years
ammonia (NH_3)	6 ppb	3–4 months
nitrous oxide (N_2O)	320 ppb	114 years
nitrogen dioxide (NO_2)	2 ppb	1–2 months
nitric acid (HNO_3)	unknown; low	2–3 weeks
oxygen (as O_2)	23.1%	7 million years
carbon		
methane (CH_4)	1.8 ppm	12 years
carbon dioxide (CO_2)	375 ppm	years to decades
carbon monoxide (CO)	0.1 ppm	1–2 months
sulfur		
sulfur dioxide (SO_2)	0.2 ppb	hours or days
hydrogen sulfide (H_2S)	0.2 ppb	hours
sulfuric acid (H_2SO_4)	1 ppb	several days
mercury	0.001 ppb	60 days
lead	0.003 ppb	2 weeks
CFCs (chlorofluorocarbons)	0.2–0.5 ppb	decades to centuries

Sources: Data from R. M. Garrels, F. T. Mackenzie, and C. Hunt, *Chemical Cycles and the Global Environment,* copyright © 1975 William Kaufmann Inc., Los Altos, CA, U.S. Environmental Protection Agency, and Carbon Dioxide Information Analysis Center.

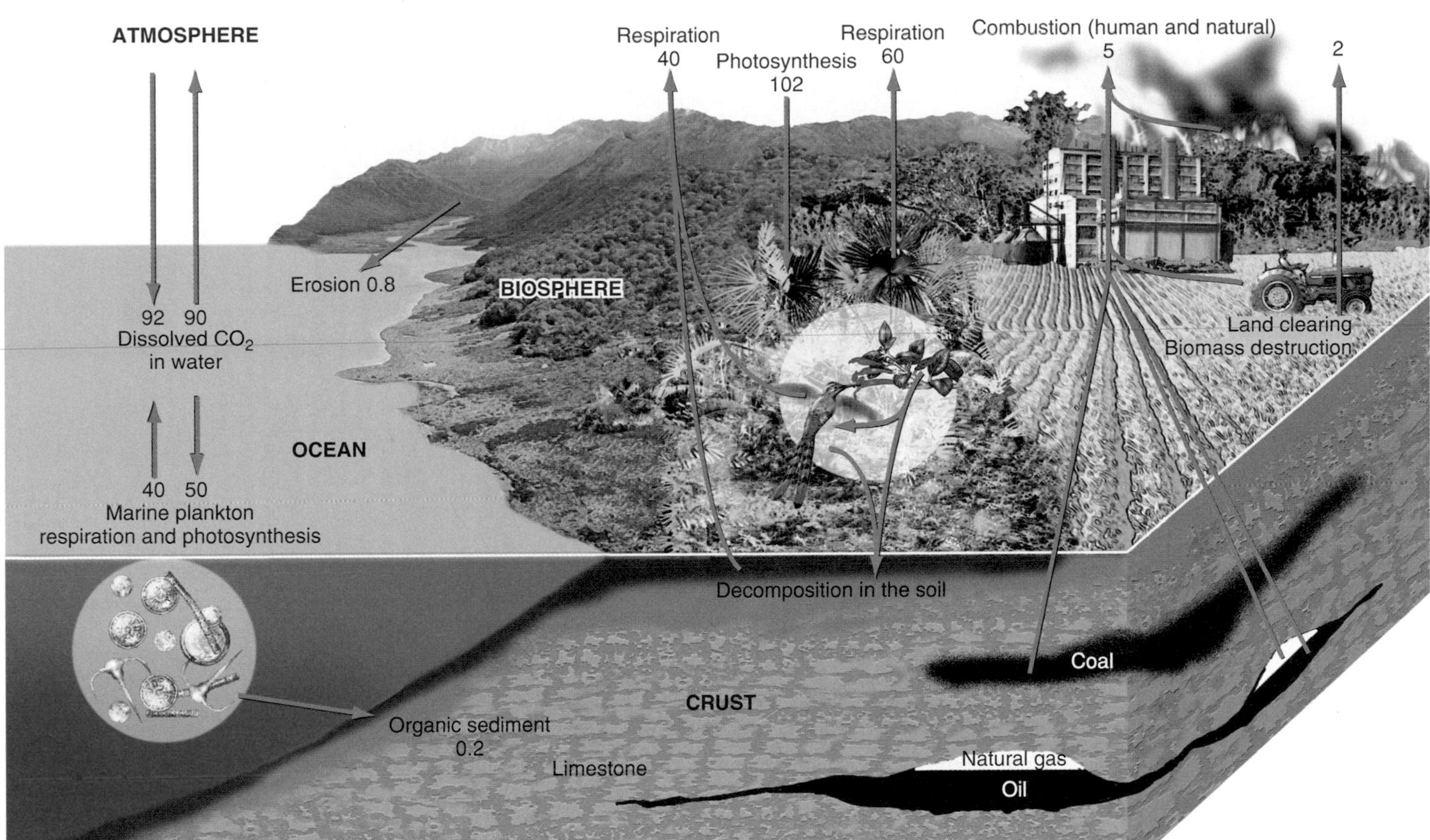

FIGURE 18.1 Reservoirs and fluxes in the global carbon cycle. Reservoirs are indicated in capital letters, fluxes from one reservoir to another by arrows. Numbers are billions of tons of carbon. While flux changes due to human activities seem relatively small, they can have a perceptible effect on atmospheric CO_2 concentrations.
Source: Intergovernmental Panel on Climate Change (IPCC), 2002.

about 375 ppm, far less than that of oxygen, and carbon dioxide has a correspondingly much shorter residence time, estimated at as little as four years. Some of the consequences of increased CO_2 content in the atmosphere were explored in chapter 10. One of the factors complicating projections of the greenhouse effect relates to imprecise knowledge of the carbon fluxes between pairs of reservoirs shown in figure 18.1. For example, consider the difficulty of estimating precisely the amount of CO_2 consumed by plants worldwide during photosynthesis. What might seem a far simpler problem, estimating net CO_2 flux between atmosphere and oceans, is not so simple after all, for the extent to which CO_2 dissolves in the oceans varies with temperature, and air and water temperatures, in turn, vary both regionally and seasonally. Still, the broad outline of the global carbon cycle, in terms of reservoirs and processes, is reasonably well understood, as shown. (However, as noted, CO_2 is not the only greenhouse gas, and climate feedbacks can be many and varied.)

The chemically inert gases, including argon and trace amounts of helium and neon, have virtually infinite residence times: they do not, by definition, react chemically with other substances, so they are not readily removed by natural processes. The geochemical cycle of nitrogen is so complex that its overall residence time in the atmosphere is not known, although residence times of individual compounds have been estimated (see table 18.1). As with many synthetic water pollutants, the fate and residence times of anthropogenic air pollutants, those added by human activity, often are also poorly known. The chlorofluorocarbon compounds, discussed later in this chapter, are just one case in point.

TYPES AND SOURCES OF AIR POLLUTION

Most air pollutants are either gases or particulates (fine, solid particles). The principal gaseous pollutants are oxides of carbon, nitrogen, and sulfur. They share some common sources but create distinctly different kinds of problems. The principal sources for each of the major pollutants are indicated in figure 18.2.

Particulates

The **particulates** include soot, smoke, and ash from fuel (mainly coal) combustion, dust released during industrial processes, and other solids from accidental and deliberate burning of

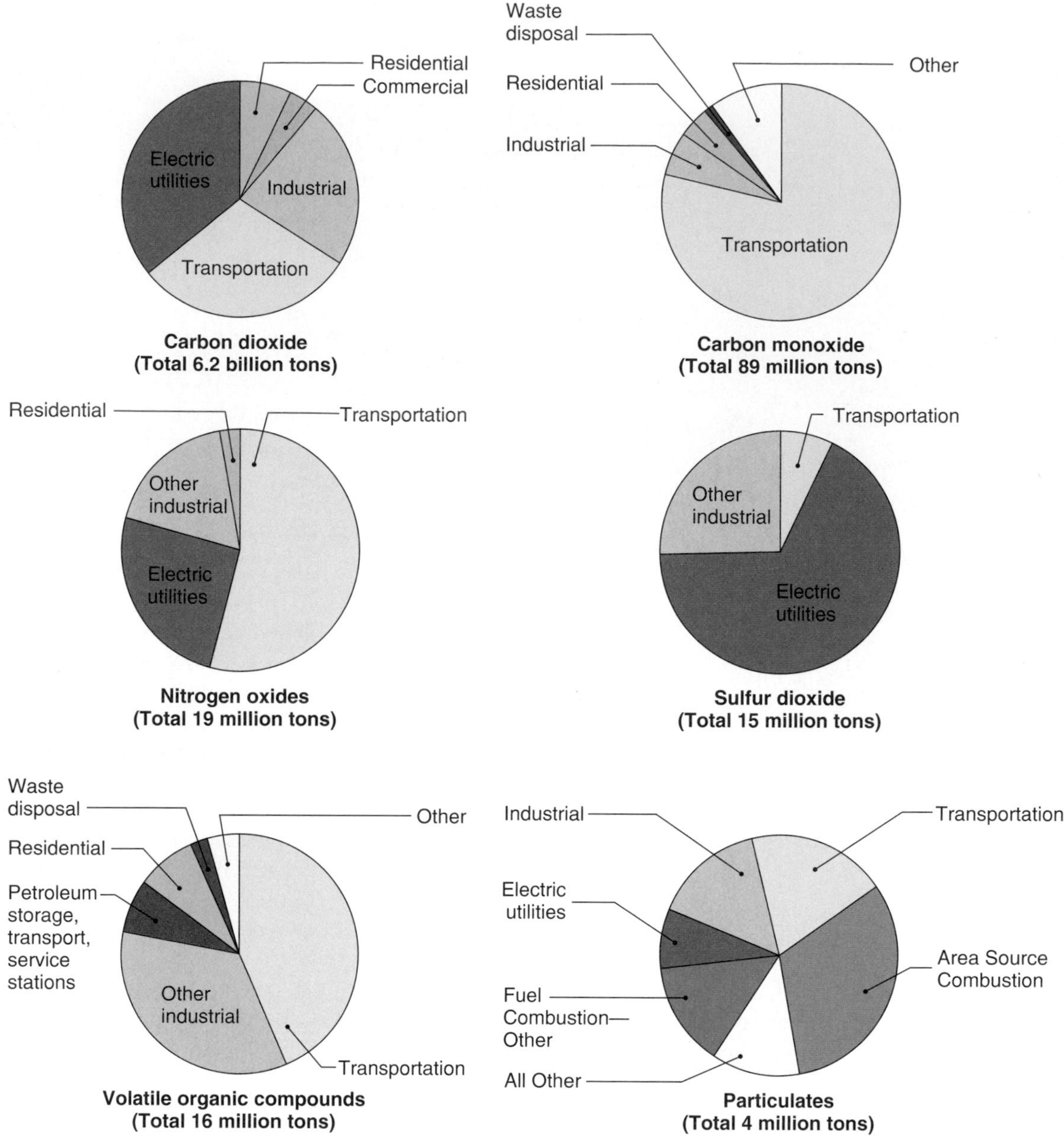

FIGURE 18.2 Principal sources of U.S. air pollutants. "Other" may include natural sources; particulates total excludes natural dust.
Source: Proportions from U.S. Environmental Protection Agency (EPA) National Emissions Trends 1990–1998; *totals, 2005 data from EPA.*

vegetation. In other words, particulate air pollutants generated by human activity are largely derived from point sources. Estimates of the magnitude of global anthropogenic contributions vary widely, from about 35 million tons/year (two-thirds from combustion) to 180 million tons/year (mostly industrial). Additional particulates are added by many natural processes, including volcanic eruptions, natural forest fires, erosion of dust by wind, and blowing salt spray off the sea surface.

Typically neither natural nor anthropogenic particulates have long residence times in the atmosphere. Generally, they are quickly removed by precipitation, usually within days or weeks. In rare cases, such as fine volcanic ash shot high into the atmosphere by violent eruptions, the finest, lightest material may stay in the air for as long as several years. Traditionally, particulate pollution has been considered a local problem, most severe close to its source and for a short time only.

The nature of the problem(s) posed by particulate pollution depends somewhat on the nature of the particulates. Certainly, dense smoke is unsightly, whatever its makeup. In areas showered with ash from a power station or industrial plant, cleaning expenses increase. Fine rock and mineral dust of many kinds has been shown to be carcinogenic when inhaled, and growing evidence suggests that the finest particulates may be especially unhealthy. Many particulates are also chemically toxic. Coal ash may contain both heavy metals and uranium stuck to the particles, for example. The dust was a significant concern after the 2001 destruction of the World Trade Center. This dust was a complex mixture of materials, including glass fibers, concrete

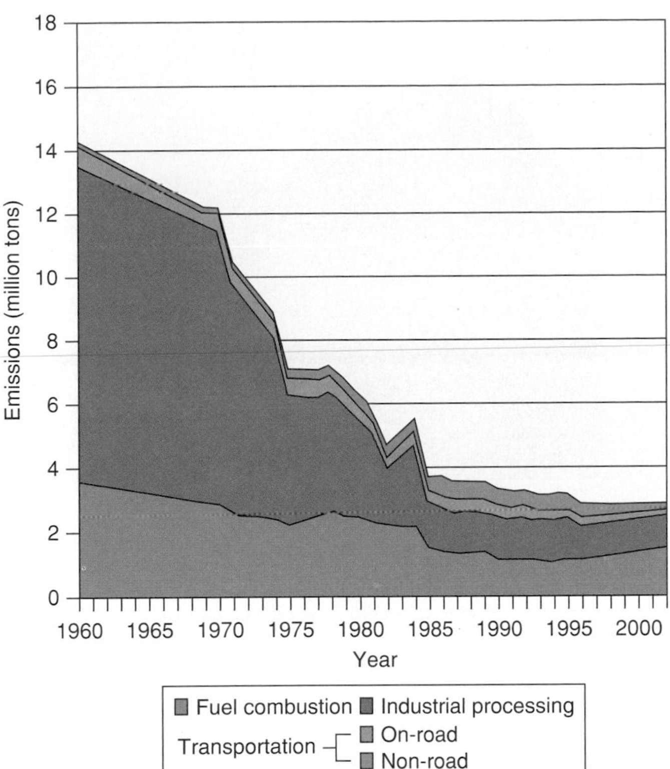

FIGURE 18.3 U.S. particulate emissions were among the first to be controlled, declining even before the various gases were addressed, but declining more sharply after the Clean Air Acts and establishment of the EPA to monitor compliance. (Note that reporting methods changed in the late 1990s.)

Source: U.S. Environmental Protection Agency Latest Findings on National Air Quality 2002 Status and Trends.

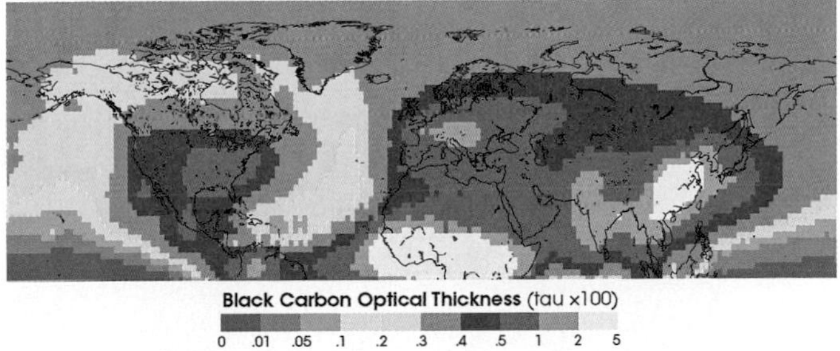

FIGURE 18.4 Distribution of black carbon (soot) in the atmosphere. "Tau" is a parameter measuring the fraction of incoming sunlight blocked from reaching the surface; the higher the tau value, the thicker the soot.

Image courtesy Dorothy Koch and James Hansen, NASA GISS.

fragments, gypsum wallboard, (chrysotile) asbestos insulation, paper, and metal-rich particles. Care had to be taken during cleanup because some of this dust could be harmful if inhaled.

In the United States, particulate emissions from many industrial and commercial sources have been sharply reduced in recent decades (figure 18.3); dust raised by such activities as agriculture, construction, and driving on unpaved roads, all remaining significant sources, is harder to control. Globally, the control of particulate pollution can be a matter of health, aesthetics, and even of climate. Worldwide, particulate pollution in the form of soot is most intense in the air above China, which burns a great deal of coal, and central Africa, where fires are commonly used in agriculture (as in slash-and-burn farming). But that soot drifts widely through the atmosphere, much of it reaching the Arctic (figure 18.4). There, as it settles out of the air to darken the surface, it can affect temperatures and weather patterns, and accelerate melting of Arctic ice.

Carbon Gases

The principal anthropogenic carbon gases are carbon monoxide and carbon dioxide. Carbon dioxide is not generally regarded as a pollutant *per se*. Some is naturally present in the atmosphere. It is essential to the life cycles of plants, and it is not unhealthful to humans, especially in the moderate concentrations in which it occurs. It is a natural end product of the complete combustion of carbon-bearing fuels:

$$\underset{\text{carbon}}{C} + \underset{\text{oxygen}}{O_2} = \underset{\text{carbon dioxide}}{CO_2}$$

As such, carbon dioxide is continually added to the atmosphere through fossil-fuel burning, as well as by natural processes, including respiration by all oxygen-breathing organisms.

At one time, it was believed that natural geochemical processes would keep the atmospheric carbon dioxide level constant, that the oceans would serve as a "sink" in which any temporary excess would dissolve to make, ultimately, more carbonate sediments. It is now apparent that this is not true. As noted in chapter 10, atmospheric carbon dioxide levels have increased about 30% in the last century and a half because of heavy fossil-fuel use, and concentrations continue to climb, spurring concerns about resultant climate change. An estimated 30 billion tons of anthropogenic carbon dioxide are added to the atmosphere each year.

Carbon monoxide (CO), while volumetrically far less important than carbon dioxide, is more immediately deadly. It is produced during the incomplete combustion of carbon-bearing materials, when less oxygen is available:

$$\underset{\text{carbon}}{2C} + \underset{\text{oxygen}}{O_2} = \underset{\text{carbon monoxide}}{2CO}$$

Anthropogenic carbon monoxide, nearly all of it from fossil-fuel burning, does not persist in the atmosphere for very long. Within a few months, carbon monoxide reacts with oxygen in the air to create more excess carbon dioxide. Actually, human activities add far less carbon monoxide to the atmosphere than do natural sources on a global basis. Nevertheless, it may constitute a serious *local* health hazard.

Carbon monoxide is an invisible, odorless, colorless, tasteless gas. Its toxicity to animals arises from the fact that it replaces oxygen in the hemoglobin in blood. The vital function of hemoglobin is to transport oxygen through the bloodstream. Carbon monoxide molecules can attach themselves to the hemoglobin molecule in the site that oxygen would normally occupy; moreover, carbon

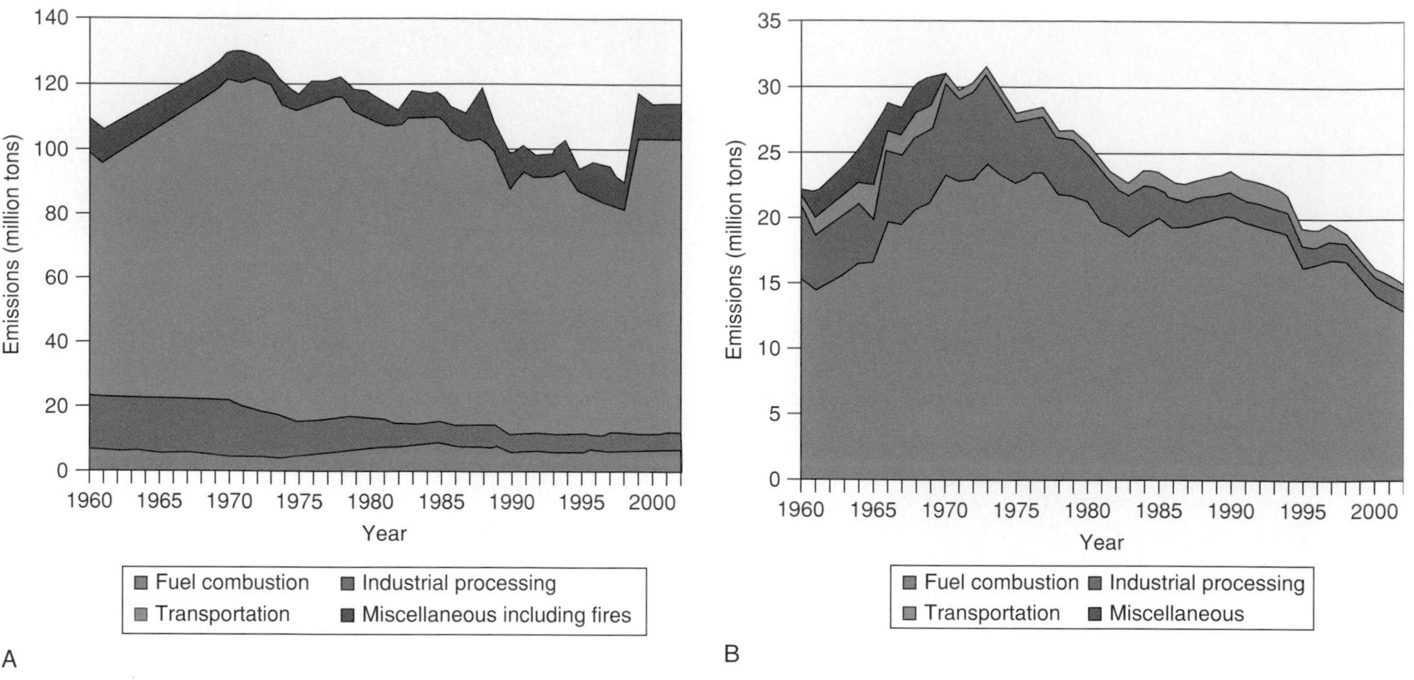

FIGURE 18.5 Emissions of carbon monoxide (A) and sulfur dioxide (B) have been reduced, primarily by regulation of auto emissions and coal-fired power plants, respectively.

Source: U.S. Environmental Protection Agency National Emissions Trends 1990–1998 *and* Latest Findings on National Air Quality 2002 Status and Trends.

monoxide bonds there more strongly. The oxygen-carrying capacity of the blood is thereby reduced. As carbon monoxide builds up in the bloodstream, cells (especially brain cells) begin to die from the lack of oxygen, and, eventually, enough cells may fail to cause the death of the whole organism. The difficulty of detecting carbon monoxide, together with the sleepiness that is an early consequence of a reduced supply of oxygen to the brain, may explain the number of accidental deaths from carbon monoxide poisoning in inadequately ventilated spaces.

Carbon monoxide does not remain in the blood indefinitely. If someone in the early stages of carbon monoxide poisoning is removed to fresh air or, if possible, given concentrated medical oxygen, the carbon monoxide is gradually released (though some irrevocable brain damage may have occurred). The problem is to recognize in time what is taking place. Carbon monoxide can build up to dangerous levels wherever combustion is concentrated and air circulation is poor.

The single largest anthropogenic source of carbon monoxide in the atmosphere, by far, is the automobile (recall figure 18.2). Nonfatal cases of carbon monoxide poisoning have been widely reported in congested urban areas with high traffic density. (Traffic police officers in Tokyo have had to take periodic "oxygen breaks" to counteract the accumulation of carbon monoxide in their blood.) This is one argument in favor of cleaning up auto emissions and making engines burn fuel as efficiently as possible. Another is simply that, if an engine is producing carbon monoxide, it is wasting energy. Burning carbon completely to make carbon dioxide releases three times the energy that is produced from incomplete combustion to carbon monoxide. When fuel is in finite supply, we should try to extract as much energy as possible from our fuels. Some progress in reducing CO emissions has been made (figure 18.5A). Even if anthropogenic production of carbon monoxide could be eliminated, however, the problems associated with carbon dioxide remain.

Sulfur Gases

The principal sulfur gas produced through human activities is sulfur dioxide, SO_2. More than 50 million tons are emitted worldwide each year. About two-thirds of this amount is released from coal combustion in factories, power-generating plants, and, in some places, home heating units. Most of the rest is released during the refining and burning of petroleum. Within a few days of its release into the atmosphere, sulfur dioxide reacts with water vapor and oxygen in the atmosphere to form sulfuric acid (H_2SO_4), which is a strong and highly corrosive acid:

$$\underset{\text{sulfur dioxide}}{SO_2} + \underset{\text{water vapor}}{H_2O} + \underset{\text{oxygen}}{\tfrac{1}{2}O_2} = \underset{\text{sulfuric acid}}{H_2SO_4}$$

Much of this is scavenged out of the atmosphere in the form of acid rain (discussed later in the chapter) to contribute to acid runoff. This, then, is another example of an air-pollution problem that becomes a water-pollution problem. As long as it remains in the air, sulfuric acid is severely irritating to lungs and eyes. Controls on power-plant emissions are the major factor in reducing SO_2 emissions (figure 18.5B).

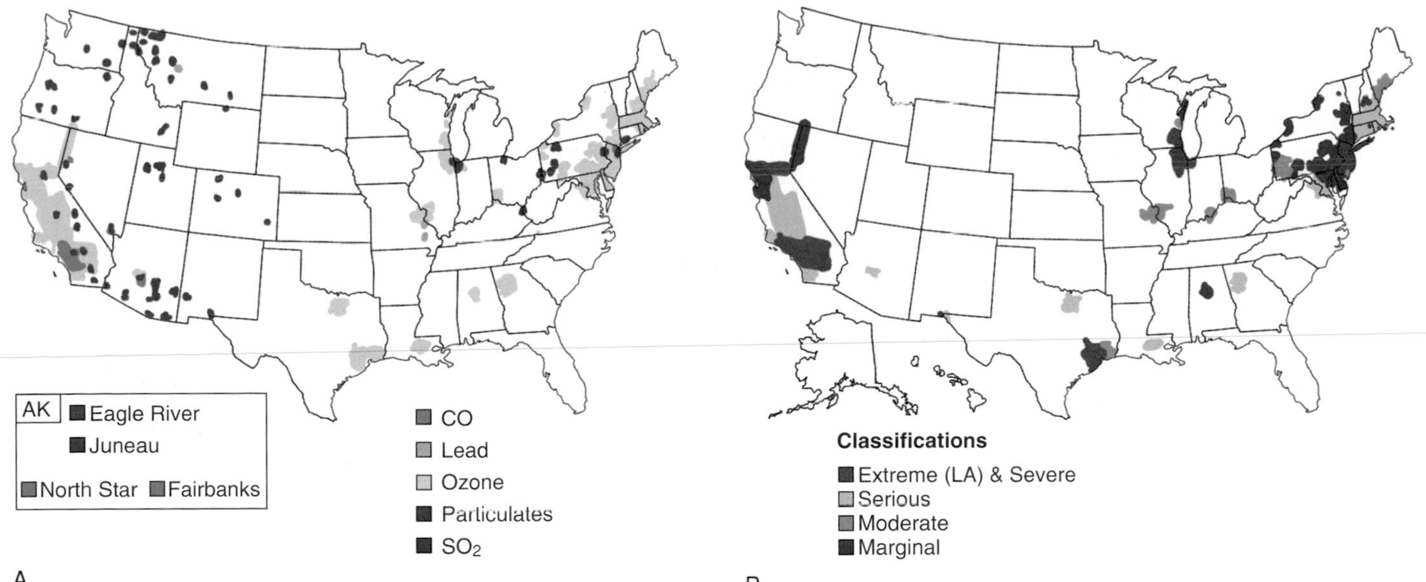

Tropospheric ozone (Dobson units)
12
55
C

FIGURE 18.6 (A) Areas identified by the EPA as having "substandard" air quality in 2002. There is an evident correlation with urbanization, except for lead, no longer present in automotive gasoline. (B) Severity of extent to which one-hour ozone levels were exceeded on September 30, 2001. (C) The higher concentration of cities and industry in the Northern Hemisphere means higher near-surface ozone levels there; once generated, the ozone is swept around the globe by westerly winds. This image is based on satellite data from 1979–2000.

(A) and (B) after U.S. Environmental Protection Agency; (C) image courtesy NASA.

Nitrogen Gases and "Smog Ozone"

The geochemistry of nitrogen oxides in the atmosphere is complex. Since nitrogen and oxygen are by far the most abundant elements in air, it is not surprising that, at the high temperatures found in engines and furnaces, they react to form nitrogen oxide compounds (principally NO and NO_2). Nitrogen monoxide (NO) can act somewhat like carbon monoxide in the bloodstream, though it rarely reaches toxic levels. In time, it reacts with oxygen to make nitrogen dioxide (NO_2). Nitrogen dioxide reacts with water vapor in air to make nitric acid (HNO_3), which is both an irritant and corrosive:

$$\underset{\text{nitrogen dioxide}}{2NO_2} + \underset{\text{water vapor}}{H_2O} + \underset{\text{oxygen}}{\tfrac{1}{2}O_2} = \underset{\text{nitric acid}}{2HNO_3}$$

Combustion of various kinds adds approximately 50 million tons of nitrogen dioxide to the air each year, which is less than 10% of the nitrogen dioxide estimated to be produced by natural biological action. However, most anthropogenic nitrogen dioxide production is strongly concentrated in urban and industrialized areas and may create serious problems in those areas. Indeed, as engines and furnaces are made more efficient to reduce emission of CO and unburned hydrocarbons, nitrogen-oxide emissions may increase.

The most damaging effect of nitrogen dioxide is its role in the production of photochemical smog, sometimes also called Los Angeles smog, after one city where it is common. Key factors in the formation of photochemical smog are high concentrations of nitrogen oxides and strong sunlight. Dozens of chemical reactions may be involved, but the critical one involving sunlight is the breakup of nitrogen dioxide (NO_2) to produce nitrogen monoxide (NO) and a free oxygen atom, which reacts with the common oxygen molecule (O_2) to make *ozone* (O_3), a somewhat unusual molecule made up of three oxygen atoms bonded together. Ozone is a strong irritant to lungs, especially dangerous to those with lung ailments or those who are exercising and breathing hard in the polluted air. Significant adverse medical effects can result from ozone concentrations below 1 ppm. Ozone also inhibits photosynthesis in plants. The dual requirement of nitrogen dioxide plus sunlight to produce ozone at ground level explains why "ozone alerts" are more often broadcast in cities with heavy traffic, and during the summertime, when sunshine is abundant and strong. Areas described as "substandard" in air quality with respect to ozone are, for the most part, urbanized areas (figure 18.6A, B). Globally, most of this near-ground ozone is concentrated in the northern hemisphere (figure 18.6C).

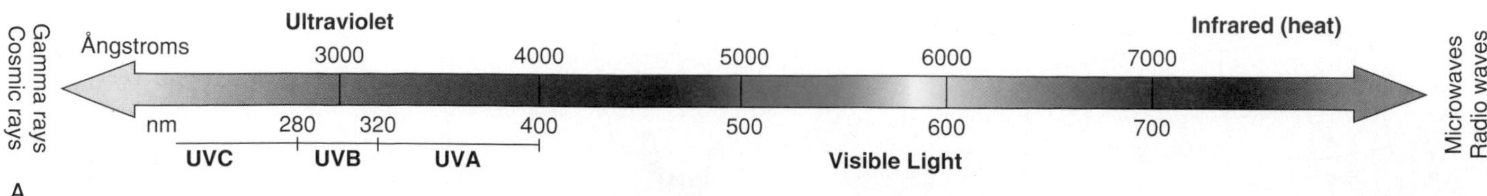

The Ozone Layer and Chlorofluorocarbons (CFCs)

If ozone is so harmful, why has so much concern been expressed over the possible destruction of the "ozone layer"? The answer is that ozone at ground level, where it can interact with animals and plants, is a good example of the chemical-out-of-place definition of a pollutant. In the stratosphere, more than about 15 kilometers above the surface, ultraviolet rays from the sun interact with ordinary oxygen to produce ozone, creating a region enriched in ozone, the so-called **ozone layer** (figure 18.7). That ozone can absorb further ultraviolet radiation, shielding the earth's surface from it. **Ultraviolet** (UV) **radiation** can cause skin cancer (it is what makes excessive tanning and sun exposure unhealthy). The presence of the ozone layer decreases that risk. Ultraviolet radiation spans a range of wavelengths shorter than those of visible light (figure 18.8A). UVA, the longest-wavelength UV, is the least damaging. UVC, the shortest, would be highly damaging, but it is largely absorbed by oxygen and water vapor in the atmosphere. It is the intermediate-wavelength UVB, also potentially damaging, that is particularly absorbed by ozone. Measurements worldwide document that the

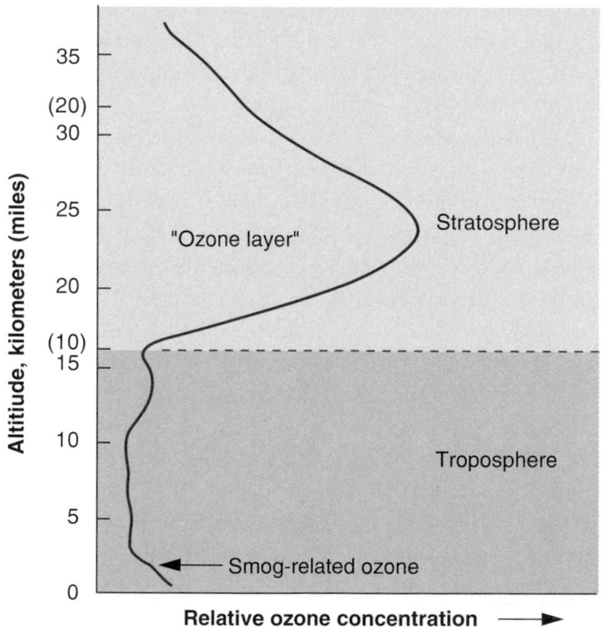

FIGURE 18.7 About 10% of atmospheric ozone is in the troposphere (the region that accounts for most weather). The other 90% is in the ozone-enriched region of the stratosphere known as the "ozone layer." Even there, ozone concentration is less than half a part per million, at most.
After National Oceanic and Atmospheric Administration.

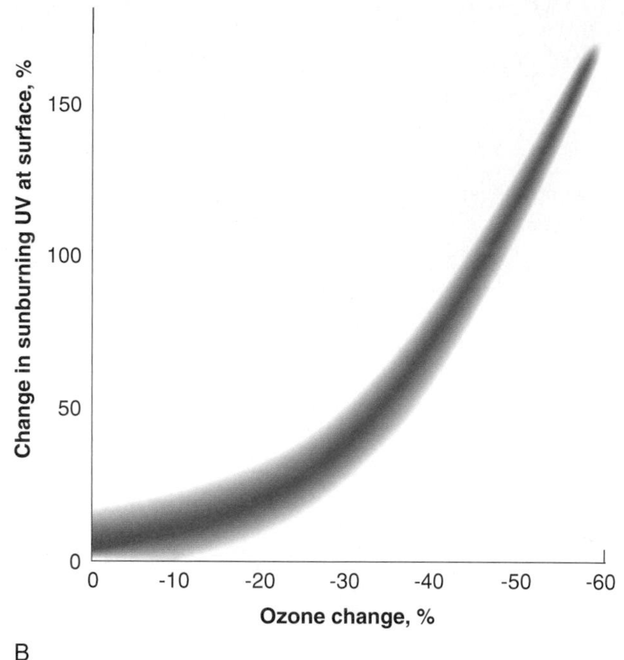

FIGURE 18.8 (A) Ultraviolet radiation is to the shorter-wavelength, higher-energy side of the visible-light spectrum. 1 nm (nanometer) = 10^{-9} meters = one-billionth of a meter = 10 Å (Ångstroms) (B) As ozone above thins, more damaging UV radiation reaches the surface.
(B) After World Meteorological Organization, UNEP, and NOAA.

amount of damaging UV reaching the surface increases as the ozone in the ozone layer decreases (figure 18.8B). Scientists of the U.N. Environment Programme estimate that each 1% reduction in stratospheric ozone could result in a 3% increase in non-melanoma skin cancers in light-skinned people, in addition to increased occurrence of melanoma, blindness related to development of cataracts, gene mutations, and immune-system damage.

Normally, oxygen (O_2) molecules in the stratosphere absorb UVC radiation, which splits them into two oxygen atoms, each of which can attach to another O_2 molecule to make O_3. The O_3 molecule, in turn, absorbs UVB radiation, which splits it into O_2 plus a free oxygen atom, and the latter, colliding with another O_3, causes regrouping back into two O_2 molecules. Over time, a balance between ozone creation and destruction develops.

The concentration of ozone in the ozone layer varies seasonally and with latitude (figure 18.9). Ozone production rates are most rapid near the equator, as a consequence of the strong sunlight there, but generally, the total ozone in a vertical column of air increases with latitude (that is, increases toward the poles) as a result of the balance between natural ozone production and destruction rates, and atmospheric circulation patterns.

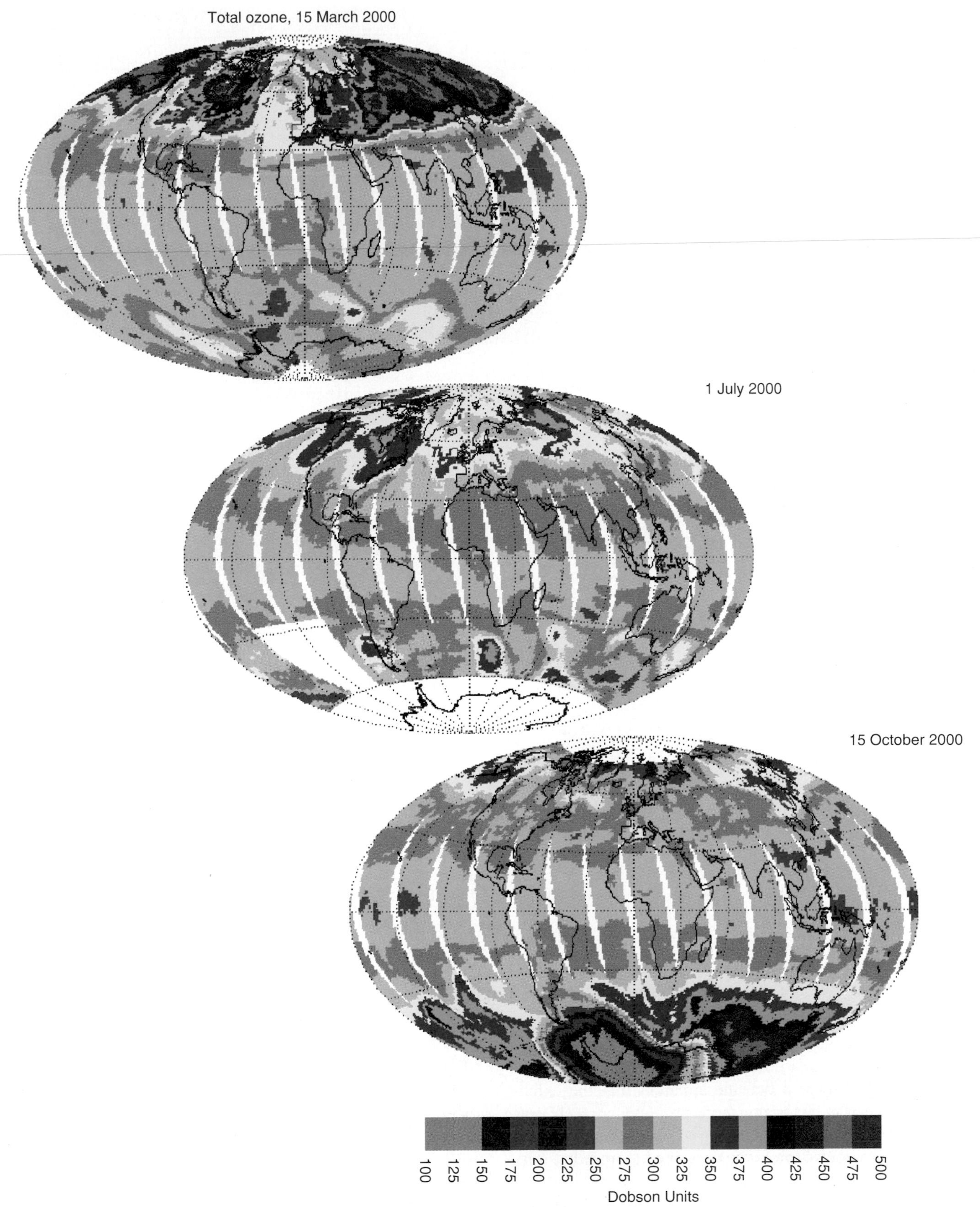

FIGURE 18.9 Typical seasonal variations in global ozone distribution as measured by NASA's Total Ozone Mapping Spectrometer. 1 Dobson unit = 0.001 cm thickness of ozone (at standard temperature and pressure) distributed over the whole vertical column of atmosphere. White areas indicate lack of data.
Images courtesy NASA.

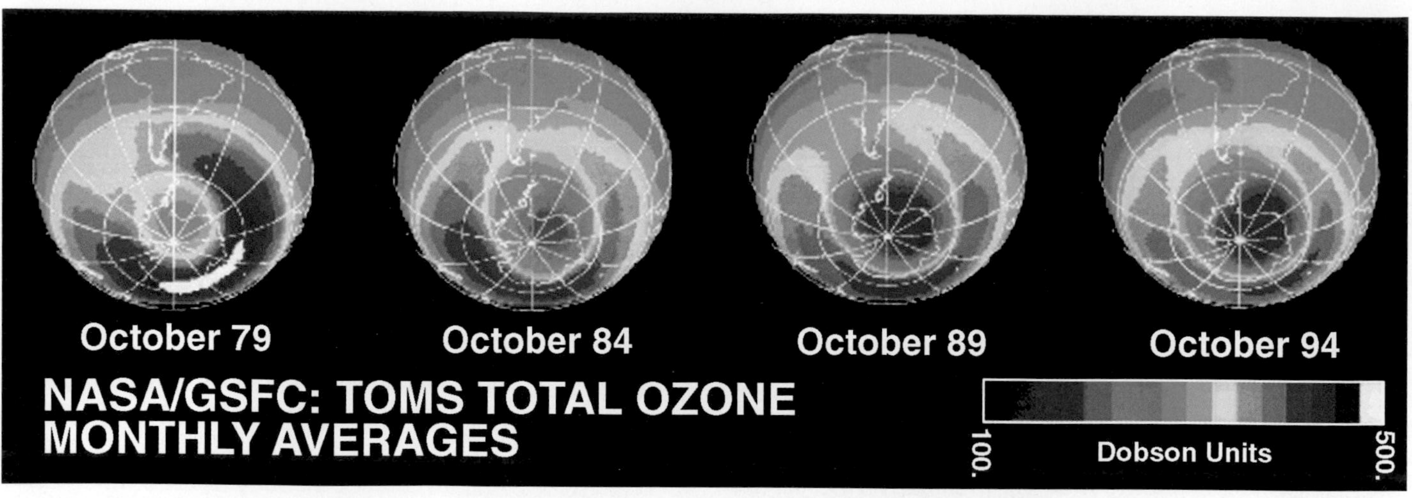

FIGURE 18.10 Map of ozone measurements over the Southern Hemisphere made by satellite each October from 1979 to 1994 revealed pronounced and deepening ozone "hole" over Antarctica—really a thinning to lower concentrations of the ozone above the south polar region.
Data from NASA Goddard Space Flight Center.

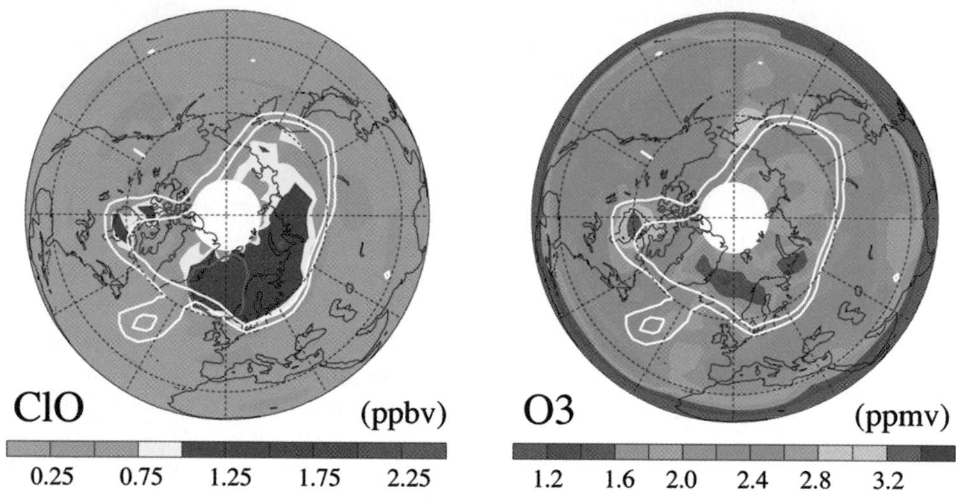

FIGURE 18.11 Higher levels of ClO correspond to lower O_3 in measurements made by satellite in the early 1990s. The probable primary source of the ClO is decomposition of anthropogenic CFCs and other chlorine compounds. Units "ppbv" and "ppmv" are parts per billion by volume and parts per million by volume, respectively.
Images courtesy UARS MLS Science Team, NASA.

It has been recognized for decades that the concentration of ozone over Antarctica decreases in the Antarctic winter. However, in the 1980s, it became apparent that the extent of the depletion was becoming more pronounced each year (figure 18.10). This extreme thinning of the protective shield of stratospheric ozone came to be described (somewhat misleadingly) as an **ozone hole.**

Measurements showed unexpectedly high concentrations of reactive chlorine compounds in the air of that region, compounds that could have been derived from anthropogenic chlorofluorocarbons (CFCs). A clear negative correlation was found between atmospheric concentrations of ClO and of ozone (O_3) (figure 18.11). While correlation does not demonstrate cause-and-effect, it was highly suggestive, and no more convincing explanation was forthcoming.

The chemistry of ozone destruction is complex. It appears that CFCs remain in the atmosphere for about a century, slowly migrating upward and breaking down over time. When chlorine atoms are freed by this CFC decomposition, which seems to occur especially readily in polar clouds during the winter, the

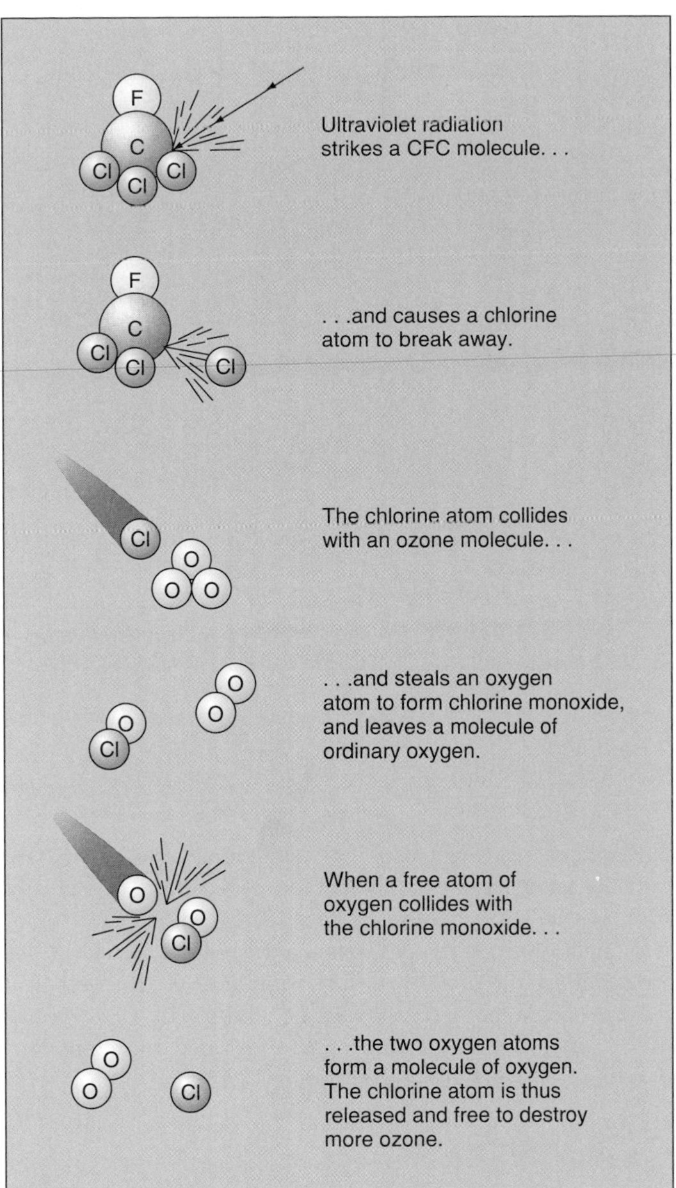

FIGURE 18.12 Schematic of the role CFCs are believed to play in ozone destruction. This would involve destruction of ozone beyond what normally occurs by interaction with ultraviolet light.
Images courtesy NASA.

chlorine reacts with ozone to form chlorine monoxide (ClO). Moreover, a single chlorine atom can destroy many ozone molecules, as seen in figure 18.12.

More recently, reduction in ozone has been detected over the northern hemisphere (figure 18.13), though the cause may not be identical in both hemispheres. The release of the reactive chlorine is catalyzed by fine ice particles in the stratosphere; thus the pronounced effect in polar regions. Other atmospheric components—including sulfate aerosols, as from Mount Pinatubo's eruption—and other human inputs such as exhaust from high-altitude aircraft like the Concorde, can also destroy ozone, but they are of shorter duration and lesser magnitude.

Recognizing the evident threat of ozone destruction by CFCs, the global community has banded together to support the Montreal Protocol on Substances that Deplete the Ozone Layer. First adopted in 1987, strengthened in 1990, and signed by 175 nations, the Montreal Protocol stipulated phaseout of production of CFCs and other ozone-depleting compounds by the year 2000. One remaining concern is that some populous developing countries have been reluctant to sign, so the overall effectiveness of the Montreal Protocol remains to be seen. Another concern is that even in developed nations, there has been some delay in the CFC phaseout; and the compounds being substituted for CFCs themselves can contribute to ozone depletion, though not as strongly. Also, the residence times of CFCs and their breakdown products are years to decades, so their effect on the atmosphere continues long after their production ceases. And old air conditioners and refrigerators continue to leak CFCs. The "ozone hole" of September 2000 was the deepest ever measured, and that of September 2003 the second most pronounced (figure 18.14). Recent projections suggest that the "ozone hole" may not recover fully until about 2065.

Lead

One air pollutant that has been greatly reduced is lead. Most of the lead once released into the atmosphere was emitted by automobile exhaust systems. While lead is not naturally a significant component of petroleum, one particular lead compound, tetraethyl lead, had been a gasoline additive since the 1940s as an antiknock agent to improve engine performance.

Lead is one of the heavy metals that accumulate in the body. It has a variety of harmful effects, including brain damage in high concentrations. Mild lead poisoning in the nervous system can cause depression, nervousness, apathy, and other psychological disorders, as well as learning difficulties. Acute lead poisoning from exhaust fumes alone has not been documented. However, children in urban areas who breathe lead-laden air and also consume chips of old lead-based paint may indeed develop high and harmful lead levels in their blood. It has been estimated that 5 to 10% of inner-city children may suffer from lead poisoning.

Lead levels allowed in most paints have been greatly reduced, and lead can be left out of gasoline. Equally good engine performance can be obtained from higher-octane gasolines at the cost of a few more cents a gallon. (In the long run, in fact, leaded gasoline could be the more costly because some lead compounds produced when leaded gasoline is burned coat engine parts and decrease their useful life.) Also, illegal use of leaded gasoline in cars equipped with catalytic converters destroys the converters' effectiveness and results in the emission of more unburned hydrocarbons, carbon monoxide, and other pollutants. Beginning in the early 1970s, the Environmental Protection Agency began to mandate reductions in the lead content of gasoline, with lead phased out almost completely from automotive fuel by 1987. Lead consumed in gasoline dropped 70% from 1975 to 1982. Concentrations of

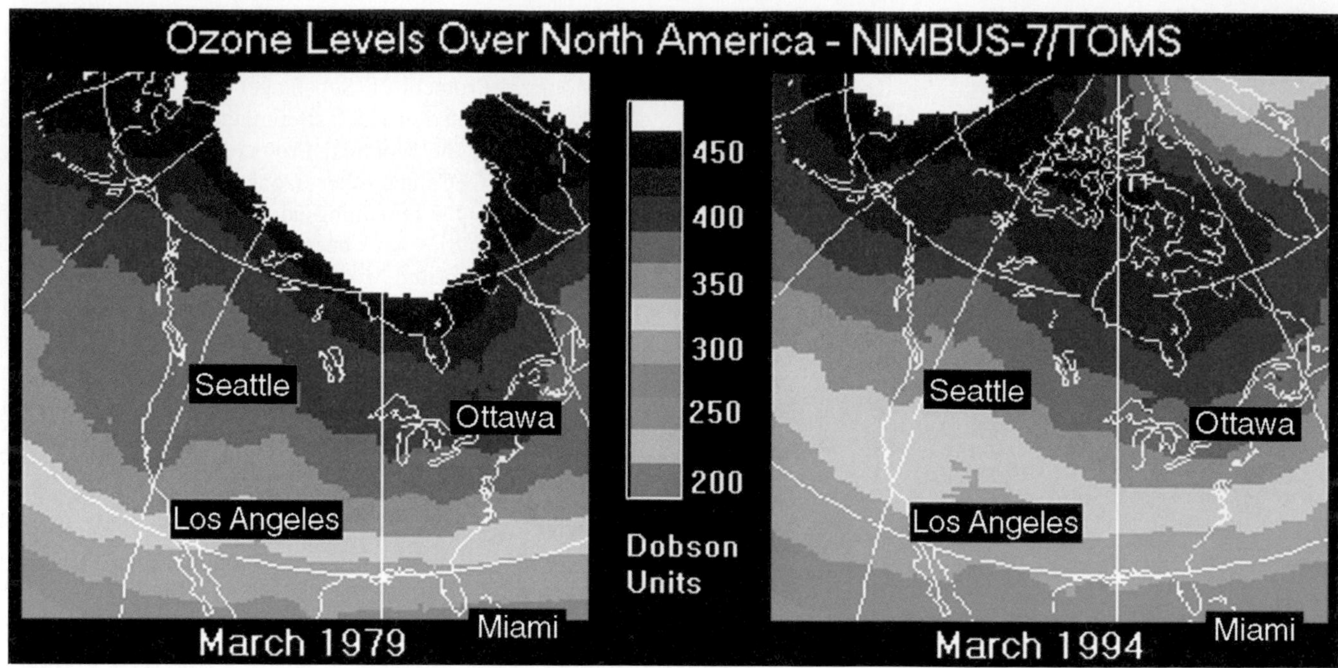

FIGURE 18.13 While the link with CFCs is more firmly established for the Antarctic "ozone hole," there is demonstrable ozone depletion over the Northern Hemisphere as well.
Data from NASA Goddard Space Flight Center.

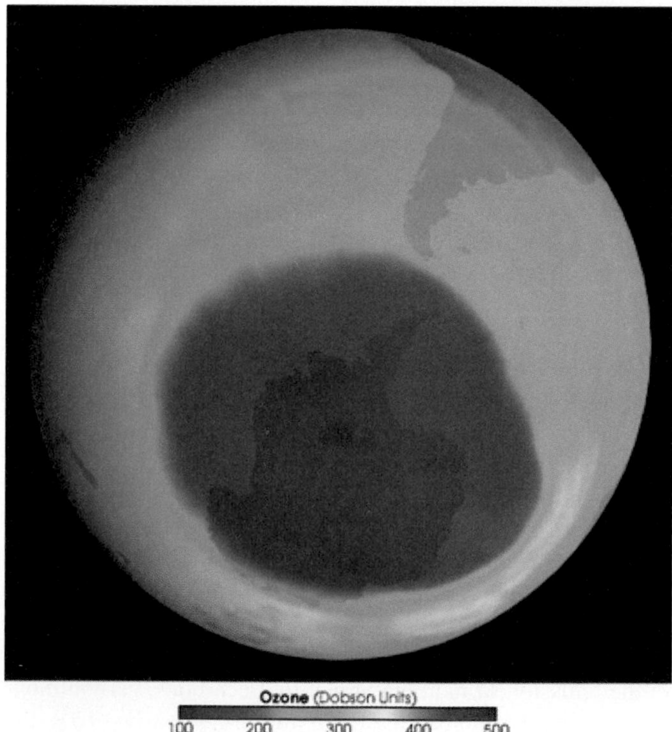

FIGURE 18.14 The ozone hole over Antarctica in September 2003 was the second "deepest" ever, meaning that the thinning of the ozone over the polar region was the second most severe measured. This is partly a function of climate, and partly a function of CFC persistence in the stratosphere.
Image courtesy NASA Goddard Space Flight Center Scientific Visualization Studio.

atmospheric lead decreased correspondingly, by about 65% over the same period, and this, in turn, was reflected in greatly reduced lead levels in blood (figure 18.15A). For a time, a limited quantity of leaded fuel remained available, primarily for the benefit of old engines in farm equipment that for technical reasons need some lead to protect their valves. By 1996, a complete ban on leaded gasoline took effect, and lead emissions have been virtually eliminated (figure 18.15B).

Other Pollutants

Volatile, easily vaporized organic compounds are a major component of air pollution. Their principal sources are (1) transportation, especially automobiles emitting unburned gasoline, and (2) industrial processes, which release a variety of organics including unburned hydrocarbons and volatile organic solvents. Unburned hydrocarbons are not themselves highly toxic, but they play a role in the formation of photochemical smog and may react to form other compounds that irritate eyes and lungs.

Other air pollutants are of more localized significance. Heavy metals other than lead, for example, can be a severe problem close to mineral-smelting operations. Such elements as mercury, lead, cadmium, zinc, and arsenic may be emitted either as vapors or attached to particulate emissions from smelters. While they are quickly removed from the atmosphere by precipitation, the metals may then accumulate in local soils, from which they are concentrated in organic matter, including growing plants. They can constitute a serious health hazard if accumulated in food or forage crops.

Finally, there are air pollutants of particular concern only indoors; see Box 18.1.

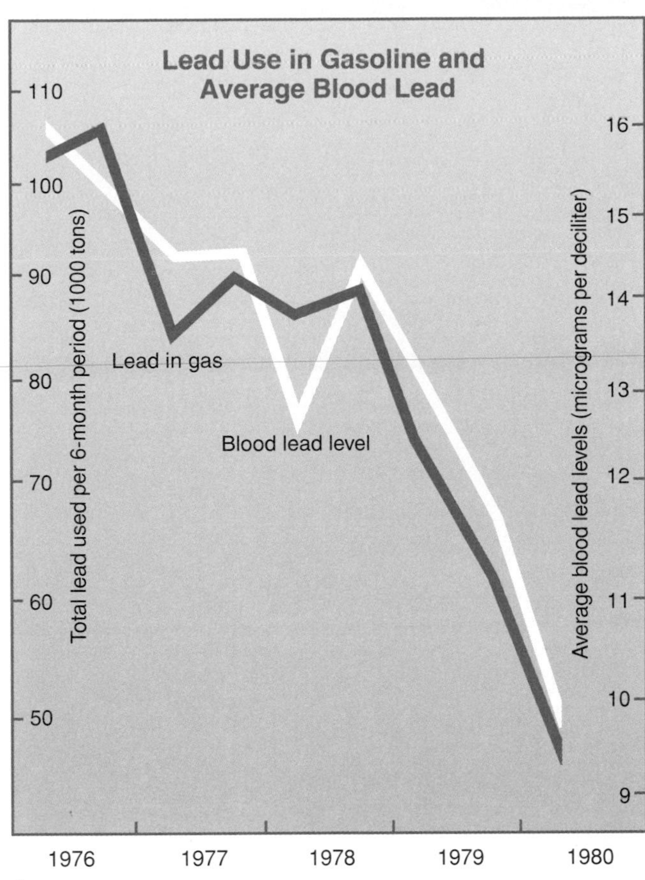

ACID RAIN

Acidity is reported on the **pH scale.** The pH of a solution is inversely proportional to the hydrogen-ion (H^+) concentration in the solution; the pH scale, like the Richter magnitude scale, is a logarithmic one (figure 18.16). Neutral liquids, such as pure water, have a pH of 7. Acid substances have pH values less than 7; the lower the number, the more acidic the solution. Typical pH values of common household acids like vinegar and lemon juice are in the range of pH 2 to 3. Alkaline solutions, like solutions of ammonia, have pH values greater than 7. All natural precipitation is actually somewhat acidic as a result of the solution of gases (for instance, CO_2) that make acids (such as carbonic acid, H_2CO_3). However, carbonic acid is what is termed a weak acid, meaning that a relatively small proportion of H_2CO_3 molecules dissociate to release the H^+ ions that contribute to acidity. **Acid rain** is rain that is more acidic than normal. While many gases in the air contribute to the acidity of rain, discussions of acid rain focus on the sulfur gases that react to form atmospheric sulfuric acid, a strong acid that dissociates extensively.

As noted in chapter 14, acid rain can contaminate water supplies. It can damage structures through its corrosive effects (figure 18.17 and chapter-opener photo). Plants may suffer defoliation, or their growth may be stunted as changes in soil-water chemistry reduce their ability to take up key nutrients such as calcium. Fish may be killed in acidified streams; or, the adult fish

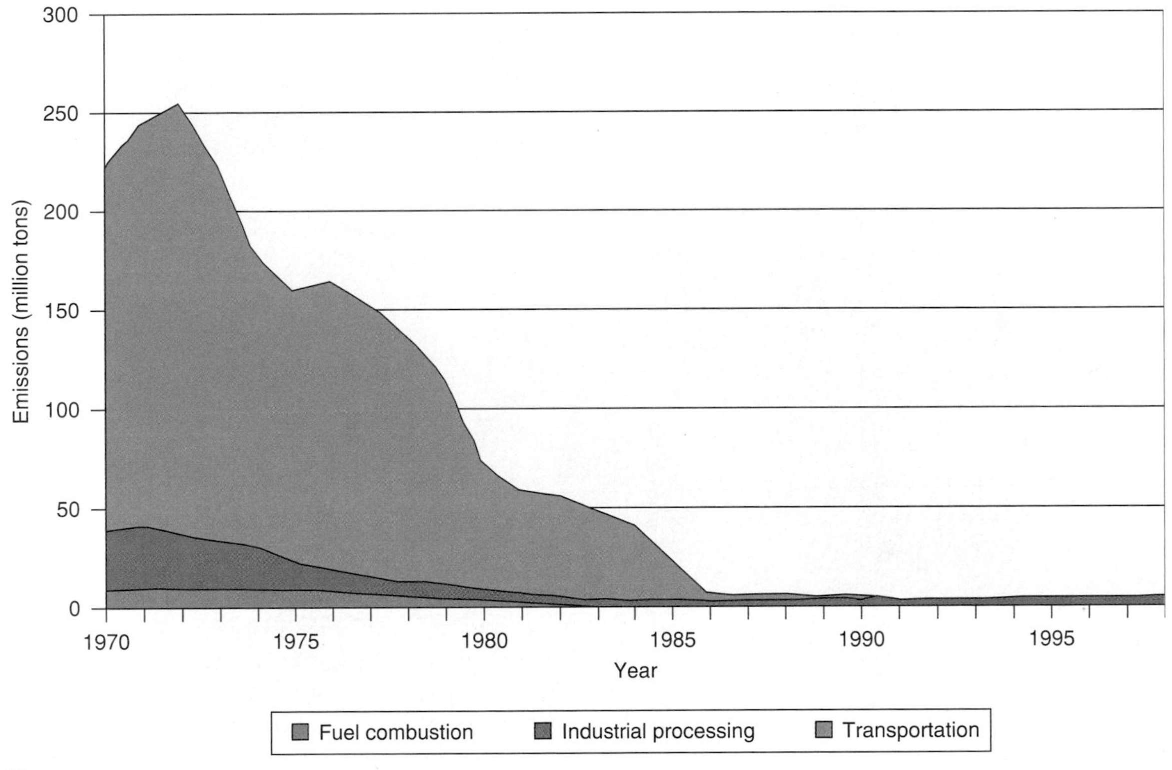

FIGURE 18.15 (A) Declines in lead used in gasoline were directly reflected in lower lead levels in blood. (B) By now, lead emissions are negligible.
(A) After L. Whiteman, "Trends to Remember: The Lead Phase Down," EPA Journal, *May/June 1992, p. 38. (B) From U.S. Environmental Protection Agency* National Emissions Trends 1990–1998.

Box 18.1 Indoor Air Pollution?

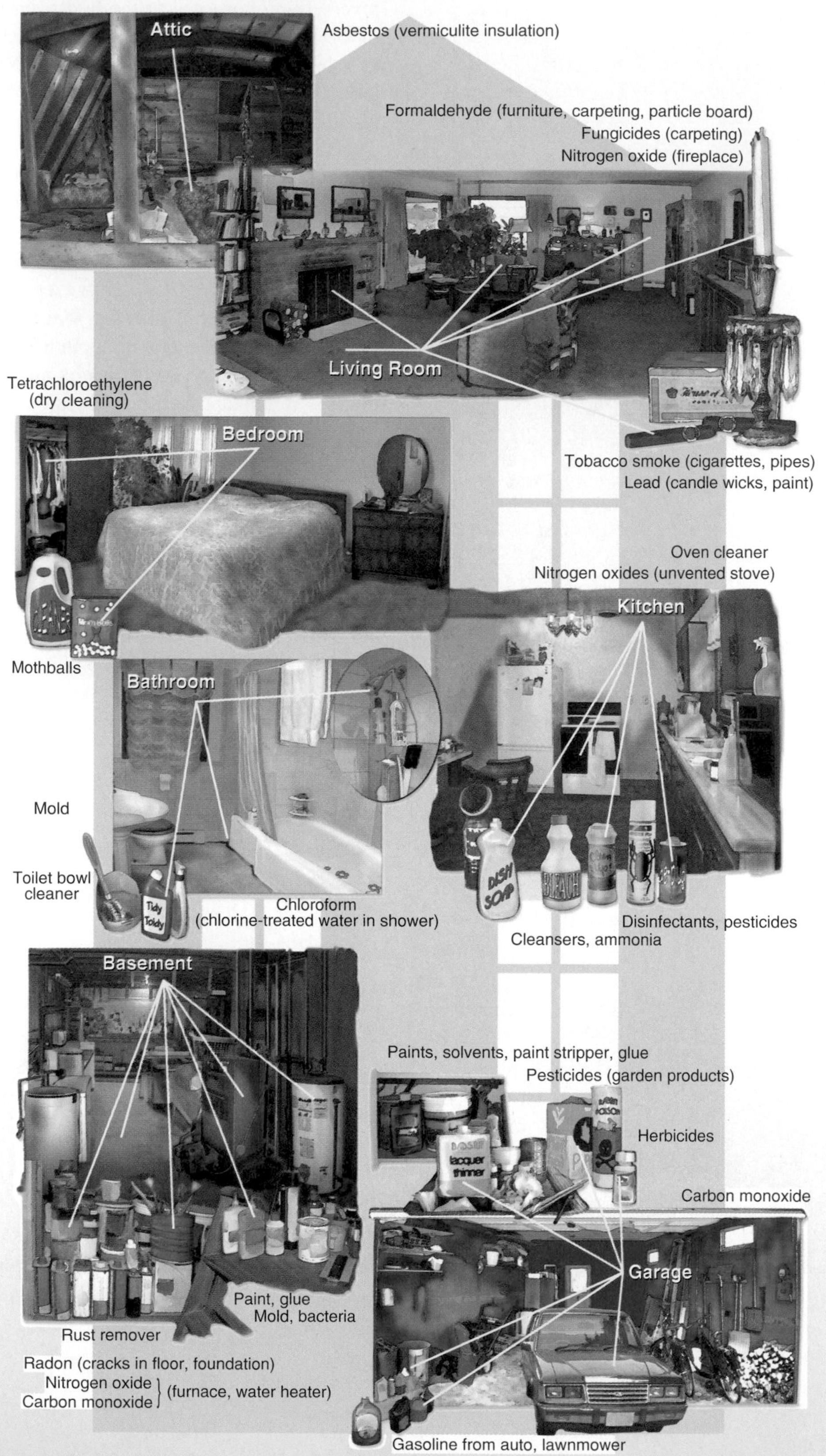

FIGURE 1 Sources of indoor air pollution are many. Sources of radon in the home include the soil, walls, and water.
Source: After U.S. Environmental Protection Agency, EPA Journal, *Sept./Oct. 1990.*

(Box 18.1 Continued)

Leslie and Kyle were looking over the new apartment they were considering renting. "Smoke detector, check. Carbon monoxide detector, too; good."

"Why do we want a carbon monoxide detector? The garage is two floors down. Car exhaust shouldn't really get up here."

"No, but there's the heater and the gas stove. If anything goes wrong, it'd be good to know it before we suffocate!"

"True. But then, what about radon?"

"*What* about radon? Where would it come in?"

Many potential hazards have been created or intensified by our search for energy efficiency in response to dwindling fuel supplies and rising costs. For example, it was learned after the fact that blown-in foam insulation releases dangerous quantities of the volatile gas formaldehyde as it cures. (Formaldehyde is the pungent liquid in which creatures are commonly preserved for dissection by biology classes.) Many homes in which this insulation had been installed had to be torn apart later to remove it. The very act of insulating tightly has meant that quantities of toxic gases that might once have escaped harmlessly through cracks and small air leaks are now being sealed in, and their concentrations are building up. Among them are smoke and carbon monoxide (from furnaces, heaters, gas appliances, and cigarettes), various volatile organic compounds (such as paint and solvent fumes), and radon (figure 1).

Radon is a colorless, odorless, tasteless gas that also happens to be radioactive. It is produced in small quantities in nature by the decay of the natural trace elements uranium and thorium. When radon is free to escape into the open atmosphere, its concentration remains low. When it is confined inside a building, the radioactivity level increases (figure 2). Tightly sealed homes may have radon levels above the levels found inside uranium mines.

How does all this radon get in? Uranium and thorium occur in most rocks and soils and in building materials made from them, like concrete or brick. Being a gas, radon can diffuse into the house through unsealed foundation walls or directly into unfinished crawl spaces from the soil. It can emanate from masonry walls. Radon seeps into ground water from aquifer rocks and thus can enter the house via the plumbing. The potential radon input from rock and soil depends on local geology; certain granites and shales are relatively rich in uranium and therefore produce relatively plentiful radon.

The risks from elevated radon in homes are disputed. As radon is a chemically inert gas, it will not remain in the lungs; one breathes it in, then out. If an atom of radon happens to decay while in the lungs, it decays to radioactive isotopes of lead, bismuth, and other metals that might lodge in the lungs and later decay there, posing a cancer risk. However, so far no correlation has been found between radon concentrations in homes and the incidence of lung cancer.

Nevertheless, given all the air pollutants that can accumulate inside the home, many experts recommend that snugly insulated homes being built or remodeled with emphasis on insulation for energy efficiency also include an air exchanger designed to provide adequate ventilation without loss of appreciable heat.

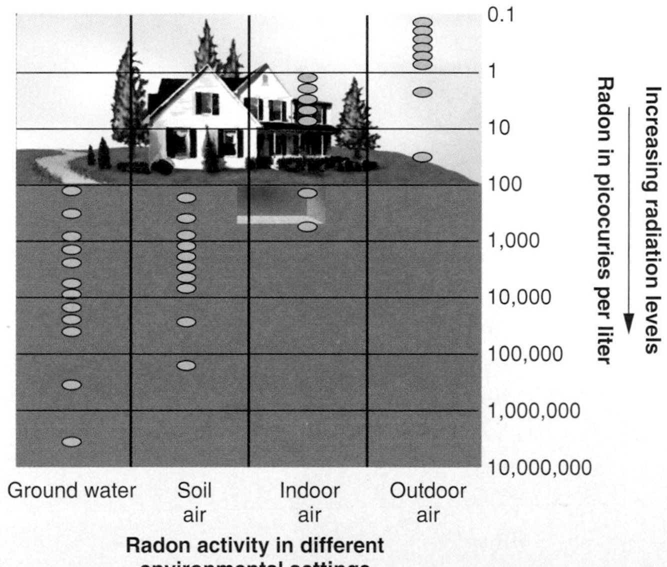

FIGURE 2 Ranges of radioactivity due to radon in natural settings—ground water, air in soil pores, outdoor air—and in the home. Note that the radioactivity scale is exponential, so radiation levels in high-radon homes may be hundreds of times what is typical in outdoor air.

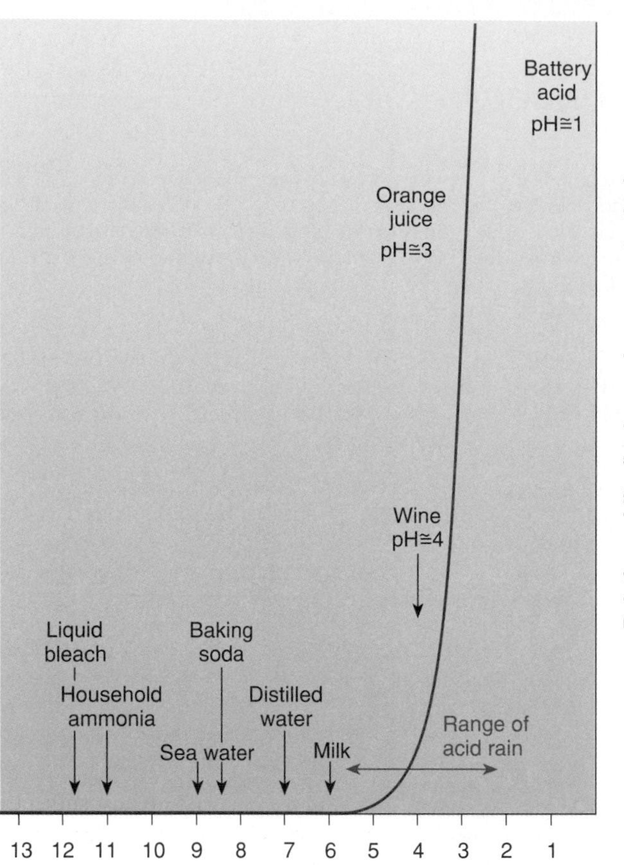

FIGURE 18.16 The pH scale is a logarithmic one; small differences in pH value translate into large differences in acidity. Range of acid precipitation is compared with typical pH values of common household substances.

may survive but their eggs may not hatch in the acidified waters. Acidic water more readily leaches potentially toxic metals from soils, transferring them to surface- or groundwater supplies, and reduces the capacity of soils to neutralize further additions of acid. Acid water leaching nutrients from lake-bottom sediments may contribute to algal bloom. Impacts of acid rain on ground water were also explored in chapter 17. Studies have found a correlation between rainfall acidity and potentially toxic levels of mercury in lakes in northern North America. Other metals that may be linked to acid rain include lead, zinc, selenium, copper, and aluminum; recall that acid mine drainage is correlated with a higher concentration of leached metals. The accumulative properties of the heavy metals, in turn, raise concerns about toxic levels in fish and consequent risks to humans and other fish-eaters.

Regional Variations in Rainfall Acidity and Impacts

Rainfall is demonstrably more acidic in regions downwind from industrial areas releasing significant amounts of sulfur among their emissions. These observations form the basis for decisions to control sulfur pollution. What is less clear is the precise role that anthropogenic sulfur plays in the sulfur cycle.

A

B

FIGURE 18.17 Cleopatra's Needle, a granite obelisk, dramatically demonstrates the effects of acid rain. It was carved around 1600 B.C., and its features remained sharply defined through 3500 years in the dry climate of Egypt (left photo). Just 75 years after it was moved to Central Park in New York City, it showed serious deterioration (right photo) due to the much more rapid weathering caused by wetter conditions compounded by acidity of rainfall.
Photograph (A) courtesy Metropolitan Museum of Art, (B) © Lee Snider/Corbis.

For one thing, there is little information about natural background levels of sulfur dioxide (SO_2) and related sulfur species. Historically, pure rainwater has been assumed to have a pH of about 5.6 in areas away from industrial sulfur sources. This is the pH to be expected as a consequence of the formation of carbonic acid due to natural carbon dioxide in the air. Recent research suggests that natural background pH may be somewhat lower as a result of other chemical reactions in the atmosphere and may vary from place to place as a consequence of localized conditions. Natural rainfall acidity may correspond to a pH as low as 5. However, air over all the globe is now polluted to some extent by human activities. Therefore, the chemistry of precipitation in wholly unpolluted air cannot be determined precisely enough to assess anthropogenic impacts fully. It does appear that, overall, more sulfate is added to the air by sea spray containing dissolved sulfate minerals than by human activities. Perhaps anthropogenic sulfate only locally aggravates an existing situation.

It may, in fact, be the case that increases in rainfall acidity due to the formation of nitric acid (another strong acid) from nitrogen oxides formed in internal-combustion engines is a much more significant problem in and near urban areas. The focus on sulfuric acid in rain partially reflects the fact that sulfur gas pollution from stationary sources is more readily controllable and partially recognizes the more long-range and global impact of anthropogenic sulfuric acid, including the political aspects of the issue (see below).

The nature of the local geology on which acid rain falls can strongly influence the severity of the acid's impact. Chemical reactions between rain and rock are complex. Farmers and gardeners have known for years that certain kinds of rock and soil react to form acidic pore waters, while others yield alkaline water. For instance, because limestone reacts to make water more alkaline, it serves, to some extent, to neutralize acid waters. The effects of acid rain falling on carbonate-rich rocks and soils are therefore moderated. Conversely, granitic rocks and the soils derived from them are commonly more acidic in character. They cannot buffer or moderate the effects of acid rain; the acid rain just makes the surface and subsurface waters more acidic in such regions. Some operators of coal-fired plants in the Midwest have argued that the predominantly granitic soils of the Northeast are principally at fault for the existence of very acid lakes and streams in the latter region. Certainly, the geology does not *reduce* the problem. However, the rainfall in the Northeast, with an average pH below 4.5, is clearly acidified (figure 18.18).

Nor is the northeastern United States the only area concerned about sulfuric acid generated in this country. The Canadian government has expressed great concern over the effects of acid rain caused by pollutants generated in the United States—especially the industrialized upper Midwest and Northeast—that subsequently drift northeast over eastern Canada. Much of that area, too, is underlain by rocks and soils highly sensitive to acid precipitation and is already subject to acid precipitation associated with Canadian sulfate sources. In Europe, acid rain produced by industrial activity in central Europe and the United Kingdom falls on Scandinavia. Acid rain, then, has become an international political concern.

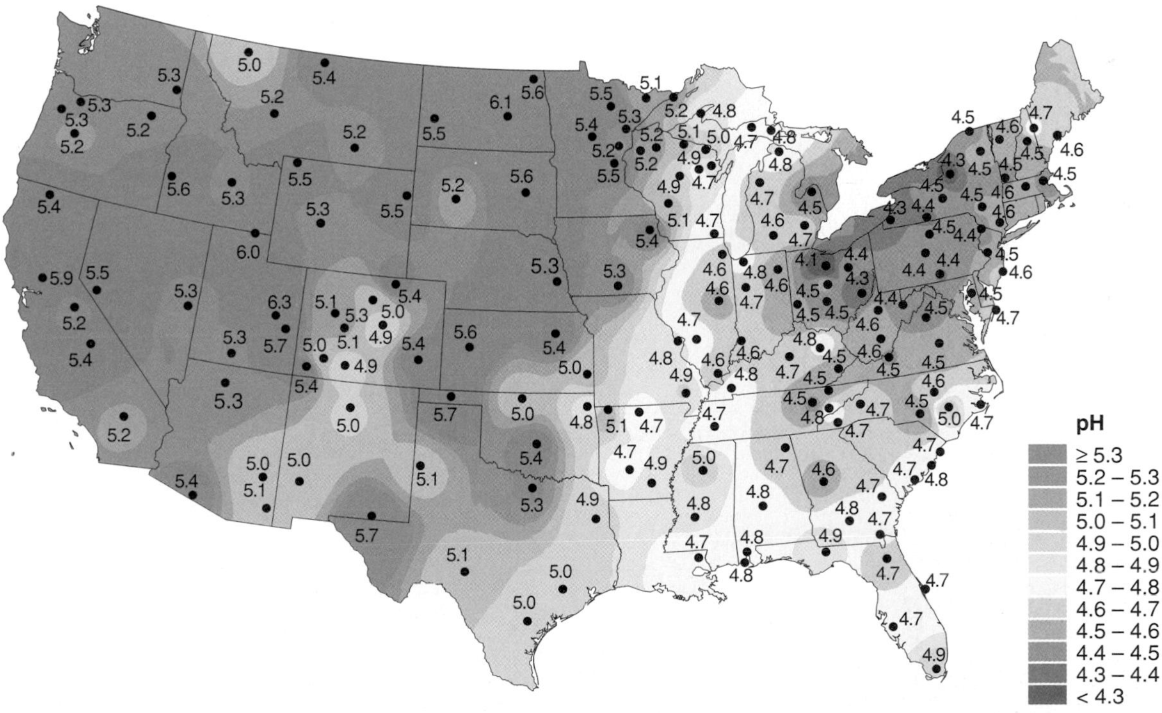

FIGURE 18.18 Rainfall is acidic everywhere, but in the United States, it is especially so in the northeast (2002 data; pH as measured in the field).
Map from National Atmospheric Deposition Program National Trends Network (2003). NADP Program Office, Illinois State Water Survey, 2204 Griffith Dr., Champaign, IL 61820.

More information is needed about natural water chemistry in various (unpolluted) geologic settings before the role of acid rain can be evaluated fully. The science of pinpointing sulfur sources is also becoming more sophisticated as means are developed to "fingerprint" those sources by the assemblage of gases present and their composition. This will allow cause-and-effect relationships to be determined more accurately.

Negative impacts of acid rain are certainly well documented. Increasingly, damage from related phenomena—acid snow and even acid fog—are being recognized. In snowy regions with acid precipitation, rapid spring runoff of acid snowmelt can create an "acid shock" to surface waters and their associated communities of organisms. Where sulfur gas release has been reduced, some recovery of water quality in lakes and streams has occurred. Better control of sulfur emissions is plainly desirable. Some strategies are discussed later in this chapter.

AIR POLLUTION AND WEATHER

Thermal Inversion

Air-pollution problems are naturally more severe when air is stagnant and pollutants are confined. Particular atmospheric conditions can contribute to acute air-pollution episodes of the kinds mentioned at the beginning of this chapter. A frequent culprit is the condition known as a **thermal inversion** (figure 18.19). Within the lower atmosphere, air temperature normally decreases as altitude increases. This is why mountaintops are cold places, even though direct sunshine is stronger there. In a thermal inversion, there is a zone of relatively warmer air at some distance above the ground. That is, going upward from the earth's surface, temperatures decrease for a time, then increase in the warmer layer (inversion of the normal pattern), then ultimately continue to decrease at still higher altitudes. Inversions may become established in a variety of ways. Warmer air moving in over an area at high altitude may move over colder air close to the ground, thus creating an inversion. Rapid cooling of near-surface air on a clear, calm night may also lead to a thermal inversion.

Most air pollutants, as they are released, are warmer than the surrounding air—exhaust fumes from automobiles, industrial smokestack gases, and so on. Warm air is less dense than colder air, so ordinarily, warm pollutant gases rise and keep rising and dispersing through progressively cooler air above. When a thermal inversion exists, a warm-air layer becomes settled over a cooler layer. The warm pollutant gases rise only until they reach the warm-air layer. At that point, the pollutants

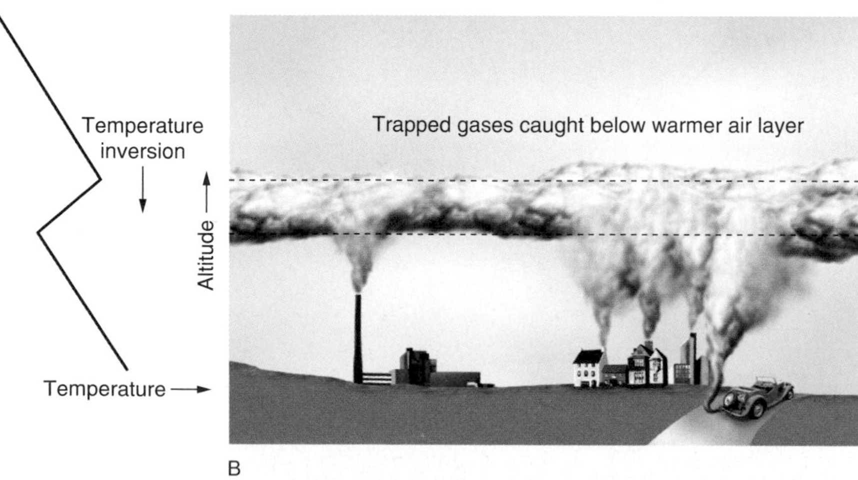

FIGURE 18.19 Effect of thermal inversion in trapping air pollution. (A) Normal conditions permit warm pollutants to rise and disperse through cooler air above. (B) Inversion traps warm pollutant gases under a stable warm-air layer.

FIGURE 18.20 Thermal inversion traps smog over an oil field near San Francisco Bay at sunset. Note drill rigs at left of center.
Photograph by Belinda Rain, courtesy of EPA/National Archives.

are no longer less dense than the air above, so they stop rising. They are effectively trapped close to the ground and simply build up in the near-surface air. Sometimes, the cold/warm air boundary is so sharp that it is visible as a planar surface above the polluted zone (figure 18.20). The Donora, Pennsylvania, episode mentioned at the beginning of the chapter was triggered by an inversion. Its effect on a freight train's smoke was described by physician R.W. Koehler: "They were firing up for the grade and the smoke was belching out, but it didn't rise…. It just spilled out over the lip of the stack like a black liquid, like ink or oil, and rolled down to the ground and lay there" (Roueché, 1953, 175).

Every one of the half-dozen major acute air-pollution episodes of the twentieth century has been associated with a thermal inversion, as have many milder ones. Certainly, air pollution can be a health hazard in the absence of a thermal inversion. The inversion, however, concentrates the pollutants more strongly. Moreover, inversions can persist for a week or longer, once established, because the denser, cooler air near the ground does not tend to rise through the lighter, warmer air above.

Certain geographic or topographic settings are particularly prone to the formation of thermal inversions (figure 18.21). Los Angeles has the misfortune to be located in such a spot. Cool air blowing across the Pacific Ocean moves over the land and is trapped by the mountains to the east beneath a layer of warmed air over the continent. Donora, Pennsylvania, lies in a valley. When a warm front moves across the hilly terrain, some pockets of cold air may remain in the valleys, establishing a thermal inversion. Germany's coal-rich Ruhr Valley suffered so from pollution trapped by a prolonged inversion during the winter of 1984–1985 that schools and many factories had to be closed and the use of private automobiles banned until the air cleared. As with most weather conditions, nothing can be done about a thermal inversion except wait it out.

Impact on Weather

While weather conditions such as thermal inversions can affect the degree of air pollution, air pollution, in turn, can affect the weather in ways other than reducing visibility, modifying air temperature, and making rain more acidic. This is particularly true when particulates are part of the pollution. Water vapor in the air condenses most readily when it has something to

FIGURE 18.21 Influence of topography on establishing and maintaining thermal inversions. (A) Cool sea breezes blow over Los Angeles and are trapped by the mountains to the east; see also Section Five opening photograph. (B) Cold air, once settled into a valley, may be hard to dislodge.

condense on. This is the principle behind cloud seeding: Fine, solid crystals are spread through wet air, and water droplets form on and around these seed crystals. Particulate pollutants can perform a similar function.

This is illustrated in figure 18.22. Southwest of Lake Michigan lies the industrialized area of Chicago, Illinois, and Gary and Hammond, Indiana. The Gary/Hammond area, especially, is a major steel milling region. At the time this satellite photograph was taken, waste gases from the mills were being blown northeast across the lake (note faint smoke plumes at bottom end of lake, just left of center). At first, they remained almost invisible, but as water vapor condensed, clouds appeared and, in this scene, snow crystals eventually formed. A large area of snow in Michigan is on a direct line with the pollution plumes from the southwest. The particulate pollutants in the exhaust gases facilitate the condensation and precipitation.

Recent studies in China and India have suggested additional effects. Climate models indicate that observed emissions of dark, sooty particulates may absorb heat, altering atmospheric temperature and circulation patterns, and thus precipitation distribution. This may account for recent trends toward more-frequent floods in south China and droughts in north China.

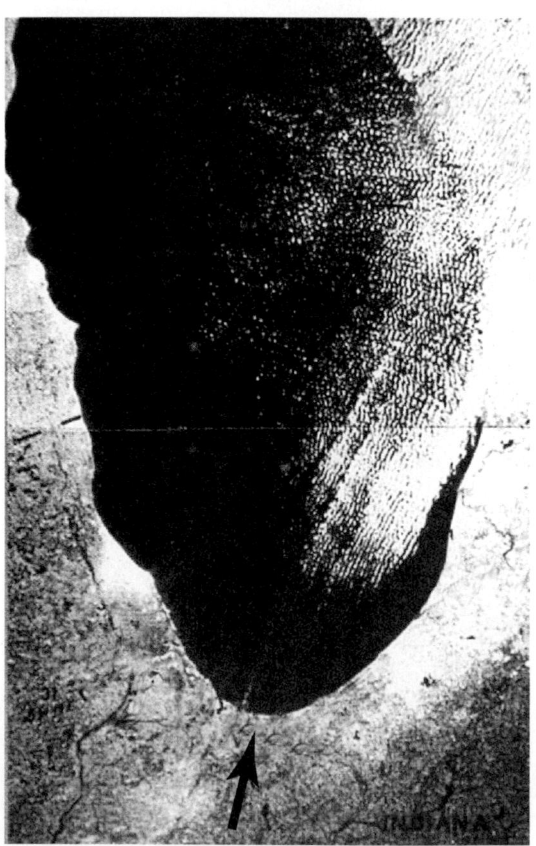

FIGURE 18.22 Air-pollution plumes from the Chicago, Illinois, and Gary/Hammond, Indiana, area bring clouds and snow to the area to the northeast. Back area is Lake Michigan; arrow points to source of pollution and initial trend.
Photograph courtesy NASA.

TOWARD AIR-POLLUTION CONTROL

Air-Quality Standards

Growing dissatisfaction with air quality in the United States led, in 1970, to the Clean Air Act Amendments, which empowered the Environmental Protection Agency to establish and enforce air-quality standards (see also chapter 19). The standards developed are designed to protect human health, clear the visible pollution from the air, and prevent crop and structural damage and other adverse effects. When meteorologists and others report on local air quality, they are referring to the Air Quality Index, developed by EPA to translate pollutant concentrations into descriptors that the general public can readily understand. The AQI is illustrated in figure 18.23. On any given day, the AQI may be different for different pollutants. Overall air quality is assigned the descriptors and color applicable to the pollutant with the highest AQI that day.

Control Methods

Air-pollutant emissions can be reduced either by trapping the pollutants at the source or by converting dangerous compounds to less harmful ones prior to effluent release. A variety of technologies for cleaning air exist.

Where particulates are the major concern, filters can be used to clear the air, with the fineness of the filters adjusted to the size range of particles being produced. Filters may be 99.9% efficient or better, but they are commonly made of paper, fiber, or other combustible material. These cannot be used with very high-temperature gases.

An alternate means of removing particulates is electrostatic precipitators or "scrubbers." In these, high voltages charge the particulates, which are then attracted to and caught on charged plates. Efficiencies are usually between 98 and 99.5%. These systems are sufficiently costly that they are practical only for larger operations. They are widely used at present in coal-fired electricity-generating plants.

Wet scrubbers are another possibility. In these scrubbers, the gas is passed through a stream or mist of clean water, which removes particulates and also dissolves out some gases. Wet chemical scrubbing, in which the gas is passed through a water/chemical slurry rather than pure water, is used particularly to clean sulfur gases out of coal-fired plant emissions. Currently, the most widely used designs use a slurry of either lime (calcium oxide, CaO) or limestone ($CaCO_3$), which reacts with and thus removes the sulfur gases. Wet scrubbing produces contaminated fluids, which present equipment-cleaning and/or waste-disposal problems.

Both carbon monoxide and the nitrogen oxides are best controlled by modifying the combustion process. Lower combustion temperatures, for example, minimize the production of nitrogen oxides and moderate the production of carbon monoxide. The use of afterburners to complete combustion is another way to reduce carbon monoxide emission, by converting it to carbon dioxide.

Air Quality Index (AQI) Value	Descriptor	Meaning
0–50	Good	Air quality is satisfactory and air pollution poses little or no risk.
51–100	Moderate	Air quality is acceptable, but a few persons unusually sensitive to the pollutant may experience negative health effects.
101–150	Unhealthy for sensitive groups	While the general public is not likely to be affected, members of sensitive groups are more likely to experience health effects. (For example, ozone at this level may be problematic for those with lung disease; those with lung or heart disease are at risk from particulates.)
151–200	Unhealthy	The population at large may experience negative health effects, and impact on sensitive groups will be more severe.
201–300	Very Unhealthy	Everyone may experience more-serious health effects.
> 300	Hazardous	This AQI level triggers health warnings of emergency conditions. The whole population is more likely to be affected.

FIGURE 18.23 For each of five major air pollutants regulated by the Clean Air Act Amendments (ground-level ozone, particulates, carbon monoxide, sulfur dioxide, and nitrogen dioxide), an air-quality standard (concentration or exposure limit) has been set. An AQI = 100 corresponds to the standard level for the particular pollutant. Various ranges of AQI are assigned different descriptors based on likely health effects, and color codes that allow the general public to understand easily the corresponding degree of health risk.
Adapted from U.S. Environmental Protection Agency.

Combustion is an effective way to destroy organic compounds, including hydrocarbons, producing carbon dioxide and water. More complex scrubbing or collection procedures for organics also exist but are not discussed in detail here. Such measures are necessary for low-temperature municipal incinerators burning materials that might include toxic organic chemicals that can only be thoroughly decomposed in the high-temperature incinerators designed for toxic-waste disposal.

Altogether, the United States spends over $100 billion a year on air-pollution abatement. We have already seen some of the results in reduced emissions. Improvements have been seen in deposition as well. Phase I of the 1990 reauthorization of the Clean Air Act Amendments targeted a number of sources of sulfur pollution, with a view to reducing acid rain. Sulfate and acidity (hydrogen ion concentration) both showed substantial reductions downwind of Phase I targets. Reductions in acid precipitation, in turn, have resulted in reduced acidity of surface waters in the northeastern United States. While concentrations of both nitrate and sulfate in precipitation remain relatively high in the region, sulfate, at least, has shown notable further declines (figure 18.24).

Automobile Emissions

As part of the efforts to reduce air pollution, the Environmental Protection Agency has imposed limits on permissible levels of carbon monoxide and hydrocarbon emissions from automobiles. Many car manufacturers use catalytic converters to meet the standard. In a catalytic converter, a catalyst (commonly platinum or another platinum-group metal) enhances the oxidation, or combustion, of the hydrocarbons and carbon monoxide to water and carbon dioxide. In this process, catalytic converters are quite effective. Unfortunately, they also promote the production of sulfuric acid from the small quantities of sulfur in gasoline, as well as the formation of nitrogen oxides. Emissions of these compounds may be *increased* by the use of catalytic converters. While the increased sulfuric acid emission by vehicles using catalytic converters is minor compared to the output from coal-fired power plants, it nevertheless makes automobile exhaust more irritating to breathe where exhaust fumes are confined. The environmental impacts of the nitrogen oxides may be more negative than previously assumed, which may dictate a need to control these more carefully as well.

Improvements in fuel economy reduce all emissions simultaneously, of course, as less fuel is burned for a given distance traveled. The EPA-mandated fuel economy standards have been increasing year by year since 1978. New-car fuel economy has actually averaged somewhat higher than the standards on occasion; however, the average mileage of *all* cars on the road in the same year is invariably lower. Since 1990, the passenger-car standard has been 27.5 mpg, but the average mileage for all cars in that year was only 20.2 mpg, rising to 22.4 mpg by 2004. Furthermore, the depressed fuel prices of the mid-1980s contributed to a rebound in popularity of larger, less-fuel-efficient cars, and some auto manufacturers began to have corresponding difficulty in meeting average new-car fuel economy standards. In addition, many of the increasingly popular vans and sport-utility vehicles are classified as light trucks rather than automobiles, so they are not currently subject to the mileage requirements of passenger cars. Worldwide, as noted earlier, fuel cost and fuel consumption are inversely correlated, and the United States has enjoyed by far the least expensive gasoline among industrialized nations, with correspondingly high consumption. Still, the stricter mileage requirements are slowly improving overall average automobile mileage, and additional limits on emissions (independent of mileage ratings) for both cars and trucks have helped bring total air pollutant emissions down.

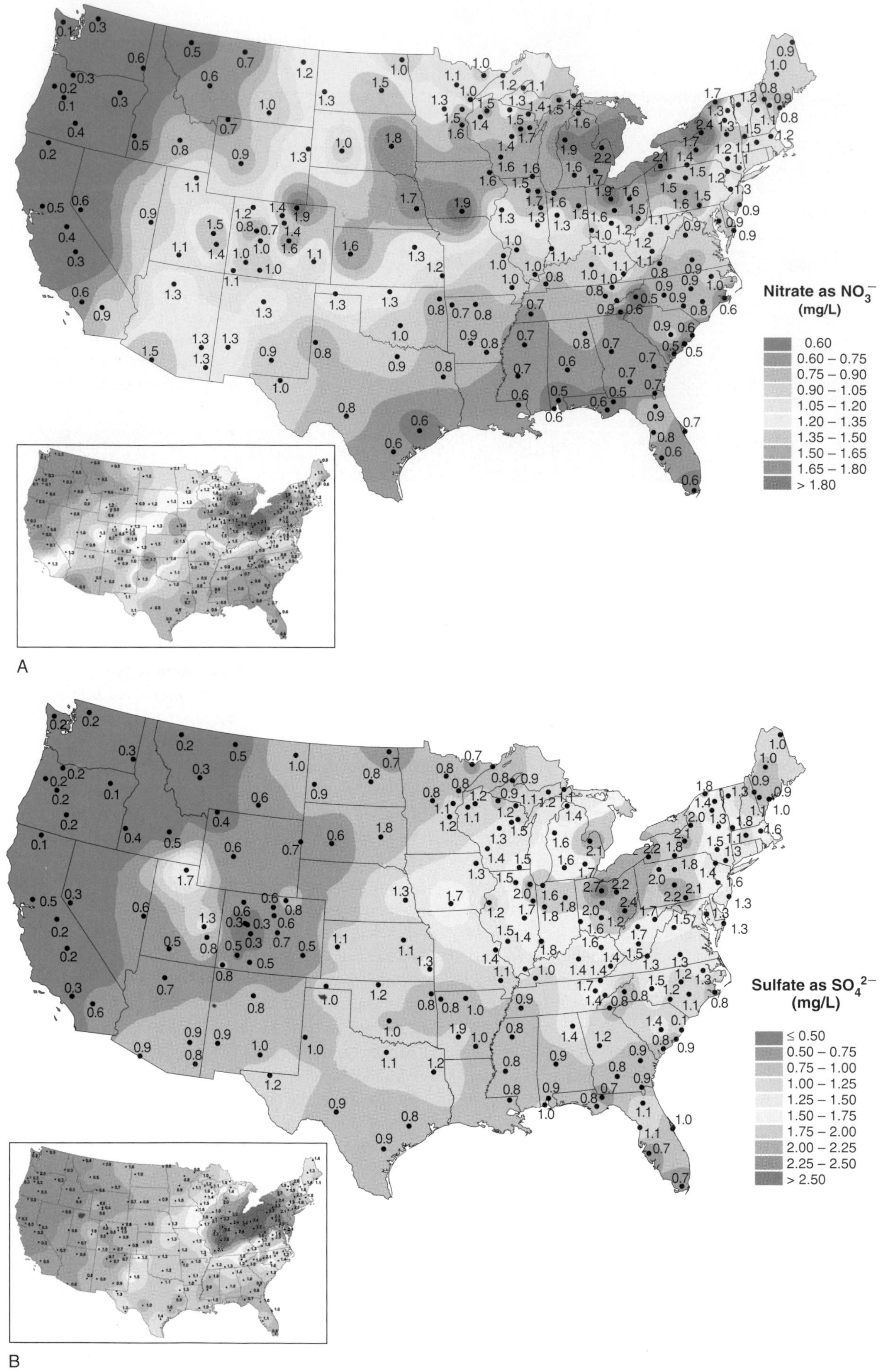

FIGURE 18.24 (A) Nitrate (NO_3^-) concentration in precipitation (mg/L) reflects particularly nitrogen-oxide emissions from urban and industrial areas upwind; (B) sulfate (SO_4^{2-}) forms from sulfur-dioxide emissions. Large maps show 2002 data; inset maps, 1994.

Maps from National Atmospheric Deposition Program National Trends Network (2003). NADP Program Office, Illinois State Water Survey, 2204 Griffith Dr., Champaign, IL 61820.

Carbon Sequestration

As we have noted, CO_2 is not a pollutant monitored by EPA for public-health reasons (though this may change in future). Clearly, however, it is a concern from a climatological point of view. Recently, there has been increased interest in various methods of *carbon sequestration,* storing carbon (often as carbon dioxide) in some reservoir isolated from atmospheric circulation. A number of geological reservoirs have been considered (figure 18.25A). One might pump CO_2 into coal seams too thin or low-quality to mine, perhaps extracting useable coal-bed methane in the process. The CO_2 could be pumped into depleted petroleum reservoirs, or into ground water too salty, too high in dissolved minerals, to be of interest as a water source. Experiments and theoretical calculations have been done on more-exotic possibilities, such as reacting CO_2 with suitable minerals to produce carbonate minerals that would be geologically stable and thus keep the carbon "locked up" for a very long time.

Some carbon-sequestration operations are already in operation (figure 18.25B). For example, natural gas from the Sleipner field offshore from Norway is high in associated CO_2. But Norway has a high carbon-emission tax. So it is economically advantageous for Statoil, the company operating the field, to separate the CO_2 and inject it back into a permeable sandstone below the sea floor, rather than releasing it into the atmosphere. Such sequestration activities will become more common if the economics are right.

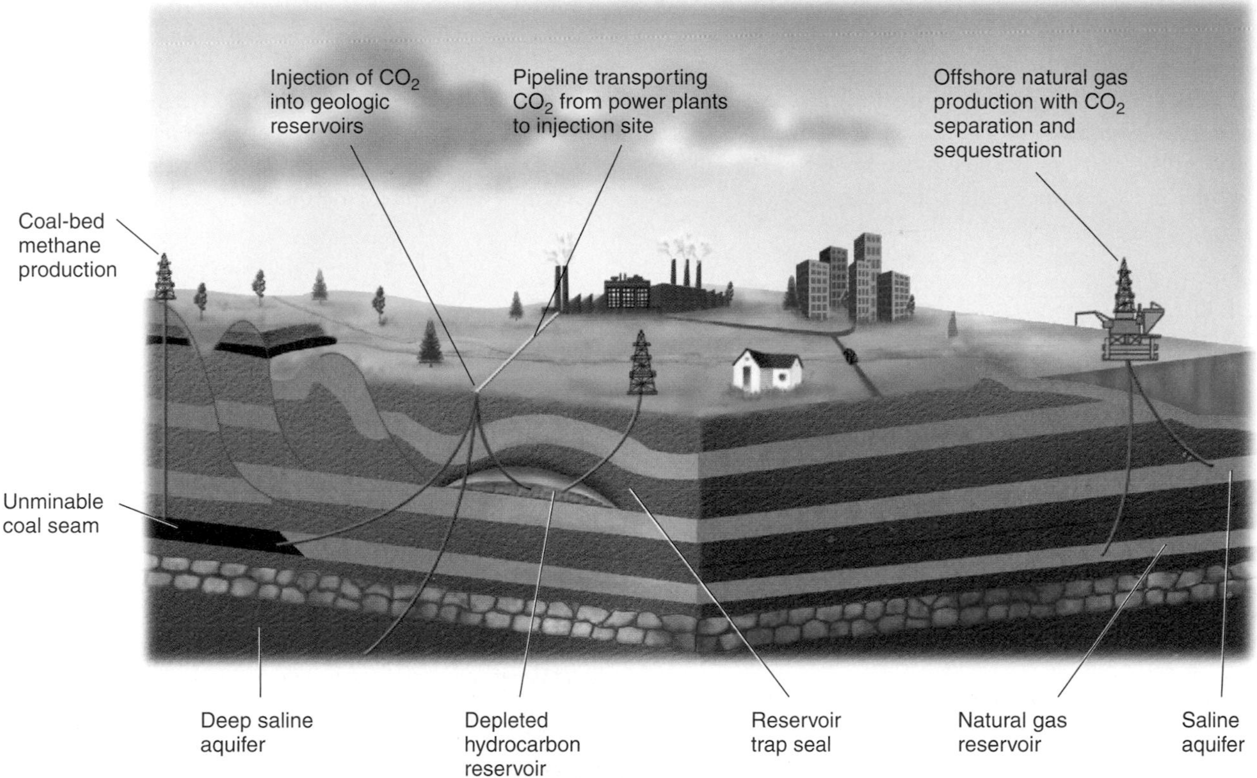

FIGURE 18.25 (A) Examples of geological methods of carbon sequestration. (B) Natural-gas extraction and carbon sequestration occur together at the Sleipner field off the coast of Norway.
(A) After USGS Fact Sheet FS-026-03; (B) Photograph © Øyvind Hagen/Statoil

SUMMARY

The principal air pollutants produced by human activities are particulates, gaseous oxides of carbon, sulfur, or nitrogen, and organic compounds. The largest source of this pollution is combustion, and especially transportation. For some pollutants, anthropogenic contributions are far less than natural ones overall, but human inputs tend to be spatially concentrated. Control of air pollutants is complicated because so many of them are gaseous substances generated in immense volumes that are difficult to contain; once released, they may disperse widely and rapidly in three dimensions in the atmosphere, thwarting efforts to recover or remove them. Climatic conditions can influence the severity of air pollution: Strong sunlight in urban areas produces photochemical smog, and temperature inversions in the atmosphere trap and concentrate pollutants. Air pollutants, in turn, may influence weather conditions by contributing to acid rain and by providing particles on which water or ice can condense. Ozone in photochemical smog can be a serious health threat, especially to those with respiratory problems. Yet in the "ozone layer" in the stratosphere, it provides important protection against damaging UV radiation. Imposition of air-quality and emissions standards by the Environmental Protection Agency for pollutants with potential health consequences has led to measurable improvements in air quality over the last four decades, particularly with respect to pollutants for which anthropogenic sources are relatively important, such as lead. However, even when pollutants are controlled, the environment may be slow to recover. For example, despite international agreements curtailing the release of the chlorofluorocarbons that contribute to the destruction of stratospheric ozone, CFC levels remain high as a consequence of the compounds' decades-long residence times, so ozone concentrations have not yet recovered proportionately. A relatively new approach to air-pollution abatement, carbon sequestration, may help to reduce CO_2 pollution, which, while not directly a health issue, is certainly an ongoing environmental concern.

Key Terms and Concepts

acid rain 449
ozone hole 446
ozone layer 444
particulates 439
pH scale 449
thermal inversion 454
ultraviolet radiation 444

Exercises

Questions for Review

1. Describe the principal sources and sinks for atmospheric carbon dioxide.
2. Carbon monoxide is a pollutant of local, rather than global, concern. Explain.
3. What is the origin of the various nitrogen oxides that contribute to photochemical smog?
4. What is photochemical smog, and under what circumstances is this problem most severe?
5. Ozone is an excellent example of the chemical-out-of-place definition of a pollutant. Explain, contrasting the effects of ozone in the ozone layer with ozone at ground level.
6. What evidence links CFC production with ozone-layer destruction? What is the role of UV radiation in forming and destroying ozone, thus protecting us from UV rays from the sun?
7. What radiation hazard is associated with indoor air pollution, and why has it only recently become a subject of concern?
8. What pollutant species is believed to be primarily responsible for acid rain? What is the principal source of this pollutant?
9. Explain briefly why the seriousness of the problems posed by acid rain may vary with local geology.
10. Describe the phenomenon of a thermal inversion. Give an example of a geographic setting that might be especially conducive to the development of an inversion. Outline the role of thermal inversion in intensifying air-pollution episodes.
11. Federal emissions-control regulations for automobiles have led to improvements in air quality with respect to some but not all pollutants in auto exhaust systems. Explain briefly.
12. What is "carbon sequestration"? Describe a geological example.

For Further Thought

1. Investigate in some detail the technology and economics associated with the control of pollutants from burning coal, and compare the apparent cost-effectiveness of different technologies and the energy requirements of each.
2. Find out who, if anyone, monitors air quality or issues air-pollution hazard warnings in your area or, perhaps, in a nearby urban area.

Keep track, over a period of months, of the weather conditions on days when warnings are issued, and see what patterns emerge.

3. Check the latest information on stratospheric ozone depletion. (The National Aeronautics and Space Administration or National Oceanic and Atmospheric Administration may be good sources.) Investigate the validity of claims, made by those opposing CFC restrictions, that natural chlorine sources (sea spray, volcanic gases) can account for ozone depletion. (See NetNotes for information sources.)

Suggested Readings/References

Akimoto, H. 2003. Global air quality and pollution. *Science* 302 (5 December): 1716–23.

Bartlett, K. 2003. Demonstrating carbon sequestration. *Geotimes* (March): 22–24.

Bell, J. N. B., and Treshow, M. (eds.). 2002. *Air pollution and plant life* (2d ed.). New York: John Wiley & Sons.

Berner, R. A. 1999. A new look at the long-term carbon cycle. *GSA Today* 9 (November): 1–6.

Clow, D. W., and Mast, M. A. 1999. *Trends in precipitation and stream-water chemistry in the northeastern United States, water years 1984–96.* U.S. Geological Survey Fact Sheet 117–99.

Friedman, S. J. 2003. Storing carbon in earth. *Geotimes* (March): 16–20.

Gates, A. E., and Gundersen, C. S., eds. 1992. *Geologic controls on radon.* Geol. Soc. Amer. Special Paper 271.

Heinsohn, R. J., and Kabel, R. 1999. *Sources and control of air pollution.* New York: Prentice-Hall.

Herzog, H., Eliasson, B., and Kaarstad, O. 2000. Capturing greenhouse gases. *Scientific American* (February): 72–79.

Indoor Air. 1993. *EPA Journal* 19 (October–December). This whole issue is devoted to indoor air pollution issues.

Jackson, R. B., et al. 2005. Trading water for carbon with biological carbon sequestration. *Science* 310 (23 December): 1944–47.

Lovins, A. B. 2005. More profit with less carbon. Scientific American (September): 74–83.

Lynch, J. A., Bowersox, V. C., and Grimm, J. W. 1996. *Trends in precipitation chemistry in the United States, 1983–94: An analysis of the effects in 1995 of Phase I of the Clean Air Act Amendments of 1990, Title IV.* USGS Open-File Report 96–0346.

Menon, S., Hansen, J., Nazarenko, L., and Luo, Y. 2002. Climate effects of black carbon aerosols in China and India. *Science* 297 (27 September): 2250–53.

Otton, J. K. *The geology of radon.* U.S. Geological Survey.

U.S. Geological Survey. 2002. *USGS environmental studies of the World Trade Center Area, New York City, after September 11, 2001.* USGS Fact Sheet 050-02.

Woestman, J. T., and Logothetis, E. M. 1995. Controlling automotive emissions. *The Industrial Physicist:* 20–24.

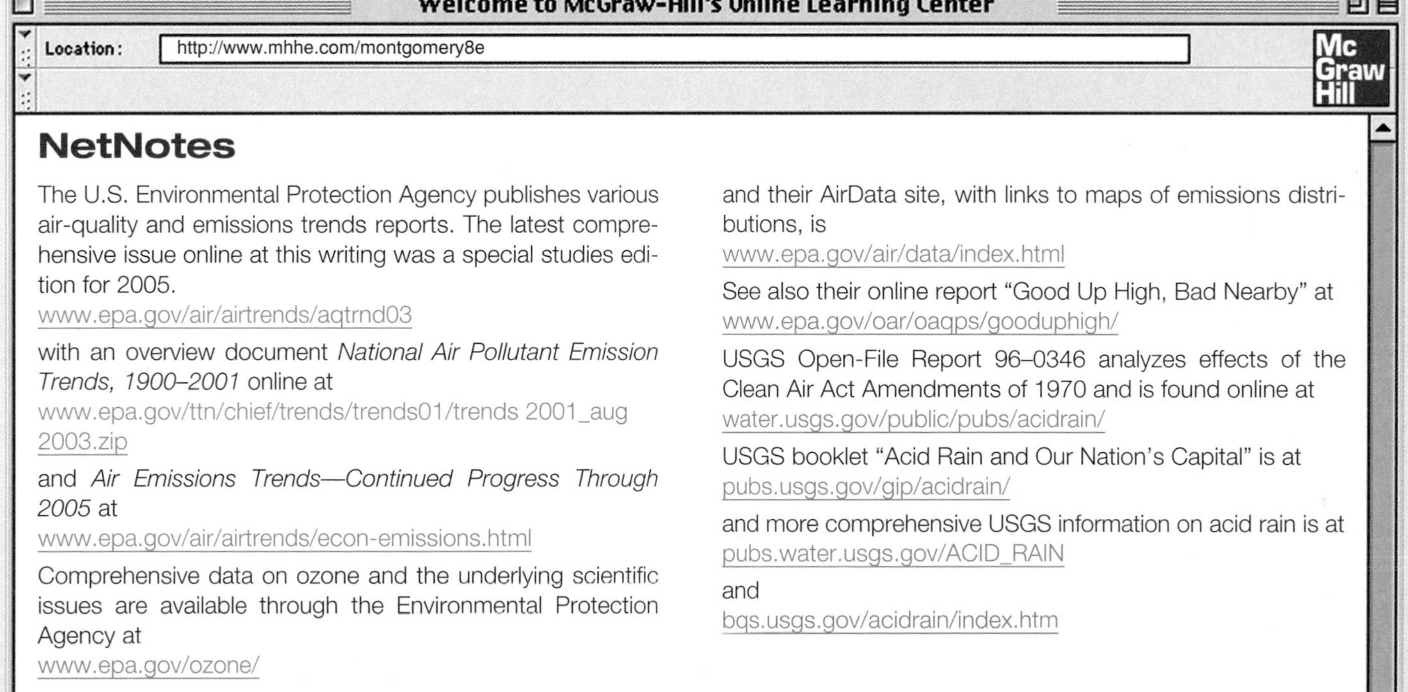

NetNotes

The U.S. Environmental Protection Agency publishes various air-quality and emissions trends reports. The latest comprehensive issue online at this writing was a special studies edition for 2005.

www.epa.gov/air/airtrends/aqtrnd03

with an overview document *National Air Pollutant Emission Trends, 1900–2001* online at

www.epa.gov/ttn/chief/trends/trends01/trends 2001_aug 2003.zip

and *Air Emissions Trends—Continued Progress Through 2005* at

www.epa.gov/air/airtrends/econ-emissions.html

Comprehensive data on ozone and the underlying scientific issues are available through the Environmental Protection Agency at

www.epa.gov/ozone/

and their AirData site, with links to maps of emissions distributions, is

www.epa.gov/air/data/index.html

See also their online report "Good Up High, Bad Nearby" at

www.epa.gov/oar/oaqps/gooduphigh/

USGS Open-File Report 96–0346 analyzes effects of the Clean Air Act Amendments of 1970 and is found online at

water.usgs.gov/public/pubs/acidrain/

USGS booklet "Acid Rain and Our Nation's Capital" is at

pubs.usgs.gov/gip/acidrain/

and more comprehensive USGS information on acid rain is at

pubs.water.usgs.gov/ACID_RAIN

and

bqs.usgs.gov/acidrain/index.htm

A recent USGS fact sheet on carbon sequestration, "Geologic sequestration of carbon dioxide—An energy resource perspective" is online at
pubs.usgs.gov/fs/fs026-03/fs026-03.html

Links to a variety of ozone-related information are available from NASA via
gcmd.gsfc.nasa.gov/Resources/FAQs/ozone.html

and maps of satellite ozone measurements for specific dates are available at
jwocky.gsfc.nasa.gov/eptoms/ep.html

The 1995 Nobel Prize in Chemistry was awarded to scientists Paul Crutzen, Mario Molnia, and F. Sherwood Rowland for their researches in atmospheric chemistry particularly as related to ozone. The citation can be viewed online at
nobelprizes.com/nobel/nobel.html

For real-time maps of UV dosage (regional or global) and many ozone-related sites, see
sedac.ciesin.columbia.edu/ozone/maps/eptpage.html

Links to laws relating to acid rain, including transboundary issues, are available at
www.epa.gov/airmarkets/

An international group including representatives of the World Meteorological Organization (WMO), United Nations Environment Programme (UNEP), NOAA, and NASA reviews the ozone situation every four years; see *Scientific Assessment of Ozone Depletion 2002* at
www.al.noaa.gov/WWWHD/pubdocs/Assessment02.html

Recently, the WMO, UNEP, and Intergovernmental Panel on Climate Change have produced a publication entitled *Carbon dioxide capture and storage,* a summary of which is online at
www.unep.ch/ipcc/activity/ccsspm.pdf

OTHER RELATED TOPICS

In this section, we move further afield from the more traditional areas of geology, and also address some topics that integrate material from many of the more specific chapters in earlier sections of the book. The subjects of this section emphasize the close relationships between geology and human affairs.

Volumes could be, and have been, written about environmental law. In chapter 19, we will very briefly examine several types of laws closely related to geologic subjects dealt with earlier: resources, pollution, and geologic hazards. Increasingly, such laws transcend national boundaries, as nations realize the global impact of geologic processes and the need for cooperation in minimizing the negative effects of human activities on those processes. Chapter 20 is concerned with the related areas of land-use planning and engineering geology. Sound land-use planning takes geology into account in trying to make the best use of each parcel of land. Sensible engineering of buildings, dams, bridges, and other structures likewise generally requires consideration of the constraints imposed by a variety of geologic phenomena and processes.

Even the creation of a national park or monument can require an environmental-impact statement these days. In the case of Mount Rushmore, creating and maintaining these massive rock sculptures in fractured granite is an exercise in engineering geology as well as art.

19

Environmental Law and Policy

Environmental laws of one sort or another have a long history. Many centuries ago, English kings restricted the burning of coal in London because the air pollution had become choking, and the penalty for violators could be execution! Penalties for breaking environmental laws in the United States today are more often financial than physical; sometimes, they are nonexistent. The number of such laws and the areas they regulate continue to increase. Through time, the philosophy of appropriate approaches to resource development, pollution control, and land use has changed, and the focus of environmental laws and the policies on which they are based has shifted in response. A comprehensive treatment of environmental legislation is well beyond the scope of this book, but in this chapter, we will look at common themes of and problems with some environmental laws related to geologic matters.

The passage of the legislation that created Superfund provided the impetus for reclaiming old dumps and mineral mining and processing sites. The Old Works golf course in Anaconda, MT, was a Superfund project founded on a sprawling abandoned smelting operation. To preserve some of the historic flavor of the site, designer Jack Nicklaus kept rows of brick smelters along one side of the course, and filled the bunkers with crushed black slag rather than sand.

RESOURCE LAW: WATER

Given the importance of water as a resource and the increasing scarcity of high-quality water, it is not surprising that laws specifying water rights are necessary. What is perhaps unexpected is that the basic principles governing water rights vary not only from nation to nation, but also from place to place within a single country. Also, quite different principles may be applied to surface-water rights and groundwater rights, even though surface and subsurface waters are inevitably linked through the hydrologic cycle.

Surface-Water Law

Navigable streams are treated as community property, accessible to all, things that cannot be "appropriated" or assigned to individuals. To the extent that they can be viewed as "owned," ownership lies with the state, on behalf of the public. However, while one cannot own the stream, one may have rights to use its water.

The two principal approaches to surface-water rights in the United States are the *Riparian Doctrine* and the *Doctrine of Prior Appropriation.* The term *riparian* is derived from the Latin word for "bank," as in riverbank, and sums up the essence of the **Riparian Doctrine.** Whoever owns land adjacent to a body of surface water (lake, stream) has a right to use that water, and all those bordering on a given body of water have an equal right to that water. Provision is also generally made to the effect that the water must be used for "natural purposes" or "beneficial uses" and returned to the body of water from which it came in essentially the same amount and quality as when it was removed. The Riparian Doctrine, long the basis for English surface-water law, likewise became the basis for assigning surface-water rights in the eastern United States, where moisture is generally ample. Often implicit or explicit in surface-water laws based on the Riparian Doctrine is the concept that no one user's water use should substantially interfere with others' use—but where water is relatively plentiful, this is typically not a problem. On the other hand, strictly speaking, *only* landowners bordering the water body have any rights to its water.

In the western United States, where surface water is in shorter supply, the prevailing doctrine is that of **Prior Appropriation.** Under this scheme, whoever is first, historically, to use water from a given surface-water source has top-priority rights to that water. Users who begin to draw on the same water source later have subordinate rights, in the order in which they begin to use the water. One need not own any land at all to have water rights under this scheme. One's water rights under this doctrine are established and maintained by continuing to divert the water for "beneficial use." Future use is to be proportional to the quantity used when the rights were established. An advantage of this system is that it makes clearer who is entitled to what in times of shortage. Ideally, it ought also to discourage excessive settlement or development in arid regions, because latecomers have no guarantee of surface-water rights if and when demand exceeds supply. In practice, however, it does not appear to have served as such a deterrent. Moreover, because failure to use the water for a period of time causes one to lose water rights under the appropriation doctrine, it may unintentionally lead to waste of water, to water use simply for the sake of preserving those rights rather than out of genuine need—not, perhaps, a wise arrangement in regions where water supplies are inadequate.

What constitutes "beneficial use" of water also varies regionally. Domestic water use is commonly accepted as beneficial. So is hydropower generation, which, in any event, does not appreciably consume water. Irrigation that is critical to agriculture may be regarded as a legitimate "beneficial use," but watering one's lawn might not be. Industrial water use is often a low-priority use. In some areas, a hierarchy of beneficial uses has been established. A typical sequence, from highest to lowest priority, might be domestic water use, municipal supply, irrigation, watering livestock, diversion of water for power generation, use for mining or drilling for oil or gas, navigation, and recreation.

One measure of the degree of benefit that could be used in balancing competing industrial and agricultural interests would be the dollar value of the return expected on the proposed water investment. The resulting priority rankings would then depend on the specific industrial activity, crops, and so forth being compared.

Such prioritizing of usage complicates water-rights questions. An additional complication is that several states, usually those with both water-rich and water-poor areas, have tried to combine both the riparian and prior-appropriation principles (see figure 19.1). The result is generally described as the "California Doctrine," although each state's particular scheme is different. Some states—for example, California—have extended the

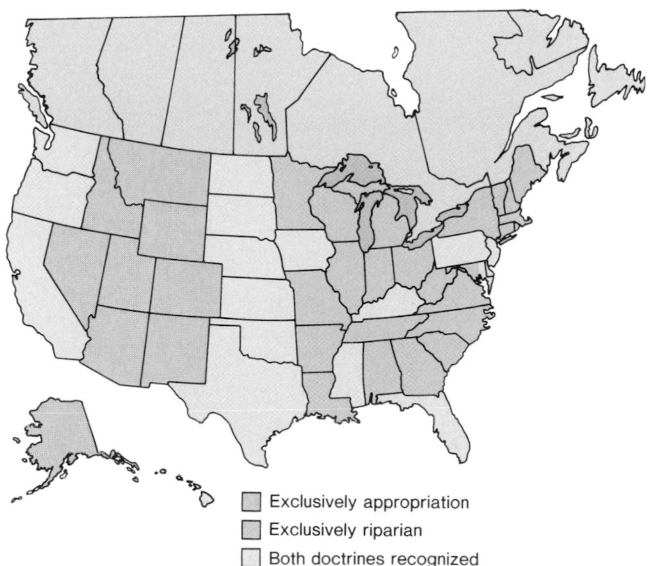

FIGURE 19.1 Distribution of underlying principles of surface-water law applied in the United States.
Source: Data from New Mexico Bureau of Mines Circular 95, 1968.

principle of appropriation to give the *state* the right to appropriate surface water and to transfer it from one region to another or otherwise redistribute it to maximize its beneficial use.

Groundwater Law

Groundwater law can be an even more confusing problem, in part because groundwater law was, to a great extent, developed before there was general appreciation of the hydrologic cycle, the movement of ground water, or its relationship to surface water. As with surface-water rights, too, different principles may govern groundwater rights in different jurisdictions, producing a similar patchwork distribution of water-rights laws across the country (figure 19.2). Two distinctly different principles have been applied. The so-called English Rule, or "Rule of Capture," gives property owners the right to all the ground water they can extract from under their own land. This is sometimes considered analogous to riparian surface-water rights. However, it ignores the reality of lateral movement of ground water. Ground water does not recognize property lines, of course, so heavy use by one landowner may deprive adjacent property owners of water, as ground water flows toward well sites. The American Rule, or "Rule of Reasonable Use," limits a property owner's groundwater use to beneficial use in connection with the land above, and it includes the idea that one person's water use should not be so great that it deprives neighboring property owners of water. Unfortunately, as with surface-water laws, there is no consistent definition of what constitutes "beneficial use." These rules have then been variously combined with appropriation or riparian philosophies of water rights.

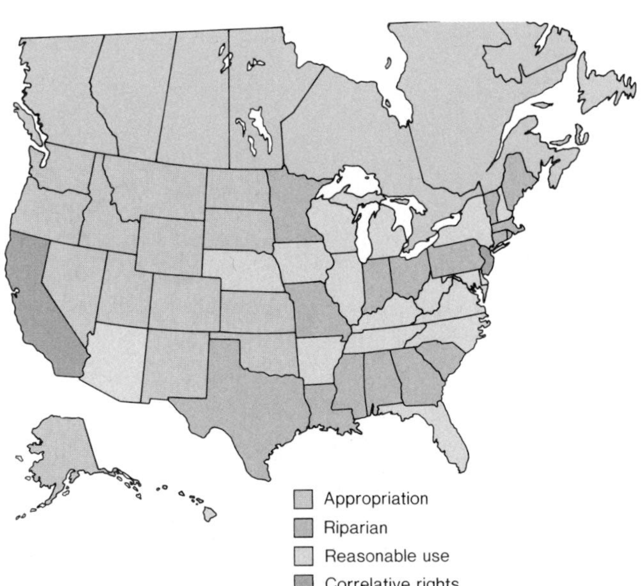

FIGURE 19.2 Distribution of principles of groundwater law applied in the United States. "Riparian" here refers to the English Rule or Rule of Capture described in the text, water rights being related to owning land overlying an aquifer rather than bordering a stream or lake.
Source: Data from New Mexico Bureau of Mines Circular 95, 1968.

Some states, notably California, have specifically addressed the problem of several landowners whose properties overlie the same body of ground water (aquifer system), specifying that each is entitled to a share of the water that is proportional to his or her share of the overlying land (*correlative rights*). Many other states have introduced a prior-appropriation principle into the determination of groundwater rights.

One factor that complicates the allocation of ground water is its comparative invisibility. The water in a lake or stream can be seen and measured. The extent, distribution, movement, and quantity of water in a given aquifer system may be so imperfectly known that it is difficult to recognize at what level one individual's water use may deprive others who draw on the same water source.

As an aside, note that similar problems arise with other reservoirs of fluids underground—specifically, oil and gas. Rights to petroleum from an oil field underlying the Iraq-Kuwait border were a point of contention between those nations prior to the 1990 invasion of Kuwait by Iraq.

RESOURCE LAW: MINERALS AND FUELS

Mineral Rights

For more than a century, U.S. federal laws have included provisions for the development of mineral and fuel resources. The fundamental underlying legislation, even today, is still the 1872 Mining Law. Its intent was to encourage exploitation of mineral resources by granting mineral rights and often land title to anyone who located a mineral deposit on federal land and invested some time and money in developing that deposit. Federal land could be converted to private ownership for $2.50 to $5 an acre—and as those prices were set in the legislation, they would remain in effect for any minerals still subject to the 1872 law, even decades later when inflation had raised land values. The individual (or company) did not necessarily even have to notify the government of the exact location of the deposit, nor was the developer required to pay royalties on the materials mined. The style of mining was not regulated, and there was no provision for reclamation or restoration of the land when mining operations were completed. What this amounted to was a colossal giveaway of minerals on public lands (and land too) in the interest of resource development.

Subsequent laws, particularly the Mineral Leasing Act of 1920 and the Mineral Leasing Act for Acquired Lands of 1947, began to restrict this giveaway. Certain materials, such as oil, gas, coal, potash, and phosphate deposits, were singled out for different treatment. Identification of such deposits on federally controlled land would not carry with it unlimited rights of exploitation. Instead, the extraction rights would be leased from the government for a finite period, and royalties had to be paid, compensating the public in some measure for the mineral or fuel wealth extracted from the public lands. Still, no requirements concerning the mining or any land reclamation were imposed, and the so-called hard-rock minerals (including iron, copper, gold, silver, and others) were exempted. The Outer Continental Shelf Lands Act of 1953 extended similar leasing provisions to offshore regions under U.S. jurisdiction.

Sparked by a 1989 General Accounting Office report recommending reform of some provisions of the 1872 Mining Law, intense debate has been taking place in Congress about possible reform of that law. Under reform proposals in a 1995 bill passed by the House of Representatives, rights to ore deposits still governed by the 1872 law would not be so freely granted, but would be treated more like the rights to oil, coal, and other materials already explicitly removed from the jurisdiction of the 1872 law. Fairer (nearer market) prices for land would have to be paid, as would royalties, and strict mine-reclamation standards would have to be met. Again, an obvious motive is to assure the public, in whose behalf federal lands are presumably held, fair value for mineral rights granted, a return on minerals extracted, and environmental protection. Industry advocated lower royalty rates, deduction of certain mineral-extraction expenses from the amount on which royalties would be paid, and other differences from the House-passed bill, backing a less radical reform passed by the Senate in 1996. Both bills stalled in committee. Subsequent attempts to revive the reform efforts failed. From time to time during the late 1990s, Congress did pass moratoria on the issuing of new mineral rights. Until a revision is passed, however, the 1872 Mining Law, for all its perceived flaws, remains in effect.

Mine Reclamation

Growing dissatisfaction with the condition of many mined lands eventually led to the imposition of some restrictions on mining and post-mining activities in the form of the Surface Mining Control and Reclamation Act of 1977. This act applies only to coal, whether surface-mined or deep-mined, and attempts to minimize the long-term impact of coal mining on the land surface. Where farmland is disturbed by mining, for example, the land's productivity should be restored afterward. In principle, the approximate surface topography should also be restored. However, in practice, where a thick coal bed has been removed over a large area, there may be no nearby source of sufficient fill material to make this possible, so this provision may be disputed in specific areas. Similarly, the act stipulates that groundwater recharge capacity be restored, but particularly if an aquifer has been removed during mining, perfect restoration may not be possible. Still, even when a complete return to pre-mining conditions is not possible, this legislation and other laws patterned after it go a long way toward preserving the long-term quality of public lands and preventing them from becoming barren wastes. The National Environmental Policy Act (discussed later in the chapter) has provided a vehicle for controlling mining practices for other kinds of mineral and fuel deposits.

Many states impose laws similar to or stricter than the federal resource and/or environmental laws on lands under state control. However, the impact of the federal laws alone should not be underestimated. Nearly one-third of the land in the United States, most of it in the west, is under federal control (figure 19.3). Indeed, part of the pressure for reforming the 1872 Mining Law comes from state land-management agents in the western states, where the great disparities between mineral-rights and mineral-development laws on federal and on state land create some tensions. For example, it may be economically advantageous under current laws for a company to seek mineral rights on, and mine, federal rather than adjacent state-controlled lands; but the states (like the federal government) receive no returns from mineral extraction on federal lands within their borders, while they do have to deal with problems such as air and water pollution associated with the mining and mineral processing.

INTERNATIONAL RESOURCE DISPUTES

As noted in chapter 13, additional legal problems develop when it is unclear in whose jurisdiction valuable resources fall. International disputes over marine mineral rights have intensified as land-based reserves have been depleted and as improved technology has made it possible to contemplate developing underwater mineral deposits.

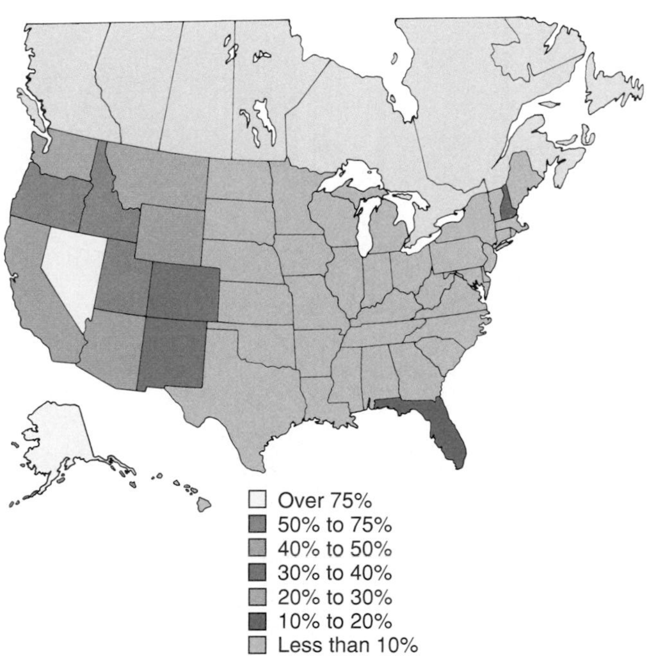

FIGURE 19.3 Federal land ownership in the United States. Of the federal lands, about 40% is under the control of the Bureau of Land Management, 30% is the national forest system, and about 15% each, the national park system and the national wildlife refuge system.
Source: 1997 National Resources Inventory, revised December 2000.

Traditionally, nations bordering the sea claimed territorial limits of 3 miles (5 kilometers) outward from their coastlines. Realization that valuable minerals might be found on the continental shelves led some individual nations to assert their rights to the whole extent of the continental shelf. That turns out to be a highly variable extent, even for individual countries. Off the east coast of the United States, for example, the continental shelf may extend beyond 150 kilometers (about 100 miles) offshore. The continental shelf off the west coast is less than one-tenth as wide. Nations bordered by narrow shelves quite naturally found this approach unfair. Some consideration was then given to equalizing claims by extending territorial limits out to 200 miles offshore. That would generally include all of the continental shelves and a considerable area of sea floor as well. The biggest beneficiaries of such a scheme would be countries that were well separated from other nations and had long expanses of coastline. Closely packed island nations, at the other extreme, might not realize any effective increase in jurisdiction over the old 3-mile limit. Landlocked nations, of course, would continue to have no seabed territorial claims at all.

Law of the Sea and Exclusive Economic Zones

In 1982, after eight years of intermittent negotiations collectively described as the "Third U.N. Conference on the Law of the Sea," an elaborate convention on the Law of the Sea emerged that attempted to bring some order out of the chaos. Resources are only one area addressed by the treaty. Territorial waters are extended to a 12-mile limit, and various navigational, fishing, and other rights are defined. **Exclusive Economic Zones** (EEZs) extending up to 200 nautical miles from the shorelines of coastal nations are established, within which those nations have exclusive rights to mineral-resource exploitation. These areas are not restricted to continental shelves only. Some deep-sea manganese nodules fall within Mexico's Exclusive Economic Zone; Saudi Arabia and the Sudan have the rights to the metal-rich muds on the floor of the Red Sea. Included in the EEZ off the west coast of the United States is a part of the East Pacific Rise spreading-ridge system, the Juan de Fuca rise, along which sulfide mineral deposits are actively forming. Areas of the ocean basins outside EEZs are under the jurisdiction of an International Seabed Authority. Exploitation of mineral resources in areas under the Authority's control requires payment of royalties to it, and the treaty includes some provision for financial and technological assistance to developing countries wishing to share in these resources.

The United States was one of only four nations (out of 141) to vote against the convention, at the time objecting principally to some of the provisions concerning deep-seabed mining. However, the United States has tended to abide by most of the treaty's provisions. In March 1983, President Reagan unilaterally proclaimed a U.S. Exclusive Economic Zone that extends to the 200-mile limit offshore. This move expands the undersea territory under U.S. jurisdiction to some 3.9 billion acres, more than 1½ times the land area of the United States and its territories (figure 19.4).

So far, exploration of this new domain has been minimal. However, certain kinds of mineral and fuel potential are already recognized. For example, off the Atlantic coast, there is potential for finding oil and gas deposits in the thick sediment deposited

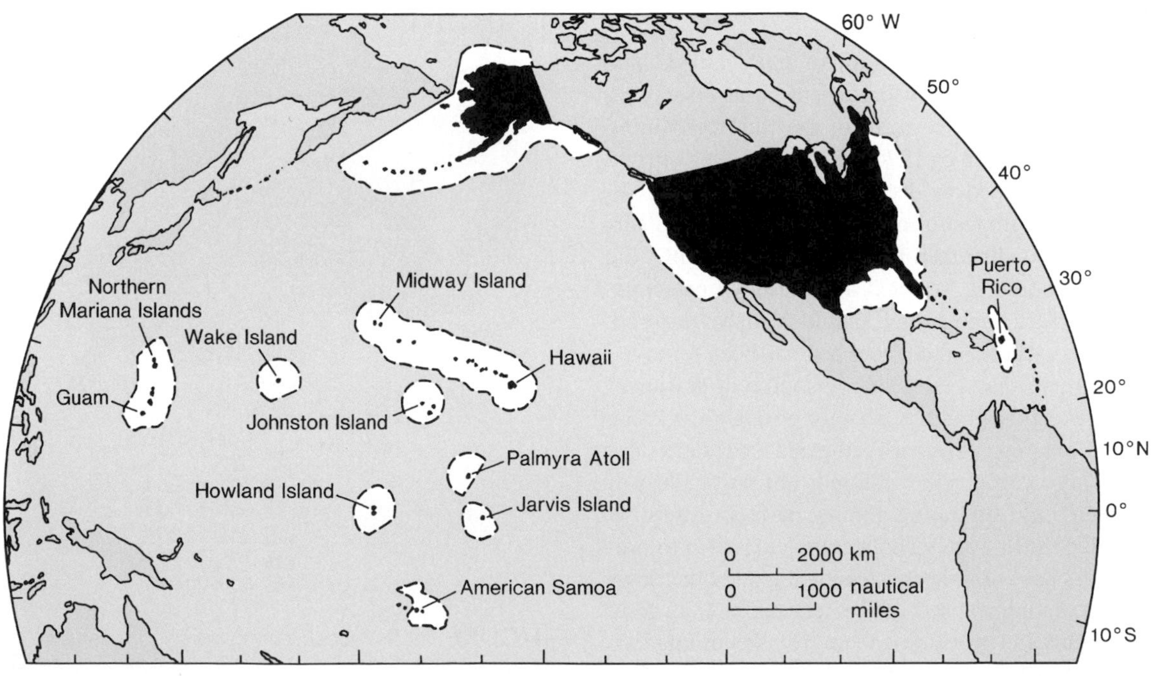

FIGURE 19.4 Exclusive Economic Zone of the United States.
Source: U.S. Geological Survey Annual Report 1983.

as the Atlantic Ocean formed—though there is also considerable concern about environmental impacts of oil and gas exploration on the continental shelf. "Black smoker" hydrothermal vents and related hydrothermal activity on the Pacific continental margin have produced sulfide and other metal deposits. Sulfides may also be associated with the Pacific island margins (consider the islands' volcanic origins), and on the flanks of those islands, in many places, mineral crusts rich in cobalt and manganese have been formed. The western and northern margins of Alaska are, like the Atlantic coast, thickly blanketed with sediment hosting known and potential petroleum deposits. Altogether, quite a variety of resources may be recoverable from the EEZ of the United States (figure 19.5).

A provision of the Convention on the Law of the Sea that has yet to take effect is prompting a flurry of seafloor mapping and seems likely to lead to some conflicts. The 200-mile limit does encompass the whole width of the continental shelf in many cases, but not all. And the provision in question allows a nation to claim rights beyond the 200-mile limit, extending its EEZ, if it can prove that the area in question represents a "natural prolongation" of its landmass. So, for instance, a nation with a 300-mile-wide continental shelf might claim the full width of that shelf. The Convention creates a Commission on the Limits of the Continental Shelf that will review such claims, which must be supported by detailed mapping of seafloor topography and sediment thickness. The first applications are due to the panel in 2009; in some places, claims out to 350 miles offshore might be made; and as with the 200-mile limit, there is plenty of potential for overlapping or conflicting claims by different nations bordering the same shelf. It will be interesting to see how these conflicts are resolved.

Antarctica

Antarctica has presented a jurisdictional problem not unlike that of the deep-sea floor. Prior to 1960, seven countries—Argentina, Australia, Chile, France, New Zealand, Norway, and the United Kingdom—had laid claim to portions of Antarctica, either on the grounds of their exploration efforts there or on the basis of geographic proximity (see figure 19.6). (Note the British claim based in part on the location of the Falkland Islands, site of a

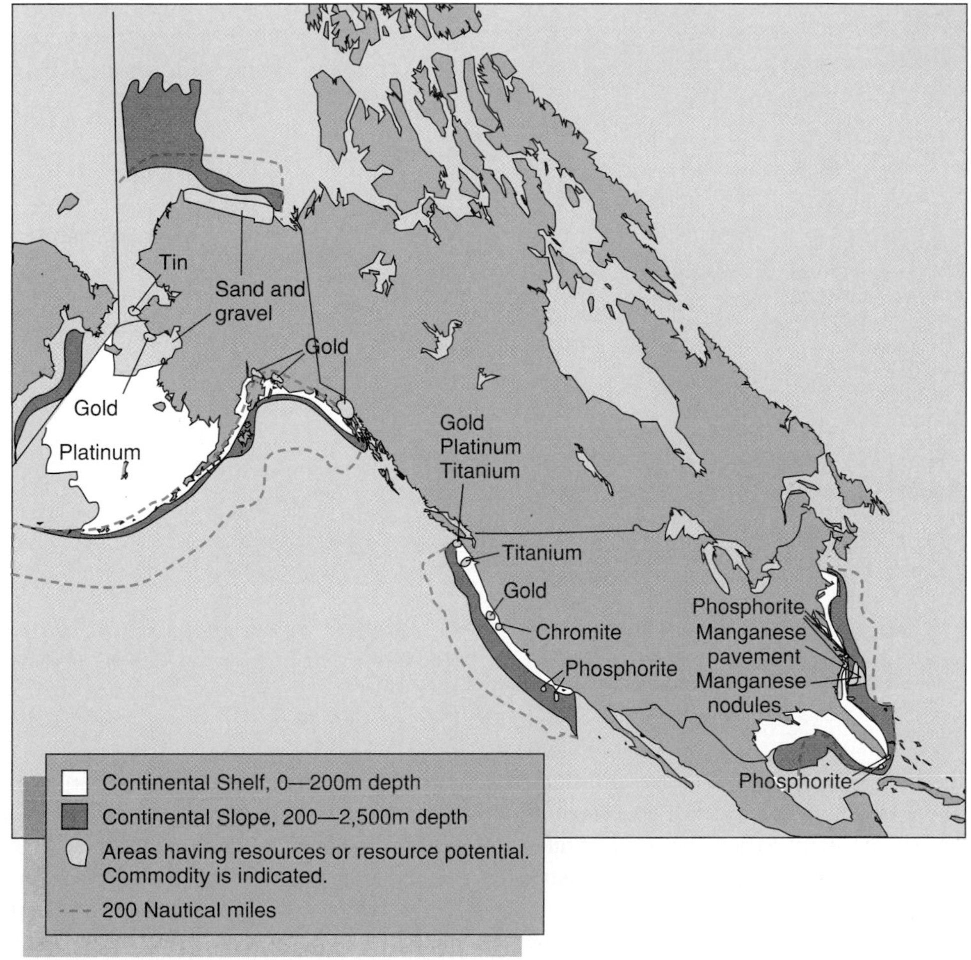

FIGURE 19.5 Possible mineral and rock resources in the North American EEZ of the United States.
Source: Adapted from B. A. McGregor and T. W. Oldfield, The Exclusive Economic Zone: An Exciting New Frontier, *U.S. Geological Survey, p. 7.*

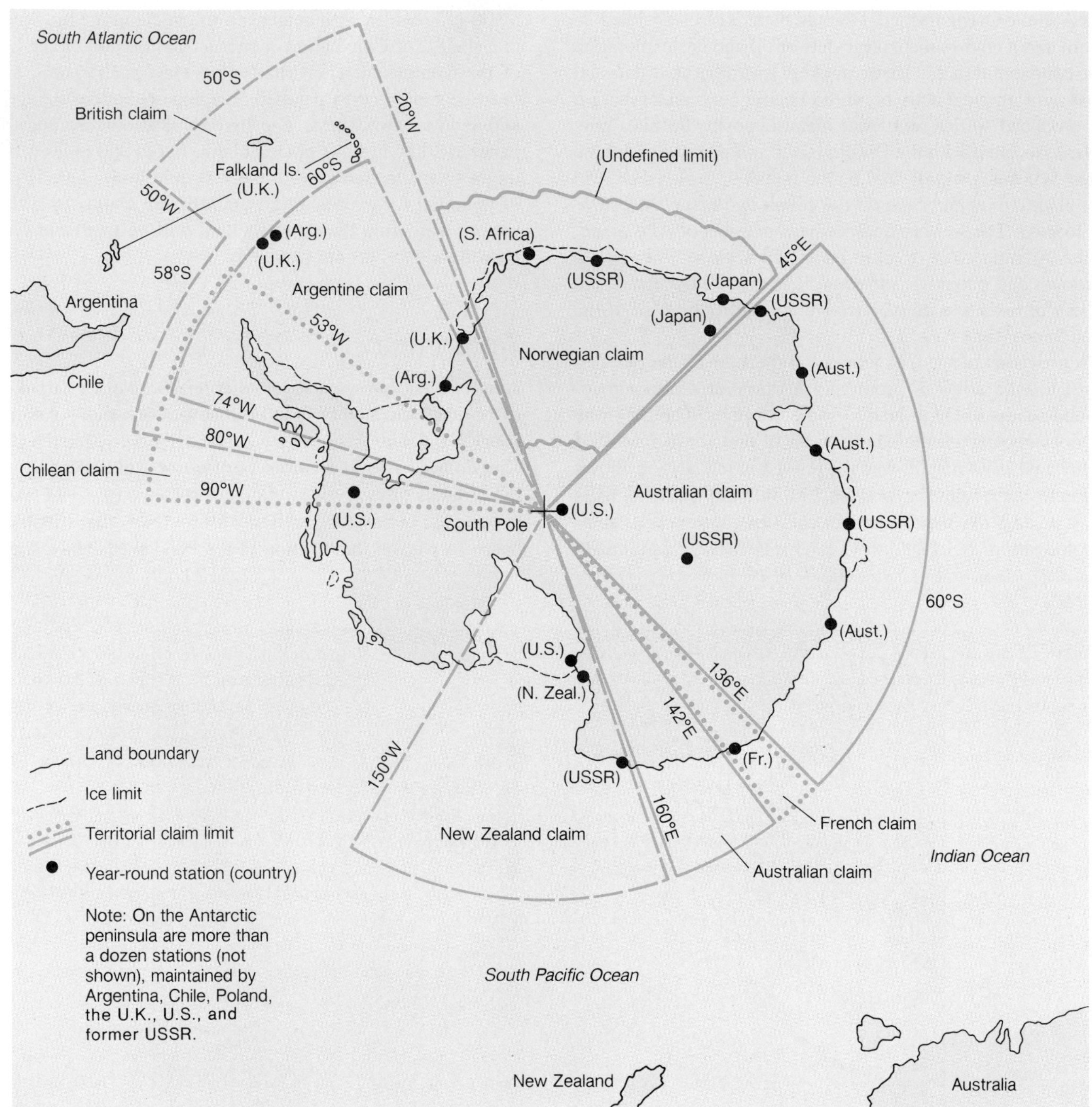

FIGURE 19.6 Conflicting land claims in Antarctica prior to the 1961 treaty. Note especially the overlapping claims.
Source: Data from C. Craddock, et al., *Geological Maps of Antarctica,* Antarctic Map Folio Series, *Folio 12, copyright 1970 by the American Geographical Society; and J. H. Zumberge, "Mineral Resources and Geopolitics in Antarctica,"* American Scientist 67, *p. 68–79, 1979.*

brief "war" between the U.K. and Argentina.) Several of these claims overlapped. None was recognized by most other nations, including five (Belgium, Japan, South Africa, the United States, and the former Soviet Union) that had themselves conducted considerable scientific research in Antarctica.

After a history of disagreement punctuated by occasional violence, these twelve nations signed a treaty in 1961 that set aside all territorial claims to Antarctica. Various additional points were agreed upon: (1) the whole continent should remain open; (2) military activities, weapons testing, and nuclear-waste disposal should be banned there; and (3) every effort should be made to preserve the distinctive Antarctic flora and fauna. Other nations subsequently signed also; there are now twenty nations that are "consultative parties" to the Antarctic treaty.

The treaty did not address the question of mineral resources, however, in part because the extent to which minerals and petroleum might occur there was not realized, and in part because their exploitation in Antarctica was then economically quite impractical anyway. Not until 1972 was the question of minerals even raised. The issue continued to be a source of contention for some years. Finally, in 1988, after six years of negotiations, the consultative parties (including the United States) produced a convention to regulate mineral-resource development in Antarctica: the Convention on the Regulation of Antarctic Mineral Resource Activities (CRAMRA). Under this convention, no such activity—prospecting, exploration, or development—could occur without prior environmental-impact assessment; significant adverse impacts would not be permitted. Actual mineral-resource development would require the unanimous consent of a supervisory commission composed mainly of representatives of the consultative parties.

In the years after CRAMRA's formulation, debate has continued. Sixteen of the twenty consultative parties must approve the treaty in order for it to take effect. Part of the reason for resistance to ratification was a concern that CRAMRA, by establishing guidelines for mineral-resource development, was effectively opening up the continent to such development (and possible resultant negative environmental impacts, safeguards notwithstanding). In the United States, this negative reaction to CRAMRA led to the passage of the Antarctic Protection Act of 1990. Briefly, the Act prohibits Antarctic mineral prospecting and development by U.S. citizens and companies and urges other nations to join in an indefinite ban on such activities and to consider banning them permanently. The 1991 Protocol on Environmental Protection to the Antarctic Treaty indeed involved an indefinite prohibition on activities related to development of geologic resources. In 2001–2002, the U.S. EPA issued a rule on environmental impact statements (described in a later section) for other nongovernmental activities—including tourism and scientific study—to be undertaken in Antarctica. It remains to be seen to what extent the consultative parties will follow this lead.

POLLUTION AND ITS CONTROL

At the outset of a discussion of laws regulating pollution, one might legitimately ask what legal basis or justification exists in the United States for antipollution legislation. Have we, in fact, any legal right to a clean environment? The primary Constitutional justification depends on an interpretation of the Ninth Amendment: "The enumeration in the Constitution of certain rights shall not be construed to deny or disparage others retained by the people." The argument made is that the right to a clean, healthful environment is one of those individual rights so basic as not to have required explicit protection under the Constitution.

This interpretation generally has not been upheld by the courts. However, those harmed or threatened by pollution have often sought relief through lawsuits alleging nuisance or negligence on the part of the polluter. Nuisance and negligence are varieties of *torts* (literally, "wrongs") that are violations of individual personal or property rights, punishable under civil law, and distinct from violations of basic public rights. The government has also exercised its authority to promote the general welfare by enacting legislation to restrict the spread of potentially toxic or harmful substances (including pollutants).

Water Pollution

Water pollution ought not to have been a significant problem in the United States since the start of the twentieth century, when the Refuse Act of 1899 prohibited dumping or discharging refuse into any body of navigable water. Because most streams drain into larger streams, which ultimately become navigable, or into large lakes, this act presumably banned the pollution of most lakes and streams. Unfortunately, the enforcement of the law left much to be desired.

In succeeding decades, stricter and more specific anti-water-pollution laws have been enacted. The Federal Water Pollution Control Act of 1956 focused particularly on municipal sewage-treatment facilities. The much broader Water Quality Improvement Act of 1970 and the subsequent amendments in the Clean Water Act of 1977 also address oil spills and various chemical pollutants, from both point and nonpoint sources. One of the objectives was to eliminate the discharge of pollutants into navigable waters of the United States by 1985—something that was, in theory, mandated eighty-six years earlier. The Clean Water Act and amendments nominally required all municipal sewage-treatment facilities to undertake secondary treatment by 1983, although practice lagged somewhat behind mandate; only about two-thirds of municipalities actually met that deadline. Another provision was that "best available technologies" to control pollutant releases by twenty-nine categories of industries were to be in place by July 1984. (This concept of "best available" treatment is discussed further later in the chapter.)

The fundamental underlying objective is "to restore and maintain the chemical, physical, and biological integrity of the Nation's waters." To this end, the Environmental Protection Agency (EPA) is directed to establish water-quality criteria for various types of water uses and to monitor compliance with water-quality standards (table 19.1). This is an increasingly complex task as the number of substances of concern grows.

Locally, significant improvements in surface-water quality resulting from improved pollution control have certainly occurred, but as is evident from chapter 17, more remains to be done. A major limitation of water-pollution legislation generally is that it treats surface waters only, although there is growing realization of the urgency of groundwater protection as well as of the links between surface and ground water. Recently, in fact, polluters have been prosecuted successfully, if indirectly, for groundwater pollution by toxic wastes after the ground water seeped into and contaminated associated surface waters.

The fate of water-quality legislation in the near future has become increasingly cloudy. Since 1991, authorization and

TABLE 19.1

EPA's Primary Safe Drinking-Water Standards for Selected Chemical Components

Contaminants	Health Effects	Max. Allowed (ppm)	Sources
Inorganic Chemicals (Total of 16 regulated)			
arsenic	skin, circulatory system damage	0.01	geological; pesticide runoff from orchards; smelter operations; glass and electronics production waste
barium	increased blood pressure	2	
cadmium	kidney damage	0.005	geological, mining and smelting
chromium	allergic dermatitis	0.1	
fluoride	skeletal damage	4	geological; additive to drinking water, toothpaste, foods processed with fluorinated water; discharge from fertilizer, aluminum processing
lead	central and peripheral nervous system damage, kidney effects, highly toxic to infants and pregnant women	0.015*	leaches from lead pipes and lead-based solder pipe joints
mercury (inorganic)	central nervous system disorders, kidney damage	0.002	used in manufacture of paint, paper, vinyl chloride; used in fungicides; geological
nitrate	methemoglobinemia ("blue-baby syndrome")	10	fertilizer, sewage, feedlots; geological
selenium	gastrointestinal effects, circulatory problems, hair or fingernail loss	0.05	geological, mining, petroleum refineries
Organic Chemicals (Total of 53 regulated)			
alachlor	eye, liver, kidney, spleen problems; anemia; increased cancer risk	0.002	herbicide runoff from row crops
benzene	anemia; increased cancer risk	0.005	factory discharge; leaching from gas storage tanks and landfills
chlordane	liver, nervous-system problems; cancer risk	0.002	residue of banned termiticide
2,4-D	liver, kidney, or adrenal-gland problems	0.07	herbicide runoff from row crops
dichloromethane	liver problems, cancer risk	0.005	discharge from drug, chemical factories
dioxin	reproductive difficulties; cancer risk	0.00000003	emission from waste incineration, combustion; discharge from chemical factories
diquat	cataracts	0.02	herbicide runoff
ethylbenzene	liver or kidney problems	0.7	discharge from petroleum refineries
methoxychlor	reproductive difficulties	0.04	runoff/leaching from insecticide use on fruits, vegetables, alfalfa, livestock
polychlorinated biphenyls (PCBs)	skin changes; thymus-gland problems; immune deficiencies; reproductive or nervous-system difficulties; increased cancer risk	0.0005	runoff from landfills; discharge of waste chemicals
styrene	liver, kidney, circulatory-system problems	0.1	discharge from rubber or plastic factories; leaching from landfills
1, 2, 4-trichlorobenzene	changes in adrenal glands	0.07	discharge from textile-finishing factories
trichloroethylene	liver problems; increased cancer risk	0.005	discharge from metal degreasing sites and other factories
vinyl chloride	increased cancer risk	0.002	leaching from PVC pipes; discharge from plastic factories

*Not a concentration limit, but an "action level" that requires municipalities to control the corrosiveness of their water; see lead sources for the reason.

Source: U.S. Environmental Protection Agency.

funding for many programs of the Clean Water Act and its later reauthorizations have lapsed. Repeated efforts since 1992 to enact further reauthorizations have failed. Recent versions of such bills, introduced in 1995 and 1996, included controversial wetlands-protection provisions that were opposed by some private-property owners. Reauthorization continues to be problematic—and enforcement of environmental-quality standards requires funding.

Air Pollution

In the United States, air pollution was not addressed at all on the federal level until 1955, when Congress allocated $5 million to study the problem. The Clean Air Act of 1963 first empowered agencies to undertake air-pollution-control efforts. It was amended in 1965 to allow national regulation of motor-vehicle emissions. The Air Quality Act of 1965 required the establishment of air-quality standards to be based on the known harmful effects of various air pollutants. An overriding objective of this act and a series of 1970 amendments was to "protect and enhance the quality of the Nation's air resources so as to promote the public health and welfare and the productive capacity of its population" (Shaw 1976, 340), though most of the details of how this was to be accomplished were unspecified. Nevertheless, substantial progress has been made with respect to many pollutants, as noted in chapter 18.

The Clean Air Act was reauthorized and amended in 1990. Various provisions of these Clean Air Act Amendments (CAAA) deal with many areas of air quality. The CAAA goes beyond the Montreal Protocol (chapter 18) in reducing production of chlorofluorocarbons and other compounds believed harmful to the ozone layer: It provides for reductions in industrial emissions of nearly 200 airborne pollutants; requires further reduction of vehicle emissions; sets standards for improved air quality with respect to ozone, carbon monoxide, and particulates in urban areas; and provides for reduction of SO_2 emissions from power plants. The SO_2/acid-rain provisions have an interesting economic component: Beginning in 1995, electricity-generating utility plants were assigned SO_2 allowances (one allowance = 1 ton SO_2 emission per year) based on power output, and allowances can be traded between plants or utility companies for cash. Utilities that reduce their emissions below their allowed output can sell their excess SO_2 allowances, which is an economic incentive to reduce pollution. In recent years, over ten million allowances per year have been traded, at an average of about $200 each.

The impact of the Phase I regulations on sulfate and acid rain and certain other pollutants was noted in chapter 18. Significant further air-quality improvements were projected to result. As will be seen later in the chapter, however, some air-pollutant emissions continue to rise, despite international agreements in principle to reduce them. There are also trade-offs: It was efforts to meet air-quality and emissions standards that caused the addition to gasoline of the MTBE that has now become a water-pollution concern.

Waste Disposal

The Solid Waste Disposal Act of 1965 was intended mainly to help state and local governments to dispose of municipal solid wastes. Gradually, the emphasis shifted as it was recognized that the most serious problem was hazardous or toxic wastes. The Resource Conservation and Recovery Act of 1976 includes provisions for assisting state and local governments in developing solid-waste management plans, but it also regulates waste-disposal facilities and the handling of hazardous waste and establishes cooperative efforts among governments at all levels to recover valuable materials and energy from solid waste. Hazardous wastes are defined and guidelines issued for their transportation and disposal.

The Environmental Protection Agency monitors compliance with the act's provisions. Development of waste-handling standards, issuance of permits to approved disposal facilities, and inspections and prosecutions for noncompliance with regulations are ongoing activities. However, there are tens of thousands of identified producers and transporters of hazardous waste in the United States subject to RCRA. In part because of this, the Environmental Protection Agency has delegated many of its responsibilities under the Resource Conservation and Recovery Act to the states.

Amendments to the Resource Conservation and Recovery Act in 1984 considerably expanded the scope of the original legislation. Approximately 100,000 firms generating only small amounts of hazardous waste (less than 1000 kilograms per month) are now subject to regulation under the act; originally, these small generators were exempt. Before 1984, a waste producer could request "delisting" as a generator of hazardous waste by demonstrating that its wastes no longer contained whatever hazardous chemicals (as identified by the EPA) had originally caused the operation to be listed; now, delisting requires that the waste producer demonstrate essentially that its wastes contain no identified hazardous chemicals at all. A whole new set of regulations has been imposed on landfills and underground storage tanks containing hazardous substances, which in many cases require expensive retrofitting of the disposal site to meet higher standards; landfill disposal of toxic liquid waste is banned. These and many other provisions clearly reflect growing recognition of the seriousness of the toxic-waste problem.

The U.S. Environmental Protection Agency

In 1970, the U.S. Environmental Protection Agency was established in conjunction with the reorganization of many federal agencies. Among its primary responsibilities were the establishment and enforcement of air- and water-quality standards. Under the Toxic Substances Control Act of 1976, the EPA was given the authority to require toxicity testing of all chemical substances entering the environment and to regulate them as necessary. The EPA also has a responsibility to encourage research into environmental quality and to develop a body of personnel to carry out environmental monitoring and improvement. As the largest and best-funded of the pollution-control agencies, it tends to have the most clout in the areas of regulation and enforcement. However, the scope of its responsibilities is broad, and expanding. The number and variety of situations with which the Environmental Protection Agency is expected to deal, restrictions imposed by limited budgets, and the varying emphases placed on different aspects of environmental quality by different EPA administrators are all factors that have contributed to limiting the agency's effectiveness. Legislation in the late 1980s and early 1990s tended to strengthen EPA's enforcement ability by establishing clear penalties for noncompliance with regulations. The CAAA, for example, allows EPA to impose penalties of up to $200,000 for violations and even to pursue criminal sanctions against serious, knowing violations of CAAA. It is also true that efforts of the U.S. EPA are often supplemented by activities of state EPAs. But enforcement also requires staff and budget for investigations, legal actions, and so on. Sharp recent budget cuts for EPA, reflecting reduced priority in Congress for environmental-protection activities in general, will certainly reduce the EPA's ability to carry out its various mandates and monitoring and enforcement of activities.

A Supreme Court decision in early 2007 has the potential to expand EPA's responsibilities under the CAAA still further. Since 1999, a dozen states and many cities had been urging EPA to add carbon dioxide to the list of air pollutants that it regulates, particularly in the context of automobile emissions. The arguments related to negative health effects of global warming. The EPA declined, claiming both lack of legal authority and lack of a definitive connection between greenhouse gases and global warming. The states, cities, and other concerned groups sued, and the issue made its way to the Supreme Court. In a split decision, the Court ruled that EPA *does* have statutory authority to regulate greenhouse-gas emissions such as CO_2 from automobiles—that given the harm these gases cause, they are clearly pollutants under the CAAA, and thus subject to such regulation. The ruling does not mean that EPA must immediately issue CO_2 emissions limits, but the pressure to begin moving in that direction has sharply increased, and the agency can no longer claim that regulation of greenhouse gases is, in effect, not in its job description.

Defining Limits of Pollution

Enforcement of antipollution regulations remains a major stumbling block, in part, because of the terms of antipollution laws themselves. Consider, for example, an objective of "zero pollutant discharge" by a certain year. At first inspection, "zero" might seem a well-defined quantity. However, from the standpoint of practical science, it is not. Water quality must be measured by instruments, and each instrument has its own detection limits for each chemical. No commonly used chemical-analysis techniques for air or water can detect the occasional atom or molecule of a pollutant. A given instrument might, for example, be able to detect manganese down to a concentration of 1 ppm, lead down to 500 ppb (0.5 ppm), mercury to 150 ppb. Discharges of these elements below the instrument's detection limit will register as "zero" when the actual concentrations in the wastewater might be hundreds of parts per billion. How close one must actually come to zero discharge, then, depends strongly on the sensitivity of the particular analytical techniques and instruments used.

The alternative approach is to specify a (detectable) maximum permissible concentration for pollutants discharged in wastewater or exhaust gases. But pollutants vary so widely in toxicity that one may be harmless at a concentration that for another is fatal. That means a need to specify *different* permissible maxima for individual chemicals, as in table 19.1. Given the number and variety of potentially harmful chemicals, a staggering amount of research would be needed to determine scientifically the appropriate safe upper limit for each. Thus, regulations are limited to a moderate number of toxins known to pose health risks. Even so, extensive data are needed to set "safe" exposure limits. With less comprehensive data, the Environmental Protection Agency and other agencies may set standards that later may be found to be too liberal or unnecessarily conservative. For some naturally occurring toxic elements, they have occasionally set standards stricter than the natural ambient air or water quality in an area. To meet the discharge standards, then, a company might be required to release wastewater cleaner than the water it took in! For instance, suppose the EPA guidelines specify a mercury concentration of 10 ppb or below in wastewater released. In an area with abundant mercury ore deposits, streams might naturally contain 100 ppb of mercury dissolved or leached from the ores. A company using that water, whether it added any mercury or not, could be required to remove some of the natural mercury to comply with the standards. A great deal of additional information, both about natural air and water quality and about specific harmful effects of individual chemicals, is needed before safe and realistic limits can be set on all pollutant discharges.

No treatment process is perfectly efficient, and emissions-control standards must necessarily recognize this fact. For instance, 1982 incinerator standards established under the Resource Conservation and Recovery Act specify 99.99% destruction and/or removal efficiency for certain specified toxic organic compounds. While this is a high level of destruction, if 1 million gallons of some such material is incinerated in a given facility in a year, the permitted 0.01% would amount to a total of 100 gallons of the toxic waste released.

Imprecision in terminology remains a concern. The CAAA, for example, includes such expressions as "lowest achievable level" (of emissions; in the context of CFC emissions) and "nonessential" applications of CFCs and other controlled ozone-threatening substances (use in such applications banned beginning in 1992; but what is an "essential" use of CFCs?). In some instances, the introduction of economic considerations further muddies the regulatory waters, both in the framing of laws and regulations and in their implementation; see "Cost-Benefit Analysis" later in the chapter.

International Initiatives

The latter half of the twentieth century saw a rapid rise in international treaties relating to the environment (figure 19.7). Some especially notable steps were taken toward international cooperation and collective action to address problems of global impact and concern during the 1990s. Some of these initiatives relate to problems that affect localized areas, widely distributed around the world—for example, the U.N. Convention to Combat Desertification, which came into force in late 1996. Others, however, recognize that all nations share some resources—the atmosphere especially—and no nation individually causes or can solve problems associated with pollution of such a common global resource (though to be fair, it is certainly true that some nations have historically contributed a disproportionately large share to that pollution).

One important outgrowth of the Earth Summit in Rio de Janeiro in 1992 was the United Nations Framework Convention on Climate Change. This Convention recognizes the role of atmospheric CO_2 and other greenhouse gases in global warming and other possible climatic consequences. It calls for nations to report on their releases of greenhouse gases and efforts to reduce these and to strive for sound management and conservation of greenhouse-gas "sinks" (plants, oceans), and it calls upon developed nations to assist developing nations in such efforts. This Convention came into force, following its ratification by over 50 nations, in March 1994. Its very existence is significant, as is the promptness of ratification. The Montreal Protocol on Substances that Deplete the Ozone Layer, discussed in chapter 18, is another example of such international commitments to pollution control for the common good.

These accords reflect a new concept in international law and diplomacy, the **precautionary principle.** Historically, nations' activities were not restricted or prohibited unless a direct causal link to some specific damage had been established. However, there are cases in which, by the time scientific certainty has been achieved, serious and perhaps irreversible damage may have been done. Under the precautionary principle, if there is reasonable likelihood that certain actions may result in serious harm, they may be restrained before great damage is demonstrated. So, given the known heat-trapping behavior of the greenhouse gases and documented increases in their atmospheric concentrations, nations agree to limit releases of greenhouse gases *before* substantial global warming or climate change has occurred; given the links between ozone depletion and increased UVB transmission, and between UV radiation and certain health risks such as skin cancer, they agree to stop releasing ozone-depleting compounds *before* significant negative health impacts have been documented.

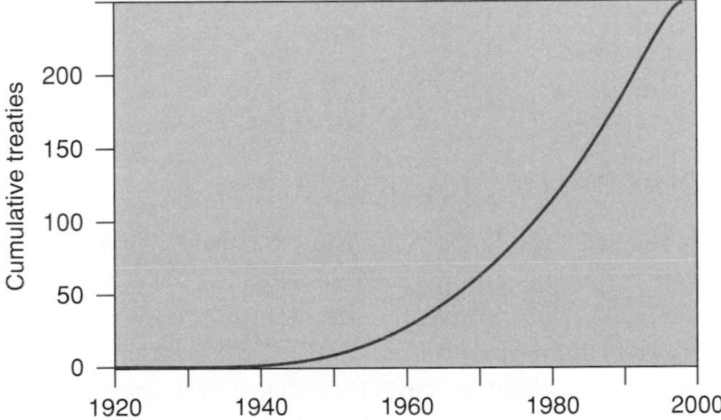

FIGURE 19.7 Increasing environmental awareness has led to increasing numbers of international environmental treaties.
Source: Data from U.N. Environment Programme.

This can be important to minimizing harm in cases in which there may be a significant time lag between exposure to a hazard and manifestation of the adverse health effects or those in which natural systems' response to changes in human inputs are slow. The CFCs are an example of the latter. Global emissions have been sharply reduced under the Montreal Protocol, but atmospheric concentrations have only leveled off (figure 19.8). This means that we will continue to be exposed to higher UVB levels for some decades. (A recent estimate by NASA scientists, considering both the long residence times of CFCs in the atmosphere and the fact that old air conditioners, refrigerators, etc. will continue to leak CFCs for some time, suggests that the ozone layer may not recover fully until 2065.) Still, the longer nations waited to take action, the further into the future the problem would have persisted.

It should also be noted that agreement in principle is one thing, implementation another, and enforcement yet another. Following the Rio summit, years of negotiations on specific targets and strategies produced the Kyoto Protocol of 1997. A major component of the Kyoto Protocol was a commitment by industrialized nations to reduce their greenhouse-gas emissions collectively to 5.2% below 1990 levels by 2008–2012. In practice, the focus must be on CO_2, overwhelmingly the principal greenhouse gas emitted in terms of quantity. But global CO_2 emissions, meanwhile, had been continuing to climb since Rio, so reductions relative to then-current CO_2 emissions would have had to be much more than 5% in many cases. The United States accounts for over 20% of global CO_2 emissions, and those emissions have been rising more rapidly than those of most other nations. For the United States, CO_2 emissions in 2004 were 20% above 1990 levels; to get to 5% below 1990 levels would have meant reducing CO_2 emissions by about 21% from 2004 levels. Concerns over the economic impact of such sharp reductions in CO_2 emissions led to the inclusion of certain flexibility provisions in the Kyoto Protocol: allowing nations to meet a portion of their commitment by enhancing CO_2 sinks (such as forests) rather than reducing emissions; permitting industrialized nations some credits for providing developing nations with low-emissions technologies; and providing for the trading of emissions credits so that one nation could acquire the right to the unused portion of the emission allowance of another.

Another principle of international environmental laws reflected in the Kyoto Protocol (as well as the Montreal Protocol and others) is the **common but differentiated responsibility of states.** (*States,* in U.N. usage, essentially means

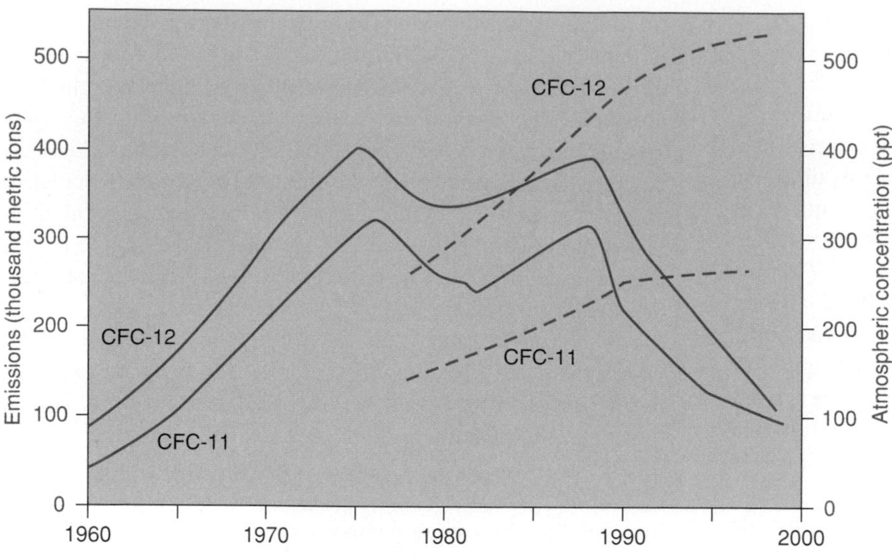

FIGURE 19.8 Though global CFC emissions have been sharply curtailed, atmospheric concentrations are just now leveling off. (Dip in 1980s is due mainly to economic conditions.) CFC-11 and CFC-12 are two of the more widely used CFCs, with chemical formulas $CFCl_3$ and CCl_2F_2, respectively.
Source: Data from U.S. Council on Environmental Quality.

nations.) The idea is that we all share the global environment, and the responsibility to protect its quality, but some nations contribute more to pollution and other negative impacts (see, for example, figure 19.9), while some have more financial resources to develop less-harmful alternatives than other nations have. Those nations that create more of the problem and can better afford the costs of addressing it are expected to do more toward a solution. So in principle, industrialized nations that produce more CO_2 and other greenhouse gases (or CFCs, or other pollutants) and can more readily pay to develop less-polluting alternatives are expected to reduce their emissions more sharply, and sooner. (Indeed, developing nations such as India and China are not subject to emissions limits under the Kyoto Protocol, and their greenhouse-gas emissions have been rising sharply since 1990—China's by 50%, India's by 70%!) Consequently, those developed nations on whom the greater financial and compliance burdens fall are not always eager to embrace what other nations see as their fair share of responsibility. This contributes to the difficulty of achieving consensus on the details of such international agreements.

In late 2000, the parties to the Kyoto Protocol met again to try to reach a binding agreement on implementation. A major point of contention involved objections by negotiators from the European Union to U.S. attempts to use CO_2 sinks for a large fraction of the target emissions reductions and, particularly, to count *existing* sinks (forestland and rangeland), rather than new or enhanced sinks, toward that total. Another significant set of issues involved questions of whether existing low-emissions technologies such as hydropower and nuclear power should be allowed to count toward "clean development mechanism" credits. After weeks of discussion, the negotiations broke up with no agreement reached.

Not until late 2004, with acceptance by the Russian Federation, did the Kyoto Protocol even achieve the critical mass of signatories to come into force. (Some observers note that by then, thanks to a depressed economy, the Russian Federation's greenhouse-gas emissions had already dropped below its Kyoto target, so not only did it have no sacrifices to make; it had excess emissions credits to sell!) By then, looking at emissions data, it was becoming increasingly evident that the Kyoto targets were unlikely to be reached by around 2010 as originally hoped, even if all of the signatory nations held to their commitments in principle. It is simply proving too difficult, economically and technically, to achieve the desired reductions fast enough in many developed nations. The United States, meanwhile, has opted out of Kyoto altogether, at least for the time being, and this will further limit achievement of reduced greenhouse-gas levels globally, though U.S. companies operating in countries bound by the Kyoto Protocol are affected there. The unrestrained growth in emissions in the developing countries is increasingly a significant factor, and a global concern. Finally, it should be noted that, like many international environmental accords, the Kyoto Protocol does not actually include enforcement provisions or penalties for failing to meet one's obligations, which certainly reduces its practical effectiveness.

COST-BENEFIT ANALYSIS

Problems of Quantification

Antipollution laws that specify using the "best available (treatment) technology economically achievable," or contain some similar clause that introduces the idea of economic feasibility, result in additional conflicts and enforcement problems. As noted in earlier chapters, costs for exhaust or wastewater treatment rise exponentially as cleaner and cleaner air and water are achieved. These costs can be defined and calculated with reasonable precision for a given mine, manufacturing plant,

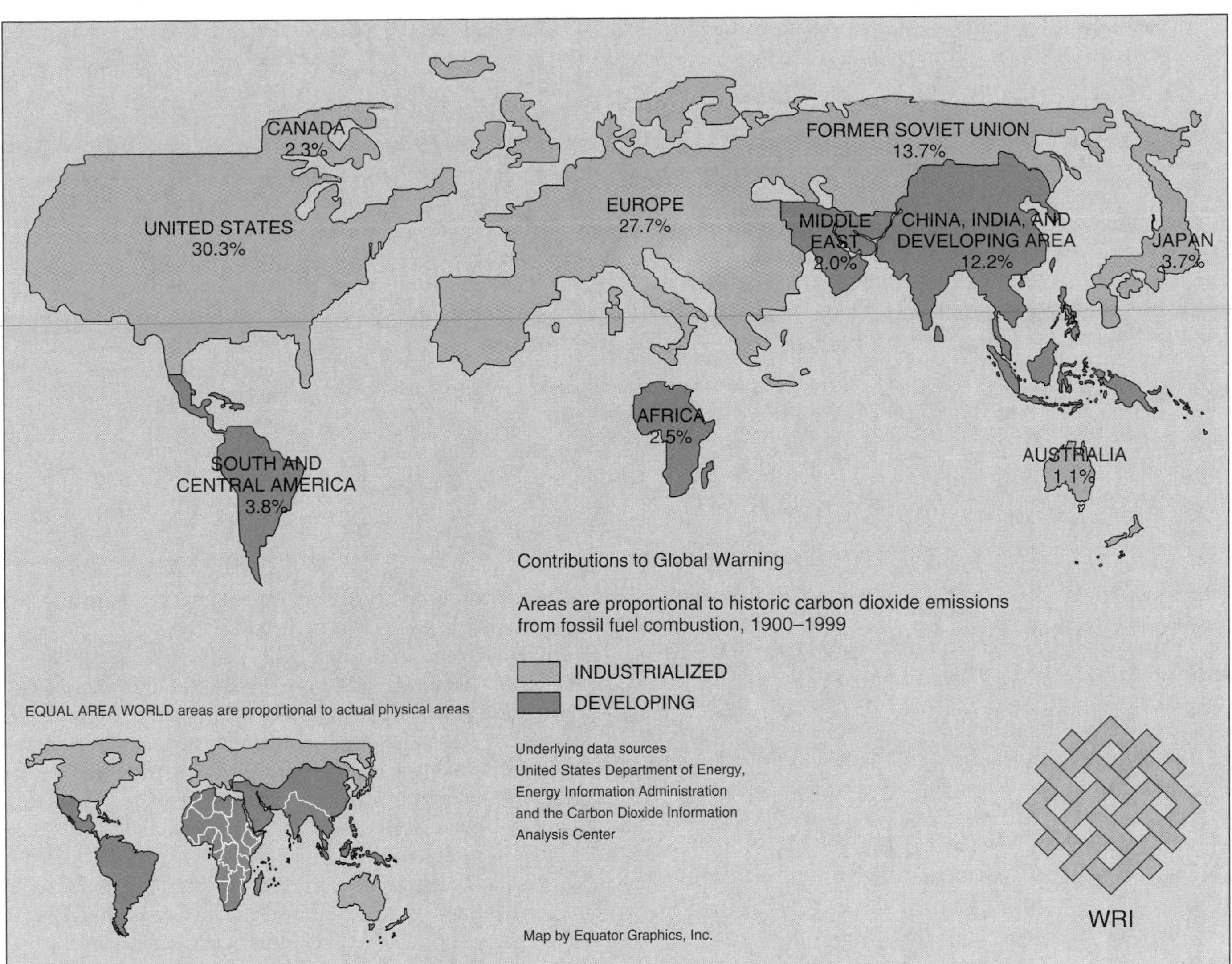

FIGURE 19.9 Inset map is a "normal," equal-area world geographic map; larger map shows sizes of countries in proportion to their contributions to global CO_2 emissions over the twentieth century.
After World Resources Institute "Contributions to Global Warming" map.

automobile, or whatever. Balanced against these cleanup costs are the adverse health effects or general environmental degradation caused by the pollutants. The latter are far harder to quantify and, sometimes, even to prove.

For example, the harmful effects of acid rain are recognized, and increased acidity of precipitation results, in part, from increased sulfur gas emissions. It is not possible, however, to say that, for each ton of sulfur dioxide released, x billion dollars worth of damage will be done to limestone and marble structures from acid rain, y trees will have their growth stunted, or z fish will die in an acidified lake. Even if such precise analysis were possible, what is the dollar value of a tree, a fish, or a statue? Conversely, scientists cannot say exactly how much environmental benefit would be derived from closing down a coal-fired power plant releasing large quantities of sulfur, and many people could suffer financial and practical hardship from the loss of that facility.

Similar debates arise in many areas. Allowing development on a barrier island may increase tourism and expand the local tax base, helping schools and allowing municipal improvements—but at what risk to the lives and property of those who would live on the vulnerable land, and cost to taxpayers providing disaster assistance? The dangers of increased atmospheric CO_2 are widely accepted, and fossil-fuel extraction may cause environmental disturbances of various sorts—but our energy consumption continues to grow; clearly we tend to decide, implicitly at least, that the benefits outweigh the risks, perhaps, again, because the benefits are readily perceived, while the risks are very hard to quantify (or adequate alternatives have not been found).

In the realm of pollution control, companies frequently contest strict regulation of their pollutant discharges on the grounds that meeting the standards is economically impossible and/or unreasonable. Each individual, each judge, and each jury has somewhat different ideas—of what is "reasonable"; of the value of clean air and water; of the importance of preserving wildlife, vegetation, natural topography, or geology; and of what constitutes an acceptably low health risk (assuming that that health risk can be quantified fairly precisely in the first place; see figure 19.10). Enforcement thus is very uneven and inconsistent. Virtually everyone agrees that, in principle, it is desirable to maintain a high-quality environment. But just what does that mean? What should it cost? And by whom should those costs be borne?

Cost-Benefit Analysis and the Federal Government

A month after assuming office in 1981, President Reagan issued Executive Order 12291, which decrees, in part, that a (pollution-control) regulation may be put forth (by an agency such as EPA) only if the potential benefits to society outweigh the potential costs. A Regulatory Impact Analysis must be performed on any proposed new "major" regulation, which is, in turn, defined as one that is likely to result in either (1) an impact on the economy of $100 million or more per year; (2) a major cost increase to consumers, individual industries, governmental agencies, or geographic regions; or (3) significant adverse effects on competition, employment, investment, productivity, or the ability of U.S.-based corporations to compete with foreign-based concerns.

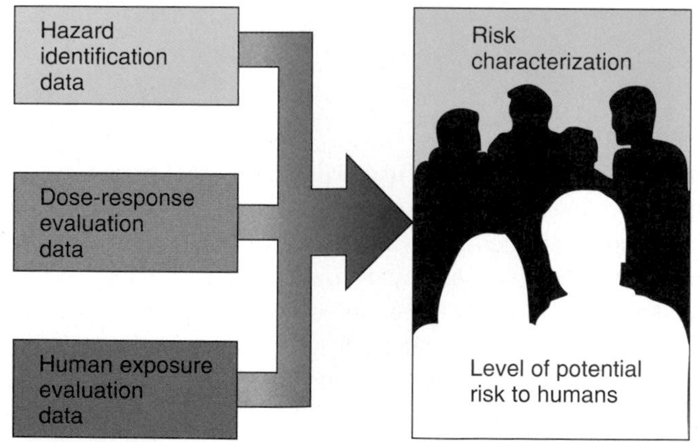

FIGURE 19.10 Components of EPA's risk-assessment process for potential health hazards. "Dose-response" refers to the relationship between exposure to or consumption of a substance and the effect on humans or other organisms, as described at the beginning of Section Five.

Source: After D. E. Pattor, "The ABCs of Risk Assessment," EPA Journal 19, *volume 1, 1993, U.S. Environmental Protection Agency.*

This represented a major departure from previous policy. The Clean Water Act orders the Environmental Protection Agency to be sensitive to costs in establishing standards, but it does not require a cost-benefit analysis; the Clean Air Act prohibits considering costs in setting health standards. Cost-benefit analysis is still not the decisive factor in regulation, but it is at least taken into account.

Cost-benefit analysis has been a key part of the debate surrounding the president's proposed "Clear Skies Act," which would cap emissions of sulfur and nitrogen oxides and of mercury by electricity-generating plants. The EPA, asked to do such an analysis in 2003, projected the costs of compliance as rising to $6.3 billion per year. Set against those costs was an impressive array of benefits, including: reduction in acidification of lakes and streams in the United States from acid rain; reduced mercury in the environment; over $3 billion per year worth of benefit from improved visibility in national parks and wilderness areas; an estimated 30,000 fewer visits to hospitals and emergency rooms, and 12.5 million fewer days with respiratory symptoms (meaning also, fewer days of school or work missed) per year by 2020; and, depending on the model used for the calculation, from 8400 premature deaths prevented, with $21 billion per year in health benefits, to 14,000 premature deaths prevented, with $110 billion per year in health benefits. Certainly, in this case, the projected benefits far outweigh the anticipated costs. However, it is necessarily also true that it is far easier to calculate the approximate cost of specified effluent reductions from existing power plants using known technology than to be equally precise about, for instance, the reductions in days of respiratory symptoms Americans would suffer, or in premature deaths, to say nothing of assigning dollar values to such benefits. And this has been a factor in ensuing years of congressional debate on this measure, which has not yet passed as this is being written.

LAWS RELATING TO GEOLOGIC HAZARDS

Laws or zoning ordinances restricting construction or establishing standards for construction in areas of known geologic hazards are a more modern development than antipollution legislation. Their typical objective is to protect persons from injury or loss through geologic processes, even when the individuals themselves may be unaware of the existence or magnitude of the risk. While such an objective sounds benign and practical enough, such laws may be vigorously opposed (see especially chapters 4 and 6). Opposition does not come only from real-estate investors who wish to maximize profits from land development in areas of questionable geologic safety. Individual homeowners also often oppose these laws designed to protect them, perhaps fearing the costs imposed by compliance with the laws. For example, engineering surveys may be required before construction can begin, or special building codes may have to be followed in places vulnerable to earthquakes or floods. People may also feel that it is their right to live where they like, accepting any natural risks. And some, of course, simply do not believe that whatever-it-is could possibly happen to *them*.

Occasionally, lawsuits have been filed in opposition to zoning or land-use restrictions, on the grounds that such restrictions amount to a "taking"—depriving the property owner of the right to use that property freely—without compensation. Such action is prohibited by the Fifth Amendment to the Constitution. However, the courts have generally held that such restrictions are, instead, simply the government's exercise of its police power on behalf of the public good and that compensation is thus unwarranted.

Construction Controls

Restrictive building codes can be strikingly successful in reducing damage or loss of life. After a major earthquake near Long Beach, California, in 1933, the California legislature passed a regulation known as the Field Act, which created improved construction standards for school buildings. Many of those standards are still being followed. Their effectiveness was demonstrated in subsequent earthquakes, notably the major one in San Fernando in 1971. It was noted afterward that damage to school buildings from the earthquake was almost wholly limited either to schools built prior to 1933 or to those later buildings whose engineers had been permitted to deviate from strict compliance with the Field Act. The 1995 Kobe earthquake further illustrated the value of earthquake-resistant design (figure 19.11).

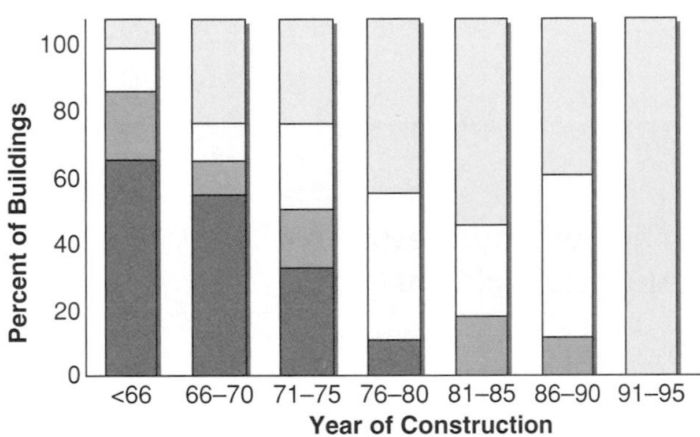

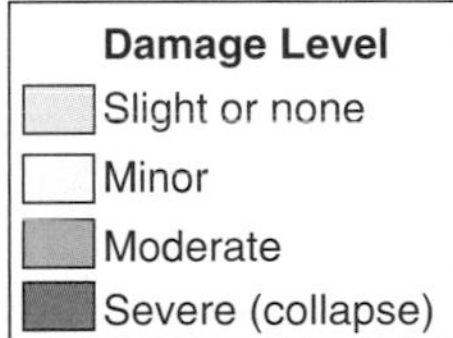

FIGURE 19.11 Effects of improving building codes for earthquake resistance are clearly shown in the aftermath of the 1995 Kobe earthquake: Newer buildings suffered far less damage than older ones.

The problem of preexisting structures remains a major problem rarely addressed even by "good" building codes aimed at hazard reduction, as was illustrated yet again in the 1989 Loma Prieta earthquake and the 1994 Northridge quake. There had been talk of "retrofitting" some older buildings, freeways, and bridges to improve safety in accordance with newer design specifications, but due to cost and inertia, many modifications went unmade, and structural failures resulted. Retrofitting of privately owned buildings tends to be even rarer (figure 19.12).

Perhaps because the state is subject to so many geologic hazards, California and its municipalities have been among the leaders in passage of legislation designed to minimize the damage from geologic hazards. One successful effort in this

A

B

FIGURE 19.12 Hazard-mitigation legislation rarely requires privately owned buildings to be modified for greater resistance; the results can be unfortunate. (A) The first story of this two-story house completely collapsed in the Loma Prieta earthquake. (B) This apartment complex was severely damaged in the 1994 Northridge quake, in part because the carport on the ground level was not as strongly built as the living quarters above.

(A) Photograph by D. Schwartz, (B) photograph by E. V. Leyendecker, both courtesy of U.S. Geological Survey.

regard has been the imposition of much-more-stringent regulations governing construction on potentially unstable, landslide-prone hillsides near Los Angeles. Over a ten-year period of rapid development (1952–1962), increasingly strict requirements for slope and soil stability studies, site grading, and construction engineering were developed. The resulting reduction in the rate of landslide damage has been dramatic (table 19.2). Although the costs of structural damage to those homes actually damaged by landslides or soil failure has continued to be high (as the result of escalating property values), the number and percentage of individual units damaged have both dropped sharply, especially since 1963. Plainly, at least where the geologic hazards are moderate, the risk of damage from the hazards can be minimized by taking them into account and "designing around" them. This is a large part of engineering geology (see chapter 20).

Other Responses to Earthquake Hazards

Not all hazard-mitigation legislation is equally effective, as evidenced by a few California examples prompted by the 1971 San Fernando earthquake, described briefly below. Each of these bills is well-intentioned and represents a positive step toward reducing earthquake damage. Collectively, however, they also illustrate a number of weaknesses that limit their effectiveness.

The Seismic Safety Element Bill (1971) requires that communities take seismic and related hazards (ground shaking, tsunamis, and so forth) into account in planning and regulating development. This is plainly sensible in a high-seismic-risk area. But the bill does not prohibit or regulate construction explicitly; public officials are left to decide what constitutes acceptable risk in siting and building. The qualifications of the person(s) evaluating the seismic risks are not stipulated either, and there is no penalty if the community actually ignores the bill altogether.

The Dam Safety Bill (1972), despite its name, has nothing directly to do with the engineering adequacy of dams. It requires dam owners to prepare maps showing the areas that would be flooded in the event of dam failure, and it requires local authorities to prepare evacuation procedures for those areas threatened. Building in the areas at risk is not prohibited, and those living there may not necessarily be made aware of the dangers.

The Alquist-Priolo Geologic Hazards Act (1972; amended 1975) is a more specific and detailed bill. Under its provisions, the state geologist identifies "Special Studies Zones" along active faults. Proposed construction in these zones requires review by local authorities, who might, for instance, require that buildings not be placed within a certain distance (25, 50, or 100 meters, perhaps) of the fault trace proper. The principal objective is to prevent damage through fault offset. (Of course, in the case of a major fault such as the San Andreas, which can cause substantial damage tens or hundreds of kilometers from the epicenter of a large earthquake, it would be of little use to worry about whether one's house was 10 or 100 meters from the fault line, for ground shaking might well level the building even if it was not torn apart directly by the rupture of the fault zone.) Local officials are expected to consider the size of earthquakes and the extent of damage anticipated along each particular fault in deciding where or whether to allow building. Engineering and site-evaluation studies can be costly, and this prompted the 1975 amendment that *exempted* individual single-family homes. (Housing developments, apartments, and commercial buildings are still covered.) Once again, economic considerations vie with scientific ones in selecting a course of action.

TABLE 19.2

Comparison of Landslide Damage with and without Restrictive Construction Guidelines in Los Angeles, California

	Prior to 1952	1952–1962	1962–1967
controls pertaining to landslide hazards	none	limited grading codes in place	much stricter requirements for engineering and soil studies; modern grading procedures required
sites constructed (approx.)	10,000	27,000	11,000
sites damaged (%)	1040 (10.4%)	350 (1.3%)	17 (0.2%)
total damage to residences (approx.)	$3,300,000	$2,800,000	$80,000
average damage per site built	$330	$100	$7
average cost per site damaged	$3200	$7900	$4700

Source: Data from J. F. Slosson, *The Role of Engineering Geology in Urban Planning,* Colorado Geological Survey Special Publication 1, 1969.

Flood Hazards, Flood Insurance

When geologic catastrophes occur, federal disaster-relief funds are often part of the rescue/recovery/rebuilding operations. In other words, all taxpayers pay for the damage suffered by those who, through ignorance or by choice, have been living in areas of high geologic risks. This has struck some as inequitable and was part of the motivation behind the Flood Insurance Act (1968) and its successor, the Federal Flood Disaster Protection Act (1973). These measures provide for federally subsidized flood insurance for property owners in identified flood-hazard areas, whether in stream floodplains or in flood-prone coastal regions. Those most at risk, then, pay for insurance against their possible flood losses. The idea is that, in time, the flood insurance will replace after-the-fact disaster assistance in flood-prone areas. To "encourage" property owners to purchase the insurance, the legislation provides that it be required of those obtaining federally insured mortgages or mortgages through federally affiliated banks and lending institutions. For a community to remain eligible for the insurance program, it must, in turn, enact strict floodplain-zoning regulations. Initially, all proposed new construction in the floodplain must be carefully evaluated and structures designed so as to minimize potential flood damage. Later, after the Department of Housing and Urban Development has provided detailed floodplain maps, still stricter provisions are enacted that require floodproofing or elevation above the 100-year flood level of new structures in the floodplain. The community may wish to include additional provisions, such as one requiring that any water-storage volume within the 100-year floodplain eliminated or filled in by the emplacement of new structures be compensated for by the creation of at least an equal volume of additional floodwater storage elsewhere.

The requisite flood-hazard maps can be slow in coming. As mentioned in chapter 6, detailed records collected over long periods, and careful analysis, are required to construct accurate maps indicating precisely the areas to be affected by 50-year, 100-year, or 200-year floods. In many communities, detailed maps will not be available for years, and, meanwhile, development goes on. The laws do not completely ban further building in the flood-prone area, and as already noted, if more buildings—"floodproof" or not—are placed in a floodplain, their very presence may in some measure increase flood stages.

An unintended and unfortunate side effect of the availability of subsidized flood insurance has been to encourage people to rebuild, sometimes several times, in severely flood-prone areas and, in a sense, the insurance has encouraged continued development in such areas. The costs to the government can then far exceed the premiums collected. In some unstable coastal zones, this has happened to such an extent that Congress has voted to remove the most severely threatened areas from the program (figure 19.13; see also chapter 7). Following the devastation of Hurricane Katrina, in fact, there has been discussion of eliminating federal subsidies for flood insurance everywhere.

Problems with Geologic-Hazard Mitigation Laws

A pervasive weakness in all kinds of geologic-hazard mitigation laws, including all of the examples discussed in this section, is the fact that they typically apply only to *new* structures. Only new schools in California need conform to Field Act building codes; only new hillside homes need be built using modern slope-stabilization methods; only new federally insured mortgages on flood-prone properties mandate that the property owner carry flood insurance; and so on. Where considerable development has preceded the legislation, those affected by the new law in the near term may be a very small percentage of those at risk: In the 1993 Mississippi River basin floods, many thousands of victims were uninsured, aware of neither the need for nor the availability of flood insurance. Where new construction is banned in very-high-risk areas, older structures nevertheless persist. A small move toward eliminating past poor practice would be a regulation requiring that, if an older structure in a geologically high-risk area were destroyed—whether by geologic processes or another means, such as fire—it could be rebuilt only if the new structure conformed to the newer laws for floodproofing, fault-trace setback, slope grading, and so forth. Even modest proposals of this kind, however, commonly meet with vocal citizen opposition.

Thus, laws designed to reduce the risk of damage from geologic hazards have several problems in common. The basic scientific information on which to base sensible regulations may be lacking. Laws may be poorly written, failing to specify fully what must be done and by whom, or they may be weakened by exceptions or omissions that exempt many from their

FIGURE 19.13 Development on barrier islands is especially vulnerable to storm damage. Note low relief and seawall backing minimal beach; person descending stairs (foreground) illustrates that a storm surge of only a few meters will overwash this island. Sandy Key Beach, Pinellas County, Florida.

Photograph by Tony Santana, courtesy U.S. Army Corps of Engineers.

Box 19.1 Floodplain? What Floodplain?

The Werners were at the closing for their new townhouse, signing what seemed an endless stack of checks. Loan fees, title search, property taxes, flood insurance …

"Flood insurance? We didn't ask for any flood insurance!" exclaimed Mr. Werner in some surprise.

"You didn't have to," replied Ms. Lupanu, the loan officer. "Actually, you have no choice. That was in your mortgage contract."

"But why? We're two lots over from the creek!"

"Yes, but you're still in the floodplain. That's the law; we didn't have any choice but to require the flood insurance, in fact."

"I don't understand," frowned Mr. Werner. "I saw the developer's maps. He *assured* me our unit was above the flood level."

"Well, I looked into that, and you're partly right," observed Ms. Lupanu. "It seems that the first floor is above hundred-year-flood level, and so is the ground surface right around the building. However, the basement isn't, and it's partly finished, potential living space. So technically, you *are* in the floodplain, and we have to require the insurance. Look at it this way: it *is* for your own good, to protect your investment."

With a resigned mumble, Mr. Werner signed another check …

With the federal flood-insurance program, federally insured lenders (virtually all banks and savings and loan institutions, for a start) must require flood insurance when they issue loans on properties identified as being in the flood hazard area under the program. Ms. Lupanu's bank has detailed maps of the county, by which it identifies properties affected. Prospective homeowners, too, can obtain information from a variety of sources. Such maps (figure 1) can be useful in understanding both the limits of the floodplain and the anticipated water depth in the event of flooding. The government's objective in making the insurance mandatory is not simply to compel homeowners to protect their own property or to make them aware that they are buying in a floodplain, but to broaden the premium base that provides the funds to pay off claims when flooding does occur. In other words, you can buy properties located in the floodplain, but you will pay for insurance against potential damage from the identified hazard.

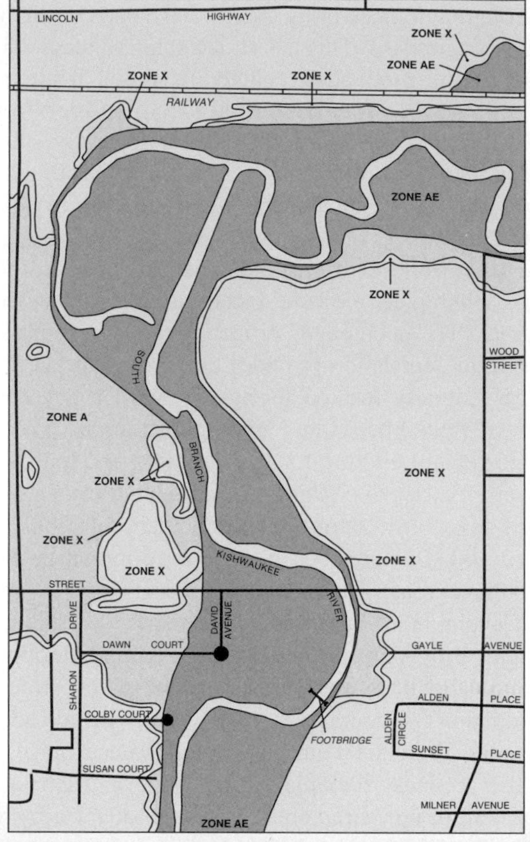

FIGURE 1 Sample section of flood insurance rate map. Zone A is estimated 100-year floodplain. In zone AE, height of 100-year flood is determined by detailed analysis of records. Zone X includes a variety of lower-risk situations: areas where the 100-year flood depth is estimated at less than 1 foot (narrow zone adjacent to zone A), areas protected by levees, and areas outside the 100-year but in the 500-year floodplain. Note cut-off meander, the result of earlier U.S. Army Corps of Engineers flood-hazard-reduction efforts.

provisions. A major omission, and a very common one, is that existing structures in areas at risk may continue unaffected (figure 19.14).

THE NATIONAL ENVIRONMENTAL POLICY ACT (1969)

The (U.S.) National Environmental Policy Act (NEPA) established environmental protection as an important national priority and provided for the creation of the Council on Environmental Quality in the Executive Office of the President. The **environmental impact statement** (EIS) is the most visible outgrowth of the NEPA. In this section, we will consider briefly what goes into an EIS, what the point is, and how well the legislation's objectives are being met.

The NEPA actually pertains only to federal agencies and their actions. Whenever a federal agency proposes legislation or "other major Federal action" (which can include anything from policy changes to construction projects that are wholly or partially supported by federal funds) that can significantly affect the quality of the human environment, an EIS must be prepared. Many states have adopted similar legislation related to projects involving funding or approval by state agencies, so the scope of the environmental impact statement process has been considerably broadened. The preparation of environmental impact statements should ensure that, before an agency acts, it considers as fully as possible the potential environmental

FIGURE 19.14 Landslide in Potrero Canyon, Pacific Palisades area, California, after 7 inches of rain in January 1956. Later building codes made newer structures safer but left old ones in areas of known risk.
Photograph by J. T. McGill, U.S. Geological Survey.

ramifications of its action, in order to make the best possible decisions. The discussion that follows is restricted to federally mandated EISs.

The NEPA specifies that an environmental impact statement should include

1. A description of the proposed action, its purpose, and why it is needed
2. A discussion of various alternatives (including the proposed action)
3. An indication of the environment to be affected and the environmental consequences anticipated
4. Lists of preparers of the statement and those agencies, organizations, or persons to whom copies of the statement are being sent

Additional supplementary material may also be included. The statement is expected to "rigorously explore and objectively evaluate all reasonable alternatives" (an analysis that presumably will ultimately supply appropriate justification for the particular course of action preferred by the agency). Possible environmental consequences might include not only those that are in some sense geologic (for example, altering natural processes like stream flow or runoff, causing air or water pollution, inevitable consumption of energy or other resources), but also biological ones (loss of habitat, destruction of organisms) and social ones (affecting urban quality, employment, historic or cultural resources). Cost-benefit analyses of various alternatives may be included if the agency is taking them into account.

After a draft environmental impact statement is prepared, a variety of others, ranging from other federal agencies to the general public, are invited to comment. In fact, comments should be actively solicited from all persons or organizations that are particularly likely to be interested or affected. The agency preparing the EIS is then expected to respond to those comments, which might mean modifying the proposed action or the analysis in the EIS or considering additional alternatives.

Over the first decade following passage of the NEPA, an average of more than 1000 environmental impact statements each year were prepared by federal agencies. Lawsuits were filed contesting about 10% of this number of projects. In many such cases, it was alleged that the EIS was inadequate, incomplete, or inaccurate. In other cases, the charge was that an EIS ought to have been prepared but was not (the responsible agency felt that no significant environmental impact was involved). Close to half the lawsuits involved citizen or environmental groups as plaintiffs. The number of environmental impact statements filed annually began to decline in the late 1970s (figure 19.15), but the number of lawsuits rose, peaking at 157 in 1982; by the mid-1980s, the number of cases filed began to drop off sharply in turn.

As might be expected, the number of cases filed against a particular agency is broadly proportional to the number of EISs that it prepares. A small number of agencies account for most of the EISs. In 2005, for example, over 80% of EISs were filed by just four agencies (figure 19.16): the Department of Agriculture

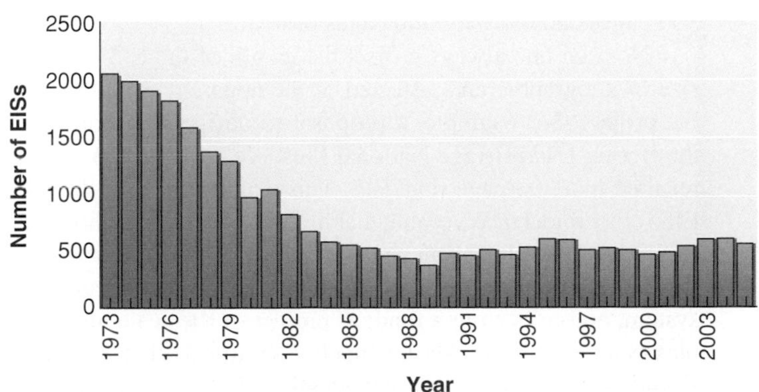

FIGURE 19.15 Number of EISs published by all federal agencies, 1973–2005.
Source: U.S. Council on Environmental Quality.

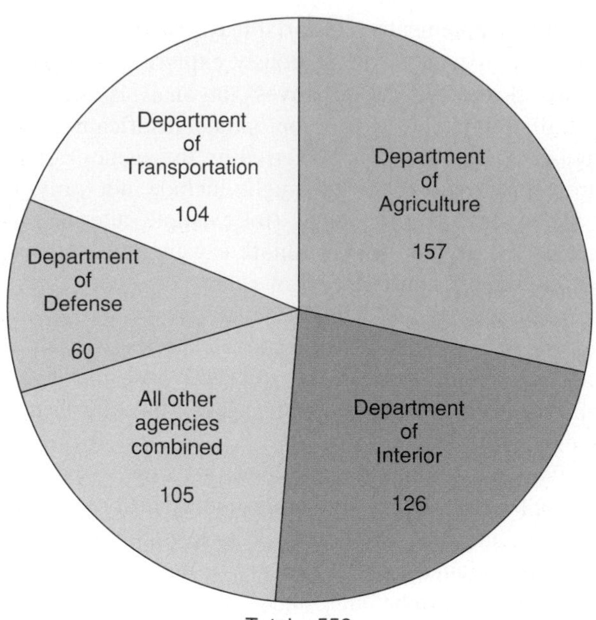

FIGURE 19.16 Distribution of EIS filings with the U.S. EPA in 1994, by agency.
Data from U.S. Council on Environmental Quality, 1995.

(especially in connection with forestry and range management, watershed protection/flood control, and use of pesticides and herbicides); Department of the Interior (especially relative to recreation areas, resource management, and mining and oil/gas drilling activities); Department of Transportation (mostly relating to road construction), and Department of Defense (especially in connection with U.S. Army Corps of Engineers and other projects related to navigation, dredging/filling, and watershed protection/flood control). By comparison, in 2004, 80 of 150 lawsuits filed in connection with NEPA involved the Department of Agriculture as defendant; the Department of the Interior (31 cases) accounted for the largest share of the rest, with 13 to 15 cases each for the Departments of Transportation and Commerce and the Corps of Engineers.

FIGURE 19.17 The size of the Trans-Alaska pipeline project, and character of much of the land it crossed, made extensive impact analysis necessary.
© Tom and Pat Leeson/Photo Researchers, Inc.

The usefulness of environmental impact statements has been limited by a number of factors. We have already noted in the context of antipollution laws the difficulty of applying cost-benefit analysis to situations where not all factors have well-defined costs associated with them. Another potential problem is the impartiality (or lack of it) of EIS preparers or proposers. An agency preparing an EIS for a project of its own has already selected a proposed course of action, and it might, consciously or otherwise, be inclined to present that action in an unduly favorable light. Where the proposed action is to be carried out by an individual or corporation—such as a company needing a federal permit to mine on federal lands—the EIS may, in part, draw on information supplied by the proposer, which could again be biased. There are also only general statutory requirements that the analysis in an EIS be thorough, which is no guarantee that all the appropriate kinds of specialists will be involved in the preparation of a given EIS. Audits of the accuracy with which EIS preparers predicted environmental impacts have often shown, in retrospect, rather low accuracy.

A major purpose of the National Environmental Policy Act is to allow the public to be informed about and involved in decisionmaking related to the environment. To prevent environmental impact statements from becoming prohibitively long and verbose, the NEPA stipulates that the length should normally be 150 pages or less, or up to 300 pages for especially complex projects. However, "tiering" is allowed, whereby one EIS may be written for each level of decision-making of a large or multistage project. One might then have to read through several statements of 150–300 pages each to comment intelligently on a particular project. Many people do not have the time to do that much reading, so they may not become involved at all. (For that matter, not all the relevant employees of the federal agencies involved may do their EIS homework adequately either, for the same reason.) An analysis of EISs filed in 1996 showed that the 243 draft EISs ranged in length from 55 to 1622 pages, with 20% over 300 pages of text; the 270 final EISs ranged from 12 to 1638 pages, averaging 204 pages, with 24% over 300 pages and 6% over 500 pages of text.

Nor can one always project the length of the EIS from the size of geographic area affected or the apparent complexity of the project. For example, a proposal to designate a portion of the Bering Land Bridge National Preserve as a wilderness area resulted in a 300-page final EIS. This was longer than the final EIS for a project in Wyoming that involved building a phosphate fertilizer plant, phosphate slurry pipeline, railroad spur, microwave communication system, and electrical power transmission system, and relocating a road! A project of the scale and complexity of the Trans-Alaska Pipeline (figure 19.17), however, would be expected to involve a lengthy EIS, and it did.

Impact Analysis and the Trans-Alaska Pipeline

The discovery of economically important reserves of oil along Alaska's North Slope raised the immediate question of how to transport that oil to refineries in the lower forty-eight states. It was proposed that a pipeline be constructed to carry the oil south to the port of Valdez, from which tankers could move the oil farther south. The planned route (figure 19.18A) crossed over 1000 kilometers of federal lands, which necessitated a permit from the government; hence the requirement of an environmental impact statement.

Various possible alternatives to the proposed route included alternate pipeline routes across Alaska; transportation by some combination of pipeline, rail, and highway across both Alaska and Canada; and tanker transport directly from the North Slope (figure 19.18B). Transport exclusively by tanker was ultimately deemed impractical, for the climate of northern Alaska is harsh, and the harbors and surrounding seas are blocked by ice much of the year. Schemes involving pipelines across Alaska and subsequent tanker transport posed threats to both terrestrial and marine environments; routes involving only pipelines through Alaska and Canada would leave marine life intact but would disturb far more land. A pipeline through Canada would also necessarily be under the jurisdiction of a foreign government, although generally good U.S.-Canada relations did not suggest that this would be a likely source of problems.

The possible environmental impacts were many and varied. Construction and operation of overland pipelines could disturb the ground, water, vegetation, fish, and wildlife. Land must be committed to the project, which might cost wildlife habitats or inhibit the migration of land animals. Warm oil in an underground pipeline could melt frozen ground, with possible subsidence of and damage to the pipeline in the resulting mud. There would be some thermal pollution from the heat released by the warm oil and a possibility of oil spills onto land or into freshwater lakes. The pipeline would cross major fault zones; southern Alaska, in particular, is subject to relatively frequent earthquakes, and as recently shown, severe earthquakes that might rupture a pipeline also occur on the Denali Fault. The use of tankers would mean possible marine oil spills also, and heavy tanker traffic could affect commercial fishing operations. On the plus side, pipeline construction would create jobs, besides providing access to major oil reserves. The influx of money and work crews and the impacts of a pipeline on the physical environment, however, would also significantly affect the lives of Alaskan natives, perhaps negatively.

On balance, the original proposed route was deemed to have the least overall adverse impact. Principal unavoidable impacts were disturbances of terrain, fish, wildlife, and some human environments; some pollution at the port of Valdez from oil discharge from treatment of tanker ballast water; and secondary effects from the influx of more people into the region. (We now know that tanker accidents can also be a significant problem.)

The analysis was a large and complex task, reflected in the size of the final environmental impact statement, which was six volumes long! However, such detailed analysis allowed

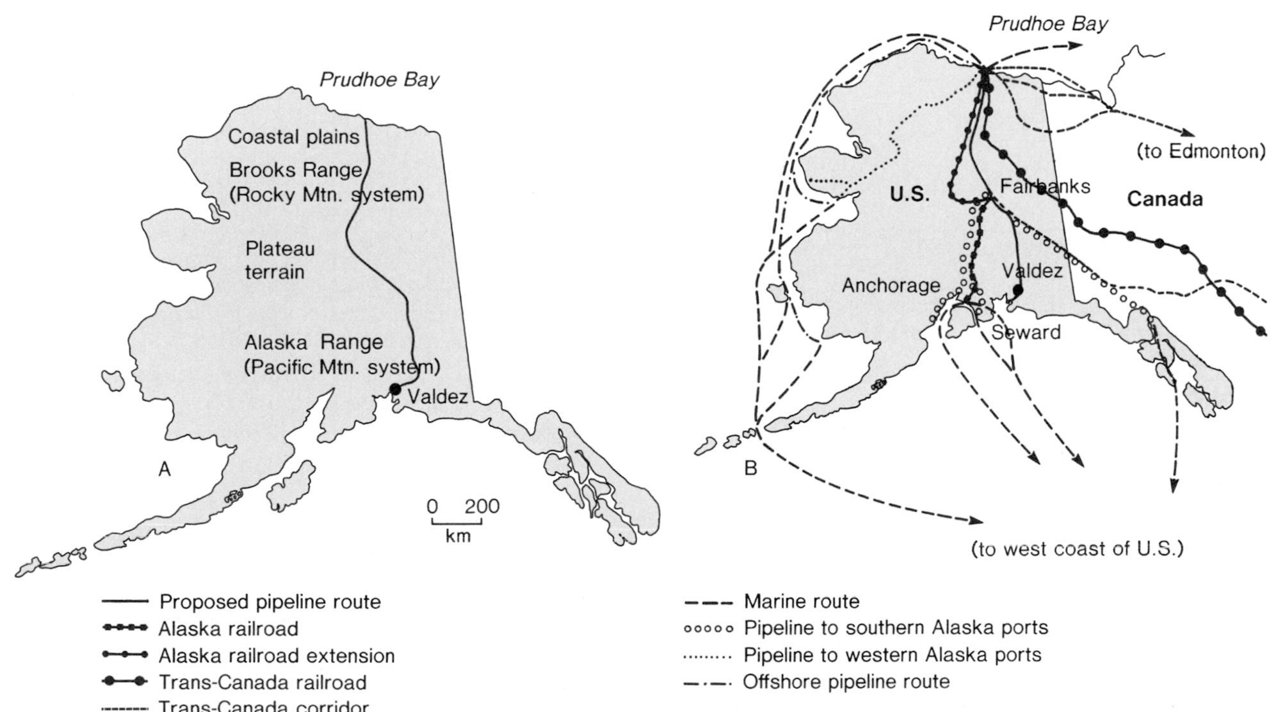

FIGURE 19.18 (A) Route of the Trans-Alaska pipeline. (B) Proposed alternative means for transport of the oil.
Source: After D. A. Brew, Environmental Impact Analysis: The Example of the Trans-Alaska Pipeline, *U.S. Geological Survey Circular 695, 1974.*

the anticipation and minimization of many potential problems and the reduction or elimination of a variety of negative environmental effects, in part through thoughtful engineering, as explored further in chapter 20.

EIS—When?

A major loophole of the NEPA, in the eyes of many environmental groups, is that statements must be prepared only when the anticipated environmental impact of a proposed project is "significant," where significance is evaluated by the federal agency involved in proposing or permitting the action in question. Clearly, *significant* is an inexact term. For each project for which an EIS is filed, many more have been deemed by the agency not to warrant one, and others may dispute that conclusion. In short, as with other kinds of environmental laws, the NEPA has some very desirable objectives, but practice has often fallen short of the ideal.

SUMMARY

Environmental laws and policies relating to geologic matters include some relating to the right to exploit certain resources, some designed to limit pollution and other kinds of environmental deterioration, and some intended to force individuals and agencies to take geologic hazards into account in development and construction. They may fall short of their goals for a variety of reasons. Terms may be inadequately defined or scientifically meaningless, making consistent enforcement difficult. Substantial economic pressure may force pursuit of a course that may be less desirable in purely scientific terms, or a rigorous cost-benefit analysis may be impossible. There may be individual or institutional resistance or bias to combat, or a particular law may recognize so many exceptions that the majority of cases ultimately end up unregulated. These and other difficulties, many without straightforward solutions, will continue to complicate efforts to develop environmental legislation. The future may see increasing development of intergovernmental agreements on critical global environmental issues, but as those agreements commonly lack strong enforcement provisions, their effectiveness may be limited.

Key Terms and Concepts

common but differentiated responsibility of states 475
environmental impact statement 482
Exclusive Economic Zone 468
precautionary principle 475
Prior Appropriation 465
Riparian Doctrine 465

Exercises

Questions for Review

1. Compare and contrast the basic concepts of the riparian and appropriation doctrines underlying much surface-water law.
2. Why are groundwater rights inherently somewhat more difficult to define than surface-water rights?
3. What was the principal objective behind early (nineteenth-century) federal mineral-resource laws? How has the emphasis shifted over the last century?
4. What are exclusive economic zones? Give two examples of types of mineral resources they might encompass.
5. Discuss some of the difficulties of defining and achieving "zero pollutant discharge."
6. Recent changes in federal policy have changed the degree of emphasis put on cost-benefit analysis in setting pollution-control standards. Explain briefly.
7. Cite two common problems with construction or zoning restrictions that limit their effectiveness, particularly in densely populated areas.
8. Briefly explain the *precautionary principle* and the concept of the *common but differentiated responsibility of states* and how they relate to the development of international environmental-protection agreements.
9. Summarize the kinds of information included in an environmental impact statement.
10. Cite and explain at least two features of the EIS process that may limit its effectiveness.

For Further Thought

1. Seek out the environmental impact statement for a local project (offices of the Army Corps of Engineers, Department of the Interior, state highway department, or other governmental agency involved in development/construction projects may be able to suggest sources of information). Note what kinds of impacts are projected and how their importance is evaluated. Is the EIS comprehensible to you as an interested individual? Would you suggest any areas in which more thorough analysis seems to be needed?

2. Consider what restrictions, if any, you as a legislator might place on development in a floodplain, on landslide-prone hills, or in active fault zones. If your community is subject to such geologic hazards, look up existing ordinances to see how extensive their restrictions are, and inquire about how routinely variances and exceptions are granted.
3. Pick an item of environmental legislation, and investigate its current status. You might choose the reauthorization of the Clean Water Act, Clean Air Act Amendments, or CERCLA, or revisions of the 1872 Mining Law; the review of the Canadian Environmental Protection Act; or the ratification status of one of the U.N. climate conventions. See NetNotes for some possible sources of current data with which to start.
4. The president pledges to increase the country's energy efficiency (the energy used per dollar of gross national product) as a way to reduce CO_2 emissions. Will enhanced energy efficiency necessarily have that result? Why or why not?

Suggested Readings/References

Evans, J. E. 1994. A course in geology and public policy. *Journal of Geological Education* 42: 10–16. (Includes references to many information sources.)

Foster, K. R., Vecchia, P., and Repachioli, M. H. 2000. Science and the precautionary principle. *Science* 288 (12 May): 979–81.

French, H., and Mastny, L. 2001. Controlling international environmental crime. Chapter 9 in *State of the World 2001.* New York: W. W. Norton and Co.

Gardner, J. V., Mayer, L. A., and Armstrong, A. 2006. Mapping supports potential submission to U.N. Law of the Sea. *EOS* 87 (18 April): 157, 160.

Kautz, S. 2005. Floods, technology, and ramifications on insurance coverage. *The Professional Geologist* (January/February): 11–14.

Keating, M. 1993. *The Earth Summit's agenda for change.* Geneva, Switzerland: Centre for Our Common Future.

Kerr, R. A. 1998. Acid rain control: Success on the cheap. *Science* 282 (6 November): 1024–27. (Discusses use of tradeable SO_2 emissions credits in the United States.)

Kintisch, E., 2006. Along the road from Kyoto. *Science* 311 (24 March): 1702–3.

Kubasek, N., and Silverman, G. 2002. *Environmental Law.* 4th ed. New York: Prentice-Hall.

McGregor, B. A., and Offield, T. W. 1987. *The Exclusive Economic Zone: An exciting new frontier.* U.S. Geological Survey.

Office of Surface Mining. 2003. *Surface coal mining reclamation: 25 years of progress, 1977–2002.* Washington, D.C.: U.S. Department of the Interior.

Reed, C., and Pinsker, L. M. 2002. New U.S. support for the Law of the Sea. *Geotimes* (January): 6–7.

Schulze, E., Wirth, C., and Heimann, M. 2000. Managing forests after Kyoto. *Science* 289 (22 September): 2058–59.

Tank, R.W. 1983. *Legal aspects of geology.* New York: Plenum Press.

NetNotes

The American Geological Institute's governmental-affairs arm tracks U.S. environmental legislation and provides links to other information sources. See AGI's home page at
www.agiweb.org

The Council on Environmental Quality (CEQ) issues extensive information on environment indicators, many relevant to pollution control legislation and agreements discussed in this chapter:
www.whitehouse.gov/ceq/

CEQ also maintains the NEPAnet site, with information on NEPA, EIS filings, and NEPA-related lawsuits:
ceq.eh.doe.gov/nepa/nepanet.htm

To obtain an EIS, see
www.epa.gov/compliance/nepa/obtaineis/index.html

and for access to information on dozens of civil and criminal cases in which EPA has prosecuted alleged polluters, 1998 to the present, see
www.epa.gov/compliance/resources/cases/index.html

The complete EPA drinking-water standards are found at
www.epa.gov/waterscience/drinking/standards/dwstandards.pdf

A discussion of the SO_2 allowances aspect of the Acid Rain Program is found at
www.epa.gov/airmarkets/arp/allfact.html

Links to acid-rain laws, including the U.S.-Canada Transboundary Agreement, are found at
www.epa.gov/airmarkets/

Cost-benefit analysis of the "Clear Skies Act" is at
www.epa.gov/air/clearskies/fact2003.html

A "Fast Facts" summary of U.S. greenhouse-gas emissions and sinks that includes information on global warming potentials of a number of greenhouse gases is at
yosemite.epa.gov/oar/globalwarming.nsf/UniqueKeyLookup/RAMR6P5MSM/$File/06FastFacts.pdf

A 2003 study, "Response of Surface Water Chemistry to the Clean Air Act Amendments of 1990," is at
www.epa.gov/ord/htm/CAAA-2002-report-2col-rev-4.pdf

The Bureau of Land Management is the agency that controls the largest fraction of federal land; its home page is
www.blm.gov/

The USGS Water Resources Division maintains "Environmental Impact Analysis Data Links" on a variety of topics; see at
water.usgs.gov/public/eap/env_data.html

Its Environmental Affairs Program (EAP) includes information on NEPA generally, and EISs referred to USGS for comment specifically, accessible via the EAP home page at
water.usgs.gov/public/eap/eappage.html

The U.N. Unit for Conventions is at
www.unep.ch/conventions/

Comprehensive information on the U.N. Framework Convention on climate change is at
unfccc.int/2860.php

and a "Beginner's Guide to the Convention" online at
www.unfccc.int/files/essential_background/application/pdf/beginner_02_en.pdf

Information on the U.N. convention on Long-Range Transboundary Air Pollution is at
www.unece.org/env/lrtap/

and on the Stockholm Convention on Persistent Organic Pollutants, at
www.pops.int

A wealth of information on the Montreal Protocol specifically and ozone generally is available via the Ozone Secretariat of the U.N. Environment Programme at
www.unep.ch/ozone/index.shtml

and answers to "Frequently Asked Questions" about the ozone layer are maintained by that Secretariat at
www.unep.ch/ozone/faqs.shtml

History of and current information on the status of the U.N. Convention to Combat Desertification can be found via
www.iisd.ca/linkages/desert.html

Materials from the Canadian Environmental Assessment Agency can be found via
www.ceaa.gc.ca/index_@.htm

This site offers links to comprehensive discussions of the Canadian Environmental Assessment Act.

20

Land-Use Planning and Engineering Geology

The purpose of land-use planning is to make the best, most sensible, practical, safe, efficient use of each parcel of land. Because such decisions are based, in part, on geologic considerations, land-use planning and engineering geology overlap. In general, however, land-use planning involves a much larger body of geologic and non-geologic facts and concerns. These may, for instance, include potential economic or practical benefits from a given use of the land and possible negative environmental or aesthetic impacts. Often, land-use planning takes the form of assessing the suitability of a particular parcel of land for a particular purpose, and proceeds somewhat like an environmental-impact assessment.

A frequent problem in land-use planning, as in other areas such as cost-benefit analysis in pollution control or evaluation of energy options, is that individual judgments about the relative importance or value of different considerations are involved. A strip-mine may be beautiful to an unemployed miner, an uninterrupted river may be more precious than a dam and hydroelectric plant to a camper, and so on. This chapter does not attempt to make such judgments, but instead reviews briefly some principles and practices of land-use planning.

Civil engineering, in some sense, goes back to the oldest human communities, to the irrigation canals of Egypt dug in 2400 B.C., to the pyramids, to the Great Wall of China, and to the aqueducts and roads of ancient Rome. The ancestral puebloans at Mesa Verde built not just dwellings, but also water reservoirs to collect and store that precious resource. The building of many ancient structures required some understanding of earth materials. Mistakes resulted from gaps in that understanding. The Leaning Tower of Pisa leans, not by the design of its builders and not because of any structural flaw within

The need for land-use planning may be especially obvious where people are densely packed, as in this Las Vegas neighborhood.
Courtesy of Lynn Betts/NRCS/USDA

the edifice proper, but because it was built on unstable soils, some of which flowed out from under it; the tilt has increased over the centuries.

Over the last 200 years, it has become increasingly possible to incorporate geologic principles and considerations into construction plans. The paramount concern of the modern engineering geologist is to take a site's geology fully into account in designing a structure so that the structure will be safe and stable. Increasingly, engineering geologists also consider the impact that the building or structure, in turn, will have on the geologic environment.

LAND-USE PLANNING—WHY?

One motive behind land-use planning is safety: Some land is unstable and unsuitable for certain kinds of structures, so the geology has restricted the possible uses to which that land can safely be put. However, much of the motivation for land-use planning arises out of the reality that a large and growing population occupies a fixed expanse of real estate. When the population was much less, it mattered less, in a sense, whether a particular parcel of land was misused. If a farm's soil eroded away or became infertile or baked hard, if a stream dried up or became polluted, one could move on to a new, pristine spot and begin again.

Centuries ago, there was ample space for all. That is becoming less true all the time. From 1997 to 2001, 2.2 million acres of rural land were converted to developed uses each year just in the United States.

Admittedly, 2.2 million acres may not seem like a lot compared to the 2300 million acres of land in the United States. Year after year, however, it adds up. From 1982 to 2001, the area developed—34 million acres—was altogether about the size of Illinois. The whole 2300 million acres in the United States are not ultimately available for conversion, either, for it is on much of that rural land that food crops and livestock (and food for export) are produced (figure 20.1). Moreover, the rate of development is accelerating: During the period 1992–2001, the rate of development was more than 50% higher than that of the previous decade and well above rates earlier in the century. From 1982 to 2003, the amount of developed land in the United States increased by over 48%.

It is becoming increasingly important to consider not only what can safely be done with a given piece of land, but the optimum uses to which the land can be put. Many of the considerations involved are geologic ones similar to those outlined at the beginning of this chapter. Additional geologic constraints might include the resource potential of a particular piece of land or its susceptibility to pollution, erosion, or other disruption if certain kinds of land uses are permitted. Again,

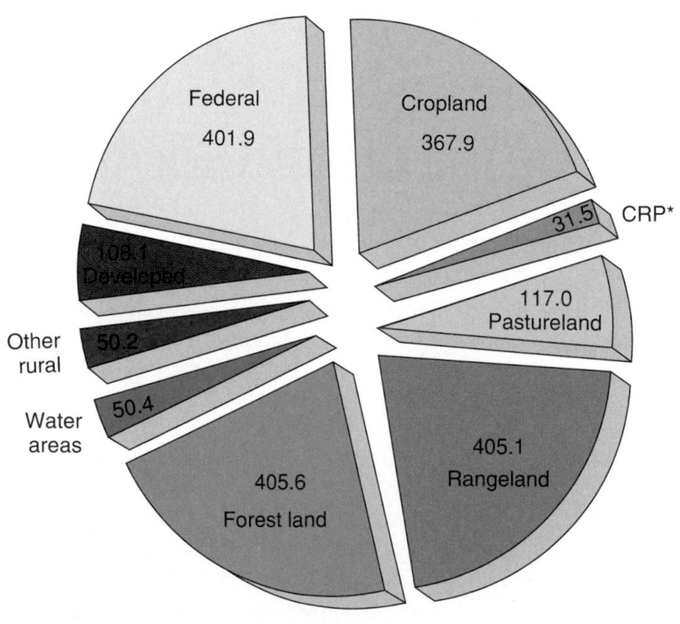

*Conservation Reserve Program Land

A

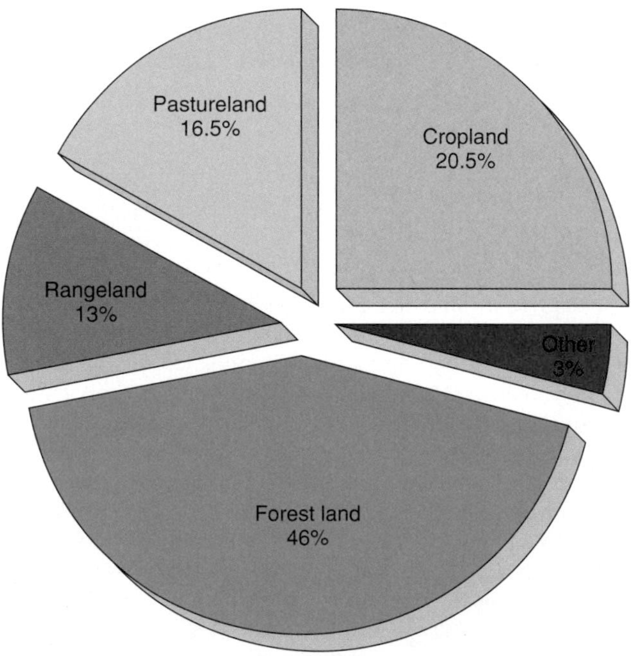

B

FIGURE 20.1 (A) Land cover/use in the United States, 2003, in millions of acres. (B) Land converted to developed land, 1997–2001, by prior land use.
From USDA Natural Resources Conservation Service National Resources Inventories 2001 and 2003.

biological/ecological, economic, and political factors may enter into the decisions ultimately made. So may aesthetic factors. For example, as noted in chapter 15, Yellowstone National Park represents a tremendous geothermal resource. However, it is also a great scenic resource and important wildlife habitat, and these considerations are deemed important enough to override the potential gain from exploiting the geothermal energy. In early oil-boom days, on the other hand, oil was so sought-after that considerations of whether or not one might want a derrick in one's front yard were often overlooked (figure 20.2). The Bingham Canyon copper mine in Utah, one of the world's largest open-pit mines, has irrevocably altered some 7500 acres of undeveloped western land, but it has yielded, on average, over $9 million per acre in metallic minerals.

FIGURE 20.2 Drilling for oil, Long Beach, California, 1901.
Photograph by C. W. Hayes, USGS Photo Library, Denver, CO.

LAND-USE OPTIONS

Particularly in densely populated areas, there is growing interest not only in making the best use of each piece of land but, if possible, in using the same parcel of land for several different purposes. The strategies involved are called *multiple use* and *sequential use.*

Multiple use, as the name implies, means using the same land for two or more purposes at the same time. Two examples are shown in figure 20.3. At first glance, the area in figure 20.3A appears just to be a ball field. However, note that it is built into a basin. When it rains—when the field would not be used as a ball field anyway—the field acts as a recharge basin, catching fresh rainwater and allowing it to infiltrate slowly down to the seriously depleted aquifers under Long Island. So the same land

A

B

FIGURE 20.3 Multiple land uses (A) This ball field doubles as a recharge basin for Long Island aquifers. (B) Switchgrass (for biomass fuel) can be grown and harvested while this wind farm in southwestern Minnesota generates electricity.
Photograph (A) courtesy of USGS Photo Library, Denver, CO, (B) by Warren Gretz, courtesy National Renewable Energy Laboratory.

serves two purposes: one recreational, the other directed toward conserving/ increasing groundwater resources. Other examples of multiple land use would be the generation of wind power from wind-generator arrays in fields used simultaneously for farmland or grazing land (or even residential areas, as is done in Denmark), or underground mining deep below the surface while urban development proceeds above. Figure 20.3B shows such multiple use with both uses directed, ultimately, to energy production.

The alternative approach of **sequential use** involves using land for two or more different purposes, one after another. Because the different land uses need not be compatible, a greater variety of combinations is possible. Abandoned underground mines, if dry and adequately ventilated, can be used for warehouse space, manufacturing, or even office space. This is being done, for example, in the Kansas City area, in abandoned underground limestone quarries. The old quarries now include nearly 1 million cubic meters of frozen-food storage space (about one-tenth of the U.S. total), and the rock's insulating properties make the practice very energy-efficient. The great strength of rock floors makes possible the use of heavy manufacturing equipment; the controlled humidity underground is an advantage to print shops and a sailboat factory located there. Rock walls are, of course, fireproof, and they also contain noise well. On a smaller scale, abandoned mine space at Wampum, Pennsylvania, has been converted to office and laboratory space for the mining company. South Dakota is hoping to convert the nearly mile-deep Homestake gold mine to a major underground scientific research laboratory.

Not all abandoned underground mines are equally suitable for subsequent occupation. The rock structure must be sufficiently strong for safety and not prone to deterioration through slow weathering with time. Even if above the water table, the space must be protected by overlying impermeable rock from infiltration of subsurface water from above. (The deepest 2000 feet of the Homestake mine is currently flooded, since the mine's owners shut off the pumps that had been keeping it dry.) Flat-lying rock strata facilitate conversion to occupied space. There must be no likelihood of quantities of dangerous gases present; old limestone mines would be safe in this respect, while old coal mines, with possible associated methane present, probably would not be.

Alternatively, abandoned mines could be used for waste disposal (recall the proposal in chapter 16 to place high-level radioactive wastes in abandoned salt mines). Strip mines have been suggested as possible landfill sites when mining is completed; this would follow resource extraction with waste disposal, after which the land could be covered, regraded, and put to another use. Denver, Colorado, has carried out a scheme of this sort: Old gravel pits were used for sanitary landfill, and then, after they had been filled, the Denver Coliseum and its parking lots were constructed over the top. (Using such permeable materials for landfill may or may not be advisable in a given situation.)

An alternative post-landfill use might be the extraction of methane gas for energy. An abandoned quarry can be flooded to make a recreational lake. Many such sequences of activities can be imagined; another example is shown in figure 20.4. For that matter, each time a Superfund site is reclaimed and converted to a productive purpose, that, too, is sequential land use. The advantage all sequential-use arrangements share is land conservation. If one piece of land can be made to do double or triple duty, that much less land must be used for or disturbed by human activities overall. Further, it becomes easier to avoid feeling forced to use the least stable or most vulnerable areas at all.

A

B

FIGURE 20.4 Aerial photographs of an area of St. Louis Park, Hennepin County, Minnesota (A) pre-1973, during operation of a coal-tar-distillation and wood-preserving facility, and (B) during subsequent redevelopment. In such a case, care must be taken to monitor possible toxic-waste hazards.
Photographs courtesy of USGS Photo Library, Denver, CO.

THE FEDERAL GOVERNMENT AND LAND-USE PLANNING

The federal government controls approximately a third of the land in the United States. The proportion, however, varies widely among states, from less than 1% in many states of the northeast and midcontinent to about 67% of Utah and Alaska, and 88% of Nevada (figure 20.5). The impact of federal land-use policies, like other federal resource-related laws, is therefore felt in very different degrees in different states.

Federal land-use policies have changed through time. As with mineral-resource laws (chapter 19), federal emphasis was initially on resource development in preference to preservation. Early preservation efforts were limited. The first of the national parks, Yellowstone, was established in 1872; the first national forests were created two decades later. Still, resource development continued to be a high priority until about the middle of the twentieth century.

Federal lands can be broadly divided into two types: those intended primarily for preservation (including national parks and wilderness areas) and those on which multiple and, one hopes, compatible land uses can be allowed (for example, national forests). On the latter lands, additional uses beyond recreation or habitat preservation might include livestock grazing, mining, logging, and exploration and drilling for petroleum. Problems arise when the multiple uses allowed turn out not to be compatible after all. Livestock may outcompete wildlife for limited food; overgrazing, careless timbering, or even just too many tourists passing through may accelerate soil erosion and loss. Because the preservation function has sometimes been less successful when multiple land uses are allowed, many groups have tried to pressure the federal government to put more of its lands into the highly protected categories.

In 1978, President Carter did that with 107 million acres of Alaska. Under the Alaska Lands Act, that land—almost one third of the state—was designated as national monuments, wilderness areas, or similarly protected land. Partially as a result of that action, mining and petroleum development now are prohibited or severely restricted on 40% of Alaska's total land. Some hailed the decision as a forward-looking move to preserve dwindling wilderness. Oil and mining interests were very unhappy at the reduction in land available for exploration and warned of possible future resource shortages. State residents who had hoped to benefit from more jobs or from taxes on those companies' profits from resource exploitation were likewise not pleased. The state, in turn, wanted more control

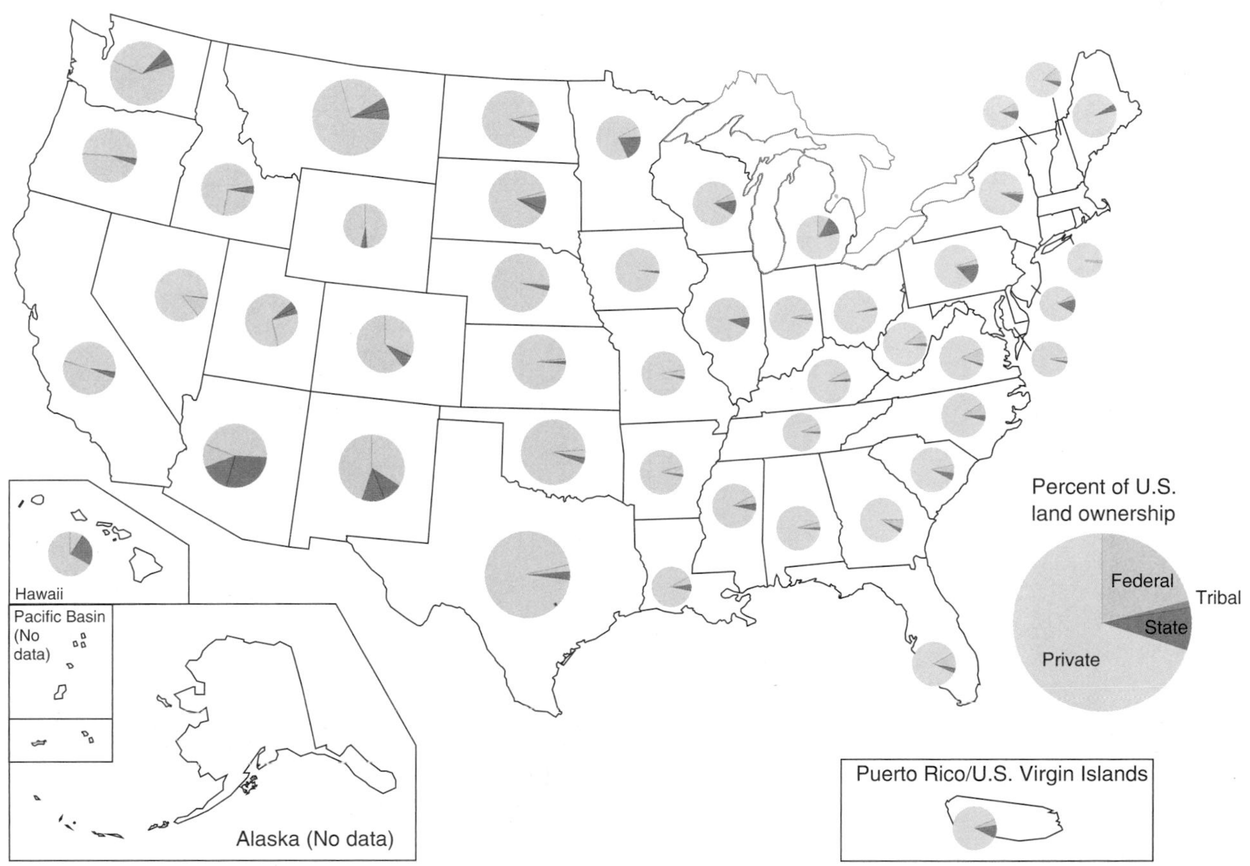

FIGURE 20.5 Land ownership by state, 1997. Note that overall pie chart excludes Alaska, where a high proportion of land is federally owned. *From USDA Natural Resources Conservation Service 1997 National Resources Inventory, revised December 2000.*

over what was to be done with its lands. Recently, proposals to open parts of the Arctic National Wildlife Refuge to petroleum exploration have intensified the debate.

Often, controversies arising in connection with the EIS process reflect significantly different opinions about appropriate or optimal land uses for the area in question. So, what's a government to do? What *does* constitute managing the public lands in everyone's best interests?

MAPS AS A PLANNING TOOL

Many kinds of information, geologic and otherwise, go into comprehensive land-use planning. Much of this information can most quickly and clearly be examined in map form. Any geological property or process that varies from place to place, including topography or steepness of slopes, bedrock geology, surficial materials or soil types, depth to water table, rates of cliff erosion, and so on, can be represented. Maps can also show locations of hazards past or present, such as fault zones, floodplains, and landslides. Figure 20.6 illustrates some examples. Nongeologic factors—vegetation, present population density, or present land use, for instance—may also lend themselves to representation in map form (figure 20.7).

A land-use planner can then seek sites for particular land uses based on whatever set of criteria seems most appropriate. Maps make it possible to see quickly where several different conditions are satisfied. In siting a major interstate highway, for example, a planner might seek gently sloping terrain not underlain by expansive clay soils or active fault zones, where the present population density is low. A survey of an appropriate set of maps permits the planner to find potentially suitable sites swiftly. Conversely, maps can aid in long-term land-use planning even when conversion of rural land is not immediately contemplated. If the location of a stream's 100-year floodplain or a major fault zone is known, restrictive zoning ordinances can prevent unwise development before it occurs, or special construction requirements for those areas can be imposed.

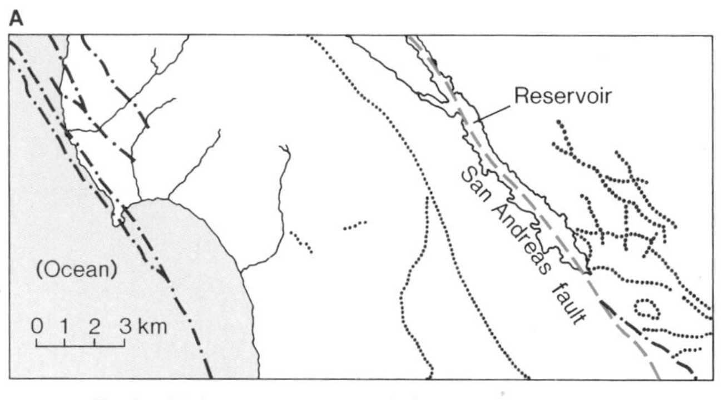

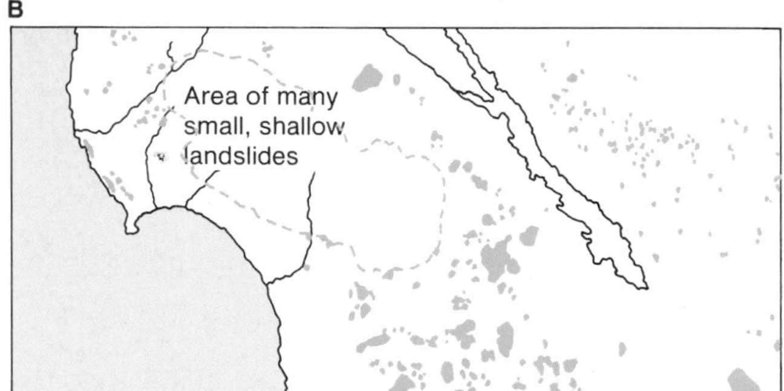

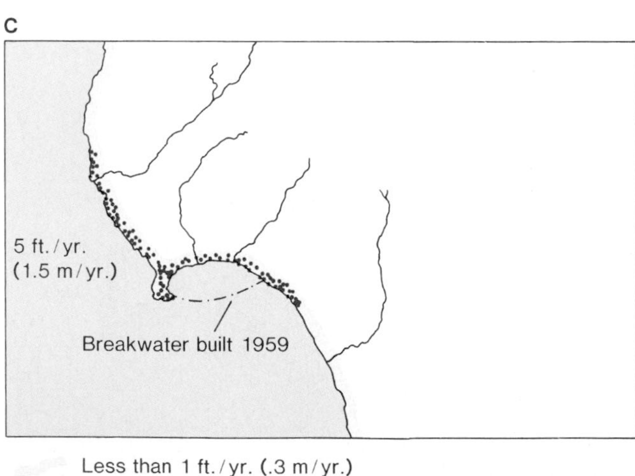

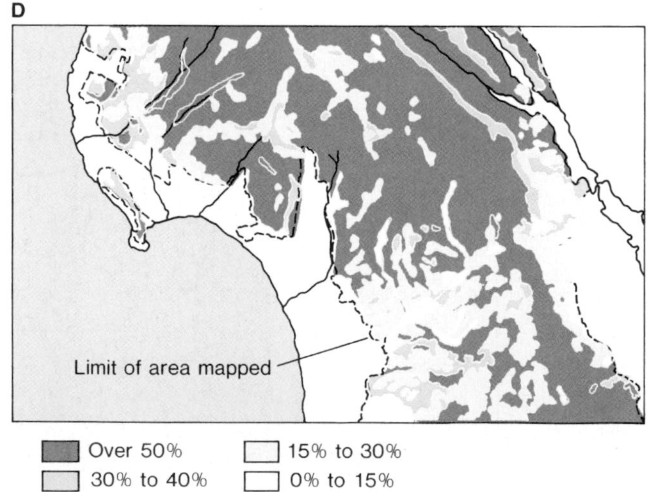

FIGURE 20.6 Map representation of several kinds of geologic considerations. (A) Fault zones. (B) Landslides (shaded). (C) Historic coastal cliff erosion rates. (D) Slope steepness.

Source: Modified from B. Atwater, "Land-Use Controls Arising from Erosion of Seacliffs, Landsliding, and Fault Movement," in U.S. Geological Survey Professional Paper 950, 1979.

LAND USE SHOWN ON MAP	CRITERIA FOR DETERMININING LAND USE (Land cover, in percent; crop type, in percent; population density, in persons per square mile; total water withdrawals for domestic use, in million gallons per day)
AGRICULTURE Wheat	Land cover: >40 crop and pasture, <40 forest, <10 urban Crop type: >50 wheat, <20 corn and soybeans
Corn and soybeans	Crop type: >50 corn and soybeans, <20 wheat
Mixed	Crop type: <50 wheat and corn and soybeans
URBAN	Land cover: <30 crop and pasture Population: >100 Total water withdrawals for domestic use: >6
FOREST	Land cover: >50 forest, <40 agriculture, <10 urban
RANGE	Land cover: >50 range and barren land, <40 forest, <40 agriculture, <10 urban
OTHER	All land cover not meeting any of the above criteria

FIGURE 20.7 U.S. land-use classifications combine elements of nature of cover, urbanization, and water use. Note regional variations in dominant land use.
From U.S. Geological Survey Water Supply Paper 2400.

Of course, the suitability of a particular area for a specific land use cannot always be determined in a clear-cut, yes-or-no fashion. A land-use planner may instead need to consider a number of alternative sites for a project, each of which is less than fully desirable in some different way. Or there may be degrees of suitability for some purpose, such as housing developments, considering a variable like slope stability. In the latter case, the planner might stipulate different levels of intensity of the same land use in different areas. A higher density of homes might be permitted on gently sloping terrain than on steeper hillsides, for instance.

In recent decades, computers have played an increasingly important role in the land-use planning process because of their capacity to manipulate large volumes of quantitative information. A map can be broken down into a grid of numbers, each data point representing a particular property (slope, soil type, and so forth) over some area (1 square kilometer, 10 square kilometers, or whatever) (figure 20.8). The computer can then combine as many kinds of information as desired, each represented on a separate array, and produce a composite measure for each point of the grid to indicate the overall suitability of that area for the land use under consideration, which is most often urbanization/housing development (figure 20.9). The same basic geologic or nongeologic data can be combined in different ways for different purposes by weighting various factors differently. In mapping the suitability of land for wildlife habitat or refuge, for example,

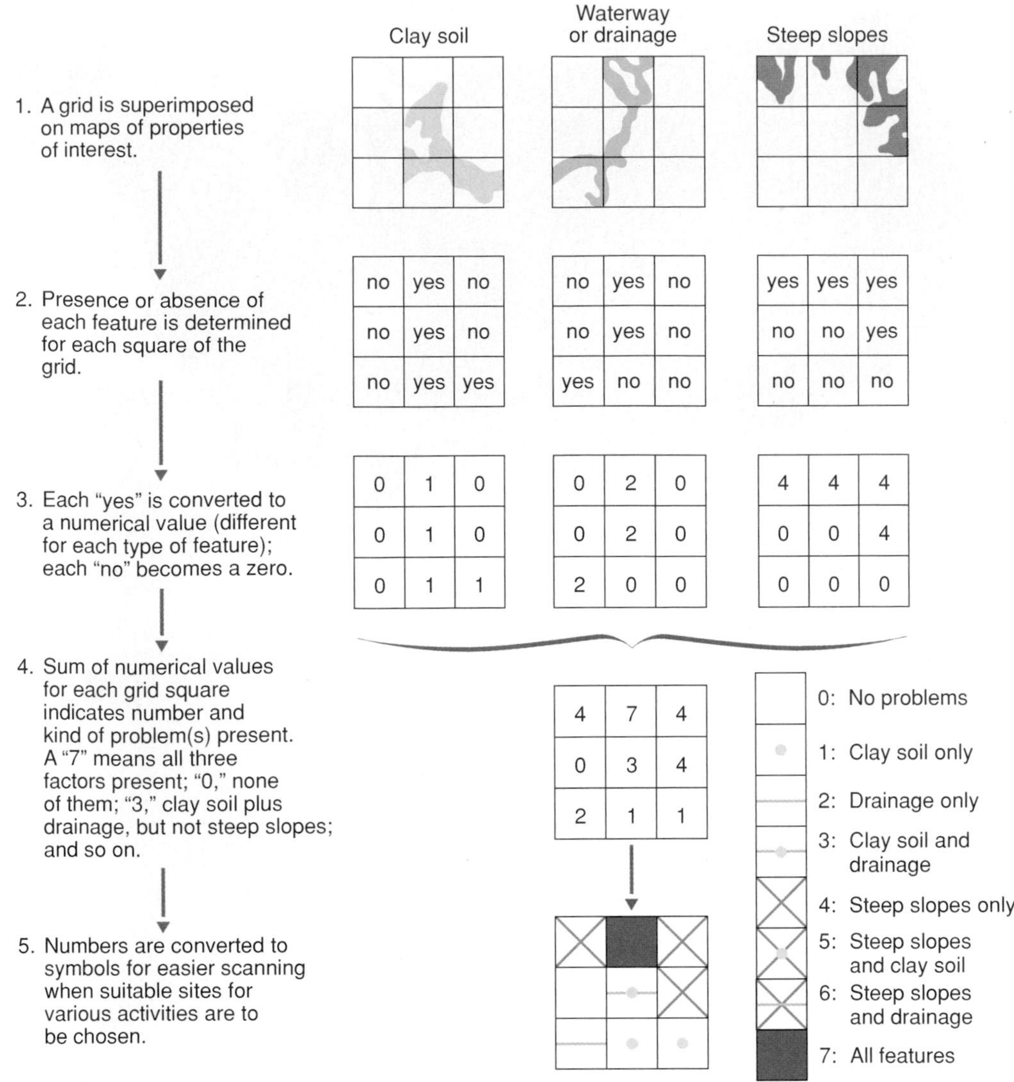

FIGURE 20.8 Digitized maps can represent data in a form that both human planners and computers can use. In this example, for each parameter (clay soil, water drainage, steep slopes), each point (area) has one of two values: Either the area is underlain by clay or it is not; either the slopes are steeper than a certain grade or they are not; and so on. In a more complex scheme, there might be several possible values for each parameter.
Source: A. J. Froelich et al., *"Planning a New Community in an Urban Setting: Lehigh," U.S. Geological Survey Professional Paper 950, pp. 69–89, 1979.*

abundant surface and near-surface water might be a positive factor and the presence of expansive clay soils largely irrelevant. A land-use planner looking at the same region with high-density housing in mind might well regard both factors as negative. The data-processing programs can be adjusted accordingly to produce summary data tailored to particular objectives.

As computers have become more powerful and their programming more sophisticated, they have increasingly been used for **geographic information systems** (GIS). A GIS is any computer-based system for storing and manipulating mappable data—in other words, data that can be associated with or plotted on points on a map. The data are often stored and displayed in so-called *layers,* where each layer is a distinct type of information (figure 20.10). The results can be examined visually and/or analyzed by computer, which can combine data from multiple layers.

Maps as tools have limitations, too. One is the matter of scale: The information presented must be given in sufficient detail that features of interest will, in fact, show up in the data. On a map on which 1 centimeter represents 1 kilometer of distance on the earth's surface (1:100,000 scale), all kinds of relatively small features—cliffs, sinkholes, small streams, and so on—might not appear at all; yet these features could be of great concern to a prospective home owner. (The astronauts had the same problem with early lunar landings. Images of the moon made from earth, covering very large expanses of the moon's surface, were used to direct them to "fairly flat" areas. But when those proposed landing areas could be seen close up on a

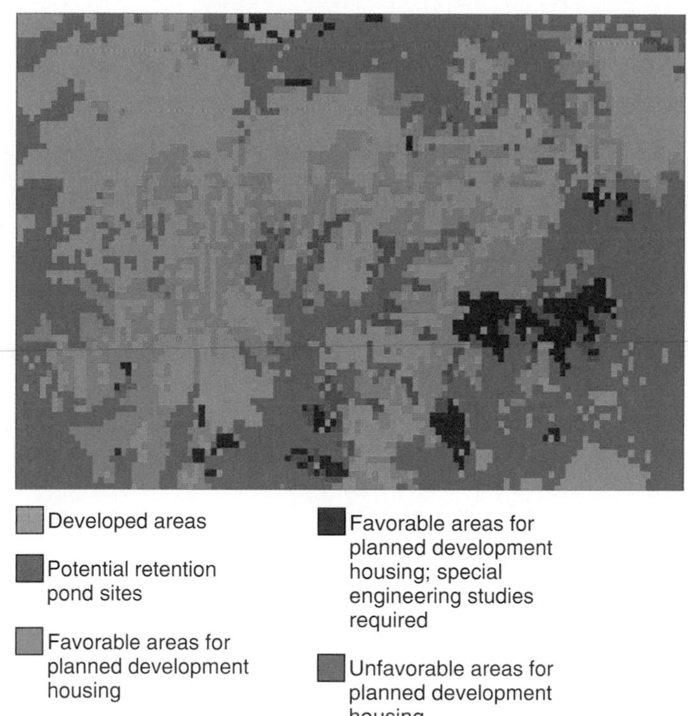

FIGURE 20.9 The computer can rapidly combine digitized data into a composite map for land-use planning, showing more- and less-suitable areas for development or any other intended purpose.

Source: A. J. Froelich et al., "Planning a New Community in an Urban Setting: Lehigh," U.S. Geological Survey Professional Paper 950, pp. 69–89, 1979.

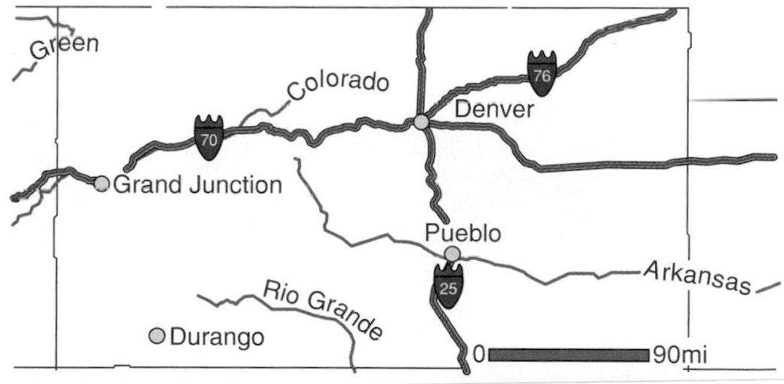

A

B

C

FIGURE 20.10 Sample GIS representations from USGS National Map Viewer. A roadmap-style map of this region near Denver (A) shows only major rivers, interstate highways, and cities. Adding the topography layer (B) or land cover (C) adds much more information.

Images generated by ArcIMS Viewer, courtesy U.S. Geological Survey.

finer scale, many were found to be strewn with boulders several meters across, which definitely presented a hazard to smooth and safe landings!) At the same time, superimposing too much detail can obscure information (figure 20.11).

Sometimes, the difficulty is that the information itself is available on only a gross scale. Maybe the only maps of vegetation or surficial geology were made from satellite images, for instance. Sometimes, the data are simply unavailable. Producing many kinds of topographic, geologic, or other earth-science maps means compiling many observations or measurements, which takes time, personnel, and funds. Even in the United States, which is rather thoroughly mapped by world standards, not all types of information have been obtained for all areas. In these cases, land-use decisions may have to be based on incomplete data, perhaps supplemented by broad, reconnaissance-type surveys of the area under study. Sometimes, too, the problem is not so much lack of data as conflict among competing options. The pressure to develop areas where risks are high or incompletely assessed is particularly acute where development is rapid, especially if population density is already high so land available for new development is scarce (compare figure 20.12 with figure 20.7).

For information on additional kinds of imagery that can be applied to land-use and other problems, see Appendix B.

ENGINEERING GEOLOGY—SOME CONSIDERATIONS

Many of the geologic factors that go into site evaluation have been noted in earlier chapters. Obviously, not every geologic factor or process is equally important to every project, but a major construction project—for example, a housing development—might require consideration of a very broad range of geologic matters. Some are outlined below.

What rock types are present? Are they strong enough to support the proposed structure(s)? Do they fracture easily? Are there structural features within the rocks—folds, faults, bedding planes—that make the rocks' properties nonuniform and that should be taken into account? How porous and permeable

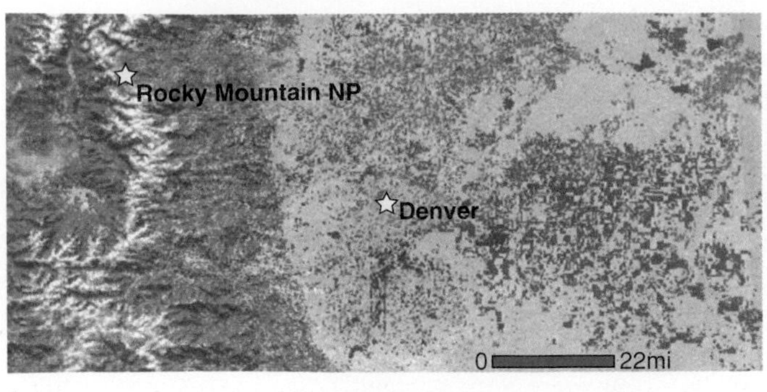

A

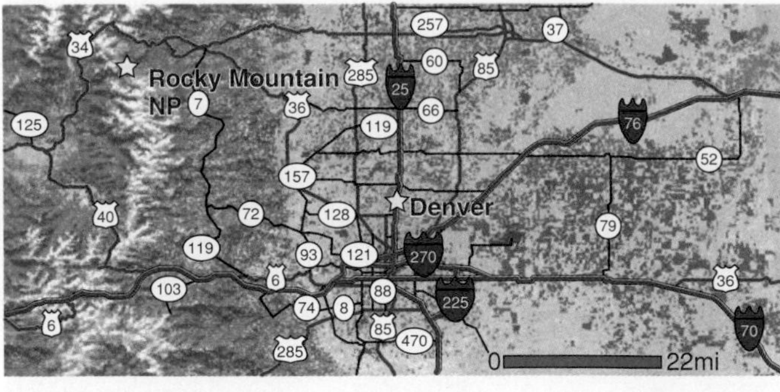

B

FIGURE 20.11 When topography and land cover are mapped on a finer scale, more information can be displayed; compare (A) with figure 20.10C. But adding the highways layer (B) obliterates some detail.

Images generated by ArcIMS Viewer, courtesy U.S. Geological Survey.

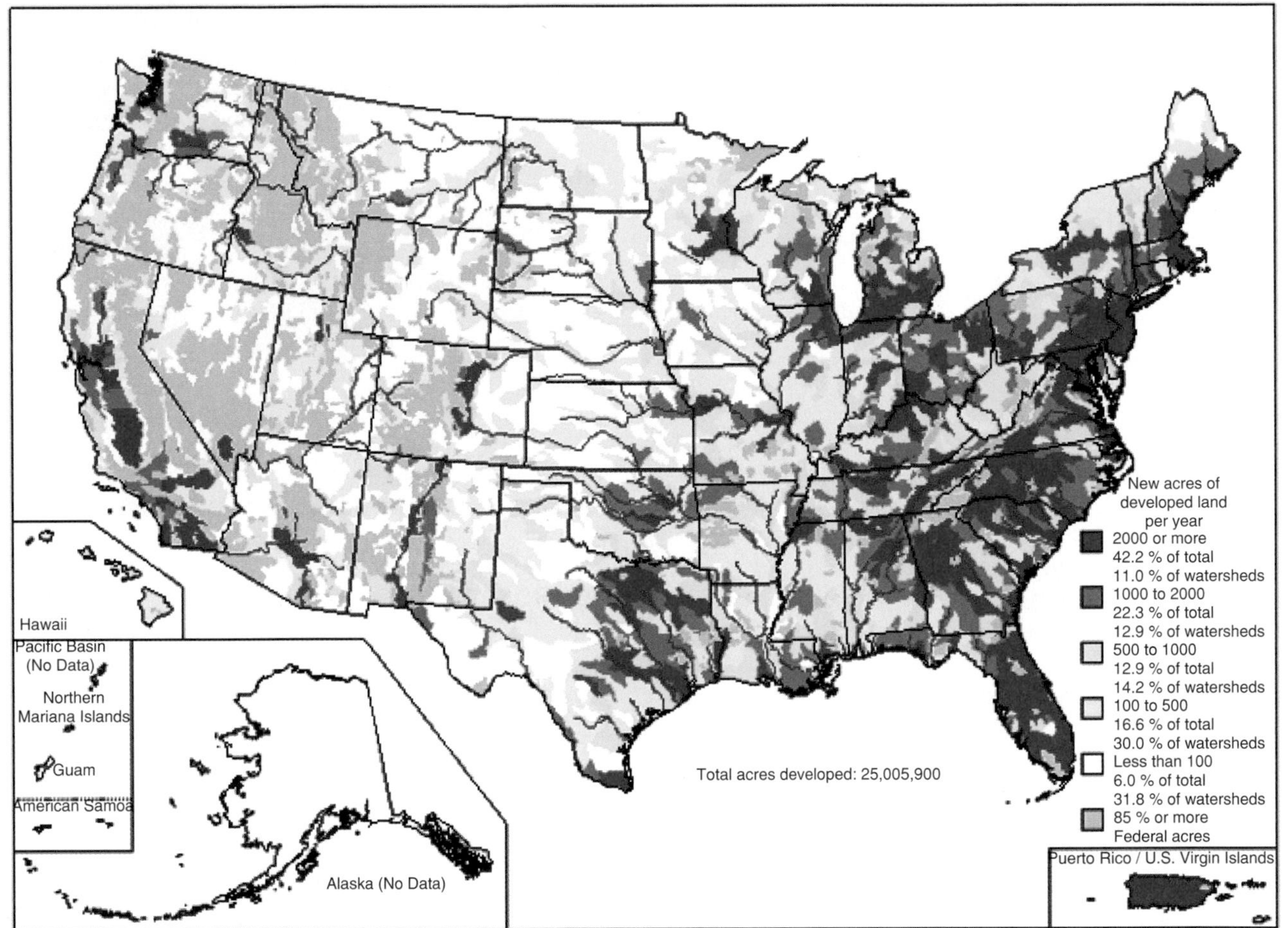

FIGURE 20.12 Annual rate of development, 1982–1997. Many areas of rapid urbanization/development in the United States are areas of high population density and/or geologic hazards (e.g., coastal areas) or limitations (e.g., of water supply).

Map from USDA Natural Resources Conservation Service 1997 National Resources Inventory, revised December 2000.

are the rocks? Over the longer term, are they unusually prone to erosion or weathering?

Many of the same questions about porosity, permeability, strength, and stability might be asked about the soils. Are any slopes likely to give way to landslides? Does the construction itself have the potential to trigger slides? How cohesive is the soil? How compressible and prone to settling? How elastic? Do the soils tend to expand and contract as moisture content varies? Are they subject to hydrocompaction? This problem of expansive clay soils as a construction hazard is not a trivial one (figure 20.13). Their failures take few, if any, lives, by contrast with more obvious hazards like floods and earthquakes, but the total structural damage to homes, commercial buildings, pavements, and utilities in the United States each year amounts to over $2.2 billion, approximately equalling the total costs of all other geologic hazards combined.

What about water? What quality and quantity of surface water or ground water is available? What are the surface runoff patterns? Does part of the site now serve as a recharge area for an aquifer system? If septic tanks are planned, are soil properties and topography appropriate for them?

Then there are the possible catastrophic hazards. Earthquakes and volcanoes are significant hazards in relatively few places, but where they are, they need to be taken into account in building siting and/or design. Landslides and floods potentially affect many more areas. Do sinkholes occur in the area? Has there ever been underground mining in the area?

Long as this list of geologic concerns is, it does not exhaust the possibilities. Consider the case of the Trans-Alaska Pipeline, discussed in chapter 19. At 1300 kilometers long, it spans a variety of geologic settings and potential problems. Much of the terrain it crosses is rugged, including two major mountain ranges (the Brooks Range and the Alaska Range). It also crosses about 800 streams, the largest of which is the Yukon River. All of those streams freeze in winter, and some thaw partially in summer, complicating the engineering. A number of faults crisscross the area, and the southern end of the pipeline, the city of Valdez, is close to the epicenter of the 1964 Alaskan earthquake. The pipeline, therefore, was designed to take the stress not only of thermal expansion and contraction, but also of earthquakes and ground displacement (figure 20.14); near the Valdez end and along the Denali Fault, it was built to withstand earthquakes up to Richter magnitude 7.5.

A further construction headache peculiar to very cold climates is permafrost, described in Chapter 10. As long as the permafrost stays frozen, it makes a fairly solid base for structures. If it is disturbed and warmed, as by construction, some of the ice melts. The deeper soil is still frozen, however, so depending on the topography, the water may drain slowly or not at all. The result is a mucky, sodden mass of saturated soil that is difficult to work in or with and that is structurally weak (figure 20.15). The work crews on the Trans-Alaska Pipeline were faced with trying to install, under very unstable conditions, structural members that would be quite adequately, solidly supported *after* construction was done and the melted permafrost had refrozen. Even the oil in the pipeline posed a problem. The oil is warm (about 60°C, or 140°F) as it comes up from deep in the earth, and its temperature stays well above freezing as

A

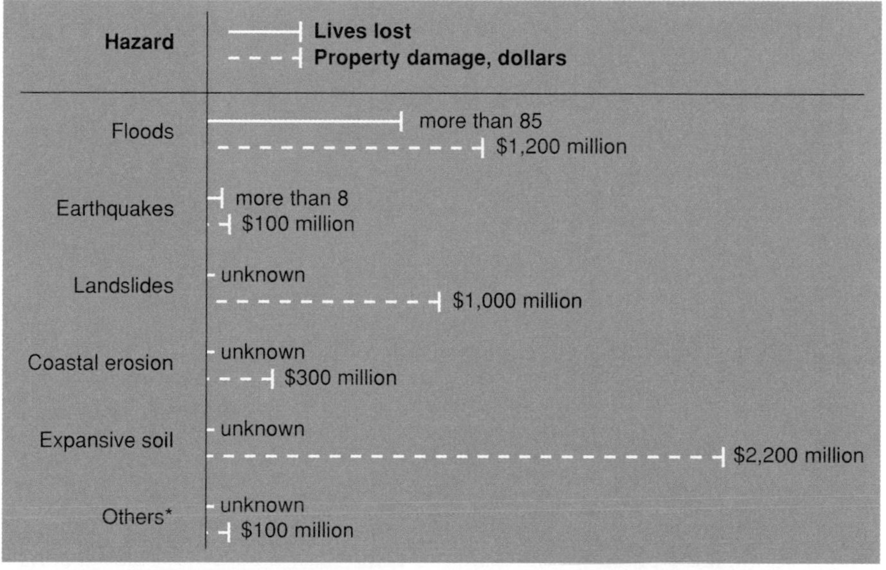

*Includes volcanic eruptions, tsunamis, subsidence, creep, and other phenomena.

B

FIGURE 20.13 Natural hazards take lives and destroy property. (A) Failure of masonry structure on unstable soil. (B) Relative magnitudes of loss of life and property damage in the United States each year from various geologic hazards. Note that expansive soils cause nearly as much property damage as all other listed hazards combined.

(A) Photograph courtesy of USDA Soil Conservation Service. (B) Source: U.S. Geological Survey Professional Paper 950, 1979.

A

B

FIGURE 20.14 Trans-Alaska pipeline design takes into account the need to allow for both thermal expansion and contraction as temperatures rise and fall, and possible displacement resulting from earthquakes. (A) "Kinks" in the pipeline, and loose coupling to supports (see also figure 20.16), help accomplish this. (B) The result is that in places, the pipeline zigzags across the land; here, it also dives under the river in the foreground.
(A) photograph courtesy U.S. Geological Survey.

A

B

FIGURE 20.15 Permafrost affects construction work and transportation systems. (A) Melted permafrost produces a sodden mass of waterlogged soil like that surrounding this tractor. (B) Differential subsidence of railroad tracks due to partial thawing of permafrost, Copper River region, Alaska. This track had to be abandoned in 1938.
Photograph (A) by T. L. Péwé, (B) by L. A. Yehle, both courtesy of O. J. Ferrians, USGS Photo Library, Denver, CO.

it flows through the pipeline. The warm oil could thaw a zone within the permafrost immediately around the pipeline, and the sagging of the pipeline in the resultant mud could lead to pipeline rupture. Ultimately, buried sections of the Trans-Alaska Pipeline were actually refrigerated in places to keep the permafrost frozen! Elsewhere, cooling systems were installed to cool the supports of aboveground sections of pipeline, to keep them firmly in place (figure 20.16).

Of course, the pipeline was engineered for the conditions at the time it was built. We have already noted (chapter 10) that warming seems to be particularly pronounced in the Arctic, and this includes much of Alaska. One consequence is extensive

FIGURE 20.16 Close-up of pipeline support system shows how pipeline can shift from side to side between "bumpers" on support posts. Fins on tops of posts are part of a system designed to cool the underground sections of the posts. Coolant fluid circulates vertically, and the fins expose maximum surface area to the cool air—working much as fins in a radiator do—to counteract the warming effects of oil or of sunlight heating the ground.

melting of permafrost—including permafrost in which pipeline supports had been anchored. A recent study has suggested that up to one-third of these supports may be becoming unstable as a result, jeopardizing the integrity of the pipeline. Replacing these supports, some of which extend as much as 70 feet below the surface, could cost an estimated $85,000 *each.*

Even where some permafrost remains at depth, warmer temperatures mean later freezing at the surface in the fall, and earlier melting in the spring. But the heavy equipment used in oil exploration and drilling cannot navigate the soft muck of the melted permafrost; that equipment can only be used in permafrost areas while the ground is solidly frozen. In areas of the Alaskan tundra, the number of days that this equipment can be deployed has shrunk dramatically over the past three decades, from 200 or more per year to only about 100. To the extent that the dwindling of the permafrost is related to warming caused by CO_2 from the burning of fossil fuels, some see real irony here.

THE ROLE OF TESTING AND SCALE MODELING

Relatively modern additions to the arsenal of tools available to the geological engineer are the use of scale models of natural systems, and testing—actual or theoretical—of the behavior of natural and construction materials.

Many scale-model experiments take place in the laboratory. Nowadays, indeed, much "testing" is done using sophisticated computer models rather than by building physical models or working with actual samples of materials. However, it is often still useful to conduct physical experiments and observe the results.

The photographs in figure 20.17 show a 1:500-scale model of the Swift Dam and Reservoir on the south flank of Mount St. Helens. The model was built by Western Canada Hydraulic Laboratories for the dam's owners, Pacific Power and Light Company of Portland, Oregon.

High-density muds like those used in drilling oil wells simulated volcanic mudflows entering the reservoir from the creeks that feed it. Figures 20.17B and 20.17C show the progress of one such flow, from the model Swift Creek, into the reservoir. With each test, maximum wave height at the model dam and overall reservoir response to mudflow input were closely observed. At the request of the State of Washington, maximum mudflow volumes corresponding to about half the reservoir volume were tested. Under these conditions, waves from single flows, and the final reservoir height, nearly reached the dam crest.

After the 1994 Northridge earthquake, engineers were rudely surprised to discover the response—and failure—of some structures believed adequately designed. The Federal Emergency Management Agency and California Office of Emergency Services committed $2.3 million to studies to address some of the problems (figure 20.18). After examining a number of failures, the engineers considered various modifications, investigated computer simulations of behavior of different designs, and conducted stress tests on full-size models of steel beams and joints. Following these initial studies, the engineers proposed interim guidelines for the design of new buildings and repair/strengthening of existing structures. Several more years of work are planned to refine the data and recommendations.

Experiments such as these can play a key role in engineering safer structures in high-risk areas.

A

B

C

FIGURE 20.17 Completed model reservoir being filled with water to normal depth prior to test runs (A). Mudflow scaled to correspond to 44,000 acre-feet is released from model Swift Creek into reservoir (B); as flow progresses into reservoir (C), water height increases and waves lap at the dam.

Photographs by J. E. Peterson, USGS Photo Library, Denver, CO.

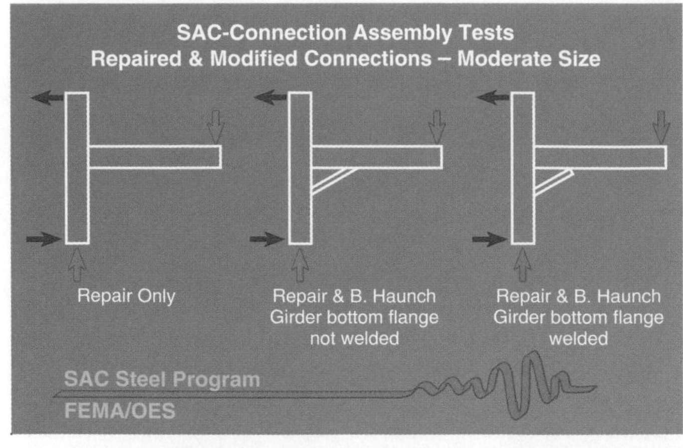

A

B

C

FIGURE 20.18 After examination revealed damage from the Northridge earthquake, engineers considered design alternatives (A), computed theoretical responses (B), and tested the modified designs (C).

Photographs courtesy Federal Emergency Management Agency and the SAC Joint Venture. Reprinted by permission.

CASE HISTORIES, OLD AND NEW

This section briefly outlines several case histories, in partial illustration of the range of problems encountered in engineering geology and some approaches to their solution.

The Leaning Tower of Pisa

The Leaning Tower of Pisa was built in several phases between A.D. 1173 and 1370. It began to tilt even before it was completed. The tilt has been self-reinforcing—that is, as the structure began to lean, more and more pressure was concentrated on the lower side, causing more flow of the unstable clay layers and more tilt. The 55-meter-high tower now leans more than 6 meters out of plumb. Over the first half of the twentieth century, the tilt increased about 0.15%, corresponding to an average rate of 165 millimeters (6.5 inches) per century. More disturbing were the early 1970s measurements that suggested that the tilt rate had increased to nearly twice that amount.

A nearby cathedral had also settled in the soft clays and had suffered enough structural damage that it had to be torn down and rebuilt. Because the Leaning Tower itself is still intact, there was hope that it could be saved if ways could be found to halt (or better, reduce) the leaning. Suggestions included physical methods, such as boring and selectively removing some material from the north (high) side of the foundation to reduce the tilt, and chemical methods, such as treating the clays to stiffen them and prevent further sinking of the settled south side.

In 1993, 900 tons of lead were placed on the high side of the Tower's base to anchor and counterweight it. In 1999, a careful plan of selective soil extraction was implemented to reduce the tilt a bit and stabilize the tower. By early 2001, the Leaning Tower, though still leaning, was pronounced stabilized, and was opened to visitors again.

The Panama Canal

The idea of a canal across the Isthmus of Panama was suggested as early as 1528 by one of Cortez's men, but it was not until 1880 that a French company began excavation. The initial design, patterned after the Suez Canal, was for a sea-level cut from Atlantic to Pacific, with no locks—this being the preference not of the engineers, but of those financing and directing the project. The Suez Canal, however, was excavated mainly through sand and soil, not rock, and in a climate with much less rainfall to promote sliding. Also, because of the higher elevation of the Isthmus of Panama, a much larger volume of material would have had to be excavated for a sea-level canal across Panama.

Little or no investigation of the geology of the Panama Canal site was made before excavation began. Such investigation would have revealed layers of young volcanic rocks, lava flows, and pyroclastic deposits, interbedded with some shale and sandstone. Because the rocks dip in many places toward the canal, excavation removed the support from some of these rock layers, which then tended to slide (figure 20.19A). Sliding was facilitated by very high rainfall, which averages 215 centimeters (85 inches) a year. Elsewhere, as the canal was dug, the weight of the rocks on the side caused flow and buckling of the rocks at the bottom of the excavation, which might rise 10 meters or more, requiring redigging to open the canal and removal of material from the sides to relieve some of the pressure (figure 20.19B).

Added to the geologic obstacles were heat, humidity, and rampant diseases, such as the mosquito-spread yellow fever, which caused many deaths among the workers. After seven frustrating years, the French company was bankrupt. A five-year lull in excavation was followed by renewed efforts by a second French firm, but it, too, later abandoned the attempt.

The United States acquired the rights to take over the project in 1903. At that point, the whole project design was reconsidered, and instead of a sea-level canal, a more complex design was chosen. A river was dammed to create a lake in mid-isthmus, and locks were to be built on either side to carry ships, in several stages, from sea level up to lake level and back down to sea level on the other side. One technical advantage of the lock design was that it would reduce the total volume of

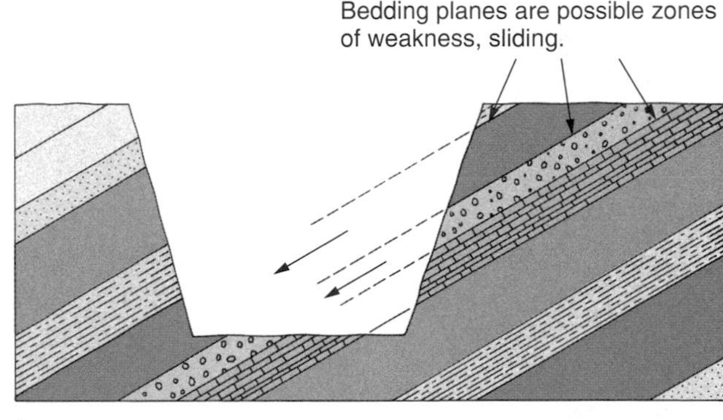

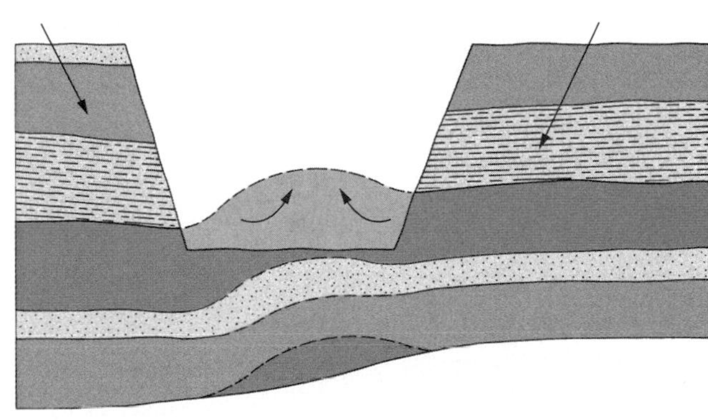

FIGURE 20.19 Geologic factors complicated construction of the Panama Canal. (A) Dipping beds sliding into a canal reduce the canal's capacity, requiring more excavation. (B) Removing the weight of the rocks in the canal may allow buckling of rocks below, which are under pressure from the rocks on the canal's sides.

excavation required. An ecological benefit was that the lock design would better separate the Atlantic and Pacific oceans; a concern with a sea-level canal and associated free flow of water and organisms between oceans was the possible impact on various ecosystems.

Even so, finishing the canal wasn't easy (figure 20.20). A geologist was not invited to examine the excavation and associated landslides until 1910, by which time many large slides were already beyond control. Sliding and excavation, in fact, continued long past the time of the canal's nominal completion and substantially increased its costs. Early consideration of the geology of the canal zone might not have eliminated the sliding problems, especially considering the lesser understanding of rock mechanics at the time of construction. However, some of the instability problems could have been anticipated and reduced, and the costs could certainly have been projected more accurately. As it was, the canal was not completed until 1914, more than thirty years after it was begun.

At that, it has continued to evolve. Parts were widened in the mid-1900s, and a new phase of widening was begun in 1992. A new set of locks is now under construction, scheduled for completion in 2015. The geology remains a challenge, as seen in figure 20.20B; better understanding of it, coupled with better equipment, makes modern excavation more manageable.

Boston's "Big Dig"

Like many cities, Boston experienced increasing traffic congestion over the last half of the twentieth century. The traffic load on the six-lane, elevated Central Artery nearly tripled from 1959 to the end of the century, creating estimated costs of half a billion dollars a year in wasted fuel in idling gridlocked cars, late delivery charges, and elevated accident rates. In a bold move, the city embarked on a massive transportation overhaul. The two major components of this Central Artery/Tunnel

A

B

FIGURE 20.20 (A) Excavation of the Panama Canal, 1907. (B) Landslides remain a threat today.

(A) Courtesy Panama Canal Commission; photograph (B) from USGS Open-File Report OFR 01-0276.

Project (quickly christened the "Big Dig") were an eight-to-ten-lane underground expressway to replace the Central Artery, and an extension of the Massachusetts Turnpike across Boston Harbor to Logan Airport via a four-lane tunnel under the harbor. The Central Artery had to be kept open throughout (it will be demolished when the replacement expressway is fully open), and various underground structures pass beneath trains and subways that had to remain operational. Logistics aside, the local geology was a challenge. The city is underlain by a complex assortment of glacial sediments and weakly metamorphosed mudstone; strong bedrock is over 120 feet deep in places. It is also densely built, with many historic buildings whose foundations could be destabilized by careless construction.

To keep the excavation walls stable while digging down to bedrock, the project relied on extensive use of *slurry walls* (figure 20.21). As trenches were dug, a clay-water slurry was poured in to replace the earth removed. The slurry is dense and stable enough to hold up the excavation walls until steel reinforcing beams can be sunk into place and concrete poured in—displacing the slurry—to form the permanent wall. In other places, the ground was purposely frozen to keep it firm during construction (just as is often done in permafrost areas).

FIGURE 20.21 Boston's "Big Dig" has posed some major geological-engineering challenges. Slurry walls keep excavations from collapsing and nearby building foundations from failing.
Photograph courtesy Massachusetts Turnpike Authority.

A different approach was taken for the tunnel under Boston Harbor. Tunnel segments were completed on land, towed to their intended positions, sunk into place, joined, and then pumped dry. (A similar technique was used for the construction of the Rotterdam Subway in Holland in 1964; there, the problem was a very shallow water table that made conventional tunnel construction in place problematic.)

The Big Dig hasn't been cheap. The size and complexity of the project have resulted in a cost of over $14 billion. But the objectives were achieved, with minimal disruption of the city's operation along the way, and the project, begun in 1991, was essentially complete in spring 2006, with all roads and tunnels open. (In July of that same year, some bolts holding up tunnel ceiling panels unexpectedly failed, and a motorist was killed. The investigation is ongoing as this is written, but the failure seems not to be geology-related.)

DAMS, FAILURES, AND CONSEQUENCES

When a building cracks and crumbles due to settling or subsidence, the repairs may be costly, but the toll rarely includes lives. Even when a bridge or tunnel collapses, only those few persons on or in it at the time are affected. A catastrophic dam failure, on the other hand, can destroy whole towns and take thousands of lives in a matter of minutes; one dam can impound a tremendous volume of water (figure 20.22). The motivation for intelligent and careful design and siting of dams is thus particularly strong. Unfortunately, past practice has frequently fallen short of the ideal. In chapter 8, we examined the Vaiont Reservoir disaster. In this section, we will look briefly at another well-documented dam disaster. Neither need have occurred; careful geologic investigation beforehand would have shown that neither site was suitable.

The St. Francis Dam

The St. Francis Dam was built in California, about 70 kilometers north of Los Angeles. The reservoir it impounded was principally intended for a water supply. The dam, 60 meters high and over 150 meters wide, was completed in 1926. The valley walls on one side of the dam were made of coarse sandstones and other sedimentary rocks. On the other side, the rocks were schists, mica-rich metamorphic rocks that tended to break along parallel planes that sloped toward the dam and reservoir. The contact between the two rock types, over which the dam was built, was a fault. The presence of that fault was known—and even mapped—before the dam was built.

A

B

FIGURE 20.22 Some 21 million acre-feet of water are impounded in Lake Mead behind Hoover Dam. (A) Close-up of the dam. (B) Regional overview gives a better appreciation of the volume of water impounded in the reservoir. Dam is at lower left (arrow); reservoir includes all the water visible above it and at right in photograph.
Photograph (A) © Corbis R.F.

On 12 March 1928, the dam abruptly failed. Several hundred people were drowned; about $10 million in damage was also done. Possible reasons for the failure became clear from subsequent laboratory tests, and the fault zone did not actually appear to have been a significant factor. The rocks of the valley walls had seemed to be sufficiently strong when dry. But the sedimentary rocks on one side of the dam were rich in gypsum, which is very soluble. When a fist-sized sample of one of the sedimentary rocks was placed in water, it bubbled out air, soaked up water, and disintegrated, in less than an hour, into a heap of sand and clay. This was hardly an appropriate rock type with which to contain a large reservoir of water! Reinvestigation of the site later also revealed a huge old landslide complex, a mile and a half long, within the schists. Recent studies and modeling now suggest that the filling of the reservoir reactivated the slide— somewhat as at Vaiont, Italy—which put pressure on the dam, uplifted its base, and promoted water seepage underneath. Apparently, the side of the dam destabilized and damaged by the sliding (which added great pressure on the structure, too) failed first, and a rapid, muddy washout began. This reduced the support for the rest of the system. As the water began to pour out, it further eroded the base of the dam. The other side of the dam then collapsed. Somewhat astonishingly, the central section of the dam remained standing (figure 20.23).

As with the Vaiont Dam case, a little more careful investigation in advance could have averted disaster. In the case of the St. Francis Dam, it seems that testing the reaction of the wall rocks to immersion in water and looking hard enough at the slopes to recognize the old landslide complex should have told the engineers everything they needed to know about site suitability!

FIGURE 20.23 Failure of the St. Francis Dam in California: The dam after its collapse. Note the waterline on the surrounding hills, marking the water level in the reservoir before failure.
Photograph by H. T. Stearns, USGS Photo Library, Denver, CO.

Other Examples and Construction Issues

These two dam disasters are not, alas, the only ones ever to have occurred. Between 1864 and 1876, some 100 dams in the United States failed because reservoir waters undermined their foundations. In 1900, the Austin, Texas, dam failed when water seeping through cracks in the foundation rocks lubricated clays and shales, and a 150-meter length of dam broke loose and slid 20 meters downstream. A Spanish dam failed in 1959 because of differential settling of supporting buttresses. In the same year, a French dam collapsed when the foundations slipped due to fractures in the gneissic rocks below. The list could go on and on.

Dams, more dramatically than most other structures, thus underscore the need for careful application of the principles of engineering geology before and during construction. They may also underscore the limits of our understanding. The Baldwin Hills reservoir was built near Los Angeles, California, between 1947 and 1951. The geologic setting was thoroughly studied and (so it was believed) adequately taken into account. The reservoir was underlain by an active fault zone; the design incorporated a seismic safety factor double that used by engineers in other earthquake-prone areas. Drainage was provided to prevent saturation of the foundations. The design was conservative, and provision was made for monitoring after construction. Nevertheless, on 14 December 1963, a differential slip of more than 10 centimeters occurred along a fault, water began to scour its way out, and two hours later, the dam was abruptly breached. Clearly, the designers did not understand the geology as well as they thought.

Geology and geography may, in fact, lead to the building of dams on faults. The zone of weakness created by a major fault zone may become a topographic low along which streams flow and lakes accumulate. This is a natural site, then, for a dam and reservoir, but damming such a stream necessarily involves placing a dam on or very close to the fault. The San Andreas is only one of many faults marked, in part, by the presence of natural sag ponds and artificial reservoirs (figure 20.24).

The majority of the approximately 50,000 dams built in the United States over the last hundred years were not built on faults, but they may nevertheless suffer from inadequate design. All dams are designed to survive any imaginable flood load, typically through the provision of a *spillway* that allows water to flow over or around the dam unimpeded when the water level in the reservoir becomes too high. The worst-case scenario for which a spillway is constructed is commonly a 1000-year flood. Still, as noted in chapter 6, it may be difficult to project just what volume of water that (or any other) flood represents, and changing land-use patterns may be altering the flood-frequency curves also. The 1993 flooding produced some unpleasant surprises in this regard in the Mississippi River basin. Dam failures have occurred, too, from dam washout as the sides or face of the spillway were scoured away by floodwaters. In 1981, the U.S. Army Corps of Engineers surveyed dams in this country. They identified 8639 of these dams as ones whose failure would cause serious property damage or loss of life. Of that number, one-third (2884) were identified as unsafe, and in over 80% of these cases, the problem was an inadequate or poorly designed spillway.

Thus, dam construction becomes increasingly complex and sophisticated (figure 20.25), aided by computers, laboratory tests, field studies, and scale-model experiments.

Three Gorges Dam (It's Not Only about Safety)

High-population developing nations such as China have large energy needs. Historically, China has relied on fossil fuels—especially coal—and the implications for greenhouse-gas emissions if per-capita energy consumption rises sharply in this nation of a billion-plus are sobering. Even now, particulate pollution from coal burning is a serious local health problem. Before energy became an issue, flooding was a major concern on the Yangtze River, the world's third largest. It has been estimated that in the twentieth century alone, half a million people died in floods of the Yangtze.

As long ago as 1919, damming the Yangtze was proposed as a means to control the flooding and enhance navigation. Structurally, a logical place was the Three Gorges area, a section of the river edged with soaring cliffs up to 1500 meters (about a mile) high and surrounded by high mountains. In 1953, Mao Zedong advocated construction of a large dam in the Three Gorges area, with a view also to development of a major hydropower facility. Finally, in 1992, the Chinese government announced a commitment to proceed, and construction began

FIGURE 20.24 Sag ponds and reservoirs abound along the San Andreas Fault; here, San Andreas and Crystal Springs reservoir. *Photograph by R. E. Wallace, USGS Photo Library, Denver, CO.*

A

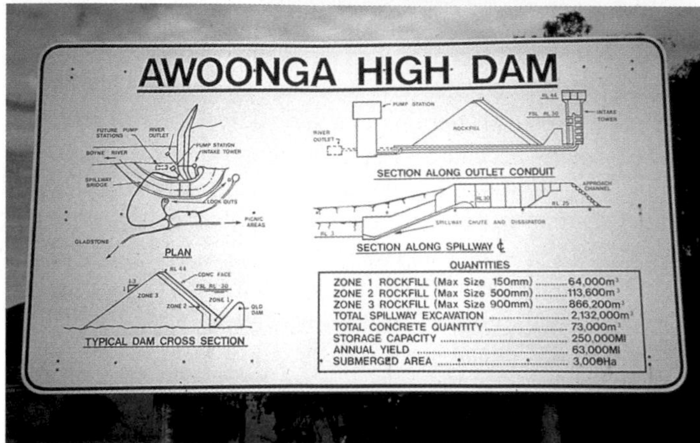

B

FIGURE 20.25 Awoonga High Dam in Australia reveals modern construction methods. (A) View of dam and reservoir; spillway in foreground. (B) Nearby billboard illustrates dam structure and presents some pertinent data.

in 1993, on a site within the Three Gorges area that offers solid granite bedrock for miles around.

The Three Gorges Dam (figure 20.26) is the largest in the world, 185 meters tall, and 2.3 kilometers (about a mile and a half) long. When the last of its planned 26 hydropower generators are installed in 2009, the complex will have a collective generating capacity of about 18,000 megawatts, making it also the world's largest hydropower project. (For comparison, a typical large nuclear-power plant has a capacity of about 1000 megawatts.) The complex could generate 10% of China's energy. Damage and loss of life from flooding should be greatly reduced. Impoundment of water upriver will make the Yangtze navigable by much larger ships farther upstream than before, and a system of locks to get the ships past the dam is part of the project.

As with all large projects of this kind, however, there are negative consequences, and there has been vocal opposition to the Three Gorges Dam since long before construction began. The reservoir behind the dam will ultimately stretch 600 km (about 375 miles) long, and it will submerge approximately 125,000 acres of prime farmland. Cities along the banks in the area the reservoir will occupy have to be abandoned and their

A

B

FIGURE 20.26 Three Gorges Dam site. (A) With the dam only partly complete, the sediment-laden Yangtze (gray) still flows past the site (left to right in this image); clear channel is probably an independent lock system for boats. (B) The dam was completed in May 2006. Most of the sediment is now trapped far upstream where the river enters the still-growing reservoir.

Images by Jesse Allen, Earth Observatory, using ASTER data made available by NASA/GSFC/MITI/ERSDAC/JAROS, and U.S./Japan ASTER Science Team.

residents moved elsewhere; an estimated one million people are being thus displaced. Cultural and archaeological features and sites that cannot be moved will be lost, and the scenic qualities of the Three Gorges area will be significantly altered. Ecologists fear the impact of habitat loss, especially on rare or endangered species that live in that region—for example, the Chinese river dolphin, which lives only in the Yangtze River. And safety concerns are, perhaps, natural with a project of such unprecedented scale.

Preliminary studies indicate some of the environmental consequences of the Three Gorges Dam project. The Yangtze River Basin is the largest in South Asia, and thousands of dams within it had already considerably reduced the sediment load reaching the East China Sea. Within two years after the Three Gorges Dam first began impounding water (2003), that sediment load was cut again by almost half. It is now not much over 100 million tons per year, compared to nearly 500 million tons per year in the mid-twentieth century. Observed results include increased channel erosion and erosion of the Yangtze delta. Future coastal erosion may threaten such important areas as Shanghai. Further, the large and growing reservoir behind the dam, with its broad surface subject to evaporation, is affecting local weather—the evaporation process cooling the air and altering circulation/wind patterns, with demonstrable changes in the distribution of rainfall not just locally but regionally. Monitoring the longer-term impacts should prove interesting.

Global concern now is focusing increasingly on the rapid proliferation of dams in some developing countries. In 2000, the World Commission on Dams issued a comprehensive review of large dams worldwide. For this purpose, "large dams" were defined as those over 15 meters high and those 5 to 15 meters high with reservoirs over 3 million cubic meters. Altogether, there are more than 45,000 such dams. Nearly half of these (22,000) are in China, and most of China's large dams have been built since 1950, which means an average of about one large dam *per day* built since that time. The reasons historically have been primarily flood control and irrigation, with increased hydropower-generating capacity a growing motive. In addition to such issues as displacement of people and flooding of habitat and cultural resources (figure 20.27), the World Commission noted serious safety concerns that have emerged as so many large dams have been built so quickly.

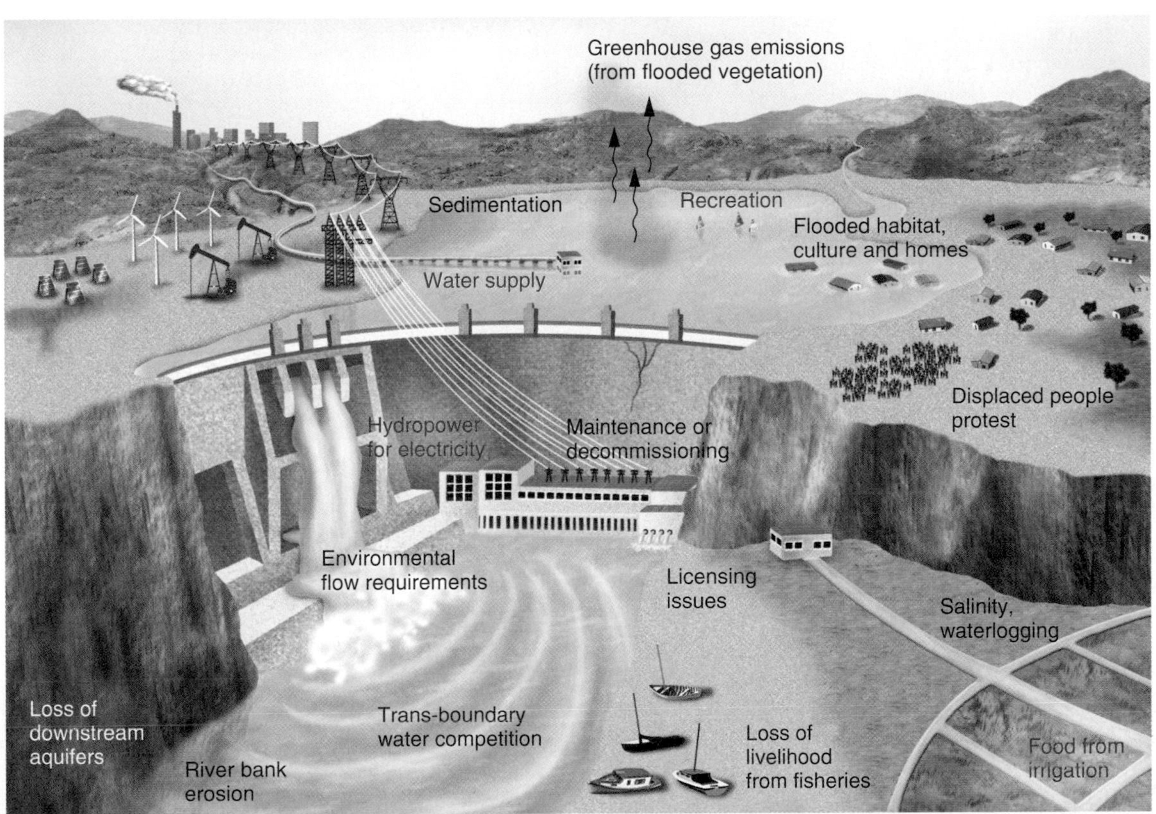

FIGURE 20.27 Benefits and issues involved in dam construction are many. Safety is an additional, very basic one, which can arise even with new dams, especially if design and construction are hurried.
After World Commission on Dams report, 2000.

Box 20.1 How Green Is My—Golf Course?

Jan, Henri, Ravi, and Judd stood on the first tee and squinted toward the green far below. Then Henri glanced past the green. "What the heck is that?" he asked. "Looks like a construction site."

"Not exactly; it's a landfill," said Jan. "In fact, you're standing on landfill right now." (See figure 1.)

"I'm *what*? Golfing on garbage?"

"Sure. It's a way to reuse the part of the landfill they've finished and bring in some revenue for the city, too. Don't worry," she added, as he wrinkled his nose. "It doesn't get smelly when the day warms up, if that's what you're expecting. Modern landfills are a lot cleaner than you think. And by the way," she added with a grin, pointing at the plastic-lumber bench Ravi and Judd were sitting on, "I think you have some of my old milk jugs there!"

The day worked out much better than the dubious Henri had expected. Except for a little muffled truck clatter when the wind blew from the active part of the operation and they were on that side of the course, and a cautionary sign or two about not fishing in the water hazards, he'd never have noticed the landfill, really. So he didn't even wince when Ravi said, at the end of the round, "Okay, next week let's go play in a floodplain!" (See figure 2.)

Historically, golf courses (of which the United States has more than 15,000) have often been roundly criticized as environmentally unfriendly land uses. By nature, they sprawl over a good deal of real estate (and as equipment has improved and even amateurs hit the ball farther, designers of new courses have tended to build them ever longer); a typical new golf course occupies about 150 acres. Natural habitat has often been cleared to make way for vast expanses of tidy close-mown grass, kept lush by lavish applications of water and fertilizer and protected by herbicides and pesticides. The water use was especially criticized in deserts and other dry settings where a golf course can consume a million gallons of water in a day; chemical runoff could pose pollution problems. However, things are changing for the better as a result of increased environmental sensitivity, reinforced by laws requiring environmental impact assessment of new construction.

Intensive research is producing turf grasses that require minimal use of pesticides and other chemical additives and need less water. Course designers are reducing the amount of area planted to short turf, leaving cactus-studded "waste areas" in desert courses in places the golfers shouldn't be hitting anyway, or letting much of the deep rough become very deep indeed (figure 3), perhaps even engaging in prairie restoration. Preservation of wetland or wooded habitat around the course can actually provide sanctuary for wildlife in an area otherwise being overrun by urban sprawl, while enhancing the course visually.

As our golfing friends discovered, building courses on landfills is increasingly common practice. The first such proposal was, in fact, put forth in 1930 in New York; now there are several dozen examples nationwide, a third of them in California, and more under development. The newest twist on this sequential-use concept is the true ugly-duckling transformation previously mentioned, beginning with a Superfund site near Butte, Montana, site of accumulated toxic wastes from an old copper-smelting operation in the town of Anaconda. The hazardous material has been carefully covered up, sealed with clay and plastic, and—with the help of Jack Nicklaus as designer, more than $40 million from ARCO (the site's last owner when the smelting ceased, faced with even higher potential costs with conventional site cleanup), and the collaboration and blessing of the EPA—the site has been transformed into an attractive, wildflower-decked golf course. Once completed, it was deeded to the local residents, so the operation not only eliminated a hazard, but now offers a new source of revenue to the region, which suffered economically from the cessation of mining and smelting activities. (See the "after" photograph, the opening photo of chapter 19.)

FIGURE 1 Golf course west of Chicago, Illinois, is built on finished section of landfill; ongoing landfill operation continues on adjacent land (in distance and at right).

FIGURE 2 This course is unusable when the river floods, but there is minimal damage cost—effective use of floodplain land. River Heights golf course, DeKalb, IL.

FIGURE 3 It adds to the challenge to have long expanses of unmanicured land on each hole, but is generally kinder to the environment.

SUMMARY

Engineering geology is concerned with making structures as safe and stable as possible, given various kinds of geologic hazards and potential problems. Land-use planning encompasses the same concerns as engineering geology, combined with additional geologic and nongeologic considerations. The aim of the land-use planner is to make the best possible use of limited land, taking all these factors into account. Using the same land for several purposes, either simultaneously or sequentially, is one approach to conserving the land resource. The success of both geological-engineering and land-use-planning efforts is heavily dependent on the accuracy and completeness of the data available to the individuals carrying out these tasks. As with pollution control, relative costs and benefits may have to be weighed in deciding how or whether to undertake a particular construction project.

Key Terms and Concepts

geographic information system 496
multiple use 491
sequential use 492

Exercises

Questions for Review

1. Describe the concepts of *multiple use* and *sequential use,* and give an example of each.
2. Maps of geologic or other factors are useful tools in land-use planning, but their usefulness can be limited by practical problems. Describe two such possible limitations.
3. How can the computer's ability to process large volumes of data assist in the planning process? How can programs be adjusted for different planning objectives?
4. Cite at least ten geologic considerations that might be important in siting an apartment building. How many of the same considerations would be relevant to siting a parking lot?
5. What is permafrost, and why is construction made more difficult by its presence?
6. How can engineers often minimize problems posed by unstable clays in near-surface rock and soil layers?
7. Dams are not infrequently built over faults. Why?
8. Identify and explain any five concerns or negative consequences associated with large dam projects.

For Further Thought

1. Most city or county planning offices have long-range plans for development in undeveloped areas. Seek out such plans for your area, if available, and investigate what kinds of considerations (geologic or otherwise) went into those plans.
2. Make a walking tour of your neighborhood or of another area to look for signs of careless or thoughtless engineering practice—buildings showing severe cracking or other structural damage from ground failure beneath, serious erosion of steep slopes, large puddles accumulated after rain because roads or buildings dam surface runoff, and so on. Is any particular problem especially common?
3. Explore the Internet for an example of geographic representation of geologic, cultural, or other data relevant to land-use decisions, and examine data at different scales, if available. See NetNotes for some places you might start; see the SAST site or Federal Geographic Data Committee site, or sites relating to geographic information systems (GIS).
4. Check the progress/current status of a major project such as the Big Dig (see NetNotes) or Three Gorges Dam via the Internet.

Suggested Readings/References

Allen, F. 1996. Inside the Panama Canal. *Invention and Technology* (Fall): 9–24.

Allison, M., et al. 2006. Hurricanes and the U.S. Gulf Coast: Science and sustainable rebuilding. *EOS* 87 (20 June): 245, 250.

Avery, B. 1996. Garbage in, golf out. *Golf Journal* (March/April): 12–17.

Covington, R. 2001. The Leaning Tower straightens up. *Smithsonian* (June): 40–47.

Gramling, C. 2006. China's massive dam alters weather. *Geotimes* (September): 14–15.

Hapgood, F. 2004. The underground cutting edge. *Invention and Technology* (Fall): 42–48. (Describes high-tech tunnel excavation.)

Howe, B. R. 2002. The plan to nuke Panama. *Invention and Technology* (Spring): 62–63. (Describes a plan to excavate a new canal with nuclear explosions.)

Löw, Simon, and Einstein, H. 2003. Building the world's longest railroad. *Geotimes* (February): 14–17.

Miller, B. A. 2002. Digging up Boston. *Geotimes* (October): 16–19.

Nixon, S. W. 2004. The artificial Nile. *American Scientist* 92 (March–April): 158–65.

Page, R. A., Boore, D. M., and Yerkes, R. F. 1995. *The Los Angeles Dam Story.* USGS Fact Sheet 096-95.

Pattison, K. 1998. Why did the dam burst? *Invention and Technology* (Summer): 22–31. (Describes the St. Francis Dam story, including modern re-analysis of the causes of failure.)

Petroski, H. 1993. The Panama Canal. *American Scientist* 81 (January/February): 13–18.

———. 2001. China Journal I. *American Scientist* 89 (May–June): 198–203. (Describes a visit to the Three Gorges Dam Project.)

———. 2003. St. Francis Dam. *American Scientist* 91 (March–April): 114–18.

———. 2004. Big Dig, big bridge. *American Scientist* 92 (July–August): 316–20.

Stuller, J. 1997. Golf gets back to nature, inviting everyone to play. *Smithsonian* (April): 56–67.

Xu, K., Milliman, J. D., Yang, Z, and Wang, H. 2006. Yangtze sediment decline partly from Three Gorges Dam. *EOS* 87 (9 May): 185, 190–91.

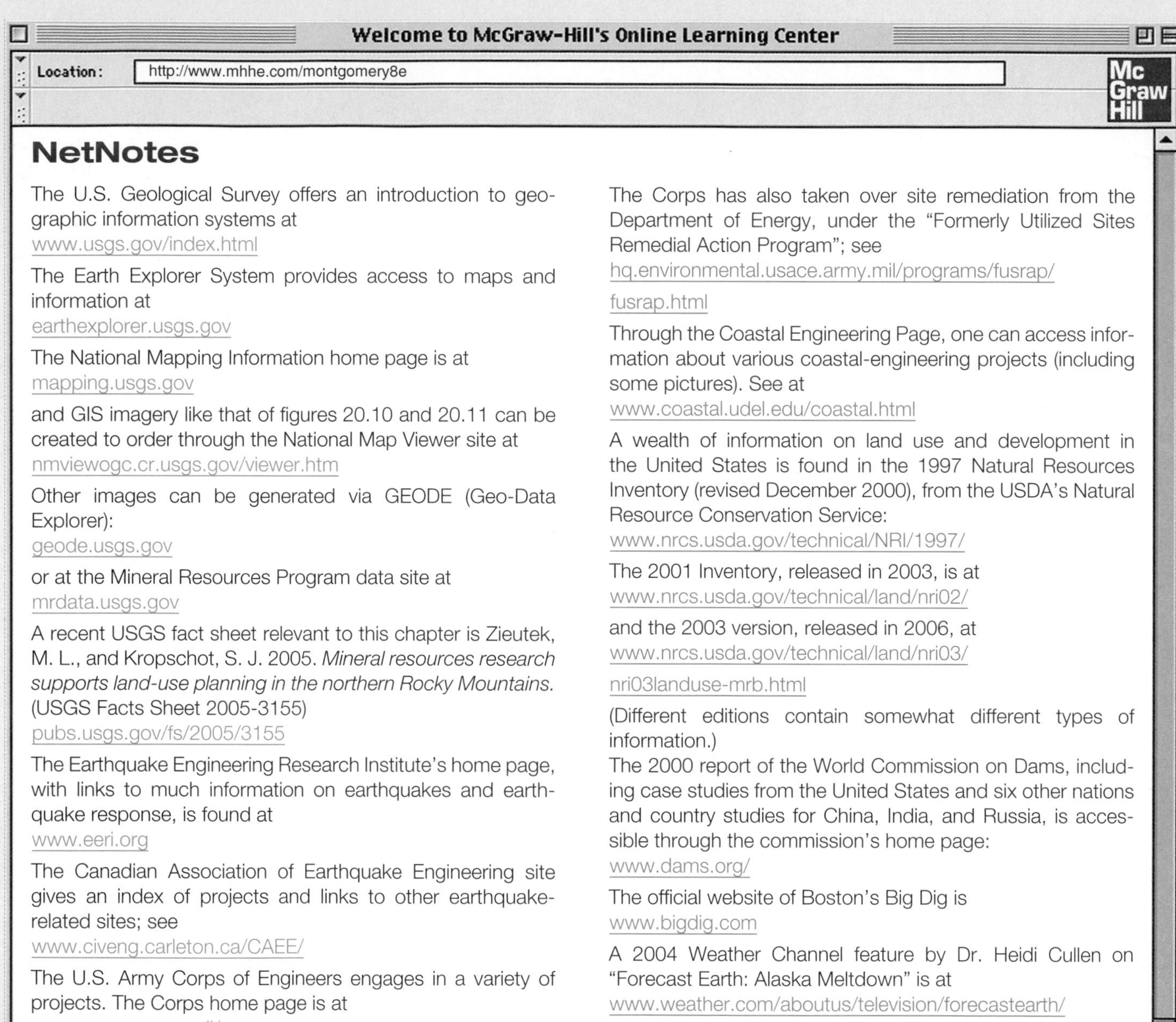

NetNotes

The U.S. Geological Survey offers an introduction to geographic information systems at
www.usgs.gov/index.html

The Earth Explorer System provides access to maps and information at
earthexplorer.usgs.gov

The National Mapping Information home page is at
mapping.usgs.gov

and GIS imagery like that of figures 20.10 and 20.11 can be created to order through the National Map Viewer site at
nmviewogc.cr.usgs.gov/viewer.htm

Other images can be generated via GEODE (Geo-Data Explorer):
geode.usgs.gov

or at the Mineral Resources Program data site at
mrdata.usgs.gov

A recent USGS fact sheet relevant to this chapter is Zieutek, M. L., and Kropschot, S. J. 2005. *Mineral resources research supports land-use planning in the northern Rocky Mountains.* (USGS Facts Sheet 2005-3155)
pubs.usgs.gov/fs/2005/3155

The Earthquake Engineering Research Institute's home page, with links to much information on earthquakes and earthquake response, is found at
www.eeri.org

The Canadian Association of Earthquake Engineering site gives an index of projects and links to other earthquake-related sites; see
www.civeng.carleton.ca/CAEE/

The U.S. Army Corps of Engineers engages in a variety of projects. The Corps home page is at
www.usace.army.mil/

The Corps has also taken over site remediation from the Department of Energy, under the "Formerly Utilized Sites Remedial Action Program"; see
hq.environmental.usace.army.mil/programs/fusrap/fusrap.html

Through the Coastal Engineering Page, one can access information about various coastal-engineering projects (including some pictures). See at
www.coastal.udel.edu/coastal.html

A wealth of information on land use and development in the United States is found in the 1997 Natural Resources Inventory (revised December 2000), from the USDA's Natural Resource Conservation Service:
www.nrcs.usda.gov/technical/NRI/1997/

The 2001 Inventory, released in 2003, is at
www.nrcs.usda.gov/technical/land/nri02/

and the 2003 version, released in 2006, at
www.nrcs.usda.gov/technical/land/nri03/nri03landuse-mrb.html

(Different editions contain somewhat different types of information.)

The 2000 report of the World Commission on Dams, including case studies from the United States and six other nations and country studies for China, India, and Russia, is accessible through the commission's home page:
www.dams.org/

The official website of Boston's Big Dig is
www.bigdig.com

A 2004 Weather Channel feature by Dr. Heidi Cullen on "Forecast Earth: Alaska Meltdown" is at
www.weather.com/aboutus/television/forecastearth/alaska.html

APPENDIX A

Geologic Time, Geologic Process Rates

INTRODUCTION

Much of the understanding of geologic processes, including the rates at which they occur, and therefore the kinds of impacts they may have on human activities, is made possible through the development of various means of "telling time" in geologic systems. In this appendix, we will explore several of these methods. A final section examines some applications of geologic age determinations to the study of process rates.

RELATIVE DATING

Arranging Events in Order

Before any ways of establishing numerical ages of rocks or geologic events were known, it was sometimes at least possible to place a sequence of events in the proper order. Among the earliest recognized efforts in this direction were those of Nicolaus Steno. In 1669, he set forth two very basic principles that could be applied to sedimentary rocks. The first, the **Principle of Superposition** (figure A.1), pointed out that, in an undisturbed pile of sediments or sedimentary rocks, those on the bottom were deposited first, followed in succession by the layers above them, ending with the youngest on top. (Today, this idea may seem so obvious as not to be worth stating, but at the time it represented a real step forward in thinking logically about rocks.) The second, the **Principle of Original Horizontality,** was based on the observation that sediments are deposited in approximately horizontal, flat-lying layers. Therefore, if one comes upon sedimentary rocks in which the layers are folded or dipping steeply, they must have been displaced or deformed after deposition and solidification into rock (figure A.2).

In later centuries, these ideas were extended to work igneous rocks into such sequences (figure A.3). If an igneous rock cuts across layers of sedimentary rocks, the sedimentary rocks must have been there first, the igneous rock introduced later. This is sometimes called the *Principle* (or *Law*) *of Crosscutting Relationships,* and in fact it applies whether the rocks that are

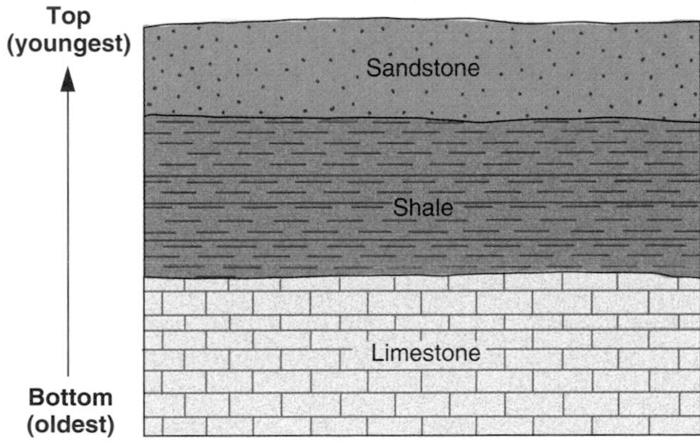

FIGURE A.1 The Principle of Superposition: The rocks on the bottom of a sequence of undisturbed sedimentary rocks were deposited first, and the depositional ages become younger higher in the sequence.

FIGURE A.2 The Principle of Superposition and Principle of Original Horizontality together indicate that the sequence of events shown here in the Grand Canyon was: deposition of the lower layers (dark red and gray); tilting and erosion; deposition of the still-flat-lying upper layers (green, cream, and light red).
© The McGraw-Hill Companies, Inc./Doug Sherman, photographer.

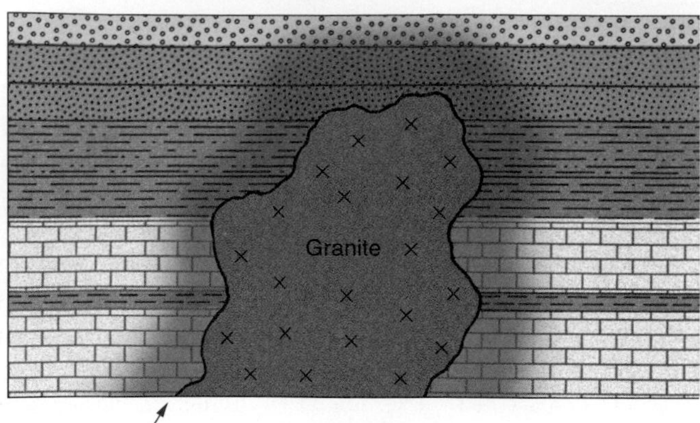

FIGURE A.3 An igneous rock crosscutting a sedimentary sequence or other rock must postdate the rocks it cuts across.

FIGURE A.4 The darker rock is a schist that was invaded by the magma that formed the (lighter-colored) granite, which picked up pieces of schist. The schist is thus older than the granite.

crosscut are sedimentary or not, though the relationships may be easier to recognize if they are. Often, there is a further clue to the correct sequence: The hotter igneous rock may have "baked," or metamorphosed, preexisting rocks immediately adjacent to it, so again the igneous rock must have come second. If an igneous rock contains pieces of other rock types (figure A.4), those pieces must have been picked up as solid chunks by the invading magma, so the rocks making up the inclusions must predate the host rock (*Principle of Inclusion*).

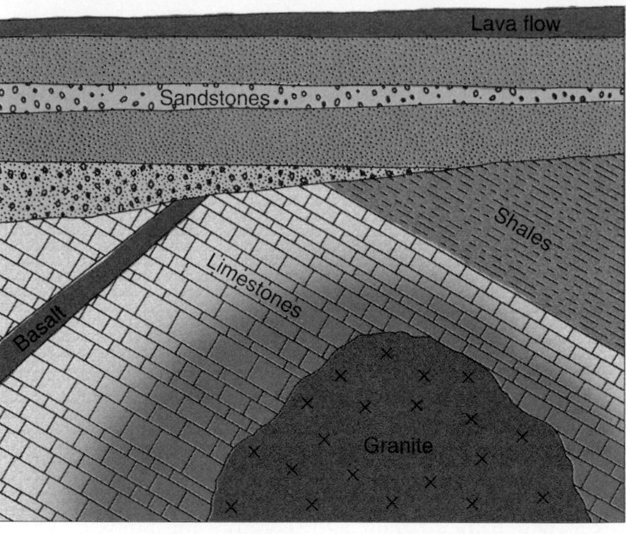

FIGURE A.5 Sorting out relative ages of rock in an outcrop: Of the sedimentary rocks, the limestones must be the oldest (Principle of Superposition), followed by the shales. The granite intrusion and basalt must both be younger than the limestone they crosscut; the granite has also metamorphosed the surrounding limestone. It is not possible to tell whether the igneous rocks predate or postdate the shales (because they are not crosscut or metamorphosed by the igneous rocks), or to determine whether the sedimentary rocks were tilted before or afer the igneous rocks were emplaced. After the limestones and shales were tilted and the basalt injected, they were eroded, and then the sandstones were deposited on top. Finally, the lava flow covered the entire sequence.

Such geologic common sense can be applied in many ways. If a strongly metamorphosed sedimentary rock is overlain by a completely unmetamorphosed one, for instance, the metamorphism must have occurred after the first sedimentary rock formed but before the second. Quite complex sequences of geologic events can be unraveled by taking into account such principles (see figure A.5 for an example).

Correlation

Fossils play a role in the determination of relative ages, too. The concept that fossils could be the remains of older life-forms dates back at least to the ancient Greeks, but for some time it fell out of favor. The idea was seriously revived in the 1700s, and around the year 1800, William Smith put forth the **Law of Faunal Succession.** The basic principle involved is that, through time, life-forms change, old ones disappear from and new ones appear in the fossil record, but the same form is never exactly duplicated at two different times in history. (The same concept could be applied to other contexts, such as car styles in a junkyard with layer upon layer of debris: Car models come and go, and while stylistic features from an earlier era may be revived in later models, each model, as a whole, corresponds to a unique period of manufacture.) This principle, in turn, implies that when one finds exactly the same type of fossil preserved in two rocks, even if the rocks are quite different compositionally and geographically widely separated, they should be the same age. Smith's law thus allowed age *correlation* between rock units exposed in different places (figure A.6). A limitation on its usefulness is that it can be applied only to rocks in which fossils are preserved, which are almost exclusively sedimentary rocks.

The foregoing ideas were all useful in clarifying age relationships among rock units. They did not, however, help answer questions like: How old is this granite? How long did it take to deposit this limestone? How recently, and over how long a period, did this apparently extinct volcano erupt? Has this fault been active in modern times?

Uniformitarianism

Geologic processes were the focus of another early worker—the physician, farmer, and part-time geologist James Hutton. Hutton is widely credited with developing and popularizing the concept of **uniformitarianism.** Uniformitarianism is sometimes described briefly by the phrase "the present is the key to the past." What Hutton meant by this is that because the same physical laws have operated throughout the earth's history, by studying present, active geologic processes and their products, geologists can infer much about how rocks were formed and changed in the past. Today, a volcanic eruption and the hardening of flowing magma into basaltic rock suggest the source of an ancient basalt flow even if geologists cannot now find the volcano from which it erupted. A broad, white sandy beach or a windswept desert might offer some useful insights about the origins of sandstones. Retreating ice sheets in Greenland and Antarctica leave moraines behind, which help geologists to understand the source and significance of the moraines found in the midwestern United States and in Canada.

Uniformitarianism has sometimes been misunderstood to mean that the *rates* of geologic processes have been the same through time as well, that by observing processes like erosion, sedimentation, and seafloor spreading as they now occur, one can know the rate at which they occurred in the past. Unfortunately, this is not true, and was not really part of uniformitarianism as originally conceived. Volcanism may have involved the same types of physical and chemical processes throughout geologic history, but it might have been much more active early in the earth's history, before the earth cooled appreciably, than it is now; rates of chemical sedimentation, as well as the proportions of different minerals in the sediments, could have been very different in the past, when the temperature and chemistry of the oceans were also different; and so on. In short, even uniformitarianism does not allow numerical answers to questions about geologic process rates in the past.

HOW OLD IS THE EARTH?

Geologists and nongeologists alike have been fascinated for centuries with the very basic question of the earth's age. Many have attempted to answer it, but with little success until the twentieth century.

Early Efforts

One of the earliest widely publicized determinations of the age of the earth was the seventeenth-century work of Archbishop Ussher of Ireland. He painstakingly and literally counted up the generations in the Bible and arrived at a date of 4004 B.C. for the formation of the earth. This very young age was impossible for most geologists to accept; the complex geology of the modern earth seemed to require far longer to develop.

Even less satisfactory from that point of view was the estimate of the philosopher Immanuel Kant. He tried to find a maximum possible age for the earth, assuming that the sun

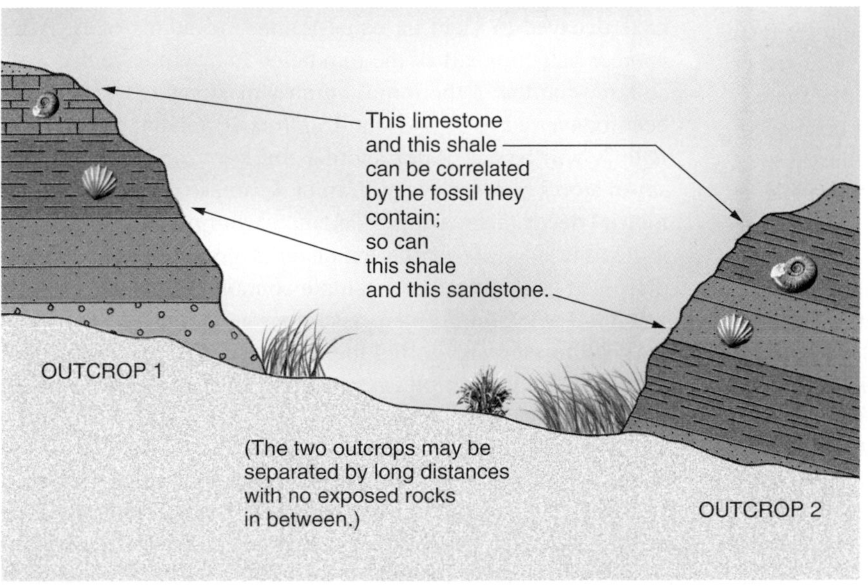

FIGURE A.6 Similarity of fossils suggests similarity of ages even in different and quite widely separated rocks.

had always shone down on earth and that the sun's tremendous energy output was due to the burning of some sort of conventional fuel. But a mass of fuel the size of the sun would burn up in only 1000 years, given the rate at which the energy is released, plainly an impossible result in view of several millennia of recorded human history. Kant, of course, knew nothing of the nuclear fusion that actually powers the sun.

Nineteenth-Century Views

About 1800, Georges L.L. de Buffon attacked the problem from another angle. He assumed that the earth was initially molten, modeled it as a ball of iron, and calculated how long it would take this quantity of iron to cool to present surface temperatures: 75,000 years. To most geologists, this still was not nearly long enough.

Around 1850, physicist Hermann L.F. von Helmholtz took the approach of supposing that the sun's luminosity was due to infall of particles into its center, converting gravitational potential energy to heat and light. This gave a maximum age for the sun (and presumably earth) of 20 to 40 million years. Again, his assumptions were wrong, so his answer was also.

In the late 1800s, Lord William T. Kelvin reworked Buffon's calculations, modeling the earth more realistically in terms of rock properties rather than those of metallic iron. Interestingly, he, too, arrived at an age of 20 to 40 million years for the earth. What he did not take into account, because natural radioactivity was then unknown, was that some heat is continually being *added* to the earth's interior through radioactive decay, so it has taken longer to cool down to its present temperature regime.

The calculations went on, and something was wrong with each. In 1893, U.S. Geological Survey geologist Charles D. Walcott tried to compute the total thickness of the sedimentary rock record throughout geologic history and, dividing by typical modern sedimentation rates, to estimate how long such a pile of sediments would have taken to accumulate. His answer was 75 million years. Walcott was hampered in his efforts by several factors, including the gaps in many sedimentary rock sequences (periods during which no sedimentation occurred or some sediments were eroded away), and the reality that there is no one spot on earth where sediments have always been accumulating. He also had no good idea of the total thickness of sediments deposited in the time before organisms capable of preservation as fossils became widespread, which turns out to be most of earth's history.

In 1899, physicist John Joly published calculations based on the salinity of the oceans. Taking the total dissolved load of salts delivered to the seas by rivers, and assuming that the ocean was initially pure water, he determined that it would take about 100 million years for the present concentrations of salts to be reached. Aside from Joly's assumption that rates of weathering and salt input into the oceans throughout earth's history were constant, he did not consider that the buildup of salts is slowed by the removal of some of the dissolved material—for example, as chemical sediments.

So the debate continued, imaginative and frequently heated, until the discovery of radioactivity provided a much more powerful and accurate tool with which to undertake a solution.

RADIOMETRIC DATING

Henri Becquerel did not set out to solve any geologic problems nor even to discover radioactivity. He was interested in a curious property of some uranium salts: If exposed to light, the salts continued to emit light for a while afterward even in a dark room (phosphoresce). He had put a vial of uranium salts away in a drawer on top of some photographic plates well wrapped in black paper. When he next examined the photographic plates, they were fogged with a faint image of the vial, as if they had somehow been exposed to light right through the paper. The uranium was emitting something that could pass through light-opaque materials. We now know that uranium is one of several substances that are naturally radioactive, that undergo spontaneous decay. Once the phenomenon of radioactive decay was reasonably well understood, physicists and geoscientists began to realize that it could be a very useful tool for investigating earth's history.

Radioactive Decay and Dating

As noted in chapter 16, one key property of any particular radioisotope is that it decays at a constant, characteristic rate, with a distinct half-life, which can be determined in the laboratory. One can then, in principle, use the relative amounts of a decaying isotope (parent) and the product isotope into which it decays (daughter) to find the age of the sample. Figure A.7 illustrates the fundamental idea. Suppose that parent isotope A decays to daughter B with a half-life of 1 million years. In a rock that contained some A and no B when it formed, A will gradually decay and B will accumulate. After 1 million years (one half-life of A), half of the atoms of A initially present will have decayed to yield an equal number of atoms of B. After another half-life, half of the remaining half will have decayed, so three-fourths of the initial number of atoms of A will have been converted to B and one-fourth will remain; the ratio of B to A will be 3:1. After another million years, there will be seven atoms of B for every atom of A, and so on. This is **exponential decay,** decreasing abundance over time in a pattern that is just the opposite of the exponential growth of populations discussed in chapter 1. If a sample contains 24,000 atoms of B and 8000 atoms of A, then, assuming no B was present when the rock formed and noting the 3:1 ratio of B to A, we would conclude that the sample was two half-lives of A (in this case, 2 million years) old.

The foregoing is, of course, an oversimplification of the dating of natural geologic samples. Many samples contain some of the daughter isotope as well as the parent at the time of formation, and correction for this must be made. Many other samples have not remained chemically closed throughout their

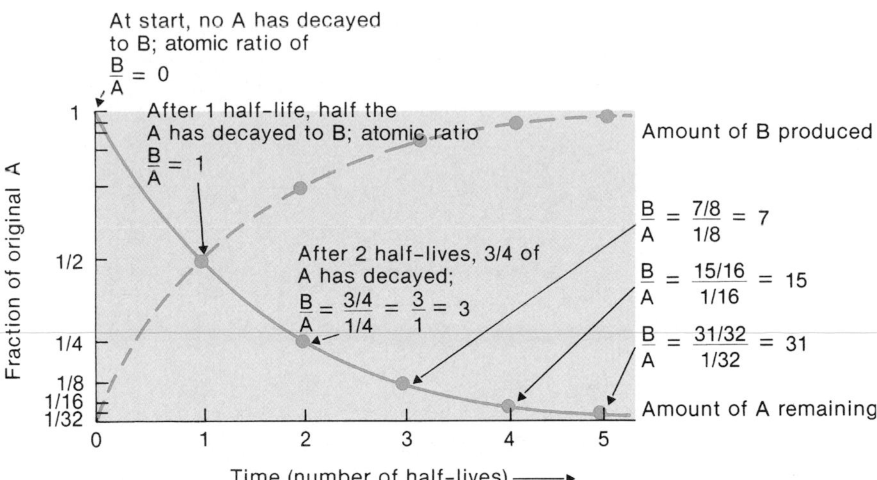

FIGURE A.7 Decay of radioactive parent A proceeds with complementary accumulation of daughter B. All ratios represent relative numbers of atoms.

histories; they have gained or lost atoms of the parent or daughter isotope of interest, which would make the apparent date incorrect. In the case of samples with a complex history—an igneous rock that has been metamorphosed once or twice and also weathered, for example—it can be difficult both to obtain a date and to decide which event, if any, is being dated! Fortunately, there are ways to recognize, and in many cases correct for, disturbances in isotopic systems. For a somewhat more thorough discussion of the complexities of dating of geologic systems, refer to pertinent texts in the reference list at the end of this appendix.

Choice of an Isotopic System

Several conditions must be satisfied by an isotopic system to be used for dating geologic materials. The parent isotope chosen must be abundant enough in the sample for its quantity to be measurable. Either the daughter isotope must not normally be incorporated into the sample initially, or there must be a means of correcting for the amount initially present. The half-life of the parent must be appropriate to the age of the event being dated: long enough that some parent atoms are still present but short enough that some appreciable decay and accumulation of daughter atoms have occurred. A radioisotope with a half-life of ten days would be of no use in dating a million-year-old rock; the parent isotope would have decayed away completely long ago, and there would be no way to tell how long ago. Conversely, a radioisotope with a half-life of 10 billion years would be useless for dating a fresh lava flow because the atoms of the daughter isotope that would have accumulated in the rock would be too few to be measurable. Only about half a dozen isotopes are widely used in dating geologic samples. They include both uranium-235 and uranium-238 (with half-lives of 700 million years and 4.5 billion years, respectively), potassium-40 (1.3 billion years), rubidium-87 (49 billion years), and carbon-14 (5730 years, meaning that carbon-14 is useful only for quite young samples, such as remains from prehistoric civilizations).

Radiometric and Relative Ages Combined

Radiometric dates (dates determined using radioisotopic methods) are sometimes imprecisely called "absolute" ages to distinguish them from the relative ages determined as described earlier. For various reasons, accurate radiometric dates cannot be determined for many rocks and fossils, so "absolute" and relative dating methods are often used in conjunction (figure A.8). Undateable sedimentary rocks crosscut by an igneous intrusion must at least be older than the igneous rock, which may be dateable; a fossil found in a rock sandwiched between two dateable lava flows has its age bracketed by the ages of the flows; and so on.

Radiometric dating has proven an extremely powerful tool for studying the earth's history. Because the earth is geologically still very active, no rocks have been preserved unchanged since its formation. However, the dating of meteorites and moon rocks, coupled with geochemical evidence that the whole solar system formed at the same time, has led to the determination of an age of about 4.5 billion years for the earth. The oldest samples from the continents are close to 4 billion years old. Sea floor is much more readily destroyed/recycled by plate tectonics; the oldest seafloor samples recovered are only about 200 million years old.

THE GEOLOGIC TIME SCALE

Particularly in the days before radiometric dating, it was necessary to establish some subdivisions of earth history by which particular intervals of time could be indicated. This initially was done using those rocks with abundant fossil remains. The resultant time scale was subsequently refined with the aid of radiometric dates.

The Law of Faunal Succession and the appearance and disappearance of particular fossils in the sedimentary rock record were first used to mark the boundaries of the time units. The

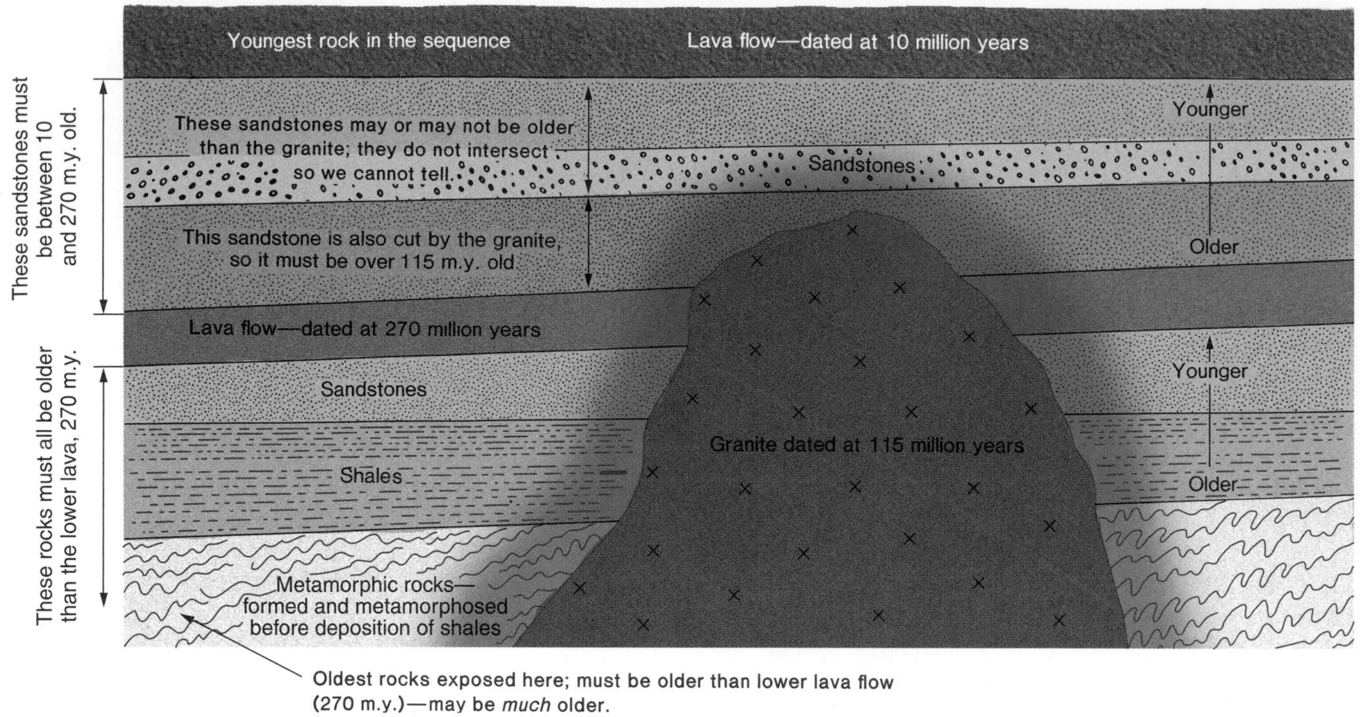

FIGURE A.8 "Absolute" and relative dating together can put constraints on ages of units not dateable directly.

principal divisions were the *eras*—**Paleozoic, Mesozoic,** and **Cenozoic,** meaning, respectively, "ancient life," "intermediate life," and "recent life." The eras were subdivided into *periods* and the periods into *epochs.* The names of these smaller time units were assigned in various ways. Often, they were named for the location of the *type section,* the place where rocks of that age were well exposed and that time unit was first defined. For example, the Jurassic period is named for the Jura Mountains of France. Other periods are named on the basis of some characteristic of the type rocks. The Cretaceous period derives its name from the Latin *creta* ("chalk"), for the chalky strata of that age in southern England and northern Europe; rocks of Carboniferous age commonly include coal beds. All of the relatively unfossiliferous rocks that seemed to be older (lower in the sequence) than the Cambrian period were lumped together as pre-Cambrian, later formally named **Precambrian.**

With the advent of radiometric dating, numbers could be attached to the units and boundaries of the time scale. It became apparent that, indeed, geologic history spanned long periods of time and that the most detailed part of the scale, the **Phanerozoic** (Cambrian and later), was by far the shortest, comprising less than 15% of earth history. The approximate time framework of the Phanerozoic is shown in table A.1.

The Precambrian has continued to pose something of a problem. A unit that spans 4 billion years of time seems to demand subdivision—but, in the virtual absence of fossils, on what basis? The principal division that has been made splits the Precambrian into two nearly equal halves—the *Archean* ("ancient") and *Proterozoic* ("pre-life"). The Archean spans the time from the earth's formation to 2500 million years ago, the Proterozoic from 2500 million years ago to the start of the Cambrian (570 million years ago). Universal agreement has not yet been reached on how to subdivide these 2-billion-year blocks of time further.

GEOLOGIC PROCESS RATES

The rates at which geologic processes occur can be estimated in a variety of ways, many of which rely on radiometric dating techniques. For example, the rate of seafloor spreading away from a particular spreading ridge can be found by dividing the age of a sample of sea floor into its present distance from the ridge. If a 10-million-year-old seafloor sample is now 400 kilometers from the ridge, then, on average, it has been moved away from the ridge at a rate of

$$\frac{400 \text{ km}}{10{,}000{,}000 \text{ yr}} \times \frac{1000 \text{ m}}{1 \text{ km}} \times \frac{100 \text{ cm}}{1 \text{ m}} = \frac{4 \text{ cm}}{\text{yr}}$$

The minimum rate of uplift of rocks in a mountain range might be estimated from the age of marine sedimentary rocks in the

TABLE A.1

The Phanerozoic Time Scale

Era	Period	Epoch	Start of Interval*	Distinctive Life-Forms
		Holocene	0.1	modern humans
	Quaternary	Pleistocene	2	Stone-Age humans
		Pliocene	5	
		Miocene	24	flowering plants common
		Oligocene	37	ancestral pigs, apes
		Eocene	58	ancestral horses, cattle
Cenozoic	Tertiary	Paleocene	66	
	Cretaceous		144	dinosaurs become extinct; flowering plants appear
	Jurassic		208	birds, mammals appear
Mesozoic	Triassic		245	dinosaurs, first modern corals appear
	Permian		286	rise of reptiles, amphibians
	Carboniferous		360	coal forests; first reptiles, winged insects
	Devonian		408	first amphibians, trees
	Silurian		438	first land plants, coral reefs
	Ordovician		505	first fishlike vertebrates
Paleozoic	Cambrian		570	first widespread fossils

*Dates, in millions of years before the present, from compilation by the Geological Society of America for the Decade of North American Geology

mountains, once deposited under water, now high above sea level. Such uplift rates are typically 1 centimeter per year or less, often much less. Beaches formed on Scandinavian coastlines during the last Ice Age have been rising since the ice sheets melted and their mass was removed from the land. From the beach deposits' ages and present elevation, uplift rates can be approximated. Typical rates of this postglacial rebound are on the order of 1 centimeter per year. Radiometric dating has shown that a small volcano may stay active for over 100,000 years, a major volcanic center for 1 to 10 million years. To build a large mountain range may take 100 million years.

Rates of continental erosion due to weathering can be deduced from the loads of major rivers draining the continents, dividing the volume of rock those loads represent by the surface area of the corresponding drainage basin(s). The North American continent is being leveled by erosion at an average rate of 0.03 millimeters per year, or about a tenth of an inch per century. The larger the area over which such measurements are made, however, the greater the potential for large local deviations from the average due to special local soil or weather conditions.

Finally, one must also be somewhat cautious about extrapolating present process rates too far into the past or future. This point is admirably illustrated by the following passage from Mark Twain's *Life on the Mississippi,* in which he speculates on the implications of the shortening of the Mississippi River by the cutoff of meanders:

> In the space of one hundred and seventy-six years the lower Mississippi has shortened itself two hundred and forty-two miles. This is an average of a trifle over one mile and a third per year. Therefore, any calm person, who is not blind or idiotic, can see that in the Old Oolitic Silurian Period just a million years ago next November, the Lower Mississippi River was upwards of one million three hundred thousand miles long, and stuck out over the Gulf of Mexico like a fishing rod. And by the same token any person can see that seven hundred and forty-two years from now the Lower Mississippi will be only a mile and three-quarters long, and Cairo and New Orleans will have joined their streets together, and be plodding along comfortably under a single mayor and a mutual board of aldermen. There is something fascinating about science. One gets such wholesale returns of conjecture out of such a trifling investment of fact.

SUMMARY

Before the discovery of natural radioactivity, only relative age determinations for rocks and geologic events were possible. Field relationships and fossil correlations were the principal methods used for this purpose. Radiometric dating has made possible quantitative age measurements of many geologic materials and events, assignment of dates to the units of the geologic time scale, and determination of the age of the earth. Radiometric dates have also been used to explore the rates at which different kinds of geologic processes occur and have considerably advanced understanding of earth's development. However, while observations of present geologic processes can be used to understand past geologic history, it cannot be assumed that all of those processes have proceeded at rates comparable to those presently observed.

Key Terms and Concepts

Cenozoic 518
exponential decay 516
Law of Faunal Succession 514
Mesozoic 518
Paleozoic 518
Phanerozoic 518
Precambrian 518
Principle of Original Horizontality 513
Principle of Superposition 513
uniformitarianism 515

Suggested Readings/References

Cloud, P. 1988. *Oasis in space: Earth history from the beginning.* New York: W.W. Norton.

Dickin, A. P. 2005. *Radiogenic isotope geology.* 2d ed. New York: Cambridge University Press.

Faure, G., and Mensing, T. M. 2005. *Isotopes: Principles and Applications.* 3d. ed. New York: John Wiley & Sons.

Hurley, P. M. 1959. *How old is the Earth?* Garden City, N.Y.: Doubleday. (This is a classic early work tracing the development of geochronology.)

Johns, R. B., ed. 1986. *Biological markers in the sedimentary record.* New York: Elsevier.

Noller, J. S., Sowres, J. M., and Lettis, W. R., eds. 2002. *Quaternary geochronology: Methods and applications.* Washington, D.C.: American Geophysical Union.

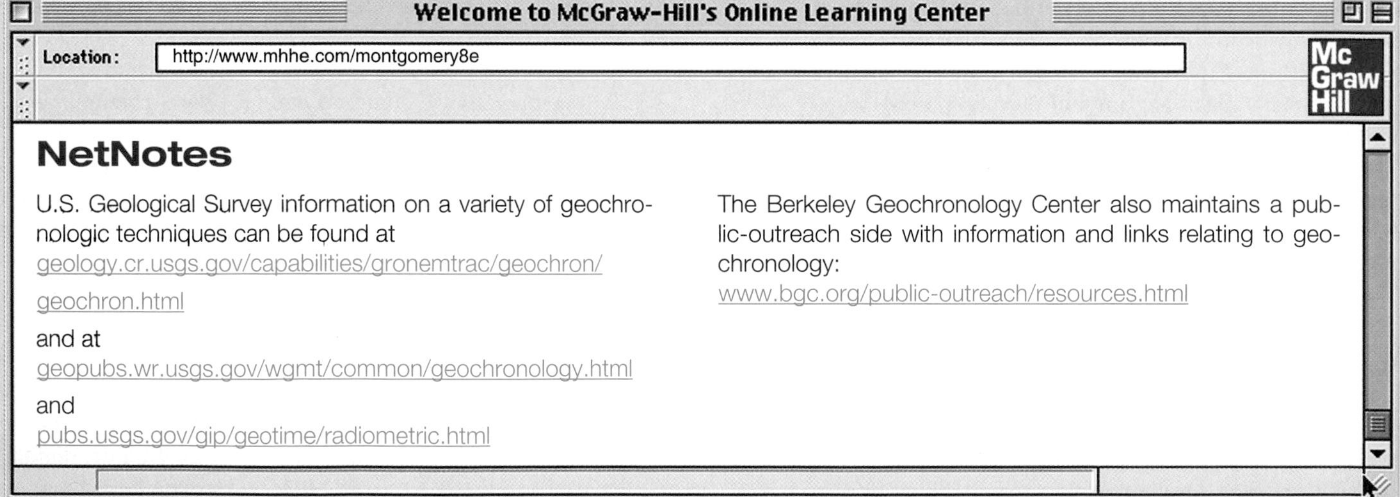

NetNotes

U.S. Geological Survey information on a variety of geochronologic techniques can be found at
geology.cr.usgs.gov/capabilities/gronemtrac/geochron/geochron.html
and at
geopubs.wr.usgs.gov/wgmt/common/geochronology.html
and
pubs.usgs.gov/gip/geotime/radiometric.html

The Berkeley Geochronology Center also maintains a public-outreach side with information and links relating to geochronology:
www.bgc.org/public-outreach/resources.html

Introduction to Topographic and Geologic Maps and Remotely Sensed Imagery

MAPS AND SCALE

Many kinds of information can be presented in map form. Topographic and geologic maps are the two kinds most frequently used by geologists. This section presents a brief introduction to both types of maps.

A point to note at the outset is that, in the United States, topographic and geologic maps commonly carry English, rather than metric, units. Many of these maps were drawn before there was any move toward the adoption of metric units in the United States. The task of redrawing maps to convert to metric units is formidable, particularly with topographic maps. (A metric map series is being prepared by the U.S. Geological Survey, but it will be many years before its completion.)

A basic feature of any map is the map **scale,** a measure of the size of the area represented by the map. Map scales are reported as ratios—1:250,000, 1:62,500, and so on; or equivalently in words, for example, "one to two hundred fifty thousand." On a 1:250,000 scale map, a distance of 1 inch on the map represents 250,000 inches (almost 4 miles) in reality; on a 1:62,500 scale map, 1 inch equals about 1 mile of actual distance. The larger the scale factor, the more real distance is represented by a given distance on the map. As a result, the fineness of detail that can be represented on a map is reduced as the scale factor is increased. The choice of scale factor often involves a compromise between minimizing the size of the map (for convenience of use) or the number of maps required to cover the area, and maximizing the amount of detail that can be shown.

TOPOGRAPHIC MAPS

Topographic maps primarily represent the form of the earth's surface. Selected other features, both natural and artificial, may be included for information. Once one becomes accustomed to reading them, topographic maps can make excellent navigational aids.

Contour Lines, Contour Intervals

The problem of representing three-dimensional features on a two-dimensional map is addressed through the use of **contour lines.** A contour line is a line connecting points of equal elevation, measured in feet or meters above or below sea level. The **contour interval** of a map is the difference in elevation between successive contour lines. For instance, if the contour interval is 10 feet, contour lines are drawn at elevations of 1000 feet, 1010 feet, 1020 feet, 1030 feet, and so on (in whatever range of elevations is appropriate to that map). If the contour interval is 20 feet, contours would be drawn at elevations of 1000 feet, 1020 feet, 1040 feet, and so on. The contour interval chosen for a particular map depends on the overall amount of relief in the map area. If the terrain is very flat, as in a midwestern floodplain, a 10-foot or even 5-foot contour interval may be appropriate to depict what relief there is. In a rugged mountain terrain, like the Rockies, the Cascades, or the Sierras, with several thousand feet of vertical relief, using a 10-foot contour interval would make for a very cluttered map, thick with contour lines; a 50-foot or even 100-foot contour interval may be used in such a case.

The relationship between the spacing of contours and the steepness of a slope can be seen in figure B.1. A comparison of the actual relief, as shown in figure B.1A, and the resultant arrangement of contour lines on the corresponding topographic map, shown in figure B.1B, illustrates that on a given map, the more closely spaced the contours, the steeper the corresponding terrain. In other words, if the map shows many closely spaced contours, this means that there is a great deal of vertical relief over a limited horizontal distance. Note also that contours run across the face of a slope; the upslope and downslope directions are perpendicular to the contours.

Contours do close, although they may not do so on any one given map. A series of concentric closed contours indicates a hill. If there is a local depression, so that the same contour is encountered twice, once going up in elevation and again within the depression, the repetition of the contour within the depression is marked by a hachured contour, as shown in figure B.2.

Where contour lines cross a stream valley, they point upstream, toward higher elevations (figure B.3). Contour lines corresponding to different elevations should not ordinarily cross each other (that would imply that the same point has multiple elevations). The only exception to this would occur in the case of an overhanging cliff, where a range in elevations exists at one spot on the map.

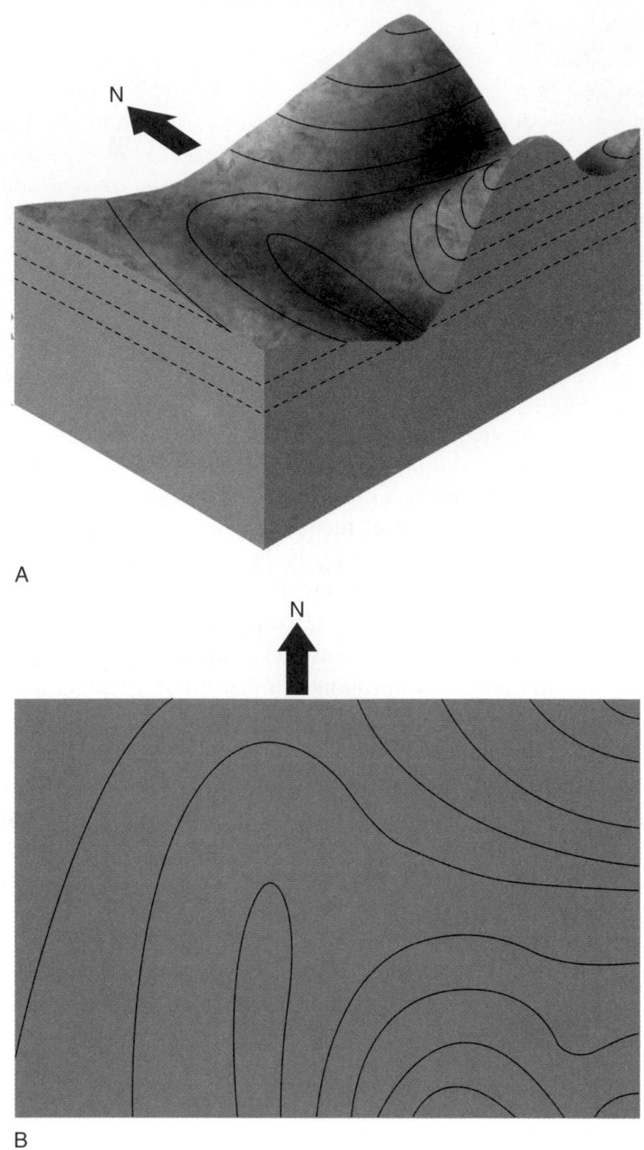

FIGURE B.1 Contours and relief. (A) Perspective view of a hilly area, with contours superimposed for reference. (B) The same contours as they would appear on a topographic map.

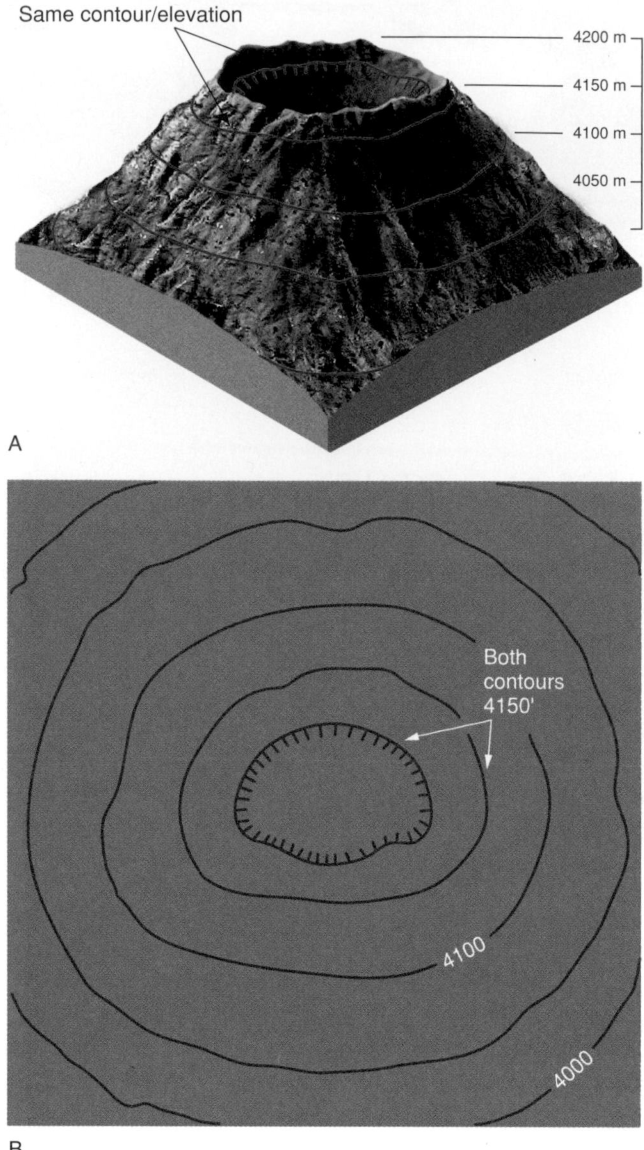

FIGURE B.2 Volcanic caldera with summit depression. (A) Actual relief. (B) As the volcano would appear on a topographic map.

Other Features on Standard Topographic Maps

The most extensive topographic maps of the United States have been compiled by the U.S. Geological Survey (USGS). The USGS has adopted a uniform set of symbols for various kinds of features, which facilitates the reading of topographic maps. Some of these symbols are illustrated by figure B.4.

Contour lines are drawn in brown. Major contours (typically, those contours corresponding to multiples of 100 feet, except on maps with large contour intervals) are drawn more boldly and labeled with the corresponding elevation. This is especially convenient in rugged terrain, where it would otherwise be easy to lose count of numerous closely spaced contours. Roads are drawn in black and/or red. Town limits are shown in light red, and the names of towns, cities, and mountains, or any other labels, are printed in black. Watery features are drawn in blue: lakes and perennial streams in solid blue, intermittent streams in dashed blue lines, swampy areas represented by blue symbols resembling tufts of low grasses. The background color for wooded areas is green; for open fields, brushy areas, deserts, bare rock, or any area lacking overhead vegetative cover, the background color is white.

GEOLOGIC MAPS

Topographic maps show the form of the land surface; geologic maps are one way of representing the underlying geology. The most common kind of geologic map is a map of **bedrock**

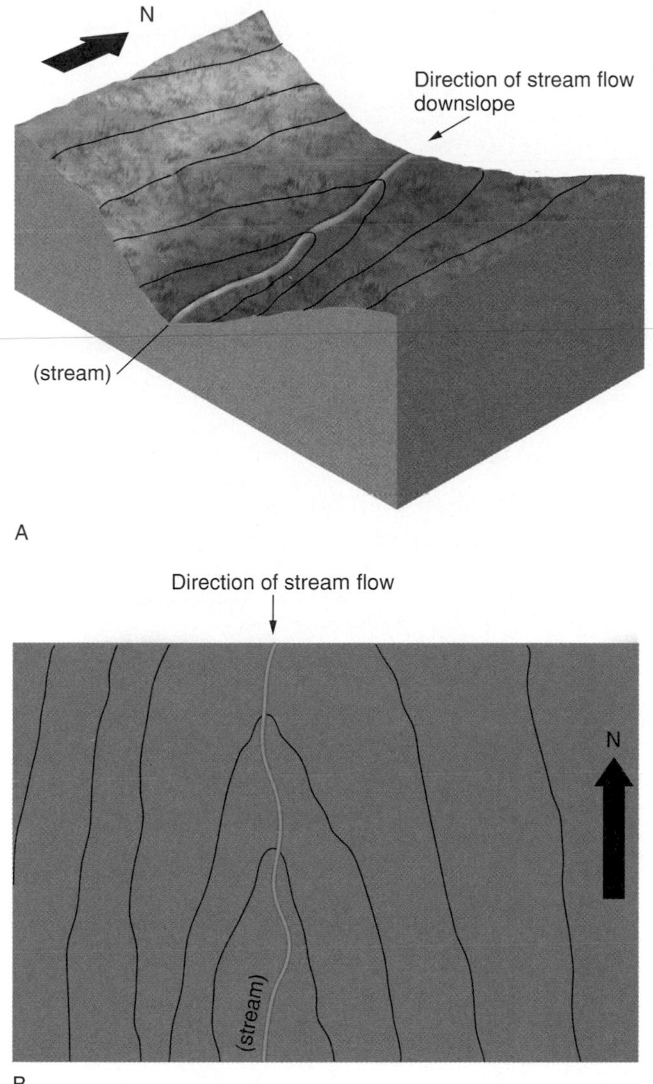

FIGURE B.3 Contours crossing stream valleys. (A) Relief with superimposed contours. (B) On topographic map, the contours point upstream.

geology, which shows the geology as it would appear with soil stripped away. (Other maps may show distribution of different soil types, glacial deposits, or other features of surface geology.) From a well-prepared geologic map, aspects of the subsurface structure can often be deduced.

Basics of Geologic Maps

Fundamental to making a geologic map is identifying a suitable set of **map units.** These may, for example, be individual sedimentary rock formations, distinguishable lava flows, or metamorphic rock units. The main requirement for a map unit is that it be identifiable by the presence or absence of some characteristic(s), and thus distinct from other map units chosen. The mapper then marks which of the map units are found at each place where the rocks are exposed.

Additional information, such as the orientation of beds or the location of contacts between map units, may also be recorded. Where obvious, contacts are drawn as solid lines; where only inferred, they are drawn as dashed lines. (If one finds granite at point *A* and limestone at point *B* a short distance away, one can infer a contact between the granite and limestone somewhere between *A* and *B,* even if the exact spot is covered by soil.)

How easy it is to produce an accurate and complete geologic map depends on several factors (aside from the competence of the mapper!). Bedrock exposure is one. Thick soil, water or swamps, and vegetation can all obscure what lies below. In some areas, the geology is well exposed only along stream valleys; in others, the rocks are completely exposed, and mapping is greatly simplified. In glaciated areas, too, one need not only deal with the cover of glacial till (assuming that is not the material of interest) but also be cautious about mistakenly identifying a large, partially buried bit of glacial debris as a small exposure of bedrock. And some areas are simply much more complicated, geologically, than others.

On a geologic map, different units are represented in different colors for clarity (figure B.5). Units of similar age may be shown in different shades of the same color. The map is accompanied by a key showing all the map units, arranged in chronological order (insofar as their ages are known), with the youngest at the top. Ordinarily, a brief description of each map unit is given; alternatively, standard patterns may be used to indicate the general rock type (figure B.6). Each unit is also assigned a symbol. The first part of the symbol consists of one or two letters corresponding to the unit's age (generally, the geologic era or period; see appendix A). This is usually followed by one to three lowercase letters corresponding to the rock type or unit's name (if any). For example, "p€qm" might be used for an unnamed Precambrian quartz monzonite, "Dl" for the Devonian Littleton Formation, "Qa" for Quaternary alluvium along a stream channel. These symbols are useful for distinguishing units mapped in similar colors, as well as for providing a general indication of the age of each unit directly on the map.

Interpretation of geologic maps likewise varies in difficulty, depending on the fundamental complexity of the geology and the extent of exposure (completeness of map). Sometimes, too, how much of the geology can be seen depends not only on cover or lack of it, but on topography. That is, in rugged terrain with considerable vertical relief, one has more of a three-dimensional look at the geology than in flat terrain. For example, in the vicinity of the Grand Canyon, the Paleozoic sedimentary sequence is quite flat-lying. So, for the most part, is the land surface. Outside the canyon, only the Kaibab limestone is exposed, forming a nearly level plateau and giving no evidence of what lies below. The cutting of the canyon has exposed many more rock units, down to the Precambrian. This is apparent on a geologic map (figure B.7).

The Grand Canyon has a regular, layer-cake geology that has a straightforward interpretation. Where the geology is more complex, patterns of repetition of units and orientation of beds must be

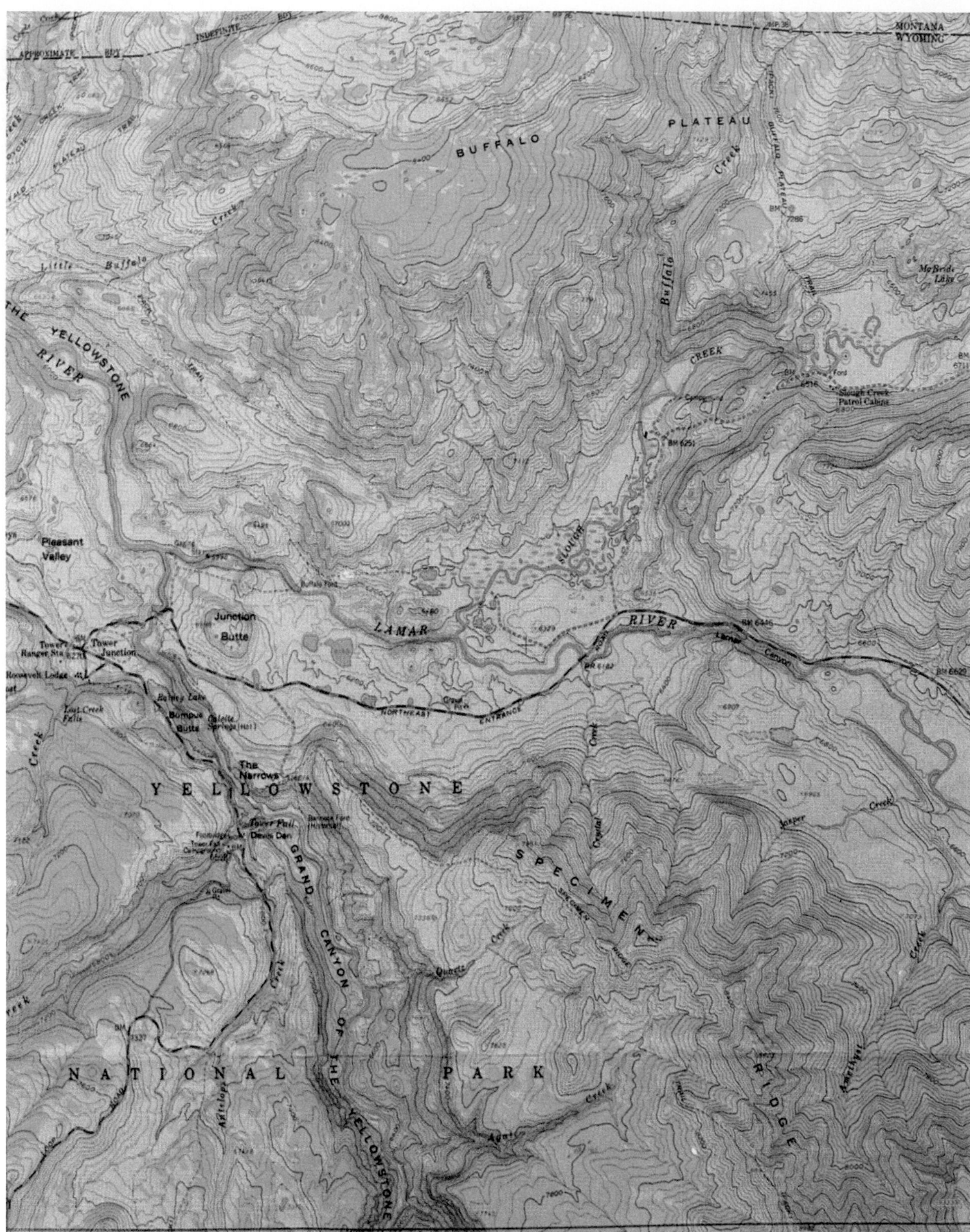

FIGURE B.4 Representative section of standard topographic map: a portion of Yellowstone National Park. Note the closely spaced contours along the Grand Canyon of the Yellowstone and edges of plateaus, the shapes of contours that cross streams, the closed contours around hills (for example, Junction Butte), and the swampy flat area (along Slough Creek).

Source: U.S. Geological Survey.

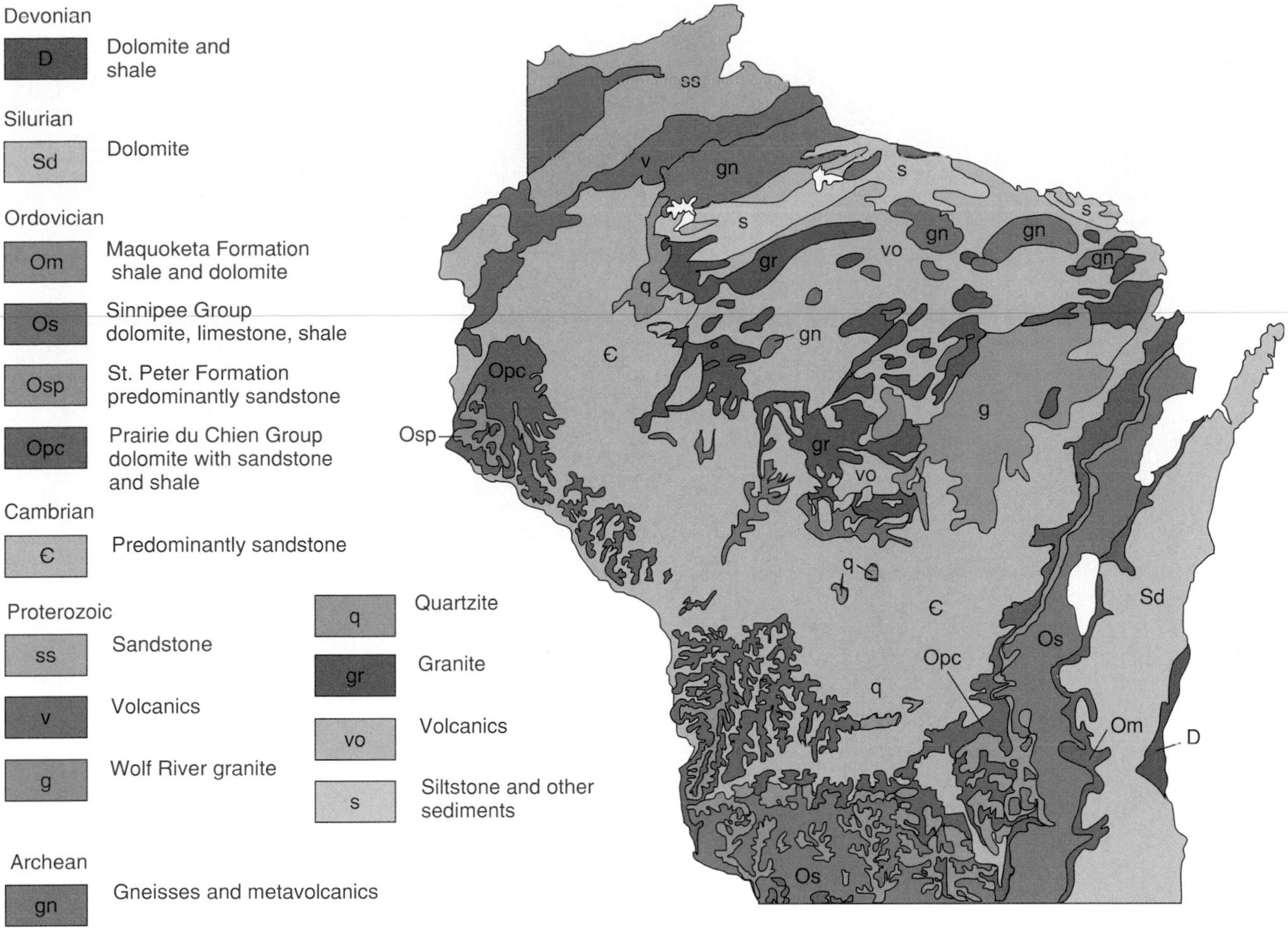

Devonian
D
Dolomite and shale
Silurian
Sd
Dolomite
Ordovician
Om
Maquoketa Formation shale and dolomite
Os
Sinnipee Group dolomite, limestone, shale
Osp
St. Peter Formation predominantly sandstone
Opc
Prairie du Chien Group dolomite with sandstone and shale
Cambrian
Є
Predominantly sandstone
Proterozoic
ss
Sandstone
v
Volcanics
g
Wolf River granite
q
Quartzite
gr
Granite
vo
Volcanics
s
Siltstone and other sediments
Archean
gn
Gneisses and metavolcanics
ss
v
gn
s
vo
gr
q
Є
Opc
Osp
g
Sd
Os
Om
D

FIGURE B.5 Sample of geologic map with key: Simplified bedrock geology of Wisconsin.
Source: After M. E. Ostrom, Wisconsin Geological and Natural History Survey, April 1981.

Limestone
Conglomerate
Dolomite
Granite
Sandstone
Basalt flows
Shale
Metamorphic rocks

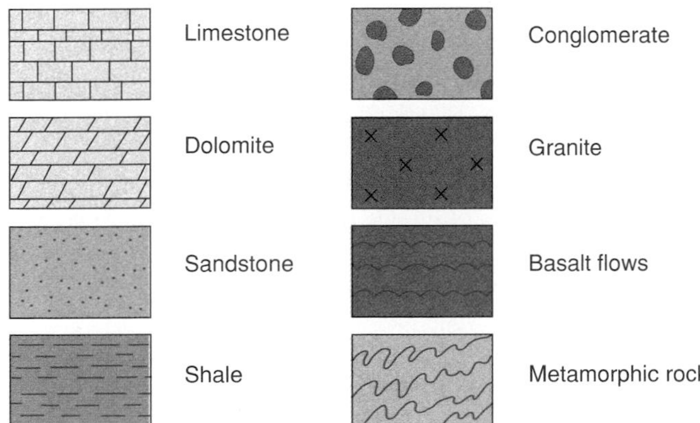

FIGURE B.6 Selected standard map symbols for various rock types.

used to interpret the structure. For example, when sets of rocks are repeated, this may indicate the presence of a fault. In the map pattern shown in figure B.8, the same sequence of units is repeated; moreover, all are dipping in the same direction. It is difficult to create such a pattern by folding. A fault, however, can account for the result (figure B.8). Offset of features that are otherwise continuous is another sign of possible faulting (figure B.9).

Cross Sections

Interpreting structure from geologic map patterns and rock orientations requires some practice and the ability to visualize in three dimensions. Often, the maker of a geologic map assists the map reader by supplying one or more geologic **cross sections.** A cross section is a three-dimensional interpretation of the geology seen at the surface. The line along which the cross section is drawn is indicated on the map. The cross section uses the same map units and symbols as the map proper and attempts to show the geometric relationships inferred to exist among those units—folds, faults, crosscutting relationships, and so on.

A cross section is drawn by starting with a topographic profile along the chosen line and marking on it the geology as seen

FIGURE B.7 The presence of the Grand Canyon allows much better knowledge of the area's subsurface geology. Geologic map shows different rock units in different colors. Note monotony outside canyon.
Used with permission Grand Canyon Natural History Association.

from the surface (figure B.10A). A structural interpretation that is consistent with all the known data is then devised (figure B.10B). Depending on the complexity of the geology and the completeness of exposure on which the original geologic map has been based, it may or may not be possible to develop a unique structural interpretation for the observed map pattern. If it is not, several plausible alternatives might be presented. Cross sections can be very valuable in evaluating site suitability for construction or other purposes.

REMOTE SENSING AND SATELLITE IMAGERY

Remote-sensing methods encompass all of those means of examining planetary features that do not involve direct contact. Instead, these methods rely on detection, recording, and analysis of wave-transmitted energy—visible light, infrared radiation, and others. Aerial photography, in which standard photographs of relatively large regions of the earth are taken from aircraft, is one example. Radar mapping of surface topography, using airplanes or spacecraft, is another. Still another involves analyzing the light reflected from the surface of a body. In the case of many planets, remotely sensed data may be the only kind readily available. In the case of the earth, remote sensing, especially using satellites, is a quick and efficient way to scan broad areas, to examine regions having such rugged topography or hostile climate that they cannot easily be explored on foot or with surface-based vehicles, and to view areas to which ground access is limited for political reasons. Probably the best-known and most comprehensive earth satellite imaging system is the one initiated in 1972, known as Landsat.

The Landsat satellites orbit the earth in such a way that images can be made of each part of the earth. Each orbit is slightly offset from the previous one, with the areas viewed on one orbit overlapping the scenes of the previous orbit. Each satellite makes 14 orbits each day; complete coverage of the earth takes 18 days. Therefore, images of any given area should be available every 18 days, although in practice, shifting distributions of clouds obscure the surface some part of the time at any point. Seven Landsat satellites have been launched; two are still functioning.

Satellite Images and Applications

The sensors in the Landsat satellites do not detect all wavelengths of energy reflected from the surface (figure B.11). They do not take photographs in the conventional sense. They are particularly sensitive to selected green and red wavelengths in the visible light spectrum and to a portion of the infrared (invisible heat radiation, with wavelengths somewhat longer than those of red light). These wavelengths were chosen particularly because plants reflect light most strongly in the green and the

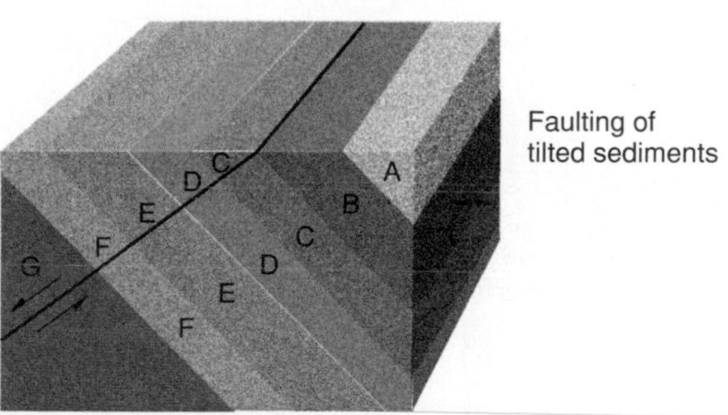

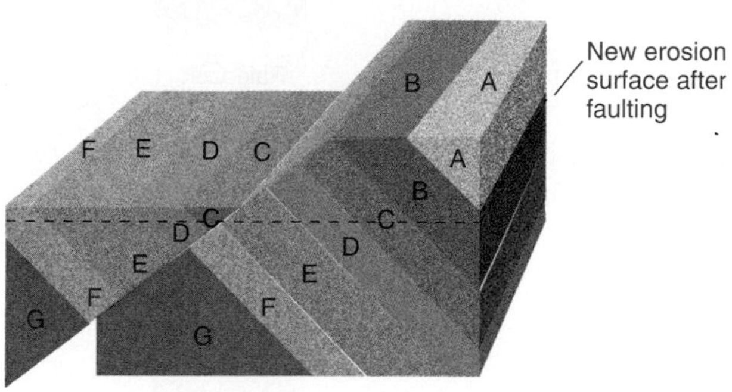

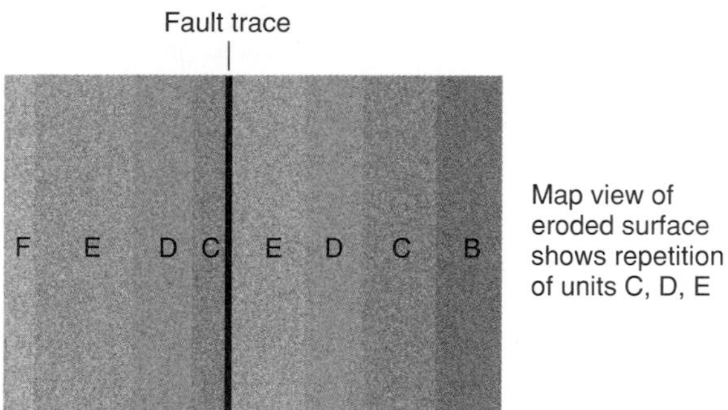

FIGURE B.8 Faulting can produce repetition of map units. Top surface of top block is the original surface exposure of units; bottom diagram is map result after faulting and erosion to the dashed surface of center diagram.

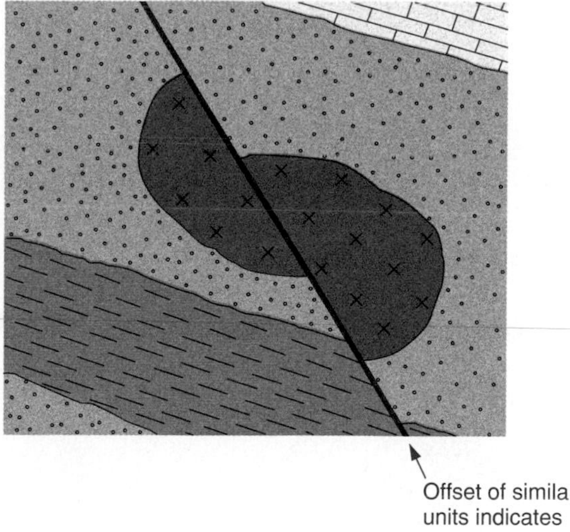

FIGURE B.9 Igneous body and surrounding rocks offset by faulting.

infrared. Later Landsat satellites (4 and 5) included sensors sensitive to more wavelengths in the visible part of the electromagnetic spectrum. Different plants, rocks, and soils reflect different proportions of radiation of different wavelengths. Even the same feature may produce a somewhat different image under different conditions. Wet soil differs from dry; sediment-laden water looks different from clear; a given variety of plant may reflect a different spectrum of radiation depending on what trace elements it has concentrated from the underlying soil or how vigorously it is growing. Landsat images can be powerful mapping tools.

A common format for Landsat imagery is photographic prints at 1:1,000,000 scale. At that scale, a 23-centimeter (9 inch) print covers 34,225 square kilometers (13,225 square miles). The smallest features that can be distinguished in the image are about 30 meters (100 feet) in size, which gives some idea of the quality of the resolution. Multiple images can be joined into mosaics covering whole countries or continents.

Images are commonly presented either in black and white or as *false-color composites.* The latter are produced by projecting the data for individual spectral regions through colored filters and superimposing the results. The false-color images are so named because the resulting pictures, though superficially resembling color photographs, do not present all features in the colors they would appear to the human eye. The most striking difference is in vegetation, which appears in shades of red, not green. Rock and soil usually show as white, blue, yellow, or brown, depending on composition. Water is blue to bluish black; snow and ice are white. Examples of false-color Landsat images have been used throughout this text. Landsat image data can also be further processed by computer to produce images in more "realistic" (expected) colors or to enhance particular features by emphasizing certain wavelengths of radiation.

Dozens of applications of Landsat and other space imagery exist in the natural sciences—for example, basic geologic mapping, identification of geologic structures, and resource exploration. It is helpful in identifying patterns of land use and in monitoring the progress of crops and the extent of damage to vegetation from fires, insects, or disease. Because satellites scan the same area repeatedly over time, seasonal or long-term changes, the progress and extent of occasional events such as flooding, and changes related to human activities can be observed (figure B.12). Satellite imagery also can be used to monitor the development of and changes in surface features such as stream channels and currents. As imaging technology has become more sophisticated, ever-better images have become available. Crews of the International Space Station have provided excellent images.

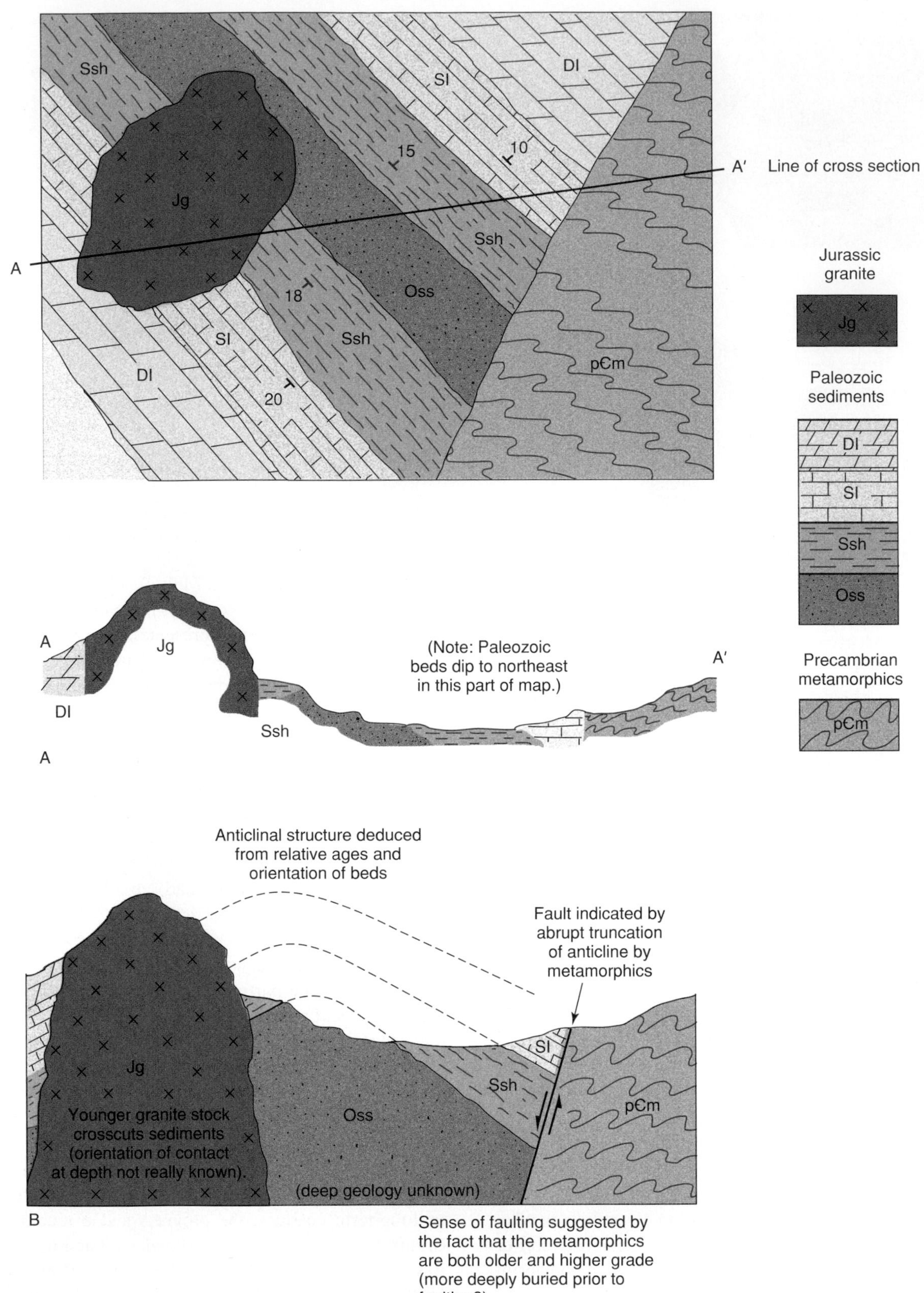

FIGURE B.10 Constructing a cross section. (A) Geology as seen at surface is sketched onto a topographic profile. (B) The pattern is interpreted, and a set of structures consistent with the pattern seen at the surface is sketched in.

A wide spectrum of solar energy reflects off the earth, with some wavelengths measured by Landsat sensors.

Three bands are shown in one image, by assigning each band a primary color (red, green, blue) and mixing them.

Different surfaces reflect differently. Vegetation looks red, water black, bare earth bright, clouds and ice white.

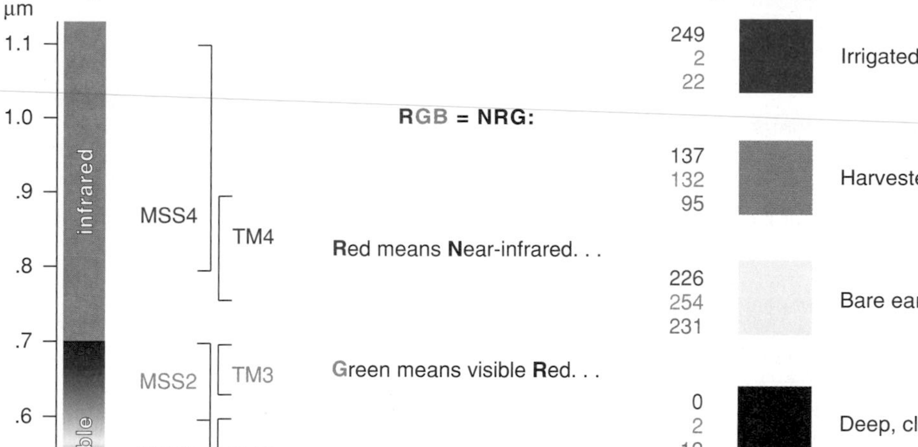

FIGURE B.11 Selected wavelengths of visible light or infrared radiation detected in three selected MultiSpectral Scanner (MSS) or Thematic Mapper (TM) bands are converted to the three primary colors and combined for a false-color composite, based on relative amounts of radiation detected in each band.

After U.S. Geological Survey.

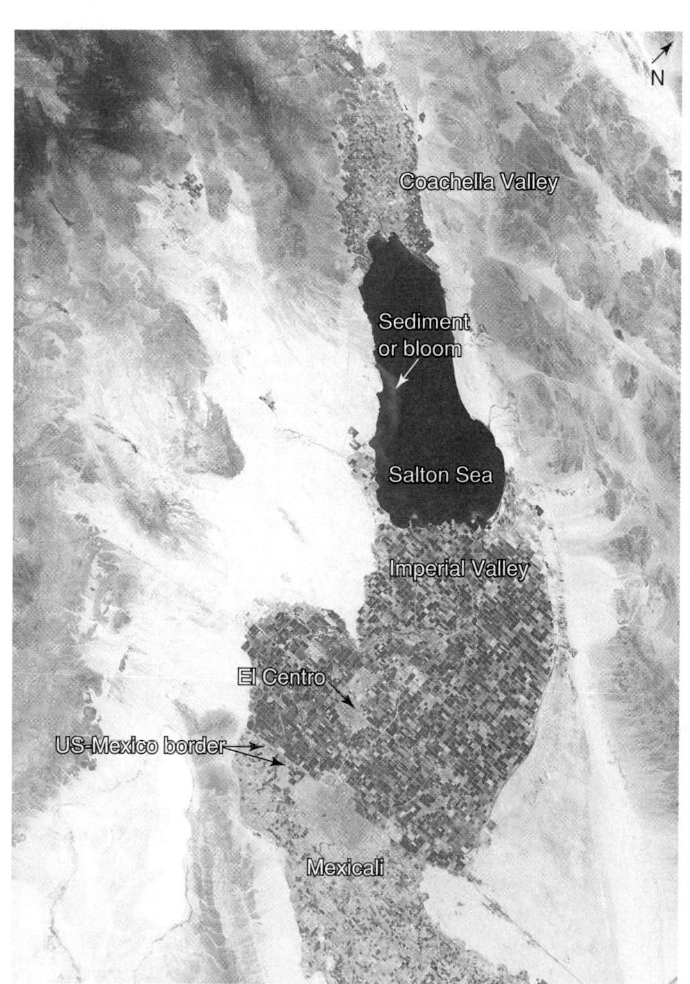

A

B

FIGURE B.12 (A) In this image, taken from the International Space Station in 2002, the importance of water from the Salton Sea to agriculture is obvious. Note how the U.S./Mexico border stands out as a result of lusher vegetation on the U.S. side. (B) Digital photograph of deforestation in eastern Bolivia, taken from the International Space Station. Agricultural development radiates outward from each of a series of small communities established by resettlement for the purpose of growing soybeans.

Images courtesy of Earth Sciences and Image Analysis Laboratory, NASA Johnson Space Center.

In all cases, imagery taken from space is especially useful when some *ground truth* can be obtained. Ground truth is information gathered by direct surface examination (best done at the time of imaging, if vegetation is involved), which can provide critical confirmation of interpretations based on remotely sensed data.

Other Airborne Imaging Techniques

A still newer technique, or set of techniques, involves the use of imaging radar. This has been likened to taking a photograph with a flash camera. Just as such a camera generates light that illuminates the pictures it takes, so imaging radar sends pulses of microwaves that bounce off the surface or scene being imaged. The techniques of processing the returned signal have become increasingly refined since NASA and the Jet Propulsion Laboratory initiated the SEASAT imaging radar system in 1978 (figure B.13). Some radar-imaging devices are flown on jets, and some have been carried on the space shuttle. One of the advantages of radar imaging (by comparison with visible-light images, for example) is that microwave radiation is penetrating enough that it is unaffected by cloud cover, heavy rain, or other such conditions that can obscure a visual image. Therefore, it can be used to monitor such phenomena as shifting lahars on Mount Pinatubo during monsoon season (figure B.14).

FIGURE B.13 Spaceborne radar image of a part of Guangdong province, China. This is a mineral-rich area, and radar imagery may assist in prospecting for additional deposits.
Image courtesy NASA.

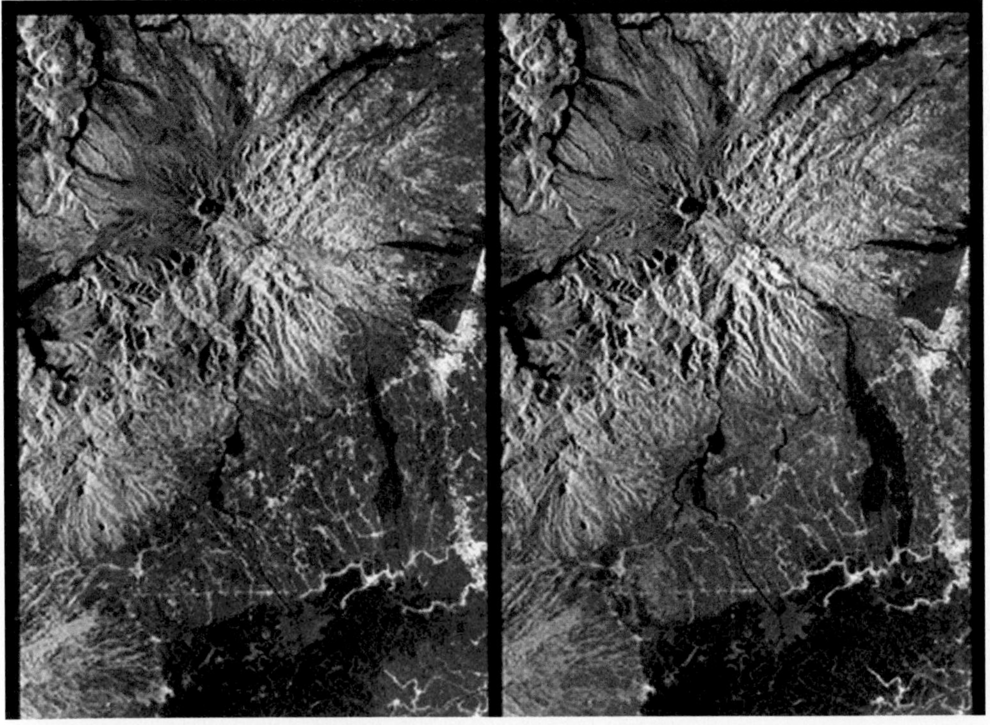

FIGURE B.14 Years after the 1991 explosive eruptions, lahars can still be triggered by monsoon rains on Mount Pinatubo. Ash from those eruptions is shown in red, lahars in black. Comparison of April (left) and October (right) 1994 images shows evolution of lahars and their deposits, and helps in hazard evaluation.
Images courtesy NASA.

Imaging spectroscopy is similar to Landsat imaging in that it relies on detection of a number of different wavelengths of radiation in the UV, visible-light, and infrared parts of the spectrum. It has become increasingly sophisticated as instrumentation has been refined. Now scans can involve detection of hundreds of individual specific wavelengths, not just a few broad bands, and certain wavelengths can serve as "fingerprints," indicating the presence of individual minerals whose crystal structures absorb energy at those precise wavelengths. The instrumentation can also be deployed from low-flying aircraft. The resultant detailed, high-resolution images can be much more useful than other types of images for certain applications, as was shown by the Leadville, Colorado, example of chapter 17.

New ways are still being found to use visible light in remote sensing, too. LIDAR (Light Detection and Ranging) is an aircraft-based laser-altimetry system for examining topography. A major application of LIDAR has been in monitoring coastal change (figure B.15). Precise assessment of coastal response to major hurricanes as well as everyday geologic processes can be important to land-use planning in such dynamic regions.

So new tools are continually developed to expand the ways we can explore and understand our planet.

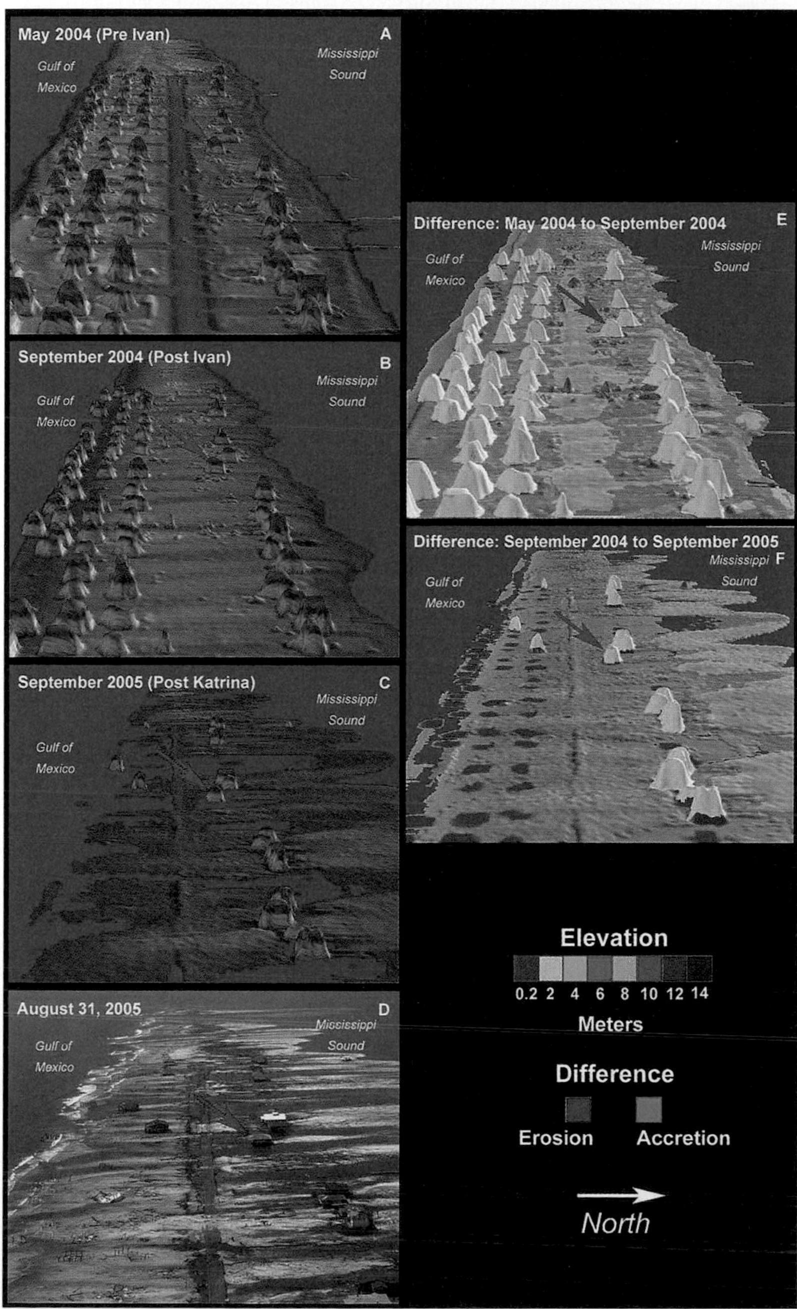

FIGURE B.15 LIDAR images (A–C) provide more precise topographic data than aerial photographs (D) and can be analyzed so that elevation changes can be mapped (E, F). Dauphin Island, AL, a barrier island.

Images courtesy U.S. Geological Survey Coastal and Marine Geology Program.

Key Terms and Concepts

bedrock geology 522
contour interval 521
contour lines 521
cross section 525
map unit 523
remote-sensing 526
scale (map) 521

Suggested Readings/References

Brock, J., and Sallenger, A. 2001. *Airborne topographic LIDAR mapping for coastal science and resource management.* USGS Open-File Report 01-46.

Gardner, J. V., Dartnell, P., Gibbons, H., and MacMillan, D. 2000. *Exposing the sea floor: High-resolution multibeam mapping along the U.S. Pacific Coast.* USGS Fact Sheet 013-00.

Holz, R. K. 1985. *The surveillant science: Remote sensing of the environment.* New York: John Wiley & Sons.

Short, N. M., Lowman, P. D., Jr., Freden, S. C., and Firch, W. A., Jr. 1976. *Mission to Earth: Landsat views the world.* National Aeronautics and Space Administration. Washington, D.C.: U.S. Government Printing Office.

Slaney, V. R. 1981. *Landsat images of Canada—A geological appraisal.* Geological Survey of Canada Special Paper 80–15.

Thompson, M. M. 1987. *Maps for America.* U.S. Geological Survey. Washington, D.C.: U.S. Government Printing Office.

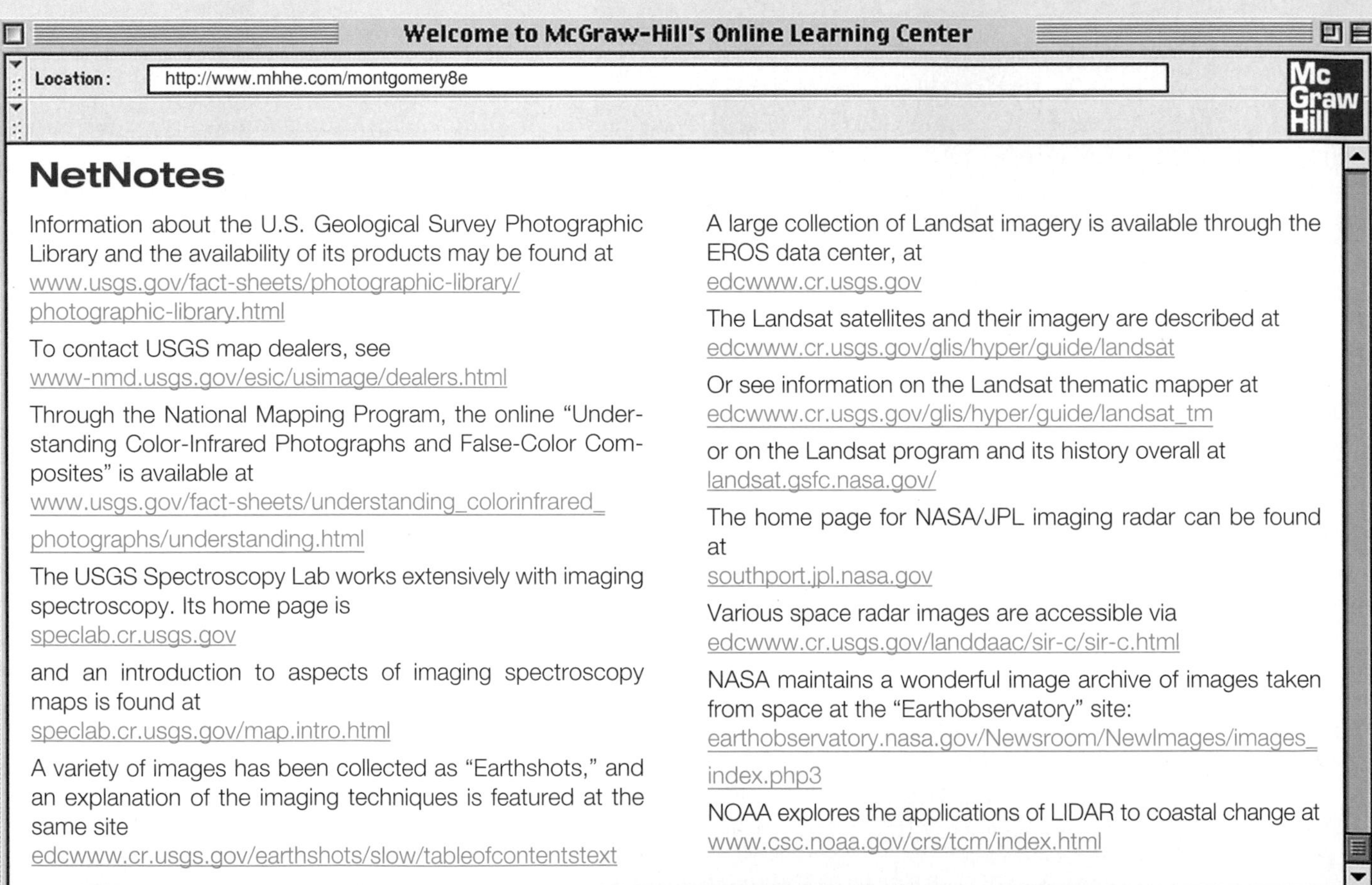

NetNotes

Information about the U.S. Geological Survey Photographic Library and the availability of its products may be found at
www.usgs.gov/fact-sheets/photographic-library/photographic-library.html

To contact USGS map dealers, see
www-nmd.usgs.gov/esic/usimage/dealers.html

Through the National Mapping Program, the online "Understanding Color-Infrared Photographs and False-Color Composites" is available at
www.usgs.gov/fact-sheets/understanding_colorinfrared_photographs/understanding.html

The USGS Spectroscopy Lab works extensively with imaging spectroscopy. Its home page is
speclab.cr.usgs.gov

and an introduction to aspects of imaging spectroscopy maps is found at
speclab.cr.usgs.gov/map.intro.html

A variety of images has been collected as "Earthshots," and an explanation of the imaging techniques is featured at the same site
edcwww.cr.usgs.gov/earthshots/slow/tableofcontentstext

A large collection of Landsat imagery is available through the EROS data center, at
edcwww.cr.usgs.gov

The Landsat satellites and their imagery are described at
edcwww.cr.usgs.gov/glis/hyper/guide/landsat

Or see information on the Landsat thematic mapper at
edcwww.cr.usgs.gov/glis/hyper/guide/landsat_tm

or on the Landsat program and its history overall at
landsat.gsfc.nasa.gov/

The home page for NASA/JPL imaging radar can be found at
southport.jpl.nasa.gov

Various space radar images are accessible via
edcwww.cr.usgs.gov/landdaac/sir-c/sir-c.html

NASA maintains a wonderful image archive of images taken from space at the "Earthobservatory" site:
earthobservatory.nasa.gov/Newsroom/NewImages/images_index.php3

NOAA explores the applications of LIDAR to coastal change at
www.csc.noaa.gov/crs/tcm/index.html

APPENDIX C

Mineral and Rock Identification

MINERAL IDENTIFICATION

Table C.1 lists many of the more common minerals. Representative chemical formulas are provided for reference. Some appear complex because of opportunities for solid solution; some have been simplified by limiting the range of compositions represented, although additional elemental substitutions are possible.

A few general identification guidelines and comments:

Hardness is an approximate measure of how readily a mineral scratches, or is scratched by, other minerals. Values of hardness range from 1 (softest) to 10 (hardest) and are measured against the ten reference minerals of the Mohs hardness scale, shown in table 2.1. For example, a mineral that scratches quartz and is scratched by topaz would have a hardness of 7½.

Luster describes the surface sheen of a mineral sample, and the terms (such as metallic, pearly, earthy) are self-explanatory. *Cleavage* is a mineral's tendency to break preferentially along certain planes of weakness in the crystal structure.

Minerals showing metallic luster are usually sulfides (or native metals, but these are much rarer). Native metals have been omitted from table C.1. Those few that are likely to be encountered, such as native copper or silver, may be identified by their resemblance to household examples of the same metals.

Of the nonmetals, the silicates are generally systematically harder than the nonsilicates. Hardnesses of silicates are typically over 5, with exceptions principally among the sheet silicates; many of the nonsilicates, such as sulfates and carbonates, are much softer.

Distinctive luster, cleavage, or other identifying properties are listed under the column "Other Characteristics." In a few cases, this column notes restrictions on the occurrence of certain minerals as a possible clue to identification; for example, "found only in metamorphic rocks," or "often found in pegmatites."

A Note on Mineral Formulas

Each chemical element is denoted by a one- or two-letter symbol. Many of these symbols make sense in terms of the English name for the element—O for oxygen, He for helium, Si for silicon, and so on. Other symbols reflect the fact that, in earlier centuries, scientists were generally versed in Latin or Greek: The symbols Fe for iron and Pb for lead, for example, are derived from *ferrum* and *plumbum,* respectively, the Latin names for these elements.

The chemical symbols for the elements can express the compositions of substances very precisely. Subscripts after a symbol indicate the number of atoms/ions of one element present in proportion to the other elements in the formula. For example, the formula $Fe_3Al_2Si_3O_{12}$ represents a compound in which, for every twelve oxygen atoms, there are three iron atoms, two aluminum atoms, and three silicon atoms. (This happens to be a variety of the mineral garnet.) The chemical formula is much briefer than describing the composition in words. It is also more exact than the mineral name "garnet," for there are several compositions of garnets with the same basic kind of formula and crystal structure: Other examples include a calcium-aluminum garnet with the formula $Ca_3Al_2Si_3O_{12}$ and a calcium-chromium garnet, $Ca_3Cr_2Si_3O_{12}$. Moreover, chemical formulas are understood by all scientists, while mineral names are known primarily to geologists.

Formulas can become very complex, especially when different elements can substitute for each other in the same site in the crystal structure. Iron and magnesium often do this in silicates. Biotite, a common, dark-colored mica, may be rich in iron and have a formula of $KFe_3AlSi_4O_{10}(OH)_2$, or it may be rich in magnesium and have a formula of $KMg_3AlSi_4O_{10}(OH)_2$, or, more commonly, it may contain some iron and some magnesium, which together total three atoms per formula. The generalized formula is then $K(Fe,Mg)_3AlSi_4O_{10}(OH)_2$, as it appears in table C.1.

ROCK IDENTIFICATION

One approach to rock identification is to decide whether the sample is igneous, sedimentary, or metamorphic and then look at the detailed descriptions in the corresponding chapter. How does one identify the basic rock type? Here are some general guidelines:

1. Glassy rocks or rocks containing bubbles are volcanic.
2. Coarse-grained rocks with tightly interlocking crystals are likely to be plutonic, especially if they lack foliation.

TABLE C.1

A Brief Mineral Identification Key

Mineral	Formula	Color
amphibole (e.g., hornblende)	$(Na,Ca)_2(Mg,Fe,Al)_5Si_8O_{22}(OH)_2$	green, blue, brown, black
apatite	$Ca_5(PO_4)_3(F,Cl,OH)$	usually yellowish
azurite	$Cu_3(CO_3)_2(OH)_2$	vivid blue
barite	$BaSO_4$	colorless
beryl	$Be_3Al_2Si_6O_{18}$	aqua to green
biotite (a mica)	$K(Mg,Fe)_3AlSi_3O_{10}(OH)_2$	black
bornite	Cu_5FeS_4	iridescent blue, purple
calcite	$CaCO_3$	variable; colorless if pure
chalcopyrite	$CuFeS_2$	brassy yellow
chlorite	$(Mg,Fe)_3(Si,Al)_4O_{10}(OH)_2$	light green
cinnabar	HgS	red
corundum	Al_2O_3	variable; colorless in pure form
covellite	CuS	blue
dolomite	$CaMg(CO_3)_2$	white or pink
epidote	$Ca_2FeAl_2Si_3O_{12}(OH)$	green
fluorite	CaF_2	variable; often green or purple
galena	PbS	silver-gray
garnet	$(Ca,Mg,Fe)_3(Fe,Al)_2Si_3O_{12}$	variable; often dark red
graphite	C	dark gray
gypsum	$CaSO_4 \cdot 2H_2O$	colorless
halite	$NaCl$	colorless
hematite	Fe_2O_3	red or dark gray
kaolinite	$Al_2Si_2O_5(OH)_4$	white
kyanite	Al_2SiO_5	blue
limonite	$Fe_4O_3(OH)_6$	yellow-brown
magnetite	Fe_3O_4	black
malachite	$Cu_2CO_3(OH)_2$	green
molybdenite	MoS_2	dark gray
muscovite (a mica)	$KAl_3Si_3O_{10}(OH)_2$	colorless
olivine	$(Fe,Mg)_2SiO_4$	yellow-green
phlogopite (a mica)	$KMg_3AlSi_3O_{10}(OH)_2$	brown
plagioclase feldspar	$(Na,Ca)(Al,Si)_2Si_2O_8$	white to gray
potassium feldspar	$KAlSi_3O_8$	white; often stained pink or aqua
pyrite	FeS_2	yellow
pyroxene (e.g., augite)	$(Na,Ca,Mg,Fe,Al)_2Si_2O_6$	usually green or black
quartz	SiO_2	variable; commonly colorless or white
serpentine	$Mg_3Si_2O_5(OH)_4$	green to yellow
sillimanite	Al_2SiO_5	white
sphalerite	ZnS	yellow-brown
staurolite	$Fe_2Al_9Si_4O_{20}(OH)_2$	brown
sulfur	S	yellow
sylvite	KCl	colorless
talc	$Mg_3Si_4O_{10}(OH)_2$	white to green
tourmaline	$(Na,Ca),(Li,Mg,Al)(Al,Fe,Mn)_6(BO_3)_3Si_6O_{18}(OH)_4$	black, red, green

Hardness	Other Characteristics
5 to 6	often forms needlelike crystals; two good cleavages forming 120-degree angle
5	crystals hexagonal in cross section
3½ to 4	often associated with malachite
3 to 3½	high specific gravity, 4.5 (denser than most silicates)
7½ to 8	usually found in pegmatites
5½	excellent cleavage into thin sheets
3	metallic luster
3	effervesces (fizzes) in weak acid
3½ to 4	
2 to 2½	cleaves into small flakes
2½	earthy luster; may show silver flecks
9	most readily identified by its hardness
1½ to 2	metallic luster
3½ to 4	powdered mineral effervesces in acid
6 to 7	glassy luster
4	cleaves into octahedral fragments; may fluoresce in ultraviolet light
2½	metallic luster; cleaves into cubes
7	glassy luster
1 to 2	streaks like pencil lead
2	
2½	salty taste; cleaves into cubes
5½ to 6½	red-brown streak regardless of color
2	earthy luster
5 to 7	found in high-pressure metamorphic rocks; often forms bladelike crystals
2 to 3	earthy luster; yellow-brown streak
6	strongly magnetic
3½ to 4	often forms in concentric rings of light and dark green
1 to 1½	cleaves into flakes; more metallic luster than graphite
2 to 2½	excellent cleavage into thin sheets
6½ to 7	glassy luster
2½ to 3	closely resembles biotite, but less black
6	may show fine parallel striations on cleavage surfaces
6	two good cleavages forming a 90-degree angle; no striations
6 to 6½	metallic luster; black streak
5 to 7	two good cleavages forming a 90-degree angle
7	glassy luster; conchoidal fracture
3 to 5	waxy or silky luster; may be fibrous (asbestos)
6 to 7	occurs only in metamorphic rocks; often forms needlelike crystals
3½ to 4	glassy luster
7 to 7½	found in metamorphic rocks; elongated crystals may have crosslike form
1½ to 2½	sulfurous odor
2	cleaves into cubes; salty taste, but more bitter than halite
1	greasy or slippery to the touch
7 to 7½	elongated crystals, triangular in cross section; conchoidal fracture

3. Coarse-grained sedimentary rocks differ from plutonic rocks in that the grains in the sedimentary rocks tend to be more rounded and to interlock less closely. A breccia does have angular fragments, but the fragments in a breccia are typically rock fragments, not individual mineral crystals.
4. Rocks that are not very cohesive, that crumble apart easily into individual grains, are generally clastic sedimentary rocks. One other possibility would be a poorly consolidated volcanic ash, but this should be recognizable by the nature of the grains, many of which will be glassy shards. (Note, however, that extensive weathering can make even a granite crumble.)
5. More-cohesive, fine-grained sedimentary rocks may be distinguished from fine-grained volcanics because sedimentary rocks are generally softer and more likely to show a tendency to break along bedding planes. Phenocrysts, of course, indicate a (porphyritic) volcanic rock.
6. Foliated metamorphic rocks are distinguished by their foliation (schistosity, compositional banding). Also, rocks containing abundant mica, garnet, or amphibole are commonly metamorphic rocks.
7. Nonfoliated metamorphic rocks, like quartzite and marble, resemble their sedimentary parents but are harder, denser, and more compact. They may also have a shiny or glittery appearance on broken surfaces, due to recrystallization during metamorphism.

Once a preliminary determination of category (igneous/sedimentary/metamorphic) has been made, table C.2 can be used in conjunction with the appropriate text chapter to identify the rock type. (Keep in mind, however, that the key and text focus on relatively common rock types.)

TABLE C.2

A Key to Aid In Rock Identification

Igneous Rocks

I. Extremely coarse-grained: Rock is a pegmatite. (Most pegmatites are granitic, with or without exotic minerals.)

II. Phaneritic (coarse enough that all grains are visible to the naked eye).

- A. Significant quartz visible; only minor mafic minerals: *granite.*
- B. No obvious quartz; feldspar (light-colored) and mafic minerals (dark) in similar amounts: *diorite.*
- C. No quartz; rock consists mostly of mafic minerals: *gabbro.*
- D. No visible quartz *or* feldspar: Rock is ultramafic.

III. Porphyritic with fine-grained groundmass: Go to Part IV to describe groundmass (using phenocryst compositions to assist); adjective "porphyritic" will preface rock name.

IV. Aphanitic (grains too fine to distinguish easily with the naked eye).

- A. Quartz is visible or rock is light in color (white, cream, pink): probably *rhyolite.*
- B. No visible quartz; medium tone (commonly gray or green); if phenocrysts are present, commonly plagioclase, pyroxene, or amphibole: probably *andesite.*
- C. Rock is dark, commonly black; any phenocrysts are olivine or pyroxene: *basalt*
- D. Rock is glassy and massive: *obsidian* (regardless of composition).
- E. Rock consists of gritty mineral grains, ash and glass shards: *ignimbrite* (welded tuff).

Sedimentary Rocks

I. Rock consists of visible shell fragments or of oolites: *limestone.*

II. Rock consists of interlocking grains with texture somewhat like that of igneous rock and is light in color: probable chemical sedimentary rock.

- A. Tastes like table salt: *halite.*
- B. No marked taste; hardness of 2 (if grains are large enough to scratch); does not effervesce: *gypsum.*
- C. Effervesces in weak HCl: *limestone* (calcite).
- D. Effervesces weakly in HCl, only if scratched: *dolomite.*

III. Rock consists of grains apparently cemented or compacted together: probable clastic sedimentary rock.

- A. Coarse grains (several millimeters or more in diameter), perhaps with a finer matrix: *conglomerate* if the grains are rounded, *breccia* if they are angular.
- B. Sand-sized grains; gritty feel: sandstone. If predominantly quartz grains, *quartz sandstone;* if roughly equal proportions of quartz and feldspar, *arkose;* if many rock fragments, and perhaps a fine-grained matrix, *greywacke.*
- C. Grains too fine to see readily with the naked eye: *mudstone.* If rock shows lamination, and a tendency to part along parallel planes, *shale.*

IV. Relatively dense; compact, dark, no visible grains; massive texture, conchoidal fracture: *chert* (silica).

Metamorphic Rocks

I. Nonfoliated; compact texture with interlocking grains: identified by predominant mineral(s).

- A. If quartz-rich, perhaps with a sugary appearance: *quartzite.*
- B. If calcite or dolomite (identified by effervescence, hardness): *marble.*
- C. Rock consists predominantly of amphiboles: *amphibolite.*

II. Foliated: classified mainly by texture.

- A. Very fine-grained; pronounced rock cleavage along parallel planes, to resemble flagstones: *slate.*
- B. Fine-grained; slatelike, but with glossy cleavage surfaces: *phyllite.*
- C. Coarser grains; obvious foliation, commonly defined by prominent mica flakes, sometimes by elongated crystals like amphiboles: *schist.*
- D. Compositional or textural banding, especially with alternating light (quartz, feldspar) and dark (ferromagnesian) bands: *gneiss.*

GLOSSARY

A

ablation The loss of glacier ice by melting or evaporation.

abrasion Erosion by wind-transported sediment or by the scraping of rock fragments frozen in glacial ice.

absorption field *See* leaching field.

acid rain Rain that is more acidic (has lower pH) than normal precipitation.

active margin A continental margin at which there is significant volcanic and earthquake activity; commonly, a convergent plate margin.

active volcano A volcano with a record of eruption within recent history.

aerobic decomposition Decomposition using or consuming oxygen.

aftershocks Earthquakes that follow the main shock when a fault has slipped; of magnitude equal to or lower than the main shock.

A horizon Usually top zone in soil profile, consisting of mix of mineral and organic material.

albedo The fraction of incoming radiation that is reflected back by a surface.

algal bloom An overly exuberant growth of algae in eutrophic water.

alluvial fan A wedge-shaped sediment deposit left where a tributary flows into a more slowly flowing stream, or where a mountain stream flows into a desert.

alpine (valley) glacier A glacier occupying a valley in mountainous terrain.

anaerobic decomposition Describes decomposition that occurs without using, or in the absence of, oxygen.

andesite Volcanic rock intermediate in composition between basalt and rhyolite.

angle of repose The maximum slope angle at which a given unconsolidated material is stable.

anion An ion with a net negative charge.

anthracite The hardest of naturally occurring coals.

anthropogenic Produced by human activity.

aquifer Rock that is sufficiently porous and permeable to be useful as a source of water.

aquitard Rock of low permeability, through which water flows very slowly.

arête Sharp-spined ridge created by erosion by valley glaciers flowing along either side of the ridge.

artesian system A confined aquifer system in which ground water can rise above its aquifer under its own pressure.

asthenosphere Partially molten, "weak" zone within the upper mantle immediately below the lithosphere.

atom The smallest particle into which a chemical element can be subdivided.

atomic mass number The sum of the number of protons and the number of neutrons in an atomic nucleus.

atomic number The number of protons in an atomic nucleus; characteristic of a particular element.

B

banded iron formation A sedimentary rock consisting of alternating iron-rich and iron-poor bands, found in Precambrian rocks, that may serve as an ore of iron.

barrier islands Long, low, narrow islands parallel to a coastline that protect the coastline somewhat from wave action.

basalt A volcanic rock rich in ferromagnesian minerals; relatively low in silica.

base level The lowest elevation to which a stream can cut down; for most streams, this is the level of the body of water into which they flow, such as another stream, lake, or ocean.

beach Gently sloping shoreline area washed over by waves.

beach face That part of the beach along the water that is regularly washed by waves.

bed load Material moved along a stream bed by flowing water.

bedrock geology The geology as it would appear with the overlying soil and vegetative cover stripped away.

Benioff zone A dipping plane of progressively deeper earthquake faci at a convergent plate boundary associated with a subducting plate.

B horizon Soil layer found below the soil's A and E horizons; also known as the zone of accumulation.

biochemical oxygen demand (BOD) Quantity of oxygen required for aerobic decomposition of organic matter in a system.

Biodiesel Fuel derived from vegetable oil or animal fat that can be used in some diesel-powered vehicles, or a blend of regular diesel with such fuels.

biogas Methane derived from decaying organic matter.

biomagnification Process through which the concentration of a harmful substance, such as a heavy metal, increases in organisms as it moves up a food chain.

biomass (energy) Energy derived from living organisms or from organic matter.

bioremediation Reduction of a pollution hazard by use of organisms—for example, breakdown of toxic organic compounds by microorganisms or fixation of toxic elements in insoluble or immobile forms.

biosphere The sum of all living things on earth.

bituminous A form of coal that is softer than anthracite but harder than lignite.

BOD *See* biochemical oxygen demand.

body waves Seismic waves that pass through the earth's interior; includes P waves and S waves.

braided stream A stream with multiple channels that divide and rejoin.

breeder reactor A reactor in which new fissionable material is produced in quantity at the same time as energy is generated.

brittle Describes materials that tend to rupture before appreciable plastic deformation has occurred.

C

caldera A large, bowl-shaped summit depression in a volcano; may be formed by explosion or collapse.

calving The formation of icebergs by the breakup of a glacier flowing out over water.

capacity (stream) The load that a stream can carry.

carbonate Nonsilicate mineral containing carbonate groups (CO_3), carbon and oxygen in the proportions of one atom of carbon to three atoms of oxygen.

carbon sequestration Isolation of carbon in some reservoir, from which it does not contribute to atmospheric CO_2.

carrying capacity The ability of a system (or the whole earth) to sustain its population in reasonably healthy and comfortable conditions.

catalyst A substance that promotes chemical reactions.

cation An ion with a net positive charge.

Cenozoic Geologic era spanning the time from 66 million years ago to the present.

chain reaction (nuclear) The process during which fission of one nucleus triggers fission of others, which, in turn, induces fission in others, and so on.

channelization The modification of a stream channel, such as deepening or straightening of the channel, usually with the objective of reducing flood hazards.

chemical sediment Sediment formed at low temperature by direct precipitation from solution.

chemical weathering The breakdown of minerals by chemical reaction with water, with other chemicals dissolved in water, or with gases in the air.

C horizon Soil layer found directly below the soil's B horizon; consists of coarsely broken bedrock.

cinder cone A volcano built of cinders and other pyroclastics piled up around the volcanic vent.

cirque Bowl-shaped depression formed at the head of an alpine glacier.

clastic Broken or fragmented; describes sediments or sedimentary rocks that are formed from fragments of preexisting rocks or minerals.

cleavage (mineral) The tendency of a mineral to break preferentially along planes in certain directions in the crystal structure.

coal A solid, carbon-rich fuel formed from the remains of land plants through the effects of heat and pressure in the earth's crust.

coal-bed methane Methane associated with, and extracted from, coal deposits.

common but differentiated responsibility of states The concept that all nations share responsibility for protecting the global environment, but individual nations' responsibilities differ with their differing contributions to the problems and resources available to address them.

composite volcano *See* stratovolcano.

composting A method of handling (nontoxic) organic waste matter by which it is converted into a beneficial soil additive through controlled decay.

compound A chemical combination of two or more elements, in specific proportions, having a distinctive set of resultant physical properties.

compressive stress Stress tending to compress an object.

concentration factor (ore) The concentration of a metal in a given ore deposit divided by its average concentration in the continental crust.

conditional resources *See* subeconomic resources.

cone of depression A broadly conical depression of the water table or potentiometric surface caused by pumped groundwater withdrawal.

confined aquifer An aquifer overlain by an aquitard or aquiclude.

contact metamorphism Local metamorphism adjacent to a cooling magma body.

contaminant plume A tongue of contaminant-rich water extending away from a point source of groundwater pollution in the direction of groundwater flow.

continental drift The concept that the continents have moved about over the earth's surface.

continental glacier A large glacier covering extensive land area; also known as an *ice cap* or *ice sheet;* may be several kilometers thick.

contour interval The difference in elevation between successive contour lines on a map.

contour line A line joining points of equal elevation on a map.

contour plowing Plowing so that rows run horizontally around the curve of a slope, rather than up and downslope.

convection cells Circulating masses of material driven by temperature differences (hot material rises, then moves laterally, cools, sinks, and is reheated to rise again).

convergent plate boundary Plate boundary at which lithospheric plates are moving toward each other; for example, a subduction zone or continental collision zone.

core The innermost zone of the earth; composed largely of iron.

core meltdown A possible nuclear reactor accident resulting from loss of core coolant and subsequent overheating.

covalent bonding Bonding involving sharing of electrons between atoms.

creep (fault) The slow, gradual slip along a fault zone without major, damaging earthquakes.

creep (rock or soil) Slow, gradual downslope movement of unstable surficial materials (as contrasted with more abrupt landslides).

crest The maximum stage reached during a flood event.

cross section An interpretation of geology and structure in the third (vertical) dimension commonly based on rock exposures and attitudes at the surface; drawn in a particular vertical plane.

crust The outermost compositional zone of the earth; composed predominantly of relatively low-density silicate minerals.

crystalline Describes materials possessing a regular, repeating internal arrangement of atoms.

cumulative reserves The total reserves, including materials already consumed or exploited.

Curie temperature The temperature above which a magnetic material loses its magnetism; different for each such material.

cut bank Steep stream bank being eroded by lateral migration of meanders.

D

Darcy's Law Relationship of groundwater flow rate between two points to the difference in hydraulic head between them.

debris avalanche A mixed flow of rock, soil, vegetation, or other materials.

decommissioning (nuclear plant) The shutdown of a nuclear reactor at the end of its safe, useful life; includes the disposal of radioactive parts.

deep-focus earthquakes Those with focal depths over 100 km.

deep-well disposal A method for the disposal of toxic liquid waste by injection into deep, porous, permeable strata confined by impermeable rocks.

deflation The wholesale removal of loose sediment by wind erosion.

delta A fan-shaped deposit of sediment formed at a stream's mouth.

desert A barren region incapable of supporting appreciable life.

desertification The process by which marginally habitable arid lands are converted to desert; typically accelerated by human activities.

desert pavement A desert surface produced by the combined effects of wind erosion and overland surface-water runoff, in which the larger rocks are exposed by the selective removal of fine sediment; these rocks, in turn, protect finer material below from erosion.

differentiation (earth) Process by which zones of different composition developed from an initially homogeneous earth.

dip-slip fault A fault with predominantly vertical displacement.

discharge (stream) The amount of water flowing past a given point per unit time.

dissolved load Sum of dissolved material transported by a stream.

divergent plate boundary A boundary along which lithospheric plates are moving apart; for example, seafloor spreading ridges and continental rift zones.

divide Topographic high separating drainage basins of different streams; also, plane in a groundwater system from which ground water flows away in different directions on either side.

dome (lava) A compact, steep-sided structure built of very viscous, silicic lava emitted from a central pipe or vent.

dormant volcano A volcano with no recent eruptive history but that still looks relatively fresh and unweathered; may become active again in future.

dose-response curve A graph illustrating the relative benefit or harm from a trace element or other substance as a function of the dosage received or amount consumed by a person or organism.

doubling time The length of time required for a population to double in size.

downstream flood A flood affecting a large area of drainage basin or a large stream system; typically caused by prolonged rain or rapid regional snowmelt.

drainage basin The region from which surface water drains into a particular stream.

dredge spoils Sediment dredged from waterways to improve navigation or to increase water capacity.

drift Sediment transported and deposited by a glacier; *see also* till, outwash.

drowned valley Along a coastline, a stream valley that is partially flooded by seawater as a consequence of land sinking and/or sea level rising.

ductile Describes material that undergoes extensive plastic deformation without rupturing.

dune A low mound or ridge of sediment (usually sand) deposited by wind.

dust bowl Any region severely affected by drought and wind erosion; when capitalized (Dust Bowl), refers specifically to a large region of the central United States so affected in the early 1930s.

E

E85 A liquid fuel consisting of 85% ethanol, 15% gasoline.

earthquake Ground displacement and energy release associated with the sudden motion of rocks along a fault.

earthquake cycle The concept that there is a periodic quality about the occurrence of major earthquakes on a given fault zone, with repeated cycles of stress buildup, rupture, and relaxation of stress through smaller aftershocks.

EEZ *See* Exclusive Economic Zone.

E horizon Layer typically found between A and B soil horizons; also known as zone of leaching.

EIS *See* environmental impact statement.

elastic deformation Deformation proportional to applied stress, from which the affected material will return to its original size and shape when the stress is removed.

elastic limit The stress above which a material will cease to deform elastically.

elastic rebound Phenomenon whereby stressed rocks snap back elastically after an earthquake to their pre-stress condition.

electron A subatomic particle with an electrical charge of −1; generally found orbiting an atomic nucleus.

end moraine A ridge of till accumulated at the end of a glacier.

enhanced recovery Any method used to increase the amount of oil or gas recovered from a petroleum reservoir.

environmental geology The study of the interactions between humans and their geologic environment.

environmental impact statement (EIS) An analysis of the environmental impacts to be anticipated from a proposed action and its alternatives; mandated by the National Environmental Policy Act and legislation patterned after it.

eolian Deposited or shaped by wind action.

epicenter The point on the earth's surface directly above the focus of an earthquake.

equilibrium line Line on the surface of a glacier at which accumulation just equals ablation.

estuary A body of water along a coastline that contains a mix of fresh and salt water.

eutrophication The development of high nutrient levels (especially, high concentrations of nitrates and phosphates) in water; may lead to algal bloom.

evaporite A sedimentary mineral deposit formed when shallow or inland seas dry up; also, the minerals commonly deposited in such an environment.

evapotranspiration Movement of water vapor from earth's surface into air by a combination of evaporation from rocks and transpiration through plants.

Exclusive Economic Zone (EEZ) A zone extending to 200 miles offshore from a nation's coast, within which the 1982 Law of the Sea Treaty recognizes that nation's exclusive right to resource exploitation.

exponential decay Breakdown of a fixed percentage or fraction of a substance per unit time.

exponential growth Growth characterized by a constant percentage increase per unit time.

extinct volcano A volcano that has no recent eruptive history and appears very weathered in appearance; not expected to erupt again.

F

fall (rock) Mass wasting by free-fall of material not always in contact with the ground underneath.

fault Planar break in rock along which one side has moved relative to the other.

ferromagnesian A term describing silicates containing significant amounts of iron and/or magnesium; these minerals are usually dark-colored.

firn Dense, coarsely crystalline snow partially converted to ice.

fission (atomic) The process by which atomic nuclei are split into smaller fragments.

fissure eruption The eruption of lava from a crack in the lithosphere, rather than from a central vent.

flood Condition in which stream stage is above channel bank height.

flood-frequency curve A graph of stream stage or discharge as a function of recurrence interval (or annual probability of occurrence).

floodplain A flat region or valley floor surrounding a stream channel, formed by meandering and sediment deposition, into which the stream overflows during flooding.

flow Mass wasting in which materials move in chaotic fashion.

focus The point of first break on a fault during an earthquake.

foliation Parallel alignment of linear or platy minerals in a rock.

fossil fuels Hydrocarbon fuels formed from organic matter.

fusion (nuclear) The process by which atomic nuclei combine to produce larger nuclei.

G

gasification Any process by which coal is converted to a gaseous hydrocarbon fuel.

geographic information system (GIS) A computer-based system for storing, manipulating, and analyzing data associated with particular geographic locations.

geomedicine Study of geographic patterns of health and disease and the relationships between health and substances in the geological environment.

geopressurized natural gas Gas dissolved in deep pore waters.

geopressurized zones Deep aquifers under unusually high pressure, exceeding normal hydrostatic (fluid) pressure.

geothermal energy Energy derived from the internal heat of the earth; its use usually requires a near-surface heat source, such as young igneous rock, and nearby circulating subsurface water.

geothermal gradient The rate of increase of temperature with depth in the earth.

GIS *See* geographic information system.

glacier A mass of ice that moves or flows over land under its own weight.

glass Solid (especially silicate) lacking a regular internal crystal structure.

gradient The slope (steepness) of a stream channel along its length.

granite Coarse-grained plutonic igneous rock, typically rich in quartz and feldspars.

greenhouse effect The warming of the atmosphere due to trapping of infrared rays by atmospheric gases, especially as due to the increased concentration of carbon dioxide derived from the burning of fossil fuels.

ground water Water in the zone of saturation, below the water table.

H

half-life The length of time required for half of an initial quantity of a radioactive isotope to decay.

halide Compound of one or more metals plus a halogen element (fluorine, chlorine, iodine, or bromine).

hardness (mineral) The ability of a mineral to resist scratching.

hard water Water containing substantial quantities of dissolved calcium, magnesium, and/or iron.

harmonic tremors A distinctive type of seismic signal, rhythmic and continuous, often associated with magma movement in or beneath a volcano.

heavy metals A group of dense metals, including mercury, lead, cadmium, plutonium, and others, that share the characteristic of being accumulative in organisms and tending to become increasingly concentrated in organisms higher up a food chain.

high-level (radioactive) waste Waste sufficiently radioactive to require special handling in disposal.

horn A peak formed by headwall erosion by several alpine glaciers diverging from the same topographic high.

hot-dry-rock Geothermal resource area in which geothermal gradients are high but indigenous ground water is lacking.

hot spot An isolated center of volcanic activity; often not associated with a plate boundary.

hydraulic gradient The difference in hydraulic head between two points, divided by the distance between them.

hydraulic head Potential energy of water above a given point, reflected in the height of the water surface (above ground), water table in an unconfined aquifer, or potentiometric surface in a confined aquifer.

hydrocarbon Any compound consisting of carbon and hydrogen.

hydrocompaction The process by which a sediment settles, cracks, and becomes compacted when wetted.

hydrograph A graph of stream stage or discharge against time.

hydrologic cycle The cycle through which water in the hydrosphere moves; includes such processes as evaporation, precipitation, and surface and groundwater runoff.

hydrosphere All water at and near the earth's surface that is not chemically bound in rocks.

hydrothermal ores The ores deposited by circulating warm fluids in the earth's crust.

hypothesis A conceptual model or explanation for a set of data, measurements, or observations.

hypothetical resources That quantity of a resource material expected to be found in areas in which like deposits are known to exist.

I

ice age Any period of extensive continental glaciation; when capitalized (Ice Age), the phrase refers to the most recent such episode, from 2 million to 10,000 years ago.

igneous rock A rock formed or crystallized from a magma.

infiltration The process by which surface water sinks into the ground.

infrared Electromagnetic radiation just to the long-wavelength side of the visible light spectrum; heat radiation.

intensity (earthquake) A measure of the damaging effects of an earthquake at a particular spot; commonly reported on the Modified Mercalli Scale.

ion Atom that has gained or lost electrons, so it has a net electrical charge.

ionic bonding Bonding due to attraction between oppositely charged ions.

island arc Chain of volcanic islands formed parallel to a subduction zone, on the overriding plate.

isotopes Atoms of a given chemical element having the same atomic number but different atomic mass numbers.

K

karst Terrain characterized by abundant formation of underground solution cavities and sinkholes; commonly underlain by limestone.

kerogen A waxy solid hydrocarbon in oil shale.

kimberlites Igneous rocks that occur as pipelike intrusive bodies that probably originated in the mantle.

L

lahar A volcanic mudflow deposit formed from hot ash and water, the latter often derived from melting snow on a snow-capped or glaciated volcano.

landslide A general term applied to a rapid mass-wasting event.

laterite An extreme variety of pedalfer soil that is highly leached; common in tropical climates.

lava Magma that flows out at the earth's surface.

Law of Faunal Succession The concept that life-forms change through time and that therefore each given fossil organism corresponds to only one interval of time.

leachate Water containing dissolved chemicals; applied particularly to fluids escaping from waste disposal sites.

leaching The removal of elements or compounds by dissolution.

leaching field A network of porous pipes and surrounding soil from which septic-tank effluent is slowly released.

levees Raised banks along a stream channel that tend to contain the water during high-discharge events.

lignite The softest of the coals.

liquefaction (coal) Any process by which coal is converted into a liquid hydrocarbon fuel.

liquefaction (soil) A quicksand condition arising in wet soil shaken by seismic waves; soil loses its strength as particles lose contact with each other.

lithification Conversion of unconsolidated sediment into cohesive rock.

lithosphere The solid, outermost zone of the earth, including the crust and a portion of the upper mantle; approximately 50 kilometers thick under the oceans and commonly more than 100 kilometers thick beneath the continents.

littoral drift Sand movement along the length of a beach.

load (stream) The total quantity of material transported by a stream: sum of bed load, suspended load, and dissolved load.

locked fault A section of fault along which friction prevents creep in response to stress.

loess Wind-deposited sediment composed of fine particles (typically 0.01 to 0.06 millimeters in diameter).

longitudinal profile Diagram of elevation of a stream bed along its length.

longshore current Net movement of water parallel to a coastline, arising when waves and currents approach the shore at an oblique angle.

low-level (radioactive) waste Wastes that are sufficiently low in radioactivity that they can be released safely into the environment or disposed of with minimal precautions.

M

magma A naturally occurring silicate melt, which may also contain mineral crystals, dissolved water, or gases.

magmatic deposits (ore) Those deposits associated with magma emplacement or crystallization.

magnitude (earthquake) Measure of earthquake size; often reported using the Richter magnitude scale.

manganese nodules Lumps of manganese and iron oxides and hydroxides, with other metals, found on the sea floor.

mantle The zone of the earth's interior between crust and core; rich in ferromagnesian silicates.

map unit A distinct, identifiable rock unit used in preparing a geologic map.

mass movement *See* mass wasting.

mass wasting The downslope movement of material due to gravity; also known as mass movement.

meanders The curves or bends in a stream channel.

mechanical weathering The physical breakup of rock or mineral grains by surface processes.

Mesozoic Geologic era from 245 to 66 million years ago.

metamorphic rock Rock that is changed in form (deformed and/or recrystallized) through the effects of heat and/or pressure.

methane hydrate A crystalline solid of natural gas and water molecules, found in arctic regions and marine sediments.

Milankovitch cycles Cyclic variations in the amount of sunlight reaching a given latitude, caused by variations in earth's orbit and tilt on its axis.

milling Erosion by the grinding action of sand-laden waves on a coast.

mineral A naturally occurring, inorganic, solid element or compound with a definite composition or range in composition, usually having a regular internal crystal structure.

moment magnitude A measure of earthquake magnitude directly proportional to energy release, calculated from the area of fault plane broken, amount of fault displacement, and shear strength of the rock; represented by M_w.

moraine Landform made of till.

mouth (stream) The point where a stream ends; where it reaches its base level.

mrem One-thousandth of a rem.

multiple barrier concept Waste-disposal approach that involves several mechanisms or materials for isolating the waste from the environment; often used in the context of high-level radioactive wastes.

multiple use The land-use practice in which a given piece of land is used for two or more purposes simultaneously.

N

native element Nonsilicate mineral consisting of a single chemical element.

natural gas Gaseous hydrocarbons, especially methane (CH_4).

neutron An electrically neutral subatomic particle with a mass approximately equal to one atomic mass unit; generally found within an atomic nucleus.

nonpoint source A diffuse source of pollutants, such as runoff from farmland or drainage from a strip-mine.

nonrenewable Not being replenished or formed at any significant rate on a human timescale.

nucleus The center of an atom, containing protons and neutrons.

O

ocean thermal energy conversion (OTEC) Power generation making use of the temperature difference between deep, cold seawater and warmer near-surface water.

oil Any of various liquid hydrocarbon compounds.

oil shale A sedimentary rock containing the waxy solid hydrocarbon kerogen.

ore A rock in which a valuable or useful metal occurs at a concentration sufficiently high to make it economically practical to mine.

organic sediments Carbon-rich sediments derived from the remains of living organisms.

OTEC *See* ocean thermal energy conversion.

outwash Glacial sediment moved and redeposited by meltwater.

oxbows Old meanders now cut off or abandoned by a stream.

oxide Nonsilicate mineral containing oxygen combined with one or more metals.

oxygen sag curve A graph depicting oxygen depletion followed by reoxygenation in a stream system below a source of organic waste matter; caused by aerobic decay of the organic matter.

ozone hole Area over which the ozone layer is thinner (containing a lower ozone concentration) than over surrounding areas; commonly develops annually over Antarctica.

ozone layer An ozone-rich layer within the stratosphere, between about 15 and 35 km above earth's surface; absorbs potentially harmful ultraviolet radiation.

P

paleomagnetism The "fossil magnetism" preserved in rocks formed in the past.

Paleozoic Geologic era from 570 to 245 million years ago.

Pangaea The name given to the ancient supercontinent that existed some 200 million years ago.

particulates (pollution) Solid particles suspended in air; includes soot, ash, and dust.

passive margin A geologically quiet continental margin, lacking significant volcanic or seismic activity.

pathogenic Disease-causing; often applied to harmful microorganisms.

pedalfer A moderately leached soil rich in residual iron and aluminum oxide minerals.

pedocal A soil in which calcium carbonate and other readily soluble minerals are retained; characteristic of drier climates.

pegmatite A very coarsely crystalline igneous rock.

percolation Movement of subsurface water through rock or soil under its own pressure.

periodic table The regular arrangement of chemical elements in a chart that reflects patterns of chemical behavior related to the electronic structure of atoms.

permafrost A condition found in cold climates, wherein ground remains frozen year-round at some depth below the surface.

permafrost table The top of the permanently frozen (permafrost) zone.

permeability A measure of how readily fluid can flow through a rock, sediment, or soil.

petroleum Liquid hydrocarbons derived from organic matter and used as fuel.

Phanerozoic The time from 570 million years ago to the present.

photovoltaic cells (solar cells) Devices that convert solar radiation directly to electricity.

phreatic eruption A violent, explosive volcanic eruption like a steam-boiler explosion, occurring when subsurface water is heated and converted to steam by hot magma underground.

phreatic zone *See* saturated zone.

pH scale Scale for reporting acidic or alkaline quality of a liquid.

placers The ores concentrated by stream or wave action on the basis of mineral densities and/or resistance to weathering.

plastic deformation Permanent strain in material stressed beyond the elastic limit; the material will not return to its original dimensions when the stress is removed.

plate tectonics The theory that holds that the rigid lithosphere is broken up into a series of movable plates.

plutonic Describes an igneous rock crystallized well below the earth's surface; typically coarse-grained.

point bar A sedimentary feature built in a stream channel, on the inside of a meander, or anywhere the water slows.

point source A single, concentrated, identifiable source of pollutants, such as a sewer outfall or factory smokestack.

polar-wander curve A plot of apparent magnetic pole positions at various times in the past relative to a continent, assuming the continent's position to have been fixed on the earth.

pore pressure Pressure of fluid filling cracks and pores in rock or soil.

porosity Proportion of void space in rock, sediment, or soil.

potentiometric surface A feature analogous to a water table but applied to confined aquifers; indicates the height to which the water's pressure would raise the water if the water were unconfined.

Precambrian The time from the formation of the earth to 570 million years ago (the start of the Cambrian period).

precautionary principle A concept in international law/diplomacy under which nations' activities may be restricted if there is reasonable likelihood that those activities may be harmful; significant damage (as to the environment) need not already have occurred.

precursor phenomena Phenomena that precede an earthquake, volcanic eruption, or other natural event, which may be used to predict the upcoming event.

Principle of Original Horizontality The concept that sediments are deposited in approximately horizontal, flat-lying layers.

Principle of Superposition The concept that, in an undisturbed pile of sedimentary rocks, those on the bottom were deposited first and those on top, last.

Prior Appropriation The principle of surface-water law by which users of water from a given source have priority rights to it on the basis of relative time of first use.

proton Subatomic particle with a charge of +1 and a mass of approximately one atomic mass unit; generally found within an atomic nucleus.

pump-and-treat Approach to groundwater purification whereby the water is extracted prior to treatment.

P waves Compressional seismic body waves.

pyroclastic flow Denser-than-air flow of hot gas and ash from a volcano.

pyroclastics The hot fragments of rock and magma emitted during an explosive volcanic eruption.

Q

quick clay The sediment formed from glacial rock flour deposited in a marine setting, weakened by subsequent flushing with fresh pore water.

R

rain shadow A dry zone landward of a mountain range, which is caused by loss of moisture from air passing over the mountains.

recharge The processes of infiltration and migration by which ground water is replenished.

recrystallization Atomic rearrangement to form new crystals in a rock while it remains in the solid state; often, the grain sizes in the rock after recrystallization are coarser than in the rock originally.

recurrence interval The average length of time between floods of a given size along a particular stream.

recycling The recovery and reprocessing or reuse of materials already used previously in construction, manufacturing, and so on; a method of conserving resources by minimizing the need for the development of new resources.

regional metamorphism Metamorphism on a large scale, involving increased heat and pressure; often associated with mountain building.

regolith Surficial sediment deposit formed in place, whether or not capable of supporting life. (*See also* soil.)

rem Short for "radiation equivalent man"; a unit of radiation exposure that takes into account the type and energy of radiation involved and the effect of that particular kind of radiation on the human body.

remote sensing Investigation without direct contact, as by using aerial or satellite photography, radar, and so on.

renewable Capable of replenishment or regeneration on a human timescale.

reprocessing (nuclear fuel) The treatment of used fuel elements to extract plutonium or other fissionable material to make new fuel elements.

reserves That quantity of a (resource) material that has been found and is recoverable economically with existing technology.

residence time The average length of time a substance persists in a system; may also be defined as (capacity)/(rate of influx).

resources Those things important or necessary to human life and civilization; also, the total quantity of a given (resource) commodity on earth, discovered and undiscovered, that might ultimately become part of the reserves.

retention pond A large basin designed to catch surface runoff to prevent its flow directly into a stream.

rhyolite Silica-rich volcanic rock; the volcanic compositional equivalent of granite.

Richter magnitude scale A logarithmic scale used for reporting the size of earthquakes; Richter magnitude (M_L) is proportional to amount of ground shaking (displacement) at the epicenter and is computed on the basis of the amplitude (height) of the largest seismic wave on a seismogram.

Riparian Doctrine The principle of surface-water law by which all landowners bordering a body of water have equal rights to it.

rock A solid, cohesive aggregate of one or more minerals.

rock cycle The concept that rocks are continually subject to change and that any rock may be transformed into another type of rock through an appropriate geologic process.

rockfall *See* fall (rock).

rock flour A fine sediment of pulverized rock produced by glacial erosion.

rupture Breakage or failure of material under stress.

S

saltation The process by which particles are moved in short jumps over the ground surface or stream bed by wind or water.

saltwater intrusion A process by which salt water replaces fresh ground water when the fresh water is being used more rapidly than it is being recharged; especially common in coastal areas.

sanitary landfill A disposal site for solid or contained liquid waste; in simplest form, a dump site at which wastes are covered with layers of earth daily or more often.

saturated zone The region of rock or soil in which pore spaces are completely filled with liquid; also known as the phreatic zone.

scale (map) The ratio of a unit of length on the map to the corresponding actual horizontal distance represented.

scientific method Means of discovering scientific principles by formulating hypotheses, making predictions from them, and testing the predictions.

scrubbers (electrostatic precipitators) Devices that remove pollutant gases and particulates from exhaust gases of power or manufacturing plants.

seafloor spreading The process by which new lithosphere is created at spreading ridges as plates of oceanic lithosphere move apart.

secure landfill A sanitary landfill designed to contain toxic chemical wastes; typically includes one or more impermeable liners and often is monitored by nearby wells.

sediment Surface accumulation of loose, unconsolidated mineral or rock particles.

sedimentary rock A rock formed from sediments at low temperature.

seismic gap A section of an active fault along which few earthquakes are occurring, in contrast to adjacent fault segments; presumably, a locked section of fault.

seismic shadow zone An effect of the earth's liquid outer core, which blocks S waves and thus casts a "seismic shadow" on the opposite side of the earth from the site of a major earthquake.

seismic tomography Technique using velocity variations of seismic body waves to map regions of relatively higher and lower temperature/density within the earth.

seismic waves The form in which most energy is released during earthquakes; divided into body waves and surface waves.

seismograph An instrument used to measure ground motion caused by seismic waves (and other disturbances).

sensitive clay A material similar in behavior to *quick clay,* but derived from different materials, such as volcanic ash.

septic system A sewage-disposal device involving the slow dispersal of wastes into soil for natural aerobic breakdown.

sequential use The land-use principle by which the same land is used for two or more different purposes, one after another.

settling pond A pond in which sediment-laden surface runoff is impounded to allow sediment to settle out before the water is released.

shallow-focus earthquakes Those with focal depths less than 100 km.

shearing stress Stress that tends to cause different parts of an object to slide past each other across a plane; with respect to mass movements, stress tending to pull material downslope.

shear strength The ability of a material to resist shearing stress.

shield volcano A low, flat, gently sloping volcano built from many flows of fluid, low-viscosity basaltic lava.

silicate Mineral containing silicon and oxygen and usually one or more additional elements.

sinkhole A circular depression in the ground surface commonly caused by collapse into an underground cavern formed by solution.

slide A form of mass wasting in which a relatively coherent mass of material moves downslope along a well-defined surface.

slip face The downwind side of a dune, on which material tends to assume the slope of the angle of repose of the sediment of the dune.

slump A slide moved only a short distance, often with a rotational component to the movement.

soil The accumulation of unconsolidated rock and mineral fragments and organic matter formed in place at the earth's surface; capable of supporting life.

soil moisture The water in the soil in the zone of aeration, or vadose zone.

solifluction Creep or very slow flow in saturated soil.

sorting A measure of the uniformity of a sediment with respect to particle size and/or density.

source (stream) The place where the stream originates.

source separation The sorting of (waste) material by type prior to collection, usually to facilitate the recycling of individual materials or to ready material for a particular disposal strategy, such as incineration.

speculative resources That quantity of a resource material that may be found in areas where similar deposits are not already known to exist.

spoil banks Piles of waste rock and soil left behind by surface mining, especially strip-mining.

stage (stream) The height (elevation) of a stream surface at a given point along the stream's length; usually expressed as elevation above sea level.

strain Deformation resulting from the application of stress.

stratovolcano A composite volcanic cone built of interlayered lava flows and pyroclastic materials.

stream A body of flowing water confined within a channel.

stress A force applied to an object.

striations Parallel grooves in a rock surface cut when a glacier containing rock debris flows over that rock.

strike-slip fault A fault with predominantly horizontal displacement.

strip mining The surface-mining technique in which the cover rock or soil is removed from a near-surface mineral or fuel deposit, and the deposit is then extracted by open excavation, in a series of parallel strips.

subduction zone A convergent plate boundary at which oceanic lithosphere is being pushed beneath another plate (continental or oceanic) into the asthenosphere.

subeconomic resources Those resources already found that cannot be exploited economically with existing technology; also known as conditional resources.

subsurface water The water in the hydrosphere below the ground surface.

sulfate Nonsilicate mineral containing sulfate (SO_4) groups, each made up of one atom of sulfur and four atoms of oxygen.

sulfide Nonsilicate mineral containing sulfur but no oxygen.

surface processes The geologic processes acting principally at the earth's surface; they involve the effects of wind, water, ice, and gravity.

surface water Liquid water above the ground surface.

surface waves The seismic waves that travel along the earth's surface.

surge (storm) Localized increase in water level of an ocean or large lake; caused by extreme low pressure and high winds associated with major storms.

suspended load (stream) The material that is light or fine enough to be moved along suspended in the stream, supported by the flowing water.

sustainable development Development that causes neither serious environmental damage nor such acute resource depletion as to imperil future development or quality of life.

S waves Shear seismic body waves; do not propagate through liquids.

T

tailings The piles of crushed waste rock created as a by-product of mineral processing.

talus Accumulated debris from rockfalls and rockslides.

tar sand Sedimentary rock, usually sand, containing thick, tarlike, heavy oils.

tectonics The study of large-scale movement and deformation of the earth's crust.

tensile stress Stress tending to pull an object apart.

terminal moraine The end moraine marking the farthest advance of a glacier.

terrane A geologically distinct region that can be distinguished from adjacent regions of different geology and/or history; commonly bounded by faults.

theory A generally accepted explanation for a set of data or observations; its validity has usually been tested by the scientific method.

thermal inversion The condition in which air temperature increases, rather than decreases, with increasing altitude; the overlying layer of warmer air may then trap warm, rising, pollutant-laden gases.

thermohaline circulation Major ocean circulation pattern, driven by winds and by differences in temperature and salinity of water masses.

thrust fault A dip-slip fault in which rocks on one side are pushed up and over those on the other, reflecting compressional stresses.

till Poorly sorted sediment deposited by melting glacial ice.

topsoil The topmost portion of soil, richest in organic matter.

trace elements Those elements present in a system at very low concentrations, typically 100 ppm or less.

transform fault A fault between offset segments of a spreading ridge, along which two plates move horizontally in opposite directions.

transuranic Describes elements with atomic numbers higher than that of uranium, none of which occur naturally on earth, and all of which are radioactive.

triple junction Point at which three plates (and three plate boundaries) meet.

tsunami A seismic sea wave, generated by a major earthquake in or near an ocean basin; sometimes incorrectly called a "tidal wave."

turbidity current A denser-than-water suspension of sediment in water that flows rapidly downslope; develops in ocean basins as a result of submarine landslides or earthquakes.

U

ultraviolet Electromagnetic radiation just to the short-wavelength side of the visible light spectrum; biologically hazardous.

unconfined aquifer An aquifer not overlain directly by an aquitard or aquiclude.

uniformitarianism The concept that the same basic physical laws have operated throughout the earth's history, and therefore, by studying present geologic processes and their products, we can infer much about geologic processes operative in the past.

unsaturated zone A partly saturated region of rock or soil, above the water table; also known as the vadose zone.

upstream flood Flood affecting only localized sections of a stream system; caused by such events as an intense, local cloudburst or a dam failure; typically brief in duration.

V

vadose zone *See* unsaturated zone.

ventifact Literally, "wind-made" rock, shaped by wind erosion.

viscosity Resistance to flow.

volcanic Describes an igneous rock formed at or near the earth's surface.

Volcanic Explasivity Index A scale for reporting the size of an explosive volcanic eruption.

W

watershed Synonym for *drainage basin.*

water table The top of the zone of saturation.

wave base Depth at which the movement of water molecules in waves becomes negligible.

wave-cut platform Steplike surface cut in rock by wave action at sea or lake level.

wave refraction The deflection of waves as they approach shore.

weathering The physical and/or chemical breakdown of rocks and minerals.

well sorted Describes sediments displaying uniform particle size and/or density.

Z

zone of accumulation *See* B horizon.

zone of leaching *See* E horizon.

INDEX

A

D

E

F

G

H

I

N

O

Y

Z

ARCTIC OCEAN
CANADA BASIN
12,468 ft.
BANKS I.
QUEEN ELIZABETH ISLANDS
GREENLAND
CHUKCHI SEA
Pt. Barrow
BEAUFORT SEA
VICTORIA ISLAND
7,008 ft.
Baffin Bay
Gunnbjørn Fjeld 12,139 ft.
BROOKS RANGE
BAFFIN ISLAND
Davis Strait
Great Bear Lake
NORTH AMERICA
Great Slave Lake
Hudson Bay
988 ft.
LABRADOR SEA
BERING SEA
Bering Strait
Yukon R.
Mt. McKinley 20,320 ft.
Mt. Logan 19,551 ft.
Gulf of Alaska
Mt. Fairweather 15,300 ft.
Mt. Waddington 13,163 ft.
Mt. Robson 12,972 ft.
Mt. Columbia 12,294 ft.
NORTHERN LOWLAND
LAURENTIAN PLATEAU
REYKJANES RIDGE
9,869 ft.
11,195 ft.
509 ft.
ALEUTIAN ISLANDS
25,576 ft.
ALEUTIAN TRENCH
17,248 ft.
L. Winnipeg
Strait of Belle Isle
NEWFOUNDLAND
Mt. Rainier 14,410 ft.
Borah Pk. 12,662 ft.
GREAT PLAINS
Granite Pk. 12,799 ft.
Columbia R.
Gannett Pk. 13,804 ft.
Cloud Pk. 13,167 ft.
The Great Lakes
St. Lawrence R.
15,398 ft.
Cape Sable
20,700 ft.
Cape Cod
Kings Pk. 13,528 ft.
Uncompahgre Pk. 14,309 ft.
Grays Pk. 14,270 ft.
Boundary Pk. 13,140 ft.
Blanca Pk. 14,345 ft.
Mt. Whitney 14,494 ft.
Colorado R.
Wheeler Pk. 13,161 ft.
APPALACHIAN MTS.
Ohio R.
Mississippi R.
CENTRAL LOWLAND
NORTH ATLANTIC OCEAN
AZORES
20,664 ft.
NORTH PACIFIC OCEAN
Cape Hatteras
MADEIRA
Jebel Toubkal 13,665 ft.
Humphreys Pk. 12,633 ft.
Mt. Elbert 14,433 ft.
Rio Grande
BERMUDA
Pico de Teide 12,188 ft.
MIDWAY IS.
5,758 ft.
17,921 ft.
22,950 ft.
CANARY ISLANDS
Gulf of Mexico 13,199 ft.
NORTH AMERICAN BASIN
MID-ATLANTIC RIDGE
HAWAIIAN RIDGE
HAWAIIAN ISLANDS
Tropic of Cancer
CAPE VERDE PLAIN
SIERRA MADRE
Cabo San Lucas
Pico de Orizaba 18,406 ft.
WEST INDIES
Mauna Kea 13,796 ft.
3,468 ft.
25,218 ft.
CAPE VERDE ISLANDS
Cape Verde
15,778 ft.
Tajumulco (Vol.) 13,845 ft.
CARIBBEAN SEA
17,783 ft.
23,925 ft.
Niger R.
18,767 ft.
Chirripo (Vol.) 12,530 ft.
Pico Bolivar 16,427 ft.
COCOS RIDGE
Nevado de Huila 18,865 ft.
GUIANA HIGHLANDS
LLANOS
13,783 ft.
Cape Palmas
Equator
Chimborazo 20,702 ft.
17,101 ft.
Rio Negro
Amazon
GALAPAGOS ISLANDS
AMAZON BASIN
SELVAS
21,448 ft.
ASCENSION
PHOENIX ISLANDS
17,996 ft.
15,877 ft.
ANDES MOUNTAINS
SOUTH AMERICA
CAMPOS
BRAZILIAN HIGHLANDS
Nevado Huascarán 22,133 ft.
WESTERN SAMOA
TUAMOTU
PACIFIC RISE
PLATEAU OF MATO GROSSO
Rio São Francisco
19,850 ft.
TONGA IS.
TONGA
SOCIETY IS.
ARCHIPELAGO
NAZCA RIDGE
ATACAMA TRENCH
26,465 ft.
Rio Paraguay
Paraná
35,704 ft.
Tropic of Capricorn
1,079 ft.
KERMADEC
18,879 ft.
RIO GRANDE RISE
Nevado Ojos del Salado 22,615 ft.
33,292 ft.
KERMADEC IS.
TRENCH
SOUTHWEST PACIFIC BASIN
Cerro Aconcagua 22,831 ft.
PAMPAS
SOUTH PACIFIC OCEAN
EAST
16,963 ft.
Río de la Plata
SOUTH ATLANTIC OCEAN
17,278 ft.
CHATHAM ISLANDS
13,049 ft.
15,998 ft.
CHONOS ARCH.
20,382 ft.
Strait of Magellan
FALKLAND ISLANDS
SOUTH GEORGIA
SOUTH SANDWICH TRENCH
PACIFIC-ANTARCTIC RIDGE
Cape Horn
Drake Passage
BELLINGSHAUSEN PLAIN
17,192 ft.
1,641 ft.
Antarctic Circle
WEDDELL PLA
ANTARCTIC PENINSULA
15,847 ft.
WEDDELL SEA
ROSS SEA
Vinson Massif 16,066 ft.

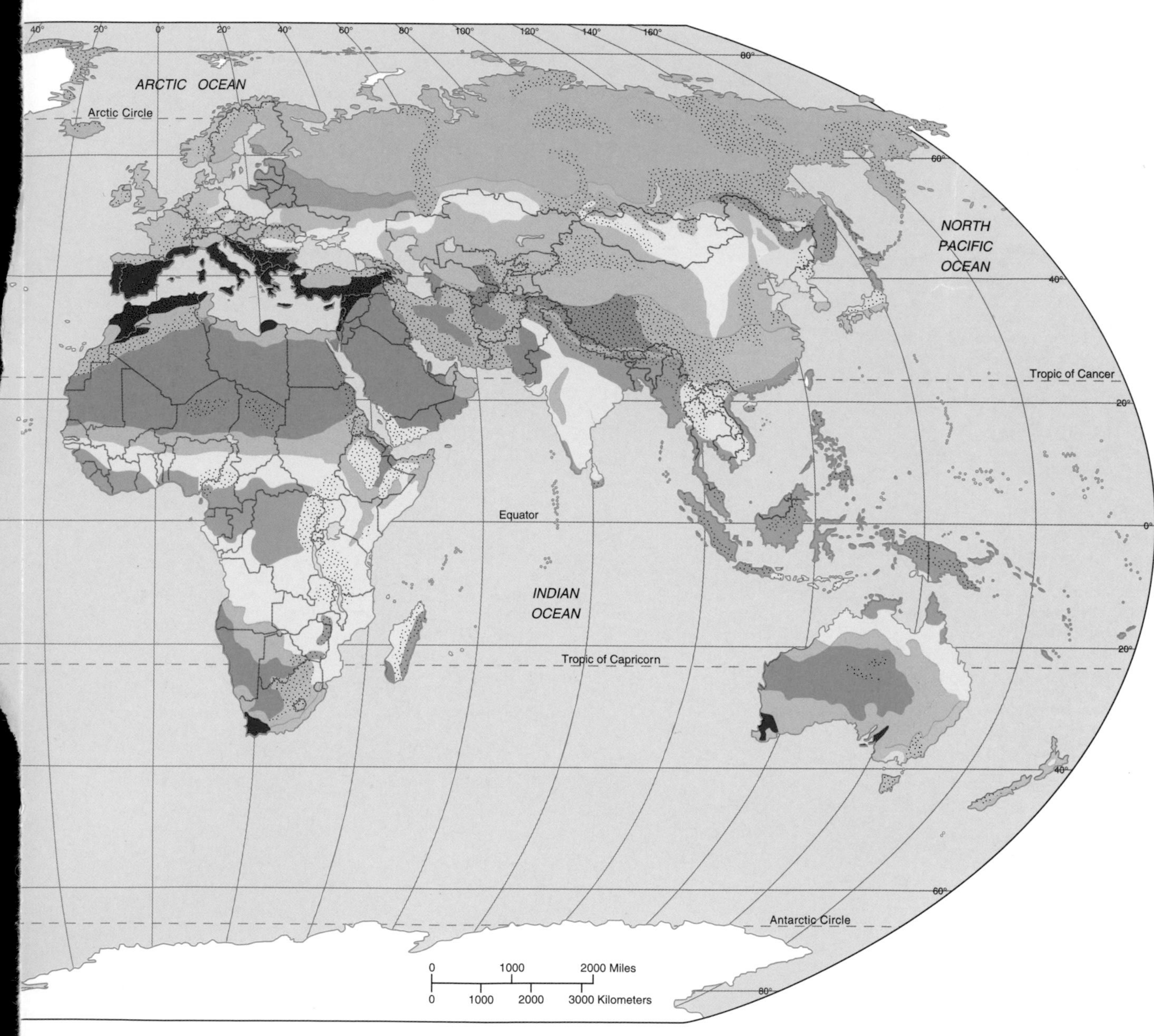

ARCTIC OCEAN
Arctic Circle
NORTH PACIFIC OCEAN
Tropic of Cancer
Equator
INDIAN OCEAN
Tropic of Capricorn
Antarctic Circle
0 1000 2000 Miles
0 1000 2000 3000 Kilometers

World Ecological Regions

Arctic and Subarctic Zone

- Ice Cap
- Tundra Province: moss-grass and moss-lichen tundra
- Tundra Altitudinal Zone: polar desert (no vegetation)
- Subarctic Province: evergreen forest, needleleaf taiga; mixed coniferous and small-leafed forest
- Subarctic Altitudinal Zone: open woodland; wooded tundra

Humid Temperate Zone

- Moderate Continental Province: mixed coniferous and broadleaf forest
- Moderate Continental Altitudinal Zone: coastal and alpine forest; open woodland
- Warm Continental Province: broadleaf deciduous forest
- Warm Continental Altitudinal Zone: upland broadleaf and alpine needleleaf forest
- Marine Province: lowland, west-coastal humid forest
- Marine Altitudinal Zone: humid coastal and alpine coniferous forest
- Humid Subtropical Province: broadleaf evergreen and broadleaf deciduous forest
- Humid Subtropical Altitudinal Zone: upland, subtropical broadleaf forest
- Prairie Province: tallgrass and mixed prairie
- Prairie Altitudinal Zone: upland mixed prairie and woodland
- Mediterranean Province: sclerophyll woodland, shrub, and steppe grass
- Mediterranean Altitudinal Zone: upland shrub and steppe

Humid Tropical Zone

- Savanna Province: seasonally dry forest; open woodland; tallgrass savanna
- Savanna Altitudinal Zone: open woodland steppe
- Rainforest Province: constantly humid, broadleaf evergreen forest
- Rainforest Altitudinal Zone: broadleaf evergreen and subtropical deciduous forest

Arid and Semiarid Zone

- Tropical/Subtropical Steppe Province: dry steppe (short grass), desert shrub, semidesert savanna
- Tropical/Subtropical Steppe Altitudinal Zone: upland steppe (short grass) and desert shrub
- Tropical/Subtropical Desert Province: hot, lowland desert in subtropical and coastal locations; xerophytic vegetation
- Tropical/Subtropical Desert Altitudinal Zone: desert shrub
- Temperate Steppe Province: medium to shortgrass prairie
- Temperate Steppe Altitudinal Zone: alpine meadow and coniferous woodland
- Temperate Desert Province: midlatitude rainshadow desert; desert shrub
- Temperate Desert Altitudinal Zone: extreme continental desert steppe; desert shrub, xerophytic vegetation, shortgrass steppe

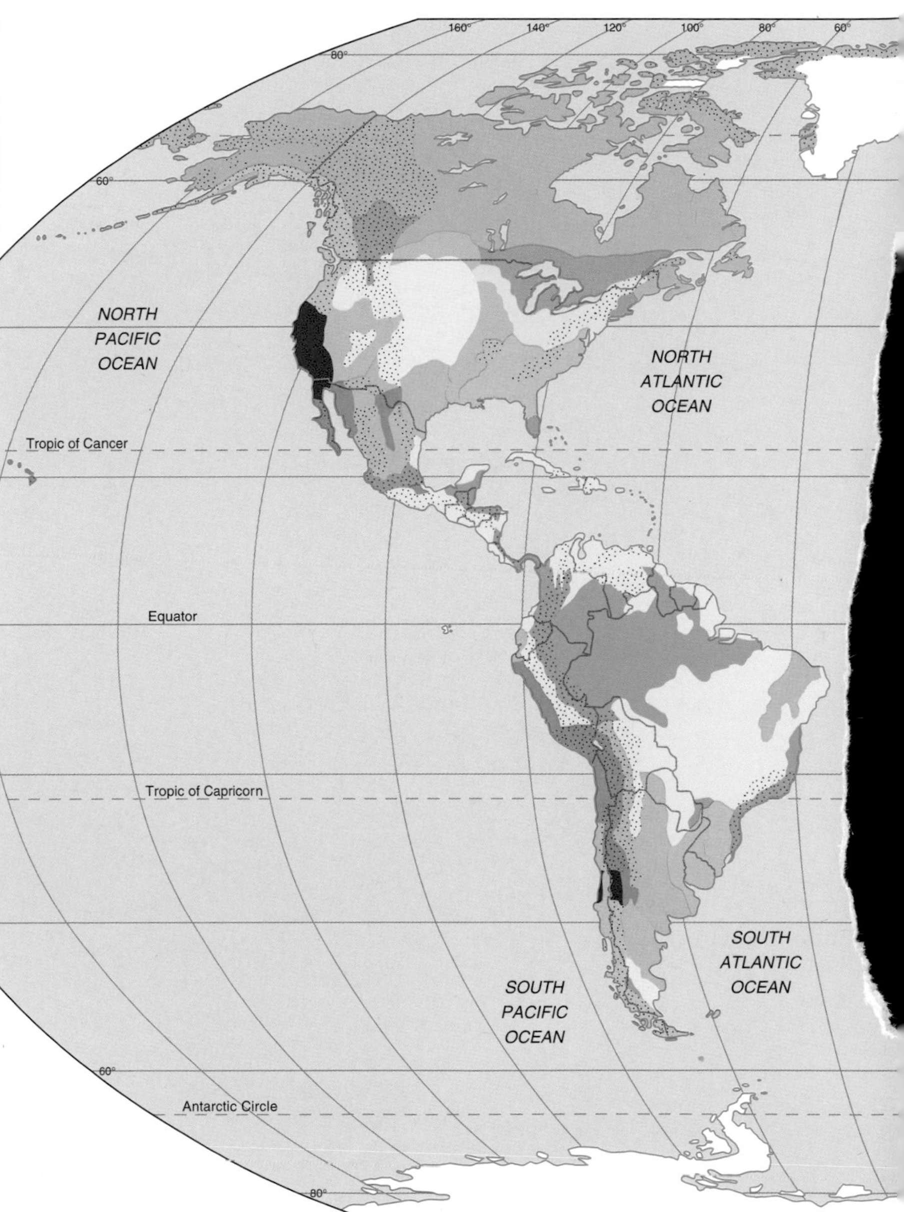

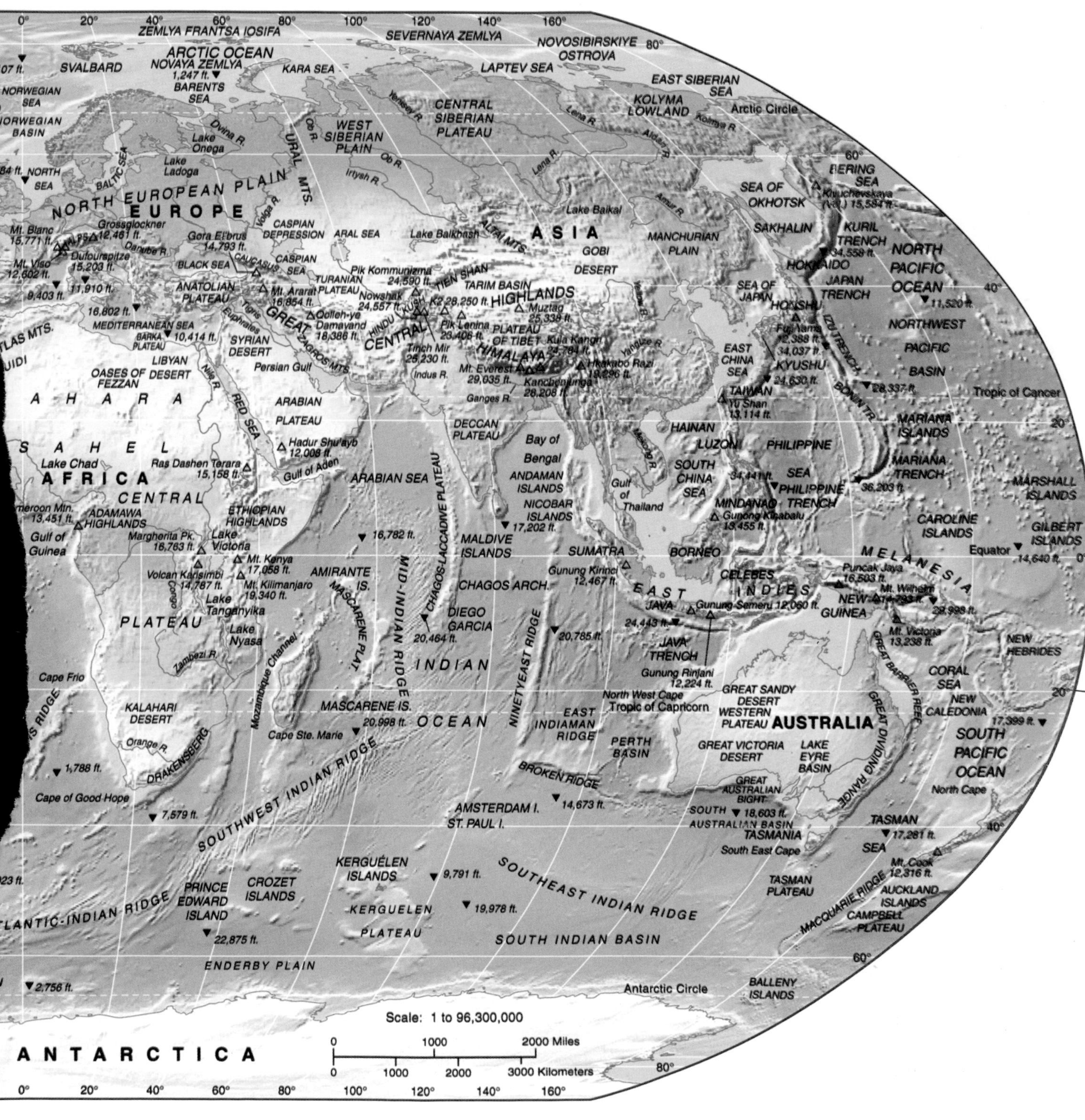

ZEMLYA FRANTSA IOSIFA
SEVERNAYA ZEMLYA
NOVOSIBIRSKIYE OSTROVA
ARCTIC OCEAN
SVALBARD
NOVAYA ZEMLYA
1,247 ft.
KARA SEA
LAPTEV SEA
NORWEGIAN SEA
BARENTS SEA
EAST SIBERIAN SEA
NORWEGIAN BASIN
KOLYMA LOWLAND
Arctic Circle
CENTRAL SIBERIAN PLATEAU
WEST SIBERIAN PLAIN
Lake Onega
Lake Ladoga
URAL MTS.
NORTH SEA
BALTIC SEA
BERING SEA
SEA OF OKHOTSK
NORTH EUROPEAN PLAIN
EUROPE
Lake Baikal
Mt. Blanc 15,771 ft.
Grossglockner 12,461 ft.
Gora El'brus 18,510 ft.
CASPIAN DEPRESSION
ARAL SEA
Lake Balkhash
ALTAI MTS.
ASIA
GOBI DESERT
MANCHURIAN PLAIN
SAKHALIN
KURIL TRENCH
NORTH PACIFIC OCEAN
Mt. Viso 12,602 ft.
Dufourspitze 15,203 ft.
BLACK SEA
CASPIAN SEA
Pik Kommunizma 24,590 ft.
TIEN SHAN
TARIM BASIN
HOKKAIDO
SEA OF JAPAN
JAPAN TRENCH
ANATOLIAN PLATEAU
TURANIAN PLATEAU
Mt. Ararat 16,854 ft.
Nowshak 24,557 ft.
K2 28,250 ft.
HIGHLANDS
Muztag 25,338 ft.
HONSHU
NORTHWEST PACIFIC BASIN
MEDITERRANEAN SEA
Qolleh-ye Damavand 18,386 ft.
CENTRAL
PLATEAU OF TIBET
Pik Lenina 23,406 ft.
Kula Kangri 24,784 ft.
Fuji Yama 12,388 ft.
SYRIAN DESERT
ATLAS MTS.
LIBYAN DESERT
Persian Gulf
ZAGROS MTS.
Tirich Mir 25,230 ft.
HIMALAYA
Mt. Everest 29,035 ft.
Hkakabo Razi 19,296 ft.
EAST CHINA SEA
KYUSHU
OASES OF FEZZAN
SAHARA
Indus R.
Ganges R.
Kanchenjunga 28,208 ft.
TAIWAN
Yu Shan 13,114 ft.
Tropic of Cancer
ARABIAN PLATEAU
RED SEA
DECCAN PLATEAU
HAINAN
MARIANA ISLANDS
SAHEL
Hadur Shu'ayb 12,008 ft.
Bay of Bengal
LUZON
PHILIPPINE SEA
Lake Chad
Ras Dashen Terara 15,158 ft.
Gulf of Aden
ARABIAN SEA
ANDAMAN ISLANDS
SOUTH CHINA SEA
Gulf of Thailand
MARIANA TRENCH
MARSHALL ISLANDS
AFRICA
CENTRAL
PHILIPPINE TRENCH
Cameroon Mtn. 13,451 ft.
ADAMAWA HIGHLANDS
ETHIOPIAN HIGHLANDS
NICOBAR ISLANDS
MINDANAO
Gunong Kinabalu 13,455 ft.
CAROLINE ISLANDS
GILBERT ISLANDS
Gulf of Guinea
Margherita Pk. 16,763 ft.
Lake Victoria
MALDIVE ISLANDS
SUMATRA
BORNEO
MELANESIA
Equator
Mt. Kenya 17,058 ft.
Volcan Karisimbi 14,787 ft.
Mt. Kilimanjaro 19,340 ft.
AMIRANTE IS.
MID-INDIAN RIDGE
CHAGOS-LACCADIVE PLATEAU
Gunung Kirinci 12,467 ft.
CELEBES
Puncak Jaya 16,503 ft.
Congo
Lake Tanganyika
MASCARENE PLAT.
CHAGOS ARCH.
EAST INDIES
JAVA
Gunung Semeru 12,060 ft.
NEW GUINEA
Mt. Wilhelm 14,793 ft.
PLATEAU
Lake Nyasa
DIEGO GARCIA
Mt. Victoria 13,238 ft.
Zambezi R.
Mozambique Channel
INDIAN
NINETYEAST RIDGE
JAVA TRENCH
Gunung Rinjani 12,224 ft.
GREAT BARRIER REEF
NEW HEBRIDES
Cape Frio
CORAL SEA
MASCARENE IS.
North West Cape
Tropic of Capricorn
GREAT SANDY DESERT
WESTERN PLATEAU
KALAHARI DESERT
OCEAN
EAST INDIAMAN RIDGE
AUSTRALIA
NEW CALEDONIA
Cape Ste. Marie
PERTH BASIN
GREAT VICTORIA DESERT
LAKE EYRE BASIN
GREAT DIVIDING RANGE
SOUTH PACIFIC OCEAN
Orange R.
DRAKENSBERG
BROKEN RIDGE
North Cape
Cape of Good Hope
SOUTHWEST INDIAN RIDGE
AMSTERDAM I.
ST. PAUL I.
GREAT AUSTRALIAN BIGHT
SOUTH AUSTRALIAN BASIN
TASMAN SEA
TASMANIA
South East Cape
KERGUELEN ISLANDS
SOUTHEAST INDIAN RIDGE
Mt. Cook 12,316 ft.
PRINCE EDWARD ISLAND
CROZET ISLANDS
TASMAN PLATEAU
AUCKLAND ISLANDS
ATLANTIC-INDIAN RIDGE
KERGUELEN PLATEAU
MACQUARIE RIDGE
CAMPBELL PLATEAU
SOUTH INDIAN BASIN
ENDERBY PLAIN
Antarctic Circle
BALLENY ISLANDS
Scale: 1 to 96,300,000
0 1000 2000 Miles
0 1000 2000 3000 Kilometers
ANTARCTICA
0° 20° 40° 60° 80° 100° 120° 140° 160°